D1064841

MYOPATHIES

HANDBOOK OF
CLINICAL NEUROLOGY

Editors

PIERRE J. VINKEN   GEORGE W. BRUYN
HAROLD L. KLAWANS

Editorial Advisory Board

R.D. ADAMS, S.H. APPEL, E.P. BHARUCHA,
M. CRITCHLEY, C.D. MARSDEN, H. NARABAYASHI,
A. RASCOL, L.P. ROWLAND, F. SEITELBERGER

VOLUME 62

ELSEVIER SCIENCE PUBLISHERS

AMSTERDAM · LONDON · NEW YORK · TOKYO

MYOPATHIES

Editors

PIERRE J. VINKEN GEORGE W. BRUYN
HAROLD L. KLAWANS

This volume has been co-edited by

L.P. ROWLAND and S. DIMAURO

REVISED SERIES 18

ELSEVIER SCIENCE PUBLISHERS

AMSTERDAM · LONDON · NEW YORK · TOKYO

ELSEVIER SCIENCE PUBLISHERS B.V.

P.O. BOX 1527

1000 BM AMSTERDAM

ISBN *for the complete series:* 0 444 90404 2

ISBN *for this volume:* 0 444 81281 4

PRINTED ON ACID-FREE PAPER

PRINTED IN THE NETHERLANDS

Foreword

The Foreword to volumes 40 and 41 ('Diseases of Muscle' - Parts I and II) in the first series of the Handbook published in 1979 reflected a rather bleak view of the state of affairs then relevant in the field of myopathies. On that occasion, we observed that the clinical descriptions of the muscular dystrophies, although refined and reclassified, had remained virtually unchanged since the first part of the nineteenth century when Charcot pointed out to his students, 'Duchenne discovered a disease that probably existed in the time of Hippocrates'. We went on to say that, sadly enough, despite the advances in our understanding of the molecular biology of myogenesis, the cause and treatment of the majority of neuromuscular diseases remained elusive, and although we were better able to detect the disease in its earliest stages and to define more accurately the pattern of inheritance, we did not fully comprehend the mechanisms of muscle failure nor effective methods of treatment. What a remarkably different panorama is revealed to the reader as he comes to consult the present work. Engulfed by the colossal intellectual and technical changes that have taken place in the intervening period, our contributors could not merely update original chapters. On the contrary, a new approach was required for each chapter.

What has happened since 1979? Without doubt, the most revolutionary force has been the advent of molecular genetics and, for reasons not entirely clear, neuromuscular diseases have been at the forefront. As a shining example, the first triumph of 'reverse genetics' - finding a gene without prior knowledge of the gene product and then deducing the protein from the nucleotide sequence - may have been chronic granulomatous disease, but that came as fall-out from the planned attack on Duchenne muscular dystrophy.

There are other examples. Molecular genetics has transformed concepts of previously known diseases, such as myotonic muscular dystrophy. Arguments about 'anticipation' have evaporated, because we do not know the molecular basis for trinucleotide amplification in the gene. Arguments about the identity of similar, but not identical, syndromes have evaporated as we have come to understand the allelic pairing of Duchenne and Becker dystrophies, or hyperkalemic periodic paralysis and paramyotonia congenita.

Naturally, molecular genetics has allowed the genetic counsellor to become more effective in his advice and guidance, and it has led to new approaches to therapy. Whether or not myoblast implantation ever becomes an effective treatment, it would surely have been unthinkable in the late 1970s. But it has also led to new concepts of diagnosis. Who would have thought that recurrent myoglobinuria or quadriceps myopathy could be allelic variations of Duchenne dystrophy? How that comes about is only one of many new mysteries.

Molecular genetics has also opened up the fascinating world of mitochondrial diseases, with a new form of human inheritance, one called 'maternal inheritance'. This has provided basic understanding of previously obscure clinical syndromes that were identified by a gibberish of acronyms: KSS, MELAS, MERRF, MNGIE, and NARP (all explained later, in the chapters devoted to metabolic myopathies and progressive external ophthalmoplegia). Here, too, discoveries have led to fresh challenges: we now have to determine how similar mutations of mitochondrial DNA lead to totally different clinical syndromes.

DNA analysis has also clarified the metabolic myopathies, so that we now have a glimmer of understanding about the differences between those forms that totally lack the affected enzyme and those that have an enzymatically inactive, but still immunoreactive, protein. As more and more clinical syndromes have been linked to specific genes, the diagnosis of limb-girdle muscular dystrophy - always known to be heterogeneous - has been whittled down to fewer and fewer cases.

Of course, molecular genetics is not the whole story. Molecular cell biology has led to new concepts of transmitters, receptors, ion channels, and surface membranes. Information about the cytoskeleton and sarcolemma provides the background for understanding the role of dystrophin in normal and Duchenne muscle, though there is still much to be learned. Identification of the gene for the sodium channel enabled us to achieve a better understanding of the function of muscle membranes and also provided a candidate gene-product that helped to localize the gene for hyperkalemic periodic paralysis. Technical advances have stimulated electrophysiology, microscopy, and tissue culture.

There have been sweeping changes in immunology. The role of T and B cells has been clarified, and the role of antigen-presenting cells has opened new vistas for myasthenia gravis and several forms of inflammatory myopathies. The molecular biology of antibody formation and the use of transgenic mice provide the basis for new theories of pathogenesis. Myasthenia gravis is now the prototypal human autoimmune disease.

Yet, myasthenia illustrates the challenges that remain. We do not understand how the disease arises, nor how the autoimmune process is initiated. Unfortunately, there has been no real advance in therapy for decades. Thymectomy is the basis for long-range treatment, but it does not cure all patients, and there continues to be debate about the proper immunosuppressive therapy for the others. For two decades, plasmapheresis prospered as a treatment for myasthenia and other immune disorders, but it now seems to have been displaced by intravenous immunoglobulin therapy, which is equally expensive and transient in effect, but more convenient and almost totally enigmatic in its mode of action. Better treatment for myasthenia is certainly needed. The treatment of inflammatory myopathies - dermatomyositis, polymyositis of several forms, and inclusion body myositis - is even more unsatisfactory.

There are other research tasks, amongst which are explanations for the structurally specific congenital myopathies. Perhaps most urgent is the question of linking molecular biology data to the clinical syndromes - and, thereby, to the development of rational therapy. Answers will undoubtedly come one day. In the mean time, the present volume reflects the state of affairs in 1992. We are most grateful to all concerned for their expert contributions.

P.J.V.
G.W.B.
H.L.K.
L.P.R.
S.DiM.

November 1992

List of contributors

V. Askanas
Neuromuscular Center, Department of Neurology, University of Southern California School of Medicine, 637 South Lucas Avenue, Los Angeles, CA 90017, U.S.A.   85

R. L. Barchi
David Mahoney Institute of Neurological Sciences, University of Pennsylvania School of Medicine, Philadelphia, PA 19104-6074, U.S.A.   261

A. S. Buchman
Department of Neurological Sciences, Rush-Presbyterian-St Lukes Medical Center, 1653 West Congress Parkway, Chicago, IL 60612, U.S.A.   197

O. J. S. Buruma
Department of Cardiology and Neurology, University Hospital, State University Leiden, Rijnsburgerweg 10, 2300 RC Leiden, The Netherlands   457

S. Carpenter
Montreal Neurological Institute, McGill University, 3801 University Street, Montreal, Quebec H3A 2B4, Canada   1

E. J. Cochran
Departments of Pathology and Neurological Sciences, Rush-Presbyterian-St Lukes Medical Center, 1653 West Congress Parkway, Chicago, IL 60612, U.S.A.   197

M.C. Dalakas
Neuromuscular Diseases Section, National Institute of Neurological Disorders and Stroke, National Institutes of Health, Bethesda, MD, U.S.A.   369

S. DiMauro
4-420 College of Physicians and Surgeons, Columbia-Presbyterian Medical Center, 630 West 168th Street, New York, NY 10032, U.S.A.   479, 553

A. G. Engel
*Department of Neurology and Neuromuscular Research Laboratory, Mayo Clinic,
Rochester, MN 55905, U.S.A.*   391

W. K. Engel
*Neuromuscular Center, Department of Neurology, University of Southern California
School of Medicine, 637 South Lucas Avenue, Los Angeles, CA 90017, U.S.A.*   85

K. H. Fischbeck
*Department of Neurology, Hospital of the University of Pennsylvania, 422 Curie Boul-
evard, Philadelphia, PA 19104, U.S.A.*   117

H. H. Goebel
*Division of Neuropathology, University of Mainz, Langenbeckstrasse 1, D-6500
Mainz, Germany*   331

R. C. Griggs
*Department of Neurology, Strong Memorial Hospital, University of Rochester School
of Medicine and Dentistry, 601 Elmwood Avenue, Box 673, Rochester, NY 14642,
U.S.A.*   117

L. C. Hopkins
*Department of Neurology, Emory University School of Medicine, 1365 Clifton Road,
NE, Atlanta, GA 30322, U.S.A.*   145

F. Jerusalem
*Department of Neurology, University of Bonn, Sigmund Freud Strasse 25, 5300 Bonn
1, Germany*   179

G. Karpati
*Montreal Neurological Institute, McGill University, 3801 University Street, Montreal,
Quebec H3A 2B4, Canada*   1

J. T. Kissel
*Division of Neuromuscular Disease, Department of Neurology, Ohio State University,
410 West 10th Avenue, Colombus, OH 43210-1228, U.S.A.*   527

H. G. Lenard
Department of Pediatrics, University of Düsseldorf, Düsseldorf, Germany   331

F. L. Mastaglia
*Department of Medicine, University of Western Australia and Department of Neurol-
ogy, Queen Elizabeth II Medical Centre, Nedlands, Perth, Western Australia 6009,
Australia*   595

J. R. Mendell
*Division of Neuromuscular Disease, Department of Neurology, Ohio State University,
410 West 10th Avenue, Columbus, OH 43210-1228, U.S.A.*   527

R. T. Moxley III
Neuromuscular Disease Center, Department of Neurology, University of Rochester Medical Center, 601 Elmwood Avenue, Box 673, Rochester, NY 14642, U.S.A.   209

T. L. Munsat
Neuromuscular Research Unit, Department of Neurology, Tufts-New England Medical Center, 750 Washington Street, Boston, MA 02111, U.S.A.   161

L. P. Rowland
Neurological Institute, Columbia-Presbyterian Medical Center, 710 West 168th Street, New York, NY 10032–2603, U.S.A.   287, 553

J. J. Schipperheyn
Department of Cardiology and Neurology, University Hospital, State University Leiden, Rijnsburgerweg 10, 2300 RC Leiden, The Netherlands   457

G. Serratrice
Department of Neurology, University of Aix-Marseille, Marseille, France   161

S. Servidei
Department of Neurology, Universitá Cattolica, Rome, Italy   479

J. P. Sieb
Department of Neurology, University of Bonn, Sigmund Freud Strasse 25, 5300 Bonn 1, Germany   179

E. Stålberg
Department of Clinical Neurophysiology, University Hospital, S-751 85 Uppsala, Sweden   49

I. Tein
Division of Neurology, The Hospital for Sick Children, 555 University Avenue, Toronto, Ontario M5G 1X8, Canada   553

P. Tonin
Department of Neurology, University of Verona, Verona, Italy   479

J. Trontelj
University Institute of Clinical Neurophysiology, University Medical Center, Ljubljana, Slovenia   49

S. T. Warren
Department of Biochemistry, Howard Hughes Medical Institute of Emory University, 4035 Rollins Research Center, 1510 Clifton Road, N.E., Atlanta, GA 30322, U.S.A.   145

A. R. Wintzen
Department of Cardiology and Neurology, University Hospital, State University Leiden, Rijnsburgerweg 10, 2300 RC Leiden, The Netherlands   457

Contents

Foreword   v

List of Contributors   vii

Chapter 1.   *Skeletal muscle pathology in neuromuscular diseases –* G. Karpati and S. Carpenter   1

Chapter 2.   *Clinical neurophysiology: the motor unit in myopathy –* E. Stålberg and J. Trontelj   49

Chapter 3.   *Cultured normal and genetically abnormal human muscle –* V. Askanas and W. K. Engel   85

Chapter 4.   *X-linked muscular dystrophies –* R. C. Griggs and K. H. Fischbeck   117

Chapter 5.   *Emery-Dreifuss muscular dystrophy –* L. C. Hopkins and S. T. Warren   145

Chapter 6.   *Facioscapulohumeral and scapuloperoneal syndromes –* T. L. Munsat and G. Serratrice   161

Chapter 7.   *The limb girdle syndromes –* F. Jerusalem and J. P. Sieb   179

Chapter 8.   *Distal myopathies –* A. S. Buchman and E. J. Cochran   197

Chapter 9.   *Myotonic muscular dystrophy –* R. T. Moxley III   209

Chapter 10.   *The nondystrophic myotonic syndromes –* R. L. Barchi   261

Chapter 11.   *Progressive external ophthalmoplegia and ocular myopathies –* L. P. Rowland   287

Chapter 12.   *Congenital myopathies –* H. H. Goebel and H. G. Lenard   331

Chapter 13.   *Inflammatory myopathies* – M. C. Dalakas   369

Chapter 14.   *Myasthenia gravis and myasthenic syndromes* – A. G. Engel   391

Chapter 15.   *Periodic paralysis* – J. J. Schipperheyn, A. R. Wintzen and
               O. J. S. Buruma   457

Chapter 16.   *Metabolic myopathies* – S. DiMauro, P. Tonin and
               S. Servidei   479

Chapter 17.   *The endocrine myopathies* – J. T. Kissel and J. R. Mendell   527

Chapter 18.   *Myoglobinuria* – I. Tein, S. DiMauro and L. P. Rowland   553

Chapter 19.   *Toxic myopathies* – F. L. Mastaglia   595

Index   623

Handbook of Clinical Neurology, Vol. 18 (62): Myopathies
L.P. Rowland and S. DiMauro, editors

CHAPTER 1

Skeletal muscle pathology in neuromuscular diseases

GEORGE KARPATI and STIRLING CARPENTER

Montreal Neurological Institute, McGill University, Montreal, Quebec, Canada

This chapter is designed to highlight key aspects of the progress in the microscopic study of skeletal muscle biopsies in neuromuscular diseases that has occurred since the last edition of Diseases of Muscle in the Handbook of Neurology in 1979 (Carpenter and Karpati 1984; Engel and Banker 1986; Walton 1988). The application of new techniques in both light and electron microscopy (such as immunocytochemistry, in-situ hybridization, tracer techniques, and automated morphometry) and the discovery of important new molecules and their isoforms have led to recognition of new diseases and new pathological processes, and improved diagnostic efficiency. Of particular importance is the identification of dystrophin (Hoffman et al. 1987), the lack or abnormality of which causes Duchenne or Becker dystrophy (Arahata et al. 1988; Bonilla et al. 1988b; Zubrinzycka-Gaarn et al. 1988). Antibodies can demonstrate the presence or absence of dystrophy in tissue sections (Nicholson et al. 1990). Another example of progress is the ability to display microscopically specific molecules, such as cytochrome oxidase in mitochondrial myopathies (Johnson et al. 1983).

Information has accumulated about the normal molecules or constituents of other tissue elements in muscle, including collagen, blood vessels and intramuscular nerves. We now have improved methods to study inflammatory cells and major histocompatibility complex proteins (Arahata and Engel 1984, 1988a,b; Karpati et al. 1988a).

In the 1979 edition, the principles of the light and electron microscopic pathology of skeletal muscle biopsies were discussed in separate chapters. Now, light and electron microscopic features of pathological muscles are integrated into one chapter, so that correlations can be made.

This review emphasizes the most important general pathological reactions that are seen in muscle biopsies. In subsequent chapters, in which specific diseases are dealt with, more extensive reference is made to pathologic features that are relevant to a particular disease.

Methods of tissue removal and processing

Diagnostic muscle samples may be obtained by either open or needle biopsies (Heckmatt et al. 1984). Both techniques have advantages and drawbacks. Advantages of the open methods include: ample sample size, possibility of removing fascicles in clamps to keep muscle fibers at resting tension (which is essential for optimal preservation of fine structure), visual control over the selection of appropriate fascicles (particularly important in advanced muscle disease) and the possibility of concomitant sural nerve biopsy when indicated. Open biopsy is technically more involved and more costly than needle biopsy, and in young children it usually requires general anesthesia. Needle biopsy may be technically simpler, faster and cheaper, but sample selection and size are limited, and isometric maintenance of muscle fibers is not possible. In young children sedation is still necessary. For most diagnostic purposes, we prefer open biopsy to needle biopsy. In fact, it may even be more cost-effective. Needle biopsy may be preferred when repeated samples are required to study a changing biochemical measurement in the tissue.

Methods of microscopic display of structures or molecules or constituents of muscle fibers and other elements of skeletal muscles

For microscopic study a portion of the biopsy is frozen and processed to obtain transverse and longitudinal cryostat sections. For routine procedures, 10 µm thick sections are adequate, whereas immunocytochemistry requires 4-6 µm sections.

Another portion of the biopsy is removed in an isometric clamp and fixed immediately in chilled 2% glutaraldehyde in cacodylate buffer. The fixed tissue is embedded in epoxy resin and from this, semi-thin (approximately 1 µm thick) sections, both transverse and longitudinal, are prepared and stained. Toluidine blue may be used, although paraphenylene diamine combined with phase microscopy gives finer resolution. If indicated, the resin blocks are then trimmed so that an area of special interest can be viewed in ultra-thin sections with an electron microscope.

In selected cases a sample of the biopsy is used for other purposes (e.g. tissue culture, DNA and RNA analysis, biochemical determinations of enzymes or mitochondrial metabolism).

The following list summarizes the general categories of cytochemical reactions that may be applied to the microscopic examination of diagnostic muscle biopsies at the light and electron microscopy level (Karpati et al. 1981):
1. Traditional cytochemistry displaying an enzyme activity.
2. Traditional marker techniques for the display of simple and complex carbohydrates or lipids, ions, or proteins.
3. Immunocytochemistry with monospecific polyclonal or monoclonal antibodies for the display of almost any immunogenic molecule. Direct or indi-

rect methods can be applied using either fluorochrome or horseradish peroxidase (HRP)-labeled antibodies, and the peroxidase-antiperoxidase technique or the biotin-avidin system. For electron microscopy colloidal gold linked to protein A or protein G is used as a postembedding procedure to display antibody binding. Different antigens vary markedly in their sensitivity to fixation or embedding.
4. Specific affinity cytochemistry, e.g. labeled alpha-bungarotoxin for acetylcholine receptors, or lectins for specific carbohydrate moieties of glycoconjugates (Pena et al. 1981), or radioactively labeled complementary deoxyribonucleic acid (DNA) for *in situ* display of a specific mRNA (Shoubridge et al. 1990) or of foreign nucleic acids such as those of viruses, or staphylococcal proteins for the display of gammaglobulins.
5. Tracer methods in fresh slabs of muscle, e.g. radiolabeled ligand for displaying a receptor, as well as receptor-mediated endocytosis of a specific receptor of HRP, or heavy metals to display the lumen of the T-tubules (Karpati et al. 1981).

As a guiding principle, the best cytochemical techniques are those that have the following characteristics:
a. precise localization (corresponding to the *in vivo* state) of the determinant molecule which the test is designed to display;
b. high degree of specificity (verified by inhibitors, as well as by intrinsic technical controls);
c. quantifiability of the reaction product;
d. suitability for both light microscopic and ultrastructure application.

Table 1 summarizes the major cytochemical features of normal muscle fibers according to various subcellular sites. In every case, the endogenous determinant molecules of structures are indicated along with the display techniques. The table includes many molecules that were identified after the 1979 edition.

MUSCLE FIBER TYPES

The traditional light microscopic and electron microscopic criteria of histochemical fiber types, key physiological features and their designation in various nomenclatures are found in Table 2. For prac-

TABLE 1

Organellar distribution of principal normal molecules of skeletal muscle fibers and methods of their microscopic display.

Organelle or cell structure	Endogenous determinant molecule	Markers	Mode of display
1. *Sarcolemma*			
a. Basal lamina	• Laminin	AB	ICC
	• Collagen IV	AB	ICC
	• Glycoconjugates	PAS or lectins	PAS or binding
	• Fibronectin	AB	ICC
b. Plasmalemma	• Na$^+$ channel	AB or TTX	ICC or binding
	• MHC	AB	ICC
	• Dystrophin-binding glycoprotein	AB	ICC
	• Neural adhesion molecule	AB	ICC
	• Glycoproteins	Lectin	Lectin binding
c. Subplasmalemmal region	• Dystrophin	AB	ICC
	• Dystrophin-related protein	AB	ICC
	• Spectrin	AB	ICC
	• Vinculin	AB	ICC
d. Motor endplate	• Acetylcholinesterase	SCC	SCC
	• Acetylcholine receptor	AB or BTX	ICC or binding
	• Glycoprotein	Lectin	Binding
2. *T-tubule*	• Ca^{2+} channel	AB	ICC
	• Lumen	Horseradish peroxidase	ICC
	• Dihydropyridine receptor	AB	ICC
3. *Sarcoplasmic reticulum*	• Ca^{2+} ATPase	AB	ICC
	• Calsequestrin	AB	ICC
	• Ryanodine receptor	Ryanodine	Binding
4. *Mitochondria*	• Succinic dehydrogenase	SCC	SCC
	• Cytochrome oxidase	SCC	SCC
	• Complex I, III, IV subunits	AB	ICC
	• NADH-dehydrogenase	SCC	SCC
	• mtDNA	AB or cDNA probe	ICC or In-situ hybridization
5. *Myofibrils*	• Myosin heavy-chains	ATPase	ATPase
	• Myosin heavy-chains (isoforms)	AB	ICC
	• Myosin light-chains	AB	ICC
	• Troponin I	AB	ICC
	• Tropomyosin	AB	ICC
	• α-actinin	AB	ICC
	• α-actin	AB	ICC
	• M line protein	AB	ICC
	• C protein	AB	ICC
	• Myomesin	AB	ICC
6. *Cytoskeleton*	• β-actin	AB	ICC
	• Nebulin	AB	ICC
	• Titin	AB	ICC
	• Desmin	AB	ICC
	• Vimentin	AB	ICC
	• Tubulin	AB	ICC

TABLE 1 *(continued)*

Organelle or cell structure	Endogenous determinant molecule	Markers	Mode of display
7. *Lysosomes*	• Acid phosphatase	SCC	SCC
	• Peroxidases	SCC	SCC
	• Cathepsins	AB	ICC
8. *Nuclei*	• DNA	AB	ICC
	• DNA	cDNA	In-situ hybridization
	• mRNA	cDNA	In-situ hybridization
	• DNA	Methyl green	SCC
9. *Ribosomes*	• rRNA	Pyronin	SCC
	• rRNA	cDNA	In-situ hybridization
10. *Cytosol*	• Parvalbumin	AB	ICC
	• Carbonic anhydrase	SCC	SCC
	• Myoglobin	AB	ICC
	• Creatine kinase	AB	ICC
	• Triglycerides	SCC	SCC
	• Glycogen	PAS	SCC

AB   = Antibody
ICC = Immunocytochemistry
SCC = Specific cytochemistry

tical purposes, in human muscle the identification of four principal histochemical fiber types (type 1, type 2A, 2B, 2C) is still sufficient (Brooke and Kaiser 1970), notwithstanding the increasing complexity of muscle fiber types recognized in other species. Electron microscopic criteria of fiber types without specific cell markers are not absolutely reliable. For practical purposes, the most generally used technique for fiber typing is still the myofibrillar ATPase, in which the reaction product accumulates as a result of the ATP-splitting capacity of the myosin heavy chain. By alkali and acid preincubations, the three basic histochemical fiber types can be clearly delineated and a fourth fiber type (type 2C) can be seen in immature and some pathological muscle fibers.

Immunocytochemical techniques, using specific monoclonal antibodies to isoforms of the myosin heavy chain, have also been used for fiber typing. Antibodies against so-called 'fast' mature myosin identify type 2A and type 2B fibers, whereas antibodies to 'slow' myosin heavy chain are used to mark type 1 fibers (Sawchak et al. 1985). Embryonic and regenerating fibers react with either an embryonic or neonatal form of myosin heavy chain (Schiaffino et al. 1986).

For routine evaluation of muscle biopsies, the identification of the three basic histochemical fiber types may be important and useful because there may be preferential or selective involvement of a specific fiber type in a pathological process. This may take the form of numerical prevalence of one fiber type, abnormal distribution of fiber types, or a specific cytochemical or cyto-architectural change in one fiber type (Table 3).

The commonest form of abnormal distribution of histochemical fiber types is *type grouping*. Clusters of the same histochemical fiber type in type grouping may be small or large. Large type groupings almost always imply reinnervation after denervation. Small type groups, however, can arise either from reinnervation after denervation or from forking of muscle fibers after regeneration (myopathic type grouping) (Schmalbruch 1984). It is sometimes difficult to decide when clustering of a histochemical fiber type is abnormal. Semi-automated image analysis may help in distinguishing pathological type grouping in borderline cases.

TABLE 2

Characteristics of major mammalian skeletal muscle fiber types*.

	Type 2A	Type 2B	Type 1
General characteristics			
Natural color	Dark	Pale	Dark
Surrounding capillary density	High	Low	High
Relative cytochemical activity			
Myosin heavy-chain	Fast	Fast	Slow
Myofibrillar ATPase pH 9.4	Strong	Strong	Weak
Myofibrillar ATPase pH 4.6	Weak	Strong	Strong
Myofibrillar ATPase pH 4.3	Weak	Weak	Strong
NADH-Tetrazolium reductase	Strong	Weak	Strong
Succinic dehydrogenase	Strong	Weak	Weak
Cytochrome oxidase	Weak	Weak	Strong
Glycogen phosphorylase	Strong	Strong	Weak
Glycogen content	High	High	Low
Myoglobin content	High	Low	High
Lipid globules	Numerous	Few	Numerous
Electron microscopy			
Mitochondria	Many, large	Few, small	Many, small
Z-disc	Wide	Narrow	Intermediate
Physiological features			
Twitch speed	Fast twitch	Fast twitch	Slow twitch
Fatiguability	Resistant	Susceptible	Resistant
Other designations			
	C	A	B
	Red	White	Intermediate
	FR	FF	S
	FOG	FG	SO

FR = fast, resistant; FF = fast, fatiguable; S = slow; FOG = fast, oxidative glycolytic; FG = fast glycolytic; SO = slow oxidative.
* Basic designation of types 2A, 2B, and 1 is after Brooke and Kaiser (1970).

TABLE 3

Predilection of pathologic reactions for histochemical muscle fiber types.

Type 1 fibers	Type 2 fibers
Targets	Tubular aggregates
Rods in nemaline myopathy	Sarcoplasmic masses
Central cores	Cylindrical spirals
Mitochondrial masses	Polyglucosan storage in Lafora's disease
Lipid globule excess in carnitine deficiency	Heteropolysaccharide storage in cardioskeletal myopathy
Central nucleation in some forms	Atrophy in:
of centronuclear neuromuscular diseases	– disuse
Glycogen excess in the adult form of	– cachexia
acid maltase deficiency	– upper motor neuron lesion
Atrophy in:	– malignant disease
– fiber type disproportion	– corticosteroid therapy
– Werdnig-Hoffmann disease	– Cushing's syndrome
– myotonic dystrophy	– myasthenia gravis
– certain centronuclear diseases	– primary hyperparathyroidism
	– Schwartz-Jampel syndrome

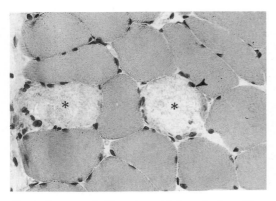

Fig. 1. Two muscle fibers (asterisks) are in the earliest phase of necrosis. They show rounded cross-sectional outline and featureless pale staining. These muscle fibers presumably became necrotic less than 6 hours prior to removal of the muscle biopsy, since macrophages have not yet invaded them. A few nuclei associated with the necrotic fibers (an example is marked by arrowhead) are presumed to be nuclei of myosatellite cells since myonuclei disintegrate very early during the necrotic cascade of events. Acute alcoholic myopathy, hematoxylin and eosin, ×350. (Figure has been reduced for printing purposes.)

MAJOR GENERAL PATHOLOGICAL REACTIONS OF MUSCLE FIBERS

Necrosis and phagocytosis

Necrosis means cell death, which may develop in response to any of numerous noxious factors. Such a factor may or may not have caused recognizable cellular abnormalities before necrosis sets in. During the process of necrosis, the muscle fiber is gradually disintegrating until it is fully converted to debris, which is eventually phagocytosed and removed by macrophages. The process of cell disintegration during necrosis gives rise to a fairly stereotyped series of microscopic changes that may provide clues to the length of time elapsed since the muscle fiber passed the point of no return. The vague term of 'degeneration', which is sometimes used to describe necrosis, is discouraged because it has also been used to refer to changes in non-necrotic fibers.

On cryostat sections, the earliest features of necrosis include the following: rounded cross-sectional outline instead of polygonality, featureless ground-glass appearance of the cytoplasm with loss of the intermyofibrillar network, and loss of

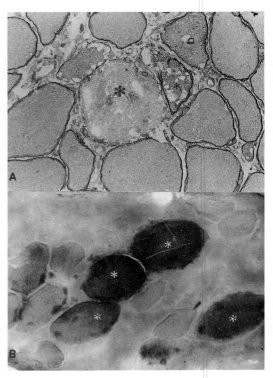

Fig. 2. (A) Dystrophin immunostaining of the surface membrane is abolished in an early necrotic fiber (large asterisk). On the other hand, in a regenerating fiber (small asterisk) dystrophin appears overexpressed in relation to the remainder of the fibers. Dermatomyositis, immunoperoxidase using a polyclonal antibody to the C-terminus of dystrophin plus biotin-streptavidin peroxidase display system, ×350. (B) Several early necrotic muscle fibers show heavy deposition of precipitated calcium salts involving the entire or partial cross-sectional area (asterisks). Duchenne dystrophy, Alizarin-red S, ×350. (Figures have been reduced for printing purposes.)

glycogen phosphorylase activity (Fig. 1). At this stage, loss of dystrophin and presence of precipitated calcium (Fig. 2) may be displayed by appropriate methods (Bodensteiner and Engel 1978). Precipitated calcium may be formed in necrotic fibers due to any cause. Immunocytochemical evidence of the third and ninth component of complement also appears early in necrotic fibers (Engel and Biesecker 1982) and may lead to binding of IgG to the necrotic debris (Fig. 3). Activation of the complement system may be the chemotactic signal that attracts macrophages into necrotic fibers.

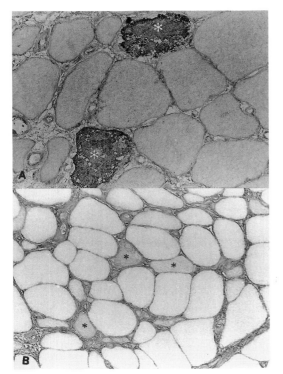

Fig. 3. (A) Asterisks mark 2 necrotic fibers in which C9 complement is present in the necrotic pulp. Dermatomyositis, immunoperoxidase using a monoclonal antibody to human C9 complement plus biotin-streptavidin peroxidase display system, ×350. (B) Three necrotic fibers (asterisks) show diffuse cytoplasmic immunostaining for IgG. This presumably is due to the presence of activated C9 component in the necrotic debris to which IgG binds non-specifically. The IgG binding in the interstitial space is a feature of normal muscle. Duchenne dystrophy, human IgG plus biotinylated rabbit antibody to human IgG and avidin peroxidase display system, ×350. (Figure has been reduced for printing purposes.)

On epoxy resin sections, necrotic fibers in early stages of dissolution usually appear paler than normal, with scattered dark dots representing abnormal mitochondria (Fig. 4). The Z-disc is the first myofibrillar component to disappear; this may occur within 2 hours after onset of necrosis. Myofibrils may also undergo hypercontraction. When this occurs in a necrotic fiber, myofibrils tend to shred or to break and slump sideways in a manner that is not seen in the hypercontraction that results from breakage at the time of biopsy.

By electron microscopy an early necrotic fiber is characterized by prominent gaps in the plasma membrane, loss of Z-disc, and mitochondrial changes which include: rounded outline, fluffy matrix densities, and linear densities in the intracristal space (Fig. 5). The rounded mitochondria tend to line up in rows between myofibrils. There is also early disappearance of triads. The basal lamina persists throughout the necrotic disintegration of the cell, although there are changes in its molecular composition (Gulati et al. 1983).

Macrophages first appear in necrotic fibers 10–12 hours after the point of no return has been passed, as indicated by experimental studies in which segmental necrosis was induced by puncturing muscle cells with a fine tungsten wire (Carpenter and Karpati 1989). Macrophages may invade earlier if they were already present in the muscle in response to a previous insult. The function of macrophages in removing the necrotic debris is probably important in optimizing regeneration. In ischemic infarcts, as in dermatomyositis, there seems to be a relative paucity of macrophages in necrotic muscle fibers, presumably due to the reduced blood flow (Fig. 6).

Because muscle fibers are multinucleated, long cylindrical cells, necrosis is usually limited to a segment, while other portions of the fiber survive. The precise length of the necrotic segment is difficult to determine in routine microscopic sections. In the rare instances when a necrotic segment can be seen from end to end, it may be as short as 300 μm or as long as 5000 μm. Necrosis may start in a limited portion of the muscle fiber and extend laterally; in experimental micropuncture, the necrotic wave seems to spread from the site of the puncture along the axis of the muscle fiber (Carpenter and Karpati 1989).

Eventually, the necrotic segment is isolated from the surviving stumps by a limiting membrane that forms, on the average, 6 hours after necrosis is triggered (Fig. 7). The mechanism of formation of this delimiting membrane is not fully understood. Aggregation of phospholipids to form a primitive bilayer occasioned by an appropriate physical-chemical milieu in the necrotic pulp seems likely, because the trajectory of the interface between the necrotic segment and the surviving stumps is usually irregular. Finger-like extensions of the stump may give the erroneous impression of 'central necrosis' or 'partial necrosis' (Fig. 8).

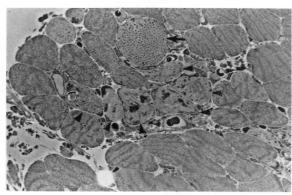

Fig. 4. In this semi-thin resin section, an enlarged rounded fiber with pale cytoplasm studded with small dark mitochondria is in an early stage of necrosis (arrow). Next to it is a group of necrotic fibers which are undergoing phagocytosis (arrowheads). The nuclei within them are either those of phagocytes or those of satellite cells. Paraphenylene diamine, phase optics, ×550. (Figure has been reduced for printing purposes.)

A phenomenon that must be distinguished from necrosis is partial invasion of non-necrotic muscle fibers by macrophages and lymphocytes (Figs. 9 and 10) in polymyositis and inclusion body myositis. On superficial evaluation of cross-sections, partially invaded muscle fibers may appear as 'incipient' necrosis, but the inflammatory cells are found under the basal lamina of fibers that are not necrotic (see Cellular Infiltrates). This process can apparently lead to complete loss of continuity of a fiber.

Muscle fiber necrosis may be seen in many forms of muscle disease. The most prominent necrosis is usually found in early and mid-phases of Duchenne dystrophy, in active polymyositis and dermatomyositis, in malignant hyperthermia crisis, and in metabolic diseases of muscles such as alcoholic myopathy, glycolytic defects and CPT

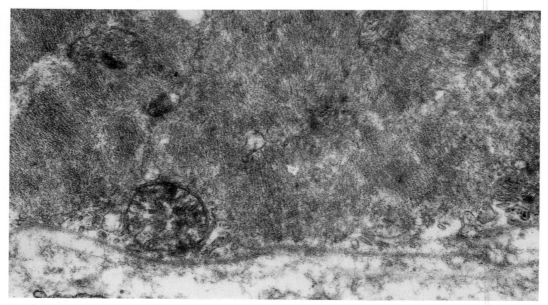

Fig. 5. Electron micrograph of part of a muscle fiber from a patient with Duchenne dystrophy in an early stage of necrosis. Note the absent plasma membrane, persistent basal lamina, lack of Z-disc, poor definition of myofibrils, and rounded mitochondrion containing fluffy matrix densities. ×28,000.

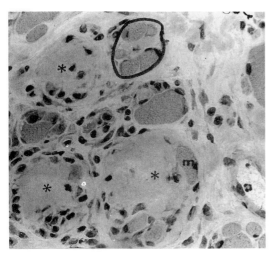

Fig. 6. Asterisks mark three necrotic fibers whose cytoplasm has been converted into featureless debris but barely invaded by macrophages, despite advanced regeneration indicated by a small regenerant daughter fiber (m). On the top there is a necrotic-regenerating fiber complex (circled). In the center it still contains some undigested necrotic debris despite the presence of at least 5 regenerant daughter fibers. These will give rise to forking. Adult dermatomyositis, hematoxylin and eosin, ×350. (Figure has been reduced for printing purposes.)

deficiency. In Duchenne dystrophy, necrotic fibers tend to occur in small clusters. The cause of this is still not clear. In some cases of dermatomyositis, there may be large groups of necrotic fibers which represent infarcts.

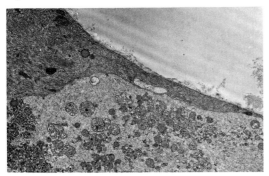

Fig. 7. Electron micrograph of a muscle fiber which has suffered experimental micropuncture. It shows part of the surviving stump and the pulp in which there are numerous mitochondria. The stump is separated from the necrotic pulp by a somewhat irregular membrane. In the center of this fiber the membrane was incomplete. ×7,000. (Figure has been reduced for printing purposes.)

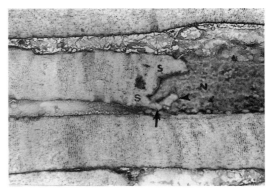

Fig. 8. A surviving portion (S) and a necrotic segment (N) of the same muscle fiber are separated by a delimiting membrane (arrowhead) which appears to be continuous with the plasmalemma of the surviving stump. The interface between the necrotic segment and surviving stump is very irregular; if a transverse section had been made at the level indicated by the arrow, a spurious picture of 'central necrosis' would have been produced. The segmental necrosis here was induced by an in-situ micropuncture of adult rat muscle fibers with a fine-tipped tungsten wire 6 hours prior to removal of the specimen. Biotinylated concanavalin-A and avidin peroxidase display system, ×425. (Figure has been reduced for printing purposes.)

Regeneration

Necrosis of muscle fibers is usually followed by regeneration, which establishes a new muscle fiber or fiber segment. Satellite cells attached to the per-

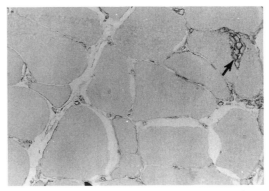

Fig. 9. 'Partial invasion': an otherwise normal-appearing muscle fiber is deeply invaginated by a group of CD8+ antigen bearing lymphocytes (arrow). Similar lymphocytes are also scattered between muscle fibers. Polymyositis, immunoperoxidase using an anti-CD8 primary antibody and biotin-streptavidin peroxidase display system, ×350. (Figure has been reduced for printing purposes.)

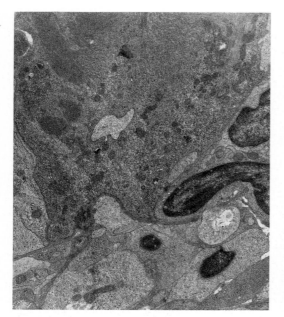

Fig. 10. Electron micrograph of a muscle fiber surrounded by inflammatory cells; although the fiber is
not necrotic, processes of these cells penetrate deeply
into its interior. The plasma membrane of the muscle
fiber remains intact, although there is extensive Z-disc
streaming. Polymyositis, ×18,000. (Figure has been
reduced for printing purposes.)

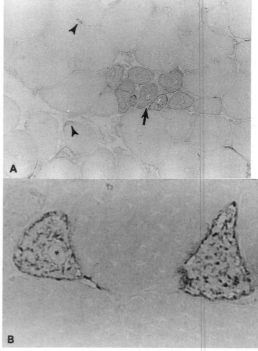

Fig. 11. (A) A cluster of small-caliber centrally nucle-
ated regenerating muscle fibers (arrow) show expression of neural cell adhesion molecule (N-CAM) both
at the cell surface and in the cytoplasm. In addition,
the surface membrane of satellite cells (two examples
marked by arrowheads) is positive. Duchenne dystrophy, immunoperoxidase using a monoclonal anti-N-
CAM primary antibody and biotin-streptavidin display dystem, ×350. (B) Higher power view of N-CAM
positive regenerating muscle fibers. Prominent cytoplasmic granules presumably represent uninserted N-
CAM molecules. Asterisk shows an activated large
nucleus. Limb-girdle muscular dystrophy, same technique as in 11(A), ×560. (Figures have been reduced
for printing purposes.)

sisting basal lamina not only survive muscle fiber
necrosis, but are stimulated to proliferate, migrate,
and eventually fuse to form myotubes (Carpenter
1990). Satellite cells throughout the length of a
muscle fiber apparently respond by thymidine incorporation to a *focal injury* (Schultz et al. 1985).
The precise molecular factors that trigger the activation of satellite cells have not yet been determined (Bischoff 1990; Grounds 1990). Proliferating satellite cells and the subsequent regenerative
myotubes (several of which form in a necrotic segment) tend to adhere to the residual basal lamina
of the muscle fiber. As the girth of the regenerating
myotubes grows, they eventually touch each other
and fuse laterally if necrotic debris has been effectively removed by macrophages. Phagocytosis and
regeneration thus progress pari passu; this has
sometimes led to the erroneous belief that regenerating fibers can become necrotic in any disease. As
a result of lateral fusion of the regenerating myotubes, a new muscle fiber segment is formed, which
eventually fuses with the surviving stumps if necro-

sis was originally segmental. An intact blood supply is needed for both phagocytosis and satellite
cell proliferation, but phagocytosis seems to be
more sensitive to ischemia.

On cryostat sections regenerating muscle fibers
are characterized by small caliber, large nuclei and
prominent nucleoli (presumably reflecting vigorous activity of gene transcription). High RNA content (as displayed by pyronine staining or acridine
orange-induced fluorescence) is attributed to
abundance of ribosomes. Other evidence of regeneration includes: enhanced oxidative enzyme activ-

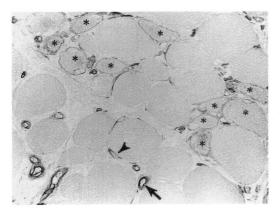

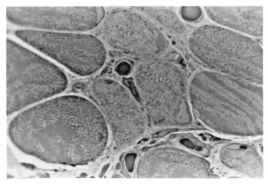

Fig. 12. Class 1 major histocompatibility complex (MHC) determinants (non-variable region) are expressed only at the surface of regenerating muscle fibers (asterisk), myosatellite cells (arrowheads) and blood vessels (one example marked by an arrow). Duchenne dystrophy, primary antibody against class 1 MHC and biotin-streptavidin display system, ×350. (Figure has been reduced for printing purposes.)

Fig. 13. Semi-thin resin section showing three regenerating muscle fiber segments. The myofibrils appear pale because they are small and separated. The nuclei are vesicular. Duchenne dystrophy, paraphenylene diamine, phase optics, ×550. (Figure has been reduced for printing purposes.)

ity forming an irregular intermyofibrillar network; lakes of prominent PAS staining glycogen; alkaline phosphatase activity; high desmin content; and the presence of an immature myosin isoform that may give a type 2C ATPase activity profile (Schiaffino et al. 1986). Regenerating fibers also express neural cell adhesion molecule (N- CAM) in the plasmalemma and sometimes in the cytoplasm as well (Fig. 11) (Cashman et al. 1987). Class 1 major histocompatibility complex proteins (MHCP) are also expressed in the plasmalemma of regenerating fibers (Fig. 12) (Karpati et al. 1988a). MyoD, a master regulatory gene product in muscle cells, is also expressed at early stages of regeneration (Beilharz 1992). Vimentin and desmin tend to be unusually abundant in regenerating fibers. When a muscle fiber completes its regenerative activity, these features disappear (usually asynchronously), although the precise time at which a muscle fiber completes its regenerative activity is not easy to recognize. In some species, but not in man, myonuclei of regenerated fibers seem to remain permanently internal (Karpati et al. 1988b).

In resin sections, regenerating fibers have a pale cytoplasm with separated small myofibrils (Fig. 13). Nuclei are relatively large and pale and have prominent nucleoli.

By electron microscopy the proliferating regenerative myoblasts adjacent to the basal lamina can be readily identified by their simple contours, in contrast to the complex contours of macrophages that are engaged in phagocytosis. The myofibrils of regenerating myotubes are small and irregular in cross-sectional outline (Fig. 14). These cells are rich in intermediate filaments and microtubules. The regenerated fibers eventually lay down their own basal lamina while the old one resorbs over a few weeks. The lamina may thus be doubled in some places around regenerating fibers. Regeneration ideally will re-establish a muscle fiber or a segment whose structure and function are indistinguishable from those of the original fiber. Sometimes, however, the results of regeneration may fall short of this ideal outcome (Fig. 15).

A regenerated muscle fiber segment may not reach the girth of the prenecrotic fiber. For example, if the number of myonuclei in the regenerated segment is not sufficient to maintain the obligatory nuclear/cytoplasmic ratio, the fiber girth will remain smaller than normal. This seems to be the principal cause of the abundant small-diameter fibers in Duchenne dystrophy and in polymyositis; in these situations, the term 'atrophy' should not be used.

Failures of lateral fusion of regenerating myotubes may produce *forked fibers* (Schmalbruch 1984). As many as 15 small-caliber muscle fibers may form by this process in place of a single fiber.

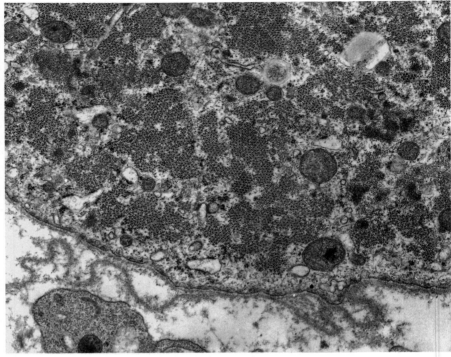

Fig. 14. In this regenerating muscle fiber the myofibrils are irregular and more widely separated than in mature fibers. Intermediate filaments are present in the subsarcolemmal area and between myofibrils. The basal lamina of the old previous fiber in places coincides with the basal lamina of the regenerated fiber and in places does not. This muscle suffered an ischemic episode 12 days earlier. ×25,000.

Since all the forked fibers are of the same histochemical type, small type groups may result (myopathic type grouping) (Fig. 16).

Failure of fusion of the regenerated segment with the surviving stumps may result in independent regenerating fibers separated by intercalated portions of collagen. If some of these segments do not become reinnervated, they will behave as denervated fibers.

The most calamitous aberration of regeneration is complete failure. As a result, the basal lamina tube of the necrotic segment remains empty, eventually collapses, and in a few weeks disappears. Empty complete skeins of basal lamina are a reliable index of regeneration failure (Carpenter and Karpati 1979). This can be seen with particularly high frequency in the early phases of Duchenne dystrophy, and less often, in other necrotic muscle diseases. The causes of regeneration failure are not clearly understood. Actual numerical depletion of satellite cells is probably rare. The number of mi-

totic cycles which a satellite cell can go through while remaining myogenic is apparently limited. Satellite cells in moderately advanced cases of Duchenne dystrophy are capable of few mitotic cycles, presumably because they have already spent most of their proliferative capacity (Blau et al. 1983).

Regenerating muscle fibers can be found in almost any muscle disease in which necrosis occurs. They are usually numerous in early stages of Duchenne dystrophy and in the active stages of some inflammatory myopathies. In Duchenne dystrophy, regenerating muscle fibers are conspicuously resistant to necrosis until they have matured. (As mentioned previously, the presence of macrophages among the myotubes during the early phase of regeneration does not indicate necrosis in regenerating fibers.) The cellular or molecular basis of the resistance of these regenerating fibers to necrosis is unclear; it may well be related to their small caliber (Karpati and Carpenter 1986).

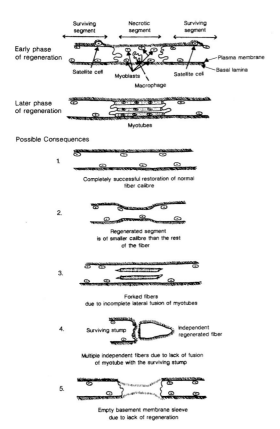

Fig. 15. Schematical illustration of skeletal muscle fiber regeneration after segmental necrosis and its possible consequences.

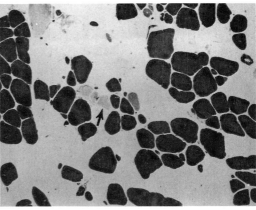

Fig. 16. Clusters of muscle fibers of the same histochemical type showing complementary contours form 'myopathic' type grouping. These groups of fibers are probably the result of forking of regenerated fibers. Some of the small fibers show type 2C myosin ATPase activity (arrow), suggesting the expression of the embryonic isoform of myosin heavy chain, which implies that these fibers are still regenerating. Duchenne dystrophy, myofibrillar ATPase, pH 4.3 preincubation, ×350. (Figure has been reduced for printing purposes.)

Atrophy and smallness

Atrophy means a significant reduction of the girth of a muscle fiber from a previously normal cross-sectional size. Abnormally small-caliber fibers that never reached their expected girth during development or regeneration, strictly speaking, should not be called atrophic. Differentiation of atrophic fibers from fibers that failed to grow to normal girth is usually not difficult (Table 4). Fibers that have shrunk are usually angulated or rounded, whereas fibers that fail to attain normal size are polygonal. By electron microscopy, rapidly shrinking fibers in denervation show redundant basal lamina sleeves.

Since myofibrils constitute the major bulk of muscle fibers, atrophy entails myofilamentous loss and myofibrils of reduced diameter. The normal equilibrium between synthesis and degradation of myofibrillar proteins is lost, with reduced synthesis, increased degradation, or both (Kelly et al. 1986). The major causes of atrophy include: denervation, disuse, chronic ischemia, corticosteroid excess, myasthenia gravis, thyrotoxicosis, myotonic dystrophy, and inclusion body myositis. The major causes of muscle fiber smallness due to lack of proper growth include: congenital myopathies of various sorts and necrotic muscle diseases where regeneration becomes limited, such as Duchenne dystrophy.

Denervation atrophy of muscle fibers is one of the most common pathological changes seen in neuromuscular diseases. The precise molecular events that lead to the atrophy are not clearly understood. Downregulation of ribosomes presumably results in reduced synthesis of the major myofibrillar proteins (Metafora et al. 1980) and enhanced proteolytic activity accelerates myofilament degradation. Disassembly of myosin monomers from thick myofilaments may precede degradation (Massa et al. 1990).

With advancing atrophy of muscle fibers, the muscle capillaries eventually adjust to the reduced

TABLE 4

Important distinguishing features of small-caliber muscle fibers in diagnostic muscle biopsies.

Type of small-caliber fibers	Cross-sectional outline	Oxidative enzyme content	Cytosolic esterase reaction	Sarcolemmal N-CAM immunostaining	Fiber type specificity	Pathophysiology	Examples of diseases in which they occur
1. Failure of natural growth	Polygonal	Normal	Normal	Negative	Possible (type 1 or type 2 smallness)	Varied	Congenital myopathies
2. Failure of regenerating fibers to attain normal size	Polygonal or round	Normal	Normal	Usually negative	None, but myopathic type grouping may be associated	Not enough satellite cells participated in the regeneration	Dystrophies
3. Small fibers due to denervation	Angular or bizarre-shaped (rounded or polygonal in infants)	Higher than normal	Increased	Positive	None, but type 2 fibers may predominate	Increased catabolism and reduced anabolism of myofibrillar proteins	Axonal neuropathies Motor neuron diseases
4. Small fibers due to disuse or cachexia	Angular	Normal	Normal	Negative	Type 2 fibers (2B > 2A)	Increased catabolism and reduced anabolism of myofibrillar proteins	Protein malnutrition Reduced mechanical activity, upper motor neuron involvement
5. Small fibers due to nuclear disintegration	Angular, polygonal or rounded	Normal	Normal	Negative	None	Readjustment to proper nuclear cytoplasmic ratio	Inclusion body myositis, infantile polymyositis
6. Miscellaneous small fibers	Varied	Normal	Normal	Negative	Type 2	Unknown	Myasthenia gravis

muscle mass. The number of capillaries per muscle fiber is reduced in longstanding severe denervation atrophy, but the number of capillaries related to cross-sectional fiber area is increased. Capillary attrition occurs by necrosis of capillaries, and in denervated muscle the empty basal laminae of vanished capillaries can often be found (Carpenter and Karpati 1982).

Small-caliber muscle fibers are resistant to necrosis as seen in extraocular muscles, denervation or disuse and pituitary insufficiency in Duchenne dystrophy, hamster dystrophy, and the mdx murine dystrophy (Karpati et al. 1988b). The basis for this is not known.

Hypertrophy

Hypertrophy indicates that the cross-sectional size of a muscle fiber is greater than the established norm for a given fiber type in a given muscle at a given age, sex and level of physical activity. In hypertrophied fibers, the number and size of myofibrils is greater than normal for reasons uncertain. Hypertrophied fibers usually lose their polygonal cross-sectional outline and become rounded.

Prominently hypertrophied fibers often show internally situated myonuclei and longitudinal splitting. Marked hypertrophy in chronic denervating conditions is not infrequently associated with an

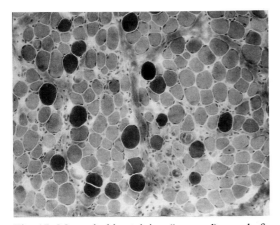

Fig. 17. Many darkly staining ('opaque') muscle fibers represent segments in which myofibrils have undergone extreme hypercontraction. These fibers show no genuine stigmata of necrosis. Duchenne dystrophy, modified trichrome, ×140. (Figure has been reduced for printing purposes.)

occasional necrotic or regenerating fiber. Hypertrophic fibers in pathological muscles may be seen in chronic partial denervation, in many forms of dystrophy, especially Duchenne dystrophy, and, contrary to common belief, also in polymyositis and inclusion body myositis. Prominent hypertrophic fibers are present in infantile and juvenile spinal muscular atrophies. In some situations (i.e. chronic partial denervation), type 2 fibers are more prone to hypertrophy than type 1. In juvenile spinal muscular atrophy, on the other hand, type 1 fibers are the only ones that hypertrophy.

True hypertrophic fibers should be distinguished from other forms of large-caliber fibers, such as those caused by hypercontraction, storage, and the early stage of necrosis when increased fluid content of the fiber transiently enlarges the caliber.

Hypercontraction

In hypercontracted segments of muscle fibers, the sarcomeres show excessive shortening. In extreme forms, the individual sarcomeres cannot be resolved even by electron microscopy. A hypercontracted segment may extend longitudinally over a distance of up to 2 mm. Adjacent to a hypercontracted segment, sarcomeres are usually overstretched, and they can tear. When this occurs, there are usually associated tears in the plasma membrane.

On cryostat sections, hypercontracted fibers are characterized by larger than normal girth, rounded cross-sectional outline, and featureless ground-glass-like cytoplasm which stains darker than normal with practically all methods ('opaque fibers') (Fig. 17). The myofibrillar tears that may be associated with the hypercontraction often appear as triangular gaps to which the term 'delta lesion' has been given (Mokri and Engel 1975). In some hypercontracted fibers precipitated calcium can be displayed.

On transverse resin sections, the intermyofibrillar network of hypercontracted fibers is not only visible but darker than usual. This contrasts with contracted segments of necrotic fibers, where only the punctate pattern of dark rounded mitochondria is seen between myofibrils. Longitudinal sections often display hypercontracted and stretched or ruptured segments together. In stretched seg-

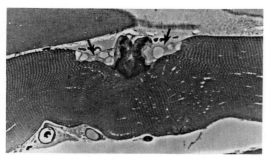

Fig. 18. Semi-thin resin section showing a muscle fiber which has undergone experimental micropuncture. The puncture tract entered the superior part of the fiber where it is surrounded by a collar of intense hypercontraction. This hypercontraction is usually absent in the depths of the fiber. On the lower side of the fiber, the tract is well outside the plane of section. Note that the hypercontraction has given rise to small tears in the fiber (arrows). ×1370. (Figure has been reduced for printing purposes.)

ments, the sarcomeres tend to be arched with their convexity pointing towards the hypercontracted segment.

The pathogenesis of hypercontracted fibers is controversial. Two points favor the view that they develop during biopsy as a result of trauma: first, they are commonly seen at the edges of normal biopsies; second, hypercontracted fibers do not show features of necrosis, which should be present if they had developed *in vivo*.

Hypercontraction may be seen at the edges of perfectly normal biopsies and occasionally within the sample, especially in children. The prevalence of hypercontraction is particularly great in Duchenne dystrophy, perhaps because lack of dystrophin makes the sarcolemma more fragile, so that they break. However, it is not clear whether the primary break occurs in a stretched or a hypercontracted segment. Bordering experimental micropuncture lesions, there is always focal hypercontraction (Fig. 18) with precipitated calcium (Carpenter and Karpati 1989). Here, the hypercontraction is probably related to the massive influx of calcium-rich extracellular fluid through the micropuncture hole. It is not clear whether or how often this mechanism is responsible for the hypercontracted muscle fibers in human biopsies. Experimentally, massive and extensive hypercontraction can be created by immersion of a rat limb in water

at 40°C for 1 hour (Personal unpublished observations).

Split fibers

On transverse view, split fibers are characterized by one or more fissures that extend from the periphery towards the interior. In the fissures, connective tissue and even capillaries may be found. Myonuclei and mitochondria may be present along the fissure within the muscle fiber. The pathogenesis of split fibers is controversial. Five possibilities have been suggested, all of which might conceivably operate in different situations:
1. actual cleavage in hypertrophic fibers;
2. activation of satellite cells in non-necrotic fibers which, by not fusing with their parent fiber, form separate fiber segments under the parent fiber's basal lamina;
3. reinnervation of denervated atrophic and bizarre shaped fibers;
4. forking of muscle fibers that may develop as an aberration of regeneration (see Regeneration);
5. autophagic vacuoles becoming connected with the extracellular space.

Split fibers occur with particular frequency in biopsies where hypertrophied fibers abound, as in chronic partial denervation or limb-girdle dystrophy. They are rare in Duchenne dystrophy.

Storage

Storage is characterized by excessive accumulation of one or more types of molecules in muscle fibers. Storage may be lysosomal or extralysosomal (Carpenter and Karpati 1986). In most instances, storage results from a deficiency of a catabolic enzyme, leading to excessive accumulation of the corresponding undegraded natural substrate. Occasionally, in addition to the specific storage material, non-specific storage substances may also accumulate (Table 5). Among storage diseases, acid maltase deficiency causes the severest neuromuscular impairment. We have seen several patients with microscopic evidence of prominent lysosomal storage disease of skeletal muscle, but they did not fit into any of the known categories and their enzyme defects remain unknown. The tentative diagnosis of lysosomal storage disease of muscle can be

TABLE 5

Features of well-defined storage diseases with storage in skeletal muscle cells.

Disease and subtypes	Deficient enzymes	Cells involved in muscle	Ultrastructural pattern of lysosomal contents in muscle cells
a. *Lysosomal*			
Acid maltase deficiency			
Infantile	α-1,4-glucosidase	Muscle cells, satellite cells, endothelium, pericytes	Glycogen, amorphous dense material
Childhood	α-1,4-glucosidase	Muscle cells	Glycogen, amorphous dense material
Adult	α-1,4-glucosidase	Muscle cells	Glycogen, amorphous dense material
Lysosomal glycogen storage with normal acid maltase activity	Unknown	Muscle cells	Glycogen, amorphous dense material
Batten's disease			
Infantile	Unknown	Muscle cells, endothelium, fibroblasts, satellite cells	Granular osmiophilic deposits (GROD)
Late infantile	Unknown	Muscle cells satellite cells, endothelium, fibroblasts	Curvilinear bodies
Juvenile	Unknown	Muscle cells, satellite cells, endothelium	Rectilinear bodies
Juvenile subgroup	Unknown	Muscle cells, endothelium, satellite cells	GROD
Adult (Kufs' disease)	Unknown	Occasional muscle cells	Rectilinear bodies
Fabry's disease type I	α-galactosidase	Muscle cells, endothelium, smooth muscle	Straight lamellae with periodicity of 5.5 nm
Mannosidosis	α-mannosidase	Muscle cells, fibroblasts, smooth muscle	Reticulofibrillar material or lucent space with sparse granules
Mucolipidosis IV	Ganglioside sialidase	Muscle cells, endothelial cells, pericytes	Concentric membranous bodies
b. *Non-lysosomal*			
Lafora's disease	Unknown	Muscle cells	Membrane-bound spaces with granulofilamentous material or linked glycogen particles
Polysaccharide storage	Unknown	Muscle cells, axons	Granulofilamentous
Phosphofructokinase deficiency (Tauri's disease)	Phosphofructokinase	Muscle cells	Glycogen particles, sometimes granulofilamentous material
Branching enzyme deficiency (Glycogenosis IV)	α-1,4-glucan: α-1,5-glucan: 6-glycosyl transferase	Muscle cells	Granulofilamentous
Debranching enzyme deficiency (Glycogenosis III)	1,6-glucosidase	Muscle cells	Glycogen particles
Myophosphorylase deficiency	Myophosphorylase	Muscle cells	Glycogen particles

made on cryostat sections by observing abundant acid phosphatase positive sites in muscle fibers. Structures such as lipofuscin, reactive lysosomes, and autophagic vacuoles must be differentiated from true storage.

As long as it is not highly soluble, the storage material may also be displayed cytochemically (i.e. glycogen), or by a characteristic electron microscopic appearance (i.e. glycogen or the various profiles in Batten's disease). In most instances, bio-

TABLE 6

Types of non-lysosomal cytoplasmic vacuolar spaces that may suggest storage with general tissue stains.

Nature of space	Disease association	Contents
Cytoplasmic space (lined vacuoles)	Inclusion body myositis	Membranous whorls, abnormal filaments
Cytoplasmic space	Non-lysosomal glycogen storage diseases	Glycogen
Cytoplasmic space	Adult dominant centronuclear myopathy	Nuclei, lipofuscin
Lipid globule	Incidental lipid excess and specific lipid storage myopathies	Lipid
Peroxisome	Lafora's disease	Polyglucosans
Various membrane-bound spaces	Periodic paralysis	Usually lucent space
Sarcoplasmic reticulum	Sarcotubular myopathy	Lucent space
Cytoplasmic space resulting from myofibrillar loss	Dermatomyositis	Glycogen, mitochondria
T-tubules	Miscellaneous, including surviving stumps of fiber with recent segmental necrosis	Lucent space

chemical assay of the deficient enzyme is required for precise diagnosis.

In some biopsies, muscle fibers show prominent vacuolar spaces that raise the question of true storage (Table 6). Sometimes vacuolization is due to storage of glycogen or lipid which has been dissolved away in fixation or processing. In other instances, however, extensive vacuolization is not a sign of storage, just as excessive accumulation of a molecule does not always indicate true storage. For instance, there may be glycogen lakes in non-specifically damaged muscle fibers, or an excessive amount of desmin in some myopathies.

Exocytosis

Exocytosis is a process by which a living cell releases substances into the extracellular space. This implies fusion of a membrane-bound space (containing the exocytosable material) with the surface membrane. In normal muscle cells, exocytosis is probably negligible. In pathological muscle fibers, exocytosis of whorls of cytomembranes or other cell debris has been seen in experimental myopathies induced by chloroquine, chlorphenterine, or ε-aminocaproic acid. In human muscle, we have seen exocytosis of dark bodies principally in polymyositis and occasionally in adult dermatomyositis (Fig. 19).

Exocytosis needs to be distinguished from mas-

sive exteriorization of non-membrane-bound cell debris, which we have seen in a 26-year-old patient with an unidentified myopathy.

Calcification

Calcium deposition in muscle fibers occurs when microscopically detectable insoluble calcium salts (usually calcium hydrophosphates) are deposited in muscle fibers. This presumably occurs when the normally micromolar range of cytosolic calcium concentration rises to the millimolar range. Insoluble calcium salts usually precipitate on hypercontracted myofibrils or in mitochondria or both. Precipitated calcium is detectable by light microscopy using complexing dyes, such as glyoxal-bis-2-hydroxyanyl (GBHA) or alizarin red. By electron microscopy, precipitated calcium can be amorphous or arranged in spicules of high electron density. By using sodium pyroantimonate, even non-precipitated calcium ions have been reportedly seen with the electron microscope. Calcification, defined as sufficient calcium deposition to form semipermanent masses that stain blue on hematoxylin and eosin, is rarely seen in human muscle, but is common in hamster dystrophy.

Calcium can be detected histochemically in early necrotic fibers of almost any cause. We have seen it with particular frequency in Duchenne dystrophy (Fig. 20), in hamster dystrophy, and in one

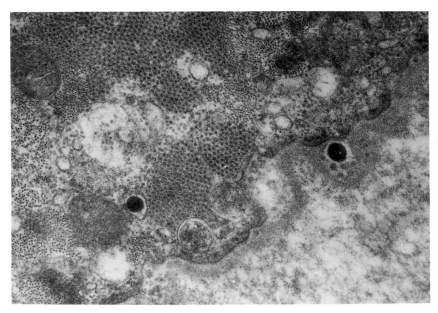

Fig. 19. Some rounded dense material is seen beneath the basal lamina of the muscle fiber but exterior to the plasma membrane. There is similar material within the fiber in a small membrane-bound space. This illustrates exocytosis in polymyositis. ×50,000.

case of acute alcoholic myopathy. Experimentally, prominent focal deposition of insoluble calcium salts on hypercontracted myofibers may follow experimental micropuncture (Fig. 21). In the 'experi-

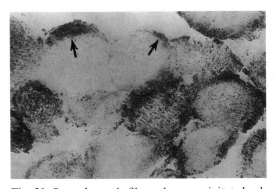

Fig. 20. Several muscle fibers show precipitated calcium, especially in their superficial portions. Most of these fibers are in the earliest phase of the necrotic cascade. In two fibers only discrete superficial crescentic regions of calcium deposition are present (arrows). These depositions may have resulted from influx of Ca^{2+} through focal breaches of the plasma membrane in this area. Duchenne dystrophy, glyoxalbis (2-hydroxyanil), ×350. (Figure has been reduced for printing purposes.)

mental calcium paradox', withdrawal of extracellular calcium damages the sarcolemmal voltage-regulated calcium channels; this is followed by unimpeded calcium influx into the muscle fibers when the normal concentration of extracellular Ca^{2+} is re-established (Soza et al. 1986).

Cellular repair

Cellular repair includes processes by which muscle fibers rectify a particular damage caused by injury. It must be distinguished from regeneration, which occurs when a necrotic muscle fiber segment is reconstituted through proliferation and fusion of satellite cells. We have observed examples of repair involving the sarcolemma, t-tubules, myofibrils, mitochondria and nuclei (Table 7).

ALTERATIONS OF COMPONENTS OR ORGANELLES OF SKELETAL MUSCLE CELLS

Cytoskeleton

Cytoskeletal components of muscle fibers (that are not associated with the sarcolemma) include inter-

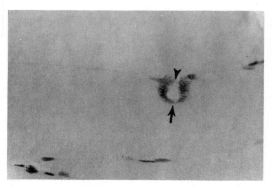

Fig. 21. Arrow indicates a U-shaped deposition of precipitated calcium salts presumably on hypercontracted myofibrils bordering a micropuncture cavity. The arrowhead indicates the presumed point and direction where the micropuncture needle pierced this rat gastrocnemius muscle fiber only a few minutes before removal of the specimen. Glyoxal-bis (2-hydroxyanil) with faint methylene blue counter stain for the demonstration of myonuclei, ×350.

mediate filaments, thin filaments, microfilaments, titin and nebulin (elastic filaments), and microtubules. The major constituent molecules of intermediate filaments are desmin and vimentin, of microfilaments actin, and of microtubules tubulin. Immunocytochemical techniques using antibodies against these molecules can be used for microscopic display (Bonilla et al. 1988a). Normally, intermediate filaments form a cage around individual myofibrils, although they are not visible in normal muscle fibers by routine electron microscopy. In early regenerating fibers they are often prominent and recognizable by their 10 nm size and random orientation. They may also be seen in some fibers in dermatomyositis where both injury and repair are going on. Microfilaments, which are 4.7 nm in diameter, may form a feltwork under the plasma membrane.

The only specific pathological reaction of cytoskeletal elements involves desmin accumulation (Fig. 22). The cases are clinically rather heterogeneous (Fardeau et al. 1978; Pellisier et al. 1989). Internalization of myonuclei in some forms of centronuclear myopathy (Sarnat 1990), or abnormal distribution of mitochondria in a variety of injured muscle fibers, and a disturbed organization of myofibrils in some cases have been attributed to a cytoskeletal abnormality, but this has not been proven.

Nebulin tends to be disproportionately depleted in Duchenne muscle fibers (Bonilla et al. 1988a).

Glycogen

Glycogen is an important source of metabolic fuel, especially in type 2 muscle fibers, and thus, its overall concentration can vary considerably in normal muscle. Normally, there is more glycogen in type 2 fibers than in type 1. Glycogen gives a magenta staining with PAS, which is eliminated by prior treatment of sections with α-amylase. Glycogen phosphorylase activity is usually co-extensive with the PAS (periodic acid Schiff) staining of glycogen. By electron microscopy glycogen appears as rounded granules of 25 nm diameter.

Pathological reactions of glycogen include (focal or diffuse) decrease or loss or (focal or diffuse) increase.

Focal decrease occurs in the centralmost portion of targets, targetoids and central cores, as well

TABLE 7

Repair of skeletal muscle fibers.

Surface membrane
  Formation of delimiting membrane between necrotic segment and surviving stumps (micropuncture)
  Possible sealing of breaches of surface membrane (micropuncture and DMD)
  Reduplication of basal lamina

T-tubules
  Restoration of normal T-tubular caliber from a dilated state in surviving stumps (micropuncture)

Myofibrils
  Re-establishment of myofibrils in the zonal lesions of tenotomized rat soleus fibers
  Reformation of myofibrils after their focal breakdown in dermatomyositis
  Restoration of A-band after thick filament depletion secondary to steroids plus denervation

Nuclei
  Extrusion of filamentous masses produced by experimental administration of 2,4-D
  Exocytosis of cell debris
  Reinnervation-induced repair of denervated muscle fibers

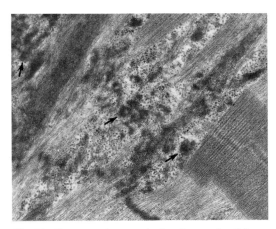

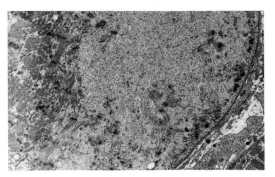

Fig. 23. Electron micrograph from a case of childhood dermatomyositis, showing an area filled with glycogen where myofibrils have been lost. ×10,000. (Figure has been reduced for printing purposes.)

Fig. 22. Electron micrograph showing cords of dense material (arrows) in the space between myofibrils. Similar deposits have been shown to be positive for antidesmin antibodies. There is also some Z-disc streaming within myofibrils. ×50,000. (Figure has been reduced for printing purposes.)

as in moth-eaten zones of moth-eaten fibers. Diffuse decrease or loss is seen in early necrosis, after denervation, and in fibers severely damaged from any cause.

Focal increase of glycogen may be seen in subsarcolemmal regions of injured fibers when glycogen appears to fill focal areas of myofibrillar loss (Fig. 23). In regenerated fibers, glycogen lakes are often present.

Focal or diffuse increase occurs in lysosomal or non-lysosomal glycogen storage diseases. Artifactual enlargement of extracellular spaces that gives a strong PAS staining must be differentiated from true increase of glycogen (Fig. 24). In polysaccharide storage disease, which affects skeletal and often cardiac muscle, a fibrillar polysaccharide, resembling amylopectin on electron microscopy, accumulates in muscle fibers (Fig. 25) (Thompson et al. 1988). In these fibers, typical glycogen is usually lacking. Fibrillar glycogen is also seen in type 4 glycogenesis (deficiency of branching enzyme) and in some cases of phosphofructokinase deficiency.

The Golgi system

The Golgi system consists of parallel stacks of cisternae located in the vicinity of nuclei. Its main function is the processing (especially glycosyla-

tion) and sorting of protein molecules delivered to the cis compartment of the Golgi structure from the endoplasmic reticulum. From the other side of the Golgi system (the trans compartment), coated vesicles arise in which specific membrane-associated processed molecules are delivered to the cell surface or to an intracellular destination such as a lysosome.

Golgi cisterns are scarce in normal skeletal muscle fibers. Thiamine pyrophosphatase activity, which is the most commonly used marker of the

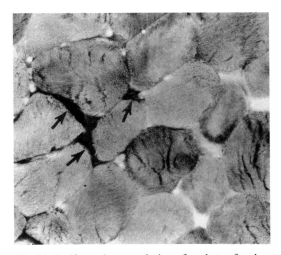

Fig. 24. Artifactual accumulation of pockets of carbohydrate-rich PAS positive extracellular fluid (arrows), which should not be confused with abnormal glycogen storage. Normal muscle, periodic acid Schiff, ×350. (Figure has been reduced for printing purposes.)

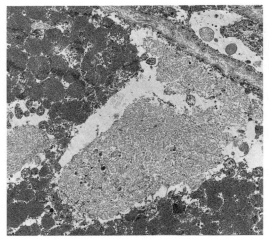

Fig. 25. Deposits of filamentous polysaccharide displace myofibrils in polysaccharide storage disease. ×10,000. (Figure has been reduced for printing purposes.)

Golgi system, is therefore inconspicuous in normal muscle fibers.

Lipid globules

Lipid globules mainly contain triglycerides. The activity of triglyceride lipase releases fatty acids, which are important metabolic fuels in myofibers.

The number and size of lipid globules vary widely in normal muscle fibers. In general, they are more numerous in type 1 and 2A fibers than in 2B fibers.

Excessively large and numerous lipid globules accumulate in type 1 and type 2A fibers in fasting guinea pig muscles and sometimes in ischemic human muscles. Even more pronounced neutral lipid storage occurs in both systemic and muscle forms of carnitine deficiency (Fig. 26) and in some forms of mitochondrial myopathy, particularly in the ragged red fibers (Karpati et al. 1991). Excess lipid does not usually accumulate in carnitine palmitoyl transferase (CPT) deficiency. The lipid globules associated with physiological lipofuscin should be differentiated from lipid storage.

The activity of triglyceride lipase can be displayed using endogenous triglyceride as substrate (Karpati et al. 1981).

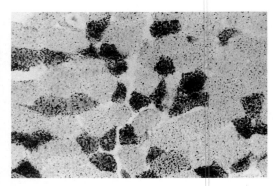

Fig. 26. There is massive increase of coarse lipid globules in type 1 and type 2A muscle fibers in a case of systemic carnitine deficiency, presumably due to a deficiency of long chain fatty acid dehydrogenase. Oilred O, ×350. (Figure has been reduced for printing purposes.)

Lysosomes

Lysosomes are membrane-bound spherical organelles of various size in which degradation of proteins, complex lipids and carbohydrates takes place. A large number of specific enzymes are located in the lysosomal membrane or matrix where they are active at an acid pH, which is maintained by a proton pump in the lysosomal membrane.

Histochemically or electron microscopically observable lysosomes are rare in normal skeletal muscle fibers. The best characterized pathological reactions of lysosomes occur in diseases associated with a genetically determined deficiency of a lysosomal enzyme. The types of lysosomal storage diseases that affect skeletal muscle fibers have been summarized under 'Storage'.

Pathological reactions of skeletal muscle lysosomes also occur in conditions other than storage diseases. In infantile myotonic dystrophy, small tubules with dense contents accumulate near nuclei. These react positively with lysosomal marker enzymes (Fig. 27). In polymyositis, some muscle fibers may have numerous small lysosomal bodies that contain membranous debris and glycogen. In an adult form of centronuclear myopathy, we have seen prominent lysosomes accumulating in a concentric pattern in the mid-portion of muscle fibers (Fig. 28). In Duchenne dystrophy, there is no evidence of activation of the lysosomes of muscle fi-

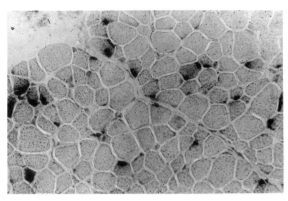

Fig. 27. Conspicuous acid phosphatase-positive sites at the periphery of numerous muscle fibers correspond to peculiar lysosomes ('dense core tubules') which accumulate in the vicinity of myonuclei. Infantile myotonic dystrophy, acid phosphatase, ×350. (Figure has been reduced for printing purposes.)

bers. Muscle fiber lysosomes should be differentiated from lysosomes of macrophages invading muscle fibers during necrosis or in partial invasion. Physiological lipofuscin accumulates in lysosomes and should also be distinguished from pathological storage (Fig. 29) (Karpati et al. 1987).

Autophagic vacuoles represent a special type of lysosome. They usually contain disintegrating remnants of cell components and whorls of cytomembranes. The membrane of autophagic vacuoles probably derives from the T-tubules. Auto-

phagic vacuoles are particularly abundant in chloroquine myopathy (Fig. 30), in one type of oculopharyngeal dystrophy, in some cases of periodic paralysis and, to some extent, in inclusion body myositis.

Mitochondria

The major role of mitochondria is to generate ATP by special oxidative processes. Fatty acids and pyruvate are the main substrates. The passage of electrons along four multiple subunit-containing complexes results in the release of free energy that is captured in the terminal phosphate bond of ATP. Mitochondria are more numerous in type 1

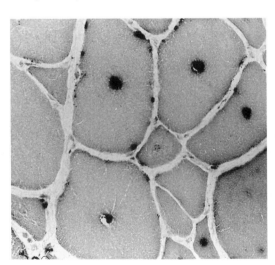

Fig. 28. Several muscle fibers show prominent central acid phosphatase-positive sites where clusters of myonuclei and abundant lysosome-bound lipofuscin is present. Adult centronuclear myopathy, acid phosphatase, ×350. (Figure has been reduced for printing purposes.)

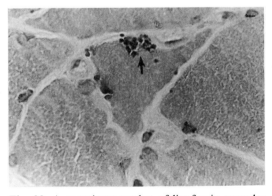

Fig. 29. A prominent pocket of lipofuscin granules (arrow) in an atrophic angulated (denervated) muscle fiber could mimic a lysosomal storage disease. Peripheral neuropathy, acid phosphatase with hematoxylin and eosin counterstain, ×520. (Figure has been reduced for printing purposes.)

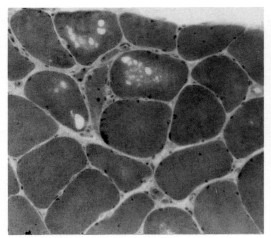

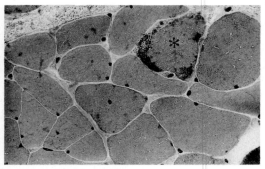

Fig. 31. Asterisk indicates a typical ragged red fiber. Late onset mitochondrial myopathy, modified trichrome stain, ×350. (Figure has been reduced for printing purposes.)

Fig. 30. Three muscle fibers at the top left show multiple irregular vacuoles. Chloroquine myopathy, hematoxylin and eosin, ×350. (Figure has been reduced for printing purposes.)

and type 2A fibers than in type 2B. Mitochondria normally occur in four general locations in muscle fibers: subsarcolemmal region, adjacent to the Z-discs, adjacent to nuclei and at the endplate region.

The major consequence of mitochondrial dysfunction is reduced generation of ATP, which can compromise energy-dependent cell functions. Mitochondrial dysfunction may result from either a primary mitochondrial defect or other disturbed cell metabolism. Primary mitochondrial defects

may result from abnormality in mitochondrial DNA, nuclear DNA, or both (DiMauro 1991).

Primary mitochondrial defects include: a. reduced availability of oxidative substrates within mitochondria (transport defects, i.e. carnitine deficiency or CPT deficiency); b. defects affecting enzymes of the fatty acid oxidation pathway or the Krebs cycle; c. defects affecting one or more complexes in the electron transfer chain; or d. uncoupling of oxidation from phosphorylation. Some of these conditions induce reactive or adaptive changes in mitochondria, such as an increase in size or number, that can be recognized microscopically. The excessive mitochondrial mass may be

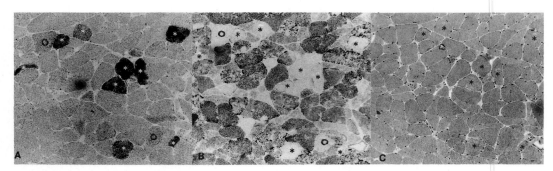

Fig. 32. On serial sections several examples of abnormal fibers are marked by asterisks. Late onset mitochondrial myopathy, ×140. (A) 'Ragged red equivalent' fibers show markedly augmented activity of phenazine methosulphate-linked succinic dehydrogenase (PMS-SH). (B) The same fibers lack cytochrome oxidase activity; however, some cytochrome oxidase-negative fibers (examples marked by O) do not appear grossly abnormal in 32A. (C) None of the marked fibers appear 'ragged red' on the modified trichrome stain, indicating that the modified trichrome is a relatively insensitive method for the study of muscle biopsies in suspected mitochondrial diseases. The discrepancy of the mentioned staining characteristics in the three panels is not due to the segmental nature of involvement. (Figure has been reduced for printing purposes.)

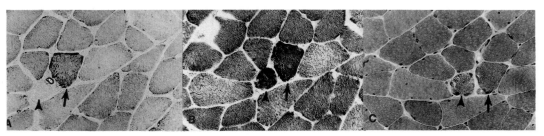

Fig. 33. Serial sections indicate discordant staining characteristics of ragged red fibers or equivalents. The fiber marked by an arrow has high cytochrome oxidase activity (A) and high PMS-SH activity (B), but does not appear 'ragged red' on the modified trichrome (C). On the other hand, the fiber marked by an arrowhead lacks cytochrome oxidase activity (A), but has abnormally high and irregular staining with PMS-SH (B) and appears 'ragged red' on the modified trichrome (C). Late onset mitochondrial myopathy, ×350. (Figure has been reduced for printing purposes.)

enough to give a 'ragged' appearance to the muscle fiber when viewed with light microscopy. On cryostat sections stained with the modified trichrome stain, the mitochondria stain red. Fibers with this ragged appearance have been termed 'ragged red fibers' (Fig. 31). Their presence usually implies a form of mitochondrial dysfunction that has led to a substantial increase of mitochondrial mass.

There are two other useful reactions in assessment of mitochondrial disease. The phenazine methosulphate-linked succinic dehydrogenase reaction (PMS-SH) makes ragged red fibers and equivalents stand out (Fig. 32). The PMS-SH can indicate abnormal mitochondrial excess in a muscle fiber even when the modified trichrome does not show a ragged red appearance. These fibers can be called 'ragged red equivalents'. Cytochrome oxidase is lacking in most, but not all, ragged red fibers (Fig. 33).

Microscopic evidence of mitochondrial abnormality includes focal or diffuse increase or decrease of number, malorientation, increase in size and architectural abnormalities (Table 8). Some of these changes can be appreciated by light microscopy, such as changes in number or distribution; others, such as structural abnormalities and enlargement, are only seen by electron microscopy.

Genuine abnormalities of mitochondria must be distinguished from artifactitious changes induced by poor fixation, in particular swelling and pallor of the matrix.

Prominent microscopic abnormalities of mitochondria are most often seen in mitochondrial myopathies. These diseases have been classified on

the basis of clinical or biochemical features (Morgan-Hughes 1986; DiMauro 1991; DiMauro et al. 1991; Karpati et al. 1991). Precise classification based on etiology and pathogenesis is not yet possible. In some of these diseases, in addition to skeletal muscles, other tissues such as myocardium, kidneys or brain may be affected. In the latter group, also called *mitochondrial encephalomyopathies*, four clinically distinct entities have been identified: Kearns-Sayre syndrome (KSS),

TABLE 8

Microscopic features of mitochondrial disorders in skeletal muscle.

Light microscopy:	• Segmental excess of mitochondria ('ragged red fibers' or RRF) demonstrable by modified trichrome stain or the phenazine-methosulphate-linked succinic dehydrogenase reaction
	• In RRF, there may be excess of lipid globules and often there is an absence of cytochrome oxidase (COX) activity
	• Absent COX activity in some non-RRF
	• Immunocytochemical demonstration of subunits of respiratory complexes by specific antibodies
	• Demonstration of excess mtRNA by appropriate DNA probes with in-situ hybridization
Electron microscopy:	• Abnormal shape and size of mitochondria
	• Intracristal crystalline inclusions
	• Matrix inclusions
	• Abnormal configuration of cristae

MELAS (mitochondrial encephalopathy, lactic acidosis, strokes), MERRF (myoclonus epilepsy with ragged red fibers), and Leigh's disease. Other forms of mitochondrial encephalomyopathy do not seem to fall into these categories or seem to be overlap cases (DiMauro et al. 1991).

In some mitochondrial diseases, there are characteristic alterations of the mitochondrial DNA (mtDNA) (Sato 1991), including deletions (in KSS and other external ophthalmoplegic syndromes) or point mutations (in MELAS and MERRF). In MELAS and MERRF, there is maternal transmission of the disorder. Muscle fiber segments that contain a large number of mutant mtDNA copies are liable to become ragged red (or equivalent). By *in situ* hybridization, using appropriate DNA probes that recognize only mutant cDNA or mRNA copies, a quantitative assessment of the mutant cDNA copy number can be made on microscopic sections (Fig. 34). In the presence of deleted mtDNA copies in a mitochondrion, the ability of the wild-type mtDNA copies to generate molecules, such as complex IV subunits, is compromised (Shoubridge et al. 1990). For this reason, most ragged red fibers or equivalents in KSS do not stain for cytochrome oxidase (Fig. 35).

The basis of the differential vulnerability of different cell types in mitochondrial diseases is not clearly understood. The higher the oxidative metabolism in a cell type, the more vulnerable it is to mitochondrial dysfunction. Subunits that make up the various complexes in the electron transfer chains may be tissue-specific, so that a defect affecting a particular subunit might lead to mitochondrial dysfunction only in the cells that share that particular species of subunit (DiMauro et al. 1991). In mtDNA mutations, ageing and other non-specific influences tend to give rise to a gradual increase and expansion of the mutated mtDNA copies, which may explain a progression of the clinical and pathological phenotypes.

Myofibrils

The major components of myofibrils are thick filaments (located in the A band), thin filaments (extending from the I band into the A band) and Z-discs. Thick filaments contain myosin heavy chain and myosin light chains; associated with them are

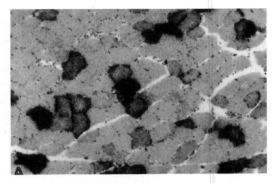

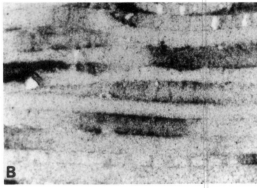

Fig. 34. (A) Muscle fibers with high grain count indicate an overexpression of a mitochondrial mRNA (related to an ND_2 mitochondrial gene) due to an excess of mitochondria in such fibers. Only some of these fibers were 'ragged red' on serial section, indicating that this in-situ hybridization technique with an ND_2 cDNA probe is much more sensitive for the identification of pathological fibers than the modified trichrome stain. Mitochondrial myopathy with external ophthalmoparesis, ×140. (B) The mitochondrial excess is segmental as shown by the same technique as used in A, ×140. (Figure has been reduced for printing purposes.)

M protein, C protein and myomesin. Thin filaments are composed of actin, tropomyosin and troponin. The Z-discs contain amorphous and filamentous components; chemically they are rich in actin and α-actinin. Titin and nebulin also appear to be within myofibrils, but their precise localization has not been settled (Horowits et al. 1986).

Myosin heavy chain occurs in multiple molecular forms that can be displayed by immunocytochemistry using specific antibodies (Sawchak et al. 1985, 1989) (see Fiber Types). The ATPase activity of myosin heavy chain, which is differentially sensitive to pH in different fiber types, can be dis-

TABLE 9

Pathological abnormalities of myofibrils.

1. *Abnormal size or shape*
   Smallness — In many types of disease ranging from denervation to dermatomyositis
   Irregular shape — In regenerating fibers

2. *Abnormal orientation*
   a. Rings — In myotonic dystrophy, some cases of limb-girdle dystrophy, some obscure cases of myopathy, some patients with rheumatoid arthritis
   b. Random disorientation — Especially in denervation

3. *Selective loss of a myofilamentous component*
   a. Thick filament loss — In dermatomyositis, in patients paralyzed with blocking agents (for assisted respiration) and given high dose steroids, in some patients with HIV infection, in rare fibers with cytochrome oxidase deficiency
   b. I band loss — In rare biopsies with denervation
   c. Z-disc loss — Typical of early necrosis when not obscured by hypercontraction

4. *Disorganization of myofilaments*
   a. Z-disc streaming — In target fibers of denervation, in dermatomyositis, in some central cores, in multicore disease, in minor degree in many other conditions
   b. Cytoplasmic bodies — Generally non-specific sign of damage. Some cases are called cytoplasmic body myopathy because of the frequency of the lesion
   c. Nemaline bodies — In nemaline myopathy. Small nemaline bodies may occur when there is selective myosin loss. Some nemaline bodies can be present in normal myotendinous junctions
   d. Spheroid bodies — Reported in one familial myopathy. Probably a form of cytoplasmic body
   e. Sarcoplasmic masses — In myotonic dystrophy
   f. Caps — In a case of congenital myopathy
   g. Miscellaneous disorganization — Non-specific. Must be distinguished from contraction artifacts

5. *Myofibrillar loss* — May be proportional to fiber shrinkage, as usually in denervation, or may precede shrinkage as in dermatomyositis

played cytochemically. This reaction is still the most convenient for identifying the histochemical type of a normal or pathological fiber (Brooke and Kaiser 1970).

Pathological abnormalities may affect the myofibrils as a whole or only certain elements (Table 9). The pathogenesis of most myofibrillar change is poorly understood.

In childhood nemaline myopathy, we observed evolution of microscopical changes in a patient who had muscle biopsies at the ages of 6 months and 13 years. At 6 months almost all fibers contained rods. At 13 years there was a remarkable failure of growth of a subpopulation of type 1 muscle fibers, some of which contained a few rods (Fig. 36A). The other type 1 fibers and the few type 2 fibers had apparently grown normally or had become hypertrophied. By lectin cytochemistry, impairment of binding of two lectins was observed in

all muscle fibers. Therefore, in childhood nemaline disease, some muscle fibers may fail to respond to a putative growth factor whose receptors could be located in the surface membrane. When nemaline myopathy is symptomatic in the neonatal period, many fibers may have extensive areas that lack thick filaments (Fig. 36b,c). This variant of nemaline myopathy has a bad prognosis (Schmalbruch et al. 1987).

We have observed abundant spheroid bodies representing the major pathological change in a muscle biopsy of a 68-year-old woman who had a 10-year history of bilateral foot drop and a distal myopathy.

A peculiar form of selective thick filament depletion follows denervation in patients who also receive large doses of glucocorticoids (Fig. 37). This reversible process has been observed, for example, in patients with status asthmaticus, who are iatro-

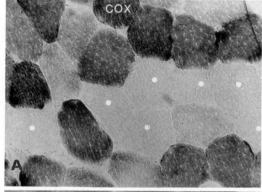

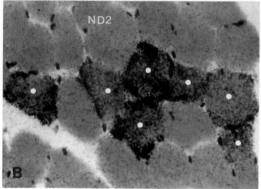

Fig. 35. (A) Dots indicate muscle fibers lacking cyto-chrome oxidase activity, ×350. (B) The corresponding fibers show heavy expression of ND$_2$ mRNA, imply-ing a marked excess of mitochondria (see Fig. 34), ×350. Both panels represent a mitochondrial myopa-thy with external ophthalmoparesis. (Figure has been reduced for printing purposes.)

genically paralyzed for a few days for mechanical ventilation while receiving high doses of glucocor-ticoids (Danon and Carpenter 1991).

Motor endplates

Motor endplates are relatively rare in routine mus-cle biopsies, unless special efforts are made to ob-tain specimens from the region of the motor point.

Pathological reactions of the neuromuscular junction may involve the terminal nerves or the motor endplates proper. Most of these changes can only be observed by electron microscopy.

Pathologic changes of the motor-nerve terminal include: terminal or ultraterminal sprouting (ana-tomical or functional denervation); absence of

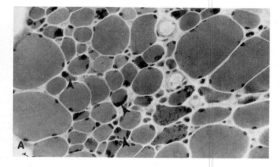

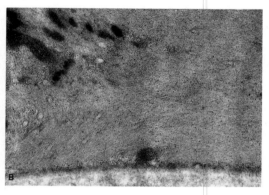

Fig. 36. (A) Only the small-diameter fibers contain nemeline rods (three examples marked by arrow-heads) in this biceps muscle biopsy of a 13-year-old patient. In his biceps biopsy at age 6 moths, practi-cally all fibers contained nemaline rods. This suggests that failure of fiber growth and nemaline rod forma-tion are linked. Modified trichrome, ×350. (B) Neona-tal nemaline myopathy. Some fibers show extensive areas without cross striations. These areas are filled with thin filaments, ×1370. (C) Electron micrograph from a case of neonatal nemaline myopathy showing a fiber in which large portions of cytoplasm are filled with thin myofilaments in the absence of Z-discs or thick filaments. ×25,000. (Figures have been reduced for printing purposes.)

nerve terminals (amyotrophic lateral sclerosis, spi-nal muscular atrophy); small caliber of nerve ter-

Fig. 37. The myofibrils have lost almost all their thick filaments, although the Z-discs and thin filaments are rather well preserved. The patient was given a neuromuscular blocking agent for several days during mechanical respiration and was treated with large doses of intravenous corticosteroids. ×25,000. (Figure has been reduced for printing purposes.)

minals associated with increased density of synaptic vesicles (a form of congenital myasthenic syndrome); dystrophic changes (neuro-axonal dystrophy); Wallerian degeneration; and non-specific changes, such as focal variation of caliber in various nerve and muscle diseases.

The most commonly observed changes of endplates occur in myasthenia gravis. By light or electron microscopic-specific affinity cytochemistry, a depletion of acetylcholine receptors can be demonstrated by impaired binding of labeled α-bungarotoxin. Binding of C9 (lytic) component of comple-

ment and of IgG (anti-AchR antibodies) can also be demonstrated by immunocytochemistry. By electron microscopy, shallow primary synaptic clefts and widened and shortened synaptic folds can be seen. In the Lambert-Eaton syndrome, there may be marked proliferation of secondary synaptic clefts, associated with abnormalities of the nerve terminals.

Myonuclei

Myonuclei in normal mature muscle fibers are situated beneath the surface membrane, hence the term sarcolemmal nuclei. How myonuclei are maintained in this position is not known; anchoring cytoplasmic elements may play a role. Longitudinal movement of myonuclei along the fibers probably does not occur. Myonuclei that occur in the motor endplate region tend to be larger and contain coarser chromatin than the extrajunctional myonuclei. Myonuclei should be distinguished from nuclei of myosatellite cells, but this can only be reliably achieved on plastic embedded semi-thin sections or by electron microscopy or with the help of anti-N-CAM antibodies, which outline the cytoplasm of the satellite cell (Fig. 38).

Pathological changes of nuclei include abnormal number, internalized location, increase of size, abnormal shape, inclusions, and disintegration. The number of myonuclei in relation to the cytoplasmic volume in normal fibers is a set value for a given age, greater in children than in adults. When fibers

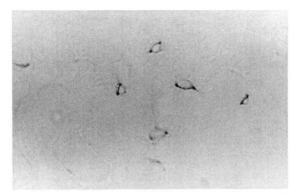

Fig. 38. The cytoplasm of several myosatellite cells in different muscle fibers is outlined by neural cell adhesion molecule (N-CAM) situated in the plasmalemma of these cells. N-CAM is not present at the surface of muscle fibers. Normal child's muscle, immunoperoxidase using a monoclonal antibody to N-CAM and biotin-streptavidin peroxidase display system, ×520. (Figure has been reduced for printing purposes.)

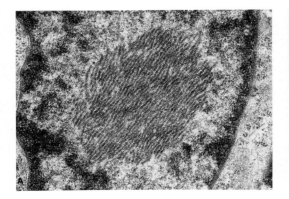

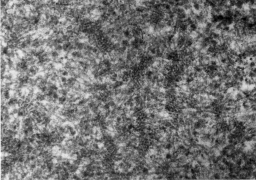

Fig. 40. In oculopharyngeal dystrophy a small percentage of the myonuclei contain inclusions made of fine tubules. ×100,000. (Figure has been reduced for printing purposes.)

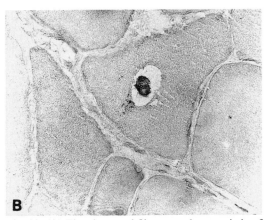

Fig. 39. (A) The abnormal filaments characteristic of inclusion body myositis (IBM) are seen here within the nucleus of a muscle cell. ×50,000. (B) A cytoplasmic filamentous mass in a vacuole is ubiquitinated in IBM. Immunoperoxidase using a monoclonal antibody to ubiquitin and biotin-streptavidin peroxidase display system, ×350. (Figures have been reduced for printing purposes.)

atrophy, significant myonuclear attrition often does not occur; in atrophic fibers, particularly after denervation, the relative number of myonuclei to the cytoplasmic volume is increased. In an extreme form, the entire muscle fiber may be converted into a bag of myonuclei (*nuclear clumps*).

The commonest nuclear abnormality is central displacement, a non-specific finding that may be seen in either myopathies or chronic neuropathies, especially in hypertrophied fibers. A peculiar plethora of internal nuclei occurs in myotonic dystrophy. In a group of diseases called *centronuclear myopathies*, the key abnormality is a chain of centrally situated nuclei. The different forms of cen-

tronuclear myopathy include an X-linked recessive infantile variety, an autosomal recessive or dominant juvenile variety, and a usually dominant adult form. In some childhood forms of centronuclear myopathies, the change is restricted to type 1 fibers which are small. In infantile centronuclear myopathy, an abnormal amount of persistent vimentin and desmin has been found in muscle fibers (Sarnat 1990).

In regenerated fibers of rodents, myonuclei remain permanently centrally situated. In humans, the central nucleation of regenerating fibers is usually temporary. Regenerating fibers, however, show large vesicular nuclei with prominent nucleoli, reflecting active gene transcription.

There are several kinds of *nuclear inclusions*. True nuclear inclusions must be differentiated from intranuclear invagination of the cytoplasm. In inclusion body myositis, abnormal tubular filamentous masses of 15–18 nm diameter can be observed in rare myonuclei (Fig. 39a). These filamentous masses contain ubiquitin (Fig. 39b) (Askanas et al. 1991). Formation of these filaments tends to lead to nuclear disintegration with release of the filaments into the cytoplasm. Familial oculopharyngeal muscular dystrophy, particularly of the French Canadian variety, is associated with a different kind of characteristic intranuclear inclusion in a very small percentage of nuclei (Fig. 40). In a peculiar form of inflammatory myopathy of infants, numerous non-specific nuclear inclusions,

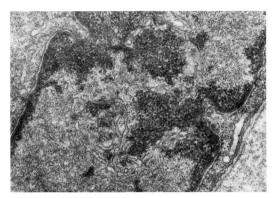

Fig. 41. This myonucleus contains numerous membranous profiles, one of the non-specific nuclear changes seen in infantile polymyositis. ×50,000. (Figure has been reduced for printing purposes.)

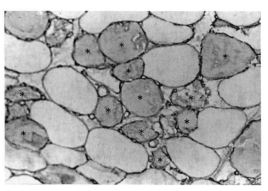

Fig. 43. There is strong immunostaining of the surface of all muscle fibers, even the necrotic ones (asterisks), for collagen IV. The necrotic fibers are recognized by diffuse immunostaining as a result of non-specific binding of the primary antibody (an IgG) to activated C9 complement (see Fig. 3). Duchenne dystrophy, monoclonal antibody to collagen IV plus immunoperoxidase using biotin-streptavidin peroxidase display system, ×350.

including mitotubules, membranes, and actin-like filaments, are present (Fig. 41).

Sarcolemma

The basal lamina. The basal lamina is apposed to the external surface of the plasma membrane of muscle cells (skeletal, smooth, and cardiac), as well as to Schwann cells, adipocytes, between epithelia and underlying connective tissue, and around vessels. The basal lamina contains heparan sulfate proteoglycans and several proteins, including laminin, entactin (nidogen), type IV collagen, and (in muscle and Schwann cells) merosin (Yurchenco

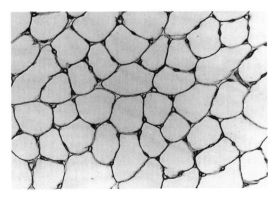

Fig. 42. Strong immunostaining of the muscle fiber surface membranes and endomysial capillaries for laminin is present in a normal adult muscle. Immunoperoxidase using a monoclonal antibody to laminin and biotin-streptavidin peroxidase display system, ×350.

and Schittny 1990). Extramembranous domains of glycoproteins embedded in the outer leaflet of the plasma membrane project into the basal lamina.

For light microscopy, the basal lamina is best displayed by antibodies to one of its proteins, such as laminin (Fig. 42) or collagen IV (Fig. 43), although the complex carbohydrates also stain with PAS. By electron microscopy a lucent band of 30 nm (lamina lucida) separates the fuzzy, moderately osmiophillic band (lamina densa) from the underlying plasma membrane. The basal lamina may partly serve as a mechanical support for the plasma membrane, but individual proteins in it have additional functions.

The basal lamina remains structurally relatively intact when a muscle fiber undergoes necrosis, although the chemical composition is altered (Gulati et al. 1983). It probably serves as an adherent platform to orient migrating and proliferating myoblasts during regeneration. A regenerated fiber segment often fills out the space within the old basal lamina, but it generates a stretch of new basal lamina wherever it does not touch the old one. A sign of failure of regeneration is the presence of long basal lamina skeins with no regenerative cell within them (Fig. 44). These must be distinguished from the empty basal lamina circles that can be produced by hypercontraction and fiber tearing

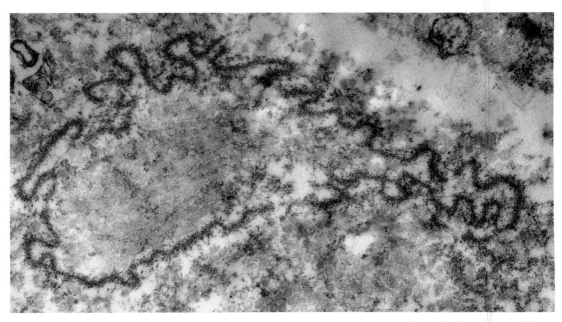

Fig. 44. Electron micrograph showing a basal lamina skein which contains collagen fibrils but no cellular component. This marks the site where a muscle fiber has failed to regenerate. ×42,000.

(Fig. 45). In such artifacts, a small amount of membranous material is usually left behind within the basal lamina.

Normally the basal lamina follows the contours of the plasma membrane closely. A muscle fiber atrophied as a result of denervation usually displays 2 or 3 empty sleeves of basal lamina protruding from the surface, but continuity is retained with the basal lamina over the rest of the fiber. In contrast, in Duchenne dystrophy, detached seg-

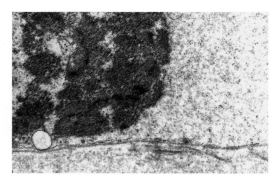

Fig. 45. Artifactitious hypercontraction of myofibrils has resulted in an empty basal lamina tube in which there are a few small residual fragments of plasma membrane and vesicles. Tay-Sachs disease, ×25,000. (Figure has been reduced for printing purposes.)

ments of basal lamina are often seen around normal-appearing fibers. Two, three or even four layers may be found in places (Fig. 46). It is not clear whether this is a consequence of episodes of necrosis and regeneration or results from a failure of anchoring of the basal lamina to the plasma membrane. Occasional short stretches of plasma membrane are sometimes seen to be denuded of basal lamina. Heaped up or reduplicated basal lamina is also seen in most cases of polymyositis, where it sometimes surrounds small, dense spheres of material probably derived from the surface of muscle fibers (Fig. 47). Some reduplication of basal lamina is also seen around damaged fibers in myotonic dystrophy.

The plasma membrane. The plasma membrane is a lipid bilayer with numerous proteins, some of which traverse the bilayer (ion channels in particular), while others are embedded in either the inner or outer leaflet (see Table 1). Plasma membrane itself is too thin a structure to be visualized by light microscopy separately from the adherent basal lamina, although some proteins in the plasma membrane can be displayed by immunocytochemistry.

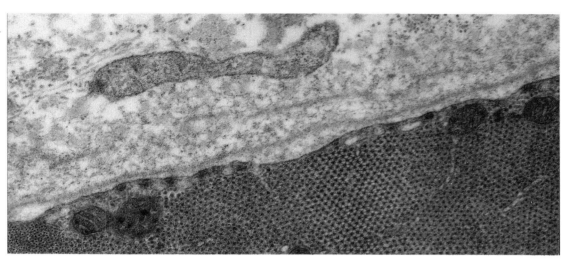

Fig. 46. The basal lamina of this muscle fiber is lifted away from the plasma membrane and a second layer of detached basal lamina is seen beyond that. Detached segments of basal lamina like this are characteristically present around numerous fibers in biopsies from Duchenne dystrophy. ×50,000.

Type 1 major histocompatibility (MHC) antigens are normally expressed on the surface of the plasma membrane of myoblasts, myotubes, and regenerating muscle fibers but not in normal muscle fibers (Fig. 48) (Karpati et al. 1988a). In inflammatory myopathies, MHC type 1 expression may be present on the surface of many fibers (Fig. 49). This can be diagnostically useful. In polymyositis the surfaces of virtually all fibers tend to be positive, while in dermatomyositis positivity is limited mainly to areas of fiber damage. In polymyositis

and inclusion body myositis, fibers that are not necrotic but are partially invaded by inflammatory cells show unusually strong positivity of MHC type 1 on their surface and even some in their interior (Fig. 50). In inclusion body myositis, positivity is virtually limited to these fibers.

Neural cell adhesion molecule (N-CAM) is another protein that is expressed on the surface of denervated, regenerating or damaged muscle fi-

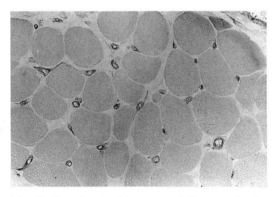

Fig. 48. Class 1 major histocompatibility complex determinants (MHC, non-variable region) are not expressed at the surface of normal-appearing muscle fibers but expressed in endothelial cells of endomysial capillaries. Duchenne dystrophy, immunoperoxidase using a monoclonal antibody to class 1 MHC plus biotin-streptavidin peroxidase display system, ×350.

(Figure has been reduced for printing purposes.)

Fig. 47. Electron micrograph from a case of polymyositis showing a dense round profile surrounded by basal lamina near the fiber's surface. ×100,000. (Figure has been reduced for printing purposes.)

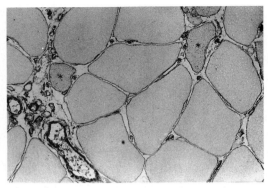

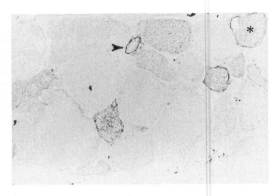

Fig. 49. All otherwise normal-appearing muscle fibers (as well as blood vessels and inflammatory cells) express class 1 MHC in their surface membrane. Two regenerating fibers (asterisks) show particularly strong surface, as well as diffuse cytoplasmic immunostaining. None of the fibers showed class 2 MHC expression while blood vessels and inflammatory cells did. Polymyositis, immunoperoxidase using a monoclonal antibody to class 1 MHC plus biotin-streptavidin peroxidase display system, ×350. (Figure has been reduced for printing purposes.)

Fig. 51. Only regenerating muscle fibers express strong sarcolemmal immunoreactivity for N-CAM. Young regenerating fibers (arrowhead) show the strongest reactivity while presumably fully regenerated fibers (example marked by asterisk) show much less reactivity. Polymyositis, immunoperoxidase using a primary monoclonal antibody to N-CAM and biotin-streptavidin peroxidase display system, ×350. (Figure has been reduced for printing purposes.)

bers, but not on normal mature fibers (Figs. 51 and 52) (Cashman et al. 1987). Satellite cells are beautifully outlined by anti-N-CAM antibodies (Fig. 52).

Acetylcholine receptors, normally limited to motor endplates, spread out over the plasma membrane in recently denervated muscle fibers. They

can be demonstrated by the binding of labeled α-bungarotoxin (Fig. 53). An intraplasmalemmal glycoprotein complex, which serves as an anchor of dystrophin to the plasma membrane, can be visualized by specific antibodies (Ervasti and

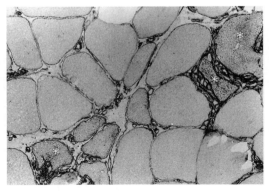

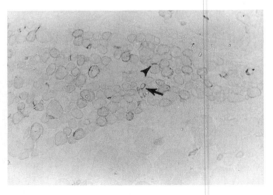

Fig. 50. Three muscle fibers partially invaded by CD8+ lymphocytes and macrophages (asterisks) show particularly strong surface and cytoplasmic expression of class 1 MHC. Polymyositis, immunoperoxidase using a monoclonal antibody to class 1 MHC and biotin-streptavidin peroxidase display system, ×350. (Figure has been reduced for printing purposes.)

Fig. 52. Approximately 60% of the muscle fibers show variably strong circumferential sarcolemmal immunoreactivity to N-CAM. In addition, stronger immunoreactivity marks the outline of satellite cells (arrow) and motor endplates (arrowhead). In a neighboring fascicle below, made up mainly of hypertrophied fibers, sarcolemmal N-CAM reactivity is completely absent. Infantile spinal muscular atrophy, immunoperoxidase technique as described in Fig. 51, ×350. (Figure has been reduced for printing purposes.)

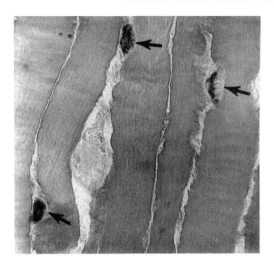

Fig. 53. Side view of 3 motor endplates on 3 different normal muscle fibers. The endplates are visualized by horseradish-peroxidase-labeled α-bungarotoxin bound to acetylcholine receptors (arrows). In the bottom endplate the localization of the label to the crowns of the junctional folds is evident. Graham-Karnovsky stain, ×350.

Campbell 1991); it is reduced in amount in dystrophin deficiency states (Ervasti et al. 1990)

The submembranous cytoskeleton. Awareness of a cytoskeletal system in intimate relationship to the cytoplasmic aspect of the plasmalemma has been roused by the discovery of dystrophin and its localization to the immediate submembranous area (Arahata et al. 1988; Bonilla et al. 1988b; Zubrzycka-Gaarn et al. 1988). By light microscopic immunocytochemistry the staining of dystrophin is indistinguishable from plasma membrane staining (Fig. 54). The ultrastructural localization, using colloidal gold electron immunocytochemistry (Cullen et al. 1990), shows the modal distance of a probe for the rod portion of the molecule to be 15 nm beneath the plasmalemma. α-Actinin and spectrin are also found in this region. The amino acid sequence of the first domain of the dystrophin molecule suggests that it can bind to actin, whereas its carboxy terminus probably binds to a plasmalemmal glycoprotein complex (Ervasti and Campbell 1991).

By light microscopic immunocytochemistry, using peroxidase or a fluochrome as a marker, the dystrophin co-localizes with the plasma membrane

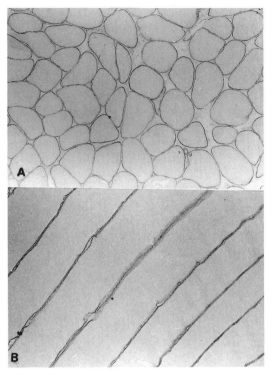

Fig. 54. Immunoreactive dystrophin appears as a thin continuous line at the surface of normal human muscle fibers. There is no staining difference between fiber types. A: transverse view. B: longitudinal view. Immunoperoxidase using a primary polyclonal antibody to the C-terminus and biotin-streptavidin peroxidase display system, ×350.

of normal skeletal muscle fibers. Some staining of the membrane of vascular smooth muscle can also be seen. Any other staining indicates an error of technique or an antibody cross-reacting with other molecules. True cross-reactivity can occur between dystrophin and an analogous protein, called dystrophin-related protein (DRP), whose gene is autosomal (chromosome 6), and which may be limited to the motor endplate region in normal muscle fibers (Ohlendieck et al. 1991). DRP, however seems to have diffuse subsarcolemmal distribution in Duchenne muscle fibers (Fig. 55). This staining may be responsible for the 'residual' sarcolemmal immunostaining of Duchenne muscle fibers that is seen with antidystrophin antibodies cross-reacting with DRP.

Dystrophin staining is generally absent in biopsies from Duchenne dystrophy, but some biopsies

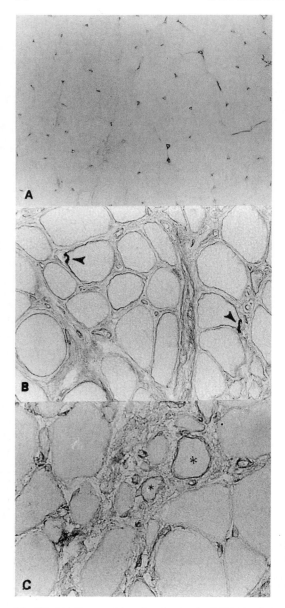

A

B

C

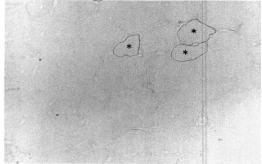

Fig. 56. There is complete absence of dystrophin of the surface of the vast majority of muscle fibers in Duchenne dystrophy. The scattered dystrophin-positive muscle fibers (asterisks) are so-called 'revertants'. Immunoperoxidase using a polyclonal antibody to the C-terminus of dystrophin, ×350. (Figure has been reduced for printing purposes.)

show isolated or small groups of fibers with normal dystrophin positivity (Fig. 56). In exceptional biopsies, there may be up to 100 in a group. In some biopsies, by chance, up to 30% of the fibers may show this phenomenon (Burrow et al. 1991). It is thought that, in some undetermined way, they have resulted from a further somatic mutation in some muscle progenitor cells, which has reversed the effects of the basic germline mutation (Burrow et al. 1991). In heterozygote carriers, a mosaic of dystrophin-positive and negative fibers may be present (Fig. 57).

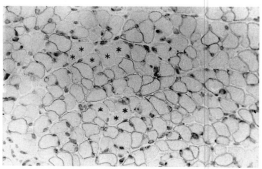

Fig. 55. (A) Dystrophin-related protein (DRP) is absent in normal human muscle fibers except at the endplate region (not shown). Endomysial capillaries, however, show staining. (B) In Duchenne dystrophy, all muscle fibers show a variable degree of circumferential sarcolemmal staining, which appears to be strongest in some regenerating fibers (example marked by asterisk). Arrowheads show very strong staining at motor endplates. (C) In dermatomyositis, some regenerating fibers show strong circumferential surface staining (asterisks). Other muscle fibers do not show staining, but blood vessels do. All three panels represent immunoperoxidase using a monoclonal antibody to the midportion of the DRP molecule and biotin-streptavidin peroxidase display system, ×350. (Antibody is kind gift of Dr. G. Morris.)

Fig. 57. In some heterozygote carriers of Duchenne dystrophy, groups of dystrophin-negative fibers are intermingled with dystrophin-positive ones ('mosaicism'). Two groups of dystrophin-negative fibers are marked with asterisks. Immunoperoxidase using a polyclonal antibody to the C-terminus of dystrophin and biotin- streptavidin peroxidase display system, plus a weak hematoxylin counterstain to display nuclei, ×140. (Figure has been reduced for printing purposes.)

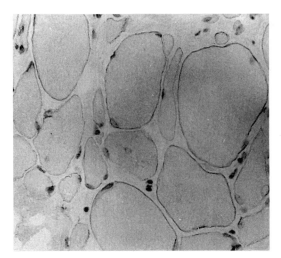

Fig. 58. Dystrophin immunostaining with a C-terminus antibody is of variable intensity from fiber to fiber and sometimes even in the same fiber. A serial section showed no immunostaining at all with an N-terminus antibody which recognizes epitopes that correspond to an in-frame deletion of the dystrophin gene. Becker dystrophy, immunoperoxidase using a C-terminus antibody to dystrophin and biotin-streptavidin peroxidase display system, plus a weak hematoxylin counterstain to demonstrate nuclei, ×350.

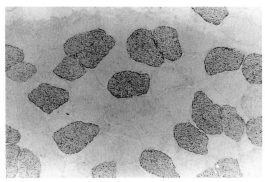

Fig. 59. Immunoreactive sarcoplasmic reticulum (SR) ATPase is abnormally absent in type 2 fibers. The primary monoclonal antibody used in this preparation should recognize both the fast (type 2) and the slow (type 1) SR ATPase. Brody's disease, immunoperoxidase plus biotin-streptavidin peroxidase display system, ×350. (Figure has been reduced for printing purposes.)

In Becker dystrophy, dystrophin is reduced in amount, and the molecule is usually abnormally small. With antibodies to the C-terminus, most fibers have an irregularly interrupted and weaker than normal stain (Fig. 58), while antibodies to the midportion of the molecule may or may not show any dystrophin.

Sarcoplasmic reticulum (SR)

The SR comprises the lateral sacs of the triads and the longitudinal or tubular component, which forms a network around the myofibrils. Release of calcium through a special channel (containing a ryanodine-binding protein) in response to a T-tubule signal and the subsequent ATP-dependent uptake of calcium by SR constitute a major link in excitation-contraction coupling of muscle fibers (Campbell et al. 1987). A genetically determined abnormality of the ryanodine-binding protein is the probable molecular basis of malignant hyperthermia (MacLennan et al. 1990). The normal SR is submicroscopic in size and there is no reliable

histochemical marker for it. Antibodies against calsequestrin or SR-ATPase show the extent of normal SR. These methods show the larger amount of SR in Type 2 fibers. Patients with Brody's disease (Karpati et al. 1986) show a genetically determined specific depletion of SR-ATPase from the type 2 fibers (Fig. 59).

The most striking pathological abnormality of SR consists of tubular aggregates (Danon et al. 1989), which appear basophilic with H & E and are red on the modified trichrome stain. They are positive with the NADH-tetrazolium reductase reaction (Fig. 60A), and with antibodies to the SR-ATPase. By electron microscopy, they generally appear as arrays of fairly regular parallel tubules 50-80 nm in diameter, sometimes containing an inner tubule. Connections can be found between the tubules of the aggregates and lateral cisterns of SR. They may be associated with many diseases, such as periodic paralysis, and appear to occur in abundance as the major abnormality in occasional patients with prominent cramps. Cylindrical spirals, which are probably also derived from SR, occur relatively rarely in muscle biopsies (Figs. 60B and D).

T-tubules. T-tubules are invaginations of the plasma membrane which serve to carry the action potential to the SR triads. Light microscopic meth-

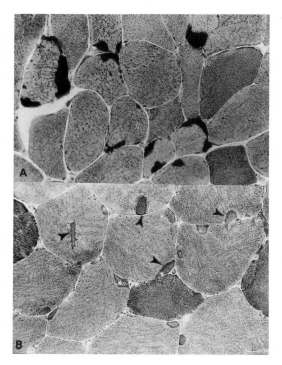

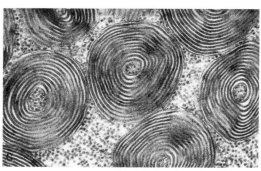

Fig. 60. (A) Dense (dark blue staining in original preparation) diformazan deposits in the periphery of several muscle fibers mark the sites of tubular aggregates. Myopathy with exercise pain and cramps, NADH tetrazolium reductase, ×350. (B) Peripheral monoformazan-positive masses are present at the periphery of several muscle fibers. They stained brown in the original preparations (examples are marked by arrowheads). These sites correspond to cylindrical spirals. Their appearance is clearly distinguishable from that of tubular aggregates (see Fig. 60A). Exercise-induced pain and cramps, NADH tetrazolium reductase, ×350. (C) Electron micrograph showing a group of cylindrical spirals. ×50,000. (Figures have been reduced for printing purposes.)

ods for their display are not available, although their lumen can be filled by horseradish peroxidase which, in turn, can be cytochemically displayed. By electron microscopy, they are best identified as the central element of triads. Outside of the triads there may be difficulty distinguishing them from the tubular component of the SR.

Proliferation of T-tubules into honeycombs is a non-specific ultrastructural reaction of muscle fibers to injury that is seen in many diseases. Large vacuoles are often found in the stumps of muscle fibers that have undergone segmental necrosis in experimental animals or in human disease, except for Duchenne dystrophy. These spaces result from marked focal swelling of T-tubules (Fig. 61) and can be explained as follows (Casademont et al. 1988): with segmental necrosis the plasma membrane of a muscle fiber breaks down, allowing a massive influx of sodium from the extracellular space. The sodium can diffuse rapidly into the adjacent non-necrotic segments, stimulating the Na^+K^+ ATPase in the T-tubular membrane to a maximum. Since this enzyme expels 3 molecules of Na^+ for every 2 of K^+ that it brings in, the ionic concentration in the T-tubules increases, resulting in the osmotic drawing of water out of the fiber into the T-tubule lumen, from where it cannot readily escape to the extracellular space, The extent of dilatation may induce formation of new T-tubule membranes from the intracellular phospholipid pool. The presence of such 'stump vacuoles' is merely a sign that segmental necrosis has occurred and should not be construed as indicating a 'vacuolar myopathy'.

Miscellaneous components of muscles

Blood vessels. Skeletal muscles are richly endowed by arteries, veins, arterioles, venules and capillaries. On transverse view, normal type 1 fibers are surrounded by more capillary lumina (4.1/fiber) than type 2 fibers (2.3/fiber). In children, endomysial capillary density is significantly less than in adults. On cryostat sections, capillaries are best visualized by a histochemical reaction for class 1 or class 2 major histocompatibility complex determinants (Fig. 62) or by immunostaining for basal lamina molecules such as laminin or collagen IV. The capillarity of muscle can also be studied on transverse plastic embedded semi-thin sections.

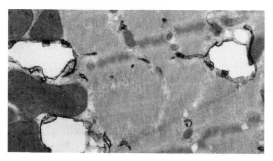

Fig. 61. Immense swelling of T-tubules has occurred in a muscle fiber which was transected and then incubated *in vitro* for 20 minutes. Lanthanum has been used as a tracer to identify the T-tubule system. ×25,000. (Figure has been reduced for printing purposes.)

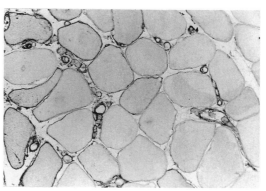

Fig. 62. Endomysial capillaries, some with unusually large lumen, stand out with strongly positive class I MHC immunostaining of their walls. In a region in the right lower corner, there is a paucity of endomysial capillaries. The staining of the surface of the muscle fibers is abnormal. Adult dermatomyositis, immunoperoxidase using a monoclonal antibody to class I MHC determinants (non-variable region) and biotin-streptavidin peroxidase display system, ×350. (Figure has been reduced for printing purposes.)

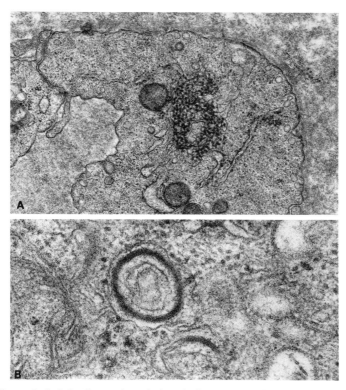

Fig. 63. (A) A swollen endothelial cell contains tubuloreticular structures in its smooth endoplasmic reticulum. Adult dermatomyositis, ×30,000. (B) Cylindrical confronting cisternae, as seen here in an endothelial cell, have the same general significance as tubuloreticular structures, with which they are often associated. Dermatomyositis, ×100,000.

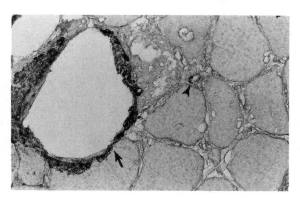

Fig. 64. In the walls of a large vein (arrow) and an endomysial capillary (arrowhead), membrane attack complex deposition is shown. Dermatomyositis, immunoperoxidase using a monoclonal antibody to human C9 complement and biotin-streptavidin peroxidase display system, ×520. (Figure has been reduced for printing purposes.)

A reduction of capillary numbers per muscle fiber and per transverse fiber area occurs in dermatomyositis, particularly in perifascicular distribution (Griggs and Karpati 1991). In these instances, necrotic capillaries may be seen on resin sections and by electron microscopy (Emslie-Smith and Engel 1990). Electron microscopy will also regularly show some tubuloreticular structures or cylindrical confronting cisternae in some endothelial cells or inflammatory cells (Fig. 63). After capillary necrosis in dermatomyositis, deposits of membrane attack complex can be demonstrated by immunocytochemistry in vessel walls (Fig. 64) (Kissel et al. 1991b). In denervation, as muscle fibers atrophy, the number of capillaries per transverse fiber area increases, but a counteracting process may result in the elimination of some capillaries (Carpenter and Karpati 1982). Therefore, the number of capillaries per fiber decreases, and empty capillary basal lamina loops may be found among the atrophic fibers. In diabetes and other unidentified diseases the thickness of the capillary basal lamina can be markedly increased (Fig. 65). A necrotizing myopathy with greatly thickened capillary basal laminae and membrane attack complex in capillary walls has been described in 3 patients (Emslie-Smith and Engel 1991). Amyloid deposits can also increase the thickness of capillary walls (Fig. 66). Hemosiderin deposits, demonstrated by the Prussian Blue reaction in pericytes, can result from prior diapedesis of erythrocytes through damaged capillary walls in certain capillary angiopathies (Fig. 67).

Increased capillary density usually occurs around ragged red fibers; in inclusion body myositis it may occur diffusely.

Arteries and arterioles may show necrosis of the vessel wall in periarteritis nodosa, Wegener's granulomatosis and leucoclastic vasculitis, and deposition of IgG may be demonstrable. These arteries may be thrombosed, and sometimes infarcts may be found in the vicinity. More often the muscle only shows denervation secondary to ischemic lesions of nerve. In other forms of vasculitis, the lumen is patent and the vessel wall is not necrotic but infiltrated with inflammatory cells. When the cell infiltrates are limited to the surrounding tissue, a perivasculitis is present (Fig. 68). This may be encountered in a variety of collagen vascular diseases.

Connective tissue. The amount of connective tissue between muscle fibers (endomysial) is scanty in normal muscle. Fascicles are separated by connective tissue septa, whose thickness can vary without implying a pathological reaction. Endomysial and septal connective tissues contain collagen fibrils (collagen I), elastic fibers, fibronectin and glycosaminoglycans. Fibrocytes are scanty. Endomysial connective tissue, particularly collagen, is increased either diffusely or focally in many neuromuscular diseases. The most dramatic example occurs in Duchenne muscular dystrophy, where it is an early phenomenon of uncertain origin; it is probably a sequel to muscle cell destruction and certainly not the cause of it. In Duchenne dystro-

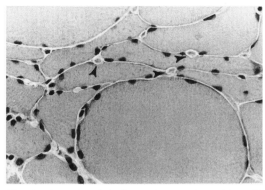

Fig. 65. Arrowheads indicate three endomysial capillaries whose walls are widened due to basal laminal thickening. The muscle is partially denervated due to diabetic peripheral neuropathy. Hematoxylin and eosin, ×350. (Figure has been reduced for printing purposes.)

Fig. 67. Prussian blue-positive hemosiderin granules fill the cytoplasm of presumed pericytes adjacent to endomysial capillaries (arrowheads). Peripheral neuropathy with an unidentified capillary angiopathy, Prussian blue, ×520. (Figure has been reduced for printing purposes.)

phy, excessive collagen is present in characteristically distinct bundles parallel to muscle fibers (Fig. 69). Focal increase of endomysial loose randomly oriented collagen occurs in inclusion body myositis and in infantile polymyositis in the midst of small diameter fibers (Fig. 70). The septal and endomysial connective tissue may be infiltrated with inflammatory cells (*vide infra*) in inflammatory myopathies and collagen vascular diseases, as well as in certain infections, including HIV and HTLV-I (Dalakas 1991).

In the eosinophilia-myalgia syndrome, inflam-

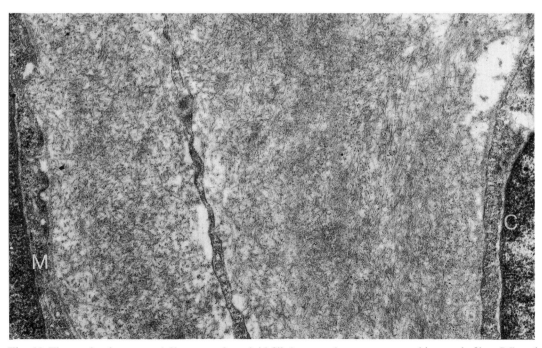

Fig. 66. The randomly arranged filaments of amyloid fill the space between an atrophic muscle fiber (M) and a capillary (C). ×25,000.

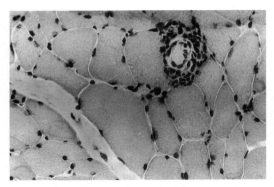

Fig. 68. A medium-sized vessel with a patent lumen and an apparently intact wall is surrounded by mononuclear inflammatory cells, mainly lymphocytes and macrophages. Mixed collagen vascular disease, hematoxylin and eosin, ×350. (Figure has been reduced for printing purposes.)

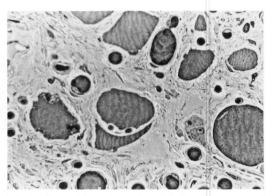

Fig. 70. Semi-thin resin section showing the pattern of loose poorly oriented connective tissue deposition in inclusion body myositis where there has been extensive loss of muscle cells. Paraphenylene diamine, phase optics, ×1370. (Figure has been reduced for printing purposes.)

mation tends to be largely limited to the connective tissues, although there may be some infiltration between muscle fibers near septa (Seidman et al. 1991).

Adipocytes are infrequent in normal muscle. Clusters or islands of adipocytes may replace destroyed muscle in Duchenne muscular dystrophy, in other myopathies and in advanced denervation. It is suspected that the endomysial histiocytes or pericytes transform into adipocytes under these circumstances.

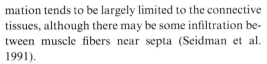

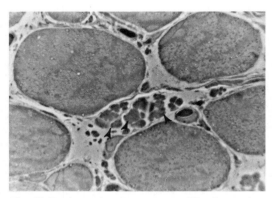

Fig. 69. Resin section from a case of Duchenne dystrophy showing discrete bundles of collagen (arrowheads) oriented with their long axis parallel to the muscle fibers. This is characteristic of fibrosis in Duchenne dystrophy. Paraphenylene diamine, phase optics, ×1370. (Figure has been reduced for printing purposes.)

Inflammatory cells. Macrophages that invade and phagocytose necrotic muscle fibers do not represent true inflammation. These cells can be demonstrated by their strong acid phosphatase positivity. Lymphocytes and plasma cells may accumulate in the septal and endomysial connective tissue in inflammatory myopathies (Arahata and Engel 1984, 1988a), but this may not always be demonstrable in a single biopsy. In dermatomyositis, the inflammatory infiltrates are situated mainly in interfascicular septa. Inflammatory cells are mainly the helper-inducer lymphocytes as well as B-lymphocytes, which can be demonstrated by special marker antibodies that recognize CD4 and CD22 antigens respectively (Fig. 71). In polymyositis and inclusion body myositis, the inflammatory cells are usually endomysial and the prominent cell type is a CD8[+] cytotoxic T-lymphocyte (Arahata and Engel 1988b). In these two diseases, such cells traverse the basal lamina and indent non-necrotic muscle fibers ('partial invasion') along with macrophages (Fig. 72). It is assumed that muscle fibers partially invaded by inflammatory cells eventually become damaged by compression or through the deleterious action of lymphokines such as perphorin. Polymorphonuclear infiltration of muscle occurs in acute trichinella myositis, but is otherwise rare.

In some cases of Duchenne dystrophy, an apparent autoimmune reaction leads to focal infiltration of the muscle with CD8[+] lymphocytes, which

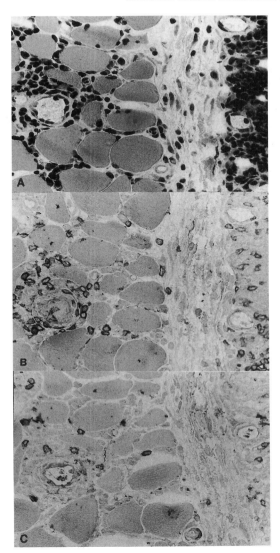

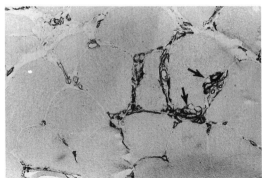

Fig. 72. An otherwise normal muscle fiber is partially invaded by CD8-antigen-bearing cells (presumably cytotoxic lymphocytes) at 2 points (arrows). Polymyositis, immunoperoxidase using a monoclonal antibody for the CD8 antigen and biotin-streptavidin peroxidase display system, ×350. (Figure has been reduced for printing purposes.)

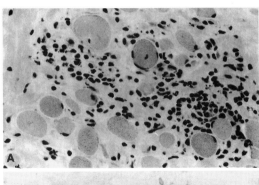

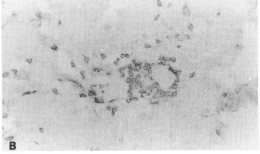

Fig. 71. Serial sections of mononuclear inflammatory infiltrates around a venule (left) and in an interfascicular septum (right) in a case of adult dermatomyositis, ×350. (A) Hematoxylin and eosin. (B) Immunoperoxidase for CD4 antigen. About 50% of the inflammatory cells are positive (helper/inducer T lymphocytes). (C) Immunoperoxidase for CD8 antigen. Only a few scattered inflammatory cells are positive (cytotoxic/suppressor T lymphocytes). (Figure has been reduced for printing purposes.)

can theoretically contribute to muscle fiber damage (Fig. 73) (Kissel et al. 1991a).

Activated lymphocytes can also express interleukin I which can be demonstrated by specific immunocytochemistry.

Fig. 73. (A) Profuse interstitial mononuclear inflammatory cell infiltrates (lymphocytes and macrophages) are shown without clear evidence of muscle fiber necrosis or partial invasion at this level. This degree of inflammatory response is highly unusual in Duchenne dystrophy. Hematoxylin and eosin, ×350. (B) The same field as in Fig. 73A showing that about 40% of the infiltrating cells carry the CD8 surface antigen. Immunoperoxidase staining, ×350. (Figure has been reduced for printing purposes.)

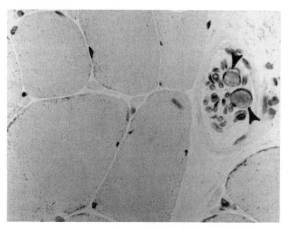

Fig. 74. Two axons in an intramuscular nerve bundle are distended with homogeneous material that stained purple with hematoxylin and eosin (arrowheads) and is positive for PAS. They represent intra-axonal polyglucosan bodies. Normal muscle from an older man, hematoxylin and eosin, ×350.

Intramuscular nerves. Intramuscular nerves may show depletion of myelinated axons or Wallerian ovoids in neuropathies and in motor neuron diseases. There may be evidence of storage in Schwann cells in some leukodystrophies, especially the metachromatic variety. Axonal swelling due to neurofilamentous masses may be seen in giant axonal neuropathy. Occasional intra-axonal polyglucosan bodies in intramuscular nerves in older people are probably not pathological (Fig. 74).

Miscellaneous pathological features of skeletal muscle fibers

The precise derivation of certain pathological alterations of muscle fibers is not certain and, therefore, these structures have not yet been discussed. These alterations include reducing bodies (Fig. 75) (Carpenter et al. 1985), fingerprint bodies, caps, and sarcoplasmic masses. Only reducing bodies are a disease-specific change; the others are non-specific.

PRACTICAL AND USEFUL POINTS IN THE
INTERPRETATION OF DIAGNOSTIC MUSCLE BIOPSIES

The myopathologist is expected to provide the clinician with diagnostically useful information. In some instances, the pathological report merely confirms a highly suspected clinical diagnosis,

while in other situations, it is the key factor in the overall diagnosis.

For the optimization of pathological diagnosis the following points are of paramount importance: good choice of muscle, artifact-free sections and judicious use of appropriate techniques. In addition to cryostat sections or resin-embedded semithin sections, electron microscopic examination may be indispensable in individual cases. Furthermore, several additional items must be considered in the pathological evaluation, such as age, sex, physical activity, nutritional status, as well as the clinical, electrophysiological and biochemical characteristics of the patient.

Taking into consideration the foregoing factors, the myopathologist must first determine if a biopsy is normal or abnormal for that patient. If it is abnormal, the relevance of the abnormality to the clinical picture must be evaluated critically. In case of relevant pathology, the pathologist must establish if the abnormalities are specific or non-specific. In some cases, they are pathognomonic, such as abundant nemaline rods in childhood nemaline myopathy, or reducing bodies in reducing body myopathy, or a constellation of features in inclusion body myositis or dermatomyositis, as well as in mitochondrial diseases. In other instances, only lesser degrees of specificity may be established, e.g. myopathic versus neuropathic process, or inflammatory versus non-inflammatory disease. If ab-

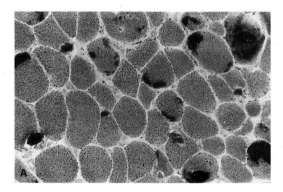
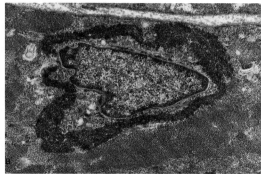

Fig. 75. (A) Numerous formazan-positive masses correspond to reducing bodies. Reducing body myopathy, nitro blue tetrazolium stain, ×350. (B) The dark material forming a ring around the nucleus is made up of the tubular filaments, which are the basic elements of reducing bodies by electron microscopy. ×25,000. (Figures have been reduced for printing purposes.)

normalities are non-specific, the clinician must use the pathological information judiciously.

We emphasize that a single normal muscle biopsy does not necessarily exclude the presence of a neuromuscular disease. Normal muscle biopsies can result from random skip of a pathological region, and a repeat biopsy may turn out to be informative.

Muscle biopsy may be helpful not only in the diagnosis of neuromuscular diseases but of other systemic diseases, even without clinically overt neuromuscular abnormality. Examples include Batten's disease, Lafora's disease, and α-fucosidosis.

REFERENCES

ARAHATA, K. and A.G. ENGEL: Monoclonal antibody analysis of mononuclear cells in myopathies. I. Quantitation of subsets according to diagnosis and sites of accumulation and demonstration and counts of muscle fibers invaded by T-cells. Ann. Neurol. (1984) 193–208.

ARAHATA, K. and A.G. ENGEL: Monoclonal antibody analysis of mononuclear cells in myopathies. V. Identification and quantitation of T8$^+$ cytotoxic and T8$^+$ suppressor cells. Ann. Neurol. 23 (1988a) 493–499.

ARAHATA, K. and A.G. ENGEL: Monoclonal antibody analysis of mononuclear cells in myopathies. IV. Cell-mediated cytotoxicity and muscle fiber necrosis. Ann. Neurol. 23 (1988b) 168–173.

ARAHATA, K., S. ISHIURA, T. TSUKAHARA, Y. SUHARA, C.

EGUCHI, T. ISHIRANA, I. NONAKA, E. OZAWA and C. SUGITA: Immunostaining of skeletal and cardiac muscle surface membrane with antibody against Duchenne muscular dystrophy peptide. Nature (London) (1988) 861–866.

ASKANAS, V., P. SERDAROGLOU, W.K. ENGEL and R.B. ALVAREZ: Immunolocalization of ubiquitin in muscle biopsies of patients with inclusion body myositis and oculopharyngeal muscular dystrophy. Neurosci. Letters 130 (1991) 73–76.

BEILHARZ, M.W.: Genetic probes for tracking muscle precursor cells in vivo: technical aspects. In: B.A. Kakulas, J.McC. Howell and A.D. Roses (Eds.), Models for Duchenne Muscular Dystrophy and Genetic Manipulation. New York, Raven Press (1992) 165–174.

BISCHOFF, R.: Control of satellite cell proliferation. Adv. Exp. Med. Biol. 280 (1990) 147–158.

BLAU, H.M., C. WEBSTER and G.K. PAVLATH: Defective myoblasts identified in Duchenne muscular dystrophy. Proc. Natl. Acad. Sci. (USA) 80 (1983) 4856–4860.

BODENSTEINER, J.B. and A.G. ENGEL: Intracellular calcium accumulation in Duchenne dystrophy and other myopathies: a study of 567,000 muscle fibers in 114 biopsies. Neurology 28 (1978) 439–446.

BONILLA, E., A.F. MIRANDA and A. PRELLE: Immunocytochemical study of nebulin in Duchenne muscular dystrophy. Neurology 38 (1988a) 1600–1603.

BONILLA, E., C.E. SAMITT, A.F. MIRANDA, A.P. HAYS, G. SALVIATI, S. DIMAURO, L.M. KUNKEL, E.P. HOFFMANN and L.P. ROWLAND: Duchenne muscular dystrophy: deficiency of dystrophin of the muscle cell surface. Cell 54 (1988b) 447–452.

BROOKE, M.H. and K.K. KAISER: Muscle fiber types: how many and what kind? Arch. Neurol. 23 (1970) 369–379.

BURROW, K.L., D.D. COUVERT, C.J. KLEIN, D.E. BULMAN, J.T. KISSEL, K.W. RAMMOHAN, A.H.M. BURGHES,

J.R.MENDELL and THE CIDD STUDY GROUP: Dystrophin expression and somatic reversion in prednisone-treated and untreated Duchenne muscular dystrophy. Neurology 41 (1991) 661–666.

CAMPBELL, K.P., C.M. KNUDSON, T. IMAGAWA, A.T. LEUNG, J.L. SUTKO, S.D. KAHL, C.R. RAAB and L. MADSON: Identification and characterization of the high affinity [^{3}H] receptor of the junctional sarcoplasmic reticulum Ca^{2+} release channel. J. Biol. Chem. 262 (1987) 6460–6463.

CARPENTER, S.: Regeneration of muscle fibers after necrosis. Adv. Exp. Med. Biol. 28 (1990) 13–15.

CARPENTER, S. and G. KARPATI: Duchenne muscular dystrophy: plasma membrane loss initiates muscle cell necrosis unless it is repaired. Brain 102 (1979) 147–161.

CARPENTER, S. and G. KARPATI: Necrosis of capillaries in denervation atrophy of human skeletal muscle. Muscle Nerve 5 (1982) 250–254.

CARPENTER, S. and G. KARPATI: Pathology of Skeletal Muscle. New York, Churchill-Livingstone (1984) 1–754.

CARPENTER, S. and G. KARPATI: Lysosomal storage in human skeletal muscle. Hum. Pathol. 17 (1986) 683–703.

CARPENTER, S. and G. KARPATI: Segmental necrosis and its demarcation in experimental micropuncture injury of skeletal muscle fibers. J. Neuropathol. Exp. Neurol. 48 (1989) 154–170.

CARPENTER, S., G. KARPATI and P. HOLLAND: New observations in reducing body myopathy. Neurology 35 (1985) 818–827.

CASADEMONT, J., S. CARPENTER and G. KARPATI: Vacuolization of muscle fibers near sarcolemmal breaks represents T-tubule dilatation secondary to enhanced sodium pump activity. J. Neuropathol. Exp. Neurol. 47 (1988) 618–628.

CASHMAN, N.R., J. COVAULT, R.L. WOLLMAN and J.R. SANES: Neural cell adhesion molecule in normal, denervated and myopathic human muscle. Ann. Neurol. 21 (1987) 481–489.

CULLEN, M.J., J. WALSH, L.V.B. NICHOLSON and J.B. HARRIS: Ultrastructural localization of dystrophin in human muscle by using gold immunolabelling. Proc. R. Soc. London (B) 240 (1990) 197–210.

DALAKAS, M.C.: Polymyositis, dermatomyositis and inclusion body myositis. N. Engl. J. Med. 325 (1991) 1487–1498.

DANON, M.J. and S. CARPENTER: Myopathy with thick filament (myosin) loss folllowing prolonged paralysis with vecuronium during steroid treatment. Muscle Nerve 14 (1991) 1131–1139.

DANON, M.J., S. CARPENTER and Y. HARATI: Muscle pain associated with tubular aggregates and structures resembling cylindrical spirals. Muscle Nerve 12 (1989) 265–272.

DIMAURO, S.: The metabolic myopathies. Curr. Opin. Neurol. Neurosurg. 4 (1990) 668–676.

DIMAURO, S., C.T. MORAES and E.A. SCHON: Mitochondrial encephalomyopathies: problems of classification. In: T. Sato and S. DiMauro (Eds.), Progress in Neuropathology, Vol. 7. New York, Raven Press (1991) 113–127.

EMSLIE-SMITH, A.M. and A.G. ENGEL: Microvascular changes in early and advanced dermatomyositis: a quantitative study. Ann. Neurol. 27 (1990) 343–356.

EMSLIE-SMITH, A.M. and A.G. ENGEL: Necrotizing myopathy with pipestem capillaries, microvascular deposition of the complement membrane attack complex (MAC) and minimal cellular infiltration. Neurology 41 (1991) 936–939.

ENGEL, A.G. and B.L. BANKER, (Eds.): Myology. New York, McGraw-Hill (1986) 1–2159.

ENGEL, A.G. and C. BIESESCKER: Complement activation in muscle fiber necrosis: demonstration of the membrane attack complex of complement in necrotic fibers. Ann. Neurol. 12 (1982) 289–296.

ERVASTI, J.M. and K.P. CAMPBELL: Membrane organization of the dystrophin-glycoprotein complex. Cell 66 (1991) 1121–1131.

ERVASTI, J.M., K. OHLENDIECK, S.D. DAHL, M.G. GAVER and K.P. CAMPBELL: Deficiency of a glycoprotein component of the dystrophin complex in dystrophic muscle. Nature (London) 345 (1990) 315–319.

FARDEAU, M., J. GODET-GUILLAIN, F.M.S. TOME, H. COLLIN, S. GAUDEAU, U. BOFFETY and P. VERVANT: Une nouvelle affection musculaire familiale, définie par accumulation intra-sarco-plasmique d'un matériel granulo-filamentaire dense en microscopie électronique. Rev. Neurol. (Paris) 134 (1978) 411–425.

GRIGGS, R.C. and G. KARPATI: The pathogenesis of dermatomyositis. Arch. Neurol. 48 (1991) 21–22.

GROUNDS, M.: The proliferation and fusion of myoblasts in vivo. Adv. Exp. Med. Biol. (1990) 101–106.

GULATI, A.K., A.H. REDDI and A.A. ZALEWSKA: Changes in the basement membrane zone components during skeletal muscle fiber degeneration and regeneration. J. Cell Biol. 97 (1983) 957–962.

HECKMATT, J.Z., A. MOOSA, C. HUTSON, C.A. MAUNDER-SEWRY and V. DUBOWITZ: Diagnostic needle biopsy: a practical and reliable alternative to open biopsy. Arch. Dis. Childh. 59 (1984) 528–532.

HOFFMAN, E.P., R.H. BROWN and L.M. KUNKEL: Dystrophin: the protein product of the Duchenne muscular dystrophy locus. Cell 51 (1987) 919–928.

HOROWITS, R., E.S. KEMPNER, M.E. BISHER and R.J. PODOLSKY: A physiological role for titin and nebulin in skeletal muscle. Nature (London) 323 (1986) 160–164.

JOHNSON, M.A., D.M. TURNBULL, D.J. DICK and H.S.A. SHERATT: A partial deficiency of cytochrome C oxidase in chronic progressive external ophthalmoplegia. J. Neurol. Sci. 60 (1983) 31–53.

KARPATI, G. and S. CARPENTER: Small-caliber skeletal muscle fibers do not suffer deleterious consequences of dystrophic gene expression. Am. J. Med. Genet. 25 (1986) 653–658.

KARPATI, G., S. CARPENTER and S. PENA: Tracer and marker techniques in the microscopic study of skeletal muscles. Meth. Achiev. Exp. Pathol. 10 (1981) 101–137.

KARPATI, G., J. CHARUK, S. CARPENTER, C. JABLECKI and P. HOLLAND: Myopathy caused by a deficiency of Ca^{2+} adenosine triphosphatase in sarcoplasmic reticulum (Brody's disease). Ann. Neurol. 20 (1986) 38–49.

KARPATI, G., S. CARPENTER and L.S. WOLFE: Clinical and experimental studies on lipofuscin in skeletal muscle fibers. In: I. Zs-Nagy (Ed.), Lipofuscin — 1987. State of the Art. Budapest, Academiai Kiado (1987) 227–249.

KARPATI, G., Y. POULIOT and S. CARPENTER: Expression of immunoreactive major histocompatibility complex products in human skeletal muscles. Ann. Neurol. 23 (1988a) 64–72.

KARPATI, G., S. CARPENTER and S. PRESCOTT: Small-caliber skeletal muscle fibers do not suffer necrosis in mdx mouse dystrophy. Muscle Nerve 11 (1988b) 795–803.

KARPATI, G., D. ARNOLD, P. MATTHEWS, S. CARPENTER, F. ANDERMANN and E. SHOUBRIDGE: Correlative multidisciplinary approach to the study of mitochondrial encephalomyopathies. Rev. Neurol. (Paris) 147 (1991) 455–461.

KELLY, F.J., J.A. MCGRATH, D.J. GOLDSPINK and M.J. CULLEN: A morphological/biochemical study on the action of corticosteroids on rat skeletal muscle. Muscle Nerve 9 (1986) 1–10.

KISSEL, J.T., K.L. BURROW, K.W. RAMMOHAN, J.R. MENDELL and THE CIDD STUDY GROUP: Mononuclear cell analysis of muscle biopsies in prednisone-treated and untreated Duchenne muscular dystrophy. Neurology 41 (1991a) 667–672.

KISSEL, J.T., R.K. HALTERMAN, K.W. RAMMOHAN and J.R. MENDELL: The relationship of complement-mediated microvasculopathy to the histologic features and clinical duration of disease in dermatomyositis. Arch. Neurol. 48 (1991b) 26–30.

MACLENNAN, O.H., C. DUFF, F. ZORZATO, J. FUJII, M. PHILLIPS, R.G. KORNELUK, W. FRODIS, B.A. BRITT and R.G. WORTON: Ryanodine receptor gene is a candidate for predisposition to malignant hyperthermia. Nature (London) 313 (1990) 559–569.

MASSA, R., S. CARPENTER and G. KARPATI: Preferential loss of myosin heavy-chain rod portion in denervated soleus muscles of rats receiving large dose glucocorticoids (Abstr.). Neurology 40 (Suppl. 1) (1990) 121.

METAFORA, S., A. FELSANI, R. COTRUFO, G.F. TAJAWA, G. DILORIO, A. DELRIO and P.P. DEPRISCO: Neural control of gene expression in skeletal muscle fibers: the nature of the lesion in the muscular protein-synthesizing machinery following denervation. Proc. R. Soc. London (B) 209 (1980) 239–255.

MOKRI, B. and A.G. ENGEL: Duchenne dystrophy: electron microscopic findings pointing to a basic or early abnormality in the plasma membrane of the muscle fiber. Neurology 25 (1975) 1111–1120.

MORGAN-HUGHES, J.: Mitochondrial diseases. Trends Neurosci. 9 (1986) 15–19.

NICHOLSON, L.V.B., M.A. JOHNSON, D. GARDNER-MEDWIN, S. BHATTACHARYA and J.B. HARRIS: Heterogeneity of dystrophin expression in patients with Duchenne and Becker muscular dystrophy. Acta Neuropathol. 80 (1990) 239–250.

OHLENDIECK, K., J.M. ERVASTI, K. MATSUMURA, S.D. DAHL, C.J. LEVEILLE and K.P. CAMPBELL: Dystrophin-related protein is localized to neuromuscular junctions of adult skeletal muscle. Neuron (1991) 499–508.

PELLISIER, J.F., J. POUGET, C. CHARPIN and D. FIGARELLA, : Myopathy associated with desmin type intermediate filaments: an immunoelectron microscopic study. J. Neurol. Sci. 89 (1989) 49–61.

PENA, S.D.J., B.B. GORDON, G. KARPATI and S. CARPENTER: Lectin histochemistry of human skeletal muscle. J. Histochem. Cytochem. 29 (1981) 542–546.

SARNAT, H.D.: Myotubular myopathy: arrest of morphogenesis of myofibres associated with persistence of fetal vimentin and desmin: four cases compared with fetal and neonatal muscle. J. Neurol. Sci. 17 (1990) 109–123.

SATO, T.: An overview of mitochondrial encephalomyopathies. In: T. Sato and S. DiMauro (Eds.), Progress in Neurobiology, Vol. 7. New York, Raven Press (1991) 1–8.

SAWCHAK, J.A., B. LEUNG and S.A. SHAFIQ: Characterization of a monoclonal antibody ot myosin specific for mammalian and human type II muscle fiber. J. Neurol. Sci. 69 (1985) 247–254.

SAWCHAK, J.A., S. LEWIS and S.A. SHAFIQ: Co-expression of myosin isoforms in muscle of patients with neurogenic disease. Muscle Nerve 12 (1989) 679–689.

SCHIAFFINO, S., L. GORZA, I. DONES, F. CORNELIO and S. SARTORE: Fetal myosin immunoreactivity in human dystrophic muscle. Muscle Nerve 9 (1986) 51–58.

SCHMALBRUCH, H.: Regenerated muscle fibers in Duchenne muscular dystrophy: a serial section study. Neurology 34 (1984) 60–65.

SCHMALBRUCH, H., Z. KAMIENIECKA and M. ARRE: Early fetal nemaline myopathy: case report and review. Dev. Med. Child Neurol. 29 (1987) 784–804.

SCHULTZ, F., D.L. JARYSZAK and C.R. VALLIERE: Response of satellite cells to focal skeletal muscle injury. Muscle Nerve 8 (1985) 217–222.

SEIDMAN, R.J., L.D. KAUFMAN, L. SOKOLOFF, L.F. MILLER, A. ILIYA and N.S. PERESS: The neuromuscular pathology of the eosinophilia-myalgia syndrome. J. Neuropathol. Exp. Neurol. 50 (1991) 49–62.

SHOUBRIDGE, E., G. KARPATI and K. HASTINGS: Deletion mutants are functionally dominant over wild-type mitochondrial genomes in skeletal muscle fiber segments in mitochondrial disease. Cell 82 (1990) 43–49.

SOZA, M., G. KARPATI and S. CARPENTER: Calcium paradox in skeletal muscles: physiologic and microscopic observations. Muscle Nerve 9 (1986) 222–232.

THOMPSON, A.J., M. SURESH, E.L. COX, D.A. INGRAM, A. GRAY and M.S. SCHWARTZ: Polysaccharide storage myopathy. Muscle Nerve 11 (1988) 349–355.

WALTON, SIR JOHN (Ed.), Disorders of Voluntary Muscles. Edinburgh, Churchill-Livingstone (1988) 1–1166.

YURCHENKO, P.D. and J.C. SCHITTNY: Molecular architecture of basement membranes. FASEB J. 4 (1990) 1577–1590.

ZUBRZYCKA-GAARN, E.E., D.E. BULMAN, G. KARPATI, A.H.M. BURGHES, B. BELFALL, H.J. KLAMUT, H.J. TALBOT, R.S. HODGES, P.N. RAY and R.G. WORTON,: The Duchenne muscular dystrophy gene product is localized in sarcolemma of human skeletal muscle. Nature (London) 333 (1988) 466–469.

Handbook of Clinical Neurology, Vol. 18 (62): Myopathies
L.P. Rowland and S. DiMauro, editors

CHAPTER 2

Clinical neurophysiology: the motor unit in myopathy

ERIK STÅLBERG[1] and JOZE TRONTELJ[2]

[1]*Department of Clinical Neurophysiology, University Hospital, Uppsala, Sweden and*
[2]*University Institute of Clinical Neurophysiology, University Medical Center, Ljubljana, Slovenia*

The diagnosis of muscle diseases is traditionally based upon clinical symptoms and signs, various blood tests, muscle biopsy, electromyography (EMG) and, more recently, molecular genetics. EMG and nerve conduction studies detect and characterize abnormalities at different levels of the motor unit, thereby also helping in differentiation between myopathic and neurogenic conditions. In addition to conventional EMG, other methods including single-fiber EMG (SFEMG), which helps in the analysis of disturbances of neuromuscular transmission and abnormal motor unit configuration, assess reinnervation and myopathy. Macro EMG, Scanning EMG, quantitative analysis of the interference pattern, and motor unit counting have opened additional possibilities for neurophysiological evaluation (Stålberg 1991). These tests may increase the sensitivity and specificity of the findings with the standard methods: they improve the understanding of structural motor unit changes in pathology, and give information about pathophysiology.

We here review advances in clinical EMG techniques and also new concepts in motor unit physiology. The focus will be on primary myopathies and disorders of neuromuscular transmission. However, pertinent examples from neurogenic disorders will also be covered. For practical details of EMG and findings in individual neuromuscular diseases, the reader is referred to comprehensive

publications (Buchthal 1973; McComas 1977; Ludin 1980; Notermans 1984; Oh 1984; Kimura 1989). Information about computer quantification of EMG may be found in several reviews (Desmedt 1983; Hausmanowa-Petrusewic and Kopec 1983; deWeerd 1984; Stålberg and Boon 1986; Stålberg and Stålberg 1989).

GENERAL COMMENTS ABOUT EMG METHODS

Conventional EMG

Conventional EMG is performed with concentric or monopolar needle electrodes (Fig. 1). In the concentric EMG method, the cannula housing the recording electrode is used as a reference electrode and is within the motor unit under study. In the monopolar EMG method, the reference electrode is outside the motor unit. The concentric electrode is less influenced by distant activity; it is slightly more selective. The normal values of the motor unit potential (MUP) differ for the two methods. We use concentric needles, and the EMG features discussed in this chapter refer mostly to these electrodes.

Routine investigation usually includes motor and sensory nerve conduction studies and EMG, sometimes complemented by repetitive nerve stimulation and evoked potential studies. EMG in-

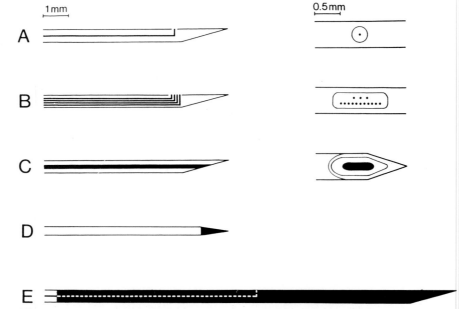

Fig. 1. Different EMG electrodes: a. SFEMG electrode; b. multielectrode for SFEMG recordings; c. concentric needle electrode; d. monopolar electrode; e. macro EMG electrode with an SFEMG electrode 7.5 mm from the tip of a 15 mm bare cannula, otherwise insulated.

cludes the study of muscle at rest, at slight activation, and at strong contraction.

Findings. In the normal muscle at rest, no activity is recorded except for a short burst of activity during and immediately after insertion of the electrode. This is called *insertional activity.* When the electrode is in the endplate region, *endplate noise* may be recorded as monophasic negative waves with amplitudes of less than 100 μV and a duration of 1–3 ms, corresponding to spontaneous quantal release of acetylcholine, insufficient in amount to trigger an action potential. When recorded with intracellular electrodes, this activity is termed *miniature endplate potentials*, which fire irregularly at a rate of about 1 Hz. The high frequency of the endplate noise indicates that it is generated by several end-plates in the vicinity of the electrode. In the same region, brief spikes with an initial negative deflection may sometimes be recorded. These are probably muscle fiber action potentials activated by the mechanical stimulation of a nerve terminal. Occasional *fasciculation discharges* may also be recorded in normal muscle, representing summated action potentials from a number of muscle fibers. They discharge with an irregular frequency.

In neuromuscular disorders, spontaneous activity may be recorded, including fibrillation potentials, positive sharp waves, myotonic discharges, complex repetitive discharges, myokymic discharges, extra-discharges of various types (Stålberg and Trontelj 1982), or fasciculations. These different forms of spontaneous activity are not specific for a given disease, but one type may be more common in one disorder than in another. They will be discussed further below.

During slight voluntary activation, the MUP is recorded as the summated activity of individual muscle fibers within the motor unit (Fig. 2). Because of volume conduction, the fibers closest to the recording electrode make the largest contribution to the recorded MUP. The MUP consists of slow initial and terminal parts and a central spike (Stålberg et al. 1986). The time between the start of the initial slow part and the end of the terminal slow part is called the MUP *duration.* Simulation studies (Nandedkar et al. 1988) have been used to analyse the relationship between motor unit characteristics and the MUP, but simulation, although useful in studying the relationship between the biological structures and MUP features, does not give an absolutely true picture of the generators. MUP duration reflects activity from a large area of the motor unit. However, 80% of the fibers that contribute to the signal lie within 2.5 mm of the electrode's recording surface. Duration therefore depends on the number of fibers of the motor unit within this radius. The endplate scatter and differences in conduction time along terminal nerve twigs and muscle fibers also contribute to the duration, more so in pathological conditions. The area under the waveform on both the positive and the negative side of the baseline depends on the number of muscle fibers and the distance from the closest fibers. Fibers within a 1.5–2 mm radius from the electrode's recording surface contribute up to 80% of the MUP area (Nandedkar et al. 1985, 1988), influenced by the distance to the closest few fibers and by the size of these fibers (Nandedkar et al. 1988). Area is easier to quantify than duration. Typically, area is correlated positively to duration and amplitude (Falck 1983). There is only a weak correlation between duration and amplitude in normal muscle.

The spike of the MUP arises mainly from the closest fibers. Simulation studies (Nandedkar et al. 1988) indicate that the peak-to-peak amplitude depends on fibers within 0.5 mm of the electrode's recording surface. Fibers at a greater distance contribute less than 20%. A semicircle with a 0.5 mm radius may contain up to 8 muscle fibers in a normal motor unit, usually 1–3 fibers. Experimental data show that the spike component (not just the component responsible for the peak-to-peak amplitude) is determined by 2–12 fibers (Thiele and Böhle 1978). Therefore, amplitude does not correlate to the total number of fibers in a motor unit but rather to the distance and the number of fibers closest to the electrode's recording surface, as well as synchronicity of the action potentials, and the diameter of these fibers.

The time between the first and last spike in the MUP, the *spike duration*, increases when the normally slight asynchronicity among muscle fiber action potentials is increased. This may happen when conduction in some terminal nerve branches is abnormally slow, when the motor endplates are abnormally dispersed, e.g. after reinnervation, and when differences in conduction time along individual muscle fibers increase (e.g. due to abnormal variation of fiber diameter as seen in myopathies).

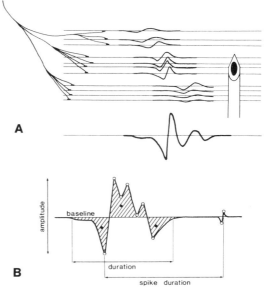

Fig. 2. (A) Schematic presentation of the generation of a MUP recorded with a concentric needle electrode. Distant fibers contribute less to the spiky part. The initial slow wave is already recorded when the arriving depolarization is still far away, and the terminal slow wave still lasts for a while after the impulse has passed. At those two stages the distance of the impulse to the recording electrode is similar for all fibers and therefore all of them contribute to a similar degree to the slow components. The spiky part, however, is generated when the depolarization is closest to the electrode tip and now the distance and the contribution is very different for the adjacent and the remote fibers. (B) Definition of MUP parameters (Stålberg et al. 1986). ● = phase ○ = turn. The late component is called a satellite potential since it occurs outside the limits for the duration (slow part).

Sometimes, individual spike components may be so delayed that they occur outside the main part of the MUP, even after the terminal slow phase. These are called *satellite potentials*. When the MUP is desynchronized because of temporal dispersion or when there is loss of individual muscle fibers, the normally triphasic MUP changes in shape and acquires more turns and phases. A *turn* is a peak in the action potential with an amplitude 20 μV or more than the previous turn and with a succeeding peak of more than 20 μV. When more than five turns occur, the MUP is called *complex* or *serrated*. A *phase* is the part of the signal between two baseline crossings. If more than four phases are present, the MUP is *polyphasic*. These

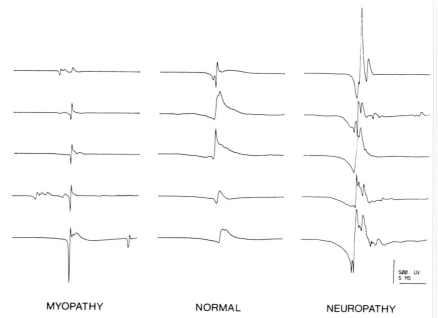

500 UV
5 MS

MYOPATHY   NORMAL   NEUROPATHY

Fig. 3. Examples of averaged MUPs (concentric needle EMG) from brachial biceps muscle in a patient with limb-girdle dystrophy, in a normal subject, and in a patient with a polyneuropathy.

two measures reflect the same type of abnormality of the motor unit, but polyphasia indicates more pronounced changes.

Thus, individual MUP measurements demonstrate characteristics such as the number and size of fibers at various distances from the electrode tip and the temporal dispersion of action potentials. These characteristics differ between muscles, and also with age and temperature (Buchthal et al. 1954; Falck 1983). It is important to standardize methods for quantification (Stålberg et al. 1986) and to collect control values with the method used in daily routine investigations.

It is important to realize that the motor units recruited at slight to moderate degrees of contraction (and only those can normally be analyzed individually in an EMG tracing) belong to low-threshold motor units, corresponding to type I motor units in histochemical classification.

If the motor unit topography is altered, resulting in an abnormal number and a changed spatial and temporal distribution of individual muscle fiber action potentials, the shape of the MUP varies in a way that differs for myopathies and neuropathies (Fig. 3). This is the quintessential use of EMG in the diagnosis of neuromuscular disorders.

The EMG pattern during strong voluntary contraction is the summated activity of many motor units within the electrode uptake area. With increasing voluntary effort, the firing rate of active units increases from low (5–8 Hz) to high (20–70 Hz) discharge frequencies, the *rate coding* (Milner-Brown et al. 1973a; Desmedt 1981). New units of higher threshold are activated starting at a low discharge frequency; this is called *recruitment*. The order of events is reproducible (Henneman et al. 1965; Grimby and Hannerz 1981) in consecutive trials of the same type of activation (e.g. tonic contraction). If another type of activation is used (e.g. phasic contraction), the recruitment order may change. The relation between rate coding and recruitment varies from muscle to muscle (for a review see Desmedt 1981). With strong contraction, the motor units discharge at high frequencies and mainly independently of each other. It is possible to see a tendency to time-lock between discharges from different motor units in non-fatigued normal muscle, probably the effect of common excitatory and inhibitory drives to many motoneurons. With fatigue, increasing synchronicity gives rise to *tremor*, which is more pronounced in neurogenic weakness.

The normal EMG during strong activity is called *full interference pattern* or, more appropriately, *full recruitment pattern*. It is not possible to discern an individual MUP from such a pattern except with computer techniques that allow decomposition of the signal (LeFever and DeLuca 1982). Individual peaks in the interference pattern are higher than the peak-to-peak amplitude of MUPs recorded during slight contraction. This has been attributed to the occasional summation of different MUPs. Simulation studies, however, indicate that this is not the main reason (Nandedkar et al. 1986a). When recorded with a concentric needle electrode, the duration of the main positive-negative slope of the MUP is brief, usually less than 0.5 ms. In order to give a summation effect on the peak-to-peak amplitude, the spike part of two MUPs must be almost exactly in phase. If they are misaligned by 1 ms or even less, the amplitude may not change or may even become reduced if a negative and a positive peak coincide. Therefore, the most likely cause of successively higher peaks in the interference pattern with increasing force, is the recruitment of larger MUPs. When defined by twitch tension, the later-recruited motor units are larger than low-threshold ones (Milner-Brown et al. 1973b). This is also seen in the Macro EMG, where larger MUPs have higher recruitment thresholds. It is also reflected in MUPs recorded with a concentric needle electrode, although the relationship of MUP amplitude to motor unit size is less consistent than with Macro EMG. The analysis of the interference pattern should aim at detecting different characteristics of the motor units such as their recruitment behavior and firing rate in relationship to force, the relationship between the number of active units and firing rates, the number of active motor units (i.e. fullness of the recruitment pattern), and characteristics of the participating motor units (frequency content).

In denervation, the recruitment pattern is reduced due to loss of active motor units but the peak-to-peak amplitude may increase due to reinnervation that results in larger motor units which are more dense. In myopathies of moderate severity there is a full recruitment pattern with fast firing rates and early recruitment of motor units in relation to the force developed. Frequency analysis

shows an increased content in the high-frequency part of the spectrum; this can be seen and heard during the recording with sharp waveforms and a high-pitched EMG sound. The patterns of different EMG findings at rest and at slight to strong voluntary contraction differ in different types of neuromuscular pathology.

Single-fiber EMG

SFEMG provides information about the physiological properties of individual muscle fibers, motor endplates, terminal nerve branches, and motor unit topography, described in detail by Stålberg and Trontelj (1979). As a refinement of the needle EMG, SFEMG has shed new light on the microphysiology of the motor unit.

Technical aspects. The recording is made with electrodes that have 1–14 recording surfaces, each 25 μm in diameter and exposed in a sideport 3 mm from the electrode tip (see Fig. 1). The cannula, 0.45–0.55 mm in diameter, is the reference. The electrode recording surface is small enough to sample action potentials from a single muscle fiber. By setting the high pass filter to 500 Hz or even higher, the contribution from distant fibers is further attenuated.

Action potentials recorded with these electrodes have a duration of about 1 ms, a peak-to-peak rise time of 100–150 μs, and a total amplitude of up to 20 mV, typically below 5 mV. The shape remains constant over consecutive discharges.

Findings. Propagation velocity in a single muscle fiber (Stålberg 1966) is determined from recordings made with a needle multielectrode (see Fig. 1). This has two parallel rows of electrodes along its long axis separated by about 200 μm. The propagation velocity varies between 1.5 and 5 m/s for different fibers, positively related to muscle fiber diameter (Håkansson 1956).

During continuous activity the propagation velocity decreases in most low-threshold muscle fibers, more so at higher innervation rates (Stålberg 1966). At the same time, the positive-negative rise-time increases proportionally. These two phenomena seem to explain the change in the power spectrum of the EMG signal seen during activity. Lindström and Petersén (1981), using surface EMG recordings, defined a 'fatigue index' from this shift in mean frequency, but the relationship to fatigue, defined as inability to maintain muscle power, has not been demonstrated.

Another change in propagation velocity is related to activity. When a muscle fiber is stimulated (indirectly through the nerve or directly) with two stimuli, the propagation velocity of the second (test) impulse differs from the first (conditioning) impulse (Stålberg 1966; Mihelin 1983). When the interval is less than 2–4 ms, the muscle fiber is refractory and no second impulse is transmitted. During the relative refractory period (3–10 ms), the velocity is decreased by up to 20% for the shortest interval. With longer intervals, the velocity of the test impulse is increased by up to 25% for intervals varying between 8 and 50 ms for different fibers. There is still some supernormal velocity after 500 ms (Mihelin et al. 1991). The curve describing interpotential interval and relative test pulse velocity is called the *velocity recovery function* (VRF) and is attributed to changes in membrane properties, possibly the accumulation of potassium in the sarcoplasmic reticulum after passage of an impulse. The effect on the propagation of a previous impulse is added to that of succeeding activity and therefore the propagation velocity may increase considerably during high-frequency stimulation. This is seen as changes in latency between stimulus and response. With voluntary activation giving a somewhat irregular innervation rate, the propagation velocity changes according to the accumulated effects of activity during the preceding 500 ms. In addition to double stimulation, the VRF can be analyzed from recordings during voluntary activity, using propagation velocity values of each impulse and the time from the previous impulse. The VRF can thus be deduced with high accuracy. The supernormal phase disappears during ischemia or low temperature and is exaggerated in muscular dystrophy or myotonic disorders. VRF is not measured in routine EMG.

Neuromuscular jitter is measured by placing the electrode to record activity from at least two muscle fibers of the same motor unit (Stålberg and Trontelj 1979). The interpotential interval between the two fibers' action potentials varies from discharge to discharge (Fig. 4). This variability, the jitter, can be expressed statistically as the standard deviation around the mean, or better, as the mean value of consecutive differences of the interpotential interval (MCD). It is attributed to variability of the neuromuscular transmission time of each motor endplate, and is a measure of the safety factor at the endplate (Stålberg et al. 1975).

Typical jitter is of the order of 5–50 µs and varies slightly in different muscles. A study is considered normal if no more than one recording out of 20 has a jitter exceeding the upper limit of normal (even if in that recording there is some blocking). A jitter below 5 µs probably represents impulses from split muscle fibers (Ekstedt and Stålberg 1969).

Jitter increases when neuromuscular transmission is disturbed, as seen after small doses of curare, with ischemia, or in myasthenia gravis (Fig. 4c). When the disturbance is more pronounced, there may be partial (intermittent) or total failure of impulse transmission. Impaired transmission is detected as increased jitter before blocking is seen. This offers an advantage over other electrophysiological and clinical tests of neuromuscular transmission which become abnormal only when there is blocking as reflected in the decremental response to repetitive nerve stimulation. SFEMG may detect a subclinical neuromuscular disorder and is useful in routine diagnostic work.

Increased jitter is seen in all disorders of neuromuscular transmission: myasthenia gravis, Lambert-Eaton myasthenic syndrome (LEMS), botulism, or intoxication with cholinesterase inhibitors. It is also abnormal with reinnervation after denervation due to functional insufficiency of new motor endplates (Fig. 4d). In muscular dystrophies (Fig. 4b), jitter may also be increased but to a lesser extent. It is usually increased in the active stage of polymyositis (see below).

To measure the local *fiber density* (FD) within motor units, the electrode is placed close to an active muscle fiber so that its action potential is recorded at maximal amplitude. Time-locked action potentials of other fibers may now be seen occurring before or after the triggering one, i.e. activity from fibers belonging to the same motor unit. If the amplitude of a potential exceeds 200 µV and the peak-to-peak rise-time is less than 300 µs, the fiber is typically within 300 µm of the recording electrode surface and is included in the calcula-

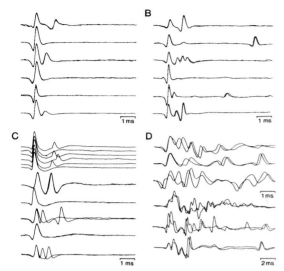

Fig. 4. SFEMG from brachial biceps with two sweeps superimposed on each trace, except in upper part of C. (A) normal subject; (B) limb-girdle dystrophy; (C) myasthenia gravis showing 6 consecutive sweeps demonstrating impulse blocking and superimpositions showing: one normal pair, single potential, a pair with increased jitter and blocking, a single potential and a pair with increased jitter and blocking; (D) ALS, showing complex recordings with instability. Note different sweep speed in the last 3 recordings.

tions. The number of time-locked action potentials fulfilling these criteria are counted and a mean of 20 recordings from different sites in the muscle is calculated. Since the triggering fiber is included in this number, the lowest value is 1. Normal values range between 1.3 and 1.7, varying slightly for different muscles. To express a relative difference in number of fibers within the uptake area from FD values, the triggering fiber should be subtracted (Gath and Stålberg 1982). An increase in FD from 1.5 to 2.5 should correspond to an average increase in number of fibers within the recording area of the electrode by a factor of 3 ($(2.5-1)/(1.5-1)$). Theoretically, this calculation is justified only when the fibers are randomly distributed over the recording area, as in the normal muscle. This is not true if there is abnormal fiber distribution (e.g. in reinnervation or in myopathies with muscle fiber splitting and regeneration); in these cases the uncorrected FD values should be used. FD increases with abnormal motor unit topography, typically after reinnervation.

Macro EMG

Macro EMG provides information about the electrical activity of the entire motor unit. The large recording surface of the electrode measures action potentials from a larger proportion of the motor unit than conventional EMG electrodes, and provides more information about the motor unit than monopolar, concentric or SFEMG recordings. The technique has been described in detail elsewhere (Stålberg 1980; Stålberg and Fawcett 1982).

Technical aspects. The recording electrode consists of a modified SFEMG electrode with a 25 μm diameter platinum wire exposed in a side-port 7.5 mm proximal from the tip (see Fig. 1). The 0.55 mm diameter steel cannula is insulated to within 15 mm of the tip. Recordings are made on two channels. One channel records the signal between the electrode shaft and a reference electrode placed remotely from the investigated muscle. The other channel displays an SFEMG signal recorded between the platinum wire and the shaft of the same electrode.

The SFEMG recording is used to trigger the oscilloscope. In this way, the action potentials of the synchronously firing motor unit are extracted. The resultant averaged potential is called the Macro MUP. Concomitantly, the number of single fiber components time-locked to the triggering action potential and exceeding 200 μV is counted and the FD is thus obtained.

Simulation studies (Nandedkar and Stålberg 1983) have shown a linear correlation between the number of muscle fibers and the amplitude and area of the Macro MUP. Mean fiber diameter is also positively related to the Macro MUP. If there is atrophy of the muscle fibers, the Macro MUP should become reduced. This effect is counteracted, however, by the shrinkage of the motor unit which reduces the distance between its fibers and the electrode, increasing the Macro MUP. The balance between these two factors for an individual motor unit, together with its number of muscle fibers, determines the final size of the MUP.

Findings. The Macro EMG is particularly useful to assess the size of the motor unit. In normal mus-

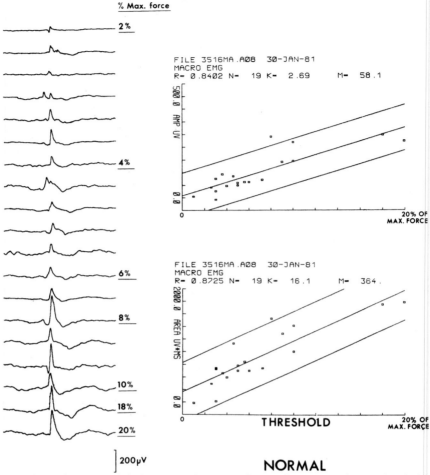

Fig. 5. Macro MUPs from brachial biceps muscle arranged according to recruitment threshold. Note that progressively larger Macro MUPs (both in amplitude and area) are recruited at higher thresholds, in the range between 0 and 20% of maximal force (Stålberg and Fawcett, unpublished).

cle, the Macro MUP shape differs from one muscle to another. In the anterior tibial muscle, the potentials often have two or more separate peaks whereas in brachial biceps muscle the potentials usually have a simple configuration with one or two negative peaks. There is a great scatter in individual Macro MUP amplitudes in the normal muscle. The largest Macro MUP may be up to 10 times larger than the smallest in brachial biceps muscle for individuals under the age of 60 and up to 20 times in those over 60 years. Larger scatter in values can be expected if high threshold motor units are included. The average Macro MUP amplitude tends to increase with age, an effect more pro-

nounced in the anterior tibial muscle than in the brachial biceps or vastus lateralis.

With increasing recruitment threshold, the Macro MUP amplitude increases (Fig. 5). The mean Macro MUP amplitude may increase fivefold for motor units recruited at 20% of maximal force as compared to those recruited at lower force, corresponding to orderly recruitment as judged from their twitches (Milner-Brown et al. 1973b).

Macro EMG has been used in neuromuscular disorders (Fig. 6). It can detect signs of reinnervation (Stålberg 1990), and quantify reinnervation capacity. The Macro MUP is typically increased in

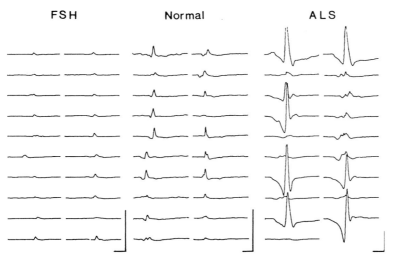

Fig. 6. Macro MUPs from brachial biceps muscle in a patient with FSH dystrophy, in a normal subject, and in a patient with ALS. Calibration 1 mV; 5 ms.

amplitude and area in neurogenic conditions. In myopathies, the electrical activity of the entire motor unit, as reflected in the Macro MUP, is reduced but not always. Macro EMG, therefore, may not be sensitive enough to detect early myopathic changes.

The combination of FD values obtained during the Macro EMG study and the Macro MUP value is, however, a useful indicator of the type of lesion: increased FD and large Macro MUPs suggest neurogenic lesions; increased FD and normal or reduced Macro MUP amplitude suggest a primary myogenic lesion; normal FD and large Macro MUPs suggest loss of small motor units as in some early cases of Guillain-Barré syndrome.

Scanning EMG

Macro EMG gives information about the total number of muscle fibers within a motor unit but it cannot tell much about the spatial distribution of the fibers; i.e. the motor unit territory is not assessed. One way to obtain this information is by using multielectrode recordings (Buchthal et al. 1955; Stålberg et al. 1976a). Another way is to move the electrode in defined steps through the muscle to map the distribution of electrical activity within a motor unit. This is the Scanning EMG (Stålberg and Antoni 1980).

Technical aspects. A single fiber electrode is inserted into a muscle which is slightly voluntarily activated and a position is sought where a single-fiber action potential is recorded. This is used to trigger the oscilloscope. A few centimeters away along the presumed direction of muscle fibers another recording electrode is inserted, e.g. a concentric needle electrode. When this electrode is recording activity time-locked to the triggering one, i.e. occurring at a constant position for each discharge along the oscilloscope sweep, the two electrodes are recording activity from the same motor unit. The concentric electrode is positioned so that a MUP with sharp components is recorded. This indicates that the concentric electrode is close to activated muscle fibers. The electrode is now pushed deeper into the muscle until no time-locked activity occurs. A step-motor is connected to the concentric electrode and via the computer, controlled to pull with a given step length, 50 μm (or multiples thereof) after each discharge. The concentric electrode is recording activity from each such recording position and the signal is stored in the computer. When it passes through the motor unit, the maximal amplitude of the MUP is recorded. The recording continues until no more time-locked activity is obtained. The recording can be displayed in different ways and analyzed (Fig. 7). The distance between the extreme positions of

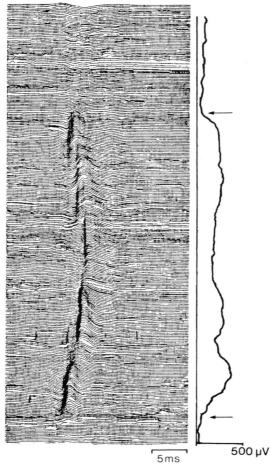

5 ms   500 µV

Fig. 7. Scanning EMG from a normal anterior tibial muscle. To the right is the profile of maximal peak-to-peak amplitude from each sweep. Arrows indicate distance between positions where electrode recorded MUPs greater than 50 µV in amplitude.

the electrode where MUPs with an amplitude exceeding 50 µV is obtained. The time dispersion between activity in different portions of the motor unit is also analyzed.

Findings. Scanning EMG gives information about the size of the motor unit territory (at least a submaximal value, since the electrode does not necessarily pass through the center or the extreme boundaries of the motor unit), as well as the distribution of activity within the motor unit. In the normal muscle the total length varies between 3 and 15 mm with differences between muscles. Clearly sep-

arated areas of maximal activity can be seen along the electrode's path interspersed with 'silent areas'. These may also be separated in time. A separate area of maximal activity within the motor unit is called a 'motor unit fraction'. Usually, a motor unit contains 1–4 fractions, fewer in the brachial biceps than in the anterior tibial muscle. The mean number of silent areas in the scan is less than one in normal muscle.

The motor unit becomes denser after reinnervation but the territory is essentially unchanged. In myopathies, silent areas within the motor unit are more commonly seen as compared to normal muscle (Stålberg and Dioszeghy 1991).

Intramuscular stimulation

To obtain a standardized pattern of activity, intramuscular stimulation can be used (Stålberg and Trontelj 1979; Troni et al. 1983; Troni 1984; Trontelj et al. 1986; Stålberg et al. 1992). With this technique, the muscle fibers or intramuscular nerve twigs can be stimulated (Fig. 8). Muscle stimulation is indicated by a low jitter between the stimulus and the recorded response.

Teflon-coated monopolar needle electrodes with a 1 mm bare tip are used for stimulation. The cathode is introduced into the muscle near the motor point and the anode is placed subcutaneously. The stimulus pulse has a typical duration of 0.05 ms with a variable amplitude. For single-fiber EMG recording, the electrode is placed into the twitching part of the muscle 10–40 mm distal or proximal to the stimulating electrode.

With slight electrical stimulation, the single-fiber electrode is positioned so that single-fiber action potentials are recorded. Measurements of jitter are made between the stimulus and the single-fiber action potential or between the different action potentials, depending on the aim of the investigation. Often, more than one spike component is obtained (Fig. 9). One possibility is that a single axon is stimulated and the recording is made from several fibers in one motor unit. This is characterized by the simultaneous disappearance of all spike components when decreasing the stimulus strength. A second possibility is that the complex recording is obtained from different motor units by stimulation of a bundle of axons. When decreasing the

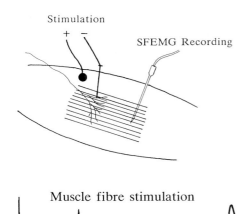

Stimulation

SFEMG Recording

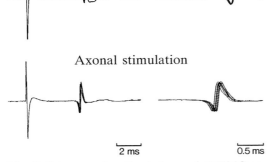

Muscle fibre stimulation

Axonal stimulation

2 ms 0.5 ms

Fig. 8. Intramuscular stimulation and SFEMG recording. Muscle fiber stimulation may be either directly (upper trace), indicated by jitter of less than 5 μs, or via its nerve (lower trace), indicated by latency jitter of more than 5 μs. Stimulus artefacts are shown at the beginning of traces.

stimulus strength, the spike components disappear individually or in multiples. Stimulation of motor axons is likely when jitter between stimulus and response exceeds 5 μs, when multiple spikes show an all or none occurrence phenomenon (and there is a jitter between these components, excluding a recording from a split muscle fiber), when axon reflexes occur (sudden latency jumps on changing stimulus strength), or when F- responses are seen (Fig. 9 a–c). A third possibility is that a group of muscle fibers is stimulated directly. Such a recording is identified by two criteria; the jitter between the stimulus and the action potentials is less than 5 μs, and the individual action potentials disappear and reappear independently upon varying the stimulus strength (Fig. 9d). Furthermore, on threshold stimulation they tend to show larger jitter.

Neuromuscular transmission is reflected in the jitter (between stimulus and the single-fiber action potential) and its dependence on low and high-frequency stimulation can be assessed. With this technique, the jitter is measured between the stimulus and the action potential of the responding fiber, hence only a single motor endplate is studied. This is different from the method in voluntarily activated muscle, which requires recording from a pair of muscle fibers, one serving as time reference while the combined jitter of both is displayed and measured on the other. The upper normal limit is therefore slightly different; for intramuscular stimulation in extensor digitorum communis muscle, for example, it is 40 μs for individual muscle fibers compared to 55 μs for voluntarily activated muscle. These values are only slightly above those theoretically expected from voluntary activation, perhaps because electrical stimulation activates both small and large motor units, in contrast to mainly small units recruited with voluntary activation.

One of the advantages of this technique is the regularity of the imposed discharge rate, which eliminates the interdischarge-interval-dependent 'myogenic' jitter that is due to propagation velocity changes in the muscle fibers. In this way it is possible to estimate the extent of muscle fiber splitting in myopathies. Low jitter (less then 5 μs) is expected between two branches of a split fiber, but it may be obscured by the addition of the 'myogenic' jitter due to uneven discharge rate during voluntary contraction.

With the described technique, it is also easy to stimulate muscle fibers directly, i.e. not through the motor axon, and the threshold is near that of the intramuscular motor axons. The responding muscle fibers have low jitter, less than 5 μs (mean 2.8 μs in the extensor digitorum communis). This also offers another possibility of estimating conduction velocity in individual muscle fibers (Buchthal et al. 1955; Troni 1984; Trontelj et al. 1990), apart from the original method of measurements across the multi-electrode (Stålberg 1966). The range of values can be used as an index of muscle fiber size variation, which may be a sensitive indicator of mild myopathy. Propagation velocity recovery function is conveniently studied by double pulse stimulation, at intervals varying from the absolute refractory period, at about 2 ms, to 1 s. Var-

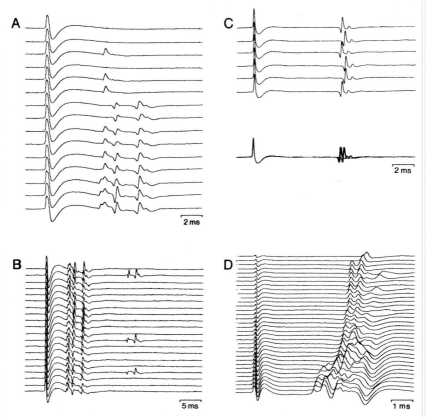

Fig. 9. Intramuscular stimulation in a normal brachial biceps muscle. (A) Upon increasing stimulus strength (traces are delayed to show stimulus artefact), one, three and then more components are recruited indicating stimulation of additional axons. The jitter is >10 μs for all components. (B) Continuation of the same recording as in A but with a slower sweep speed to show F-responses in the components recruited starting at trace 7 in A, proof of nerve stimulation. (C) At a given stimulus strength, this response shows two alternating latencies, i.e. an axon reflex. The nerve terminal to the fibers is stimulated directly (shorter latency) or via an antidromic impulse from another branch in the same motor unit (longer latency). (D) Stimulation electrodes placed distally in the muscle. With increasing stimulus strength individual fibers are recruited. At constant stimulation strength above recruitment threshold (not shown here) the jitter is about 5 μs for the spikes that could be analysed, indicating direct muscle stimulation.

ious degrees and time courses of subnormality and supernormality can be observed, with clear differences for example between normal and denervated muscle fibers (Mihelin 1983; Mihelin et al. 1991).

Similarly, it is possible to plot the action potential amplitude recovery function, which may also pass through a period of supernormality. This function shows wide variation even in normal muscle fibers (Mihelin 1983; Mihelin et al. 1991). Another way to study abnormal membrane repolarization is to measure decrement of single muscle fiber action potentials during repetitive stimulation at different rates. Quite abnormal decrements may be obtained in myotonic disorders (see below).

Motor unit counting techniques

McComas et al. (1971a) described the increment-counting technique to estimate the number of functioning motor units in a muscle. The nerve to a muscle is stimulated through surface electrodes. The muscle response is recorded with a pair of surface electrodes, usually two silver strips. The stimulus intensity is gradually increased and the response amplitude increases in discrete increments which is thought to signify the excitation of additional axons. From the first increments a mean motor unit amplitude is calculated. The M-response obtained at maximal nerve stimulation is then divided by the mean amplitude to estimate the number of excited axons (motor units). A number of factors make this estimate a rough approximate, as discussed by McComas (1977). One problem is the fact that the addition of one motor unit of different latency from the main peak may not give any increase in the amplitude of the response. Panayiotopoulos et al. (1974) used a photographic technique to demonstrate this. In modification introduced by Ballantyne and Hansen (1974a), the previous M-response is subtracted from the present, giving the shape of the response added by the last increase in stimulus intensity. This method has been used to study mean motor unit size and number of motor units in normal controls and in nerve-muscle disorders (Ballantyne and Hansen 1974a,b). Stålberg (1976) used SFEMG recordings to trigger an averager to which the surface recorded signal from silver strip electrodes was fed, another way to determine the electrical size of the motor unit. Brown et al. (1988) used this recording technique to estimate the number of motor units from the maximal M-wave amplitude and the average motor unit size.

EMG IN NEUROMUSCULAR DISORDERS

Spontaneous activity

Fibrillation potentials are spontaneously firing single-fiber action potentials and are seen in denervated muscle fibers. The firing rate is typically 1–30 Hz (Buchthal and Rosenfalck 1966). Sometimes two fibrillation potentials may be time-locked, which is assumed to be due to ephaptic driving of one fiber by the other (Trontelj and Stålberg 1983a). The rhythm is usually regular; sometimes, however, an irregular pattern is seen. Intracellular recordings have shown that rhythmical fibrillations are due to diphasic membrane potential oscillations and are preferentially generated in the former endplate region of the denervated muscle fiber (Thesleff et al. 1974; Thesleff and Ward 1975; Thesleff 1982). Critical for this type of fibrillation potentials is a fall in the resting membrane potential by 10–15 mV due to a reduction in the electrogenic activity of the sodium pumping mechanism of the membrane, with decreased resting potassium permeability (Purves and Sakmann 1974; Thesleff and Sellin 1980). The irregular fibrillation potentials, on the other hand depend on discrete, randomly occurring focal depolarizations with a sharp rising phase followed by exponential decay. Fibrillation is triggered when these local non-propagated potentials are high enough to reach the firing threshold, for which summation of several potentials may be required. The irregular fibrillations are assumed to be started from small foci in the transverse tubular system (Thesleff 1982), presumably due to increased sodium conductance in its membranes (Purves and Sakmann 1974). In contrast to the rhythmical fibrillations, the irregular fibrillations do not show the self-inhibiting behavior. They fire at a mean rate of 0.1–25 Hz, having a cluster of interdischarge intervals between 100 and 250 ms and numerous longer intervals, with pauses up to 7 seconds mixed among these basic intervals (Partanen 1987). Warming the muscle increases the amount of fibrillations. Positive sharp waves are seen in muscles with more severe denervation. Fibrillations and positive sharp waves can be seen in non-innervated muscle fibers, both neurogenic and myopathic. They may be seen in denervated muscle fibers (e.g. in motor neuron disease, peripheral nerve lesions or axonal neuropathies), in myasthenia gravis with motor endplate destruction, after functional denervation (e.g. in botulism (Cornblath et al. 1983)), and in myopathies. Fibrillation potentials are also seen in hyperkalemic periodic paralysis, probably as a sign of hyperexcitability.

Myotonic discharges are spontaneous muscle fiber discharges with a frequency of 40–100 Hz with waxing and waning amplitude and frequency. They are typically seen in myotonic disorders, but may also be seen in polymyositis, acid maltase deficiency and hyperkalemic periodic paralysis. They can occur in the absence of clinical myotonia. The frequency within the repetitive discharges can be up to 100 Hz in myotonia congenita and 60 Hz in myotonic dystrophy, but is usually considerably lower, about 1 Hz, in paramyotonia congenita. The duration of bursts tends to be shorter in myotonia congenita than in myotonic dystrophy

(Rüdel and Lehmann-Horn 1985). In a typical discharge in myotonia congenita, the frequency increases and the spike amplitude decreases, followed occasionally by a reversed sequence, which gives rise to a peculiar acoustic effect ('dive bomber'). In myotonic dystrophy, the typical discharges are longer and the frequency and amplitude tend to fall more slowly or not at all. Occasional myotonia-like discharges may be seen in early denervation and in a number of other neuromuscular disorders.

Complex repetitive discharges may be mistaken for myotonic discharges but differ on scrutiny. They consist of cyclical, time-locked activity of groups of single muscle fibers. They start and stop abruptly and have a frequency between 3 and 150 Hz, i.e. there are both low and high-frequency types. The high-frequency type has highly regular repetition rates, often slowly declining, while the low-frequency type may be less regular. SFEMG investigations have shown (Trontelj and Stålberg 1983b) that jitter between the individual single-fiber components is often low, suggesting ephaptic activation. Occasionally, single fibers or groups of fibers start to show larger jitter and may then disappear. Occasionally, groups of fibers disappear from the complex permanently or for shorter repetitive periods. These discharges are attributed to ephaptic transmission between adjacent hyperexcitable, probably denervated, muscle fibers that were initially triggered from a fibrillating muscle fiber. The high repetition rate is thought to be due to ephaptic re-excitation of the fiber which initiated the discharge, resulting in a closed circuit loop. The loop may be open in the low-frequency type. Sometimes, a complex repetitive discharge is started in a voluntarily firing motor unit. In this case the discharge rate increases sharply and becomes highly regular, the interpotential intervals and sequence of the individual muscle fibers in the discharge change and other muscle fibers may be recruited. Complex repetitive discharges are seen in both myopathic and neurogenic disorders, including polymyositis, Duchenne dystrophy, Schwartz-Jampel syndrome (Jablecki and Schultz 1982) and in chronic neurogenic diseases such as spinal muscular atrophy or Charcot-Marie-Tooth disease. Emeryk et al. (1974) studied these discharges in neuromuscular disorders.

Spontaneous activity of muscle fibers and motor units is also seen in tetanic cramps, at one stage taking the form of irregularly occurring doublets, triplets and multiplets. Voluntary discharges of motor units may be followed by repetitive extra-discharges. Other types of spontaneous activity also occur and show different EMG patterns such as fasciculations, myokymia, neuromyotonia, hemifacial spasm and muscle cramps. Fasciculations are considered to represent spontaneous firing of single motor units. They are characterized by highly irregular interdischarge intervals, with a mean of about 5 s in one study, though with frequent shorter intervals below 300 ms, interpreted as extra-discharges (Janko et al. 1989). They may be an important early sign of motoneuron disease. Most, if not all, fasciculations are believed to be generated distally in the axonal tree, perhaps as distal as the presynaptic portion of the motor endplate (Conradi et al. 1982). Some single fiber observations drew attention to differences between fasciculations in motoneuron disease and those in healthy persons. The latter may not represent activity of individual motor units but may be based on ephaptic activation of a bundle of adjacent muscle fibers, driven by an irregularly fibrillating muscle fiber (Stålberg and Trontelj 1982). This as well as other types of spontaneous activity, such as neuromyotonia, myokymia and muscle cramps, are not a feature of primary muscle disease and will not be further discussed here.

The motor unit in denervation and reinnervation

In cases of denervation fibrillation potentials and positive sharp waves start to appear after 7–21 days, depending on the distance between the nerve lesion and the muscle. Several weeks after complete denervation, fibrillation potentials may become so profuse as to result in a full interference pattern. Later, particularly when significant reinnervation (either by nerve regeneration or collateral sprouting) has taken place, fibrillation activity becomes progressively diminished and may finally, after years, cease altogether, whether reinnervation has been successful or not. With electrical stimulation in the denervated muscle, the jitter of single-fiber responses may be extremely large when the stimulus strength is near threshold, but de-

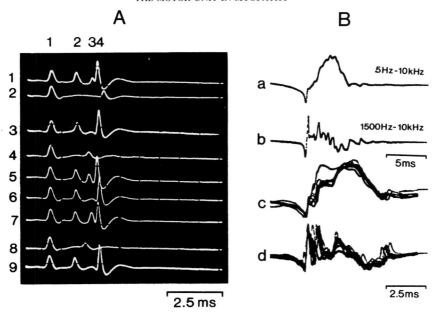

Fig. 10. EMG in reinnervation (from a patient with ALS). SFEMG (A) shows individual blocking in component 3 (traces 3,6,9) and concomitant blocking in components 2 and 4 (traces 2,4,8). Concentric needle EMG (B) shows unstable MUP with conventional filter settings 5 Hz and 10 kHz (a=single sweep, c=superimposed sweeps). The instability is more clearly seen with high pass filtering, 1500 Hz and 10 kHz (b=single sweep, d=superimposed sweeps) (from Stålberg 1986).

creases when the stimulus is well above threshold, to the order of 1–5 μs, the same jitter as seen in normal muscles when the muscle fiber is directly stimulated. On increasing the stimulus strength, the latency to the response usually decreases to some extent, and it does so in a continuous mode. Occasionally, an increasing stimulus strength, even when well above the threshold, results in a sudden stepwise latency change in some muscle fibers. This is interpreted as being due to the existence of discrete low-threshold sites along the denervated muscle fiber. Another possible explanation would be the stimulation of alternative branches of a split muscle fiber. A further phenomenon in the denervated muscle is the occurrence of an extra-discharge in the same muscle fiber after a single stimulus. It is probably elicited at a low-threshold site along the muscle fiber during a critical period of hyperexcitability following the passage of the first action potential. Such sites may also participate in ephaptic driving of one muscle fiber by another which is electrically stimulated or spontaneously firing. A denervated muscle fiber can follow stimulation rates up to 100 Hz. At these frequencies

there is usually a progressive increase in latency often associated with the change in the potential shape. Double pulse stimulation shows that the denervated fibers have a less pronounced supernormal phase of the velocity recovery function with a significantly delayed onset as compared to normal (Mihelin 1983; Mihelin et al. 1991).

In summary, most of the phenomena observed during stimulation of denervated muscle indicate the existence of discrete sites of hyperexcitability scattered along the denervated muscle fibers. These are probably the starting points for spontaneous or externally induced depolarization giving rise to fibrillation potentials, extra-discharges, ephaptic transmission between muscles fibers and complex repetitive discharges.

In *single-fiber EMG*, ongoing reinnervation is characterized by very unstable complex potentials showing large jitter and intermittent blocking (Fig. 10). This has been assumed to reflect immaturity in newly formed motor endplates and impaired conduction along nerve twigs. The evidence for failing axonal conduction is in concomitant blocking where two or more muscle fibers drop out simultaneously and reappear in the recording together again. This is due to intermittent blocking in the axonal branch common to the dropping-out muscle fibers. Theoretically, concomitant blocking can also be due to neuromuscular blocking in a single motor endplate on a split muscle fiber. The two possible explanations can be differentiated since the latter should show abnormally low jitter (<5 µs) between the individual components. In a study of reinnervation in transplanted muscle, impulse transmission became progressively more stable during the first six months (Hakelius and Stålberg 1974). In cases of single acute neurogenic lesions (e.g. nerve injury), after six months most of the recorded MUPs show only slightly increased jitter. This also applies to other essentially non-progressive and chronic neurogenic conditions. Reinnervation causes fiber type grouping (seen in histochemically stained muscle biopsies (Karpati and Engel 1968)), because surviving or regenerated axons innervate many neighboring muscle fibers. Therefore, there is a greater chance to record activity from many fibers in the same motor unit within the restricted uptake area of an SFEMG electrode. Fiber density (FD) is increased in all situations of reinnervation. SFEMG seems to be a more sensitive tool to detect moderately abnormal motor unit organization than muscle biopsy, at least without detailed morphometric analysis.

The combination of jitter and FD measurements helps to characterize the status of the motor unit during reinnervation. The amount of jitter and blocking indicates the stage of maturation of reinnervating structures, an initially large jitter becomes progressively more normal with ongoing reinnervation. FD indicates the extent of reinnervation, increasing during the active process and then usually remaining high.

In *conventional EMG*, reinnervation is reflected in most of the MUP characteristics. After *complete denervation* (e.g. nerve transection) the reinnervating motor units initially contain only a few muscle fibers. The MUPs are polyphasic, and may be of short duration and low amplitude. There is a pronounced variation of the shape on consecutive discharges due to large jitter and intermittent blocking of individual components. As more fibers become innervated, the amplitude and duration increase. As long as reinnervation progresses, there

is an increased variability in the shape of the MUP corresponding to the increased jitter in SFEMG recordings. The end-stage is characterized by high amplitude, stable MUPs of moderately or markedly prolonged duration. In *partial denervation* (e.g. partial nerve lesion) with collateral sprouting, late and unstable components may be added to the MUP, the original part of which can show a normal shape. A discrepancy between normal-sized fibers and often atrophic fibers undergoing reinnervation may give rise to very late satellite components. Their delayed occurrence may be attributed to several factors: the conduction velocity along newly grown nerve sprouts is slow, the new endplates may be located outside of the original endplate zone, and a recently reinnervated muscle fiber may have a reduced fiber diameter and hence lower conduction velocity compared to normal fibers in the same motor unit. This increased dispersion in time between the different single-fiber components is seen as an increased spike duration of the MUP.

Often the MUP shape varies with repeated discharges because of the jitter between individual muscle fiber components. This feature is visually enhanced in the recording if the high-pass filter of the amplifier is set to 500 Hz or above, up to 3000 Hz at 12 dB per octave (Fig. 10b) (Payan 1978). At later stages of reinnervation there is a higher degree of synchronicity, which results in progressive reduction of spike duration and increase in its amplitude. Because of the increased number of muscle fibers in the motor unit, the duration as well as the amplitude is increased. Due to the above mentioned maturation with improved synchronicity between individual fiber components, the amplitude may continue to grow even if no additional fibers are reinnervated.

Thus, in conventional EMG, active reinnervation is characterized by an increased number of polyphasic MUPs and great variability in the shape of the MUP over consecutive discharges. In the chronic stage, the MUPs are more stable, duration is long and amplitude high. The polyphasic MUPs after reinnervation typically have long duration of the slow initial and terminal phases (satellites not included) which differentiates them from the polyphasic MUPs of myopathy. Sometimes, however, the differentiation between neurogenic

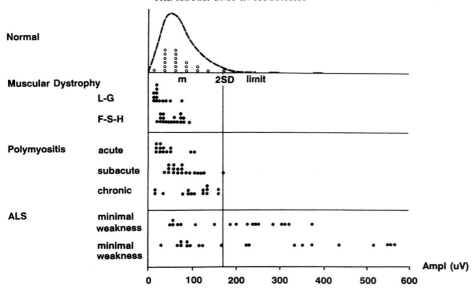

Fig. 11. Amplitudes of individual Macro MUPs in patients with different neuromuscular disorders compared to age-matched controls.

and myopathic changes in MUPs may be difficult. For example, small MUPs in the early stages of reinnervation after severe denervation may be misinterpreted to indicate myopathy. In a distal neuropathy with transmission failure in terminal nerve branches, there may be a loss of individual fibers in a motor unit, which gives rise to short-duration, low-amplitude MUPs. The end-stage of a chronic neurogenic disease with atrophy and fibrosis may show polyphasic short duration MUPs. Therefore, diagnostic EMG should not be performed in a severely affected muscle.

Long-duration, high-amplitude MUP does not necessarily suggest an enlarged motor unit: it just indicates an increased number of fibers within the uptake area of the electrode. Exceptionally, as in late stages of a neurogenic disorder, discrete clusters of fibers in a motor unit may give rise to large MUPs, but with an overall loss of fibers in other areas of the motor unit. The total number of fibers may even be less than normal. Concentric needle EMG does not supply information about the total number of fibers in a motor unit; Macro EMG reflects much more of the total motor unit activity than other EMG recordings.

In *Macro EMG* the number and size of the muscle fibers in a motor unit is reflected in the amplitude and area of the Macro MUP. This method

can be used to follow reinnervation quantitatively (Stålberg 1986, 1990). Some MUPs may be within normal amplitude limits in muscles where other MUPs are abnormal (Fig. 11), indicating a variable involvement of individual motor units. It is not yet clear whether this is due to random distribution of the pathology or reflects a principal difference in reinnervation capacity between different units.

Often the individual Macro MUPs in reinnervation have an amplitude exceeding the normal limit by a factor of 10. In patients with old polio, individual MUPs may be increased 20 times. In amyotrophic lateral sclerosis (ALS), the situation is more complex (Stålberg and Sanders 1984). In some patients with rapid progression, the Macro MUPs are increased only slightly and FD is only moderately increased. In cases of slow progression, the Macro MUPs increase much more, with individual Macro MUPs 10–20 times higher than the upper normal limit, still in parallel with the increase in FD, indicating a homogeneous and effective reinnervation. In later ALS, the average Macro MUP amplitude may start to decline although the FD is still high. This has been interpreted as either fragmentation of large motor units or an effect of selective dropout of the largest motor units, leaving the smaller ones preserved.

Thus, a small Macro MUP does not imply that

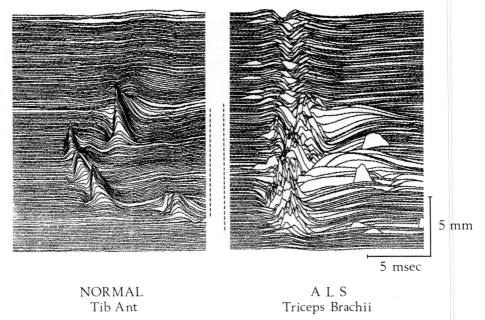

NORMAL
Tib Ant

A L S
Triceps Brachii

Fig. 12. Scanning EMG from a normal subject and a patient with ALS. The territories are of equal size in this example. In the normal muscle 3 fractions are seen. In the abnormal muscle the potentials show a great variability in shape.

the motor unit has not been involved in reinnervation. Conversely, a large Macro MUP does not necessarily indicate pathology in that unit. In early stages of Guillain-Barré syndrome, large MUPs may be recorded both with concentric needle electrodes and with Macro EMG (Stålberg 1986). The concomitant finding of a normal FD shows that the recorded units represent normal large units. These are usually activated at a higher threshold and are not studied under normal conditions with a slight degree of voluntary activation of the muscle. With the selective loss of small units, the average size of the recorded MUPs will increase because large normal motor units are recruited early.

Scanning EMG has been used to map the reinnervating motor unit. No dramatic increase in the maximal topographical size has been observed in various neurogenic conditions (Fig. 12). These findings seem to indicate that a particular motor unit does not support collateral reinnervation outside the fascicles where it is located, a suggestion first put forward by Kugelberg et al. (1970) in their studies on reinnervation in the rat. The reinnervated motor unit is certainly much denser than in

normal muscle and therefore its MUP can be recorded over a larger distance with an EMG electrode or multielectrode (Erminio et al. 1959). On a few occasions in scanning EMG in ALS patients, it has been observed that only one part of the scan shows a great shape variability in consecutive discharges. Here the motor unit is actively involved in reinnervation in one part of its territory where it happens to overlap with a denervated motor unit.

In summary, typical neurogenic atrophy can rather easily be distinguished from myopathic conditions. In cases of very slight involvement, the increased FD is the first sign of reinnervation; however, an increased FD may also be seen in myopathies. The additional findings in neurogenic conditions of prolonged MUP duration in concentric needle EMG and increased Macro MUP amplitudes may help to differentiate between these two conditions. The combination of SFEMG and Macro EMG techniques may help in quantitating ongoing reinnervation and in monitoring dynamic changes in the motor unit, as in ALS, post-polio syndrome and other progressive neurogenic disorders.

The motor unit in myopathies

Conventional EMG. The EMG abnormalities vary in degree of severity and localization for different myopathies in accordance with the clinical distribution of involvement and progression. Although some types show rather specific EMG features, it is usually not possible to distinguish between different myopathies on an individual EMG recording. Therefore, this description will mainly concern general findings. The classical EMG findings in muscular dystrophies will be reviewed first and then short comments will be given about some other myopathies.

Spontaneous activity: Fibrillation potentials and *positive sharp waves* are not exceptional in myopathies, particularly in the acquired forms. They are relatively common in Duchenne dystrophy, and are quite frequently seen in myotonic dystrophy, distal hereditary myopathy and polymyositis. This type of activity is rarely seen in congenital and metabolic myopathies. Fibrillation potentials may be generated in regenerating muscle fibers, in sequestered muscle fiber segments after focal necrosis, and in split muscle fibers. Regenerating muscle fibers are abundant in dystrophic muscles, and some of those never become innervated (Schmalbruch 1979). Whether fibrillations really occur in experimentally cut muscle fibers is still under debate. Desmedt and Borenstein (1975) found fibrillation activity after division of fascicles of fibers in the distal part of the brachial biceps muscle in the baboon. Kereshi et al. (1983) failed to detect fibrillation potentials after crush, ligation or transection of rat muscle, or after myotomy in human brachioradialis muscle. They only found fibrillation potentials after nerve section. They suggested that the fibrillation potentials recorded in the previous studies might have been due to inadvertent nerve section. Whether immature regenerating fibers, split fibers, or fibers damaged by inflammatory processes can give fibrillation activity without axonal involvement, has not been experimentally tested. Hence, the presence of fibrillation potentials and positive sharp waves should not be taken as evidence to exclude the diagnosis of what is conventionally classified as primary myopathy.

Myotonic discharges, as described above, are typically seen in myotonic disorders such as myotonic dystrophy, myotonia congenita, paramyotonia and also in acid maltase deficiency and hyperkalemic periodic paralysis. They increase with cooling and with mechanical stimulation.

Complex repetitive discharges may also be seen in myopathies (e.g. Duchenne dystrophy, limb-girdle dystrophy, Schwartz-Jampel syndrome, acid maltase deficiency, distal hereditary myopathy and polymyositis). SFEMG evidence has been presented to suggest that normally innervated muscle fibers may participate in such discharges (Trontelj and Stålberg 1983b).

Extra discharges of MUPs, i.e. initiated in the axon or motoneuron, are rather typical for myotonic dystrophy but are also seen in other myopathies, particularly polymyositis (Fig. 13).

The Motor Unit Potential: The classical EMG findings in myopathy are those of short-duration, low- amplitude MUPs (see Fig. 3) and an increased proportion of MUPs that are complex (with more than 5 turns) or polyphasic (with more than 4 phases).

The reduced duration of the MUP may be due to loss of muscle fibers, decrease in fiber diameter or fibrosis. The amplitude of the MUP in myopathies is usually reduced, probably for the same reasons. The finding of reduced amplitude was made from recordings, in which measurements were made of all MUPs in a given sweep with an amplitude exceeding 50 µV. However, many MUPs of low amplitude may be distant from the recording electrode. If the recording position is standardized for each MUP so that only MUPs with a rise time of less than 500 µs are used (which means that at least some of their fibers are very close to the electrode's recording surface), the amplitude may be normal or even increased (due to fiber hypertrophy seen in myopathies). Great variation in MUP amplitude is seen with slight electrode movement in Scanning EMG in myopathy. The amplitude drop with distance seems to be more pronounced in myopathy than in normal muscle. This may explain the nearly normal mean amplitudes of selected nearby MUPs versus the low mean amplitude when all MUPs within a certain radius from the electrode are included. This suggests that MUP amplitude is a rather insensitive parameter for reflecting myopathic changes. The complex shape with increased numbers of turns and phases can be

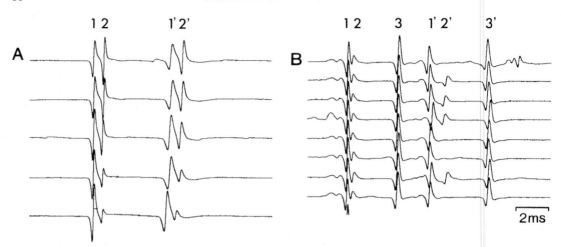

Fig. 13. Extra discharges in an SFEMG recording (extensor digitorum communis muscle) from a patient with limb-girdle muscular dystrophy. Note the change in shape in the extra discharge (in A: 1′,2′, and in B: 1′,2′,3′) depending on the relative refractoriness of the muscle fibers. In A the electrode was moved intentionally to demonstrate the parallel shape change in the original and extra-discharge and thus support its identification. In A the rise-time is slightly prolonged and the interval increased. In B three components are seen with occasional failure to produce extra-discharges in component 2, probably due to refractoriness.

due to partial loss of fibers. However, an even more important cause of the complex MUP shape may be the increased variation of fiber diameters and hence, increased variation of propagation velocities, giving rise to temporal dispersion between individual spike components. The changed MUP shape could also be partly due to loss of fibers. Individual components may be seen as satellite potentials. These could be due to atrophic or regenerating muscle fibers (Desmedt and Borenstein 1973, 1976; Stålberg et al. 1974; Stålberg 1977), or else branches of split muscle fibers (see below). SFEMG has shown that the satellite potentials have lower muscle fiber conduction velocity than early components (Stålberg 1977), indicating that late occurrence is due mainly to small muscle fiber diameter rather than aberrant motor endplates or other factors.

The MUP, including the satellite potentials, shows only slightly increased jitter, which is distinctly different from the polyphasic MUPs in early reinnervation, as best seen in SFEMG.

Just as in neurogenic conditions, there are many exceptions to the classical EMG picture in myopathy. Individual MUPs may have a normal or even high amplitude, particularly in mild cases or in clinically unaffected muscles, e.g. if the electrode is close to normal or hypertrophic muscle fibers. This may be due to compensatory hypertrophy of individual muscle fibers early in the disease. Furthermore, as indicated above, the spike duration, i.e. the time from the first to the last spike component (not the MUP duration which includes the slow early and late components but the satellites) is sometimes extremely prolonged in severely affected muscles, up to 50 ms, due to late satellite potentials. Thus, if duration is given as the time between first and last spike component (rather than between the beginning and the end of the slow components), the values may be considerably prolonged in the polyphasic MUPs. In Duchenne dystrophy, polyphasic MUPs with prolonged spike duration are rather typical. Sometimes the end of a long MUP is difficult to detect since it may be hidden in the next discharge from the same motor unit.

EMG at full effort: The interference pattern in moderately affected muscles is typically dense and contains an increased number of sharp components, but in severely atrophic muscles a loss of motor units may be seen. The average amplitude may be reduced even though a large number of motor units contribute to the signal. Individual MUPs become reduced, due to fiber necrosis, loss of fibers and fibrosis, particularly at a distance from the electrode. Automatic methods for analysis are used to quantify these features (see below).

A common observation in muscular dystrophy is the high innervation rate with early recruitment of new motor units in relation to force. A question has been raised whether mismatch between innervation rate and force generated is of pathogenic importance. Vrbová (1983) suggested that immature muscle fibers may be sensitive to high activation frequency and therefore undergo degeneration. Hilton-Brown and Stålberg (1986) found that the mean firing rate of individual motor units was the same in muscular dystrophies as in healthy muscles when the test was made at a given relative force of maximal value. However, a higher innervation rate will be used by each motor unit to produce a given absolute force. There was no indication of mismatch between force and innervation rate in clinically normal muscles in their patients. The impression of a more dense interference pattern than normal at maximal voluntary contraction is due partly to the fact that the MUPs in muscular dystrophy are complex and sometimes of long duration. This gives rise to an increased number of phases and turns which could then be misinterpreted as an increased firing rate. These changes in MUP are the basis for the increase in the high frequency content of the interference pattern.

The general description of EMG in myopathy is contingent upon the severity and the type of disease.

Congenital myopathies may show pictures of scanty fibrillation potentials and short duration MUPs in many cases. Often the abnormalities are of a slight degree compared to the clinical weakness. A normal EMG does not exclude a congenital myopathy. Myotubular (centronuclear) myopathy seems to give the most pronounced changes in spontaneous activity (fibrillation potentials, positive sharp waves, complex repetitive discharges and myotonic discharges); sometimes the pattern is similar to myotonic dystrophy. Excessive fiber type disproportion, seen in some congenital myopathies, may lead to suspicion of fiber type grouping. FD measurements do not show a corresponding increase in local fiber concentration within individual motor units but rather a normal scatter of fibers.

Endocrine myopathies may show classical myopathic picture on EMG. In hyperthyroidism, fasciculations or myokymia may be the only spontaneous activity. The myopathic picture, not directly related to the severity of the hyperthyroidism, becomes normal with treatment. In hypothyroidism, there is a low incidence of EMG abnormalities (Puvanendran et al. 1979). The slow relaxation after tendon tap is electrically silent (Salick and Pearsson 1967). Patients with hyperparathyroidism and weakness or subjective fatiguability usually have normal EMG. SFEMG shows slight abnormalities (Ljunghall et al. 1984). Nerve conduction may be minimally slow, but no clinical signs of neuropathy are seen. In hypoparathyroidism, hypocalcemic tetany may be seen as double or multiple extra-discharges following the MUP, as well as spontaneously occurring discharges.

In familial (hypokalemic) periodic paralysis, the EMG is often normal between attacks unless there is a progressive myopathy. During an attack the number of active motor units is reduced and a low muscle response follows nerve stimulation. In hyperkalemic periodic paralysis there is spontaneous single-fiber activity (fibrillation potentials). Sometimes myotonic and complex repetitive discharges are found between attacks. MUPs are of short duration during an attack. In the normokalemic form, there is no spontaneous activity in muscle. Diseases of the adrenal or the pituitary glands as well as steroid-induced myopathy show type II fiber atrophy without degenerative changes. Since EMG mainly reflects activity in type I fibers, these muscle abnormalities do not give much EMG change, particularly when clinical involvement is slight. SFEMG study of primary Cushing disease showed only slight abnormality, with mild elevation of fiber density and borderline jitter values.

Metabolic myopathies vary in the EMG. In acid maltase deficiency, spontaneous activity such as fibrillation potentials, myotonic discharges and complex repetitive discharges is seen in addition to abnormal MUPs. Other metabolic myopathies also may show abnormally short and low MUPs, but in some the EMG is normal. For example, in phosphorylase deficiency the voluntary EMG is usually normal but repetitive nerve stimulation induces a slowly developing decrement (Brandt et al.

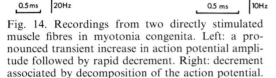

0.5 ms ┆ 20Hz   0.5 ms ┆ 10Hz

Fig. 14. Recordings from two directly stimulated muscle fibres in myotonia congenita. Left: a pronounced transient increase in action potential amplitude followed by rapid decrement. Right: decrement associated by decomposition of the action potential. A positive afterpotential (arrows) appears, which gradually drifts away during stimulation.

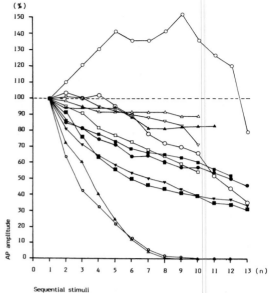

Fig. 15. Action potential amplitude changes in 12 muscle fibers of four patients with myotonia congenita at stimulation rate of 10 Hz.

1977). The muscle is electrically silent during the painful cramps (contractures). In a large family with mitochondrial cytopathy, myopathic changes were seen in early stages. In late stages, EMG signs of neuropathy were seen in addition (Torbergsen et al. 1991b). In some patients with Marinesco-Sjögren's syndrome, EMG showed myopathic changes and some other electrophysiologic phenomena indicating a membrane disturbance (Torbergsen et al. 1991a).

Patients with chronic progressive external ophthalmoplegia, a mitochondrial myopathy, are sometimes referred for EMG investigation to exclude myasthenia gravis. 17 such patients were studied by Krendel et al. (1985). The concentric EMG showed slight MUP changes indicating myopathy in 13. SFEMG showed increased jitter in at least one muscle in 13. Therefore increased jitter should not be taken as evidence against the diagnosis of progressive external ophthalmoplegia.

In *myotonia congenita*, typical spontaneous activity is recorded, as described above. The MUPs recorded with coaxial needle EMG on full effort show rapidly declining amplitude of the interference pattern, e.g. to about one-third within the first second (Rüdel and Lehmann-Horn 1985), with an increase in low frequency spectrum. After-dis-

charges occur on termination of short voluntary activation, and this is associated with transient weakness. On repetitive voluntary contractions, 'warm-up' is paralleled by a recovery of amplitude in the interference pattern, progressive shortening of after-discharges, and increased force. Responses of single muscle fibers to repetitive direct or axonal electrical stimulation showed pronounced decrement of the action potential amplitude, in extreme cases to zero within 10 seconds of stimulation at 2 Hz (Trontelj et al. 1992). In that study there was an average decrement at 10 Hz of 25–60% in the first second, usually with increase in rise time and deformed shape. Decrement was occasionally preceded by an initial increment (Figs. 14 and 15). These changes may be due to abnormal tubular accumulation of potassium.

There is a similar surface-recorded decrement on repetitive nerve stimulation, in this case not due to neuromuscular transmission block. On repeated testing, the decrement becomes progressively smaller, related to the clinical 'warm-up' phenomenon. Similar, but less pronounced abnormalities are recorded in myotonic dystrophy.

In myotonic dystrophy, concentric needle EMG

differs little from that seen in other muscular dystrophies, except for the myotonic discharges.

Acute polymyositis is characterized by abundant fibrillation potentials, positive sharp waves, complex repetitive discharges and MUPs of short duration and low amplitude. In a study of 153 patients with polymyositis or dermatomyositis, Bohan et al. (1977) found fibrillation potentials, positive sharp waves and increased insertional activity in 74%, complex repetitive discharges in 38% and abnormal MUPs in 90%. 10% showed a normal EMG. In another study of polymyositis (Devere and Bradley 1975), 11% had normal EMG. Serial EMG investigations during treatment show a reduction in amount of fibrillation potentials, which may be an early sign of improvement. With treatment, the amplitude and duration of the MUPs increase from abnormally low and short to enlarged and the number of polyphasic MUPs decreases. The frequency spectrum of the interference pattern initially contains increased amounts of high-frequency components, giving a high mean frequency that shifts to lower values with improvement (Sandstedt 1981). On repetitive nerve stimulation, a decrementing, and exceptionally an incrementing response may be obtained, signifying disturbed neuromuscular transmission (Simpson 1966).

On the usefulness of conventional needle EMG in the diagnosis of myopathy. Since the first descriptions of abnormal EMG in myopathy (Kugelberg 1949; Pinelli and Buchthal 1953; Buchthal and Rosenfalck 1964), the method has become progressively more sensitive in diagnosis and more useful in quantification of change in motor unit microanatomy and physiology over time e.g. for follow-up studies (for review see Wilbourn 1987). This is the result of a better understanding of the relationship between morphology and physiology, accumulated experience of EMG findings in normal and diseased muscle, improvement in recording equipment and the introduction of computers into neurophysiology in the last 10 years. Computer-aided EMG analysis is now used in the quantification of nearly all measurements mentioned in this chapter. It has been used for characteristics that were difficult to quantify, and to define new measurements of abnormality in the motor unit. Of par-

amount importance, computers make all of these measurements in a standardized way, so improving the accuracy. These methods are not described here but references are given in the introduction and elsewhere in the text. In many areas the development is so fast that it is impossible to keep textbooks up-to- date.

A few examples of the usefulness of EMG in the diagnosis of myopathy will be given. (This summary does not include the results from SFEMG or techniques other than conventional concentric needle EMG.) The figures on diagnostic yield given in the literature are minimal values for one or a few measures. As soon as additional features are measured, the yield improves. Although figures may show a high incidence of a given abnormality (i.e. high sensitivity to detect deviation from normal), this may not necessarily reflect accuracy in differentiating neuropathy and myopathy and, even less, the ability to discriminate different types of myopathy (diagnostic specificity). For the individual patient, a combination of findings must be used to arrive at a specific certain diagnosis. Fibrillation potentials provide such an example. They may exceptionally be seen in normal muscle but always in acute denervation. They are also seen in about one third of patients with muscular dystrophy, in at least 50% of patients with polymyositis, and are equally common in distal hereditary myopathy. Thus, although this finding indicates pathology in the individual case, it is seen in both neurogenic and myopathic conditions and is not specific for either. Therefore, additional electrophysiological information is always required. For diagnosis clinical information is essential and often histological and histochemical guidelines or DNA analysis is necessary.

Individual motor unit measures have a different diagnostic impact. The results should not be used for comparison between techniques or authors because of differences in patients. Duration shows a decreased mean value in muscular dystrophy. In Buchthal's study (1977), the MUP duration became even more sensitive when long duration polyphasic potentials, usually with late components, were excluded. This is why we suggest (see above under 'General comments about EMG methods') separate measurements of MUP duration, including the slow components but excluding the satellites, and of spike duration, including satellites. When results from many muscles were considered, short-duration MUPs (simple and short polyphasic) were found in 100% of patients with muscular dystrophy and in 80–90% in other myopathies (Buchthal 1977). In other studies, duration was decreased in all patients with polymyositis, in 50% of patients with other collagen myopathies (Vilppula 1972) and in all patients with thyrotoxic myopathy (Ramsay 1965). In contrast to the short-duration MUPs of muscular dystrophies is the nearly obligatory findings of prolonged MUPs in neurogenic conditions. Duration is thus of differential diagnostic importance.

The incidence of polyphasic potentials is increased to 20–30% of all recordings in muscular dystrophies, compared to 10–20% in most normal muscles. As seen in Table 1, this increase is seen in a large proportion of investigated muscles and is the single most sensitive measurement in muscular dystrophies but not in all myopathies. For example, in one study of polymyositis, there was an increased incidence of polyphasic potentials in 62% of all patients even though the duration of the MUPs was decreased in 100% (Vilppula 1972).

The incidence of polyphasic MUPs is also increased in neurogenic conditions. The polyphasic MUPs are typically of short duration in myopathy which distinguishes them from the long-duration polyphasic MUPs in neurogenic conditions.

The normal range of MUP amplitude is rather wide and this parameter is often normal in cases of myopathy. When abnormal, the amplitude may either be increased or decreased. A decrease, most typical for myopathy, may also be seen in neurogenic conditions (e.g. early reinnervation). In the individual case, the amplitude does not differentiate myopathy from neuropathy.

The analysis of the EMG at full effort has been difficult before the advent of automatic methods. With Willison's measurement of turns and amplitudes (Rose and Willison 1967) and modifications thereof (Fuglsang-Frederiksen and Månsson 1975; Fuglsang-Frederiksen et al. 1976; Stålberg et al. 1983; Nandedkar et al. 1986a), an abnormal pattern is found in 80–90% of patients with muscular dystrophy. In neurogenic conditions, the turns and amplitude analysis has shown abnormal findings in 70–80% of the patients, typically with increased mean amplitudes and fewer turns (Hayward and Willison 1973; Fuglsang-Frederiksen et al. 1977; Stålberg et al. 1983; Nandedkar et al. 1986b).

Other features of needle EMG include shape variability of the MUP, spike duration/total duration index and firing pattern. Myotonic discharges, complex repetitive discharges, and the anatomic distribution of abnormalities are also important. M-response amplitudes are usually available from the routine nerve conduction studies. All these measurements are included in the final interpretation.

Apart from these standard techniques, other methods may be applied to give further insight into motor unit physiology. These are more sensitive in detecting specific abnormalities and will be reviewed briefly.

Single-fiber EMG: SFEMG in myopathy shows an increased fiber density, with up to 10 individual components of a single motor unit at one recording site and a long interval between the first and last spike components (see Fig. 4b). In one study FD in the brachial biceps was increased in 22 of 32 investigations in 20 patients with limb-girdle or facioscapulohumeral (FSH) dystrophy (Hilton-Brown and Stålberg 1983b). There was no significant difference in the FD between limb-girdle and FSH patients. Even higher FD values were seen in 10 extensor digitorum communis (EDC) muscles in 10 patients with Duchenne dystrophy. Among 37 patients with limb-girdle, FSH or Duchenne dystrophy, the jitter was increased above normal in 33 of 49 brachial biceps and EDC muscles studied, or in 12–40% of the recordings. Abnormally low jitter, implying split muscle fibers, was seen in 5–16% of the recordings. The jitter abnormalities, both increased and decreased, were most pronounced in Duchenne dystrophy and least pronounced in FSH. In fact, even weak muscles in FSH may show rather normal jitter (Trontelj et al. 1988). In polymyositis, increased FD, abnormal jitter and intermittent blocking are seen particularly in acute stages (Henriksson and Stålberg 1978). If with corticosteroid treatment the patient becomes weaker, SFEMG may differentiate between relapse of the disease and steroid myopathy. The latter does not produce abnormal jitter and blocking.

The supernormal part of velocity recovery function was exaggerated in some studies of myopathies, particularly muscular dystrophies and myotonia congenita, and was taken as an indication of abnormal muscle fiber membrane or sarcoplasmic reticulum. Except for the spontaneous activity, principally the same findings are obtained in myotonic dystrophy. In myotonia congenita, however, the FD was normal, with only mild jitter abnormalities.

Could the observed increase in FD in many myopathies be due to atrophy and shrinkage of the motor unit? This is not likely because small muscle fibers are poor electrical generators and therefore the electrode uptake area is reduced. The increased FD is most likely due to abnormal fiber distribution in the dystrophic motor unit. Non-uniform atrophy of muscle fibers may lead to increased FD because of closer approximation of non-atrophic fibers. SFEMG studies showing a greater scatter of single-fiber conduction velocities in dystrophic muscle compared to normal also suggest non-uniform atrophy (Stålberg 1977). The morphological findings of increased variation of muscle fiber diameter supports these conclusions. Another finding related to increased scatter of fiber diameter and muscle fiber propagation velocity is an increased mean interspike interval (MISI) seen in nearly all cases of myopathy. Values above 0.8 ms and as high as 2 ms and greater are not specific of myopathies; increased MISI may also be seen in myasthenia gravis, where it can be due to non-uniformly prolonged neuromuscular transmission time, and in muscles with recent reinnervation. However, increased MISI in association with mildly abnormal jitter is typical of myopathy and can be used to support the diagnosis. This finding demonstrates that the increased fiber size variation occurs within the same motor unit rather than between different motor units. The effect on FD of this non-uniform atrophy is probably limited, particularly when packing is counteracted by increased interstitial connective tissue. Furthermore, the concomitant findings of abnormal jitter, either low or high, are not explained by this mechanism. There are other possible reasons for the increased FD. *Splitting of muscle fibers* is frequently seen in muscle biopsies in muscular dystrophy, which could give rise to fibers generating several action potentials with low jitter between each other. Branching of *regenerating fibers* could also cause increased FD. Furthermore, *focal necrosis* may produce fiber segments without innervation. These segments could later be reinnervated, leading to a reorganization of the motor unit with increased FD. In the early stage of reinnervation, large jitter could be caused by the immature motor endplates and axon twigs.

An increased FD not associated with fiber type grouping in biopsy could be due to *ephaptic transmission* between muscle fibers of different motor units. This phenomenon occurs only rarely and is revealed by time-locking between individual muscle fiber action potentials which is dependent upon firing rate. A single or exceptionally complex potential may be recruited into the potential under study when the innervation rate is voluntarily increased. When the rate is reduced, the extra components disappear. Finally, secondary (or primary) *neurogenic involvement* in myopathies could lead to reinnervation with type grouping and increased FD. This mechanism may be more important in late stages with fibrosis and possible involvement of intramuscular nerves.

In summary, SFEMG will detect early myopathic reorganization of the motor unit as reflected by increased FD and MISI. The jitter is only slightly abnormal and is, per se, not particularly useful as a diagnostic criterion. In combination with increased FD and MISI, the nearly normal jitter points strongly to myopathy. However, SFEMG does not give information about the total size of the motor unit or its general topography. Macro EMG, representing non-selective recording, may give additional information on this.

Macro EMG. Macro EMG has been studied in 35 muscles from patients with limb-girdle or FSH dystrophy (Hilton-Brown and Stålberg 1983a). On average the group showed a reduced mean Macro MUP amplitude (Fig. 16). Among individual patients, 15 showed an abnormally low mean amplitude (on average reduced by about 52%), 16 were normal and 4 had abnormally high amplitudes (on average increased by 120%). FD was increased in 23 of the muscles independent of the Macro EMG findings. The relatively slight reduction in Macro EMG amplitude may be due to the pronounced

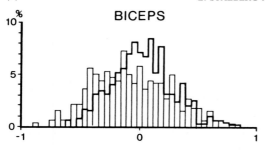

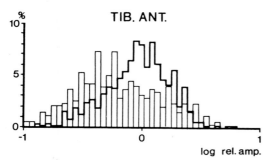

Fig. 16. Macro EMG in brachial biceps and anterior tibial muscles from a group of 18 patients with FSH and limb-girdle muscular dystrophy (columns). Heavy lines are from control material. The values are given as logarithms of the ratio of recorded amplitude value/normal median amplitude for age. (From Hilton-Brown and Stålberg 1983a).

regenerative capacity where fiber splitting, regeneration from satellite cells and reinnervation of sequestered muscle segments could give rise to action potentials. Mechanical abnormalities or a primary defect in the contractile properties may explain the discrepancy between the reduced force and the well-preserved electrical activity.

The discrepancy between the small concentric needle EMG potentials and the nearly normal Macro EMG potentials has been difficult to explain. One reason might be the difference between the two recording electrodes. Fibrosis of dystrophic muscle may restrict volume conduction between fascicles. The concentric needle electrode, with its small leading-off surface, may record from a restricted compartment. The Macro EMG electrode, on the other hand, physically penetrates a wide area of muscle and is therefore less susceptible to the possible shielding effect of fascicular walls within the muscle.

A large mean Macro MUP amplitude was found in 3 patients with FSH and one with limb-girdle

dystrophy with slight to no clinical involvement; this may indicate a compensatory hypertrophy, as seen also by others (Dubowitz and Brooke 1973; McComas 1977). The mild changes in Macro EMG in muscular dystrophy are not specific, but the dual finding of increased FD and normal or slightly reduced Macro MUP amplitude is strong evidence for myopathic changes within a motor unit, a combination not seen in neurogenic lesions.

Scanning EMG. Scanning EMG has been performed only in a few myopathy patients (Fig. 17) (Hilton-Brown and Stålberg 1983a). The mean total length of the scan on motor unit territory is not significantly different from that in normal muscle. Buchthal et al. (1960) found a 30% reduction of the territory in 18 patients with muscular dystrophy using multielectrodes. In Scanning EMG of the brachial biceps muscle, the mean diameter was 7.9 mm (SD=2.3) in limb-girdle and FSH dystrophies, compared with 6.0 (SD=3.9) in normal controls. In anterior tibial muscles, the values were 8.1 (SD=3.1) and 7.9 (SD=1.3) respectively. The maximal length was about 15 mm in both muscles in the patient group, similar to findings in controls. In the limb-girdle and FSH groups, the mean number of silent areas is increased, both in the brachial biceps (1.0 compared to 0.5 in the normal muscle) and in the anterior tibial muscle (1.2 compared to 0.2 in the normal muscle). The silent areas probably correspond to loss of muscle fibers in certain parts of the motor unit territory. This finding is distinctly different from that in neurogenic conditions where the motor unit is typically more dense.

Intramuscular stimulation. Electrical stimulation in conjunction with SFEMG was performed in patients with limb-girdle, FSH or Duchenne dystrophy (Hilton-Brown et al. 1985). Extensor digitorum communis and brachial biceps muscles were investigated. Only recordings with a jitter between stimulus artifact and response greater than 10 μs were accepted. This ensured that the nerve, rather than the muscle itself, was stimulated. For analysis, the jitter was measured between individual spike components. Among these, 853 muscle fiber action potentials were analyzed. The jitter was increased in 35%, normal in 48%, and abnor-

NORMAL   MUSCULAR DYSTROPHY

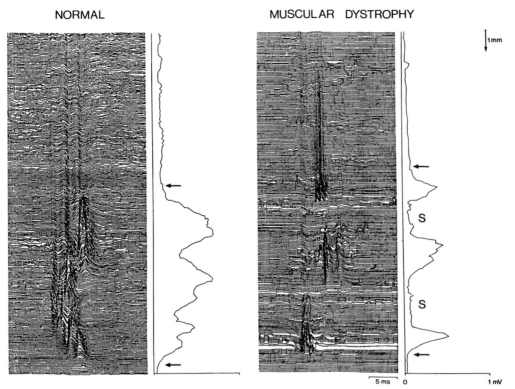

Fig. 17. Scanning EMG from anterior tibial muscle: normal subject (left); FSH (right). The peak-to-peak amplitude profile is shown on the right. Note that in this example the motor unit in the dystrophic muscle has a slightly larger total territory (arrows) than in the normal muscle, and also shows 2 silent areas (S). (From Hilton-Brown and Stålberg 1983a.)

mally low in 27% of the recordings. The proportion of abnormally large and low jitter was highest in limb-girdle and Duchenne patients. When abnormally low jitter occurred, this was often seen between more than two of the components, up to 4 or 5. Abnormally low jitter is attributed to split muscle fibers. Increased jitter is taken as a sign of uncertain transmission in nerve-twigs and motor endplates, probably in regenerating and reinnervating structures. Upon increasing the stimulation rate, new muscle fiber action potentials were sometimes recruited. When the rate was again lowered, they dropped out. These muscle fibers may have been recruited from neighboring silent motor units by ephaptic transmission, as has been observed during voluntary activation of muscle. Occasionally the opposite was seen, with components only occurring at low stimulation rates, suggesting pro-

longed subnormality of conduction across a segment along such fibers.

In another set of recordings in muscular dystrophies, direct muscle fiber stimulation was studied and 51 action potentials were analyzed. In 91% the jitter was low (less than 10 μs including the tape recorder's jitter of 5 μs as in normal muscle, and only slightly increased (10–16 μs) in 9%. This indicates that the increased jitter seen in these patients during voluntary activity or during nerve stimulation is not due to disturbed impulse propagation along the muscle fiber.

Direct stimulation of muscle fibers in muscular dystrophies demonstrates an increased range of values for propagation velocity. The multiple potentials recorded 25 mm away from the stimulating cathode might be dispersed over more than 30 ms, indicating that the propagation velocities of the

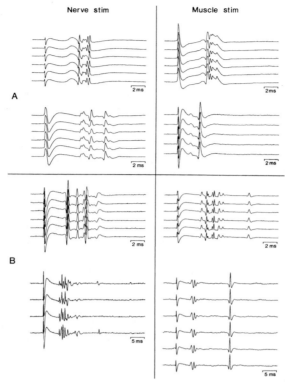

Fig. 18. Intramuscular stimulation and SFEMG recording in brachial biceps muscle in a healthy subject (A) and in a patient with limb-girdle dystrophy (B). Both indirect (nerve) and direct (muscle) stimulation is shown. Note the long intervals between extreme components in the dystrophic muscle, indicating a greater variation in fiber diameters in muscular dystrophy. In each recording the stimulus artefact is seen to the left.

slowest fibers are below 1 m/s (Fig. 18). However, some of the late potentials may be derived from branches of split fibers with a longer impulse pathway or represent extra-discharges. Troni (1984) has used direct intramuscular stimulation in patients with neuromuscular disorders. He found a reduced propagation velocity in patients with denervation but in 13 patients with limb-girdle and FSH dystrophies he found normal (c.f. Stålberg 1977) or increased velocities, a finding of interest in the light of normal or occasionally increased Macro MUP amplitudes. Further studies on more patients are necessary to establish the potential usefulness of this technique. Recordings obtained with nerve stimulation cannot be used for studies of propagation velocity since factors such as nerve conduction time, neuromuscular transmission time and motor endplate localization are unknown. Similar to the finding during voluntary ac-

tivity, propagation velocity recovery function obtained during direct double pulse stimulation shows differences from the normal, although not as pronounced as with denervated muscle fibers. The main trend is towards a more pronounced supernormal part of the curve (Mihelin 1983; Mihelin et al. 1991).

Motor unit counting. McComas et al. (1971a,b) studied extensor digitorum brevis, thenar, hypothenar and soleus muscles in 41 boys with Duchenne dystrophy. They found a reduced number of motor units and approximately normal amplitudes in most patients. These results and other findings formed the basis for the so-called sick motoneuron hypothesis, which created a debate about the pathogenesis of the muscular dystrophies, focusing interest on nerve-muscle interaction both in normal and pathological conditions. This question is

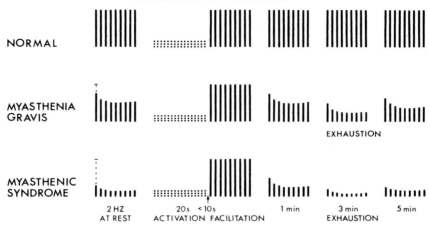

Fig. 19. Protocol for repetitive nerve stimulation in testing the neuromuscular junction. Bars indicate amplitude of obtained muscle responses using surface electrodes.

not yet solved physiologically. Ballantyne and Hansen (1974b), however, found a normal number of motor units in Duchenne, limb-girdle and FSH dystrophies, although the number was reduced in myotonic muscular dystrophy. Panayiotopoulos et al. (1974) found normal motor unit counts in Duchenne dystrophy. Molecular genetics, however, indicates that the fundamental disorder in Duchenne dystrophy is myopathic.

In summary these special EMG techniques have revealed reorganization of muscle fibers within the motor unit with focal grouping and silent areas, sometimes with a rather normal number of muscle fibers in individual motor units. The conduction in nerve and transmission across motor endplates is disturbed. Muscle membrane characteristics are changed and ephaptic transmission occurs. Electrical nerve stimulation has in some studies also shown a reduced number of excitable axons which is taken as evidence for abnormalities outside the muscle fibers, at least in myotonic dystrophy where similar results have been reported by different investigators.

Myasthenia gravis

Conventional needle EMG: In some cases fibrillation potentials may be seen, perhaps a sign of denervation of muscle fibers caused by destruction of motor endplates. Other spontaneous activity is usually absent in the untreated myasthenic patient. MUPs recorded with concentric or monopolar electrodes often show clear variability in shape over consecutive discharges (Lindsley 1935) due to increased jitter and intermittent blocking of individual components. The amplitude may decrease during continuous activity in pronounced cases producing small MUPs of complex shape that are similar to those in myopathies. Immediately after a pause, however, the MUP configuration reverts towards normal. The interference pattern at maximal voluntary contraction is dense.

The most important reason for performing conventional EMG is to exclude other diseases that resemble, or occur concomitantly with, myasthenia, such as polymyositis, thyroid myopathy, or steroid myopathy. In steroid myopathy the changes are minimal, whereas in the former conditions the MUPs have short duration, polyphasic appearance and rather stable shape over consecutive discharges. It should be remembered, however, that the MUPs usually show some degree of shape variability in acute polymyositis.

Repetitive nerve stimulation: The classical test for myasthenia is repetitive nerve stimulation with 4–10 pulses at a frequency of 2–10 Hz (Fig. 19). The difference in amplitude and area of the M-response between the first and fourth or fifth impulse is usually the parameter used to quantify a decrementing response. The test is also performed immediately (within 10–20 s) after a short period of maximal voluntary activity or high frequency stimulation. In myasthenia gravis, facilitation is often seen in two ways: the amplitude of the first response increases or the decrement becomes less prominent compared to the test prior to activity. These phenomena may occur independently. The initial M-response amplitude in myasthenia gravis is usually not less than 5 mV in the rested hand muscle and the amplitude increment after activation is usually less than 100%. In contrast, the amplitude in the Lambert-Eaton syndrome is initially low and increases often by more than 100%.

After prolonged stimulation or voluntary activation, posttetanic exhaustion or postactivation exhaustion may be seen after a rest of 3–10 minutes, depending on the length of activation and the clinical severity of the disease. This appears as a decreasing initial amplitude or more pronounced decrement. The decrement may be enhanced by warming, ischemia, or regional injection of curare.

Single-fiber EMG: SFEMG is a sensitive test of neuromuscular transmission and is therefore helpful in the diagnosis of myasthenia gravis and other disorders of neuromuscular transmission. When studying fiber pairs, a mixture of findings is observed: normal jitter, increased jitter, or increased jitter with intermittent blocking (see Fig. 4c). The proportion of recordings showing blocking increases with severity of the disease. In muscles with no blocking but only increased jitter there is no weakness and the decrement is typically normal. The fact that SFEMG detects subclinical changes makes it valuable in early diagnosis. In 304 myasthenia gravis patients the following results were obtained (Stålberg and Sanders 1981). In ocular myasthenia gravis, the detection rate was 85% if recordings from orbicularis oculi and EDC were combined. In generalized myasthenia gravis, the detection rate was better than 96%. Patients in complete remission had abnormal SFEMG findings in 69%. Increased jitter and blocking are not specific for myasthenia gravis but indicate disturbed neuromuscular transmission. Hence, this may also be seen in early reinnervation.

In 71 patients (Hilton-Brown et al. 1982), the FD was significantly increased compared with normal controls. The increase was not related to severity or duration of symptoms. It was more pronounced in patients treated with cholinesterase inhibitors. The reorganization of the motor unit in untreated patients may be due to destruction of individual motor endplates giving rise to denerva-

tion and subsequent reinnervation. The additional changes seen in treated patients were attributed to neurotoxic effects of cholinesterase inhibitors (Engel et al. 1973).

SFEMG has also shown slight neuromuscular disturbances in relatives of patients with myasthenia gravis, suggesting a genetic factor (Stålberg et al. 1976b; Hokkanen 1978).

Single fiber recording used with intramuscular electrical stimulation of motor axons may also prove valuable, as it can help differentiate typical myasthenic disturbance of neuromuscular transmission from that responsible for Lambert-Eaton myasthenic syndrome (LEMS) at the level of single motor endplates. When tested at different stimulation frequencies, most motor endplates show an increase in jitter when the rate is raised from 0.5–1 Hz to 3–5 Hz. However, many show a reduced jitter and degree of blocking when the rate is further increased to 10–20 Hz. This is obviously due to the prevailing effect of tetanic facilitation. On the other hand, in LEMS the jitter and degree of blocking become dramatically reduced when the stimulation frequency is increased from below 5 Hz to 10 Hz or more (Trontelj and Stålberg 1990).

Stimulation SFEMG is also helpful in studying very weak muscles in which the patients find it difficult to maintain steady contraction. Finally, it may be easier to perform than voluntary activation in patients who have a considerable proportion of motor endplates in permanent transmission block, making it difficult to collect a sufficient number of recordings from paired potentials.

Other tests such as the stapedius reflex and intracellular recordings also have high detection rates. Tonography and nystagmography have a lower yield in cases of mild myasthenia gravis.

Myasthenic syndrome

Several other disorders of neuromuscular transmission have been identified and classified as myasthenic syndromes. The first described is the most common, the Lambert-Eaton myasthenic syndrome (Eaton and Lambert 1957), which may be associated with carcinoma, especially small-cell bronchogenic carcinoma. Lambert (1986) reported that 52% of 155 patients with the syndrome had an underlying tumor. Although, in typical cases, the clinical findings differ from those of myasthenia gravis, in some situations the diagnosis may be difficult. Electrophysiologic tests are essential because the condition is defined by the response to repetitive stimulation.

Repetitive nerve stimulation: Typical findings with this test are the following: there is a small initial M-response amplitude (less than 5 mV in hand muscles) after a single-nerve stimulus, associated with normal evoked sensory nerve action potentials. With low-frequency repetitive nerve stimulation up to 10 Hz, the M-response shows a decrement similar to that in myasthenia gravis. In contrast, a high stimulation rate (above 10 Hz) or a preceding short voluntary contraction for 10–20 seconds produces a pronounced increase in the M-response amplitude, typically by more than 100%. Posttetanic or postactivation exhaustion then follows, as in myasthenia gravis. In LEMS, both distal and proximal muscles are involved. Changes can also be found in muscles without clinical symptoms.

Conventional EMG usually shows no spontaneous activity. During slight voluntary contraction, the shape of individual MUPs varies considerably with repetitive firing, indicating jitter and intermittent blocking of individual fiber components. During maximal voluntary contraction, one can see an incrementing amplitude, but this is not a reliable sign since it cannot be clearly distinguished from a normal recruitment. When the patient has a malignancy, other secondary neurogenic changes may be superimposed.

SFEMG shows increased jitter in all patients. A pronounced increase in jitter with blocking is often seen. At increasing innervation rates (or stimulus frequency when using stimulation SFEMG), the degree of blocking may decrease, contrary to the finding in myasthenia gravis. The difference between myasthenia gravis and LEMS is most pronounced when the stimulation rate is raised from 0.1 or 1 Hz to 2 or 5 Hz (Trontelj and Stålberg 1990). With similar degree of muscular weakness and fatigue, the SFEMG findings are usually much more abnormal in LEMS than in myasthenia gravis. This is because of excessive variation in the amount of transmitter released on successive nerve terminal depolarizations in LEMS, giving rise to large variation in the EPP amplitude.

Atypical cases

There are also patients with abnormalities that suggest both myasthenia gravis and LEMS in the same patient. In one group, there is a clinical picture of myasthenia gravis, with no malignancy, abnormal titers of antibodies against acetylcholine receptors and a facilitating response. In another group are patients with lung cancer and clinical myasthenic syndrome but EMG findings characteristic of myasthenia gravis (Sanders and Stålberg 1987). Other myasthenic syndromes are described in detail elsewhere in this book.

Neonatal myasthenia

In approximately 12% of infants born to myasthenic mothers, transient signs of myasthenia are seen. There is a transplacental transfer of cholinergic receptor antibodies from myasthenic mothers to their children. In addition, affected children show a transient synthesis of autoreceptor antibodies, which seem to be a factor in the pathogenesis of this condition (Lefvert and Osterman 1983). Neurophysiological abnormalities are comparable to myasthenia gravis with a decrementing response particularly at higher stimulation rates. These patients can be conveniently studied with stimulation SFEMG. Usually the clinical signs disappear within 6 weeks. These children do not show an increased tendency to develop myasthenia gravis later in life.

SUMMARY

The findings in EMG may be succinctly described in terms of hyperexcitability, reorganization of the motor unit and changes in the number of motor units. This permits an interpretation of motor unit anatomy and physiology in the normal and pathological states. Analysis of these findings can usually localize the abnormality to the central or peripheral nervous systems and in the latter case to a certain level within the motor unit (i.e. ventral horn cell, axon, neuromuscular junction or muscle fiber). In some cases, more specific diagnostic considerations may be deduced from the EMG data. Based on the recorded signals, it is important for the electromyographer to describe and interpret the structural and functional status of the motor unit as completely as possible. The EMG and other neurophysiological tests (e.g. nerve conduction studies, twitch measurements, etc.) can be combined with muscle biopsy and biochemical tests to complement and elucidate the clinical findings and ultimately lead to an exact diagnosis.

Acknowledgements

The investigations were supported by the Swedish Medical Research Council, Grant 135 (ES) and Research Community of Slovenia, Grant C3-0558- 306 (JT).

REFERENCES

BALLANTYNE, J.P. and S. HANSEN: Computer method for the analysis of evoked motor unit potentials. 1. Control subjects and patients with myasthenia gravis. J. Neurol. Neurosurg. Psychiatry 37 (1974a) 1187–1194.

BALLANTYNE, J.P. and S. HANSEN: New method for the estimation of the number of motor units in a muscle. 2. Duchenne, limb-girdle, facioscapulohumeral and myotonic muscular dystrophies. J. Neurol. Neurosurg. Psychiatry 37 (1974b) 1195–1201.

BOHAN, A., J.B. PETER, R.L. BOWMAN and C.M. PEARSON: A computer-assisted analysis of 153 patients with polymyositis and dermatomyositis. Medicine 56 (1977) 255–286.

BRANDT, N.J., F. BUCHTHAL, F. EBBESEN, Z. KAMIENIECKA and C. KRARUP: Post-tetanic mechanical tension and evoked potentials in McArdle's disease. J. Neurol. Neurosurg. Psychiatry 40 (1977) 920–925.

BROWN, W.F., M.J. STRONG and R. SNOW: Methods for estimating numbers of motor units in biceps-brachialis muscles and losses of motor units with aging. Muscle Nerve 11 (1988) 423–432.

BUCHTHAL, F.: Electromyography (Part B). In: A. Remond (Ed.), Handbook of Electroencephalography and Clinical Neurophysiology. Amsterdam, Elsevier (1973) 65–97.

BUCHTHAL, F.: Diagnostic significance of the myopathic EMG. In: L. Rowland (Ed.), Pathogenesis of human muscular dystrophies. Amsterdam, Excerpta Medica (1977) 205–218.

BUCHTHAL, F. and P. ROSENFALCK: Electrophysiological aspects of myopathy with particular reference to progressive muscular dystrophy. In: G.H. Bourne and M.N. Golarz (Eds.), Muscular Dystrophy in Man and Animals. New York, Hafner (1963) 193– 262.

BUCHTHAL, F. and P. ROSENFALCK: Spontaneous electrical activity of human muscle. Electroenceph. Clin. Neurophysiol. 20 (1966) 321–336.

BUCHTHAL, F., P. PINELLI and P. ROSENFALCK: Action potential parameters in normal human muscle and their physiological determinants. Acta Physiol. Scand. 32 (1954) 219–229.

BUCHTHAL, F., C. GULD and P. ROSENFALCK: Propagation velocity in electrically activated muscle fibers in man. Acta Physiol. Scand. 34 (1955) 75–89.

BUCHTHAL, F., P. ROSENFALCK and P. ERMINIO: Motor unit territory and fiber density in myopathies. Neurology (Minneap) 10 (1960) 398–408.

CONRADI, S., L. GRIMBY and G. LUNDEMO: Pathophysiology of fasciculations in ALS as studied by electromyography of single motor units. Muscle Nerve 5 (1982) 202–208.

CORNBLATH, D.R., J.T. SLADKY and A. SUMNER: Clinical electrophysiology of infantile botulism. Muscle Nerve 6 (1983) 448–452.

DESMEDT, J.E.: Motor unit types, recruitment and plasticity in health and disease. In: Progress in Clinical Neurophysiology, vol. 9. Basel, Karger (1981).

DESMEDT, J.E.: Computer aided electromyography. In: Progress in Clinical Neurophysiology, vol. 10. Basel, Karger (1983).

DESMEDT, J.E. and S. BORENSTEIN: Collateral reinnervation of muscle fibre by motor axons of dystrophic motor units. Nature (London) 246 (1973) 500–501.

DESMEDT, J.E. and S. BORENSTEIN: Relationship of spontaneous fibrillation potentials to muscle fibre segmentation in human muscular dystrophy. Nature (London) 258 (1975) 531–534.

DESMEDT, J.E. and S. BORENSTEIN: Regeneration in Duchenne muscular dystrophy. Arch. Neurol. 33 (1976) 642–650.

DEVERE, R. and W.G. BRADLEY: Polymyositis: its presentation, morbidity and mortality. Brain 98 (1975) 637–666.

DEWEERD, J.P.C.: Quantitative methods in electromyography and electroneurography. In: S.L.H. Notermans (Ed.), Current Practice of Clinical Electromyography. Amsterdam, Elsevier (1984) 479–522.

DUBOWITZ, V. and M.H. BROOKE: Muscle Biopsy, A Modern Approach. London, Saunders (1973).

EATON, L.M. and E.H. LAMBERT: Electromyography and electric stimulation of nerves in diseases of motor unit: observations on the myasthenic syndrome associated with malignant tumors. J. Am. Med. Assoc. 163 (1957) 1117–1124.

EKSTEDT, J. and E. STÅLBERG: Abnormal connections between skeletal muscle fibers. Electroencephalogr. Clin. Neurophysiol. 27 (1969) 607–609.

EMERYK, B., J. HAUSMANOWA-PETRUSEWICZ and T. NOWAK: Spontaneous volleys of bizarre high frequency potentials (b.h.f.p) in neuromuscular diseases. I. Occurrence of spontaneous volleys of b.h.f.p. in neuromuscular diseases. Electromyogr. Clin. Neurophysiol. 14 (1974) 303–312.

ENGEL, A.G., E.H. LAMBERT and T. SANTA: Study of long-term anticholinesterase therapy: effects on neuromuscular transmission and on motor endplate fine structure. Neurology (Minneap) 23 (1973) 1273–1281.

ERMINIO, F., F. BUCHTHAL and P. ROSENFALCK: Motor

unit territory and muscle fiber concentration in paresis due to peripheral nerve injury and anterior horn cell involvement. Neurology (Minneap) 9 (1959) 657–671.

FALCK, B.: Automatic analysis of individual motor unit potentials with a special two channel electrode. Turku, Academic thesis Turku (1983).

FUGLSANG-FREDERIKSEN, A. and A. MÅNSSON: Analysis of electrical activity of normal muscle in man at different degrees of voluntary effort. J. Neurol. Neurosurg. Psychiatry 38 (1975) 683–694.

FUGLSANG-FREDERIKSEN, A., U. SCHEEL and F. BUCHTHAL: Diagnostic yield of analysis of the pattern of electrical activity of muscle and of individual motor unit potentials in myopathy. J. Neurol. Neurosurg. Psychiatry 39 (1976) 742–750.

FUGLSANG-FREDERIKSEN, A., U. SCHEEL and F. BUCHTHAL: Diagnostic yield of the analysis of the pattern of electrical activity of muscle and of individual motor unit potentials in neurogenic involvement. J. Neurol. Neurosurg. Psychiatry 40 (1977) 544–554.

GATH, I. and E. STÅLBERG: On the measurement of fibre density in human muscle. Electroencephalogr. Clin. Neurophysiol. 54 (1982) 699–706.

GRIMBY, L. and J. HANNERTZ: Flexibility of recruitment order of continuously and intermittently discharging motor units in voluntary contraction. In: J.E. Desmedt (Ed.), Motor Unit Types, Recruitment and Plasticity in Health and Disease. Progress in Clinical Neurophysiology, vol. 9. Basel, Karger (1981) 201–211.

HAKELIUS, L. and E. STÅLBERG: Electromyographical studies of free autogenous muscle transplants in man. Scand. J. Plast. Reconstr. Surg. 8 (1974) 211–219.

HAUSMANOWA-PETRUSEWICZ, I. and J. KOPEC: Quantitative EMG and its automation. In: J.E. Desmedt (Ed.), Computer-Aided Electromyography. Basel, Karger (1983) 164–185.

HAYWARD, M. and R.G. WILLISON: The recognition of myogenic and neurogenic lesions by quantitative EMG. In: J.E. Desmedt (Ed.), New Developments in Electromyography and Clinical Neurophysiology. Basel, Karger (1973) 448–453.

HENNEMAN, E., G. SOMJEN and D.O. CARPENTER: Functional significance of cell size in spinal motor neurone. J. Neurophysiol. 28 (1965) 560–580.

HENRIKSSON, K.G. and E. STÅLBERG: The terminal innervation pattern in polymyositis: a histochemical and SFEMG study. Muscle Nerve 1 (1978) 3–13.

HILTON-BROWN, P. and E. STÅLBERG: The motor unit in muscular dystrophy, a single fibre EMG and scanning EMG study. J. Neurol. Neurosurg. Psychiatry 46 (1983a) 981–995.

HILTON-BROWN, P. and E. STÅLBERG: Motor unit size in muscular dystrophy; a Macro EMG and Scanning EMG study. J. Neurol. Neurosurg. Psychiatry 46 (1983b) 996–1005.

HILTON-BROWN, P. and E. STÅLBERG: Size of motor units and firing rate in muscular dystrophy. In: Vrbová and M. Dimitrijevic (Eds.), Recent

Achievements in Restorative Neurology. 2. Progressive Neuromuscular Diseases. Basel, Karger (1986) 289–304.

HILTON-BROWN, P., E. STÅLBERG and P.O. OSTERMAN: Signs of reinnervation in myasthenia gravis. Muscle Nerve 5 (1982) 215–221.

HILTON-BROWN, P., E. STÅLBERG, J.V. TRONTELJ and M. MIHELIN: Causes of the increased fibre density in muscular dystrophies studied with single fibre EMG during electrical stimulation. Muscle Nerve 8 (1985) 383–388.

HOKKANEN, E., B. EMERYK-SZAJEWSKA and K. ROWINSKA-MARCINSKA: Evaluation of the jitter phenomenon in myasthenic patients and their relatives. J. Neurol. 219 (1978) 73–82.

HÅKANSSON, C.H.: Conduction velocity and amplitude of the action potential as related to the circumference in the isolated fibre of frog muscle. Acta Physiol. Scand. 37 (1956) 14–34.

JABLECKI, C. and P. SCHULTZ: Single muscle fibre recordings in the Schwartz-Jampel syndrome. Muscle Nerve 5 (9S) (1982) S64–S69.

JANKO, M., J.V. TRONTELJ and K. GERSÅK: Fasciculations in motor neuron disease; discharge rate reflects extent and recency of collateral sprouting. J. Neurol. Neurosurg. Psychiatry 52 (1989) 1375–1381.

KARPATI, G. and W.K. ENGEL: Type grouping in skeletal muscle after experimental reinnervation. Neurology (Minneap) 18 (1968) 447–455.

KERESHI, S., A. MCCOMAS, N. KOWALUCHUK and A. STURART: Absence of fibrillation after muscle injury. Exp. Neurol. 80 (1983) 645–651.

KIMURA, J.: Electrodiagnosis in Diseases of Nerve and Muscle: Principles and Practice. Philadelphia, F.A. Davis (1989).

KRENDEL, D., D.B. SANDERS and J.M. MASSEY: Single fiber electromyography in chronic external opthalmoplegia. 32nd AAEE Annual Meeting, Las Vegas (1985).

KUGELBERG, E.: Electromyography in muscular dystrophy. J. Neurol. Neurosurg. Psychiatry 12 (1949) 129–136.

KUGELBERG, E., L. EDSTRÖM and M. ABRUZZESE: Mapping of motor units in experimentally reinnervated rat muscle. J. Neurol. Neurosurg. Psychiatry 33 (1970) 319–329.

LAMBERT, E.H.: The Lambert-Eaton Myasthenic Syndrome; Clinical Features and Pathophysiology. 7th International Conference on Myasthenia Gravis, The New York Academy of Sciences (1986).

LEFEVER, R.S. and C.J. DELUCA: A procedure for decomposing the myoelectric signal into its constituent action potentials. I. Technique, theory and implementation. IEEE Trans. Biomed. Eng. 29 (1982) 149–157.

LEFVERT, A.K. and P.O. OSTERMAN: Newborn infants to myasthenic mothers: a clinical study and an investigation of acetylcholine receptor antibodies in 17 children. Neurology (Minneap) 33 (1983) 133–138.

LINDSLEY, D.B.: Electrical activity of human motor

units during voluntary contraction. Am. J. Physiol. 114 (1935/36) 90–99.

LINDSTRÖM, L. and I. PETERSÉN: Power spectra of myoelectric signals: motor unit activity and muscle fatigue. In: E. Stålberg and R.R. Young (Eds.), Clinical Neurophysiology. London, Butterworths (1981) 66–87.

LJUNGHALL, S., G. ÅKERSTRÖM, G. JOHANSSON, Y. OLSSON and E. STÅLBERG: Neuromuscular involvement in primary hyperparathyroidism. J. Neurol. 231 (1984) 263–265.

LUDIN, H.P.: Electromyography in Practice. Stuttgart-New York, Georg Thieme Verlag (1980).

MCCOMAS, A.J.: Neuromuscular Function and Disorders. London, Butterworth & Co. (1977).

MCCOMAS, A.J., P.R.W. FAWCETT, M.J. CAMPBELL and R.E.P. SICA: Electrophysiological estimation of the number of motor units within a human muscle. J. Neurol. Neurosurg. Psychiatry 34 (1971a) 121–131.

MCCOMAS, A.J., R.EP. SICA and S. CURRIE: An electrophysiological study of Duchenne dystrophy. J. Neurol. Neurosurg. Psychiatry 34 (1971b) 461–468.

MIHELIN, M.: Automatic Recognition and Diagnostics in Single Fiber EMG (Thesis, in Slovene). Ljubljana, University of Ljubljana (1983).

MIHELIN, M., J.V. TRONTELJ and E. STÅLBERG: Muscle fiber recovery functions studied with double pulse stimulation. Muscle Nerve 14 (1991) 739–747.

MILNER-BROWN, H.S., R.B. STEIN and R. YEMM: The orderly recruitment of human motor units during voluntary isometric contractions. J. Physiol. (London) 230 (1973a) 359–370.

MILNER-BROWN, H.S., R.B. STEIN and R. YEMM: Changes in firing rate of human motor units during linearly changing voluntary contractions. J. Physiol. (London) 230 (1973b) 371–390.

NANDEDKAR, S. and E. STÅLBERG: Simulation of macro EMG motor unit potentials. EEG Clin. Neurophysiol. 56 (1983) 52–62.

NANDEDKAR, S.D., E. STÅLBERG and D.B. SANDERS: Simulation techniques in electromyography. IEEE Trans. Biomed. Eng. 32 (1985) 775–785.

NANDEDKAR, S., D.B. SANDERS and E.V. STÅLBERG: Simulation and analysis of the electromyographic interference pattern in normal muscle. I. Turns and amplitude measurement. Muscle Nerve 9 (1986a) 423–430.

NANDEDKAR, S, D.B. SANDERS and E.V. STÅLBERG: Automatic analysis of the electromyographic interference pattern. II. Findings in control subjects and in some neuromuscular diseases. Muscle Nerve 9 (1986b) 491–500.

NANDEDKAR, S.D., D.B. SANDERS, E. STÅLBERG and S. ANDREASSEN: Simulation of concentric needle EMG motor unit action potentials. Muscle Nerve 11 (1988) 151–159.

NOTERMANS, S.L.H.: Current Practice of Clinical Electromyography. Amsterdam, Elsevier (1984).

OH, S.J.: Clinical Electromyography. Nerve Conduction Studies. Baltimore, University Park Press (1984).

PANAYIOTOPOULOS, C.P., S. SCARPALEZOS and T. PAPAPETROPOULOS: Electrophysiologic estimation of motor units in Duchenne muscular dystrophy. J. Neurol. Sci. 23 (1974) 89–98.

PARTANEN, J.: Fibrillation potentials and end-plate spikes. Electroencephalogr. Clin. Neurophysiol. 66 (1987) S78.

PAYAN, J.: The blanket principle: a technical note. Muscle Nerve 1 (1978) 423–426.

PINELLI, P. and F. BUCHTHAL: Muscle action potentials in myopathies with special regard to progressive muscular dystrophy. Neurology (Minneap) 3 (1953) 347.

PURVES, D. and B. SAKMANN: Membrane properties underlying spontaneous activity of denervated muscle fibres. J. Physiol. 239 (1974) 125.

PUVANENDRAN, K., J.S. CHEAH, N. NAGANATHAN, P.P.B. YEO and P.K. WONG: Thyrotoxic myopathy: a clinical and quantitative analytic electromyographic study. J. Neurol. Sci. 42 (1979) 441–451.

RAMSAY, I.D.: Electromyography in thyreotoxicosis. Q. J. Med. 34 (1965) 255–267.

ROSE, A.L. and R.G. WILLISON: Quantitative electromyography using automatic analysis: studies in healthy subjects and patients with primary muscle disease. J. Neurol. Neurosurg. Psychiatry 30 (1967) 403–410.

RÜDEL, R. and LEHMANN-HORN: Membrane changes in cells from myotonia patients. Physiol. Rev. 65 (1985) 310–356.

SALICK, A.I. and C.M. PEARSON: Electrical silence of myoedema. Neurology (Minneap) 17 (1967) 899.

SANDERS, D.B. and E. STÅLBERG: The overlap between myasthenia gravis and Lambert-Eaton myasthenic syndrome. In: D.B. Drachman (Ed.), Myasthenia Gravis: Biology and Treatment. Ann. N.Y. Acad. Sci. 505 (1987) 864–865.

SANDSTEDT, P.: Quantitative Examination in Neuromuscular Disorders Studied by Muscle Biopsy and Electromyography. Linköping, Academic thesis Linköping (1981).

SCHMALBRUCH, H.: Manifestations of regeneration in myopatic muscle. In: A. Mauro (Ed.), Muscle Regeneration. New York, Raven Press (1979) 201–232.

SIMPSON, J.A.: Disorders of neuromuscular transmission. Proc. R. Soc. Med. 59 (1966) 993.

STÅLBERG, E.: Propagation velocity in human muscle fibres in situ. Uppsala, Acta Physiol. Scand. 70 (1966) 1–112.

STÅLBERG, E.: Electrogenesis in human dystrophic muscle. In: L.P. Rowland (Ed.), Pathogenesis of Human Muscular Dystrophies. Amsterdam-Oxford, Excerpta Medica (1977) 570–587.

STÅLBERG, E.: Macro EMG, a new recording technique. J. Neurol. Neurosurg. Psychiatry 43 (1980) 475–482.

STÅLBERG, E.: Single fiber EMG, macro EMG, and scanning EMG: new ways of looking at the motor unit. Crit. Rev. Clin. Neurobiol. 2 (1986) 125–167.

STÅLBERG, E.: Use of SFEMG and Macro EMG in

study of reinnervation. Muscle Nerve 13 (1990) 804–813.

STÅLBERG, E.: Electrodiagnostic assessment and monitoring of motor unit changes in disease (Lambert lecture 1989). Muscle Nerve 14 (1991) 293–303.

STÅLBERG, E. and L. ANTONI: Electrophysiological cross section of the motor unit. J. Neurol. Neurosurg. Psychiatry 43 (1980) 469–474.

STÅLBERG, E. and K. BOON: Electromyography. In: L. Da Silva (Ed.), Handbook of Clinical Neurophysiology. Amsterdam, Elsevier Science Publishers (1986) 449–497.

STÅLBERG, E. and P. DIOSZEGHY: Scanning EMG in normal muscle and in neuromuscular disorders. Electroencephalogr. Clin. Neurophysiol. 81 (1991) 403–416.

STÅLBERG, E. and P.R.W. FAWCETT: Macro EMG changes in healthy subjects of different ages. J. Neurol. Neurosurg. Psychiatry 45 (1982) 870–878.

STÅLBERG, E. and D.B. SANDERS: Electrophysiological tests of neuromuscular transmission. In: E. Stålberg and R. Young (Eds.), Clinical Neurophysiology. London, Butterworth (1981) 88–116.

STÅLBERG, E. and D.B. SANDERS: The motor unit in ALS studied with different neurophysiological techniques. In: C.F. Rose (Ed.), Research Progress in Motor Neurone Disease. London, Pitman Books (1984) 105–122.

STÅLBERG, E. and S. STÅLBERG: The use of small computers in the EMG lab. In: J.E. Desmedt (Ed.), Computer-Aided Electromyography and Expert Systems. Amsterdam, Elsevier (1989) 1–32.

STÅLBERG, E. and J. TRONTELJ: Single Fibre Electromyography. Old Woking, Surrey, Mirvalle Press (1979).

STÅLBERG, E. and J.V. TRONTELJ: Abnormal discharges generated within the motor unit as observed with single fibre electromyography. In: W.J. Culp and J. Ochoa (Eds.), Abnormal Nerves and Muscles as Impulse Generators. Oxford, Oxford University Press (1982) 443–474.

STÅLBERG, E., J.V. TRONTELJ and M. JANKO: Single fibre EMG findings in muscular dystrophy. In: I. Hausmanowa-Petrusewics and H. Jedrzejowska (Eds.), Structure and Function of Normal and Diseased Muscle and Peripheral Nerve. Warsaw, Polish Medical Publishers (1974) 185–190.

STÅLBERG, E., H.H. SCHILLER and M.S. SCHWARTZ: Safety factor in single human motor end-plates studied in vivo with single fibre electromyography. J. Neurol. Neurosurg. Psychiatry 38 (1975) 799–804.

STÅLBERG, E., M.S. SCHWARTZ, B. THIELE and H. SCHILLER: The normal motor unit in man. J. Neurol. Sci. 27 (1976a) 291–301.

STÅLBERG, E., J.V. TRONTELJ and M.S. SCHWARTZ: Single muscle fibre recording of the jitter phenomenon in patients with myasthenia gravis and in members of their families. Ann. N.Y. Acad. Sci. 274 (1976b) 189–202.

STÅLBERG, E., J. CHU, V. BRIL, S. NANDEDKAR, S.

STÅLBERG and M. ERICSSON: Automatic EMG analysis of EMG interference pattern. Electroencephalogr. Clin. Neurophysiol. 56 (1983) 672–681.

STÅLBERG, E., S. ANDREASSEN, B. FALCK, H. LANG, A. ROSENFALCK and W. TROJABORG: Quantitative analysis of individual motor unit potentials — a proposition for standardized terminology and criteria for measurement. J. Clin. Neurophysiol. 3 (1986) 313–348.

STÅLBERG, E., J.V. TRONTELJ and M. MIHELIN: Electrical microstimulation with single fibre EMG: a useful method to study physiology of the motor unit. J. Clin. Neurophysiol. 9(1) (1992) 105–119.

THESLEFF, S.: Fibrillation in denervated mammalian skeletal muscle. In: W.J. Culp and J. Ochoa (Eds.), Abnormal Nerves and Muscles as Impulse Generators. Oxford, Oxford University Press (1982) 678–694.

THESLEFF, S. and L.C. SELLIN: Denervation supersensitivity. Trends Neurosci. 4 (1980) 122.

THESLEFF, S. and M.R. WARD: Studies on the mechanism of fibrillation potentials in denervated muscle. J. Physiol. (London) 244 (1975) 313–323.

THESLEFF, S., F. VISKOCIL and M.R. WARD: The action potential in end-plate extrajunctional regions of rat skeletal muscle. Acta Physiol. Scand. 91 (1974) 196–202.

THIELE, B. and A. BOEHLE: Anzahl der Spike-Komponenten im Motor Unit Potential. EEG-EMG 9 (1978) 125–130.

TORBERGSEN, T., E. STÅLBERG, J. AASLY and S. LINDAL: Myopathy in Marinesco-Sjögren syndrome: an electrophysiological study. Acta Neurol. Scand. 84 (1991a) 132–138.

TORBERGSEN, T., E. STÅLBERG and J. BLESS: Nerve-muscle involvement in a large family with mitochondrial cytopathy. Muscle Nerve 14 (1991b) 35–41.

TRONI, W.: Human muscle fiber conduction velocity in clinical application. In: P. Pinelli, C. Pasetti and G. Mora (Eds.), Neurophysiological Contributions for Assessing Rehabilitation in Lower and Upper Motorneuron Diseases. Padova, Liviana Editrice (1984) 77–84.

TRONI, W., R. CANTELLA and I. RAINERO: Conduction velocity along human muscle fibers in situ. Neurology (Cleveland) 33 (1983) 1453–1459.

TRONTELJ, J.V. and E. STÅLBERG: Responses to electrical stimulation of denervated human muscle fibre recorded with single fibre EMG. J. Neurol. Neurosurg. Psychiatry 46 (1983a) 305–309.

TRONTELJ, J.V. and E. STÅLBERG: Bizarre repetitive discharges recorded with single fibre EMG. J. Neurol. Neurosurg. Psychiatry 46 (1983b) 310–316.

TRONTELJ, J.V. and E. STÅLBERG: Single motor end-plates in myasthenia gravis and LEMS at different firing rates. Muscle Nerve 14 (1990) 226–232.

TRONTELJ, J.V., M. MIHELIN, J.M. FERNANDEZ and E. STÅLBERG: Axonal stimulation for end-plate jitter studies. J. Neurol. Neurosurg. Psychiatry 49 (1986) 677–685.

TRONTELJ, J., J. ZIDAR, M. DENISLIC, D.B. VODUSEK and

M. MIHELIN: Facioscapulohumeral dystrophy: jitter in facial muscles. J. Neurol. Neurosurg. Psychiatry 51 (1988) 1988.

TRONTELJ, J., E. STÅLBERG and M. MIHELIN: Jitter in the muscle fiber. J. Neurol. Neurosurg. Psychiatry 53 (1990) 49–54.

TRONTELJ, J.V., E. STÅLBERG and M. MIHELIN: Extracellularly recorded single muscle fibre responses to electrical stimulation in myotonia congenita. (1992) in press.

VILPPULA, A.: Muscular disorders in some collagen diseases. Acta Med. Scand. Suppl. 540 (1972) 1–47.

VRBOV, G.: Duchenne dystrophy viewed as a disturbance of nerve-muscle interactions. Muscle Nerve 6 (1983) 671–675.

WILBOURN, A.: The EMG examination with myopathies. Proc. Am. Assoc. Electromyography Electrodiagnosis, Rochester, U.S.A. (1987) 7–20.

Handbook of Clinical Neurology, Vol. 18 (62): Myopathies
L.P. Rowland and S. DiMauro, editors

CHAPTER 3

Cultured normal and genetically abnormal human muscle

VALERIE ASKANAS and W. KING ENGEL

University of Southern California School of Medicine, USC Neuromuscular Center, Los Angeles, CA, USA

Tissue culture of human muscle provides an unusual opportunity to study the development of normal and genetically abnormal human muscle. In contrast to studies related to animal muscle, which in addition to tissue culture, can also be performed in vivo, both during embryonic development and under various experimental conditions in an adult animal, studies related to human muscle rely mainly on tissue culture approach, since invivo material is not available for experimentation.

In normal human muscle, tissue culture provides possibilities for the studies of: 1) developmental expression of human muscle-specific genes and their products; 2) mechanisms regulating human muscle-specific genes and their products; 3) intracellular and extracellular environmental requirements for development of muscle phenotype; 4) influencing 1, 2, and 3 by: (a) trophic factors; (b) hormones; (c) neuronal factors; and (d) muscle contractile activity; 5) genetic manipulations; and 6) models of immunologic human muscle diseases, i.e. polymyositis and myasthenia gravis, by using patients' lymphocytes or immunoglobulins.

In genetically abnormal human muscle, tissue culture provides possibilities for the studies of: 1) expression of an abnormal gene, or lack of it; 2) developmental expression of an abnormal gene product, or lack of it; 3) developmental expression of a pathologic phenotype as a consequence of an abnormal gene, or lack of it; 4) conditions required for expression of a pathologic phenotype; and 5) 'treating' a manifested abnormality by: (a) nonspecific means, such as trophic factors, hormones, etc.; (b) specific means, such as transferring of a normal gene; and (c) enhancing a normal 'minor gene' governing a parallel function to the abnormal gene.

During the last 2 decades tissue culture of normal human muscle has provided important information regarding both the developmental expression of muscle-specific gene products and the factors regulating their expression in normal human muscle. Tissue culture of genetically abnormal adult human muscle has provided important information regarding the regulatory mechanisms of the manifestation, or lack of it, of abnormal or missing genes and their products in abnormal muscle.

In this article, we will attempt to describe and critique various methods of culturing of human muscle, briefly review the results regarding tissue culture of normal and genetically abnormal human muscle, and outline future goals. The scope of this article does not permit the review of all studies regarding cultured human muscle, therefore, a possible omission of some important communications may occur. Previous reviews may be helpful (Askanas and Engel 1977, 1979; Askanas 1984; Miranda and Mongini 1986; Witkowski 1986a,b; Askanas et al. 1989).

NORMAL MUSCLE

Aneural cultures

Methodological considerations

It is now well-established that adult human muscle can develop aneurally in tissue culture. Aneural cultures are initiated from satellite cells, which after being liberated from an original muscle specimen are capable of dividing and subsequently initiating muscle growth in vitro. Several methods to establish tissue cultures from satellite cells of adult human muscle have been used.

Each of the methods has its advantages and disadvantages, which we will try to address. In 1975, we described an explant–re-explantation method to culture human muscle (Askanas and Engel 1975). According to this method, monolayer cultures of human muscle are initiated from the microexplants of a biopsied muscle and can be established from patients of any age, including those 80 years and older. In this method, after satisfactory growth of myoblasts has emerged from the explants, the explants are removed and the myoblasts are allowed to fuse and to develop in monolayer. At the time of their removal, the original microexplants are re-explanted into another culture dish and after mononucleated cell growth is again obtained, the original explant can be either re-explanted again or it is discarded. (This decision depends on the number of myoblasts still emerging after the second re-explantation.) After re-explantation, the amount of myoblasts is substantially enriched and the multinucleated myotubes grow and develop with only a minimal number of fibroblasts between them (Fig. 1D) (Askanas and Engel 1975). Because the original muscle explants are removed, the monolayer muscle cultures do not contain any remnants of the original muscle tissue. Using this method, the amount of cultured muscle obtained from a small fraction of diagnostic muscle biopsy is very large (35–40 35-nm Petri dishes from each explantation). Therefore, a variety of multidisciplinary studies, including extensive morphological and biochemical investigations, can be performed simultaneously (Askanas and Engel 1979; Askanas 1984). In addition, even after cultures have been performed, there are always muscle biopsy fragments remaining, which can be vitally frozen for future culturing (Carter and Askanas 1981).

Depending on the composition of medium, human muscle cultured by the explant–re-explantation method survives for up to 2 months can achieve very good biochemical and morphologic differentiation (Askanas and Engel 1975; Meinhofer et al. 1977; Askanas and Gallez-Hawkins 1985; Askanas et al. 1985b), and it can be successfully innervated (see below). Several of the studies demonstrated that this culture method is very suitable for the investigations related to genetically abnormal human muscle, since (a) cultures can be established from individuals of any age and (b) long survival and relatively advanced maturation of cultured muscle enables the expression of the abnormalities in several genetic muscle diseases (reviewed in Askanas and Engel 1977, 1979; Askanas 1984; Askanas et al. 1989).

Another way of establishing a monolayer culture is to trypsinize the cells outgrowing from the muscle microexplants (after the removal of all original explants) and to plate them at desirable densities (Askanas et al. 1985b; Pegolo et al. 1990; and reviewed in Miranda and Mongini 1986 and in Witkowsky 1986a). When monolayer muscle cultures are established using this method, the amount of fibroblasts may be reduced by 'preplating', as described for rat muscle cells (Yaffe 1968). In our laboratory, both mass and clonal human muscle cultures are being established by this method (Fig. 1A,B,C,E) and the number of myoblasts is determined by desmin immunoreactivity (McFerrin et al. 1990). Direct dissociation of an original muscle biopsy fragment has also been used by many investigators to initiate either a mass monolayer or clonal human muscle culture (Yasin et al. 1977; Blau and Webster 1981; Sklar and Brown 1991). In 1977, Yasin et al. described a quantitative method for muscle mass-monolayer culture obtained from muscle biopsies of adult patients. The authors later modified this technique in order to develop clonal muscle cultures (Yasin et al. 1977, 1981). This method is now often used to study clonal growth of human muscle (Sklar and Brown 1991). Purified mass cultures of human muscle cells are being established either from pooled muscle clones (Sklar and Brown 1991) or by a fluorescent cell sorter (Webster et al. 1988).

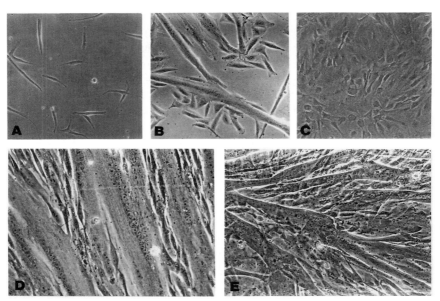

Fig. 1. Phase-contrast inverted microscopy of living human muscle cultured aneurally according to different methods. (A,B) Myoblast cloning performed after population of myoblasts has been enriched by selective preplating. (A) Single cell stage (× 180). (B) Early fusion (× 290). (C) Monolayer of fibroblasts remaining after myoblasts have been removed (× 180). (D) 3-week-old muscle cultured according to the explant–re-explantation technique (× 300). (E) 5-week-old muscle cultured in primary mass monolayer, after cells outgrowing from the explant have been dissociated and myoblasts have been enriched (× 400).

Successful clonal cultures of fetal human muscle have been also performed (Hauschka 1974).

Each culture model has its advantages and disadvantages. The advantages of a primary, monolayer human muscle culture are as follows. 1. Cultures can be easily established from patients regardless of their age and disease process. 2. Since cultured muscle fibers survive for up to 6–8 weeks, it is possible to study many aspects of aneural development and aneural maturation of human muscle.

We and others have shown that the degree of biochemical and morphologic maturation of human muscle cultured aneurally in primary monolayer culture depends on the duration of culture. For example, according to our studies (Pegolo et al. 1990), the percent of muscle-specific isozyme of creatine kinase (CK) in cultured human muscle increases approximately 50% between 1 and 6 weeks of culture, reaching 67% of total CK activity. The percent of muscle-specific isozyme of phosphorylase increases 20%, reaching 18% of the total phosphorylase activity (Pegolo et al. 1990). Previously, muscle-specific isozyme of phosphorylase

could not be demonstrated immediately after fusion, or in early culture, in either normal or phosphorylase-deficient cultured human muscle (Sato et al. 1977; DiMauro et al. 1978), but it was demonstrated in human muscle cultured aneurally for several weeks (Meinhofer et al. 1977; Martinuzzi et al. 1986, 1991; Pegolo et al. 1990). Carnitine uptake into aneurally cultured human muscle is also regulated by muscle differentiation (Avigan et al. 1983; Martinuzzi et al. 1991). Since in several genetic muscle diseases, advanced maturation of cultured human muscle is absolutely required for the genetic defect to manifest itself (Askanas et al. 1989, and see below), it is important to establish culture conditions which would allow a high degree of differentiation.

In addition, because the primary monolayer muscle cultures also contain some fibroblasts, this culture system resembles more the in-vivo situation in which, both during the development and in the adult state, muscle fibers are surrounded by and co-exist with nonmuscle cells. For example, it has been demonstrated that fibroblasts participate in the deposition of type-IV collagen in the basal lamina of myotubes and that animal myotubes cultured in clonal cultures do not synthesize basal lamina, while the myotubes cultured in the presence of fibroblasts do (Kühl et al. 1984; Sanderson et al. 1986; Lipton 1977).

Therefore, the primary monolayer culture system is very suitable when the aim is to study growth and trophic factors which exert their beneficial influence on human muscle, either directly or through their action on nonmuscle cells. Moreover, muscle trophic factors identified in this culture system should be more suitable for their potential future therapeutic application in patients, than those exerting the influence on muscle cell lines or on young myotubes cultured by cloning method.

On the other hand, because of the presence of nonmuscle cells, the mixed human muscle-fibroblast culture is not suitable for studies related to (a) identification of mitogenic factors acting directly on satellite cells or on replicating myoblasts, (b) trophic factors affecting muscle through their direct influence on muscle cell, and (c) gene transfer into myoblasts.

The above obstacles of mixed human primary muscle cultures are overcome by cloning of muscle cells and their subsequent culturing either in individual clones, or in pure mass muscle culture obtained either from pooled clones, or through fluorescent cell sorter (Blau and Webster 1981; Yasin et al. 1981; Miranda et al. 1988b; Webster et al. 1988; Sklar and Brown 1991). Purified muscle culture enables studies related to the mitotic cycle of normal and abnormal satellite cells and to various molecular aspects of early human myogenesis, including the influence of mitogens, hormones, and other factors influencing satellite cell proliferation and fusion. Pure muscle cultures provide important information regarding early human myogenesis in vitro (Webster and Blau 1990; Sklar and Brown 1991). However, there are several disadvantages of cultures obtained from either cloned or purified satellite cells. 1) The potential of cloned-satellite cells to divide decreases with age of a donor (Webster and Blau 1990), making it difficult to establish cultures from older patients. 2) Fused myotubes survive in culture for a very short time, i.e. 50% of myotubes were reported to die either after 4 days of initiation of fusion (Sklar and

Brown 1991) or to detach from the bottom of a petri dish after 6–8 days (Kaplan et al. 1990). 3) The degree of muscle maturation is rather poor, i.e., among CK isozymes, the highest percentage belongs to MB isozyme, even at the peak of muscle differentiation (between 4–6 days after initiation of fusion) (Blau et al. 1983), or to BB isozyme (Yasin et al. 1983). 4) Following the peak of muscle differentiation, which occurs very early (4–6 days after fusion), total CK activity and the number of acetylcholine receptors (AChRs) decline (Blau et al. 1983). 5) There is a poor degree of innervation by fetal rat spinal cord neurons, when compared to primary human muscle culture (V. Askanas, unpublished observation).

Similar differences between cultures obtained by various methods have been described for animal muscle (Florini 1987; Florini et al. 1991). For example, the properties and response of myogenic cells obtained from donors of various ages as well as those cultured as cell clones versus primary culture have been emphasized previously (Florini 1987).

In summary, depending on the final aim to be achieved, different methods to culture human muscle should be used. However, besides the method used to initiate cultures of human muscle, other factors, including media, quality of serum, culture substrata, the source of muscle, etc., will also substantially influence the results, and may be responsible for often diametrically opposite and seemingly controversial data reported by different laboratories. Therefore, because various systems of culturing of human muscle differ considerably, caution must be exercised, not only in the interpretation of the results, but also when the results obtained by various laboratories are compared.

Factors influencing the development of muscle phenotype in cultured human muscle

Fibroblast growth factor (FGF), epidermal growth factor (EGF) and insulin. FGF first discovered in pituitary extracts (Gospodarowicz and Moran 1976) exerts a variety of effects on mesoderm-derived cells. FGF has been shown to inhibit the terminal muscle differentiation in clonal cultures established from embryonic myoblasts and satellite cells of rats and humans (Linkhart et al. 1980). In primary culture of adult human muscle, which is

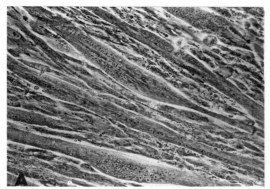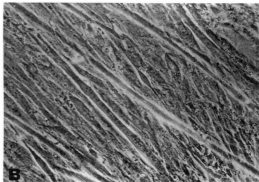

Fig. 2. Phase-contrast inverted microscopy of living human muscle cultured for 28 days. (A) Culture treated continuously with control medium containing insulin, epidermal growth factor, and fibroblast growth factor (IEF) for 28 days. (B) Culture treated with control medium plus IEF for 14 days and then for another 14 days with control medium alone. Note thinning of cultured muscle fibers in B ($\times$ 200). (Reproduced from Askanas and Gallez-Hawkins (1985) (by courtesy of the Editors of *Arch. Neurol.*)

mixed to some degree with fibroblasts, FGF delays, but does not inhibit, myoblast fusion (V. Askanas, unpublished observation) and, when introduced to culture medium after myoblast fusion has been completed, it increases specific activity of creatine kinase and causes AChR clustering (Askanas and Gallez-Hawkins 1985; Askanas et al. 1985b). However, the expression of muscle-specific isozyme of creatine kinase, CK-MM, is not influenced by FGF. The removal of FGF from culture medium for several days results in muscle fiber thinning and its accelerated degeneration (Fig. 2A,B) (Askanas and Gallez-Hawkins 1985). Therefore, it appears that in human muscle, FGF exerts both proliferative and trophic influence.

Insulin is the most universal factor for growth and maturation of virtually all cells in tissue culture (reviewed in Florini 1987; Florini et al. 1991). The beneficial influence of supraphysiologic doses of insulin on differentiation of muscle cells of various animal species has been reported by many investigators (Mandel and Pearson 1979; Sandra and Przybylski 1979) and insulin appears to be an essential component of virtually all serum-free media (reviewed by Florini 1987). We were the first to demonstrate that insulin exerts several important influences on cultured human muscle. Even though insulin alone does not influence total CK activity, addition of insulin to the media containing both FGF and EGF causes synergistic increase of CK activity and significant increase of CK-MM

(Askanas and Gallez-Hawkins 1985). In addition to stimulating the expression of CK in cultured human muscle, insulin causes a significant increase of the total number of AChRs, without having any influence on AChR clustering (Askanas et al. 1985b). Interestingly, addition of insulin to the medium containing EGF, which by itself does not influence either the total number of AChRs or AChR clustering, results in a higher number of AChRs and AChR clustering as compared to the treatment with insulin alone (Askanas et al. 1985b). However, addition of insulin to the medium containing FGF, which alone does not influence the total number of AChRs but significantly increases AChR clustering, does not potentiate the influence of FGF on AChR aggregation. However, combination of all 3 factors has an additive effect on both the number of AChRs and AChR clustering (Askanas et al. 1985a).

EGF alone, which was reported to be the most important factor in inducing muscle differentiation in cloned human satellite cells (Ham et al. 1990), does not influence either the expression of CK or AChRs in primary cultures of human muscle.

In all our studies regarding the influence of the above factors on the expression of CK and AChRs, the factors were added to culture medium after myoblast fusion had been completed (1 week after initiation of cultures), and at least 2 weeks of treatment was required for the maximum effect to be exerted. Therefore, under these experimental

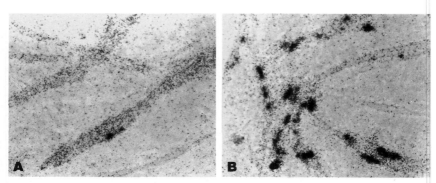

Fig. 3. Autoradiography of ^{125}I-α-bungarotoxin in 28-day-old human muscle cultured in serum-free medium without (A) and with (B) addition of 50 μM hydrocortisone. Note prominent acetylcholine receptor aggregates due to hydrocortisone treatment (× 400).

conditions, the action of all 3 factors should be interpreted as trophic for cultured human muscle, since the increased levels of CK and AChRs do not reflect the increased muscle mass.

In addition, long-term treatment (3 weeks) of aneurally cultured human muscle with insulin also caused a significant increase of both the voltage-dependent calcium channels and Ca^{2+} uptake (Desnuelle et al. 1987). FGF and EGF did not exert any influence on the voltage-dependent calcium channel.

Recently, we have shown that in human muscle cultured in serum-free medium, insulin stimulates both RNA synthesis (Ibrahim et al. 1991) and the phosphorylation of ribosomal S6 protein (Greenlee et al. unpublished observation 1992). It is presently not known whether insulin exerts those metabolic effects on cultured human muscle through its own receptors or through the receptors for insulin-growth factor (IGF-1). Both receptors exist on human muscle grown in clonal cultures (Shimizu et al. 1986) and in primary cultures (V. Askanas and D. Greenlee, unpublished observation 1992).

It is of importance to emphasize that besides influencing a variety of significant metabolic responses in cultured human muscle, pretreatment of aneurally cultured human muscle with FGF and insulin appears to be very important for obtaining their successful innervation by fetal rat spinal cord neurons (below). Even though the reason for this requirement is not known, one can speculate that FGF and insulin induce muscle membrane properties that prepare the muscle to accept innervation (or makes the muscle attractive for a motor neuron

to innervate). Whether this involves membrane properties other than those described above awaits further exploration.

Glucocorticoids. Glucocorticoids induce a variety of effects on cells in vivo and in culture (reviewed in Florini 1987; Florini et al. 1991). Dexamethasone has been shown to stimulate proliferation and differentiation of muscle cells from a variety of species (reviewed in Florini 1987), including clonal human muscle cell lines (Ham et al. 1988).

We have shown that glucocorticoids also exert trophic effects on human muscle cultured aneurally either in serum-free or in serum-supplemented medium. Two-week treatment with glucocorticoids (hydrocortisone (HC) and dexamethasone) initiated 8 days after culture have been established and at the time when fusion of myoblasts was completed increased both the number of AChRs and AChR clustering in human muscle (Figs. 3A,B and 4A,B) (Askanas et al. 1986). The influence of glucocorticoid was dependent on the duration of treatment (Fig. 4A). Enzymatic activity of CK and its isozymes were not influenced by HC treatment (Fig. 4B) (Askanas et al. 1986). Table 1 illustrates our results of the influence of insulin, EGF, FGF and HC on AChR of cultured human muscle.

Subsequently, it has been reported that dexamethasone introduced at the beginning of myoblast fusion and continued for 6 days increased the number of AChRs in clonal human muscle cultures (Kaplan et al. 1990). Other parameters of muscle differentiation, i.e. fusion index, myotube

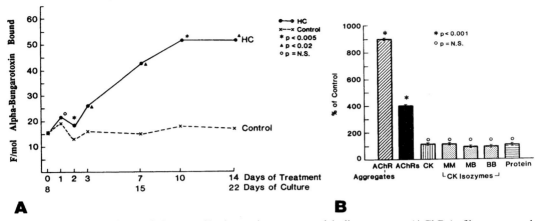

Fig. 4. Graphs illustrating the influence of hydrocortisone on acetylcholine receptors (AChRs) of human muscle cultured aneurally in serum-free medium. (A) Time-related influence on the total number of AChRs as measured by α-bungarotoxin binding. (B) AChR aggregates, total number of AChRs, total creatine kinase (CK) activity and CK isozymes in human muscle cultured for 4 weeks in serum-free medium with the addition of hydrocortisone and expressed as percent of control (human muscle cultured for 4 weeks in serum-free medium without hydrocortisone).

survival, expression of actin, and total CK activity, were reported not to be influenced by 6-day dexamethasone treatment, except that after 8 days in culture, control myotubes were coming off the dish, while dexamethasone-treated cultures remained attached (Kaplan et al. 1990). In apparent contrast, recent studies from another laboratory demonstrated that treatment with glucocorticoid methylprednisolone, also initiated at the beginning of myoblast fusion, inhibited myotube death as early as 4 days in fusion medium, resulting in increased levels of muscle-specific proteins, dystrophin and myosin (Sklar and Brown 1991).

Since in both studies glucocorticoids were introduced to culture medium at the beginning of myoblast fusion, the reason for this discrepancy is not clear, unless it may be explained by the variations in culture techniques and by the differences between various clones in their response to glucocorticoids (Dr. Robert Sklar, personal communication 1992). In clonal human muscle culture, glucocorticoids inhibit myotube death only when introduced at the very beginning of myoblast fusion, but not when introduced at the later stage of muscle differentiation (Sklar and Brown 1991). Similarly, in primary cultures of human muscle, treatment with glucocorticoids initiated after the completion of myoblast fusion does not influence the number of myotubes (J. McFerrin et al., unpub-

lished observation 1991). Therefore, when studying trophic influence of glucocorticoids on the expression of muscle-specific proteins, we prefer to initiate treatment after myoblast fusion is completed, since, in that case, the influence of glucocorticoids on either muscle cell proliferation or myotube death does not interfere with the interpretation of results.

Recently, we demonstrated that several days of treatment with hydrocortisone introduced to 7–10-day-old cultured human muscle increased the voltage-dependent calcium channels 2.3-fold ($p<0.025$) and the K^+-induced, nifedipine-inhibitable Ca^{2+} uptake 1.4-fold ($p < 0.05$) (Braun et al. 1991a).

TABLE 1

Factors influencing AChRs of aneurally cultured human muscle

	Total number	Aggregates
Insulin	↑	(-)
Epidermal growth factor (EGF)	(↑)	(-)
Fibroblast growth factor (FGF)	(-)	↑
EGF + insulin	↑	↑
FGF + insulin + EGF	↑↑	↑↑
Hydrocortisone	↑	↑↑

In summary, our studies regarding the stimulatory influence of glucocorticoids on AChRs and voltage-dependent calcium channels indicate that glucocorticoids exert trophic influence on aneurally cultured human muscle.

Other factors influencing properties of aneurally cultured human muscle

Five days of treatment with recombinant human interferon-γ has been shown to induce the expression of HLA-class II on human muscle cells in culture (Hohlfeld and Engel 1990). The induction of HLA-DR and HLA-DP has also been reported (Mantegazza et al. 1991). In both papers, the authors discuss their results in relation to the potential use of cultured myoblasts for transplantation and raise caution regarding possible myoblast rejection.

Recently, tumor necrosis factor-α (TNF) and interleukin 1 α (IL-1) were shown to induce ferritin and ferritin-H mRNA in cultured human muscle cells (Wei et al. 1990). Interferon-β and -γ did not have any influence on ferritin (Wei et al. 1990).

Innervated cultured human muscle

In spite of many methods available to aneurally culture human muscle, none has yet proved to be successful in inducing advanced expression of adult muscle phenotype. Even though above mentioned growth factors and hormones exert trophic effects on aneurally cultured human muscle, most of the studies have been hindered by the fact that adult human muscle fibers cultured aneurally are still quite immature in their ontogenic development and several phenotypic characteristics of innervated adult normal human muscle are not expressed in them. We have speculated that this pronounced immaturity may be responsible for the fact that morphologic or biochemical 'pathophenotypic' aspects of several genetic muscle diseases have not yet been reproducibly demonstrated in aneurally cultured muscle (Askanas 1984).

The development of muscle fiber specialization and its phenotypic characteristics is presumably based on an orderly expression of muscle-specific genes, culminating with those controlling molecules characteristic of mature muscle fibers. Because in vitro the maturation of muscle fibers is

governed to a great degree by innervation, innervation of cultured human muscle seemed to be a logical step in inducing its further maturation.

Described below are the comments regarding 2 most commonly used methods of innervation of cultured human muscle by fetal rodent spinal cord.

Comments on the method of innervation of human muscle cultured in organotypic culture

Crain and Peterson were the first to establish, in what they termed 'organotypic' culture, that human muscle fibers in vitro can be innervated by rodent spinal cord neurons (Crain et al. 1970; Peterson and Crain 1979). Innervation of human muscle fibers in that organ-culture system provides a very good in-vitro model of in-situ regeneration of muscle fibers and their innervation/reinnervation, and important information has been obtained using this model (Crain et al. 1970; Peterson and Crain 1979; Ecob-Prince et al. 1989). However, in that system one is not able to study easily the completely de-novo formation of neuromuscular junctions (NMJs), because in organ cultures, although the original explanted muscle fibers are damaged, their basal lamina survive. Therefore, de-novo formation of the basal lamina and the basal lamina specialization due to innervation are impossible to evaluate in that system. Moreover, in organ culture, new muscle fibers form within the original basal lamina of the explanted muscle fibers; they can sometimes be innervated at the original synaptic sites on the basal lamina (Ecob 1984), as is known to be the usual occurrence during muscle fiber regeneration in vivo (Marshall et al. 1977; Sanes et al. 1978; Burden et al. 1979). Four other important problems with the organ-culture technique are that: (a) apparently it has been very difficult (Peterson et al. 1986) or impossible (Ecob 1983) to obtain regeneration and innervation of organotypic muscle explants obtained from biopsies of persons more than 3 years of age; (b) the possible presence of residual degenerating muscle fibers within the organotypic explants can complicate interpretation of biochemical studies; (c) only a few muscle fibers in each culture are innervated, severely limiting the amount of muscle fibers available for various special analyses (Peterson et al. 1986); and (d) it is impossible to compare innervated and non-innervated muscle fibers, grown

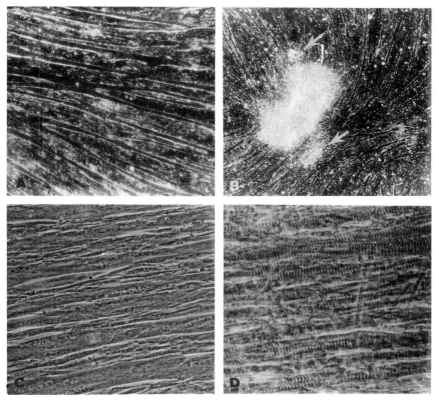

Fig. 5. Dark-field, low-power micrographs (A,B) and phase-contrast micrographs of living cultured human muscle (C,D). (A) Human muscle, aneurally cultured in monolayer, photographed 1 day before co-culture with fetal rat spinal cord. Note the abundance of muscle fibers (× 140). (B) Micrograph obtained 1 day after fetal rat spinal cord with dorsal root ganglia (arrows) was placed on top of the monolayer cultured human muscle (× 75). (C) Aneurally cultured human muscle fibers, 1 month old, are not cross-striated (× 600). (D) Human muscle fibers, well cross-striated and contacted by neurites, 16 days after co-culture with rat spinal cord (× 1200). (Reproduced from Askanas et al. (1978b) by courtesy of the Editors of J. Neurocytol.)

under the same conditions, because non-innervated muscle fibers do not regenerate in this culture system (Peterson and Crain 1979).

Comments on the method of innervation of human muscle cultured in monolayer

Monolayer primary cultures of normal and diseased human muscle can be obtained from persons of any age in contrast to organ cultures, in monolayer cultures new muscle fibers are formed from outgrowing myoblasts without remnants of the original basal lamina surrounding them. Therefore, this culture model appears very suitable to study different stages of innervation and its influence of human muscle fibers.

In the innervation model developed by us, explants of the spinal cord with dorsal root ganglia attached, from 12–14-day-old rat embryos, are placed on the top of a monolayer of the newly-fused myotubes (Fig. 5A,B) (Askanas et al. 1987b). This monolayer consists of abundant myotubes and a small degree of fibroblasts dispersed between them. Before innervation, muscle cultures are pretreated for 1 week with FGF, insulin, and EGF (Askanas et al. 1985b). From the moment of muscle-spinal cord co-culture, FGF and EGF are removed from the medium and cultures are maintained in medium supplemented with 10% fetal bovine serum and insulin only (Askanas et al. 1987b; Kobayashi et al. 1987). After 4 weeks of innervation, serum is decreased to 5% (Askanas et al. 1987b).

In living cultures successful innervation is manifested by: (a) contractions in nearly continuous rhythm of large groups of muscle fibers, which are reversibly blocked by 1 mM d-tubocurarine (Askanas et al. 1985a); (b) well-developed cross-striations (Fig. 5D); and (c) survival of muscle fibers for 5 months or longer. When the muscle fibers are not pretreated with FGF, EGF, and insulin, the degree of innervation, as measured by the number of contracting muscle fibers, is greatly decreased (see above for possible explanation).

In contrast to human muscle grown in primary culture, successful innervation of human muscle grown in clonal culture is much more difficult to achieve (V. Askanas, unpublished observation). Whether much less satisfactory innervation of clonal human muscle culture results from the absence of fibroblasts, which are synthesizing some of the components of synaptic basal lamina (Kühl et al. 1984; Sanderson et al. 1986) and are promoting neurite outgrowth (V. Askanas, unpublished observation), or it depends on other factors remains to be elucidated. Since the innervation of human muscle grown in clonal culture is desirable for a variety of studies, most likely this method will receive considerable attention in the future.

Neuromuscular junction (NMJ) formation on human muscle cultured in monolayer and innervated by fetal rat spinal cord

In innervated cultured human muscle, AChR and acetylcholinesterase (AChE) accumulate at the NMJs (Kobayashi and Askanas 1985; Kobayashi et al. 1987). Junctional AChRs have different metabolic properties than extrajunctional AChRs, as reflected by their increased stability (Braun et al. 1992). For example, in our culture system, half-life of junctional AChRs is 3.5 days, whereas that of extrajunctional AChRs is 1.3 days (Braun et al. 1992).

Ultrastructurally, in early innervation (7–10 days), contacts of nerve terminals with cultured human muscle fibers are superficial and unorganized, and there is no basal lamina-like material between the nerve terminals and muscle fibers (Askanas et al. 1987b). At 2–3 weeks of innervation, shallow 'beds' are formed on the muscle fibers just beneath nerve terminals, and occasionally there are irregular minuscule fragments of basal lamina-like material in the cleft. No Schwann cell apposes

the nerve terminal at this stage. After 4–5 weeks of innervation, there is more definite basal lamina material in the cleft and suggestive postsynaptic plasmalemmal densities and invaginations are present; however, no Schwann cell apposes the nerve terminal. At 6–8 weeks of innervation, deep postsynaptic folds are present, a Schwann cell apposes the nerve terminal (Fig. 6A,B), and ultrastructural cytochemistry of α-bungarotoxin binding reveals that AChRs are located exclusively at the NMJ (Fig. 6C) (Askanas et al. 1987b).

Histoenzymatic differentiation of innervated cultured human muscle fibers

Cultured human muscle fibers innervated for 60–90 days show an advanced degree of morphologic and histoenzymatic maturation. The innervated muscle fibers contain a well-developed intermyofibrillar network with NADH tetrazolium reductase (NADH-TR) and succinate dehydrogenase (SDH) reactions. Although most of the innervated cultured muscle fibers are not fully differentiated into 2 histochemical fiber types (because they have strong ATPase activity after both alkaline and acid preincubation), a few fibers have ATPase profile similar to type 2 fibers in human adult muscle and also have reciprocal staining with phosphorylase and NADH-TR reactions. Therefore, there is evidence of neuronally induced differentiation into different histochemical fiber types of cultured human muscle (Vita et al. 1988).

Neuronally induced expression of muscle-specific isozymes (MSIs) in cultured human muscle

MSIs of creatine kinase (CK) and glycogen phosphorylase. Isozymes of CK and phosphorylase are excellent markers of skeletal muscle fiber maturation. In adult innervated muscle, only MSIs are expressed, whereas aneurally cultured human muscle has the fetal pattern of isozymes (Miranda et al. 1979; Martinuzzi et al. 1986, 1987, 1988). In innervated cultured muscle fibers, the expression of MSIs is significantly increased and the increase directly correlates with the duration of innervation (Martinuzzi et al. 1988).

MSI of phosphoglycerate mutase (PGAM). The electrophoretic pattern of PGAM of adult normal

Fig. 6. (A,B) Transmission electron microscopy of organized neuromuscular junctions. Nerve-muscle contacts 6 weeks after co-culture. (A) Distinct postsynaptic folds, cross-striations, and basal lamina extending around the muscle fibers are present. Cross-section of a part of a nerve trunk is seen in the upper left (× 3780). (B) Nerve-muscle contact 8 weeks after co-culture. Similar postsynaptic organization as in (A), but plasmalemma at the tips of postsynaptic folds is thicker and more darkly stained than at the bottom of the folds. A Schwann cell is apposing the nerve terminal. Many clear and several dense-core vesicles are present in the nerve terminal (×29,400). (C) Ultrastructural cytochemistry of HRP-α-bungarotoxin at the nerve-muscle contact. The reaction product is confined to the organized nerve-muscle contact (× 4590). (Reproduced from Askanas et al. (1987b) by courtesy of the Editors of *J. Neurocytol.*)

human muscle is composed predominantly of the muscle-specific isozyme of PGAM, whereas the pattern of aneurally cultured human muscle fibers is composed virtually only of the brain-specific isozyme (Martinuzzi et al. 1987). In muscle fibers cultured in monolayer and innervated for 20–83 days, the total activity of PGAM was increased and the muscle-specific isozyme was expressed. The latter directly correlated with duration of innervation, suggesting that its gene is neurally regulated. Because muscle-specific isozyme of PGAM accumulated only in muscle fibers actually contacted by neurites (i.e., innervated and contracting muscle fibers) but not in the non-innervated muscle fibers from the same culture dish that were

equally exposed to diffusible neuronal factors, physical nerve-muscle contact seems required for full expression of the muscle-specific PGAM gene (Martinuzzi et al. 1987). The induction of muscle-specific isozymes CK and PGAM was also demonstrated in human muscle innervated in organotypic culture (Miranda et al. 1988a). It is of interest that MSI of PGAM is virtually never expressed in human muscle cultured aneurally even if the muscle is cultured in our serum-free medium, which greatly induces the expression of other MSIs (Pegolo et al. 1990).

Therefore, de-novo innervation of human muscle fibers cultured in monolayer exerts the maturational influence on their biochemical parameters and physical nerve-muscle contact, not simply diffusible factors, is required to produce this biochemical maturation of cultured human muscle fibers.

Neuronal regulation of electrical properties of cultured human muscle

We have shown that innervated cultured muscle fibers have higher resting membrane potentials (63.8 ± 4.1 mV) than aneurally cultured muscle fibers (52.1 ± 2.2 mV). Action potential amplitude in innervated muscle fibers is also higher, 95 ± 24 mV versus 89 ± 2 mV. The rate of rise of the action potential is 147 ± 6.6 V/sec in innervated muscle fibers versus 111 ± 1 V/sec in aneurally cultured human muscle fibers. Therefore, innervation induces electrical maturational properties of cultured human muscle fibers (Saito et al. 1990).

Recently, early events of chemical transmission have been studied in this culture system and have demonstrated to resemble immature nerve-muscle synapses in vivo (Michikawa et al. 1991).

Table 2 summarizes different aspects of neuronally induced adult muscle phenotype in cultured human muscle.

Influence of neuronal factors versus muscle contractile activity on innervated cultured human muscle

Influence of innervation versus muscle contractile activity and paralysis on morphologic characteris-

tics. To determine whether the expression of the muscle morphologic phenotype is induced only by neural factors generated from the spinal cord explants or also by their frequent contractile activity, we paralyzed innervated cultured human muscle fibers with 2 μM tetrodotoxin (TTX) for 4 weeks, either from the first day of muscle contractions or following 4 weeks of muscle contractions (Park et al. 1990; Park-Matsumoto et al. 1992). In both experimental designs using light microscopy, TTX paralysis abolished cross-striations and caused prominent internalization of muscle nuclei; however, it did not influence the intensity of AChE staining at the NMJs. Using electron microscopy, there was no difference between paralyzed and contracting muscle fibers in development of t-tubules, basal lamina, and postsynaptic folds. Therefore, in cultured human muscle, contractile activity (a) regulates peripheral migration of nuclei and development of cross-striations and (b) does not influence development of the NMJ, basal lamina, and t-tubules, which are mainly regulated by neural influences (Park et al. 1990; Park-Matsumoto et al. 1992). This culture model should be useful for studying detailed mechanisms of the influence of muscle contractile activity on human muscle fiber development and structural abnormalities in human neuromuscular diseases.

Influence of innervation versus muscle contractile activity and paralysis on MSIs. As described above, innervation greatly enhances the accumula-

---

TABLE 2

Neuronal induction of phenotypic expression in cultured human muscle fibers (CHMFs).

Aneurally CHMFs have	Innervated CHMFs have
1. Limited cross-striations	1. Well-organized cross-striations
2. Poorly developed t-tubules	2. Well-developed t-tubules
3. Lack of basal lamina	3. Basal lamina surrounding muscle fibers
4. AChRs distributed over the plasmalemma	4. AChRs localized postsynaptically at NMJ
5. Negative AChE stain	5. AChE localized at NMJ
6. Negative staining with Dolichos biflorus agglutinin	6. Dolichos biflorus agglutinin localized at NMJ
7. One histochemical fiber type	7. Tendency toward MF differentiation
8. Nuclei localized internally	8. Most nuclei localized at the periphery
9. Limited expression of muscle-specific isozymes (MSIs), predominant expression of fetal isozymes	9. Increased expression of MSIs and decreased expression of fetal isozymes
10. Lower resting membrane potential (mean 50 mV) and action potential (mean 70 mV)	10. Higher resting membrane potential (mean 64 mV) and action potential (mean 76 mV)

tion of MSIs. Since innervated cultured human muscle fibers also continuously contract while sister non-innervated fibers do not, we studied the influence of muscle contractile activity on developmental accumulation of MSIs in innervated-contracting cultured human muscle fibers vs. 30-day paralyzed innervated muscle fibers and vs. non-innervated muscle fibers that had never contracted. In our system, (a) neuronal influence seems to be important beyond its induction of muscle contractile activity for accumulation (possibly reflecting genetic expression) of MSIs of CK, GP, LDH, and PGAM in cultured human muscle fibers and (b) only MSIs of CK and LDH, but not of PG and PGAM, were enhanced by muscle activity per se, indicating that accumulations of MSIs of different enzymes are differently regulated (possibly due to differential gene regulation) (Martinuzzi et al. 1990a).

Influence of glucocorticoids on molecular and morphologic properties of NMJs of innervated cultured human muscle

The glucocorticoid prednisone has been widely used in high doses for about 2 decades for treating myasthenia gravis (MG) (Engel and Warmolts 1971; Brunner et al. 1972; Seybold and Drachman 1974). Its mode of benefit has been attributed to an anti-dysimmune function. Whether prolonged treatment with prednisone has an additional direct beneficial influence on the postsynaptic properties of the NMJ has not been known.

Although reports vary, short-term treatment (hours to a few days) with glucocorticoid seems to have caused facilitation of NMJ transmission due to stimulation of presynaptic mechanisms (acetylcholine release and synthesis) (Hall 1983; Veldesma-Currie et al. 1984). Inhibition by glucocorticoid of NMJ transmission in vivo, in animals and humans, has also been reported (Hall 1983). Intravenous glucocorticoid in animals acutely inhibits stimulus-evoked repetitive discharges from the nerve terminal and the muscle contraction increase produced by administration of cholinesterase inhibitors (Patten et al. 1974; Dengler et al. 1979; Noble et al. 1979). These inhibitory effects were considered to possibly underline the initial worsening of myasthenic symptoms that sometimes occur

at the beginning of treatment of MG patients with prednisone, especially when they are concurrently receiving an oral anticholinesterase, such as pyridostigmine (Patten et al. 1974).

Described below is the influence of glucocorticoids on various postsynaptic properties of the NMJs of innervated cultured human muscle.

Influence on AChE at the NMJ. Under our standard culture conditions, all AChE positive sites are linear and thin between 5 and 21 days of innervation (non-innervated cultured human muscle does not have AChE positive sites). Between 4 and 8 weeks of innervation, 37% of them become complexly organized (Kobayashi et al. 1987). HC treatment, initiated after 4 weeks of innervation and continued for 3 weeks, strikingly increased complexity and intensity of AChE staining at the NMJs (Fig. 7A,B). As quantitated by a computerized video- image analysis system (RAS, Amersham), HC increased the size (in pixels) of AChE sites 2.7-fold and their optical density 1.1-fold (both $p < 0.05$). Biochemically measured total AChE activity was increased 1.5-fold, and endplate-specific 16S fraction of AChE was increased 2.7-fold ($p < 0.001$) (Fig. 8) (Askanas et al. 1992). Therefore, (a) HC exerts significant influence on AChE at the NMJs; (b) increased staining intensity corresponds to increase of 16S- AChE; and (c) the AChE increased by HC should be a consideration in the treatment of myasthenia gravis, in which both glucocorticoid and AChE inhibitors are commonly used.

Junctional AChRs. We studied the influence of HC on AChRs at the NMJs formed on cultured human muscle fibers that have been innervated by fetal rat spinal cord neurons (Braun et al. 1991b). HC treatment was begun 4 weeks after co-culture and continued for up to 3 weeks thereafter. AChRs were detected by autoradiography of ^{125}I-α-bungarotoxin and quantified by computerized video-image analysis. HC increased the size of AChR clusters (reflecting size of NMJs) 2.3-fold ($p < 0.05$) and their optical density (reflecting relative number of AChRs) 1.6-fold ($p < 0.001$). The same results were obtained with dexamethasone and methylprednisolone. Therefore our studies indicate that prolonged treatment with HC increases

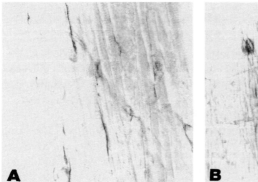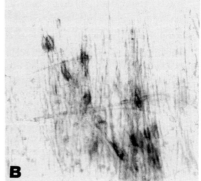

Fig. 7. Bright-field microscopy of AChE staining of (A) control and (B) hydrocortisone-treated innervated cultured human muscle. Muscle was innervated for 8 weeks and treated with hydrocortisone during the last 4 weeks of innervation (×700). In this example, quantitative video-image analysis indicated that the average size of AChE-positive sites in the hydrocortisone-treated culture was 4.5 times larger and 2 times darker than in the control. (Reproduced from Askanas et al. (1992) by courtesy of the Editor of *Exp. Neurol.*)

the number of AChRs at human NMJs (Braun et al. 1991b; Braun et al., unpublished observation 1992).

Postsynaptic membrane organization. We have studied morphometrically (a) the length of membrane constituting the postsynaptic folds, (b) the depth of the secondary synaptic clefts, (c) the midpoint width of the largest secondary synaptic cleft, and (d) the number of postsynaptic folds per junction in control and hydrocortisone-treated NMJs on innervated cultured human muscle fibers (Park et al. 1990; Askanas et al. 1992).

Three weeks of treatment with HC-induced organization of the postsynaptic membrane of cultured muscle fibers (Fig. 9A,B). This was evidenced by increase of (a) the length of postsynaptic

membrane (1.6-fold), (b) the depth of secondary synaptic clefts (2-fold), and (c) the number of postsynaptic folds (2-fold), ($p < 0.001$ for all 3 parameters) (Park et al. 1990; Askanas et al. 1992).

In patients with myasthenia gravis, flattening of the postsynaptic folds at the NMJ commonly occurs (Zacks et al. 1962; Fardeau et al. 1972; Engel 1986). The increased organization of the postsynaptic folds and the increased AChE we have found seem to be salutary effects of glucocorticoid. Even though it is difficult to relate the dose of glucocorticoid active in our culture system to the therapeutic level of glucocorticoid in myasthenia gravis patients, these, or other, influences of glucocorticoid on the NMJ might contribute to the long-term beneficial effect of prednisone in those patients, in

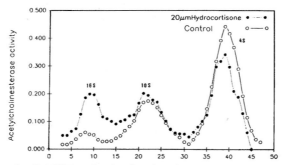

Fig. 8. Representative graph of AChE molecular fractions in control and hydrociortisone-treated cultured human muscle fibers innervated for 8 weeks. Hydrocortisone treatment was begun after 4 weeks of innervation. Note increased 16S and decreased 4S molecular fractions in the hydrocortisone-treated cultured muscle. (Reproduced from Askanas et al. (1992) by courtesy of the Editors of *Exp. Neurol.*)

Fig. 9. Electron-microscopy of the AChE reaction at the neuromuscular junctions of (A) control and (B) hydrocortisone-treated cultured human muscle fibers (×6500). The AChE reaction was performed after 2.5-minutes incubation of the hydrocortisone-treated culture and after 25-minutes incubation of the control. The organization of postsynaptic folds and the intensity of the AChE reaction are greater in the hydrocortisone-treated specimen (B). (Reproduced from Askanas et al. (1992), by courtesy of the Editors of *Exp. Neurol.*)

addition to its anti-dysimmune benefit. By demonstrating that chronic treatment with glucocorticoid increases junctional AChRs, AChE, and postsynaptic-membrane organization in cultured innervated human muscle, we have identified a novel action of glucocorticoid.

This culture model should provide a very useful system for studying the influence of various substances on the postsynaptic membrane of human muscle.

GENETICALLY ABNORMAL HUMAN MUSCLE

Tissue culture of human biopsied muscle fibers which carry a defect in their genome should enable precise developmental studies of the expression of the genetic defect. Several different types of abnormalities caused by an abnormal or missing gene can be manifested in tissue culture (Table 3). The full phenotypic manifestation of a genetic abnor-

mality will depend on the degree of maturity achieved by cultured muscle fibers, the longevity of cultured muscle, and the degree of transcription and translantion of a given gene in normal cultured human muscle. Accordingly, if a given gene product is either weakly expressed or not of biological importance in a normal developing cultured human muscle, its deficiency may not be manifested in culture. Likewise, since some genetic defects in vivo require years for their full expression, it is not likely that they will be fully expressed in cultured muscle, which rarely survives longer than 5–6 months (Askanas et al. 1989).

Naturally, the ultimate goal is to develop culture conditions, which would enable full phenotypic manifestation of an abnormality in all genetic muscle diseases. Such culture model would be ideal for the studies related to gene therapy, since it would allow both the transfection of a normal gene and the analysis of its ability to ameliorate the abnormality.

To date there have been only a few genetic muscle diseases in which specific-disease phenotype, either morphologic or biochemical or both, has been manifested in culture. Those diseases include acid maltase deficiency (Askanas et al. 1976; Askanas and Engel 1977), glycogen debrancher enzyme deficiency (Miranda et al. 1981) in aneural cultures, and infantile rod (nemaline) disease in innervated cultured muscle (Askanas et al. 1988b). Even though the biochemical phenotype, dystrophin deficiency, exists in cultured muscle of Duchenne muscular dystrophy patients (Hurko et al. 1989;

TABLE 3

Types of abnormalities in cultured human muscle caused by an abnormal (or missing) gene.

A. Disease-specific morphologic abnormality (even though the underlying pathogenic mechanism or an abnormal gene, or its product, are not known)
B. Disease-specific biochemical abnormality
C. Disease-specific electrophysiologic abnormality
D. Disease-characteristic (but not disease-specific) morphologic, biochemical or electrophysiologic abnormalities

Miranda et al. 1988b), the morphologic phenotype is not explained.

In this article we will give a short summary of just a few genetic muscle diseases in which the consequences of their genetic defects took various forms depending either on the degree of muscle maturity in culture, and on whether muscle had been cultured aneurally or had been innervated. Finally, we will discuss a genetic muscle disease in which the defect appears to be spontaneously ameliorated in cultured muscle.

For all the other diseases, including mitochondrial myopathies, the readers are referred to previous reviews and to some of the most current publications (Askanas and Engel 1977, 1979; Miranda and Mongini 1986; Askanas et al. 1989; Moraes et al. 1989a,b; DiMauro et al. Chapter 16 this volume).

Duchenne muscular dystrophy (DMD)

Muscle of patients with DMD has been studied in tissue culture probably more intensively than that from any other genetic muscle disease. In spite of those extensive investigations, early studies failed to demonstrate reproducible abnormalities (reviewed in Miranda and Mongini 1986). More recent studies demonstrated that both dystrophin, the missing gene product in DMD (Hoffman et al. 1987), and its mRNA are not present in cultured DMD muscle (Lev et al. 1987; Miranda et al. 1988b; Ecob-Prince et al. 1989; Hurko et al. 1989; Sklar et al. 1990). In addition, because DMD is an X-linked recessive disorder in which 1 of the 2 X chromosomes in females undergoes somatic inactivation, dystrophin-positive and dystrophin-negative muscle clones were demonstrated in culture (Miranda et al. 1989).

It is of interest that in spite of dystrophin deficiency, the only abnormality reproducibly demonstrated in aneurally cultured DMD muscle is that of abnormally increased calcium (Mongini et al. 1988; Fong et al. 1990). All other characteristics of an abnormal DMD phenotype in vivo, i.e. muscle necrosis, regeneration, and creatine kinase abnormalities are not present in aneurally cultured DMD muscle. Moreover, none of the morphologic and biochemical parameters of aneural cultured muscle differentiation and survival differ between

control and DMD muscle (reviewed in Miranda and Mongini 1986; Askanas et al. 1987a). A decreased proliferative potential of DMD satellite cells in culture (Blau et al. 1983; Webster and Blau 1990) and their decreased myogenic response to stimulation by glucocorticoid (Hardiman et al. 1992) were concluded to result from their increased division in a patient and not from a primary genetic defect (Webster et al. 1986; Hardiman et al. 1992).

We have postulated that the lack of apparent differences between normal and DMD muscle in aneural culture may be due to the fact that dystrophin (postulated role of which is to preserve integrity of the membrane of a muscle fiber (reviewed in Arahata and Sugita 1989)), does not yet play an important role in aneurally cultured muscle, because aneurally cultured muscle fibers are quite immature, survive for a very short time only (6–8 weeks maximally), and do not contract. Under such conditions, lack of the appropriate membrane stability may not be of biological significance, even though it is severe enough to cause the increased accumulation of calcium. Therefore, we were interested to see whether long-term innervation of DMD-cultured muscle and their subsequent contraction will enable better manifestation of the consequences of the genetic defect (Askanas et al. 1987a).

Aneurally cultured DMD muscle grew and developed indistinguishably from control (Fig. 10A,B). DMD-cultured muscle fibers became innervated and contracted virtually continuously, also indistinguishably from our cultured control-innervated human muscle fibers. Observation of the living cultures by phase-contrast light microscopy did not reveal any obvious differences between DMD- and control-innervated cultured human muscle fibers. There was no apparent difference between DMD- and control-innervated cultured muscle fibers regarding acceptance of innervation as evidenced by: neuronally-driven, virtually continuous, muscle-fiber contractions; characteristic muscle-fiber organization by phase-contrast microscopy (Fig. 10C,D); and increased longevity of the innervated muscle fibers (both control and DMD muscle survived over 6 months).

Comparing DMD non-innervated cultured muscle fibers and control non-innervated cultured

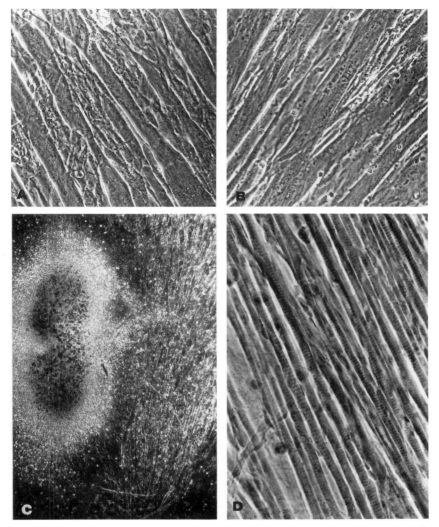

Fig. 10. Phase-contrast microscopy of living cultured muscle of control (A) and Duchenne muscular dystrophy (DMD) patients (B–D). (A,B) 4-week-old living aneurally cultured muscle (×400). There are no morphologic differences between control- and DMD-cultured muscle. (C,D) Living 1-month innervated cultured muscle of a DMD patient. (C) Lower power illustrating a large area of dense packed muscle fibers parallel to each other, in proximity to spinal cord explant, which were continuously contracting (×59). (D) Higher power illustrating well cross-striated muscle fibers (a nerve branch is emerging from the lower left) (×582). (C and D reproduced from Askanas et al. (1987a) by courtesy of the Editors of *Life Sci.*)

muscle fibers, there was no difference in accumulation of muscle-specific isozymes (MSIs) (Askanas et al. 1987a). In DMD-innervated muscle fibers, accumulation of MSIs of phosphorylase and lactose dehydrogenase (LDH) were significantly increased compared to DMD non-innervated cultured muscle fibers (Askanas et al. 1987a). The PGAM-MSI was present in both DMD-inner-

vated and control-innervated cultured muscle fibers, but not in DMD- non-innervated or control-non-innervated cultured muscle fibers (Askanas et al. 1987a).

Interestingly, in DMD-innervated cultured muscle fibers, the percent of CK-MM isozyme (in relation to total CK) was significantly lower compared to that of control-innervated muscle fibers and

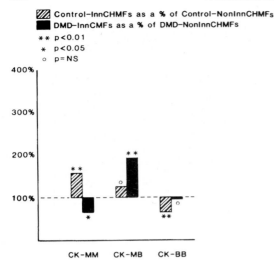

Fig. 11. CK-MM, CK-MB, and CK-BB in innervated cultured muscle fibers of DMD and control patients expressed as a percent of their respective noninnervated sister-cultured muscle fibers. P values are comparison between innervated and noninnervated muscle fibers of each isoenzyme. (Reproduced from Askanas et al. (1987a) by courtesy of the Editors of *Life Sci.*)

DMD-non-innervated cultured muscle fibers (Fig. 11), and the percent of CK-MB was higher compared to control-innervated muscle fibers.

It has been suggested that dystrophin binds cytoplasmic contractile filaments to an internal membrane system of the muscle cell and that it is associated with t-tubule system (Hoffman et al. 1987; reviewed in Arahata and Sugita 1989). An intriguing question is whether impaired accumulation of CK-MM in DMD-innervated cultured muscle fibers can be related to the putative fragility of cytoplasmic contractile filaments in DMD muscle. It is of interest that the decreased accumulation of CK-MM was only present in DMD-innervated cultured muscle fibers. DMD-non-innervated cultured muscle fibers did not differ from control. Besides being much more mature than non-innervated muscle fibers, in respect to biochemical and morphologic criteria (Askanas et al. 1988a), innervated cultured muscles also virtually continuously contract (Askanas et al. 1985a), and are almost 4 times longer than their non-innervated counterparts (Kobayashi et al. 1987). Possibly, because of their lack of dystrophin, innervated DMD-cultured muscle fibers are more

fragile than innervated control-cultured muscle fibers (as was previously postulated for innervated DMD muscle fibers in vivo (Engel 1977b)). If so, with the repeated vigorous contractions resulting from their in-culture innervation, some DMD muscle fibers would be more easily damaged than the putatively hardier innervated control muscle fibers. Another possibility is that the abnormality observed by us in DMD-innervated cultured muscle fibers is due to both muscle fiber growth and their contractile activity. Therefore, it is quite possible that lack of dystrophin may not be important for the function of inactive, aneurally cultured DMD muscle fibers, but it becomes significant when DMD-cultured muscle fibers are continuously contracting, and requiring a strong cytoskeleton. This is in accord with a previously postulated theory 'that some sort of physiological or biochemical threshold must be crossed before a dystrophin-deficient myofiber is subject to necrosis' (Hoffman et al. 1989). The increased percent of CK-MB isozyme may indicate that more DMD- than control-cultured muscle fibers are in a regenerative stage, perhaps resulting from increased degeneration and subsequent regeneration.

In conclusion, our studies demonstrated that innervation and subsequent increased maturity and contractile activity of DMD-cultured muscle fibers has enabled the demonstration of creatine-kinase abnormality in them. This culture system in which innervated rather mature cultured human muscle fibers can be obtained in large quantity should provide an excellent experimental model to study other possible abnormalities associated with genetic lack of dystrophin.

Oculopharyngeal muscular dystrophy (OPMD)

OPMD is inherited, usually as an autosomal dominant trait with strong gene penetrance (for review see Tomé and Fardeau 1986). It typically manifests itself in the 5th–6th decade, with eyelid ptosis and dysphagia. Limb muscles also become weak. The most characteristic EM changes are intranuclear tubular filaments of 8.5-nm outer and 3-nm inner diameter described for the first time by Tomé and Fardeau (1980). The inclusions occur only in muscle fiber nuclei and not in nuclei of any other cell type, including muscle satellite cells (Tomé and

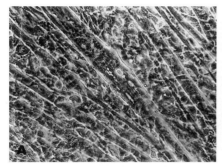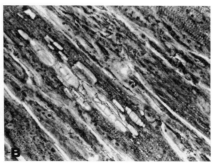

Fig. 12. Phase-contrast microscopy of living cultured OPMD muscle fibers. (A) 4-week-old aneurally cultured muscle fibers. Note vacuoles in the fiber in the center (×750). (B) 4-week-innervated muscle fibers. Note cross-striations and prominent vacuoles (×1200). (Reproduced from Tomé et al. (1989) by courtesy of the Editors of *Neurology.*)

Fardeau 1986). The nature of the nuclear inclusions is not known.

The 'Tomé bodies' represent the morphological marker of OPMD, as they have not been described in any other normal or pathologic condition.

We studied OPMD-muscle fibers cultured aneurally and innervated by fetal rat spinal cord from 3 patients with OPMD, in order to see whether the characteristic nuclear inclusions will be also present in their cultured muscle (Tomé et al. 1989). Of the aneurally cultured muscle fibers, most had normal morphology, but some were vacuolated (Fig. 12A) and had enlarged mitochondria by EM. Innervated muscle fibers contracted vigorously, and the majority contained large vacuoles (Fig. 12B). They were breaking easily, and could not be maintained longer than 2 months, versus a routine 3–6-month survival of control muscle fibers. Using EM, the patients' innervated muscle fibers had abnormally dilated sarcoplasmic reticulum forming balloon-like vacuoles, abnormal lateral sacs, and some mitochondrial abnormalities (resembling those found occasionally in the biopsied OPMD muscle fibers). The most striking abnormalities were the unusual nuclear inclusions which we have not found to date in aneurally cultured muscle fibers of OPMD or in any innervated normal- and disease-control human muscle fibers. The nuclear inclusions in the cultured muscle fibers seem similar, but not identical, to those found in biopsied muscle fibers. In culture they were usually smaller and did not contain the distinct tubules characteristic of the nuclear inclusions in biopsied muscle fibers, but the diameter and size were identical

(Fig. 13). Recently, we have found identical abnormalities to those described above, in innervated cultured muscle fibers from 3 OPMD patients from Uruguay (Medici et al. 1990).

Because OPMD-characteristic nuclear inclusions are not found in muscle biopsy satellite cells or in aneural or very immature innervated cultured muscle fibers, it is logical to assume that the neural influence or advanced maturation, or both, may be

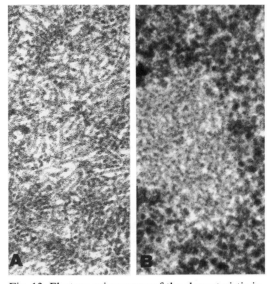

Fig. 13. Electron microscopy of the characteristic intranuclear inclusions in muscle biopsy (A) and in innervated cultured muscle fibers (B) from OPMD patients, to compare their ultrastructural similarities (×72,000). (Reproduced from Tomé et al. (1989) by courtesy of the Editors of *Neurology.*)

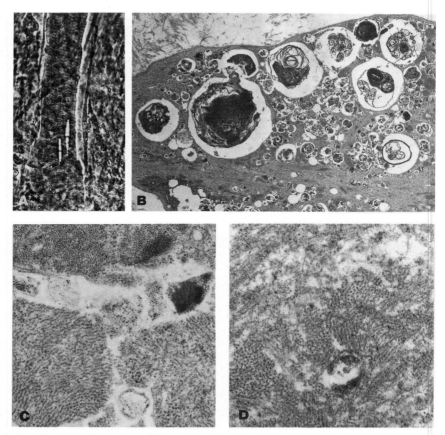

Fig. 14. Innervated cultured muscle fibers from a patient with hereditary IBM. (A) Phase-contrast light micros-copy of a 12-week-innervated cultured muscle fiber illustrating cross-striations and long, well-demarcated vacuoles (×1200). (B) Electron microscopy of a vacuolated area. There is a collection of large and small membranous whorls (×3360). (C,D) Characteristic cytoplasmic tubulofilaments in muscle biopsy (C) and inner-vated-cultured muscle (D). (C. ×38,000; D. ×42,300). (Patient was diagnosed by Dr. Michel Fardeau, Fig. C provided by Dr. Fernando Tomé, both at INSERM-U, Paris.)

necessary for the formation of the inclusions in muscle fiber nuclei in OPMD. Neither the genetic defect responsible for OPMD nor the chemical composition and the cause of intranuclear inclu-sion are presently known.

In general, the protein composing the abnormal nuclear tubules in OPMD might reflect: (a)(i) over-production or (ii) under-catabolism of either a nor-mal nuclear component or a normal cytoplasmic component moving aberrantly through nuclear pores into nucleoplasm; or (b) synthesis of an ab-normal molecule. Further, culturing and inner-vating muscle of OPMD patients should provide a model for studying developmental stages of typical adult intranuclear inclusions and more abundant

material for chemical characterization and molec-ular studies.

Inclusion-body myositis (IBM)

Inclusion-body myositis (IBM) is diagnosed by a combination of clinical and pathologic features. Light-microscopic pathologic features include: various degrees of inflammation; muscle fibers with rimmed vacuoles; and atrophic muscle fibers, indicative of lower motor neuron involvement. Using EM, abnormal muscle fibers contain CTFs, 15–21-nm external diameter and 3–6-nm internal diameter; these are a diagnostic criterion of IBM (Chou 1967; Yunis and Samaha 1971; Carpenter et al. 1978; Lotz et al. 1989; Tomé et al. 1981). Autosomal recessive 'hereditary IBM' designates rare patients with progressive muscle weakness and CTFs in vacuoles of abnormal muscle fibers (Cole et al. 1988; Fardeau et al. 1990; Massa et al. 1991).

Recently, we have described that vacuolated muscle fibers in both sporadic and hereditary IBM contain darkly immunoreactive ubiquitinated inclusions (Askanas et al. 1991a, 1992a). Using EM, ubiquitin immunoreactivity was localized to CTFs (Askanas et al. 1991a, 1992a). Our most recent studies also demonstrate an abnormal accumulation of Alzheimer β-amyloid protein in vacuolated fibers of IBM (Askanas et al. 1992b,c).

We have previously studied aneural and innervated muscle cultured from muscle biopsies of 2 brothers, ages 42 and 29 with progressive proximal and distal muscle weakness since early teens (Alvarez et al. 1990; Askanas, Alvarez and Fardeau, to be published). In their muscle biposies, rimmed vacuoles and eosinophilic inclusions were present histochemically. Cytoplasmic inclusions, consisting of 16–18-nm filaments and myelin-like bodies, were present ultrastructurally (Fardeau et al. 1990). Aneurally cultured muscle fibers were normal. Cultured muscle fibers that had been innervated and contracting for 4–18 weeks had vacuoles with light microscopy, which with electron microscopy contained many myelin-like bodies (Fig. 14A,B). Also ultrastructurally evident were cytoplasmic inclusions, consisting of 15–20-nm filaments, identical in the organization to those in the muscle biopsy (Fig. 14C,D). Thus, this study demonstrated that the reproduction of the typical inclusions in cultured muscle fibers reflects a genetic defect and the innervation and/or advanced maturity or contractile activity is required for the expression of the genetic defect.

Recently, we have seen the same type of abnormalities in muscle fibers cultured from a 42-year-old patient from another family with hereditary IBM. In addition, our preliminary studies demonstrated that this patient's innervated muscle fibers also had darkly stained ubiquitinated inclusions (Fig. 15).

Although the origin and nature of the pathologic CTFs in sporadic and hereditary IBM are not known, the fact that they are ubiquitinated places their protein in the ubiquitin-mediated turnover

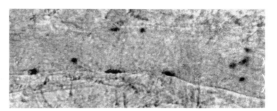

Fig. 15. Bright-field microscopy of 16-week-innervated cultured muscle fiber from a patient with hereditary IBM illustrating dark ubiquitinated inclusions. Immunoreactive ubiquitin is identified with PAP reaction (×1600).

pathway, providing a focus for further studies to identify their composition and pathogenesis. Because ubiquitin is covalently bound to proteins, affinity, purification, and molecular analysis of ubiquitinated CTFs should be possible. Since CTFs are present in cultured hereditary IBM, this source should provide material for purification studies.

Myotonic atrophy/dystrophy (MA)

MA is an autosomal dominant disease of unknown pathogenesis. Recently, a CTG triplet that undergoes expansion in MA has been identified on chromosome 19 (Brook et al. 1992). The CTG fragment is transcribed in the mRNA, which encodes a polypeptide that is a member of the protein kinase family (Brook et al. 1992). Whether the myopathogenic mechanism reflects a genetic defect expressed in the muscle fiber or is consequent to a defect expressed in a non-muscle cell, or both, is uncertain. The typical muscle pathology in vivo is atrophy of muscle fibers, often resembling denervation atrophy (Engel and Brooke 1966). In the severe neonatal form of MA, there is muscle fiber hypotrophy (Engel 1977b). Nerve endings on muscle fibers can be abnormal, sometimes extending along the muscle fibers instead of being limited to a discrete junctional region (Coers and Woolf 1959; McDermot 1961; Coers et al. 1973). Recently, peripheral nerve involvement was decribed in MA (Cros et al. 1988). Because of the above, we favor the name of myotonic atrophy instead of myotonic dystrophy (Engel 1979).

Recently, we have reported electrophysiological and morphological studies in MA muscle fibers cultured aneurally and innervated by fetal rat spinal cord (Kobayashi et al. 1990b). Aneural muscle was cultured in our new medium containing fibroblast growth factor (FGF), epidermal growth factor (EGF), and insulin (Askanas and Gallez-Hawkins 1985). At the moment of innervation, FGF and EGF were removed, and muscle continued to be cultured in the medium containing only insulin (10 μg/ml) and 5% fetal bovine serum (Askanas et al. 1987b; Kobayashi et al. 1990b).

There were no morphological differences between control and MA muscle fibers cultured aneurally. On the innervated MA muscle fibers, 96% of acetylcholinesterase-stained NMJs were linear, and only 4% appeared as complicated, pretzel-like, more mature-looking structures; on control innervated fibers, 37% of acetylcholinesterase-stained NMJs had the mature appearance (Kobayashi et al. 1990b). The normal trend from multifocal innervation toward unifocal innervation described by us previously (Kobayashi et al. 1987) was decreased in innervated MA muscle fibers, i.e. for every 600-μm length of contracting muscle fibers, MA-innervated fibers had 2.3±0.3 (mean SEM) AChE-denoted NMJs and 2.3±0.3 AChR clusters, compared with 0.5±0.061 of AChE-positive NMJs and 0.6±0.06 of AChR clusters (p<0.5 for both) in control-innervated fibers (Kobayashi et al. 1987, 1990b).

In addition, the degree of successful innervation in MA-cultured muscle was lower than that of cultured control muscle fibers. Whereas 75% of spinal cord explants successfully innervated control muscle fibers, as manifested by virtually continuous contractions of large areas of cross-striated muscle fibers (reversibly inhibitable by d-tubocurarine (Askanas et al. 1985a)), in MA muscle fibers only 30% of spinal cord explants induced the innervation-driven contractions (Kobayashi et al. 1990b). However, MA muscle fibers that were successfully innervated were well cross-striated and contracted similarly to the control fibers.

Electrophysiologic properties of cultured MA muscle fibers were described in detail previously (Kobayashi et al. 1990b). Briefly, the mean resting membrane potentials of the 2 groups (aneurally cultured MA muscle fibers and innervated-contracting cultured MA muscle fibers) were 8 and 9 mV lower, respectively, than those of their counterpart controls. The mean amplitude of action po-

tentials, the maximum rate of rise of action potentials in innervated MA muscle fibers, and the action potential amplitude in aneural MA muscle fibers were significantly smaller than in corresponding control fibers. These results differ from those demonstrated by us in aneurally cultured muscle fibers from acid maltase deficiency and paramyotonia congenita patients (Tahmoush et al. 1984; Kobayashi et al. 1990a).

Even though in the past a decreased resting membrane potential (RMP) in aneurally cultured muscle fibers from MA patients has been reported (Merickel et al. 1981), previous studies by our group were not able to detect any significant electrical abnormalities in MA muscle fibers cultured aneurally in our previous, nonsupplemented standard medium (Tahmoush et al. 1983). It is of interest that in our present studies, when both MA and control muscle fibers were cultured in medium supplemented with growth factors, including insulin, the RMP of MA muscle fibers was as in our previous studies (Tahmoush et al. 1983), but the RMP of control muscle fibers was considerably increased. Whether this difference reflects a defective response of MA muscle to insulin, proposed previously in MA patients (Moxley et al. 1984), will have to be elucidated.

Defective innervation of MA muscle fibers in vivo, due to the muscle fiber not accepting innervation, has been hypothesized previously and termed muscle dysreception to innervation, an idea based partly on denervation-like atrophy histochemically and elongated motor nerve endings on muscle fibers (Engel 1979). This hypothesis is supported by the present results demonstrating an abnormal innervation response of cultured MA muscle fibers innervated by normal motor neurons from fetal rats. Recently, extrajunctionally located neural cell adhesion molecule was found in muscle biopsy specimens of patients with MA, as occurs in noninnervated (Covault and Sanes 1985; Moore and Walsh 1985; Walsh et al. 1988) or denervated (Covault and Sanes 1985) muscle fibers.

In conclusion, we have demonstrated that compared with controls, MA muscle fibers (a) when cultured aneurally or innervated, had plasmalemmal abnormalities regarding RMP, AP amplitude and dv/dt; (b) had no difference in electrically-evoked repetitive discharges; (c) had more spontaneous firing in aneural cultures; and (d) when innervated by fetal rat spinal cord did not form as mature-looking AChE-denoted NMJs (Kobayashi et al. 1990b). Our results indicate endogenous muscle-cell abnormalities, some of which suggest plasmalemmal unresponsiveness to innervation. It will be of interest to determine how the abnormalities found by us relate to the newly identified genetic defect in MA.

Infantile rod (nemaline) disease

Rod disease is known to be present in 4 clinical forms: (a) congenital, rapidly progressive, and fatal; (b) congenital, non- or slowly progressive; (c) adult-onset; and (d) asymptomatic, recognizable only by muscle biopsy (usually a parent of an affected child) (reviews in Engel 1977a; Banker 1986). The pathogenesis(es) of rod disease is unknown.

The characteristic pathologic feature in muscle fibers are rodlike structures resembling Z-disk material and seeming to be lateral aggrandizements of Z-disks. Rods have the same ultrastructure as normal Z-disks, and both contain axial actin filaments and a large amount of α-actinin.

In all forms, rods are in the cytoplasm. Similar rods within muscle fiber nuclei are fairly common in the adult-onset form but rare in the congenital forms (Engel 1977a). Motor end-plate areas have not been extensively studied; there is disagreement whether they are abnormal (Banker 1986). Since muscle fibers of patients with congenital rod disease are predominantly type I (type-II fiber paucity) and the myosin light-chain pattern is practically pure slow-myosin type (Banker 1986), abnormality of neural influence, either neurogenic or myogenic, has been postulated (Engel 1977a). Finding rods in tenotomized rat muscle fibers has provoked the suggestion that they might be caused (at least partly) by inappropriate muscle fiber activity and/or muscle fiber length-adaptive remodeling (Engel 1977a).

We have demonstrated abundant rods in innervated muscle fibers cultured from a biopsy of a 25-day-old floppy male infant, who had been on a respirator since birth (his sibling had died neonatally with rod disease) (Askanas et al. 1988b). His biopsied muscle fibers were uniformly small, poorly differentiated, and cross-striated, and they

contained numerous rods. Aneurally cultured muscle fibers were flat (Fig. 16A) and degenerated early (after 2–3 weeks of growth vs. 6–7 weeks for normal aneurally cultured muscle fibers). By EM they were poorly differentiated and had numerous autophagic lysosomal inclusions, but no rods. Only a minority of muscle fibers (30% vs. 75% seen in normal muscle cultures) responded to innervation, as evidenced by muscle fiber contractions. Innervated muscle fibers were very thin, lacked well-developed cross-striation (Fig. 16B), and survived only up to 2 months (vs. 3–5-month routine survival of innervated normal cultured human muscle fibers). By EM, innervated-contracting muscle fibers had disorganized myofibrils and smearing of Z-disks (Fig. 16C). Many fibers contained rods, closely resembling those present in the patients' muscle biopsy (Fig. 16E). By demonstrating that rods can be reproduced in muscle fibers cultured from a patient but innervated by normal rat motor neurons, our studies provide evidence of an intrinsically myogenic component of this patient's disease. By demonstrating an inability of muscle fibers to become properly innervated, our studies raise a possibility of a 'dysreception' at the level of the plasmalemma or a defective step further along the 'trophic pathways' in the muscle fiber (e.g., messengers, translation, transcription, protein synthesis).

This infant's mother's muscle biopsy was normal by light and electron microscopy, and there were no rods present in her cultured muscle. His father's muscle biopsy was normal by light microscopy, but with electron microscopy, 1 atrophic muscle containing numerous rods was found after an extended and long search. In contrast, many of the father's cultured muscle fibers contained numerous rods, which could be easily identified. Therefore, based on tissue culture results, it appears that in this case, the father is a probable carrier of the disease.

Those studies indicate that under some circumstances, tissue culture of genetically abnormal human muscle may provide information regarding carrier detection.

Autosomal dominant neuromuscular disease with cylindrical spirals

Cylindrical spirals (CS) reported in muscle biop-

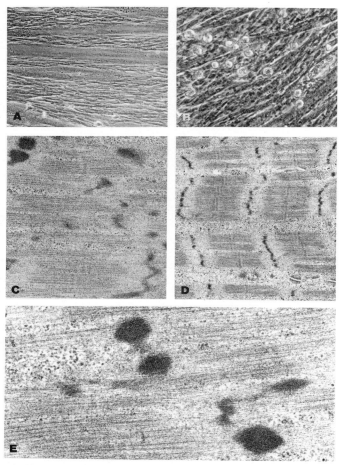

Fig. 16. Cultured muscle fibers from a patient with infantile rod disease. (A,B) Phase-contrast light microscopy of living cultured muscle fibers. (A) 10-day-old aneurally cultured muscle has normal appearance (×460). (B) 4-week-innervated cultured muscle fibers are thin and are lacking cross-striations (×665). (C–E) Electron microscopy of 4-week-innervated cultured muscle fibers from patient with rod disease (C,E) and control (D). There is lack of cross-striations, smearing of Z-disks in (C), and typical rods in (E). In contrast, there are well-defined cross-striations in (D) (C. ×17,000; D. ×9,800; E. ×34,600).

sies are of unknown origin (Gibbels et al. 1983). We studied in culture (Askanas et al. 1991b) muscle biopsy containing numerous CSs from a 50-year-old man with lower limb weakness from age 46. His mother's muscle biopsy also had CSs; she and several other family members, in an autosomal dominant pattern, have progressive muscle weakness (Taratuto et al. 1990). Aneurally cultured muscle fibers had very excessive, longitudinally oriented, long lanthanum-filled dilated t-tubules (Fig. 17A). Well-organized CSs were not present. Cultured muscle fibers innervated for 8–12 weeks

by embryonic rat spinal cord neurons had numerous CSs, 0.2–1.5 µm diameter, composed of 3–20 concentric lamellae 10–15-nm thick separation by 5–20-nm clefts (Fig. 17B). CSs were located within lateral sacs, with one edge of each lamella attached to the intra-sac end of a t-tubule junctional bridge. Masses of tubular aggregates (Fig. 17C) and various disfigurations of t-tubules were also present. This study demonstrated that 1) CSs originate in the lateral sacs, 2) in this family, because the CSs develop in cultured muscle fibers, they result from a genetic defect, which affects the lateral sac and

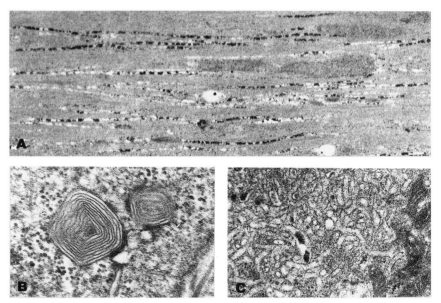

Fig. 17. Electron microscopy of innervated cultured muscle fibers from a patient with familial neuromuscular disease with cylindrical spirals (CSs). (A) Electron microscopy of lanthanum staining of aneurally cultured muscle fiber illustrating abundant longitudinally oriented tubules filled with lanthanum (×13,000). (B) Two cylindrical spirals adjacent to t-tubules in innervated muscle fiber (×42,700). (C) Aggregate of tubules in innervated muscle fiber (×25,700).

t-tubule system, and 3) innervation and/or advanced maturity of cultured muscle fiber is required for the defect to be demonstrable. This tissue culture system should provide an excellent experimental model for future studies regarding molecular events leading to CS formation.

Table 4 summarizes the expression of abnormalities in innervated muscle from patients with genetic muscle disorders.

TABLE 4

Expression of abnormalities in innervated cultured muscle fibers.

A. *Presumably due to more advanced maturation and/ or to contractile activity of muscle fibers*
1. Duchenne muscular dystrophy
2. Oculopharyngeal muscular dystrophy
3. Familial inclusion-body myositis (vacuolar myopathy)

B. *Presumably due to defective response of muscle fibers to normal innervation – putative 'dysreception'*
1. Myotonic atrophy (dystrophy)
2. Infantile rod (nemaline) disease

Myophosphorylase deficiency (McArdle's myopathy)

In muscle biopsies of patients with myophosphorylase deficiency, muscle glycogen phosphorylase (GP) and its mRNA are absent or greatly reduced (Mommaerts et al. 1959; Gautron et al. 1987; Servidei et al. 1988; McConchie et al. 1991). Even though major deletions in M-GP gene could not be demonstrated (Anderson et al. 1986; Gautron et al. 1987; Lebo et al. 1990), point mutations or small alternation of the flanking gene close to the region of the M-GP gene were not excluded.

In human tissues GP exists in 3 different isoforms: muscle-type (M-GP); liver-type (L-GP); and brain type (BP) (Proux and Dreyfus 1973). Each of the isoforms is encoded by a separate gene (Lebo et al. 1984; Newgard et al. 1987, 1988).

Fetal muscle contains L- and B-GP, which are during maturation replaced by M-GP. The change of the isoenzymatic pattern of GP isozymes occurs also during differentiation in culture. In human cultured muscle a percent of M-GP gradually increases during maturation and it is influenced both by media components and neuronal factors (Martinuzzi et al. 1986, 1988; Pegolo et al. 1990). In young immature cultured human myotubes, the percent of M-GP is very low, approximately 5%; however, in 2-month-old innervated cultured muscle, the percent of M-GP reaches 35% (Martinuzzi et al. 1986, 1988, 1990b).

We have previously demonstrated that aneurally cultured human muscle from normal and myophosphorylase-deficient patients does not differ regarding their M-GP expression (Meinhofer et al. 1977). In the more recent studies, using much more mature and contracting innervated cultured muscle of myophosphorylase-deficiency patients both the expression of M-GP and its reactivity with anti-M-GP antibody were also indistinguishable from control-cultured muscle fibers, which were of the same age and had the same degree of innervation (Martinuzzi et al. 1990b, unpublished observation 1992).

In muscle biopsies of patients from which cultures were established, there was no detectable phosphorylase activity, no M-GP mRNA, and no reactivity with anti-M-GP antiserum (Martinuzzi et al. 1990b, unpublished observation 1992). Table 5 summarizes the findings in the biopsied and cultured muscle of myophosphorylase-deficiency patients.

Therefore, our study indicates that in cultured muscle of myophosphorylase-deficiency patients, the M-GP gene can be effectively transcribed and translated. The promoter sequence of the human M-GP gene has not yet been fully characterized and the possibility of a mutation in this region or the existence of other active transcription sites active in cultured muscle cannot be ruled out. Therefore, the following hypothetical reasons for lack of transcription of the M-GP gene in the adult but not in the developing muscle may be considered. 1. Genetic defect is in a transcriptional promoter that plays a role in the adult but not in the developing muscle. 2. Two promoters exist for the M-GP, 'fetal' and a 'mature' form, and the genetic defect is in the 'mature promoter'. 3. Genetic defect is in transcriptional enhancer that plays a role in the adult but not in the developing muscle. 4. Genetic defect is in 'pre-mRNA' processing (splicing) in the adult but not in the developing muscle.

If the factors influencing transcription of the M-GP gene could be identified, the pathogenesis of myophosphorylase deficiency and therapeutic approaches to this disease might be able to be elucidated.

FUTURE POSSIBILITIES

Culture of normal human muscle offers the opportunity to study molecular details of development and maintenance of the human muscle fiber. For example, identification of DNA regulatory sequences and factors that influence transcription of muscle-specific genes is of great recent interest (reviewed in Emerson 1990 and Tapscott et al. 1990). The identification of a family of human muscle-specific regulatory factors, that have some homology with the myc oncogene (Wright et al. 1989; Braun et al. 1990a,b), opened the field for studying events leading to the expression of muscle-specific genes. With continuing development of molecular biology techniques, tissue culture of normal human muscle should serve as an excellent model for studying (a) factors regulating transcription and translation of various muscle-specific genes, (b) regulatory genes of myogenesis, including genes of the MyoD family, (c) functional regulation of expression of muscle-specific genes by the binding of MyoD, or members of its family, to enhancers of muscle-specific proteins, and (d) identification of new human myogenic determination factors.

TABLE 5

Myophosphorylase deficiency.

Detection of phosphorylase	Cultured muscle (aneural and inner-vated)	Biopsied muscle	
		Mature fibers	Regenerating fibers
Hystoenzymatic reaction	+	0	+
Biochemical enzymatic reaction	+	0	?
Immunoblot, muscle-specific type	+	0	?
mRNA, muscle-specific type	?	0	?
Gene, muscle-specific type	?	+	?

Tissue culture of genetically abnormal human muscle should become even more valuable for identifying: (a) genetic defects; (b) abnormalities of gene regulation; and (c) the step-by-step manifestations of an abnormal gene. Moreover, studies of genetically abnormal cultured human muscle that expresses the characteristic disease phenotype can serve as an excellent model for therapeutic experimentation on the patients' own living muscle, including drug trials and gene transfection.

Acknowledgements

Studies described in this article were supported in part by the Muscular Dystrophy Association, National Institutes of Health, March of Dimes Foundation, Myasthenia Gravis Foundation, and the Sheldon Katz, Norma Bard, and Tim Murrell Research Funds. The authors are grateful to the following postdoctoral fellows and collaborators with whom they have had the pleasure to work in recent years regarding cultured human muscle: A. Martinuzzi, T. Kobayashi, C. Desnuelle, G. Vita, S. Braun, G. Pegolo, Y. Park, K. Saito, F. Tomé, M. Fardeau, A.L. Taratuto, and M. Medici. Special thanks are due to Janis McFerrin and Renate B. Alvarez for their years of dedication and excellence. Maggie Baburyan provided excellent assistance in photography and Sharon Jaffe in preparation of this article.

REFERENCES

ALVAREZ, R.B., M. FARDEAU, V. ASKANAS, W.K. ENGEL, J. MCFERRIN and F.M.S. TOMÉ: Characteristic filamentous inclusions reproduced in cultured innervated muscle fibers from patients with familial 'inclusion body myositis' (FIBM). J. Neurol. Sci. 98 (1990) 178.

ANDERSON, L., R.J. FLETTERICK, S. DIMAURO, P. HWANG, F. GORIN and R. LEBO: Restriction enzyme analysis of McArdle's syndrome gene locus. Muscle Nerve 9 (1986) 231 (abstract).

ARAHATA, K. and H. SUGITA: Dystrophin and the membrane hypothesis of muscular dystrophy. Trends Pharmacol. Sci. 10 (1989) 437–439.

ASKANAS, V.: Human muscle and Schwann cells in tissue culture as a tool in studying pathogenesis and treatment of neuromuscular disorders. In: G. Serratrice, D. Cros, C. Desnuelle, J.-L. Gastaut, J.F. Pellissier, J. Pouget and A. Schiano (Eds), Neuromuscular Diseases. New York, Raven Press (1984) 373–380.

ASKANAS, V. and W.K. ENGEL: New program for investigating adult human skeletal muscle grown aneurally in tissue culture. Neurology 25 (1975) 58–67.

ASKANAS, V. and W.K. ENGEL: Diseased human muscle in tissue culture. A new approach to the pathogenesis of human neuromuscular disorders. In: L.P. Rowland (Ed.), Pathogenesis of the Human Muscular Dystrophies. New York, Elsevier (1977) 856–871.

ASKANAS, V. and W.K. ENGEL: Normal and diseased human muscle in tissue culture. In: P.J. Vinken and G.W. Bryun (Eds), Handbook of Clinical Neurology, Vol. 40. New York, North-Holland (1979) 183–196.

ASKANAS, V. and G. GALLEZ-HAWKINS: Synergistic influence of polypeptide growth factors on cultured human muscle. Arch. Neurol. 42 (1985) 749–752.

ASKANAS, V., W.K. ENGEL, S. DIMAURO, B.R. BROOKS and M. MEHLER: Adult-onset acid maltase deficiency. N. Engl. J. Med. 294 (1976) 573–578.

ASKANAS, V., W.K. ENGEL and T. KOBAYASHI: TRH enhances motor-neuron-evoked contractions of cultured human muscle. Ann. Neurol. 18 (1985a) 716–719.

ASKANAS, V., S. CAVE, G. GALLEZ-HAWKINS and W.K. ENGEL: Fibroblast growth factor, epidermal growth factor and insulin exert neuronal-like influence on acetylcholine receptors in aneurally cultured human muscle. Neurosci. Lett. 61 (1985b) 213–219.

ASKANAS, V., S. CAVE, A. MARTINUZZI and W.K. ENGEL: Glucocorticoids increase number of acetylcholine receptors (AChRs) and AChR aggregates in human muscle cultured in serum free hormonally-chemically defined medium. Neurology 36 (1986) 241.

ASKANAS, V., A. MARTINUZZI, W.K. ENGEL, T. KOBAYASHI, L.Z. STERN and J.D. STERN: Accumulation of CK-MM is impaired in innervated and contracting cultured muscle fibers of Duchenne muscular dystrophy patients. Life Sci. 41 (1987a) 927–933.

ASKANAS, V., H. KWAN, R.B. ALVAREZ, W.K. ENGEL, T. KOBAYASHI, A. MARTINUZZI and E.F. HAWKINS: De novo neuromuscular junction formation on human muscle fibers cultured in monolayer and innervated by fetal rat spinal cord: ultrastructural and ultrastructural-cytochemical studies. J. Neurocytol. 16 (1987b) 523–537.

ASKANAS, V., W.K. ENGEL, T. KOBAYASHI, R.B. ALVAREZ, A. MARTINUZZI, K. SAITO and G. VITA: Phenotypic expressions of cultured human muscle induced by neuronal influence via nerve-muscle contact. Neurology 38 (1988a) 160–161.

ASKANAS, V., W.K. ENGEL, J. MCFERRIN, R.B. ALVAREZ, E. GANGITANO and C. IMBUS: Rod inclusions in innervated cultured muscle fibers from a patient with fatal infantile rod (Nemaline) disease. Ann. Neurol. 24 (1988b) 138.

ASKANAS, V., W.K. ENGEL, T. KOBAYASHI, F.M.S. TOMÉ, R.B. ALVAREZ, A. MARTINUZZI, C.S. LEE and J. MCFERRIN: Aneural and neural induction of abnormalities in cultured diseased human muscle fibers. In: G. Serratrice, J.-F. Pellissier, C. Desnuelle and J. Pouget (Eds), Myélopathies, Neuropathies et Myopathies. Advances in Neuromuscular Diseases. Paris, Expansion Scientifique Francaise (1989) 24–43.

ASKANAS, V., P. SERDAROGLU, W.K. ENGEL and R.B. AL-VAREZ: Immunolocalization of ubiquitin in muscle biopsies of patients with inclusion body myositis and oculopharyngeal muscular dystrophy. Neurosci. Lett. 130 (1991) 73–76.

ASKANAS, V., R.B. ALVAREZ, A.L. TARATUTO, J. MCFER-RIN and W.K. ENGEL: Cylindrical spirals (CSs) expressed in innervated cultured muscle from a patient with autosomal dominant neuromuscular disease. Neurology 41 (1991b) 422.

ASKANAS, V., J. MCFERRIN, Y.C. PARK-MATSUMOTO, C.S. LEE and W.K. ENGEL: Glucocorticoid increases acetylcholinesterase and organization of the postsynaptic membrane in innervated cultured human muscle. Exp. Neurol. 115 (1992a) 368–375.

ASKANAS, V., W.K. ENGEL and R.B. ALVAREZ: Light- and electronmicroscopic localization of β-amyloid protein in muscle biopsies of patients with inclusion-body myositis. Am. J. Pathol. 141 (1992b) 36.

ASKANAS, V., W.K. ENGEL, R.B. ALVAREZ and G.G. GLEN-NER: β-Amyloid protein immunoreactivity in muscle biopsies of patients with inclusion-body myositis. Lancet 339 (1992c) 560–561.

AVIGAN, J., V. ASKANAS and W.K. ENGEL: Muscle carnitine deficiency: fatty acid metabolism in cultured fibroblasts and muscle cells. Neurology 33 (1983) 1021–1026.

BANKER, B.Q.: Congenital myopathies. In: A.G. Engel and B.Q. Banker (Eds), Myology. New York, McGraw-Hill Book Co. (1986) 1527–1581.

BLAU, H.M. and C. WEBSTER: Isolation and characterization of human muscle cells. Proc. Natl. Acad. Sci. USA 78 (1981) 5623–5627.

BLAU, H.M., C. WEBSTER, C.P. CHIU, S. GUTTMAN and F. CHANDLER: Differentiation properties of pure populations of human dystrophic muscle cells. Exp. Cell Res. 144 (1983) 495–503.

BRAUN, S., V. ASKANAS and W.K. ENGEL: Glucocorticoid increases the expression of voltage-dependent Ca^{2+} channels of human muscle in culture. Soc. Neurosci. Abstr. 17 (1991a) 68.

BRAUN, S., V. ASKANAS, E. IBRAHIM and W.K. ENGEL: Long- term treatment with hydrocortisone (HC) increases accumulation of acetylcholine receptors (AChRs) at human neuromuscular junctions (NMJs) in culture. Neurology 41 (1991b) 154.

BRAUN, S., V. ASKANAS and W.K. ENGEL: Different degradation rates of junctional and extrajunctional acetylcholine receptors of human muscle cultured in monolayer and innervated by fetal rat spinal cord neurons. Int. J. Dev. Neurosci. 10 (1992) 37–44.

BRAUN, T., B. WINTER, E. BOBER and H.H. ARNOLD: Transcriptional activation domain of the muscle-specific gene-regulatory protein myf5. Nature (London) 346 (1990a) 663–665.

BRAUN, T., E. BOBER, B. WINTER, N. ROSENTHAL and H.H. ARNOLD: Myf-6, a new member of the human gene family of myogenic determination factors: evidence for a gene cluster on chromosome 12. Eur. Mol. Biol. Organ. J. 9 (1990b) 821–831.

BROOK, J.D., M.E. MCCURRACH, H.G. HARLEY, A.J. BUCK-LER, D. CHURCH, H. ABURATANI, K. HUNTER, V.P. STANTON, J.P. THIRION, T. HUDSON, R. SOHN, B. ZE-MELMAN, R.G. SNELL, S.A. RUNDLE, S. CROW, J. DA-VIES, P. SHELBOURNE, J. BUXTON, C. JONES, V. JU-VONEN, K. JOHNSON, P.S. HARPER, D.J. SHAW and D.E. HOUSMAN: Molecular basis of myotonic dystrophy: expansion of a trinucleotide (CTG) repeat at the 3′ end of a transcript encoding a protein kinase family member. Cell 58 (1992) 799–808.

BRUNNER, N.G., T. NAMBA and D. GROB: Corticosteroids management of severe generalized myasthenia gravis. Neurology 22 (1972) 603—610.

BURDEN, S.J., P.B. SARGENT and U.J. MCMAHAN: Acetylcholine receptors in regenerating muscle accumulate at original synaptic sites in the absence of the nerve. J. Cell Biol. 182 (1979) 412–425.

CARPENTER, S., G. KARPATI, I. HELLER and A. EISEN: Inclusion body myositis: a distinct variety of idiopathic inflammatory myopathy. Neurology 28 (1978) 8– 17.

CARTER, L. and V. ASKANAS: Vital freezing of cultured human muscle cells. Muscle Nerve 4 (1981) 367–369.

CHOU, S.M.: Myxovirus-like structures in a case of human chronic polymyositis. Science 158 (1967) 1453–1455.

COERS, C. and A.L. WOOLF: The Innervation of Muscle. Oxford, Blackwell Science (1959).

COERS, C., N. TELERMAN-TOPPET and J.M. GERARAD: Terminal innervation ratio in neuromuscular disease. Arch. Neurol. 29 (1973) 215–222.

COLE, A.J., R. KUZNIECKY, G. KARPATI, S. CARPENTER, E. ANDERMANN and F. ANDERMANN: Familial myopathy with changes resembling inclusion body myositis and periventricular leukoencephalopathy. Brain 111 (1988) 1025–1037.

COVAULT, J. and J.R. SANES: Neural cell adhesion molecule (N-CAM) accumulates in denervated and paralyzed skeletal muscle. Proc. Natl. Acad. Sci. USA 82 (1985) 4544–4548.

CRAIN, S.M., L. ALFEI and E.R. PETERSON: Neuromuscular transmission in cultures of adult human and rodent skeletel muscle after innervation in vitro by fetal rodent spinal cord. J. Neurobiol. 1 (1970) 471–489.

CROS, D., P. HARNDEN, J. POUGET, J.F. PELLISSIER, J.L. GASTAUT and G. SERRATRICE: Peripheral neuropathy in myotonic dystrophy: a nerve biopsy study. Annu. Neurol 23 (1988) 470–476.

DENGLER, R., R. RUDEL, J. WARELAS and K.L. BIRN-BERGER: Corticosteroids and neuromuscular transmission: electrophysiological investigation of the effects of prednisone on normal and anticholinesterase-treated neuromuscular junction. Pfluegers Arch. Eur. J. Physiol. 380 (1979) 145–151.

DESNUELLE, C., V. ASKANAS and W.K. ENGEL: Insulin enhances development of functional voltage-dependent Ca^{2+} channels in aneurally cultured human muscle. J. Neurochem. 49 (1987) 1133–1138.

DIMAURO, S., S. ARNOLD, A. MIRANDA and L.P. ROW-LAND: McArdle disease: the mystery of reappearing phosphorylase activity in muscle culture. A fetal isozyme. Annu. Neurol. 3 (1978) 60–66.

ECOB, M.S.: The application of organotypic muscle nerve cultures to problems in neurology with special reference to their potential use in research into neuromuscular diseases. J. Neurol. Sci. 58 (1983) 1–15.

ECOB, M.: The location of neuromuscular junctions on regenerating adult mouse muscle in culture. J. Neurol. Sci. 64 (1984) 175–182.

ECOB-PRINCE, M.S., M.A. HILL and A.E. BROWN: Localization of dystrophin in cultures of human muscle. Muscle Nerve 12 (1989) 594–597.

EMERSON, C.P.: Myogenesis and developmental control genes. Curr. Opin. Cell Biol. 2 (6) (1990) 1065–1075.

ENGEL, A.G.: The neuromuscular junction. In: A.G. Engel and B.G. Banker (Eds), Myology, New York, McGraw-Hill (1986) 209–253.

ENGEL, W.K.: Rod (nemaline) disease. In: E.S. Goldensohn and S.H. Appel (Eds), Scientific Approaches to Clinical Neurology. Philadelphia, Lea Febiger (1977a) 1667–1691.

ENGEL, W.K.: Integrative histochemical approach to the defect of Duchenne muscular dystrophy. In: L.P. Rowland (Ed.), Pathogenesis of Human Muscular Dystrophies. Amsterdam, Excerpta Medica (1977b) 277.

ENGEL, W.K.: Dagen des Oordeels: mechanisms and molecular message (a personal view). Arch. Neurol. 36 (1979) 329–339.

ENGEL, W.K. and M.H. BROOKE: Histochemistry of the myotonic disorders. In: E. Kuhn (Ed.), Symposium über Progressive Muskeldystrophie, Myotonie, Myasthenie. New York, Springer-Verlag Inc. (1966) 203–222.

ENGEL, W.K. and J.R. WARMOLTS: Myasthenia gravis: a new hypothesis of the pathogenesis and a new form of treatment. Ann. N.Y. Acad. Sci. 183 (1971) 72–87.

FARDEAU, M., J. GODET-GUILLAIN and M. CHEVALLAY: Modifications ultrastructurales des plaques motrices dans la myasthénie et les syndromes myasthéniques: la transmission cholinergique de l'excitation. INSERM-COLLOQUET (1972) 247–256.

FARDEAU, M., V. ASKANAS, F.M.S. TOMÉ, W.K. ENGEL, R. ALVAREZ, J. MCFERRIN and M. CHEVALLAY: Hereditary neuromuscular disorder with inclusion body myositis-like filamentous inclusions: clinical, pathological, and tissue culture studies. Neurology 40 (1990) 120.

FLORINI, J.R.: Hormonal control of muscle growth. Muscle Nerve 10 (1987) 577–598.

FLORINI, J.R., D.Z. EWTON and K.A. MAGRI: Hormones, growth factors, and myogenic differentiation. Annu. Rev. Physiol. 53 (1991) 201–216.

FONG, P., P.R. TURNER, W.F. DENETCLAW and R.A. STEINHARDT: Increased activity of calcium leak chan-nels in myotubes of Duchenne human and mdx mouse origin. Science 250 (1990) 673–675.

GAUTRON, S., D. DAEGELEN, F. MENNECIER, D., DUBOCQ, A. KAHN and J.C. DREYFUS: Molecular mechanisms of McArdle's disease (muscle glycogen phosphorylase deficiency). J. Clin. Invest. 79 (1987) 275–281.

GIBBELS, E., U. HENKE, H.J. SCHADLICH, W.F. HAUPT and W. FIEHN: Cylindrical spirals in skeletal muscle: a further observation with clinical, morphological, and biochemical analysis. Muscle Nerve 6 (1983) 646–655.

GOSPODAROWICZ, D. and J.S. MORAN: Growth factors in mammalian cell culture. Annu. Rev. Biochem. 45 (1976) 431–558.

HALL, E.D.: Direct effects of glucocorticoids on neuromuscular function. Clin. Neuropharmacol. 6 (1983) 169–183.

HAM, R.G., J.A. ST CLAIR, C. WEBSTER and H.M. BLAU: Improved media for normal human muscle satellite cells; serum-free clonal growth and enhanced growth with low serum. In Vitro Cell Dev. Biol. 24 (1988) 833–844.

HAM, R.G., J.A. ST CLAIR and S.D. MEYER: Improved media for rapid clonal growth of normal human skeletal muscle satellite cells. Adv. Exp. Med. Biol. 280 (1990) 193–199.

HARDIMAN, O., R.H. BROWN JR., A.H. BEGGS, L. SPECHT and R.M. SKLAR: Differential glucocorticoid effects on the fusion of Duchenne/Becker and control muscle cultures: pharmacological detection of accelerated aging in dystrophic muscle. Neurology 42 (1992) 1085–1091.

HAUSCHKA, S.D.: Clonal analysis of vertebrate myogenesis. II. Environmental influences on human muscle differentiation. Dev. Biol. 37 (1974) 239–344.

HOFFMAN, E.P., B.H. BROWN JR. and L.M. KUNKEL: Dystrophin: the protein product of the Duchenne muscular dystrophy gene. Cell 51 (1987) 919–928.

HOFFMAN, E.P., C. BERTELSON and L.M. KUNKEL: Alterations of dystrophin quality and quantity: the genetic and biochemical basis of Duchenne and Becker muscular dystrophies in cellular and molecular biology of muscle development. In: F. Stockdale and L. Kedes (Eds), Cellular and Molecular Biology of Muscle Development. New York, Alan R. Liss (1989) 917–935.

HOHFELD, R. and A.G. ENGEL: Induction of HLA-DR expression on human myoblasts with interferon-gamma. Am. J. Pathol. 136(3) (1990) 503–508.

HURKO, O., E.P. HOFFMAN, L. MCKEE, D.R. JOHNS and L.M. KUNKEL: Dystrophin analysis in clonal myoblasts derived from a Duchenne muscular dystrophy carrier. Am. J. Hum. Genet. 44 (1989) 820–826.

IBRAHIM, E.N., V. ASKANAS, S. BRAUN and W.K. ENGEL: Insulin increases RNA accumulation in human muscle cultured in a chemically-hormonally-defined medium. Ann. Neurol 30 (1991) 262.

KAPLAN, I., B.T. BLAKELY, G.K. PAVLATH, M. TRAVIS and H.M. BLAU: Steroids induce acetylcholine receptors on cultured human muscle: implications for myasthenia gravis. Proc. Natl. Acad. Sci. USA 87 (1990) 8100–8104.

KOBAYASHI, T. and V. ASKANAS: Acetylcholine receptors and acetylcholinesterase accumulate at the nerve-muscle contacts of de novo-grown human monolayer muscle co-cultured with fetal rat spinal cord. Exp. Neurol. 88 (1985) 327–335.

KOBAYASHI, T., V. ASKANAS and W.K. ENGEL: Human muscle cultured in monolayer and co-cultured with fetal rat spinal cord: importance of dorsal root ganglia for achieving successful functional innervation. J. Neurosci. 7 (1987) 3131–3141.

KOBAYASHI, T., V. ASKANAS and W.K. ENGEL: High excitability and remained slow calcium repolarization components of action potentials in cultured human paramyotonia congenita muscle cells. J. Biomed. Res. 11 (1990a) 299–305.

KOBAYASHI, T., V. ASKANAS, K. SAITO, W.K. ENGEL and K. ISHIKAWA: Abnormalities of aneural and innervated cultured muscle fibers from patients with myotonic atrophy (dystrophy). Arch. Neurol. 47 (1990b) 893–896.

KÜHL, U., M. OCALAN, R. TIMPL, R. MAYNE, E. HAY and K. VON DER MARK: Role of muscle fibroblasts in the deposition of type-IV collagen in the basal lamina of myotubes. Differentiation 28 (1984) 164–172.

LEBO, R.V., F. GORIN, R.J. FLETTERICK, F.T. KAO, M.C. CHEUNG, B.D. BRUCE and Y.W. KAN: High resolution chromosome sorting and DNA spot analysis assign McArdle's syndrome to chromosome 11. Science 225 (1984) 57–59.

LEBO, R.V., L.A. ANDERSON, S. DIMAURO, E. LYNCH, P. HWANG and R. FLETTERICK: Rare McArdle disease locus polymorphic site on 11q13 contains CpG sequence. Hum. Genet. 86 (1990) 17–24.

LEV, A.A., C.C. FEENER, L.M. KUNKEL and R.H. BROWN: Expression of the Duchenne's muscular dystrophy gene in cultured muscle cells. J. Biol. Chem. 262 (1987) 15817–15820.

LINKHART, T.A., C.H. CLEGG and S.D. HAUSCHKA: Control of mouse commitment to terminal differentiation by mitogens. J. Supramol. Struct. 14 (1980) 483–498.

LIPTON, B.H.: Collagen synthesis by normal and bromodeoxyuridine-modulated cells in myogenic cultures. Dev. Biol. 61 (1977) 153–165.

LOTZ, B.P., A.G. ENGEL, H. NISHINO, J.C. STEVENS and W.J. LITCHY: Inclusion body myosits: observations in 40 patients. Brain 112 (1989) 727–747.

MANDEL, J.L. and M.L. PEARSON: Insulin stimulates myogenesis of rat myoblast line. Nature (London) 251 (1979) 618–620.

MANTEGAZZA, R., S.M. HUGHES, D. MITCHELL, M. TRAVIS, H.M. BLAU and L. STEINMANN: Modulation of MHC class II antigen expression in human myoblasts after treatment with IFN-gamma. Neurology 41 (1991) 1128–1132.

MARSHALL, L.M., J.R. SANES and U.J. MCMAHAN: Reinnervation of original synaptic sites on muscle fiber basement membrane after disruption of the muscle

cells. Proc. Natl. Acad. Sci USA 74 (1977) 3073–3077.

MARTINUZZI, A., V. ASKANAS, T. KOBAYASHI, W.K. ENGEL and S. DIMAURO: Expression of muscle-gene specific isoenzymes of phosphorylase and creatine kinase in innervated cultured human muscle. J. Cell Biol. 103 (1986) 1423–1429.

MARTINUZZI, A., V. ASKANAS, T. KOBAYASHI, W.K. ENGEL and J. GORSKY: Developmental expression of the muscle- specific isozyme of phosphoglycerate mutase in human muscle cultured in monolayer and innervated by fetal rat spinal cord. Exp. Neurol. 96 (1987) 365–372.

MARTINUZZI, A., V. ASKANAS, T. KOBAYASHI and W.K. ENGEL: Asynchronous regulation of muscle specific isozymes of creatine kinase, glycogen phosphorylase, lactic dehydrogenase and phosphoglycerate mutase in innervated and non-innervated cultured human muscle. Neurosci. Lett. 89 (1988) 216–222.

MARTINUZZI, A., V. ASKANAS and W.K. ENGEL: Paralysis if innervated cultured human muscle fibers affect enzymes differentially. J. Neurochem. 54 (1990a) 223–229.

MARTINUZZI, A., L. VERGANI, M. FANIN, M. ROSA, R. CARROZZO, C. ANGELINI, V. ASKANAS and W.K. ENGEL: Glycogen phosphorylase in normal and McArdle's innervated cultured muscle. J. Neurol. Sci. 98 (1990b) 60.

MARTINUZZI, A., L. VERGANI, M. ROSA and C. ANGELINI: L-Carnitine uptake in differentiating human cultured muscle. Biochim. Biophys. Acta 1095 (1991) 217–222.

MASSA, R., B. WELLER, G. KARPATI, E. SHONBRIDGE and S. CARPENTER: Familial inclusion body myosites among Kurdish-Iranian Jews. Arch. Neurol. 48 (1991) 519–522.

MCCONCHIE, S.M., J. COAKLEY, R.H.T. EDWARDS and R.J. BEYNON: Molecular heterogeneity in McArdle's disease. Biochim. Biophys. Acta 1096 (1991) 26–32.

MCDERMOT, V.: The histology of the neuromuscular junction in dystrophia myotonica. Brain 84 (1961) 75–84.

MCFERRIN, J., V. ASKANAS and W.K. ENGEL: Expression of desmin in human muscle cultured aneurally and innervated by fetal rat spinal cord neurons. J. Neurol. Sci. 98 (1990) 193.

MEDICI, M., F.M.S. TOMÉ, V. ASKANAS, J. MCFERRIN and W.K. ENGEL: Characteristic nuclear inclusions present in cultured innervated muscle fibers (MFs) from two Uruguayan patients with oculopharyngeal muscular dystrophy (OPMD). J. Neurol. Sci. 98 (1990) 190–191.

MEINHOFER, M.C., V. ASKANAS, D. PROUX-DAEGELEN, J.C. DREYFUS and W.K. ENGEL: Muscle-type phosphorylase activity present in muscle cells cultured from three patients with myophosphorylase deficiency. Arch. Neurol. 34 (1977) 779–781.

MERICKEL, M., R. GRAY, P. CHAUVIN and S. APPEL: Cultured muscle from myotonic muscular dystrophy patients; altered membrane electrical properties. Proc. Natl. Acad. Sci. USA 78 (1981) 648–652.

MICHIKAWA, M., T. KOBAYASHI and H. TSUKAGOSHI: Early events of chemical transmission of newly formed neuromuscular junctions in monolayers of human muscle cells co-cultured with fetal rat spinal cord explants. Brain Res. 538 (1991) 79–85.

MIRANDA, A.F. and T. MONGINI: Diseased muscle in tissue culture. In: A.G. Engel and B.Q. Banker (Eds), Myology. New York, McGraw-Hill Book Co. (1986) 1123–1150.

MIRANDA, A.F., H. SOMER and S. DIMAURO: Isoenzymes as markers of differentiation. In: A. Mauro (Ed.), Muscle Regeneration. New York, Raven Press (1979) 453–473.

MIRANDA, A.F., S. DIMAURO, A. ANTLER, L.Z. STERN and L.P. ROWLAND: Glycogen debrancher deficiency is reproduced in muscle culture. Ann. Neurol. 9 (1981) 283–288.

MIRANDA, A.F., E.R. PETERSON and E.B. MASUROVSKY: Differential expression of creatine kinase and phosphoglycerate mutase isozymes during development in aneural and innervated human muscle culture. Tissue Cell 20 (1988a) 179–191.

MIRANDA, A.F., E. BONILLA, G. MARTUCCI, C.T. MORAES, A.P. HAYS and S. DIMAURO: Immunocytochemical study of dystrophin in muscle cultures from patients with Duchenne muscular dystrophy and unaffected control patients. Am. J. Pathol. 132 (1988b) 410–416.

MIRANDA, A.F., U. FRANCKE, E. BONILLA, G. MARTUCCI, B. SCHMIDT, G. SALVIATI and M. RUBIN: Dystrophin immunocytochemistry in muscle culture: detection of a carrier of Duchenne muscular dystrophy. Am. J. Med. Genet. 32 (1989) 268–273.

MOMMAERTS, W.F.H.M., B. ILLINGWORTH, C.M. PEARSON, R.J. GUILLORY and K. SERAYDARIAN: A functional disorder of muscle associated with the absence of phosphorylase. Proc. Natl. Acad. Sci. USA 45 (1959) 791–797.

MONGINI, T., D. GHIGO, C. DORIGUZZI, F. BUSSOLINO, G. PESCARMONA, B. POLLO, D. SCHIFFER and A. BOSIA: Free cytoplasmic Ca^{++} at rest after cholinergic stimulus is increased in cultured muscle cells from Duchenne muscular dystrophy patients. Neurology 38 (1988) 476–480.

MOORE, S.E. and F.S. WALSH: Specific regulation of N-CAM/D2-CAM cell adhesion molecule during skeletal muscle development. Eur. Mol. Biol. Organ. J. 4 (1985) 623–630.

MORAES, C.T., E.A. SCHON, S. DIMAURO and A.F. MIRANDA: Heteroplasmy of mitochondrial genomes in clonal cultures from patients with Kearns-Sayre syndrome. Biochem. Biophys. Res. Commun. 160 (1989a) 765–771.

MORAES, C.T., S. DIMAURO, M. ZEVIANI, A. LOMBES, S. SCHANSKE, A.F. MIRANDA, H. NAKASE, E. BONILLA, L.C. WERNECK, S. SERVIDEI, I. NONAKA, Y. KOGA, A.J. SPIRO, A.K.W. BROWNELL, B. SCHMIDT, D.L. SCHOTLAND, M. ZUPANG, D.C. DEVIVO, E.A. SCHON and L.P. ROWLAND: Mitochondrial DNA deletions in progressive external ophthalmoplegia and Kearns-Sayre syndrome. N. Engl. J. Med. 320 (1989b) 1293–1299.

MOXLEY, R.T., A.J. CORBETT, K.L. MINAKER and J.W. ROWE: Whole body insulin resistance in myotonic dystrophy. Ann. Neurol. 15 (1984) 157–162.

NEWGARD, C.B., R.J. FLETTERICK, L.A. ANDERSON and R.V. LEBO: The polymorphic locus for glycogen storage disease VI (liver glycogen phosphorylase) maps to chromosome 14. Am. J. Hum. Genet. 40 (1987) 351–364.

NEWGARD, C.B., D.R. LITTMAN, C. VAN GENDEREN, M. SMITH and R.J. FLETTERICK: Human brain glycogen phosphorylase: cloning, sequence analysis, chromosomal mapping, tissue expression and comparison with the human liver and muscle isozymes. J. Biol. Chem. 263 (1988) 3850–3857.

NOBLE, M.D., J.H. PEACOCK, J.A. LACHER and W.W. HOFMANN: Prednisone-neostigmine interactions at cholinergic junctions. Muscle Nerve 2 (1979) 155–157.

PARK, Y.C., V. ASKANAS and W.K. ENGEL: Influence of hydrocortisone (HC) on the development of neuromuscular junctions (NMJs) in cultured human muscles. J. Neurol. Sci. 98 (1990) 181–182.

PARK-MATSUMOTO, Y.C., V. ASKANAS and W.K. ENGEL: The influence of muscle contractile activity versus neural factors on morphologic properties of innervated cultured human muscle. J. Neurocytol. 21 (1992) 329–340.

PATTEN, B.M., K.L. OLIVER and W.K. ENGEL: Adverse interaction between steroid hormones and anticholinesterase drugs. Neurology 24 (1974) 442–449.

PEGOLO, G., V. ASKANAS and W.K. ENGEL: Expression of muscle-specific isozymes of phosphorylase and creatine kinase in human muscle fibers cultured aneurally in serum-free hormonally/chemically enriched medium. Int. J. Dev. Neurosci. 8 (1990) 299–308.

PETERSON, E.R. and S.M. CRAIN: Maturation of human muscle after innervation by fetal mouse spinal cord explants in long-term cultures. In: A. Mauro (Ed.), Regeneration. New York, Raven Press (1979) 429–441.

PETERSON, E.R., E.B. MASUROVSKY, A.J. SPIRO and S.M. CRAIN: Duchenne dystrophic muscle develops lesions in long-term co-culture with mouse spinal cord. Muscle Nerve 9 (1986) 797–808.

PROUX, D. and J.C. DREYFUS: Phosphorylase isoenzymes in tissues: prevalence of the liver type in man. Clin. Chim. Acta 48 (1973) 167–172.

SAITO, K., T. KOBAYASHI, V. ASKANAS, W.K. ENGEL and E. ISHIKAWA: Electrical properties of human muscle cultured in monolayer aneurally and co-cultured with fetal rat spinal cord. Biomed. Res. 11 (1990) 19–28.

SANDERSON, R.D., J.M. FITCH, T.R. LINSENMAYER and R. MAYNE: Fibroblasts promote the formation of a continuous basal lamina during myogenesis in vitro. J. Cell Biol. 102 (1986) 740–747.

SANDRA, A. and R.Y. PRZYBYLSKI: Antogeny of insulin binding during chick skeletal myogenesis in vitro. Dev. Biol. 68 (1979) 546–556.

SANES, J.R., L.M. MARSHALL and U.J. MCMAHAN: Reinnervation of muscle fiber basal lamina after removal of myofibers. Differentiation of regenerating axons at original synaptic sites. J. Cell Biol. 78 (1978) 176–198.

SATO, K., F. IMAI, I. HATAYAMA and R.I. ROELOFS: Characterization of glycogen phosphorylase isoenzymes present in cultured skeletal muscle from patients with McArdle's disease. Biochem. Biophys. Res. Commun. 78 (1977) 663–668.

SERVIDEI, S., S. SHANSKE, M. ZEVIANI, R. LEBO, R. FLETTERICK and S. DIMAURO: McArdle's disease: biochemical and molecular genetic studies. Ann. Neurol. 24 (1988) 774–781.

SEYBOLD, M.D. and D.B. DRACHMANN: Gradually increasing doses of prednisone in myasthenia gravis: reducing the hazards of treatment. N. Engl. J. Med. 290 (1974) 81–84.

SHIMIZU, M., C. WEBSTER, D.O. MORGAN, H.M. BLAU and R.A. ROTH: Insulin and insulin like growth factor receptors and responses in cultured human muscle cells. Am. J. Physiol. 251 (1986) E611–E615.

SKLAR, R.M. and R.H. BROWN JR: Methylprednisolone increases dystrophin levels by inhibiting myotube death during myogenesis of normal human muscle in vitro. J. Neurol. Sci. 101 (1991) 73–81.

SKLAR, R.M., A.H. BEGGS, A.A. LEV, L. SPECHT, F. SHAPIRO and R.H. BROWN JR: Defective dystrophin in Duchenne and Becker dystrophy myotubes in cell culture. Neurology 40 (1990) 1854–1858.

TAHMOUSH, A.K., V. ASKANAS, P.G. NELSON and W.K. ENGEL: Electrophysiologic properties of aneurally cultured muscle from patients with myotonic muscular atrophy. Neurology 33 (1983) 311–316.

TAHMOUSH, A.J., V. ASKANAS, P.G. NELSON and W.K. ENGEL: Adult-onset acid maltase deficiency. Electrophysiological properties of aneurally cultured muscle. Arch. Neurol. 41 (1984) 1190–1192.

TAPSCOTT, S.J., R.L. DAVIS, A.B. LASSAR and H. WEINTRAUB: MyoD: a regulatory gene of skeletal myogenesis. Adv. Exp. Med. Biol. 280 (1990) 3–5.

TARATUTO, A.L., M. MATTEUCCI and C. BARREIRO: Dominant autosomal neuromuscular disease with cylindrical spirals (CS). J. Neurol. Sci. 98 (1990) 336.

TOMÉ, F.M.S. and M. FARDEAU: Nuclear inclusions in oculopharyngeal dystrophy. Acta Neuropathol. 49 (1980) 85–89.

TOMÉ, F.M.S. and M. FARDEAU: Ocular myopathies. In: A.G. Engel and B.Q. Banker (Eds), Myology. New York, McGraw-Hill Book Co. (1986) 1327–1347.

TOMÉ, F.M.S., M. FARDEAU, P. LEBON and M. CHEVALLAY: Inclusion body myositis. Acta Neuropathol. 7 (1981) 287–291.

TOMÉ, F.M.S., V. ASKANAS, W.K. ENGEL, R.B. ALVAREZ and C.S. LEE: Nuclear inclusion in innervated cultured muscle fibers from patients with oculopharyngeal muscular dystrophy. Neurology 39 (1989) 926–932.

VELDESMA-CURRIE, R.D., H.V. VAN WILGENBURG, W.T. LABRUYERE and M.W.E. LANGEMEIJER: Presynaptic, facilitatory effects of the corticosteroid dexamethasone on rat diaphragm: modulation by beta-bungarotoxin. Brain Res. 294 (1984) 315–325.

VITA, G., V. ASKANAS, A. MARTINUZZI and W.K. ENGEL: Histoenzymatic profile of human muscle cultured in monolayer and innervated de novo by fetal rat spinal cord. Muscle Nerve 11 (1988) 1–9.

WALSH, F.S., S.E. MOORE and J.G. DICKSON: Expression of membrane antigens in myotonic dystrophy. J. Neurol. Neurosurg. Psychiatr. 51 (1988) 136–138.

WEBSTER, C. and H.M. BLAU: Accelerated age-related decline in replicative life-span of Duchenne muscular dystrophy myoblasts: implications for cell and gene therapy. Somatic Cell Mol. Genet. 16 (1990) 557–565.

WEBSTER, C., G. FILIPPI, A. RINALDI, C. MASTROPAOLO, M. TONDI, M. SINISCALCO and H.M. BLAU: The myoblast defects identified in Duchenne muscular dystrophy is not a primary expression of the DMD mutation. Hum. Genet. 74 (1986) 74–80.

WEBSTER, C., G.K. PAVLATH, D.R. PARKS, F.S. WALSH and H.M. BLAU: Isolation of human myoblasts with the fluorescence-activated cell sorter. Exp. Cell Res. 174 (1988) 252–265.

WEI, Y., S.C. MILLER, Y. TSUJI, S.V. TORTI and F.M. TORTI: Interleukin 1 induces ferritin heavy chain in human muscle cells. Biochem. Biophys. Res. Commun. 169 (1990) 289–296.

WITKOWSKI, J.A.: Tissue culture studies of muscle disorders. Part 1. Techniques, cell growth, morphology, cell surface. Muscle Nerve 9 (1986a) 191–207.

WITKOWSKI, J.A.: Tissue culture studies of muscle disorders. Part 2. Biochemical studies, nerve-muscle culture, metabolic myopathies and animal models. Muscle Nerve 9 (1986b) 283–298.

WRIGHT, W.E., D.A. SASSOON and V.K. LIN: Myogenin, a factor regulating myogenesis, has a domain homologous to MyoD. Cell 56(4) (1989) 607–617.

YAFFE, D.: Retention of differentiation potentialities during prolonged cultivation of myogenic cells. Proc. Natl. Acad. Sci. USA 61 (1968) 477–483.

YASIN, R., G. VAN BEERS, K.C.E. NURSE, S. EL-AMI, D.N. LANDON and E.S. THOMPSON: A quantitative technique for growing human adult skeletal muscle in culture starting from mononucleated cells. J. Neurol. Sci. 32 (1977) 347–360.

YASIN, R., D. KUNDU and E.J. THOMSON: Growth of adult human cells in culture at clonal densities. Cell Differ. 10 (1981) 131–137.

YASIN, R., F.S. WALSH, D.N. LANDON and E.J. THOMPSON: New approaches to the study of human dystrophic muscle cells in culture. J. Neurol. Sci. 58 (1983) 315–334.

YUNIS, E.J. and F.J. SAMAHA: Inclusion body myositis. Lab. Invest. 25 (1971) 240–248.

ZACKS, S.I., W.C. BAUER and J.M. BLUMBERG: The fine structure of myasthenic neuromuscular junction. J. Neuropathol. Exp. Neurol. 21 (1962) 334–347.

Handbook of Clinical Neurology, Vol. 18 (62): Myopathies
L.P. Rowland and S. DiMauro, editors
© 1992 Elsevier Science Publishers B.V. All rights reserved

CHAPTER 4

X-linked muscular dystrophies

ROBERT C. GRIGGS[1] and KENNETH H. FISCHBECK[2]

[1]*University of Rochester School of Medicine and Dentistry, Department of Neurology, Rochester, NY, USA and*
[2]*Hospital of the University of Pennsylvania, Department of Neurology, Philadelphia, PA, USA*

This chapter considers X-linked myopathies, emphasizing Duchenne and Becker dystrophies; it updates the previous edition of the chapter by Rowland and Layzer (1979). The accompanying chapter by Hopkins and Warner (Chapter 5) discusses Emery-Dreifuss and Barth syndrome. At least six neuromuscular diseases have been localized to the X chromosome. Four of these disorders are traditionally considered myopathies. One disorder, bulbar spinal muscular atrophy (Kennedy syndrome), is an anterior horn cell disorder (La Spada et al. 1991). Another X chromosome neuromuscular disease includes one or more of the forms of Charcot-Marie-Tooth disease, a peripheral neuropathy. These two neurogenic conditions are not considered further in this chapter. The gene lesions and absent or abnormal protein products for Duchenne and Becker muscular dystrophy have been defined (Monaco et al. 1986). The clinical features of Duchenne and Becker muscular dystrophy as well as a number of intermediate and less severe variants can be at least partially explained by the genetic and resulting biochemical heterogeneity (discussed below).

DUCHENNE MUSCULAR DYSTROPHY

Duchenne muscular dystrophy (DMD) is an X-linked disease characterized by progressive weakness of skeletal muscle and caused by total lack of dystrophin in skeletal muscle. It is present at birth, causes weakness apparent before age 5 and progresses to loss of ambulation before age 15.

History

In 1868 Duchenne reported 13 patients, the majority of which clearly had the disease that bears his name and which he termed *paralysie musculaire pseudohypertrophique* (Duchenne 1868). Duchenne cited earlier cases and in fact Meryon had the earliest published case that clearly had the same disease (Meryon 1852). Gowers' (1879) description of the clinical features of the disease includes the still important sign of proximal weakness that bears the name *Gowers' sign*: the way a boy with proximal weakness rises from lying supine on the floor that is a landmark in the recognition of the disease. Gowers clearly described the predilection of the disease for males and noted that both parents were characteristically normal. He also noted that male relatives on the mother's side were affected with a similar illness. It was not until the 1950s that the classification of the muscular dystrophies on clinical and genetic grounds could pave the way for research into the specific biochemical and genetic defects of DMD and other muscular dystrophies. Stephens and Tyler (1951), Stevenson (1953), and Walton and Nattrass (1954) developed the classification of muscular dystrophy into X-linked recessive, autosomal recessive, and autosomal dominant diseases. The DMD phenotype was noted to be X-linked in a majority of cases (Walton 1960, 1964), although an identical phenotype was considered to be due to an autosomal recessive disease in some cases (Walton 1964). Particularly helpful in defining DMD was the observation that the blood creatine kinase level was invariably elevated to extremely high values (Ebashi et al. 1959; Walton 1960). (Initially termed creatine phosphokinase [CPK] the preferred term is now creatine kinase [CK]). Earlier studies by Sibley and Lehniger (1949) and Schapira et al. (1953) identified marked elevations in aldolase in DMD. Subsequent studies have shown that the aldolase is less specific and not as useful in the diagnosis of muscle disease as the CK (Munsat et al. 1973). It was not until 1973 that serial observations of CK demonstrated convincingly that enzyme levels could return to normal late in the course of the disease (Munsat et al. 1973).

The histopathologic findings of DMD were described by Duchenne (1868) and earlier by Griesinger (1865). Autopsy studies of a patient with DMD demonstrated the normality of the central and peripheral nervous system (Griesinger 1865). The relatively specific muscle histopathologic criteria for DMD were first identified in the studies of Hathaway et al. (1970) and Dubowitz and Brooke (1973): groups of basophilic, regenerating fibers within a muscle that shows signs of chronic myopathy and type 1 fiber predominance (Dubowitz and Brooke 1973). The presence of hyaline, 'opaque' fibers had earlier been identified (Pearce and Walton 1962) but was often viewed as a non-specific finding (Boxler and Jerusalem 1978; Nonaka and Sugita 1980). Numerous early reports of electron microscopy of muscle demonstrated largely non-specific findings. It remained for Mokri and Engel (1975) to identify small membrane lesions in relatively normal fibers as an early and characteristic feature of the illness.

The heterogeneity of X-linked muscular dystrophy was first established when Becker and Kiener (1955) clearly identified a less severe, 'benign sex-linked recessive muscular dystrophy' that now bears the name Becker dystrophy. The careful clinical definition of the disease (Walton 1960) made

possible sequential, quantitative clinical assessment of a large population of Duchenne patients (Brooke et al. 1983). These studies first characterized the relative homogeneity of the severe group of patients and described the existence of a milder 'outlier' group of intermediate severity. It was also possible to identify intrafamily homogeneity in comparison to a population of patients (Hyser et al. 1987). These studies paved the way for subsequent clinical-gene lesion correlations (see below).

Prevalence and incidence

The incidence of DMD is approximately 1 in 3500 male births (Emery 1991). The disease occurs in all populations that have been studied carefully and differing incidences may reflect early confusion as to disease definition and disease ascertainment (Rowland and Layzer 1979). Roughly one-third of cases appear to arise *de novo* with two-thirds of cases having inherited the gene from the carrier mother (Emery 1980). There is no evidence for an increased incidence of mutations in any specific population. Because DMD shortens lifespan, prevalence in the population is much lower, approximately 1 in 18,000 males (Gardner-Medwin and Sharples 1989).

Clinical manifestations

Age of onset. The majority of patients with DMD are normal at birth, and sit and stand at a normal age. Some patients will walk at a normal age but in over half of patients walking is delayed to 15 months or more (Gardner-Medwin et al. 1978). A minority of cases may appear abnormal and hypotonic even at birth (Firth et al. 1983). Most parents that have normal children for comparison identify abnormalities in patients with DMD before age 3 including difficulty in walking, running, and jumping (Brooke 1986). 'Toe walking' and enlarged calves are often noted before age 3 (Emery and Skinner 1976). By definition, all patients have an onset before age 5 (Brooke et al. 1983). Patients with a later onset, in all likelihood, are either Becker dystrophy or outliers. Neck flexor weakness that is so severe that the patient is unable to fully lift the head against gravity is present in all patients with DMD from birth onward and never improves as the normal infant does (Brooke et al. 1983). In situations where family history indicates an infant is at risk, DMD can be diagnosed at birth on the basis of either an elevated blood CK level or an abnormal muscle biopsy. Mass screening of blood CK can identify all cases at birth (Zellweger and Antonik 1975).

Characteristic features. The disease is usually first evident to parents and physicians because of a broad-based waddling gait with an exaggerated lumbar lordosis. When patients rise from the floor, they demonstrate the typical Gowers' sign in which the boys turn to face the floor, spread their legs, elevate their butt, and climb up their thighs using their hands (Gowers 1879). Patients never run or jump normally and fall behind playmates in sports. Muscles often feel rubbery and firm and there may be enlargement of the calves, quadriceps, gluteals and deltoids. Occasionally this muscle enlargement is extreme, while in other patients it may not be easy to distinguish from a normal child. Initially, the enlargement of muscles includes true muscle fiber hypertrophy. As the disease progresses, muscle fibers are replaced by connective tissue and fat and the term 'pseudohypertrophy', which gave the disease its early name, is justified (Dubowitz and Brooke 1973).

Clinical progression and natural history. The progression of muscle weakness follows a stereotyped pattern in the vast majority of patients. The fact that neck flexors are severely weak and cannot oppose gravity often goes unnoticed early in the course of the disease. Proximal lower extremity muscles are involved earlier than distal, and proximal arm strength earlier than distal. Evident leg weakness is present before age 5 but arm strength may appear normal until age 7 or 8. Ankle plantar flexion and ankle inversion remain remarkably strong throughout the entire disease. Cranial nerve-innervated muscles remain relatively unaffected until late in the course of the disease; tongue musculature is often enlarged but symptomatic tongue weakness uncommon. Functional ability follows a relatively predictable pattern of progression (Dubowitz and Heckmatt 1980; Allsop and Ziter 1981; Brooke et al. 1989). Importantly, boys between ages 3 and 6 often appear to transiently 'improve' in the performance of motor tasks such as climbing stairs, running, or arising from the floor (Brooke 1986; Brooke et al. 1989). By age 8, patients begin to have difficulty in climbing stairs and by age 10 patients have difficulty in walking. Most patients become dependent on long leg braces by age 10 and without pharmacologic intervention become wheelchair-dependent by around age 12.

Manual muscle testing, functional assessment, and measurement of contractures have been carefully characterized by several groups of investigators (Allsop and Ziter 1981; Scott et al. 1982; Brooke et al. 1989; Smith et al. 1991). Longitudinal studies of patients examined at 3 month intervals were carried out by Brooke et al. (1989) and characterized 283 patients with DMD. Over 100 were followed for five years or more. This population allows certain conclusions as to the age at which various functions are lost. Using functional grades for arms and shoulders (Brooke et al. 1981b) and hips and legs modified from Swinyard et al. (1957), Table 1 and Fig. 1 show the ages at which sequential functional grades are past and various functions lost.

Other manifestations.

Scoliosis does not occur before age 11 but in the majority of Duchenne patients begins while the patient is still able to walk or to stand an hour a day (Brooke et al. 1989). Three-quarters of patients eventually develop some element of scoliosis (>10 degrees) and nearly 40% develop severe scoliosis (>70 degrees).

Contractures. In most patients contractures are detectable by age 6, including the iliotibial bands, hip joints and heel cords (Scott et al. 1982; Brooke et al. 1989). Iliotibial bands severely limit hip flexion because of the fibrosis connecting the ilium with the tibia. The 'toe walking' observed early in childhood is the result of heel cord contractures. Knee, elbow, and wrist extensor contractures also occur but only after age 8. Contractures, particularly those at the elbows and the knees, correlate with difficulty with ambulation and accelerate once patients are confined to a wheelchair. Shoulder joint contractures do not occur until very late in the illness.

Cardiac involvement. As many as 90% of Duch-

TABLE 1

Functional grades.

Arms and shoulders	Hips and legs
1. Starting with arms at the sides, patient can abduct the arms in a full circle until they touch above the head	1. Walks and climbs stairs with assistance
2. Can raise arms above head only by flexing the elbow (i.e. shortening the circumference of the movement) or by using accessory muscles	2. Walks and climbs stairs with aid of railing (<12 sec)
3. Cannot raise hands above head but can raise an 8 oz. glass of water to mouth (using both hands if necessary)	3. Climbs stairs slowly (>12 sec)
4. Can raise hands to mouth but cannot raise an 8 oz. glass of water to mouth	4. Walks unassisted and rises from chair but cannot climb stairs
5. Cannot raise hand to mouth but can use hands to hold pen or pick up pennies from table	5. Walks unassisted but cannot rise from chair or climb stairs
6. Cannot raise hands to mouth and has no useful function of hands	6. Walks only with assistance or walks independently with long leg braces
	7. Walks in long leg braces but requires assistance for balance
	8. Stands in long leg braces but unable to walk even with assistance
	9. Is in wheelchair
	10. Is confined to bed

From Brooke et al. 1989.

enne patients demonstrate electrocardiographic abnormalities (Griggs et al. 1977; Sanyal and Johnson 1982). The presence of tall, right precordial R-waves with an increased R-S amplitude ratio in V_1, and deep, narrow Q-waves in the left precordial leads represent the most commonly observed and distinctive changes (Perloff et al. 1967) (Fig. 2). Common patterns of conduction system involvement include intraatrial defects which are more common than ventricular and infra-nodal disturbances. Heart block seldom occurs. Most patients have a labile or persistent sinus tachycardia or other sinus arrhythmia (Sanyal and Johnson 1982). Ectopic beats and tachyarrhythmias are rare. DMD results in replacement myocardial fibrosis, particularly in the posterobasal and the surrounding region of the left ventricular wall. These areas are the initial and most extensive site of myocardial fibrosis. The ventricular septum, the right ventricle, and atrial myocardium are characteristically only minimally involved. Despite the evidence for left ventricular cardiac muscle fibrosis, most patients remain free of cardiovascular symptoms. Congestive heart failure and cardiac arrhythmias occur as a rule only in late stages of

the illness. These manifestations of cardiac dysfunction are much more common when there is intercurrent a pulmonary infection. Rare patients have overt signs of congestive heart failure and may die of cardiac failure with relative sparing of respiratory muscle function. In patients with congestive cardiomyopathy due to DMD, pulmonary emboli and systemic emboli can occur. In the longitudinal natural history study of Brooke et al. (1989), patients who had more severe muscle weakness were more likely to die from a pulmonary than cardiac cause whereas those in a stronger group were more likely to die of cardiac dysfunction.

Gastrointestinal manifestations. Clinical and pathologic involvement of smooth muscle of the gastrointestinal tract is often an important feature of DMD. The syndrome of acute gastric dilatation (Crowe 1961; Barohn et al. 1988) produces 'intestinal pseudo-obstruction'. The clinical manifestations are sudden onset of vomiting associated with abdominal pain and distention. This syndrome can lead to hypotension and death. Acute gastric dilatation may result from degeneration of smooth muscle within the stomach where dystrophin has

DUCHENNE MUSCULAR DYSTROPHY

AGES AT WHICH 'MILESTONES' ARE PASSED

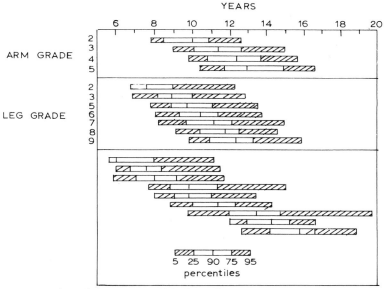

Fig. 1. Graphic representation of the ages at which the patients cross various milestones. (Reproduced with permission from Brooke et al. (1989) *Neurology*.)

been shown to be deficient. Quantitative studies of gastric function indicate that subclinical evidence for impaired contractility is common even in asymptomatic patients with DMD (Barohn et al. 1988).

Central nervous system. Mental retardation occurs in approximately one third of Duchenne patients (Zellweger and Niedermeyer 1965). In DMD the average IQ falls approximately one standard deviation below the mean (Leibowitz and Dubowitz 1981). This impairment of intellectual function is non-progressive and affects verbal more than performance IQ (Leibowitz and Dubowitz 1981). The mental deficiency is not the result of the physi-

cal limitations of a disabled child since control populations disabled by spinal muscular atrophy or other disease do not show a similar degree of deficiency (Marsh and Munsat 1974). Despite a suggestion of abnormal brain structure in DMD (Rosman and Kakulas 1966), there have been no consistent CNS histopathologic abnormalities to account for the mental retardation (Dubowitz and Crome 1969). The presence of dystrophin in brain and its absence in DMD (Hoffman et al. 1988b) has not been shown to be related to the mental deficiency. Since brain dystrophin is structurally different from muscle dystrophin, it is possible that truncated portions of the dystrophin gene can still

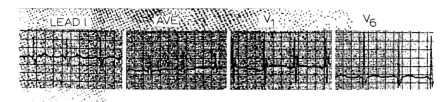

Fig. 2. The electrocardiogram in Duchenne dystrophy shows a distinctive RSR' (or RSR'S'R'') pattern as well as deep Q-waves in lateral precordial leads. (Reproduced by permission from Griggs, Reeves and Moxley (1977) *Pathogenesis of Human Muscular Dystrophies*, Amsterdam, Excerpta Medica.)

code for relatively normal brain dystrophin in patients with DMD.

Respiratory failure. The majority, if not all Duchenne patients have decreased respiratory function reflected in a low forced vital capacity at each age measured. Longitudinal analysis of pulmonary function of individual patients shows, however, a steady increase in absolute forced vital capacity in a majority of patients between ages 5 and 15 (Brooke et al. 1983) (Fig. 3). Even when related to expected norms, a quarter or more of patients show improvement (Brooke et al. 1983). Respiratory failure and difficulty clearing oropharyngeal and bronchial secretions does not occur until patients have been confined to wheelchair for at least two years (Brooke et al. 1989). Some patients, however, never develop a vital capacity of over 1 liter. Signs of impending respiratory failure include glossopharyngeal breathing (frog breathing) and signs of cor pulmonale: hepatic congestion, distended neck veins, pulmonary congestion, ankle edema, and other signs of right heart failure. Early morning headache signifying hypercapnia, and progressive lethargy are signs of impending ventilatory failure. In rare instances hypercapnic patients develop papilledema; distended retinal veins are commonly seen. Such patients

typically begin to lose weight, develop erythremia, and evidence other signs of chronic hypoxia (Griggs and Sutton 1992).

Differential diagnosis

The ability to define Duchenne and Becker dystrophy by genetic and biochemical criteria have actually broadened the differential diagnosis by making it clear that a number of patients with similar features do *not* have a dystrophin-related disease. DMD is by far the commonest neuromuscular disease presenting with proximal weakness in young boys between ages 2 and 5. The other relatively common disorder, juvenile spinal muscular atrophy, is occasionally misdiagnosed as DMD because of the presence of similar progressive proximal weakness and enlarged calves. Patients with spinal muscular atrophy, however, usually have tremor and fasciculations and are typically either of earlier onset and much more severe than DMD, or of later onset and less severe (Dubowitz 1991). Patients with spinal muscular atrophy can have moderate elevations in CK (up to 5-fold) (Munsat et al. 1973) but never reach the 20-fold or greater levels characteristic of the young patient with DMD (Brooke et al. 1981a,b). Other conditions that occasionally resemble typical DMD are childhood polymyositis, or dermatomyositis with evanescent rash, myasthenia gravis and the congenital myasthenias, chronic inflammatory demyelinating polyneuritis, Emery-Dreifuss and limb-girdle dystrophies. The one disorder that cannot be easily distinguished by clinical examination alone or by routine blood studies are the milder variant forms of X-linked dystrophy. If the patient has antigravity neck flexor strength, Becker dystrophy or 'outlier' is likely. CK values are usually 20-fold elevated early in the course of all dystrophinopathies.

Atypical presentations

Unusually severe cases of DMD may present with hypotonia and severe weakness at birth (Iannacone et al. 1988). Such cases may be mistaken for various congenital myopathies, particularly congenital muscular dystrophy. The extremely high CK values (20-fold elevated or higher) found in such severe cases serves to exclude other congeni-

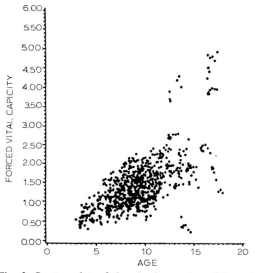

Fig. 3. Scatter plot of the absolute value of forced vital capacity against the age of the patient. (Reproduced with permission from Brooke et al. (1983) *Muscle and Nerve*.)

tal myopathies. The severe mental retardation seen in occasional patients may overshadow muscle weakness.

Patients with coincidental infantile illnesses such as febrile seizures resulting from ear or other infections often have routine blood tests that include liver function studies. In the patient with DMD, AST and LDH values are invariably elevated and may lead to a diagnosis of hepatitis. Liver biopsy is sometimes inappropriately performed if weakness is not evident or is overlooked. CK testing, once obtained, indicates that the enzyme abnormalities result from myopathy rather than liver disease. Patients with DMD are susceptible to malignant hyperthermia. In rare instances, cases have presented with malignant hyperthermia during correction of strabismus or hernia done before the age when weakness is recognized (Rosenberg and Heiman-Patterson 1983). Rare instances of large deletions of Xp21 have first come to medical attention for glycerol kinase deficiency, retinitis pigmentosa, chronic granulomatous disease, or other severe disorders located in the vicinity of the DMD locus which, because of their severity, have overshadowed DMD (Kousseff 1981; Dunger et al. 1986; Love et al. 1990).

Histology

The histochemical features of DMD have been well described by Carpenter and Karpati (1984) and by Engel (1986). Muscle biopsies characteristically show necrotic or degenerating muscle fibers. Although not necessarily evident in fetal muscle samples, this finding is present after birth from the earliest stages of the disease, even before weakness is clinically apparent. The degenerating fibers often occur in clusters, and on longitudinal section the process may be seen to be segmental. The fibers may be somewhat enlarged and hyalinized early in the necrotic process. As the degeneration progresses, necrotic vacuoles are seen. The muscle fiber subsequently falls apart, with the debris invaded by macrophages and inflammatory cells. The muscle has the capability to regenerate: satellite cells become stimulated and differentiate into myoblasts, fuse to form myotubes, and then develop into regenerating muscle fibers. These regenerating fibers are recognized in muscle biopsy specimens as small, round, basophilic fibers with large, pale nuclei containing prominent nucleoli (Dubowitz and Brooke 1973). The histochemical appearance of these fibers reflects the high levels of mRNA and associated protein synthesis that is taking place. The regenerating fibers also appear in clusters. Late after regeneration, these fibers may be indistinguishable from normal fibers, except that the nuclei are centrally placed rather than in the normal subsarcolemmal position. DMD muscle may go through repeated cycles of muscle fiber degeneration and regeneration before the regenerative capacity of the muscle is exhausted. Since muscle fibers (as other cells) have a limited regenerative capacity (Blau et al. 1990), regeneration may fail. Then, the muscle gradually becomes replaced with fat and connective tissue. Increased endomysial fibrosis is thus a common feature in DMD muscle biopsies. In the late stages of the disease, the muscle may become completely replaced with connective tissue; little recognizable muscle tissue remains.

While this pattern of muscle fiber degeneration and regeneration with endomysial fibrosis may be characteristic of DMD, it is not entirely specific for the disease, Similar changes are found in other myopathies. In particular, the biopsy findings in some forms of autosomal recessive limb-girdle dystrophy may resemble those seen in DMD, except that they tend to be less severe (McGuire and Fischbeck 1991). In Becker dystrophy, also, the range of biopsy findings is similar to but less severe than those seen in DMD. Becker dystrophy biopsies may also show small, atrophic fibers such as are seen in denervated muscle (Bradley et al. 1978; Kaido et al. 1991). This finding, which – as noted for DMD –may be a result of segmental muscle fiber necrosis, can occasionally lead to the mistaken diagnosis of spinal muscular atrophy.

More specific structural alterations in DMD muscle relate to the primary defect in the disease, deficiency of the structural protein dystrophin, which is normally adherent to the muscle plasma membrane. Well before dystrophin was identified, the effects of its alteration were noticed by Mokri and Engel (1975) as disruption of the plasma membrane in fibers that had not yet become necrotic. They termed these minute tears 'delta' lesions after their appearance on electron microscopy.

Electrophysiological studies

Nerve conduction velocities and terminal latencies are normal and repetitive nerve stimulation demonstrates normal neuromuscular junction transmission. Electromyography (EMG) demonstrates findings typical of myopathy: motor unit potentials are of short duration and low amplitude; as many as 50% of motor unit potentials are polyphasic (vs <10% of potentials in normals); early recruitment of motor units is observed; and a full recruitment pattern is observed until late in the course of disease (Desmedt and Borenstein 1976). Fibrillations and positive sharp waves, as well as complex repetitive discharges are often present and presumably reflect the separation of muscle fibers from their innervation as the sequel to segmental necrosis and subsequent regeneration. This reinnervation of muscle fibers by regeneration of intramuscular nerve axonal sprouting results in late components, so-called satellite potentials that occur late (15–25 msec) after the beginning of the motor unit potential (Desmedt and Borenstein 1975). Late in the course of disease, muscle becomes replaced by connective tissue and areas of muscle become electrically silent.

Motor unit counts by the studies of McComas et al. (1971, 1977) have indicated a reduced number of units in the majority of boys with DMD. The technique is based upon dividing the amplitude of the compound muscle action potential evoked by supramaximal nerve stimulation by the mean amplitude of the increments in the muscle action potential induced by gradually increasing nerve stimulation. Other workers have not found such a decrease (Ballantyne and Hansen 1974; Panayiotopoulos et al. 1974) and there was no reduction in numbers of limb anterior horn cells in postmortem study of a Duchenne patient (Tomlinson et al. 1974). The recent finding of dystrophin-like protein in nerve, presumably the product of alternatively spliced dystrophin (Feener et al. 1989) again raises the possibility that abnormal nerve function will prove to be part of the illness.

Biochemical abnormalities of blood and urine. The serum level of CK is markedly elevated in DMD (Ebashi et al. 1959; Walton 1960) and this enzyme is the most useful of the serum markers for DMD.

The CK is elevated to diagnostically high levels at birth, continues to rise until age 3 and then decreases by about 20% per year thereafter (Munsat et al. 1973; Brooke et al. 1983). Levels eventually decline to normal levels (Munsat et al. 1973). Other serum enzyme levels are also elevated, including alanine aminotransferase, aspartate aminotransferase, aldolase (Thomson et al. 1960), carbonic anhydrase III (Heath et al. 1982), lactic dehydrogenase, and pyruvate kinase (Harano et al. 1973). The serum CK in DMD originates from muscle sarcoplasm (Rowland 1984) and consists predominantly of the muscle-specific (M) isoenzyme (Dawson and Fine 1967). In muscular dystrophies such as DMD, however, the skeletal muscle content of the MB isoenzyme increases such that serum MB activity also rises (Somer et al. 1973). The level of the MB isozyme in serum does not correlate with nor indicates that cardiac disease is present (Silverman and Mendell 1974). Serum levels of myoglobin are also elevated (Ando et al. 1978; Kagen et al. 1980) to an extent that roughly correlates with CK levels (Florence et al. 1985). The levels of both increase with exercise (Florence et al. 1985).

Urinary excretion of creatinine is diminished in DMD (Milhorat 1952; Fitch and Sinton 1964) and decreases progressively with age (Griggs et al. 1983). The urinary excretion of creatinine, if carefully quantitated in 3 or more consecutive 24-hour periods, provides an accurate estimate of muscle mass in DMD (Griggs et al. 1983). Urine collections must be made while patients are on a diet free of animal muscle. Urinary excretion of creatine, the biological precursor of creatinine, is elevated in DMD (Milhorat 1952). Creatine synthesis begins in the kidney and is completed in the liver. It is then transported to muscle where it is stored either free (20%) or phosphorylated to phosphocreatine (80%) (Fitch and Sinton 1964). In DMD and other muscle-wasting diseases, muscle creatine stores are reduced such that the ratio of creatine to creatinine in the urine is markedly increased (Milhorat 1952). Urinary excretion of a number of other amines is increased (relative to creatinine): putrescine, spermine, and spermidine (Russell and Stern 1981); carnosine (Hartlage et al. 1983); and 3-methylhistidine (McKeran et al. 1977; Griggs et al. 1980). The 3-methylhistidine to creatinine ratio is particularly useful as an indicator of the rate of muscle breakdown since histidine is methylated post-translationally and is excreted quantitatively in the urine after muscle protein is degraded. Since most 3-methylhistidine is confined to muscle actin and myosin protein, its excretion (in subjects on a flesh-free diet) reflects skeletal muscle protein degradation (Griggs et al. 1980).

Genetics

Finding the Duchenne dystrophy gene. The isolation of the gene which is defective in DMD is one of the great accomplishments of molecular biology. This was the first time the biochemical cause of a hereditary human disease was identified through isolation of the DNA which is altered. The story of this research will be told briefly here. Other reviews have appeared elsewhere (Witkowski and Caskey 1988; Worton and Thompson 1988; Gutmann and Fischbeck 1989).

Because DMD follows an X-linked recessive pattern of inheritance, the defective gene has long been known to be located on the X chromosome. Several patients were identified in the early 1980s who had cytologically detectable alterations in the X chromosome (translocation or deletion); in each case the damage occurred at the same spot, at a band near the middle of the short arm of the chromosome designated Xp21 (Verellen-Dumoulin et al. 1984; Francke et al. 1985). DNA fragments isolated from the Xp21 region identified DNA sequence polymorphisms (restriction fragment length polymorphisms, or RFLPs) which tended to track through DMD families together with the disease (Davies et al. 1983), indicating that patients without visible alteration of the chromosome also had defects at Xp21.

DNA segments within the gene were isolated by two independent approaches. Worton and his colleagues in Toronto and Philadelphia were able to clone the breakpoint or junction fragment from a patient with an Xp21 translocation (Ray et al. 1985), and Kunkel and his colleagues in Boston were able to isolate pieces of DNA deleted from the X chromosome of a patient with an Xp21 deletion causing DMD and other diseases due to disruption of nearby genes on the X chromosome (Kunkel et al. 1985). The DNA segments isolated by both groups turned out to be within the same gene, although some distance, more than 200 kilobases (kb) apart. Chromosomal DNA in this region was often found to be deleted in patients with DMD. Kunkel's group was able to isolate complementary DNA (cDNA) from muscle RNA corresponding to the deleted genomic segment (Monaco et al. 1986). The cDNA probes identified a messenger RNA in skeletal muscle which is 14 kb long. The sequence of nucleotides in this message predicted a protein of 3,685 amino acids which had not been previously identified (Koenig et al. 1988). This protein was given the name *dystrophin*. Portions of the gene were incorporated into expression vectors, producing fusion proteins which were used to raise antibodies against dystrophin (Hoffman et al. 1987). The antibodies identified a protein of about 430 kilodaltons which was present in normal skeletal muscle and absent in DMD muscle (Hoffman et al. 1988a).

Dystrophin structure and function. Immunohistochemical studies done with antibodies against dystrophin fusion peptides showed that the protein is located at or near the muscle plasma membrane (Zubrzycka-Gaarn et al. 1988). Dystrophin is also present in cardiac muscle and, to a lesser extent, in smooth muscle and brain (Hoffman et al. 1988b). In brain, it is present within neurons rather than glia, and appears to be localized to postsynaptic membranes (Lidov et al. 1990). In muscle, dystrophin appears when myoblasts fuse to form myotubes (Lev et al. 1987). Differential splicing of the gene transcript produces different isoforms in different tissues: it appears that smooth muscle dystrophin has a different sequence at the 3' (carboxy-terminal) end (Feener et al. 1989), and that brain dystrophin has a different promoter and 5' (amino-terminal) end from that found in skeletal muscle (Nudel et al. 1989). Recently, a 6.5 kb transcript containing sequences from the 3' half of the dystrophin gene has been found in a variety of tissues, including liver, testis, lung, and kidney, but not skeletal muscle (Bar et al. 1990). The dystrophin protein sequence is well conserved between human and rodent muscle, and even the chicken has a dystrophin gene with a similar sequence, particularly at the 3' end (Lemaire et al. 1988).

Analysis of the amino acid sequence of dystrophin predicts a rod-shaped cytoskeletal protein (Koenig et al. 1988). The protein can be divided into four domains. The first 240 amino acids at the amino-terminal end have homology to the actin-binding domain of *alpha-actinin* (Hammond 1987). The second and largest portion of the protein consists of 26 repeated units of about 100 amino acids each. These repeated units consist of triple helical segments similar to those found in the cytoskeletal protein *spectrin*. Following the long, presumably rod-shaped repeated domain is a domain of 280 amino acids which is rich in cysteine and bears sequence homology to the carboxy-terminal end of alpha-actinin. There is also some immunologic cross-reactivity to the spectrin-binding protein *ankyrin* in this region. The last 325 amino acids, which are highly conserved across species (Lemaire et al. 1988), have little resemblance to any previously described protein. Since the discovery of dystrophin, however, another gene expressed in skeletal muscle and other tissues has been identified, which has a similar sequence in this region (Love et al. 1989, 1991). This suggests that dystrophin may be one member of a family of related large structural proteins in muscle.

Dystrophin has no apparent transmembrane segment, but both cytochemical and immunocytochemical studies point toward it being tightly membrane-associated, probably through adherence to one or more integral membrane glycoproteins (Watkins et al. 1988; Campbell and Kahl 1989). At least one of these dystrophin-associated glycoproteins also binds to laminin in the basal lamina sheath on the external surface of the muscle fiber (Ibraghimov-Breskrovnaya et al. 1992). By analogy with spectrin, dystrophin may self-associate in antiparallel dimers, with the rod domains side-by-side and the actin-binding and other active domains at either end, forming a network of filaments on the cytoplasmic surface of the plasma membrane. The evidence therefore indicates that dystrophin is a part of the muscle fiber cytoskeleton bound to actin and to membrane proteins in a lattice-work which serves to support the muscle plasma membrane and lends it stability as the fiber contracts and relaxes.

Duchenne dystrophy genotype and phenotype. DMD has a characteristic clinical course,

with onset in early childhood, loss of ambulation by age 10 to 12, and death by the early twenties. Becker muscular dystrophy (BMD), a milder form of X-linked muscular dystrophy which was long thought to be a separate disease, is now known to be caused by defects in the same gene. In DMD, dystrophin is undetectable or present at very low levels in the muscle, presumably because the protein is severely truncated and rapidly degraded (Hoffman et al. 1988a). Although messenger RNA corresponding to the dystrophin gene may be present in normal or even increased amounts (Scott et al. 1988), translation of the RNA into protein is usually disrupted by mutations which shift the reading frame of three nucleotides for each amino acid (Monaco et al. 1988). In BMD, identifiable mutations usually do not shift the translational reading frame; the protein which is produced may be altered in size (smaller or larger) or reduced in quantity, but not altogether lost from the muscle (Hoffman et al. 1988a). In general, the severity of the disease is inversely related to the amount of dystrophin remaining in the muscle (Hoffman et al. 1989).

Since the dystrophin gene was first isolated, it has been apparent that the distribution of mutations within the gene is not uniform (Koenig et al. 1987, 1989; Den Dunnen et al. 1989). About 60% of DMD and BMD patients have detectable deletions or partial duplications of the gene; the others presumably have point mutations or other minor alterations of the gene which are more difficult to detect by standard technique (Roberts et al. 1991). The vast majority of detectable mutations fall in two clusters or 'hot spots', one in the first 4 kb of the 14 kb messenger RNA (the first 20 exons of the 79-exon gene), the other in the central 2 kb of the RNA (between exons 43 and 54 of the gene). The stretch between these hot spots (between 20 exons and 43) is a 'cold spot' with relatively few deletions in DMD and BMD patients, and mutations affecting the last 5 kb of the RNA (beyond exon 55 of the gene) are rarely detected. The reason for these cold spots is not clear. It may be that there is less susceptibility to mutation in these regions or that mutations here do not produce a disorder recognizable as DMD or BMD. The latter possibility is supported by a report of a family with a particularly mild phenotype (myalgia and increased serum

CK but no weakness) caused by a deletion in the cold spot in the first third of the gene (Gospe et al. 1989). This deletion, which removed 15% of the gene, would be expected to excise several of the repeat units in the rod domain of dystrophin, presumably producing a protein which is shorter but still functional enough to prevent widespread muscle degeneration. Other patients with very mild symptoms have had defects in this portion of the gene (England et al. 1990).

At the other end of the clinical spectrum associated with defects in the dystrophin gene is at least one report of absent dystrophin in a patient with the Fukuyama type of congenital muscular dystrophy (Arahata et al. 1989a). This suggests that some dystrophin gene defects produce an unusually severe, early-onset disease. The majority of Fukuyama patients have normal dystrophin, however, indicating that this congenital muscular dystrophin is in fact a different disorder.

At least two X-linked myopathies in animals have been found to be associated with loss of dystrophin: canine X-linked muscular dystrophy (Cooper et al. 1988) and the *mdx* mouse (Torres and Duchen 1987). The canine disorder has a phenotype similar to DMD, although there is substantial variation in severity within the pedigree. The *mdx* mice, which have a point mutation producing a stop codon within the gene (Sicinski et al. 1989), have an increased serum CK and mild histopathological changes of muscle degeneration, but no weakness. It is not clear why dystrophin loss in *mdx* mice does not produce the severe, progressive weakness which occurs in dogs and humans.

Glycerol kinase deficiency and McLeod myopathy. Mild myopathic syndromes may be caused by defects near the dystrophin gene, either distal (towards the telomere) and associated with glycerol kinase deficiency, or proximal (towards the centromere) with McLeod's syndrome.

Glycerol kinase deficiency had been reported earlier, but it was not until the report by Guggenheim et al. in 1980 that it was noted to be associated with myopathy. In this paper, two brothers were reported with increased glycerol in the urine due to glycerol kinase deficiency, and with associated findings of developmental delay, mental retardation, congenital adrenal hypoplasia, and a mild,

non-specific myopathy. All of these disorders (adrenal hypoplasia, glycerol kinase deficiency, and myopathy) were recognized to be X-linked. Over the next few years they were reported to occur either independently or in various associations: adrenal hypoplasia with glycerol kinase deficiency, glycerol kinase deficiency with myopathy, or all three with mental retardation. Cytogenetic and Southern blot studies of patients with the combined syndromes showed deletions of Xp21 (Dunger et al. 1986). In one case deletion extended across the entire Duchenne locus to include the binding site of probe p754 (Dunger et al. 1986); in this case the myopathy was not mild, as originally reported, but severe as is more typical of DMD. Analysis with first genomic and later with dystrophin cDNA probes showed all glycerol kinase and Duchenne deletions to extend distally (3′) from the dystrophin gene, with a variable extent (Darras and Francke 1988). The extent of involvement of the dystrophin gene correlated with the severity of the muscular dystrophy. The original patient described by Guggenheim et al. (1980), 'CM', had only deletion of the last three kilobases of the dystrophin cDNA. Patients with glycerol kinase deficiency or adrenal hypoplasia who had no muscle involvement had no deletion of the dystrophin cDNA, and those with deletions extending further into the dystrophin gene had typical DMD. Recently, the cloned breakpoint from a patient with DMD and glycerol kinase deficiency without adrenal hypoplasia was found to fall between the glycerol kinase and adrenal hypoplasia genes in a segment of at least four megabases of DNA (by pulse field gel electrophoresis) which separates the 3′ end of the dystrophin gene from the binding site of probe C7, the next closest marker to be previously identified (Love et al. 1990). Thus, deletions must be large to cause both glycerol kinase deficiency and myopathy. In summary, glycerol kinase deficiency is rarely associated with mild myopathy due to involvement of the distal end of the dystrophin gene. Only one such patient has been reported to date, and he had an unusual phenotype with mental retardation and developmental delay.

In 1961, a patient named Hugh McLeod was reported by Allen et al. to have acanthocytes and loss of the Kell red blood cell antigen. Since then a number of such patients have been reported. Most are asymptomatic blood donors with compensated hemolysis. Increased serum CK is common. The disorder is X-linked and may be associated with chronic granulomatous disease. It maps proximal to the DMD locus. The well-known patient 'BB' from the Seattle area who was reported by Francke et al. (1985) had DMD and McLeod's syndrome with chronic granulomatous disease and a deletion of the X chromosome at Xp21. DNA from this patient was used to derive the PERT clones within the dystrophin gene and also was instrumental in identifying the defect in chronic granulomatous disease. Other patients with McLeod's red cell phenotype may have mild myopathic symptoms and findings. Do they have dystrophin gene involvement similar to that seen in the glycerol kinase deficiency patients? A paper by Bertelson et al. (1988) suggested that McLeod myopathy is in fact a separate locus. Although BB had a large Xp21 deletion with McLeod's syndrome and DMD, two members of another family had mild myopathy with McLeod's syndrome and a deletion that fell short of the dystrophin gene. The possibility that there is a remote effect on dystrophin expression in McLeod's syndrome has been addressed by studies of dystrophin in patients with McLeod's syndrome and mild myopathy (without weakness). Muscle biopsies showed mild myopathic changes, and the serum CK was increased. Dystrophin cDNA analysis, immunohistochemistry, and Western blot are all normal (Carter et al. 1990; Danek et al. 1990). Thus, there is evidence for a separate McLeod's locus which may be associated with mild myopathy at Xp21 but does not involve the dystrophin gene.

Clinical testing for the gene lesion and protein product. Deletions and duplications in the dystrophin gene can readily be identified in DNA obtained from the white blood cells in a single tube of blood, either by Southern blot with dystrophin cDNA probes (Southern 1979) or by polymerase chain reaction amplification of frequently deleted exons (Chamberlain et al. 1988). Since the majority of DMD and BMD patients have genetic defects which can be identified in this way, and no control sample has yet been found to have a deletion in the dystrophin gene, this test provides a fairly sensitive and highly specific means for confirming the clini-

cal diagnosis of DMD and BMD. If a patient is shown to have a defect in the dystrophin gene, then the diagnosis is proven and there is no need for EMG or muscle biopsy.

The same test can be done on samples from female relatives for carrier detection and on fetal samples obtained by chorionic villus sampling or amniocentesis for prenatal diagnosis. If no abnormality can be identified in the dystrophin gene (and the diagnosis is established by other means such as clinical criteria, 20-fold CK elevation, and muscle biopsy), carrier detection and prenatal diagnosis can be offered through the use of genetic markers (simple sequence repeats and restriction fragment length polymorphisms) within and around the dystrophin gene. Because the dystrophin gene is large enough for intragenic recombination to occur, it is necessary to use a set of informative markers and to obtain samples from as many family members as possible to establish a *haplotype* associated with the disease allele in any given family (Witkowski and Caskey 1988; Gutmann and Fischbeck 1989).

The diagnosis of DMD or BMD can also be confirmed by assay of the protein dystrophin in a muscle sample by Western blot or immunohistochemistry (Hoffman et al. 1988a, 1989; Arahata et al. 1989a). Western blot testing, which can be done with as little as 10 mg of frozen muscle, is more sensitive than DNA analysis of the dystrophin gene, showing abnormalities in over 95% of patients with DMD. When properly done, it is also highly specific. Furthermore, dystrophin protein analysis can differentiate DMD from BMD and thus help to determine the prognosis in an early symptomatic patient (Hoffman et al. 1988a, 1989), which cannot be done reliably by DNA analysis alone (Monaco et al. 1988; Baumbach et al. 1989; Fischbeck 1989). Analysis of dystrophin in the muscle by immunohistochemistry may also be useful in DMD carrier detection (Bonilla et al. 1988; Arahata et al. 1989b), although the sensitivity and specificity of this type of testing has not been well established.

In conclusion, clinical testing of the dystrophin gene and its product can provide sensitive and accurate confirmation of the clinical diagnosis of DMD or BMD and reliable genetic counseling in families affected by these disorders.

Treatment

The discovery of the gene defect in DMD has led to a number of approaches that are attempting to reverse the defect by cell or gene transfer (see pp. 131). Prior to the discovery of the gene lesion, various strategies for medical intervention led to the trial of a large number of agents (Dubowitz and Heckmatt 1980). Therapeutic trial literature prior to 1983 was usually based on inadequate statistical methodology (Brooke et al. 1983). The establishment of multicenter studies of DMD (Brooke et al. 1981a,b; Heckmatt et al 1989) has characterized the natural history of the disorder and permitted definitive trials of various agents. Most have been negative (Table 2). Those trials with corticosteroids have demonstrated improvements.

Methodology. The design of a controlled treatment trial in DMD as in any disease is predicated on defining the natural history of one or more clinical measurements. Muscle strength as assessed by manual muscle testing (MMT) is the most pertinent clinical measure. MMT has been shown in DMD patients age 5–15 to decline by 0.4 unit/year (on a 10-point, expanded MRC scale (Brooke et al. 1983)). The variability of MMT is 0.39 unit (one standard deviation). In order for a trial to have adequate power to *exclude* a beneficial effect of a putative therapeutic agent that slows progression

TABLE 2

Randomized controlled treatment trials* in Duchenne dystrophy.

No benefit	
leucine	(Mendell et al. 1984)
allopurinol	(Bertorini et al. 1985)
nifedipine	(Moxley et al. 1987)
penicillamine and vitamin E	(Fenichel et al. 1988)
diltiazem	(Bertorini et al. 1988)
mazindol	(Griggs et al. 1990)
azathioprine	(Griggs et al. 1991b)
dantrolene	(Bertorini et al. 1991)
Benefit	
prednisone	(Mendell et al. 1989; Griggs et al. 1991; Fenichel et al. 1991a,b)
deflazacort	(Mesa et al. 1991; Angelini et al. 1991)

* Includes only those with adequate statistical power.

of disease by 75%, 80 patients (40 in each of 2 groups, placebo vs agent) must be studied for one year (Brooke et al. 1983).

Other groups have noted a comparable rate of decline (Scott et al. 1982). Various quantitative systems such as handheld myometry (Edwards et al. 1987) and quantitative myometry (Brussock et al. 1992) might, in theory, provide a more sensitive measure for quantitating strength. Handheld myometry is, however, no better than MMT (Griggs et al. 1993). Whether other systems will be as good as or better then MMT, remains unknown. Pulmonary function testing, particularly the forced vital capacity, may be the most precise measurement for detecting a change in strength (Griggs 1990). While forced vital capacity *improves* in absolute terms between ages 5 and 15 (Brooke et al. 1983, 1989), small changes between treated and untreated groups can be readily appreciated (Griggs 1990).

Corticosteroids. An uncontrolled trial of prednisone by Drachman et al. (1974) suggested that slight clinical improvement and CK level decline resulted from treatment. Subsequently, a single center, double-blind trial of prednisone (alternate day) failed to confirm efficacy (Siegel et al. 1974). The first controlled study of daily prednisone in DMD (Brooke et al. 1987) indicated an increase in muscle strength after six months of treatment. Subsequently, a randomized, double-blind trial of prednisone confirmed improvement (Mendell et al. 1989). Subsequently, additional randomized control trials have confirmed the observation (Griggs et al. 1991)(Figs. 4,5). These two trials have shown that 0.75 mg/kg is the minimum dose necessary to improve strength to the maximum extent possible in boys with DMD.

Daily prednisone is more effective than alternate day (Fenichel et al. 1991b), presumably accounting for the early negative controlled trial of Siegel et al. (1974). Subsequent studies of 93 boys taking prednisone for 3.25 years or more have shown a significant and sustained beneficial effect of prednisone in slowing the decline of muscle strength and functional grade in boys with DMD (Fenichel et al. 1991a). Patients treated with a dose of 0.6 mg/kg/d or more had not declined significantly from their initial muscle scores (Fenichel et al. 1991a). Major

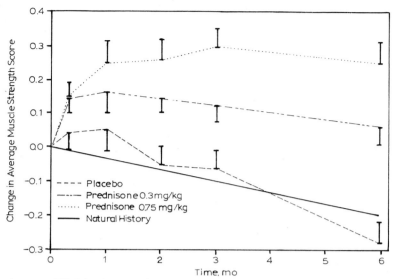

Fig. 4. Change (mean ± SEM) in the score for average muscle strength in the placebo and prednisone groups after the initiation of their regimens. Solid line ('natural history') represents the values for change observed in 177 patients with Duchenne dystrophy who received no treatment. (Reproduced with permission from Griggs et al. (1991a) *Archives of Neurology.*)

side effects occur with prednisone treatment (Table 3). In addition marked weight gain and slowing of growth occur (Griggs et al. 1993).

Mechanism of prednisone-induced improvement. Prednisone has numerous effects which could contribute to the improvement in muscle strength. While DMD is not ordinarily considered

TABLE 3

Number of boys reporting side effects during the first year and during the maintenance phase.

Side effect	Only first year	Only mainte- nance	Both	Never
Negative behavior*	10	16	53	13
Cataracts	0	8	2	82
Easy bruising	8	16	2	66
Cushingoid	0	32	59	1
GI symptoms	8	29	48	7
Glycosuria	1	4	5	82
Excess hair	12	15	39	26
Skin rash	9	20	29	34

* Insomnia, hyperactivity, aggression, emotional lability.

an immunologic disorder, there is evidence that both humoral and cellular immune responses participate in the pathology of the disease. Complement activation with membrane attack complex deposition is observed on necrotic muscle fibers (Engel and Biesecker 1982). Cytotoxic lymphocytes are present in and around necrotic fibers (Arahata and Engel 1984). Muscle biopsies from prednisone-treated DMD patients have been shown to contain fewer cytotoxic lymphocytes than placebo-treated controls, supporting an anti-inflammatory effect of prednisone (Kissel et al. 1991). Prednisone did not alter the number of necrotic fibers nor did it produce an increase in dystrophin by Western blots or by immunofluorescence microscopy (Burrow et al. 1991). Recently, a trial of azathioprine alone or in combination with prednisone (Griggs et al. 1993) did not show any improvement attributable to azathioprine (Fig. 5). Moreover, the fact that the biopsies of azathioprine-treated patients showed a reduction in cytotoxic lymphocytes comparable to that seen in prednisone-treated patients (Kissel et al. 1993) suggests that the improvement in strength induced by prednisone is not the result of immunosuppression.

Muscle mass improves with prednisone (Mendell et al. 1989; Griggs et al. 1991). In vivo studies of muscle protein synthesis and breakdown indicate that prednisone increases muscle mass by decreasing muscle protein *breakdown* without significantly altering muscle protein synthesis (Moxley et al. 1990; Rifai et al. 1992).

Other agents.

Deflazacort is an oxazoline derivative of prednisolone that is effective in the treatment of a variety of steroid-responsive disorders such as rheumatoid arthritis and may have fewer side effects than prednisone. In particular, it may have anabolic effects. Two small, randomized trials of deflazacort have shown benefit in DMD. In one trial, significant changes in functional scores and a trend towards improvement in manual muscle strength testing were observed (Angelini et al. 1991). In another, both myometric and functional improvement occurred (Mesa et al. 1991). Further studies with this less catabolic steroid are in progress.

Cyclosporine. The use of cyclosporine as an immunosuppressant agent in patients receiving experimental myoblast transfer has been associated with improvement of strength (Sharma et al. 1991). Improvement of strength has not occurred during cyclophosphamide (Karpati et al. 1991) or azathioprine treatment (Griggs et al. 1993).

Myoblast transfer and other experimental approaches to gene therapy

The availability of animals with disorders homologous to DMD has allowed trials of definitive therapy for the disease. The objective of this therapy is clear: replacement of the defective dystrophin gene and protein in the muscle. A number of obstacles prevent this objective from being attained at present, however.

In 1989, Partridge et al. reported that injection of cultured normal myoblasts into *mdx* mouse muscle can lead to replacement of the muscle with up to 30–40% dystrophin-positive fibers. Subsequently, similar treatment has been attempted in patients with DMD. The efficacy of this treatment must be judged against the background rate of dystrophin-positive fibers in DMD muscle, which

are presumably due to spontaneous reversion of the genetic defect by somatic mutation. A recent study has shown that myoblasts from normal first-degree relatives injected into DMD muscle can lead to dystrophin-positive fibers, with evidence by polymerase chain reaction of expression of the normal (rather than revertant) dystrophin gene (Gussoni et al. 1992). However, the frequency of dystrophin-positive fibers was quite low, less than 1%, and no better than the frequency of spontaneous revertant fibers, certainly not enough to prevent the progressive weakness of the disease.

Another approach to definitive treatment of DMD is direct gene injection. This approach was suggested by the surprising observation of Wolff et al. in 1990 that injection of naked DNA for various marker genes directly into muscle can lead to stable incorporation and expression of the injected gene for up to two months. Application of this technique to DMD and dystrophin-deficient animals first required preparation of a DNA construct encoding the entire full-length dystrophin message. This was done successfully by Lee et al. in 1991, who also demonstrated that the construct could be used to transfect cultured kidney cells that do not normally make dystrophin. The transfected cells not only made dystrophin which was apparently normal by Western blot analysis, they also had dystrophin localized to the cell periphery near the plasma membrane. Subsequently, Acsadi et al. (1991) used both full length and truncated dystrophin cDNA constructs to produce dystrophin-positive muscle fibers by direct injection into *mdx* mouse muscle in vivo. Here, again, the frequency of dystrophin-positive fibers was low, 1% or less.

For gene replacement therapy to be effective as a treatment for DMD the efficiency of gene transfer must be increased. The factors which limit the efficiency of these procedures include the fibrosis and limited regenerative capacity of DMD muscle, and the limited proliferation and spread of the cells and plasma vectors which have been used to deliver the normal gene to the muscle. Viral vectors may serve as better delivery systems, if current problems with virulence, tissue specificity, cloning capability, and immune response can be overcome. Thus, while effective gene therapy is beyond the reach of currently available methodology, there is

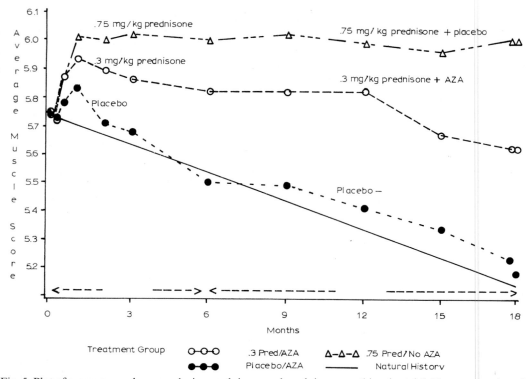

Fig. 5. Plot of average muscle scores during prednisone and prednisone-azathioprine trial. There was no benefit of azathioprine either alone or in combination with prednisone. Prednisone 0.75 mg/kg/d was more effective than prednisone 0.3 mg/kg/d. (From Griggs et al. (1993) *Neurology*.)

a reasonable hope for definitive treatment in the future.

Supportive therapies

Bracing and physical therapy.

Thoracolumbosacral orthoses (body jackets) are used in an attempt to control the scoliosis of DMD. While this approach seems sensible, patient compliance is often poor. Obese patients do not receive sufficient support and thin patients may develop pressure sores. Longitudinal studies of patients using body jackets have failed to demonstrate a beneficial effect (Brooke et al. 1989) but weaknesses in study design leave the question open for further study.

Physical therapy. Stretching exercises are frequently recommended with the goal of preventing contractures (Scott et al. 1981). Night splints are often used to maintain ankle position. In longitudi-

nal studies (Brooke et al. 1989) no correlation could be detected between the use of passive joint stretching and prevention of joint contractures. On the other hand, leg braces appear to prevent contractures of heel cords, knee extensors and iliotibial bands. Night splints have had a marked effect in reducing contractures of heel cords (Brooke et al. 1989). Although adequate comparative data are lacking, the relatively aggressive use of braces in one population (Brooke et al. 1989) appeared to prolong ambulation by three years in comparison with another group (DeSilva et al. 1987).

Surgery.

Spine stabilization. Scoliosis reduces lung compliance (McCool et al. 1986) and predisposes patients to atelectasis and pneumonia. It is also disfiguring and painful. Maintaining ambulation delays the onset of scoliosis (Brooke et al. 1989) and the use of a standing table may also delay scoliosis development. Proper use of a wheelchair may also retard scoliosis by promoting an extensor posture through the use of a reclining back rest and neck extension support (Brooke et al. 1989).

Major advances in spine stabilization procedures permit the correction of scoliosis (Luque 1977; Siegel 1982; Rideau et al. 1984). The Luque (1977) technique of segmental spinal stabilization is now widely used for correction of scoliosis. Patients with a progressive scoliosis of over 35 degrees should be considered surgical candidates (Siegel 1982). The postoperative course is much more benign if the vital capacity is greater than 35% of predicted. Patients with a lower vital capacity are likely to have complications including pneumonia or difficulty in extubation and weaning from a ventilator. There is evidence that spine stabilization improves and delays the deterioration of pulmonary function (Rideau 1984; Brooke et al. 1989; Pandya et al. 1990).

Surgical release of contractures. Heel cords and iliotibial bands often require surgical release in order to allow patients to be fitted with long leg braces (calipers) (Siegel 1975). The availability of light weight, plastic, molded braces now make possible three or more years of additional ambulation and, at times, an even longer period of standing with support (Heckmatt et al. 1985). Since such management delays the development of scoliosis and bone demineralization, most patients benefit from bracing (Brooke et al. 1989). Following surgical release, long-leg braces are often necessary so that most patients scheduled for surgery should be fitted with braces prior to surgery (Siegel 1975).

Respiratory failure. Patients with late-stage DMD in whom the vital capacity is less then 30% of normal, require careful attention to the prevention of respiratory complications. Both the patient and family must be told of the risk of both acute and chronic ventilatory failure, and the management of acute and chronic respiratory failure should be discussed (Griggs and Donahoe 1982; Gilgoff et al. 1989). Health care providers, patient, and family members may have differing opinions as to the appropriateness of ventilatory support for the Duchenne patient. It is imperative that the opinions and wishes of the patient and family be discussed and documented in writing and that a plan of action (or a plan for inaction) be developed (Gilgoff et al. 1989). If such discussions are not held, ambulance crews and emergency room physicians often make the decision to support the patient's ventilation regardless of his wishes and of the plans of family, and of the health care providers who have been supervising the care of the patient for years.

Certain respiratory complications can be prevented. Respiratory exercises have been recommended (DiMarco et al. 1985) but are not of benefit (Smith et al. 1988b). Intermittent positive pressure breathing devices have also been recommended (Adams and Chandler 1974) but do not help (McCool et al. 1986). Patients should receive prophylactic immunization for pneumococcal infections (Immunization Practices Advisory Committee 1982) and annual immunization for influenza viral infections (Immunization Practices Advisory Committee 1985). The sudden appearance of respiratory symptoms in DMD usually results from acute, frequently-reversible illness (O'Donohue et al. 1976). The appearance of acute respiratory symptoms is seldom due to muscle weakness alone. Atelectasis is the commonest cause of acute deterioration and typically occurs in the setting of an upper respiratory infection. Other causes of acute ventilatory failure in the patient with DMD include pneumothorax, pulmonary emboli, aspiration, congestive heart failure, asthma, and pneumonitis. Chest roentgenograms in a DMD patient with acute respiratory symptoms from atelectasis or pneumonitis often fail to show any change, particularly if obtained shortly after the onset of symptoms. Hypercapnia ($pCO_2 > 60$) usually requires emergency endotracheal intubation. If hypoxia alone is present and the patient is alert, low flow oxygen (at rates less than 1 l/min) improves hypoxia without causing hypercapnia. Special equipment is usually necessary to provide such low flow rates safely. Blood gases or oximetry should be monitored to determine the safety of treatment. Oxygen administration in too high of a concentration will depress ventilation by lessening hypoxic drive and result in progressive hypercapnia (Griggs and Sutton 1992). Patients developing respiratory failure who have a vital capacity at baseline of less than 30% often require tracheostomy (Griggs and Donahoe 1982). The majority of patients can be extubated and return to independent respiratory status following their *first* episode of ventilatory failure (O'Donohue et al. 1976).

Chronic ventilatory failure. Late-stage DMD patients should be offered overnight respiratory monitoring with oximetry. Such monitoring can be performed at home (Carroll et al. 1991) and often demonstrates hypoxemia during sleep; such hypoxemia may result in cor pulmonale and can be treated with low flow oxygen (Smith et al. 1989a). Protriptyline may also improve oxygenation but has undesirable anticholinergic side effects (Smith et al. 1989b). Once chronic ventilatory failure has occurred, patients can be easily maintained on a home ventilatory support system. Positive pressure ventilation using nasal systems such as BiPap can prevent cor pulmonale and maintain normal blood oxygen for a number of years prior to total loss of respiratory muscles (Bach et al. 1987; Heckmatt et al. 1990). Once total ventilatory failure has developed, particularly if atelectasis, pneumonitis or aspiration have occurred, it is usually necessary to perform a tracheostomy. Patients with DMD can be maintained for many years with normal ability to talk and independent function on a ventilatory system. Some patients tolerate this approach extremely well and can attend school, work and otherwise remain productive (Bregman 1980). Such an approach usually requires a well-motivated, intelligent family or other health care delivery personnel (Griggs and Donahoe 1982).

Cardiac failure. Baseline electrocardiography and chest roentgenogram are indicated in patients with DMD. There is convincing evidence for myocardial fibrosis in all patients at autopsy. On the other hand, cardiac failure occurs only in those patients in whom muscle strength is disproportionally preserved or who are receiving prolonged ventilatory support (Brooke et al. 1989). When cardiomegaly and signs of congestive heart failure develop, patients are usually hypoxic and have cor pulmonale. Patients with evidence for congestive heart failure in whom oxygenation is well maintained, probably have primary myocardial failure. No treatment is particularly successful. If hypoxia is present, its correction often improves heart failure. Patients with marked impairment of ventricular ejection

fraction merit anticoagulation. Diuretic therapy and strategies for afterload reduction are often associated with symptomatic improvement. The long-term benefit of such treatment remains to be established. Cardiac arrhythmias are infrequent in DMD and may correlate with hypoxemia (Carroll et al. 1991).

Gastrointestinal manifestations. Bowel hypomotility results in severe obstipation in many late-stage Duchenne dystrophy patients. Stool softeners, high fiber diets, and agents that increase intestinal fluid are necessary in many patients. Cathartics, suppositories, and enemas are occasionally indicated. Manual disimpaction is usually needed only if other measures have been neglected. Gastric dilatation is a life-threatening complication (see pp. 120 and 121 above) (Barohn et al. 1988). Anecdotal evidence suggests that metaclopamide may prevent recurrent gastric dilatation. Aspiration is a problem in late-stage DMD and may culminate in pneumonia. It also contributes to inanition. A feeding procedure is occasionally necessary.

BECKER MUSCULAR DYSTROPHY

Becker muscular dystrophy (BMD) is a disease caused by incomplete deficiency of dystrophin in skeletal muscles that results in progressive weakness that is less severe than DMD. Ambulation is generally possible until after age 15 years. There are many different gene lesions which produce a variety of phenotypes.

History

Becker and Kiener (1955) first proposed that a benign form of muscular dystrophy occurring as an X-linked recessive disease existed and could be separated from DMD. Other families with similar clinical features were subsequently described (Blythe and Pugh 1959; Becker 1962; Mabry et al. 1965; Zellweger and Hanson 1967; Markand et al. 1969; Conomy 1970; Bradley et al. 1978). The heterogeneity of the syndrome was identified even in early reports (Becker 1962). The biochemical and genetic basis for BMD was established by dystrophin analysis (Hoffman et al. 1988a). The distinction between autosomal recessive limb-girdle muscular dystrophy and BMD occurring in sporadic cases has proved difficult. As many as one third of cases termed limb-girdle dystrophy in the past prove on dystrophin analysis to have BMD (Hoffman et al. 1989). Dystrophin studies have clarified the diagnosis and shown that there is a major overlap in clinical features between BMD and limb-girdle dystrophy (Hoffman et al. 1989).

Incidence and prevalence

Since many disease phenotypes exist, precise incidence figures await the full characterization of the frequency of incomplete deficiency of dystrophin in various populations. A figure for BMD of 10% of all cases with X-linked dystrophies has long been quoted (Walton 1964). More recent data suggest that milder X-linked dystrophinopathies have an incidence that is approximately 1/20,000 (Emery 1991). Since survival of patients with BMD is much greater than that of DMD patients, the relative prevalence is higher at approximately 1 in 40,000 (Emery 1991).

Clinical manifestations

Table 4 lists a spectrum of disorders caused by abnormalities of, rather than total or near total lack, muscle dystrophin. Patients with reduced or abnormal dystrophin present findings ranging from severe weakness only slightly less severe than DMD to completely normal strength, and includes patients with various symptoms such as exertional myalgias (Gospe et al. 1989). It is reasonable to reserve the term BMD for patients whose clinical picture is similar to DMD but simply of later onset and slower progression.

The pattern of muscle wasting seen in typical BMD closely resembles that seen in DMD. Prominent involvement of pelvic, girdle and thigh muscles occurs with anterior tibial and peroneal muscles less severely involved. Upper extremity weakness in proximal muscles appears well after the onset of leg weakness. Initially, neck flexors are strong relative to DMD and are in the 4 MRC grade range. This distinction permits early recognition of BMD as opposed to DMD. As the disease progresses, however, neck flexors become weak. Neck extensors remain strong. Forearm muscles, hand intrinsics, and ankle plantar flexors are normal early in the disease and remain strong until very late in the illness. Little or no facial muscle weakness occurs. Calf muscle enlargement is present early in most patients. Toe walking is an early feature and the severity of heel cord contractures is one of the few features that may distinguish patients with BMD from limb-girdle dystrophy clinically (Ringel et al. 1977). Scoliosis is much less frequent in BMD than DMD but in severely weak BMD patients may develop after wheelchair confinement.

Age at onset. As emphasized in Table 4, patients with partial deficiency of dystrophin have extreme variation in the age of presentation. The majority of patients with BMD will, however, experience symptoms between ages 5 and 15. Patients with relatively severe abnormalities of dystrophin have a reduced life expectancy but the majority of BMD patients survive into the 4th and 5th decades. Mental retardation occurs in BMD but much less frequently than in DMD (Ringel et al. 1977). Cardiac involvement occurs in BMD but the distinctive ECG findings of DMD are less frequent in BMD (Emery and Skinner 1976); patients may, however, develop cor pulmonale from respiratory failure, and severe cardiomyopathy has been identified in some instances even in patients who remain ambulatory.

Differential diagnosis

The wide variation in clinical phenotypes resulting from partial dystrophin deficiency or abnormal dystrophin leads to an extremely large differential diagnosis list. Suffice it to say, dystrophin testing is

TABLE 4

Clinical phenotypes of incomplete dystrophin deficiency.

Becker's dystrophy –benign Duchenne. Varying age of onset and rate of clinical progression
Focal myopathies (e.g. quadriceps myopathy)
Exertional myalgias
Myoglobinuria (with or without myalgias)
Cardiomyopathy (with minimal-moderate weakness of limb muscles)
Malignant hyperthermia susceptibility

indicated in most patients with the presentations listed in Table 4.

Treatment

Complications. When patients with BMD reach the late stages of illness, they can develop a variety of complications similar to those of late-stage DMD. There is much less data upon which to base decisions than for DMD but presumably respiratory, cardiac, gastrointestinal, and other complications merit similar diagnostic and treatment strategies.

Medication trials. The number of therapeutic trials including BMD patients is limited. A small number of patients with BMD have been included in studies of the effects of prednisone and appear to respond to at least the extent that DMD patients do (Rifai et al. 1992).

REFERENCES

ADAMS, M.S. and L.S. CHANDLER: Effects of physical therapy program on vital capacity of patients with muscular dystrophy. Phys. Ther. 54 (1974) 494–496.

ACSADI, G., G. DICKSON, D.R. LOVE, A. JANI, F.S. WALSH, A. GURUSINGHE, J.A. WOLFF and K.E. DAVIES: Human dystrophin expression in mdx mice after intramuscular injection of DNA constructs. Nature 352 (1991) 815–818.

ALLEN, F.H., S.M.R. KRABBE and P.A. CORCORAN: A new phenotype (McLeod) in the Kell Blood-group system. Vox Sang 6 (1961) 555–560.

ALLSOP, K.G. and F.A. ZITER: Loss of strength and functional decline in Duchenne's dystrophy. Arch. Neurol. 38 (1981) 406–411.

ANDO, T., T. SHIMIZU, T. KATO, M. OHSAWA and Y. FUKUYAMA: Myoglobinemia in children with progressive muscular dystrophy. Clin. Chim. Acta 85 (1978) 17–22

ANGELINI, C., E. PEGORARO, F. PERINI, E. TURELLA, M. INTINO, A. PINI, A. OTTOLINI and C. COSTA: A trial with a new steroid in Duchenne muscular dystrophy. Muscular Dystrophy Research. In: C. Angeli, G.A. Danieli and D. Fontanari (Eds.), Molecular Diagnosis Foward Therapy. Excerpta Medica International Congress Series (1991) 173–179.

ARAHATA, K. and A.G. ENGEL: Monoclonal antibody analysis of mononuclear cells in myopathies. I. Quantitation of subsets according to diagnosis and sites of accumulation and demonstration and counts of muscle fibers invaded by T cells. Ann. Neurol. 16 (1984) 193–208.

ARAHATA, K., E.P. HOFFMAN, L.M. KUNKEL, S. ISHIURA, T. TSUKAHARA, T. ISHIHARA, N. SUNOHARA, I. NONAKA, E. OZAWA and H. SUGITA: Dystrophin diagnosis: comparison of dystrophin abnormalities by immunofluorescence and immunoblot analyses. Proc. Natl. Acad. Sci. USA 86 (1989a) 7154–7158.

ARAHATA, K., T. ISHIHARA, K. KAMKURA, T. TSUKAHARA, S. ISHIURA, C. BABA, T. MATSUMOTO, I. NONAKA and H. SUGITA: Mosaic expression of dystrophin in symptomatic carriers of Duchenne's muscular dystrophy. N. Engl. J. Med. 320 (1989b) 138–142.

BACH, J.R., A. ALBA, R. MOSHER and A. DELAUBIER: Intermittent positive pressure ventilation via nasal access in the management of respiratory insufficiency. Chest 92 (1987) 168–170.

BALLANTYNE, J.P. and S. HANSEN: New method for the estimation of the number of motor units in a muscle. II. Duchenne, limb-girdle and facioscapulohumeral and myotonic muscular dystrophies. J. Neurol. Neurosurg. Psychiatry 37 (1974) 1195–1201.

BAR, S., E. BARNEA, Z. LEVY, S. NEUMAN, D. YAFFE and U. NUDEL: A novel product of the Duchenne muscular dystrophy gene which greatly differs from the known isoforms in its structure and tissue distribution. Biochem. J. 272 (1990) 557–560.

BAROHN, R.J., E.J. LEVINE, J.O. OLSON and J.R. MENDELL: Gastric hypomotility in Duchenne's muscular dystrophy. N. Engl. J. Med. 319 (1988) 15–18.

BAUMBACH, L.L., J.S. CHAMBERLAIN, P.A. WARD, N.J. FARWELL and C.T. CASKEY: Molecular and clinical correlations of deletions leading to Duchenne and Becker muscular dystrophies. Neurology 39 (1989) 465–474.

BECKER, P.E.: Two new families with benign sex-linked recessive muscular dystrophy. Rev. Can. Biol. 21 (1962) 551–566.

BECKER, P.E. and F. KIENER: Eine neue X-chromosomal Muskeldystrophie. Arch. Psychiatr. Nervenkr. 193 (1955) 427–448.

BERTELSON, C.J., A.O. POGO, A. CHAUDHURI, W.L. MARSH, C.M. REDMAN, D. BANERJEE, W.A. SYMMANS, T. SIMON, D. FREY and L.M. KUNKEL: Localization of the McLeod locus (XK) within Xp21 by deletion analysis. Am. J. Hum. Genet. 42 (1988) 703–711.

BERTORINI, T.E., G.M.A. PALMIERI, J. GRIFFIN, C. CHESNEY, D. PIFER, L. VERLING, D. AIROZO and I.H. FOX: Chronic allopurinol and adenine therapy in Duchenne muscular dystrophy: effects on muscle function, nucleotide degradation, and muscle ATP and ADP content. Neurology 35 (1985) 61–65.

BERTORINI, T.E., G.M.A. PALMIERI, J.W. GRIFFIN, M. IGARASHI, J. MCGEE, R. BROWN, D.F. NUTTING, A.B. HINTON and J.G. KARAS: Effect of chronic treatment with the calcium antagonist diltiazem in Duchenne muscular dystrophy. Neurology 38 (1988) 609–613.

BERTORINI, T.E., G.M. PALMIERI, J. GRIFFIN, M. IGARASHI, A. HINTON and J.G. KARAS: Effect of dantrolene in Duchenne muscular dystrophy. Muscle Nerve 14 (1991) 503–507.

BLAU, H.M., G.K. PAVLATH, K. RICH and S.G. WEBSTER:

Localization of muscle gene products in nuclear domains: does this constitute a problem for myoblast therapy? In: R.C. Griggs and G. Karpati (Eds.), Myoblast Transfer Therapy. New York, Plenum Press (1990) 167–172.

BLYTHE, H. and R.J. PUGH: Muscular dystrophy in childhood. The genetic aspects: a field study in the Leeds region of clinical types and their inheritance. Ann. Hum. Genet. 23 (1959) 127–163.

BONILLA, E., B. SCHMIDT, C.E. SAMITT, A.F. MIRANDA, A.P. HAYS, A.B. DE OLIVEIRA, H.W. CHANG, S. SERVIDEI, E. RICCI and D.S. YOUNGER: Normal and dystrophin-deficient muscle fibers in carriers of the gene for Duchenne muscular dystrophy. Am. J. Pathol. 133 (1988) 440–445.

BOXLER, K. and F. JERUSALEM: Hyperreactive (hyaline, opaque, dark) muscle fibers in Duchenne dystrophy: a biopsy study of 16 dystrophy and 205 other neuromuscular disease cases and controls. J. Neurol. 219 (1978) 63–72.

BRADLEY, W.G., M.Z. JONES, J.M. MUSSINI and P.R.W. FAWCETT: Becker-type muscular dystrophy. Muscle Nerve 1 (1978) 111–132.

BREGMAN, A.M.: Living with progressive childhood illness: parental management of neuromuscular disease. Soc. Work Health Care 5 (1980) 387–408.

BROOKE, M.H.: A Clinician's View of Neuromuscular Diseases, second edit. Baltimore, Williams and Wilkins (1986) 117–120.

BROOKE, M.H., R.C. GRIGGS, J.R. MENDELL, G.M. FENICHEL, J.B. SHUMATE and the CIDD GROUP: The natural history of Duchenne muscular dystrophy (DMD): a caveat for therapeutic trials. Trans. Am. Neurol. Assoc. 106 (1981a) 195–199.

BROOKE, M.H., R.C. GRIGGS, J.R. MENDELL, G.M. FENICHEL, J.B. SHUMATE and R.J. PELLEGRINO: Clinical trials in Duchenne muscular dystrophy: 1. The design of the protocol. Muscle Nerve 4 (1981b) 186–197.

BROOKE, M.H., G.M. FENICHEL, R.C. GRIGGS, J.R. MENDELL, R.T. MOXLEY, J.P. MILLER, M.A. PROVINCE and the CIDD GROUP: Clinical investigation in Duchenne dystrophy: 2. Determination of the 'power' of therapeutic trials based on the natural history. Muscle Nerve 6 (1983) 91–103.

BROOKE, M.H., G.M. FENICHEL, R.C. GRIGGS, J.R. MENDELL, R.T. MOXLEY, J.P. MILLER, K.K. KAISER, J.M. FLORENCE, S. PANDYA and L. SIGNORE: Clinical investigation of Duchenne muscular dystrophy: interesting results in a trial of prednisone. Arch. Neurol. 44 (1987) 812–817.

BROOKE, M.H., G.M. FENICHEL, R.C. GRIGGS, J.R. MENDELL, R.T. MOXLEY, J.P. MILLER and the CIDD GROUP: Duchenne muscular dystrophy: patterns of clinical progression and effects of supportive therapy. Neurology 39 (1989) 475–481.

BRUSSOCK, C.M., S.M. HALEY, T.L. MUNSAT and D.B. BERNHARDT: Measurement of isometric force in children with and without Duchenne's muscular dystrophy. Phys. Ther. 72 (1992) 105–114.

BURROW, K.L., D.D. COOVERT, C.J. KLEIN, D.E. BULMAN, J.T. KISSEL, K.W. RAMMOHAN, A.H. BURGHES, J.R. MENDELL and the CIDD GROUP: Dystrophin expression and somatic reversion in prednisone treated and untreated Duchenne dystrophy. Neurology 41 (1991) 661–666.

CAMPBELL, K.P. and S.D. KAHL: Association of dystrophin and an integral membrane glycoprotein. Nature (London) 338 (1989) 259–262.

CARPENTER, S. and G. KARPATI: Pathology of Skeletal Muscle. New York, Churchill Livingstone (1984) 484–505.

CARROLL, N., R.J.I. BAIN, P.E.M. SMITH, S. SALTISSI, R.H.T. EDWARDS and P.M.A. CALVERLEY: Domiciliary investigation of sleep-related hypoxaemia in Duchenne muscular dystrophy. Eur. Resp. J. 4 (1991) 434–440.

CARTER, N.D., J.E. MORGAN, A.P. MONACO, M.S. SCHWARTZ and S. JEFFERY: Dystrophin expression and genotypic analysis of two cases of benign X-linked myopathy (McLeod's syndrome). J. Med. Genet. 27 (1990) 345–347.

CHAMBERLAIN, J.S., R.A. GIBBS, J.E. RANIER, P.N. NGUYEN and C.T. CASKEY: Deletion screening of the Duchenne muscular dystrophy locus via multiplex DNA amplification. Nucleic Acids Res. 16 (1988) 11141–11156.

CONOMY, J.P.: Late-onset slowly progressive sex-linked muscular dystrophy. Mil. Med. 135 (1970) 471–475.

COOPER, B.J., N.J. WINAND, H. STEDMAN, B.A. VALENTINE, E.P. HOFFMAN, L.M. KUNKEL, M.O. SCOTT, K.H. FISCHBECK, J.N. KORNEGAY and R.J. AVERY: The homologue of the Duchenne locus is defective in X-linked muscular dystrophy of dogs. Nature (London) 334 (1988) 154–156.

CROWE, C.G.: Acute dilatation of stomach as a complication of muscular dystrophy. Br. Med. J. 1 (1961) 1371.

DANEK, A., T.N. WITT, H.B. STOCKMANN, B.J. WEISS, D.L. SCHOTLAND and K.H. FISCHBECK: Normal dystrophin in McLeod myopathy. Ann. Neurol. 28 (1990) 720–722.

DARRAS, B.T. and U. FRANCKE: Myopathy in complex glycerol kinase deficiency patients is due to 3′ deletions of the dystrophin gene. Am. J. Hum. Genet. 43 (1988) 126–130.

DAVIES, K.E., P.L. PEARSON, P.S. HARPER, J.M. MURRAY, T. O'BRIEN, M. SARFARAZI and R. WILLIAMSON: Linkage analysis of two cloned DNA sequences flanking the Duchenne muscular dystrophy locus on the short arm of the human X chromosome. Nucleic Acids Res. 11 (1983) 2303–2312.

DAWSON, D.M. and I.H. FINE: Creatine kinase in human tissues. Arch. Neurol. 16 (1967) 175–180.

DEN DUNNEN, J.T., P.M. GROOTSCHOLTEN, E. BAKKER, L.A. BLONDEN, H.B. GINJAAR, M.C. WAPENAAR, H.M. VAN PAASSEN, C. VAN BROECKHOVEN, P.L. PEARSON and G.J. VAN OMMEN: Topography of the Duchenne muscular dystrophy (DMD) gene: FIGE and cDNA analysis of 194 cases reveals 115 deletions and 13 duplications. Am. J. Hum. Genet. 45 (1989) 835–847.

DESILVA, S., D.B. DRACHMAN, D. MELLITS and R.W. KUNCL: Prednisone treatment in Duchenne muscular dystrophy. Arch. Neurol. 44 (1987) 818–822.

DESMEDT, J.E. and S. BORENSTEIN: Relationship of spontaneous fibrillation potentials to muscle fibre segmentation in human muscular dystrophy. Nature (London) 258 (1975) 531–534.

DESMEDT, J.E. and S. BORENSTEIN: Regeneration in Duchenne muscular dystrophy: electromyographic evidence. Arch. Neurol. 33 (1976) 642–650.

DIMARCO, A.F., J.S. KELLING, M.S. DIMARCO, I. JACOBS, R. SHIELDS and M.D. ALTOSE: The effects of inspiratory resistive training on respiratory muscle function in patients with muscular dystrophy. Muscle Nerve 8 (1985) 284–290.

DRACHMAN, D.B., K.V. TOYKA and E. MYER: Prednisone in Duchenne muscular dystrophy. Lancet ii (1974) 1409–1412.

DUBOWITZ, V.: Chaos in classification of the spinal muscular atrophies of childhood. Neuromusc. Disord. 1 (1991) 77–80.

DUBOWITZ, V. and M.H. BROOKE: Muscle Biopsy. A Modern Approach. London, Saunders (1973).

DUBOWITZ, V. and L. CROME: The central nervous system in Duchenne muscular dystrophy. Brain 92 (1969) 805–808

DUBOWITZ, V. and J. HECKMATT: Management of muscular dystrophy: pharmacological and physical aspects. Br. Med. Bull. 36 (1980) 139–144.

DUCHENNE, G.B.: Recherches sur la paralysie musculaire pseudohypertrophique, ou paralysie myosclérosique. Arch. Gén. Méd (6 Ser.) 11 (1868) 5–25, 179–209, 305–321, 421–443. 552–588, 868.

DUNGER, D.B., K.E. DAVIES, M. PEMBREY, B. LAKE, P. PEARSON, D. WILLIAMS, A. WHITFIELD and M.J. DILLON: Deletion on the X chromosome detected by direct DNA analysis in one of two unrelated boys with glycerol kinase deficiency, adrenal hypoplasia, and Duchenne muscular dystrophy. Lancet i (1986) 585–587.

EBASHI, S., Y. TOYOKURA, H. MOMOI and H. SUGITA: High creatine phosphokinase activity of sera of progressive muscular dystrophy patients. J. Biochem. 46 (1959) 103–104.

EDWARDS, R.H.T., S.J. CHAPMAN, D.J. NEWHAM and D.A. JONES: Practical analysis of variability of muscle function measurements in Duchenne muscular dystrophy. Muscle Nerve 10 (1987) 6–14.

EMERY, A.E.: Duchenne muscular dystrophy: genetic aspects, carrier detection and antenatal diagnosis. Br. Med. Bull. 36 (1980) 117–122.

EMERY, A.E.H.: Population frequencies of inherited neuromuscular diseases – a world survey. Neuromusc. Disord. 1 (1991) 19–29.

EMERY, A.E.H. and R. SKINNER: Clinical studies in begin (Becker type) X-linked muscular dystrophy. Clin. Genet. 10 (1976) 189–201.

ENGEL, A.G.: Duchenne dystrophy. In: A.G. ENGEL and

B.Q. BANKER (Eds.), Myology. New York, McGraw-Hill (1986) 1185–1240.

ENGEL, A.G. and G. BIESECKER: Complement activation in muscle fiber necrosis: demonstration of the membrane attack complex of complement in necrotic fibers. Ann. Neurol. 12 (1982) 289–296.

ENGLAND, S., L.V. NICHOLSON, M.A. JOHNSON, S.M. FORREST, D.R. LOVE, E.E. ZUBRZYCKA-GAARN, D.E. BULMAN, J.B. HARRIS and K.E. DAVIES: Very mild muscular dystrophy associated with the deletion of 46% of dystrophin. Nature (London) 343 (1990) 180–182.

FEENER, C.A., M. KOENIG and L.M. KUNKEL: Alternative splicing of human dystrophin mRNA generates isoforms at the carboxy terminus. Nature (London) 338 (1989) 509–511.

FENICHEL, G.M., M.H. BROOKE, R.C. GRIGGS, J.R. MENDELL, J.P. MILLER, R.T. MOXLEY, J.H. PARK, M.A. PROVINCE, J. FLORENCE, K.K. KAISER, W.M. KING, S. PANDYA, J. ROBISON and L. SIGNORE: Clinical investigation in Duchenne muscular dystrophy: penicillamine and vitamin E. Muscle Nerve 11 (1988) 1164–1168.

FENICHEL, G.M., J.M. FLORENCE, A. PESTRONK, J.R. MENDELL, R.T. MOXLEY III, R.C. GRIGGS, M.H. BROOKE, J.P. MILLER, J. ROBISON, W. KING, L. SIGNORE, S. PANDYA, J. SCHIERBECKER and B. WILSON: Long-term benefit from prednisone therapy in Duchenne muscular dystrophy. Neurology 41 (1991a) 1874–1877.

FENICHEL, G.M., J.R. MENDELL, R.T. MOXLEY, R.C. GRIGGS, M.H. BROOKE, J.P. MILLER, A. PESTRONK, J. ROBISON, W. KING, L. SIGNORE, S. PANDYA, J. FLORENCE, J. SCHIERBECKER and B. WILSON: A comparison of daily and alternate-day prednisone therapy in the treatment of Duchenne muscular dystrophy. Arch. Neurol. 48 (1991b) 575–579.

FIRTH, M., D. GARDNER-MEDWIN, G. HOSKING and E. WILKINSON: Interviews with parents of boys suffering from Duchenne muscular dystrophy. Dev. Med. Child Neurol. 25 (1983) 466–471.

FISCHBECK, K.H.: The difference between Duchenne and Becker dystrophies. Neurology 39 (1989) 584–585.

FITCH, C.D. and D.W. SINTON: A study of creatine metabolism in diseases causing muscle wasting. J. Clin. Invest. 43 (1964) 444–451.

FLORENCE, J.M., P.T. FOX, J. PLANER and M.H. BROOKE: Activity, creatine kinase, and myoglobin in Duchenne muscular dystrophy: a clue to etiology? Neurology 35 (1985) 758–761.

FRANCKE, U., H.D. OCHS, B. DEMARTINVILLE, J. GIACALONE, V. LINDGREN, C. DISTECHE, R.A. PAGON, M.H. HOFKER, G.J. VAN OMMEN and P.L. PEARSON: Minor Xp21 chromosome deletion in a male associated with expression of Duchenne muscular dystrophy, chronic granulomatous disease, retinitis pigmentosum and McLeod syndrome. Am. J. Hum. Genet. 37 (1985) 250–267.

GARDNER-MEDWIN, D. and P. SHARPLES: Some studies of the Duchenne and autosomal recessive types of muscular dystrophy. Brain Dev. 11 (1989) 91–97.

GARDNER-MEDWIN, D., S. BUNDEY and S. GREEN: Early diagnosis of Duchenne muscular dystrophy. Lancet i (1978) 1102.

GILGOFF, I., W. PRENTICE and A. BAYDUR: Patient and family participation in the management of respiratory failure in Duchenne's muscular dystrophy. Chest 95 (1989) 519–524.

GOSPE, S.M., R.P. LAZARO, N.S. LAVA, P.M. GROOT-SCHOLTEN, M.O. SCOTT and K.H. FISCHBECK: Familial X-linked myalgia and cramps: a nonprogressive myopathy associated with a deletion in the dystrophin gene. Neurology 39 (1989) 1277–1280.

GOWERS, W.R.: Pseudohypertrophic Muscular Paralysis: A Clinical Lecture. London, Churchill (1879).

GRIESINGER, W.: Über Muskelhypertrophie. Arch. Heilkd. 6 (1865) 1–13.

GRIGGS, R.C.: The use of pulmonary function testing as a quantitative measurement for therapeutic trial. Muscle Nerve (Suppl.) (1990) S30–S34.

GRIGGS, R.C. and K.M. DONOHOE: Recognition and management of respiratory insufficiency in neuromuscular disease. J. Chron. Dis. 35 (1982) 497–500.

GRIGGS, R.C. and J.R. SUTTON: Neurologic manifestations of respiratory disease. In: A.K. Asbury, G.M. McKhann and W.I. McDonald (Eds), Diseases of the Nervous System (2nd edit.). Philadelphia, W.B. Saunders (1992) 1432–1441.

GRIGGS, R.C., W. REEVES and R.T. MOXLEY III: The heart in Duchenne dystrophy. In: L.P. Rowland (Ed.), Pathogenesis of Human Muscular Dystrophies. Amsterdam, Excerpta Medica (1977) 661–671.

GRIGGS, R.C., R.T. MOXLEY and G.B. FORBES: 3-Methylhistidine excretion in myotonic dystrophy. Neurology 30 (1980) 1262–1267.

GRIGGS, R.C., G. FORBES, R.T. MOXLEY and B.E. HERR: The assessment of muscle mass in progressive neuromuscular disease. Neurology 33 (1983) 158–165.

GRIGGS, R.C., R.T. MOXLEY, J.R. MENDELL, G.M. FENICHEL, M.H. BROOKE, P.J. MILLER, S. MANDEL, J. FLORENCE, J. SCHIERBECKER, K.K. KAISER, W. KING, S. PANDYA, J. ROBISON and L. SIGNORE: Randomized double-blind trial of mazindol in Duchenne dystrophy. Muscle Nerve 13 (1990) 1169–1173.

GRIGGS, R.C., R.T. MOXLEY, J.R. MENDELL, G.M. FENICHEL, M.H. BROOKE, A. PESTRONK, J.P. MILLER and the CIDD GROUP: Prednisone in Duchenne dystrophy: a randomized, controlled trial defining the time course and dose response. Arch. Neurol. 48 (1991a) 383–388.

GRIGGS, R.C., R.T. MOXLEY, J.R. MENDELL, J.P. MILLER, A. PESTRONK, J. FENICHEL, M. BROOKE and THE CIDD GROUP: Randomized, controlled trial of prednisone and azathioprine in Duchenne dystrophy. Neurology 41 (1991b) 166.

GRIGGS, R.C., R.T. MOXLEY, J.R. MENDELL, G.M. FENICHEL, M.H. BROOKE, A. PESTRONK, J.P. MILLER, S. PANDYA, J. ROBISON, W. KING, L. SIGNORE, J. SCHIERBECKER, J. FLORENCE, N. MATHESON-BURDEN and B.

WILSON: Duchenne dystrophy: randomized, controlled trial of prednisone (18 months) and azathioprine (12 months). Neurology (1993) in press.

GUGGENHEIM, M.A., E.R.B. MCCABE, M. ROIG, S.I. GOODMAN, G.M. LUM, W.W. BULLEN and S.P. RINGEL: Glycerol kinase deficiency with neuromuscular, skeletal, and adrenal abnormalities. Ann. Neurol. 7 (1980) 441–449.

GUSSONI, E., G.K. PAVLATH, A.M. LANCTOT, K.R. SHARMA, R.G. MILLER, L. STEINMAN and H.M. BLAU: Normal dystrophin transcripts in Duchenne muscular dystrophy patients after myoblast transplantation. Nature (London) 356 (1992) 435–438.

GUTMANN, D.H. and K.H. FISCHBECK: Molecular biology of Duchenne and Becker's muscular dystrophy: clinical applications. Ann. Neurol. 26 (1989) 189–194.

HAMMOND, R.G.: Protein sequence of DMD gene is related to actin-binding domain of alpha-actinin. Cell 51 (1987) 1.

HARANO, Y., R. ADAIR, P.J. VIGNOS, M. MILLER and K. KOWAL: Pyruvate kinase isoenzymes in progressive muscular dystrophy and in acute myocardial infarction. Metab. Clin. Exp. 22 (1973) 493–501.

HARTLAGE, P.L., A. HARMATZ, E. MOBLEY, P.R. BLANKENSHIP and R.A. ROESEL: Carnosine excretion in muscular dystrophy. Fed. Proc. 42 (1983) 791.

HATHAWAY, P.W., W.K. ENGEL and H. ZELLWEGER: Experimental myopathy after microarterial embolization: comparison with childhood X-linked pseudohypertrophic muscular dystrophy. Arch. Neurol. 22 (1970) 365–378.

HEATH, R., S. JEFFREY and N. CARTER: Radioimmunoassay of human muscle carbonic anhydrase III in dystrophic states. Clin. Chim. Acta 119 (1982) 299–305.

HECKMATT, J.Z., V. DUBOWITZ, S.A. HYDE, J. FLORENCE, A.C. GABAIN and N. THOMPSON: Prolongation of walking in Duchenne muscular dystrophy with lightweight orthoses: review of 57 cases. Dev. Med. Child Neurol. 27 (1985) 149–154.

HECKMATT, J.Z., E. RODILLO and V. DUBOWITZ: Management of children: pharmacological and physical. Br. Med. Bull. 45 (1989) 788–801.

HECKMATT, J.Z., L. LOH and V. DUBOWITZ: Night-time nasal ventilation in neuromuscular disease. Lancet 335 (1990) 579–582.

HOFFMAN, E.P., R.H. BROWN and L.M. KUNKEL: Dystrophin: the protein product of the Duchenne muscular dystrophy locus. Cell 51 (1987) 919–928.

HOFFMAN, E.P., K.H. FISCHBECK, R.H. BROWN, M. JOHNSON, R. MEDORI, J.D. LOIKE, J.B. HARRIS, R. WATERSTON, M. BROOKE and L. SPECHT: Characterization of dystrophin in muscle biopsy specimens from patients with Duchenne's or Becker's muscular dystrophy. N. Engl. J. Med. 38 (1988a) 1363–1368.

HOFFMAN, E.P., M.S. HUDECKI, P.A. ROSENBERG, C.M. POLLINA and L.M. KUNKEL: Cell and fiber-type distribution of dystrophin. Neuron 1 (1988b) 411–420.

HOFFMAN, E.P., L.M. KUNKEL, C. ANGELINI, A. CLARKE,

M. JOHNSON and J.B. HARRIS: Improved diagnosis of Becker muscular dystrophy via dystrophin testing. Neurology 39 (1989) 1011–1017.

HYSER, C.L., M. PROVINCE, R.C. GRIGGS, J.R. MENDELL, G.M. FENICHEL, M.H. BROOKE, J.P. MILLER, R. PO-LAKOWSKA, R.A. DOHERTY, S. QUIRK and the CIDD GROUP: Genetic heterogeneity in Duchenne dystrophy. Ann. Neurol. 22 (1987) 553–555.

IANNACCONE, S.T., K. BOVE, R. TOWBIN, L.M. FORD and P. KEEBLER: Congenital muscular dystrophy. Neurology 38 (Suppl. 1) (1988) 187.

IBRAGHIMOV-BRESKROVNAYA, O., J.M. ERVASTI, C.J. LEV-EILLE, C.A. SLAUGHTER, S.W. SERNETT and K.P. CAMPBELL: Primary structure of dystrophin-associated glycoproteins linking dystrophin to the extracellular matrix. Nature (London) 355 (1992) 696–702.

IMMUNIZATION PRACTICES ADVISORY COMMITTEE, Centers for Disease Control, Atlanta, Georgia: Pneumococcal polysaccharide vaccine. Ann. Intern. Med. 96 (1982) 203–205.

IMMUNIZATION PRACTICES ADVISORY COMMITTEE, Centers for Disease Control, Department of Health and Human Services, Atlanta, Georgia: Prevention and control of influenza. Ann. Intern. Med. 103 (1985) 560–565.

KAGEN, L.J., S. MOUSSAVI, S.L. MILLER and P. TSAIRIS: Serum myoglobin in muscular dystrophy. Muscle Nerve 3 (1980) 221–226.

KAIDO, M., K. ARAHATA, E.P. HOFFMAN, I. NONAKA and H. SUGITA: Muscle histology in Becker muscular dystrophy. Muscle Nerve 14 (1991) 1067–1073.

KARPATI, G, AJDUKOVIC, D. ARNOLD, S. CARPENTER, R. GLEDHILL, R. GUTTMANN, P. HOLLAND, P. KOCH, D. SPENCE, M. VANASSE and G.W. WATTERS: Normal myoblast transfer in Duchenne muscular dystrophy. Neurology 41 (Suppl. 1 (1991) 165.

KISSEL, J.T., K.L. BURROW, K.W. RAMMOHAN, J.R. MENDELL and the CIDD GROUP: Mononuclear cell analysis of muscle biopsies in prednisone-treated and untreated Duchenne muscular dystrophy. Neurology 41 (1991) 667–672.

KISSEL, J.T., J. LYNN, K. RAMMOHAN, R.C. GRIGGS, R.T. MOXLEY, M.H. BROOKE, V. CWIK and J.R. MENDELL: Mononuclear cell analysis of muscle biopsies in azathioprine-treated Duchenne dystrophy. Neurology (1993) in press.

KOENIG, M., E.P. HOFFMAN, C.J. BERTELSON, A.P. MONACO, C. FEENER and L.M. KUNKEL: Complete cloning of the Duchenne muscular dystrophy (DMD) cDNA and preliminary genomic organization of the DMD gene in mouse and affected individuals. Cell 50 (1987) 509–517.

KOENIG, M., A.P. MONACO and L.M. KUNKEL: The complete sequence of dystrophin predicts a rod-shaped cytoskeletal protein. Cell 53 (1988) 219–228.

KOENIG, M., A.H. BEGGS, M. MOYER, S. SCHERPF, K. HEINDRICH, T. BETTECKEN, G. MENG, C.R. MULLER, M. LINDLOF and H. KAARIANEN: The molecular basis for Duchenne versus Becker muscular dystrophy: cor-

relation of severity with type of deletion. Am. J. Hum. Genet. 45 (1989) 498–506.

KOUSSEFF, B.: Linkage between chronic granulomatous disease and Duchenne's muscular dystrophy? Am. J. Dis. Child. 135 (1981) 1149.

KUNKEL, L.M., A.P. MONACO, W. MIDDLESWORTH, H.D. OCHS and S.A. LATT: Specific cloning of DNA fragments absent from the DNA of a male patient with an X chromosome deletion. Proc. Natl. Acad. Sci. USA 82 (1985) 4778–4782.

LA SPADA, A.R., E.M. WILSON, D.B. LUBAHN, A.E. HARDING and K. FISCHBECK: Androgen receptor gene mutations in X-linked spinal and bulbar muscular atrophy. Nature (London) 352 (1991) 77–79.

LEE, C.L., J.A. PEARLMAN, J.S. CHAMBERLAIN and C.T. CASKEY: Expression of recombinant dystrophin and its localization to the cell membrane. Nature (London) 349 (1991) 334–336.

LEIBOWITZ, D. and V. DUBOWITZ: Intellect and behaviour in Duchenne muscular dystrophy. Dev. Med. Child Neurol. 23 (1981) 577–590.

LEMAIRE, C., R. HEILIG and J.L. MANDEL: The chicken dystrophin cDNA: striking conservation of the C-terminal coding and 3" untranslated regions between man and chicken. Embo J. 7 (1988) 4157–4162.

LEV, A.A., C.C. FEENER, L.M. KUNKEL and R.H. BROWN: Expression of the Duchenne's muscular dystrophy gene in cultured muscle cells. J. Biol. Chem. 262 (1987) 15817–15820.

LIDOV, H.G.W., T.J. BYERS, S.C. WATKINS and L.M. KUNKEL: Localization of dystrophin to postsynaptic regions of central nervous system cortical neurons. Nature (London) 348 (1990) 725–728.

LOVE, D.R., D.F. HILL, G. DICKSON, N.K. SPURR, B.C. BYTH, R.F. MARSDEN, F.S. WALSH, Y.H. EDWARDS and K.E. DAVIES: An autosomal transcript in skeletal muscle with homology to dystrophin. Nature (London) 339 (1989) 55–58.

LOVE, D.R., J.F. BLOOMFIELD, S.J. KENWRICK, J.R. YATES and K.E. DAVIES: Physical mapping distal to the DMD locus. Genomics 8 (1990) 106–112.

LOVE, D.R., G.E. MORRIS, J.M. ELLIS, U. FAIRBROTHER, R.F. MARSDEN, J.F. BLOOMFIELD, Y.H. EDWARDS, C.P. SLATER, D.J. PARRY and K.E. DAVIES: Tissue distribution of the dystrophin-related gene product and expression in the mdx and dy mouse. Proc. Natl. Acad. Sci USA 88 (1991) 3243–3247.

LUQUE, E.R.: Segmental correction of scoliosis with rigid internal fixation: preliminary report in orthopaedic transactions. J. Bone Joint Surg. 2 (1977) 136.

MABRY, C.C., I.E. ROCKEL, R.L. MUNICH and D. ROBERTSON: X-linked pseudohypertrophic muscular dystrophy with a late onset and slow progression. N. Engl. J. Med. 273 (1965) 1062–1070.

MARKAND, O.N., R.R. NORTH, A.N. DAGOSTINO and D.D. DALY: Benign sex-linked muscular dystrophy. Neurology 19 (1969) 617–633.

MARSH, G.G. and T.L. MUNSAT: Evidence for early impairment of verbal intelligence in Duchenne muscular dystrophy. Arch. Dis. Child. 49 (1974) 118–122.

MCCOMAS, A.J., R.E.P. SICA and S. CURRIE: An electrophysiological study of Duchenne dystrophy. J. Neurol. Neurosurg. Psychiatry 34 (1971) 461–468.

MCCOMAS, A.J., R.E.P. SICA and M.E. BRANDSTATER: Further motor unit studies in Duchenne muscular dystrophy. J. Neurol. Neurosurg. Psychiatry 40 (1977) 1147–1151.

MCCOOL, F.D., R.F. MAYEWSKI, D.S. SHAYNE, C.J. GIBSON, R.C. GRIGGS and R.W. HYDE: Intermittent positive pressure breathing in patients with respiratory muscle weakness: alterations in total respiratory system compliance. Chest 90 (1986) 546–552.

MCGUIRE, S.A. and K.H. FISCHBECK: Autosomal recessive Duchenne-like muscular dystrophy: molecular and histochemical results. Muscle Nerve 14 (1991) 1209–1212.

MCKERAN, R.O., D. HALLIDAY and P. PURKISS: Increased myofibrillar protein catabolism in Duchenne muscular dystrophy measured by 3-methylhistidine excretion in urine. J. Neurol. Neurosurg. Psychiatry 40 (1977) 979–981.

MENDELL, J.R., R.C. GRIGGS, R.T. MOXLEY, G.M. FENICHEL, M.H. BROOKE, J.P. MILLER, M.A. PROVINCE, W.D. DODSON and THE CIDD GROUP: Clinical trial in Duchenne dystrophy: IV. double-blind controlled trial of leucine. Muscle Nerve 7 (1984) 535–541.

MENDELL, J.R., R.T. MOXLEY, R.C. GRIGGS, M.H. BROOKE, G.M. FENICHEL, J.P. MILLER, W. KINGER, L. SIGNORE, S. PANDYA and J. FLORENCE: Randomized double-blind six-month trial of prednisone in Duchenne's muscular dystrophy. N. Engl. J. Med. 320 (1989) 1592–1597.

MERYON, E.: On the granular and fatty degeneration of the voluntary muscles. Med. Chir. Trans. 35 (1852) 73–85.

MESA, L.E., A.L. DUBROVSKY, J. CORDERI, P. MARCO and D. FLORES: Steroids in Duchenne muscular dystrophy – deflazacort trial. Neuromusc. Disord. 1 (1991) 261–266.

MILHORAT, A.T.: Creatine and creatinine metabolism and diseases of the neuromuscular system. Res. Nerv. Ment. Dis. 32 (1952) 400.

MOKRI, B. and A.G. ENGEL: Duchenne dystrophy: electron microscopic findings pointing to a basic or early abnormality in the plasma membrane of the muscle fiber. Neurology 25 (1975) 1111.

MONACO, A.P., R.L. NEVE, C. COLLETTI-FEENER, C.J. BERTELSON, D.M. KURNIT and L.M. KUNKEL: Isolation of candidate cDNAs for portions of the Duchenne muscular dystrophy gene. Nature (London) 323 (1986) 646–650.

MONACO, A.P., C.J. BERTELSON, S. LIECHTI-GIALATI, H. MOSER and L.M. KUNKEL: An explanation for the phenotypic differences between patients bearing partial deletions of the DMD locus. Genomics 2 (1988) 90–95.

MOXLEY, R.T., M.H. BROOKE, G.M. FENICHEL, J.R. MENDELL, R.C. GRIGGS, J.P. MILLER, M.A. PROVINCE, V.

PATTERSON and THE CIDD GROUP: Clinical investigation in Duchenne dystrophy: VI. double-blind controlled trial of nifedipine. Muscle Nerve 10 (1987) 22–33.

MOXLEY, R.T., M. LORENSON, R.C. GRIGGS, J.R. MENDELL, G.M. FENICHEL, M.H. BROOKE J.P. MILLER and the CIDD Group: Decreased breakdown of muscle protein after prednisone therapy in Duchenne dystrophy. J. Neurol. Sci. 98 (1990) 419.

MUNSAT, T.L., R. BALOH, C.M. PEARSON and W. FOWLER JR: Serum enzyme alterations in neuromuscular disorders. J. Am. Med. Assoc. 226 (1973) 1536–1543.

NONAKA, I. and H. SUGITA: Muscle pathology of Duchenne dystrophy, with particular reference to 'opaque fibers'. Adv. Neurol. Sci. 24 (1980) 718.

NUDEL, U., D. ZUK, P. EINAT, E. ZEELON, Z. LEVY, S. NEUMAN and D. YAFFE: Duchenne muscular dystrophy gene product is not identical in muscle and brain. Nature (London) 337 (1989) 76–67.

O'DONOHUE, W.J., J.P. BAKER, G.M. BELL, O. MUREN, C.L. PARKER and J.L. PATTERSON: Respiratory failure in neuromuscular disease: management in a respiratory intensive care unit. J. Am. Med. Assoc. 235 (1976) 733–735.

PANAYIOTOPOULOS, C.P., S. SCARPALEZOS and T. PAPAPETROPOULOS: Electrophysiological estimation of motor units in Duchenne muscular dystrophy. J. Neurol. Sci. 23 (1974) 89–98.

PANDYA, S., R. MOXLEY, R. GRIGGS, B. WILSON, P. MILLER and the CIDD Group: Pulmonary function in Duchenne dystrophy. J. Neurol. Sci. 98 (1990) 421.

PARTRIDGE, T.A., J.E. MORGAN, G.R. COULTON, E.P. HOFFMAN and L.M. KUNKEL: Conversion of mdx myofibres from dystrophin-negative to positive by injection of normal myoblasts. Nature (London) 337 (1989) 176–179.

PEARCE, G.W. and J.N. WALTON: Progressive muscular dystrophy: the histopathologic changes in skeletal muscle obtained by biopsy. J. Pathol. Bacteriol. 83 (1962) 535–550.

PERLOFF, J.K., W.C. ROBERTS, A.C. DE LEON and D. O'DOHERTY: The distinctive electrocardiogram of Duchenne's progressive muscular dystrophy: an electrocardiographic-pathologic correlative study. Am. J. Med. 42 (1967) 179–188.

RAY, P.N., B. BELFALL, C. DUFF, C. LOGAN, V. KEAN, M.W. THOMPSON, J.E. SYLBESTER, J.L. GORSKI, R.D. SCHMICKEL and R.G. WORTON: Cloning of the breakpoint of an Xp21 translocation associated with Duchenne muscular dystrophy. Nature (London) 318 (1985) 672–675.

RIDEAU, Y., B. GLORION, A. DELAUBIER, O. TARLE and J. BACH: The treatment of scoliosis in Duchenne muscular dystrophy. Muscle Nerve 7 (1984) 281–286.

RIFAI, Z., S. WELLE, R.T. MOXLEY, M. LORENSON and R.C. GRIGGS: Mechanism of action of prednisone in Duchenne dystrophy. Neurology 42 (1992) 1428.

RINGEL, S.P., J.E. CARROLL and S.C. SCHOLD: The spectrum of mild X-linked recessive muscular dystrophy. Arch. Neurol. 34 (1977) 408–416.

ROBERTS, R., T. BARBY, M. BOBROW and D. BENTLEY:

Direct detection of dystrophin gene mutations by analysis of dystrophin mRNA in peripheral blood lymphocytes. Am. J. Hum. Genet. 49 (suppl.) (1991) 202.

ROSENBERG, H. and T. HEIMAN-PATTERSON: Duchenne's muscular dystrophy and malignant hyperthermia: another warning. Anesthesiology 59 (1983) 362.

ROSMAN, N.P. and B.A. KAKULAS: Mental deficiency associated with muscular dystrophy: a neuropathological study. Brain 89 (1966) 769–788.

ROWLAND, L.P.: The membrane theory of Duchenne dystrophy: where is it? Ital. J. Neurol. Sci. 5 (suppl. 3) (1984) 13.

ROWLAND, L.P. and R. LAYZER: X-linked muscular dystrophies. In: P.J. Vinken and G.W. Bruyn (Eds.), Handbook of Clinical Neurology, Vol. 40: Diseases of Muscle, Part. 1. Elsevier/North Holland Biomedical Press (1979) 349–414.

RUSSELL, D.H. and L.Z. STERN: Altered polyamine excretion in Duchenne muscular dystrophy. Neurology 31 (1981) 80–83.

SANYAL, S.K. and W.W. JOHNSON: Cardiac conduction abnormalities in children with Duchenne's progressive muscular dystrophy: electrocardiographic features and morphologic correlates. Circulation 66 (1982) 853–863.

SCHAPIRA, G., J.C. DREYFUS and F. SCHAPIRA: L'élévation du taux de l'aldolase sérique: test biochimique de myopathies. Sem. Hôp. Paris 29 (1953) 1917–1920.

SCHOTLAND, D.L., E. BONILLA and Y. WAKAYAMA: Freeze fracture of muscle plasma membrane in human muscular dystrophy. Acta Neuropathol. 54 (1981) 189–197.

SCOTT, M.O., S.A. HYDE, C. GODDARD and V. DUBOWITZ: Prevention of deformity in Duchenne muscular dystrophy: a prospective study of passive stretching and splintage. Physiotherapy 67 (1981) 177–180.

SCOTT, M.O., S.A. HYDE, C. GODDARD and V. DUBOWITZ: Quantitation of muscle function in children: a prospective study in Duchenne muscular dystrophy. Muscle Nerve 5 (1982) 291–301.

SCOTT, M.O., J.E. SYLVESTER, T. HEIMAN-PATTERSON, Y.J. SHI, W. FIELDS, H. STEDMAN, A. BURGHES, P. RAY, R. WORTON and K.H. FISCHBECK: Duchenne muscular dystrophy gene expression in normal and diseased human muscle. Science 239 (1988) 1418–1420.

SHARMA, K., P. YU and R.G. MILLER: Effect of cyclosporin upon muscle force generation in Duchenne muscular dystrophy. Neurology 41 (Suppl. 1) (1991) 166.

SIBLEY, J.A. and A.L. LEHNIGER: Aldolase in the serum and tissues of tumor-bearing animals. J. Natl. Cancer Inst. 9 (1949) 303–309.

SICINSKI, P., Y. GENG, A.S. RYDER-COOK, E.A. BARNARD, N.G. DARLISON and P.J. BARNARD: The molecular basis of muscular dystrophy in the mdx mouse: a point mutation. Science 244 (1989) 1578–1580.

SIEGEL, I.M.: Plastic-molded knee-ankle-foot orthoses in the treatment of Duchenne muscular dystrophy. Arch. Phys. Med. Rehab. 56 (1975) 322.

SIEGEL, I.M.: Spinal stabilization in Duchenne muscular dystrophy: rationale and method. Muscle Nerve 5 (1982) 417–418.

SIEGEL, I.M., J.E. MILLER and R.D. RAY: Failure of corticosteroid in the treatment of Duchenne (pseudohypertrophic) muscular dystrophy: report of a clinically matched three year double blind study. Illinois Med. J. 145 (1974) 32–36.

SILVERMAN, L.M. and J.R. MENDELL: Creatine kinase isoenzymes in muscular dystrophy. Clin. Chem. 20 (1974) 865.

SMITH, P.E.M., P.M.A. CALVERLEY and R.H.T. EDWARDS: Hypoxemia during sleep in Duchenne muscular dystrophy. Am. Rev. Respir. Dis. 137 (1988a) 884–888.

SMITH, P.E.M., J.H. COAKLEY and R.H.T. EDWARDS: Respiratory muscle training in Duchenne muscular dystrophy. Muscle Nerve 11 (1988b) 784–785.

SMITH, P.E.M., R.H.T. EDWARDS and P.M.A. CALVERLEY: Oxygen treatment of sleep hypoxaemia in Duchenne muscular dystrophy. Thorax 44 (1989a) 997–1001.

SMITH, P.E.M., R.H.T. EDWARDS and P.M.A. CALVERLEY: Protriptyline treatment of sleep hypoxaemia in Duchenne muscular dystrophy. Thorax 44 (1989b) 1002–1005.

SMITH, R.A., R.G. NEWCOMBE, J.R. SIBERT and P.S. HARPER: Assessment of locomotor function in young boys with Duchenne muscular dystrophy. Muscle Nerve 14 (1991) 462–469.

SOMER, H., M. DONNER, J. MURROS and A. KONTTINEN: A serum isozyme study in muscular dystrophy: particular reference to creatine kinase, aspartate aminotransferase, and lactic acid dehydrogenase isozymes. Arch. Neurol. 29 (1973) 343.

SOUTHERN, E.: Gel electrophoresis of restriction fragments. Methods Enzymol. 68 (1979) 152–176.

STEPHENS, F.E. and F.H. TYLER: Studies in disorders of muscle. V. The inheritance of childhood progressive muscular dystrophy in 33 kindreds. Am. J. Hum. Genet. 3 (1951) 111–125.

STEVENSON, A.C.: Muscular dystrophy in Northern Ireland: an account of the condition in 51 families. Ann. Eugenet. 18 (1953) 50–93.

SWINYARD, C.A., G.G. DEAVER and L. GREENSPAN: Gradients of functional ability of importance in rehabilitation of patients with progressive muscular and neuromuscular disease. Arch. Phys. Med. Rehab. 38 (1957) 574.

THOMSON, W.H.S., P. LEYBURN and J.N. WALTON: Serum enzyme activity in muscular dystrophy. Br. Med. J. 2 (1960) 1276–1281.

TOMLINSON, B.E., J.N. WALTON and D. IRVING: Spinal cord limb motor neurones in muscular dystrophy. J. Neurol. Sci. 22 (1974) 305–327.

TORRES, L.F.B. and L.W. DUCHEN: The mutant mdx: inherited myopathy in the mouse. Brain 110 (1987) 296–299.

VERELLEN-DUMOULIN, C., M. FREUND, R. DEMEYER, C. LATERRE, J. FREDERIC, M.W. THOMPSON, V.D. MARKOVIC and R.G. WORTON: Expression of an X-linked muscular dystrophy in a female due to translocation involving Xp21 and non-random inactivation of the normal X chromosome. Hum. Genet. 67 (1984) 115–119.

WALTON, J.N.: Muscular dystrophy and its relation to the other myopathies. Res. Publ. Ass. Nerv. Ment. Dis. 38 (1960) 378–421.

WALTON, J.N.: Muscular dystrophy: some recent advances in knowledge. Br. Med. J. 1 (1964) 1271–1274, 1344–1348.

WALTON, J.N. and F.J. NATTRASS: On the classification, natural history and treatment of the myopathies. Brain 77 (1954) 169–231.

WATKINS, S.C., E.P. HOFFMAN, H.S. SLAYER and L.M. KUNKEL: Immunoelectron microscope localization of dystrophin in myofibres. Nature (London) 333 (1988) 863– 866.

WITKOWSKI, J.A. and C.T. CASKEY: Duchenne muscular dystrophy: DNA diagnosis in practice. Curr. Neurol. 8 (1988) 1–36.

WOLFF, J., R.W. MALONE, P. WILLIAMS, W. CONG, G. ACSADI, A. JANI and P.L. FELGNER: Direct gene transfer into mouse muscle in vivo. Science 247 (1990) 1465–1468.

WORTON, R.G. and M.W. THOMPSON: Genetics of Duchenne muscular dystrophy. Annu. Rev. Genet. 22 (1988) 601–629.

ZELLWEGER, H. and A. ANTONIK: Newborn screening for Duchenne muscular dystrophy. Pediatrics 55 (1975) 30–34.

ZELLWEGER, H. and J.W. HANSON: Slowly progressive X-linked recessive muscular dystrophy (Type IIIb). Arch. Intern. Med. 120 (1967) 525–535.

ZELLWEGER, H. and E. NIEDERMEYER: Central nervous system manifestations in childhood muscular dystrophy. I. Psychometric and electroencephalographic findings. Ann. Paediatr. 205 (1965) 25–42.

ZUBRZYCKA-GAARN, E.E., D.E. BULMAN, G. KARPATI, A.H. BURGHES, B. BELFALL, H.J. KLAMUT, J. TALBOT, R.S. HODGES, P.N. RAY and R.G. WORTON: The Duchenne muscular dystrophy gene product is localized in the sarcolemma of human skeletal muscle. Nature (London) 333 (1988) 466–469.

Handbook of Clinical Neurology, Vol. 18 (62): Myopathies
L.P. Rowland and S. DiMauro, editors
© 1992 Elsevier Science Publishers B.V. All rights reserved

CHAPTER 5

Emery-Dreifuss muscular dystrophy

LINTON C. HOPKINS and STEPHEN T. WARREN

*Department of Neurology and the Howard Hughes Medical Institute, Emory University School of Medicine,
Atlanta, GA, USA*

Emery-Dreifuss muscular dystrophy (EDMD) is the third most frequent X-linked recessive dystrophy after Duchenne (DMD) and Becker (BMD). EDMD is a primary myopathy and, like Becker, is slowly progressive. However, in every other way, typical EDMD patients can easily be distinguished from patients with these more common diseases. Whether the patients are analyzed from a neuromuscular, cardiologic, or molecular point of view, their biology sets them apart.

The major purpose of this Chapter is to describe the features that distinguish EDMD from more familiar diseases. There is considerable variation in severity, and mild and severe 'atypical' EDMD patients will be described. Also, the unique cardiac disorder will be emphasized. Finally, we will discuss other rare syndromes that resemble EDMD but do not fulfill the diagnostic criteria.

We will first describe the phenotype, emphasizing the unique neuromuscular and cardiac features. This will be followed by discussion of genetic aspects. In this Chapter, following the guidelines proposed at the European Workshop on EDMD in Baarn (Yates 1991), EDMD is defined as a progressive myopathy with the following features/criteria:
1. early contractures of the Achilles tendons, elbows, and spine;
2. slowly progressive muscle wasting and weakness with a predominantly humeral (upper arm) and peroneal (lower leg) distribution, bilateral and approximately symmetrical;
3. cardiac conduction defect or other evidence of cardiomyopathy;
4. muscle biopsy showing myopathic features or overt muscular dystrophy; and
5. pedigree consistent with unequivocal X-linked inheritance.

HISTORICAL REVIEW

Dreifuss and Hogan (1961) described an Appalachian Virginia family with an X-linked recessive muscular dystrophy and emphasized that the prolonged survival and less prominent calf hypertrophy differed from DMD. Although the unusual contractures were not emphasized in the initial report, the authors noted the cardiac manifestations; atrial flutter with 4-to-1 block was seen in 1 patient.

Five years later, the same family was restudied and the important phenotypic features were described in 8 affected males in 3 generations (Emery and Dreifuss 1966). The unusual elbow and spine contractures, the proximal arm and distal leg pattern of weakness, and the essential cardiac features were all first described in this report.

In 1972, an analysis of a German family with a disease identical to the Virginia family emphasized the high incidence of sudden death, which occurred in 9 of 17 patients (ages 37–59). The electro-cardiogram (EKG) showed partial or complete atrioventricular block in 5 patients. Electro-myography (EMG) and muscle pathology were in-dicative of myopathy (Rotthauwe et al. 1972). The disease first appeared in Becker's (1972) classifica-tion of muscular dystrophies. A third family was described as 'X-linked scapuloperoneal syn-drome', and attention was drawn to linkage of the disease to color blindness, suggesting a gene local-ization different from DMD and BMD (Thomas et al. 1972).

In 1975, the name 'X-linked humeroperoneal neuromuscular disease' was given to the disorder of 2 Appalachian families from North Georgia and Alabama, in which there were 37 affected males. The heart disease evolved from atrial arrhythmias to permanent atrial paralysis with a slow junc-tional pacemaker. Permanent atrial paralysis was demonstrated by inability to evoke a response from the atria, even by using intracavitary elec-trodes in 4 patients (Waters et al. 1975). In 1979, a fourth Appalachian family, primarily from West Virginia, was described, including 12 affected males. The subtle neuromuscular findings were noted to be external markers of the potentially le-thal heart disease (Hassan et al. 1979).

The term *Emery-Dreifuss muscular dystrophy* was first suggested by Rowland et al. (1979), who also noted the often ambiguous EMG changes, with features of both myopathy and neuropathy in the same muscle. Hopkins et al. (1981) restudied the Georgia and Alabama families, emphasized the EMG and muscle pathology, and suggested that slowly progressive myopathies commonly produce EMG and muscle pathology features that resemble neurogenic changes.

Subsequently, Emery and Skinner (1976) sepa-rated the disease from BMD, restudied the Vir-ginia and West Virginia families (Emery 1987; Bialer et al. 1991), emphasized the common feature of sudden death (Merlini et al. 1986; Bialer et al. 1991), and defined an *Emery-Dreifuss syndrome* to

include the neurogenic and dominantly inherited disorders that give a similar appearance (Emery 1989).

DESCRIPTION

History and physical examination

Patients with the EDMD gene are usually thought by their parents to be normal at birth and in early infancy. Most patients show mild weakness in childhood and elbow contractures appear as mini-mal limitation of full arm extension in the first dec-ade, progressing over 5–10 years to a maximum of 90°. Although the fully developed picture of humeroperoneal muscle wasting and cardiac con-duction defect is not fully developed until the end of the second decade or beginning of the third, the usual patient is thinly muscled as a youth but can perform the activities of daily living (Fig. 1) (Dickey et al. 1984). Only when challenged by diffi-cult tasks requiring effort of affected muscles does the patient recognize a problem. For instance, many patients report a lifelong inability to pull or push their body weight against gravity (chin-up or push-up exercises).

Early in the second decade the inability to fully extend the elbows limits the ability to reach out in front or above the head (Fig. 2). Some wasting of the humeral muscles, biceps, and triceps can usu-ally be noted by the time contracture of the biceps muscles begin to affect routine activities. Wasting is initially less severe in more proximal deltoid and spinati than in biceps and triceps, so the appear-ance of the arm differs from that of DMD, BMD, or other proximal myopathies (in which muscles that move the shoulder are more affected than those that move the elbow). A common appear-ance in EDMD on front view shows relatively pre-served, rounded deltoids proximal to the obviously wasted biceps and triceps (Fig. 2). However, scap-ular muscles are not spared and scapular winging is common (Hausmanowa-Petrusewicz 1988).

→

Fig. 1. (A,B,C) A patient at age 6. Note muscle atrophy in upper trunk and proximal arms, scapular winging, lordotic stance, and absence of elbow contractures at this age. (D,E,F) Same patient at age 10. Note mild elbow contractures, but no spine contracture.

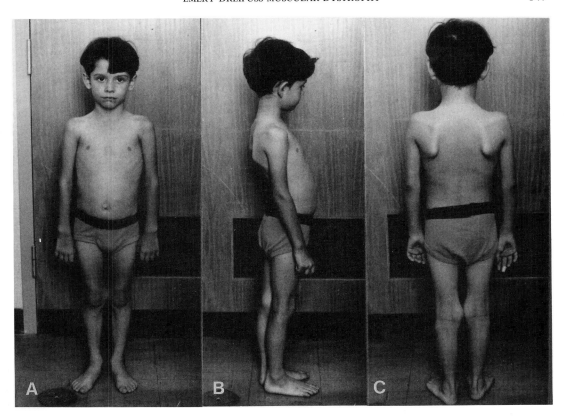

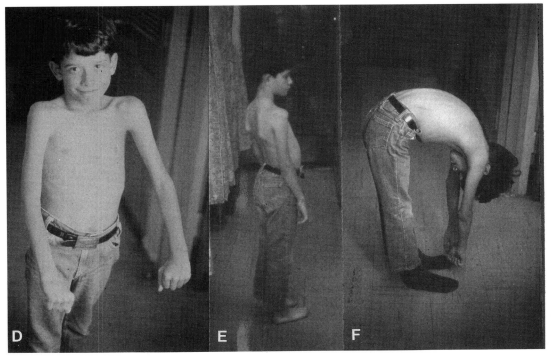

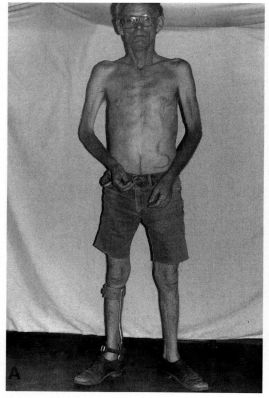
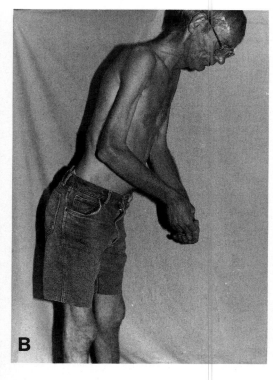

Fig. 2. (A) General adult appearance. Note relative preservation of proximal deltoid muscles, atrophy of humoral muscles, elbow contractures, and pacemaker. Brace is needed because of superimposed mild right hemiparesis due to previous stroke. (B) Maximum neck flexion.

There is some variability of distribution within families and even in the same individual over time. In the same family, the wasting may seem more 'scapuloperoneal' in some, more 'humeroperoneal' in others (Thomas et al. 1972). Also, the distribution of weakness may evolve. EDMD may be 'humeroperoneal' early in the course and, later, more 'scapulohumeropelvic-peroneal' (Emery 1987).

Initially, leg weakness is also mild and difficulty climbing stairs or walking is usually delayed until the late second or early third decade. However, the parents frequently note mild abnormality of gait or stance in the first decade. In patients with early leg symptoms, Achilles contractures usually cause most of these problems early in the disease. Early symptoms are toe walking, slow running, and loss of balance on uneven ground. Later, as in any generalized myopathy, when the disease progresses, patients experience problems rising from the seated position. However, the distribution of weakness in EDMD modifies the pattern of rising that is characteristic of more proximal myopathies like DMD and BMD.

In EDMD, the act of rising from a chair is modified by the elbow contractures and humeral weakness (especially triceps) which combine to limit the usual push from the chair seat or action of thigh muscles that are effective in patients with purely proximal myopathies. To compensate, patients must hold onto a higher stable point than the seat, such as the back of the chair or a long staff (Fig. 3). In walking, a similar adjustment is required: a cane of conventional length is not effective; a staff or extended cane is needed.

On palpation, affected muscles feel indurated, and the biceps becomes taut as the arm is maximally extended. There is no limitation of flexion at the elbow (Shapiro and Specht 1991).

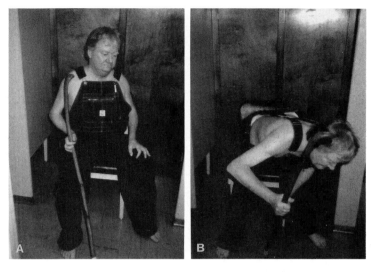

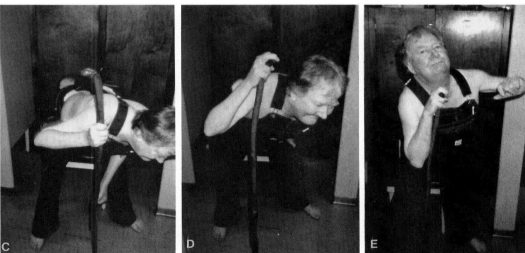

Fig. 3. (A–E) Rising from the seated position using a staff. Elbow contractures and humoral weakness limit the push with the arms, but the preserved grip allows the patient to pull himself to an erect position.

Limitation of flexion of the spine is prominent. It is not seen in the youngest patients (Fig. 1), but is obvious in most patients by the third decade. Spinal contracture is not limited to the posterior cervical muscles, but eventually involves the entire spine (Fig. 4). Rarely, there is contracture of the lower spine without any cervical involvement (1 patient, observed by author). Hip- and knee-flexion contractures are not usually present while the patient can still walk, but may appear later.

Additional neuromuscular features that distinguish EDMD from BMD and DMD are early hy-poreflexia or areflexia, absence of calf pseudohy-pertrophy, and lack of marked increase of serum creatinephosphokinase (CK).

Mild EDMD

Some affected patients are asymptomatic (Fig. 5). However, even in mildly involved patients, the cardiac lesion may be fully expressed with potentially lethal complete heart block, so attention to the subtle manifestations of the disease is important (Hassan et al. 1979; Voit et al. 1988). Minimal

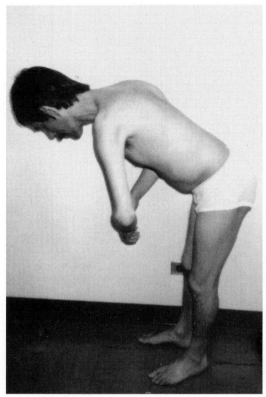

Fig. 4. Maximum spine flexion.

Clinical neurophysiology

Conventional needle EMG usually shows some features of myopathy, but others suggest reinnervation (Hopkins et al. 1981). Quantitative EMG and single fiber EMG have suggested myopathy, but there is increased mean amplitude of voluntary muscle action potentials, a feature that is usually considered neurogenic. This pattern has also been reported in other long-standing slowly progressive myopathies (Martinez and Du Theil 1989).

Because of the nonspecific EMG pattern, no particular EMG finding is required for diagnosis. However, EMG is important in the evaluation of sporadic patients because some findings, such as myotonic discharges or clear evidence of axonal loss, would exclude EDMD. Nerve conduction studies are normal.

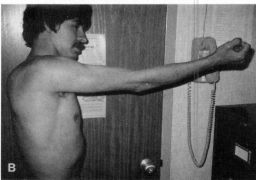

Fig. 5. Mildly affected patient. (A) General appearance is normal. Affected brother is on right. Note pacemaker. (B) Full elbow extension: note mild elbow contracture.

elbow or spine contracture without weakness may be seen in mildly affected patients. EMG and muscle biopsy may be abnormal in patients with no weakness or wasting (Hopkins et al. 1981). Typically, severity varies in different members of each family. Large families show the full spectrum of the disease, including both mild and severely affected patients in the same sibship.

Severe EDMD

Severely affected patients (Fig. 6) may lose the ability to run by age 10 and require a wheelchair by age 25–30. The spine becomes rigidly extended, and although further passive extension is still possible, forward neck flexion is limited by contracture. Contractures are found throughout, but variations occur; there may be limitation of either wrist flexion or extension by forearm contractures (Figs. 6C, 7).

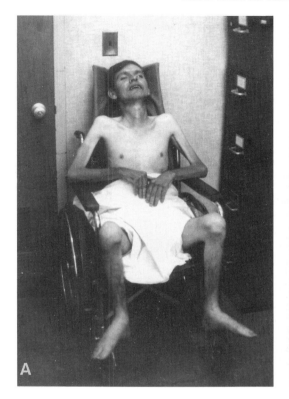
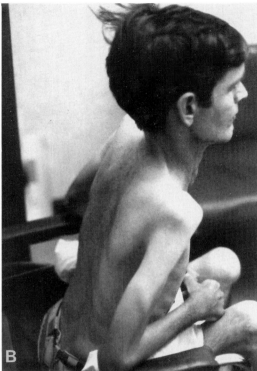
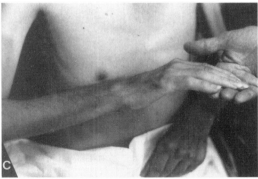
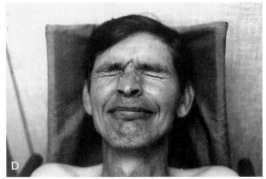

Fig. 6. Severely affected patient. (A) Note head in extension requiring posterior cervical support and generalized muscle wasting. Surgery was done on ankle and foot contractures. (B) Complete spine rigidity, maximum neck flexion. (C) Limitation of finger extension by forearm flexion contractures. (D) Little or no facial weakness.

Muscle pathology

Muscle biopsies in EDMD show a spectrum of changes from mild fiber size variation and occasional single fiber atrophy in minimally affected muscles, such as quadriceps, to marked type I fiber atrophy, fibrosis, and scattered necrosis in more severely affected biceps muscles (Hopkins et al. 1981; Hausmanowa-Petrusewicz 1988). The grouped necrosis lesion, so commonly seen in DMD, has not been described. There may be type I muscle fiber atrophy and either type II (Hopkins et al. 1981) or type I (Merlini et al. 1986) predominance. There is usually an increase of connective tissue and number of sarcolemmal nuclei and muscle fiber atrophy. Usually fiber necrosis is less prominent than in DMD or BMD (Fig. 8).

Only 1 autopsy has been reported in a patient

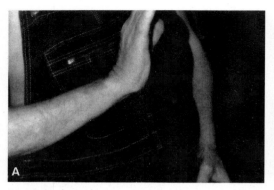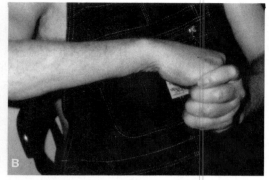

Fig. 7. (A) Mild forearm flexion contracture. (B) Contracture of wrist extensors. The patient is attempting to forcefully flex the wrist with his left hand.

who met all of the 5 criteria given earlier for the diagnosis of EDMD (Thomas et al. 1972). The spinal cord was not examined, but the sciatic nerve was normal, excluding anterior horn cell disease as the cause of the calf muscle weakness. The spinal cord was normal at autopsy of another patient who had no family history but met all

other diagnostic criteria for EDMD (Hara et al. 1987).

The heart in EDMD

EDMD is not the only dystrophy with cardiomyopathy. Autosomal-dominant myotonic muscular

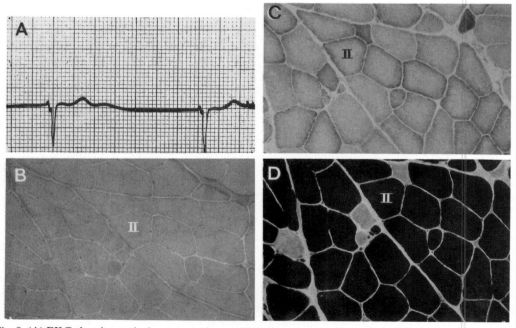

Fig. 8. (A) EKG showing typical pattern and rate. (B) Trichrome stain of muscle biopsy shows mild myopathy: fiber size variation, mild increase in endomysial connective tissue and sarcolemmal nuclear proliferation. (C) and (D) Oxidative enzyme stain (NADH) and myosin ATPase stain (pH 9.4) show type I fiber atrophy and type II predominance.

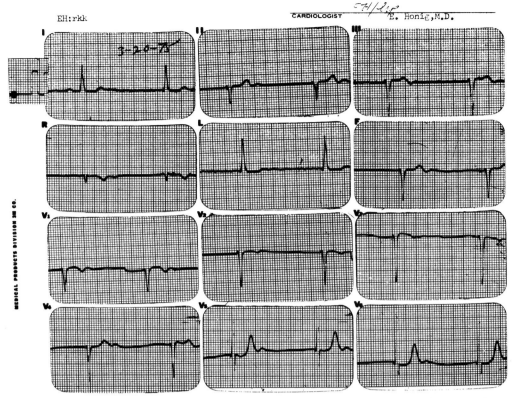

Fig. 9. When fully evolved, the EKG in EDMD shows a slow rate, absent P waves, and either a regular nodal rhythm (shown) or atrial fibrillation/flutter.

dystrophy (which has distal muscular weakness and may superficially resemble EDMD) and DMD affect the heart; and myotonic and Duchenne dystrophies may have atrial arrhythmias (Jozefowicz and Griggs 1988). However, atrial paralysis occurs only in EDMD. One patient originally thought to have facioscapulohumeral muscular dystrophy (FSH) with atrial paralysis probably had EDMD (Baldwin et al. 1973; Rowland et al. 1979).

There is a wide spectrum of cardiac problems in EDMD. In patients with no cardiac symptoms, the earliest EKG change consists of low amplitude P waves and first degree heart block, while the most advanced state consists of 4-chamber dilated cardiomyopathy with complete heart block and ventricular arrhythmias (Emery and Dreifuss 1966; Rotthauwe et al. 1972; Thomas et al. 1972; Waters et al. 1975; Oswald et al. 1987; Yoshioka et al.

1989). Initial depolarization (tall R waves in lead V_1 and Q waves in inferolateral leads), so characteristic of DMD, is not a feature of EDMD (Hassan et al. 1979) (Fig. 9).

Atrial arrhythmia usually appears prior to complete heart block. Reported features include first degree heart block, followed by Wenckebach phenomenon and then, complete atrioventricular dissociation (Hassan et al. 1979) and atrial fibrillation or flutter with progressive slowing of the rate (Emery and Dreifuss 1966; Waters et al. 1975; Voit et al. 1988). Syncope or near-syncope commonly occurs late in the second decade or early in the third (Waters et al. 1975). Subsequently, with more prolonged survival, a generalized cardiomyopathy appears (Thomas et al. 1972; Voit et al. 1988). Some patients develop generalized left ventricular hypokinesis, perfusion defects in the cardiac muscle, and tricuspid and mitral valve regurgitation (Yoshioka et al. 1989).

Although all affected patients have heart disease, there is considerable variation in the evolution, as illustrated by specific patients. One had a normal EKG at age 16 and only first degree AV block with a normal rate at ages 40 and 43. Another, normal at 13, had first degree AV block with a rate of 70 at age 37; continuous EKG monitoring during sleep showed wandering atrial pacemaker, frequent premature atrial contractions, and periods of complete heart block with a rate of 20. Another, in atrial fibrillation at age 28, had atrial fibrillation with tachycardia and wide QRS complexes at age 52, an exception to the expected progression to bradycardia and complete heart block (Bialer et al. 1991).

The full spectrum of symptoms due to the heart disease in EDMD includes palpitations due to underlying arrhythmia, near-syncope and syncope, sudden death, poor exercise tolerance, congestive heart failure, and cerebral emboli.

There is no correlation between the severity of the cardiac abnormalities and the severity of limb weakness (Pinelli et al. 1987). For this reason, EDMD should be considered whenever a young man is found to have an atrial arrhythmia and strongly considered if there is a slow nodal rhythm (Hassan et al. 1979).

Prognosis and suggestions concerning management

When the cardiac symptoms were all ascribed to the atrial conduction abnormality, it was generally thought that ventricular pacing, life-saving in the short term, would also be effective long-term management. However, a pacemaker only treats the bradycardia and other life-threatening symptoms may occur.

One patient had a pacemaker for 14 years and then had 2 transient ischemic attacks (TIAs). Also, sudden death in patients with presumably well-functioning pacemakers was noted in the Virginia and West Virginia families (Bialer et al. 1991). In the Georgia family, 1 death with pacemaker was apparently cardiac without progressive heart failure; another was related to recurrent cerebral emboli, with death in status epilepticus (L.H., 1991,

unpublished observations). EDMD is a muscular dystrophy which may cause TIA and stroke, presumably due to cerebral emboli from atrial thrombus (Hopkins et al. 1981) (Fig. 2A).

In the autopsied patient with EDMD, all chambers of the heart were diffusely dilated without much hypertrophy of the ventricular walls, and there was diffuse myocardial fibrosis (Thomas et al. 1972). With or without a pacemaker, patients with EDMD usually die suddenly while they can still walk. This differs from DMD and BMD, in which death usually occurs in bed-ridden patients and is due to progressive hypoventilation from weakness of respiratory muscles (Waters et al. 1975; Oswald et al. 1987).

Management should include 24-hour EKG monitoring to study nocturnal heart rate. Pacemaker insertion is suggested when rates drop below 50 beats per minute. Ultimately, heart transplantation should be considered (Witt et al. 1988; Merchut et al. 1990), especially in patients with mild weakness, whose life expectancy should be normal if the cardiac threat were removed. Because of the risk of stroke due to embolus from the heart, antiplatelet therapy and anticoagulation should be considered.

Carriers

Carriers of EDMD are usually free of muscle or spinal manifestations, but they may have arrhythmias and a pacemaker may be needed (Mawatari and Katayama 1973; Dickey et al. 1984; Emery 1987; Pinelli et al. 1987; Bialer et al. 1991). In the original Virginia family, 6 of 34 carrier females (18%) had arrhythmia, 2 had pacemakers, and the frequency of heart disease increased with age. All five carriers over age 60 had an abnormal EKG (Bialer et al. 1991).

DISEASES SIMILAR TO X-LINKED RECESSIVE EDMD

Emery-Dreifuss syndrome (Criteria 1, 2, and 3 only)

Emery-Dreifuss syndrome is a useful term, encouraging physicians to recognize that the distinctive clinical appearance is associated with a lethal cardiomyopathy. The term includes patients with both neurogenic and myopathic features and allows different modes of inheritance (Emery 1989). The key triad is elbow contracture, humeroperoneal weakness, and cardiomyopathy with heart block.

It might be thought that the combination of limb muscle weakness and cardiomyopathy implies a myopathic, not a neurogenic, disorder. In fact, most reported patients with clearly neurogenic scapuloperoneal atrophy do not have cardiomyopathy (Emery et al. 1968; Mercelis et al. 1980).

Two reports, however, take exception to the theory that severe cardiomyopathy with contractures and muscle weakness always means primary muscle disease. Kudo et al. (1982) described abnormal spinal cord pathology with reduced number of anterior horn cells in a 15-year-old boy with no family history. Witt et al. (1988) described a family with a dominantly inherited disease characterized by fiber-type grouping that suggested a neurogenic disorder, but the spinal cord was not studied. Concentric and single-fiber EMG studies were interpreted as neurogenic. However, early recruitment of voluntary potentials and type I muscle fiber atrophy were present, indicating a resemblance to EDMD.

In some autosomal-dominant families, male-to-male transmission effectively excluded X-linked recessive transmission (Chakrabarti and Pearce 1981; Serratrice and Pouget 1986; Witt et al. 1988). Muscle wasting was described as scapulohumeral-peroneal, a pattern resembling EDMD.

Early ventricular cardiomyopathy may be a feature of the dominantly inherited disease (Gilchrist and Leshner 1986) and the dominantly inherited form may be more rapidly progressive (Chakrabarti and Pearce 1981). In the German family, 1 patient had generalized cardiomyopathy and a heart transplant, but it was uncertain whether ventricular dysfunction followed atrial arrhythmia. Pathologic features resembled EDMD, and there was type I atrophy. Another patient had some fiber-type grouping, suggesting a neurogenic disorder and single-fiber EMG suggested neurogenic factors. The authors indicated that the pattern was nonspecific and could occur in either neurogenic or myopathic disease, but they favored a neurogenic cause in the case presented (Witt et al. 1988). Given the complex nature of EMG changes seen in

chronic dystrophy and the similarity of the biopsy to other reported cases, these patients may have had a primary myopathy with neurogenic features, a combination typical of EDMD.

One disadvantage of the term 'Emery-Dreifuss syndrome' is that it includes diseases that are transmitted by genes on different chromosomes and which may be neurogenic instead of myopathic. In the future the gene products of these disorders may be found to be different.

Myopathies with early contractures but without cardiomyopathy (Criteria 1 and 4 only)

The *rigid spine syndrome* (Dubowitz 1971; Seay et al. 1977) refers to patients who are normal at birth but have early onset of spine, elbow, and other contractures with mild weakness and wasting and a benign course. Males and females have been affected and the condition is either sporadic or autosomal recessive. There are myopathic changes on muscle biopsy (Van Munster et al. 1986), but cardiomyopathy is not a feature (Poewe et al. 1985; Goto et al. 1986). The rigid spine syndrome is distinguished from EDMD by the absence of cardiac conduction defects and lack of documented X-linked recessive inheritance.

Bethlem myopathy is another slowly progressive disease with early contractures without cardiomyopathy (Bethlem and Van Wijngaarden 1976; Arts et al. 1978; Mohire et al. 1988). It differs from EDMD by the absence of cardiac disorder and by autosomal-dominant inheritance. Spine contractures are not emphasized in reported patients but congenital torticollis and weakness of head flexion have been described. There are usually contractures of the long finger flexors.

The rigid spine syndrome and Bethlem myopathy are generalized, proximal and distal, myopathies. In these syndromes, there is no scapuloperoneal or humeroperoneal accentuation.

Facioscapulohumeral muscular dystrophy (FSH)

Though FSH is autosomal-dominant, there is a superficial resemblance to EDMD in some patients, in that humeral weakness and wasting in association with peroneal weakness are common in both disorders. Facial weakness may be mild in FSH but elbow contractures and posterior cervical contractures do not occur, and the cardiomyopathy, if present at all, does not dominate the clinical picture as it does in EDMD. There is no well-documented case of FSH dystrophy with heart block.

Stevenson et al. (1990) studied 13 females and 17 males with FSH. They excluded patients with elbow contractures or X-linked recessive inheritance. P wave abnormalities were noted in 60% of the 30 patients and 24% had infranodal conduction defects. Echocardiograms were normal in 15 patients, and there were no disturbances of rhythm or conduction on 24-hour ambulatory ECGs, although 5 patients did have sinus bradycardia at rest. Program-atrial stimulation showed 'high susceptibility to induced atrial flutter or fib'. This study suggested that the advanced atrial and ventricular abnormality of EDMD might have a much milder counterpart in the heart of patients with FSH.

Other conditions that resemble EDMD but have distinguishing features

1. One family had dominantly inherited cardiomyopathy that resembled EDMD with slow atrial fibrillation and humeral weakness, but with no elbow or spine contractures and leg weakness was more severe proximally than distally (Fenichel et al. 1982).
2. One family had X-linked recessive scapuloperoneal weakness, hypertrophic cardiomyopathy, mild serum CK elevation, and severe mental retardation (Bergia et al. 1986). Like EDMD, there was no pseudohypertrophy and progression was slow, but there were no flexion contractures.
3. One family had autosomal-dominant inheritance and muscle pathology that included abnormal mitochondrial accumulation and ragged red fibers (Tanaka et al. 1989). The pedigree was compatible with, but not diagnostic of, maternal transmission.
4. An EDMD-like patient without cardiomyopathy had a vacuolar muscle biopsy (Petty et al. 1986).
5. A woman with contracture and cardiomyopathy had pseudohypertrophy and probable autosomal-dominant inheritance (Miller et al. 1985).
6. In 1 unusual family with 4 affected females, including 1 pair of identical twins, there was myopa-

thy, cardiomyopathy, and short stature (Orstavik et al. 1990).
7. Although BMD is usually clinically distinct from EDMD, 1 patient with early elbow contractures and no pseudohypertrophy had brothers with typical BMD (Wadia et al. 1976).
8. Two different types of fatal congenital myopathy have now been mapped by linkage analysis to Xq28. However, both are probably localized to proximal Xq28 and are distinct from the more distal EDMD gene. In the first syndrome, 'X-linked myotubular myopathy with fatal neonatal asphyxia', muscle biopsy showed centronuclear myopathy and death was due to respiratory muscle failure. There was no cardiomyopathy or mitochondrial abnormality (Barth et al. 1975; Liechti-Gallati et al. 1991). Also, in another family with severe centronuclear myopathy, the mother had a very different and milder lifelong disease with both central and peripheral nervous symptom signs (Torres et al. 1985).

In the second syndrome, 'X-linked cardioskeletal myopathy with neutropenia and abnormal mitochondria', dilated cardiomyopathy was universal, both mitochondrial structure and function were abnormal, and centronuclear myopathy was absent (Barth et al. 1983; Bolhuis et al. 1991). Of considerable interest is that 1 of the patients with mitochondrial abnormalities had low muscle carnitine content, was treated with oral carnitine, and had a more indolent course and longer survival than other affected patients.

Though each of these myopathies is different from the other, and both are quite different from EDMD, the fact that their genes map to the proximal part of Xq28 raises the question that they are allelic and suggests that different mutations in the same gene might be able to produce very different but related clinical syndromes.

GENETICS

EDMD is inherited in X-linked fashion, with affected sons of carrier women and no male-to-male transmission. Penetrance is 100% in males by the second or third decade of life. Typical of the X-linked recessive pattern, most heterozygous women show no clinically evident skeletal muscle disorder. However, bradycardia, arrhythmias, and prolonged PR intervals are occasionally found in carrier women and could be due to nonrandom inactivation of the X chromosomes in cardiac tissue.

By pedigree inspection, transmission of EDMD is similar to that of both DMD and BMD. However, DMD and BMD are allelic at Xp21, and the EDMD gene is remote from that site. Linkage analysis with X-linked polymorphic loci demonstrated frequent recombination between EDMD and those marker loci tightly linked to DMD and BMD (Hodgson et al. 1986; Thomas et al. 1986; Yates et al. 1986). EDMD was then linked to distal Xq markers (Romeo et al. 1988; Saviranta et al. 1988) and definitively mapped to the distal portion of Xq28 (Consalez et al. 1991). Therefore, EDMD is genetically distinct from DMD and BMD, as long suspected clinically.

Linkage of EDMD to polymorphic marker loci allows the diagnostic prediction of both presymptomatic boys at risk and female carriers, as well as for prenatal diagnosis. The disease can be ruled out with a high degree of accuracy in family members at risk, easing anxiety and reproductive concerns.

Two considerations of marker loci are important for diagnosis by linkage. First, heterozygosity frequency indicates how often a given family will be informative; that is, the carrier mother is heterozygous at that locus. Second, the genetic distance, Θ, of the marker locus from the disease locus denotes how often the marker locus recombines with the EDMD gene (Table 1). One polymorphism that is detected by the gene encoding coagulation factor 8 seems to be closest to EDMD,

with no recombinants yet identified (Consalez et al. 1991). Within the confidence interval stated, there is an approximate 6% probability of recombination between F8C and EDMD. Therefore, in a family segregating EDMD that is informative for the F8C polymorphism, the diagnosis can be made or excluded in an at-risk individual with a maximum likelihood of 94%.

An example using the F8C polymorphism in an EDMD family is shown in Fig. 10. Digestion of DNA with the restriction endonuclease BclI revealed 2 alleles in this family following Southern blotting and hybridization with the probe F8. One allele was observed as a 1.2-kb band and the other as a 0.9-kb band. All clinically affected males exhibited the 1.2-kb band, showing that that allele was associated with the EDMD mutation in this family. VI-1 was a boy with an a priori risk of 50% for eventually being affected (from the left, lane 6). Individual VI-1 was a 4-year-old boy who inherited the 1.2-kb allele from his mother; she was an obligate carrier of EDMD and heterozygous for the F8 polymorphism, so that particular marker locus would not be informative in the study of any children she may have.

When the F8C polymorphism is not informative, other nearby loci may also be used (Table 1). The marker locus DXS52 is particularly useful since it is informative in nearly 75% of families examined and, although slightly farther away from EDMD, can predict genotype in an appropriate family to within 92% probability.

Many of the polymorphisms listed in Table 1 can be assessed simultaneously, allowing a compiled risk estimate that may significantly modify

TABLE 1

Marker loci exhibiting genetic linkage with Emery-Dreifuss muscular dystrophy.

Locus	Probe	Enzyme	Heterozygosity	Z_{max}	Θ_{max}	95% Confidence interval
DXS52	St14	Taq I	0.74	15.7	0.02	$0.001 \rightarrow 0.075$
DXS15	Dx13	Bgl II	0.36	11.3	0.02	$0.001 \rightarrow 0.097$
R/GCP	Hs7	Sst I	0.50	9.6	0	$0 \rightarrow 0.082$
F8C	F8	Bcl I Bgl I Msp I	0.33	10.8	0	$0 \rightarrow 0.062$

Data compiled from Consalez et al. (1991).

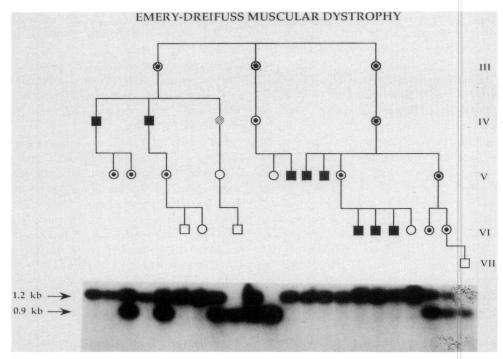

Fig. 10. Results of Southern blot analysis of DNA from a portion of a large family segregating EDMD. DNA was digested with BclI, separated by electrophoresis, blotted onto a nylon filter, and hybridized with the radiolabelled probe F8 prior to autoradiography. Closed symbols are clinically affected and partially closed circles are carrier females. Open symbols are clinically normal and hatched symbols are individuals not studied in this example. The lane showing the F8C alleles for an individual are immediately below the individual's pedigree symbol.

the risk estimates. Because molecular genetics is changing so rapidly, EDMD families should be referred to a laboratory experienced in these linkage studies. Also, because linkage studies have narrowed the location of the EDMD gene to perhaps less than 2000 kb of DNA (Consalez et al. 1991; Poustka et al. 1991), the gene itself may be isolated in the near future, allowing even more precise gene diagnosis and hope that the gene product will be identified.

REFERENCES

ARTS, W.F., J. BETHLEM and W.S. VOLKERS: Further investigations on benign myopathy with autosomal dominant inheritance. J. Neurol. 217 (1978) 201–206.

BALDWIN, B.J., R.C. TALLEY, C. JOHNSON and D.O. NUTTER: Permanent paralysis of the atrium in a patient with facioscapulohumeral muscular dystrophy. Am. J. Cardiol. 31 (1973) 649–653.

BARTH, P.G., G.K. VAN WIJNGAARDEN and J. BETHLEM: X-linked myotubular myopathy with fatal neonatal asphyxia. Neurology 25 (1975) 531–536.

BARTH, P.G., H.R. SCHOLTE, J.A. BERDEN, J.M. VAN DER KLEI-VAN MOORSEL, I.E.M. LUYT-HOUWEN, E.TH. VAN T VEER-KORTHOP, J.J. VAN DER HARTEN and M.A. SOBOTKA-PLOJHAR: An X-linked mitochondrial disease affecting cardiac muscle, skeletal muscle and neutrophil leucocytes. J. Neurol. Sci. 62 (1983) 327–355.

BECKER, P.E.: Neues zur Genetik und Klassifikation der Muskeldystrophien. Humangenetik 17 (1972) 1–22.

BERGIA, B., H.D. SYBERS and I.J. BUTLER: Familial lethal cardiomyopathy with mental retardation and scapuloperoneal muscular dystrophy. J. Neurol. Neurosurg. Psychiatr. 49 (1986) 1423–1426.

BETHLEM, J.A.A.P. and G.K. VAN WIJNGAARDEN: Benign myopathy with autosomal dominant inheritance. Brain 99 (1976) 91–100.

BIALER, M.G., N.L. MCDANIEL and T.E. KELLY: Progression of cardiac disease in Emery-Dreifuss muscular dystrophy. Clin. Cardiol. 14 (1991) 411–416.

BOLHUIS, P.A., G.W. HENSELS, T.J.M. HULSEBOS, F. BASS and P.G. BARTH: Mapping of the locus for X-linked cardioskeletal myopathy with neutropenia and abnormal mitochondria (Barth syndrome) to Xq28. Am. J. Hum. Genet. 48 (1991) 481–485.

CHAKRABARTI, A. and J.M.S. PEARCE: Scapuloperoneal syndrome with cardiomyopathy: report of a family with autosomal dominant inheritance and unusual features. J. Neurol. Neurosurg. Psychiatr. 44 (1981) 1146–1152.

CONSALEZ, G.G., N.S.T. THOMAS, C.L. STAYTON, S.J.L. KNIGHT, M. JOHNSON, L.C. HOPKINS, P.S. HARPER, L.J. ELSAS and S.T. WARREN: Assignment of Emery-Dreifuss muscular dystrophy to the distal region of Xq28: the results of a collaborative study. Am. J. Hum. Genet. 48 (1991) 468–480.

DICKEY, R.P., F.A. ZITER and R.A. SMITH: Emery-Dreifuss muscular dystrophy. J. Pediatr. 104 (1984) 555– 559.

DREIFUSS, F.E. and G.R. HOGAN: Survival in X-chromosomal muscular dystrophy. Neurology 11 (1961) 734– 741.

DUBOWITZ, V.: Recent advances in neuromuscular disorders. Rheumatol. Phys. Med. 11 (1971) 126–133.

EMERY, A.E.H.: X-linked muscular dystrophy with early contractures and cardiomyopathy (Emery-Dreifuss type). Clin. Genet. 32 (1987) 360–367.

EMERY, A.E.H.: Emery-Dreifuss syndrome. J. Med. Genet. 26 (1989) 637–641.

EMERY, A.E.H. and F.E. DREIFUSS: Unusual type of benign X-linked muscular dystrophy. J. Neurol. Neurosurg. Psychiatr. 29 (1966) 338–342.

EMERY, A.E.H. and R. SKINNER: Clinical studies in benign (Becker type) X-linked muscular dystrophy. Clin. Genet. 10 (1976) 189–201.

EMERY, E.S., G.M. FENICHEL and G. ENG: A spinal muscular atrophy with scapuloperoneal distribution. Arch. Neurol. 18 (1968) 129–133.

FENICHEL, G.M., Y.C. SUL, A.W. KILROY and R. BLOUIN: An autosomal-dominant dystrophy with humeropelvic distribution and cardiomyopathy. Neurology 32 (1982) 1399–1401.

GILCHRIST, J.M. and R.T. LESHNER: Autosomal dominant humeroperoneal myopathy. Arch. Neurol. 43 (1986) 734–735.

GOTO, I., S. ISHIMOTO, T. YAMADA, H. HARA and Y. KUROIWA: The rigid spine syndrome and Emery-Dreifuss muscular dystrophy. Clin. Neurol. Neurosurg. 88 (1986) 293–298.

HARA, H., H. NAGARA, S. MAWATARI, A. KONDO and H. SATO: Emery-Dreifuss muscular dystrophy. An autopsy case. J. Neurol. Sci. 79 (1987) 23–31.

HASSAN, Z.U., C.P. FASTABEND, P.K. MOHANTY and E.R. ISAACS: Atrioventricular block and supraventricular arrhythmias with X-linked muscular dystrophy. Circulation 60 (1979) 1365–1369.

HAUSMANOWA-PETRUSEWICZ, I.: The Emery-Dreifuss disease. Neuropathol. Pol. 26 (1988) 278–281.

HODGSON, S., E. BOSWINKEL, C. COLE, A. WALTER, V. DUBOWITZ, C. GRANATA, L. MERLINI and M. BOBROW: A linkage study of Emery-Dreifuss muscular dystrophy. Hum. Genet. 74 (1986) 409–416.

HOPKINS, L.C., J.A. JACKSON and L.J. ELSAS: Emery-Dreifuss humeroperoneal muscular dystrophy: an X-linked myopathy with unusual contractures and bradycardia. Ann. Neurol. 10 (1981) 230–237.

JOZEFOWICZ, R.F. and R.C. GRIGGS: Myotonic dystrophy. Neurol. Clin. 6 (1988) 455–472.

KUDO, H., S. YANO, K. SAITO, M. YANAGISAWA and S. KAMOSHITA: Spinal cord involvement in Emery-Dreifuss muscular atrophy (X-linked recessive scapuloperoneal muscular atrophy). Brain Dev. (Japan) 14 (1982) 363–369.

LIECHTI-GALLATI, S., B. MULLER, T. GRIMM, W. KRESS, C. MULLER, E. BOLTSHAUSER, H. MOSER and S. BRAGA: X-linked centronuclear myopathy: mapping the gene to Xq28. Neuromusc. Disord. 1 (1991) 239–245.

MARTINEZ, A.C. and L.A. DU THEIL: Electrophysiologic evaluation of Emery-Dreifuss muscular dystrophy. A single fiber and quantitative EMG study. Electromyogr. Clin. Neurophysiol. 29 (1989) 99–103.

MAWATARI, S. and K. KATAYAMA: Scapuloperoneal muscular atrophy with cardiopathy. An X-linked recessive trait. Arch. Neurol. 28 (1973) 55–59.

MERCELIS, R., J. DEMEESTER and J.J. MARTIN: Neurogenic scapuloperoneal syndrome in childhood. J. Neurol. Neurosurg. Psychiatr. 43 (1980) 888–896.

MERCHUT, M.P., D. ZDONCZYK and M. GUJRATI: Cardiac transplantation in female Emery-Dreifuss muscular dystrophy. J. Neurol. 237 (1990) 316–319.

MERLINI, L., C. GRANATA, P. DOMINICI and S. BONFIGLIOLI: Emery-Dreifuss muscular dystrophy: report of five cases in a family and review of the literature. Muscle Nerve 9 (1986) 481–485.

MILLER, R.G., R.B. LAYZER, M.A. MELLENTHIN, M. GOLABI, R.A. FRANCOZ and J.C. MALL: Emery-Dreifuss muscular dystrophy with autosomal dominant transmission. Neurology 35 (1985) 1230–1233.

MOHIRE, M.D., R. TANDAN, T.J. FRIES, B.W. LITTLE, W.W. PENDLEBURY and W.G. BRADLEY: Early-onset benign autosomal dominant limb-girdle myopathy with contractures (Bethlem myopathy). Neurology 38 (1988) 573–580.

ORSTAVIK, K.H., R. KLOSTER, C. LIPPESTAD, L. RODE, T. HOVIG and K.N. FUGLSETH: Emery-Dreifuss syndrome in three generations of females, including identical twins. Clin. Genet. 38 (1990) 447–451.

OSWALD, A.H., J. GOLDBLATT, A.R. HORAK and P. BEIGHTON: Lethal cardiac conduction defects in Emery-Dreifuss muscular dystrophy. S. Afr. Med. J. 72 (1987) 567–570.

PETTY, R.K.H., P.K. THOMAS and D.N. LANDON: Emery-Dreifuss syndrome. J. Neurol. 233 (1986) 108–114.

PINELLI, G., P. DOMINICI, L. MERLINI, G. DI PASQUALE, C. GRANATA and S. BONFIGLIOLI: Valutazione cardiologica in una famiglia affetta da distrofia muscolare di Emery-Dreifuss. G. Ital. Cardiol. 17 (1987) 589–593.

POEWE, W., H. WILLEIT, E. SLUGA and U. MAYR: The rigid spine syndrome: a myopathy of uncertain nosological position. J. Neurol. Neurosurg. Psychiatr. 48 (1985) 887–893.

POUSTKA, A., A. DIETRICH, G. LANGENSTEIN, D. TONIOLO, S.T. WARREN and H. LEHRACH: Physical map of human Xq27-qter: localizing the region of the fragile X mutation. Proc. Natl. Acad. Sci. USA 88 (1991) 8302–8306.

ROMEO, G., L. RONCUZZI, S. SANGIORGI, M. GIACANELLI, M. LIQUORI, D. TESSAROLO and M. ROCCHI: Mapping of the Emery-Dreifuss gene through reconstruction of crossover points in two Italian pedigrees. Hum. Genet. 80 (1988) 59–62.

ROTTHAUWE, H.W., W. MORTIER and H. BEYER: Neuer Typ einer recessiv X-chromosomal vererbten Muskeldystrophie: scapulohumerodistale Muskeldystrophie mit fruhzeitigen Kontrakturen und Herzrhythmusstorungen. Humangenetik 16 (1972) 181–200.

ROWLAND, L.P., M. FETELL, M. OLARTE, A. HAYS, N. SINGH and F.E. WANAT: Emery-Dreifuss muscular dystrophy. Ann. Neurol. 5 (1979) 111–117.

SAVIRANTA, P., M. LINDLOF, A.E. LEHESJOKI, H. KALIMO, H. LANG, V. SONNINEN, M.L. SAVONTAUS and A. DE LA CHAPELLE: Linkage studies in a new X-linked myopathy, suggesting exclusion of DMD locus and tentative assignment to distal Xq. Am. J. Hum. Genet. 42 (1988) 84–88.

SEAY, A.R., F.A. ZITER and J.H. PETAJAN: Rigid spine syndrome. Arch. Neurol. 34 (1977) 119–122.

SERRATRICE, G. and J. POUGET: Maladie d'Emery-Dreifuss ou syndrome d'amyotrophie avec rétractions précoces et troubles secondaires de la conduction cardiaque d'hérétité variable. Rev. Neurol. (Paris) 142 (1986) 766–770.

SHAPIRO, F. and L. SPECHT: Orthopedic deformities in Emery-Dreifuss muscular dystrophy. J. Pediatr. Orthopaed. 11 (1991) 336–340.

STEVENSON, W.G., J.K. PERLOFF, J.N. WEISS and T.L. ANDERSON: Facioscapulohumeral muscular dystrophy: evidence for selective, genetic electrophysiologic cardiac involvement. J. Am. Coll. Cardiol. 15 (1990) 292–299.

TANAKA, K., T. YOSHIMURA, H. MURATANI, J. KIRA, Y. ITOYAMA and I. GOTO: Familial myopathy with scapulohumeral distribution, rigid spine, cardiopathy and mitochondrial abnormality. J. Neurol. 236 (1989) 52–54.

THOMAS, N.S.T., H. WILLIAMS, L.J. ELSAS, L.C. HOPKINS, M. SARFARAZI and P.S. HARPER: Localization of the gene for Emery-Dreifuss muscular dystrophy to the distal long arm of the X chromosome. J. Med. Genet. 23 (1986) 596–598.

THOMAS, P.K., D.B. CALNE and C.F. ELLIOTT: X-linked scapuloperoneal syndrome. J. Neurol. Neurosurg. Psychiatr. 35 (1972) 208–215.

TORRES, C.F., R.C. GRIGGS and J.P. GOETZI: Severe neonatal centronuclear myopathy with autosomal dominant inheritance. Arch. Neurol. 42 (1985) 1011–1014.

VAN MUNSTER, E.T.L., E.M.G. JOOSTEN, M.A.M. VAN MUNSTER-UIJTDEHAAGE, H.J.A. KRULS and H.J. TER LAAK: The rigid spine syndrome. J. Neurol. Neurosurg. Psychiatr. 49 (1986) 1292–1297.

VOIT, T., O. KROGMANN and H.G. LENARD: Emery-Dreifuss muscular dystrophy: disease spectrum and differential diagnosis. Neuropediatrics 19 (1988) 62–71.

WADIA, R.S., S.U. WADGAONKAR, R.B. AMIN and H.V. SARDESAI: An unusual family of benign 'X' linked muscular dystrophy with cardiac involvement. J. Med. Genet. 13 (1976) 352–356.

WATERS, D.D., D.O. NUTTER, L.C. HOPKINS and E.R. DORNEY: Cardiac features of an unusual X-linked humeroperoneal neuromuscular disease. N. Engl. J. Med. 293 (1975) 1017–1022.

WITT, T.N., C.G. GARNER, D. PONGRATZ and X. BAUR: Autosomal dominant Emery-Dreifuss syndrome: evidence of a neurogenic variant of the disease. Eur. Arch. Psychiatr. Neurol. Sci. 237 (1988) 230–236.

YATES, J.R.W.: European Workshop on Emery-Dreifuss Muscular Dystrophy 1991. Neuromusc. Disord. 1 (1991) 393–396.

YATES, J.R.W., N.A. AFFARA, D.M. JAMIESON, M.A. FERGUSON-SMITH, I. HAUSMANOVA-PETRUSEWICZ, J. ZAREMBA, J. BORKOWSKA, A.W. JOHNSTON and K.KELLY: Emery-Dreifuss muscular dystrophy: localization to Xq27.3-qter confirmed by linkage to the Factor VIII gene. J. Med. Genet. 23 (1986) 587–590.

YOSHIOKA, M., K. SAIDA, Y. ITAGAKI and T. KAMIYA: Follow up study of cardiac involvement in Emery-Dreifuss muscular dystrophy. Arch. Dis. Child. 64 (1989) 713–715.

Handbook of Clinical Neurology, Vol. 18 (62): Myopathies
L.P. Rowland and S. DiMauro, editors
© 1992 Elsevier Science Publishers B.V. All rights reserved

CHAPTER 6

Facioscapulohumeral and scapuloperoneal syndromes

THEODORE L. MUNSAT[1] and GEORGES SERRATRICE[2]

[1]*Department of Neurology, Tufts-New England Medical Center, Boston, MA, USA and*
[2]*Department of Neurology, University of Aix-Marseille, Marseille, France*

FACIOSCAPULOHUMERAL SYNDROME

Although Duchenne's series of papers in the Archives of General Medicine (1868) is usually cited for the X-linked dystrophy that bears his name, they also describe families with facioscapulohumeral (FSH) distribution of weakness. However, the first identification of FSH disease (FSHD) is appropriately attributed to Landouzy and Dejerine (1884). Dawidenkow (1919, 1929, 1930, 1939) wrote extensively about restricted forms of neuromuscular disease and pointed out that they are often benign and limited in distribution. Bell (1942) attempted unsuccessfully to define subgroups of neuromuscular disease according to the presence or absence of facial involvement.

In 1950, Tyler and Stephens described an extraordinary family from Utah. Among 1249 individuals in 6 generations stemming from a single ancestor, they found that 159 were known to have the disease. Fifty-eight were examined. This unique study revealed the frequent 'subclinical' involvement and the characteristic distribution of muscle weakness now identified with FSHD. Chung and Morton (1959), utilizing genetic criteria, noted the autosomal-dominant pattern.

It was the study of Walton and Nattrass (1954) that truly established FSHD as 1 of the 3 major forms of muscular dystrophy. Their classification was based primarily on clinical and genetic observations. Although electrophysiologic studies and muscle biopsies were available, they did not use them to create their diagnostic categories. They defined FSH 'dystrophy' as a disease that began any time in the first 30 years of life and had a slowly progressive course, with arrest of progression common. The face was initially involved, then the shoulder girdle, and finally the pelvic muscles. Asymmetry of weakness was not unusual. Lifespan was not significantly affected.

Recent linkage of FSHD to the end of chromosome 4 (Wijmenga 1990) provided the first step in unraveling the etiology and pathogenesis of this disorder.

Clinical features

A full understanding of the manifestations of FSHD is hampered by a paucity of longitudinal studies. Time of onset is difficult to determine, and the muscles involved have complex and overlapping functions that resist analysis by traditional neurologic examination. However, a composite picture can be pieced together. In addition to standard texts on muscle disorders (Dubowitz 1978; Walton 1981) and specific reviews (Munsat 1986), the thesis by Padberg (1982) is particularly recommended.

It is clear from examining presymptomatic family members that the disease exists many years before a patient seeks medical attention. Age of onset has been estimated as 7–20 (Tyler and Stephens 1950), 7–27 (Walton and Nattrass 1954), and 3–44 (Padberg 1982). Infantile onset has been described by Carroll and Brooke (1979) and was noted by both Duchenne (1868) and Landouzy and Dejerine (1884) in their original reports. Hanson and Rowland (1971) described 3 cases of Möbius' syndrome in which congenital facial weakness was associated with later development of FSH weakness. However, onset in infancy is relatively uncommon.

The initial weakness affects the facial muscles, especially orbicularis oculi, zygomaticus, and orbicularis oris. Masseters, temporalis, extraocular, and pharyngeal muscles are characteristically spared. The onset is so slow and insidious that patients are usually completely unaware of the deficit, and many are reluctant to accept the examining physician's observations. Involvement is often asymmetrical, causing unusual facial expressions. Early, the face may appear uninvolved, but eventually it loses the normal lines and wrinkles and assumes an expressionless appearance easily mistaken for depression. Later, progressive loss of muscle results in drooping and increased tissue laxity.

Typically, patients have never been able to whistle or drink through a straw. Upon smiling, the mouth moves transversely and often asymmetrically. With more forceful smiling, characteristic dimples can be observed at the corners of the mouth. With eyelid closure, the lids are buried incompletely or, in the later stages, fall short of complete closure and leave a rim of exposed sclera. Even very early, forceful closure of the lids can be easily overcome by the examiner's fingers. Although cases are seen where facial weakness is either equivocal or absent, these are extremely rare.

The scapular fixators are weak at onset (Fig. 1). Involvement of latissimus dorsi, the lower portion of trapezius, the rhomboids, and serratus anterior results in a characteristic shoulder posture. At rest the scapula is positioned more laterally than normal. It rides upwards and forwards, giving the shoulders a forward-sloped appearance. This often gives the mistaken impression that the patient is athletic and well-muscled. With arm elevation, the upper border of the scapula rises into the substance of trapezius, giving it a hypertrophied ap-

pearance. The clavicle becomes more horizontal and eventually may angle downwards.

Scapular winging may or may not be observed at rest. However, it becomes more apparent with abduction and especially forward movement of the arms. If deltoid function is preserved, as it often is in early stages, scapular elevation will be further accentuated by the upward pull of this muscle. Winging can be partially corrected by mechanically holding the scapula against the rib cage as the arm is elevated.

The supraspinatus and infraspinatus muscles are relatively normal in the early stages, but pectoralis major, particularly the sternocostal head, is almost always weak and atrophic. This results in a characteristic upward slope of the anterior axillary fold (Fig. 2) (in contrast to the normal diagonal) and a flattened appearance to the anterior chest wall. Pectus excavatum can be seen in the more extreme instances.

It is the functional impairment of these shoulder girdle muscles that usually brings the patient to seek medical advice. In Padberg's series (1982), 82% first sought care because of shoulder weakness.

The first 25° of arm elevation are performed by supraspinatus. Movement to the horizontal is performed by the deltoid. Bringing the arm from the horizontal to the face and head for facial and hair care and reaching for objects on high shelves is performed by scapular rotation (Fig. 3). The scapula must be fixed to the chest wall for appropriate mechanical advantage. When it is not, and the above skills are impaired, patients seek help. Meanwhile, most of them learn to achieve over-the-head elevation by literally throwing the arm into the air with a quick jerking motion and holding it there with the other arm or hand.

Functional shoulder weakness is usually accompanied by weakness of the leg anterior compartment muscles. Involvement of tibialis anterior, as can be shown by muscle CT-scan (Fig. 4), is highly characteristic. Few patients can sustain heel-walking; many have a frank foot-drop gait and a small number will present with a gait disorder. The posterior leg muscles are spared, even in advanced stages.

The muscle weakness in FSH syndrome has a tendency to progress downward. Facial involve-

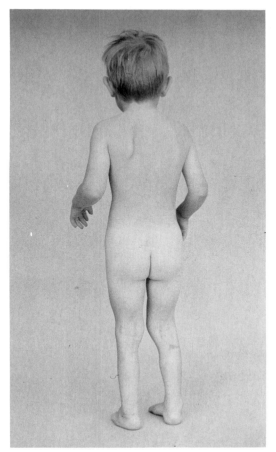

Fig. 1. Early scapular and periscapular atrophy may be difficult to differentiate from normal. This child demonstrates very early scapular winging.

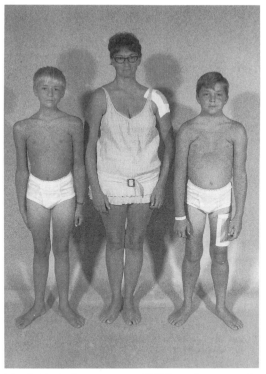

Fig. 2. Typical appearance of FSH syndrome are the 'flattened' facial appearance, asymmetrical shoulder girdle atrophy, and upward sloping of the axillary folds.

ment is followed by weakness of the shoulder girdle, then biceps and triceps, and finally the pelvic girdle. More than any other 'dystrophy', FSHD may show prolonged periods of apparent arrest. These can be several years to a few decades in duration, or even permanent. In about half of those affected, the disease never involves the pelvic girdle and remains arrested in the shoulders and arms. These patients remain ambulatory all their lives.

Although very slow progression is most typical, periods of acceleration may be observed. This may be more apparent than real and actually represents the loss of only a small number of motor units or muscle fibers that have been functionally important.

Although occasional patients are severely af-fected, wheelchair confinement is the exception and life-span is not significantly reduced. Padberg (1982) observed a mean age at death of 64.2 years in males and 70.5 years in females, compared to the control expectation of 65.6 and 67.3, respectively.

Inheritance and genetics

Autosomal-dominant inheritance is evident in the vast majority of cases. There are no well-studied families with recessive or sex-linked patterns of inheritance. Sporadic cases must remain suspect until all primary relatives are carefully examined by an experienced clinician. One should be most wary of accepting a negative family history from even the most cooperative and intelligent patients. Detailed examination of all family members at risk is essential.

Developments in DNA analysis will soon provide more precise genetic diagnosis. Meanwhile,

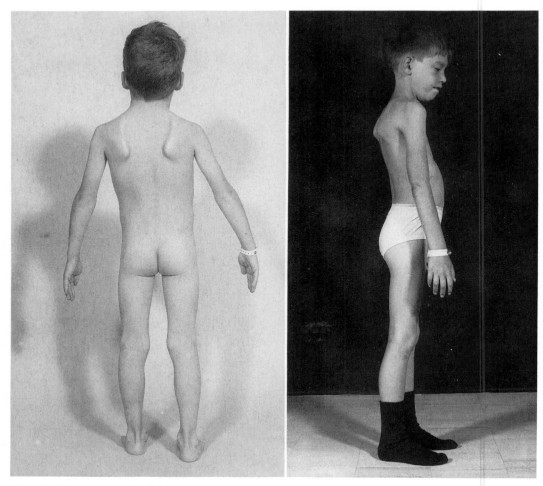

Fig. 3. In the mid-stage of the disease, shoulder girdle atrophy and weakness are unmistakable, especially upon arm abduction.

clinical examination is the only way to detect early or preclinical manifestations. In the early stages, clinical findings may be equivocal, and serum enzyme tests, electromyography (EMG), and muscle biopsy are not of specific help. The following screening examination, which takes only a few minutes, will detect most affected persons: (1) reviewing old family photographs; (2) observing the face during attempts to smile, wrinkling the nose, and 'blow out the cheeks'; (3) checking resistance to eye closure; (4) observing arm elevation and abduction from the front, side, and back and evaluating scapular winging; and (5) asking the patient to walk on his heels.

The propensity of affected family members to remain asymptomatic and undetected is most striking. Padberg (1982) stated that, in his experience 30% of affected people were asymptomatic and our observations support that figure. Zundel and Tyler (1965) observed that 24 of 58 patients, 13 over age 20, were unaware that they had disease.

The reasons for this are several. The muscles initially involved are not functionally important. Although the disease most likely begins in early childhood, damage to proximal arm muscles does not cause concern until the late teens or, frequently, much later. Early facial, shoulder, or gait involvement is frequently interpreted as a benign family

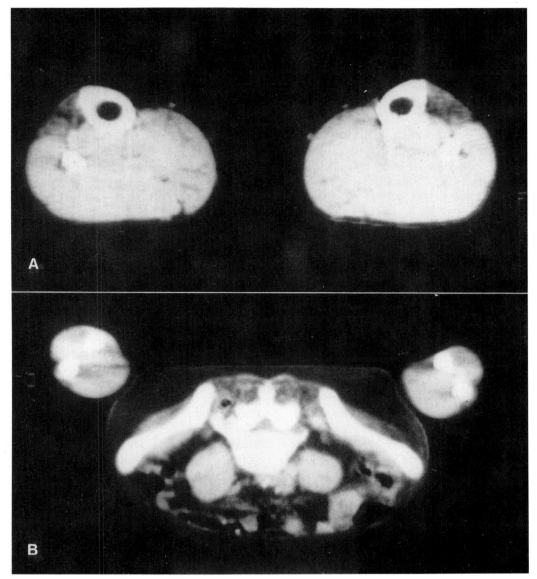

Fig. 4. Typical CT scan of patient with FSH syndrome. (A) The destructive process in the anterior compartment muscles is represented by decreased density. (B) Hypertrophy of the psoas muscle is often seen in these patients.

trait of no importance. Even when 1 familial disorder is known, there is a tendency not to accept the presence of an inherited disease.

Penetrance of FSH is remarkably complete (Lunt et al. 1989). When families are examined by experienced clinicians, the ratio of affected to nonaffected siblings is nearly 1:1 (Kazakov et al. 1977; Padberg 1982). Reports of skipped generations are probably examples of incomplete case de-termination. The prevalence rate is about 1 per 100,000 population (Morton and Chung 1959) and fertility is unaffected.

It is difficult to generalize about intrafamilial variation. The disease often runs a remarkably similar course in the affected individuals of a family, but we have seen families in which some members have infantile onset of moderate severity and others remain asymptomatic into late adult years.

Recent mapping of the FSHD gene to the end of the long arm of chromosome 4 (Wijmenga et al. 1990) is the first step in understanding the true nature of FSHD. Patients and scientists alike hope that cloning of the gene and identification of the defective gene product will follow shortly. Indeed progress is already being made in this task (Wijmenga et al. 1991). To date, although there is clinical evidence for infrequent spontaneous mutations (Lunt and Harper 1991), all families have mapped to the same location and there is no evidence for genetic heterogeneity.

Laboratory features

Specific tests diagnostic of FSHD are not available. Changes detected in the laboratory are those common to any slowly evolving neuromuscular disease. Routine blood and urine studies are normal. Electrocardiography and echocardiography are usually unremarkable in autosomal-dominant cases. Most patients have an elevated serum creatine kinase level. Munsat et al. (1973) found an elevation in 50% while Hughes (1971) observed abnormal levels in 66% of women and 80% of men. Serum enzyme activity is not age-related in the early years, but levels decrease after age 55. Fewer patients have elevation of LDH (20%), aldolase (15%), or SGPT (5%) (Kaeser 1965). EMG shows polyphasic motor units of low amplitude and short duration, with increased interference pattern.

Pathology

Results of muscle biopsy vary considerably, depending on the site selected for sampling. FSHD is characterized by regional muscle involvement, often with considerable asymmetry, and unevenness of the disease process extends even to single muscles. Parts of a muscle may be severely affected while adjacent fascicles are normal. Because of the slow evolution, most involved muscles will show fairly advanced histologic changes, with portions being replaced by adipose and connective tissue that can obscure more useful pathologic features. It is therefore advisable to biopsy muscles that are minimally affected or to biopsy those partially affected at the border zone. Most biopsies will then demonstrate those changes felt to be consistent

with a defect of the muscle fiber itself. These myopathic changes consist of random variation in fiber diameter, centrally placed nuclei, occasional necrosis of muscle fibers, and increased endomysial and perimysial tissue. Occasional fibers are lobulated or partitioned by aberrant myofibrils (Bethlem et al. 1973). Within each fasciculus, small rounded fibers are mixed with large fibers as well as normal-sized fibers. Regenerating fibers are occasionally present, as are 'moth-eaten' fibers demonstrating focal decrease in oxidative enzyme activity. Collections of mononuclear cells may be seen in perivascular or perimysial distribution (Munsat et al. 1972).

In 1967, Fenichel and colleagues reported a 17-year-old woman with a 4 year history of progressive facial and shoulder girdle weakness and atrophy. Her mother had a similar problem, and the family history suggested involvement of prior generations. Sensation was normal. Serum CPK was modestly elevated. Electrophysiologic studies revealed that the muscle was 'irritable'. Muscle biopsy (in the daughter only) showed occasional small, angulated fibers of both fiber types. The authors thought this was a form of spinal muscular atrophy.

Subsequent case reports provide a clinical picture similar to that of FSH 'dystrophy', but with electrophysiologic or pathologic evidence of denervation. In one case of Furukawa et al. (1969), the patient had clinical fasciculations and the EMG showed a reduced interference pattern with increased motor unit potentials. The muscle biopsy demonstrated grouped atrophy. These data suggested denervation. However, in a later case studied by Furukawa and Toyokura (1976), the histologic examination was inadequate and the EMG pattern was mixed. In a case of Kazakov et al. (1977), histologic confirmation was lacking and the diagnosis was based on the EMG alone. Twins described by Iwamoto et al. (1979) had histologic evidence of denervation, but the EMG was myopathic and there was an unusual clinical presentation.

Inflammatory changes have been reported in several typical FSHD families. It is nevertheless unlikely that they had a restricted form of polymyositis. Polymyositis and dermatomyositis rarely cause weakness in a facioscapulohumeral distribution. Polymyositis may progress slowly and arrest spontaneously, as does FSHD, but any facial involvement is minor. It is rarely, if ever, early or prominent. In FSHD, the pattern is quite the opposite. Moreover, the degree of muscle weakness tends to be less marked than expected for the amount of wasting, while in polymyositis the reverse often is true. Serologic findings and clinical features typical of other autoimmune disorders are sometimes found in association with polymyositis, but not with FSHD.

In occasional FSHD patients, inflammatory changes are found with other myopathic changes, making the differentiation from polymyositis even more difficult. An early report (Munsat et al. 1972) suggested that FSH patients with inflammatory changes in muscle improved with steroid treatment. However, the same patients subsequently continued to worsen (Munsat and Bradley 1977) (Fig. 5). Other cases have been unresponsive to steroids (Wulff et al. 1982). Steroid unresponsiveness may argue for the differences between polymyositis and FSH with inflammatory changes, although many patients with polymyositis do not respond to steroids either. Lastly polymyositis is not inherited, while FSH syndrome is.

It has been suggested that the inflammatory changes in FSHD are transient, similar to those observed in Duchenne dystrophy. However, on the contrary, in the patient reported by Wulff et al. (1982), 3 muscle biopsies over a period of 13 years showed prominent inflammatory changes. In 2 of our cases (Munsat et al. 1972), repeat biopsies 6 and 7 years later found the prominent inflammatory changes unaltered. Such evidence strongly suggests that the inflammatory reaction is an intrinsic part of the disease. Hudgson et al. (1972) described a family with 12 affected individuals in 4 generations with FSH weakness. Muscle biopsy revealed abnormal lipid accumulation associated with giant, bizarre mitochondria. The mitochondria had distorted cristae and electron-dense paracrystalline inclusions, suggesting a defect of oxidative metabolism. These changes were similar to those of 'mitochondrial myopathy', which usually presents with weakness in limb-girdle distribution. Other cases witih FSH distribution have been reported more recently by Rowland et al. (1991).

An unusual association of FSH weakness with hearing loss and retinal disease has been reported in recent years (Small 1968; Hudgson et al. 1972; Carroll and Brooke 1979; Taylor et al. 1982; Wulff et al. 1982). Known as Coats' syndrome, the myo-

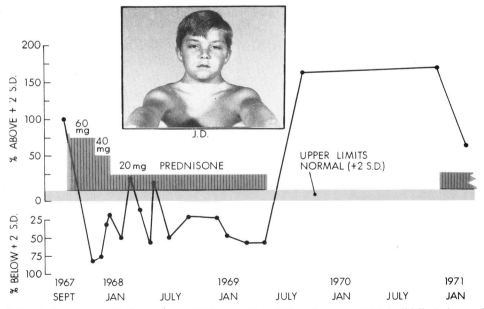

Fig. 5. Typical example of inflammatory FSH myopathy. Although serum CPK (solid line) dropped with steroid therapy, clinical improvement did not occur.

pathy begins in infancy and is rapidly progressive. One patient (Taylor et al. 1982) had pelvic girdle weakness by age 6 and was in a wheelchair by age 15. CK (creatine kinase) activity was moderately elevated, and muscle biopsy showed myopathic changes.

Hearing loss may be unilateral but more common is bilateral and sensorineural in character. Brainstem auditory-evoked potentials are normal. The retinopathy is characteristic of Coats' syndrome, with retinal telangiectasia, exudation of proteinaceous fluid, and retinal detachment. Photocoagulation, if done early, can be quite successful. The inheritance is autosomal-dominant with incomplete penetrance.

Van Wijngaarden and Bethlem (1973) observed a FSH distribution of muscular weakness in 17 patients. Five had myotubular myopathy, 1 each had nemaline bodies and central cores, and 1 had mitochondrial abnormalities. Such disorders are not usually associated with weakness in FSH distribution. Two other patients had myasthenia gravis and developed severe atrophy and weakness of facial and shoulder girdle muscles.

Therapy

Medical therapy is limited to the use of steroids in patients with FSH syndrome who have inflammatory changes. However, as noted above, although steroids will depress CK levels and result in some improvement in functional strength, the benefits are short-lasting, and the progressive nature of the disease eventually becomes apparent.

Appropriately timed surgery to stabilize the scapulae in a more functional position can be effective in carefully selected patients. Since glenohumeral function is usually intact (first 90° of abduction), elevation above the horizontal can be achieved by fixing the scapula to the rib cage at about 20° of external rotation. This prevents winging and lateral movement. It also allows an additional 20° of range above the horizontal, which can be important functionally for the care of the head and face and for reaching high objects.

Muscle transplantations have not been useful. Although physical therapy is of unknown value, it is recommended on empirical grounds with the provision that overextension of partially weak

muscles is avoided. Posterior plastic ankle supports can be most helpful in correcting the footdrop, if it is not too severe. In more advanced cases, spring-loaded braces are necessary.

THE SCAPULOPERONEAL SYNDROME

The nosology of neuromuscular cases reported under the heading of scapuloperoneal atrophy remains uncertain. They were initially thought to be a variety of Charcot-Marie-Tooth disease (Dawidenkow 1929, 1939; Stark 1958). Other authors used the designation scapuloperoneal 'myopathy' and classified it as a muscular dystrophy or as a variant of FSH (Oransky 1927; Seitz 1957; Hausmanowa-Petrusewicz and Zielinska 1962; Serratrice et al. 1969; Thomas et al. 1972). Several cases were considered as variants of spinal muscular atrophy (Kaeser 1965; Emery et al. 1968; Ricker et al. 1968; Schumann 1970; Serratrice et al. 1976).

In most cases the nature of the pathologic change is unclear. Several genetic patterns are present and muscle biopsies show both myopathic and neuropathic changes (Takahashi et al. 1971; Mercelis et al. 1980). Weakness or wasting may involve facial or pelvic girdle muscles.

It has been suggested that the scapuloperoneal syndrome is a stage in the course of other neuromuscular diseases (Bethlem 1979). As the disease progresses, it may affect other muscle groups, thereby losing its characteristic scapuloperoneal features. In fact, several neuromuscular diseases initially present with a scapuloperoneal topography including myotonic dystrophy, amyotrophic lateral sclerosis, some cases of hereditary ataxia, phosphofructokinase deficiency, and myotubular myopathy. If these secondary cases are excluded, however, 3 main groups may be defined using clinical, electrophysiological, and pathological data. The most frequently encountered is scapuloperoneal dystrophy.

Scapuloperoneal dystrophy

The scapuloperoneal dystrophies are usually of autosomal-dominant inheritance or sporadic (Oransky 1927; Seitz 1957; Hausmanowa-Petrusewicz and Zielenska 1962; Ricker et al. 1968; Feigenbaum and Munsat 1970; Thomas et al. 1975; Delwaide and Schoenen 1976; Serratrice et al. 1982a). Onset occurs in childhood or young adulthood. Symptoms usually begin with the peroneal muscles but sometimes with those of the shoulder girdle. The clinical features are characteristic (Fig. 6). Wasting and weakness involve the peroneal muscles, especially anterior tibialis. CT shows reduced density of these muscles. Asymmetric cases are often misdiagnosed as peroneal nerve compression. Later, bilateral foot-drop occurs. The extensor digitorum brevis may be hypertrophic. Ankle contractures are frequent and prominent, limiting dorsiflexion of the foot. Reflexes are present and there are no sensory disturbances. Wasting of shoulder girdle muscles is characteristic, with winged scapulae. Humeral muscles are often spared. Serum CK activity is occasionally slightly elevated. EMG shows low amplitude and short duration potentials in distal or proximal muscles. Histological changes are variable, often with type I fiber atrophy. The course is slow, with periods of

stabilization, and wheelchair confinement is the exception. Life-span is not significantly reduced.

Facial weakness in 21 of 31 cases (Serratrice et al. 1969) suggested a relationship between FSHD and scapuloperoneal syndrome. There appears to be a continuum with 4 clinical forms: pure FSHD; FSHD with peroneal weakness (22 of 40 patients, Serratrice et al. 1982a); scapuloperoneal dystrophy with facial weakness (in these cases, humeral muscles are frequently spared and ankle contractures are constant); and pure scapuloperoneal dystrophy.

Spinal scapuloperoneal atrophy

In 1965, Kaeser described a family with neurogenic scapuloperoneal atrophy. In 1 case, autopsy showed degeneration of anterior horn motoneurons. The inheritance was autosomal-dominant, which is rare in typical chronic spinal atrophies. Later identical cases were reported as Stark-Kaeser syndrome (1 sporadic case: Fotopoulos and

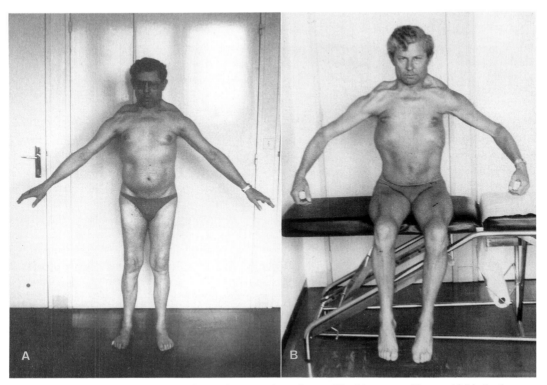

Fig. 6. (A, B) Two typical examples of scapuloperoneal syndrome. The biopsy was 'dystrophic' in both cases. The clinical involvement was similar to FSH disease, but no facial weakness could ever be detected.

Schultz 1966; 1 familial case: Gori et al. 1971; 2 sporadic cases: Andre et al. 1972; 3 familial cases: Negri et al. 1973; 1 case: Rozhold et al. 1975; 5 familial and sporadic cases: Thomas et al. 1975; 10 cases with 6 familial cases: Serratrice et al. 1976). The main features of the so-called Stark-Kaeser syndrome are the following.

Muscle wasting is moderate with respect to weakness and extends to all the muscles of the leg, not only to the peroneal muscles (shoulder-shank syndrome). However, the peroneal muscles are the first to be affected, while the intrinsic foot muscles are spared. The peroneal wasting is more pronounced than that in the shoulder girdle atrophy, which remains minor without winged scapulae. The tongue is sometimes involved. Fasciculations are not uncommon. Areflexia is constant and there are no sensory disturbances. The course is protracted. Weakness of the proximal legs may occur later. The EMG is neuropathic, but motor and sensory conduction velocities are normal. Muscle biopsy is neuropathic and the nerve biopsy is normal. Thus, the clinical features suggest a chronic spinal disorder.

There is a clinical subtype with recessive inheritance, rather heterogeneous, with infantile muscle weakness, a more severe course, and foot deformities such as pes equinovarus or clubfoot (Emery et al. 1968; Munsat 1968; Zellweger and McCormick 1968).

Several additional cases reported in the literature are questionable because of the distal extension of atrophy (Meadows and Marsden 1969; Schumann 1970) or atypical EMG or muscle biopsy (Ricker et al. 1968).

Scapuloperoneal neuropathy

Scapuloperoneal neuropathy, sometimes called Dawidenkow type, is probably a different subgroup. Few cases have been reported in detail (Dawidenkow 1919, 1929, 1930, 1939; Tohgi et al. 1971; Schwartz and Swash 1975; Serratrice et al. 1982a). The main features of these cases are autosomal-dominant inheritance, scapuloperoneal atrophy with glove and stocking sensory disturbances, pes cavus, areflexia, slow nerve motor conduction velocities, demyelination, and onion bulbs on nerve biopsy. However, without necropsy it is impossible to assess whether the syndrome is restricted to peripheral nerve, and the peripheral nerves are not hypertrophic in all cases (Schwartz and Swash 1975; Serratrice et al. 1976). The hypertrophic and axonal subtypes are in fact very close to the hypertrophic and neuronal types of Charcot-Marie-Tooth disease, according to the Dyck and Lambert classification (1968).

The Dawidenkow syndrome may be a special form of HSMN (hereditary sensory motor neuropathy) I and II. Except for the unusual scapular girdle amyotrophy, the main features are identical: autosomal-dominant inheritance; distal sensory disturbances; areflexia; findings on EMG and nerve biopsy; and a very slow course. Moreover, there is distal wasting and weakness in the upper limbs in some cases (Dawidenkow 1929; Schwartz and Swash 1975; Serratrice et al. 1982a).

Inflammatory scapuloperoneal syndrome

In rare cases, inflammatory infiltrates have been reported in scapuloperoneal dystrophy (Serratrice et al. 1982a). They suggest the inflammatory form of FSHD (Munsat et al. 1972) and are not polymyositis.

THE SCAPULOHUMERAL SYNDROME

Scapulohumeral dystrophy

The juvenile muscular dystrophy of Erb (1884) is similar in many respects to FSHD but without facial weakness. It is slowly progressive, with wasting restricted to the shoulder girdle or to the humeral muscles. Weakness is most prominent in the biceps, triceps, trapezius, rhomboid, and serratus anterior muscles. There are winged scapulae and the deltoid is spared, as in FSH. Walton and Gardner-Medwin (1981) consider it a subgroup of limb-girdle dystrophy. This myopathy is uncommon. It is probably not a separate entity, and there is no reason to differentiate this syndrome from FSHD.

Scapulohumeral spinal atrophy

This condition initially reported by Vulpian (1886) is also rare. The amyotrophy is restricted to the shoulder girdle, with frequent neck and proximal

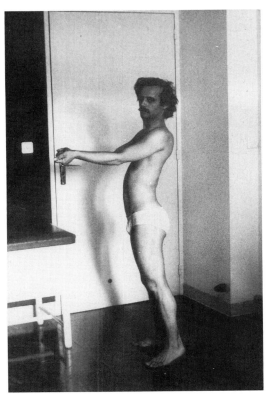

Fig. 7. Emery-Dreifuss disease. This patient had elbow and ankle contractures, progressive scapular winging, and cardiomyopathy.

arm weakness. The EMG and muscle biopsy are neurogenic. Some cases are unilateral in the early stages (Kaeser et al. 1983)

Emery-Dreifuss disease

Several cases of focal neuromuscular disease have been reported with 3 features in common; early contractures; progressive wasting and weakness in childhood with scapulohumeroperoneal distribution; and death caused by cardiac arrhythmia and atrial paralysis (Figs 7, 8).

In 1966, Emery and Dreifuss reported a family of 7 affected males in 3 generations. This family was examined again by McKusik (1971). The patients had joint contractures at elbows and ankles, with toe-walking; proximal weakness, especially in biceps and triceps brachii; areflexia; and cardiopathy. Out of the 7, 3 had a nodal rhythm and another died suddenly of a 'heart attack'.

A number of similar X-linked cases have been reported. In 1972, a family including 17 males in 3 generations was reported by Rotthauwe and co-workers. Contractures occurred before the age of 10. Muscle weakness began in the biceps brachii and tibialis anterior with a scapulohumeral distal distribution. Serum muscle enzymes were high. The EMG was myopathic, but with fibrillations. Muscle biopsy showed marked fibrosis. Nine patients died suddenly.

Thomas et al. (1972) reported an X-linked disease affecting 5 males in 2 generations. The main features were contractures with a rigid neck in 1; slowly progressive scapuloperoneal weakness with femoral weakness in 1 case, and areflexia. Three cases had a cardiopathy with atrial fibrillation, nodal rhythm, atrioventricular conduction defect, and marked cardiac enlargement. Unfortunately, no muscle biopsy was performed, but post-mortem

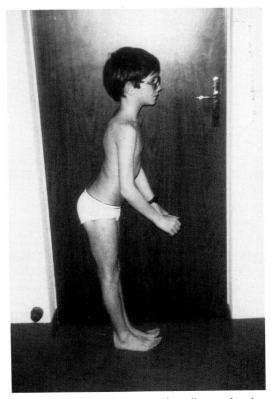

Fig. 8. Early onset Emery-Dreifuss disease showing pelvic and elbow contractures and periscapular atrophy and weakness. (Courtesy of Shinji Ishimoto, M.D.)

examination showed pronounced cardiac fibrosis. All striated muscles showed myopathy.

The 5 cases of Mawatari and Katayama (1973) were very similar. There was X-linked inheritance with contractures, and scapulohumeroperoneal wasting and weakness with areflexia. The EKG showed a complete heart block in 2 and an incomplete right bundle-branch block in another. Muscle biopsy, performed in 1 case only, showed increased connective tissue with small atrophic groups. EMG suggested denervation. The authors concluded that their patients had an unusual form of spinal muscular atrophy, but their cases are very similar to those above, as are the cases of Cammann et al. (1974) and Waters et al. (1975), which had elbow and neck contractures, areflexia, and cardiopathy. The EMG was both neuropathic and myopathic. Muscle biopsy showed a marked increase in connective tissue and type I fiber atrophy. Rowland et al. (1979) reported a sporadic case and suggested the eponym, 'Emery-Dreifuss dystrophy'. Their patient had weakness of humeroperoneal distribution associated with limited range of motion of the cervical spine and elbows. Later he developed permanent atrial paralysis. The EMG and muscle biopsy showed a mixed neuropathic and myopathic pattern. These authors suggested that the clinical criteria determined a distinct form of muscular dystrophy. They discussed the possibility of 2 genetic disorders, 1 neuropathic and the other myopathic, but felt that the multisystem clinical pattern was sufficient to define the disease.

Hopkins and Karp (1976) reported the 2 largest familial series, with 35 affected males. Contractures and cardiac arrhythmia were observed, but the pattern of muscle wasting was mainly humeral (biceps and triceps) and peroneal. Muscle biopsy, performed in 8, showed type I fiber atrophy. Several patients had an EMG with both myopathic and neuropathic findings.

Autosomal-dominant inheritance was reported by Chakrabarti and Pearce (1981) in 4 members of a family with scapuloperoneal weakness and features of Emery-Dreifuss syndrome. They believed their cases to represent an autosomal-dominant myopathic scapuloperoneal syndrome, with unusual symptoms similar to cases reported by Emery and Dreifuss. Serratrice et al. (1986) reported 6 additional cases, 2 sporadic and 4 X-linked. All had early contractures, scapulohumeroperoneal atrophy, cardiac conduction disorders, and neuropathic and myopathic EMGs and biopsies. They pointed out the prominence of contractures and myofibrosis. The authors discussed the relationship with myosclerosis, and suggested that fibrosis could be the common feature of both the cardiac and striated muscle lesion. They questioned the wisdom of considering Emery-Dreifuss a separate disease, noting that the atrophy and weakness were not scapulohumeral and peroneal in Emery and Dreifuss' original cases (1966). Serratrice et al. (1982b) preferred the term, 'X-linked scapulohumeralperoneal amyotrophy with early contractures and cardiac conduction abnormalities'.

Similar to the cases of Chakrabarti and Pearce (1981), those of Fenichel et al. (1982) were of autosomal-dominant inheritance. The proband was a man with shortening of the Achilles tendons, humeropelvic weakness, atrial fibrillation, neuropathic EMG, and myopathic muscle biopsy. His father and sister were similarly affected with contractures and cardiopathy. There are no important clinical differences between the autosomal-dominant and X-linked cases.

In a recent paper, Takamoto et al. (1984) described an autosomal-recessive variant with early joint contractures, humeropelvic weakness, and permanent atrial paralysis in a 38-year-old woman. The disease was assumed to be inherited through an autosomal-recessive trait because of parental consanguinity.

Miller et al. (1985) reported a third case with autosomal-dominant inheritance. The woman had ankle and elbow contractures, limitation of neck flexion, atrial fibrillation, atrophy of biceps, brachii, triceps, brachioradialis, legs and peroneal muscles, with areflexia. Both EMG and muscle biopsy were myopathic. The woman's father had similar muscular and cardiac defects. Another female with autosomal dominant inheritance was recently described by Serratrice and Pouget (1986).

In summary, Emery-Dreifuss dystrophy consists of 4 essential components.
1. When inherited, the pattern is X-linked.
2. Contractures at ankles, elbow, and neck occur early in childhood.
3. The distribution of weakness involves the biceps

and triceps more than scapular muscles. In the legs, the distal muscles, especially peroneal, are affected earlier than proximal. The progression is very slow. Areflexia is the rule.

4. Cardiac conduction abnormalities are always observed and lead ultimately to atrial paralysis and arrest. Sudden death before 50 is not uncommon.

Some unclear cases are found in the cardiology literature. Bensaid et al. (1972) reviewed 24 published cases of permanent atrial paralysis. Among them, 7 were associated with a neuromuscular disease but often incorrectly classified. Four were diagnosed as facioscapulohumeral dystrophy (Bloomfield and Sinclair-Smith 1965; Baldwin et al. 1973) in spite of photos showing contractures and features compatible with Emery-Dreifuss disease. Other cases were classified as Charcot-Marie-Tooth disease (Bensaid et al. 1972).

Recently, Serratrice et al. (1982b) reported 2 brothers with rigid spine syndrome and cardiac abnormalities. In addition, the muscle biopsy showed an increase of connective tissue (such cases are similar to those described as myosclerosis, a heterogeneous and poorly documented entity). The EMG was unusual, with myopathic and neuropathic changes in the same muscle. There was an increased number of potentials, reduced mean duration, fibrillations, and increased potential amplitude.

A clear understanding is difficult in these cases. Few authors report careful EMG evaluation (Rotthauwe et al. 1972; Thomas et al. 1972; Cammann et al. 1974; Hopkins and Karp 1976; Rowland et al. 1979; Serratrice et al. 1982a,b). Spontaneous activity with fibrillations is frequent. The number of motor unit potentials is mildly to moderately reduced. They are often polyphasic, decreased in amplitude and duration, suggesting both a neuropathic and myopathic disorder. In the quantitative study of Rowland et al. (1979), the mean duration of motor unit potentials was reduced while the amplitude was normal. Fibrillations were seen with sharp positive waves of 5-msec duration and bizarre high frequency responses, and the number of units was reduced. Similar findings were reported by Rotthauwe et al. (1972), Waters et al. (1975), Hopkins and Karp (1976) and Serratrice et al. (1982a,b). However, in some cases, EMG was only neuropathic (Mawatari and Katayama 1973; Fenichel et al. 1982) or only myopathic (Cammann et al. 1974).

The same ambiguities are found in the histological changes. Some cases are myopathic (Chakrabarti and Pearce 1981; Fenichel et al. 1982) and some neuropathic (Mawatari and Katayama 1973). Most often, both changes are reported (Rowland et al. 1979; Serratrice et al. 1982b). In many cases fibrosis is very prominent (Rotthauwe et al. 1972; Thomas et al. 1972; Mawatari and Katayama 1973; Waters et al. 1975). This is of considerable interest. The proliferation of connective tissue could explain the mixed pattern in EMG and biopsy. The same mixed findings are reported in several types of muscle fibrosis including congenital muscle dystrophy, arthrogryposis, and rigid spine syndrome. It could also explain the atrial paralysis. In the case of Thomas et al. (1972), postmortem examination showed extensive fibrous replacement in the myocardium. The question arises whether such cases are true muscular dystrophies or whether the primary lesion is a proliferation of connective tissue.

Myosclerosis is an unclear term. Unequivocal and well-documented cases are rare. Isolated cases have been reported as 'dystrophia muscularis progressiva' or 'heredofamilial myosclerosis'. There may be several forms of progressive muscle fibrosis, some generalized (heredofamilial myosclerosis) or even localized (rigid spine syndrome), with or without cardiac conduction abnormalities. Classification of these cases is most difficult, and they are genetically different. Several types are present. The first group, corresponding to the X-linked cases reported by Emery and Dreifuss (1966), is most frequently seen (Rotthauwe et al. 1972; Thomas et al. 1972; Mawatari and Katayama 1973; Cammann et al. 1974; Waters et al. 1975; Hopkins and Karp 1976). An autosomal-recessive inheritance was suggested in 1 woman (Takamoto et al. 1984). An autosomal-dominant inheritance has been observed in 4 members of a family (Chakrabarti and Pearce 1981), in 3 members of another family (Fenichel et al. 1982), and in 2 members of a third family (Miller et al. 1985). There are no significant clinical differences between the X-linked and autosomal-dominant inherited cases.

This genetic heterogeneity underlines the fact that Emery-Dreifuss dystrophy is more likely a syndrome than a disease. In their review, Rowland et al. (1979) indicate that several cases similar to the Emery-Dreifuss dystrophy were different in essential features. For instance, some patients had limb weakness and cardiac conduction disorders but not atrial paralysis or early contractures. Some cases were classified as myopathy and others as spinal muscular atrophy. Some patients had early contractures but lacked cardiac conduction abnormalities. Some had an autosomal dominant inheritance, while others had an X-linked inheritance apparently without contractures and weakness.

As with other inherited neuromuscular disorders, recently developed technology which allows gene identification will soon permit us to correlate these varying phenotypes with specific genotypes. This will enormously enhance our understanding of these various clinical conditions. The defective gene causing Emery-Dreifuss dystrophy has recently been mapped to the distal portion of the long arm of the X chromosome (Consalez et al. 1991).

Quadriceps myopathy

The designation 'quadriceps myopathy' is controversial. In 1922, Bramwell reported 2 cases which were followed by others who described quite dissimilar cases. They were classified as muscular dystrophy (Walton 1956; Mumenthaler et al. 1958; Van Wijngaarden et al. 1968; Espir and Matthews 1973; Serratrice et al. 1983), localized types of polymyositis (Denny-Brown 1939; Turner and Heathfield 1961), or spinal amyotrophies (Boddie and Stewart-Wyne 1974; Furukawa et al. 1977; Serratrice et al. 1985). The main characteristics of these patients are predominance of men, adult onset, atrophy of the quadriceps muscles, and occasional extension to brachioradialis. Knee jerks are present or absent. Some of the above cases had a high creatine kinase, myogenic EMG, and muscle biopsy. In several cases, however, there were neuropathic or inflammatory changes.

The reported cases are not homogeneous, and the quadriceps myopathies would be better classified under the heading of 'quadriceps amyotrophic syndrome'. They might be divided into myopathic types with 2 subgroups: (1) true quadriceps myopathies (sporadic or family history) and (2) quad-

riceps myopathy 'plus' (for example, the case of Bramwell with atrophy of the brachioradialis muscles). The other cases of quadriceps amyotrophy indicate glycogenosis, inclusion body myositis, restricted polymyositis, or chronic spinal muscular atrophy.

REFERENCES

ANDRE, J.M., M. ANDRE, I. FLOQUET, P. TRIDON and G. ARNOULD: Les amyotrophies scapulo-peroneales. (A propos de deux cas de syndrome de Stark-Kaeser). Ann. Med. Nancy 2 (1972) 925–938.

BALDWIN, B.J., R.D. TALLEY, C. JOHNSON and D.O. NUTTER: Permanent paralysis of the atrium in a patient with facioscapulohumeral muscular dystrophy. Am. J. Cardiol. 31 (1973) 649–653

BELL, J.: On age of onset and age of death in hereditary dystrophy with some observation bearing on the question of antedating. Annals of antedating. Ann. Eugen. 11 (1942) 272–289.

BENSAID, J., J.M. GILGENKRANTZ, F. FERNANDEZ ET AL.: Paralysie suriculaire permanente familiale en rapport probable avec une maladie genetique du type Charcot-Marie. Arch. Mal Coeur 65 (1972) 935–952.

BETHLEM, J.: The scapulo-peroneal syndrome. In: G. Serratrice and H. Roux (Eds). Peroneal Atrophies and Related disorders. Paris, Masson (1979) 225–229.

BETHLEM, J., G.K. VAN WIJNGAARDEN and J. DE JONG: The incidence of lobulated fibers in the facioscapulohumeral type of muscular dystrophy and the limb-girdle syndrome. J. Neurol. Sci. 18 (1973) 351–358.

BLOOMFIELD, D.A. and B.C. SINCLAIR-SMITH: Persistent atrial standstill. Am. J. Med. 39 (1965) 335–340.

BODDIE, H.G. and E.G. STEWART-WYNNE: Quadriceps myopathy–entity or syndrome? Arch. Neurol. 31 (1974) 60–62.

CAMMANN, R., T. VEHRERCHILD and K. ERNST: Eine neue Sippe von x-chromosomaler benigner Muskeldystrophie mit Fruhkontrakturen (Emery-Dreifuss). Psychiatr. Neurol. Med. Psychol. (Leipzig) 26 (1974) 431–438.

CARROLL, J.E. and M.J. BROOKE: Infantile facioscapulohumeral dystrophy. In: G. Serratrice and H. Roux (Eds.), Peroneal Atrophies and Related Disorders. New York, Masson (1979).

CHAKRABARTI, A. and J.M.S. PEARCE: Scapuloperoneal syndrome with cardiomyopathy: report of a family with autosomal dominant inheritance and unusual features. J. Neurol. Neurosurg. Psychiatr. 44 (1981) 1146–1152.

CHUNG, C.S. and N.E. MORTON: Discrimination of genetic entities in muscular dystrophy. Am. J. Hum. Genet. 11 (1959) 339–359.

CONSALEZ, G.G., N.S. THOMAS, C.L. STAYTON ET AL.: Assignment of Emery-Dreifuss muscular dystrophy to the distal region of xq28: the results of a collaborative study. Am. J. Hum. Genet. 48 (1991) 468–480.

DAWIDENKOW, S.: Scapulo-peroneal amyotrophy. Arch. Neurol. Psychatr. 1 (1919) 694–701.

DAWIDENKOW, S.: Uber die scapulo-peroneale Amyotrophie (Die Familie 'Z'). Z. Gesamte Neurol. Psychiatr. 122 (1929) 628–650.

DAWIDENKOW, S.: Uber die Verehbung der dystrophia musculorum progressiva und uber Unterforemen. Arch. Rassenbiol. 18 (1930) 22.

DAWIDENKOW, S.: Scapuloperoneal amyotrophy. Arch. Neurol. 41 (1939) 694–701.

DELWAIDE, P.J. and J. SCHOENEN: Atrophie scapuloperoniere sporadique d'origine myogene. Rev. Neurol. (Paris) 132 (1976) 424–429.

DENNY-BROWN, D.: Myopathic weakness of quadriceps. Proc. R. Soc. Med. 32 (1939) 867–869.

DUBOWITZ, V.: Muscle Disorders in Childhood. Philadelphia, W.B. Saunders (1978).

DUCHENNE, G.B.: Recherches sur la paralysis musculaire pseudo-hypertrophique, ou paralysis myosclerosique. Arch. Gen. Med. 11 (1868) 5–25, 179–209, 305–321, 421–443, 552–588.

DYCK, P.J. and E.H. LAMBERT: Lower motor and primary sensory neuron diseases with peroneal muscular atrophy. Neurologic, genetic and electrophysiologic findings in hereditary polyneuropathies. Arch. Neurol. (Chicago) 18 (1968) 603–618.

EMERY, A.E.H. and E.E. DREIFUSS: Unusual type of benign X-linked muscular dystrophy. J. Neurol. Neurosurg. Psychiatr. 29 (1966) 338–342.

EMERY, E.S., G.M. FENICHEL and G. ENG: A spinal muscular atrophy with scapuloperoneal distribution. Arch. Neurol. (Chicago) 18 (1968) 129–133.

ERB, W.H.: Uber die juvenile Form der progressiven Muskelatrophie und ihre Beziehungen zur sogenannten Pseudohypertrophie der Musckeln. Dtsch. Arch. Klin. Med. 34 (1884) 467–519.

ESPIR, M.L.E. and W.B. MATTHEWS: Hereditary quadriceps myopathy. J. Neurol. Neurosurg. Psychiatr. 36 (1973) 1041–1045.

FEIGENBAUM, J.A. and T.L. MUNSAT: A neuromuscular syndrome of scapuloperoneal distribution. Bull. Los Angeles Neurol. Soc. 35 (1970) 47–57.

FENICHEL, G.M., E.S. EMERY and P. HUNT: Neurogenic atrophy simulating facioscapulohumeral dystrophy: a dominant form. Arch. Neurol. 17 (1967) 257–260.

FENICHEL, G.M., Y.C. SUE, A.W. KILROY and R. BLOVIN: An autosomal-dominant dystrophy with humeropelvic distribution and cardiomyopathy. Neurology 32 (1982) 1399–1401.

FOTOPOULOS, D. and H. SCHULTZ: Beitrag zur Pathogenese des scapulo-peronealen Syndroms. Psychiatr. Neurol. Med. Psychol. (Leipzig) 18 (1966) 129–136.

FURUKAWA, J. and Y. TOYOKURA: Chronic spinal muscular atrophy of facioscapulohumeral type. J. Med. Genet. 13 (1976) 285–289.

FURUKAWA, T., N. AKAGAMI and S. MARUYAMA: Chronic neurogenic quadriceps amyotrophy. Ann. Neurol. 2 (1977) 528–530.

FURUKAWA, T., H. TSUKAGOSHI, H. SUGITA and Y. TOYOKURA: Neurogenic muscular atrophy simulating facioscapulohumeral muscular dystrophy with particular reference to the heterogeneity of Kugelberg-Welander disease. J. Neurol. Sci. 9 (1969) 389–397.

GORI, Z., B. GANDINI and R. ROSSI: Un case di amiotrofia scapulo-peroneale. Studio al microscoio eiettronico. Acta Neurol. (Napoli) 26 (1971) 381–385.

HANSON, P.A. and L.P. ROWLAND: Mobius syndrome and facioscapulohumeral muscular dystrophy. Arch. Neurol. 24 (1971) 31–39.

HAUSMANOWA-PETRUSEWICZ, I. and S. ZIELINKSA: Zur nosologischen Stellung des scapulo-peronealen Syndrome. Dtsch. Z. Nervenheilkd. 183 (1962) 377–382.

HOPKINS, L.C. and H.R. KARP: X-linked humero-peroneal neuromuscular disease associated with atrial paralysis and sudden death. Trans. Am. Neurol. Assoc. 101 (1976) 84–86.

HUDGSON, P., W.G. BRADLEY and M. JENKINSON: Familial mitochondrial myopathy. Part I. Clinical, electrophysiological and pathological findings. J. Neurol. Sci. 16 (1972) 343–370.

HUGHES, B.P.: Creatine phosphokinase in facioscapulohumeral muscular dystrophy. Br. Med. J. 3 (1971) 464–465.

IWAMOTO, N., C.S. KOH, K. OGUCHI, N. YANAGISAWA and H. TSUKAGOSHI: Chronic spinal muscular atrophy of facioscapulohumeral type with congenital facial palsy in twins. Clin. Neurol. 19 (1979) 301–307.

KAESER, H.E.: Scapulo-peroneal muscular atrophy. Brain 88 (1965) 407–418.

KAESER, H.E., R. FEINSTEIN and W. TACKMANN: Unilateral scapulohumeral muscular atrophy. Eur. Neurol. 22 (1983) 70–77.

KAZAKOV, V.M., T.M. KOVALENKO, A.A. SKOROMETZ and E.P. MIKHAILOV: Chronic spinal muscular atrophy simulating facioscapulohumeral type and limbgirdly type of muscular dystrophy. Report of two cases. Eur. Neurol. 16 (1977) 90–98.

LANDOUZY, L. and J. DEJERINE: De la myopathie atrophique progressive (myopathie héréditaire debutant dans l'enface, par la face sans alteration du systeme nerveaux). C.R. Acad. Sci. (Paris) 98 (1884) 53.

LUNT, P.W. and P.S. HARPER: Genetic counselling in facioscapulohumeral muscular dystrophy. J. Med. Genet. 28 (1991) 655–664.

LUNT, P.W., D.A. COMPSTON and P.S. HARPER: Estimation of age dependent penetrance in facioscapulohumeral muscular dystrophy by minimizing ascertainment bias. J. Med. Genet. 26 (1989) 755–769.

MAWATARI, S. and K. KATAYAMA: Scapuloperoneal muscular atrophy with cardiopathy. Arch. Neurol. (Chicago) 28 (1973) 55–59.

MCKUSIK, V.A.: X-linked muscular dystrophy: benign

form with contractures. Birth Defects Orig. Artic. Sci. 7 (1971) 113–115.

MEADOWS, J.C. and C.D. MARSDEN: Scapuloperoneal amyotrophy. Arch. Neurol. (Chicago) 20 (1969) 9–12.

MERCELIS, R., J. DEMEESTER and J.J. MARTIN: Neurogenic scapuloperoneal syndrome in childhood. J. Neurol. Neurosurg. Psychiatr. 43 (1980) 888–896.

MILLER, R.G., R.B. LAYZER, M.A. MELLENTHIN, M. GOLABI, R.A. FRANCOZ and J.C. MALL: Emery-Dreifuss muscular dystrophy with autosomal dominant transmission. Neurology 35 (1985) 1230–1233.

MORTON, N.E. and C.S. CHUNG: Formal genetics of muscular dystrophy. Am. J. Hum. Genet. 11 (1959) 360–379.

MUMENTHALER, M., T. BOSCH, E. KATLENSTEIN and E. LEHNER: Uber den isolienten befall des M. quadriceps femoris bei der dystrophia Muscularum progressive. Confin. Neurol. 18 (1958) 416–441.

MUNSAT, T.L.: Infantile scapulo-peroneal muscular atrophy. Neurology (Minneapolis) 18 (1968) 285.

MUNSAT, T.L.: Facioscapulohumeral dystrophy and the scapuloperoneal syndrome. In: A.G. Engel and B.Q. Banker (Eds.), Myology, Vol. 1. New York, McGraw-Hill (1986) 1251–1266.

MUNSAT, T.L. and W.G. BRADLEY: Serum creatine phosphokinase levels and prednisone treated muscle weakness. Neurology 27 (1977) 96–97.

MUNSAT, T.L., D. PIPER, P. CANCILLA and J. MEDNICK: Inflammatory myopathy with facioscapulohumeral distribution. Neurology 22 (1972) 335–347.

MUNSAT, T.L., R. BALOH, C.M. PEARSON and W. FOWLER: Serum enzyme alterations in neuromuscular disorders. J. Am. Med. Assoc. 226 (1973) 1536–1543.

NEGRI, S., T. CARACENI and F. CORNELIO: A case of scapulo-tibio-peroneal syndrome. Electromyographic and histoenzymologic considerations. Eur. Neurol. 10 (1973) 31–40.

ORANSKY, W.: Uber einen heriditaren Typus progressiver Muskel-dystrophie. Dtsch. Z. Nervenheilkd. 99 (1927) 147–155.

PADBERG, G.: Facioscapulohumeral disease. Doctoral Thesis. University of Leiden, The Netherlands, Intercontinental Graphics (1982).

RICKER, K., H.G. MERTENS and K. SCHIMRIGK: The neurogenic scapulo-peroneal syndrome. Eur. Neurol. 1 (1968) 257–274.

ROTTHAUWE, H.W., W. MORTIER and H. BEYER: Neuer Type einen recessiv X-chromosomal vererbten Muskeldystrophie: Scapulo-humero-distale Muskeldystrophie mit fruhzeitigen Kontrakturen und Herzrhythmusstorungen. Humangenetik 16 (1972) 181–200.

ROWLAND, L.P., M. FETELL, M. OLARTE, A. HAYS, N. SINGH and F.E. WANAT: Emery-Dreifuss muscular dystrophy. Ann. Neurol. 5 (1979) 111–117.

ROWLAND, L.P., D.M. BLADE, M. HIRATO ET AL.: Clinical syndromes associated with ragged red fibers. Rev. Neurol. (Paris) 147 (1991) 467–473.

ROZHOLD, O., L. STEIDI and J. HROMADA: Histopatholo-gische und elektromyographische Befunde beim skapuloperonealen Syndrom. Acta Univ. Palacki. Olomuc. Fac. Med. 73 (1975) 291–299.

SCHUMANN, L.: Spinal muscular atrophy of the scapulo-peroneal type. Z. Kinderheilkd. 109 (1970) 118–123.

SCHWARTZ, M.S. and M. SWASH: Scapulo-peroneal atrophy with sensory involvement: Davidenkow's syndrome. J. Neurol. Neurosurg. Psychiatr. 38 (1975) 1063–1067.

SEITZ, D.: Zur nosologischen Stellung des sogenannten scapulo-peronealen Syndrome. Dtsch. Z. Nervenheilkd. 175 (1957) 547–552.

SERRATRICE, G. and J. POUGET: Maladie d'Emery-Dreifuss ou syndrome d'amyotrophie avec retractions precoces et troubles secondaires de la conduction cardiaque d'heredite variable. Rev. Neurol. (Paris), 1986 (in press).

SERRATRICE, G., H. ROUX, R. AQUARON, D. GAMBARELLI and J. BARET: Myopathies scapulo-peronieres. A propos de 14 observations dont 8 avec atteinte faciale. Sem. Hop. (Paris) 45 (1969) 2678–2683.

SERRATRICE, G., J.L. GASTAUT, J.F. PELLISSIER and J. POUGET: Amyotrophies scapulo-peronieres chroniques de type Stark-Kaeser (A propos de 10 observations). Rev. Neurol. (Paris) 132 (1976) 823–832.

SERRATRICE, G., J.F. PELLISSIER, J. POUGET, J.L. GASTAUT and D. CROS: Les syndromes scapulo-peroniers. Rev. Neurol. (Paris) 138 (1982a) 691–711.

SERRATRICE, G., J. POUGET, J.F. PELLISSIER, J.L. GASTAUT and D. CROS: Les atteintes musculaires à transmission autosomique récessive liées a l'X avec rétractions musculaires précoces et troubles de la conduction cardiaque. Rev. Neurol. (Paris) 138 (1982b) 713–724.

SERRATRICE, G., J.L. GASTAUT, J.F. PELLISSIER, D. CROS and J. POUGET: Myopathie quadricipitale ou syndrome amyotrophique quadricipital. Etude nosologique a propos de 10 observations. Rev. Neurol. (Paris) 139 (1983) 367–373.

SERRATRICE, G., A. POU-SERRADEL, J.F. PELLISSIER, H. ROUX, J. LAMARCO-CIVRA and J. POUGET: Chronic neurogenic quadriceps amyotrophy. J. Neurol. 232 (1985) 150–153.

SMALL, R.G.: Coats' disease and muscular dystrophy. Trans. Am. Acad. Opthalmol. Otolaryngol. 72 (1968) 225–231.

STARK, D.: Etude cliniquie et genetique d'une famille atteinte d'atrophie musculaire progressive neurale. J. Genet. Hum. 7 (1958) 1–32.

TAKAHASHI, K., H. NAKAMURA, K. TANAKA and T. MUTSUMURA: Neurogenic scapulo-peroneal amyotrophy associated with dystrophic changes. Clin. Neurol. (Tokyo) 2 (1971) 650–658.

TAKAMOTO, K., K. HIROSE, M. UNO and I. NONAKA: A genetic variant of Emery-Dreifuss disease. Muscular Dystrophy with humerpelvic distribution on early joint contracture and permanent paralysis. Arch. Neurol. 41 (1984) 1292–1293.

TAYLOR, D.A., J.E. CARROLL, M.E. SMITH, M.O. JOHNSON,

G.P. JOHNSON and M.H. BROOKE: Facioscapulohumeral dystrophy associated with hearing loss and Coats' syndrome. Ann. Neurol. 12 (1982) 395–398.

THOMAS, P.K., D.R. CAINE and C.E. ELLIOTT: X-linked scapuloperoneal syndrome. J. Neurol. Neurosurg. Psychiatr. 35 (1972) 208–215.

THOMAS, P.K., G.D. SCHOTT and J.A. MORGAN-HUGHES: Adult onset scapulo-peroneal myopathy. J. Neurol. Neurosurg. Psychiatr. 38 (1975) 1008–1015.

TOHGI, H., H. TSUKAGOSHI and Y. TOYOKURA: Neurogenic scapulo-peroneal syndrome with autosomal recessive inheritance (Japonais). Clin. Neurol. 2 (1971) 215–220.

TURNER, J.W.A. and K.W.C. HEATHFIELD: Quadriceps myopathy occurring in middle age. J. Neurol. Neurosurg. Psychiatr. 24 (1961) 18–21.

TYLER, F.H. and F.E. STEPHENS: Studies in disorders of muscles. II. Clinical manifestations and inheritance of facioscapulohumeral dystrophy in a large family. Ann. Intern. Med. 32 (1950) 640–660.

VAN WIJNGAARDEN, G.K. and J. BETHLEM: The facioscapulohumeral syndrome. Internal Congress Series #295. In: B.A. Kakulas (Ed.), Clinical Studies in Myology. Amsterdam, Excerpta Medica (1973) 498–501.

VAN WIJNGAARDEN, G.K., C.J. HAGEN, J. BETHLEM and A.G.F. MEIJER: Myopathy of the quadriceps muscles. J. Neurol. Sci. 7 (1968) 201–206.

VULPIAN, A.: Maladies du Systeme Nerveux. Paris, Doin Ed. (1886).

WALTON, J.N.: Two cases of myopathy limited to the quadriceps. J. Neurol. Neurosurg. Psychiatr. 19 (1956) 106–108.

WALTON, J.N.: Disorders of Voluntary Muscle. Edinburgh, Churchill Livingstone, 4th Ed. (1981) 502–505.

WALTON, J.N. and D. GARDNER-MEDWIN: Progressive muscular dystrophy and the myotonic disorders. In: J.N. Walton (Ed.), Disorders of Voluntary Muscle. Edinburgh, Churchill Livingstone, 4th Ed. (1981).

WALTON, J.N. and F.J. NATTRASS: On classification, natural history and treatment of the myopathies. Brain 77 (1954) 169–231.

WATERS, D.D., D.D. NUTTER, L.C. HOPKINS and E.R. DORNEY: Cardiac features of an unusual X-linked humeroperoneal neuromuscular disease. N. Engl. J. Med. 293 (1975) 1017–1022.

WIJMENGA, C., R.R. FRANTS, O.F. BROUWER, P. MOERER, J.L. WEBER and G.W. PADBERG: Location of facioscapulohumeral muscular dystrophy gene on chromosome 4. Lancet 336 (1990) 651–653.

WIJMENGA, C., G.W. PADBERG, P. MOERER ET AL.: Mapping of facioscapulohumeral muscular dystrophy gene to chromosome 4q35-qter but multipoint linkage analysis and in situ hybridization. Genomics 9 (1991) 570–575.

WULFF, J.D., J.T. LIN and J.J. KEPES: Inflammatory facioscapulohumeral muscular dystrophy and Coats' syndrome. Ann. Neurol. 12 (1982) 398–401.

ZELLWEGER, H. and W.F. MCCORMICK: Scapuloperoneal dystrophy and scapuloperoneal atrophy. Helv. Paediat. Acta 23 (1968) 643–649.

ZUNDEL, W.S. and F.H. TYLER: The muscular dystrophies. Parts I and II. N.Engl. J. Med. 273 (1965) 537–543, 596–601.

Handbook of Clinical Neurology, Vol. 18 (62): Myopathies
L.P. Rowland and S. DiMauro, editors

CHAPTER 7

The limb girdle syndromes

FELIX JERUSALEM and JÖRN P. SIEB

Neurologische Klinik der Universität, Bonn, Germany

Knowledge of genetics and pathophysiology of muscle diseases is increasing rapidly, particularly because of the application of modern molecular biology. Nosological classifications have to be revised; others are being defined as terms of molecular biology. Any classification that is not based on etiology in each disease can be regarded only as an interim arrangement.

Of all the muscular dystrophies, limb girdle muscular dystrophy has always been the most difficult to establish as an entity. Many diseases cause a similar picture of slowly proximal muscle weakness. Therefore, this Chapter is entitled 'The limb girdle syndromes'. Several investigators have asked whether limb girdle muscular dystrophy exists at all (Sica and McComas 1971; Munsat 1977; Gardner-Medwin 1980; Walton and Gardner-Medwin 1988).

At a time when knowledge is increasing so rapidly, it is difficult to review this inadequately defined group of diseases. The main objectives of our review are to define and describe limb girdle muscular dystrophy itself, based on the earlier chapter by Bradley (1979) in Volume 40 of the original series of this handbook.

Cases described as 'limb girdle muscular dystrophy' have varied from author to author. Walton and Nattrass (1954) popularized the term, applying it to forms of progressive muscular dystrophy that were more benign than Duchenne muscular dystrophy, did not involve facial muscles, and began with either upper or lower limb weakness. They included syndromes earlier designated as juvenile scapulohumeral type of muscular dystrophy ascribed to Erb (1884), the atrophic pelvifemoral type of Leyden (1876) and Möbius (1879), and the late adult type of Nevin (1936). Autosomal-recessive inheritance separated it from facioscapulohumeral muscular dystrophy, Duchenne muscular dystrophy, and myotonic dystrophy.

Thus limb girdle muscular dystrophy might be defined as a progressive primary degeneration of skeletal muscle, relatively benign in course, sparing facial muscles until advanced stages, and transmitted as an autosomal-recessive trait in 59% of cases and being sporadic in 41% (Morton and Chung 1959). However, as new diagnostic methods have implied, the concept has been eroded; the term 'limb girdle syndromes' is therefore more suitable (Shields 1986).

HISTORY OF CLASSIFICATION (TABLE 1)

Earlier descriptions of patients with progressive muscle wasting included examples of a relatively benign course (see reviews by Bell 1943; Levison 1951; Becker 1953; Stevenson 1953, 1955; Walton and Nattrass 1954). Meryon (1852) described 2 groups of siblings of 4 and 2 brothers with the pseudohypertrophic muscular dystrophy later denoted by Duchenne. However, he also mentioned 3 brothers with leg weakness that began about the age of 12; they were disabled by about 25 years. This could have been either limb girdle muscular dystrophy or benign X-linked muscular dystrophy, characterized by Becker and Kiener in 1955.

Duchenne (1868) gave a full description of the childhood dystrophy that now bears his name. He also described both adult and infantile forms of 'l'atrophie musculaire graisseuse progressive' or progressive muscular fatty atrophy. The acute form of this condition in childhood was probably acute anterior poliomyelitis, because it usually began with fever, and autopsy studies indicated the presence of spinal cord disease. The progressive infantile form was probably facioscapulohumeral muscular dystrophy, later more fully described by Landouzy and Dejerine (1885, 1886). However, the adult cases of 'l'atrophie musculaire graisseuse progressive' mentioned by Duchenne probably included limb girdle syndromes.

Leyden (1876) described a familial form of atrophy of the pelvic girdle musculature, and Möbius (1879) grouped this form of pelvic girdle atrophy with Duchenne dystrophy.

In 1884 Erb described a juvenile scapulohumeral form of progressive muscular atrophy, with the onset from early childhood to adult life. In an 1891 review, Erb distinguished this form, as well as facioscapulohumeral dystrophy of Landouzy and Dejerine and Duchenne dystrophy as primary diseases of muscle without observed central or peripheral neural abnormalities, based largely on clinicopathological correlation. He applied the title of 'dystrophia muscularis progressiva' to the overall group. Autopsy studies demonstrated that conditions now termed acute anterior poliomyelitis or progressive muscular atrophy were due to anterior horn cell damage, and that some diseases with progressive muscular wasting had normal spinal cord and peripheral nerves at autopsy (Eulenberg and Cohnheim 1866; Landouzy and Dejerine 1885).

From the beginning of the 20th century, an increasing number of upper or lower limb girdle cases were described, with attempts to classify the diverse syndromes. Later, Levison (1951) reviewed 123 personal cases from Denmark; 30 had been investigated by muscle biopsy and 40 by electro-

TABLE 1

History of the description and classification of limb girdle dystrophies.

Duchenne (1868)
  Childhood form of pseudohypertrophic muscular dystrophy
Leyden (1876) Möbius (1879)
  Atrophic pelvifemoral type of muscular dystrophy
Erb (1884)
  Juvenile scapulohumeral type of muscular dystrophy
Landouzy and Dejerine (1885, 1886)
  Facioscapulohumeral muscular dystrophy
Bramwell (1922)
  Quadriceps myopathy
Nevin (1936)
  Late adult type of muscular dystrophy
Walton and Nattrass (1954)
  Limb girdle muscular dystrophy: definition as nosological entity
Becker and Kiener (1955)
  A new X-chromosomal muscular dystrophy
Emery and Dreifuss (1966)
  Benign X-linked muscular dystrophy with early-onset of contractures
Bethlem and Van Wijngaarden (1976)
  Benign myopathy with autosomal-dominant inheritance
Hoffman et al. (1987)
  Identification of Duchenne/Becker gene product as dystrophin

myography (EMG). He grouped the cases into juvenile scapulohumeral form of Erb, facioscapulohumeral muscular dystrophy, and the lower extremity type which included the forms that would now be called Duchenne dystrophy, Becker dystrophy, and limb girdle muscular dystrophy.

Stevenson (1953, 1955) investigated 60 families with muscular dystrophy in Northern Ireland. He divided the nonmyotonic dystrophies into a Duchenne type with X-linked inheritance and a non Duchenne type of autosomal-recessive inheritance. He suggested the term 'autosomal limb girdle dystrophy' for the latter group of patients with upper or lower limb girdle weakness, whether the face was involved or not. He believed that these cases were all due to the same gene mutation.

Walton and Nattrass (1954) investigated 105 cases of muscular dystrophy from the northeast of England, propounding the most widely accepted classification of the muscular dystrophies. They separated facioscapulohumeral muscular dystrophy and grouped the pelvic atrophic form of Leyden and Möbius, the juvenile scapulohumeral form of Erb, and late-onset cases of Nevin (1936) as limb girdle muscular dystrophy. They stated that 'their observations have not indicated any significant differences in natural history and mode of inheritance in these cases which could be related to the pattern of muscle affection and the age of onset'.

In the view of Walton and Nattrass (1954), limb girdle muscular dystrophy was characterized by the following features:
1. expression in either males or females;
2. onset usually late in the first decade but also in the second or third decades and even in middle age;
3. transmission usually as an autosomal-recessive but occasionally as a dominant trait;
4. primary involvement of either the shoulder girdle or pelvic girdle muscles, frequently asymmetric when upper limbs were first involved;
5. spread from the upper to the lower limbs or vice versa;
6. uncommon occurrence of abortive cases or muscle pseudohypertrophy;
7. variable rate of progression;
8. severe disability leading to inability to walk within 20–30 years of onset; and
9. muscular contractures uncommon, except in the late stages.

It is important to recognize that the status of limb girdle muscular dystrophy as a nosological entity has been questioned (Munsat 1977; Gardner-Medwin 1980). Referring to most of the above-quoted reports and to our own experience, we believe it is useful to retain the term limb girdle muscular dystrophy, although there are no absolute diagnostic criteria (Fowler and Nayak 1983). Recent findings point to a high proportion of dystrophinopathy cases within the limb-girdle syndromes, and to the heterogeneity of the latter (Arikawa et al. 1991)

Although Walton and Gardner-Medwin (1988) omitted the term from their most recent classification, they included 4 subtypes of limb girdle muscular dystrophy: scapulohumeral muscular dystrophy, myopathy confined to the quadriceps, childhood muscular dystrophy with autosomal-recessive inheritance, and dominant late-onset proximal muscular dystrophy.

Panegyres et al. (1990a) suggested a classification that is phenomenological and does not take the modus of inheritance into account. They differentiate the following:
1. limb girdle muscular dystrophy (autosomal-dominant or -recessive)
(a) proximal limb and limb girdle,
(b) with facial involvement,
(c) with distal limb involvement;
2. limb girdle syndromes with myopathic and neurogenic features; and
3. limb girdle syndromes secondary to other causes.

We shall adopt the classification of Shields (1986) (Table 2), adding the autosomal-dominant early-onset type with contractures (Bethlem myopathy; Bethlem and Van Wijngaarden 1976).

GENETIC ASPECTS

The evidence from several investigations on the frequency of affected siblings and offspring of consanguineous marriages indicates that limb girdle muscular dystrophy is generally inherited as an autosomal-recessive trait (Stevenson 1953; Walton and Nattrass 1954). There have been a few cases of autosomal-dominant inheritance (see below).

The observation that many of the reported familial and sporadic cases have occurred in inbred communities is additional evidence of autosomal-recessive inheritance. Large kinships of limb girdle muscular dystrophy have been reported in Amish communities of Swiss and French extraction in the midwest of America (Jackson and Carey 1961;

TABLE 2

Classification of limb girdle syndromes.

A. Autosomal-recessive muscular dystrophy of childhood
    Pelvifemoral type (Leyden-Möbius)
    Scapulohumeral type (Erb)
B. Autosomal-dominant early-onset type with contractures (Bethlem myopathy)
    Autosomal-dominant late-onset type (Nevin)
    Quadriceps myopathy

C. Limb girdle syndromes secondary to other causes (see Table 3)

Jackson and Strehler 1968). An increased number of cases with autosomal-recessive limb girdle muscular dystrophy were observed in Switzerland (Moser et al. 1966), Tunisia (Ben Hamida and Fardeau 1980), and Libya (Radhakrishnan et al. 1987). In these areas the high frequency of autosomal-recessive limb girdle muscular dystrophy is a direct result of large family sizes (in the Arabic countries) and a high rate of consanguineous marriages.

The incidence and prevalence of limb girdle muscular dystrophy has been estimated by Morton and Chung (1959). There are always some reservations in accepting these figures, because diagnostic criteria were too imprecise for a reliable epidemiological survey; conversely, incomplete registration would lead to incorrectly low estimates. These authors applied the genetic method of segregation analysis to separate autosomal-recessive cases of limb girdle muscular dystrophy from sporadic cases. A total of 59% were autosomal-recessive, while 41% were sporadic. In the former the prevalence was approximately 12 per 10^6 living individuals, and the incidence about 38 per 10^6 births. The combined figures indicate a prevalence of about 20 per 10^6 living individuals and an incidence of about 65 per 10^6 births. The frequency of heterozygous carriers of the gene was calculated at approximately 16 per 10^3 persons.

A more reliable study conducted in Denmark indicated a higher frequency (Leth et al. 1985). The prevalence was 36.5 per 10^6 inhabitants. The incidence, calculated from the prevalence and the mean age at onset and at death, was 108 per 10^6 newborns. There was evidence of excess mortality. Complete registration of adult-onset limb girdle muscular dystrophy in the Lothian Region of Scotland (Edinburgh and environs: 750,000 inhabitants) was attempted by Yates and Emery (1985). In 10 families they found 11 affected subjects, significantly less than expected for autosomal-recessive inheritance.

CLINICAL FEATURES

The disease is expressed in either sex, usually in the second or third decade, with a progressive weakness affecting the limb girdle muscles in a specific distribution. In about half of the cases, weakness begins in the pelvic girdle musculature (the Leyden-Möbius type), and in the other half, it begins in the scapulohumeral musculature (the Erb type). Gradually, the weakness spreads and comes to involve the rest of the limb girdle musculature. The average interval before symptoms appear in the latter is about 5–10 years, although it can frequently be longer.

The pattern of muscle involvement seems to be similar in both forms, but the severity of individual muscle damage differs in the 2 types. The serratus anterior, pectoralis major (especially sternal head), trapezius, supraspinatus, infraspinatus, latissimus dorsi, biceps, brachioradialis, triceps, sacrospinalis, quadriceps femoris, hamstring muscles, and the hip flexors, extensors, abductors and adductors, are especially involved. Eventually, all muscles in the body are affected, but it is the tendency for certain muscles to become weak and wasted early in the disease, with relative sparing of others, which produces the characteristic appearance of these patients. They usually show a drooped shoulder girdle, although uptilting of the inner angle of the scapula may produce a false appearance of square muscular shoulders. They develop winging of the scapula, lumbar hyperlordosis, and a waddling gait. The characteristic anterior axillary fold appears with wasting of the sternal head of the pectoralis major (see Fig. 1). The biceps and brachioradialis muscles are affected before and more severely than the triceps. The deltoid muscle is relatively spared, but it is a difficult muscle to test because the subject cannot fix the scapula.

Pseudohypertrophy is seen in about 20% of the cases, and is more prominent in males than in females in the same family. Tendon reflexes are preserved when the muscle weakness first appears, but are lost as the weakness becomes more severe. The ankle jerk is often preserved when all other reflexes are lost.

Facial involvement occurs either not at all or only when the patient is severely disabled from limb weakness. In families with undoubted facioscapulohumeral muscular dystrophy, occasional patients have no facial weakness but only scapular or scapuloperoneal weakness.

The progressive muscle weakness eventually causes loss of the ability to walk, occurring earlier when the disease begins in the pelvic girdle. The

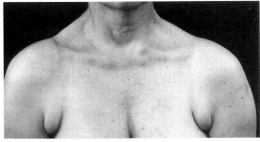

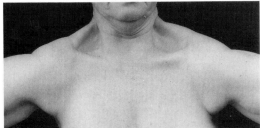

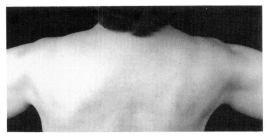

Fig. 1. 45-year-old woman with limb girdle muscular dystrophy. Note the characteristic anterior axillary fold due to wasting of the sternal head of the pectoralis major. The other muscles of the shoulder girdle are also visibly involved.

interval between onset of symptoms and inability to walk varies from 10 to 20 years. The disease may ultimately produce severe disability with muscle contractures. Some patients have respiratory muscle impairment, the muscles of forced expiration being involved earlier than are those of inspiration (see below). Pneumonia and respiratory failure may shorten the lifespan, with death in the forties and fifties. Some patients, however, remain mobile for the rest of their life.

Autosomal-recessive muscular dystrophy of childhood

For examples of autosomal-recessive muscular dystrophy of childhood, we refer to 37 patients in 14 families reported by Jackson and Strehler (1968) and 49 patients described by Ben Hamida and Fardeau (1980). Onset occurs in childhood or adolescence. Males and females are affected equally. Weakness affects proximal muscles of the legs before and more severely than those of the shoulder. Muscle hypertrophy may be present. Facial muscles are spared and electrocardiographic abnormalities are absent. Serum creatine kinase is markedly elevated. Muscle biopsy reveals a myopathic pattern, similar to but less marked than seen in Duchenne dystrophy. The progression of autosomal-recessive muscular dystrophy of childhood varies; some children cannot walk before they reach puberty, while others are only slightly handicapped and walk well into the fifth decade. An unusual variety of muscular dystrophy in 2 sisters with earlier onset and a more progressive course than in the autosomal-recessive limb girdle form was described by Wharton (1965).

Pelvifemoral type (Leyden-Möbius)

The pelvifemoral type (Leyden-Möbius) resembles the autosomal-recessive muscular dystrophy of childhood in all respects except for later onset and a more benign course. Age at onset varies from the second to the sixth decade. Many cases are sporadic. The course is slowly progressive, with some patients showing periods of apparent clinical arrest. In some siblings age at onset varies between childhood and early adulthood (Moser et al. 1966; Jackson and Strehler 1968); thus, an overlap between the pelvifemoral type (Leyden-Möbius) and the childhood variety can occur.

Scapulohumeral type (Erb)

The scapulohumeral type (Erb) is characterized by the distribution of muscle weakness and wasting, which may be asymmetrical in the early stages of the disease. The disorder is usually inherited as an autosomal-recessive trait, but a large number of sporadic cases have been observed. Age at onset varies from childhood to the fourth decade. The course is slowly progressive. Muscle hypertrophy and facial weakness are uncommon, and when the latter occurs it is always mild.

Autosomal-dominant early-onset type with contractures (Bethlem myopathy)

This condition was described by Bethlem and Van Wijngaarden (1976) in 3 unrelated Dutch families, to which they applied the term 'benign myopathy with autosomal-dominant inheritance'. Some authors now apply the term 'Bethlem myopathy' to this condition. The salient clinical features were: (a) onset of limb girdle weakness in infancy or childhood; (b) slow progression with periods of arrest for several decades; (c) contractures of fingers, elbows, and ankles; (d) autosomal-dominant inheritance; (e) mild weakness with preserved ability to work until old age; (f) normal life expectancy, and (g) absence of cardiac involvement. Serum creatine kinase was either normal or slightly raised, and electromyography suggested myopathy. Muscle biopsy revealed nonspecific features of a myopathy without fiber necrosis or regeneration.

A similar disorder was described in a family of Polish descent by Arts et al. (1978), in a large French-Canadian kinship by Mohire et al. (1988), in 2 French families by Serratrice and Pellisier (1988), in 2 members of a Japanese family by Tachi et al. (1989), and in 2 Italian families by Ballestrazzi et al. (1990).

Autosomal-dominant late-onset type

There are few reports of families with the late-onset dominant type of muscular dystrophy. In one pedigree, onset of slowly progressive proximal weakness was in the third decade (Schneidermann et al. 1969). Wasting and weakness affected the limb girdles, being more pronounced in the legs than in the arms. Facial muscles were spared. Muscle biopsy revealed a myopathic pattern. There was a suggested link to the Pelger-Huët anomaly (bilobal polymorphonuclear leukocytes).

Chutkow et al. (1986) described 1 kinship with a similar autosomal-dominant myopathy limited to the limb girdle muscles, beginning insidiously any time from the late second decade through to the sixth decade and being slowly progressive. Pelvifemoral preceded scapulohumeral weakness. Other reports have appeared from Gilchrist et al. (1988) and from Somer et al. (1989). In another family the distribution of muscle involvement varied. Some members showed weakness of the shoulder girdle and arms, while others demonstrated an equal involvement of the arms and legs or only of the legs. Age at onset was the second and third decade and the inheritance pattern was clearly dominant (Bacon and Smith 1971). In the 1 patient adequately described, a specific distribution of proximal muscle weakness somewhat similar to that seen in limb girdle muscular dystrophy was present, but the deltoid muscles were wasted, the triceps was more involved than the biceps, and there was anterior tibial muscle weakness. The muscle showed various myopathic changes on biopsy. No modern electromyography or histological techniques were used.

A large pedigree in which 9 males were affected over 4 generations with male-to-male transmission was reported by De Coster et al. (1974). The onset was in the fifth decade. Weakness and wasting were confined to the pelvic girdle. A mutation of the dystrophin gene cannot be excluded in this family, as dystrophin had not at that time been discovered.

Henson et al. (1967) described a family with a clinical picture of the limb girdle type restricted to females, 8 being affected in 2 generations of the family. The disease started in the legs and ascended. The muscles were affected in a relatively specific distribution, and included the medial head of the gastrocnemius and the deltoid muscles. Serum creatine kinase activity was up to 10 times its normal level. Muscle biopsy showed marked myopathic features with necrosis, phagocytosis and regeneration. Electromyographic studies showed a reduced interference pattern with myopathic potentials. Heyck and Laudahn (1969) as well as Hertrich (1957) similarly reported cases of families in which only females suffered from limb girdle dystrophy.

Quadriceps myopathy

This disorder is characterized by chronic, slowly progressive weakness and wasting confined to the quadriceps muscles. It was first described in 2 isolated cases by Bramwell (1922). His second case had congenital absence of the brachioradialis muscle, which may well indicate spread of the disease.

Further cases have been reported by Denny-

Brown (1939), Walton (1956), Mumenthaler et al. (1958), Turner and Heathfield (1961), Boddie and Stewart-Wynne (1974), Bradley et al. (1978), and Swash and Heathfield (1983). One sibship included 2 affected brothers (Van Wijngaarden et al. 1968). Autosomal-dominant pedigrees are also known. In 1 family, a man, his brother, and 3 of his daughters suffered from quadriceps myopathy with onset in adult life (Espir and Matthews 1973).

In spite of these observations in familial cases, it is not certain whether quadriceps myopathy represents an independent (genetic) entity. One has to consider that some patients with similar clinical features may have suffered from a neuropathic or myositic disease (Turner and Heathfield 1961; Thage 1965; Boddie and Stewart-Wynne 1974; Furukawa et al. 1977; Serratrice et al. 1985). A forme fruste of Becker muscular dystrophy can also be present (Sunohara et al. 1990; Wada et al. 1990).

PULMONARY AND CARDIAC INVOLVEMENT

Inspiratory muscle weakness limits the expansion of the chest wall and can be considered a special form of restrictive lung disease. Chronic alveolar hypoventilation can arise, particularly when the diaphragm is involved. Symptoms are morning headaches, hypersomnia, and impaired mental function.

Pulmonary function should be tested in all patients with muscular dystrophies. Restrictive lung disease can be detected when the patient is awake. However, it is especially important to record respiratory parameters continuously during nocturnal sleep, for instance the blood oxygen saturation. A respiratory impairment is sometimes only detectable during nocturnal sleep (Newsom-Davis 1980; Rideau et al. 1981; George 1989).

Systematic examinations have however not yet been carried out in larger patient groups. Skatrud et al. (1980) studied a single patient with limb girdle muscular dystrophy, bilateral diaphragmatic paralysis, chronic hypercapnic respiratory failure, and hypersomnolence. During NREM (non rapid eye movement) sleep, breathing rate was decreased and the thoracic and abdominal excursions were reduced, although oxygen saturation remained

high. However, during REM (rapid eye movement) sleep dramatic and repetitive oxygen desaturations occurred followed by arousals. Respiratory insufficiency is seldom so marked that a nocturnal assisted ventilation is indicated (Heckmatt et al. 1990).

In addition, evidence of cardiomyopathy and conduction disturbance can be seen in patients with limb girdle muscular dystrophy, though cardiac involvement is much less common than it is in Duchenne dystrophy. The characteristic electrocardiographic alterations of Duchenne dystrophy, i.e. tall R waves in the right precordial and deep Q waves in the left precordial and limb leads, are not encountered in limb girdle muscular dystrophy (Jackson and Carey 1961; Kovick et al. 1975; Hunter 1980).

Clinically apparent cardiac manifestations have been described only rarely. Syncope may result from underlying arrhythmia (Lambert and Fairfax 1976; Antonio et al. 1978; Hoshio et al. 1987). Features of congestive heart failure are less commonly reported (Welsh et al. 1963; Perloff et al. 1966; Kawashima et al. 1990).

SERUM LEVELS OF MUSCLE ENZYMES

Assays of serum creatine kinase (CK) and pyruvate kinase (PK) are widely used as the most sensitive biochemical tests for the detection of muscular dystrophy (Harano et al. 1973; Zatz et al. 1978; Gardner-Medwin 1980; Pennington 1980). Usually these enzymes are mildly to moderately elevated, but in some cases CK can be increased as much as in Duchenne dystrophy. CK levels show a tendency to decrease with the duration of muscular dystrophy. In addition to that of CK and PK, the determination in serum of aldolase, lactic dehydrogenase, and myoglobin is used but is less sensitive (Kagen et al. 1980). Familially elevated serum CK in the absence of symptoms can be the phenotypic expression of the heterozygous state of autosomal-recessive limb girdle muscular dystrophy, as described in 2 families by Sabatelli et al. (1986). Creatinuria, hypocreatinuria and decreased urinary excretions of methylhistidine may reflect the decrease of muscle mass (Röthig et al. 1984).

ELECTROMYOGRAPHY (EMG)

EMG findings in proximal muscles are generally described as being myopathic (Hayward 1980), but these alterations are not specific. It is noteworthy that there is a discordance between EMG and biopsy results in a certain percentage of neuromuscular disorders. Therefore, the diagnostic classification of limb girdle syndromes cannot be based purely on electrophysiological criteria. Twelve to 21% of cases with chronic nuclear atrophies have small amplitude and short duration potentials in the EMG, while biopsy reveals a neurogenic pattern (Mastaglia and Walton 1971; Fewings et al. 1977; Brenni et al. 1981).

Increased mechanical resistance to insertion of the needle electrode due to muscle fibrosis may occur late in the disease. Some cases show decreased and some increased insertional activity, and also fibrillation potentials, positive sharp waves, and motor units with late components (Desmedt and Borenstein 1975; Lang and Partanen 1976).

Single-fiber electromyography (SFEMG) reveals lower fiber density, jitter, and a percentage of abnormal pairing consistent with myopathy (Shields 1984). The opposite findings, i.e. higher values of fiber density and jitter in a small percentage of muscle fibers, were observed by Stalberg (1974). Conduction velocities and potential amplitudes of both motor and sensory nerve fibers, as well as motor nerve conduction velocities along the proximal segment (F wave) and terminal latencies, are normal in the limb girdle dystrophies (Panayiotopoulos and Scarpalezos 1977; Hayward 1980). Electrophysiological estimates of the number of motor units are reported to be normal in limb girdle muscular dystrophy (Ballantyne and Hansen 1974; Panayiotoupoulos and Scarpalezos 1975).

The contractile properties of ankle dorsiflexor and plantarflexor muscles in 20 patients with limb girdle muscular dystrophy have been compared with those in matched controls (Belanger and McComas 1985). Twitch and voluntary torques were significantly smaller in the patients and in 9 patients it was impossible to record a twitch from the tibialis anterior. The disease seemed more rapid in the tibialis anterior than in plantar flexor muscles. Surviving fibers in dorsiflexor and plantar flexor muscles did not reveal evidence of excitation-contraction uncoupling and they exhibited normal post-activation potentiation and fatigue properties. Some patients were initially incapable of exciting motor units maximally during voluntary contractions. A finding of possible pathogenetic significance was that 1 patient, with prominent calves, developed exceptionally large voluntary torque in his plantarflexor muscles (Belanger and McComas 1985), implying that muscle enlargement may sometimes reflect an increase in excitable muscle mass rather than hyperplasia of connective tissue and fat.

MUSCLE BIOPSY

Ultrasound imaging (Heckmatt and Dubowitz 1985), computed tomography (Jiddane et al. 1983), and magnetic resonance imaging (Murphy et al. 1986) may be helpful in selected cases in order to detect focal involvement in limb girdle syndromes and they could be helpful in choosing a biopsy site.

The changes found in muscle biopsy detectable with conventional histological methods depend on the stage of the disease. They are non specific and do not enable one to differentiate the limb girdle forms from other types of muscular dystrophy and myopathies (Fareau 1970; Escourolle and Fardeau 1973; Schmalbruch 1982; Dubowitz et al. 1985). Mild to moderate changes include pathological caliber variation of cross-sectional muscle fibers, a few degenerative changes of muscle fibers, to some extent hyaline fibers, splitting of fibers, ring fibers, lobulated fibers, whorled fibers, moth-eaten fibers, an increased number of internal nuclei, and a slight increase of fat and connective tissue (Fig. 2). Small basophilic muscle fibers with large vesicular nuclei indicating regeneration may be present. Not all of these changes are regularly detectable in every case of limb girdle dystrophy. Advanced stages show a marked variation in the cross-sectional diameter with hyper- and atrophic muscle fibers and a wide spectrum of muscle fiber degeneration, while all the other morphological features mentioned may become more prominent. A few biopsies demonstrate small mononuclear cell infiltrates in the vicinity of degenerating muscle fibers. The fiber type distribution and the size of fibers of differing histochemical types may be changed but vary between different biopsies and do not have a diagnostic significance. Type 1 fiber predominance and a relative deficiency of type 2B fibers were observed (Dubowitz et al. 1985). In several affected muscles, the pronounced loss of muscle parenchyma is replaced by broad strands and clusters of connective tissue and fat.

There is normal expression of dystrophin in limb girdle dystrophy. Proof of the normal presence of dystrophin, immunocytochemically or by means of immunoblot analysis, allows the segregation of XP21.2-linked muscular dystrophies, as is also possible at the DNA level (Patel et al. 1988; Norman et al. 1989).

Degradation of connectin, a myofibrillar elastic filament also called titin, in Duchenne dystrophy has been shown immunochemically. Connectin is only minimally degraded in limb girdle dystrophy, even when the biopsied muscles exhibit a similar degree of weakness as those in Duchenne dystrophy (Matsumara et al. 1989).

DIFFERENTIAL DIAGNOSIS

The more common diseases that have to be ruled out before the diagnosis of limb girdle muscular dystrophy can be posed are listed in Table 2. It is impossible to discuss each of these here, and most are fully described elsewhere in this Volume. However, an outline of the salient features which enable us to differentiate these diseases from limb girdle muscular dystrophy are given (Table 3).

Polymyositis heads the list of diseases that need to be considered, because it is treatable (DeVere and Bradley 1975; Plotz et al. 1989) and is still misdiagnosed as muscular dystrophy with regrettable frequency. Often, muscle pains, dysphagia, neck weakness, Raynaud's phenomenon, and skin rash are clinical pointers to the diagnosis of polymyositis or dermatomyositis. Generally, there is a rapid evolution of proximal, or occasionally diffuse, muscle weakness which differs from the slowly progressive weakness and selective distribution of limb girdle muscular dystrophy. The raised sedimentation rate, characteristic muscle biopsy, and EMG findings help to confirm the diagnosis, but sometimes all of these findings may be normal. In atypical cases corticosteroid therapy over a 3-

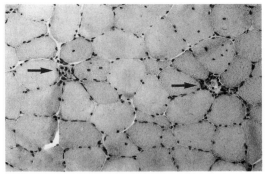

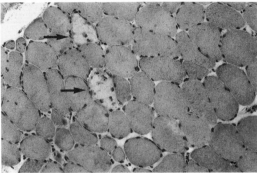

Fig. 2. Muscle biopsy of quadriceps from a 45-year-old woman with limb girdle dystrophy. The hematoxylin-eosin-stained muscle section shows a pathological variation of caliber, an increased degree of internal nuclei, and occasional fiber degeneration with phagocytosis (arrow).

month period may help to differentiate polymyositis from muscular dystrophy.

Inclusion body myositis may appear clinically as a chronic limb girdle syndrome. Riggs et al. (1989) described a 9-year-old boy with proximal muscle weakness that was diagnosed as limb girdle dystrophy. After 30 years, inclusion body myositis was found.

Infectious agents can also cause slowly progressive proximal limb weakness. Parasitic organisms such as *Trichinella*, *Schistosoma*, *Cysticercus*, *Sarcocystis*, and *Trypanosoma* are more important in underdeveloped countries.

Toxic causes include many drugs, particularly the fluorinated corticosteroids, vincristine, chloroquine, and cholesterol synthesis-inhibiting agents. In acquired immune deficiency syndrome, muscular weakness can also occur as a result of zidovudine therapy (Panegyres et al. 1990b). Chronic alcoholics commonly have diffuse muscle wasting, sometimes more proximally marked. This is usually found to be due to an alcoholic neuropathy, although an acute or chronic alcoholic myopathy may exist (Ekbom et al. 1964). Most toxic neuropathies usually cause a non selective distal muscle weakness.

A number of other *acquired disorders* of skeletal muscle should be recognized, as it is possible to treat them. The endocrine myopathies include those due to hyperthyroidism and hypothyroidism, Cushing's syndrome, hyperparathyroidism, and hyperaldosteronism. They generally cause a nonselective proximal weakness, and diagnosis depends upon the appropriate investigation.

Vitamin deficiency is an uncommon cause of proximal muscle weakness. Vitamin E deficiency in animals is associated with severe skeletal muscle degeneration, but this has never been described with certainty in man. Vitamin D deficiency, resulting from various causes including malabsorption, malnutrition, renal disease, and anticonvulsant therapy, may produce a characteristic picture of proximal or diffuse skeletal muscle weakness and wasting with marked pains at rest and on movement.

Carcinoma may occasionally cause polymyositis or the Eaton-Lambert syndrome, with a characteristic incremental response to repetitive nerve action. More commonly, the diffuse muscle wasting and weakness associated with carcinomatous cachexia is due to an underlying neuropathy. A muscle weakness is seldom the result of an ectopic hormone production by the carcinoma.

In a small proportion of cases it might be difficult to distinguish brachial plexus neuropathy, which affects both shoulders and arms. Finally, it should be noted that a cervical neurofibroma may occasionally cause shoulder girdle wasting without disturbance of sensibility and pain.

A large number of *inherited, congenital, or at least endogenous conditions* of the skeletal muscle should be considered in the differential diagnosis of limb girdle muscular dystrophy.

Dystrophia myotonica is usually easy to diagnose on the basis of the characteristic face, ptosis, other clinical signs, and EMG myotonia.

In some cases it can be difficult to separate *pseudomyopathic spinal muscular atrophies* from the muscular dystrophies as a whole, and in particular

TABLE 3

Differential diagnosis of the limb girdle syndromes.

Acquired
Inflammatory myopathies: polymyositis, dermatomyositis, inclusion body myositis, sarcoidosis, infectious agents
Toxins and drugs: chloroquine, fluorinated corticosteroids, vincristine, cholesterol synthesis inhibitors, alcohol
Endocrine disorders: hyper- and hypothyroidism, Cushing syndrome, hyperparathyroidism, hyperaldosteronism, diabetes mellitus
Vitamin deficiencies: Vitamin D due to malabsorption, anticonvulsants, etc.
Paraneoplastic: Eaton-Lambert syndrome, ectopic hormone production, polymyositis
Brachial plexus neuropathy
Inherited
Dystrophia myotonica
Spinal muscular atrophies
Motor neuron disease (sporadic and familial)
Duchenne muscular dystrophy
Manifesting carrier of Duchenne muscular dystrophy
Becker muscular dystrophy
Emery-Dreifuss muscular dystrophy
Facioscapulohumeral muscular dystrophy without facial involvement
Scapuloperoneal myopathy
Congenital myopathies: centronuclear myopathy, central core myopathy, nemaline myopathy, myopathy with tubular aggregates, myopathy with cytoplasmic bodies
Metabolic disorders: glycogen storage diseases (McArdle disease, acid maltase deficiency), lipid storage diseases (carnitine deficiency), mitochondrial diseases
Periodic paralysis myopathy

Kugelberg-Welander pseudomyopathic spinal muscular atrophy from limb girdle muscular dystrophy. Clearly, there are many different forms of spinal muscular atrophy, just as there are many different types of primary muscle disease (Bouwsma and Leschot 1986).

The difficulties of classification are highlighted by reports such as that by Kazakov et al. (1976) on the 'K' kinship, which has been the subject of extensive studies spanning more than 50 years, and which various authorities have diagnosed as scapuloperoneal muscular atrophy, facioscapulohumeral muscular dystrophy, Erb scapulohumeral muscular dystrophy, or the spinal muscular atrophy form of the scapuloperoneal syndrome.

Many of the spinal muscular atrophies either cause diffuse muscle weakness and wasting, or at an early stage, although only partially, affect distal muscles selectively, such as the small muscles of the hands; these conditions are thereby easily differentiated from limb girdle muscular dystrophy. Occasionally, there is a tendency for spinal muscular atrophy to involve proximal muscle in an unusual pattern; for instance, the triceps may be involved more than the biceps and the brachioradialis muscles. The presence of fasciculations on initial examinations in half of the cases or following the injection of neostigmine may indicate motor nerve involvement. EMG and muscle biopsy classically point to denervation, and serum CK is generally normal or only slightly increased. These features allow the distinction of the majority of cases of spinal muscular atrophy from limb girdle muscular dystrophy. However, there are some cases of spinal muscular atrophy where the diagnosis is very difficult. Histological findings in patients with proximal muscle weakness may be inconclusive (Coërs and Telerman-Toppet 1979); there may be a discordance between EMG and biopsy, or between mixed features of myopathy and neurogenic syndrome by either examination (Mastaglia and Walton 1971; Shields 1984). Three patients from a large pedigree with an autosomal-dominant spinal muscular atrophy were described (Jansen et al. 1986). This disease became manifest between the end of the fourth and the sixth decade, and progressed rapidly without evidence of corticospinal tract impairment; within 3 years the patients died of respiratory failure. A rare differential diagnosis is, for example, the Fotopoulos syndrome (chronic spinal amyotrophy of shoulder girdle and chronic chorea; Serratrice et al. 1984).

Modern immunohistological and molecular biological methods enable a definite segregation of limb girdle muscular dystrophy from muscle dystrophies that arise as a result of a mutation in the chromosome region Xp21.2. The expression of dystrophin remains unaltered in limb girdle dystrophy. It may be difficult to distinguish clinically between Becker muscular dystrophy and carriers of Duchenne dystrophy (Patel et al. 1988; Defesche et al. 1989; Norman et al. 1989). The presence of affected females is incompatible with a diagnosis of X chromosome-linked dystrophy, but the differential diagnosis is difficult in sporadic male cases or where more than 1 male in a group of siblings is affected. Furthermore, in the absence of a male relative with Duchenne muscular dystrophy, the clinical distinction between carriers of Duchenne dystrophy and cases of autosomal-recessive limb girdle dystrophy may not be possible (Barkhaus and Gilchrist 1989). Richards et al. (1990) reported on monozygotic twins, 1 of whom, the sister, was suffering from muscular dystrophy. As she was a female a limb girdle dystrophy was suspected. Using DNA probes complementary for the dystrophin gene, a gene deletion was detected in the twins and their mother. An uneven lyonization (X chromosome inactivation) was suggested as the underlying mechanism for disease expression in the affected female.

The serum activity of CK in the early stages of Becker and limb girdle muscular dystrophies may be similar. The finding of increased serum CK in the mother of the patient or in female siblings suggests an X-linked inheritance, but this finding is not absolutely reliable.

The observation of pseudohypertrophy is of little help, as this is present in almost all cases of Becker muscular dystrophy (Bradley et al. 1978) and also occurs in about 20% of patients with limb girdle muscular dystrophy.

A separate form of X-linked muscular dystrophy with an early onset of contractures, the absence of pseudohypertrophy, and the frequent presence of cardiac involvement has been described by Emery and Dreifuss (1966). Here the muscles of the upper arms and lower legs are most affected, a so-called humeroperoneal pattern. In particular, the early contractures are different from the usual pattern of limb girdle muscular dystrophy.

Facioscapulohumeral muscular dystrophy may occasionally be abortive, without significant facial involvement, as mentioned previously. The finding of other cases with classical facial involvement in the family may be of diagnostic help.

An increasing number of congenital myopathies with specific pathological features have been recognized. Usually these diseases are present from birth, although symptoms may not arise until later in life. The muscle involvement is often diffuse and nonselective, but in some the distribution may be restricted to the limb girdle; often they show dysmorphic features. Their pattern of inheritance is variable (Engel 1966; Harriman and Haleem 1972; Bonnette et al. 1974; Brownell et al. 1978). Moreover, there is an increasing number of disorders in which the inborn biochemical error is recognized. These include glycogen storage diseases such as McArdle disease (myophosphorylase deficiency; Engel et al. 1963), 1→4-α-glucosidase deficiency (Pompe infantile and adult-onset forms; Engel 1970); and the adult form of Anderson disease (deficiency of branching enzyme, Ferguson et al. 1983). Also disorders of lipid metabolism, including deficiencies of carnitine and abnormalities of the mitochondrial function, have been characterized. In some examples, such as adult-onset 1→4-α-glucosidase deficiency and mitochondrial myopathy, the pattern of muscle involvement may be selective. A family with benign, prednisone-responsive, lipid mitochondrial myopathy presenting as a limb girdle syndrome, has been described (Heiman-Patterson et al. 1985). Myoadenylate deaminase deficiency may be present in patients with facial and limb girdle myopathy (Mercelis et al. 1981).

Patients with familial hypokalemic periodic paralysis may develop progressive proximal vacuolar myopathy (Goldflam myopathy) in later life. A rather similar picture has been observed in families with hyperkalemic periodic paralysis, though this is much less common than is the hypokalemic form. In addition, the sporadic or familial 'limb girdle myasthenia syndromes' should be mentioned. They show varying muscle pathology, often with tubular aggregates, a typical decremental response to repetitive stimulation and clear and sustained improvement with anticholinesterase treatment (McQuillen 1966; Dobkin and Verity 1978; Dalakas and Engel 1987; Engel 1990). These diseases were probably originally classified as rather atypical examples of limb girdle muscular dystrophy, and for this reason it is conceivable that other distinct disease entities produce the limb girdle syndrome. Patients at present classified as having limb girdle muscular dystrophy undoubtedly show some heterogeneity, and this suggests the existence of a number of different disease etiologies.

ETIOLOGY AND TREATMENT

In recent years great advances have been attained by research on the pathogenesis of the X chromosome-linked muscular dystrophies. The search for the etiology of the muscular dystrophies has concentrated to a very large extent on Duchenne muscular dystrophy. This is mainly because it is the most frequently encountered muscular dystrophy, the most rapidly lethal, and the most distressing.

The Duchenne and Becker muscular dystrophies are allelic disorders due to a mutation at the chromosome region of Xp21.2, and the gene product affected, dystrophin, has been identified. Although the precise role of dystrophin in the muscle fiber is still not known, an attractive postulate is that it acts as a link between the cell membrane and the underlying cytoskeleton and that its deficiency weakens the membrane, rendering it susceptible to physical damage during muscle contraction (for review see Kakulas and Mastaglia 1990).

It is not hard to imagine a comparable pathologic mechanism for the other muscular dystrophies. However, hitherto there are few indications of such a defect with regard to limb girdle dystrophy. Mechler et al. (1989) measured the intracellular free calcium concentration $(Ca^{2+})i$ in lymphocytes of patients suffering from various muscular dystrophies. The $(Ca^{2+})i$ level was significantly increased in facioscapulohumeral and limb girdle dystrophy, while it was decreased in Duchenne dystrophy. In addition by using polyvalent immunoglobin and concanavalin-A as ligands, a decreased proportion of capped lymphocytes could be demonstrated in limb girdle muscular dystrophy patients and in carriers (Bader et al. 1982).

After almost 100 years of research on muscular dystrophy, hope has now become stronger that through molecular genetic advances we will have an effective therapy in the foreseeable future. For example, it is quite conceivable to bypass the genetic defect by gene transfer or to introduce normogenomic myonuclei into dystrophic muscle fibers. Purely empirical attempts to cure the disease have become increasingly uncertain (Wolf and Lewis 1985).

Although no specific therapy is available at present, the physician's role in helping the patient with any muscular disease may be summarized as follows:

1. diagnosis and advice concerning the nature and prognosis of the disease;

2. rehabilitation and physiotherapy to prevent contractures or deformities, and to maintain independence for as long as possible;

3. provision of aids that enable the patient to continue actions of daily life and employment (e.g., wheelchairs, ramps, hoists, and assisted ventilation);

4. regular supervision to allow cardiopulmonary complications to be diagnosed and treated in good time; and

5. comforting the patient with an optimistic and positive attitude.

REFERENCES

ANTONIO, J.H., M.C. DINIZ and D. MIRANDA: Persistent atrial standstill with limb girdle muscular dystrophy. Cardiology 63 (1978) 39–46.

ARIKAWA, E., E.P. HOFFMAN, M. KAIDO, I. NONAKA, H. SUGITA and X. ARAHAT: The frequency of patients with dystrophin abnormalities in a limb-girdle patient population. Neurology 41 (1991) 1491–1496.

ARTS, W.F., J. BETHLEM and W.S. VOLKERS: Further investigations on benign myopathy with autosomal dominant inheritance. J. Neurol. 217 (1978) 201–206.

BACON, P.A. and B. SMITH: Familial muscular dystrophy of late onset. J. Neurol. Neurosurg. Psychiatr. 34 (1971) 93–97.

BADER, P.I., C.J. BENDER, M.T. CREASON, P.S. CONN and D.T. TOWNSEND: Lymphocyte capping in limb-girdle muscular dystrophy: patients and carriers in an Amish isolate. Am. J. Med. Genet. 12 (1982) 255–269.

BALLANTYNE, J.P. and S. HANSEN: New method for the estimation of the number of motor units in a muscle. 2. Duchenne, limb-girdle and fascioscapulohumeral, and myotonic muscular dystrophies. J. Neurol. Neurosurg. Psychiatr. 37 (1974) 1195–1201.

BALLESTRAZZI, A., C. GRANATA, U. TRENTIN, A. STAGNI, T. CAPELLI, L. MORANDI and L. MERLINI: Bethlem myopathy: clinical data of 14 patients in 2 families. J. Neurol. Sci. 98 (Suppl.) (1990) 461.

BARKHAUS, P.E. and J.M. GILCHRIST: Duchenne muscular dystrophy manifesting carriers. Arch. Neurol. 46 (1989) 673–675.

BECKER, P.E.: Dystrophia Musculorum Progressiva. Stuttgart, G. Thieme Verlag (1953).

BECKER, P.E. and F. KIENER: Eine neue X-chromosomale Muskeldystrophie. Arch. Psychiatr. Nervenkr. 193 (1955) 427–448.

BELANGER, A.Y. and A.J. MCCOMAS: Neuromuscular function in limb girdle dystrophy. J Neurol. Neurosurg. Psychiatr. 48 (1985) 1253–1258.

BELL, J.: The Treasury of Human Inheritance, Vol. IV. Nervous Diseases and Muscular Dystrophies. Cambridge, University Press (1943).

BEN HAMIDA, M. and M. FARDEAU: Severe autosomal recessive limb-girdle muscular dystrophies frequent in Tunisia. In: C. Angelini and G.A. Danieli (Eds), Muscular Dystrophy Research: Advances and New Trends. Amsterdam, Excerpta Medica (1980) 143–146.

BETHLEM, J. and G.K. VAN WIJNGAARDEN: Benign myopathy with autosomal dominant inheritance, a report of three pedigrees. Brain 99 (1976) 91–100.

BODDIE, H.G. and E.G. STEWART-WYNNE: Quadriceps myopathy: entity or syndrome? Arch. Neurol. 31 (1974) 60–62.

BONNETTE, H., R. ROELOFS and W.H. OLSON: Multicore disease. Report of a case with onset in middle age. Neurology 24 (1974) 1039–1044.

BOUWSMA, G. and N.J. LESCHOT: Unusual pedigree patterns in seven families with spinal muscular atrophy; further evidence for the allelic model hypothesis. Clin. Genet. 30 (1986) 145–149.

BRADLEY, W.G.: The limb-girdle syndromes. In: P.J. Vinken and G.W. Bruyn (Eds), Handbook of Clinical Neurology, Vol. 40. Diseases of Muscle, Part I. Amsterdam, New York, Oxford, North-Holland Publishing Company (1979) 433–469.

BRADLEY, W.G., M.Z. JONES, J.M. MUSSINI and P.R. FAWCETT: Progressive X-linked neuromuscular disease: Becker-type muscular dystrophy. Muscle Nerve 1 (1978) 111–132.

BRAMWELL, E.: Observations on myopathy. Proc. R. Soc. Med. 16 (1922) 1–12.

BRENNI, G., F. JERUSALEM and H. SCHILLER: Myopathologie chronischer Denervationsprozesse. Nervenarzt 52 (1981) 692–702.

BROWNELL, A.K.W., J.J. GILBERT, D.T. SHAW, B. GARCIA, G.F. WENKEBACH and A.K.S. LAM: Adult onset nemaline myopathy. Neurology 28 (1978) 1306–1309.

CHUTKOW, J.G., R.R. HEFFNER JR, A.A. KRAMER and J.J. EDWARDS: Adult-onset autosomal dominant limb-girdle muscular dystrophy. Ann. Neurol. 20 (1986) 240–248.

COERS, C. and N. TELERMAN-TOPPET: Differential diagnosis of limb-girdle muscular dystrophy and spinal muscular atrophy. Neurology 29 (1979) 957–972.

DALAKAS, M.C. and W.K. ENGEL: Prednisone-responsive limb-girdle syndrome: a special disorder? Neuropediatrics 18 (1987) 88–90.

DE COSTER, W., J. DE REUCK and E. THIERY: A late autosomal dominant form of limb girdle muscular dystrophy. Eur. Neurol. 12 (1974) 159–172.

DEFESCHE, J.C., M. DE VISSAR, E. BAKKER, G. BOUWSMA, J.J. DE VIJLDER and P.A. BOLHUIS: DNA restriction fragment polymorphisms in differential diagnosis of genetic disease: application in neuromuscular diseases. Hum. Genet. 82 (1989) 55–58.

DENNY-BROWN, D.: Myopathic weakness of quadriceps. Proc. R. Soc. Med. 32 (1939) 867–869.

DESMEDT, J.E. and S. BORENSTEIN: Relationship of spontaneous fibrillation potentials to muscle fibre segmentation in human muscular dystrophy. Nature (London) 258 20(1975) 531–534.

DE VERE, R. and W.G. BRADLEY: Polymyositis: its presentation, morbidity and mortality. Brain 98 (1975) 637–666.

DOBKIN, B.H. and M.A. VERITY: Familial neuromuscular disease with type I fiber hypoplasia, tubular aggregates, cardiomyopathy and myasthenic features. Neurology 28 (1978) 1135–1140.

DUBOWITZ, V., C.A. SEWRY and R.B. FITZSIMONS: Muscle Biopsy. A Practical Approach. London, Philadelphia, Toronto, Mexico City, Rio de Janeiro, Sydney, Tokyo, Hong Kong, Baillière Tindall (1985) 344–357.

DUCHENNE, G.B.A.: Recherches sur la paralysie musculaire pseudohypertrophique ou paralysie myosclérosique. Arch. Génér. Méd. 11 (1868) 5–25, 179–209, 305–321, 421–443, 552–588.

EKBOM, K., R. HED, L. KIRSTEIN and K.E. ASTROM: Muscular affections in chronic alcoholism. Arch. Neurol. 10 (1964) 449–458.

EMERY, A.E.H. and F.E. DREIFUSS: Unusual type of benign X-linked muscular dystrophy. J. Neurol. Neurosurg. Psychiatr. 19 (1966) 338–342.

ENGEL, A.G.: Late-onset rod myopathy (a new syndrome)?: light and electron microscopic observations in two cases. Mayo Clin. Proc. 41 (1966) 713–741.

ENGEL, A.G.: Acid maltase deficiency in adults: studies in four cases of syndrome which may mimic muscular dystrophy or other myopathies. Brain 93 (1970) 599–616.

ENGEL, A.G.: Congenital disorders of neuromuscular transmission. Semin. Neurol. 10 (1990) 12–26.

ENGEL, W.K., E.L. EYERMAN and H.E. WILLIAMS: Late-onset type of skeletal-muscle phosphorylase deficiency, a new familial variety with completely or partially affected subjects. N. Engl. J. Med. 268 (1963) 135.

ERB, W.: Ueber die 'juvenile Form' der progressiven Muskelatrophie und ihre Beziehungen zur sogenannten Pseudohypertrophie des Muskeln. Dtsch. Arch. Klin. Med. 34 (1884) 467–519.

ERB, W.: Dystrophia muscularis progressiva. Klinische und pathologisch-anatomische Studien. Dtsch. Z. Nervenheilkd. 1 (1891) 13–94, 173–261.

ESCOUROLLE, R.M. and M. FARDEAU: Aspects histopathologiques des dystrophies musculaires progressives. La place des techniques histologiques classiques dans leur diagnostic. Ann. Anat. Pathol. (Paris) 18 (1973) 109–137.

ESPIR, M.L.E. and W.B. MATTHEWS: Hereditary quadriceps myopathy. J. Neurol. Neurosurg. Psychiatr. 36 (1973) 1041–1045.

EULENBERG, A. and J.C. COHNHEIM: Ergebnis der anatomischen Untersuchung eines Falles von sog. Muskelhypertrophie. Verh. Berl. Med. Ges. (1866) 2. Hft. p. 191.

FAREAU, M.: Ultrastructural lesions in progressive muscular dystrophies. A critical study of their specificity. In: J.N. Walton, N. Canal and G. Scarlato (Eds), Muscle Diseases. Amsterdam, Excerpta Medica (1970) 98–108.

FERGUSON, I.T., M. MAHON and W.J.K. CUMMING: An adult case of Anderson's disease: type IV glycogenosis. A clinical, histochemical, ultrastructural and biochemical study. J. Neurol. Sci. 60 (1983) 337–353.

FEWINGS, J.D., J.B. HARRIS, M.A. JOHNSON and W.G. BRADLEY: Progressive denervation of skeletal muscle induced by spinal irradiation in rats. Brain 100 (1977) 157–183.

FOWLER, W.M. and N.N. NAYAK: Slowly progressive proximal weakness: limb-girdle syndromes. Arch. Phys. Med. Rehabil. 64 (1983) 527–538.

FURUKAWA, T., N. AKAGAMI and S. MARUYAMA: Chronic neurogenic quadriceps amyotrophy. Ann. Neurol. 2 (1977) 528–530.

GARDNER-MEDWIN, D.: Clinical features and classification of the muscular dystrophies. Br. Med. Bull. 36 (1980) 109–115.

GEORGE, C.F.P.: Sleep in neuromuscular diseases. In: M.H. Kryger, T. Roth and W.C. Dement (Eds), Principles and Practice of Sleep Medicine. Philadelphia, London, Toronto, Montreal, Sydney, Tokyo, W.B. Saunders Company (1989) 630–632.

GILCHRIST, J.M., M. PERICAK-VANCE, L. SILVERMAN and A.D. ROSES: Clinical and genetic investigation in autosomal dominant limb-girdle muscular dystrophy. Neurology 38 (1988) 5–9.

HARANO, Y., R. ADAIR, P.J. VIGNOS, M. MILLER and J. KOWAL: Pyruvate-kinase isoenzymes in progressive muscular dystrophy and in acute myocardial infarction. Metabolism 22 (1973) 493–501.

HARRIMAN, D.G.F. and M.A. HALEEM: Centronuclear myopathy in old age. J. Pathol. 108 (1972) 237–248.

HAYWARD, M.: Electrodiagnosis of the muscular dystrophies. Br. Med. Bull. 36 (1980) 127–132.

HECKMATT, J.Z. and V. DUBOWITZ: Diagnostic advantage of needle muscle biopsy and ultrasound imaging in the detection of focal pathology in a girl with limb girdle dystrophy. Muscle Nerve 8 (1985) 705–709

HECKMATT, J.Z., L. LOH and V. DUBOWITZ: Night-time nasal ventilation in neuromuscular disease. Lancet 335 (1990) 579–582.

HEIMAN-PATTERSON, T.D., A.J. TAHMOUSH, E.B. BONILLA and S. DI MAURO: Familial prednisone-responsive mitochondrial myopathy presenting as a limb-girdle dystrophy syndrome. Neurology 35 (1985) 302 (abstract).

HENSON, T.L., J. MULLER and W.E. DEMYER: Hereditary myopathy limited to females. Arch. Neurol. (Chicago) 17 (1967) 238–247.

HERTRICH, O.: Kasuistische Mittelung über eine Sippe

mit dominant vererblicher, wahrscheinlich weiblich geschlechtsgebundener progressiver Muskeldystrophie des Schulötergürteltyps. Nervenarzt 28 (1957) 325–327.

HEYCK, H. and G. LAUDAHN: Die progressiv-dystrophischen Myopathien. Berlin, Springer Verlag (1969) 58–60.

HOFFMAN, E.P., R.H. BROWN and L.M. KINKEL: Dystrophin: the protein product of the Duchenne muscular dystrophy locus. Cell 51 (1987) 919–928.

HOSHIO, A., H. KOTAKE, M. SAITOH, K. OGINO, Y. FUJIMOTO, J. HASEGAWA, T. KOSAKA and H. MASHIBA: Cardiac involvement in a patient with limb-girdle muscular dystrophy. Heart Lung 16 (1987) 439–441.

HUNTER, S.: The heart in muscular dystrophy. Br. Med. Bull. 36 (1980) 133–134.

JACKSON, C.E. and J.H. CAREY: Progressive muscular dystrophy: autosomal recessive type. Pediatrics 7 (1961) 77–84.

JACKSON, C.E. and D.A. STREHLER: Limb-girdle muscular dystrophy: clinical manifestations and detections of preclinical disease. Pediatrics 41 (1968) 495–502.

JANSEN, P.H.P., E.M.G. JOOSTEN, H.H. JASPAR and H.M. VINGERHOETS: A rapidly progressive autosomal dominant scapulohumeral form of spinal muscular atrophy. Ann. Neurol. 20 (1986) 538–540.

JIDDANE, M., J.L. GASTAUT, J.F. PELLISIER, J. POUGET, G. SERRATRICE and G. SALAMON: CT of primary muscle diseases. Am. J. Neuroradiol. 4 (1983) 773–776.

KAGEN, L.J., S. MOUSSAVI, S.L. MILLER and P. TSAIRIS: Serum myoglobin in muscular dystrophy. Muscle Nerve 3 (1980) 221–226.

KAKULAS, B.A. and F.L. MASTAGLIA (EDS): Pathogenesis and Therapy of Duchenne and Becker Muscular Dystrophy. New York, Raven Press (1990)

KAWASHIMA, S., M. UENO, T. KONDO, J. YAMAMOTO and T. IWASAKI: Marked cardiac involvement in limb-girdle muscular dystrophy. Am. J. Med. Sci. 299 (1990) 411–414.

KAZAKOV, V.M., D.K. BOGORODINSKY and A.A. SKOROMETZ: The myogenic scapuloperoneal syndrome. Muscular dystrophy in the K. kindred: clinical study and genetics. Clin. Genet. 10 (1976) 41–50.

KOVICK, R.B., A.M. FOGELMAN, A.S. ABBASI, J.B. PETER and M.L. PEARCE: Echocardiographic evaluation of posterior left ventricular wall motion in muscular dystrophy. Circulation 52 (1975) 447–454.

LAMBERT, C.D. and A.J. FAIRFAX: Neurological associations of chronic heart block J. Neurol. Neurosurg. Psychiatr. 39 (1976) 571–575.

LANDOUZY, L. and J. DEJERINE: De la myopathie atrophique progressive. Rev. Méd. 5 (1885) 81–117; 253–366.

LANDOUZY, L. and J. DEJERINE: Myopathie atrophique progressive. Rev. Méd. 6 (1886) 977–1027.

LANG, A. and V.S.J. PARTANEN: 'Satellite potentials' and the duration of motor unit potentials in normal, neuropathic and myopathic muscles. J. Neurol. Sci. 27 (1976) 513–524.

LETH, A., K. WULFF, M. CORFITSEN and J. ELMGREEN: Progressive muscular dystrophy in Denmark. Natural history, prevalence and incidence. Acta Paediatr. Scand. 74 (1985) 881–885.

LEVISON, H.: Dystrophia musculorum progressiva. Clinical and diagnostic criteria, inheritance. Acta Psychiat. Scand. (Suppl.) 76 (1951) 7–175.

LEYDEN, E.: Klinik der Rückenmarkskrankheiten. Vol. 2. Berlin, Verlag A. Hirschwald (1876) 525–540.

MASTAGLIA, F.L. and J.N. WALTON: Histological and histochemical changes in skeletal muscle from cases of chronic juvenile and early adult spinal muscular atrophy (the Kugelberg-Welander syndrome). J. Neurol. Sci. 2 (1971) 15–44.

MATSUMURA, K., T. SHIMIZU, I. NONAKA and T. MANNEN: Immunochemical study of connectin (titin) in neuromuscular diseases using monoclonal antibody: connectin is degraded extensively in Duchenne muscular dystrophy. J. Neurol. Sci. 93 (1989) 147–156.

MCQUILLEN, M.P.: Familial limb-girdle myasthenia. Brain 89 (1966) 873–880.

MECHLER, F., M. MOLNAR, M. BALAZS and J. MATKO: Intracellular free calcium concentration in lymphocytes of patients with muscular dystrophies. Biochim. Biophys. Acta 1012 (1989) 227–230.

MERCELIS, R., J.J. MARTIN, I. DEHAENE, T. DE BARSY and G. VAN DEN BERGHE: Myoadenylate deaminase deficiency in a patient with facial and limb girdle myopathy. J. Neurol. 225 (1981) 157–166.

MERYON, E.: On granular and fatty degeneration of the voluntary muscles. Med.-Chir. Trans. (1852) 73–85.

MÖBIUS, P.J.: Ueber die hereditären Nervenkrankheiten. Volkmanns Samml. Klin. 171 (1879) 1501–1531.

MOHIRE, M.D., R. TANDAN, T.J. FRIES, B.W. LITTLE, W.W. PENDLEBURY and W.G. BRADLEY: Early-onset benign autosomal dominant limb-girdle myopathy with contractures (Bethlem myopathy). Neurology 38 (1988) 573–580.

MORTON, N.E. and C.S. CHUNG: Formal genetics of muscular dystrophy. Am. J. Hum. Genet. 11 (1959) 360–379.

MOSER, H., U. WIESMANN, R. RICHTERICH and E. ROSSI: Progressive Muskeldystrophie, VIII Häufigkeit, Klinik und Genetik der Typen I und II. Schweiz. Med. Wochenschr. 96 (1966) 169–174, 205–211.

MUMENTHALER, V.M., T. BOSCH, E. KATZENSTEIN and F. LEHNER: Über den isolierten Befall des M. quadriceps femoris bei der Dystrophia musculorm progressiva. Confin. Neurol. 18 (1958) 416–441.

MUNSAT, T.L.: The classification of human dystrophies. In: L.P. Rowland (Ed.), Pathogenesis of Human Muscular Dystrophies. Amsterdam, Excerpta Medica, (1977) 21–31.

MURPHY, W.A., W.G. TOTTY and J.E. CARROLL: MRI of normal and pathologic skeletal muscle. Am. J. Roentgenol. 146 (1986) 565–574.

NEVIN, S.: Two cases of muscular degeneration occurring in late adult life, with a review of the recorded cases of late progressive muscular dystrophy (late progressive myopathy). Q. J. Med. 17 (1936) 51–68.

NEWSOM-DAVIS, J.: The respiratory system in muscular dystrophy. Br. Med. Bull. 36 (1980) 135–138.

NORMAN, A., N. THOMAS, J. COAKLEY and P. HARPER: Distinction of Becker from limb-girdle muscular dystrophy by means of dystrophin cDNA probes. Lancet i (1989) 466–468.

PANAYIOTOPOULOS, C.P. and S. SCARPALEZOS: Electrophysiological estimation of motor units in limb-girdle muscular dystrophy and chronic spinal muscular atrophy. J. Neurol. Sci. 24 (1975) 95–107.

PANAYIOTOPOULOS, C.P. and S. SCARPALEZOS: F-wave studies on the deep peroneal nerve. Part 2. 1. Chronic renal failure 2. Limb-girdle muscular dystrophy. J. Neurol. Sci. 31 (1977) 331–341.

PANEGYRES, P.K., F.L. MASTAGLIA and B.A. KAKULAS: Limb girdle syndromes. Clinical, morphological and electrophysiological studies. J. Neurol. Sci. 95 (1990a) 201–218.

PANEGYRES, P.K., J.M. PAPDIMITRIOU, P.N. HOLLINGWORTH, J.A. ARMSTRONG and B.A. KAKULAS: Vesicular changes in the myopathies of AIDS. Ultrastructural observations and their relationship to zidovudine treatment. J. Neurol. Neurosurg. Psychiatr. 53 (1990b) 649–655.

PATEL, K., T. VOIT, M.J. DUNN, P.N. STRONG and V. DUBOWITZ: Dystrophin and nebulin in the muscular dystrophies. J. Neurol. Sci. 87 (1988) 315–326.

PENNINGTON, R.J.T.: Clinical biochemistry of muscular dystrophy. Br. Med. Bull. 36 (1980) 123–126.

PERLOFF, J.K., A.C. DE LEON and D. ODOHERTY: The cardiomyopathy of progressive muscular dystrophy. Circulation 36 (1966) 625–648.

PLOTZ, P.H., M. DALAKAS, R.L. LEFF, L.A. LOVE, F.W. MILLER and M.E. CRONIN: Current conceps in the idiopathic inflammatory myopathies: polymyositis, dermatomyositis, and related disorders. Ann. Intern. Med. 111 (1989) 143–157.

RADHAKRISHNAN, K., M.A. EL-MANGOUSH and S.E. GERRYO: Descriptive epidemiology of selected neuromuscular disorders in Benghazi, Libya. Acta Neurol. Scand. 75 (1987) 95–100.

RICHARDS, C.S., S.C. WATKINS, E.P. HOFFMAN, N.R. SCHNEIDER, I.W. MILSARK, K.S. KATZ, J.D. COOK, L.M. KUNKEL and J.M. CORTADA: Skewed X inactivation in a female MZ twin results in Duchenne muscular dystrophy. Am. J. Hum. Genet. 46 (1990) 672–681.

RIDEAU, Y., L.W. JANKOWSKI and J. GRELLET: Respiratory function in muscular dystrophies. Muscle Nerve 4 (1981) 155–164.

RIGGS, J.E., S.S. SCHOCHET, JR., L. GUTMANN and S.C. LERFALD: Childhood onset inclusion body myositis mimicking limb-girdle muscular dystrophy. J. Child Neurol. 4 (1989) 283–285.

RÖTHIG, H.J., W. BERNHARDT and E.G. AFTING: Excretion of total and muscular N-methylhistidine and creatinine in muscle diseases. Muscle Nerve 7 (1984) 374–379.

SABATELLI, M., G. GALLUZZI, M. LO MONACO, A. UNCINI and P. TONALI: Familiar idiopathic hyperckemia associated with limb-girdle muscular dystrophy. Muscle Nerve 9 (1986) 210 (abstract).

SCHMALBRUCH, H.: The muscular dystrophies. In: F.L. Mastaglia and J. Walton (Eds), Skeletal Muscle Pathology. Edinburgh, London, Melbourne, New York, Churchill Livingstone (1982) 235–265.

SCHNEIDERMAN, L.J., W.I. SAMPSON, W.C. SCHOENE and G.B. HAYDON: Genetic studies of a family with two unusual autosomal dominant conditions: muscular dystrophy and Pelger-Huet anomaly. Clinical, pathologic and linkage considerations. Am. J. Med. 46 (1969) 380–393.

SERRATRICE, G. and J.F. PELLISIER: Deux familles de myopathies bénignes prédominant sur les ceintures d'hérédité autosomique dominante. Rev. Neurol. (Paris) 144 (1988) 43–46.

SERRATRICE, G., G. CREMIUX, J.F. PELLISIER and J. POUGET: Deux cas de syndrome de Fotopoulos (amyotrophie spinale chronique de la ceinture scapulaire et choree chronique). Presse Med. 13 (1984) 1274 (letter).

SERRATRICE, G., A. POU-SERRADEL, J.F. PELLISIER, H. ROUX, J. LAMARCO-CIVRO and J. POUGET: Chronic neurogenic quadriceps amyotrophies. J. Neurol. 232 (1985) 150–153.

SHIELDS JR., R.W.: Single fiber electromyography in the differential diagnosis of myopathic limb girdle syndromes and chronic spinal muscular atrophy. Muscle Nerve 7 (1984) 265–272.

SHIELDS JR., R.W.: Limb girdle syndromes. In: A.G. Engel and B.Q. Banker (Eds), Myology. Vol. 2. New York, McGraw-Hill Book Company (1986) 1349–1365.

SICA, R.E.P. and A.J. MCCOMAS: An electrophysiological investigation of limb-girdle and facioscapulohumeral dystrophy. J. Neurol. Neurosurg. Psychiatr. 34 (1971) 469–474.

SKATRUD, J., C. IBER and W. MCHUGH: Determinants of hypoventilation during sleep in diaphragmatic paralysis. Am. Rev. Respir. Dis. 121 (1980) 587–593.

SOMER, H., V. LAULUMAA, L. PALJÄRVI, J. PARTANEN and M. HALTIA: Adult onset limb-girdle muscular dystrophy with autosomal inheritance. Prog. Clin. Biol. Res. 306 (1989) 69–71.

STALBERG, E.: Single fibre electromyography in muscular dystrophy. In: I. Hausmanova-Petrusewicz and H. Jedrezejowska (Eds), Structure and Function of Normal and Diseased Muscle and Peripheral Nerve. Warsaw, Polish Medical Publishers (1974) 185–190.

STEVENSON, A.C.: Muscular dystrophy in Northern Ireland. I. An account of the condition in fifty-one families. Ann. Eugen. (London) 18 (1953) 50–93.

STEVENSON, A.C.: Muscular dystrophy in Northern Ireland. II. An account of nine additional families. Ann. Hum. Genet. 19 (1955) 159–164.

SUNOHARA, N., K. ARAHATA, E.P. HOFFMAN, H. YAMADA, J. NISHIMIYA, E. ARIKAWA, M. KAIDO, I. NONAKA and

H. SUGITA: Quadriceps myopathy: forme fruste of Becker muscular dystrophy. Ann. Neurol. 28 (1990) 634–639.

SWASH, M. and K.W. HEATHFIELD: Quadriceps myopathy. A variant of the limb-girdle dystrophy syndrome. J. Neurol. Neurosurg. Psychiatr. 46 (1983) 355–357.

TACHI, N., M. TACHI, K. SASAKI and S. IMAMURA: Early-onset benign autosomal-dominant limb-girdle myopathy with contractures (Bethlem myopathy). Pediatr. Neurol. 5 (1989) 232–236.

THAGE, O.: The 'Quadriceps syndrome', an electromyographic and histological evaluation. Acta Neurol. Scand. 41 (Suppl.) 13 (1965) 245–249.

TURNER, J.W.A. and K.W.G. HEATHFIELD: Quadriceps myopathy occurring in middle age. J. Neurol. Neurosurg. Psychiatr. 18 (1961) 18–21.

VAN WIJNGAARDEN, G.K., C.J. HAGEN, J. BETHLEM and A.E.F.H. MEIJER: Myopathy of the quadriceps muscles. J. Neurol. Sci. 7 (1968) 201–206.

WADA, Y., Y. ITOH, T. FURIKAWA, H. TSUKAGOSHI and K. ARAHATA: 'Quadriceps myopathy': a clinical variant form of Becker muscular dystrophy. J. Neurol. 237 (1990) 310–312.

WALTON, J.N.: Two cases of myopathy limited to the quadriceps. J. Neurol. Neurosurg. Psychiatr. 19 (1956) 106–108.

WALTON, J.N. and D. GARDNER-MEDWIN: The muscular dystrophies. In: J.N. Walton (Ed.), Disorders of Voluntary Muscle. Edinburgh, London, Melbourne, New York, Churchill Livingstone, 5th Ed. (1988) 519–568.

WALTON, J.N. and F.J. NATTRASS: On the classification, natural history and treatment of the myopathies. Brain 77 (1954) 170–231.

WELSH, J.D., T.N. LYNN JR. and G.R. HAAS: Cardiac findings in 73 patients with muscular dystrophy. Arch. Intern. Med. 112 (1963) 199–206.

WHARTON, B.A.: An unusual variety of muscular dystrophy. Lancet i (1965) 248–249.

WOLF, S. and N.J. LEWIS: Clinical trials in chronic neuromuscular diseases. Muscle Nerve 8 (1985) 453–492.

YATES, J.R. and A.E. EMERY: A population study of adult onset limb-girdle muscular dystrophy. J. Med. Genet. 22 (1985) 250–257.

ZATZ, M., L.J. SHAPIRO, D.S. CAMPION, E. ODA and M.M. KABACK: Serum pyruvate-kinase (PK) and creatinine phosphokinase (CPK) in progressive muscular dystrophies. J. Neurol. Sci. 36 (1978) 349–362.

Handbook of Clinical Neurology, Vol. 18 (62): Myopathies
L.P. Rowland and S. DiMauro, editors
© 1992 Elsevier Science Publishers B.V. All rights reserved

CHAPTER 8

Distal myopathies

ARON S. BUCHMAN[1] and ELIZABETH J. COCHRAN[1,2]

[1]*Department of Neurological Sciences and*
[2]*Department of Pathology, Rush-Presbyterian-St. Lukes Medical Center, Chicago, IL, USA*

The first description of distal myopathy is commonly credited to Gowers (1902). He described a boy who had trouble walking at about 12 years of age, followed shortly thereafter by weakness of the hands. At 18, ankle flexion was weaker than ankle dorsiflexion and his calves were hypertrophic. Facial and neck weakness with tongue atrophy were also noted. There is still controversy as to whether this patient had distal myopathy or myotonic dystrophy which had not been well described at that time (Kratz and Brooke 1979; Furukawa et al. 1984).

Subsequent reports of distal myopathy may have been spinal muscular atrophy, hereditary sensorimotor neuropathy, or myotonic dystrophy. Milhorat and Wolff (1943) probably described a family with distal myopathy, but the disorder was not firmly established until 1951, after the introduction of electromyography, when Welander reported 249 patients with distal muscle weakness which she named myopathia distalis tarda hereditaria (Welander 1951). Additional types of distal myopathy have been proposed based on reports of individual families or sporadic cases which differ from Welander's cases. This chapter will focus on the reports which have followed Welander's description because earlier reports lack essential details necessary for definite classification. The earlier reports have been analyzed by Kratz and Brooke (1979) and Markesbery and Griggs (1986).

Traditional classification of myopathic disorders based on clinical, electrophysiologic, and histologic findings are limited because of the nonspecific nature of these abnormalities. Remarkable advances in molecular genetics have been successfully applied to elucidating the underlying genetic defect of Duchenne muscular dystrophy (Rowland 1988). This disorder can now be classified as a myopathy with mutation at Xp21, a specific locus on the short arm of the X chromosome with a known abnormal gene product, dystrophin. Application of these advances to patients that formerly would have been classified as having different types of muscular dystrophy, such as classic Duchenne muscular dystrophy, muscular dystrophy of the Becker type, or limb-girdle dystrophy, can now be shown to share the same genetic defect. As in Duchenne muscular dystrophy, future application of molecular genetic techniques to distal myopathy should facilitate unraveling the pathophysiology and provide a more accurate basis for diagnosis, classification, and possible treatment.

DIAGNOSIS

Distal myopathy is a descriptive term that comprises a heterogeneous group of syndromes in which weakness is first noticed in the distal extremities. At the present time there is no laboratory marker that can definitely identify distal myopathy. The diagnosis of distal myopathy can only be made after proving that distal weakness is caused by a myopathic process and excluding other neuropathic and myopathic disorders that can cause distal weakness (Table 1). Therefore, the appropriate use of magnetic resonance imaging or computed tomography (CT) scans of the spine, nerve conduction studies, electromyography, muscle enzymes, and muscle biopsy are necessary for evaluation of distal weakness (Table 2).

A number of distinct myopathic syndromes may present with distal weakness. Myotonic dystrophy can be identified by the presence of percussion myotonia, cataracts, facial and ocular weakness, as well the presence of myotonic discharges on electromyography (Harper 1986). The scapuloperoneal syndrome which begins with only lower extremity weakness may be difficult to distinguish from distal myopathy until prominent shoulder girdle weakness becomes clinically evident (Munsat 1986). Mild facial weakness may help distinguish scapuloperoneal syndrome from distal myopathy. Fascioscapulohumeral dystrophy is distinguished from distal myopathy by the early prominent facial and shoulder girdle weakness.

TABLE 1

Differential diagnosis of distal myopathy.

Structural diseases
Cervical spondylosis
Cervical rib
Cervical cord pathology

Anterior horn cell diseases
Amyotrophic lateral sclerosis
Spinal muscular atrophy
Distal spinal muscular atrophy

Neuropathies
Hereditary motor sensory neuropathies
Motor neuropathies

Myopathic disorders
Dystrophic
    Myotonic dystrophy
    Fascioscapulohumoral dystrophy
    Scapulohumeroperoneal dystrophy
Inflammatory
    Inclusion body myositis
    Polymyositis
Metabolic
    Debrancher disease

TABLE 2

Criteria for diagnosis of distal myopathy.

Clinical examination
Weakness preferentially involving the distal muscles
    of the upper or lower extremities.
No significant clinical sensory disturbance.
No evidence of central nervous system dysfunction
    such as increased deep tendon reflexes and extensor
    plantar responses.
No bowel, bladder, or respiratory dysfunction.
No weakness of extraocular muscles.

Electrophysiologic testing
No evidence for polyneuropathy or motor neuro-
    pathy with routine nerve conduction studies.
Needle electromyographic evidence of myopathy in
    clinically weak muscles.
No electromyographic evidence of myotonia.

Muscle biopsy
Myopathic abnormalities on muscle biopsy without
    significant inflammatory changes or evidence
    suggesting a metabolic myopathy.

Needle electromyographic studies will demonstrate myopathy rather than the widespread neurogenic changes characteristic of anterior horn cell disorders. Nerve conduction studies should be performed in both upper and lower extremities to exclude a sensorimotor or motor neuropathy.

Muscle biopsy is necessary to document myopathy and exclude an inflammatory myopathy which can occasionally present with distal weakness (Bates et al. 1973; Van Kasteren 1979). Muscle biopsy and electron microscopy are helpful to exclude inclusion body myositis which can be difficult to distinguish from distal myopathy (see below) (Krendel et al. 1988). Muscle biopsy may be useful for identifying Debrancher disease which can present with preferential distal weakness (DiMauro et al. 1979).

CLASSIFICATION OF DISTAL MYOPATHY

The present classification of distal myopathy is based on: (a) age at onset of weakness; (b) clinical signs; (c) mode of inheritance; (d) histopathology; and (e) serum levels of sarcoplasmic enzymes (Tables 3, 4). Patients with autosomal-dominant inheritance with weakness beginning after age 40

TABLE 3

Classification of distal myopathy.

Autosomal-dominant inheritance
Infantile distal myopathy (IDM)
Juvenile distal myopathy (JDM)
Myopathia distalis tarda hereditaria of Welander
(MDTH)

Autosomal recessive or sporadic inheritance
Distal muscular dystrophy (dMD)
Distal myopathy with rimmed vacuole formation
(DMRV)

---

years form 1 category that seems to be rare outside of Sweden. Reports to suggest the possibility of an infantile or juvenile form of distal myopathy are tentative and require additional confirmation (Biemond 1955, 1966; Magee and Dejong 1965; Van der Does de Willebois et al. 1968; Bautista et al. 1978).

Most recent reports are of distal myopathy with sporadic or autosomal-recessive inheritance with weakness beginning in the legs. These patients can be divided into 2 categories, distal muscular dystrophy or distal myopathy with rimmed vacuoles, based on (a) whether there is preferential extensor of flexor leg weakness, (b) serum creatine kinase (CK) elevation greater or less than 10 times normal, and (c) muscle biopsy demonstrating dystrophic changes without rimmed vacuoles or biopsy showing myopathic changes with rimmed vacuoles.

AUTOSOMAL-DOMINANT DISTAL MYOPATHY

Myopathia distalis tarda hereditaria of Welander

Welander's description of 249 patients from 72 Swedish families still constitutes the vast majority of all reported cases of distal myopathy (Welander 1951, 1957; Barrows and Duemler 1962; Borg et al. 1987a,b, 1989). Autosomal- dominant inheritance was documented in these families with a preponderance of males, 1.5:1. The mean age at onset of weakness was 50.4±10.7 years (34–82 years).

In most patients (>90%) weakness began in the hands. Weakness started in the thumb or index finger and was first noticed as clumsiness in fine movements, such as fastening buttons, sewing, or typing. Weakness gradually spread to other fingers and later to wrist mucles. Asymmetry of weakness was occasionally noted. About 10% reported that weakness began in the legs or simultaneously in the hands and legs. Some patients noted weakness of proximal limb muscles, trunk, cranial, and neck muscles. Weakness of the anterior tibial muscles made it difficult for 58% of patients to stand on heels, but only 7% had difficulty standing on their toes. In most weakness was confined to the extensor and intrinsic muscles distal to the elbow and knees throughout the course of the disease. Hand and toe flexors were affected with the increasing duration of disease but proximal muscles and feet and finger flexors were spared. The course was slowly progressive, with a normal lifespan.

A minority of patients were 'moderately' or 'grossly atypical' because of early weakness of proximal limb muscles and the long finger and feet flexors. They were also distinguished by the common onset of symptoms in the legs and more rapid course. Both parents of 'atypical' patients were more likely to have been affected with myopathia distalis tarda hereditaria. 'Typical' cases were thought to be heterozygous and 'atypical' cases were thought to be homozygous accounting for more severe manifestations.

Welander (1957) described 1 family in which some siblings were severely affected and others only mildly. Five of 7 siblings had 'typical' myopathia distalis tarda hereditaria; 2 were 'grossly atypical'. This disparity of expression is common in most autosomal-dominant diseases.

With increasing duration there was mild distal vibratory loss in the legs while cutaneous sensation was normal. Tendon reflexes were usually normal, after more than 25 years, about 50% of cases demonstrated loss of ankle jerks.

Autonomic testing has demonstrated normal vagal and sympathetic function (Borg et al. 1987a). Mild sensory dysfunction may be demonstrated by sensitive laboratory tests (Borg et al. 1987b) and sural nerve abnormalities have been found (Thornell et al. 1984; Borg et al. 1989). If these findings are confirmed, myopathia distalis tarda hereditaria may in fact be more widespread, as is the case of myotonic dystrophy in which peripheral nerve can also be affected (Jamal et al. 1986).

TABLE 4

Classification of distal myopathy (DM).*

DM	type				
	IDM	JDM	MDTH	dMD	DMRV
Inheritance	AD	AD	AD	AR/sporadic	AR/sporadic
Age of onset (year)	2–5	5–15	>40	<40	<40
First symptom	LE	Both	UE	LE	LE
Weakness	E<F	E=F	E<F	E>F	E<F
Course	Stable	SP	SP	PROG	PROG
CK	?	?	<10X	>10X	<10X
Biopsy	?MYOP	?MYOP	MYOP	MYOP/DYS	MYOP/RV

* Modified from Miyoshi et al. (1986).
IDM, infantile distal myopathy; JDM, juvenile distal myopathy; MDTH, myopathia distalis tarda hereditaria of Welander; dMD, distal muscular dystrophy; DMRV, distal myopathy with rimmed vacuole formation; AD, autosomal-dominant; AR, autosomal-recessive; UE, upper extremity; LE, lower extremity; E, extensor; F, flexor; SP, slow progression; PROG, progressive; CK, MYOP, myopathic; DYS, dystrophic; RV, rimmed vacuoles.

Pathology of myopathia distalis tarda hereditaria. Welander examined 3 patients at autopsy and muscle biopsies from 26 additional cases. At autopsy no abnormalities were seen in spinal cord, sympathetic ganglia, or peripheral nerves. The most severely involved muscles were the distal long extensors of the hands and feet; 26 of 37 samples showed advanced abnormalities. The distal long flexors were also severely involved. In 8 samples of proximal muscles, however, only 1 showed definite abnormality.

The severely involved muscles showed marked fibrosis and great variation in fiber size with many atrophic or hypertrophic fibers. Fiber-splitting and centrally-placed nuclei were seen. Vacuoles were occasionally noted and myophages were abundant. Intramuscular nerves were normal. Welander described several muscle biopsies that showed early changes of interstitial fibrosis and variation in fiber size with central nuclei.

Borg et al. (1989) confirmed Welander's description but also noted rimmed vacuoles with autophagic characteristics which were confirmed by electron microscopy. These findings were similar to those described by Nonaka et al. (1981, 1985). Of 5 sural nerve biopsies, 2 showed moderate loss of predominantly small myelinated nerve fibers, but blinded control observations were not described (Borg et al. 1989).

Other reports of autosomal-dominant late-onset distal myopathy. The predilection for hand weakness in 'typical' myopathia distalis tarda hereditaria has led others to classify non-Swedish reports of distal myopathy which began in the legs into separate categories. Patients whose weakness began in the feet were labeled as 'atypical' by Welander and both 'typical' and 'atypical' cases were included in 1 catagory in her original report (Welander 1951). Therefore, our classification would include distal myopathy beginning in the leg or arm with autosomal-dominant inheritance and age of onset later than 40 years of age without significant elevation of muscle enzymes in the category of myopathia distalis tarda hereditaria. This category would include the family described by Markesbery et al. (1974, 1977). That family showed autosomal-dominant distal myopathy starting at age 46–51 with leg weakness. This report raised the possibility of distal myopathy causing cardiomyopathy, which has been rarely reported in distal myopathy but is described in other myopathic disorders.

Edstrom et al. (1980) described a family with autosomal-dominant distal myopathy of late onset with more severe weakness of hand flexors than extensors, in contrast to myopathia distalis tarda hereditaria.

Sumner et al. (1971) reported a non-Swedish

family with autosomal-dominant distal myopathy in which men were more severely affected at an early age, while women had milder symptoms later in life. The differences in the expression of disease in that family were similar to Welander's description of presumed homozygous expression (Welander 1957).

Juvenile distal myopathy

Biemond (1955, 1966) described 19 members of a family from Holland with autosomal-dominant inheritance of distal myopathy. He called this disorder myopathia distalis juvenilis hereditaria. In contrast to myopathia distalis tarda hereditaria, weakness began between ages 5 and 15 years in 18 of 19 patients. Weakness was slowly progressive until age 40–50 years and then plateaued. In addition, symmetrical weakness began simultaneously in hands and feet, and distal flexors and extensors were equally affected. No sensory loss was noted, but all patients lacked ankle jerks. EMG in 1 patient showed 'decreased amplitude only of the action potentials'. Autopsy examination (Biemond 1966) showed both myopathic and mild neurogenic histologic features in the muscle. The spinal cord including dorsal and ventral roots was normal. Increased connective tissue was noted in nerve. Histopathology EMG and clinical descriptions are needed to firmly establish this form of distal myopathy.

Infantile distal myopathy

Magee and Dejong (1965) described infantile onset of autosomal-dominant distal myopathy in 18 family members. Only 2 of the family members were personally examined in this report. EMG and muscle biopsy in 1 patient showed evidence of myopathy. Bilateral foot-drop developed or was noticed at about age 2. Mild finger extension weakness and hand weakness was noticed in some members later in childhood. Long periods without progression were also reported. Van der Does de Willebois et al. (1968) described a family in which foot-drop was noted when those affected children began to walk at about 2 years. The father, aged 38, had mild finger and hand extension weakness. All 3 affected individuals had hypertrophic calves.

It is not clear whether this disorder is actually progressive or is a static congenital myopathy. Bautista et al. (1978) described a family with autosomal-dominant distal myopathy beginning in infancy. EMG exams were compatible with myopathy; however, their patients had more diffuse weakness, pes cavus, scoliosis, and selective type I fiber atrophy and predominance.

AUTOSOMAL-RECESSIVE OR SPORADIC DISTAL MYOPATHY

Distal muscular dystrophy

Miyoshi et al. (1986) described autosomal-recessive distal myopathy which began in young Japanese adults usually in the 2nd or 3rd decade. Consanguinity was common, onset symptoms included difficulty running and a tendency to fall, and there was more severe weakness of gastrocnemius and soleus muscles than anterior tibial and peroneal muscles. Consequently, these patients demonstrated more severe difficulty with toe walking as compared to heel walking, an important distinguishing feature from other forms of distal myopathy. Despite the initial distal onset, progressive proximal lower extremity weakness was common. However, once standing, patients could still walk and 16 of 17 cases were not wheelchair-dependent. Mild forearm weakness and wasting were noted, but intrinsic hand muscle atropy was not as severe as in autosomal-dominant distal myopathy. More diffuse muscle wasting was observed with increasing duration of disease including neck, sternomastoid, pharyngeal, facial, or trunk muscles. Respiratory weakness or cranial nerve symptoms were not noted.

An important distinguishing feature of distal muscular dystrophy from other forms of distal myopathy was marked elevated serum CK levels to 20–100 times normal. There was a suggestion of decreased CK levels with increasing age since the symptomatic 56-year-old father of 1 patient had only modest elevation of CK. Three asymptomatic family members who later developed weakness had marked CK elevation and were presumed to be homozygotes. Four other asymptomatic family members with only mild CK elevation were presumed heterozygote carriers for this disease.

Electromyography showed typical myopathic findings without significant spontaneous activity, although later reports described mild fibrillation potentials (Orrico et al. 1987; Cavaletti et al. 1988).

Pathology of distal muscular dystrophy. Miyoshi et al. (1986) described muscle biopsies in 9 patients from 6 different families. The findings were similar in all patients. Abnormalities were severe in the gastrocnemius, but only mild changes were noted in the quadriceps and gluteus maximus. In the gastrocnemius, there were decreased numbers of muscle fibers, extensive fatty infiltration, and fibrosis. The remaining muscle fibers showed marked variation in size with rounding, hypertrophy, and fiber-splitting. Internal nuclei and nuclear clumps were noted. Necrotic fibers associated with macrophages and regenerating fibers were present (Fig. 1). The atrophy and degeneration involved both type 1 and 2 fibers; many type 2C fibers were noted without fiber-type predominance or grouping. On electron microscopic examination, the degenerating fibers showed no other specific findings. Similar but less conspicuous abnormalities were found in the quadriceps and gluteus maximus.

In 1 autopsy exam, generalized muscle wasting was present which more severely affected the lower extremities. The gastrocnemius and soleus muscles were almost entirely replaced by fat and the tibialis anterior muscle was less severely involved. The thigh, arm, limb girdle, and trunk muscles were atrophied but showed little fat or connective tissue infiltration. Muscle histologic findings were similar to those findings described above. In all of the muscle tissue examined, there was evidence of severe myopathy without signs of denervation atrophy or nerve degeneration. Electron microscopy of the diaphragm showed numerous lysosomes and rod-like bodies. The brain, spinal cord, spinal ganglia, nerve roots, peripheral nerves, and heart were unremarkable.

Non-Japanese distal muscular dystrophy. Additional reports of autosomal-recessive and sporadic distal myopathy share many of the features of distal muscular dystrophy (Kuhn and Schroder 1981; Galassi et al. 1987; Cavaletti et al. 1988; Isaacs et al. 1988, cases 2–4; Orrico et al. 1987). The sporadic cases could be isolated expressions of autosomal-recessive inheritance.

Atypical distal muscular dystrophy. Scoppetta et al. (1984) described 2 sisters with all the features of distal muscular dystrophy except that they had preferential anterior tibial and peroneal muscle weakness. Based on clinical features and elevated CKs, a number of additional cases probably represent distal muscular dystrophy. Muscle biopsies in these cases demonstrated a variety of findings, indicating the variability and non-specificity of histologic changes in some cases (Markesbery et al. 1977, cases 1, 2; Miller et al. 1979, cases 1,2; De Visser 1983, cases 1, 2).

Distal myopathy with rimmed vacuole formation

Onset of weakness in this form of distal myopathy is usually during the 2nd or 3rd decade, as in distal muscular dystrophy, with a female predominance of 1.6:1 (Miller et al. 1979, case 3; Nonaka et al. 1981, 1985; Vacario et al. 1981; Kumamoto et al. 1982; Matsubara and Tanabe 1982; Furukawa et al. 1984; Mizusawa et al. 1987a,b; Isaacs et al. 1988, case 1; Sunohara et al. 1989). Most patients with distal myopathy with rimmed vacuoles have been Japanese, although some recent reports in non-Japanese share similar features (see review in Sunohara et al. 1989). A family history of muscle weakness was reported in about 35% of cases with consanguinity frequently noted in other patients despite a lack of familial weakness. The first complaints were lower extremity weakness or gait difficulties in more than 90% of patients. The clinical features of distal myopathy with rimmed vacuoles are similar to distal muscular dystrophy except that lower extremity weakness preferentially affects tibialis anterior and peroneal muscles rather than gastrocnemius and soleus. Consequently, these patients have more difficulties with heel walking, while toe walking is relatively normal.

More diffuse and severe weakness developed with increasing duration of disease. Steppage and waddling gait developed within 1–4 years of onset with patients noticing difficulties climbing stairs. Within 2–7 years patients were unable to lift their head when in the supine position and began to notice upper extremity weakness. Ten to 13 years

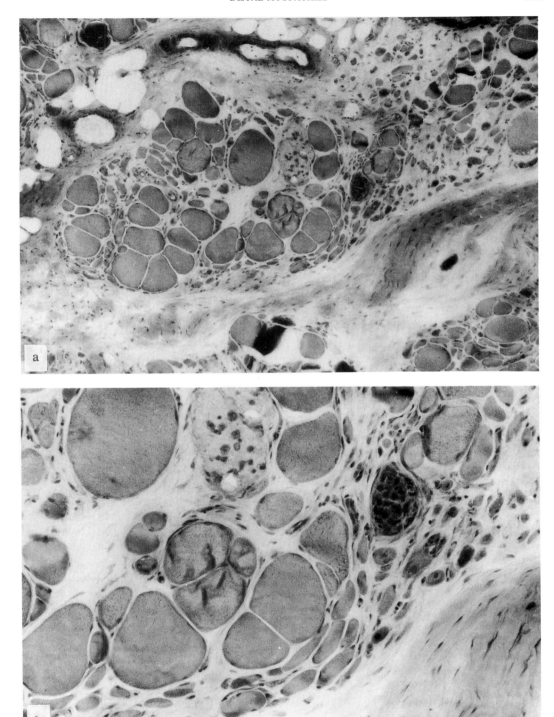

Fig. 1. Light micrograph of gastrocnemius muscle from a young woman with distal myopathy and marked CK elevation showing fiber hypertrophy, atrophy, and degeneration, typical of distal muscular dystrophy. Trichrome, original magnifications (A) ×50, (B) ×80. (Courtesy of Scott Heller, M.D.)

after onset of disease, patients could no longer stand by themselves and proximal upper extremity weakness was more severe than distal weakness. By 15 years after the onset of disease, patients showed severe generalized muscle wasting without involvement of facial, extraocular, bulbar, or intercostal muscles or diaphragm. Patients could not turn over or sit from the supine position because of axial muscle weakness.

In contrast to distal muscular dystrophy, CK in distal myopathy with rimmed vacuoles was generally normal or less than 5 times normal, although occasionally as high as 10 times normal. Electromyography showed myopathic abnormalities. Fibrillation potentials were reported in more than 50% of cases. Although some patients complained of cold hands and feet, clinical sensory exam and nerve conduction studies were normal.

Pathology of distal myopathy with rimmed vacuoles. The most prominent finding on muscle biopsies reported by Nonaka et al. (1981, 1985) was the presence of rimmed vacuoles in both type I and II muscle fibers. The vacuoles were rimmed with basophilic granules which were acid phosphatase-positive but Oil Red O- and PAS-negative. On electron microscopy, the fibers contained numerous intracytoplasmic vacuoles with lamellar bodies resembling myelinated fibers; autophagic vacuoles were also noted. Marked variation of muscle fiber size was present with a type 1 fiber predominance and an increased number of type 2C muscle fibers. In most biopsies, moderate endomysial fibrosis and adipose replacement was present. In general there was no necrosis, phagocytosis, or regeneration of muscle fibers (Fig. 2). One biopsy (Case 3, Nonaka et al. 1981) contained 2 myonuclei with filamentous inclusion bodies. There was no evidence of inflammation in any of these biopsies. The lack of inflammation in distal myopathy with rimmed vacuoles can be used to distinguish this disorder from inclusion body myositis in which more than 90% of biopsies demonstrate endomysial inflammatory exudates (Ringel et al. 1987; Lotz et al. 1989).

Rimmed vacuoles in neuromuscular diseases. There are a variety of neuromuscular diseases which exhibit vacuoles on muscle biopsy. The type of vacu-

ole, its histochemical characteristics, overall muscle pathology, and the clinical picture must be considered in determining the significance of this pathological finding. Autophagic vacuoles, whether rimmed or not, are usually membrane-bound, contain degradation products, and are acid phosphatase-positive (Banker and Engel 1986). Fukuhara et al. (1980) published an excellent description of the morphology and histochemistry of rimmed vacuoles. They reviewed 12 muscle biopsies from a variety of disorders including distal myopathy, Kugelberg-Welander disease, and polymyositis, which all showed rimmed vacuoles on light microscopy. The rimmed vacuole was centrally or peripherally placed in the muscle fiber and was surrounded by a basophilic rim on hematoxylin and eosin, with an eosinophilic rim on Gomori's trichrome stain. One or 2 vacuoles per muscle fiber were noted in 0.2–14.4% of the fibers of the muscle biopsies. Vacuoles were more commonly found in the type 1 muscle fibers, but were occasionally seen in the type 2 fibers as well. Histochemically, the rim of the vacuoles was almost always acid phosphatase-positive, and some were PAS-, Oil Red O- and Alcian blue (pH 2.6)-positive. Ultrastructurally, the vacuoles were multilaminated membranous structures with a mixture of glycogen granules, small myelin-like figures, dense bodies, and amorphous granular or fibrillar material. A membrane delineating the vacuole was not often found.

Inclusion body myositis or distal myopathy with rimmed vacuoles? No single clinical or histological feature distinguishes distal myopathy with rimmed vacuoles from inclusion body myositis. However, when all clinical and histological features are considered, it is possible to place most patients into one or another category. In contrast to distal myopathy with rimmed vacuoles, weakness in inclusion body myositis usually begins after age 50 and there is a preponderance of men (Ringel et al. 1987; Lotz et al. 1989). Later onset has been noted in recent studies that have not confirmed earlier reports of a bimodal age distribution of inclusion body myositis (Eisen et al. 1983). Weakness is more likely to affect proximal leg muscles (Ringel et al. 1987; Lotz et al. 1989). In contrast to distal myopathy with rimmed vacuoles, there is slow but relentless progression of weakness with

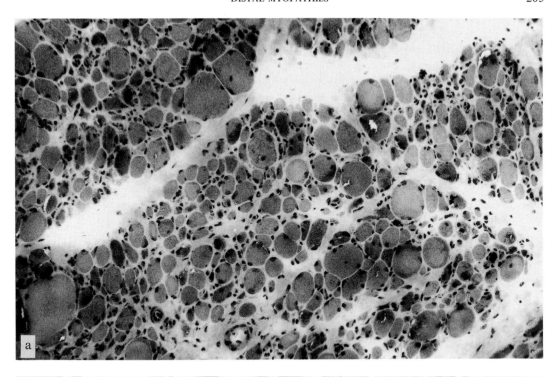

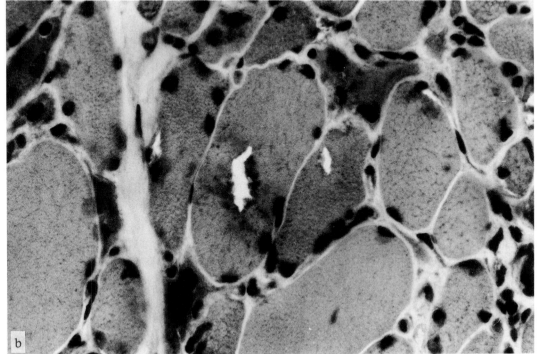

Fig. 2. Light micrograph of gastrocnemius muscle of young woman with distal myopathy and slight elevation of CK showing marked variation in fiber size with numerous rimmed vacuoles typical of distal myopathy with rimmed vacuoles. Hematoxylin and eosin, original magnifications (A) ×33, (B) ×132.
(Courtesy of Scott Heller, M.D.)

selectively severe involvement of quadriceps, iliopsoas, tibialis anterior, biceps, and triceps with relatively early disappearance of the knee jerks. Although distal weakness is common in inclusion body myositis, it is usually not more severe than proximal weakness. Dysphagia, not seen in distal myopathy with rimmed vacuoles, is reported in up to 40% of patients with inclusion body myositis.

The histologic findings in inclusion body myositis overlap with those of distal myopathy with rimmed vacuoles. As described by Carpenter et al. (1978), the typical findings in inclusion body myositis are vacuoles rimmed with basophilic granules, inflammatory infiltrates, atrophic fibers with a variable number of angulated fibers, fibrosis, type 1 and 2 muscle fiber involvement, and the demonstration of filamentous inclusions in the cytoplasm or nuclei of the muscle fibers. These inclusions range from 10–25 nanometers in diameter and are straight or curve slightly (Engel and Banker 1986).

The presence of rimmed vacuoles and filamentous inclusions, although necessary for the diagnosis of inclusion body myositis, does not alone appear sufficient (Carpenter et al. 1978). A patient with a clinical diagnosis of Kugelberg-Welander disease showed rimmed vacuoles with a mass of filaments of 16–19 nanometers in diameter without inflammation on muscle biopsy (Fukuhara et al. 1980). Similarly, Matsubara and Tanabe (1982) reported a case of hereditary distal myopathy with filamentous inclusions without inflammatory infiltrates. Several reports of inclusion body myositis with typical clinical history and filamentous inclusions described only sparse or a total lack of inflammatory infiltrates (Carpenter et al. 1978; Julien et al. 1982).

Lotz et al. (1989) stated that, with adequate sampling of the muscle biopsy, the presence of rimmed vacuoles, at least 1 group of atrophic fibers, and endomysial inflammation predicted the presence of filamentous inclusions in 93% of the 48 muscle specimens examined. More than 90% of patients with inclusion body myositis have some evidence of inflammation on their muscle biopsy and some studies of inclusion body myositis have used inflammation as an inclusion criteria (Ringel et al. 1987; Lotz et al. 1989). These reports underscore the lack of specificity of any individual feature seen on muscle biopsy. The muscle biopsy must be reviewed in the context of the other clinical and laboratory data to allow accurate classification of an individual patient.

TREATMENT

There is no known therapy for distal myopathy. A number of patients treated with steroids empirically have shown decreases in elevated serum CK without demonstrable improvement (Miyoshi et al. 1986). Supportive therapy including the use of orthotic devices should be used to increase function and safety. Clinical and laboratory evaluation of other family members is suggested to allow for accurate genetic counseling.

REFERENCES

BANKER, B.Q. and A.G. ENGEL: Basic reactions of muscle. In: A.G. Engel and B.Q. Banker (Eds.), Clinical Myology. New York, McGraw Hill (1986) 872.

BARROWS, M.S. and L.P. DUEMLER: Late distal myopathy. Report of a case. Neurology 12 (1962) 547–550.

BATES, D., J.C. STEVENS and P. HUDGSON: 'Polymyositis' with involvement of facial and distal musculature: one form of facioscapulohumeral syndrome? J. Neurol. Sci. 19 (1973) 105–108.

BAUTISTA J., E. RAFEL, J.M. CASTILLA and R. ALBERCA: Hereditary distal myopathy with onset in early infancy. J. Neurol. Sci. 37 (1978) 149–158.

BIEMOND, A.: Myopathia distalis juvenilis hereditaria. Acta Psychiatr. Neurol. Scand. 30 (1955) 25–38.

BIEMOND, A.: Myopathia distalis juvenilis. In: E. Kuhn (Ed.), Symposion über progressive Muskeldystrophien. Berlin, Springer (1966) 95–100.

BORG, K., L. SACHS and L. KAIJSER: Autonomic cardiovascular responses in distal myopathy (Welander). Acta Neurol. Scand. 76 (1987a) 261–266.

BORG, K., J. BORG and U. LINDBLOM: Sensory involvement in distal myopathy (Welander). J. Neurol. Sci. 80 (1987b) 323–332.

BORG, K., G. SOLDERS, J. BORG, L. EDSTROM and K. KRISTENSSON: Neurogenic involvement in distal myopathy (Welander). Histochemical and morphological observations on muscle and nerve biopsies. J. Neurol. Sci. 91 (1989) 53–70.

CARPENTER, S., G. KARPATI, I. HELLER and A. EISEN: Inclusion body myositis: a distinct variety of idiopathic inflammatory myopathy. Neurology 28 (1978) 8–17.

CAVALETTI, G., G. BOGLIUN and P. APALE: Sporadic adult onset distal myopathy. J. Neurol. Neurosurg. Psychiatr. 51 (1988) 486–469.

DE VISSER, M.: Computed tomographic findings of the skeletal musculature in sporadic distal myopathy with early adult onset. J. Neurol. Sci. 59 (1983) 331–339.

DIMAURO, S., G.B. HARTWIG, A. HAYS, A.B. EASTWOOD, R. FRANCO, M. OLARTE, M. CHANG, A.D. ROSES, M. FETELL, R.S. SCHOENFELDT and L.Z. STERN: Debrancher deficiency: Neuromuscular disorder in 5 adults. Ann. Neurol. 5 (1979) 422–436.

EDSTROM, L., L.E. THORNELL and A. ERIKSSON: A new type of hereditary distal myopathy with characteristic sarcoplasmic bodies and intermediate (skeletin) filaments. J. Neurol. Sci. 47 (1980) 171–190.

EISEN, A., K. BERRY and G. GIBSON: Inclusion body myositis (IBM): Myopathy or neuropathy? Neurology 33 (1983) 1109–1114.

ENGEL, A.G. and B.Q. BANKER: Ultrastructural changes in diseased muscle. In: A.G. Engel and B.Q. Banker (Eds.), Clinical Myology. New York, McGraw Hill (1986) 969.

FUKUHARA, N., T. KUMAMOTO and T. TSUBAKI: Rimmed vacuoles. Acta Neuropathol. 51 (1980) 229–235.

FURUKAWA, T., N. ODAJIMA, S. WATABIKI and H. TSUKAGOSH: Distal myopathy of Gowers: a reappraisal. Eur. Neurol. 23 (1984) 144–147.

GALASSI, G., L.P. ROWLAND, A.P. HAYS, L.C. HOPKINS and S. DIMAURO: High serum levels of creatine kinase: asymptomatic prelude to distal myopathy. Muscle Nerve 10 (1987) 346–350.

GOWERS, W.R.: A lecture on myopathy and a distal form. Lancet ii (1902) 89–92.

HARPER, P.: Myotonic Disorders. In: A.G. Engel and B.Q. Banker (Eds), Clinical Myology, New York, McGraw Hill (1986) 1267–1296.

ISAACS, H., M. BADENHORST and T. WHISTLER: Autosomal recessive distal myopathy. J. Clin. Pathol. 41 (1988) 188–194.

JAMAL, G.A., A.I. WEIR, S. HANSEN and J.P. BALLANTYNE: Myotonic dystrophy: a reassessment by conventional and more recently introduced neurophysiological techniques. Brain 109 (1986) 1279–1296.

JULIEN, J., C.L. VITAL, J.M. VALLAT, A. LAGUENNY and A. SAPINA: Inclusion body myositis: clinical, biological and ultrastructural study. J. Neurol. Sci. 55 (1982) 15–24.

KRATZ, R. and M.H. BROOKE: Distal myopathy. In: P.J. Vinken and G.W. Bruyn (Eds). Handbook of Clinical Neurology, Vol. 40. Amsterdam, North-Holland, Publ. Co. (1979) 471–483.

KRENDEL, D.A., J.M. GILCHRIST and E.H. BOSSES: Distal vacuolar myopathy with complete heart block. Arch. Neurol. 45 (1988) 698–699.

KUHN, E. and J.M. SCHRODER: A new type of distal myopathy in two brothers. J. Neurol. 226 (1981) 181–185.

KUMAMOTO, T., N. FUKUHARA, M. NAGASHIMA, T. KANDA and M. WAKABAYASHI: Distal myopathy: histochemical and ultrastructural studies. Arch. Neurol. 39 (1982) 367–371.

LOTZ, B.P., A.G. ENGEL, H. NISHINO, J.C. STEVENS and W.J. LITCHY: Inclusion body myositis: observation in 40 patients. Brain 112 (1989) 727–747.

MAGEE, K.E. and R.N. DEJONG: Hereditary distal myopathy with onset in infancy. Arch. Neurol. 13 (1965) 387–390.

MARKESBERY, W.R., R.C. GRIGGS, R.P. LEACH and L.W. LAPHAM: Late onset hereditary distal myopathy. Neurology 24 (1974) 127–134.

MARKESBERY, W.R., R.C. GRIGGS and B. HERR: Distal myopathy: electron microscopic and histochemical studies. Neurology 27 (1977) 727–735.

MARKESBERY, W.R. and R.C. GRIGGS: Distal myopathies. In: A.G. Engel and B.Q. Banker (Eds), Clinical Myology. New York, McGraw Hill (1986) 1313–1325.

MATSUBARA, S. and H. TANABE: Hereditary distal myopathy with filamentous inclusions. Acta Neurol. Scand. 65 (1982) 363–368.

MILHORAT, A.T. and H.G. WOLFF: Studies in diseases of muscle. XIII. Progressive muscular dystrophy of atrophic distal type; report on a family; report of autopsy. Arch. Neurol. Psychiatr. 49 (1943) 655–664.

MILLER, R.G., N.K. BLANK and R.B. LAYZER: Sporadic distal myopathy with early adult onset. Ann. Neurol. 5 (1979) 220–227.

MIYOSHI, K., H. KAWAI, M. IWASA, K. KUSAKA and H. NISHINO: Autosomal recessive distal muscular dystrophy as a new type of progressive muscular dystrophy. Brain 109 (1986) 31–54.

MIZUSAWA, H., H. KURIASAKI, M. TAKATSU, K. INOUE, Y. TOYOKURA and T. NAKANISHI: Rimmed vacuolar distal myopathy: a clinical, electrophysiological, histopathological and computed tomographic study of seven cases. J. Neurol. 234 (1987a) 129–136.

MIZUSAWA, H., H. KURIASAKI, M. TAKATSU, K. INOUE, Y. TOYOKURA and T. NAKANISHI: Rimmed vacuolar distal myopathy. An ultrastructural study. J. Neurol. 234 (1987b) 137–145.

MUNSAT, T.L.: Facioscapulohumeral dystrophy and the scapuloperoneal syndrome. In: A.G. Engel and B.Q. Banker (Eds), Clinical Myology. New York, McGraw Hill (1986) 1251–1266.

NONAKA, I., N. SUNOHARA, S. ISHIURA and E. SATOYOSHI: Familial distal myopathy with rimmed vacuole and lamellar (myeloid) body formation. J. Neurol. Sci. 51 (1981) 141–155.

NONAKA, I., N. SUNOHARA, E. SATOYOSHI, K. TERASAWA and K. YONEMOTO: Autosomal recessive distal muscular dystrophy: a comparative study with distal myopathy with rimmed vacuole formation. Ann. Neurol. 17 (1985) 51–59.

ORRICO, D., G. TOMELLERI, D.D.E. GRANDIS, E. FINCATI and A. FIASCHI: Sporadic adult onset distal myopathy. J. Neurol. Neurosurg. Psychiatr. 50 (1987) 107–108.

RINGEL, S.P., C.E. KENNY, H.E. NEVILLE et al.: Spectrum of inclusion body myositis. Arch. Neurol. 44 (1987) 1154–1157.

ROWLAND, L.P.: Clinical concepts of Duchenne muscular dystrophy. The impact of molecular genetics. Brain 111 (1988) 479–495.

SCOPPETTA, C., M.L. VACCARIO, C. CASALI, G.D. TRAPANI

and G. MENNUNI: Distal muscular dystrophy with autosomal recessive inheritance. Muscle Nerve 7 (1984) 478–481.

SUMNER, D., M.D'A. CRAWFORD and D.C.F. HARRIMAN: Distal muscular dystrophy in an English family. Brain 94 (1971) 51–60.

SUNOHARA, N., I. NONAKA, N. KAMEI and E. SATOYOSHI: Distal myopathy with rimmed vacuole formation. A follow-up study. Brain 112 (1989) 65–83.

THORNELL, L.E., L. EDSTROM, R. BILLETER, G.S. BUTLER-BROWNE, U. KJÖRELL and R.G. WHALEN: Muscle fibre type composition in distal myopathy (Welander): an analysis with enzyme- and immuno-histo-chemical, gel-electrophoretic and ultrastructural techniques. J. Neurol. Sci. 65 (1984) 269–292.

VACCARIO, M.L., C. SCOPPETTA, R. BRACAGLIA and A. UNCINI: Sporadic distal myopathy. J. Neurol. 224 (1981) 291–295.

VAN DER DOES DE WILLEBOIS, A.E.M., J. BETHLEM, A.E.F.H. MEYER and J.R. SIMONS: Distal myopathy with onset in early infancy. Neurology 18 (1968) 383–390.

VAN KASTEREN, B.J.: Polymyositis presenting with progressive distal muscular weakness. J. Neurol. Sci. 41 (1979) 307–310.

WELANDER, L.: Myopathia distalis tarda hereditaria. Acta Med. Scand. 141, Suppl. 265 (1951) 1–124.

WELANDER, L.: Homozygous appearance of distal myopathy. Acta Genet. 7 (1957) 321–325.

Handbook of Clinical Neurology, Vol. 18 (62): Myopathies
L.P. Rowland and S. DiMauro, editors
© 1992 Elsevier Science Publishers B.V. All rights reserved

CHAPTER 9

Myotonic muscular dystrophy

RICHARD T. MOXLEY III

Neuromuscular Disease Center, Department of Neurology, University of Rochester Medical Center, Rochester, NY, USA

Myotonic dystrophy (dystrophia myotonica, myotonia atrophica, Curschmann-Steinert disease) is a multisystem autosomal dominant disease, the most common form of adult muscular dystrophy (Harper 1989). The symptoms and signs arise in several tissues and organs; throughout the course of the condition the manifestations vary in different patients. Ultimately, in patients, the major problems arise from damage to muscle (skeletal, cardiac, and smooth muscle).

The primary challenge for clinicians presented with an undiagnosed patient with myotonic dystrophy is to consider the disease. The diagnosis becomes apparent if the clinician searches for myotonia, and identifies one specific pattern of weakness: the face, long flexors of the fingers, and dorsiflexors of the feet. However, undiagnosed patients may also be seen first by a neonatologist or pediatrician, with neonatal complications; an obstetrician who faces a failed labor or placenta previa; a general surgeon or anesthesiologist who notes prolonged apnea after cholecystectomy; a cardiologist because of arrhythmia or heart block; an internist or gastroenterologist because of dysphagia, abdominal discomfort and constipation; a pulmonologist because of recurrent atelectatic pneumonitis or sleep apnea; a school counselor or psychologist for evaluation of behavior disorders or mental retardation; an orthopedist for talipes; an oral surgeon for recurrent dislocation

of the mandible; or an ophthalmologist for cataracts.

These undiagnosed patients may not have the severe muscle wasting and weakness often seen in the typical adult form, and often these patients have no complaints about muscle problems. Sometimes, the patient is brought by others (interested family members, spouse, or friends) because of their concern about tripping or weak hands. The neurologist eventually learns about myotonia after persistent questioning or on examination. Only infrequently is myotonia the first symptom. The clinician must recognize that this diversity of clinical problems is characteristic of myotonic dystrophy.

The first portion of the text focuses on the genetics of myotonic dystrophy and the molecular biology that have transformed our concepts of this disorder. The latter portion of this chapter discusses the diagnosis and management and devotes an entire section to possible future treatments and therapeutic trials.

GENETICS OF MYOTONIC DYSTROPHY

Discovery of the gene

The earlier history and formal genesis of myotonic dystrophy have been reviewed in detail by Harper (1989) and in the original series of the Handbook by Roses et al. (1979), and will not be covered here.

The present discussion focuses primarily on genetic studies since 1989.

The gene lesion in myotonic dystrophy (DM) is due to an abnormal expansion of a trinucleotide repeat, CTG, in the q13.3 region of the long arm of chromosome 19. This discovery is the most important finding since the original chapter was published in Vol. 40 of the Handbook (Roses et al. 1979). Three simultaneous reports (Aslandis et al. 1992; Buxton et al. 1992; Harley et al. 1992) summarized the investigations that led to the identification of the affected DM gene. Then, three other reports (Brook et al. 1992; Fu et al. 1992; Mahadevan et al. 1992) provided the description of the isolation (Fig. 1; Table 1).

In his textbook Harper reviewed the historic advances in understanding the genetic basis of myotonic dystrophy. In 1971 the DM gene was linked closely to a locus responsible for the secretion of Lewis red blood cell antigens (secretion of ABH blood group substances) and to the Lutheran blood group locus (Harper 1989). In 1982 the gene was localized to chromosome 19, based upon linkage to the third component (Whitehead et al. 1982). In the mid-1980s the gene was further localized to the long arm of chromosome 19 (Harper 1989). Then the powerful techniques of molecular biology were used to isolate the segment of DNA containing the DM gene. Investigators use three complementary approaches (Harley et al. 1991).

One approach required the construction and characterization of somatic cell hybrid lines that contained contiguous human chromosome 19 portions translocated onto either another human chromosome or a rodent chromosome. These cell lines were used to localize new and existing DNA, protein, and cell surface markers to small regions of chromosome 19, providing a consensus map of the whole chromosome.

Another approach was used simultaneously, the construction of DNA libraries in lambda phage, yeast artificial chromosomes, and cosmids to identify new polymorphic DNA markers linked to the DM locus. This produced a genetic map of the relevant region of chromosome 19 that contains the gene. A third approach involved the application of pulsed-field gel electrophoresis to produce fine-structure maps in the region of DNA markers that lie within 2–3 cM of the DM gene.

Using these different approaches researchers developed both genetic and physical maps of chromosome 19, identifying flanking markers that surround the DM gene (MacKenzie et al. 1989; Johnson et al. 1990; Nokelainen et al. 1990; Yamaoka et al. 1990; Bailly et al. 1991; Brook et al. 1991; Harley et al. 1991a,b; Shutler et al. 1991; Tsilfidis et al. 1991; Aslanidis et al. 1992; Brook et al. 1992; Buxton et al. 1992; Fu et al. 1992; Harley et al. 1992; Jansen et al. 1992; Mahadevan et al. 1992) (see Fig. 1). Before February 1992, several candidate genes for DM were excluded. One example involved insulin resistance, which had been thought to cause the muscle wasting and had been thought to be involved in pathogenesis (Moxley 1983; Moxley et al. 1978, 1980, 1984; Hudson et al. 1987). The gene for the insulin receptor is located on chromosome 19. The insulin resistance could have been due to an alteration in the structure of the insulin receptor caused by the DM gene lesion. However, the gene for the insulin receptor was localized to the short arm of chromosome 19, a long distance from the DM gene. Other candidate genes, such as the gene for hormone-sensitive lipase (an important enzyme in lipid metabolism), the poliovirus receptor gene, the carcino-embryonic antigen gene family, and the kallikrein super family of genes (Schonk et al. 1990) were also excluded. The closest gene markers are those for apolipoprotein C2, for creatine kinase (skeletal muscle type) and for two excision repair genes ERCC1 and ERCC2 (Fig. 1). Other DNA probes are positioned even closer to the DM gene (Fig. 1) but their specific functions require clarification.

The final localization and discovery of the DM gene occurred with remarkable speed, in large measure due to the sharing of gene probes and information prior to publication by a consortium of international laboratories. This coordinated effort was facilitated by the Muscular Dystrophy Association of America and the Piton Foundation who brought together the groups of investigators (Aslanidis et al. 1992; Brook et al. 1992; Buxton et al. 1992; Fu et al. 1992; Harley et al. 1992; Mahadevan et al. 1992). In 1988 the area of chromosome 19 containing the DM gene was 6 million base pairs in length. By 1990 it was 5 million base pairs; by October 1991, 300,000 base pairs; and by February 1992 8,500 base pairs and identification

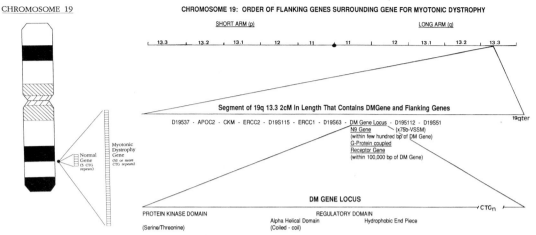

Fig. 1. The left hand panel is a pictorial representation of the normal and abnormally expanded CTG trinucleotide repeats that occur in the region of the DM gene lesion in chromosome 19 for normals and patients with myotonic dystrophy. The right hand panel is a schematic which shows the order of flanking genes that surround the DM gene and there is an enlargement of the DNA segment containing the DM gene as well as an amplified picture of the DM gene locus. Probes D19S37, D19S115 (pEO.8), D19S63 (pD10), N9GENE, D19S112 (pX75b), and D19S51 (p134C) are anonymous DNA probes without an identified gene product. APOC2 represents the gene coding for apolipoprotein C2. CKM is the gene coding for the skeletal muscle form of creatine kinase. ERCC1 and ERCC2 are DNA excision repeat genes. The estimated distance between the segment from ERCD1 to D19S1 ranges from 1 to 1.5 Mb of DNA (Jansen et al. 1992). The X75b-VSSM is a 12-allele $(TG)_n$ variable simple sequence motif (VSSM) gene which is a highly informative marker for the DM gene. Within the DM gene locus there is coding region for a protein kinase which is possibly a serine/threonine kinase (Mahadevan et al. 1992). The gene lesion lies in the regulatory, untranslated region, at the 3' end (Brook et al. 1992; Mahadevan et al. 1992; Fu et al. 1992).

of the DM gene. Future coordinated research efforts should bring together the critical mass of researchers needed to answer important questions in an effective, timely fashion.

Figure 1 denotes the segment of DNA that contains the DM gene locus and the surrounding markers. It also shows the amplification of the DM gene locus, the abnormal expansion of trinucleotide (CTG) repeats. This abnormally expanded region of the gene lies in the regulatory domain at the 3' end, a region that is not involved in the direct translation of the gene product. The flanking segments of the DM gene that surround the expanded CTG repeat have a normal base pair sequence, and there is no evidence of a deletion or other damage to the gene. The only gene alteration that has been identified is the expansion of the CTG repeat (Brook et al. 1990; Fu et al. 1992; Mahadevan et al. 1992).

The normal alleles differ from the alleles observed in patients with DM (Table 1). Normal individuals have a range of expansion for the trinu-

cleotide CTG repeat which varies between 5 and 30 repeats. 70–80% of normal people are heterozygous for this gene; the two most frequent alleles in normal individuals contain either 5 CTG repeats or 13 repeats. In Caucasians the 5 CTG repeat is the most frequent, but in the Japanese it is the 13 CTG repeat allele (Davies et al. 1992). In contrast, patients with DM have a CTG trinucleotide expansion that exceeds 50 repeats. In the vast majority of patients there is a strong correlation between the genotype (the magnitude of the CTG expansion) and the phenotype (the clinical severity). The greater the expansion of the gene, the more severe the symptoms (Aslandis et al. 1992; Brook et al. 1992; Buxton et al. 1992; Fu et al. 1992; Harley et al. 1992; Mahadevan et al. 1992; Tsilfidis et al. 1992).

The unstable region of DNA in the DM gene is similar to the gene lesions described in both the fragile X syndrome (Oberle et al. 1991; Rousseau et al. 1991; Sutherland et al. 1991; Verkerk et al. 1991; Yu et al. 1991) and the X-linked spinal and

TABLE 1

Allelic expansion of the CTG trinucleotide repeat sequence in the myotonic dystrophy gene locus on chromosome 19: comparison of leukocyte DNA analysis in normals to patients with myotonic dystrophy.

	Normal alleles	Myotonic dystrophy alleles
Brook et al. *Cell*, 1992	Analysis of 282 normal alleles shows CTG repeats ranging from 5 to 27 with peaks at 5 (over 40%) and 10 to 16 repeats (over 50%). Over 75% of normals are heterozygous at this locus.	Analysis of 3 minimally affected (cataracts only) grandparents of typical patients reveals CTG expansions ranging from 50 to 61 repeats. Southern DNA analysis in 12 other patients shows an expansion of 2 kb or more. These patients had more severe clinical symptoms and signs.
Mahadevan et al. *Science*, 1992	Analysis of 124 normal alleles shows CTG repeats ranging from 5 to 30 with 5 (35%) and 13 (19%) being the most common. Overall heterozygosity among normals is 81%.	Analysis of 258 alleles from patients in 98 separate families reveals an abnormal expansion of the CTG repeat in 253 out of the 258. Patients from only 2 out of the 98 families have failed to show expansion of the CTG repeat on combined Southern blot DNA analysis and PCR-based oligonucleotide analysis.
Fu et al. *Science*, 1992	Analysis of 40 normals shows CTG repeats ranging from 5 to 30 with 5 repeats being the most common. Heterozygote frequency is 75%.	Analysis with Southern blot DNA methodology reveals expansion of the CTG repeat in 9 out of 9 patients with congenital myotonic dystrophy and in 14 out of 16 classically affected adults.
Tsilfidis, MacKenzie, Mettler, Barcelo and Korneluk *Nature*, 1992	See Mahadevan et al. above.	Analysis of CTG repeat length in 272 affected individuals reveals no visible expansion in 27% on Southern blot; 29% have expansion of 0 to 1.5 kb corresponding to 0–500 CTG repeats; 29% have expansion ranging 1.5 to 3.0 kb (up to 1000 repeats); 13% have expansions of 3.0–4.5 kb (up to 1500 repeats); 2% have expansions greater than 4.5 kb (over 2000 repeats) majority being 4.5–6.0 kb.

bulbar muscular atrophy (LaSpada 1991), which are also due to expansions of trinucleotide repeats. In 1991 investigators described an abnormal expansion of a CGG trinucleotide repeat in patients with fragile X syndrome. By evaluating different family members and monitoring the expansion of the CGG repeat, the researchers found that unaffected carriers had an expansion smaller than that of symptomatic individuals. This 'premutation' was asymptomatic in contrast to the full mutation that leads to full expression of the disease.

An analogy holds with patients having myotonic dystrophy. Patients with very mild signs of DM (cataracts, baldness, electrical myotonia only) may have a slight expansion of the CTG repeat (approximately two times more than the maximum number of repeats of normal individuals (Table 1). Patients with typical DM have a larger number of CTG repeats, and patients with the severe congenital form of DM frequently have the greatest number of repeats, often exceeding 2000. Patients with congenital DM are born in most cases to mothers having CTG repeats exceeding 1500 (Tsilifidis et al. 1992). If expansion of the gene is not seen in Southern blot DNA analysis, no child of a DM mother has yet been found to have congenital DM (Tsilfidis et al. 1992).

However, a word of caution is necessary. Patients from 2 of 98 families (Mahadevan et al. 1992) failed to show a significant expansion of the

CTG region of the DM gene, even though the individuals had clinical signs (Table 1). These cases, in which there is a lack of direct correlation between the size of the CTG expansion and clinical symptoms, indicate the need for more study to clarify the mechanisms responsible for the exceptions.

One other issue requires emphasis. The assay for the CTG expansion uses DNA isolated from circulating white blood cells. However, tissues are affected to different degrees and there may be mosaicism of the gene lesion in different tissues. Varying expansion of the CTG repeat may be related to changes that occur in early somatic cell division following fertilization (personal communication, Roses 1992; Weringa and Jansen 1992) and may thereby produce an expansion of CTG repeats that is not the same in DNA of skeletal muscle, cardiac muscle, brain, or circulating white blood cells. Analysis of the number of CTG repeats in DNA of different tissues in DM patients is needed. The findings should clarify the relationship between the gene lesion, tissue specificity, and severity of symptoms in individual tissues.

The gene product and gene instability

When this chapter was written, little was known about the product protein of the DM gene. The region of the CTG expansion is in the regulatory domain of the gene near the 3′ prime end (Fig. 1). Northern blot analysis shows that messenger RNA is expressed in high levels in heart as well as in skeletal muscle, with smaller amounts of message in brain (Brook et al. 1992). Another recent report (Roses 1992) shows mRNA expression for the DM gene in frontal lobe regions as well as skeletal and cardiac muscle. It seems likely that the gene product is a protein kinase (perhaps either a cAMP dependent kinase or a serine/threonine kinase), an enzyme involved in regulatory signalling reactions (Brook et al. 1992; Fu et al. 1992; Mahadevan et al. 1992; personal communication, Weringa and Jansen 1992; Perryman 1992). It is not known how the gene lesion affecting this protein kinase could be expressed in an autosomal dominant fashion. Perhaps, analogous to the gene lesion in the fragile X syndrome, the CTG expansion in DM causes a loss of expression of the allele carrying the expanded repeat (Brook et al. 1992). If this is respon-

sible for the development of symptoms, the cell response to the DM gene must be very sensitive to gene dosage, because the range of gene expression in the presence of the normal allele ranges only between 50 and 100% of normal.

Another possible mechanism for the dominant effect of the gene lesion is that overproduction of the kinase leads to symptoms. Expansion of the CTG repeat may impair or block the normal cellular control mechanisms that constrain expression of the DM gene. If the gene lesion leads to overproduction of the kinase, there might be a dominant-negative effect on cell metabolism (Brook et al. 1992; personal communication, Weringa and Jansen 1992). Loss of a function by overproduction of a protein kinase has been described (Ahringer et al. 1991; Wightman et al. 1991). Some phosphorylation reactions that result from overproduction of the DM protein kinase might inactivate other enzymes involved in second messenger functions. The gene-related mechanism that leads to the overexpression of DM protein kinase might be caused by the expansion of the CTG repeat so great that normal binding of a gene regulatory element (one that exerts a negative control to this gene site) can no longer be maintained. The protein kinase associated with the DM gene might also exert a dominant-negative effect if the gene lesion leads to overexpression of an abnormal isoform of the kinase. There are normally at least three or more isoforms of this enzyme (personal communication, Weringa and Jansen 1992). If an abnormal isoform were produced in a specific region of the muscle cell, due to an impairment of alternative splicing to produce the desired isoform, an abnormal abundance of an abnormal isoform of the kinase might have a negative effect.

The instability of DNA containing the abnormal CTG repeat is not understood. It seems unlikely that the DM gene lesion is due to frequent mutations at this site. Instead a specific group of people are likely to develop the abnormal CTG expansion in this portion of the gene. For example, there is clear evidence of linkage disequilibrium between the gene for DM and polymorphisms at nearby gene loci in several populations (Mathieu et al. 1990; Nokelainen et al. 1990; Harley et al. 1991, 1992; Tsilifidis et al. 1991; Yamagata et al. 1992). This observation suggests that there have been few mutations, perhaps only a single ancestral event. Alternatively, specific nearby polymorphisms may have predisposed individuals to mutations in the DM gene (Brook et al. 1992). However, it is difficult to imagine how multiple polymorphisms could predispose to the DM gene mutation.

On the other hand, it is possible that the DM gene is a stable, longstanding, mutation. If, in fact, there have been few new mutations, there must be a mechanism that maintains the DM disease allele in the population. This would be particularly necessary in view of the genetic endpoint of severely affected individuals. There is probably a pool of clinically unrecognized individuals who carry an abnormal CTG expansion of the DM gene locus (Brook et al. 1992). In future screening of patients who have elective cataract surgery, it may be possible to identify people with a small but abnormal expansion of CTG repeats (Brook et al. 1992). These people might be viewed as having a premutation similar to that proposed for patients who carry, but do not express, symptoms of the fragile X syndrome (Rousseau et al. 1991). In one family, there was stable transmission of a mildly expanded DM gene in individuals who had few if any clinical signs of disease (Brook et al. 1992). Also, one DM patient with a mild expansion of the CTG repeat had a child who had a reversion of the CTG repeat back into the normal range of repeats (Brunner et al. 1992). Therefore, there is a stable population of individuals who carry the DM gene and can transmit the gene to subsequent generations but may have children with normal DNA on the DM gene.

Another observation supports the hypothesis that DM derives from only a few or one ancestral mutation. The distribution of CTG repeats in specific alleles differs in normal populations and patients with DM. The smallest number of CTG repeats in patients with myotonic dystrophy is 50. A doubling or tripling in the CTG repeat number may have been the ancestral event that predisposed this allele at the DM locus to be susceptible to further expansion up to the allele associated with complete clinical expression (Brook et al. 1992).

The original 'premutation' responsible for DM may have involved a lengthening of the CTG repeat above a critical threshold, perhaps due to a rare event, such as unequal crossing over (Tsilfidis et al. 1992). Then, the metastable lengthened CTG repeat could be at risk for further amplification, and, like the fragile X syndrome, the expansion might become completely unstable after a second threshold of lengthening. The human telomeres consist of tandem repeats of nucleotides rich in guanine. Expansion of the CTG repeat might involve an enzyme such as telomerase (Tsilfidis et al. 1992), which adds repeat nucleotide units to the ends of broken chromosomes (Blackburn 1991). An enzyme of this kind might mediate expansion of the CTG repeat in DM.

The ethnic distribution of DM also supports the hypothesis that DM arises from a small number of mutations. In the Saguenay-Lac-Saint-Jean area of Quebec, Canada, DM occurs 30–60 times more than the world prevalence. These people have been traced back to an affected couple from Europe who settled in Quebec in 1657 (Mathieu et al. 1990). Molecular genetic studies demonstrated a specific haplotype for the DM gene in this population, indicating that a single mutation accounted for all DM patients in one area (Mathieu et al. 1990; Nokelainen et al. 1990; Harley et al. 1991; Tsilfidis et al. 1991; Yamagata et al. 1992).

There is an extremely low prevalence of DM among ethnic Africans (especially in central and southern Africa), Chinese (especially southern China), Cantonese, Thai, and probably Oceanians (Ashizawa and Epstein 1991). These observations suggest that the ethnic distribution is consistent with a proto-world theory of evolution and migration of the human species from Africa (Ashizawa and Epstein 1992). If the DM mutation originally occurred in North Eurasian groups migrating out of Africa to Europe, Southwest Asia, Korea and Japan, the pattern of distribution was compatible with a single early mutation.

Anticipation

For many years clinicians noted that in DM families, symptoms appeared at an earlier age and were more severe in successive generations (Chaughey and Myrianthopoulos 1963; Howeler 1986; Harper 1989; Howeler et al. 1989; Harper et al. 1992) (Fig. 2). This phenomenon, known as anticipation, generated debate among geneticists and clinicians. Harper et al. (1992) commented in detail about the

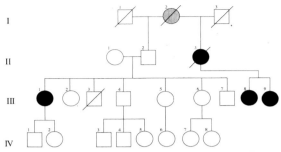

Fig. 2. Four generations in a family in which four women have definite clinical signs of myotonic dystrophy (I-2, II-3, III-1, III-8, III-9). Patient II-2 is a 76 year old man who is an obligate carrier of myotonic dystrophy who on clinical examination has no muscle weakness and no evidence of cataract formation. Electromyographic investigation of several muscles reveals no myotonia or diagnostic abnormalities. The younger patients in the 4th generation have not shown clinical signs of myotonic dystrophy. Analysis of DNA obtained from circulating leukocytes to search for evidence of the abnormal gene lesion in myotonic dystrophy, an expansion of the CTG trinucleotide repeat, is not available at present. Patient I-2, the mother of patient II-2, was suspected of having myotonic dystrophy and died suddenly at 32 years of age. She was not examined by Dr. Ricker. Special appreciation is given to Dr. Kenneth Ricker for providing the clinical information and the family tree on this family afflicted by myotonic dystrophy.

Observations that the degree of expansion of the CTG repeat correlates generally with the clinical severity (Brook et al. 1992; Fu et al. 1992; Mahadevan et al. 1992; Tsilfidis et al. 1992), have provided the molecular basis of anticipation. Instability in the DNA of the DM gene accounts for this phenomenon. The nature of the DM gene lesion has supported the clinical observations that defied explanation by genetics available for 75 years. As Harper et al. (1992) emphasize, the lesson for both investigators and clinicians is that new mechanisms have to be considered when we are faced with observations that seem implausible at first glance.

They also asked whether the phenomenon of anticipation, caused by an instability in DNA, could be more widespread. Expression of a trinucleotide sequence might be one genetic alteration that accounts for other diseases. An autosomal dominant disorder, such as Huntington's disease, might result from an instability in DNA. Such a gene lesion might explain the childhood forms and early-onset adult cases in successive generations in some Huntington's disease families.

Prenatal diagnosis

Prior to the discovery of the DM gene lesion, DNA probes had already been identified that were close to the DM gene. The use of these markers provided an opportunity for both presymptomatic and prenatal diagnosis (Ott et al. 1990; Lavedan et al. 1990, 1991; Speer et al. 1990; Milunsky et al. 1991a,b). Milunsky et al. (1991), using a variety of DNA probes in 74 members of 12 families, found valuable information concerning genetic counseling for patients in 11 of the 12 families. Now, with knowledge of the gene lesion, presymptomatic screening and prenatal diagnosis is more reliable and very feasible. The CTG repeat has been shown to be expanded in more than 96% of DM patients using a combination of Southern blotting and PCR techniques (Tsilfidis et al. 1992; Shelbourne et al. 1992). Related technology has been used for prenatal testing for the fragile X syndrome (Rousseau et al. 1991; Sutherland et al. 1991).

Tsilfidis et al. (1992) emphasized the usefulness of DM gene screening in genetic counseling. They found that women with DM who had a CTG ex-

debate, and emphasized the carefully controlled studies of Howeler (1986) and Howeler et al. (1989) in substantiating the validity of anticipation, even before the DM gene was identified. Howeler et al. provided clear support to the theory of genetic anticipation in DM. Their observations have been confirmed by reports that the severity of findings in successive generations worsens as the expansion of the gene lesion increases (Brook et al. 1990; Fu et al. 1992; Mahadevan et al. 1992; Tsilfidis et al. 1992). Harper et al. (1992) point out that, as early as 1918, Fleischer had proposed that anticipation occurred in DM. However, in 1948 Penrose developed a counter argument to this proposal, and over the years the phenomenon of anticipation was considered to be an artifact of ascertainment bias by many geneticists (Harper et al. 1992). However, Howeler et al. (1989) followed a careful clinical definition for documenting disease severity in DM; in 14 DM families there were 61 parent-child pairs. In 98% symptoms appeared earlier in the child.

pansion exceeding 1500 repeats had a higher risk of having a child with the severe congenital form. In contrast, in DM mothers who had a small CTG expansion, such that Southern DNA blot analysis failed to detect enlargement of the DNA segment containing the gene, there was a lower risk that they would have an affected child with the severe congenital form.

It will be important to define the DM gene size in the normal population and establish the variation in its size in DM patients, especially those who carry only a small abnormal expansion. Some of these individuals will never develop overt clinical signs of DM. Also, some can transmit the small expansion to the next generation who, in turn, will have few if any clinical symptoms (Brook et al. 1992). In addition there are a few cases in which the allele decreases in size on transmission (Brunner et al. 1992; Shelbourne et al. 1992). It is a challenge to determine the appropriate advice for these individuals who have this kind of 'premutation' at the DM gene locus.

CLINICAL FEATURES OF THE ADULT FORM AND CONGENITAL FORM OF MYOTONIC DYSTROPHY AND MULTISYSTEM INVOLVEMENT

Clinical features in adult forms

Modern reviews of the clinical features of DM have been published (Harper 1989; Roses 1992). These sources supplement the clinical descriptions given here.

DM commonly presents in adults as either the mild or classical form. The discussion that follows will compare the pattern of muscle weakness that is seen in classical DM with that observed in the milder adult form (Table 2). This comparison will demonstrate the increase in the severity of DM in successive generations. Figures 3, 4, 5, 6, 7 and 8 show patients with either the classical or mild forms of myotonic dystrophy. Some of the photographs depict the classical pattern of muscle wasting. However, as the companion photographs of the mild form depict, there is much variation in the severity of signs within a single family. The number of individuals with the classical pattern is less than 25% of the actual total number of affected individ-

uals (Roses et al. 1979; Roses 1992). Over the life of the patient the proportion rises to approximately 50% (Roses et al. 1979). Future studies based on DNA probes may redefine the actual proportion of affected individuals who develop the classical form of the disease, as well as those who fail to manifest clinical symptoms.

There is little or no relationship between the rate of disease progression and sex of the patient, sex of the affected parent, or age at onset (Mathieu et al. 1992). However, genotype-phenotype correlations were assessed. DM manifestations within a family may be partial or incomplete (Pryse-Phillips et al. 1982; Telerman-Toppet et al. 1984; Mathieu et al. 1989). Mathieu et al. (1989) identified 130 patients from a population of 602 people examined in 88 kindreds with a partial syndrome of facial muscle weakness and wasting who lacked both myotonia and cataracts. Of these 130, 44 were re-evaluated 2.5 years later. Thirty showed no change; 8 had progressed to typical signs of DM; and 6 no longer had abnormal clinical findings. These observations emphasize the important practical role that testing for the DM gene will have in identifying actual carriers of the gene.

The muscles most frequently involved in the classical form of adult DM are the facial muscles, elevators of the eyelids, temporalis and masseter muscles, sternocleidomastoid muscles (Figs 3 and 4), distal muscles of the forearm (especially the

TABLE 2

Classifications of myotonic dystrophy: correlations of clinical signs with the age of onset. (Guidelines of Dyken and Harper 1973)

Mildest
  Onset: middle to old age
  Major findings: cataracts, minimal or no muscle
    abnormality

Classical
  Onset: adolescence and early adult life
  Major findings: myotonia, muscle weakness
    (face, forearms, foot dorsiflexors)

Congenital
  Onset: at birth, frequent history of hydramnios and
    reduced fetal movements
  Major findings: neonatal respiratory distress,
    hypotonia, bilateral facial weakness, feeding
    difficulty, talipes, mental retardation

Fig. 3. The upper and lower panels of this figure represent serial photographs of a father and son both afflicted with myotonic dystrophy. To the right are two photographs, in the upper panel of the father (age 77 yrs) and two in the lower panel of the son (age 41 yrs) obtained in 1991. The two photographs furthest to the left were obtained when the father was 45 yrs of age and the son was 18 yrs old. Those photographs immediately adjacent to the ones obtained in 1991 show the father at age 62 yrs and the son at age 26 yrs. Both patients have had cataract surgery, right eye lens implant for the father and left eye for the son. Ptosis, bifacial weakness, wasting of the sternocleidomastoid, temporalis, and masseter muscles as well as frontal balding are apparent in the son and difficult to detect in the father.

flexor profundus in the initial stages) (Fig. 7), and dorsiflexors of the feet (Fig. 6). Later some patients show weakness of the quadriceps femoris muscle (especially the distal portion), the diaphragm and intercostal muscles, intrinsic muscles of the hands and feet, palatal and pharyngeal muscles, tongue, and extraocular muscles. Pelvic girdle muscles and the hamstrings are frequently spared, and the gastrocnemius and soleus muscles have often been affected only late in the disease.

The clinical course of patients with either the mild or classical forms varies from individual to individual, but is usually one of slow progression with relatively long periods (months to years) of stable function. Patients with the mild form are troubled mainly by cataracts and minimal weak-

ness, primarily in muscles of the forearm, dorsiflexors of the feet, or, in late stages, muscles of respiration and swallowing. These patients are only infrequently troubled by myotonia or disturbances of central nervous system (hypersomnia, sleep apnea, and personality changes). Cardiac conduction abnormalities, on the other hand, and occasional disturbances of gastrointestinal function, are not uncommon.

In contrast to those affected with the mild adult form, those with the classical type of disease have a more rapid progression of symptoms and are more troubled by the problems due to muscle weakness and myotonia as well as non-muscular complications. Patients with the classical form are less likely to live a normal life span. In midlife, they

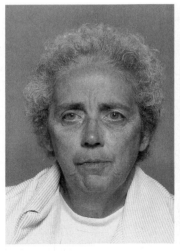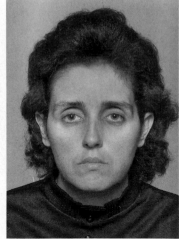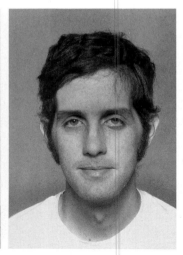

Fig. 4. A mother (58 yrs), her daughter (age 35 yrs) and son (age 31 yrs). The mother was not diagnosed as having myotonic dystrophy until she accompanied her daughter and son to their initial neurological evaluation. The mother has cataracts, grip myotonia, mild weakness of muscles in the forearms, hands, distal portions of the lower extremities and face. Both her daughter and son have symptoms of the classical type of myotonic dystrophy (see Table 2).

often face disability caused by respiratory muscle weakness and may die from hypoventilation and pneumonia. Weakness of pharyngeal and laryngeal muscles, as well as esophageal and gastric hypomotility, predispose to aspiration which leads to atelectatic collapse of the lung or pneumonia, some turning fatal. Cardiac conduction disturbances, especially heart block, can be life-threatening. Patients with the classical DM are also at greater risk of life-threatening complications of anesthesia (see Table 9). Measures to improve the quality of life for these patients are discussed in the section on *Management*.

Clinical features of a congenital myotonic dystrophy

Congenital DM is the most severe form (Table 2). The original chapter on myotonic dystrophy in the Handbook (Roses et al. 1979) presented typical photographs of patients with congenital DM and summarizes much of the historical background. In 1960 Vanier provided the first convincing description of congenital DM (Vanier 1960). Harper and Dyken (1972) suggested a maternal environmental factor to account for the clinical observation that only women affected with DM had offspring with the congenital form. Subsequent publications

(Harper 1989) firmly established maternal transmission but the explanation remains unknown. It was thought that the maternally mediated pattern of inheritance might relate to an influence of mitochondrial DNA. Mitochondria are inherited from the ovum, and any abnormality in mitochondrial DNA might have a greater influence in infants born to an affected mother. However, there is no alteration in the DNA isolated from mitochondria in children with congenital DM (Thyagarajan et al. 1991). Another possible explanation might be a circulating factor that travels transplacentally; for instance, the bile salt, chenodeoxycholate, (Tanaka et al. 1982). However, the circulating levels of this compound vary in women (Soderhall et al. 1982), making it difficult to demonstrate a direct relationship between bile salt level and congenital DM. Very recently Roses (1992) has discussed the various theories for the cause of congenital myotonic dystrophy and has emphasized the possibility that genetic imprinting may provide an explanation. The reader should review this article for further detail (Roses 1992).

The vast majority of DM women who have a child with the congenital form, there is an extremely large expansion of the CTG repeat in the maternal gene (Tsilfidis et al. 1992). A specific group of DM women seem to be at special risk. This abnormal expansion of CTG repeats has also been seen in males with DM, but they do not father offspring with the congenital form.

Before the discovery of the gene lesion, researchers had already determined that the risks of a DM woman for having a child with the congenital form seemed to be related to the mother's clinical status (Koch et al. 1991). In an affected woman who had had no previous pregnancies, her general risk was believed to be 3–9% for having a congenitally affected child in any affected offspring with her first pregnancy. If such a mother had a child with congenital DM her risk was said to have increased to 20–37%. This percentage risk may be in error due to limitations in ascertainment and the number of cases in the series. None of the women with DM who had minimal clinical signs of the disease (cataracts without muscle weakness or clinical myotonia) had had a congenitally affected child. Based upon studies identifying the size of the DM gene (Tsilfidis et al. 1992), not all affected women have equal risk of having children with the congenital form. Future genetic studies may account for the infrequent cases in which women with mild clinical signs do have a child with congenital DM.

The clinical course of congenital DM depends on the severity of the respiratory problems and brain damage that appear in the immediate perinatal period. Many of the patients need ventilator support immediately after birth, and those who have continuous ventilation for four weeks or more have a poor prognosis for survival (Rutherford et al. 1989; Nicholson et al. 1990). Rutherford et al. (1989) noted that 13 of 14 babies with congenital DM had asphyxia, 11 were premature, and 4 had intrauterine growth retardation. Chest X-rays revealed that 9 of 14 had thin ribs and 5 of the 14 had raised right hemidiaphragms. All of the babies who were ventilated for 4 weeks or longer died of respiratory complications before age 15 months. If an infant with congenital DM has only mild respiratory problems during the months after delivery, it is likely that the child will survive with mental retardation and mild to moderate muscle weakness that will not prevent participation in most forms of physical activity. Club foot deformity requires aggressive treatment, as do all childhood infections, especially those involving hearing.

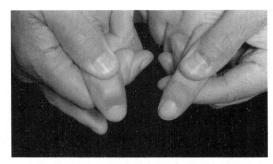

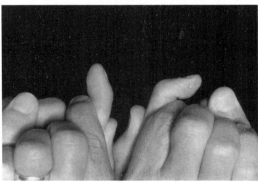

Fig. 5. The upper and lower panels of this photograph demonstrate weakness in the flexor profundus (flexion of the distal PIP joint of the index finger) in a mother 58 yrs old and her daughter 35 yrs old both of whom have myotonic dystrophy. The index fingers are being grasped by the examiner and both patients have been asked to flex the distal portion of their index finger with maximum power. In the upper panel on the left-hand side and in the lower panel on the right-hand side the examiner is holding the index finger of the mother. On the right-hand side of the upper panel and on the left-hand side of the lower panel the examiner is holding the index finger of the daughter. There is a much greater degree of flexion of the distal PIP joint by the mother compared to that observed in her daughter. This demonstrates the significant weakness that develops in the flexor profundus muscles of the forearm and emphasizes the selective pattern of muscle involvement in both mild and classical forms of the disease. Weakness of the flexor profundus often occurs early when there is little weakness of the superficial flexors of the fingers, wrist flexors or extensors.

Management of the muscular and non-muscular complications of congenital DM will be reviewed later (the section on *Management*).

Pathology of skeletal muscle

Previous authors have provided thorough reviews

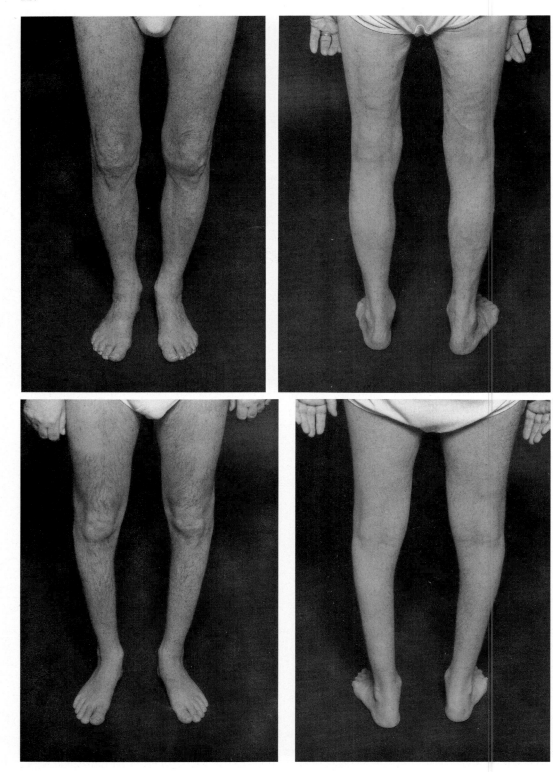

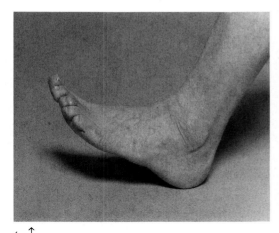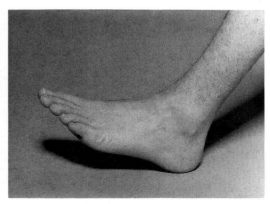

← ↑

Fig. 6. The upper and middle panels of this figure show the lower extremities in a father age 77 (left-hand side) and son age 41 (right hand side). The lower panel shows the action of the foot dorsiflexors in both the father (upper portion) and son (lower portion). The son shows more wasting of the leg musculature and more limitation in dorsiflexion of the foot. The shaved areas on the anterior lateral regions of both thighs in the son represent the sites for needle muscle biopsies that were obtained in association with a research study in which the patient participated. The small linear scar associated with the procedure is shown in greater magnification in Figure 16.

of the muscle pathology (Roses et al. 1979; Carpenter and Karpati 1984; Harper 1989). This section highlights only selected aspects of the muscle pathology (Fig. 9). The degree of structural changes seen in muscle varies greatly from patient to patient and from muscle to muscle. This reflects the tissue specific and muscle specific nature of the pathology. This variation in the degree of pathological change in skeletal muscle is one of the reasons that muscle biopsy is a secondary rather than primary diagnostic procedure (see Table 5). The structural changes in adult muscle (Fig. 9) characteristically include increased central nuclei, chains of nuclei, ringed muscle fibers, sarcoplasmic masses, type 1 fiber atrophy and predominance, increased fiber splitting in muscle spindles, and an increased arborization of nerve terminals.

Other changes such as type 2 fiber hypertrophy, small angular fibers, and moth eaten fibers, are also commonly seen. In contrast, biopsies in the first months of life, in congenital DM show smallness of muscle fibers. Fiber-type differentiation is difficult or impossible to determine based on oxidative enzyme reactions (Carpenter and Karpati 1984). Fibers usually have central nuclei and in muscle obtained during the first two months of life there is a peripheral zone in the fibers which lacks

fibrils and mitochondria (Carpenter and Karpati 1984). Numerous acid phosphatase positive sites can be found in biopsies in infants with congenital DM and may erroneously suggest a storage disease (Carpenter and Karpati 1984). These acid phos-

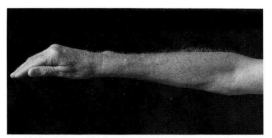

Fig. 7. Greater severity of forearm muscle wasting (forearm flexor compartment being more wasted than forearm extensors) in the son age 41 (lower panel) compared to the father (upper panel).

Fig. 8. This figure shows the 31 year old patient from Figure 4 exerting a forceful grip (left), immediately after attempting to release his grip (middle), and a few seconds after he has attempted to release his grip (right). During the phase of maximum grip there is a relative hyperextension of the right wrist due to the imbalance of muscle forces across the wrist caused by greater weakness of forearm flexors (flexors of the wrist and fingers) compared to wrist extensors. Approximately 20–30 sec was required for the patient to become fully relaxed. Following five forceful contractions the patient was able to relax his grip within 3–4 sec. He had prominent percussion myotonia in the thenar and forearm extensor muscles.

phatase positive sites gradually begin to disappear by 6 years of age. The normal acid phosphatase activity at motor inplates in infancy needs to be distinguished from this alteration. In general the structural changes in muscle obtained from children as well as adults with DM are most consistent with a mechanism that causes muscle fiber atrophy.

The specific pathophysiology of the structural changes in skeletal muscle in DM is not understood. Some authors have suspected a delay or failure in maturation of certain muscle fibers (Carpenter and Karpati 1984). Others suggest that there is an immature or incomplete interaction between the motor axon and its muscle fibers (Kobayashi et al. 1990). Biopsies often show an increased number of satellite cells, suggesting an attempt at muscle fiber regeneration (Carpenter and Karpati 1984). However, atrophy, especially of type 1 fibers, persists and there is a gradual dropout of type 2 fibers (Borg et al. 1987).

The bee venom toxin, apamin, is a polypeptide that attaches to calcium-activated potassium channels; apamin binds to muscle membranes in patients with DM (Renaud 1987). It also binds to

noninnervated rat myotubes in culture, to embryonic fetal muscle before maturation, and to denervated adult rat muscle (Renaud 1987). These findings have raised the possibility that the process in DM may alter the normal regulation of muscle development and perhaps also alter regeneration. On the other hand, there may be a failure of effective tropism mediated by the nerve. There are no histological signs of denervation, such as spread and increased number of acetylcholine receptors (Drachman et al. 1976). However, there is an extra- junctional spread of neural cell adhesion molecules in muscle biopsy samples from DM patients (Walsh et al. 1988). This change may be seen in noninnervated (Moore and Walsh 1985; Covault and Sanes 1986) or denervated (Covault and Sanes 1985) muscle, and does not necessarily indicate that classical denervation is responsible. But the possibility of an alteration in the normal trophic interaction between motor axon and muscle could account for the finding.

Neural and innervated cultured muscle fibers obtained from DM patients have been compared to normal muscle (Kobayashi et al. 1990). Studies of innervated fibers revealed that only 4% of those

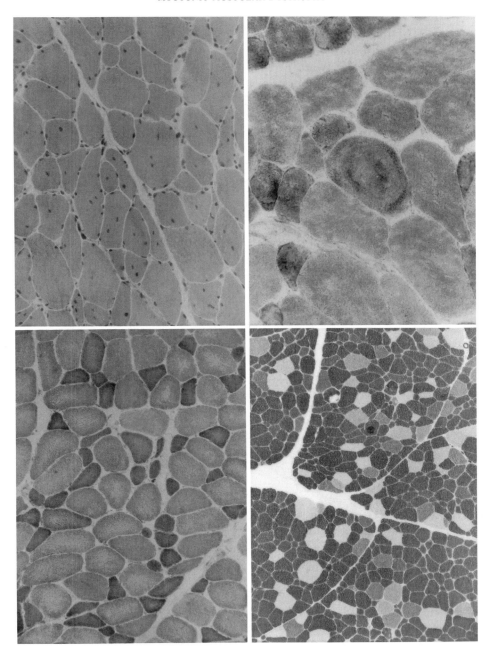

Fig. 9. The four panels in this figure show typical changes observed on muscle biopsy in patients with myotonic dystrophy. Reading from left to right, the first panel is a H & E stain magnified 375x showing prominent central nuclei, angular fibers, and pyknotic nuclear clumps. The second shows muscle stained with NADH tetrazolium reductase (magnified 750x) demonstrating a prominent ringed fiber in the center of the photograph. The third shows muscle fibers stained with NADH tetrazolium reductase (magnified 375x) with atrophy of type I, darkly staining, muscle fibers. The fourth panel shows muscle tissue stained with sodium-potassium ATPase emphasizing type I fiber predominance (darkly staining fibers). Special appreciation goes to Dr. Ziad Rifai for his preparation of these photographs.

from DM patients showed the complex, mature-appearing, staining for acetylcholine esterase. In contrast, 37% of the control innervated cultured fibers demonstrated this mature staining pattern. Electrophysiologic properties of the cultured fibers were different in DM. Both cultures of aneural muscle fibers and innervated fibers from patients showed a lower resting membrane potential than controls indicating an intrinsic myogenic defect.

Kobayashi et al. also demonstrated a possible defect in the response of cultured myotonic dystrophy muscle fibers to insulin. Cells grown in standard growth medium deficiency in insulin, in fibroblast growth factor, and epidermal growth factor had resting membrane potentials that were similar to DM fibers and normal control fibers (Tamoush et al. 1983). However, the resting membrane potential of control muscle fibers cultured aneurally in hormonally supplemented medium containing insulin was 9 mV more negative than the resting potential observed in the fibers from patients with DM. Insulin increases the total number of acetylcholine receptors in aneurally cultured normal muscle fibers and the response of normal cultured fibers to innervation was diminished when insulin was omitted from the medium (Askanas et al. 1985; Desnuelle et al. 1987). The DM gene defect may compromise the normal growth-related function of the insulin receptor, perhaps due to the altered protein kinase regulated by the DM gene.

An explanation for the atrophy of type 1 fibers and the relative loss of type 2 fibers observed in muscle biopsies from DM patients may be due to an alteration in the trophism or effective interaction between type 2 neurons and the muscle fibers in their motor unit. Borg et al. (1987) studied biopsies from the tibialis anterior muscle, and characterized the electrophysiologic response of the extensor digitorum brevis muscle in 7 DM patients (35–46 years of age). Based on the histochemical findings, there seemed to have been a successive loss of type 2 muscle fibers in the dorsiflexors of the foot. Histochemical staining of fibers from muscles showing more advanced myopathic changes revealed a scarcity of type 2A fibers (using the terminology of Brooke and Kaiser 1970). The remaining populations of fibers stained homogeneously for myofibrillar ATPase, had a high content of acid-stable ATPase, and a higher content of alkali-stable ATPase than normal type 1 muscle fibers. Immunohistochemical staining for slow myosin indicated that most fibers were positive. This combination of findings suggested that the type 1 fibers in DM are not entirely typical, and that the predominance of these fibers is related to loss of type 2 muscle fibers. There is, however, no explanation for the hypothesized dropout of type 2 fibers. One possible explanation might be overuse, because overuse of the tibialis anterior muscle in patients with poliomyelitis is associated with transformation of muscle fiber type. The changes in DM may result from an altered pattern of motor neuron firing.

Physiology of myotonia in myotonic dystrophy

Muscle relaxation is shown in classical DM (Fig. 8), and is due to electrical myotonia (Fig. 10). Roses et al. (1979) and Harper (1989) provide a detailed overview of the history of studies of myotonia. This section deals with more recent electrophysiological investigations.

The myotonia in several nondystrophic myotonic diseases (paramyotonia congenita, paramyotonia congenita with hyperkalemic periodic paralysis, hyperkalemic periodic paralysis) (Ptacek et al. 1991; Rojas et al. 1991) is associated with an abnormal sodium conductance caused by an inherited defect in the gene for the alpha subunit of the sodium channel on chromosome 17, and is covered in Chapter 15 in this volume. Similarly, an alteration in chloride conductance is observed in mice with autosomal recessive myotonia (Steinmeyer et al. 1991a,b) and in the autosomal dominant form of myotonia congenita is associated with a defect in the gene for the chloride channel located on chromosome 7, discussed in Chapter 15. Based on these discoveries, the genetic alteration responsible for myotonia in DM is distinct from those responsible for myotonia in the nondystrophic myotonic disorders. The DM gene lesion might cause an alteration in the function of the chloride channel or the sodium channel and in turn produce myotonia.

Patch clamp studies of resealed muscle fiber segments obtained from muscle biopsies of patients with DM (Frank et al. 1990) demonstrated abnormalities in the inactivation of sodium channels, both as an isolated finding, and in conjunction with a reduced chloride conductance. Late openings of the sodium channels were observed frequently in patch clamp studies of muscle fiber segments from DM patients, and these changes in sodium channel function were noted whether or not there was a diminished chloride conductance. There was also a difference in the resting membrane potential in muscle fiber segments from mildly affected patients (myotonia without weakness) compared to segments from more severely symptomatic individuals. The resting membrane potential in mildly affected patients was not significantly different from normals, whereas it was more depolarized in segments from muscle showing more dystrophic changes (−60 to −70 mV compared to −80 mV in normals). The mechanism responsible for the alteration in sodium channel function remains to be clarified. The authors indicate that nerve supply may play a role because aneurally grown muscle from patients fails to express myotonia. Indirect support for the potentially important role that effective innervation may play, has been discussed in the preceding section.

In muscle biopsies from controls and DM patients, there was a 3–6 fold reduction in ouabain-binding sites (Desnuelle et al. 1982) in patients. This reduction in the number of sodium pumps could also play a role in provoking the myotonic phenomenon by causing a rise in extracellular potassium, which itself exacerbates myotonia. There is an excessive rise in venous potassium after forearm exercise in DM (Biezenbos et al. 1990; Wevers et al. 1990). This is perhaps implying altered function of the sodium-potassium pump.

Moreover, there may be an additional disturbance in the regulation of transmembrane potassium flux caused by the presence of immature-fetal type potassium channels. Renaud (1986, 1987) points out that there are apamin binding sites observed in DM muscle that ordinarily would be seen only in immature fetal muscle or denervated muscle. He indicates that the most important property of myotonic fibers is the tendency to fire repetitively. One way to generate repetitive action potentials is to combine a decreased resting membrane potential with the value close to the threshold for the activation of the sodium channel, with the presence of apamin-sensitive calcium activated potas-

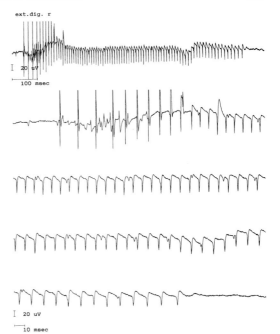

Fig. 10. A typical myotonic discharge recorded from the right extensor digitorum muscle of a patient with the classical form of myotonic dystrophy. This recording was performed in the laboratory of Dr. Kenneth Ricker at the University of Würzburg. Appreciation is given to Dr. Ricker for providing this recording.

sium channels. This combination creates an after-hyperpolarization that follows an action potential and which permits the transient reactivation of the sodium channel after the normal cycle of activation and inactivation during the action potential. These alterations could create repetitive bursts of myotonic activity. The gene lesion in myotonic dystrophy, which affects expression of a protein kinase, may involve a kinase that is intimately involved, not only in mediating signalling from the insulin receptor, but also in regulating signals responsible for the recruitment of the sodium potassium ATPase and other membrane-associated proteins required to maintain normal ion flux, such as, the slow type potassium channel (Lee and Roses 1992; Ropers et al. 1992). A derangement in this signalling could be a major factor contributing to the myotonic phenomenon.

Support for an alteration in the function of the sodium potassium ATPase comes from study of

DM patients. Fenton et al. (1991), studied the amplitude of compound muscle action potentials recorded from the abductor pollicis brevis, extensor digitorum brevis, and tibialis anterior muscles. In DH patients after maximum voluntary contraction, these muscles failed to show the normal post-exercise potentiation of the evoked compound muscle action potential. The authors, as have previous authors (Hull and Roses 1975), suggested that the sarcolemmal sodium-potassium pump has a raised threshold for activation in DM. This interpretation fits with the hypothesized alterations noted above and would be consistent with the possibility that the altered protein kinase may be involved in signalling-activation of the muscle sodium potassium ATPase.

The phenomenon of 'warm-up' (Cooper et al. 1988) and transient paresis (Zwarts and Van Weerden 1989) are two well-known findings found in some myotonic disorders. Both phenomena have been investigated in DM (Cooper et al. 1988). Electrical stimulation was designed to mimic fatiguing exercise. In DM the adductor pollicis muscle showed improvement in force and relaxation characteristics which could best be attributed to normalization of muscle membrane excitation. During this fatiguing electrical stimulation, patients developed a normal muscle twitch potentiation. The authors interpreted these findings as evidence of intact excitation-contraction coupling within the muscle fibers. The apparent normalization of muscle membrane excitation created by exercise seemed to be an important physiologic change and helped to explain the warm-up effect observed in patients. The specific cellular events responsible for the normalization of muscle membrane excitation, however, requires further study.

Transient paresis, a loss of muscle force on initial contraction, is especially prominent after a period of muscle relaxation. This is a prominent clinical symptom in autosomal recessive generalized myotonia congenita. Zwarts and Van Weerden (1989) studied the force of muscle contraction and the associated surface electromyogram in the biceps brachii muscle in three patients with generalized recessive myotonia congenita, in eight patients with DM and three normal controls. The studies were performed after 10 min of rest. Each individual was asked to carry out a maximum voluntary flexion of the elbow for 10.3 sec and given 13 sec for recovery between contractions. A series of five maximum voluntary contractions was recorded. Patients with autosomal recessive generalized myotonia congenita showed a decline in maximum voluntary contraction force of 30–52%. This decline lessened with each subsequent contraction and showed a type of warm-up phenomena. However, this type of warm-up differs from the warm-up of the slowed relaxation. In contrast to the patients with recessive generalized myotonia, patients with DM showed no transient paresis and did not differ from normals except for having a lower value for their maximum isometric contraction force. The DM patients also showed less severe fatigue in maintaining muscle power than normals, perhaps because they maintained a more effective circulation to muscle as a result of the lower absolute value for the maximum contraction force. The occlusion of blood flow that occurs during contraction is directly related to the absolute level of force generated within the muscle. The findings in this study highlight the fact that the clinically delayed muscle relaxation can have different underlying cellular alterations. Specific electrophysiological and biochemical studies are needed to distinguish the specific types of myotonias seen in the different myotonic disorders.

Cardiac abnormalities

Disturbances in cardiac conduction are a frequent finding in adults with myotonic dystrophy (Moorman et al. 1985; Fragola et al. 1987; Hiromasa et al. 1988; Olofsson et al. 1988; Fall et al. 1990; Florek et al. 1990; Hawley et al. 1991). However, conduction block and the various arrhythmias often do not correlate with the severity of symptoms. Altered cardiac conduction is also seen in congenital DM (Forsberg et al. 1990) as well as in childhood onset cases (Harris et al. 1984). Frequently, the management of these conduction disturbances requires a pacemaker (Fig. 11). Occasionally the arrhythmia demands immediate interim treatment with a transvenous pacemaker, which may lead infrequently to complications such as myocardial perforation (Harris et al. 1984). Anticipation of potentially life-threatening arrhythmias and monitoring cardiac rhythm with serial electrocardiograms are essential for the overall treatment plan for DM patients. Atrial arrhythmias and first degree heart block are the most common disturbances (Table 3).

Many patients have first degree heart block (Table 3) and are particularly at risk for subsequent development of complete heart block and sudden death. Pacemaker treatment is effective in controlling Stokes-Adams attacks and should be considered early in management. Fragola et al. (1987) and Hawley et al. (1991) emphasized the role of serial monitoring of the ECG in patients with DM and the usefulness of 24 hour monitoring. Any patients who have presyncope or syncope should be closely questioned, and all patients should have an ECG annually to check for evidence of covert conduction disturbances and arrhythmias, as described in the section on *Management*.

Severe congestive heart failure is an unusual feature of the cardiac disease in DM (Fall et al. 1990), but alterations in left ventricular function are common (Roses et al. 1979; Harper 1989). In an echocardiography assessment of 43 patients Gospavic et al. (1990) found alterations in 12 different measurements of left ventricular contractility. The severity of these alterations paralleled the clinical severity of muscle symptoms. However, other studies of cardiac function failed to reveal such a close correlation between cardiac abnormalities and clinical severity (Holt and Lambert 1964; Veneo et al. 1978; Moorman et al. 1984). This lack of close correlation between clinical signs, cardiac function, and cardiac arrhythmias is another example of variability of tissue involvement in DM; there may be some selective vulnerability of cardiac muscle, especially the Purkinje fibers.

In 12 patients with DM, Nguyen et al. (1988) found that the distribution and extent of lesions in the conduction system tended to correspond closely with antimortem ECG abnormalities, including prolongation of the PR interval, intraventricular conduction delays, and bundle branch block. The most frequently observed histopathologic abnormalities in the conduction system were fibrosis, fatty infiltration, and atrophy. The patients ranged in age from 40 days to 65 years; most were aged between 43 and 57 years. None had died from chronic cardiac disease. Four died from sud-

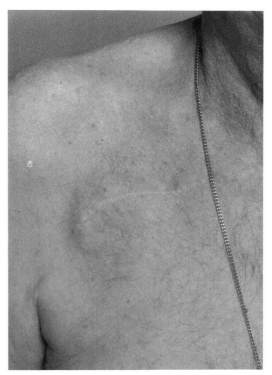

Fig. 11. The linear incision and subcutaneous bulge of a pacemaker is shown, placed to treat heart block in a 77 year old patient with myotonic dystrophy. The patient developed cataracts and conduction disturbance in his heart in his 60s and now has mild muscle weakness and minimal grip myotonia.

den death, two from pulmonary embolus, two from bronchial pneumonia, one from heart attack, one from myocarditis, and two from respiratory failure. Pathologic changes occurred in the S-A node in 6 of 12 cases and in the A-V node in 7 of 12. There was a close correlation between the ECG alterations observed prior to death with excessive fibrosis of the A-V node observed in 5 of 6 cases having first degree heart block and between the fatty infiltration or fibrosis observed in the His bundle in 6 of 6 cases having intraventricular conduction delay.

While pathological studies clearly demonstrated disease involving the myocardium, as well as pathological alterations in the fibers of the conduction system, evidence of clinical disease of the myocardium is much less frequent than that associated with damage to the conduction system. This fact

TABLE 3

Prevalence of ECG abnormalities in selected patients with myotonic dystrophy having mild to moderately severe clinical symptoms.

Reference source	Number of patients	Percentage with abnormal ECG	Percentage with AV block
DeWind and Jones 1950	98	62%	--
Fisch 1951	85	68%	48%
Payne and Greenfield 1963	47	45%	11%
Church 1967	300	86%	38%
Gearrington et al. 1964	17	70%	30%
Moorman et al. 1985	46	72%	33%
Oloffson et al. 1988			
mild symptoms	17	35%	18%
moderate symptoms	24	50%	21%
severe symptoms	24	96%	23%
Florak et al. 1990			
initial examination	45	58%	20%
during follow-up (mean 4–6 years)	45	78%	33%

along with the more severe pathological changes observed on autopsy studies of the conduction system, provides pathological and clinical evidence to support the hypothesis that the expression of the DM gene lesion in the muscle fibers of the Purkinje system may differ from that in cardiac myocytes.

Pregnancy

Women with DM require special care during pregnancy (Table 4). The factors accounting for the complications are not entirely clear. Myotonia and altered contraction power of smooth muscle probably contribute to the prolonged and often ineffective labor. There is anecdotal evidence of increase of leg weakness during pregnancy, which may be due to enlargement of the uterus. The increase in intra-abdominal pressure and the associated upward force on the diaphragm as pregnancy proceeds, may lead to increased weakness of respiration, with increased risk of aspiration and pneumonitis.

There is a 2–3 fold greater rate of spontaneous abortion in mothers with DM. The peak incidence occurs at around 3 months of gestation (Harper 1989). The cause of this increased incidence of abortion is unknown. The growing fetus may have some form of developmental abnormality or may produce a circulating factor that in turn leads to spontaneous abortion. No definite abnormalities

of ovarian function have been identified to indicate a maternal cause (Harper 1989), and it is not yet possible to identify women at increased risk of spontaneous abortion. Specific alterations of the DM gene, such as a critical size of the CTG expansion, may indicate the likelihood of specific maternal and obstetric complications, just as they predict risk of congenital DM.

Pregnant women with DM who have failed labor, or developed complications such as placenta previa or marked postpartum hemorrhage, may require emergency surgery. These women are at increased risk of the complications of anesthesia (Aldridge 1985; Chung et al. 1987; Harper 1989; Blumgart et al. 1990). Careful monitoring of patients during pregnancy and coordination of their care around the time of delivery is necessary (Table 4). The use of epidural analgesia during labor is often helpful, and caesarian section performed with regional anesthesia has been successful. Chung et al. (1987) and Blumgart et al. (1990) discuss the considerations for anesthesia and management during labor and post partum. Chung et al. (1987) described a previously undiagnosed woman; during pregnancy and delivery she experienced placenta previa, atrial fibrillation, shock following caesarian section, and the loss of two infants following two earlier deliveries. In retrospect the children who died were probably afflicted with congenital DM. The mother had classical DM and

TABLE 4

Maternal and obstetric complications in myotonic dystrophy (adapted from Harper, 1989).

Maternal complications
  Increased muscle weakness, especially respiratory
  Increased rate of spontaneous abortion
  Reduced fetal movements and hydramnios
  Prolonged and often ineffective labor
  Sensitivity to general anesthesia (arrhythmias, apnea following C-section)

Obstetric complications

Retained placenta:	All three of these complica-
Placenta praevia:	tions are several times more
Neonatal deaths:	frequent than in normal
	women

exemplifies a frequent problem; failure to have the correct diagnosis established prior to the appearance of complications of DM.

Gonadal abnormalities and alterations in regulation of pituitary hormonal release

While no alterations of ovarian function have been clearly defined, testicular atrophy is common in men with DM affecting 62–86% of individuals (Harper 1989). The past literature was reviewed in detail by Harper (1989). He noted a consensus that there is a marked elevation of follicle stimulating hormone levels in men with DM and that the pattern of secondary overproduction of pituitary gonadotrophins results from primary gonadal pathology. There is strong histological evidence of early selective damage to the seminiferous tubules in DM. The regulation of FSH secretion is thought to be controlled by a factor released by the tubules. In contrast there is only modest or no elevation of luteinizing hormone (LH) (Nappi et al. 1982; Harper 1989). The interstitial cells of the testes exert a primary control on the release of LH. In the later stages of the gonadal atrophy associated with DM there is more damage to the interstitial cells, and at this time there is a greater alteration in LH secretion. The varying degree of damage to the seminiferous tubules and interstitial cells of the testes probably account for reports of both normal and high levels of LH secretion (Harper 1989).

It is uncertain whether gonadal atrophy and decreased plasma levels of testosterone play a role in the muscle wasting of DM patients. Griggs et al. (1985) studied 22 men with DM and compared them to 16 men with other muscle-wasting diseases and 36 normal men. They found no direct correlation between low plasma testosterone levels and diminished muscle mass in either the DM patients or the controls. Weekly intramuscular injections of testosterone at a dosage that produced a 2–3 fold elevation above baseline (values well above the normal range), also increased muscle protein synthesis and muscle mass (Griggs et al. 1986; Kingston et al. 1986; Griggs et al. 1989). Testosterone seemed to stimulate the uptake of amino acids (Kingston et al. 1986) and protein synthesis (Griggs et al. 1986). These anabolic effects seemed to be independent of the effectiveness of insulin (Kingston et al. 1986). There was no improvement in insulin-stimulated glucose uptake following treatment. Despite the improved rate of muscle protein synthesis and the increased muscle mass, the strength was unchanged after one year of testosterone therapy (Griggs et al. 1989). The lack of correlation between circulating levels of testosterone and total body muscle mass (Griggs et al. 1985) and the failure of high replacement doses of testosterone to improve strength (Griggs et al. 1989), suggested that the fundamental defect in skeletal muscle anabolism does not involve intracellular lack of testosterone or testosterone-mediated protein synthesis. Other anabolic factors, perhaps growth hormone, insulin-like growth factor-1 (IGF-1), and insulin may be involved.

No consistent alteration in pituitary-hypothalamic regulation of thyroid or parathyroid function has been found (Harper 1989). Hypometabolism with a decreased basal metabolic rate suggested alterations in thyroid or other hormonal function (Harper 1989). However, Jozefowicz et al. (1987) found when basal oxygen consumption was corrected for the degree of muscle wasting, patients had a normal or somewhat elevated basal metabolic rate compared to controls. Fukazawa et al. (1990) found that levels of thyroid hormone, thyrotropic stimulating hormone, and binding globulin were normal in DM patients. They found a mild decrease in the uptake of radioactive iodine by the thyroid and a decrease in the release of thyroid-stimulating hormone after infusion of thyrotropic releasing hormone. They concluded that there was no evidence of alterations in circulating thyroid hormone, but that there was a slight functional failure of normal regulation of TSH secretion.

Altered parathyroid function has been proposed but no consistent alteration observed (Harper 1989). Yoshida et al. (1988a) suggested that parathyroid function may be increased in DM. They found a greater than normal excretion of urinary calcium and a greater increment in serum calcium following oral calcium loading. There was also a greater urinary excretion of renal derived cyclic AMP. They suggested that the apparent increase in function of the parathyroid gland might be due to an alteration in the negative feedback control of parathyroid secretion, and also an abnormality in the action of parathyroid hormone on the renal tubules.

Some investigators (Okimura et al. 1988; Harper 1989) described deficiencies in the release of growth hormone during sleep and after certain provocative stimuli in patients with DM. The mechanisms responsible for the abnormal patterns of release for growth hormone are unknown. However, a discordance between the regulation of growth hormone release in response to intravenous infusion of growth releasing hormone and that seen in response to hypoglycemia provoked by intravenous infusion of insulin was observed (Okimura et al. 1988). Seven patients with DM were given separate infusions of growth hormone releasing hormone and arginine; stimulation of plasma growth hormone response was less than normal. Following hypoglycemia provoked by intravenous insulin, however, the same patients demonstrated a normal increase in plasma growth hormone. These investigators suggested that patients with DM have abnormal receptors for growth hormone releasing hormone, and that substances such as L-dopa and arginine, which also failed to stimulate an increase in plasma growth hormone, act by increasing the release of human growth hormone releasing hormone. In contrast, hypoglycemia stimulates a rise in plasma growth hormone by inhibiting release of somatostatin. The authors point out that there are normal circulating levels of IGF-1 in patients with DM and the 24 hour secretion of growth hormone is probably not greatly impaired.

Damage to the CNS in DM may influence neurons related to the regulation of growth hormone release. Alternatively, the moderate decrease in the release of growth hormone may be related to the generalized insulin resistance (Moxley et al. 1978, 1980, 1983, 1984; Moxley 1983; Harper 1989). The release of growth hormone and its actions on peripheral tissues antagonize the actions of insulin. The late phase release of growth hormone following oral glucose loading is thought to be a normal regulatory response to prevent an inappropriate fall in circulating glucose after a period of insulin release by the pancreas. The diminished action of insulin on peripheral tissues in DM may result in a feedback on brain to diminish the normal release of growth hormone. This interpretation and other alternative explanations require further investigation. The significance of the alterations in growth hormone release and the beneficial effects observed with short term treatment remain to be clarified (see therapeutic trials in the section on *Management*).

Hyperinsulinism and insulin resistance

It is not known why there is excessive release of insulin in circulating blood following glucose challenge in DM (Huff et al. 1967; Barbosa et al. 1974; Moxley 1983; Moxley et al. 1983; Harper 1989) or why there is tissue-specific insulin resistance (Moxley et al. 1978, 1980, 1984, 1987; Hudson et al. 1987; Harper 1989). Since Huff et al. (1967) described excessive release of insulin following oral glucose, others described increased release of insulin following either oral or intravenous glucose challenge as well as after administration of other secretogogues, such as arginine and leucine (Moxley et al. 1978, 1983; Harper 1989). The insulin released is normal, biologically active, insulin, and the proportion of pro-insulin to insulin is normal (Poffenbarger et al. 1976). The excessive output of insulin following oral glucose is not related to a decrease in muscle mass (Moxley et al. 1983) but seems to be due to tissue-specific insulin resistance that is most severe in skeletal muscle, especially wasted forearm muscles (Moxley et al. 1978, 1980). Adipose tissue shows less marked resistance to insulin (Moxley et al. 1978, 1980; Hudson et al. 1987). Hepatic responsiveness to insulin is normal

(Moxley et al. 1987). Intrabrachial arterial infusions of insulin in patients with classical DM demonstrated that at low, high, and supraphysiologic elevations of insulin (10-fold higher than maximum physiologic levels) (Moxley et al. 1978, 1980) there is a 3-fold lower stimulation of glucose uptake by forearm muscle in DM patients than in comparably wasted disease controls or normals (Moxley et al. 1978). The arteriovenous concentration difference observed across subcutaneous tissues of the forearm (adipose tissue and skin) was also lower in DM patients than in disease controls or normals, but that difference was less marked than in skeletal muscle, being only 40–50% lower than normal. In vitro studies of adipose tissue from patients also documented that there is less severe insulin resistance in fat (Hudson et al. 1987). Whole body 120 min euglycemic insulin infusions using doses that increased in insulin levels from low physiologic to the supraphysiologic range have demonstrated a marked decrease in insulin-stimulated whole body glucose uptake in DM patients (Moxley et al. 1984).

In addition to the diminished uptake of glucose by peripheral tissues observed during whole body insulin infusion, there is also abnormal regulation of amino acid metabolism, especially a fall in arterial alanine as opposed to the normal slight rise (Moxley et al. 1985, 1986). It is not known whether the fall in alanine content reflects decreased production of alanine from pyruvate in muscle, or a combination of decreased production and inappropriately continued uptake of alanine by the liver. Oral glucose tolerance tests (Krentz et al. 1990) show an abnormal elevation in fasting blood levels of glycerol, pyruvate and lactate and a persistent elevation of these metabolites after oral glucose loading. The hyperinsulinemia at rest and after glucose ingestion may inhibit hepatic uptake of gluconeogenic substrates (glycerol, pyruvate, and lactate) to account for the persistent elevation of these potential precursors of glucose. Such a response by the liver would indicate normal hepatic sensitivity to insulin, because insulin normally inhibits gluconeogenesis. This interpretation agrees with previous research in DM demonstrating normal liver sensitivity to insulin (Moxley et al. 1987).

Insulin resistance in DM may in part result from an abnormality of the insulin receptor. DM patients show decreased insulin binding and fewer than normal receptors in monocytes (Harper 1989) and in adipocytes (Hudson et al. 1987) isolated after an overnight fast. There is also an abnormality in insulin binding to monocytes in blood after oral glucose (Moxley et al. 1981). Unlike monocytes from normals the cells from patients showed no increase in binding affinity for insulin. The physiological role of the increased binding affinity in circulating monocytes is not known. Glucose ingestion by normal rats rapidly increases insulin binding to fat cells, with an associated increase in insulin action in these adipocytes (Livingston and Moxley 1982). A similar phenomenon occurs in human fat cells after a glucose load (Arner et al. 1983). Oral glucose may enhance insulin action by an effect on the insulin receptor, perhaps increasing binding affinity for the hormone. This glucose-mediated enhancement of insulin action does not function properly in DM. Oral glucose, even in small amounts (15 or 25 g) enhances the action of insulin on target tissues in normals (Kingston et al. 1986). However, non-obese patients lacked the normal rapid enhancement in whole-body sensitivity to insulin after small doses of oral glucose (Moxley et al. 1987).

The underlying mechanism of the insulin resistance in DM may be due to an alteration in the signalling functions of the insulin receptor. It was once thought that the gene for DM and the gene for the insulin receptor might be one and the same, but the insulin receptor gene has been mapped to the short arm of chromosome 19. There is no structural abnormality of the receptor for insulin isolated from erythrocytes, cultured skin fibroblasts, and transformed lymphocytes (Kakehi et al. 1990). Insulin binding and autophosphorylation are also normal. Something other than a structural alteration in the insulin receptor must be responsible for the insulin resistance of DM.

For activity, insulin must bind to the alpha subunit of the receptor (located on the surface of the cell), which in turn activates the tyrosine kinase intrinsic to the beta subunit of the receptor located inside the cell (Kahn and White 1988). Activation of the tyrosine kinase then stimulates the different actions of insulin through second messenger mechanisms (reactions that in part involve phosphorylation of the beta subunit of the insulin receptor)

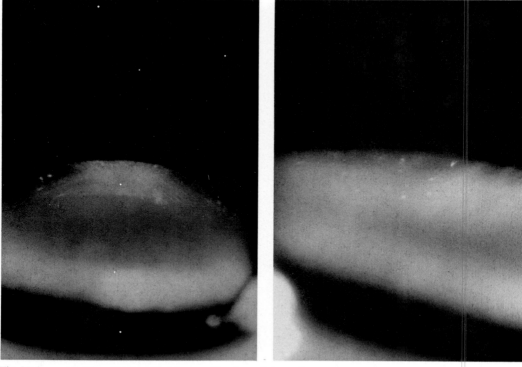

Fig. 12. Lower (left) and higher (right) magnifications of the stellate and punctate characteristics of the cataracts typically seen in patients with myotonic dystrophy. Special appreciation goes to Dr. Douglas D. Koch in the Department of Ophthalmology, Baylor College of Medicine, for providing these photographs.

(Kahn and White 1988). Effective intracellular second messenger signalling is necessary for the actions of insulin, including recruitment of the insulin stimulatable glucose transporter that mediates glucose uptake (Lienhard et al. 1992): recruitment of the sodium-potassium ATPase; activation of carriers that stimulate uptake of amino acids; and activation of processes that inhibit the rate of skeletal muscle protein breakdown. If the protein kinase associated with the DM gene is needed to facilitate all or some aspects of the intracellular signalling required for insulin action, this alteration may have a fundamental role in the insulin resistance of DM. The compromise in the normal anabolic actions of insulin may account for the muscle wasting. The severity of the gene alteration (the degree of expansion of the CTG repeat) may directly control the severity of the alteration in the protein kinase and the insulin resistance. The severity of the insulin resistance in different muscles may also be related to the differences in proportion

of different fiber types. Activation of the kinase associated with the insulin receptor is different in red and white muscle (James et al. 1986). The tyrosine kinase activity is greater in the insulin receptors in type 1 compared to type 2 muscle. Receptor from skeletal muscle of patients with non-insulin dependent diabetes mellitus (NIDDM) show defective intramolecular autoactivation of the tyrosine kinase of the insulin receptor (Obermaier-Kusser et al. 1989). Insulin resistance is characteristic of non-insulin dependent diabetes, and there may be similarities in the mechanism of insulin resistance in NIDDM and DM.

The DM gene lesion may be related to the insulin resistance and the muscle atrophy of DM. Muscle wasting and weakness develop at a rate and severity consistent with the degree of expansion of the CTG repeat. The greater the expansion, the greater the alteration in function of the protein kinase, causing a more severe derangement in intracellular signalling by the insulin receptor. The normal actions of insulin are compromised because the normal phosphorylation reactions mediated through the tyrosine kinase of the insulin receptor may be compromised by the abnormal DM protein kinase. This cellular pathology could lead to progressive muscle wasting and weakness and to increasing derangement of insulin-stimulated glucose transport.

The eye

Abnormalities of the eye in DM include cataracts, retinal degeneration, weakness of extraocular muscles, morphologic abnormalities of the iris, altered pupillary responses, and decreased intraocular pressure, as described in other reviews (Harper 1989; Ashizawa and Epstein 1992); this discussion highlights only a few recent reports. The typical iridescent lens opacities and posterior cortical lens opacities are highly specific for DM (Ashizawa and Epstein 1992) (Fig. 12). The rosette-like subcapsular opacities are still a diagnostic finding of primary value (Table 3). Cataracts are common in both the mild and classical adult forms. About 86% of adult patients have typical cataracts (Ashizawa and Epstein 1992), another example of the tissue-specific expression of DM. Children with congenital DM do not have cataracts, especially in the first few years. It is not known whether there is a specific alteration in the delayed expression of the DM gene in the lens.

Constriction of peripheral visual fields is attributed to retinal pathology (Harper 1989; Ashizawa and Epstein 1992), but abnormalities of visual evoked potential responses are probably due to lesions outside the retina, either pre- or postchiasmatic (Pinto et al. 1987; Kerty and Ganes 1989; TerBruggen et al. 1990). The abnormal visual evoked responses do not vary with age or duration of symptoms (Pinto et al. 1987; Kerty and Ganes 1989). TerBruggen et al. (1990) found slowing of the maximum velocity of visually-guided saccades in 83% of patients. EMG findings suggested that the alterations in saccadic movement were directly related to alterations in the extraocular muscles rather than neural controls. Ophthalmoparesis is seen in some DM patients, but these changes only rarely lead to visual symptoms. If there are symptoms, it is much more likely that they are caused by cataracts, or less commonly, to retinal-visual impairment.

Gastrointestinal disturbances

Hypomotility and delayed transit time are the most prominent abnormalities of gastrointestinal function in DM (Nowak et al. 1982; Eckardt and Nix 1991). Abnormal coordination of contraction in the pharyngo-esophageal area (Siciliano et al. 1990) or the anal sphincter (Eckardt and Nix 1991) are other common, troublesome problems. It is uncertain whether the decreased motility and coordination of swallowing and intestinal function only are due to intrinsic alterations in muscle tissue (dystrophy or myotonia) or whether there is an abnormality of neural control. Investigations of swallowing function in 15 patients with myotonic dystrophy (age 17–59 years) revealed alterations in the junctional pharyngo-esophageal area (Siciliano et al. 1990). The abnormalities were thought to represent selective dysfunction of striated muscle in the upper esophagus. The alterations were independent of duration or severity of other symptoms. The authors have recommended a digitalized recording technique for records of swallowing function in evaluating dysphagia, regurgitation, heartburn, or aspiration.

Abdominal cramps, distention, and constipation are also common complaints. Investigations of the colon (Yoshida et al. 1988b) and anal sphincter (Eckardt and Nix 1991) suggest an alteration in neural function, rather than abnormalities of smooth or voluntary muscle. One 32-year-old man with DM (Yoshida et al. 1988) had a hemicolectomy for treatment of megacolon. Histochemical analysis of the colonic mucosa, smooth muscle and connective tissue revealed no abnormalities, but there were markedly fewer neurons in the myenteric plexus than in normals. The authors speculated that the abnormal colonic motor function in DM may be related to a visceral neuropathy.

Patients with DM show exaggerated rebound contractions of the anal sphincter with lower resting pressures of the anal sphincter and decreased pressure during voluntary contraction (Eckardt and Nix 1991). EMG recordings from sphincter muscle revealed typical myotonic discharges. However, it is unclear whether the abnormal repetitive rebound contractions of the sphincter resulted solely from myotonia, because the EMG of myotonia was intermittent and did not coincide with inflation of the rectal balloon that was used to provoke the repetitive contractions. The authors offered an alternative explanation for these abnormal contractions, that there may be a multitude of defects in the control of anal sphincter function, a combination of neural abnormalities, myopathy, and myotonia.

Other intestinal functions, such as the digestion of fat, may be compromised by impaired contraction of the gall bladder (Nowak et al. 1982; Harper 1989). Gall stones commonly occur in DM patients (Nowak et al. 1982) and may result from abnormal contraction of smooth muscle due to myotonia and dystrophy. There have been no prospective studies of antimyotonia treatment for impaired digestion of fatty foods or gall stone formation.

Other abnormalities include bile salt metabolism (Tanaka et al. 1982; Tanaka and Takeshita 1983) and abnormal elevations of serum levels of gamma-glutamyl transferase (Rönnemaa et al. 1987), suggesting that hepatic disease is associated with DM. Tanaka et al. (1982) suggested that the altered bile salt metabolism might be the maternal factor in congenital DM. However, Soderhall et al. (1982) found that elevated circulating levels of deoxycholic acid were inconsistent. Only two of the three women studied who had children with congenital DM had high serum levels of bile salts. Two other women with children who had late onset DM had slightly elevated levels. Further research is needed to establish if in fact the gene lesion in DM is expressed in liver tissue.

Abnormalities in peripheral nerve and central nervous system

Peripheral nerve. Typically, tendon reflexes become hypoactive as DM worsens (Messina et al. 1976). Some clinicians have suspected some involvement of the peripheral nervous system in DM. Harper (1989) provided a thorough review of the sometimes conflicting findings that support and argue against the presence of peripheral neuropathy. Normal nerve conduction, normal mor-

phologic studies of teased nerve fibers, and lack of postsynaptic spread of acetylcholine receptors (Harper 1989) provide evidence against neuropathy, but other data include a decrease in the number of motor units in the extensor brevis muscle, abnormal nerve terminals on intravital methylene blue staining, reduced myelinated fiber density in sural nerve biopsies, and abnormal thresholds for both heat and cold (Hughes and Wilson 1991) that provide evidence of peripheral nerve dysfunction. On balance, there seems to be some peripheral nerve damage.

Patients in some families manifest signs of both DM and Charcot-Marie-Tooth disease (Bird et al. 1984; Spaans et al. 1986; Harper 1989; Brunner et al. 1991). One father and son had typical features of Charcot-Marie-Tooth disease, including distal sensory loss and pes cavus deformity, in combination with typical DM (Figs. 13 and 14). The son had earlier and more severe signs of DM than his father. However, the son shows less sensory loss and milder pes cavus than his father. The father said that the son's sensory disorder was similar to his own at the same age. It is not known whether the manifestations of Charcot-Marie-Tooth disease are due to separate inheritance of one of the genes for that disease, or whether some aspect of the expression of the DM gene accounts for the Charcot-Marie-Tooth-like syndrome in patients with DM. Brunner et al. (1991) used gene markers for the form of Charcot-Marie-Tooth located on chromosome 17 to evaluate a family with combined symptoms of DM and Charcot-Marie-Tooth disease. They found evidence against involvement of the locus for Charcot-Marie-Tooth disease on chromosome 17. This does not prove that the alteration in the DM gene was responsible for the co-appearance of DM and Charcot-Marie-Tooth disease.

One other possible abnormality in the peripheral nervous system has been mentioned. Wright et al. (1988) found sensorineural hearing loss in 17 of 25 DM patients. Deafness in DM may result from changes in the bony structure of the internal auditory meatus and in the bony wall of the first cochlear turn (Cornacchia et al. 1989). Computerized tomographic in DM patients show overgrowth of bone (Avrahami et al. 1987; Kawamura et al. 1987), and bony compression of the eighth nerve

ness, hearing loss can sometimes masquerade as mental retardation.

Central nervous system. Clinicians have long suspected alterations in the brain in DM because there are often changes in personality (Harper 1989; Roses 1992), hypersomnia and alterations of sleep (Coccagna et al. 1982; Harper 1989; Broughton et al. 1990), neuropsychological abnormalities (Censori et al. 1990; Perini et al. 1989; Harper 1989; Broughton et al. 1990) and an increased risk of depression (Brumback et al. 1981; Brumback and Carlson 1983; Huber et al. 1989; Brumback 1990). The diffidence and nonchalance of DM patients is remarkable (Roses 1992). Unaffected members in a family may have problems dealing with the behavior disorders of patients, especially the precocious and aggressive sexual behavior of young men. Neuropsychological abnormalities have been documented (Huber et al. 1989; Perini et al. 1989; Broughton et al. 1990; Censori et al. 1990). Intellectual impairment was identified in 24–50%, often with selective deficiency on specific tests rather than a uniform decline in capability. In one study of 41 DM patients (Huber et al. 1989) the most significant differences from normal occurred in the mini-mental status examination (largely due to problems with calculation, fluency, visual spatial testing, digit span and depression scale). Ten of the 41 patients had evidence of severe and generalized intellectual dysfunction; the others had little or no cognitive impairment. Patients with severe intellectual disturbance had an earlier onset of both weakness and myotonia and were more likely to have inherited the disease from the mother.

Broughton et al. (1990) sought to determine if these alterations correlated with sleep apnea or hypoventilation. The authors wondered if hypoxemia related to sleep apnea might damage the brain to account for the cognitive deficits. Overnight polysomnography was performed in 8 DM patients. Evidence of central type sleep apnea and other abnormalities of sleep were measured and related to results of neuropsychological tests. The 5 of 8 patients having significant cognitive loss did not have the greatest sleep disturbance or breathing problems in sleep. The authors concluded that the neuropsychological deficiencies of DM cannot be a

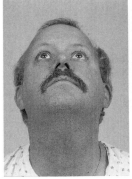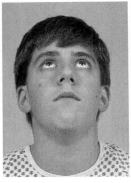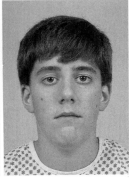

Fig. 13. This figure shows a father, age 41 yrs (top) and his son, in his mid-teens (15 yrs) (lower panel) with myotonic dystrophy and the axonal form of Charcot-Marie-Tooth disease. Clinical examination and electrodiagnostic evaluation of both patients revealed myotonia. Both patients have high-arched feet and a stocking distribution decrease in sensation to pinprick, light touch, and vibration. On electromyographic investigation of distal muscles in the upper and lower extremities, there was a decrease in the recruitment of motor unit potentials typical for chronic denervation. Nerve conduction velocities were within normal limits for both the father and son. The father has mild bifacial weakness and mild weakness of forward head flexion. The son has moderate bifacial weakness, and a mild decrease in the bulk of the temporalis and sternocleidomastoid muscles. Both father and son had readily elicitable grip myotonia. The photographs of these patients and the electrodiagnostic evaluation were kindly provided by Dr. Keshav R. Rao, Director of the EMG Laboratory at the University of Connecticut Health Center.

may be a causative factor. There appears to be increased risk of hearing loss in DM. Patients may benefit from hearing aids. It is also important to consider hearing loss when evaluating mental retardation in children or adults. Like bifacial weak-

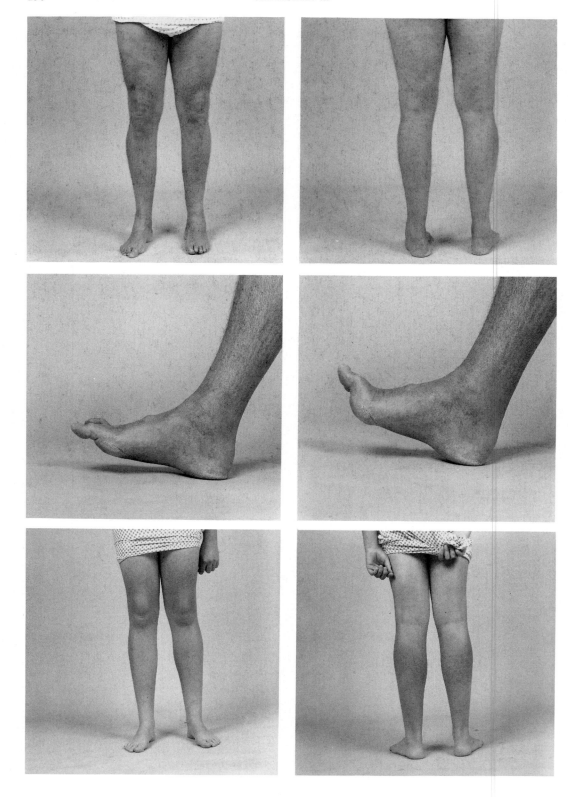

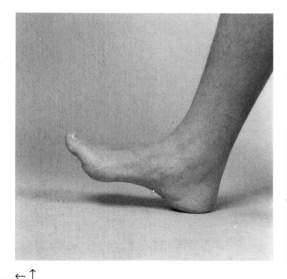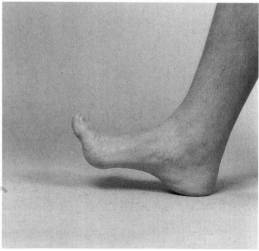

← ↑

Fig. 14. Upper and middle panels show the lower extremities in the father age 41 yrs (left-hand side) and son (right hand side) age 15 years. The lower panel shows the father (left) and the son (right) with their feet first at rest (top) and during maximum dorsiflexion of the feet (bottom). Mild wasting of the distal muscles in the legs is apparent in both patients. There is prominent high-arching of the feet in both and some trophic skin changes can be seen in the feet of the father. Special thanks go to Dr. Keshav R. Rao, Director of the EMG Laboratory, University of Connecticut Health Center, for providing us with these photographs of his patients.

secondary effect of sleep apnea or sleep disruption, but are probably due primarily to a cerebral disorder or damage within the central nervous system related to myotonic dystrophy.

Magnetic resonance imaging shows abnormalities of white matter and cerebral atrophy. Glantz et al. (1989) found ventriculomegaly and periventricular hyperintensity. Huber et al. (1989) found cerebral atrophy, focal nonspecific white matter changes similar to those of multiple sclerosis, and unique white matter abnormalities in the anterior temporal lobes but these were unlike the abnormalities of demyelinating disease (Fig. 15). The etiology of these changes is unknown.

Specific thalamic inclusions have been described in DM (Harper 1989). However, Yoshimura et al. (1990) believe that the thalamic inclusions are nonspecific. They also found Alzheimer's neurofibrillary changes in the brain of a 61-year-old DM patient who died after a cardiac arrest. The neurofibrillary changes showed a similar distribution pattern to those observed in Alzheimer's disease. However, unlike Alzheimer's there were few senile plaques. The neurofibrillary changes also involved the olfactory bulb, brainstem nuclei, and the cer-

ebral neocortex. There was gliosis and neuronal atrophy in the thalamus without Alzheimer neurofibrillary changes. The authors thought that the brainstem and cerebral lesions might play a role in the mental symptoms and altered respiration of DM. They attributed the neurofibrillary changes to increased phosphorylation or accumulation of an abnormally phosphorylated tau protein. Since the DM gene lesion involves a protein kinase and phosphorylation reactions, altered function of this kinase might be responsible for the pathological changes.

Impaired cerebral circulation might be still another mechanism for damage to brain in DM. Anderson et al. (1991) found a decrease of cerebral blood flow in the parietal lobes in 5 of the 6 patients. These alterations in circulation may precede the DM abnormalities. Fiorelli et al. (1992) used positron emission tomography (PET) and found a 20% reduction in cortical glucose utilization. They speculated that there might be an impairment of insulin-facilitated transport of glucose across the blood-brain barrier, because insulin receptors are well represented in brain capillaries. However, there is doubt that there are insulin regulatable

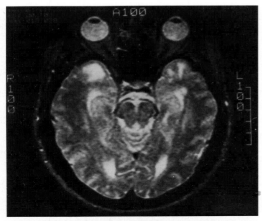

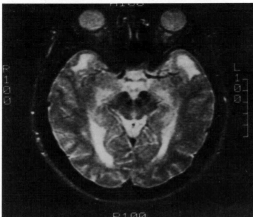

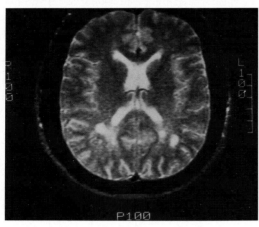

Fig. 15. Three serial MRI images of brain from the same patient. Special appreciation goes to Dr. John Kissel in the Department of Neurology, Ohio State University, for providing these photographs.

glucose transporters in endothelial cells (Lienhard et al. 1990). Nevertheless, there are insulin recep-

tors in brain of uncertain function (Baskin et al. 1987; Unger et al. 1989, 1991). If circulating insulin is important in brain metabolism, the signalling functions associated with activation of the tyrosine kinase portion of the insulin receptor may be altered by the abnormal protein kinase of the DM gene.

Respiratory system

Respiratory problems are troublesome in advanced DM. Elevation of the right hemidiaphragm and weakness of respiratory muscles in congenital DM are usually associated with life-threatening illness; and, if ventilator support is required for more than 4 weeks prognosis is extremely poor. Adults with diaphragmatic and respiratory muscle weakness are at increased risk for complications of anesthesia (see *Management*) and may develop linear atelectasis or pneumonitis, often related to the increased frequency of aspiration in patients (Harper 1989). On of the more insidious and troublesome disorders in DM is alveolar hypoventilation (Coccagna et al. 1982; Harper 1989; Broughton et al. 1990), which may be related to cerebral pathology. Altered function in the centers responsible for hypoxic drive of ventilation may be responsible for episodes of apnea in the postoperative period. Careful monitoring of respiratory function is essential in long-term care (*Management*).

Most respiratory symptoms relate either directly to dystrophic weakness of the respiratory muscles or to an alteration in the neural mechanisms responsible for respiration. Fitting and Leuenberger (1989) described a bout of severe dyspnea in a 47-year-old woman with DM; she improved promptly with tocainide treatment. The authors attributed the dyspnea to myotonia of respiratory muscles rather than weakness. They thought that the antimyotonic action of tocainide was responsible for the improvement. A controlled therapeutic trial would be needed to determine whether tocainide, mexiletine or phenytoin would be useful in preventative therapy for respiratory problems.

Early balding and epitheliomata of the skin

One hallmark finding in most men with DM is early frontal balding (Harper 1989). The mechanism is unknown, and there is no clear relationship between early baldness and testicular function. It is also not known whether DM balding is similar in any way to other forms of inherited baldness. Screening studies with the DM gene should include patients with premature balding as well as cataracts.

Another skin problem in DM is an increased frequency of a benign neoplasm, pilomatrixoma or calcifying epithelioma of Malherbe (Harper 1989). This neoplasm is thought to originate from the primitive cells of the hair matrix.

Immunologic alterations

Deranged immunoglobulin metabolism (Kuroiwa et al. 1980; Harper 1989) and altered function of polymorphonuclear leukocytes (Mege et al. 1988; Harper 1989) have been noted in DM but there is no clear evidence of increased rates of infection or other problems. Catabolism of IgG is accelerated but the mechanisms remain unknown. Regarding polymorphonuclear leukocytes, there is a reduction in the chemotactic response and an altered electrophoretic mobility, despite of normal phagocytosis and killing power (Hughes and Wilson 1991). Mege et al. (1988) found a decrease in the burst of oxygen by polymorphonuclear leukocytes after stimulation by two agents. They suggested that the principal alteration underlying this defective response affected the cell membrane, perhaps involving the activation of protein kinase-C and phospholipase C, which could be related to one protein kinase associated with the DM gene.

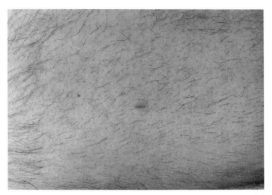

Fig. 16. Healed muscle biopsy site in a 41 year old patient.

DIFFERENTIAL DIAGNOSIS

The differential diagnosis of DM is now much simpler because DM probes are available. Diagnostic procedures can be considered of primary and secondary value (Table 5). Identification of the expansion of the DM gene will provide definite diagnosis. Clinical examination may suffice to identify early manifestations of DM in skeletal muscle: weakness of flexor profundus of the fingers, intrinsic hand muscles, dorsiflexors of the feet, facial muscles, sternocleidomastoid, temporalis and masseter muscles. EMG may detect myotonia even if this is not apparent on percussion or grip testing. Slit lamp is essential to search for the characteristic cataracts. The creatine kinase activity may be normal or slightly increased and in a non-specific range. Muscle biopsy (Figs. 9 and 16) can provide useful supporting evidence. However, structural pathology in muscle is not always seen in mild DM and biopsy is less sensitive than the test given primary value (Table 5).

In adults and children several conditions are considered (Table 6). Clinical evaluation and EMG with quantitation of muscle strength and myotonia before and after exercise in the cold distinguishes hereditary nondystrophic myotonic disorders from DM (Ricker et al. 1990). These different diseases are discussed in Chapter 10. The genes responsible for the nondystrophic, autosomal dominant myotonic diseases (myotonia congenita, myotonia fluctuans, paramyotonia congenita with or without periodic paralysis) and the gene for the

TABLE 5

Useful diagnostic tests in identifying individuals with myotonic dystrophy.

Primary value
   Analysis for gene lesion in DNA isolated from
      white blood cells
   Clinical examination focused toward detection of
      early muscular and non-muscular alterations
   Electromyography
   Slit-lamp examination

Secondary value
   Serum creatine kinase
   Muscle biopsy

autosomal recessive form of generalized myotonia have all been identified. Myotonia fluctuans and paramyotonia congenita with or without periodic paralysis, as well as the hyperkalemic form of periodic paralysis all have gene lesions associated with the sodium channel on chromosome 17 (Ptacek et al. 1991; Rojas et al. 1991). Probes to detect these disorders should be available by the time this chapter is published. The gene probes for facioscapulohumeral dystrophy and nemaline myopathy are also likely to be known soon, and accurate diagnosis will be much easier. Oculopharyngeal dystrophy may closely resemble the classical form of DM, but these patients have no clinical myotonia and do not have the typical pattern of distal muscle weakness and wasting of DM. Facioscapulohumeral dystrophy, distal myopathy, and limb-girdle dystrophy occasionally resemble DM. EMG, slit lamp examination, and DNA analysis of DNA may be needed to establish the diagnosis. Muscle biopsy may prove helpful to distinguish distal myopathy from facioscapulohumeral and limb-girdle dystrophies. Other chapters in this volume deal specifically with these muscle diseases.

Differential diagnosis of adult or congenital DM is rarely a problem. The real difficulty in establishing the diagnosis of DM is that patients may be seen first by cardiologists, anesthesiologists, orthopedists, pediatricians, ophthalmologists, obstetricians, pulmonologists, or oral surgeons. These clinicians often do not include myotonic dystrophy in their differential diagnosis. This is especially a problem for patients with previously undiagnosed DM. In patients known to have the disease, it is imperative for them to carry a card or other information identifying the diagnosis and providing the telephone number of their physician so that appropriate consultation can be provided in case of emergency. This type of coordinated care is needed to avoid inappropriate therapeutic agents or procedures that may entail especially high risk.

MANAGEMENT OF MUSCULAR AND NON-MUSCULAR SYMPTOMS IN MYOTONIC DYSTROPHY

The opportunity for more rational, possibly curative, treatment of DM is now at hand. Once we have a better definition of the actions of the protein kinase that is affected by the gene lesion, specific treatment strategies should follow. Meanwhile, treatment continues to be supportive. The following discussion is divided into sections. The first deals with the management of complaints related to skeletal muscle in adults with DM. The second describes the management of the non-muscular complaints in adults, emphasizing preoperative and postoperative care as well as anesthesia. The third section describes problems in infancy and early childhood. The fourth section describes therapeutic trials and methods for quantitating efficacy of treatment.

Management of complaints related to skeletal muscle in adults with myotonic dystrophy

Serial examinations are helpful in anticipating problems that occur at some point during the illness (Table 7). Patients are often remarkably diffident about their problems. Clinicians must take an active role in eliciting new symptoms. Foot drop

TABLE 6

Differential diagnosis of myotonic dystrophy.

Facioscapulohumeral dystrophy	Autosomal dominant
Distal myopathy	Autosomal dominant
Oculopharyngeal dystrophy	Autosomal dominant
Limb girdle dystrophy	Autosomal recessive
Certain congenital myopathies, e.g.	
Nemaline myopathy and centronuclear myopathy	Autosomal dominant
Certain non-dystrophic hereditary myotonias, e.g.	
Myotonia congenita	Autosomal dominant
Myotonia fluctuans	Autosomal dominant
Paramyotonia congenita	Autosomal dominant
Autosomal recessive generalized myotonia	Autosomal recessive

may lead to ankle sprains, frequent falls, or even ankle fracture. Lightweight ankle/foot orthoses solve the problem.

The distal portion of the quadriceps becomes weaker and wasted, with loss of strength in knee extensors and liability to sudden collapse of the knees. The patient may hyper-extend both knees to maintain stability.

Patients occasionally develop painless effusions in the knee joint. Some patients, especially in the early stages of knee extensor weakness, benefit from new lightweight knee/ankle/foot orthoses and can continue to walk for several additional years.

Myotonia is usually most pronounced in the distal muscles, primarily in the hands. Typically patients have pronounced grip myotonia in the early stages of the classical form but do not complain about the myotonia. In the few whose occupation demands dexterous movements with their hands, a trial of phenytoin (Dilantin) in a dosage of 300–400 mg daily may be helpful. Phenytoin is safer than

some other antimyotonia drugs because it does not affect cardiac conduction (Griggs et al. 1975). If the patient is allergic to phenytoin, mexiletine is an effective alternative in doses of 300–600 mg daily. Before commencing antimyotonia treatment, it is important to obtain a resting ECG as well as blood tests to document normal hepatic function.

Some patients become wheelchair bound. Initially, they may benefit from a combination of lightweight long leg braces for standing and walking short distances. A wheelchair is necessary for longer trips. At this phase support for the head and neck may be necessary, depending on the severity of weakness of the sternocleidomastoids and other anterior neck muscles. In addition to the limb weakness, there is often weakness of the diaphragm and other muscles of respiration. More detailed comments about the evaluation and management of respiratory weakness is presented in Table 8.

Ptosis is frequent, but does not commonly lead to complaints. Occasionally lid-elevating crutches

TABLE 7

Management of complaints related to skeletal muscle in myotonic dystrophy.

Problem	Management
Foot drop	Light-weight molded ankle foot orthoses.
Knee extensor weakness (extreme hyper-extension of knee; painless effusion)	Light-weight long leg knee-ankle-foot orthoses may allow prolonged weight bearing without provoking effusion of the knee joint.
Myotonia	If patients symptoms warrant, phenytoin therapy as initial choice. Mexilitene is an effective second choice.
Generalized weakness	Wheelchair. Use intermittently with braces in earlier stages of generalized weakness (e.g. for traveling long distances within the home or trips outside).
Respiratory muscle weakness with ventilatory failure	Consider nasal ventilation, initially at night and subsequently during daytime. In very selected cases, could consider positive pressure ventilation via tracheostomy.
Neck weakness	Fitted collar; head supports in car seats and chairs.
Ptosis	Lid elevating crutches or props on eyeglasses. Rarely is lid surgery indicated.
Recurrent dislocation of the mandible due to weakness of temporalis and masseter muscles	Infrequently, mandibular surgery is recommended by an oral surgeon evaluating the patient. Guidelines for preoperative and postoperative care need to be followed carefully.

or props on eyeglasses are useful. Rarely, patients benefit from lid surgery.

Many patients with the classical adult form of DM develop a chronic mouth-open posture due to weakness of the temporalis and masseter muscles, and there may be recurrent dislocation of the mandible (Wilson et al. 1989). These patients occasionally benefit from surgery, with careful attention to postoperative care as well as of anesthesia (see Table 9).

Management of non-muscular complaints in adults with myotonic dystrophy

Non-muscular complaints are common in adults with either mild or classical forms. One of the most important is abnormal cardiac conduction. Yearly ECG helps to search for signs of covert arrhythmia or conduction block. Once the PR interval becomes greater than 0.2 sec first degree heart block has developed, and the patients are at increased risk for more severe heart block. If arrhythmias or heart block are present, cardiology consultation and coordination of care with an internist are necessary. In more severely affected patients a cardiac pacemaker should be considered to prevent sudden death. Some patients have a varying degree of heart block that is not detected on resting ECG, and serial 24 hour monitoring may be required to correlate the clinical signs with a specific abnormality in cardiac conduction. In individuals with symptoms of intermittent dizziness, poor exercise tolerance, palpitations, sudden bouts of sleepiness (in contrast to prolonged sleepiness typical of hypersomnia) cardiac conduction disturbance should be suspected. In a patient of this kind who has arrhythmias a pacemaker may be lifesaving. Once a pacemaker is in place, the use of digoxin and antiarrhythmic medications that affect the conduction system becomes safer.

Another frequent complication involves respiration. Patients are at risk for both sleep apnea and primary respiratory failure. Patients suspected of having sleep apnea, based on history or observations by the spouse, are candidates for sleep apnea monitoring and nasal continuous positive airway pressure.

There are frequent bouts of pneumonia, morning headaches, or deterioration of forced vital capacity. The patient needs formal pulmonary function testing, measurements of arterial blood gases, and overnight oximetry and capnography (Table 8). For respiratory insufficiency we often consider a trial of nasal ventilation, which does not require invasive procedures such as a tracheostomy and is often tolerated well by the patient. Only few patients with DM seem to be good candidates for tracheostomy and long-term positive pressure ventilation with a portable ventilator. Negative pressure ventilatory equipment, such as the raincoat or turtle shell, has not been tolerated well and is not suitable. The patient may sleep better with chest and head elevated, often sufficient to maintain comfortable ventilation throughout most of the illness. Sleeping with the head of the bed elevated also helps to avoid any tendency for aspiration. Patients should avoid eating large meals, especially in the evening. Fitting et al. (1989) described a bout of dyspnea that responded to treatment with tocainide, suggesting that the respiratory symptoms were due to myotonia. Controlled studies are needed to assess the efficacy of antimyotonia agents in preventing respiratory complications.

Cataracts are common and require yearly or more frequent evaluation by an ophthalmologist. Cataract surgery is well tolerated and can be carried out under local anesthesia.

Gastrointestinal disturbances are also common, primarily in patients with the classical form, and are often difficult to treat. There may be pseudo-obstruction of the large or small intestine and intermittent gastric dilatation. There may be intermittent vomiting, intermittent swelling of the abdomen, or constipation. Again, controlled studies are needed for prophylactic treatment with antimyotonia drugs. It is uncertain whether the bowel hypomotility and the gastroparesis are due to primary weakness of smooth muscle or whether, especially in the early stages, myotonia of smooth muscle accounts for these problems. Some patients benefit from the use of metoclopramide.

The increased occurrence of gall stones in DM often leads to surgery. The occurrence of gall stones relates both to myotonia of the gall bladder and to an independent derangement in bile salt metabolism. Again, no controlled study has established whether long-term antimyotonia therapy would prevent gall stones. If surgery is necessary, attention to the potential complications of anes-

TABLE 8

Management of non-muscular complaints in adult myotonic dystrophy.

Mild form of myotonic dystrophy	Classical (moderate to severe) form of myotonic dystrophy
Cardiac	
Check ECG yearly. If PR interval 70.2 sec, obtain 24 hour recording and coordinate future monitoring with cardiologist. Atrial arrhythmias require cautious monitoring as well to avoid toxic effects on conduction.	Check ECG yearly. If PR interval 0.2 sec obtain 24 hour recording and coordinate care with cardiologist. Would move toward placement of pacemaker early in management. Simplifies use of antiarrhythmia drugs and Dixogin which may have toxic effects on cardiac conduction.
Respiratory	
Check forced vital capacity FVC yearly in sitting and supine position. If value falls below 40 ml/kg, check FVC at 6 month intervals. Monitor for signs of respiratory insufficiency, sleep apnea, and start daily incentive spirometry exercises.	Check FVC at 6–12 month intervals depending upon degree of decline in FVC (30 ml/kg every 6 months; 40 ml every 12 months). Once FVC falls between 20 to 30 ml/kg, would continue? to obtain continuous overnight sleep monitoring of oxygen saturation (O_2 sat) and end tidal pCO_2 ($ETpCO_2$).
	If O_2 sat falls frequently below 85% and/or $Etpco_2$ rises above 50 mmHg indicates respiratory insufficiency during sleep and would consider need for night time ventilation. If there is a history suggesting sleep apnea, even with FVC >30 ml/kg, would obtain overnight monitoring with sleep apnea protocol.
	Routinely encourage daily incentive spirometry exercises. If frequent bouts of pneumonitis occur, obtain swallowing function studies to evaluate degree of aspiration. Sleep with head of bed elevated and avoid large meals, especially at night.
Eye	
Should be seen by ophthalmologist yearly and include a slit lamp examination	Same as for mild form.
Gastrointestinal	
Not a frequent problem. Evaluate as under classical form of disease.	If history suggests trouble swallowing, recommend reducing amount of food in each bite and having food cut into very small pieces.
	Avoid foods high in fat. Maintain bulk in diet and fluid intake (>1.5 liters/day).
	If aspiration likely, obtain swallowing function studies to assess severity, and check chest X-ray.
	If vomiting or intermittent abdominal discomfort suggests gastric retention, consider initiation of metoclopramide.
	If symptoms of gall bladder disease or ob bowel hypomotility occur, obtain consult of gastroenterologist. However, if problem is considered to require surgery, coordinate preoperative and postoperative with anesthesiologist to avoid complications of anesthesia.
Central nervous system	
Not a frequent problem. Treat as under classical form of disease.	If hypersomnolence becomes an active complaint, would consider a trial of methylphenidate.
	If depression becomes prominent, would consider trial of tricyclic antidepressant.

thesia must be followed (Table 7). Postoperative care should be wary after abdominal operations; the resulting pain and ileus magnify any predilection for respiratory insufficiency. Postoperative atelectatic pneumonitis, aspiration, and even apnea become more likely. Monitoring ventilatory function is important until the patient is ambulatory and free of pain (Table 8).

Aspiration is common in advanced DM. Patients with dysphagia must pay attention to the size of meals, the size of bites of food, and the consistency of foods to be swallowed. Serial X-rays of the chest may show evidence of covert atelectasis and aspiration. Patients with aspiration often have a pronounced morning cough and should be counseled to sleep with the head of the bed elevated and to avoid consuming large meals.

Hypersomnia (sleeping more than 10 hours a day) is common, especially in those with the classical form. Anecdotal evidence suggests that many patients benefit from methylphenidate in doses of 10– 20 mg once or twice daily. Treatment may start with a single morning dose and after 1–2 weeks a second dose may be added at midday, but no controlled studies have been done.

Many patients become depressed and tricyclic

drugs may be effective. Imipramine in doses of 100–150 mg at bedtime has proven helpful (Brumback and Carlson 1983; Brumback 1990).

Preoperative and postoperative care and complications of anesthesia in myotonic dystrophy

For many years clinicians have observed complications from anesthetic agents such as thiopental (Aldridge 1985), as well as depolarizing muscle relaxants such as suxamethonium (Orndahl and Sternberg 1962; Thiel 1967; Azar 1984; Aldridge 1985; Tanaka and Tanaka 1991) and opiates (Aldridge 1985). Gall bladder removal or bowel surgery may lead to respiratory complications during or after surgery (Aldridge 1985). Some women with DM have had anesthetic complications during emergency caesarian section (Blumgart et al. 1990; Chung et al. 1987). Cardiac arrhythmias and respiratory failure may occur during the general anesthesia for caesarian section. A 13-year-old boy with DM died during an orthopedic procedure on

TABLE 9

Preoperative and postoperative care and complications of anesthesia in myotonic dystrophy.

Preoperative evaluation
ECG, pulmonary function testing (including supine and upright forced vital capacity) and arterial blood gas measurements.

Intraoperative monitoring
Monitor ECG; measure arterial blood pressure; use a peripheral nerve stimulator to monitor blockade of peripheral muscle; monitor temperature; warm mattress; warm intravenous fluids; and maintain humidification of anesthetic gases.

Postoperative care
Retain endotracheal tube in place and ventilate if necessary on an intensive care unit; monitor respiratory efficiency with checks of oxygen saturation and pCO_2 for at least 24 hours postoperatively to avoid overlooking delayed onset apnea; use controlled flow oxygen therapy with close monitoring of ventilation in patients relying upon hypoxic drive due to chronic respiratory insufficiency; provide early physiotherapy; monitor ECG; keep patient warm; monitor swallowing closely to check for signs of aspiration; treat all infections vigorously.

Anesthetic agents
When possible use local or regional anesthesia, such as an epidural block.
Avoid suxamethonium and other depolarizing muscle relaxants.
Avoid or use only minimal doses of thiopental; for muscle relaxation use short acting agents, such as, atracurium or vecuronium.
Avoid or use only minimal doses of opiates to avoid respiratory depression.
When possible, avoid general anesthesia; if necessary, use combination of nitrous oxide/oxygen mixture with an agent, such as 0.8% enflurane or 1.0% isoflurane.
Use anticholinesterases, such as neostigmine, with care; may be preferable to ventilate the patient until residual curarization wears off.

his foot (Brahams 1989), stimulating a legal discussion in Great Britain that led to a recommendation for respiratory monitoring of patients with DM for 24 hours after surgery (Coakley and Calverley 1989). A consensus of the recommendations has been suggested to avoid or lessen the risk of anesthetic complications (Table 9). It is also important to educate patients and their families about the risks of anesthesia. Patients should carry on their person information about their diagnosis, the risks of anesthesia, and the name and telephone number of the physician for rapid coordination of care between neurologist and surgeon in the event of an emergency operation.

The gene for malignant hyperthermia is also carried on chromosome 19 (MacKenzie et al. 1990) but it is not close to the DM gene (MacKenzie et al. 1990). Moreover, there has been only one report of the occurrence of malignant hyperthermia in a patient with DM (Schellnack et al. 1976). It does not appear that patients with DM are at high risk for malignant hyperthermia. However, there may be inappropriate increase in muscle rigidity in response to suxamethonium or the use of halothane as a general anesthetic agent (Aldridge 1985). For these reasons, and because of the greatly increased risk of prolonged postoperative apnea, seen especially in patients receiving thiopental (Aldridge 1985), it is important to follow the recommended guidelines (Table 9).

Management of myotonic dystrophy in infancy and childhood

The treatment of infants with congenital DM typically requires admission to a neonatal intensive care unit and ventilatory support. The major difficulties involve respiration, feeding problems, aspiration, increased risk of hypoxia and cerebral hemorrhage (Table 10). Deformities such as club foot and elevation of the diaphragm also pose challenges. Infants requiring ventilatory support for more than 4 weeks have a poor prognosis for survival. Neonates with swallowing difficulty are also at high risk of severe pneumonia and death due to aspiration and hypoxia. Skilled nursing care with careful monitoring of respiratory and swallowing is needed to prevent irreversible pulmonary complications and brain damage.

Infants who survive the immediate perinatal period show gradual slow improvement and often do not require further treatment other than early diagnosis and correction of deformities, especially club foot.

Later in childhood it becomes increasingly important to make an accurate identification of the degree of mental retardation. It is essential to separate the muscular symptoms, such as palatal weakness and bifacial weakness, from true mental retardation. Speech problems do not necessarily parallel the severity of retardation. The child may be placed inappropriately in a special class with children of lesser capability. After careful evaluation of the child, more appropriate placement can be made in the classroom and decisions can be made about the need for vocational as opposed to normal classroom courses.

Abdominal pain and abdominal complaints are as common in children as in adults. Children with abdominal complaints often have a form of 'spastic colon'. There is no established role for anti-myotonia agents in ameliorating the abdominal pain of children. Pediatricians often manage these patients as if they had typical childhood spastic colon. Maintaining good hydration and fiber content in the diet is important.

Occasionally there is concern about the need to limit physical activity in children with either congenital or early-onset DM. These patients have only moderate limb weakness, and it is not appropriate to limit physical activity other than that required by their specific pattern of weakness. Serial examinations on a yearly basis may help to tailor the physical education program at school. On a yearly basis the ECG should be checked for covert development of arrhythmias or conduction block.

Children with DM, like affected adults, are at increased risk of anesthetic complications, and recommendations are the same as those for adults.

THERAPEUTIC TRIALS

Drug treatment and other interventions

Future therapeutic trials in myotonic dystrophy will benefit from a clarification of the mechanisms by which the involved protein kinase functions in the cell and by an understanding of the events that control expansion of the CTG repeat within the DM gene. Rational agents can then be developed to use prophylactically as well as for active treatment. Previous trials have had general applicability to many muscle wasting disorders, including testosterone, growth hormone, and exercise. To establish the efficacy of any treatment, it is necessary to define the natural history of the disease, and to identify the most appropriate measures that pro-

vide the maximum statistical power to detect a significant change.

Studies of adults between 18 and 62 years of age with classical DM have shown that the natural history of whole body muscle wasting is one of slowly progressive loss of muscle. Griggs et al. (1983) compared the use of 24 hour excretion of creatinine to measurements of total body potassium; excretion of creatinine, as determined from care-

TABLE 10

Management of myotonic dystrophy in infancy and childhood (adapted from Harper 1989).

Problem	Management
Respiratory insufficiency or failure	Neonatal intensive care; ventilator support; chest X-ray to check on elevation of diaphragm; very poor prognosis for survival if on ventilator longer than 4 weeks. Standard care as necessary for pulmonary immaturity.
Feeding problems and aspiration	Frequent monitoring of respiration and check chest X-ray. Poor swallowing frequently causes recurrent episodes of collapse of the lung. Neonatal intensive care or highly skilled nursing care needed. Evaluate esophageal function if problems persist.
Risk of anoxia and cerebral hemorrhage in infants	Neonatal intensive care. Serial cerebral ultrasound measurements helpful.
Talipes	Early diagnosis and active correction. Splinting. Surgery, if indicated.
General delay in development	Accurate evaluation of contributing factors: especially, muscular and cerebral components.
Speech problems	Anticipate early: Distinguish palatal weakness and other muscular defects from mental retardation.
Abdominal pain	A frequent and difficult problem. Treat as for 'spastic colon'. Role of smooth muscle disease and myotonia in this complaint is unknown. Response to medications variable and not established.
Surgery and anesthesia	Be aware of risks. Avoid general anesthesia if possible.
Recurrent otitis; hearing difficulty	Active management of otitis; awareness of later neural hearing loss.
School decisions	May need special classroom setting and vocational education in later school years; Requires full assessment of functional capabilities, awareness of facial immobility and hearing and speech defects which exaggerate mental retardation; physical disability rarely severe and usually is static during childhood.
Should physical activity be restricted	Not feasible or justified. Obtain yearly ECG to search for signs of cardiac conduction disturbance.

fully collected 24 hour urine samples in patients consuming a standardized meatless diet, is an accurate and reproducible measure of total body muscle mass. However, comparison of samples collected in an out-patient setting and relying on patients to control the diet was much less accurate than urine collections in a clinical research center (Griggs et al. 1980); out-patient 24 hour urine collections in patients with myotonic dystrophy are of questionable reliability.

Table 11 lists different laboratory methods to monitor changes in muscle mass and structure. MRI, CT, and ultrasound have been used. They could be used to assess the natural history of muscle wasting in DM, but no study has been reported using these techniques. It would be necessary to determine whether these expensive imaging methods are sufficiently superior in sensitivity and accuracy, as compared to 24 hour urinary excretion of creatinine, to justify the expense for use in assessing improvement or decline in muscle mass or structure.

The natural history of changes in muscle strength in DM has been defined in 20 patients followed for 12 months (Griggs et al. 1989). The patients received placebo in a double-blind randomized therapeutic trial comparing the effects of weekly intramuscular injections of placebo or testosterone enanthate (3 mg/kg/wk). Over the course of the 12 months on placebo, they showed no significant decline in an average muscle strength score that had been done by Brooke et al. (1983) as

TABLE 11

Laboratory methods to monitor changes in muscle mass and structure.

Total muscle mass
24 hour urinary excretion of creatinine (easy to perform; requires diet and accurate collection of urine).
Total body potassium (requires a counting chamber; presently a research technique).

Specific muscles and their structural changes
MRI (expensive; not routinely used except in research studies).
Ultrasound (less expensive; used in both clinical and research studies).
Muscle biopsy (commonly used; increasing use is being made of needle muscle biopsy compared to open biopsy).

well as functional tests. Changes in pulmonary function, as opposed to muscle strength in the arms and legs, did show some small decline over the 12-month period as determined from serial measurements of forced vital capacity. Improvement in the sensitivity and reliability of measurements of strength may make it possible to discern changes in the strength in individual muscles assessed by quantitative myometry to document the progression of weakness. These muscles will have to be identified in future therapeutic trials. Refinements in the measurements of grip strength, especially in the long flexors of the fingers, have already been applied to patients with other forms of myotonia (Ricker et al. 1990). These methods and others (Torres et al. 1983), including the use of maximum isometric voluntary contraction determined with specialized equipment (Andres et al. 1986; Brussock et al. 1992) should prove valuable. Ultimately, some of these quantitative methods may become routine in the out-patient monitoring of muscle strength in patients with neuromuscular disease.

The methods presently used to assess muscle strength have advantages and disadvantages (Table 12). Manual muscle strength testing is the most adaptable (flexible) technique for patients of all ages, but it is relatively insensitive in detecting small changes in the range of 3–5 (5/5 being normal). Quantitative techniques, using hand-held myometry and large more expensive instrumentation for quantitating maximum isometric voluntary contraction, have therefore gained greater usage. More expensive isokinetic devices such as the Lido and Cybex have been used to assess changes in strength in athletes and adults in training programs. However, this expensive equipment is often not adaptable or appropriate for patients with neuromuscular disease, especially those who have limitations in joint range of motion or body structure changes that do not allow them to be seated appropriately for isokinetic testing. In future therapeutic trials for DM patients the techniques listed in Table 11 may be used; their efficacy, reliability, and interrater and intrarater reliability will be determined. Based on this information, it will become easier to decide whether quantitative myometric techniques or timed functional tests or other methods are the most suitable to fol-

TABLE 12

Methods for quantitating changes in muscle strength.

1. *Quantitative testing of strength*

Manual muscle testing (using trained physical therapist)
– advantages: good interrater reliability; 'cheap', flexible.
– disadvantages: relatively insensitive to changes in strength in 3–5 range (0–5 scale); ordinal scale

Hand held myometry (using trained personnel)
– advantages: intrarater reliability good (5–25%); interrater reliability more variable
– disadvantages: not reliable in detecting changes in the 4 to 5- out of 5 range

Maximum voluntary isometric contraction force with quantitative myometry equipment
– advantages: intrarater reliability very good (5–15%)
– disadvantages: interrater reliability more variable; expensive; time consuming; inflexible in adaptability to
  patients of different body sizes and shapes

2. *Changes in function abilities*

Timed functional tests (such as travel 30 feet, climb 4 steps, arise from supine to standing, etc.)

Functional grade (such as ability to raise arms above head, walk normally on heels or toes, etc.)
– advantages: reliable; flexible; pertinent to patients
– disadvantages: relatively insensitive in detecting changes in strength; ordinal scale

low muscle strength. It is already apparent that timed functional test scores, such as the time to travel 30 feet time to climb 4 steps, and time to arise from supine to standing, were helpful in monitoring the efficacy of prednisone treatment in Duchenne dystrophy.

Therapeutic trial of testosterone

Men with DM are frequently hypogonadal, manifest by depressed plasma testosterone and by elevated plasma gonadotropin levels. They also have a decrease in the rate of protein synthesis in skeletal muscle (Griggs et al. 1986), with a normal rate of muscle breakdown as evidenced by normal ratio of 3-methylhistidine to creatinine excretion in the urine (Griggs et al. 1980), and a loss of normal anabolic response to insulin (Moxley et al. 1978, 1980, 1984, 1985). Therefore a deficiency in protein anabolism is an important contributing factor, if not a fundamental alteration, accounting for the muscle wasting and weakness in DM. In an attempt to reverse this deficiency in muscle anabolism and to increase muscle strength. Griggs et al.

(1989) carried out a double-blind randomized therapeutic trial in 40 men comparing weekly intramuscular injections of testosterone enanthate (3 mg/kg/wk) to intramuscular injections of placebo. The men were 18 to 65 years of age, ambulatory, and had the classical form of DM. Measurements were made at baseline, 3, 6, 9, and 12 months. Manual muscle strength in 34 muscles, hand-held myometric testing in selected muscles of the right arm and leg, and hand grip dynamometry were performed along with timed functional testing, assessment of pulmonary function, and measurements of forearm muscle volume. No improvement was seen in muscle strength even though plasma testosterone levels were 2–4 fold above normal. Despite the lack of improvement in strength, there was a significant increase in muscle mass. Therefore testosterone seems to enhance the synthesis of muscle proteins not involved in the contractile process. Pilot studies performed prior to the initiation of this double-blind randomized therapeutic trial (Griggs et al. 1986) had revealed a significant increase in skeletal muscle protein synthesis. In the same patients testosterone treatment did not reverse insulin resistance (Kingston et al. 1986). The anabolic effect of testosterone seems to be mediated separately from the anabolic action of insulin. At present, testosterone treatment has no established value in improving muscle strength in DM, and it is not known whether prolonged treatment would prevent the gradual loss of muscle bulk and strength.

Kolkin et al. (1987) studied 10 men with DM who were treated by transcutaneous muscle stimulation (6 hr/day) for 14 days prior to a 12-week course of weekly injections of testosterone enanthate (3 mg/kg). Transcutaneous muscle stimulation was carried out on the quadriceps femoris muscle, with sham stimulation of the opposite leg. After the testosterone treatment, there was a follow-up period of 14 days of transcutaneous muscle stimulation. They found an increase in the cross-sectional area of the quadriceps femoris muscle as assessed by CT but, again, there was no improvement in muscle strength. Thus, no beneficial effect was observed from either the 12-week course of testosterone enanthate or the two 14-day trials of transcutaneous muscle stimulation. Transcutaneous stimulation can increase muscle strength, but a longer course of transcutaneous muscle stimulation must be necessary.

Trials of growth hormone treatment

Investigators studied the effects of 14 days of treatment with intramuscular injections of human growth hormone derived from pooled human brain (Rudman et al. 1971, 1972; Chyatte et al. 1974). These short-term studies showed an enhanced sensitivity to growth hormone and a marked anabolic response. There was no significant improvement in muscle strength but, in 3 of the 4 patients, a significant shortening of the PR interval was apparent by the third day of treatment. Because of the contamination of pooled human brain derived growth hormone by the Creutzfeldt-Jakob agent (Gibbs et al. 1985) further studies of human growth hormone in patients with myotonic dystrophy had to be delayed. Now, it is possible to obtain synthetic, c-DNA derived, human growth hormone for use in therapeutic trials. Plot studies assessing the effect of daily subcutaneous injections of human growth hormone are

in progress at the University of Rochester. There may be a non-specific beneficial improvement in muscle mass and function with growth hormone therapy, comparable to the effects of growth hormone in normal elderly individuals, in athletes, and patients with acquired growth hormone deficiency (Rudman et al. 1990).

Human, c-DNA derived, insulin-like growth factor 1 (IGF-1) is becoming available for research study. Treatment with IGF-1 may have more beneficial effects and less toxic side effects than growth hormone. IGF-1 has a receptor on skeletal muscle that is homologous to the receptor for insulin and shares with insulin in possessing important anabolic actions on skeletal muscle (Froesch et al. 1985). If the muscle wasting in DM is related to the insulin resistance, the use of IGF-1 acting through its own receptor on skeletal muscle may bypass the defect in muscle anabolism that is caused by the deficiency in insulin action. This possibility will require further consideration.

Exercise and related interventions

Supervised exercise, as a treatment for DM, has received only limited investigation but provides an opportunity for therapy that can be used with other treatments, such as drugs or trophic factors. Both whole body exercise (Booth and Thomason 1991) and the intake of carbohydrate (Kingston et al. 1986; Moxley et al. 1987) enhance the action of insulin. In view of the insulin resistance in DM, exercise seems to be a feasible intervention to reverse some of the negative effects of insulin resistance and to increase muscle bulk and function. The intake of small amounts of carbohydrate by DM patients before the administration of insulin does not demonstrate the enhancement of insulin action that is seen in normals (Moxley et al. 1987). Exercise, as opposed to food intake, may be a means for bypassing the putative deficiency in anabolism caused by insulin resistance. Exercise might also slow the rate of muscle atrophy associated with non-weight bearing and inactivity. The physiologic and biochemical changes associated with exercise and with non-weight bearing have been reviewed (Booth and Thomason 1991), and animal models also support the role of regular physical activity in preventing muscular atrophy.

There may also be a beneficial role for transcutaneous stimulation of muscle (Gibson et al. 1989; Hainaut and Duchateau 1989; Booth and Thomason 1991; Russell et al. 1992; Russell and Dix 1992). Controlled studies are needed.

One therapeutic trial (Milner-Brown and Miller 1990) compared the beneficial effects of a combination of regular weight training (involving elbow flexors and extensors, knee flexors and extensors, and ankle dorsiflexors and plantar flexors) with night-time doses of amitriptyline, 50 mg. Eight patients participated, and all had previously been participating in an on-going, supervised, weight training program (Milner-Brown et al. 1986; Milner-Brown and Miller 1988). The therapeutic trial had two principal objectives, to quantify muscle weakness and myotonia, and to determine quantitatively the efficacy of amitriptyline on muscle strength and myotonia. Four of the 8 patients noted subjective improvement in their energy level, and 2 felt their myotonia was reduced. No definite change in clinical findings was observed in any of the 8 patients. Quantitative myometry was carried out over the 4–6 month trial of amitriptyline and was limited to the first dorsal interosseous muscle. This muscle was not part of the weight training program. Strength of the first dorsal interosseous muscle improved in 6 of 8 patients. Relaxation time, used to quantify the degree of grip myotonia, was also reduced by a mean factor of 2. Therefore, the combination of amitriptyline and exercise ameliorated myotonia, as well as a possible beneficial effect on muscle strength. It was not possible to distinguish a direct action of amitriptyline and the synergistic effect of exercise in mediating this improvement. Additional studies, combining exercise with amitriptyline or other agents, deserve consideration.

Future studies of exercise therapy alone in patients with myotonic dystrophy also deserve consideration. The exercise program should be supervised and individualized to the pattern of muscle weakness of each patient. In view of the cardiac conduction abnormalities and the additional stress placed on the heart by exercise, especially with static, muscle contractions, the programs should involve exercises that emphasize isotonic muscle contractions and avoid isometric contractions that are prolonged and close to maximum. In addition to an individualized resistive muscle strengthening program, some patients may be able to participate safely in a dynamic, aerobic, whole body exercise program. Because of the pattern of muscle weakness, exercise on a stationary bicycle or with an arm ergometer may be more appropriate than a program in which patients are asked to walk prolonged distances or walk on a treadmill. Whole-body aerobic exercise will need supervision, especially monitoring of the EKG for arrhythmia.

Future therapeutic trials may evaluate the efficacy of a combination of low intensity resistive exercises, whole body dynamic exercises, and treatment with growth hormone or IGF-1. A combination of therapies may have a synergistic effect to overcome the deficiency in anabolism in myotonic dystrophy. This approach might also have value in maintaining strength in other neuromuscular disorders.

EPILOGUE

Future questions related to the newly discovered gene lesion in myotonic dystrophy

The cause of the abnormal instability in DNA that leads to expansion of the CTG trinucleotide repeat in the DM gene is unknown, and the function of the associated protein kinase (presumably serine-threonine protein kinase) has to be determined. Roses and Appel (1973, 1974) had earlier found abnormal patterns of phosphorylation in membranes from erythrocytes and skeletal muscle in patients with myotonic dystrophy. Second messenger signalling, especially that mediated through the insulin receptor, also involves phosphorylation and dephosphorylation (Kahn and White 1988). Patients with DM have tissue-specific insulin resistance and it is uncertain whether the DM protein kinase has altered phosphorylation reactions that produce tissue damage; insulin resistance may be one such reaction. Whether an alteration in intracellular signalling can provide a sound pathophysiologic mechanism to account for the selective pattern of muscle wasting and the involvement of other tissues (lens, testes, smooth muscle, cardiac and perkingi muscle fibers, and neurons) requires future research. However, based on the discovery of the gene lesion (Aslandis et al. 1992; Brook et al.

1992; Buxton et al. 1992; Fu et al. 1992; Harley et al. 1992; Mahadevan et al. 1992), several questions require consideration.

Armed with the easy screening of DNA in circulating white blood cells to test for the gene lesion, clinicians can redefine the natural history of DM. Properly designed studies of the normal population and of all individuals at risk in identified families with DM are needed for information about the relation between genotype and phenotype. Will the genotype, as reflected in circulating white blood cells, be sufficient to determine disease severity or will it be necessary to assess the gene lesion (number of CTG trinucleotide repeats) in DNA in target tissues such as skeletal muscle or lens? Will it be possible to predict the clinical course in patients from DNA analysis in circulating leukocytes? If not, what studies will be necessary to provide a more accurate means of prediction in individual patients?

Other questions concern our ability to predict the clinical course from the degree of expansion of CTG repeat. What is the relationship between the severity and pattern of muscle wasting and the gene lesion? Is the CTG expansion identical in DNA isolated from muscles that are mildly and severely affected? Is there a difference in the expression of the protein kinase associated with the DM gene in different muscles? Do the metabolic alterations in patients (hyperinsulinemia, insulin resistance, altered amino acid regulation, hyperlipidemia, decreased protein synthesis) parallel the expansion of the CTG repeat as seen in DNA from circulating white cells or other tissues? Cataracts and balding are seen in patients who have no overt muscle wasting or weakness, electrical myotonia without clinically delayed muscle relaxation. In these patients, and perhaps in all DM patients, is there some specific tissue vulnerability of the lens, regulation of hair growth, or ion transport that is influenced by a relatively mild expansion of the CTG repeat in the DM gene (50–100 repeats)? Do the alterations in personality mentation and disordered sleep (sleeping more than 10 hours a day) parallel the degree of expansion of the CTG repeat? Is there a correlation between symptomatic sleep apnea and the degree of expansion? Is there some aspect of the metabolism of Purkinje muscle fibers that make them particularly susceptible to cause heart block? Could at-risk individuals be identified early enough for prophylactic therapy? Congenital myotonic dystrophy occurs only in infants born to mothers affected with the disease, even though DM fathers may also have an expansion of the CTG repeat comparable to that in affected women who have affected infants, and congenital DM is never seen in infants born to affected fathers. What is the maternal factor that is accounting for congenital DM?

Many other questions can and should be asked about the implications of the gene discovery for diagnosis and care. Three broad questions now seem particularly important. What DNA strategy should be used to detect affected individuals? What recommendations should be made for presymptomatic and prenatal screening? Are we already at a point at which we could consider designing long term prophylactic therapeutic trials once a rational therapeutic agent becomes available? Will following any specific genetic markers, including the expansion of the CTG repeat, be a useful measurement in a clinical trial? As Roses and colleagues stated in their original Handbook chapter 'When the inborn error is delineated another clinical chapter will need to be written with more defined measures for diagnosis, genetic analysis, and treatment'. The present chapter is timely and, in large measure, follows their recommendations. However, still another chapter is needed, perhaps in the near future, when we can account for the instability in the DM gene and when there is a more detailed understanding of the relation between genotype, expression of the protein kinase, and phenotype.

ACKNOWLEDGEMENTS

The author expresses appreciation to Dr. Kenneth Ricker, Dr. Alan Roses, Dr. Henry Epstein, Dr. Keith Johnson, Dr. John Kissel, and to the entire Working Group on Myotonic Dystrophy (*Neurology* 39:420–421, 1991) for helpful discussions prior to the preparation of this chapter. Thanks go to the Muscular Dystrophy Association of America, and to Dr. John Kissel, Dr. Douglas Koch, Dr. Keshav Rao, Dr. Kenneth Ricker, and Dr. Ziad Rifai for their help in preparing the illustrations. Special thanks go to Jean Ellinwood for her help in typing and organizing this chapter.

REFERENCES

ABDALLA, J.A., W.L. CASLEY, J.K. COUSIN, A.J. HUDSON, E.G. MURPHY, F.C. CORNELIS, L. HASHIMOTO and G.C. EBERS: Linkage of autosomal dominant myotonia congenita to TCRB gene locus on chromosome FQ 35. Neurology (1992) in press.

ALDRIDGE, L.M.: Anaesthetic problems in myotonic dystrophy. Br. J. Anesth. 57 (1985) 1119–1130.

AHRINGER, J. and J. KIMBLE: Control of the sperm-oocyte switch in *Caenorhabditis elegans hermaphrodites* by the Fem-33′ untranslated region. 349 (1991) 346–348.

ANDERSON, T.L., L. CHANG, R.J. GIAMBETTI, L. MILLER, K. GARRETT, J. VELLAVEREVA-MEYER and I. MENA: Abnormal cerebral blood flow in myotonic dystrophy. Neurology 41 Suppl. (1991) 420.

ANDRES, P.L., W. HEDLUND, L. FINNISM, T. CONLON, M. FELMUS and T.L. MUNSAT: Quantitative motor assessment in amyotrophic lateral sclerosis. Neurology 36 (1986) 937–941.

ARNER, P., J. BOLINDER and J. OSTMAN: Marked increase in insulin sensitivity of human fat cells one hour after glucose ingestion. J. Clin. Invest. 71 (1983) 709–714.

ASHIZAWA, T. and H.F. EPSTEIN: Ethnic distribution of the myotonic dystrophy gene. Lancet 338 (1991) 642–643.

ASHIZAWA, T., J.F. HEJTMANCIK, J. LIU, M.B. PERRYMAN, H.F. EPSTEIN and D.D. KOCH: Diagnostic value of ophthalmologic findings in myotonic dystrophy. Am. J. Med. Genet. 142 (1992) 55–60

ASKANAS V., S. CAVE, G. GALLEZ-HAWKINS and W.K. ENGEL: Fibroblasts growth factor, epidermal growth factor, and insulin exert a neuronal-like influence on acetylcholine receptors in aneurally cultured human muscle. Neurosci. Lett. 61 (1985) 213–219.

ASLANIDIS, C., G. JANSEN, C. AMEMIYA, G. SHUTLER, M. MAHADERVAN, C. TSILFIDIS, C. CHEN, J. ALLEMAN, N.G.M. WORMSKAMP, M. VOOIJS, J. BUXTON, K. JOHNSON, H.J.M. SMEETS, G.G. LENNON, A.V. CARRANSO, R.G. KORNELUK, B. WERINGA and P. DEJONG: Cloning of the essential myotonic dystrophy region and mapping of the putative defect. Nature (London) 356 (1992) 548–551.

AVRAHAMI, E., A. KATZ, A. BORNSTEIN and A.D. KORCZYN: Computed tomographic findings of brain and skull in myotonic dystrophy. J. Neurol. Neurosurg. Psychiatry 50 (1987) 435–438.

AZAR I.: The response of patients with neuromuscular disorders to muscle relaxants: A review. Anesthesiology 61 (1984) 173–187.

BAILLY, J., A.E. MACKENZIE, S. LEBLOND and R.G. KORNELUK: Assessment of a creatine kinase isoform M defect as a cause of myotonic dystrophy and the characterization of two novel CKMM polymorphisms. Hum. Genet. 86 (1991) 457–462.

BARBOSA, J., F.Q. NUTTALL, W. KENNEDY and F. GOETZ: Plasma insulin in patients with myotonic dystrophy and their relatives. Medicine 53 (1974) 307–323.

BASKIN, D.G., D.P. FIGLEWICZ, S.C. WOODS, D. PORTE JR and D.M. DORSA: Insulin in the brain. Annu. Rev. Physiol. 49 (1987) 335–347.

BIEZENBOS, J.B.M., E.M.G. JOOSTEN, R.A. WEVERS, R.A. BINKHORST, D.F. STEGEMAN and A.G. M. THEEUWES: Excessive potassium efflux after standardized ischemic forearm exercise in myotonic dystrophy. J. Neurol. Sci. 98 (1990) Suppl. 424.

BIRD, T.D., A.M. REENAN and M. PFEIFER: Autonomic nervous system function in genetic neuromuscular disorders. Hereditary motor-sensory neuropathy and myotonic dystrophy. Arch. Neurol. 41 (1984) 43–46.

BLACKBURN, C.H.: Structure and function of telomeres. Nature (London) 350 (1991) 569–573.

BLUMGART, C.H., D.G. HUGHES and N. REDFERN: Obstetric anaesthesia in dystrophia myotonica. Anaesthesia 45 (1990) 26–29.

BOOTH, F.W. and D.B. THOMASON: Molecular and cellular adaptation of muscle in response to exercise: Perspectives of various models. Physiol. Rev. 71 (1991) 541–574.

BORG, J., L. EDSTRÖM, G.S. BUTLER-BROWNE and L.-E. THORNELL: Muscle fibre type composition, motoneuron firing properties, axonal conduction velocity and refractory period for foot extensor motor units in dystrophia myotonica. J. Neurol. Neurosurg. Psychiatry 50 (1987) 1036–1044.

BRAHAMS, D.: Postoperative monitoring in patients with muscular dystrophy. Lancet (1989) 1053–1054.

BROOK, J.D., H.G. HARLEY, K.V. WALSH, S.A. RUNDLE, M.J. SICILIANO, P.S. HARPER and D.J. SHAW: Identification of new DNA markers close to the myotonic dystrophy locus. J. Med. Genet. 28 (1991) 84–88.

BROOK, J.D. M.E. MCCURRACH, H.G. HARLEY, A.J. BUCKLER, D. CHURCH, H. ABURATANI, K. HUNTER, V.P. STANTON, J-P. THIRION, T. HUDSON, R. SOHN, B. ZEMELMAN, R.G. SNELL, S.A. RUNDE, S. CROW, J. DAVIES, P. SHELBOURNE, J. BUXTON, C. JONES, V. JUVONEN, K. JOHNSON, P.S. HARPER, D.J. SHAW and D.E. HOUSMIN: Molecular basis of myotonic dystrophy: Expansion of a trinucleotide (CTG) repeat at the 3′ end of a transcript encoding a protein kinase family member. Cell 68 (1992) 799–808.

BROOKE, M.H. and K.K. KAISER: Muscle fibre types: how many and what kind? Arch. Neurol. (Chicago) 23 (1970) 369–379.

BROOKE, M.H., G.M. FENICHEL, R.C. GRIGGS, J.R. MENDELL, R.T. MOXLEY, J.P. MILLER, M.A. PROVINCE and CIDD GROUP: Clinical investigations in Duchenne dystrophy: determination of the 'power' of therapeutic trials based on the natural history. Muscle Nerve 6 (1983) 91–103.

BROOKE, M.H., E. WILLIAMSON and K.K. KAISER: The behavior of four fiber types in developing and reinnervated muscle. Arch. Neurol. (Chicago) 25 (1971) 360–366.

BROUGHTON, R., D. STUSS, M. KATES, J. ROBERTS and W. DUNHAM: Neuropsychological deficits and sleep in myotonic dystrophy. Can. J. Neurol. Sci. 17 (1990) 410–415.

BRUMBACK, R.A.: Magnetic resonance imaging and clinical correlates of intellectual impairment in myotonic dystrophy. Arch. Neurol. 47 (1990) 253.

BRUMBACK, R.A. and K.M. CARLSON: The depression of myotonic dystrophy: response to imipramine. J. Neurol. Neurosurg. Psychiatry 46 (1983) 587–588.

BRUMBACK, R.A., K.M. CARLSON, H. WILSON and R.D. STATON: Myotonic dystrophy as a disease of abnormal membrane receptors: An hypothesis of pathophysiology and a new approach to treatment Med. Hypothesis 7 (1981) 1059–1066.

BRUNNER, H.G. ET AL.: Reversions in myotonic dystrophy. N. Engl. J. Med. (1992) in press.

BRUNNER, H.G., F. SPAANS, H.J.M. SMEETS, M. COER-WINKEL-DRIESSEN, T. HULSEBOS, B. WIERINGA and H.H. ROPERS: Genetic linkage with chromosome 19 but not chromosome 17 in a family with myotonic dystrophy associated with hereditary motor and sensory neuropathy. Neurology 41 (1991) 80–84.

BRUSSOCK, C.M., S.M. HALEY, T.L. MUNSAT and D.B. BEERNHARDT: Measurement of isometric force in children with and without Duchenne dystrophy. Phys. Ther. 72 (1992) 105–114.

BUXTON, J., P. SHELBOURNE, J. DAVIES, C. JONES, T. VAN TONGEREN, C. ASLANIDIS, P. DEJONG, G. JANSEN, M. ANVRET, B. RILEY, R. WILLIAMSON and K. JOHNSON: Detection of an unstable fragment of DNA specific to individuals with myotonic dystrophy. Nature (London) 355 (1992) 547–551.

CARPENTER, S. and G. KARPATI: Pathology of Skeletal Muscle. New York, Churchill Livingstone (1984) 616–631.

CAUGHEY, J.E. and N.C. MYRIANTHOPOULOS: Dystrophica Myotonica and Related Disorders. Springfield, IL, Charles C. Thomas (1963).

CENSORI, B., M. DANNI, M. DELPESCE and L. PROVINCIALI: Neuropsychological profile in myotonic dystrophy. J. Neurol. 237 (1990) 251–256.

CHUNG, H.T., A.Y.C. TAM, V. WONG, D.F.H. LI, J.T.C. MA, C.Y. HUANG, Y.L. YU and E. WOO: Dystrophia myotonica and pregnancy — an instructive case. Postgrad. Med. J. 63 (1987) 555–557.

CHURCH, S.C.: The heart in myotonia atrophica. Arch. Int. Med. 119 (1967) 176–181.

CHYATTE, S.B., D. RUDMAN, J.H. PATTERSON, P. AHMANN and A. JORDAN: Human growth hormone in myopathy: myotonic dystrophy, Duchenne muscular dystrophy, and limb-girdle muscular dystrophy. South. Med. J. 67 (1974) 170–172.

COAKLEY, J.H. and P.M.A. CALVERLEY: Anaesthesia and myotonic dystrophy. Lancet 335 (1989) 409.

COCCAGNA, G., P. MARTINELLI and E. LUGARESI: Sleep and alveolar hypoventilation in myotonic dystrophy. Acta Neurol. Belg. 82 (1982) 185–194.

COOPER, R.G., M.J. STOKES and R.H.T. EDWARDS: Physiological characterisation of the 'warm up' effect of activity in patients with myotonic dystrophy. J. Neurol. Neurosurg. Psychiatry 51 (1988) 1134–1141.

CONACCHIA, L., R. MARINA, V. BALLARINI and G. SOZZI:

Computed tomographic findings of brain and skull in myotonic dystrophy. J. Neurol. Neurosurg. Psychiatry 51 (1988) 1463–1464.

COVAULT, J. and J.R. SANES: Distribution of NpCAM in synaptic and extrasynaptic portions of developing and adult skeletal muscle. J. Cell Biol. 102 (1986) 716–730.

COVAULT, J. and J.R. SANES: Neural cell adhesion molecule (N-CAM) accumulates in denervated and paralyzed skeletal muscle. Proc. Natl. Acad. Sci. USA 92 (1985) 4544–4548.

DAVIES, J., H. YAMAGATA, P. SHELBOURNE, J. BUXTON, T. OGIHARA, P. NOKELAMIN, M. NAKAGAWA, R. WILLIAMSON, K. JOHNSON and T. MIKI: Comparison of the myotonic associated CTG repeat in European and Japanese populations. J. Med. Genet. (1992) in press.

DESNUELLE, C., V. ASKANAS and W.K. ENGEL: Insulin enhances development of functional voltage dependent calcium channels in aneurally cultured human muscle. J. Neurochem. 49 (1987) 1133–1138.

DESNUELLE, C., A. LOMBET, G. SERRATRICE and M. LAZDUNSKI: Sodium channel and sodium pump in normal and pathologic muscles from patients with myotonic muscular dystrophy and lower motor neuron impairment. J. Clin. Invest. 69 (1982) 358–367.

DRACHMAN, D.B. and D.M. FAMBROUGH: Are muscle fibers denervated in myotonic dystrophy? Arch. Neurol. (Chicago) 33 (1986) 485–488.

ECKARDT, V.F. and W. NIX: The anal sphincter in patients with myotonic muscular dystrophy. Gastroenterology 100 (1991) 424–430.

FALL, L.H., W.W. YOUNG, J.A. POWER, C.S. FAULKNER, B.D. HETTLEMEN and J.F. ROBB: Severe congestive heart failure and cardiomyopathy as a complication of myotonic dystrophy in pregnancy. Obstet. Gynecol. 76 (1990) 481–484.

FENTON, J., S. GARNER and A.J. MCCOMAS: Abnormal M-wave responses during exercise in myotonic muscular dystrophy: A Na^+-K^+ pump defect? Muscle Nerve 14 (1991) 79–84.

FIORELLI, M., D. DUBOC, B.M. MAZOYER, J. BLIN, B. EYMARD, M. FARDEAU and Y. SAMSON: Decreased cerebral glucose utilization in myotonic dystrophy. Neurology 42 (1992) 91–94.

FITTING, J.-W. and P. LEUENBERGER: Procainamide for dyspnea in myotonic dystrophy. Amer. Rev. Resp. Dis. 140 (1989) 1442–1445.

FLOREK, R.C., D.W. TRIFFON, D.E. MANN and M.J. REITER: Electrocardiographic abnormalities in patients with myotonic dystrophy. West. J. Med. 153 (1990) 24–27.

FLORENCE, J.M., S. PANDYA, W. KING, J. ROBISON, J. BATY, J.P. MILLER, J. SCHIERBECKER and L.C. SIGNORE: Intrarater reliability of manual muscle test grades in Duchenne dystrophy. Phys. Ther. 72 (1992) 115–126.

FORSBERG, H., B.-O. OLOFSSON, A. ERIKSSON and S.

ANDERSSON: Cardiac involvement in congenital myotonic dystrophy. Br. Heart. J. 63 (1990) 119–121.

FRAGOLA, P.V., G.C. RUSCITTI, C. AUTORE, G. ANTONINI, A. CAPRIA, S. FIORITO, R. VICHI, E. PENNISI and D. CANNATA: Ambulatory electrocardiographic monitoring in myotonic dystrophy (Steinert's disease). A study of 22 patients. Cardiology 74 (1987) 362–368.

FRANKE, C. H. HATT, P.A. IAIZZO and F. LEHMANN-HORN: Characteristics of Na^+ channels and Cl^- conductance in resealed muscle fibre segments from patients with myotonic dystrophy. J. Physiol. 425 (1990) 391–405.

FROESCH, E.R., C. SCHMID, J. SCHWANDER and J. ZAPF: Actions of insulin-like growth factors. Annu. Rev. Physiol. 47 (1985) 443–467.

FU, Y.H., A. PIZZUTI, R.G. FENWICK JR, J. KING, S. RAJNARAYAN, P.W. DUNNE, J. DUBEL, G.A. NASSER, T. ASHIZAWA, P. DEJONG, B. WERINGA, R. KORNELUK, M.B. PERRYMAN, H.F. EPSTEIN and C.T. CASKEY: An unstable triplet repeat in a gene related to myotonic muscular dystrophy. Science 255 (1992) 1256–1258.

FUKAZAWA, H., T. SAKURADA, K. YOSHIDA, N. KAISE, K. KAISE, T. NOMURA, M. YAMMAMOTO, S. SAITO, S. TAKASE and K. YOSHINAGA: Thyroid function in patients with myotonic dystrophy. Clin. Endocrinol. 32 (1990) 485–590.

GIBBS, C.J., A. JOY, R. HEFFNER, M. FRANKO, M. MIYAZAKI, D.M. ASHER, J.E. PARISI, P.W. BROWN and D.C. GAJDUSEK: Clinical and pathological features and laboratory confirmation of Creutzfeldt-Jakob disease in a recipient of pituitary-derived human growth hormone. N. Engl. J. Med. 313 (1985) 734–738.

GIBSON, J.N.A., W.L. MORRISON, C.M. SCRIMGEOUR, K. SMITH, P.J. STOWARD and M.J. RENNIE: Effects of therapeutic percutaneous electrical stimulation of atrophic human quadriceps on muscle composition, protein synthesis and contractile properties. Eur. J. Clin. Invest. 19 (1989) 206–212.

GILBERT, J., H. TAYLOR, C. SCHWARTZBACH and A.D. ROSES: Myotonic dystrophy. In: B.A. Kakulas, J. McHowell and A.D. Roses (Eds.), Duchenne Muscular Dystrophy: Animal Models and Genetic Manipulation. In press.

GLANTZ, R.H., R.B. WRIGHT, M.S. HUCKMAN, D.C. GARRON and I.M. SIEGEL: Central nervous system magnetic resonance imaging findings in myotonic dystrophy. Arch. Neurol. 45 (1989) 36–37.

GOSPAVIC, J.J., V.G. KUJACIC, M.A. KUJACIC, N.D. SEMIN and C.V. BUGARSKI-PROKOPLJEV: Echocardiographic assessment of left ventricular function in myotonic dystrophy. Acta Cardiomiol. 2 (1990) 35–44.

GRIGGS, R.C., R.J. DAVIS, D.C. ANDERSON and J.T. DOVE: Cardiac conduction in myotonic dystrophy. Amer. J. Med. 59 (1975) 37–42.

GRIGGS, R.C., G. FORBES, R.T. MOXLEY and B.E. HERR: The assessment of muscle mass in progressive neuromuscular disease. Neurology 33 (1983) 158–165.

GRIGGS, R.C., D. HALLIDAY, W.J. KINGSTON and R.T.

MOXLEY: Effect of testosterone on muscle protein synthesis in myotonic dystrophy. Ann. Neurol. 20 (1986) 590–596.

GRIGGS, R.C., W.J. KINGSTON, R.F. JOZEFOWICZ, B.E. HERR, G. FORBES and D. HALLIDAY: Effect of testosterone on muscle mass and muscle protein synthesis. J. Appl. Physiol. 66 (1989) 498–503.

GRIGGS, R.C., W.J. KINGSTON, B.E. HERR, G. FORBES and R.T. MOXLEY: Lack of relationship of hypogonadism to muscle wasting in myotonic dystrophy. Arch. Neurol. 42 (1985) 881–885.

GRIGGS, R.C., R.T. MOXLEY and G.B. FORBES: 3-Methylhistidine excretion in myotonic dystrophy. Neurology 30 (1980) 1262–1267.

GRIGGS, R.C., S. PANDYA, J.M. FLORENCE, M.H. BROOKE, W.J. KINGSTON, J.P. MILLER, J. CHUTKOW, B.E. HERR and R.T. MOXLEY: Randomized controlled trial of testosterone in myotonic dystrophy. Neurology 39 (1989) 219–222.

HAINAUT, K. and J. DUCHATEAU: Muscle fatigue, effects of training and disease. Muscle Nerve 12 (1989) 660–669.

HARLEY, H.G., J.D. BROOK, J. FLOYD, S.A. RUNDLE, S. CROW, K.V. WALSH, M.C. THIBAULT, P.S. HARPER and D.J. SHAW: Detection of linkage disequilibrium between the myotonic dystrophy locus and a new polymorphic DNA marker. Amer. J. Hum. Genet. 49 (1991) 68–75.

HARLEY, H.G., J.D. BROOK, S.A. RUNDLE, S. CROW, W. REARDON, A.J. BUCKLER, P.S. HARPER, D.E. HOUSMAN and D.J. SHAW: Expansion of an unstable DNA region and phenotypic variation in myotonic dystrophy. Nature (London) 355 (1992) 545–546.

HARLEY, H.G., K.V. WALSH, S. RUNDLE, J.D. BROOK, M. SARFARAZI, M.C. KOCH, J.L. FLOYD, P.S. HARPER and D.J. SHAW: Localisation of the myotonic dystrophy locus to 19q13.2-19q13.3 and its relationship to twelve polymorphic loci on 19q. Hum. Genet. 87 (1991) 73–80.

HARPER, P.S.: Myotonic Dystrophy. London, W.B. Saunders Co. (1989).

HARPER, P.S., H.G. HARLEY, W. REARDON and D.J. SHAW: Anticipation in myotonic dystrophy — new light on an old problem. Amer. J. Hum. Genet. (1992) in press.

HARRIS, J.P., N.C. NANDA, R. MOXLEY and J.A. MANNING: Myocardial perforation due to temporary transverse pacing cathethers in pediatric patients. Catheterization Cardiovasc. Diagnosis 10 (1984) 329–333.

HAWLEY, R.J., M.R. MILNER, J.S. GOTTDIENER and A. COHEN: Myotonic heart disease — a clinical follow up. Neurology 41 (1991) 259–262.

HIROMASA, S., I. TAKAYUKI, K. KUBOTA, N. HATTORI, H. COTO, C. MALDONADO and J. KUPERSMITH: Ventricular tachycardia and sudden death in myotonic dystrophy. Amer. Heart J. 115 (1988) 914–915.

HOLT, M. and K. LAMBERT: Heart disease as a presenting feature in myotonia dystrophica. Br. Heart J. 26 (1964) 433–436.

HOWELER, C.J.: A clinical genetic study in myotonic

dystrophy. Thesis, Rotterdam. (1986).

HOWELER, C.J., H.F.M. BUSCH, J.P.M. GERAEDTS, M.F. NI-ERMEIJER and A. STAAL: Anticipation in myotonic dystrophy. Fact or fiction. Brain 112 (1989) 779–797.

HUBER, S.J., J.T. KISSEL, E.C. SHUTTLEWORTH, D.W. CHA-KERS, L.E. CLAPP and M.A. BROGAN: Magnetic resonance imaging and clinical correlates of intellectual impairment in myotonic dystrophy. Arch. Neurol. 46 (1989) 536–537.

HUDSON, A.J., M.W. HUFF, C.G. WRIGHT, M.M. SILVER, T.C.Y. LO and D. BANERJEE: The role of insulin resistance in the pathogenesis of myotonic muscular dystrophy. Brain 110 (1987) 469–488.

HUFF, T.A., E.S. HORTON and H.E. LEBOVITZ: Abnormal insulin secretion in myotonic dystrophy. N. Engl. J. Med. 297 (1967) 837–841.

HUGHES, E.F. and J. WILSON: Response to treatment with antihistamines in a family with myotonia congenita. Lancet 337 (1991) 28–30.

HULL, K.L. and A.D. ROSES: Stoichiometry of sodium and potassium transport in erythrocytes from patients with myotonic muscular dystrophy. J. Physiol. 254 (1976) 169–181.

JAMES, D.E., A. ZORZANO, M. BONI-SCHETZLER, R.A. NE-MENOFF, A. POWERS, P. PILCH and N.B. RUDEMAN: Intrinsic differences of insulin receptor kinase activity in red and white muscle. J. Biol. Chem. 261 (1986) 14939–14944.

JANSEN, G., P.J. DEJONG, C. AMEMIYA, C. ASLANIDIS, D.J. SHAW, H.G. HARLEY, J.D. BROOK, R. FENWICK, R.G. KORNELUK, C. TSILFIDIS, G. SHUTLER, R. HERMENS, N.G.M. WORMSKAMP, H.J.M. SMEETS and B. WERINGA: Physical and genetic characterization of the distal segment of the myotonic dystrophy area on 19q. Genomics (1992) in press.

JOHNSON, K., P. SHELBOURNE, J. DAVIES, J. BUXTON, E. NIMMO, M.J. SICILIANO, L.L. BACHINSKI, M. ANVRET, H. HARLEY, S. RUNDLE, T. MIKI, H. BRUNNER and R. WILLIAMSON: A new polymorphic probe which defines the region of chromosome 19 containing the myotonic dystrophy locus. Amer. J. Hum. Genet. 46 (1990) 1073–1081.

JOZOFOWICZ, R.F., S.L. WELLE, K.S. NAIR, W.J. KING-STON and R.C. GRIGGS: Basal metabolic rate in myotonic dystrophy: Evidence against hypometabolism. Neurology 37 (1987) 1021–1025.

KAHN, E.R. and M.F. WHITE: The insulin receptor and the molecular mechanism of insulin action. J. Clin. Invest. 82 (1988) 1151–1156.

KAKEHI, T., H. KUZUYA, A. KOSAKI, K. YAMADA, Y. YO-SHIMASA, M. OKAMOTO, H. NISHIMURA, H. NISHITANI, K. SAIDA, S. KUNO and H. IMURA: Binding activity and autophosphorylation of the insulin receptor from patients with myotonic dystrophy. J. Lab. Clin. Med. 115 (1990) 688–695.

KAWAMURA, T.: Computed tomographic findings of brain and skull in myotonic dystrophy. J. Neurol. Neurosurg. Psychiatry 50 (1987) 1723.

KERTY, E. and T. GANES: Clinical and electrophysiological abnormalities in the visual system in

myotonic dystrophy. Ophthalmologica 198 (1989) 95–102.

KINGSTON, W.J., J.N. LIVINGSTON and R.T. MOXLEY: Enhancement of insulin action after oral glucose ingestion. J. Clin. Invest. 77 (1986) 1153–1162.

KINGSTON, W.J., R.T. MOXLEY and R.C. GRIGGS: Effect of testosterone on whole body amino acid utilization in myotonic dystrophy. Metabolism 35 (1986) 928–932.

KOBAYASHI, T., V. ASKANAS, K. SAITO, W.K. ENGEL and K. ISHIKAWA: Abnormalities of aneural and innervated cultured muscle fibers from patients with myotonic atrophy (dystrophy). Arch. Neurol. 47 (1990) 893–896.

KOCH, M.C., T. GRIMM, H.G. HARLEY and P.S. HARPER: Genetic risks for children of women with myotonic dystrophy. Amer. J. Genet. 48 (1991) 1084–1091.

KOLKIN, S., R.J. BAROHN, K.A. KUDSK, C.L. HYSER and J.R. MENDELL: The effects of transcutaneous muscle stimulation and testosterone on strength and muscle mass in myotonic dystrophy. Neurology 37 (1987) 209.

KRENTZ, A.J., N.H. COLES, A.C. WILLIAMS and M. NAT-TRASS: Abnormal regulation of intermediary metabolism after oral glucose ingestion in myotonic dystrophy. Metabolism 39 (1990) 938–942.

KUROIWA, Y., H. SUGITA, Y. TOYOKURA, M. MIZOGUCHI, H. MATSUO and Y. NONAKA: Immunologic derangement in myotonic dystrophy. J. Neurol. Sci. 47 (1980) 231–239.

LASPADA, A.R., E.M. WILSON, D.B. LUBAHN, A.E. HAR-DING and K.H. FISHBECK: Androgen receptor gene mutation sin X-linked spinal and bulbar muscular dystrophy. Nature (London) 352 (1991) 77–80.

LAVEDAN, C., D. DUROS, D. SAVOY, S. LEBLOND, J. BAILLY, R. KORNELUK and C. JUNIEN: Direct haplotyping by double digestion of PCR-amplified creatine kinase (CKMM): Application to myotonic dystrophy diagnosis. Genomics 8 (1990) 739–740.

LAVEDAN, C., H. HOFMANN, P. SHELBOURNE, C. DUROS, D. SAVOY, K. JOHNSON and C. JUNIEN: Prenatal diagnosis of myotonic dystrophy using closely linked flanking markers. J. Med. Genet. 28 (1991) 89–91.

LIENHARD, G.E., J.W. SLOT, D.E. JAMES and M.M. MUECK-LER: How cells absorb glucose. Sci. Amer. 266 (1992) 86–91.

LIVINGSTON, J.N. and R.T. MOXLEY: Glucose ingestion mediates a rapid increase in insulin-responsiveness of rat adipocytes. Endocrinology (1982) 1749–1759.

MACKENZIE, A.E., R.G. KORNELUK, F. ZORZATO, J. FUJII, M. PHILLIPS, D. ILES, B. WIERINGA, S. LEBLOND, J. BAILLY, H.F. WILLARD, C. DUFF, R.G. WORTON and D.H. MACLENNAN: the human ryanodine receptor gene: its mapping to 19q13.3.1, placement in a chromosome 19 linkage group, and exclusion as the gene causing myotonic dystrophy. Amer. J. Hum. Genet. 46 (1990) 1082–1089.

MACKENZIE, A.E., H.L. MACLEOD, A.G.W. HUNTER and R.G. KORNELUK: Linkage analysis of the apolipoprotein C2 gene and myotonic dystrophy on human chromosome 19 reveals linkage disequilibrium in a French-Canadian population. Amer. J. Hum. Genet. 44 (1989) 140–147.

MAHADEVAN, M., C. TSILFIDIS, L. SABOURIN, G. SHULTER, C. AMENNIYA, G. JANSEN, C. NEVILLE, M. NORANG, J. BARCELO, K. O'HOY, S. LEBLOND, J. EARLE-MAC-DONALD, P.J. DEJONG, B. WERINGA and R.G. KORNE-LUK: Myotonic dystrophy mutation: an unstable CTG repeat in the 3' untranslated region of the gene. Science 255 (1992) 1253–1255.

MARINKOVIC, Z., G. PRELEVIC, R. HAN, M. WËRZBURGER and S. TODOROVIC: Prolactin secretion in myotonic dystrophy. Exp. Clin. Endocrinol. 96 (1990) 247–252.

MATHIEU, J., M. DEBRAEKELEER and C. PRÉVOST: Genealogical reconstruction of myotonic dystrophy in the Saguenay-Lac-Saint-Jean area (Quebec, Canada). Neurology 40 (1990) 839–842.

MATHIEU, J., M. DEBRACEKELEER, C. PRÉVOST and C. BOILY: Myotonic dystrophy: Clinical assessment of muscular disability in an isolated population with presumed homogeneous mutation. Neurology 42 (1992) 203–208.

MATHIEU, J., M. SIMARD, M. DEBRAEKELEER, C. BOILY and A. DESCHêNES: Partial syndrome of myotonic dystrophy: Clinical presentation and follow-up. Can J. Neurol. Sci. 16 (1989) 99–103.

MEGE, J.L., J. POUGET, C. CAPO, P. ANDRE, A.M. BENO-LIEL, G. SERRATRICE and P. BONGRAND: Myotonic dystrophy: defective oxidative burst of polymorphonuclear leukocytes. J. Leukocyte Biol. 44 (1988) 180–186.

MESSINA, C., P. TONALI and C. SCOPPETTA: The lack of deep tendon reflexes in myotonic dystrophy: A neurophysiological study. J. Neurol. Sci. 30 (1976) 303–311.

MILNER-BROWN, H.S., M. MELLENTHIN and R.G. MILLER: Quantifying human muscle strength, endurance and fatigue. Arch. Phys. Med. Rehabil. 67 (1986) 530–535.

MILNER-BROWN, H.S. and R.G. MILLER: Muscle strengthening through high-resistance weight training in patients with neuromuscular disorders. Arch. Phys. Med. Rehabil. 69 (1988) 14–19.

MILNER-BROWN, H.S. and R.G. MILLER: Myotonic dystrophy: Quantification of muscle weakness and myotonia and the effect of amitriptyline and exercise. Arch. Phys. Med. Rehabil. 71 (1990) 983–987.

MILUNSKY, J.M., J.C. SKARE and A. MILUNSKY: Presymptomatic and prenatal diagnosis of myotonic muscular dystrophy with linked DNA probes. Amer. J. Med. Sci. 301 (1991) 231–237.

MILUNSKY, J.M., J.C. SKARE, J.M. MILUNSKY, T.A. MAHER and J.A. AMOS: Prenatal diagnosis of myotonic muscular dystrophy with linked deoxyribonucleic acid probes. Amer. J. Obstet. Gynecol. 164 (1991) 751–755.

MOORE, S.E. and F.S. WALSH: Specific regulation of N-CAM/D2-CAM cell adhesion molecule during skeletal muscle development. Eur. Mol. Biol. Org. J. 4 (1985) 623–630.

MOORMAN, R., E. COLEMAN and D. PACKER: Cardiac involvement in myotonic muscular dystrophy. Medicine 64 (1985) 371–387.

MOXLEY, R.T.: Metabolic studies in muscular dystrophy and their relationship to clinical pathophysiology: A focus upon myotonic dystrophy and Duchenne muscular dystrophy. In: Seminars in Neurology. New York, Thiemie-Stratton, Inc. (1983) 308–318.

MOXLEY, R.T., A.J. CORBETT, K.L. MINAKER and J.W. ROWE: Whole body insulin resistance in myotonic dystrophy. Ann. Neurol. 15 (1984) 157–162.

MOXLEY, R.T., R.C. GRIGGS, G.B. FORBES,, D. GOLDBLATT and K. DONOHOE: Influence of muscle wasting on oral glucose tolerance testing. Clin. Sci. 64 (1983) 601–609.

MOXLEY, R.T., R.C. GRIGGS and D. GOLDBLATT: Decreased insulin sensitivity of forearm muscle in myotonic dystrophy. J. Clin. Invest. 62 (1978) 857–867.

MOXLEY, R.T., R.C. GRIGGS and D. GOLDBLATT: Muscle insulin resistance in myotonic dystrophy: effect of supraphysiologic insulinization. Neurology 30 (1980) 1077–1983.

MOXLEY, R.T., W. KINGSTON and R.C. GRIGGS: Abnormal regulation of venous alanine after glucose ingestion in myotonic dystrophy. Clin. Sci. 68 (1985) 151–157.

MOXLEY, R.T., W.J. KINGSTON, R.C. GRIGGS and J.N. LIVINGSTON: Lack of rapid enhancement of insulin action after oral glucose challenge in myotonic dystrophy. Diabetes 36 (1987) 693–701.

MOXLEY, R.T., W.J. KINGSTON, K.L. MINAKER, A.J. CORBETT and J.W. ROWE: Insulin resistance and regulation of serum amino acid levels in myotonic dystrophy. Clin. Sci. 71 (1986) 429–436.

MOXLEY, R.T., J.N. LIVINGSTON, D.H. LOCKWOOD, R.C. GRIGGS and R.L. HILL: Abnormal regulation of monocyte insulin-binding affinity after glucose ingestion in patients with myotonic dystrophy. Proc. Natl. Acad. Sci. USA 78 (1981) 2567–2571.

NAPPI, G., E. MARTIGNONI, G. SANCES, G. MURIALDO, C. ZAULI and L. MURRI: Dynamic and chronobiological changes of endocrine indices in myotonic dystrophy. Acta Neurol. Belg. 82 (1982) 168–177.

NGUYEN, H.H., J.T. WOLFE III, D.R. HOLMES and W.D. EDWARDS: Pathology of the cardiac conduction system in myotonic dystrophy: a study of 12 cases. J. Amer. Coll. Cardiol. 11 (1988) 662–671.

NICHOLSON, A., E. RIVLIN, D.G. SIMS, M.L. CHISWICK and S.W. D'SOUZA: Developmental delay in congenital myotonic dystrophy after neonatal intensive care. Early Hum. Dev. 22 (1990) 99–103.

NOKELAINEN, P., L. ALANEN-KURKI, R. WINQVIST, B. FALCK, H. SOMER, J. LEISTI, K. JOHNSON, M.L. SAVONTAUS and L. PELTONEN: Linkage disequilibrium detected between dystrophia myotonica and APOC2 locus in the Finnish population. Hum. Genet. 85 (1990) 541–545.

NOWAK, T.V., V. IONASESCU and S. ANURAS: Gastrointestinal manifestations of the muscular dystrophies. Gastroenterology 82 (1982) 800–810.

OBERLE, I., F. ROUSSEAU, D. HEITZ, C. KRETZ, D. DEVYS, A. HANAUER, J. BONE, M.F. BERTHEAS and J.L. MANDEL: Instability of a 550-base pair DNA segment and abnormal methylation in fragile X syndrome. Science 252 (1991) 1097–1102.

OBERMAIER-KUSSER, B., M.F. WHITE, D.E. PONGRATZ, Z. SU, B. ERMEL, C. MUHLBACHER and H.U. HARING: A defective intramolecular autoactivation cascade may cause the reduced kinase activity of the skeletal muscle insulin receptor from patients with non-insulin dependent diabetes mellitus. J. Biol. Chem. 264 (1989) 9467–9504.

OKIMURA, Y., K. CHIHARA, T. KITA, Y. KASHIO, M. SATA, N. KITAJIMA, H. ABE, K. TAKAHASHI and T. FUJITA: Discordance between growth hormone (GH) and insulin responses after GH releasing hormone and insulin hypoglycemia in myotonic dystrophy. J. Clin. Endocrinol. Metab. 67 (1988) 1074–1079.

OLOFSSON, B.-O., H. FORSBERG, S. ANDERSSON, P. BJERLE, A. HENRIKSSON and I. WEDIN: Electrocardiographic findings in myotonic dystrophy. Br. Heart J. 59 (1988) 47–52.

ÖRNDAHL, G. and K. STENBERG: Myotonic human musculature; stimulation with depolarizing agents. Acta Med. Scand. 172 Suppl. (1962) 389.

OTT, J., J. CAESAR, M. MACHLER, A. SCHINZEL and W. SCHMID: Presymptomatic exclusion of myotonic dystrophy in a one-generation pedigree of half-siblings. Hum. Hered. 40 (1990) 305–307.

PERINI, G.I., G. COLOMBO, M. ARMANI, A. PELLEGRINI, M. ERMANI, M. MIOTTI and C. ANGELINI: Intellectual impairment and cognitive evoked potentials in myotonic dystrophy. J. Nerv. Ment. Dis. 177 (1989) 750–754.

PERRYMAN, M.B., J.F. HEJTMANCIK, T. ASHIZAWA, R. ARMSTRONG, S.-C. LIN, R. ROBERTS and H.F. EPSTEIN: NcoI and TaqI RFLPs for human M creatine kinase (CKM). Nucleic Acids Res. 16 (1988) 8744.

PINTO, F., A. AMANTINI, G. DE SCISCIOLO, V. SCAIOLI, R. FROSINI, A. PIZZI and G. MARCONI: Electrophysiological studies of the visual system in myotonic dystrophy. Acta Neurol. Scand. 76 (1987) 351–358.

POFFENBARGER, P.L., T. POZEFSKY and J.S. SOELDNER: The direct relationship of proinsulin-insulin hypersecretion to basal serum levels of cholesterol and triglyceride in myotonic dystrophy. J. Lab. Clin. Med. 87 (1976) 384–396.

POLGAR, J.G., W.G. BRADLEY, A.R.M. UPTON, J. ANDERSON, J.M.L. HOWAT, F. PETITO, D.F. ROBERTS and J. SCOPA: The early detection of dystrophia mytonica. Brain 95 (1972) 761–776.

PRYSE-PHILIPS, W., G.J. JOHNSON and B. LARSEN: Incomplete manifestations of myotonic dystrophy in a large kinship in Laborador. Ann. Neurol. 11 (1982) 582–591.

PTACEK, L.J., J.S. TRIMMER, W.C. AGNEW, J.W. ROBERTS, J.H. PETAJAN and M. LEPPERT: Paramyotonia congenita and hyperkalemic periodic paralysis map to

the same sodium-channel gene locus Amer. J. Hum. Genet. 49 (1991) 851–854.

PTACEK, L.J., A.L. GEORGE, R.C. GRIGGS, ET AL.: Identification of a mutation in the gene causing hyperkalemic periodic paralysis. Cell 67 (1991) 1021–1027.

RENAUD, J.F.: Involvement of cation transporting systems in myotonic diseases. Biochimie 69 (1987) 407–410.

RENAUD, J.F., C. DESNUELL, H. SCHMID-ANTOMARCHI, M. HUGHES, G. SERRATRICE and M. LAZDUNSKI: Expression of apamin receptor in muscles of patients with myotonic dystrophy. Nature (London) 319 (1986) 678–680.

RICKER, K., F. LEHMANN-HORN and R.T. MOXLEY III: Myotonia fluctuans. Arch. Neurol. 47 (1990) 268–272.

ROJAS, C.V., J. WANG, L.S. SCHWARTZ, E.P. HOFFMAN, B.R. POWELL and R.H. BROWN JR: A met-to-val mutation in the skeletal muscle Na^+ channel alpha subunit in hyperkalemic periodic paralysis. Nature (London) 354 (1991) 387–389.

RÖNNEMAA, H. ALARANTA, J. VIIKARI, R. TILVIS and B. FALCK: Increased activity of serum γ-glutamyltransferase in myotonic dystrophy. Acta Med. Scand. 222 (1987) 267–273.

ROPERS, H.H., M. PERICAK-VANCE, M. SICILIANO and H. MOHRENWEISER: Report on the second international workshop on human chromosome 19. Human Genet., in press.

ROSES, A.D.: Myotonic dystrophy. Trends Genet. 8 (1992).

ROSES, A.D.: Myotonic dystrophy. In: The Molecular and Genetic Basis of Neurological Diseases (1992) in press.

ROSES, A.D. and S.H. APPEL: Muscle membrane protein kinase in myotonic muscular dystrophy. Nature (London) 250 (1974) 245–246.

ROSES, A.D. and S.H. APPEL: Protein kinase activity in erythrocyte ghosts of patients with myotonic muscular dystrophy. Proc. Natl. Acad. Sci. USA 70 (1973) 1855–1859.

ROSES, A.D., P.S. HARPER and E.H. BOSSEN: Myotonic muscular dystrophy. In: P.J. Vinken, G.W. Bruyn and S.P. Ringel (Eds.), Handbook of Clinical Neurology. Vol. 40. Diseases of Muscle, Part I. Amsterdam, North-Holland Publ. Co. (1979).

ROUSSEAU, F., D. HEITZ, V. BIANCALANA, S. BLUMENFELD, C. KRETZ, J. BOUE, N. TOMMERUP, C. VANDERHAGEN, C. DELOZIER-BLANCHET, M-F. CROQUETTE, S. GELGENKRANTZ, P. JALBOT, M-A. VOELCKEL, I. OBERLE and J-L. MANDEL: Direct diagnosis by DNA analysis of the fragile X syndrome of mental retardation. N. Engl. J. Med. 325 (1991) 1673–1681.

RUDMAN, D., S.B. CHYATTE, G.G. GERRON, I. O'BEIRNE and J. BARLOW: Hyper-responsiveness of patients with Limb-girdle dystrophy to human growth hormone. J. Clin. Endocrinol. Metab. 35 (1972) 256–260.

RUDMAN, D., S.B. CHYATTE, J.H. PATTERSON, G.G. GERRON, I. O'BEIRNE, J. BARLOW, P. AHMANN, A. JORDAN and R.C. MOSTELLER: Observations on the responsiveness of human subjects to human growth hormone. J. Clin. Invest. 50 (1971) 1941–1949.

RUDMAN, D., A.G. FELLER, H.J.S. NAGRAJ, G.A. GERGANS, P.Y. LALITHA, A.F. GOLDBERG, R.A. SCHLANKER, L. COHN, I.W. RUDMAN and D.E. MATTSON: Effects of human growth hormone in men over 60 years of age.

RUSSELL, B. and D.J. DIX: Mechanisms for intracellular distribution of mRNA: in situ hybridization studies in muscle. Amer. J. Physiol. 262 (1992) C1–C8.

RUSSELL, B., D.J. DIX, D.L. HALLER and J. JACOBS-EL: Repair of injured skeletal muscle: A molecular approach. Med. Sci. Sports Exercise 24 (1992) 189–196.

RUTHERFORD, M.A., J.Z. HECKMATT and V. DUBOWITZ: Congenital myotonic dystrophy: respiratory function at birth determines survival. Arch. Dis. Childhood 64 (1989) 191–195.

SCHELLNACK, K., G. REGLING, J. HOFFMAN, P. BUNTROCK, H. MARTIN and H. HECHT: Maligne Hyperhermie bei subklinischer myotonischer Dystrophie. Beitr. Orthop. Traumatol. 23 (1974) 537–544.

SCHONK, D., P. VAN DIJK, P. RIEGMANN, J. TRAPMAN, C. HOLM, T.C. WILLCOCKS, P. SILLEKENS, W. VAN VENROOIJ, E. WIMMER, A.G. VAN KESSEL, H-H. ROPERS and B. WIERINGA: Assignment of seven genes of distinct intervals on the midportion of human chromosome 19q surrounding the myotonin dystrophy gene region. Cytogenet. Cell Genet. 54 (1990) 15–19.

SHELBOURNE, P., J. DAVIES, J. BUXTON, M. ANVRET, E. BLENNOW, M. BONDUELLE, E. SCHMEDDING, I. GLASS, R. LINDENBAUM, R. LANE, R. WILLIAMSON and K. JOHNSON: Direct diagnosis of ambiguous cases of myotonic dystrophy using the disease-specific DNA marker. N. Engl. J. Med. (1992) in press.

SHUTLER, G., A.E. MACKENZIE, H. BRUNNER, B. WIERINGA, P. DEJONG, F.P. LOHMAN, S. LEBLOND, J. BAILLY and R.G. KORNELUK: Physical and genetic mapping of a novel chromosome 19 ERCC1 marker showing close linkage with myotonic dystrophy. Genomics 9 (1991) 500–504.

SICILIANO, G., B. ROSSI, L. FRATINI, F. FALASCHI and A. MURATORIO: Digitalized radiologic assessment of pharyngo-esophageal function in myotonic dystrophy. Acta Cardiomiol. 2 (1990) 45–53.

SODERHALL, S., J. GUSTAFSSON and I. BJORKHEIM: Deoxycholic acid in myotonic dystrophy. Lancet i (1982) 1068–1069.

SPAANS, F., F.G.I. JENNEKENS, J.F. MIRANDOLLE, J.B. BIJLSMA and G.C. DEGAST: Myotonic dystrophy associated with hereditary motor and sensory neuropathy. Brain 109 (1986) 1149–1168.

SPEER, M.C., M.A. PERICAK-VANCE, L. YAMAOKA, W.-Y. HUNG, A. ASHLEY, J.M. STAJICH and A.D. ROSES: Presymptomatic and prenatal diagnosis in myotonic dystrophy by genetic linkage studies. Neurology 40 (1990) 671–676.

STEINMEYER, K., R. KLOCKE, C. ORTLAND, M. GRONEMEIER, H. JOCKUSCH, S. GRUNDER and T.J. JENTSCH: In activation of muscle chloride channel by trans-

poson insertion in myoptonic mice. Nature (London) 354 (1991) 304–308.

STEINMEYER, K., C. ORTLAND and T.J. JENTSCH: Primary structure and function expression of a developmentally regulated skeletal muscle chloride channel. Nature (London) 354 (1991) 301–304.

SUTHERLAND, G.R., A. GIDEON, L. KORNMAN, A. DONNELLY, R.W. BYARD, J.C. MULLEY, E. KREMER, M. LYNCH, M. PRITCHARD, S. YUN and R.I. RICHARDS: Prenatal diagnosis of fragile X syndrome by direct detection of the unstable DNA sequence. N. Engl. J. Med. 325 (1991) 1720–1722.

TAMOUSH, A.J., V. ASKANAS, P.G. NELSON and W.K. ENGEL: Electrophysiologic properties of aneurally cultured muscle from patients with myotonic muscular atrophy. Neurology 33 (1983) 311–316.

TANAKA, K. and K. TAKESHITA: Analysis of serum bile acids in patients with various types of muscular dystrophy. Muscle Nerve 10 (1983) 606–607.

TANAKA, K., K. TAKESHITA and M. TAKITA: Abnormalities of bile acids in serum and bile from patients with myotonic muscular dystrophy. Clin. Sci. 62 (1982) 627–642.

TANAKA, M. and Y. TANAKA: Cardiac anaesthesia in a patient with myotonic dystrophy. Anaesthesia 46 (1991) 462–465.

TELERMAN-TOPPET, N., P. KHOUBESSERIAN, M. BACQ, J. DURDU, D. LAMBEBIN, G. LOUSBERG and C. COËRS: Unclassified family myopathy resembling Steinert disease, without myotonia. Muscle Nerve 7 (1984) 439–441.

TERBRUGGEN, J.P., L.A.K. BASTIAENSEN, C.C. TYSSEN and G. GIELEN: Disorders of eye movement in myotonic dystrophy. Brain 113 (1990) 463–473.

THIEL, R.E.: The myotonic response to suxamethonium. Br. J. Anesth. 39 (1967) 815.

THYAGARAJAN, D., E. BYRNE, S. NOER, P. LERTRIT, P. UTHANOPHOL, R. KAPSA and S. MARZUKI: Mitochondrial DNA sequence analysis in congenital myotonic dystrophy. Ann. Neurol. 30 (1991) 724–727.

TORRES, C., R.T. MOXLEY and R.C. GRIGGS: Quantitative testing of handgrip strength, myotonia, and fatigue in myotonic dystrophy. J. Neurol. Sci. 60 (1983) 157–168.

TSILFIDIS, C., A.E. MACKENZIE, G. METTLER J. BARCELO and R.G. KORNELUK: Correlation between CTG trinucleotide repeat length and frequency of severe congenital myotonic dystrophy. Nature Genet. (1992) in press.

TSILFIDIS, C., A.E. MACKENZIE, G. SHUTLER, S. LEBLOND, J. BAILLY, K. JOHNSON, R. WILLIAMSON, J. SIEGEL-BARTELT and R.G. KORNELUK: D19S51 is closely linked with maps distal to the myotonic dystrophy locus on 19q. Amer. J. Hum. Genet. 49 (1991) 961–965.

UNGER, J., T. MCNEILL, R.T. MOXLEY, M. WHITE, A. MOSS and J.N. LIVINGSTON: Distribution of insulin receptor-like immunoreactivity in the rat forebrain. Neuroscience 31 (1989) 143–157.

UNGER, J.W., A.M. MOSS and J.N. LIVINGSTON: Insulin

receptors in brain: localization and functional implications. Prog. Neurobiol. 36 (1991) 343–362.

VANIER, T.M.: Dystrophin myotonica in childhood. Br. Med. J. 29 (1960) 1284–1288.

VENEO, A.M., M. SAVIOTTI, D. BESANA, G. FINARDI and G. LANZI: Noninvasive assessment of left ventricular function in myotonic dystrophy. Br. Heart J. 40 (1978) 1262–1268.

VERKERK, A.J.M.H., M. PIERETTI, J.C. SUTCLIFFE, Y-H. FU, D.P.A. KUHL, A. PIZZUTI, O. REINO, S. RICHARDS, M.F. VITORIA, F. ZHANG, B.E. EUSSEN, G.J.B. VAN OMMEN, L.A.J. BLONDEN, G.J. RIGGINS, J.L. CHASTAIN, C.B. KUNST, H. GALJAARD, C.T. CASKEY, D.L. NELSON, B.A. DOSTRA and S.T. WARREN: Identification of a gene (FMR-1) containing a CGG repeat coincident with a breakpoint cluster region exhibiting length variation in fragile X syndrome. Cell 65 (1991) 905–914.

WALSH, F.S., S.E. MOORE and J.G. DICKSON: Expression of membrane antigens in myotonic dystrophy. J. Neurol. Neurosurg. Psychiatry 51 (1988) 136–138.

WEVERS, R., M.G. JOOSTEN, J.B.M. VAN DE BIEZENBOS, G.M. THEEUWES and J.M. VEERKAMP: Excessive potassium K+ increase after ischemic exercise in myotonic dystrophy. Muscle Nerve 13 (1990) 27–32.

WHITEHEAD, A.S., E. SOLOMON, S. CHAMBERS, W.F. BODMER, S. POVEY and G. FEY: Assignment of the structural gene for the third component to chromosome 19. Proc. Natl. Acad. Sci. USA 79 (1982) 5021–5025.

WIGHTMAN, B., T. BURGLIN, J. GATTO, P. ARASU and G. RUVKUN: Negative regulatory sequences in the lin-14 3'-untranslated region are necessary to generate a temporal switch during Caenorhabditis elegans development. Genes Dev. 5 (1991) 1813–1824.

WILSON, A., L. MACKAY and R.A. ORD: Recurrent dislocation of the mandible in a patient with myotonic dystrophy. J. Oral Maxillofacial Surg. 47 (1989) 129–1332.

WINTERS, S.J., B. SCHREINER, R.C. GRIGGS, P. ROWLEY and N.C. NANDA: Familial mitral valve prolapse and myotonic dystrophy. Ann. Intern. Med. 85 (1976) 19–22.

WRIGHT, R.B., R.H. GLANTZ and J. BUTCHER: Hearing loss in myotonic dystrophy. Ann. Neurol. 23 (1988) 202–203.

YAMAGATA, H., T. MIKI, T. OGIHARA, M. NAKAGAWA, I. HIGUCHI, M. OSAME, P. SHELBOURNE, J. DAVIES and K. JOHNSON: Expansion of an unstable DNA region in Japanese myotonic dystrophy patients. Lancet 339 (1992) in press.

YAMAOKA, L.H., M.A. PERICAK-VANCE, M.C. SPEER, P.C. GASKELL, JR., J. STAJICH, C. HAYNES, W.-Y. HUNG, C. LABERGE, M.-C. THIBAULT, J. MATHIEU, A.P. WALKER, R.J. BARTLETT and A.D. ROSES: Tight linkage of creatine kinase (CKMM) to myotonic dystrophy on chromosome 19. Neurology 40 (1990) 222–226.

YOSHIDA, H., H. OSHIMA, E. SAITO and M. KINOSHITA: Hyperparathyroidism in patients with myotonic dystrophy. J. Clin. Endocrinol. Metab. 67 (1988) 488–492.

YOSHIDA, M.M., S. KRISHNAMURTHY, D.A. WATTCHOW, J.B. FURNESS and M.D. SCHUFFLER: Megacolon in myotonic dystrophy caused by a degenerative neuropathy of the myenteric plexus. Gastroenterology 95 (1988) 820–827.

YOSHIMURA, N.: Alzheimer's neurofibrillary changes in the olfactory bulb in myotonic dystrophy. Clin. Neuropathol. 9 (1990) 240–243.

YOSHIMURA, N., M. OTAKE, K. IGARASHI, M. MATSUNAGA, K. TAKEBE and H. KUDO: Topography of Alzheimer's neurofibrillary change distribution in myotonic dystrophy. Clin. Neuropathol. 9 (1990) 234–239.

ZWARTS, M.J. and T.W. VAN WEERDEN: Transient paresis in myotonic syndromes. Brain 112 (1989) 665–680.

Handbook of Clinical Neurology, Vol. 18 (62): Myopathies
L.P. Rowland and S. DiMauro, editors

CHAPTER 10

The nondystrophic myotonic syndromes

ROBERT L. BARCHI

University of Pennsylvania School of Medicine, Philadelphia, PA, USA

Myotonia is a form of delayed relaxation of skeletal muscle after a voluntary contraction or following mechanical or electrical stimulation. This abnormal relaxation is the external correlate of sustained electrical activity that occurs at the level of the muscle membrane and causes the characteristic repetitive discharges seen on electromyography (Landau 1952; Lindsley and Curnen 1936). Myotonia is associated with several different neuromuscular diseases; it is not a disease in itself. This Chapter will discuss the diseases of muscle other than myotonic muscular dystrophy in which myotonia is a prominent feature.

Individuals with myotonia often describe the persistent contractions as a 'stiffness' of their muscles. The myotonic contractions are usually painless, although some report a feeling of soreness similar to that experienced by normal people after hard exercise. Myotonic stiffness is most likely to occur at the onset of activity; for example, with the first few steps taken after a rest. The stiffness is typically relieved by continued exercise, and affected individuals can often carry out a full range of athletics after the stiffness has abated. This *warm-up* phenomenon is not universal, however. In some myotonic disorders the stiffness does not appear immediately but worsens with exercise; this is referred to as *paradoxical myotonia*. In many forms of myotonia, symptoms are aggravated by emotional excitement, and falls due to myotonic stiffness are often reported after an affected individual has been startled or surprised.

Myotonia can be induced by voluntary muscle contraction, by electrical stimulation of motor nerves, or by direct percussion of the muscle itself. More than just a single brief contraction is required to trigger a myotonic discharge; for instance, contractions elicited with tendon reflexes usually do not exhibit slowed relaxation. Similarly, contractions must be strong; slight movements of a finger do not produce myotonia while a firm grip does.

Studies of the mechanisms underlying the repetitive electrical activity in myotonia indicate that the site of the abnormality lies in the muscle sarcolemma (Brown and Harvey 1939; Lanari 1946). Myotonic discharges recorded with intramuscular concentric needle electrodes in response to needle movement or to percussion of the muscle cannot be eliminated by nondepolarizing blockade of the neuromuscular junction or by treatment with agents such as benzodiazepines that affect synaptic transmission in the spinal cord. The discharges also persist after motor nerve section. In this regard, myotonic contractions can be differentiated from common cramps and from peripheral nerve and spinal cord disorders on one hand (Isaacs 1961; Nakanishi 1978), and from the electrically silent contractures of the glycogen-storage diseases on the other (DiMauro et al. 1984).

Although the myotonic disorders share a characteristic phenotypic feature, these disorders do not appear to be the result of a single common defect; rather, myotonia is the final common expression of an abnormality in any one of several membrane ion channels. Even within a single seemingly homogeneous disease, the precise nature of the genetic defect may differ from family to family. In most cases, the molecular basis of these fascinating diseases is only beginning to be understood.

THE MYOTONIC DISCHARGE

Characteristics of the myotonic discharge

The characteristic finding of myotonia in electromyography (EMG) is the appearance of runs of repetitive discharges at 20–80 Hz that appear after needle insertion or movement, or following percussion of the muscle belly (Fig. 1) (Landau 1952). The individual elements of the myotonic discharge usually resemble either positive sharp waves or fibrillation potentials, and represent action potentials from single muscle fibers. Although the classical description of these myotonic discharges is one of 'dive-bomber discharges' which first increase and then decrease progressively in amplitude and

Fig. 1. A myotonic discharge induced in skeletal muscle by percussion, recorded by a concentric needle electrode. The characteristic increasing and then decreasing amplitude and frequency of the muscle action potentials is evident. This particularly striking example, while similar to myotonic discharges in human disease, was obtained from the tibialis anterior muscle of a rat treated chronically with 20,25-diazacholesterol (Furman and Barchi 1981).

frequency, thus giving rise to the audio image, the discharges are, in fact, much more variable in appearance. A similar range of myotonic waveforms is seen in the various disorders described in the following section, but some features distinguish individual diseases.

The pathognomonic myotonic discharge increases in frequency and amplitude over the first few potentials. Then both frequency and amplitude decrease progressively until the discharge terminates. However, this complete pattern was seen in only ~5% of an extensive series of recordings (Ricker and Meinck 1972a). Shorter discharges, characterized by increasing frequency and decreasing amplitude, were more common in myotonia congenita. In myotonic dystrophy, the most common pattern was one of longer discharges of lower frequency with little variation in frequency or amplitude (Buchthal and Rosenfalck 1963; Marinacci 1964; Ricker and Meinck 1972b).

In paramyotonia, myotonic discharges similar to those of myotonia congenita are found at normal ambient temperatures. When the muscle is cooled, however, persistent runs of low amplitude, high frequency potentials are recorded in which the individual waveforms resemble fibrillation potentials (Haass et al. 1981). The discharges of chondrodysplastic myotonia also differ from those of myotonia congenita; the discharges have a higher frequency and show little change in frequency during bursts. The bursts themselves can be quite long and may exhibit abrupt onset and termination. The spontaneous activity in this syndrome is virtually continuous.

When different muscles are systematically surveyed, the distribution of myotonic discharges also varies among the several myotonic syndromes. In myotonia congenita of either the dominant or recessive form, myotonic discharges are easily elicited throughout the muscle belly and are found widely in proximal and distal muscles of both upper and lower limbs. In myotonic dystrophy, on the other hand, myotonic discharges are more focal in distribution within a given muscle, are most commonly found in the distal muscles of the upper extremity (finger extensors, thenar muscles) and in the orbicularis oculi (>90% of patients), and are less frequent in other muscles such as the deltoid (38% of patients) (Strieb and Sun 1983).

Repetitive electrical activity resembling myotonia

Myotonic discharges, while characteristic of the primary myotonic muscle disorders discussed in this Chapter, are not absolutely confined to them. Simple and complex repetitive discharges similar to those seen in the true myotonic disorders have also been reported in glucosidase deficiency (Pompe's disease; Manz 1980) and hypothyroid myopathy (Norris and Panner 1966) as well as in chronic denervation of muscle (Brumlik et al. 1970; Ricker and Meinck 1972b; Valenstein et al. 1978; Chida et al. 1981). In at least one study, the muscle membrane was implicated because the discharges did not disappear after curarization (Ricker and Meinck 1972b).

Other forms of electrical activity observed in muscle with EMG can be mistaken for myotonic discharges. *Neuromyotonic discharges* are bursts of repetitive motor unit action potentials that exhibit an abrupt onset and termination, and usually are at high rates (150–300 Hz). Although the frequency of the discharges may remain constant, the amplitude typically decrements. Neuromyotonic discharges are abolished by curarization but not by peripheral nerve block or anesthesia, and are thought to arise in the distal segments of motor neurons (Isaacs 1961; Black et al. 1972; Isaacs and Heffron 1974). They are most characteristic of *Isaacs' syndrome* (see below), but similar discharges are seen in other disorders of peripheral nerve. *Myokymic discharges* are grouped bursts of motor unit discharges in which single motor units fire briefly and at constant frequency. The rhythmic discharges correspond to the undulating local contractions that give myokymia its characteristic visual appearance (Norris 1977). The electrical activity of myokymia does not disappear with sleep, but is abolished by Xylocaine block of the peripheral nerve trunk (Hughes and Matthews 1969; Parry-Jones et al. 1977). *Complex repetitive discharges* are polyphasic potentials of constant amplitude and frequency (usually 50–100 Hz) that start and stop abruptly. The signals arise from a group of contiguous muscle fibers that stimulate each other through ephaptic transmission (Trontelj and Stålberg 1983). These discharges are seen in myopathies such as polymyositis and Duchenne dystrophy as well as in neuronal disorders such as

Charcot-Marie-Tooth disease and spinal muscular atrophy (Emeryk et al. 1974).

CLINICAL FEATURES OF THE MYOTONIC DISORDERS

Classification

Myotonia is a characteristic feature of myotonic muscular dystrophy (Steinert 1909), by far the most common disorder associated with the phenomenon. That disease is considered in Chapter 9, this Volume.

The nondystrophic myotonic disorders can be divided into genetic and acquired conditions (Table 1). The major genetic myotonic syndromes include the *dominantly inherited* form of *myotonia congenita* or *Thomsen's disease* (Thomsen 1876) and an uncommon syndrome with variable myotonic features, *myotonia fluctuans*. Myotonia congenita also occurs in a recessively inherited form, *recessive generalized myotonia*, whose symptoms vary in a characteristic fashion from the dominant form (Becker 1957). Other rare disorders also have myotonia as a prominent feature; these include the dominantly inherited *paramyotonia congenita* (Eulenberg 1886; Becker 1970), some forms of *hyperkalemic periodic paralysis* (Gamstorp 1956; Buchthal et al. 1958; Van der Meulen et al. 1961; Carson and Pearson 1964), and *chondrodysplastic myotonia* (Schwartz and Jampel 1962). Each of these disorders will be discussed in detail below.

Inherited forms of myotonia resembling myotonia congenita have been characterized in detail in the goat (Bryant 1979) and mouse (Rüdel 1990). These animal models have proven valuable in studying the membrane mechanisms responsible for myotonia. In addition, toxins and drugs can induce human myotonic syndromes that closely resemble the inherited myotonias. These compounds, including the insecticide 2,4-D (Pohl 1917) and the anticholesterolemic agent 20,25-diazacholesterol (Winer et al. 1966), produce myotonia in experimental animals and have also proven useful in studying disease mechanisms.

The salient clinical features of diseases associated with myotonia are given in the following paragraphs. A detailed discussion of the pathophysiology of the abnormal membrane electrical activity in each of these myotonic disorders is presented in a later section.

Myotonia congenita: Thomsen's disease

The dominantly inherited form of myotonia congenita is best exemplified by a single extended kinship that can be traced to a mutation affecting the family of Dr. Asmus Thomsen (Leyden 1874; Thomsen 1876). Through Dr. Thomsen's account of the disorder that afflicted his family, we have an unusual insight into the signs and symptoms of *dominant myotonia congenita* (DMC) (Thomsen 1876). This form of myotonia has also been extensively studied in large German kinships by Dr. P. Becker; this collected information describes a surprisingly homogeneous disorder (Becker 1977).

In DMC, the myotonic symptoms are generalized but characteristically nonprogressive. To the discerning eye, symptoms are present at birth, often as a difficulty experienced by an infant in opening its eyes after squeezing them tightly to cry. Members of the Thomsen family were said to be able to predict affected children with a high degree of accuracy in this manner. Myotonic symptoms are most prominent in the thigh and lower leg muscles. The hands are the next most frequently involved, followed by the facial muscles including the orbicularis oculi and oris. Patients usually date the onset of symptoms as far back as they can remember, and clinical records usually indicate onset between the ages of 1 and 6 years. Although myotonia of trunk muscles is seldom reported, some individuals report stiffness of the trunk after coughing or sneezing.

The myotonic features of this disease do not fluctuate much with time, remaining relatively constant from day to day. In addition, most affected individuals report little change in the severity of their symptoms with increasing age. In this respect, myotonia congenita differs consistently from the recessive form, in which progression of

TABLE 1

Human disorders associated with myotonia.

Inherited disorders
Autosomal-dominant
    Myotonic muscular dystrophy
    Dominant myotonia congenita
    Myotonia fluctuans
    Paramyotonia congenita
    Hyperkalemic periodic paralysis

Autosomal-recessive
    Recessive generalized myotonia
    Schwartz-Jampel syndrome

Acquired disorders
Aromatic carboxylic acid intoxication (2,4D; 9-AC)
Inhibitors of cholesterol synthesis
    20,25-Diazacholesterol,
    Clofibrate

symptoms is commonly reported. Despite the constancy of symptoms in a given individual, it is not uncommon for the severity to vary between members of a given family. In some people, symptoms may even be restricted to a few muscles.

Most patients with DMC do not report major changes with exposure to cold. Some experience a mild increase in stiffness, but never as much as in paramyotonia (see below). Myotonic symptoms may be aggravated by exhaustion or extreme muscular fatigue. Emotional excitement is often cited as an augmenting factor. Finally, pregnancy has been reported to induce a worsening of myotonia.

Muscular hypertrophy is common in DMC. This was described by Thomsen as a tendency toward an athletic habitus (Thomsen 1876), while Becker comments on the well-developed appearance of the muscles with obviously demarcated muscle edges and relatively little subcutaneous fat (Becker 1977) (Fig. 2).

The incidence of DMC is difficult to ascertain since many individuals with mild disease are never seen by a physician. Becker has estimated that in Germany, where the disorder is more common

$\rightarrow$

Fig. 2. Myotonia congenita. (a) and (b) are photographs of a descendent of Thomsen who expresses the dominant form of the disease. In this case, muscles are neither hypertrophied nor atrophied. In contrast, (c) shows the hypertrophic calf muscles of a patient with generalized recessive myotonia. (Reproduced from R.J. Lipicki (1979), Handbook of Clinical Neurology, Vol. 40.)

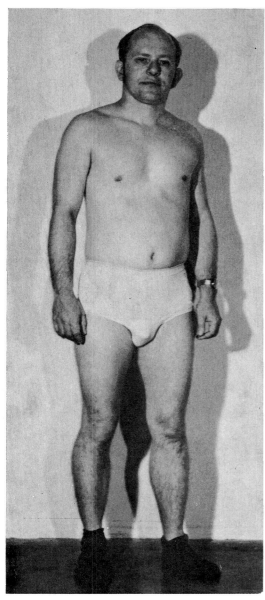

(a)

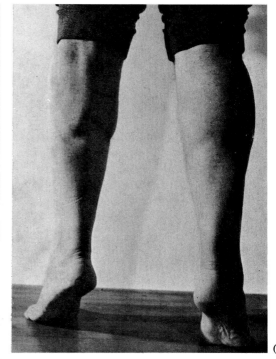

(b)

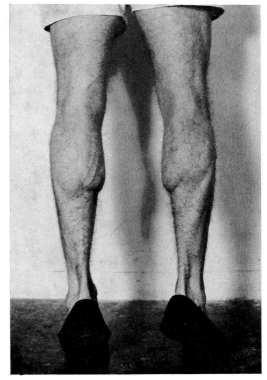

(c)

than in many other countries, the incidence probably lies between 0.25 and 4/100,000 (Becker 1977). Males and females are affected equally. Life expectancy is not altered.

Genetic studies on German families with myotonia congenita have demonstrated linkage between the expression of DMC and a region on chromosome 7 that contains the T-cell beta receptor family genes and the gene for the skeletal muscle chloride channel (Koch et al. 1992). This chloride channel gene is homologous to that already shown to be involved in the pathogenesis of myotonia in the *adr* mouse mutant model (Steinmeyer et al. 1991b).

Myotonia fluctuans

Another dominantly inherited form of myotonia can be distinguished based on the variability of symptoms reported by affected family members. This form, designated type II by Becker (1977), has been called *myotonia fluctuans* because of the characteristic wide temporal variations in severity (Ricker et al. 1990). These variations can occur over hours or days but, unlike myotonia congenita, are consistently reported by members of an affected family. However, the disease does not show the long-term temporal progression typical of the recessive form of myotonia congenita.

A second distinguishing feature of this disorder is the report of pain that accompanies the myotonic contracture. In Becker's families, 23 of 29 patients with this form of fluctuating myotonia reported muscle pain, while this complaint is distinctly unusual for members of families with Thomsen's disease.

Painful dominant myotonia

Several families have been reported with a syndrome closely resembling DMC that differs from Thomsen's disease in having consistently painful muscle contractions. In one of these families (Sanders 1976) some painful contractions were found to be electrically silent, while in the other family (Trudell et al. 1987) muscle stiffening was always associated with profuse electrical activity. Of particular interest, the latter family was unique in responding to acetazolamide while showing little response to other traditional antimyotonic agents. Muscle stiffness was provoked by fasting or by oral potassium. This latter family may prove to be a variant of the paramyotonia-periodic paralysis group of disorders since expression of the disease phenotype has been linked to the skeletal muscle sodium channel locus on chromosome 17q (Ptacek et al. 1992b).

Recessive generalized myotonia

In *recessive generalized myotonia* (RGM), the onset of symptoms is characteristically later than in the dominant form (Becker 1973, 1977). Age at onset is between 6 and 12 years. Stiffness usually first affects the legs; the upper extremities may be free from myotonic symptoms for many years. For most patients with recessively inherited myotonia, symptoms are constant from day to day but worsen slowly throughout the first 20 or 30 years of life.

In addition to the stiffness and the difficulty with force generation during the myotonic contracture itself, RGM patients often have mild muscle weakness when myotonia is absent. Such weakness was found in nearly 50% of patients with RGM in the extensive German pedigrees evaluated by Becker (1977). This weakness is also slowly progressive, although usually not debilitating. Proximal muscles of the legs and arms are usually the most severely affected. Unlike DMC, the tendon reflexes in RGM are typically depressed. Achilles tendon reflexes may be absent in 30% of affected individuals (Becker 1977).

The exacerbating factors for RGM are similar to those of the dominant form. Falls may be induced by startle. Individuals may be unable to rise for several seconds or minutes. Some people report worsening of myotonia with cold, but as with the dominant form, this is not a consistent finding.

Muscular hypertrophy is said to be more pronounced in RGM than in DMC (Becker 1977). Affected individuals are often described as having Herculean figures with prominent musculature. The muscles themselves have an abnormal firmness even at rest that is not due to persistent myotonia. In spite of this description, patients with RGM often develop thinning and atrophy of the forearm muscles and muscles of the hands later in life. This feature led to speculation in the early literature of a progression of RGM to myotonic muscular dystrophy, although experience indicates that this transition does not occur. The incidence of RGM in Germany is approximately 2/100,000, although incomplete ascertainment makes it likely that this is an underestimate (Becker 1977).

Direct measurements of membrane properties in muscle biopsies have shown that RGM is associated with an abnormal reduction in sarcolemmal chloride conductance. RGM has also been linked to the region of chromosome 7 that contains the skeletal chloride channel gene (Koch et al. 1992).

Hyperkalemic periodic paralysis with myotonia

The periodic paralyses comprise a group of disorders characterized by intermittent episodes of muscular weakness or paralysis associated with loss of excitability of the muscle surface membrane. The severity of attacks can vary between brief periods of muscle stiffness or mild weakness lasting a few minutes to periods of flaccid paralysis lasting hours or even days. Paralytic episodes are often accompanied by shifts in serum potassium concentration. These shifts formed the basis of the first classifications of these disorders; Becker observed that in affected German families serum potassium reproducibly rose (hyperkalemic periodic paralysis), fell (hypokalemic periodic paralysis), or remained unchanged (normokalemic variant) during paralytic episodes. The periodic paralyses are considered in detail in Chapter 15, this Volume.

The *hyperkalemic* form of *periodic paralysis* (HPP) can present in association with the clinical features of myotonia (Gamstorp 1956, 1963; Samaha 1965). In some families, myotonia is present only in a cold environment, while in others it is present at normal ambient temperature. The electrophysiological appearance of the myotonic potentials is indistinguishable from that of the other myotonic disorders discussed above. Most commonly, the disorder is inherited in an autosomal-dominant manner.

The most striking symptom in HPP is not myotonia, but rather the episodic weakness or paralysis. Episodes may be brief, and may occur repeatedly during the course of a day. The predominant triggering factors in HPP are rest after exercise and hypoglycemia; episodes of weakness can be evoked clinically by oral potassium loading.

Various forms of HPP appear to be due to allelic defects in the human skeletal muscle sodium channel. Linkage has been shown between expression of the disease phenotype and the gene encoding this channel on chromosome 17 (Fontaine et al. 1990), and several different mutations within the gene have been identified (Ptacek et al. 1991a; Rojas et al. 1991). A kinetic abnormality of the channel protein has been characterized at the single channel level that is consistent with the small noninactivating current reported in earlier electrophysiological studies on biopsied muscle (Ricker et al. 1989; Cannon et al. 1991). These findings are discussed in detail in the section: *Pathophysiology of myotonia*.

Paramyotonia congenita

Paramyotonia congenita (PC) is a rare disorder originally described by Eulenburg in a single family extending over 6 generations (Eulenberg 1886). Becker (1977), in his extensive studies of the myotonic disorders in Germany, identified only 157 individuals with PC; he attributed all cases to a mutation in a single individual. Other extensive families have been described (Rich 1894; LaJoie 1961). Inheritance is autosomal-dominant. Symptoms usually begin in the first decade of life.

Patients with PC exhibit myotonic symptoms that are profoundly exacerbated by cold (Drager et al. 1958) (Fig. 3). Often there is no myotonia at all in a warm environment, or there may be only mild symptoms. Exposure to cold increases the severity of the myotonic stiffness; continued exposure may lead to weakness or even paralysis of affected muscles. Becker (1977) found that only 3 of 157 individuals with PC had attacks of periodic paralysis. Both myotonia and weakness are reversed by rewarming the affected muscles, although there may be a lag of minutes to hours. In winter, stiffness of the facial muscles causes difficulty opening the eyes and restricts facial movements (Fig. 3). The disorder is not progressive; in fact, many affected individuals report that symptoms were most severe in childhood and improved with age (LaJoie 1961; DeJong et al. 1973).

The myotonic symptoms in PC are typically

Fig. 3. Paramyotonia. (A) The normal facial appearance of a patient with paramyotonia at rest in a warm environment. (B) After exposure to cold, the same patient exhibits myotonia of the lid muscles during attempted eye opening after several forceful closures of the eyelids. (Julien et al., 1971) (Reproduced from R.J. Lipicki (1979), Handbook of Clinical Neurology, Vol. 40.)

'paradoxical'; that is, the myotonia is not present at the start of motor activity but rather appears and increases with exercise. In contrast to the warm-up phenomenon seen in DMC and RGM, patients with PC usually report that myotonic stiffness increases if they continue to exercise. In addition, stiffness is often followed by periods of weakness. Indeed, paralytic episodes can be induced by exercise alone, although they are more likely to occur in cold weather than in warm.

Although PC is often considered a distinct clinical entity, there has been a controversy as to whether this syndrome is indeed distinct from the HPP or represents an extreme on a spectrum that combines the 2 descriptive entities. This debate has been fueled by the observations that some forms of HPP are associated with myotonia (Buchthal et al. 1958; Van der Meulen et al. 1961; Gamstorp 1963; Samaha 1965), that some patients with otherwise classical PC can experience attacks of weakness in the absence of myotonia (Haass et al. 1981), that cold exposure can induce local paralysis in HPP (Layzer et al. 1967), and that some patients with PC have paralytic episodes associated with hyperkalemia potassium loading (French and Kilpatrick 1957; Becker 1977; Ricker et al. 1983).

Some investigators believe that the rate at which the compound muscle action potential is lost after exercise or during cooling differentiates PC from HPP (Subramony et al. 1983). Patients with PC show a rapid loss of compound muscle action potential (CMAP) amplitude, while in HPP the decline is delayed and slower. However, De Silva et al. (1990) described a family in which individual members expressed clinical features of either PC or HPP, and in which the electrophysiological results also showed either the rapid or slowly developing pattern of loss.

Recent genetic linkage studies have now resolved the issue of whether PC and HPP are different entities or extremes along a continuum of phenotypic expression of a single pathophysiologic defect. Several groups have reported linkage between expression of the PC phenotype and a restriction fragment length polymorphism either whithin the human skeletal muscle sodium channel gene (Ptacek et al. 1991b) or adjacent to but inseparable from it (Ebers et al. 1991). In PC, specific point mutations in the sodium channel gene have also been identified, but occur at different locations than those found in HPP (McClatchey et al. 1992; Ptacek et al. 1992a).

PC and HPP are allelic disorders of the skeletal muscle sodium channel. Depending on the nature and site of the altered amino acid introduced by each mutation, expression of the resultant sodium channel protein α subunit can give rise to the extreme phenotypes reported for PC and HPP, or to an intermediate phenotype with characteristics of both. The degree to which the symptoms assumed to be characteristic of either disease are expressed in a given family would depend on the nature of the specific mutation present in that family. Some families might express symptoms of only PC or HPP, while others express a mixed picture even though the defective molecule is the same in all cases.

Chondrodystrophic myotonia

Chondrodystrophic myotonia, the *Schwartz-Jampel syndrome,* is characterized by dwarfism, multiple skeletal abnormalities, persistent skeletal muscle electrical activity, and characteristic facies (Fig. 4) (Schwartz and Jampel 1962). Although the usual inheritance pattern is autosomal-recessive, some cases of autosomal-dominant inheritance have been reported (Ferrannini et al. 1982). In most patients, symptoms are obvious at birth, and the diagnosis is typically made in the first year of life. Affected infants have narrow palpebral fissures, blepharospasm, micrognathia, and flattened facies (Farrell et al. 1987). They almost invariably have limitations of joint motion. Skeletal abnormalities include short neck, kyphosis, and pectus carinatum (Fig. 4). The muscles are typically hypertrophic and clinically stiff. Electrophysiologically, there is nearly continuous electrical activity and electrical silence is difficult to obtain.

The characteristic muscle stiffness arises from sustained muscle membrane electrical activity (Spaans et al. 1990). Discharges differ from those of myotonia congenita and RGM in that there is relatively little waxing and waning in either amplitude or frequency. Although there is some similarity to the persistent muscular activity of the stiffman syndrome or Isaacs' syndrome, the electrical activity in chondrodysplastic myotonia does not disappear with sleep and, contrary to some earlier reports (Taylor et al. 1972; Fowler et al. 1974), is not affected by curarization (Cao et al. 1978; Spaans et al. 1990). Benzodiazepines and other drugs that modify spinal cord synaptic activity also have no effect on the muscle electrical activity.

Muscle biopsy in chondrodysplastic myotonia exhibits mostly nonspecific changes (Fowler et al. 1974; Spaans et al. 1990), including vacuoles between myofibrils that are interpreted to be dilated elements of the T-tubular system, focal disarrangement and discontinuities in myofibrils, and Z-disk streaming (Spaans et al. 1990).

Although not characteristic of the disorder, about 20% of all reported cases have had mild mental retardation (Huttenlocher et al. 1969; Ferannini et al. 1982). Hyporeflexia is seen in about half of the cases, and infants are often described as having a harsh cry with laryngeal stridor. The disorder is inherited as an autosomal-recessive trait. The symptoms are not progressive in nature.

Electrophysiological studies at the single channel level have reported abnormal sodium channel kinetics in the sarcolemma of muscle obtained at biopsy from patients with the Schwartz-Jampel syndrome. Genetic analysis, however, remains to be carried out.

Myotonia induced by drugs or toxins

A myotonic syndrome can be induced in otherwise normal individuals by exposure to several drugs or toxins. One of the first described was the herbicide 2,4-dichlorphenoxyacetate (2,4-D) (Bucher 1946), an aromatic dicarboxylic acid that specifically blocks the skeletal muscle chloride conductance pathway (Palade and Barchi 1977b). Exposure to or ingestion of this chemical or related compounds may cause a syndrome in which muscle stiffness and delayed muscle relaxation are prominent

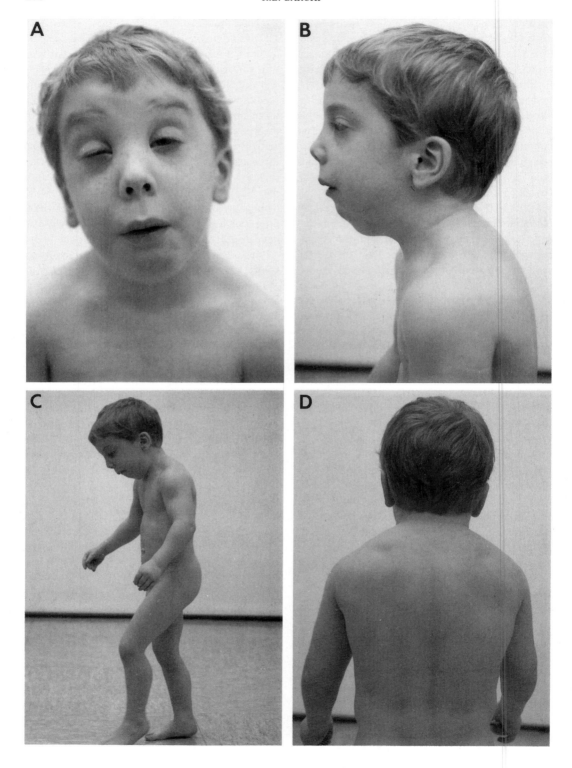

(Eyzaguirre et al. 1948). The myotonia is acute in onset, and reverses rapidly after exposure to the agent is terminated.

Other hydrocarbon- and halogen-substituted benzoic acid derivatives also produce myotonic symptoms when injected intraperitoneally into mice, rats, monkeys, and several other mammalian species (Tang et al. 1968). These aromatic carboxylic acids seem to act by blocking sarcolemmal chloride conductance (G_{Cl}). Aromatic monocarboxylic acid derivatives have been studied in rats (Palade and Barchi 1977b); many reduce membrane G_{Cl} in vitro in a dose-dependent manner. The inhibitory constant for block of (G_{Cl}) correlated directly with the physicochemical properties of the compound. The most potent was anthracene-9-carboxylic acid (9-AC), which inhibited G_{Cl} with an apparent K_i of 1×10^{-5} M (Fig. 5). These compounds partition into the lipid phase of the membrane and bind at a single class of intramembranous sites, presumably on the chloride channel itself. Interaction with the channel alters the anion selectivity and effectively reduces channel conductance to chloride ions (Palade and Barchi 1977b). When these compounds were tested for ability to induce myotonia in isolated rat muscle, there was a direct relationship between the concentration required to block G_{Cl} in vitro and that required to produce myotonia (Furman and Barchi 1978). The electrophysiological characteristics of the resultant myotonia, as recorded with intracellular microelectrodes (Furman and Barchi 1978), were identical to those found by Adrian and Bryant (1974) in isolated muscle fibers from the myotonic goat (Fig. 5).

Myotonia can also be seen in humans as a consequence of treatment with the anticholesterolemic agent 20,25-diazacholesterol (20,25-D) (Winer et al. 1966). This drug inhibits the conversion of desmosterol to cholesterol by the enzyme δ-24-reductase in the terminal step of cholesterol biosyn-

thesis. Treatment with diazacholesterol results in an increase in the desmosterol level and a decrease in the cholesterol level in plasma, erythrocyte membranes, and whole muscle (Winer et al. 1966; Chalikian and Barchi 1980b). After variable periods of treatment with the drug, humans and animals develop a syndrome characterized by delayed muscle relaxation, repetitive muscle electrical activity, cataracts, and (in rats) changes in fur texture. The gradual onset of symptoms correlates with the slow accumulation of desmosterol in cell membranes; unlike the myotonia that follows exposure to aromatic carboxylic acids, these symptoms may persist for several weeks after termination of 20,25-D treatment.

Rats treated with 20,25-D exhibit numerous biochemical abnormalities, some of which involve properties of cell surface membranes such as increased ($Na^+ + K^+$)-ATPase activity (Peter et al. 1973; Chalikian and Barchi 1980b), increased ouabain binding (Seiler et al. 1974), decreased Ca^{2+}-ATPase activity, and increased apparent membrane microviscosity (Butterfield and Watson 1977; Chalikian and Barchi 1980a). The animals do not, however, exhibit a reproducible reduction in muscle membrane G_{Cl} of the magnitude seen with the aromatic carboxylic acids (Furman and Barchi 1981). Membrane G_{Cl} decreases transiently after initiation of 20,25-D treatment but rises again with continued exposure at a time when myotonia is still present (D'Alonzo and McArdle 1982). While the pathogenesis of the repetitive activity following 20,25-D treatment remains to be defined, an abnormality in the sodium of potassium conductance pathways has been suggested.

Hereditary myotonic syndromes in animals

A hereditary myotonic syndrome in goats that has many of the clinical characteristics of human myotonic congenita was reported as early as 1904

←

Fig. 4. A 4-year-old boy with Schwartz-Jampel syndrome. The characteristic facial features include a receding chin, long philtrum, and upturned nose (A), as well as prominent epicanthal folds (B). Blepharospasm on attempted eye opening after forceful closure is also seen in (B). When walking, the patient's forefoot hits the floor before the heel; there is moderate kyphosis (C). The short neck and hypertrophic shoulder and neck musculature are obvious in (D). (Modified from Spaans et al. (1990).)

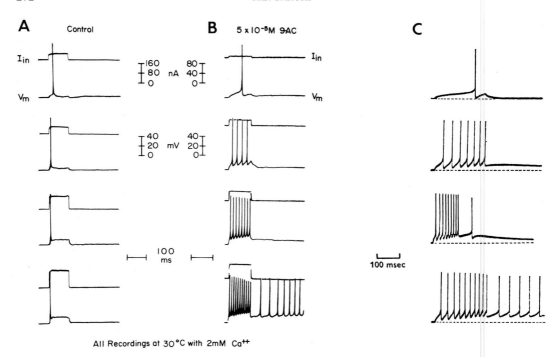

Fig. 5. Intracellular recordings from a normal rat muscle fiber (A) and from a fiber in which chloride conductance (G_{Cl}-) was blocked with anthracine-9-carboxylic acid (9-AC) (B) are compared with similar recordings from muscle fibers of a goat with congenital myotonia (C). In each series, depolarizing current pulses of increasing amplitude are applied through 1 electrode (I_{in}) and membrane voltage response recorded through a second (V_m). In normal muscle, long depolarizing pulses rarely elicit more than a single action potential (A). In muscle treated with 9-AC (B) or in congenital myotonia (C) larger depolarizations generate repetitive driven action potentials and produce an increasingly large after-depolarization that persists beyond the duration of the current pulse. Eventually this depolarization becomes large enough to independently trigger the repeated action potentials of a myotonic discharge.

(White and Plaskett 1904). This disorder is inherited in an autosomal-dominant pattern without sexual preponderance. Animals usually show hypertrophied muscles. Muscle stiffness is especially apparent when the animals are surprised and lessens with exercise. This feature is said to have led to the initial recognition of the disorder; members of a herd of goats in Tennessee were noted to fall over in rigid spasms after being startled by the whistle of a passing train, only to appear completely normal after a brief struggle to regain their feet. The train engineers were said to take great pleasure in surprising the animals on each trip through the area!

Much of our current understanding of the membrane mechanisms involved in the production of myotonic discharges is the result of research on hereditary goat myotonia (Bryant 1979). Early experiments quickly established the sarcolemma as the site of pathology in this disorder (Brown and Harvey 1939). Later studies with isolated intercostal muscle fibers in vitro provided the first intracellular recordings of myotonic discharges and the conclusive demonstration that these discharges persisted in spite of nerve section or neuromuscular blockade (Adrian and Bryant 1974). This work eventually led to the identification of a markedly reduced sarcolemmal membrane chloride conductance as the primary inherited defect (Bryant 1962).

Another syndrome that resembles human recessive generalized myotonia is seen in mouse mutants designated *mto* or *adr* (Rüdel 1990). These mutations, first described in 1982 (Heller et al. 1982; Watkins and Watts 1984), and 3 additional mutations are now thought to be allelic at a single site on mouse chromosome 6 (Davisson et al.

1989). When animals homozygous for the *adr* mutation start to walk, the hind legs become stiffly extended. The forelimbs are less severely affected, so the animals can move about in part by dragging the hind limbs along. The spasms of the hind limbs are especially prominent when the cage is shaken, startling the animals. This stiffness leads to difficulty in righting when the animals are placed on their backs, and is apparent by 10 days of age; this property accounts for the label attached to the mutant: arrested development of righting response (*adr*).

In the *mto* allele Entriken et al. (1987) confirmed delayed relaxation and myotonic discharges. The half-time for relaxation of skeletal muscle following tetanic stimulation was increased more than 25-fold over control. Subsequent studies of the general physiologic properties of the *adr* allele showed similar results (Reininghaus et al. 1988). These investigators also confirmed that the abnormal electrical activity was not influenced by nerve section or neuromuscular blockade. Studies with intracellular microelectrodes (Bryant et al. 1987; Mehrke et al. 1988) demonstrated that electrical responses in both *mto* and *adr* mutants were indistinguishable from those of goat myotonia or myotonia produced by aromatic carboxylic acids. Measurements of membrane passive properties confirmed the presence of a marked reduction (probably in excess of 80%) in membrane chloride conductance (Mehrke et al. 1988).

Muscle transplant experiments between *adr* and wild type mice indicate that the defect is intrinsic to the muscle fiber and not a reflection of the environment in which the fiber develops (Füchtbauer et al. 1988). Recent genetic analysis has confirmed an inactivating mutation in the gene encoding the skeletal muscle chloride channel caused by insertion of a transposon element (see below).

Other conditions resembling myotonia

Myotonia must be differentiated from other forms of delayed relaxation or muscle stiffness that arise in spinal cord or peripheral nerve, or that are the result of intrinsic muscle disease. In true myotonia, the *sine qua non* is repetitive electrical activity on EMG that originates *at the level of the muscle surface membrane*. This is distinct from *neuro-* *myotonia* and *myokymia*, where the repetitive activity in muscle is driven from a hyperirritable focus in the peripheral nerve or spinal cord, from *cramps*, where repetitive motor unit firing probably originating within the spinal cord produces painful muscle contraction, and from *contractures*, the persistent painful muscle shortening of metabolic myopathies that occurs in the setting of muscle electrical silence.

Several syndromes associated with muscle stiffness, delayed relaxation, and persistent muscle electrical activity can be initially confused with myotonia. In *Isaacs' syndrome* (Isaacs 1961; Isaacs and Heffron 1974; Wolter and Brauer 1977), muscle stiffness appears first as myokymia, the slow, undulating contraction of adjacent groups of muscle fibers, driven by rhythmic firing of motor nerves. Stiffness is usually apparent first in the distal muscles of the hands and feet, and later spreads more proximally. In later stages, generalized muscle stiffness worsens and the EMG exhibits continuous activity at rest, with characteristic repeating doublets and triplets similar to tetany. The delayed relaxation seen in this syndrome is associated with an after-discharge of muscle electrical activity that does not wax and wane. The stiffness and delayed relaxation of Isaacs' syndrome differ from true myotonia in being abolished by neuromuscular block. Stiffness and delayed relaxation increase with exertion in Isaacs' syndrome, contrary to true myotonia, and muscle does not exhibit percussion myotonia. Fasciculations are commonly seen in Isaacs' syndrome but not in the true myotonic disorders. Since the muscle electrical activity in this syndrome is reduced but not abolished by peripheral nerve block, but is completely eliminated by neuromuscular blockade, the site of origin is most likely in the terminal arborizations of the motor nerves.

Neuromyotonia has also been reported following peripheral nerve trauma, and in association with intrinsic cord lesions. The EMG correlate is typically a high frequency discharge of constant frequency and amplitude with an abrupt onset and termination. An unusual form of 'pseudomyotonia', characterized clinically by difficulty in opening the hand after a forceful grasp, has also been reported with chronic cervical radiculopathy. This probably represents aberrant re-innervation of finger extensor muscles by C7 root fibers normally destined for finger flexors, rather than abnormal hyperexcitability of the nerves themselves (Satoyoshi et al. 1972).

Another syndrome presenting with muscle stiffness and persistent muscle electrical activity is the *Moersch-Woltman syndrome* or *stiff-man syndrome* (Moersch and Woltman 1956; Gordon et al. 1967). Painful muscle spasms accompany the muscular rigidity during later phases of this disease. The symptoms are usually most prominent in the axial and proximal limb musculature. The abnormal electrical activity is completely abolished by neuromuscular block, differentiating it from true myotonia, but is also eliminated by peripheral nerve block and by spinal anesthesia (Gordon et al. 1967; Mertens and Ricker 1968). These observations place the site of origin more proximally than in Isaacs' syndrome, most likely in the spinal cord at the level of the internuncial neuronal pool. Although this syndrome is usually sporadic, a hereditary form of the disease has also been reported.

Muscle cramps are another form of abnormally delayed muscle relaxation associated with persistent electrical activity (Layzer and Rowland 1971; Rowland 1985). In spite of their common presentation, the exact site of hyperexcitability has yet to be defined. Most evidence points to a central location, however.

At the other extreme, painful muscle shortening can be seen in a number of metabolic diseases of muscle. These contractures are electrically silent and easily differentiated from myotonia by EMG. They form a prominent feature of phosphorylase deficiency as well as in other disorders involving defects in the enzymes of glycolysis and glycogenolysis (DiMauro et al. 1984). While initially thought to be the result of depleted energy reserves, abnormal ATP levels have yet to be confirmed during a contracture (Rowland et al. 1965; Ross et al. 1981).

MUSCLE BIOPSY IN NONDYSTROPHIC MYOTONIA

Several investigators found a relative decrease in the frequency of type II-b fibers in muscle biopsies from patients with myotonia congenita and paramyotonia (Pépin et al. 1975; Crews et al. 1976). Several explanations have been postulated, includ-

ing a genetic factor (Heene et al. 1986). However, experimental evidence indicates that the distribution of fiber types in otherwise normal muscle can be changed when chronic myotonia is induced by chemical agents that induce repetitive firing in the muscle sarcolemma (Salviati et al. 1986). 20,25-Diazacholesterol, when administered to rats, produces a chronic myotonic syndrome that has been well-characterized biochemically and electrophysiologically. Salviati et al. have found that the induction of myotonia by this chemical produces a decline in the percentage of type II-b fibers in the extensor digitorum longus (EDL) with a commensurate increase in the percentage of type II-A and type I fibers. Similarly, the *adr* homozygotes exhibit a marked shift in the distribution of fiber types in their skeletal muscle, with a striking predominance of oxidative type II fibers as compared to normal (Reininghaus et al. 1988). Chronic treatment with tocainide, which eliminates the myotonic signs in affected mice, produces a partial reversion of this fiber type pattern to normal suggesting that it is the result of the superimposed pattern of abnormal myotonic electrical activity. Apparently, the change in the pattern of electrical activity in the sarcolemma induces a change in phenotypic expression of contractile proteins that results in the alteration in fiber type distribution. Similar changes in phenotype have also been documented in regard to contractile properties and actin and myosin isoforms as a function of electrical stimulation of muscle or stimulation through transposed motor nerves (Lomo et al. 1980; Pette et al. 1984; Pette and Vrbová 1985).

TREATMENT OF MYOTONIA

The mainstay in the treatment of myotonia is the type I class of cardiac antiarrhythmic agents. These drugs act on the voltage-dependent sodium channel to reduce the rate of channel activation, thereby prolonging the effective channel refractory period. In essence, these drugs reduce the ease with which an action potential can be triggered by a preceding potential with a given interpotential latency. In the context of repetitive action potential generation in myotonia, the result is a reduction in the frequency of repetitive spikes and in the duration of spiking activity. In experimental myotonia induced by blockage of membrane G_{Cl} by 9-AC, the multiple driven action potentials seen with depolarizing current pulses are reduced in frequency by 20-μM phenytoin, and the self-sustaining repetitive activity is completely abolished (Furman and Barchi 1978) (Fig. 6).

The type 1 antiarrhythmic agents available in clinical practice include phenytoin, procainamide, quinine, mexiletine, and tocainide. Some of these agents, such as phenytoin and quinine, are relatively innocuous, while others, such as tocainide, have been associated with serious side effects. There is a serious question as to whether it is justifiable to expose a patient to the risk of a potentially fatal complication of drug therapy to treat a non-lethal complication of a chronic disease. In some cases, however, the effects of severe myotonia can be so disabling that therapy might be considered if other less risky alternatives are exhausted.

For most patients with myotonic muscular dystrophy, myotonia is so localized or mild that drug therapy is not warranted. Indeed, many patients elect to discontinue antimyotonic medication rather than deal with the inconvenience of daily dosage or the mild side effects. In myotonia con-

genita, on the other hand, the myotonic features may be severe and interfere with activities of daily life. In these cases, initial therapy with phenytoin (Munsat 1967) or quinine (Wolf 1936) is often chosen. Serum phenytoin levels higher than those usually associated with seizure control may be necessary for suppression of myotonia. In addition, patients initially controlled on phenytoin may find that the medication loses its effectiveness after a variable period.

In patients where phenytoin is not effective, a trial of procainamide may be undertaken although the risk of drug-induced systemic lupus erythematosus must be considered (Geschwind and Simpson 1955; Finlay 1982). In resistant cases, mexiletine is often found useful. This drug does not appear to carry the risk of fatal hematological complications seen with tocainide. Myotonia associated with PC or with HPP may be particularly sensitive to tocainide (Strieb et al. 1983; Strieb 1987; Ricker et al. 1980, 1983) and its use in selected cases may be necessary.

In addition to the class I antiarrhythmic agents, other drugs have been reported to be effective in controlling myotonia. Some of these reports are

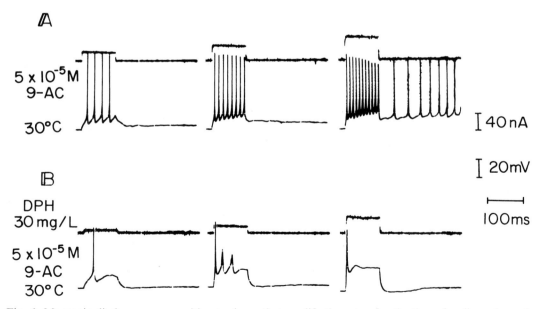

Fig. 6. Myotonic discharges are sensitive to drugs that modify the rate of activation of sodium channels, especially those that act in a use-dependent manner. Myotonic discharges recorded with intracellular microelectrodes from a muscle treated with 9-AC in vitro (A) are inhibited by exposure to phenytoin (30 μg/ml) (B) (Furman and Barchi, 1978).

uncontrolled case studies, while others involve more rigorous blinded analysis. The calcium channel blocker diltiazem has been shown to reduce the delayed relaxation in experimental 2,4-D-induced myotonia in vitro by 90% at a concentration of 5×10^{-5} M (Al-Rajeh et al. 1989). One patient with severe myotonia was treated effectively with nifedipine (Grant and Weir 1987). How calcium channel blockers act in reducing myotonia is not clear; perhaps inhibition of the calcium channel in T-tubular triadic junctions may reduce the effectiveness of excitation-contraction coupling, thereby limiting the contractile response of muscle to myotonic membrane activity without abolishing the electrical activity itself. Several reports have appeared concerning the use of the drug chromakalim (BRL 34915) as an antimyotonic agent (Spuler et al. 1989; Quasthoff et al. 1990). This drug activates an ATP-dependent K^+ channel, increasing membrane potassium conductance and stabilizing the membrane potential nearer to the potassium equilibrium potential, and speeding the repolarization process after a membrane action potential. In experiments on human muscle in vitro, using specimens obtained from patients with myotonia congenita, myotonic dystrophy, or HPP with myotonia, this drug and the related drug EMD 52962 completely abolished repetitive electrical activity, presumably by opening these ancillary potassium channels. Trials in vivo have not yet been reported.

Isolated reports have also appeared concerning the use of taurine, β-adrenergic stimulants such as metaproterenol, and tricyclic antidepressants such as imipramine and clomipramine for the treatment of myotonia. In a double-blind study, taurine reduced the severity of myotonic symptoms observed in response to intra-arterial potassium infusion (Durelli et al. 1983). The low incidence of side effects produced by this naturally occurring amino acid may warrant further clinical trials. Imipramine (Gascon et al. 1989) and clomipramine (Antonini et al. 1990) may also have beneficial effects on myotonia in cases where more traditional approaches have failed. Conversely, the β-adrenergic antagonists fenoterol and pindolol have been reported to have deleterious effects on recessive myotonia (Haass and Ricker 1979).

The myotonia of HPP or paramyotonia may respond to medications intended to control the paralytic events, such as acetazolamide and hydrochlorothiazide.

PATHOPHYSIOLOGY OF MYOTONIA

Although the clinical features of myotonia appear similar in the various syndromes outlined above, a number of different membrane mechanisms can produce the same clinical picture of membrane hyperexcitability and delayed muscle relaxation. Abnormalities involving the voltage-dependent sodium channel, the chloride channel, and conceivably the various membrane potassium and calcium channels, theoretically could give rise to self-sustaining repetitive activity. At least the first 2 pathways seem to be involved in human myotonia (Barchi 1982; Rüdel and Lehmann-Horn 1985).

Myotonia due to low chloride conductance

Physiology of chloride conductance in myotonia. Bryant (1979) and Bryant and Owenburg (1980) clearly demonstrated that the major membrane defect in the myotonic goat model was a marked reduction in the sarcolemmal conductance to chloride ions. When intercostal muscle fibers from myotonic goats were impaled with microelectrodes and depolarized with long current pulses just above threshold, a repetitive series of driven action potentials was seen, followed by a prolonged depolarizing after-potential (Adrian and Bryant 1974). The amplitude of the depolarizing after-potential was proportional to the number of driven action potentials and, if large enough, was sufficient to itself trigger a regenerative string of action potentials after the stimulus current was terminated. These phenomena were not seen in normal goat fibers, but could be induced in normal muscle if the chloride conductance was eliminated by replacement of external Cl^- ions with an impermeant anion. Low G_{Cl} is also the primary defect in the myotonic *adr* mouse mutant (Mehrke et al. 1988), and has been reported in intercostal fibers of patients with RGM (Lipicky et al. 1971; Lipicky and Bryant 1973; Franke et al. 1991), but variable results have been reported in patients with DMC (Iaizzo et al. 1991) and myotonic muscular dystrophy (Lipicky 1977).

Early computer models of muscle membrane excitation were used to test the hypothesis that a reduction in membrane G_{Cl} could produce myotonic discharges (Bretag 1973; Barchi 1975). These models confirmed that a reduction in membrane G_{Cl} to 20% or less of control values could allow repeated action potential to develop. A more elaborate computer model including simulation of an active T-tubular system supported and extended these conclusions (Adrian and Marshall 1976). The development of myotonic discharges with low G_{Cl} in these models was sensitive to a number of variables, the most prominent of which were the potassium equilibrium potential (V_k), the peak sodium conductance (G_{Na}) and the rate constants for sodium channel activation and inactivation.

Conclusions drawn from work with the myotonic goat and computer simulations were subsequently verified by experiments demonstrating that experimental blockade of membrane chloride channels with aromatic carboxylic acids (ACAs) would produce in vitro a myotonic syndrome indistinguishable from that seen in the goat in vivo (Bryant and Morales-Aquilera 1971; Furman and Barchi 1978); myotonia developed at concentrations of ACAs that produced about 80% reduction in G_{Cl}, and the resultant repetitive activity could be blocked by alteration in external potassium concentration or by drugs that modify the rate of sodium channel activation.

An explanation for the appearance of myotonia with low G_{Cl} can be found in the normal physiology of the muscle membrane. In normal mammalian skeletal muscle, G_{Cl} accounts for nearly 80% of the resting membrane conductance (Palade and Barchi 1977a); most of the remainder is attributable to various potassium conductances (G_K), while G_{Na} accounts for less than 1%. During an action potential, when there are rapid time and voltage-dependent changes in membrane G_{Na} and G_K, small quantities of Na^+ and K^+ ions move across the surface membrane. At the sarcolemma, these ionic fluxes have negligible effect of the actual concentration gradients of the 2 ions present across the membrane due to the large volumes of the cytoplasm and the extracellular space. When action potentials propagate into the narrow elements of the T-tubular system, however, the situation is quite different. The K^+ released into the small volume of

the T-tubular lumen with each passing action potential is sufficient to raise the intraluminal $[K^+]$ by about 0.3 mM (Adrian and Bryant 1974). Under normal circumstances, the depolarizing effect of this increase in tubular $[K^+]$ is not reflected in the surface membrane, because of the large shunting conductance to chloride ions that is present. When this G_{Cl} is removed, however, T-tubular K^+ accumulation will produce surface depolarization; because of the slow diffusion of K^+ out of the narrow lumen of the T-system, the effects of sequential action potentials will be additive, ultimately producing sufficient after-depolarization to re-activate sodium channels and trigger another action potential.

This role for T-tubular K^+ accumulation is supported by experimental studies in which myotonia was induced by substitution of chloride ions with an impermeant anion or by blocking G_C with aromatic carboxylic acids. Subsequent disconnection of the T-tubular system from the surface membrane by osmotic shock abolishes both the long-lasting depolarizing after-potential and the sustained spontaneous electrical activity seen in these myotonic models (Adrian and Bryant 1974).

An abnormal reduction in surface membrane G_C most likely accounts for the myotonic activity seen in RGM and DMC, in the toxic myotonic syndrome produced in man by 2,4-D and related herbicides, and in the hereditary myotonic syndromes in the goat and the *adr* mouse mutant.

Molecular pathology of chloride channel defects.

Recently the channel protein subserving G_C has been cloned, sequenced, and functionally expressed from both *Torpedo* electroplax and rat skeletal muscle (Jentsch et al. 1990; Steinmeyer et al. 1991b). These homologous proteins differ from both the voltage-sensitive ion channels and the ligand-gated ion channels. The *Torpedo* channel is 805 amino acids in length with a calculated MW of 89 kDa; the rat muscle channel is slightly larger, containing 994 residues with a MW core of 110 kDa. Both contain 12–13 putative transmembrane helical regions without evidence of internal repeat domains or amphipathic S4-type helices. When expressed in oocytes, the rat channel gene product exhibits physiological properties very similar to those previously described for macroscopic chloride currents in rat muscle in vivo (Steinmeyer et al. 1991b; Palade and Barchi 1977a).

Elucidation of the rat chloride channel sequence has allowed probes to be developed that identify the homologous gene in the mouse. Such probes can be used to investigate the role of the chloride channel in the pathophysiology of the mouse *adr* myotonic mutant. Through Southern blot analysis of DNA from affected and normal mice, Steinmeyer et al. (1991a) demonstrated a consistent defect in the genomic DNA encoding the mouse skeletal muscle chloride channel in animals expressing the *adr* phenotype. Subsequent sequencing of the defective DNA revealed that the normal coding sequence was interrupted at an exon-intron junction by the insertion of a transposable element or transposon. This insertion caused the information for most of the 3′ end of the channel message to be lost. No full-length normal transcripts for the chloride channel were found in the homozygous *adr* mice. Conversely, the insertion was never seen in any of the normal mice examined. In addition, the mouse chloride channel gene was mapped to chromosome 6, and further analysis using interspecies backcrosses showed no evidence of recombination between this gene and the *adr* locus. Taken together, these data provide very compelling evidence that the genetic defect in this mouse model of RGM is an insertional mutation in the muscle chloride channel gene.

Although one electrophysiological study of human muscle biopsies in DMC has suggested that the disease is associated with a defect in the skeletal muscle sodium channel (Iaizzo et al. 1991), no linkage has been found between the expression of this disease and the SCN4A gene encoding this protein. Recently, tight linkage has been found between a region on chromosome 7 that contains the genes encoding the T-cell beta receptor and the skeletal muscle chloride channel in both DMC and RGM located at 7q32-qter (Koch et al. 1992). The chloride channel encoded by this human gene has been partially sequenced and is about 88% identical to the previously characterized rat chloride channel at the amino acid level. In two families with DMC a phenylalanine to cysteine mutation was identified in putative transmembrane domain 8 that segregates with the disease. These results strongly suggest that both DMC and RGM are also caused by mutations in the skeletal muscle chloride channel gene.

Myotonia due to other membrane channel abnormalities

Abnormalities of voltage-dependent sodium channels. Not all the myotonic syndromes are caused by low G_{Cl} (Barchi 1982). In myotonic dystrophy and PC, for example, the G_{Cl} is inconsistently and mildly affected and is often normal. In the experimental myotonia produced by 20,25-D, abnormal G_{Cl} also does not appear to be the primary pathophysiological defect.

Abnormalities of sodium channel kinetics can also lead in themselves to repetitive action potential generation of the type seen in the myotonic disorders and repetitive activity is produced by a number of sodium channel neurotoxins (Strichartz et al. 1982). This is especially true for abnormalities which increase the rate of channel recovery from short-term inactivation; such changes can predispose the channel to reactivation before the membrane has fully repolarized after an action potential or series of action potentials. Normal sodium channels typically open once with a variable latency after depolarization and then close to a long-lasting inactivated state which persists until the membrane is repolarized. Rare late openings from the inactivated state and bursts of openings that seem to represent a transient loss of inactivation are seen in normal muscle. However, frequent bursts of channel openings are not found. Evidence of abnormal sodium channel activity has been presented for the Schwartz-Jampel syndrome; patch clamp studies found bursts of late single channel openings in response to depolarization. In the macroscopic currents, bursts of late openings such as these will produce a slowly inactivating sodium current of much longer duration than that seen in normal membranes.

Cannon et al. (1991) have made single channel measurements on myotubes developing in cultures from muscle biopsies of a family with HPP with myotonia. A class of abnormal sodium channels was identified that shifted between normal gating kinetics and an abnormal kinetic state characterized by persistent channel openings during depolarization and failure of channel inactivation. This

Regional Mapping of the Human Skeletal Muscle Sodium Channel on Chromosome 17

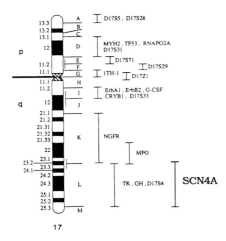

Fig. 7. The gene encoding the tetrodotoxin (TTX)-sensitive form of the adult human skeletal muscle sodium channel (designated SCN4A) has been localized to chromosome 17q23.1-25.3.

abnormal kinetic state was seen only in the presence of elevated extracellular potassium, consistent with the appearance of clinical symptoms in this form of periodic paralysis. These findings provide the single channel correlates of the small non-inactivating sodium current previously reported by Ricker et al. (1989).

Abnormal sodium channel single channel characteristics have also been reported in DMC, and may play a role in the repetitive activity seen in that disorder as well. Iaizzo et al. (1991) carried out membrane measurements and patch clamp studies of resealed single fiber preparations from 8 patients with DMC. In 7 of these patients, the G_{Cl} was said to be normal, suggesting that this disorder may differ from RGM in the primary underlying defect. The authors found abnormal sodium channel single channel activity; multiple reopening events were observed for single channels during a 10-ms depolarizing pulse, suggesting a defect in short-term channel inactivation. The result of these late reopenings is a small but very significant prolonged inward sodium current that persists for tens of milliseconds beyond the normal termination of the action potential sodium current.

Molecular pathology of disorders involving sodium channels. Work with the molecular pathophysiology of myotonias associated with defects in sodium channel function has moved rapidly during the past few years. The sodium channel protein has been purified from a number of species and tissues (see Barchi (1989) for review) and 2 isoforms of the rat skeletal muscle sodium channel have already been cloned, sequenced, and functionally expressed (Trimmer et al. 1989; Kallen et al. 1990). Using sequence information from these rat channels, the human skeletal muscle and cardiac sodium channels have now also been cloned and expressed (Gellens et al. 1992; George et al. 1992) and the genes encoding them localized to their respective chromosomes. The adult tetrodotoxin (TTX)-sensitive skeletal muscle sodium channel gene (designated SCN4A) is located on the long arm of chromosome 17 at q23.1-q25.3, adjacent to the location of the human growth hormone gene (George et al. 1991) (Fig. 7). The gene encoding the TTX-resistant sodium channel found in cardiac muscle and denervated skeletal muscle is located on chromosome 3 (George and Barchi, in preparation).

The identification of cDNAs for these channels has allowed investigators to isolate fragments of both genes and to identify characteristic RFLPs within them. Using these RFLPs, the hypothesis

TABLE 2

Muscle diseases linked to the skeletal muscle sodium channel gene at 17q23.1-25.3.

Reference	Disorder	Probe	LOD score @. 00
Fontaine et al. (1990)	HyPP w Myo	hNa2	4.00
Ptacek et al. (1991c)	HyPP w Myo	pM8	10.35
Koch et al. (1991)	HyPP w Myo	hNa2, hGH	3.06
Ebers et al. (1991)	HyPP w/o Myo	C6b, hNa2	4.09
Ebers et al. (1991)	PC	GH1	3.79
Ptacek et al. (1991b)	PC	pM8	4.43
Ptacek et al. (1992b)	MC w pain	pM8	4.19

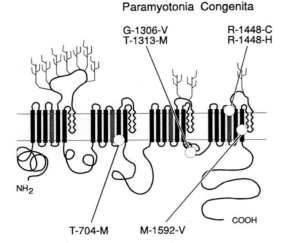

Paramyotonia Congenita

G-1306-V   R-1448-C
T-1313-M   R-1448-H

NH₂

T-704-M   M-1592-V   COOH

Hyperkalemic Periodic Paralysis

Fig. 8. Mutations in the gene encoding the adult skeletal muscle sodium channel can produce either the electrical hyperactivity seen in paramyotonia congenita or the intermittent weakness characteristic of hyperkalemic periodic paralysis. Known point mutations associated with these allelic disorders are shown, and the regions of the channel structure that are affected are identified.

that the adult skeletal muscle sodium channel is a candidate gene defect in a particular disorder can be tested. Linkage between this gene at 17q23.1-25.3 and the phenotype of HPP both with and without myotonia (Fontaine et al. 1990; Koch et al. 1991; Ptacek et al. 1991c) and of PC (Ebers et al. 1991; Ptacek et al. 1991b) has been reported (Table 2). For each disorder, LOD scores greater than 4.0 at a recombination frequency of 0.0, indicating that the diseases are linked to the sodium channel gene with an error of less than 1 in 10,000.

Analysis of either the genomic DNA encoding the skeletal muscle sodium channel or cDNA prepared from defective muscle RNA has allowed several groups to identify specific point mutations in the sodium channel protein associated with several of these diseases. In HPP, a threonine to methionine mutation (T704M) affecting a conserved residue near the cytoplasmic terminus of the S5 helix in repeat domain 2 has been identified in 2 families and in a third sporadic case as a spontaneous mutation (Ptacek et al. 1991a). This mutation is not present in any of the unaffected family members or

in any of 116 control individuals examined. Another mutation associated with the same phenotype was found in 3 other families; this mutation results in the conversion of a methionine to valine (M1592V) in the S6 helix of domain 6 (Rojas et al. 1991).

Several other mutations in the sodium channel gene have been identified in families with PC. McClatchey et al. (1992) reported 2 independent mutations affecting the cytoplasmic loop connecting repeat domains 3 and 4. This region has previously been implicated in channel inactivation. One mutation introduces a valine for a conserved glycine residue (G1306V), while the second produces a threonine to methionine transition (T1313M) 7 amino acids away in the same segment. Ptacek et al. (1992a) have identified families with PC in which different base mutations have occurred within the same codon, both producing alterations in the same absolutely conserved arginine residue in the S4 helix of domain 4. In 1 kindred the mutation results in the substitution of cysteine for arginine (R1448C), while in 2 other kinships, the result is replacement of the same arginine by histidine (R1448H). The S4 amphipathic helix is the most conserved characteristic element of voltage-dependent ion channels and is postulated to play a central role as the voltage-sensing element of the protein.

It is now clear that the various forms of HPP and PC represent allelic variants of mutations within the same sodium channel gene (Fig. 8). Single amino acid substitutions at a number of locations within this large protein can produce alterations in channel physiology that are expressed as clinically similar phenotypes. Undoubtedly, other mutations will be identified that have similar effects. This concept clarifies much of the confusion that has surrounded the nosology of these diseases, and leads to a simple classification on the basis of the underlying genetic pathology. Experience with PC and the periodic paralyses also serves to underscore the similarity of mechanisms that can be involved in producing either hyperexcitable or hypoexcitable states of the muscle membrane.

DISCUSSION

The myotonic disorders represent a diverse group

of muscle diseases which share a common pheno-typic expression of abnormal muscle surface mem-brane hyperexcitability. This in turn leads to the common external appearance of an abnormal delay in the relaxation of voluntary skeletal mus-cle, and the recording of repetitive electrical activ-ity from such muscle by electromyography. In spite of the common general appearance of the phenotype, it is clear that most of these disorders have different genotypes and that the underlying defect may involve a number of different mem-brane proteins or ion channels. The pathophysiol-ogy of the acquired myotonic disorders reflects this variability as well, with several clearly differ-ent mechanisms leading to a similar appearance at the whole animal level. The advent of single chan-nel electrophysiological methods and of modern molecular biological techniques makes these disor-ders ideal candidates for investigation. Undoubt-edly, the genetic defect for most of the inherited myotonias will be determined within the next few years, allowing a more rational approach to their classification and treatment, and providing poten-tial avenues to the correction of the defects them-selves.

REFERENCES

ADRIAN, R.H. and S.H. BRYANT: On the repetitive dis-charge in myotonic muscle fibers. J. Physiol. (Lon-don) 240 (1974) 550–515.

ADRIAN, R.H. and M.W. MARSHALL: Action potentials reconstructed in normal and myotonic muscle fi-bers J. Physiol. (London) 258 (1976) 125–143.

AL-RAJEH, S., V. IYER and W.H. OLSON: Effect of dil-tiazem on 2,4-D-induced myotonia in rats. Muscle Nerve 12 (1989) 470–472.

ANTONINI, G., R. VICHI, M.G. LEARDI, E. PENNISI, G.C. MONZA and M. MILLEFIORINI: Effect of clomipram-ine on myotonia: a placebo-controlled, double-blind, crossover trial. Neurology 40 (1990) 1473–1474.

BARCHI, R.L.: Myotonia: an evaluation of the chloride hypothesis. Arch. 32 (1975) 175–180.

BARCHI, R.L.: A mechanistic approach to the myotonic syndromes. Muscle Nerve 5, Suppl. 9 (1982) S60–S63.

BARCHI, R.L.: Probing the molecular structure of the voltage-dependent sodium channel. Annu. Rev. Neurosci. 11 (1989) 455–495.

BECKER, P.E.: Zur Frage der Heterogenie der erblichen Myotonien. Nervenarzt 28 (1957) 455–460.

BECKER, P.E.: Paramyotonia congenita (Eulenburg). In: P.E. Becker, W. Lenz, I. Vogel and G.G. Wendt (Eds), Fortschritte der allgemeinen und klinischen Human-genetik. Stuttgart, Thieme (1970) 1–131.

BECKER, P.E.: Generalized non-dystrophic myotonia. The dominant (Thomsen) type and the recently identified recessive type. In: J.E. Desmedt (Ed.), New Developments in Electromyography and Clinical Neurophysiology. Basel, Karger (1973) 407–412.

BECKER, P.E.: Myotonia Congenita and Syndromes Associated with Myotonia. Stuttgart, Thieme (1977).

BLACK, J.T., R. GARCIA-MULLIN, E. GOOD and S. BROWN: Muscle rigidity in a newborn due to continuous pe-ripheral nerve hyperactivity. Arch. Neurol. 27 (1972) 413–425.

BRETAG, A.H.: Mathematical modelling of the myotonic action potential. In: J.E. Desmedt (Ed.), New Developments in Electromyography and Clinical Neurophysiology, Vol. 1. Basel, Karger (1973).

BROWN, G.L. and A.M. HARVEY: Congenital myotonia in the goat. Brain 62 (1939) 341–363.

BRUMLIK, J., B. DRECHSLER and T.M. VANNIN: The myotonic discharge in various neurological syn-dromes: a neurophysiological analysis. Electro-myography 10 (1970) 369–383.

BRYANT, S.H.: Muscle membrane of normal and myotonic goats in normal and low external chlo-ride. Fed. Proc. 21 (1962) 312.

BRYANT, S.H.: Myotonia in the goat. Ann. N.Y. Acad. Sci. 317 (1979) 314–324.

BRYANT, S.H. and A. MORALES-AGUILERA: Chloride con-ductance in normal and myotonic muscle fibres and the action of monocarboxylic aromatic acids. J. Physiol. (London) 219 (1971) 367–383.

BRYANT, S.H. and K. OWENBURG: Characteristics of the chloride channel in skeletal-muscle fibers from myotonic and normal goats. Fed. Proc. 39 (1980) 579.

BUCHER, N.L.: Effects of 2,4-dichlorophenoxyacetic acid on experimental animals. Proc. Soc. Exp. Biol. Med. 63 (1946) 204–205.

BUCHTHAL, F. and P. ROSENFALCK: Electrophysiologi-cal aspects of myopathy with particular reference to progressive muscular dystrophy. In: G.H. Bourne and M.N. Golarz (Eds), Muscular Dystro-phy in Man and Animals. New York, Hafner Pub-lishing Company (1963) 193–262.

BUCHTHAL, F., L. ENGBAEK and I. GAMSTORP: Paresis and hyperexcitability in adynamia episodica hereditaria. Neurology 8 (1958) 347–351.

BUTTERFIELD, D.A. and W.E. WATSON: Electron spin resonance studies of an animal model of human congenital myotonia: increased erythrocyte mem-brane fluidity in rats with 20,25-diazacholesterol-induced myotonia. J. Membr. Biol. 32 (1977) 165–176.

CANNON, S.C., R.H. BROWN and D.P. COREY: A sodium channel defect in hyperkalemic periodic paralysis:

potassium–induced failure of inactivation. Neuron 6 (1991) 619–626.

CAO, A., C. CIANCHETTI, L. CALISTI, S. DEVIRGILIS, A. FERRELLI and W. TANGHERONI: Schwartz-Jampel syndrome. Clinical electrophysiological and histopathological study of a severe variant. J. Neurol. Sci. 36 (1978) 175–187.

CARSON, M.H. and C.M. PEARSON: Familial hyperkalemic periodic paralysis with myotonic features. J. Pediatr. 64 (1964) 853–865.

CHALIKIAN, D.M. and R.L. BARCHI: Fluorescent-probe analysis of erythrocyte-membranes in myotonic-dystrophy. Neurology 30 (1980a) 277–285.

CHALIKIAN, D.M. and R.L. BARCHI: 20,25-Diazacholesterol myotonia: biochemical characteristics of the sarcolemma. Neurology 30 (1980b) 423–424.

CHIDA, T., K. HIROSE, H. SHOJI, R. HANAKAGO and M. UONO: An unusual form of myotonia-like phenomenon associated with the lesions of cervical cord or roots. Clin. Neurol. 21 (1981) 1–6.

CREWS, J., K.K. KAISER and M.H. BROOKE: Muscle pathology of myotonia congenita. J. Neurol. Sci. 28 (1976) 449–457.

D'ALONZO, A.J. and J.J. MCARDLE: An evaluation of fast-and slow-twitch muscle from rats treated with 20,25-diazacholesterol. Exp. Neurol. 78 (1982) 46–66.

DAVISSON, M.T., B. HARRIS and P.W. LANE: Jackson Laboratory. Mouse News Letter 83 (1989) 167.

DE JONG, J.G.Y., J.L. SLOOFF and A.A.J.J. VAN DER EERDEN: A family with paramyotonia congenita with the report of an autopsy. Acta Neurol. Scand. 49 (1973) 480–494.

DE SILVA, S.M., R.W. KUNCL. J.W. GRIFFIN, D.R. CORNBLATH and S. CHAVOUSTIE: Paramyotonia congenita or hyperkalemic periodic paralysis? Clinical and electrophysiological features of each entity in one family. Muscle Nerve 13 (1990) 21–26.

DIMAURO, S., N. BRESOLIN and A.P. HAYS: Disorders of glycogen metabolism of muscle. CRC Crit. Rev. Clin. Neurobiol. 1 (1984) 83–116.

DRAGER, G.A., J.F. HAMMILL and G.M. SHY: Paramyotonia congenita. Arch. Neurol. Psychiatr. (Chicago) 80 (1958) 1–9.

DURELLI, L., R. MUTANI and F. FASSIO: The treatment of myotonia: evaluation of chronic oral taurine therapy. Neurology 33 (1983) 599–603.

EBERS, G., A.L. GEORGE, R.L. BARCHI, R.G. KALLEN, S.S. TING-PASSADOR, M. LATHROP, J. BECKMANN, A.F. HAHN, W.F. BROWN, R. CAMPBELL and A.J. HUDSON: Paramyotonia congenita and hyperkalemic periodic paralysis are linked to the adult muscle sodium channel gene. Ann. Neurol. 30 (1991) 810–816.

EMERYK, B., I. HAUSMANOWA-PETRUSEWICZ and T. NOWAK: Spontaneous volleys of bizarre high frequency potentials (b.h.f.p.) in neuromuscular diseases. Part I. Occurrence of spontaneous volleys of b.h.f.p. in neuromuscular diseases. Part II. An analysis of the morphology of spontaneous volleys in b.h.f.p. in neuromuscular diseases. Electro-

myogr. Clin. Neurophysiol. 14 (1974) 303–312, 339–354.

ENTRIKIN, R.K., R.T. ABRESCH, R.B. SHARMAN, D.B. LARSON and N.A. LEVINE: Contractile and EMG studies of murine myotonia (mto) and muscular dystrophy (dy/dy). Muscle Nerve 10 (1987) 293–298.

EULENBERG, A.: Über eine familiäre, durch 6 Generationen verfolgbare Form congenitaler Paramyotonie. Neurol. Zentralbl. 12 (1886) 265–272.

EYZAGUIRRE, C., B.P. FOLK, K.L. ZIERLER and J.L. LILIENTHAL: Experimental myotonia and repetitive phenomena: the veratrinic effect of 2,4-dichlorophenoxyacetate (2,4-D) in the rat. Am. J. Physiol. 155 (1948) 69–77.

FARRELL, S.A., R.G. DAVIDSON and P. THORP: Neonatal manifestations of Schwartz-Jampel syndrome. Am. J. Med. Genet. 27 (1987) 799–805.

FERRANNINI, E., T. PERNIOLA, G. KRAJEWSKA, L. SERLENGA and M. TRIZIO: Schwartz-Jampel syndrome with autosomal-dominant inheritance. Eur. Neurol. 21 (1982) 137–146.

FINLAY, M.: A comparative study of disopyramide and procainamide in the treatment of myotonia in myotonic dystrophy. J. Neurol. Neurosurg. Psychiatr. 45 (1982) 461–463.

FONTAINE, B., T.S. KHURANA, E.P. HOFFMAN, G.A.P. BRUNS, J.L. HAINES, J.A. TROFATTER, M.P. HANSON, J. RICH, H. MACFARLANE, D.M. YESEK, D. ROMANO, J.F. GUSELLA and R.H. BROWN: Hyperkalemic periodic paralysis and the adult muscle sodium channel α-subunit gene. Science 250 (1990) 1000–1002.

FOWLER JR., W.M., R.B. LAYZER, T.D. TAYLOR, E.D. EBERLE, G.E. SIMS, T.L. MUNSAT, M. PHILLIPPART and B.W. WILSON: The Schwartz-Jampel syndrome: its clinical, physiological and histological expressions. J. Neurol. Sci. 22 (1974) 127–146.

FRANKE. C., P.A. LAIZZO, H. HATT, W. SPITTELMEISTER, K. RICKER and F. LEHMANN-HORN: Altered Na$^+$ channel activity and reduced Cl$^-$ conductance cause hyperexcitability in recessive generalized myotonia. Muscle Nerve 14 (1991) 762–770.

FRENCH, E.B. and R. KILPATRICK: A variety of paramyotonia congenita. J. Neurol. Neurosurg. Psychiatr. 20 (1957) 40–46.

FÜCHTBAUER, E.M., J. REININGHAUS and H. JOCKUSCH: Developmental control of the excitability of muscle: transplantation experiments on a myotonic mouse mutant. Proc. Natl. Acad. Sci. USA 85 (1988) 3880–3884.

FURMAN, R.E. and R.L. BARCHI: The pathophysiology of myotonia produced by aromatic carboxylic acids. Ann. Neurol. 4 (1978) 357–365.

FURMAN, R.E. and R.L. BARCHI: 20,25-Diazacholesterol myotonia: an electrophysiological study. Ann. Neurol. 10 (1981) 251–260.

GAMSTORP, I.: Adynamia episodica hereditaria. Acta Paediatr. (Stockholm) Suppl 108 (1956) 1–126.

GAMSTORP, I.: Adynamia episodica hereditaria and myotonia. Acta Neurol. Scand. 39 (1963) 41–58.

GASCON, G.G., R.D. STATON, B.D. PATTERSON, P.J. KONEWKO, H. WILSON, K.M. CARLSON and R.A. BRUM-

BACK: A pilot controlled study of the use of imipramine to reduce myotonia. Am. J. Phys. Med. Rehabil. 68 (1989) 215–220.

GELLENS, M.E., A.L. GEORGE, L. CHEN, M. CHAHINE, R. HORN, R.L. BARCHI and R.G. KALLEN: Primary structure and functional expression of the human cardiac tetrodotoxin-insensitive voltage-dependent sodium channel. Proc. Natl. Acad. Sci. USA 89 (1992) 554–558.

GEORGE, A.L., D.H. LEDBETTER, R.G. KALLEN and R.L. BARCHI: Assigment of a human skeletal muscle sodium channel gene (SCN4A) to 17q23.1-25.3 Genomics 9 (1991) 555–556.

GEORGE, A.L., J. KOMISAROF, R.G. KALLEN and R.L. BARCHI: Primary structure of the adult human skeletal muscle voltage-dependent sodium channel. Ann. Neurol. 31 (1992) 131–137.

GESCHWIND, N. and J.A. SIMPSON: Procainamide in the treatment of myotonia. Brain 78 (1955) 81–91.

GORDON, E.E., D.M. JANUSKO and L. KAUFMAN: A critical survey of stiff-man syndrome. Am. J. Med. 42 (1967) 589–599.

GRANT, R. and A.I. WEIR: Severe myotonia relieved by nifedipine [letter]. Ann. Neurol. 22 (1987) 395.

HAASS, A. and K. RICKER: Deleterious effect of β-receptor stimulating and blocking drugs fenoterol and pindolol in recessive myotonia congenita (abstract). Naunyn-Schmiedeberg's Arch. Pharmacol. 307 (1979) R77.

HAASS, R., K. RICKER, R. RÜDEL, F. LEHMANN-HORN, R. BHLEN, R. DENGLER and H.G. MERTENS: Clinical study of paramyotonia congenita with and without myotonia in a warm environment. Muscle Nerve 4 (1981) 388–395.

HEENE, R., R.-R. GABRIEL, F. MANZ and K. SCHIMRIGK: Type 2B muscle fibre deficiency in myotonia and paramyotonia congenita. A genetically determined histochemical fibre type pattern? J. Neurol. Sci. 73 (1986) 23–30.

HELLER, A.H., E.M. EICHER, M. HALLET and R.L. SIDMAN: Myotonia, a new inherited muscle disease in mice. J. Neurosci. 2 (1982) 924–933.

HUGHES, R.C. and W.B. MATTHEWS: Pseudo-myotonia and myokymia. J. Neurol. Neurosurg. Phychiatr. 32 (1969) 11–14.

HUTTENLOCHER, P.R., J. LANDWIRTH, V. HANSON, B.B. GALLAGHER and K. BENSCH: Osteochondro-muscular dystrophy: a disorder manifested by multiple skeletal deformities, myotonia and dystrophic changes in muscle. Pediatrics 44 (1969) 945–958.

IAIZZO, P., C. FRANKE, H. HATT, W. SPITTELMEISTER, K. RICKER, R. RÜDEL and F. LEHMANN-HORN: Altered Na⁺ channel behavior causes myotonia in dominantly inherited myotonia congenita. Neuromusc. Dis. (1992) in press.

ISAACS, H.: A syndrome of continuous muscle-fiber activity. J. Neurol. Neurosurg. Psychiatr. 24 (1961) 319–325.

ISAACS, H. and J.J.A. HEFFRON: The syndrome of continuous muscle fibre activity cured: further studies.

J. Neurol. Neurosurg. Psychiatr. 37 (1974) 1231–1235.

JENTSCH, T.J., K. STEINMEYER and G. SCHWARZ: Primary structure of Torpedo marmorata chloride channel isolated channel by expression cloning in Xenopus oocytes. Nature (London) 348 (1990) 510–514.

KALLEN, R.G., Z. SHENG, J. YANG, L. CHEN, R. ROGART and R.L. BARCHI: Primary structure and expression of a sodium channel characteristic of denervated and immature rat skeletal muscle. Neuron 4 (1990) 233–242.

KOCH, M.C., K. RICKER, M. OTTO, T. GRIMM, E.P. HOFFMAN, R. RÜDEL, K. BENDER, B. ZOLL, P. HARPER and F. LEHMANN-HORN: Confirmation of linkage of hyperkalemic periodic paralysis to chromosome 17. J. Med. Genet. 28 (1991) 583–586.

KOCH, M.C., K. STEINMEYER, C. LORENZ, K. RICKER, F. WOLF, M. OTTO, B. ZOLL, F. LEHMANN-HORN, K.-H. GRZESCHIK and T.J. LENTSCH: The skeletal muscle chloride channel in dominant and recessive human myotonia. Science 257 (1992) 797–800.

LAJOIE, W.J.: Paramyotonia congenita, clinical features and electromyographic findings. Arch. Physiol. Med. 42 (1961) 507–512.

LANARI, A.: Mechanism of myotonic contraction. Science 104 (1946) 221–222.

LANDAU, W.M.: The essential mechanism in myotonia: an electromyographic study. Neurology 2 (1952) 369–388.

LAYZER, R.B. and L.P. ROWLAND: Cramps. N. Engl. J. Med. 285 (1971) 31–40.

LAYZER, R.B., R.E. LOVELACE and L.P. ROWLAND: Hyperkalemic periodic paralysis. Arch. Neurol. 16 (1967) 455–472.

LEYDEN, E.: Klinik der Rückenmarkskrankheiten. Vol. 1. Berlin (1874) 128.

LINDSLEY, D.B. and E.C. CURNEN: An electromyographic study of myotonia. Arch. Neurol. Psychiatr. 35 (1936) 253–269.

LIPICKY, R.J.: Studies in human myotonic dystrophy. In: L.P. Rowland (Ed.), Pathogenesis of Human Muscular Dystrophies. Amsterdam, Excerpta Medica (1977) 729–738.

LIPICKY, R.J. and S.H. BRYANT: A biophysical study of human myotonias. In: J.E. Desmedt (Ed.), New Developments in Electromyography and Clinical Neurophysiology, Vol. 1. Basel, Karger (1973) 451–463.

LIPICKY, R.J., S.H. BRYANT and J.H. SALMON: Cable parameters, sodium, potassium, chloride and water content, and potassium efflux in isolated external intercostal muscles of normal volunteers and patients with myotonia congenita. J. Clin. Invest. 50 (1971) 2091–2103.

LOMO, T., R.H. WESTGAARD and L. ENGEBRETSEN: Different stimulation patterns affect contractile properties of denervated rat soleus muscles. In: D. Pette (Ed.), Plasticity of Muscle. Berlin, W. de Gruyter (1980) 297–309.

MANZ, F.: Pseudomyotone EMG-Entladungen als unspezifischer Befund bei Muskelglykogenose Typ II (Morbus Pompe). Arch. Psychiatr. Nervenkr. 228 (1980) 45–51.

MARINACCI, A.A.: Electromyogram in muscular dystrophy. Bull. Los Angeles Neurol. Soc. 29 (1964) 7–16.

MC CLATCHEY, A.I., P.V. BERGH, M.A. PERICAK-VANCE, W. RASKIND, C. VERELLEN, D. MCKENNA-YASEK, K. RAO, J.L. HAINES, T. BIRD, R.H. BROWN and J.F. GUSELLA: Temperature-sensitive mutations in the III-IV cytoplasmic loop region of the skeletal muscle sodium channel gene in paramyotonia congenita. Cell 68 (1992) 769–774.

MEHRKE, G., H. BRINKMEIER and H. JOCKUSCH: The myotonic mouse mutant adr: electrophysiology of the muscle fiber. Muscle Nerve 11 (1988) 440–446.

MERTENS, H.G. and K. RICKER: Übererregbarkeit der γ-Motoneurone beim 'Stiff-man'-syndrome. Klin. Wochenschr. 46 (1968) 33–42.

MOERSCH, F.P. and H.W. WOLTMAN: Progressive fluctuating muscular rigidity and spasm (stiffman syndrome): report of a case with some observations in 13 other cases. Proc. Staff Meet. Mayo Clin. 31 (1956) 421–427.

MUNSAT, T.L.: Therapy of myotonia. A double-blind evaluation of diphenylhydantoin, procainamide, and placebo. Neurology 17 (1967) 359–367.

NAKANISHI, T.: Neuromyotonia. Adv. Neurol. Sci. 22 (1978) 830–832.

NORRIS, F.H.: Myokymia (correspondence). Arch. Neurol. 34 (1977) 133.

NORRIS, F.H. and B.J. PANNER: Hypothyroid myopathy: clinical, electromyographical and ultrastructural observations. Arch. Neurol. 14 (1966) 574–589.

PALADE, P.T. and R.L. BARCHI: Characteristics of the chloride conductance in muscle fibers of the rat diaphragm. J. Gen. Physiol. 69 (1977a) 325–342.

PALADE, P.T. and R.L. BARCHI: On the inhibition of muscle membrane chloride conductances by aromatic carboxylic acids. J. Gen. Physiol. 69 (1977b) 875–896.

PARRY-JONES, N.O., J.A. STEPHENS, A. TAYLOR and D.A.H. YATES: Myokymia, not myotonia. Br. Med. J. 2 (1977) 300.

PÉPIN, B., M. HAGUENAU and J. MIKOL: Observation familiale de myotonie avec hypertrophie musculaire, faiblesse corrigé par l'effort et atrophie des fibres du type II. Rev. Neurol. (Paris) 131 (1975) 285–292.

PETER, J.B., R.M. ANDIMAN, R.L. BOWMAN and T. NAGATOMO: Myotonia induced by diazacholesterol: increased (Na+K)-ATPase activity of erythrocyte ghosts and development of cataracts. Exp. Neurol. 41 (1973) 738–744.

PETTE, D. and G. VRBOVÁ: Neural control of phenotypic expression in mammalian muscle fibers. Muscle Nerve 8 (1985) 676–689.

PETTE, D., A. HEILIG, G. KLUG, H. REICHMANN, U. SEEDORF and W. WIEHRER: Alterations in phenotype expression of muscle by chronic nerve stimulation. In: R.C. Strohmann and S. Wolf (Eds.), Gene Expression in Muscle. New York, Plenum Press (1984) 169–178.

POHL, J.: Physiologische Wirkung des Hydroatophans. Z. Exp. Pathol. Ther. 19 (1917) 198–204.

PTACEK, L.J., A.L. GEORGE, R.C. GRIGGS, R. TAWIL, R.G. KALLEN. R.L. BARCHI, M. ROBERTSON and M.F. LEPPERT: Identification of a mutation in the gene causing hyperkalemic periodic paralysis. Cell 67 (1991a) 1021–1027.

PTACEK, L.J., J.S. TRIMMER, W.S. AGNEW, J.W. ROBERTS, J.H. PETAJAN and M.F. LEPPERT: Paramyotonia congenita and hyperkalemic periodic paralysis map to the same sodium channel gene locus. Am. J. Hum. Genet. 49 (1991b) 851–854.

PTACEK, L.J., F. TYLER, J.S. TRIMMER, W.S. AGNEW and M.F. LEPPERT: Analysis in a large hyperkalemic periodic paralysis pedigree supports tight linkage to a sodium channel locus. Am. J. Genet. 49 (1991c) 378–382.

PTACEK, L.J., A.F. GEORGE, R.L. BARCHI, R.C. GRIGGS, J.E. RIGGS, M. ROBERTSON and M.F. LEPPERT: Mutations in an S4 segment of the adult skeletal muscle sodium channel cause paramyotonia congenita. Neuron 8 (1992a) 891–897.

PTACEK, L.J., R. TAWIL, R.C. GRIGGS, D. STORVICK and M.F. LEPPERT: Linkage of atypical myotonia congenita to a sodium channel locus. Neurology 42 (1992b) 431–433.

QUASTHOFF, S., A. SPULER, W. SPITTELMEISTER, F. LEHMANN-HORN and P. GRAFE: K+ channel openers suppress myotonic activity in human skeletal muscle in vitro. Eur. J. Pharmacol. 186 (1990) 125–128.

REININGHAUS, J., E.M. FÜCHTBAUER, K. BERTRAM and H. JOCKUSCH: The myotonic mouse mutant adr: physiological and histochemical properties of muscle. Muscle Nerve 11 (1988) 433–439.

RICH, E.C.: A unique form of motor paralysis due to cold. Med. News (New York) 62 (1894) 210–213.

RICKER, K. and H.M. MEINCK: Vergleich myotoner Engladungen bei Myotonia congenita und Dystrophia myotonica. Zg Neurol. 201 (1972a) 62–72.

RICKER K. and H.M. MEINCK: Verlaufsdynamik und Herkunft pseudomyotoner Entladungsserien bei Denervationssyndromen. EEG EMG 3 (1972b) 170–178.

RICKER, K., R. BÖHLEN, A. HAASS, H.G. MERTENS and R. RÜDEL: Successful treatment of paramyotonia congenita (Eulenberg): muscle stiffness and weakness prevented by tocainide. J. Neurol. Neurosurg. Psychiatr. 43 (1980) 268–271.

RICKER, K., R. BÖHLEN and R. ROHKAMM: Different effectiveness of tocainide and hydrochlorothiazide in paramyotonia congenita with hyperkalemic episodic paralysis. Neurology 33 (1983) 1615–1618.

RICKER, K., L. CAMACHO, P. GRAFE, F. LEHMANN-HORN and R. RÜDEL: Adynamia episodica hereditaria: what causes the weakness? Muscle Nerve 12 (1989) 883–891.

RICKER, K., F. LEHMANN-HORN and R. MOXLEY: Myotonia fluctuans. Arch. Neurol. 47 (1990) 268–272.

ROJAS, C.V. , J. WANG, L.S. SCHWARTZ, E.P. HOFFMAN, B.R. POWELL and R.H. BROWN: A Met-to-Val mutation in the skeletal muscle sodium channel alpha subunit in hyperkalemic periodic paralysis. Nature (London) 354 (1991) 387–389.

ROSS, B.D., G.K. RADDA, D.G. GADIAN, G. ROCKER, M. ESIRI and FALCONER-SMITH: Examination of a case of suspected McArdle's syndrome by 31P nuclear magnetic resonance. N. Engl. J. Med. 304 (1981) 1338– 1342.

ROWLAND, L.P.: Cramps, spasms, and muscle stiffness. Rev. Neurol. (Paris) 141 (1985) 261–273.

ROWLAND, L.P., S. ARAKI and P. CARMEL: Contracture in McArdle's disease. Arch. Neurol. 13 (1965) 541–544.

RÜDEL, R.: The myotonic mouse: a realistic model for the study of human recessive generalized myotonia. Trends Neurosci. 13 (1990) 1–3.

RÜDEL R. and F. LEHMANN-HORN: Membrane changes in cells from myotonic patients. Physiol. Rev. 65 (1985) 310–356.

SALVIATI, G., E. BIASIA, R. BETTO and D.D. BETTO: Fast to slow transition induced by experimental myotonia in rat EDL muscle. Pflueger's Arch. Eur. J. Physiol. 406 (1986) 266–272.

SAMAHA, F.J.: Hyperkalemic periodic paralysis. A genetic study, clinical observations, and report of a new method of therapy. Arch. Neurol. (Chicaco) 12 (1965) 145–154.

SANDERS, D.B.: Myotonia congenita with painful muscle contractions. Arch. Neurol. 33 (1976) 580–582.

SATOYOSHI, E., Y. DOI and M. KINOSHITA: Pseudomyotonia in cervical root lesions with myelopathy. A sign of the misdirection of regenerating nerve. Arch. Neurol. (Chicago) 27 (1972) 307–313.

SCHWARTZ, O. and R.S. JAMPEL: Congenital blepharophimosis associated with a unique generalized myopathy. Arch. Ophthalmol. 68 (1962) 52–57.

SEILER, D., W. FIEHN and E. KUHN: Disturbances in cholesterol biosynthesis as a cause of experimental myotonia. In: W.G. Bradley (Ed.), Recent Advances in Myology. Amsterdam, Excerpta Medica (1974) 429–434.

SPAANS. F., P. THEUNISSEN, A.D. REEKERS, L. SMIT and H. VELDMAN: Schwartz-Jampel syndrome: I. Clinical, electromyographic, and histologic studies. Muscle Nerve 13 (1990) 516–527.

SPULER, A., F. LEHMANN-HORN and P. GRAFE: Chromokalim (BRL 34915) restores in vitro the membrane potential of depolarized human skeletal muscle fibers. Naunyn-Schmeideberg's Arch. Pharmacol. 339 (1989) 327–331.

STEINERT, H.: Myopathologische Beiträge. I. Über das klinische und anatomische Bild des Muskelschwunds der Myotoniker. Dtsch. Z. Nervenheilkd. 37 (1909) 58–104.

STEINMEYER, K., R. KLOCKE, C. ORTLAND, M. GRONEMEIER, H. JOCKUSCH, S. GRUNDER and T. JENTSCH: A muscle chloride channel is inactivated by transposon insertion in myotonic mice. Nature (London) 354 (1991a) 301–304.

STEINMEYER, K., C. ORTLAND and T. JENTSCH: Primary structure and functional expression of a developmentally regulated skeletal muscle chloride channel. Nature (London) 354 (1991b) 304–308.

STREIB, E.W.: Paramyotonia congenita: successful treatment with tocainide. Clinical and electrophysiologic findings in seven patients. Muscle Nerve 10 (1987) 155–162.

STREIB, E.W. and S.F. SUN: Distribution of electrical myotonic muscular dystrophy. Am. Neurol. 14 (1983) 80–82.

STREIB, E.W., S.F. SUN and M. HANSON: Paramyotonia congenita: clinical and electrophysiologic studies. Electromyogr. Clin. Neurophysiol. 23 (1983) 315–325.

STRICHARTZ, G., R. HAHIN and M. CAHALAN: Pharmacological models for sodium channels producing abnormal impulse activity. In: W. Culp and J. Ochoa (Eds.), Abnormal Nerves and Muscles as Impulse Generators. Oxford, Oxford University Press (1982) 98–129.

SUBRAMONY, S.H., C.P. MALHOTRA and S.K. MISHRA: Distinguishing paramytonia congenita and myotonia congenita by electromyography. Muscle Nerve 6 (1983) 374–379.

TANG, A.H., L.A. SCHROEDER and H.H. KEASLING: U-23,223 (3-chloro-2,5,6-trimethylbenzoic acid), a veratrinic agent selective for the skeletal muscles. Arch. Intern. Pharmacodyn. 175 (1968) 319–329.

TAYLOR, R.G., R.B. LAYZER, H.S. DAVIS and W.M. FOWLER: Continuous muscle fiber activity in the Schwartz-Jampel syndrome. Electroencephalogr. Clin. Neurophysiol. 33 (1972) 497–509.

THOMSEN, J.: Tonische Krämpfe in willkürlich beweglichen Muskeln in Folge von ererbter psychischer Disposition. Arch. Psychiatr. 6 (1876) 702–718.

TRIMMER, J.S., S.S. COOPERMAN, S.A. TOMIKO, J. ZHOU, S.M. CREAN, M.B. BOYLE, R.G. KALLEN, Z. SHENG, R.L. BARCHI, F.J. SIGWORTH, R.H. GOODMAN, W.S. AGNEW and G. MANDEL: Primary structure and functional expression of a mammalian skeletal muscle sodium channel. Neuron 3 (1989) 33–49.

TRONTELJ, J. and E. STÅLBERG: Bizarre repetitive discharges recorded with single fibre EMG. J. Neurol. Neurosurg. Psychiatr. 46 (1983) 310–316.

TRUDELL, R.G., K.K. KAISER and R.C. GRIGGS: Acetazolamide-responsive myotonia congenita. Neurology 37 (1987) 488–491.

VALENSTEIN, E., R.T. WATON and J.L. PARKER: Myokymia, muscle hypertrophy and percussion 'myotonia' in chronic recurrent polyneuropathy. Neurology 28 (1978) 1130–1134.

VAN DER MEULEN, J.P., G.J. GILBERT and C.K. KANE: Familial hyperkalemic paralysis with myotonia. N. Engl. J. Med. 264 (1961) 1–6.

WATKINS, W.J. and D. WATTS: Biological features of the new A2G-adr mouse mutant with abnormal muscle function. Lab. Anim. 18 (1984) 1–6.

WHITE, G.R. and J. PLASKETT: 'Nervous', 'stiff-legged', or 'fainting' goats. Am. Vet. Rev. 28 (1904) 556–560.

WINER, N., D.M. KLACHKO, R.D. BAER, P.L. LANGLEY and T.W. BURNS: Myotonic response induced by inhibitors of cholesterol biosynthesis. Science 153 (1966) 312–313.

WOLF, A.: Quinine: an effective form of treatment for myotonia. Arch. Neurol. Psychiatr. 36 (1936) 382–383.

WOLTER, M. and D. BRAUER: Das Neuromyotonie–syndrome. Fortschr. Neurol. Psychiatr. 45 (1977) 98–105.

Handbook of Clinical Neurology, Vol. 18 (62): Myopathies
L.P. Rowland and S. DiMauro, editors
© 1992 Elsevier Science Publishers B.V. All rights reserved

CHAPTER 11

Progressive external ophthalmoplegia and ocular myopathies

LEWIS P. ROWLAND

Neurological Institute, Columbia-Presbyterian Medical Center, New York, NY, USA

Definition: The syndrome of progressive external ophthalmoplegia (PEO) is defined in purely clinical terms, and has the following characteristics: 1. there is progressive ptosis and immobility of the eyes; 2. the condition is bilateral; 3. the muscles affected are innervated by more than one nerve; 4. the pupil is spared; 5. progression is gradual, for months or years; 6. there are no remissions or exacerbations; 7. there is no evidence of a specific disorder. Specifically, there is no evidence of thyroid disorder, myotonic muscular dystrophy, or myasthenia gravis. The last three criteria are especially arbitrary, serving only to exclude known diseases that sometimes cause ophthalmoplegia, otherwise clinically identical with PEO. Other complications of this definition are discussed later.

As defined, PEO is a syndrome of diverse causes that might be due to a disease of the brain, involving either supranuclear connections or neurons in the oculomotor nuclei. Or it might be a disease of cranial nerves, neuromuscular junction, or muscle itself. We presume that there are sets of diseases at each site, all resulting in more or less the same clinical syndrome.

Changing times: This essay is a revision and update of a chapter written for the Handbook that was first submitted in 1971 (Rowland 1975). Our intellectual world has changed since then, the assignment in 1971 was to write about 'progressive nuclear external ophthalmoplegia' for a volume on spinal and cranial muscular atrophies. The volume was entitled 'System Disorders and Atrophies'. It was assumed that there is a disease of cranial oculomotor nuclei, due to 'alterations of the neurons alone', which parallels motor neuron disease. Even then, however, there were problems with this concept, leading to the following comment: 'But the identification of this kind of disorder, if it really exists, has become a major problem for modern neurology. It is therefore necessary to consider all possible mechanisms in describing the syndrome, which may appear to be the same whatever the anatomic origin.'

The other central problem in the seventies was a question of lumping and splitting. In writing that essay, it occurred to me that the Kearns-Sayre syndrome (KSS) was an identifiable and separate disorder. There were only 25 reported cases in those days, but the clinical picture was remarkably uniform and, although the multisystem disease resembled many heritable diseases, none of the cases were familial. That set off a debate that has continued up to the present. In contrast to my own views about KSS, others believe that KSS is not unique, but is part of a spectrum of diseases that are familial.

Another problem in the mid-seventies is also still with us, but somehow seems to command less attention, determining whether the syndrome is myopathic or neurogenic. For two decades we grappled with that problem and others. But the world changed in 1988 when Holt et al. described deletions of mitochondrial DNA (mtDNA) and it was soon found that these deletions were restricted to syndromes that included ophthalmoplegia (Zeviani et al. 1988; Moraes et al. 1989; DiMauro, this volume, Ch.16). That established PEO itself and KSS as unique syndromes, and it opened a new approach to the thorny problem of the inheritance (or lack of inheritance) of KSS.

Therefore, what has changed more than anything else in this field, is the advent of molecular genetics. This has reset the central problems. We still have to identify clinical syndromes, but we also have to explain overlaps. We still have to determine how much of a syndrome is due to a muscle disorder, or one of neurons or peripheral nerves, but the challenge now is to determine how a mutation of mtDNA leads to a biochemical abnormality that causes the clinical symptoms. And we have to understand why some syndromes are inherited in mendelian patterns, while others are found in mitochondrial 'maternal' patterns of inheritance, and why KSS is sporadic in most cases. There have been basic changes in concepts since 1975.

History: The term 'progressive nuclear ophthalmoplegia' was apparently first used by Beaumont (1890). The name implies that the weakness is due to degeneration of neurons and the condition was thought to be analogous to motor neuron disease of infancy (Werdnig-Hoffmann disease) or amyotrophic lateral sclerosis. But ocular muscles are only exceptionally involved in either of these two model disorders and neither the term nor the concept has been accepted as an explanation for progressive ophthalmoplegia. The conflict is explained in the following brief statement of the history as taken from several sources (Kiloh and Nevin 1951; Drachman 1968, 1975; Rosenberg 1968; Daroff 1969).

The first recorded case is credited to Von Graefe (1868). Hutchinson (1879) also provided an early description, the first in English. By 1900, Wilbrand and Saenger collected reports of 22 cases. They thought the primary disorder was a neuronal degeneration, following an earlier suggestion of

Möbius. But, on the basis of a biopsy of the levator of the eyelid, Fuchs (1890) has already suggested that the disorder might be a restricted form of muscular dystrophy, a myopathy in which the primary disorder was in the muscle. This difference of opinion continued during the next 50 years, with eminent authorities expressing one view or the other and usually without substantial evidence one way or the other.

It was not until 1928 that Langdon and Cadwalader described the first autopsy examination, and they attributed the syndrome to neuronal abnormality. For decades, that concept of neuronal disease dominated the field, as seen in the authoritative reviews of Marburg (1936) and Hassler (1953).

However, there were also descriptions of patients with ophthalmoplegia with limb weakness in patterns resembling muscular dystrophy (Martin 1936; Elliott 1939). Sandifer (1948) provided the first biopsy of a lateral rectus muscle and found evidence of myopathy. Then, in 1951, Kiloh and Nevin discounted the changes previously described by Langdon and Cadwalader (1928) and also those described in the second autopsy examination by Jedlowski (1943). Others were similarly disdainful of the early reports (Beckett and Netsky 1953; Schwarz and Liu 1954). On the basis of biopsies of ocular and limb muscles, Kiloh and Nevin (1951) opted for the myopathic theory. In autopsy studies by Beckett and Netsky (1953) and Schwarz and Liu (1954), there seemed to be no significant changes in oculomotor nuclei. For the next two decades the myopathic theory was generally accepted and the condition was considered an 'ocular myopathy'.

Then, in 1968, there was another shift because of two reports of patients who had ophthalmoplegia in combination with evidence of neural disease in the form of peripheral neuropathy, ataxia or spinal cord syndromes. Rosenberg et al. (1968) found that 5 of 27 cases of ocular myopathy were associated with neural syndromes. Drachman (1968) described four similar cases and introduced what became a popular term, 'ophthalmoplegia-plus'. These cases suggested that the ophthalmoplegia might also be neural in origin and called into question the criteria used to determine whether a condition was neurogenic or myopathic. As pointed out by Rosenberg et al. (1968), all previous biopsies had been deemed myopathic; no one had ever made a diagnosis of denervation from a muscle biopsy. Drachman et al. (1969) then found that section of the oculomotor nerve in dogs was followed by histological changes of myopathy or myositis. His twin brother, D.B. Drachman (1967), found myopathic changes in muscles of patients surviving paralytic poliomyelitis.

In 1975, we summarized the problems generated by the use of biopsy or electromyographic examination of ocular muscles. Neither has been considered reliable because both tend to show only 'myopathic' changes in eyelids or ocular muscles. This is not merely the idiosyncratic conclusion of the author; neither test is much used any more.

Subsequently, a new criterion was introduced and gradually became a dominant force. That is, the demonstration that 'ragged red fibers' (RRF), as seen in the Gomori trichrome stain applied to a muscle biopsy, implied structurally abnormal mitochondria (Olson et al. 1972) gave rise to the concept that many cases of ocular myopathy were, in fact, 'mitochondrial myopathies'. This test was applied to biopsies of limb muscles.

The next step was the application of biochemical analysis of mitochondrial enzymes, an approach that has been sufficiently fruitful for DiMauro and his colleagues to provide a biochemically based classification of these diseases (DiMauro, this volume, Ch.16). This was followed by the study of deletions and point mutations of mtDNA in these diseases. Unfortunately, these approaches have not yet resolved the question of classification of PEO. For this chapter, we have therefore maintained a clinical-anatomical classification (Table 1).

Classification

Postmortem examination: Autopsy examination is usually considered the final arbiter in questions of this kind. But, morphology has problems of its own. First, it is inapplicable to living patients unless it is possible to extrapolate from the clinical disorder because there is an unambiguous and consistent relationship between the clinical and autopsy findings. Second, neurons may be sick without structural alteration as viewed by the light or electron microscope.

Consider the following incongruities: Daroff et al. (1966) studied a case of KSS. There were changes in oculomotor neurons but ocular muscles showed myopathic changes. In another case, Stephens et al. (1958) found ophthalmoplegia in a patient with clinical evidence of motor neuron disease. At autopsy, the oculomotor nuclei seemed normal and ocular muscles showed myopathic changes. In the spinal cord, there was degeneration of neurons in ventral horns as well as demyelination of posterior columns. Was the ophthalmoplegia neurogenic or myopathic? Similarly, Kearns and Sayre (1958) regarded the ophthalmoplegia in one of the original cases of KSS as myopathic because the oculomotor nuclei seemed normal but there was extensive vacuolation throughout the hemispheres, brain stem, and spinal cord. The patient was deaf and the cochlear nuclei appeared normal. Could oculomotor neurons malfunction but seem normal histologically?

When we reviewed the cases in 1975, there had been 16 reports of postmortem examination in PEO. In about half, the oculomotor neurons seemed normal and the syndrome was deemed to be myopathic. In the others there were neuronal changes, but many of the reports were criticized by later authors. In very few of the cases were the cranial nerves and ocular muscles available for examination.

This problem should come as no surprise in multisystem disorders such as KSS. In that condition, the central nervous system (CNS) is affected, and so is skeletal muscle (denoted by the RRF). Under those circumstances, we cannot determine whether the ophthalmoplegia is neurogenic or myopathic.

It would seem that we cannot count on anatomy to resolve this problem. As a matter of fact, it may not be important, practically or theoretically, to determine whether the ophthalmoplegia is neurogenic or myopathic.

Practical approach: What seems to be needed is a practical approach to these syndromes, one that starts with clinical observations and is then modified by laboratory tests. To do that, we can use a classification that was presented in 1975 and, with slight modification, is now brought up to date (see Table 1). As we will explain in subsequent sections,

TABLE 1

A tentative classification of progressive external ophthalmoplegia (Modified from Rowland 1975).

Anatomic site uncertain
    Ophthalmoplegia and ptosis; no other manifestations
        Clinical variations: ptosis alone; ophthalmoplegia alone.
        Onset congenital, juvenile, or later.
        Sporadic, autosomal dominant, autosomal recessive, or maternal inheritance.
        Mitochondrial DNA deletion present or absent.

Ocular myopathies
    Ocular and other cranial muscles
        Oculopharyngeal muscular dystrophy (autosomal dominant)
        Oculopharyngeal myopathy (sporadic)
    Ocular and proximal limb muscles
    Ocular and distal limb muscles
    Myotonic muscular dystrophy
    Myotubular or centronuclear myopathy
    Ophthalmoplegia, glycogen storage, and abnormal mitochondria
    Ophthalmopathy of Graves' disease
        Euthyroid, hypothyroid, hyperthyroid
    Orbital myositis (orbital pseudotumor)
    Congenital myopathic ptosis or ophthalmoplegia
        Anomalous insertion of ocular muscles
        Ocular muscle fibrosis
        With or without limb weakness
    Ptosis of aging

Disorders of neuromuscular junction
    Myasthenia gravis
    Curare-sensitive ocular myopathy

Possibly neural ophthalmoplegias (nuclear or supranuclear)
    Congenital: with facial diplegia (Möbius syndrome)
    Ophthalmoplegia with myelopathy or encephalopathy of later onset, with:
        Mental retardation, optic atrophy or other abnormalities
        Hereditary ataxias
        Hereditary spastic paraplegia
        Hereditary multisystem disease, including Joseph disease
        Dystonia musculorum deformans
        A-beta-lipoproteinemia (Bassen-Kornzweig)
        Progressive supranuclear bulbar palsy (Steele-Richardson-Olszewski)
    Ophthalmoplegia with motor neuron disease
        Infantile spinal muscular atrophy (Werdnig-Hoffmann)
        Juvenile spinal muscular atrophy (Kugelberg-Welander)
        Adult motor neuron disease (amyotrophic lateral sclerosis)
    Ophthalmoplegia with retinopathy, cardiac conduction abnormality and neural disorder (Kearns-Sayre syndrome)
    Ophthalmoplegia with mitochondrial encephalomyopathy, lactic acidosis, and stroke (MELAS)
    Ophthalmoplegia with myoclonus epilepsy and ragged red fibers (MERRF)
    Ophthalmoplegia with peripheral neuropathy in myo-, neuro-, gastro-intestinal encephalopathy (MNGIE) or ophthalmoplegia, sensorimotor peripheral neuropathy, intestinal pseudo-obstruction, and mitochondrial myopathy (OSPOM)

the classification is based on the following assumptions:

a. If the patient has PEO alone or with manifestations that might qualify for KSS, the most important studies are those directed to analysis of mitochondrial disorders. A comprehensive family history, sometimes with examination of family members, is needed to determine patterns of inheritance.

b. For PEO, alone or with limb weakness, it is necessary to evaluate several possibilities: mitochondrial myopathy, Graves' disease, myasthenia gravis, or orbital myositis.

c. For PEO with cranial and limb weakness, in addition to mitochondrial myopathies, consider other myopathies, such as myotonic muscular dystrophy, centronuclear or myotubular myopathy.

d. For PEO with retinopathy and cerebellar or corticospinal signs, consider KSS or other mitochondrial disorders, abetalipoproteinemia, spinocerebellar disorders or multisystem diseases of brain and spinal cord.

This anatomically and clinically based classification is deemed quaintly old-fashioned by DiMauro and colleagues (1991), who believe that the time has come for a genetic classification. They may be correct in the long run, but clinicians still need guidance to select cases for molecular study.

Epidemiology

There are no figures for the incidence and prevalence of PEO in any part of the world, nor for the relative frequency of uncomplicated PEO, PEO with myopathy, PEO with abnormal mitochondria, or PEO with multisystem neurological disease. Among those deemed to be ocular myopathies, Serratrice and Pellissier (1987) found the following distribution of syndromes in a personal series of 49 patients. First there was a group of 38 with abnormal mitochondria, including three subgroups: pure ocular myopathy, 14 patients; descending ocular myopathy, 14; and ocular myopathy with central abnormalities, including KSS, 10 patients. The other 11 patients were classified as oculopharyngeal muscular dystrophy.

Another way to estimate the relative frequency of different syndromes is to consider the diseases that are denoted by the presence of abnormal mitochondria in skeletal muscle, as selected by the appearance of RRF in the biopsy (Table 2) (Rowland et al. 1991).

SYNDROMES INCLUDING PEO

PEO WITHOUT EVIDENCE OF NEURAL DISEASE (OCULAR MYOPATHIES)

Progressive external ophthalmoplegia alone or with limb weakness

Definition. The purest form of PEO occurs in individuals and families with no weakness of other cranial or limb muscles, and no neural signs. A similar syndrome may occur with weakness of facial or limb muscles, and the two variations may be encountered in different members of the same family. There seems to be no basis for differentiating PEO alone or with weakness of other cranial or limb muscles, but it is uncertain how many genetic or non-familial forms there are.

This syndrome seems to be almost invariably associated with RRF and with deletions of mtDNA. For instance, in a recent review (Rowland et al. 1991) of syndromes associated with RRF, there were 8 patients with PEO and 7 of them had deletions of mtDNA; almost all had weakness of limb muscles in addition to PEO. It is not clear whether the few cases that show no deletions imply genetic heterogeneity, or whether more sensitive methods of detection would show similar abnormalities. In time, both RRF and deletion of mtDNA may become part of the definition of the syndrome.

Terminology. 'Progressive external ophthalmoplegia' is an awkward phrase but the abbreviation (PEO) is succinct, has withstood the vicissitudes of time, and is appropriately ambiguous because it has no implications about pathogenesis. It might be preferable to use PEO for all cases that include this feature. However, the definition would then have to be extended to include cases with limb weakness.

Some authors use the term 'ocular myopathy' or variants such as 'descending ocular myopathy' (Lees and Liversedge 1962) or 'progressive ophthalmoplegia with extra-ocular extension (ocular myopathy)' (Thiel 1954). These terms may be suited better for syndromes that involve limb muscles in addition to PEO because these cases are likely to show myopathic patterns in electro-

TABLE 2

Clinical syndromes associated with ragged red fibers at the Columbia-Presbyterian Medical Center (from Rowland et al. 1991).

	1975–82	1982–89	Denmark*
Number of biopsies	1,055	1,215	–
Number with RRF	20	40	22
PEO alone	7	4	10
PEO + limbs	–	11	–
PEO + exercise intolerance	–	2	–
MELAS	5	3	0
MERRF	0	2	0
KSS	3	5	2
Incomplete KSS	3	1	2
Infantile COX deficiency	1	0	0
Adult limb myopathy, no PEO	1	1	4
FSH myopathy	0	4	4
Spinal muscular atrophy	0	1	0
Polymyositis	0	0	2
Exercise intolerance	0	3	0
MNGIE	0	2	0
Borderline, not KSS	0	2	0

COX, cytochrome c oxidase; FSH, facioscapulohumeral; KSS, Kearns-Sayre syndrome; MELAS, mitochondrial encephalomyopathy with lactic acidosis and stroke; MERRF, myoclonus epilepsy and ragged red fibers; MNGIE, myo- neuro- gastrointestinal encephalomyopathy; PEO, progressive external ophthalmoplegia; RRF, ragged red fibers.
*Data of Kamienieczka and Schmalbruch (1980)

myography (EMG) and muscle biopsy. However, patients with PEO alone in clinical examination often show RRF in limb muscle biopsy, thereby invalidating any distinction between the two groups. Moreover, as will be seen, there may be neuronal changes in some patients with PEO alone. The presence of RRF may be the most important distinction from oculopharyngeal muscular dystrophy (Colombo et al. 1988); there is only one report of RRF in oculopharyngeal muscular dystrophy (Goas et al. 1991). On the other hand, RRF may be found in ocular muscles of patients with mitochondrial diseases even when there is no ophthalmoparesis (Takeda et al. 1990).

In this chapter, 'ocular myopathy' is sometimes used to mean PEO without any neural manifestations.

Clinical manifestations. There has been no recent review of the clinical manifestations of PEO. As a result, we have to use generalities rather than numbers in describing the manifestations. For the same reason, the following description is taken mostly from papers cited in the 1975 review.

Symptoms usually begin in childhood; juvenile or later onset is not infrequent. The age at onset is usually relatively constant within a family, but there are variations. In one family (Faulkner 1939), symptoms usually appeared around age 30, but one girl was affected in childhood.

Ptosis is usually the first symptom and may occur as an isolated manifestation for years. There is no diurnal variation and there are no remissions or exacerbations to suggest myasthenia gravis. Generally, there are both ptosis and ophthalmoplegia but either may occur alone. Ophthalmoplegia without ptosis is less common (Magora and Zauberman 1969; Walsh 1969). When ptosis is marked, the head is often thrown back to permit vision beneath the lowered lids. The lids themselves may or may not appear grossly thin and atrophic.

The ophthalmoparesis is usually symmetric, so diplopia is not part of the picture. Yet asymmetric weakness has been recorded often enough to make diplopia no surprise (Rowland 1975). Some patients are unaware of the ophthalmoplegia, learning of it only from physicians who examine them, or from relatives who notice that the head, not the eyes, turns for lateral gaze (Kearns 1965).

That the disease may indeed be restricted to the eyes was evident in the report of Langdon and Cadwalader (1928); symptoms began at age 43 and at age 84 there were still no weakness elsewhere. In another patient (McAuley 1956), ptosis appeared in childhood and the disorder was still ocular at age 63. Also, there are families in which manifestations are only ocular (Bradburne 1912; Faulkner 1939; Beckett and Netsky 1953).

Limb weakness is rarely severe, rarely disabling. Sometimes it is serious, however. For instance, in the family described by Chanmugam and Haerer (1970), ocular symptoms were apparent by age 4 and limb weakness followed shortly thereafter. By age 11, the children could not walk. The limb weakness is most often proximal (Martin 1936; Elliott 1939; Gartner and Billet 1949), but may be more marked distally (Schotland and Rowland 1964; Satoyoshi et al. 1965). The distal form resembles the myopathy of myotonic dystrophy, but there was no evidence of myotonia in these families. Sometimes, exercise intolerance is prominent (Driesssen-Kletter et al. 1987).

Depression of tendon jerks and focal wasting have not been consistently described, wasting may be pronounced (Elliott 1939; Gartner and Billet 1949). Teasdall et al. (1964) described two patients, still unique, with urinary and sphincter symptoms.

Weakness of limb and ocular muscles usually proceed together. In one case (Gartner and Billet 1949), limb weakness was evident in early childhood but the ophthalmoplegia did not appear until age 20, a pattern of progression that is characteristic of centronuclear myopathy. Some of the early cases had heart block and could have been examples of KSS (Sandifer 1948; Ross et al. 1970). There seems to be no danger of sudden death in patients with PEO who have no stigmata of KSS (Bril et al. 1984). Reports of subclinical retinopathy, as revealed by abnormal electroretinograms, are of uncertain significance (Berdjis et al. 1985), and cases of PEO with retinopathy may prove to be KSS (Korner and Probst 1975). Rarely, there are corneal abnormalities (Kosmorsky et al. 1989).

Endocrine abnormalities have been noted in a few cases, including gonadal failure (Lakin and Locke 1961; Lundberg 1962, 1966). Desnuelle et al. (1989) described a woman with PEO and amenorrhea at age 24.

In some cases there is clinical or laboratory evidence of peripheral neuropathy (Grisold et al. 1983; Cantello et al. 1985).

Muscle histopathology mitochondrial abnormalities: Although mitochondrial abnormalities are suspected in all cases of ocular myopathies, RRF are not found in all cases. For instance, Mussini et al. (1984) found mitochondrial changes in 6 of 10 cases; it is not clear whether the patients who do not show these changes have a different disease, or whether the RRF were missed because the limb muscle biopsy is remote from the eye and is a minuscule sample of skeletal muscle. In diagnostic studies, it is not clear how many RRF pass the threshold of normality. Pongratz et al. (1979) counted abnormal fibers per high-power field, finding that 0.3-2.5 fibers in each field contained RRF. Others have set some arbitrary percentage of fibers.

Conversely, Petty et al. (1986) found that more than half of all patients with mitochondrial myopathies (defined by the presence of RRF) had PEO of some kind. In our series (Rowland et al. 1991), 15 of 40 had ocular myopathy and 9 others had PEO in combination with other features (such as KSS). Serratrice et al. (1991) found ophthalmoparesis alone in 44 patients and PEO with multisystemic disease in 15 patients with RRF. Abnormal mitochondria may be found in asymptomatic relatives of an affected individual (Aasly et al. 1990). Similarly, deletions of mtDNA can also be found in asymptomatic relatives (Zeviani et al. 1990a). Therefore, changes in mtDNA and histopathology antedate clinical PEO, but it is still uncertain whether these people will be affected clinically.

The abnormal mitochondria are enlarged, increased in number, and lose normal structure when viewed with the electron microscope. In some cases, biochemical studies show abnormally low activity of cytochrome C oxidase (COX) but there is no consistent pattern (Aasly et al. 1990; Reichmann et al. 1991; DiMauro, this volume, Ch.16). Müller-Höcker et al. (1983) found that the enzyme was lacking in fibers with normal mitochondria but was active in fibers showing RRF, as though the enzyme activity were a sign of mito-chondrial proliferation. This pattern, with only some fibers showing no activity, might explain why muscle homogenates show no consistent level of COX activity. Yamamoto and Nonaka (1988) found that COX activity was irregularly distrib-uted along the length of a fiber, with active and inactive areas alternating.

Mita et al. (1989) related COX activity to the proportion of deleted mtDNA. The higher the pro-portion of deleted genomes, the lower the propor-tion of normal mtDNA in a fiber, and the more likely it was to be deficient in COX activity, and this was especially true of RRF. Nakamura et al. (1990) found decreased expression of the mRNA for COX. There is a correlation between the num-ber of COX-deficient fibers and the population of deleted mtDNA, and there may be more COX-deficient fibers in patients with deletions than in

those who have no deletion (Goto et al. 1990c). Shoubridge et al. (1990) suggested that the mu-tated mtDNAs were dominant over normal genomes, thus explaining why segmental areas of COX deficiency were related to deleted mtDNA even though the same area might also contain nor-mal mtDNA.

Other changes in muscle fibers are usually non-specific indicators of myopathy. Sometimes, accu-mulation of glycogen and lipid is impressive. Rarely, there are other structural abnormalities, such as cytoplasmic bodies (Sahashi et al. 1990).

Pathogenesis. Although these PEO syndromes are regarded as myopathies and may prove to be so, there are still uncertainties. Four of the re-ported autopsies were performed in patients with no manifestations other than PEO (Langdon and Cadwalader 1928; Jedlowski 1943; Beckett and Netsky 1953; Schwarz and Liu 1954), assuming

TABLE 3

Diagnostic tests for PEO syndromes.

Clinical signs	Diseases	Diagnostic test
PEO alone PEO + possible KSS	Mitochondrial diseases	Muscle biopsy for RRF Analysis mtDNA Brain CT, MRI, ECG, CSF, EMG, NCV Family history Retinal examination Audiogram
PEO alone PEO + limb weakness	Myasthenia gravis	Edrophonium test AChR antibodies Repetitive stimulation Chest CT for thymoma
	Congenital myasthenia	Endplate ultrastructure Endplate physiology
PEO alone	Graves' disease	Thyroid function tests Sensitive TSH test Orbital CT or MRI
PEO + dysphagia PEO + BKS	OPMD A-beta-lipoproteinemia	Muscle biopsy for nuclear inclusions Beta-lipoprotein Smear for acanthocytes Tests for malabsorption
PEO + MND	Infantile, juvenile SMA, ALS	EMG, NCV, chromosome-5 DNA Paraproteinemia, anti-GM1 Bone marow examination

AChR, acetylcholine receptor; ALS, amyotrophic lateral sclerosis; BKS, Bassen-Kornzweig syndrome; CSF, cerebrospinal fluid; CT, computed tomography; ECG, electrocardiogram; EMG, electromyography; KSS, Kearns-Sayre syndrome; MRI, magnetic resonance imaging; NCV, nerve conduction velocity; OPMD, oculo-pharyngeal muscular dystrophy; PEO, progressive external ophthalmoplegia; RRF, ragged red fibers; SMA, spinal muscular atrophy; TSH, thyroid stimulating hormone

that mental deterioration and aqueductal stenosis in one case were coincidental, as was the glioblastoma in another case. The two earlier investigators reported neuronal changes but these observations were deemed insignificant by the later authors who, in turn, explained away minor neuronal alterations in their own cases.

In cases with limb weakness, the EMG and biopsy changes have almost invariably shown myopathic changes. In the case of Cogan et al. (1962), the symptoms were primarily ocular but there was also some leg weakness. Autopsy examination showed myopathic changes in limb and ocular muscles, with no abnormality of oculomotor neurons. Orbital computerized tomography (CT) may show abnormalities consistent with muscular atrophy (Hansman et al. 1988). Single-fiber EMG may show non-specific abnormalities (Krendel et al. 1987). A major argument in favor of a myopathic origin was advanced by Ringel et al. (1979) who found RRF in ocular muscles in proportion to the percentage of RRF in limb muscles. Moreover, they found no extrajunction acetylcholine receptor as might be expected in denervated muscle.

Although the consensus is that PEO is an inherited condition, Calzetti et al. (1983) raised the possibility of an autoimmune disorder when they found evidence of circulating immune complexes in a family with an affected woman and four of her affected children.

It is not clear how many different forms are subsumed with this heterogeneous category, united only by the occurrence of PEO. Cases could conceivably be divided by patterns of inheritance, onset in childhood or later, with or without limb weakness, with proximal or distal limb weakness, and with or without endocrine disorder. These decisions await more specific information about DNA or biochemical abnormalities.

Genetics. Deletions of mtDNA have been found in about half of the patients with uncomplicated ocular myopathy (Holt et al. 1989; Moraes et al. 1989; Degoul et al. 1991). Conversely, almost all patients with deletions of mtDNA have had ophthalmoplegia, either as ocular myopathy or as part of KSS. However, there has been no clear maternal transmission of a deletion of mtDNA (Zeviani et al. 1990a). In some families, the inheritance pattern implies autosomal dominance. Some of these cases have had multiple deletions of mtDNA, which has been attributed trans-acting factors that are encoded in nuclear DNA and somehow cause multiple deletions in mtDNA (Zeviani et al. 1989, 1990a; Otsuka et al. 1990; Servidei et al. 1991). In one family (Ozawa et al. 1988), a mother and daughter with PEO had different deletions, a pattern that does not suggest maternal inheritance. In other families, the pattern was clinically consistent with maternal inheritance of mtDNA (Aasly et al. 1990). In the case of Desnuelle et al. (1989), the propositus was the child of consanguineous parents, but there was also an affected maternal uncle, so that it could have been maternal transmission. It seems likely that the phenotype of PEO or ocular myopathy is due to more than one genetic disorder.

There is little information about patients and families that do not show a deletion. In some cases there may be a deletion that defies current technology; in others there may be a point mutation (Lauber et al. 1991).

Oculopharyngeal muscular dystrophy

Definition. About two thirds of all PEO patients show facial weakness and many also note dysphagia (Roberts and Bamforth 1968; Rosenberg et al. 1968). Dysarthria and dysphagia are often mild but may be severe, especially in a French-Canadian form that is thought to be a unique disorder, which was first described by Taylor (1915) and is marked by later onset, severe dysphagia, and autosomal dominant inheritance (Victor et al. 1962). On clinical grounds it might be difficult to select out the severity of dysphagia needed to differentiate this type from other forms of PEO. However, Tome and Fardeau (1980) have found nuclear inclusions in almost all cases of French Canadian origin. If so, this morphological change might also be an identifying character. Additionally, RRF are not found and there are no deletions of mtDNA.

The term 'oculopharyngeal muscular dystrophy' (OPMD) might therefore be reserved for patients who meet 5 criteria: PEO; prominent dysphagia; family with documented autosomal dominant inheritance; typical inclusions; and no RRF beyond the number encountered in asymptomatic people.

Genetics. It may seem arbitrary to restrict use of the term OPMD to those with documented family history, but the autosomal dominant pattern has been seen around the world (Barbeau 1966; Schmitt and Krause 1981; Bouchard et al. 1989). In an unusual feat of medical detective work, Barbeau (1966) traced many of the cases reported from the United States to the westward migration of a family that originated in Montreal. The founder had come from France in 1634. As his migrating descendants settled in a city before resuming the trek, a new kindred arose and, in due time, was reported by an intrepid investigator who had not searched for the origin of the family.

In France, however, it was not possible to link all affected families in a study that went back to the 18th century (Brunet et al. 1990). Rather, there seems to have been more than one mutation. However, this cannot be put to test until there is information about the mutated DNA. The disorder is not restricted to the French and has been seen in other Europeans (Brunet et al. 1990).

Clinical features. In these families, symptoms are usually not apparent in childhood, with onset usually after age 40, even into the eighth decade (Amyot 1948; Victor et al. 1962; Weitzner 1969). The course is usually slowly progressive, but Weitzner (1969) described a 75-year-old man who had symptoms for only a few months before he died of carcinoma of the stomach; similar symptoms of typically chronic nature had been present in 3 generations of the family. Ptosis usually precedes the dysphagia, and proximal limb muscles are affected last. Distal muscles may also be prominently affected in some cases (Bosch et al. 1979; Fukuhara et al. 1982; Vita et al. 1983; Man et al. 1976; Klostermann et al. 1990), sometimes with neurogenic features (Jaspar et al. 1977). Ocular movements may remain normal even when the ptosis is severe (Victor et al. 1962).

Motility and pressure studies have shown impaired function of both the pharynx and esophagus (Lewis 1966; Roberts and Bamforth 1968), including the entire esophagus (Bray et al. 1965) or upper third only (Murphy and Drachman 1968). Cricopharyngeal myotomy has been reported to improve symptoms as well as functional measures

of swallowing (Fradet et al. 1988; Taillefer and Duranceau 1988).

Pathology: In autopsy examination of 6 cases, the brain, spinal cord, and peripheral nerves were essentially normal (Rebeiz et al. 1967; Weitzner 1969; Little and Perl 1982; Serratrice and Pellissier 1987). Schmitt and Krause (1981) were impressed by changes in the anterior horn cells of the spinal cord. The usual pattern in muscle is myopathic (Tome and Fardeau 1986), but sometimes there are neurogenic changes, too (Schmitt and Krause 1981; Probst et al. 1982). In limb muscles, Kozachek and Wilson (1982) found Z-line streaming and disorganization of the sarcomere. Schmitt and Krause (1981) found abnormalities throughout the body, sufficient to regard this as a generalized dystrophy, but, like the clinical symptoms and signs, the histopathologic changes were much more pronounced in ocular and oropharyngeal muscles, including the tongue. Replacement by fat and fibrous tissue was prominent in the levator and eye muscles. Little and Perl (1982) called attention to an unexplained discrepancy: the levator palpebrae are the most severely affected muscles, but there is relative sparing of the superior rectus muscles that share a common embryologic origin but differ in innervation.

The nuclear inclusions are described as 'tubulofilaments' and are accumulations of short (0.1 μm), straight, unbranched filaments of about 8.5 nm diameter (Coquet et al. 1983, 1990). They are irregularly oriented but tend to be roughly parallel. They are not membrane-bound. The central tubular portion is about 3 nm in diameter. Similar structures are seen in muscles cultured from patients with OPMD (Tome et al. 1989), but the inclusions did not appear in muscle cultured from patients unless the cultures were innervated by cocultured spinal cord. The inclusions are seen in about 5% of muscle nuclei but are thought to occur in all. The nature of the inclusions is still obscure because they do not react with antibodies to any of the constituents of other cellular filaments (Tome et al. 1989).

As recently as 1989, the inclusions were thought be 'the only ultrastructural change that is totally specific to one primary muscle disease' (Bouchard et al. 1989).

Sometimes there is a second type of filament, one with a larger diameter of 16–18 nm, which is similar to the one that has been considered characteristic of inclusion body myositis (Coquet et al. 1990). These broader inclusions are therefore specific for neither disease.

In addition to the nuclear filaments, muscles often show rimmed vacuoles, an unusual but nonspecific finding of unknown origin (Bouchard et al. 1989). There may also be non-specific mitochondrial changes (Julien et al. 1974), but RRF appear no more often than in asymptomatic people (Bouchard et al. 1989).

Occasionally, there is an inflammatory cellular response. This should not be mistaken for polymyositis if there is a clearly autosomal dominant pattern of inheritance. Sometimes, the inflammatory response is no longer seen in a later, second, biopsy (Vita et al. 1983).

Congenital myopathies: myotubular or centronuclear myopathy, and others

When PEO occurs in children, there may be evidence of what might be called a histologically specific structural abnormality (Goebel and Lenard, this volume, Ch.12). Perhaps the most common form is myotubular or centronuclear myopathy (Spiro et al. 1966; Sher et al. 1967). Were it not for the histological features, there would be no way to distinguish them from uncomplicated PEO with limb weakness (Munsat et al. 1969). Almost all patients have ptosis and almost as many have ophthalmoplegia. Facial weakness may also be evident. Goebel et al. (1984) noted ptosis of adult-onset in only 1 of 4 patients, so there is some variability in these manifestations (Bergen et al. 1980). Also, the cranial symptoms are overshadowed by delayed psychomotor development and severe limb weakness; 2 patients died in early childhood. Ortiz de Zarate and Maruffo (1970) called the disorder the 'descending ocular myopathy of early childhood'.

The congenital myopathies are discussed in more detail in Ch.12 in this volume by Goebel and Lenard. They and Owen et al. (1981) have listed the childhood myopathies that include ophthalmoplegia: centronuclear myopathy, congenital fiber type disproportion with fiber type 1 hypotrophy

(Sugie et al. 1982), muscle fiber hypotrophy (Bender and Bender 1977), microfibers (Hanson et al. 1977), nemaline myopathy (Fukunaga et al. 1980), congenital myasthenia gravis (Misulis and Fenichel 1989), tubular aggregates (Pierobon-Bormioli et al. 1985), multicore disease (Swash and Schwartz 1981), congenital myotonic muscular dystrophy, minimal change myopathy (Ohtaki et al. 1990), and mitochondrial diseases (PEO with RRF).

Congenital ptosis and congenital ophthalmoplegia

In some families, ptosis and ophthalmoplegia seem to date from birth (Bradburne 1912; Holmes 1956; Krüger and Friedreich 1963). In others, ptosis is said to have been present from birth but the ophthalmoplegia appeared later in childhood (Erdbrink 1957; Cogan 1962; Lind and Prame 1963). There have also been sporadic cases of PEO beginning in early childhood (Li 1923; Holmes 1956; Erdbrink 1957; Isaksson 1962; Slatt 1965) and congenital ptosis, without ophthalmoplegia, might be a limited form of the condition (Isaksson 1962).

Is this the same condition as PEO appearing at an unusually early age, perhaps even complete at birth? Or is it a different genetic defect? Advances in molecular genetics may resolve this problem. However, as in other heritable diseases, there have been individual families with both congenital PED in some individuals and later-onset in others (Gerber et al. 1970).

Congenital limitation of eye movements has been attributed to causes other than a myopathy. One form is attributed to congenital fibrosis of ocular muscles (Brown 1950; Laughlin 1956; Hansen 1968; Frazzetto et al. 1969; Crawford 1970). Ocular movements may also be restricted by anomalous fibrous bands in the orbits (McNeer and Jampolsky 1965; Von Noorden 1970). Forced duction tests may help to make this distinction. Cibis (1984) thought that EMG evidence of co-contraction of antagonist muscles implied some abnormality of CNS regulation of eye movements.

Congenital myasthenia gravis is another important cause of congenital ophthalmoplegia and has to be excluded before a case can be considered a congenital ocular myopathy.

Ophthalmoplegia, glycogen storage and abnormal mitochondria

In the 1975 review we thought this might have been a separate category. However, morphologic evidence of excessive glycogen deposition cannot be regarded as specific, and the abnormal mitochondria were probably more important, placing these cases in the general category of PEO. It would be desirable to test these patients for deletions of mtDNA (Zintz 1966; Zintz and Villiger 1967; Sluga and Moser 1970; DiMauro et al. 1973).

Orbital myositis

Orbital myositis is identified by soft-tissue swelling of periorbital tissues and exophthalmos in association with progressive ophthalmoplegia in individuals lacking evidence of thyroid disease. The first report is credited to Gleason (1903), but his patient, as well as many of the other reported cases, were described before the advent of the T3 suppression test or more modern methods to diagnose Graves' disease and could therefore have been due to endocrine ophthalmoplegia (Jellinek 1969; Dresner and Kennerdell 1985; De Waard et al. 1983; Glaser 1984). Some of them responded to steroid therapy, but so do some patients with Graves' disease (Bahn et al. 1988; Wiersinga et al. 1988; Kendall-Taylor 1989). On the other hand, papilledema and visual loss are said to be much less common in orbital myositis than in Graves' disease; nevertheless, visual loss has been reported (Mottow and Jakobiec 1978; Weinstein et al. 1983). The typical pattern of orbital myositis may be followed a year or more later by evidence of hyperthyroidism (Heersink et al. 1977). Both disorders may occur in childhood or adolescence (Mottow and Jakobiec 1978).

The syndrome is mentioned here because the soft tissue signs may be minimal and could conceivably be lacking, a so-called oligosymptomatic form (Papst et al. 1958, 1959). In these cases the pattern would resemble other forms of PEO. The problem is compounded by the difficulty of identifying 'euthyroid' Graves' disease, which therefore has to be considered in all patients with adult-onset PEO.

Orbital myositis is not to be confused with polymyositis of limb muscles. Ophthalmoplegia is virtually unknown in typical dermatomyositis and cases of polymyositis (lacking a rash) with ophthalmoplegia would be indistinguishable from myasthenia gravis (Rowland et al. 1960). The histopathology of myasthenia gravis can also include inflammatory lesions in skeletal muscle and myocardium (Rowland et al. 1956). There is a syndrome of myasthenia, thymoma, and giant-cell myocarditis and there have been differences of opinion; some regard this as a variant of myasthenia gravis (Rowland et al. 1973); others consider it the association of two different diseases, myasthenia and polymyositis (Namba et al. 1974). Klein et al. (1989) described a case that encompassed the ambiguity. There were clinical manifestations of myasthenia but the ophthalmoplegia was accompanied by swelling of the lid and there was CT evidence of enlargement of lateral and medial rectus muscles. Serum lactate dehydrogenase and creatine kinase levels were markedly elevated. At autopsy there was severe myocarditis. That is, the syndrome included elements of both myasthenia and polymyositis. Unfortunately, antibodies to acetylcholine receptor were not recorded nor was the patient tested for a decremental response to repetitive stimulation.

Additionally, orbital myositis may be unilateral and may therefore simulate an intraorbital mass lesion; another name for orbital myositis is 'orbital pseudotumor' (Kennerdell and Dresner 1984; Curtin 1987). Bullen and Younge (1982) attempted to differentiate orbital myositis from pseudotumor, but this now seems a fruitless task; the authors concluded that the two conditions 'cannot be entirely separated'.

The differential diagnosis includes Graves' disease, myasthenia gravis, and intraorbital or retroorbital mass lesions. Even as a diagnosis of exclusion, the same syndrome can appear in patients who have other manifestations of temporal or giant-cell arteritis (Goldberg 1983; Laidlaw et al. 1990), sarcoidosis (Cassan et al. 1970; Stannard and Spalton 1985; Atlas et al. 1987), Wegener granulomatosis (Bullen et al. 1983; Kalina et al. 1990), systemic lupus erythematosus (Grimson and Simons 1983; Weinstein et al. 1983), Hodgkin disease (Slavin and Glaser 1982), macroglobulinemia (Lossos et al. 1990), retroperitoneal sclerosis

(Schonder et al. 1985), multiple sclerosis (Bixenman and Buchsbaum 1988), and herpes zoster (Volpe and Shore 1991). When the syndrome is unilateral and there is pain, the syndrome overlaps with the equally vague concepts of Tolosa-Hunt syndrome and ophthalmoplegic migraine. When all of these other conditions are excluded, we are left with 'idiopathic orbital myositis' but, lacking clear concepts of etiology or pathogenesis, what remains is still a vague syndrome, impossible to define precisely, and dependent on clinical features that merge with other syndromes. Even what remains is undoubtedly a heterogeneous syndrome of diverse etiology.

Clinical manifestations. Most patients are adults; in a Mayo Clinic series, 13 of the 15 patients were between ages 18 and 30 (Bullen and Younge 1982). Children may be affected (Goldberg et al. 1982; Slavin and Glaser 1982). Some attacks are preceded by non-specific viral infections (Purcell and Taulbee 1981), but there are no systemic symptoms in most cases. The acute syndrome may appear within a matter of days and it may be unilateral, but most cases are bilateral, usually asymmetrically so and therefore likely to be associated with diplopia. Local pain is prominent and is usually worse on movement of the eyes. In the full syndrome, there is edema and congestion of periorbital soft tissues and conjunctiva. There may be proptosis. In contrast to cellulitis or other orbital soft-tissue inflammations, ocular muscles must be involved to be considered 'myositis'; the soft-tissue signs do not appear without impaired ocular motility (Slavin and Glaser 1982; Bach et al. 1988). Sometimes the swelling may be accompanied by orbital hemorrhage and ecchymosis of the eyelid (Reifler et al. 1989).

The attack may terminate in days or weeks, spontaneously or in response to steroid therapy. All 15 of the Mayo patients responded to prednisone therapy in a dosage of 50–80 mg daily for 7–10 days, followed by reduction of dose for the next few weeks. There were no complications of steroid therapy for the acute attack. Pain was relieved within 72 hours and eye signs disappeared within two weeks. This pattern seems typical and some authors consider a favorable response to steroids as part of the definition, but some attacks

resolve spontaneously (Keane 1977; Slavin and Glaser 1982). If there is complete resolution of symptoms, there may be a relapse at some future time; the later attacks may also seem to respond to steroids but repeated or continued use of prednisone may lead to intolerable side effects (Bullen and Younge 1982; Slavin and Glaser 1982). Sometimes, the acute phase is followed by chronic ophthalmoplegia, and sometimes there is a chronic syndrome without a prior acute attack. In the late stages, fibrosis may limit passive mobility of the eyes in the forced duction test (Bixenman and Von Noorden 1981; Abramovitz et al. 1983).

Pathology. The infiltration of ocular muscles and soft tissue has no distinguishing features. The infiltrating cells include lymphocytes, granulocytes, and eosinophils in varying numbers. Jakobiec et al. (1984) found that most of the lymphocytes were B-cells in orbital myositis and in Graves' disease. Both the acute inflammation and the chronic fibrosis can be induced in rabbits by injection of phorbol ester into the superior rectus muscle (Fries et al. 1987).

Diagnosis: relation to Graves' disease; impact of imaging. The differential diagnosis of orbital myositis involves structural lesions of the orbit and retro-orbital areas. In former days the diagnosis could be made firmly only by exploration of the orbit, often with the use of arteriography to evaluate the neighborhood of the cavernous sinus (Enzmann et al. 1976, 1979; Flanders et al. 1989). Now, CT and magnetic resonance imaging (MRI) identify the few cases that need arteriography and those that are due to tumors, carotid-cavernous or other aneurysms, or thrombosis of the cavernous sinus. MRI can detect dural arteriovenous malformations (Hirabuki et al. 1988). Imaging has also become crucial in the distinction between orbital myositis and the ophthalmopathy of Graves' disease. Bahn et al. (1988) consider MRI and CT equivalent; Atlas et al. (1987) think MRI is more specific. Sometimes it is necessary to repeat studies before the mass lesion becomes evident (Abdul-Rahim et al. 1989).

In many cases of Graves' disease, there is evidence of previous or concomitant hyperthyroidism or hypothyroidism. In the series of Gorman (1983) only 29 of 194 patients with Graves' ophthalmopathy had not been clinically hyperthyroid. In 81% of the patients with hyperthyroidism, the thyroid disorder commenced within 18 months before or after the eye disease. The cases with prior thyroid disease are not at issue. However, Graves' ophthalmopathy may occur in euthyroid people; in that case, it may be impossible to distinguish the two conditions, myositis and Graves' eye disease.

There are few clinical clues and none seem reliable in making the distinction. Graves' orbitopathy is less likely to be acute, but may be (Leonard et al. 1984). Ocular discomfort in Graves' orbitopathy is more often described as sandy or a sensation that there is a foreign material in the eye, rather than deep pain or pain on eye movement. Eyelid stare, retraction and lid-lag are thyroid signs. There may be greater likelihood of visual loss due to the crowding of swollen soft tissue structures at the orbital apex (Neigel et al. 1988). But none of these is sufficiently invariant or specific to make diagnosis reliable. Slavin and Glaser (1982) believe that response to steroids confirms the diagnosis of myositis, but Graves' disease may also improve (Bach et al. 1988; Kendall-Taylor et al. 1988; Bartalena et al. 1989; Prummel et al. 1989). Therefore, we have come to rely on two sets of tests: imaging and biochemical or immunological thyroid tests.

In the pre-CT era, ultrasonography was used to make the distinction but this has now been largely supplanted by CT and MRI. At the moment, both are used, but it seems likely that MRI will ultimately supplant CT.

In orbital myositis, CT may show a mass or densities that involve orbital soft tissues. The posterior wall of the globe may be involved and the anterior insertions of the muscle may enhance. Rarely, there may be bony destruction and intracranial extension of the process (Frohman et al. 1986). The muscles may be enlarged; sometimes only one muscle is enlarged (Keane 1977; Purcell and Taulbee 1981). In Graves' disease, according to some authors (Trokel and Hilal 1979, 1980; Trokel and Jakobiec 1981), muscle enlargement is more uniform, the process is almost invariably bilateral, and anterior herniation of orbital fat or enlargement of the optic nerve is more likely to occur. Only in myositis is there supposed to be thickening of the muscle tendon, and there may be other distinguishing features. However, Rothfus and Curtin (1984) and Dresner et al. (1984) found that none of these features was pathognomonic.

The difficulty of separating orbital myositis and Graves' disease was illustrated by a patient described by Bixenman and Von Noorden (1981). Months after an attack of left herpes zoster ophthalmicus, a 62-year-old woman suddenly had diplopia and there was ophthalmoparesis on that side, with limitation of passive upward movement in the forced duction test. Except for lid-lag on the left, the clinical and CT findings were consistent with orbital myositis, but one month later there was overt evidence of thyrotoxicosis. Tamai et al. (1980) found that more than 70% of patients with ostensibly euthyroid Graves' disease will soon develop clinical or laboratory evidence of hyperthyroidism. There is no specific antibody test to identify Graves' ophthalmopathy (Kadlubowsky et al. 1986; Hosten et al. 1989; Fleck and Toft 1990) and the limitation of movement is not always due to fibrosis as indicated by normal responses to forced duction (Hermann 1982). It is therefore necessary to determine how euthyroid Graves' disease is to be recognized (Dresner and Kennerdell 1985).

First, there are two meanings to 'euthyroid'. Some patients show no clinical evidence of hyperthyroidism but abnormal laboratory tests indicate that there is a hyperthyroid state (Gorman 1983). Second, there are patients who have no evidence of hyperthyroidism, clinically or in routine laboratory tests (Solomon et al. 1977; Rosen and Burde 1990). Werner (1961) introduced the concept of the autonomous thyroid gland as the central sign of Graves' disease. In a normal person, administration of T3 would be followed by decreased uptake of radio-iodine by the thyroid; in Graves' disease, the abnormally increased uptake is not affected by administration of T3 (Sergott and Glaser 1981).

With the advent of better blood tests and the availability of synthetic thyroid-releasing hormone (TRH), the Werner test was modified. In Graves' disease, there was no increase in thyroid function with administration of TRH (Gorman 1983; Spector and Carlisle 1987).

The modern version depends on sensitive and reliable tests for thyroid stimulating hormone (TSH) (thyrotropin) itself. After injection of synthetic TRH, blood TSH levels normally rise, do not change in hyperthyroidism, and show an exaggerated rise in hypothyroidism. If the thyroid gland is autonomous, there will be no response to TRH and the circulating levels of TSH will be immutably low (Klee and Hay 1987; Surks et al. 1990; Scott et al. 1991). Therefore, if a patient had appropriate eye signs, normal T3 and T4 levels but low TSH values, the findings are consistent with euthyroid ophthalmic Graves' disease. Lawton (1979) thought the TRH test was especially useful in the study of unilateral ophthalmopathy. Marcocci et al. (1989) estimated that 8.6% of all patients with ophthalmic Graves' disease were euthyroid. Using the sensitive test for TSH levels, Salvi et al. (1990) studied 22 apparently euthyroid patients and found that three were hyperthyroid, one hypothyroid, and 18 euthyroid. All 18 had an abnormality in at least one of several immunological tests that included antibodies to TSH receptors and cytotoxicity assays.

Peter (1986) described a woman with a relapsing orbitopathy in whom thyroid function tests were repeatedly normal for 12 years; she had a sister with hyperthyroidism. However, in the absence of some functional test that indicates abnormal thyroid function, it seems inappropriate to make the diagnosis of Graves' orbitopathy. In addition to the impossibility of distinguishing this state from non-thyroid orbital myositis, a retro-orbital mass might be missed (Reifler et al. 1986).

Diagnosis: problems with Tolosa-Hunt syndrome and ophthalmoplegic migraine. The syndrome of painful unilateral ophthalmoplegia has been called the Tolosa-Hunt syndrome, and it seems to have captured the imagination of neurologists throughout the world. The manifestations are consistent with a lesion in the region of the superior orbital fissure (Bruyn and Hoes 1986). As will be seen, however, the findings are difficult to differentiate from orbital myositis on the one hand, and ophthalmoplegic migraine on the other hand. Hunt described the clinical features (Hunt et al. 1961; Hunt 1976): pain is noted with or before the unilateral ophthalmoplegia; there is paresis of muscles innervated by all three ophthalmic nerves and, often, the sensory trigeminal is affected as well, and the pupil may be involved. The attack may last weeks or months. There may be spontaneous remission

or the symptoms may be relieved dramatically by steroids, which may be followed by recurrence. In Hunt's case, as in the original case of Tolosa (1954), the intracranial portion of the carotid artery was wrapped in granulomatous tissue (Campbell and Okazaki 1987). The early cases were diagnosed by arteriography, exploration, or autopsy. The condition was regarded as a diagnosis of exclusion (Kline 1982). Now, MRI may show abnormal soft tissue in the cavernous sinus (Goto et al. 1990a) or at the orbital apex (Ketonen et al. 1985), even when arteriography is normal (Thomas et al. 1988). Clinically, therefore, there is little difference from orbital myositis except when the trigeminal nerve or pupil is affected, and there is no proptosis or soft-tissue swelling. In fact, the diagnosis depends entirely on recognition of abnormality in the cavernous sinus.

Ophthalmoplegic migraine is similarly defined clinically. It also causes unilateral painful ophthalmoplegia, and is recognized because the subject has had other hemicranial attacks without ophthalmoplegia, or there is a family history of migraine.

However, many authors now view these concepts with skepticism. For instance, Hansen et al. (1990) wrote: 'The clinical characteristics of ophthalmoplegic migraine and Tolosa-Hunt syndrome cannot be clearly separated and it is likely that diagnosing one or the other depends more on the doctor than on the patient'. Others (Kandt and Goldstein 1985; Rapin and Echenne 1987; Hansen et al. 1990) have expressed similar views.

There seems to be little specificity to the Tolosa-Hunt syndrome. The presumably invariant response to steroid therapy has led to delayed errors in the recognition of mass lesions (Ambrosetto et al. 1986; Spector and Fiandaca 1986; Nakashima et al. 1987; Koppel 1987; Hansen et al. 1990) or recurrent cranial neuropathy (Barontini et al. 1987). Goto et al. (1989, 1990a) recognized the difficulty of separating THS from orbital myositis, indicating that it depends on the appearance of the cavernous sinus in the CT or MRI (Goadsby and Lance 1989), or the results of arteriography. The clinical picture is insufficient and even imaging may be ambiguous.

If intracavernous pathology is demonstrated, treatment with steroids should come first, but exploration and removal of some of the granulomatous tissue may be helpful (Goadsby and Lance 1989).

Familial relapsing ophthalmoplegia

According to the criteria for identifying PEO, we should not be considering intermittent ophthalmoplegia. However, if orbital myositis occasionally relapses, then the familial nature of the syndrome described by Currie (1970) warrants consideration, and may be related to familial ophthalmoplegias. Currie's cases were presumed to be neurogenic because weakness was confined to muscles innervated by the third nerve (albeit sparing the pupil and involving both eyes at different times) and because it was associated with attacks of Bell's palsy. Rowland (1975) mentioned a brother and sister who had recurrent attacks of unilateral ophthalmoplegia and soft tissue swelling, alternating sides, and resembling orbital myositis. The attacks had started at age 8 and sometimes seemed to respond to steroids. An aunt was said to have had similar symptoms. Familial orbital myositis has not been otherwise reported, and there have been no further cases like Currie's.

Senile ptosis

Senile ptosis is a familiar sight, but one that has been little studied. It is said that, in 90% of the cases, it is due to dehiscence or disinsertion of the levator aponeurosis from its tarsal insertion, and that the other 10% show degeneration of the anterior portion of the levator (Dortzbach and Sutula 1980; Collin 1986), which might be considered a limited ocular myopathy (Shore and McCord 1984). Aponeurotic ptosis can be recognized by the following features: ptosis despite good excursion of the levator, raised or absent upper skin lid crease, and thinning of the upper lid above the tarsal plate (Deady et al. 1989). It is said that the symptom is often worse at the end of the day even though it is not myasthenia. The disorder is rarely a serious concern to the subject but if it is, it can be corrected surgically.

Myotonic muscular dystrophy

For reasons that are not clear, eye muscles are spared in the conventional forms of progressive muscular dystrophy. In myotonic muscular dystrophy, however, there may be impairment of both saccadic and pursuit movements, sometimes with overt ophthalmoplegia. The problem has been studied by Ter Bruggen et al. (1990), who concluded that 83% of all patients had a decrease of the maximum velocity of visually guided saccades, and that the disorder was peripheral, not one of the CNS.

Other myopathic forms of ophthalmoplegia

Sandyk and Brennan (1983) described bilateral medial rectus weakness of uncertain nature in a 12-year-old girl who had celiac disease. Ophthalmoplegia may also be seen in temporal arteritis (Fisher 1959; Barricks et al. 1977; Dimant et al. 1980; Mehler and Rabinowich 1988). Keane (1986) analysed 60 patients with acute bilateral ophthalmoparesis; he did not include any with myositis or other myopathy. Not all pupil-sparing ophthalmoplegias are myopathic (Rowland 1975; Kissel et al. 1983; Nadeau and Trobe 1983). In comatose patients, there may be drug-induced ophthalmoplegia and the site of action may be uncertain (Beal 1982).

DISEASES OF THE NEUROMUSCULAR JUNCTION

Ocular myasthenia gravis, myasthenic ocular myopathy, and curare-sensitive ocular myopathy

Ordinarily, there is no problem in distinguishing myasthenia gravis from PEO. Myasthenia has several characteristics that are not seen in PEO; diurnal fluctuation, remissions and exacerbations, asymmetry of ocular disorder with prominent diplopia, and response to edrophonium. The presence of antibodies to acetylcholine receptor (AChR) is another identifying feature.

Problems may arise, however, if there is total ophthalmoparesis with little fluctuation. Moreover, AChR antibodies are often absent in patients with solely ocular myasthenia gravis and it may be difficult to demonstrate the response to edrophonium or other cholinergic drugs. And, in ocular myasthenia gravis, there may be no EMG abnormality in limb muscles.

Another problem concerns the reliability of

edrophonium as a diagnostic test. In my experience, botulism is the only condition other than myasthenia that responds reproducibly and unequivocally to cholinergic drugs (Rowland et al. 1960). The problem has been clouded in recent years because Dirr et al. (1989) published a photograph of an edrophonium-responsive ophthalmoplegia that was due to a third nerve palsy originating in a brainstem glioma. Then, a group of eminent neurologists and neuro-ophthalmologists (Moorthy et al. 1989) described several similar cases. These papers suggest that the test is not absolutely specific and there are 'false positive' results.

However, to avoid false positive responses, there should be absolutely unequivocal and reproducible responses. Except for the report of Dirr et al. (1989), published claims have not been documented by photographs (and could have been erroneous impressions) or could have involved slight changes in tropias (and therefore could have been erroneous impressions). Some included patients with pupillary abnormalities that could not be part of myasthenia or were patients who, on other grounds, clearly had motor neuron disease. Another source of error is failure to exclude placebo responses (in patients with fatigue states or simulated limb weakness). These are not 'false positives' but are 'false impressions'. On the other hand, myasthenia may rarely be found in a patient who also has a brain tumor (Moorthy et al. 1989); or there may be features of both myasthenia gravis and Lambert-Eaton syndrome (Fetell et al. 1978; Newsom-Davis et al. 1991; Oh and Cho 1990); this is not some other condition that responds to edrophonium but a combination of two conditions, like the combination of ocular myasthenia and Graves' ophthalmopathy. The single photograph presented by Dirr et al. (1989) tempers the absolute position, but does not diminish the clinical value of the edrophonium test under appropriate clinical conditions.

Before the advent of the antibody test, we studied a patient who had weakness in the characteristic distribution of myasthenia gravis but who failed to show a response to cholinergic drugs; thymectomy was carried out nevertheless and a thymoma was found (Black et al. 1973). There have been few other similar cases (Garcin et al. 1965; Hausmanowa-Petrusewicz et al. 1965) and there are likely to be fewer in the future, because CT of the chest would show the thymoma.

These two tests, chest CT and AChR antibodies, have also eliminated the use of the curare test in the diagnosis of atypical cases of myasthenia gravis. That may be one reason why there have been no new cases of 'curare-sensitive ocular myopathy', a syndrome we discussed with some skepticism in the 1975 Handbook review. In the series of Rosenberg et al. (1968), none of 24 patients with PEO was sensitive to curare. We continued to do that test for another 25 years, in another 50 cases or so, without finding a positive reaction. We have now stopped doing the curare test if there are no AChR antibodies, if chest CT is normal, and if there is no response to edrophonium or clear diurnal variation.

Another consideration is the possibility that a patient with hyperthyroidism may have both orbitopathy and ocular myasthenia gravis. The physical signs of proptosis and soft-tissue swelling would identify the Graves' component. Myasthenia would be identified by ptosis (not often a part of Graves' orbitopathy) and because the ptosis and ophthalmoplegia may respond to edrophonium. The photograph used to illustrate the edrophonium test in Merritt's Textbook of Neurology since the 4th edition (1967) illustrated one such patient. Alfaro et al. (1982) described a patient with hyperthyroidism and chronic ophthalmoplegia that lasted for 10 years before there were symptoms of generalized myasthenia. Reports of lid-lag on the side opposite ptosis in patients with presumed Graves' disease may not have been tested sufficiently to exclude myasthenia (Lohman et al. 1984). There have also been reports of myasthenia with Graves' ophthalmopathy and multiple sclerosis (Lo and Feasby 1983; Bixenman and Buchsbaum 1988). We have also seen one woman with these three diseases. Myasthenia has often been seen in patients with Graves' eye disease and Hashimoto's thyroiditis; in one patient, the sequence was ophthalmopathy at age 22, ocular myasthenia at age 26, and thyroid disorder manifest as hypothyroidism at age 62 (Czernobilsky and Ziegler 1985). When ocular myasthenia occurs in a patient with thyrotoxicosis, there is no need to consider the condition 'pseudomyasthenia' (Rousselle et al. 1990).

Ophthalmoplegia may be a paraneoplastic manifestation (Barontini et al. 1985), but it is then difficult to determine whether it is myasthenia gravis or an oligosymptomatic orbital myositis.

OPHTHALMOPLEGIA WITH NEURAL DISEASE

Central disorders

Nuclear and supranuclear mechanisms. Patients with clear clinical evidence of neural disease may also have syndromes of ophthalmoparesis that resemble ocular myopathy. It is then uncertain whether the ophthalmoplegia is neurogenic or myogenic. Familial diseases of this type have included Friedreich's ataxia (Heck 1964) and spastic paraplegia (Landau and Gitt 1951; Brown and Coleman 1966; Kato et al. 1986). Gaze palsies bordering on ophthalmoplegia have been described in a multisystem disease that involves pyramidal, cerebellar, sensory, and optic tracts, resembling multiple sclerosis (Ferguson and Critchley 1929; Mahloudji 1963). In these descriptions, the ocular disorder has not been emphasized because the other aspects were more constant and more devastating. In Brown's (1892) description of the cerebellar ataxia that bears his eponym, ptosis was described in several patients, slow and faulty eye movements in others, and total ophthalmoplegia in two. The presence of ophthalmoplegia was used as a major criterion to identify one form of olivopontocerebellar degeneration by Konigsmark and Weiner (1970). Ophthalmoparesis is also seen in a-beta-lipoproteinemia, the one hereditary ataxia in which there is a clue to the biochemical abnormality (Schwartz et al. 1963; Rosenberg et al. 1968).

Ophthalmoplegia has also been prominent in patients with corticospinal tract disease (Helfand 1939; Brown and Coleman 1966; Rosenberg et al. 1968) or dystonia (Rosenberg et al. 1968). Ocular movements have been impaired in other hereditary ataxias (Rowland 1975; Bastiaensen et al. 1977), sometimes with pigmentary retinopathy (Rowland 1975). Schimke et al. (1984) described an apparently X-linked syndrome of severe mental and growth retardation, chorea, ophthalmoplegia, and deafness.

Jampel et al. (1961) suggested that the disordered eye movements in most reported cases of ataxia were supranuclear in origin. In most reports there was impairment of vertical gaze. Jampel and coworkers noted that there was dissociation between the preservation of the levator when upward gaze was impaired, implication of gaze movements, and preservation of some superior rectus function, all suggesting a supranuclear disorder. Several pioneer neuro-ophthalmologists were long ago impressed with supranuclear gaze palsies that might terminate in total ophthalmoplegia (Gowers 1879; Daroff and Hoyt 1971; Holmes 1921, 1931; Collier 1927; Wilson 1941). Electromyographic studies suggested that congenital ophthalmoplegia might be due to supranuclear lesions (Gornig et al. 1967).

Supranuclear ophthalmoplegia is prominent in the syndromes of progressive supranuclear palsy (Steele et al. 1964; Maher and Lees 1986) and Joseph disease (McQuinn and Kemper 1987; Shimizu et al. 1990), and may be seen in diffuse Lewy body disease (Lewis and Gawel 1990; Fearnley et al. 1991). Sangla et al. (1990) reported improvement of eye movements on treatment with sulfamethoxazole and trimethoprim. Although tract degenerations at autopsy make it likely that the disordered eye movements are due to supranuclear lesions, there is also neuronal degeneration in some of these multisystem diseases, and this may also contribute to the ophthalmoplegia (Jampel et al. 1961; Brion and De Recondo 1967; Blumenthal and Miller 1969; Cambier et al. 1969; Woods and Schaumburg 1972; Coutinho et al. 1982; Iizuka et al. 1984; Janzer and Barontini 1985; Brunet et al. 1986; Hayashi et al. 1986; McQuinn and Kemper 1987; Shimizu et al. 1990; Staal et al. 1990; Gilden and Kleinschmidt-Demasters 1991).

Also, in a multisystem neurological disease, it is not possible to exclude the possibility that there is also a myopathy of ocular muscles, especially if there are RRF in limb muscles, as there were in a case of spastic paraplegia with ophthalmoplegia (Alonzo et al. 1984). Atsumi et al. (1980) described loss of motor neurons in a multisystem disease that included abnormal mitochondria in muscle. This problem, deciding whether the essential pathology is in nerve or muscle, is especially evident in the MNGIE syndrome to be discussed later, in which PEO and pseudo-obstruction of the bowel are the

TABLE 4

Causes of ocular muscle enlargement on CT (from Rothfus and Curtin 1984).

Diagnosis	Number cases studied	More than one muscle affected			
		Number cases	% Total	Number unilateral	Number bilateral
Graves' disease	67	62	93	7	55
Myositis	15	9	60	1	8
Other inflammatory	13	3	23	3	0
C-C fistula, AVM	7	6	86	6	0
Acromegaly	3	3	100	0	3
Tumors	23	12	25	11	1
Trauma	9	4	44	4	0
Total	137	99	–	–	–

C-C, carotid-cavernous; AVM, arteriovenous malformation.

major manifestations of a mitochondrial myopathy in which there is also a profound sensorimotor peripheral and cranial neuropathy.

Augustin et al. (1984) described optic atrophy as the only neural complication of PEO without retinopathy. Meire et al. (1985) described a similar syndrome of PEO, optic nerve atrophy and hearing loss. Both reports suggested autosomal dominant inheritance. Cantello et al. (1985) described a syndrome of PEO, leukoencephalopathy, RRF, and demyelinating neuropathy in a family that seemed also to have myotonia congenita with muscle hypertrophy and disabling myotonia. Cooles et al. (1988) described a family from the Caribbean island of Dominica with PEO, RRF, retinopathy, spastic paraparesis and cerebellar syndrome. Although the pattern of inheritance in 5 generations was thought to be autosomal dominant, it is quite consistent with maternal inheritance, as pointed out by DiMauro (personal communication, 1988).

Motor neuron degeneration. In addition to the multisystem neurological diseases with ophthalmoplegia, there are some cases of what seems clinically to be a primarily lower motor neuron disease. For instance, although ophthalmoplegia is exceedingly rare in typical amyotrophic lateral sclerosis, there have been some cases (Kushner et al. 1984).

TABLE 5

Differential diagnosis of orbital myositis and euthyroid Graves' ophthalmopathy (Modified from Bahn et al. 1988).

Condition	Diagnostic features
Primary or metastatic tumor, lymphoma	CT or MRI needed. Rarely bilateral. May have systemic disease. Biopsy may be necessary.
Vascular abnormality	CT or MRI may show mass with enlarged superior orbital vein; ocular muscles may be enlarged; arteriography may be needed.
Sarcoidosis	CT or MRI may show mass or findings of orbital myositis. May have systemic disease. Angiotensin converting enzyme blood levels may be high. Orbital biopsy may be needed.
Wegener granulomatosis	CT or MRI may be compatible with orbital myositis; usually bilateral; severe nasal symptoms, hemoptysis and hematuria in many cases. Diagnosis by biopsy of nasal mucosa or orbit.
Orbital infections	Evidence of systemic infection, immunosuppression, or diabetes in some cases. Evidence of contiguous sinus disease in many.

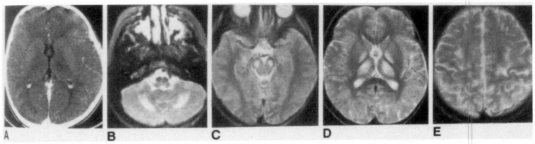

Fig. 1. Brain imaging in Kearns-Sayre syndrome. (A) Contrast-enhanced CT shows hypodensity of thalami and globus pallidus. (B–E) MRI shows areas of hypersignal in regions of dentate nucleus of cerebellum (B), superior cerebellar peduncles and substantia nigra (C), thalamus (nucleus ventralis lateralis) and pars medialis of globus pallidus (D), and white matter around central sulcus (E). The lesions correspond to areas of spongy degeneration in autopsy cases. From Demange et al. (1989).

Moreover, if patients with amyotrophic lateral sclerosis are kept alive on respirators for many years, ophthalmoplegia may appear (Mizutani et al. 1990).

Clinical evidence of a motor neuron disease with ophthalmoplegia has been seen in the Kugelberg-Welander syndrome (Aberfeld and Namba 1969;

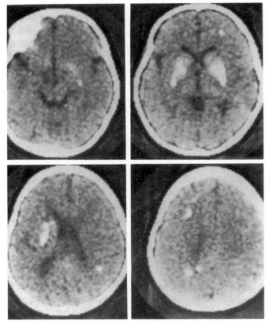

Fig. 2. CT of the brain of a 9-year-old girl with Kearns-Sayre syndrome. Hyperdensity indicates areas of calcium deposition in basal ganglia and white matter of the cerebral hemispheres on both sides. From Seigel et al. (1979).

Gruber et al. 1983), Werdnig-Hoffmann disease (Rosenberg et al. 1968), and a syndrome of distal spinal muscular atrophy with cataracts (Dubrovsky et al. 1981). There may be EMG evidence of denervations in limb muscles (Schwartzman et al. 1990). Loss of spinal neurons was prominent in the familial cases of Stephens et al. (1958) and Brown and Coleman (1966), but there was no specific loss of neurons in oculomotor nuclei. Lower motor neuron signs were clinically evident in the multisystem diseases described by Iizuka et al. (1984) and Hayashi et al. (1986). These cases were exceptional for the diseases mentioned.

Central multisystem diseases with PEO: Kearns-Sayre syndrome

Terminology. Cases have been reported under numerous titles, including Kearns-Sayre syndrome (KSS), Kearns-Sayre-Daroff syndrome, Kearns-Shy syndrome, ophthalmoplegia-plus, oculocraniosomatic neuromuscular disease, mitochondrial encephalomyopathy, and mitochondrial cytopathy (Pavlakis et al. 1988). The different terms largely reflect the views of lumpers and splitters in the days before 1988; in that year, deletions were found in almost all patients with the syndrome. Use of the eponym still seems justified because the pathogenesis of the disorder is still not known and, specifically, the essential gene product is not known. We shall therefore continue to use the term 'Kearns-Sayre syndrome', which now seems preferable to any of the others.

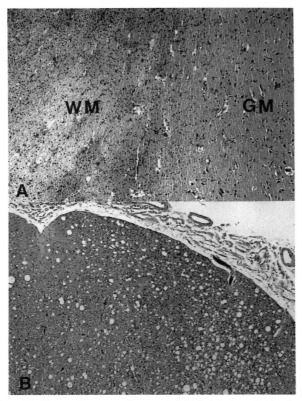

Fig. 3. Brain and spinal cord in Kearns-Sayre syndrome. A 4.9 kb deletion of mitochondrial DNA was present in all tissues analysed, including skeletal muscle and central nervous system (Shanske et al. 1990). (A) The gray matter of the cerebral cortex is normal, but the white matter shows patchy rarefaction and hypercellularity due to astrocytosis. There was little evidence of spongiform degeneration of the white matter in the cerebral hemispheres. (B) Spongiform degeneration is prominent in the posterior columns of the spinal cord. This was associated with mild astrocytosis, but there were no macrophages to suggest secondary (Wallerian) degeneration. Hematoxylin and eosin, × 70. Courtesy of Prof. A.P. Hays, Columbia-Presbyterian Medical Center, New York.

Definition. We have used a clinical definition (Berenberg et al. 1977; Rowland 1983) that includes three obligatory features, which should all be present: onset before age 20, PEO, and pigmentary retinopathy. In addition, there must be at least one of the following three: heart block, CSF protein content of about 100 mg/dl or higher, or cerebellar syndrome. Most patients have a deletion of mtDNA, which is not seen in any other multisystem neurological disease; a deletion is therefore an important diagnostic feature and would probably function as a fourth secondary criterion. An incomplete case may lack one of the obligatory or all three of the original secondary features. These cases can be identified if the missing features appear later, or if there is a deletion of mtDNA.

Clinical manifestations. In 1982 there were detailed descriptions of 70 cases (Rowland et al. 1983). There have been no large series of cases since then but there are probably over 100 reported cases by now, with no current incentive to record the details of individual cases. The most recent summary is that of Pavlakis et al. (1988); the most thoroughly documented review is that of Berenberg et al. (1977), in which reports of 35 cases were analysed. The clinical manifestations listed in Table 6 provide a reasonable estimate of the rela-

TABLE 6

Clinical manifestations of Kearns-Sayre syndrome (Modified from Rowland et al. 1983).

	1977 Berenberg et al. 1977	1982 Rowland et al. 1983
Number of cases	35	35
Retarded	14	20
Cerebellar syndrome	24	32
Babinski signs	9	9
Limb weakness	13	29
Hearing loss	19	24
Seizures	0	3
Short stature	22	25
Delayed puberty	11	–
CSF protein		
Not done	10	7
Normal	1	2
>100 mg/dl	20/24	26/28
RRF in limb muscle	5/7	28/30
CT lucencies	–	4
pontine atrophy	–	4
calcification	–	2
normal	–	4
Autopsies	5	8
Spongy degeneration	5	8

tive frequency of different signs. By definition all patients had PEO, pigmentary retinopathy and onset before age 20.

PEO. Symptoms usually begin after age 5, sometimes earlier. Ptosis and restricted eye movements are usually the very first symptoms of the disease. Both eyes are affected symmetrically and there is no ptosis. By the time of the first examination, pigmentary change in the retina is also evident. The disorder of ocular motility is identical to that of other forms of PEO.

Pigmentary retinopathy. In the earlier literature the retinal disorder was called 'atypical retinitis pigmentosa' but that term is now restricted to heritable disorders and the preferred name for the changes in KSS is 'pigmentary retinopathy' or 'salt-and-pepper retinopathy' (Mullie et al. 1985). The retinopathy of KSS differs from retinitis pigmentosa in four ways: genetics, appearance, visual loss, and optic atrophy. Retinitis pigmentosa is usually familial; arteriolar narrowing and structures that resemble 'bone corpuscles' are seen on fundoscopy; and both visual loss and optic atrophy are common. In contrast, the retinopathy of KSS is often asymptomatic; visual loss is recorded

in about half the cases but is usually mild. Serious loss of vision seems to be exceptional even when there are dramatic changes in the retina. The pigment is distributed diffusely or in widespread punctate fashion, most marked in the equatorial region. The pattern differs from the 'bone corpuscle' appearance, but that, too, may be seen in some cases. The changes are most marked in the pigment epithelium (Korner and Probst 1975; Babel 1982). Steindler et al. (1985) found a poor correlation of pigmentary change on fundoscopy, electroretinography (ERG), and fluoroangiography. Sandberg et al. (1979) also found that ERG was often normal in patients with retinal pigmentary deposition. In two cases (Chamlin and Billet 1950; Davis et al. 1981), there was an episode of optic neuritis. Retinal pathology will be described later.

Cerebral manifestations: mental retardation, dementia, seizures, cerebellar syndrome. There have been no longitudinal studies to determine the sources of disability in KSS, but the cerebral manifestations are surely prominent. Mental retardation or dementia occurs in about half of the patients. The cerebellar syndrome is somewhat more common and is often severe. It primarily affects

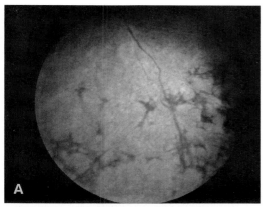

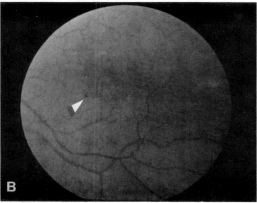

Fig. 4. (A) In 'typical retinitis pigmentosa', pigment accumulates in patterns that have been likened to 'bone spicules'. (B) In the pigmentary retinopathy of Kearns-Sayre syndrome (formerly called 'atypical retinitis pigmentosa'), the pigment is scattered in a salt-and-pepper pattern but there may be clumping at the macula (arrow) or in the periphery. Courtesy of Prof. Myles M. Behrens.

gait, but tremor and dysarthria are also part of the disorder. Seizures are uncommon, except in patients with hypoparathyroidism. Episodic coma has been attributed to both insensitivity of the respiratory center and diabetic acidosis (Coulter and Allen 1981; Barohn et al. 1990; Manni et al. 1991). Stroke has been reported in three cases (Zupanc et al. 1991). Symptomatic peripheral neuropathy has not been recorded and there has been no instance of spinal muscular atrophy. Myoclonus has been seen in a few patients with otherwise typical KSS.

Cardiac disorder. Among the early cases, syncope and sudden death were inordinately common.

Once the association of the neurological disorder and heartblock were recognized, however, patients were properly monitored and pacemakers were inserted to prevent these symptoms, which are now exceptions. Ophthalmoplegia usually antedates cardiac symptoms, so that monitoring is feasible. Congestive heart failure is exceptional (Uppal 1973; Kleber et al. 1987); the cardiac disorder is primarily one of the conduction system rather than a cardiomyopathy (Clark et al. 1975; Charles et al. 1981).

Endocrine disorders. The most common endocrine disorder is short stature; the exact prevalence is not known but it is not seen in all patients and the pathogenesis is not known. There have been no systematic studies of growth hormone in these patients; low levels of growth hormone were recorded in some. Similarly, delayed sexual maturation is often mentioned but the frequency and nature of this disorder is not known. Fertility seems to be diminished, but that could be the result of the restriction of social contacts because of the neurological disorders and limited mobility.

Diabetes mellitus may be seen in as many as 18% of patients (Boltshauser and Gauthier 1978). Episodic coma has been noted in a few patients and it has been suggested that the cerebral disease makes the patients more susceptible to metabolic acidosis; this could be a cause of death in those who already have pacemakers (Manni et al. 1991). In some cases (Curless et al. 1986; Feinsmith et al. 1988), administration of steroids was a precipitating factor.

Hypoparathyroidism has been described in 7 cases (Horwitz and Roessmann 1978; Pellock et al. 1978). This seems to be the major, perhaps the only, cause of seizures in KSS. Stridor and tetany are subordinate. All of the hypoparathyroid patients were mentally retarded.

Renal disorder may be manifest as the Lowe syndrome (Moraes et al. 1991) or tubular acidosis (Eviatar et al. 1990; Goto et al. 1990b).

Peripheral neuropathy has been recorded in some

cases (Shy et al. 1967; Berenberg et al. 1977; Groothuis et al. 1980; Mizusawa et al. 1991).

Among the systems that seem to be spared are skin, gastrointestinal tract, liver, and lungs. Also, there has not yet been a syndrome resembling motor neuron disease. There are no characteristic skeletal abnormalities or immunologic disorder.

Partial expression. If a disorder is defined by several clinical criteria and if the sequence of the appearance of those characteristics is not identical in all cases, it is likely that some patients will have some but not all of the requisite features. These are therefore 'incomplete' cases. As discussed by Berenberg et al. (1977) and Rowland et al. (1991), some patients may have an otherwise typical syndrome but lack retinopathy or ophthalmoplegia and then these features appear at a later date.

Histopathology

Brain. Spongy degeneration of the brain has been seen in all 20 reported autopsies of patients with typical KSS (Kearns and Sayre 1958; Jager et al. 1960; Kearns 1965; Daroff et al. 1966; Castaigne et al. 1971; Adachi et al. 1973; Schneck et al. 1973; Crosby and Chou 1974; Rowland and Berenberg 1974; Tridon et al. 1974; Azubuilke et al. 1975; McComish et al. 1976; Berenberg et al. 1977; Castaigne et al. 1977; Atsumi et al. 1980; Groothuis et al. 1980; Coulter and Allen 1981; Egger et al. 1982; Curless et al. 1986; Bresolin et al. 1987; Bordarier et al. 1990; Oldfors et al. 1990; McKelvie et al. 1991). The changes are usually widespread throughout the white matter of the cerebral hemispheres and basal ganglia. In two cases, changes were restricted to the brainstem (Castaigne et al. 1977; Groothuis et al. 1980), a pattern that may lead to misdiagnosis as Leigh's disease (Crosby and Chou 1974; Rowland and Berenberg 1974) or some other spongiform disorder (Rowland 1980; Boltshauser 1980; Goodhue et al. 1980).

Although the brain was said to have been normal in the case of Jager et al. (1960), the original microscopical slides of the brainstem were retrieved and reviewed by A.P. Hays and, as reported by Berenberg et al. (1977): 'Vacuolar changes were present and were restricted to the tegmentum, but preservation of the tissue was not adequate to exclude artifact.'

Perivascular calcification may be seen in the basal ganglia, not only in those cases with hypoparathyroidism. In two cases (Kearns and Sayre 1958; Oldfors et al. 1990), there was prominent deposition of iron in the basal ganglia.

Abnormalities of mitochondria have not been prominent in postmortem series, but were seen in a cerebellar biopsy (Adachi et al. 1973; Schneck et al. 1973). Sometimes there is loss of Purkinje cells (Castaigne et al. 1971; Bresolin et al. 1987) but there is usually no evidence of spinocerebellar tract degeneration to indicate the pathogenesis of the clinical cerebellar disorder. Castaigne et al. (1971) did find degeneration of cerebellar tracts.

The changes in the brainstem may be severe, and the oculomotor nuclei were affected prominently in the cases of Daroff et al. (1966) and Castaigne et al. (1971).

Skeletal muscle. RRF are found in almost all KSS patients but may be lacking in otherwise typical cases. There may be excessive deposition of lipid or glycogen. Electron microscopy shows that the mitochondria are enlarged, too numerous, distorted in structure and marked by arrays of paracrystalline inclusions. There are no major histological signs of myopathy and signs of denervation are rarely seen. The muscle findings do not differ from other forms of PEO (Niebroj-Dobosz et al. 1985; Mechler et al. 1986).

Ocular muscle. The findings in ocular muscle, including mitochondrial abnormalities, are similar to those in skeletal muscle PEO (Daroff et al. 1966; Castaigne et al. 1971; Mechler et al. 1986). Because of the findings in the oculomotor nuclei, it is not clear whether the ophthalmoplegia is myopathic. For instance, similar pathological changes are seen ocular muscles of patients with MELAS or MERRF, in which there is no ophthalmoplegia (Takeda et al. 1990).

Heart. Despite the prominence of the cardiac disorder, the autopsy may show no abnormality even in patients who have died abruptly, presumably of heart block and cardiac arrest (Kearns 1965; Jager et al. 1960; Daroff et al. 1966; McComish et al. 1976; Atsumi et al. 1980). A right ventricular endomyocardial biopsy in a living patient showed proliferation of typically abnormal mitochondria between myofibrils and beneath the sarcolemma (Charles et al. 1981). In an exceptional case of congestive heart failure, Hubner et al. (1986) found a diffuse mitochondrial cardiomyopathy. In general, the cardiac disorder is remarkably restricted to the conduction system as defined by physiological studies that are described below.

Retina. There have been at least 7 morphological examinations of the retina (Kearns and Sayre 1958; Jager et al. 1960; Daroff et al. 1966; Eagle et al. 1982; McKechnie et al. 1985). In the original case, Kearns and Sayre (1958) found that the retina appeared atrophic on transillumination and, microscopically, the posterior retina was adherent to the underlying choroid. Rods and cones were absent in some areas and there was clumping of pigment along the walls of small blood vessels. There were also circinate atrophic areas or giant pseudorosettes. In the case of Jager et al. (1960), however, no abnormality of the retina was discerned despite fundoscopic retinopathy, and in the case of Daroff et al. (1966), there was perimacular degeneration of rods and cones in a patient with good vision. In two cases (Eagle et al. 1982; McKechnie et al. 1985), the pigment epithelial cells were affected rather than the photoreceptors. Abnormal mitochondria have been seen in the pigment epithelium (Eagle et al. 1982; McKechnie et al. 1985; Mullie et al. 1985). As pointed out elsewhere, the case of Eagle et al. (1982) was atypical in several respects (Rowland 1983; Pavlakis et al. 1988). Sometimes, there is no anatomic abnormality even when the pigmentary change is evident clinically (McKechnie et al. 1985; Mullie et al. 1985).

Laboratory data. Routine blood and urine studies show no abnormality. High values for blood and cerebrospinal fluid (CSF) lactate and pyruvate are found in some but not all cases (Kuriyama et al. 1984; Robinson and Sherwood 1984). The lactic acidosis is never severe enough to be symptomatic.

The CSF cell count may be slightly abnormal, with no more than 20 cells. The protein content is about 100 mg/dl, even in childhood; in some of our cases there seems to have been a gradually rising value through adolescence. The significance of this characteristic abnormality is uncertain; there are no clues to the origin of the protein. The CSF gamma globulin content was increased in more than half of the cases thus studied (Berenberg et al.

1977), but there has been no tabulation of frequency of oligoclonal bands.

For reasons unknown, CSF folate content may be low (Allen et al. 1983; Laplane and Hasboun 1984). There is no evidence of malabsorption; plasma values are normal and there is no hematologic abnormality. Replacement therapy increases the CSF levels with no clinical improvement.

CT shows hypodense lesions in the white matter, corresponding to the spongiform leukoencephalopathy (Tridon et al. 1974; Azubilke et al. 1975; Castaigne et al. 1977; Bertorini et al. 1978; Seigel et al. 1979; Okamoto et al. 1981; Yoda et al. 1984; Dewhurst et al. 1986). Calcification of the basal ganglia may be seen in the cases with hypoparathyroidism, including one who had received replacement therapy from early childhood (Dewhurst et al. 1986). MRI would presumably show similar changes but has not been applied for patients with pacemakers that contain metal. The EEG shows non-specific and non-focal abnormalities.

The electrocardiographic (ECG) abnormality has been analysed in detail (Clark et al. 1975; Charles et al. 1981; Schwartzkopff et al. 1986; Gallestegui 1987). At first there is a left anterior fascicular block, followed by right bundle branch block and then complete heart block. In patients with bifascicular block in the ECG, His bundle recording may show trifascicular block, short P-R interval, and enhanced atrioventricular conduction.

Biochemistry and molecular genetics. These topics are analysed in greater detail in DiMauro's chapter on mitochondrial diseases (Ch.16, this volume). Suffice it to say here that there is no characteristic abnormality of any mitochondrial enzyme; in many cases there is subnormal activity of cytochrome oxidase, but other mitochondrial functions have been affected in some cases or there may be no biochemical abnormality (Berenberg et al. 1977; Johnson et al. 1983; Rivner et al. 1989; Tulinius et al. 1991; DiMauro, this volume, Ch.16). In patients with deletions of mtDNA, Moraes et al. (1989) found decreased activities of several enzymes of the respiratory chain that are normally encoded by mtDNA. DiMauro et al. (1991) conclude that, in conditions characterized by deletions of mtDNA, there is likely to be decreased activity of all mitochondrial enzymes that contain subunits encoded by mtDNA. Trounce et al. (1991) suggest that the large deletions may also interfere with nuclear-mitochondrial regulation of complex I, and the amount of mutated mtDNA may increase with time (Larsson et al. 1990).

Similarly, although deletions of mtDNA have been found in almost all cases studied, there has been no consistency of the site or size of the deletion and the biochemical abnormalities found in individual cases (Holt et al. 1989; Moraes et al. 1989; Gerbitz et al. 1990). Deleted mtDNA is found is tissues other than skeletal muscle (Bordarier et al. 1990; DiMauro, this volume, Ch.16). Nakase et al. (1990) found that deleted genomes are actively transcribed to mRNA but there seems to be defective translation from mRNA to protein. This is attributed to lack of indispensable mtDNA-encoded tRNAs for subunits of polypeptides in elements of the respiratory chain.

Therapy. Lacking a clear understanding of the pathogenesis of the syndrome, it has not been possible to develop a rational therapy. Ogasahara et al. (1986) gave coenzyme Q by mouth in doses of 120–150 mg daily and then reported improvement in oxidation of NADH, CSF protein content, lactate/pyruvate ratio, ECG abnormalities, and evoked potentials in median nerve. These findings have not yet been confirmed and other trials have been inconclusive (Bresolin et al. 1987; Nishikawa et al. 1989; Zierz et al. 1989; Scarlato et al. 1991). Other treatments of unproven value include folate, carnitine, and vitamin E.

Use of a pacemaker ultimately becomes necessary. Nitsch et al. (1990) advocate insertion of the pacemaker when there is evidence of bifascicular block and prolonged His-ventricular conduction.

Patients may be sensitive to barbiturates so that the required dosage may be smaller than normal in preparing for surgery (James 1985).

Genetics. In 1975, Rowland pointed out that all previously reported cases of KSS were simplex, with no other cases in the family. From then until 1988, when the deletions of mtDNA were found, there were two debates. The first argument was whether KSS was identifiable or merged with other disorders (as suggested by the term 'ophthalmoplegia plus' or by lumping all cases with RRF as mito-

chondrial encephalomyopathies). The second question concerned the question of inheritance, as repeated attempts were made to describe families with an autosomal dominant pattern.

We have provided details of our views of these contests in several reviews (Rowland 1983, 1988; Rowland et al. 1983, 1991; Pavlakis et al. 1988) and will not repeat them here. The contrary views of the lumpers have been similarly presented (Mullie et al. 1985; Petty et al. 1986; Holt et al. 1989; Truong et al. 1990). There seems to be no doubt now that KSS is a unique and identifiable disorder, confirming the views of the splitters. However, there are overlap cases that also have to be explained, as warned by the lumpers. For instance, a few cases of stroke (Zupanc et al. 1991) or myoclonus (Truong et al. 1990) have been noted in exceptional patients with KSS. How these overlapping manifestations between KSS, MELAS, and MERRF come about is a challenge for molecular biology.

The genetic question also remains a challenge. In 1975, noting that cases were sporadic, that some had had childhood meningitis or encephalitis, that CSF pleocytosis was seen in some cases, and that the characteristic brain pathology was spongiform degeneration, Rowland suggested that there might be a persistent viral infection. Aside from this circumstantial evidence, nothing ever came of that view.

In 1988, there came one of those embarrassing moments in personal history. The same issue of the journal *Neurology* contained two papers from our department. One (Zeviani et al. 1988) had been submitted 9 months earlier, reporting KSS in monozygotic twins, and suggesting that the syndrome might arise from an autosomal dominant lethal mutation, 'lethal' in the sense that affected individuals are not likely to reproduce. A second article (Rowland et al. 1988) was submitted from our department 6 months after the first but it was published as a 'rapid communication' because it recorded the then sensational news that deletions of mtDNA were characteristic of KSS; with expedited publication, it appeared in the same issue. Although seemingly contradictory, the two papers are in one sense complementary. That is, the deletions of mtDNA may arise in the ovum and if the diseases were genetically lethal because the patients do not reproduce, it would appear to be sporadic. The inheritance, however, could be maternal (or mitochondrial) rather than autosomal dominant.

There have been few women with KSS who have had children, but two were reported by DiMauro (this volume, Ch.16) and had no deletion. Similarly, no deletion has been found in a mother of an affected child with KSS. The best theory now seems to be a mutation in the ovum which then affects somatic cells in different organs, but which is not transmitted to the next generation (DiMauro, this volume, Ch.16).

This theory might explain the appearance of the disease in brothers (Schnitzler and Robertson 1979), or twins (Rowland et al. 1988). But how would it explain the father and son reported by Jankowicz et al. (1977), which cannot be attributed to maternal inheritance? Also, Desnuelle et al. (1989) described a patient with possible KSS who was the child of consanguineous parents and had a deletion of mtDNA. Another conundrum was the finding of the same mtDNA deletion in a KSS patient and two asymptomatic relatives (Poulton et al. 1991), suggesting a germ-line mutation, although previous studies of the parents of KSS patients had not shown deletions (Moraes et al. 1989).

It is striking that deletions of mtDNA seem always to cause PEO, either as a manifestation of KSS or as ocular myopathy. The exception seemed to have been Pearson syndrome, a usually fatal hematologic disorder of infancy. However, if those children live long enough, the manifestations of KSS may appear (McShane et al. 1991). However, not all juvenile Pearson patients manifest KSS (Blaw and Mize 1990).

Since PEO bears this constant relationship to deletions of mtDNA, KSS might be the full expression of a single disorder that encompasses both ocular myopathy and KSS, a view long espoused by Bastiaensen et al. (1978, 1982; Bastiaensen 1987). However, if that were true, there ought to be more families in which siblings are affected with one or the other disorder. That is, some siblings of patients with KSS ought to have ocular myopathy, but that has not been recorded.

This brief summary delineates the challenges to modern clinical investigation:

1. How do the deletions of mtDNA lead to the clinical syndrome? What is the affected gene product? How does an abnormality of a single gene product lead to a multisystem disease? What determines which different organ systems are involved or spared? How does the same deletion cause KSS in one person and ocular myopathy in another?
2. Why, in fact are so many cases sporadic and so few familial?
3. Why do point mutations in mtDNA lead to the syndromes of MELAS and MERRF? And why are these syndromes different from KSS? Why are there so few patients with overlapping features of the three syndromes?

Peripheral and cranial neuropathy with ophthalmoplegia

As we have seen, sensorimotor polyneuropathy, usually demyelinating, has been encountered in a few cases of KSS or ocular myopathy. PEO is also part of the syndrome of a-beta-lipoproteinemia (Schwartz et al. 1963) and may be the result of cranial neuropathy. Telerman-Toppet et al. (1969) described an adult with celiac disease, PEO, slow conductions and deafness but no retinopathy. Rarely, PEO may precede other manifestations of a chronic relapsing polyneuropathy (Brunet et al. 1986; Piccolo and Martino 1988). However, the most impressive syndrome that incorporates PEO and sensorimotor peripheral neuropathy is called MNGIE, POLIP, OSPOM (Table 1) or MEPOP, acronyms that are explained in the next section.

Mitochondrial myopathy, neuropathy, ophthalmoplegia, and gastrointestinal pseudo-obstruction. In 1983, there were two reports of patients with the unusual combination of PEO and a bowel disorder that was manifest by vomiting and diarrhea. In the propositus of the family described by Ionasescu et al. (1983) and Anuras et al. (1983), the ophthalmoplegia started at age 5; the bowel disorder first appeared at age 20. There was no evidence of malabsorption but there was a disorder of motility that was consistent with 'pseudo-obstruction' and at autopsy there was no evidence of autonomic neuropathy. Both muscle biopsy and EMG gave evidence of myopathy and mitochondria appeared to be normal. There was consanguinity in two generations and the inheritance was assumed to be autosomal recessive on the basis of the 4 affected individuals. Two of the patients were the children of an affected woman and the fourth was the daughter of an unaffected man, possibly excluding maternal inheritance. The authors called this 'oculogastrointestinal muscular dystrophy'. Anuras (1988) and Chokhavatia and Anuras (1991) classified this as familial visceral myopathy type 2, to differentiate from two types of gastromotility disorder without PEO.

Cervera et al. (1988) described a woman who had gastrointestinal symptoms starting in early childhood and ophthalmoplegia started at age 23; at age 28, hearing loss became apparent. There was no retinopathy. RRF were seen in muscle biopsy and CSF protein content was normal. No other family members were known to have been affected. Gastrointestinal studies indicated esophageal aperistalsis, gastroparesis with delayed emptying, megaduodenum, and diffuse dilatation of the rest of the intestine.

Steiner et al. (1987) described three affected siblings of one family and the propositus and maternal cousin in another family. Affected individuals showed PEO, bowel disease with malnutrition and severe sensorimotor peripheral neuropathy. No comment was made about retinopathy, hearing loss, RRF or ECG. Motor nerve conductions were slow and CSF protein content was 74 in one patient and 110 in another. Two patients died but there were no autopsies.

Bardosi et al. (1987) described an affected woman, daughter of consanguineous parents but without other family information. The onset of bowel symptoms was not explicitly stated but she 'never weighed more than 40 kg' and, at age 32, she had a bowel resection for 'idiopathic peritonitis'. PEO commenced at age 37. She had a steppage gait and had paresthesias. Nerve conduction studies showed slowing and CSF protein content was 62 mg/dl, but EMG was said to have been normal. CT showed hypodensity of cerebral white matter. No comment was made about retina, ECG, or hearing. Lactic acidosis was induced administration of glucose. She died at age 42. At autopsy, RRF were found in skeletal and ocular muscles; mitochondrial inclusions were seen on electron microscopy and biochemical studies showed a partial

defect of COX. There was evidence of a demyelinating neuropathy. The authors invented a new acronym, MNGIE syndrome, for myo-, neuro-, gastrointestinal encephalopathy.

The same acronym was used by Blake et al. (1990) in describing an affected brother and sister with the same clinical syndrome. Both had diarrhea and malnutrition from early childhood, with PEO commencing after age 20. Both were short and had severe hearing loss. Blood and CSF lactate values were increased. Motor and sensory nerve conductions were slow. CSF protein content was 63 mg/dl in the man. The EMG gave a mixed myopathic pattern in proximal muscles and denervation in distal muscle. MRI showed decreased signal in cerebral white matter. Muscle biopsies showed RRF and scattered fibers devoid of COX as well as features of denervation. Nerve biopsy showed features of both axonal and demyelinating neuropathy. Biochemical analysis showed partial defect of COX; the activity was about half normal, but immunological studies showed that the protein itself was normal. There were no deletions of mtDNA.

Blake et al. (1990) described a third patient, unrelated to these siblings.

Despite the triplet of consecutive consonants, the acronym seemed to be euphonious when spoken as 'min-gee', but it was soon challenged by a new one, POLIP, for polyneuropathy, ophthalmoplegia, leukoencephalopathy, and intestinal pseudo-obstruction, the term used by Simon et al. (1990). They described 5 affected people in 3 different families. There was no consanguinity. In each of two families, a pair of siblings was affected. The last was a simplex case. None of the parents was affected, so the pattern of inheritance was moot. The neuropathy clinically affected sensory as well motor fibers. Three of the five had hearing loss. There were no comments about retinopathy, ECG, or RRF. The CSF protein content was 56 in one patient, 70 in another, and normal in two. Nerve conductions were slow in all four studied.

In one autopsy, the oculomotor nerve showed thin myelin sheaths and endoneurial fibrosis. There was pallor of central white matter in both myelin and axonal preparations and there was vacuolation in spinal cord. Ventral and dorsal roots were affected as well as peripheral nerves, which showed massive loss of axons. There was also evidence of autonomic neuropathy.

Another case was reported in a clinico-pathological conference (Case Records of the MGH, 1990), with PEO, pseudo-obstruction, and sensorimotor neuropathy. Again, there were no other cases in the family and the possibility of mitochondrial disorder was not considered but RRF were presumably absent.

I have personally seen another patient. Limb weakness began at age 10 in 1968. At first we thought she had Kugelberg-Welander syndrome because sensation, CSF protein content, and nerve conduction studies were all normal. Later in adolescence, however, she developed ophthalmoplegia, slow conductions, severe aperistalsis of the esophagus and intestines, and intermittent tachycardia. She died at age 24 and postmortem tissues were examined by A.P. Hays. There were no RRF and the postmortem findings were consistent with a demyelinating cranial, peripheral, and autonomic neuropathy.

Although the combination of PEO and pseudo-obstruction seems to be so unique that a specific syndrome is suspected, the several reported families have not been studied in a uniform way. There seems to be a combination of sensorimotor demyelinating neuropathy and mitochondrial myopathy. Mitochondrial disorder was not specifically evaluated in cases, but the inheritance pattern was consistent with maternal inheritance in all but one family. There have been no studies of mtDNA in any of the families other than that of Blake et al. (1990) and there was no evidence of mutation or deletion in that one. Whether this syndrome has more than one etiology or pathogenesis remains to be determined. Leukoencephalopathy and hearing loss in some of the cases seem compatible with mitochondrial disorder. At any rate, it differs from the combination of PEO and diarrhea seen in a-beta-lipoproteinemia because the neurological complications of that syndrome are due to malabsorption, which is not seen in the pseudo-obstruction syndrome.

This new syndrome has appeared without a satisfactory name. Two acronyms have been proposed, but neither includes the ophthalmoplegia or the mitochondrial disorder. Without intending to add to already existing chaos, a more satisfactory acronym might be MEPOP, for *mitochondrial encephalomyopathy*, sensorimotor *polyneuropathy*, *ophthalmoplegia*, and *pseudo-obstruction*.

REFERENCES

AASLY, J., S. LINDAL, T. TORBERGSEN, O. BORUD and S.I. MELLGREN: Early mitochondrial changes in chronic progressive ocular myopathy. Eur. Neurol. 30 (1990) 314–318.

ABERFELD, D.C. and T. NAMBA: Progressive ophthalmoplegia in Kugelberg–Welander disease. Arch. Neurol. (Chic.) 20 (1969) 253–256.

ABDUL RAHIM, A.S., P.J. SAVINO, R.A. ZIMMERMAN, R.C. SERGOTT and T.M. BOSLEY: Cryptogenic oculomotor nerve palsy: the need for repeated neuroimaging studies. Arch. Ophthalmol. 107 (1989) 387–390.

ABRAMOVITZ, J.N., D.L. KASDON, F. SUTULA, K.D. POST and F.K. CHONG: Sclerosing orbital pseudotumor. Neurosurgery 12 (1983) 463–468.

ADACHI, M., J. TORII, B.W. VOLK, P. BRIET, A. WOLINTZ and L. SCHNECK: Electron microscopic and enzyme histochemical studies of cerebellum, ocular and skeletal muscles in chronic progressive ophthalmoplegia with cerebellar ataxia. Acta Neuropathol. 23 (1973) 300–312.

ALFARO, A., A. GILSANZ, M.A. CERVELLO, M.A. ANTOLIN, C. ALBEROLA and C. VILLOSLADA: [Chronic ophthalmoplegia in a case of myasthenia gravis, hyperthyroidism and thymolipoma]. Med. Clin. 79 (1982) 236–239.

ALLEN, R.J., S. DIMAURO, D.L. COULTER, A. PAPADIMITRIOU and S.P. ROTHENBERG: Kearns-Sayre syndrome with reduced plasma and cerebrospinal fluid folate. Ann. Neurol. 13 (1983) 679–682.

ALONZO, B., G. BOUDOURESQUES, A. ALI CHERIF, J.F. PELLISSIER and R. KHALIL: [Progressive external ophthalmoplegia with involvement of the central nervous system: nosological discussion apropos of 2 cases]. Rev. Otoneuroophthalmol. 56 (1984) 191–197.

AMBROSETTO, P., G. MESSEROTTI and A. BACCI: Unruptured intracavernous carotid artery aneurysm mimicking 'painful ophthalmoplegia': case report. Ital. J. Neurol. Sci. 7 (1986) 609–611.

AMYOT, R.: Hereditary, familial and acquired ptosis of late onset. Can. Med. Assoc. J. 59 (1948) 434–438.

ANURAS. S.: Intestinal pseudo-obstruction syndrome. Annu. Rev. Med. 39 (1988) 1–15.

ANURAS, S., F.A. MITROS, T.V. NOWAK, V.V. IONASESCU, N.J. GURLL, J. CHRISTENSEN and J.B. GREEN: A familial visceral myopathy with external ophthalmoplegia and autosomal recessive transmission. Gastroenterology 84 (1983) 346–353.

ATLAS, S.W., R.I. GROSSMAN, P.J. SAVINO, R.C. SERGOTT, N.J. SCHATZ, T.M. BOSLEY, D.B. HACKNEY, H.I. GOLDBERG, L.T. BILLANIUK and R.A. ZIMMERMAN: Surfacecoil MR of orbital pseudotumor. Am. J. Roentgenol. 148 (1987) 803–808.

ATSUMI, T., Y. YAMAMURA, T. SATO, and F. IKUTA: Hirano bodies in the axon of peripheral nerves in a case with progressive external ophthalmoplegia with multisystemic involvements. Acta. Neuropathol. 49 (1980) 95–100.

AUGUSTIN, P., M. DUJARDIN, N. DALUZEAU and P. DENIS: [Ophthalmoplegia plus with optic atrophy: apropos of a case]. Rev. Otoneuroophthalmol. 56 (1984) 183–189.

AZUBUIKE, J.C., F. GULLOTTA, H.C. KALLFELZ, K. GELLISSEN, S. MENDE and R. EXSS: [Juvenile spongy dystrophy of CNS with necrosis of the medulla. A. Complication of hydroxyquinoline therapy (author's transl)]. Neuropädiatrie 6 (1975) 292–306.

BABEL, J.: [The retinopathy of ophthalmoplegia plus (Kearn's syndrome)]. J. Fr. Ophthalmol. 5 (1982) 601–608.

BACH, M.C., M. KNOWLAND and W.B. SCHUYLER: Acute orbital myositis mimicking orbital cellulitis. Ann. Intern. Med. 109 (1988) 243–245.

BAHN, R.S., J.A. GARRITY, G.B. BARTLEY and C.A. GORMAN: Diagnostic evaluation of Graves' ophthalmopathy. Endocrinol. Metab. Clin. North Am. 17 (1988) 527–545.

BARBEAU, A.: The syndrome of hereditary late-onset ptosis and dysphagia in French Canada. In:. E. Kuhn (Ed.), Progressive Muskeldystrophie, Myotonie, Myasthenie. Berlin, Springer Verlag (1966) 102–109.

BARDOSI, A., W. CREUTZFELDT, S. DIMAURO, K. FELGENHAUER, R.L. FRIEDE, H.H. GOEBEL, A. KOHLSCHUTTER, G. MAYER, G. RAHLF and S. SERVIDEI: Myo-, neuro-, gastrointestinal encephalopathy (MNGIE syndrome) due to partial deficiency of cytochrome-c-oxidase: a new mitochondrial multisystem disorder. Acta Neuropathol. 74 (1987) 248–258.

BAROHN, R.J., T. CLANTON, Z. SAHENK and J.R. MENDELL: Recurrent respiratory insufficiency and depressed ventilatory drive complicating mitochondrial myopathies. Neurology 40 (1990) 103–106.

BARONTINI, F., S. MAURRI and P. MARINI: Total extrinsic ophthalmoplegia as only paraneoplastic sign two years before X-ray diagnosis of bronchial carcinoma. Ital. J. Neurol. Sci. 6 (1985) 441–445.

BARONTINI, F., S. MAURRI and E. MARRAPODI: Tolosa-Hunt syndrome versus recurrent cranial neuropathy: report of two cases with a prolonged follow-up. J. Neurol. 234 (1987) 112–115.

BARRICKS, M.E., D.B. TRAVIESA, J.S. GLASER and I.S. LEVY: Ophthalmoplegia in cranial arteritis. Brain 100 (1977) 209–221.

BARTALENA, L., C. MAROCCI, F. BOGAZZI, M. PANICUCCI, A. LEPRI and A. PINCHERA: Use of corticosteroids to prevent progression of Graves' ophthalmopathy after radioiodine therapy for hyperthyroidism. N. Engl. J. Med. 321 (1989) 1349–1352.

BASTIAENSEN, L.A.K.: Pigment changes of the retina in chronic progressive external ophthalmoplegia (CPEO). Acta Ophthalmol. Suppl. 138 (1978) 5–36.

BASTIAENSEN, L.A.K.: Ocular myopathies: syndromes with chronic progressive external ophthalmoplegia (CPEO). In: E.A.C.M. Sanders, R.J.W. de Keizer

and D.S. Zee (Eds.), Eye Movement Disorders. Dordrecht, Nijhoff/Junk (1987) 131–142.

BASTIAENSEN, L.A.K., H.H.F. JASPAR, A.M. STADHOUDERS, G.J.M. EGBERINK and J.J. KORTEN: Chronic progressive external ophthalmoplegia in a heredo-ataxia: neurogenic or myogenic? A clinical, neuropathological and submicroscopic study. Acta Neurol. Scand. 56 (1977) 483–507.

BASTIAENSEN, L.A.K., E.M. JOOSTEN, J.A. DE ROOIJ, O.R. HOMMES, A.M. STADHOUDERS, H.H. JASPAR, J.H. VEERKAMP, H. BOOKELMAN and V.W. VAN HINSBERGH: Ophthalmoplegia-plus, a real nosological entity. Acta Neurol. Scand. 58 (1978) 9–34.

BASTIAENSEN, L.A.K., S.L. NOTERMANS, C.H. RAMAEKERS, B.J. VAN DIJKE, E.M. JOOSTEN, H.H. JASPAR, A.M. STADHOUDERS and C.T. BELJAARS: Kearns syndrome or Kearns disease: further evidence of a genuine entity in a case with uncommon features. Ophthalmologia 184 (1982) 40–50.

BEAL, M.F.: Amitriptyline ophthalmoplegia. Neurology 32 (1982) 1409.

BEAUMONT, W.M.: Notes of a case of progressive nuclear ophthalmoplegia. Brain 13 (1890) 386–387.

BECKETT, R.S. and M.G. NETSKY: Familial ocular myopathy and external ophthalmoplegia. Arch. Neurol. Psychiatry 69 (1953) 64–72.

BENDER, A.N. and M.B. BENDER: Muscle fiber hypotrophy with intact neuromuscular junctions: a study of a patient with congenital neuromuscular disease and ophthalmoplegia. Neurology 27 (1977) 206–212.

BERDJIS, H., W. HEIDER and K. DEMISCH: ERG and VECP in chronic progressive external ophthalmoplegia (CPEO). Doc. Ophthalmol. 60 (1985) 427–434.

BERENBERG, R.A., J.M. PELLOCK, S. DIMAURO, D.L. SCHOTLAND, E. BONILLA, A. EASTWOOD, A. HAYS, C.T. VICALE, M. BEHRENS, A. CHUTORIAN and L.P. ROWLAND: Lumping or splitting? 'Ophthalmoplegia-plus' or Kearns-Sayre syndrome? Ann. Neurol. 1 (1977) 37–54.

BERGEN, B.J., M.P. CARRY, W.B. WILSON, M.T. BARDEN and S.P. RINGEL: Centronuclear myopathy: extraocular- and limb-muscle findings in an adult. Muscle Nerve 3 (1980) 165–171.

BERTORINI, T., W.K. ENGEL, G. DI CHIRO and M. DALAKAS: Leukoencephalopathy in oculocraniosomatic neuromuscular disease with ragged-red fibers: mitochondrial abnormalities demonstrated by computerized tomography. Arch. Neurol. 35 (1978) 643–647.

BIXENMAN, W.W. and H.W. BUCHSBAUM: Multiple sclerosis, euthyroid restrictive Graves' ophthalmopathy, and myasthenia gravis. A case report. Graefes. Arch. Clin. Exp. Ophthalmol. 226 (1988) 168–171.

BIXENMAN, W.W. and G.K. VON NOORDEN: Atypical presentation of restrictive orbital myositis. J. Pediatr. Ophthalmol. Strabismus. 18 (1981) 6–10.

BLACK, J.T., K.A. BRAIT, P.V. DEJESUS JR., R.N. HARNER and L.P. ROWLAND: Myasthenia gravis lacking response to cholinergic drugs. Neurology 23 (1973) 851–853.

BLAKE, D., A. LOMBES, C. MINETTI, E. BONILLA, A. HAYS, R.E. LOVELACE, I. BUTLER and S. DIMAURO: Mingie syndrome: report of 2 new patients. Neurology 40 (Suppl 1) (1990) 294 (Abstract).

BLAW, M.E. and C.E. MIZE: Juvenile Pearson syndrome. J. Child. Neurol. 5 (1990) 187–190.

BLUMENTHAL, H. and C. MILLER: Motor nuclear involvement in progressive supranuclear palsy. Arch. Neurol. 20 (1969) 362–367.

BOLTSHAUSER, E.: Juvenile form of spongy degeneration: an instance of Kearns-Sayre syndrome? [letter]. Arch. Neurol. 37 (1980) 256.

BOLTSHAUSER, E. and G. GAUTHIER: Diabetes mellitus in Kearns-Sayre syndrome. Am. J. Dis. Child. 132 (1978) 321–322.

BORDARIER, C., C. DUYCKAERTS, O. ROBAIN, G. PONSOT and D. LAPLANE: Kearns-Sayre syndrome: two clinico-pathological cases. Neuropediatrics 21 (1990) 106–109.

BOSCH, E.P., J.D. GOWANS and T. MUNSAT: Inflammatory myopathy in oculopharyngeal dystrophy. Muscle Nerve 2 (1979) 73–77.

BOUCHARD, J.P., F. GAGNE, F.M. TOME and D. BRUNET: Nuclear inclusions in oculopharyngeal muscular dystrophy in Quebec. Can J. Neurol. Sci. 16 (1989) 446–450.

BRADBURNE, A.A.: Hereditary ophthalmoplegia in five generations. Trans. Ophthalmol. Soc. U.K. 32 (1912) 142–153.

BRAY, G.M., M. KAARSOO and R.T. ROSS: Ocular myopathy with dysphagia. Neurology 15 (1965) 678–684.

BRESOLIN, N., M. MOGGIO, L. BET, A. GALLANTI, A. PRELLE, E. NOBILE ORAZIO, L. ADOBBATI, C. FERRANTE, G. PELLEGRINI and G. SCARLATO: Progressive cytochrome c oxidase deficiency in a case of Kearns-Sayre syndrome: morphological, immunological, and biochemical studies in muscle biopsies and autopsy tissues. Ann. Neurol. 21 (1987) 564–572.

BRIL, V., N.B. REWCASTLE and J. HUMPHREY: Oculoskeletal myopathy with abnormal mitochondria. Can. J. Neurol. Sci. 11 (1984) 390–394.

BRION, S. and J. DE RECONDO: [Progressive nulear ophthalmoplegia and hereditary spinal cerebellar degeneration: study of 1 anatomo-clinical case]. Rev. Neurol. 116 (1967) 383–400.

BROWN, H.W.: Congenital structural muscle anomalies. In: J.H. Allen (Ed.), Ophthalmic Strabismus Symposium. St. Louis, C.V. Mosby (1950) 205–236.

BROWN, S.: On hereditary ataxy, with a series of twenty-one cases. Brain 15 (1892) 250–282.

BROWN, J.W. and R.F. COLEMAN: Hereditary spastic paraplegia with ocular and extra-pyramidal signs. (A clinical pathologic study of a family.) Bull. Los Angeles. Neurol. Soc. 31 (1966) 21–34.

BRUNET, P., F. VIADER and D. HENIN: [Conference at La Salpétrière. January 1985. Encephalopathy of rapid development with supranuclear ophthalmo-

plegia and peripheral neuropathy]. Rev. Neurol. 142 (1986) 159–166.

BRUNET, G., F.M. TOME, F. SAMSON, J.M. ROBERT and M. FARDEAU: [Oculopharyngeal muscular dystrophy: a census of French families and genealogic study]. Rev. Neurol. 146 (1990) 425–429.

BRUYN, G.W. and M.J.A.J.M. HOES: The Tolosa-Hunt syndrome. In: P.J. Vinken, G.W. Bruyn and H.L. Klawans (Eds.), Handbook of Clinical Neurology, Vol. 48. Amsterdam, Elsevier Science Publishers (1986) 291–307.

BULLEN, C.L. and B.R. YOUNGE: Chronic orbital myositis. Arch. Ophthalmol. 100 (1982) 1749–1751.

BULLEN, C.L., T.J. LIESEGANG, T.J. MCDONALD and R.A. DEREMEE: Ocular complications of Wegener's granulomatosis. Ophthalmology 90 (1983) 279–290.

CALZETTI, S., F. GEMIGNANI, A. MARBINI, M. SAVI and M.M. BRAGAGLIA: Immunological abnormalities in a family with progressive external ophthalmoplegia. J. Neurol. Sci. 61 (1983) 13–20.

CAMBIER, J., R. ESCOUROLLE, M. MASSON and B. LECHEVALIER: Progressive supranuclear paralysis. Anatomical and clinical observation. Rev. Neurol. (Paris) 121 (1969) 139–154.

CAMPBELL, R.J. and H. OKAZAKI: Painful ophthalmoplegia (Tolosa-Hunt variant): autopsy findings in a patient with necrotizing intracavernous carotid vasculitis and inflammatory disease of the orbit. Mayo Clin. Proc. 62 (1987) 520–526.

CANTELLO, R., L. BERGAMINI, W. TRONI, A. RICCIO, I. CHIADO, L. PALMUCCI and M. DE MARCHI: Familial progressive external ophthalmoplegia with multisystem abnormalities: 'new' features raising nosological problems. J. Neurol. 232 (1985) 102–108.

Case records of the Massachusetts General Hospital. Weekly Clinicopathological exercises. Case 12-1990. A 21-year-old man with progressive gastrointestinal statis, hepatomegaly, and a neurologic disorder. N. Engl. J. Med. 322 (1990) 829–841.

CASSAN, S.M., M.B. DIVERTIE, R.W. HOLLENHORST and E.G. HARRISON JR.: Pseudotumor of the orbit and limited Wegener's granulomatosis. Ann. Intern. Med. 72 (1970) 687–693.

CASTAIGNE, P., D. LAPLANE, R. ESCOUROLLE, P. AUGUSTIN, J. RECONDO, G.J. MARTINEZ LAGE and J.A. VILLANUEVA EUSA: [Progressive external ophthalmoplegia with brainstem nuclei spongiosis]. Rev. Neurol. 124 (1971) 454–466.

CASTAIGNE, P., F. LHERMITTE, R. ESCOUROLLE, F. CHAIN, M. FARDEAU, J.J. HAUW, J. CURET and C. FLAVIGNY: [Anatomo-clinical study of a case of 'ophthalmoplegia plus' with analysis of muscular, central nervous, ocular, myocardial, and thyroid lesions]. Rev. Neurol. 133 (1977) 369–386.

CERVERA, R., J. BRUIX, A. BAYES, R. BLESA, I. ILLA, J. COLL and A.M. GARCIA PUGES: Chronic intestinal pseudoobstruction and ophthalmoplegia in a patient with mitochondrial myopathy. Gut 29 (1988) 544–547.

CHAMLIN, M. and E. BILLET: Ophthalmoplegia and pigmentary degeneration of the retina. Arch. Ophthalmol. 43 (1950) 217–223.

CHANMUGAM, D. and A.F. HAERER: Progressive external ophthalmoplegia of early onset followed by generalized myopathy in a Ceylonese family. J. Neurol. Sci. 10 (1970) 101–105.

CHARLES, R., S. HOLT, J.M. KAY, E.J. EPSTEIN and J.R. REES: Myocardial ultrastructure and the development of artrioventricular block in Kearns-Sayre syndrome. Circulation 63 (1981) 214–219.

CHOKHAVATIA, S. and S. ANURAS: Neuromuscular disease of the gastrointestinal tract. Am. J. Med. Sci. 301 (1991) 201–214.

CIBIS, G.W.: Congenital familial external ophthalmoplegia with co-contraction. Ophthalmic Paediatr. Genet. 4 (1984) 163–167.

CLARK, D.S., R.J. MYERBURG, A.R. MORALES, B. BEFELER, F.A. HERNANDEZ and H. GELBAND: Heart block in Kearns-Sayre syndrome: electrophysiologic-pathologic correlation. Chest 68 (1975) 727–730.

COGAN, D.G., T. KUWABARA and E.P. RICHARDSON JR.: Pathology or abiotrophic ophthalmoplegia externa. Bull. Johns Hopkins Hosp. 111 (1962) 42–56.

COLLIER, J.: Nuclear ophthalmoplegia, with special reference to retraction of the lids and ptosis, and to lesions of the posterior commissure. Brain 50 (1927) 488–498.

COLLIER, J.R.: Involutional ptosis. Aust. N.Z.J. Ophthalmol. 14 (1986) 109–112.

COLOMBO, A., E. MERELLI, P. SOLA, P. PANZETTI, D. QUAGLINO JR. and C. FORNIERI: Mitochondrial oculoskeletal myopathy: case report. Ital. J. Neurol. Sci. 9 (1988) 385–389.

COOLES, P., R. MICHAUD and P.V. BEST: A dominantly inherited progressive disease in a black family characterised by cerebellar and retinal degeneration, external ophthalmoplegia and abnormal mitochondria. J. Neurol. Sci. 87 (1988) 275–288.

COQUET, M., J.M. VALLAT, C. VITAL, M. FOURNIER, M. BARAT, J.M. ORGOGOZO, J. JULIEN and P. LOISEAU: Nuclear inclusions in oculopharyngeal dystrophy: an ultrastructural study of six cases. J. Neurol. Sci. 60 (1983) 151–156.

COQUET, M., C. VITAL and J. JULIEN: Presence of inclusion body myositis-like filaments in oculopharyngeal muscular dystrophy: ultrastructural study of 10 cases. Neuropathol. Appl. Neurobiol. 16 (1990) 393–400.

COULTER, D.L. and R.J. ALLEN: Abrupt neurological deterioration in children with Kearns-Sayre syndrome. Arch. Neurol. 38 (1981) 247–250.

COUTINHO, P., A. GUIMARAES and F. SCARAVILLI: The pathology of Machado-Joseph disease: report of a possible homozygous case. Acta Neuropathol. 58 (1982) 48–54.

CRAWFORD, J.S.: Congenital fibrosis syndrome. Can. J. Ophthalmol. 5 (1970) 331–336.

CROSBY, T.W. and S.M. CHOU: 'Ragged-red' fibers in Leigh's disease. Neurology 24 (1974) 49–54.

CURLESS, R.G., J. FLYNN, B. BACHYNSKI, J.B. GREGORIOS, P. BENKE and R. CULLEN: Fatal metabolic acidosis, hyperglycemia, and coma after steroid therapy for Kearns-Sayre syndrome. Neurology 36 (1986) 872–873.

CURRIE, S.: Familial oculomotor palsy with Bell's palsy. Brain 93 (1970) 193–198.

CURTIN, H.D.: Pseudotumor. Radiol. Clin. North Am. 25 (1987) 583–599.

CZERNOBILSKY, H. and R. ZIEGLER: Graves' ophthalmopathy, ocular myasthenia gravis and Hashimoto's thyroiditis. Isr. J. Med. Sci. 21 (1985) 377–380.

DAROFF, R.B.: Chronic progressive external ophthalmoplegia: a critical review. Arch. Ophthalmol. 82 (1969) 845–850.

DAROFF, R.B. and W.F. HOYT: Supranuclear disorders of ocular control systems in man: clinical, anatomical and physiological correlations. In: P. Bach-y Rita and C.C. Collins (Eds.), The control of Eye Movements. New York-London, Academic Press (1971) 175–236.

DAROFF, R.B., G.B. SOLITARE, J.H. PINCUS and G.H. GLASER: Spongiform encephalopathy with chronic progressive external ophthalmoplegia: central ophthalmoplegia mimicking ocular myopathy. Neurology 16 (1966) 161–169.

DAVIS, J.C., J.A. REIFFEL, M. BEHRENS, L. ROWLAND, R. MASCITELLI and A. SEPLOWITZ: Optic neuritis and heart block in Kearns-Sayre syndrome. N.Y. State J. Med. 81 (1981) 1364–1368.

DEADY, J.P., A.J. MORRELL and G.A. SUTTON: Recognising aponeurotic ptosis. J. Neurol. Neurosurg. Psychiatry 52 (1989) 996–998.

DEGOUL, F., I. NELSON, P. LESTIENNE, D. FRANCOIS, N. ROMERO, D. DUBOC, B. EYMARD, M. FARDEAU, G. PONSOT and M. PATURNEAU JOUAS: Deletions of mitochondrial DNA in Kearns-Sayre syndrome and ocular myopathies: genetic, biochemical and morphological studies. J. Neurol. Sci. 101 (1991) 168–177.

DEMANGE, P., H.P. GIA, G. KALIFA and N. SELLIER: MR of Kearns-Sayre syndrome. AJNR 10 (1989) S91.

DESNUELLE, C., J.F. PELLISSIER, G. SERRATRICE, J. POUGET and D.M. TURNBULL: [Kearns-Sayre syndrome: mitochondrial encephalomyopathy caused by deficiency of the respiratory chain]. Rev. Neurol. 145 (1989) 842–850.

DE WAARD, R., L. KOORNNEEF and B. VERBEETEN JR.: Motility disturbances in Graves' ophthalmopathy. Doc. Ophthalmol. 56 (1983) 41–47.

DEWHURST, A.G., D. HALL, M.S. SCHWARTZ and R.O. MCKERAN: Kearns-Sayre syndrome, hypoparathyroidism, and basal ganglia calcification. J. Neurol. Neurosurg. Psychiatry 49 (1986) 1323–1324.

DIMANT, J., D. GROB and N.G. BRUNNER: Ophthalmoplegia, ptosis, and miosis in temporal arteritis. Neurology 30 (1980) 1054–1058.

DIMAURO, S., D.L. SCHOTLAND and L.P. ROWLAND: Ocular myopathy, glycogen storage, and abnormal mitochondria. Neurology 21 (1973) 412 (Abstract).

DIMAURO, S., C.T. MORAES, S. SHASSKE, A. LOMBES, H.

NAKASE, S. MITA, H.J. TRITSCHLER, E. BONILLA, A.F. MIRANDA and E.A. SCHON: Mitochondrial encephalomyopathies: biochemical approach. Rev. Neurol. (Paris) 147 (1991) 443–449.

DIRR, L.Y., P.D. DONOFRIO, J.F. PATTON and B.T. TROOST: A false-positive edrophonium test in a patient with a brainstem glioma. Neurology 39 (1989) 865–867.

DORTZBACH, R.K. and F.C. SUTULA: Involutional blepharoptosis: a histopathological study. Arch. Ophthalmol. 98 (1980) 2045–2049.

DRACHMAN, D.A.: Ophthalmoplegia plus: the neurodegenerative disorders associated with progressive external ophthalmoplegia. Arch. Neurol. 18 (1968) 654–674.

DRACHMAN, D.A.: Ophthalmoplegia plus: a classification of the disorders associated with progressive external ophthalmoplegia. In: P.J. Vinken and G.W. Bruyn (Eds.), Handbook of Clinical Neurology, Vol. 22. Amsterdam, North-Holland (1975) 203–216.

DRACHMAN, D.A., N. WETZEL, M. WASSERMAN and H. NAITO: Experimental denervation of ocular muscles: a critique of the concept of 'ocular myopathy'. Arch. Neurol. 21 (1969) 170–183.

DRACHMAN, D.B., S.R. MURPHY, M.P. NIGAM and J.R. HILLS: 'Myopathic' changes in chronically denervated muscle. Arch. Neurol. 16 (1967) 14–24.

DRESNER, S.C. and J.S. KENNERDELL: Dysthyroid orbitopathy. Neurology 35 (1985) 1628–1634.

DRESNER, S.C., W.E. ROTHFUS, T.L. SLAMOVITS, J.S. KENNERDELL and H.D. CURTIN: Computed tomography of orbital myositis. Am. J. Roentgenol. 143 (1984) 671–674.

DRIESSEN-KLETTER, M.F., P.R. BAR, H.R. SCHOLTE, T.U. HOOGENRAAD and I.E.M. LUYT-HOUWEN: Striking correlation between muscle damage after exercise and mitochondrial dysfunction in patients with chronic external ophthalmoplegia. J. Inherited Metab. Dis. 10 Suppl. 2 (1987) 252–255.

DUBROVSKY, A., A.L. TARATUTO and R. MARTINO: Distal spinal muscular atrophy and ophthalmoparesis: a case with selective type 2 fiber hypotrophy. Arch. Neurol. 38 (1981) 594–596.

EAGLE, JR. R.C., T.R. HEDGES and M. YANOFF: The atypical pigmentary retinopathy of Kearns-Sayre syndrome: a light and electron microscopic study. Ophthalmology 89 (1982) 1433–1440.

EGGER, J., C.J. WYNNE WILLIAMS and M. ERDOHAZI: Mitochondrial cytopathy or Leigh's syndrome? Mitochondrial abnormalities in spongiform encephalopathies. Neuropediatrics 13 (1982) 219–224.

ELLIOTT, F.A.: Myopathic wasting associated with ptosis and external ophthalmoplegia. Proc. R. Soc. Med. 32 (1939) 876 (Abstract).

ENZMANN, D., S.S. DONALDSON, W.H. MARSHALL and J.P. KRISS: Computed tomography in orbital pseudotumor (idiopathic orbital inflammation). Radiology 120 (1976) 597–601.

ENZMANN, D.R., S.S. DONALDSON and J.P. KRISS: Appearance of Graves' disease on orbital computed tomography. J. Comput. Assist. Tomogr. 3 (1979) 815–819.

ERDBRINK, W.L.: Ocular myopathy with retinitis pigmentosa. Arch. Ophthalmol. 57 (1957) 335–338.

EVIATAR, L., S. SHANSKE, B. GAUTHIER, C. ABRAMS, J. MAYTAL, M. SLAVIN, E. VALDERRAMA and S. DIMAURO: Kearns-Sayre syndrome presenting as renal tubular acidosis. Neurology 40 (1990) 1761–1763.

FAULKNER, S.H.: Familial ptosis with ophthalmoplegia externa starting in adult life. Br. Med. J. 2 (1939) 854.

FEARNLEY, J.M., T. REVESZ, D.J. BROOKS, R.S. FRACKOWIAK and A.J. LEES: Diffuse Lewy body disease presenting with a supranuclear gaze palsy. J. Neurol. Neurosurg. Psychiatry 54 (1991) 159–161.

FEINSMITH, B.M., W.P. LIEBESMAN and P. GUIBOR: Steroid danger in Kearns-Sayre syndrome (KSS). N.J. Med. 85 (1988) 659–663.

FERGUSON, F.R. and M. CRITCHLEY: A clinical study of an heredo-familial disease resembling disseminated sclerosis. Brain 52 (1929) 203–225.

FETELL, M.R., H.S. SHIN, A.S. PERIN, R.E. LOVELACE and L.P. ROWLAND: Combined Eaton-Lambert syndrome and myasthenia gravis. Neurology 28 (1978) 398.

FISHER, C.M.: Ocular palsy in temporal arteritis. Minn. Med. 42 (1959) 1258–1268.

FLANDERS, A.E., M.F. MAFEE, V.M. RAO and K.H. CHOI: CT characteristics of orbital pseudotumors and other orbital inflammatory processes. J. Comput. Assist. Tomogr. 13 (1989) 40–47.

FLECK, B.W. and A.D. TOFT: Graves' ophthalmopathy. Br. Med. J. 300 (1990) 1352–1353.

FRADET, G., D. POULIOT, S. LAVOIE and S. ST PIERRE: Inferior constrictor myotomy in oculopharyngeal muscular dystrophy: clinical and manometric evaluation. J. Otolaryngol. 17 (1988) 68–73.

FRAZZETTO, F., E.B. STREIFF and W.G. FORSSMANN: L'ophtalmyopathie congénitale. Ann. Oculist. (Paris) 12 (1969) 1217–1253.

FRIES, P.D., D.S. FOHRMAN and D.H. CHAR: Phorbol ester-induced orbital myositis. Arch. Ophthalmol. 105 (1987) 1273–1276.

FROHMAN, L.P., M.J. KUPERSMITH, J. LANG, D. REEDE, R.T. BERGERON, S. ALEKSIC and S. TRASI: Intracranial extension and bone destruction in orbital pseudotumor. Arch. Ophthalmol. 104 (1986) 380–384.

FUCHS, E.: Ueber isolierte doppelseitige ptosis. Graefe Arch. Ophthalmol. 36(1) (1890) 234–259.

FUKUHARA, N., T. KUMAMOTO, T. TSUBAKI, T. MAYUZUMI and H. NITTA: Oculopharyngeal muscular dystrophy and distal myopathy: intrafamilial difference in the onset and distribution of muscular involvement. Acta Neurol. Scand. 65 (1982) 458–467.

FUKUNAGA, H., M. OSAME and A. IGATA: A case of nemaline myopathy with ophthalmoplegia and mitochondrial abnormalities. J. Neurol. Sci. 46 (1980) 169–177.

GALLASTEGUI, J., R.J. HARIMAN, B. HANDLER, M. LEV

and S. BHARATI: Cardiac involvement in the Kearns-Sayre syndrome. Am. J. Cardiol. 60 (1987) 385–388.

GARCIN, R., M. FARDEAU and J. GODET-GUILLAIN: A clinical and pathological study of a case of alternating and recurrent external ophthalmoplegia with amyotrophy of the limbs observed for forty-five years; discussion of the relationship of this condition with myasthenia gravis. Brain 88 (1965) 739–752.

GARTNER S. and E. BILLET: Progressive muscular dystrophy involving the extraocular muscles: report of a case. Arch. Ophthalmol. 41 (1949) 334–340.

GERBER, N., E. ESSLEN and F. REGLI: Ocular muscular dystrophy. In: J.N. Walton, N. Canal and G. Scarlato (Eds.), Muscle Diseases. Amsterdam, Excerpta Med. (1970) 603–608.

GERBITZ, K.D., B. OBERMAIER KUSSER, S. ZIERZ, D. PONGRATZ, J. MULLER HOCKER and P. LESTIENNE: Mitochondrial myopathies: divergences of genetic deletions, biochemical defects and the clinical syndromes. J. Neurol. 237 (1990) 5–10.

GILDEN, D.H. and B. KLEINSCHMIDT-DEMASTERS: A 47-year old man with ophthalmoplegia and dementia. J. Neuroimag. 1 (1991) 140–145.

GLASER, J.S.: Graves' ophthalmopathy. Arch. Ophthalmol. 102 (1984) 1448–1449.

GLEASON, J.E.: Idiopathic myositis involving the extraocular muscles. Ophthalmic Rec. 12 (1903) 471–478.

GOADSBY, P.J. and J.W. LANCE: Clinicopathological correlation in a case of painful ophthalmoplegia: Tolosa-Hunt syndrome. J. Neurol. Neurosurg. Psychiatry 52 (1989) 1290–1293.

GOAS, J.Y., J.P. LEROY, Y. MOQUARD and F. ROUHART: [A case of mitochondrial myopathy in a family with oculo-pharyngeal dystrophy]. Rev. Neurol. (Paris) 147 (1991) 536–537.

GOEBEL, H.H., H.M. MEINCK, M. REINECKE, K. SCHIMRIGK and U. MIELKE: Centronuclear myopathy with special consideration of the adult form. Eur. Neurol. 23 (1984) 425–434.

GOLDBERG, L., A. TAO and P. ROMANO: Severe exophtalmos secondary to orbital myopathy not due to Graves' disease. Br. J. Ophthalmol. 66 (1982) 392–395.

GOLDBERG, R.T.: Ocular muscle paresis and cranial arteritis – an unusual case. Ann. Ophthalmol. 15 (1983) 240–243.

GOODHUE, JR. W.W., R.D. COUCH and H. NAMIKI: Ophthalmoplegia plus or Kearns-Sayre syndrome? [letter]. Arch. Neurol. 37 (1980) 256.

GORMAN, C.A.: Temporal relationship between onset of Graves' ophthalmopathy and diagnosis of thyrotoxicosis. Mayo Clin. Proc. 58 (1983) 515–519.

GORNIG, H., W. ZETT and K. ZERNAHLE: [On the clinical picture and genesis of hereditary congenital external ophthalmoplegia]. Klin. Monatsbl. Augenheilkd. 150 (1967) 342–351.

GOTO, Y., I. GOTO and S. HOSOKAWA: Neurological and radiological studies in painful ophthalmoplegia: Tolosa-Hunt syndrome and orbital pseudotumor. J. Neurol. 236 (1989) 448–451.

GOTO, Y., S. HOSOKAWA, I. GOTO, R. HIRAKATA and K. HASUO: Abnormality in the cavernous sinus in three patients with Tolosa-Hunt syndrome: MRI and CT findings. J. Neurol. Neurosurg. Psychiatry 53 (1990a) 231–234.

GOTO, Y., N. ITAMI, N. KAJII, H. TOCHIMARU, M. ENDO and S. HORAI: Renal tubular involvement mimicking Bartter syndrome in a patient with Kearns-Sayre syndrome. J. Pediatr. 116 (1990b) 904–910.

GOTO, Y., Y. KOGA, S. HORAI and I. NONAKA: Chronic progressive external ophthalmoplegia: a correlative study of mitochondrial DNA deletions and their phenotypic expression in muscle biopsies. J. Neurol. Sci. 100 (1990c) 63–69.

GOWERS, W.R.: Note on a relax mechanism in the fixation of the eyeballs. Brain 2 (1879) 39–41.

GRIMSON, B.S. and K.B. SIMONS: Orbital inflammation, myositis, and systemic lupus erythematosus. Arch. Ophthalmol. 101 (1983) 736–738.

GRISOLD, W., K. JELLINGER and B. MAMOLI: Axonal polyneuropathy in familial progressive external ophthalmoplegia (neuromuscular mitochondriopathy). Neuropsychiatr. Clin. 2 (1983) 101–114.

GROOTHUIS, D.R., S. SCHULMAN, R. WOLLMAN, J. FREY and N.A. VICK: Demyelinating radiculopathy in the Kearns-Sayre syndrome: a clinicopathological study. Ann. Neurol. 8 (1980) 373–380.

GRUBER, H., J. ZEITHOFER, J. PRAGER and P. PILS: Complex oculomotor dysfunctions in Kugelberg-Welander disease. Neuro-Ophthalmology 3 (1983) 125–128.

HANSEN, E.: Congenital general fibrosis of the extraocular muscles. Acta Ophthalmol. 43 (1968) 439–476.

HANSEN, S.L., L. BORELLI-MØLLER, P. STRANGE, B.M. NIELSEN and J. OLESON: Ophthalmoplegic migraine: diagnostic criteria, incidence of hospitalization and possible etiology. Acta Neurol. Scand. 81 (1990) 54–60.

HANSMAN, M.L., R.G. PEYSTER, T. HEIMAN PATTERSON and V.S. GREENFIELD: CT demonstration of extraocular muscle atrophy. J. Comput. Assist. Tomogr. 12 (1988) 49–51.

HANSON, P.A., A.F. MASTRIANNI and L. POST: Neonatal ophthalmoplegia with microfibers: a reversible myopathy? Neurology 27 (1977) 974–980.

HASSLER, R.: Erkrankungen der Oblongata, der Brücke und des Mittelhirns. In: F. Bergmann and Schwiegk, (Eds.), Handbuch der inn. Medizin, Vol. 5, part 3, 4th ed. Berlin, Springer (1953) 552–619.

HAUSMANOWA-PETRUSEWICZ, I., S. FALKIEWICZOWA, H. JEDRZEJEWSKA, Z. KAMIENIECKA and A. FIDIANSKA: Descending dystrophy or advanced myasthenia. Schweiz. Arch. Neurol. Psychiatr. 95 (1965) 233–245.

HAYASHI, Y., K. NAGASHIMA, Y. URANO and M. IWATA: Spinocerebellar degeneration with prominent involvement of the motor neuron system: autopsy report of a sporadic case. Acta Neuropathol. 70 (1986) 82–85.

HECK, A.F.: A study of neural and extraneural findings in a large family with Friedreich's ataxia. J. Neurol. Sci. 1 (1964) 226–255.

HEERSINK, B., M.R. RODRIGUES and J.C. FLANAGAN: Inflammatory pseudotumor of the orbit. Ann. Ophthalmol. 9 (1977) 17–22, 25–9.

HELFAND, M.: Congenital familial external ophthalmoplegia without ptosis with a lesion of the pyramidal tract. Arch. Ophthalmol. 21 (1939) 823–827.

HERMANN, J.S.: Paretic thyroid myopathy. Ophthalmology 89 (1982) 473–478.

HIRABUKI, N., T. MIURA, M. MITOMO, K. HARADA, T. HASHIMOTO, R. KAWAI and T. KOZUKA: MR imaging of dural arteriovenous malformations with ocular signs. Neuroradiology 30 (1988) 390–394.

HOLMES, G.: Palsies of the conjugate ocular movements. Br. J. Ophthalmol. 5 (1921) 241–250.

HOLMES, G.: Observations on ocular palsies. Br. Med. J. 2 (1931) 1165–1167.

HOLMES, W.J.: Hereditary congenital ophthalmoplegia. Am. J. Ophthalmol. 41 (1956) 615–618.

HOLT, I.J., A.E. HARDING and J.A. MORGAN HUGHES: Deletions of muscle mitochondrial DNA in patients with mitochondrial myopathies. Nature (London) 331 (1988) 717–719.

HOLT, I.J., A.E. HARDING, J.M. COOPER, A.H. SCHAPIRA, A. TOSCANO, J.B. CLARK and J.A. MORGAN HUGHES: Mitochondrial myopathies: clinical and biochemical features of 30 patients with major deletions of muscle mitochondrial DNA. Ann. Neurol. 26 (1989) 699–708.

HORWITZ, S.J. and U. ROESSMANN: Kearns-Sayre syndrome with hypoparathyroidism. Ann. Neurol. 3 (1978) 513–518.

HOSTEN, N., B. SANDER, M. CORDES, C.J. SCHUBERT, W. SCHORNER and R. FELIX: Graves ophthalmopathy: MR imaging of the orbits. Radiology 172 (1989) 759–762.

HOYT, W.F. and R.B. DAROFF: Supranuclear disorders of ocular control systems in man: clinical, anatomical and physiological correlations. In: P. Bach y Rita and C.C. Collins (Eds.), The Control of Eye Movements. New York, Academic Press (1971) 175–235.

HUBNER, G., J.M. GOKEL, D. PONGRATZ, A. JOHANNES and J.W. PARK: Fatal mitochondrial cardiomyopathy in Kearns-Sayre syndrome. Virchows Arch. [A] 408 (1986) 611–621.

HUNT, W.E.: Tolosa-Hunt syndrome: one cause of painful ophthalmoplegia. J. Neurosurg. 44 (1976) 544–549.

HUNT, W.E., J.N. MEAGHER, H.E. LEFEVER and W. ZEMAN: Painful ophthalmoplegia. Neurology 11 (1961) 56–62.

HUTCHINSON, J.: On ophthalmoplegia externa or symmetrical immobility (partial) of the eyes, with ptosis. Med. Chir. Trans. (London) 62 (1879) 307–329.

IIZUKA, R., K. HIRAYAMA and K.A. MAEHARA: Dentato-rubro-pallido-luysian atrophy: a clinico-pathological study. J. Neurol. Neurosurg. Psychiatry 47 (1984) 1288–1298.

IONASESCU, V.: Oculogastrointestinal muscular dystrophy. Am. J. Med. Genet. 15 (1983) 103–112.

IONASESCU, V., S.H. THOMPSON, R. IONASESCU, C. SEARBY, S. ANURAS, J. CHRISTENSEN, F. MITROS, M. HART and P. BOSCH: Inherited ophthalmoplegia with intestinal pseudo-obstruction. J. Neurol. Sci. 59 (1983) 215–228.

IONASESCU, V.V., H.S. THOMPSON, C. ASCHENBRENER, S. ANURAS and W.S. RISK: Late-onset oculogastrointestinal muscular dystrophy. Am. J. Med. Genet. 18 (1984) 781–788.

ISAKSSON, I.: Studies on congenital genuine blepharoptosis: morphological and functional investigations of the upper eyelid. Acta Ophthalmol. Suppl. (Copenhagen) 72 (1962) 1–121.

JAGER, B.V., H.L. FRED, R.B. BUTLER and W.H. CARNES: Occurrence of retinal pigmentation. ophthalmoplegia, ataxia, deafness, and heart block: report of a case, with findings at autopsy. Am. J. Med. 29 (1960) 888–893.

JAKOBIEC, F.A., J. LEFKOWITCH and D.M. KNOWLES: B- and T-lymphocytes in ocular disease. Ophthalmology 91 (1984) 635–654.

JAMES, R.H.: Thiopentone and ophthalmoplegia plus. Anaesthesia 40 (1985) 88.

JAMES, R.H.: Induction agent sensitivity and ophthalmoplegia plus. Anaesthesia 41 (1986) 216.

JAMPEL, R.S., H. OKAZAKI and H. BERNSTEIN: Ophthalmoplegia and retinal degeneration associated with spinocerebellar ataxia. Arch. Ophthalmol. 66 (1961) 247–259.

JANKOWICZ, E., H. BERGER, S. KURASZ, W. WINOGRODZKA and L. ELJASZ: Familial progressive external ophthalmoplegia with abnormal muscle mitochondria. Eur. Neurol. 15 (1977) 318–324.

JANZER, R.C. and F. BARONTINI: An unusual association of dentato-rubral degeneration with spinal ataxia, ophthalmoplegia and multiple cranial nerve palsies. J. Neurol. 231 (1985) 319–323.

JASPAR, H.H., L.A. BASTIAENSEN, H.J. TER LAAK, E.M. JOOSTEN, M.W. HORSTINK and A.M. STADHOUDERS: Oculopharyngodistal myopathy with early onset and neurogenic features. Clin. Neurol. Neurosurg. 80 (1977) 272–282.

JEDLOWSKI, P.: Sulla oftalmoplegia esterna nucleare cronica progressiva. Riv. Oto. neuro. oftal. 20 (1943) 203–239.

JELLINEK, E.H.: The orbital pseudotumour syndrome and its differentiation from endocrine exophthalmos. Brain 92 (1969) 35–58.

JOHNSON, M.A., D.M. TURNBULL, D.J. DICK and H.S. SHERRATT: A partial deficiency of cytochrome c oxidase in chronic progressive external ophthalmoplegia. J. Neurol. Sci. 60 (1983) 31–53.

JULIEN, J., C. VITAL, J.M. VALLAT, M. VALLAT and M. LE BLANC: Oculopharyngeal muscular dystrophy: a case with abnormal mitochondria and 'fingerprint' inclusions. J. Neurol. Sci. 21 (1974) 165–169.

KADLUBOWSKI, M., W.J. IRVINE and A.C. ROWLAND: The lack of specificity of ophthalmic immunoglobulins in Graves' disease. J. Clin. Endocrinol. Metab. 63 (1986) 990–995.

KALINA, P.H., J.A. GARRITY, D.C. HERMAN, R.A. DEREMEE and U. SPECKS: Role of testing for anticytoplasmic autoantibodies in the differential diagnosis of scleritis and orbital pseudotumor. Mayo Clin. Proc. 65 (1990) 1110–1117.

KAMIENIECKA, Z. and H. SCHMALBRUCH: Neuromuscular disorders with abnormal muscle mitochondria. Int. Rev. Cytol. 65 (1980) 321–357.

KANDT, R.S. and G.W. GOLDSTEIN: Steroid-responsive ophthalmoplegia in a child: diagnostic considerations. Arch. Neurol. 42 (1985) 589–591.

KATO, T., A. HIRANO, M.N. WEINBERG and A.K. JACOBS: Spinal cord lesions in progressive supranuclear palsy: some new observations. Acta Neuropathol. 71 (1986) 11–14.

KEANE, J.R.: Alternating proptosis: a case report of acute orbital myositis defined by the computerized tomographic scan. Arch. Neurol. 34 (1977) 642–643.

KEANE, J.R.: Acute bilateral ophthalmoplegia: 60 cases. Neurology 36 (1986) 279–281.

KEARNS, T.P.: External ophthalmoplegia, pigmentary degeneration of the retina, and cardiomyopathy: a newly recognized syndrome. Trans. Am. Ophthalmol. Soc. 63 (1965) 559–625.

KEARNS, T.P. and G.P. SAYRE: Retinitis pigmentosa, external ophthalmoplegia and complete heart block: unusual syndrome with histologic study in one of two cases. Arch. Ophthalmol. 60 (1958) 280–289.

KENDALL-TAYLOR, P.: The management of Graves' ophthalmopathy. Clin. Endocrinol. 31 (1989) 747–756.

KENDALL-TAYLOR, P., A.L. CROMBIE, A.M. STEPHENSON, M. HARDWICK and K. HALL: Intravenous methylprednisolone in the treatment of Graves' ophthalmopathy. Br. Med. J. 297 (1988) 1574–1578.

KENNERDELL, J.S. and S.C. DRESNER: The nonspecific orbital inflammatory syndromes. Surv. Ophthalmol. 29 (1984) 93–103.

KETONEN, L., H. TERAVAINEN, A. PILKE and K. KATEVUO: Computed tomography in a case of ophthalmoplegia syndrome. Ann. Clin. Res. 17 (1985) 37–42.

KILOH, L.G. and S. NEVIN: Progressive dystrophy of the external ocular muscles (ocular myopathy). Brain 74 (1951) 115–143.

KISSEL, J.T., R.M. BURDE, T.G. KLINGELE and H.E. ZEIGER: Pupil-sparing oculomotor palsies with internal carotid-posterior communicating artery aneurysms. Ann. Neurol. 13 (1983) 149–154.

KLEBER, F.X., J.W. PARK, G. HUBNER, A. JOHANNES, D. PONGRATZ and E. KONIG: Congestive heart failure due to mitochondrial cardiomyopathy in Kearns-Sayre syndrome. Klin. Wochenschr. 65 (1987) 480–486.

KLEE, G.G. and I.D. HAY: Assessment of sensitive thyrotropin assays for an expanded role in thyroid function testing: proposed criteria for analytic performance and clinical utility. J. Clin. Endocrinol. Metab. 64 (1987) 461–471.

KLEIN, B.R., T.R. HEDGES III, Y. DAYAL and L.S. ADELMAN: Orbital myositis and giant cell myocarditis. Neurology 39 (1989) 988–990.

KLINE, L.B.: The Tolosa-Hunt syndrome. Surv. Ophthalmol. 27 (1982) 79–95.

KLOSTERMANN, W., K. WESSEL, E. REUSCHE, C. KESSLER and D. KOMPF: [Oculopharyngeal muscular dystrophy: clinical, electromyography and muscle biopsy findings in 2 cases]. Nervenarzt 61 (1990) 351–355.

KONIGSMARK, B.W. and L.P. WEINER: the olivopontocerebellar atrophies: a review. Medicine 49 (1970) 227–241.

KOPPEL, B.S.: Steroid-responsive painful ophthalmoplegia is not always Tolosa-Hunt. Neurology 37 (1987) 544.

KORNER, F. and I. PROBST: [Typical retinitis pigmentosa with chronic progressive external ophthalmoplegia (author's transl.)]. Albrecht von Graefes Arch. Klin. Exp. Ophthalmol. 195 (1975) 195–200.

KOSMORSKY, G.S., D.M. MEISLER, L.A. SHEELER, R.L. TOMSAK, P. SWEENEY, H. MITSUMOTO and S.M. MACRAE: Familial ophthalmoplegia-plus syndrome with corneal endothelial disorder. Neuro-Ophthalmology 9 (1989) 271–277.

KOZACHEK, J.W. and F.J. WILSON: Oculopharyngeal dystrophy: ultrastructure of muscles distinct from the primary myopathy. Acta Neuropathol. 57 (1982) 7–12.

KRENDEL, D.A., D.B. SANDERS and J.M. MASSEY: Single fiber electromyography in chronic progressive external ophthalmoplegia. Muscle Nerve 10 (1987) 299–302.

KRUEGER, K.E. and D. FRIEDREICH: [Familial congenital disorders of motility of the eye]. Klin. Monatsbl. Augenheilkd. 142 (1963) 101–117.

KURIYAMA, M., M. SUEHARA, N. MARUME, M. OSAME and A. IGATA: High CSF lactate and pyruvate content in Kearns-Sayre syndrome. Neurology 34 (1984) 253–255.

KUSHNER, M.J., M. PARRISH, A. BURKE, M. BEHRENS, A.P. HAYS, B. FRAME and L.P. ROWLAND: Nystagmus in motor neuron disease: clinicopathological study of two cases. Ann. Neurol. 16 (1984) 71–77.

LAIDLAW, D.A., P.E. SMITH and P. HUDGSON: Orbital pseudotumour secondary to giant cell arteritis: an unreported condition. Br. Med. J. 300 (1990) 784.

LAKIN, M. and S. LOCKE: Progressive ocular myopathy with ovarian insufficiency and diabetes mellitus: report of a case. Diabetes 10 (1961) 228–231.

LANDAU, W.M. and J. GITT: Hereditary spastic paraplegia and hereditary ataxia: a family demonstrating a variety of phenotypic manifestation. Arch. Neurol. Psychiatry 66 (1951) 346–354.

LANGDON, H.W. and W.B. CADWALADER: Chronic progressive external ophthalmoplegia: report of a case with necropsy. Brain 51 (1928) 321–333.

LAPLANE, D. and D. HASBOUN: [Disorders of folate metabolism in the Kearns-Sayre syndrome]. Rev. Otoneuroophthalmol. 56 (1984) 199–202.

LARSSON, N.G., E. HOLME, B. KRISTIANSSON, A. OLDFORS and M. TULINIUS: Progressive increase of the mutated mitochondrial DNA fraction in Kearns-Sayre syndrome. Pediatr. Res. 28 (1990) 131–136.

LAUBER, J., C. MARSAC, B. KADENBACH and P. SEIBEL: Mutations in mitochondrial tRNA genes: a frequent cause of neuromuscular diseases. Nucleic Acids Res. 19 (1991) 1393–1397.

LAUGHLIN, R.C.: Congenital fibrosis of the extraocular muscles; a report of six cases. Am. J. Ophthalmol. 41 (1956) 432–438.

LAWTON, N.F.: Exclusion of dysthyroid eye disease as a cause of unilateral proptosis. Trans. Ophthalmol. Soc. U.K. 99 (1979) 226–228.

LEES, F. and L.A. LIVERSEDGE: Descending ocular myopathy. Brain 85 (1962) 701–710.

LEONARD, T.J., E.M. GRAHAM, M.R. STANFORD and M.D. SANDERS: Graves' disease presenting with bilateral acute painful proptosis, ptosis, ophthalmoplegia, and visual loss. Lancet ii (1984) 431–433.

LEWIS, A.J. and M.J. GAWELL: Diffuse Lewy body disease with dementia and oculomotor dysfunction. Mov. Disord. 5 (1990) 143–147.

LEWIS, I.: Late-onset muscle dystrophy: oculopharyngoesophageal variety. Canad. Med. Ass. J. 95 (1966) 146–150.

LI, T.M.: Congenital total bilateral ophthalmoplegia. Am. J. Ophthal. 6 (1923) 816–821.

LIND, I. and P. PRAME: Chronic progressive external ophthalmoplegia and muscular dystrophy. Acta Ophthalmol. 41 (1963) 497–507.

LITTLE, B.W. and D.P. PERL: Oculopharyngeal muscular dystrophy: an autopsied case from the French-Canadian kindred. J. Neurol. Sci. 53 (1982) 145–158.

LO, R. and T.E. FEASBY: Multiple sclerosis and autoimmune diseases. Neurology 33 (1983) 97–98.

LOHMAN, L. J.A. BURNS, W.R. PENLAND and K.V. CAHILL: Unilateral eyelid retraction secondary to contralateral ptosis in dysthyroid ophthalmopathy. J. Clin. Neurol. Ophthalmol. 4 (1984) 163–166.

LOSSOS, A., L. AVERBUCH HELLER, A. RECHES and O. ABRAMSKY: Complete unilateral ophthalmoplegia as the presenting manifestation of Waldenstrom's macroglobulinemia. Neurology 40 (1990) 1801–1802.

LUNDBERG, P.O.: Ocular myopathy with hypogonadism. Acta Neurol. Scand. 38 (1962) 142–155.

LUNDBERG, P.O.: Observations on endocrine function in ocular myopathy. Acta. Neurol. Scand. 42 (1966) 39–61.

MAGORA, A. and H. ZAUBERMAN: Ocular myopathy. Arch. Neurol. 20 (1969) 1–8.

MAHER, E.R. and A.J. LEES: The clinical features and natural history of the Steele-Richardson-Olszewski syndrome (progressive supranuclear palsy). Neurology 36 (1986) 1005–1008.

MAHLOUDJI, M.: Hereditary spastic ataxia simulating disseminated sclerosis. J. Neurol. Neurosurg. Psychiatry 26 (1963) 511–513.

MAN, H.X., J. MIKOL, A. GUILLARD and G. BOUDIN: [Study of the clinical anatomy of oculo-pharyngeal myopathy]. Bull. Soc. Ophthalmol. Fr. 76 (1976) 23–30.

MANNI, R., G. PICCOLO, P. BANFI, I. CERVERI, C. BRUSCHI, C. ZOIA and A. TARTARA: Respiratory patterns during sleep in mitochondrial myopathies with ophthalmoplegia. Eur. Neurol. 31 (1991) 12–17.

MARBURG, O.: Ponto-mesencephale Form (Die Chronisch progressiven nuclearen Amyotrophien) [Chronic progressive ophthalmoplegia]. In: O. Bumke and O. Foerster (Eds.), Handbuch der Neurologie, Vol. 16. Berlin, Springer Verlag (1936) 548–550.

MARCOCCI, C., L. BARTALENA, F. BOGAZZI, M. PANICUCCI and A. PINCHERA: Studies on the occurrence of ophthalmopathy in Graves' disease. Acta Endocrinol. 120 (1989) 473–478.

MARTIN, J.P.: Progressive external ophthalmoplegia with myopathy [case history]. Proc. R. Soc. Med. 29 (1936) 383.

MCAULEY, F.D.: Progressive external ophthalmoplegia. Br. J. Ophthalmol. 40 (1956) 686–690.

MCCOMISH, M., A. COMPSTON and D. JEWITT: Cardiac abnormalities in chronic progressive external ophthalmoplegia. Br. Heart J. 38 (1976) 526–529.

MCKECHNIE, N.M., M. KING and W.R. LEE: Retinal pathology in the Kearns-Sayre syndrome. Br. J. Ophthalmol. 69 (1985) 63–75.

MCKELVIE, P.A., J.B. MORLEY, E. BYRNE and S. MARZUKI: Mitochondrial encephalomyopathies: a correlation between neuropathological findings and defects in mitochondrial DNA. J. Neurol. Sci. 102 (1991) 51–60.

MCNEER, K.W. and A. JAMPOLSKY: Double elevator palsy caused by anomalous insertion of the inferior rectus. Am. J. Ophthalmol. 59 (1965) 317–319.

MCQUINN, B.A. and T.L. KEMPER: Sporadic case resembling autosomal-dominant motor system degeneration (Azorean disease complex). Arch. Neurol. 44 (1987) 341–344.

MCSHANE, M.A., S.R. HAMMANS, M. SWEENEY, I.J. HOLT, T.J. BEATTIE, E.M. BRETT and A.E. HARDING: Pearson syndrome and mitochondrial encephalomyopathy in a patient with a deletion of mtDNA. Am. J. Hum. Genet. 48 (1991) 39–42.

MECHLER, F., F.L. MASTAGLIA, M. SERENA, M. JENKINSON, M.A. JOHNSON, P.R. FAWCETT, P. HUDGSON and J.N. WALTON: Mitochondrial myopathies: a clinico-pathological study of cases with and without extraocular muscle involvement. Aust. N.Z.J. Med. 16 (1986) 185–192.

MEHLER, M.F. and L. RABINOWICH: The clinical neuro-ophthalmologic spectrum of temporal arteritis. Am. J. Med. 85 (1988) 839–844.

MEIRE, F., J.J. DE LAEY, S. DE BIE, M. VAN STAEY and M.T. MATTON: Dominant optic nerve atrophy with progressive hearing loss and chronic progressive external ophthalmoplegia (CPEO). Ophthalm. Paediatr. Genet. 5 (1985) 91–97.

MERRITT, H.H.: A Textbook of Neurology, 4th ed. Philadelphia, Lea and Eebiger (1967) 551.

MISULIS, K.E. and G.M. FENICHEL: Genetic forms of myasthenia gravis. Pediatr. Neurol. 5 (1989) 205–210.

MITA, S., B. SCHMIDT, E.A. SCHON, S. DIMAURO and E. BONILLA: Detection of 'deleted' mitochondrial genomes in cytochrome-c oxidase-deficient muscle fibers of a patient with Kearns-Sayre syndrome. Proc. Natl. Acad. Sci. U.S.A. 86 (1989) 9509–9513.

MIZUSAWA, H., N. OHKOSHI, M. WATANABE and I. KANAZAWA: Peripheral neuropathy of mitochondrial myopathies. Rev. Neurol. (Paris) 147 (1991) 501–507.

MIZUTANI, T., M. AKI, R. SHIOZAWA, M. UNAKAMI, T. NOZAWA, K. YAJIMA, H. TANABE and M. HARA: Development of ophthalmoplegia in amyotrophic lateral sclerosis during long-term use of respirators. J. Neurol. Sci. 99 (1990) 311–319.

MOORTHY, G., M.M. BEHRENS, D.B. DRACHMAN, T.H. KIRKHAM, D.L. KNOX, N.R. MILLER, T.L. SLAMOVITZ and S.J. ZINREICH: Ocular pseudomyasthenia or ocular myasthenia 'plus': a warning to clinicians. Neurology 39 (1989) 1150–1154.

MORAES, C.T., S. DIMAURO, M. ZEVIANI, A. LOMBES, S. SHANSKE, A.F. MIRANDA, H. NAKASE, E. BONILLA, L.C. WERNECK and S. SERVIDEI: Mitochondrial DNA deletions in progressive external ophthalmoplegia and Kearns-Sayre syndrome. N. Engl. J. Med. 320 (1989) 1293–1299.

MORAES, C.T., M. ZEVIANI and E.A. SCHON: Mitochondrial DNA deletion in a girl with manifestations of Kearns-Sayre and Lowe syndromes: an example of molecular mimicry? Am. J. Med. Genet. 41 (1991) 301–305.

MOTTOW, L.S. and F.A. JAKOBIEC: Idiopathic inflammatory orbital pseudotumor in childhood. I. Clinical characteristics. Arch. Ophthalmol. 96 (1978) 1410–1417.

MÜLLER-HÖCKER, J., D. PONGRATZ and G. HUBNER: Focal deficiency of cytochrome-c-oxidase in skeletal muscle of patients with progressive external ophthalmoplegia: cytochemical-fine-structural study. Virchows Arch. [A]. 402 (1983) 61–71.

MULLIE, M.A., A.E. HARDING, R.K. PETTY, H. IKEDA, J.A. MORGAN HUGHES and M.D. SANDERS: The retinal manifestations of mitochondrial myopathy. A study of 22 cases. Arch. Ophthalmol. 103 (1985) 1825–1830.

MUNSAT, T.L., L.R. THOMPSON and R.F. COLEMAN: Centronuclear ('myotubular') myopathy. Arch. Neurol. 20 (1969) 120–131.

MURHPY, S.F. and D.A. DRACHMAN: The oculopharyngeal syndrome. J. Am. Med. Ass. 203 (1968) 1003–1008.

MUSSINI, J.M., M. FRIOL VERCELLETTO, Y. FORGEAU, J.R. FEVE and C. MAGNE: [Role of muscle biopsy in the diagnosis of ocular myopathies]. Rev. Otoneuroophthalmol. 56 (1984) 203–208.

NADEAU, S.E. and J.D. TROBE: Pupil sparing in oculo-

motor palsy: a brief review. Ann. Neurol. 13 (1983) 143–148.

NAKAMURA, S., T. SATO, H. HIRAWAKE, R. KOBAYASHI, Y. FUKUDA, J. KAWAMURA, H. UJIKE and S. HORAI: In situ hybridization of muscle mitochondrial mRNA in mitochondrial myopathies. Acta. Neuropathol. 81 (1990) 1–6.

NAKASE, H., C.T. MORAES, R. RIZZUTO, A. LOMBES, S. DIMAURO and E.A. SCHON: Transcription and translation of deleted mitochondrial genomes in Kearns-Sayre syndrome: implications for pathogenesis. Am. J. Hum. Genet. 46 (1990) 418–427.

NAKASHIMA, A., M. KAWAI, Y. TOFUKU, R. TAKEDA, S. KATSUDA and Y. OOKADA: Painful ophthalmoplegia caused by metastasis of cholangiocarcinoma of the liver. Ala. J. Med. Sci. 24 (1987) 29–30.

NAMBA, T., N.G. BRUNNER and D. GROB: Idiopathic giant cell polymyositis: report of a case and review of the syndrome. Arch. Neurol. 31 (1974) 27–30.

NEIGEL, J.M., J. ROOTMAN, R.G. ROBINSON, F.A. DURITY and R.A. NUGENT: The Tolosa-Hunt syndrome: computed tomographic changes and reversal after steroid therapy. Can. J. Ophthalmol. 21 (1986) 287–290.

NEIGEL, J.M., J. ROOTMAN, R.I. BELKIN, R.A. NUGENT, S.M. DRANCE, C.W. BEATTIE and J.A. SPINELLI: Dysthyroid optic neuropathy. The crowded orbital apex syndrome. Ophthalmology 95 (1988) 1515–1521.

NEWSOM-DAVIS, J.: Edrophonium responsiveness not necessarily diagnostic of myasthenia gravis [letter]. Muscle Nerve 13 (1990) 1186.

NEWSOM-DAVIS, J., K. LEYS, A. VINCENT, I. FERGUSON, G. MODI and K. MILLS: Immunological evidence for the co-existence of the Lambert-Eaton myasthenic syndrome and myasthenia gravis in two patients. J. Neurol. Neurosurg. Psychiatry 54 (1991) 452–453.

NIEBROJ-DOBOSZ, I., B. RYNIEWICZ, A. FIDZIANSKA and B. BADURSKA: Lipid storage myopathy in Kearns-Sayre syndrome. Neurology 35 (1985) 1582–1586.

NISHIKAWA, Y., M. TAKAHASHI, S. YORIFUJI, Y. NAKAMURA, S. UENO, S. TARUI, T. KOZUKA and T. NISHIMURA: Long-term coenzyme Q10 therapy for a mitochondrial encephalomyopathy with cytochrome c oxidase deficiency: a 31P NMR study. Neurology 39 (1989) 399–403.

NITSCH, J., S. ZIERZ, K.P. JANSSEN, W. JUNG, M. MANZ, F. JERUSALEM and B. LUDERITZ: [Indications for pacemaker therapy in ophthalmoplegia plus and Kearns-Sayre syndrome]. Z. Kardiol. 79 (1990) 60–65.

OGASAHARA, S., Y. NISHIKAWA, S. YORIFUJI, F. SOGA, Y. NAKAMURA, M. TAKAHASHI, S. HASHIMOTO, N. KONO and S. TARUI: Treatment of Kearns-Sayre syndrome with coenzyme Q10. Neurology 36 (1986) 45–53.

OH, S.J. and H.K. CHO: Edrophonium responsiveness not necessarily diagnostic of myasthenia gravis. Muscle Nerve 13 (1990) 187–191.

OHTAKI, E., Y. YAMAGUCHI, Y. YAMASHITA, T. MATSUISHI, K. TERASAWA, Y. KATAFUCHI and I. NONAKA:

Complete external ophthalmoplegia in a patient with congenital myopathy without specific features (minimal change myopathy). Brain Dev. 12 (1990) 427–430.

OKAMOTO, T., K. MIZUNO, M. IIDA, I. SOBUE and M. MUKOYAMA: Ophthalmoplegia-plus: its occurrence with periventricular diffuse low density on computed tomography scan. Arch. Neurol. 38 (1981) 423–426.

OLDFORS, A., I.M. FYHR, E. HOLME, N.G. LARSSON and M. TULINIUS: Neuropathology in Kearns-Sayre syndrome. Acta Neuropathol. 80 (1990) 541–546.

OLSON, W., W.K. ENGEL, G.O. WALSH and R. EINAUGLER: Oculocraniosomatic neuromuscular disease with 'ragged-red' fibers. Arch. Neurol. 26 (1972) 193–211.

ORTIZ DE ZARATE, J.C. and A. MARUFFO: The descending ocular myopathy of early childhood; myotubular or centronuclear myopathy. Eur. Neurol. 3 (1970) 1–12.

OTSUKA, M., K. NIIJIMA, Y. MIZUNO, M. YOSHIDA, Y. KAGAWA and S. OHTA: Marked decrease of mitochondrial DNA with multiple deletions in a patient with familial mitochondrial myopathy. Biochem. Biophys. Res. Commun. 167 (1990) 680–685.

OWEN, J.S., L.B. KLINE, S.J. OH, N.E. MILES and J.W. BENTON: Ophthalmoplegia and ptosis in congenital fiber type disproportion. J. Pediatr. Ophthalmol. Strabismus 18 (1981) 55–60.

OZAWA, T., M. YONEDA, M. TANAKA, K. OHNO, W. SATO, H. SUZUKI, M. NISHIKIMI, M. YAMAMOTO, I. NONAKA and S. HORAI: Maternal inheritance of deleted mitochondrial DNA in a family with mitochondrial myopathy. Biochem. Biophys. Res. Commun. 154 (1988) 1240–1247.

PAPST, W., H.G. MERTENS and E. ESSLEN: [Chronic ocular myositis. I. Exophthalmic ocular myositis]. Klin. Monatsbl. Augenheilkd. 133 (1958) 673–694.

PAPST, W., H.G. MERTENS and E. ESSLEN: [Chronic ocular myositis. II. Oligosymptomatic ocular myositis]. Klin. Monatsbl. Augenheilkd. 134 (1959) 374–396.

PAVLAKIS, S.G., L.P. ROWLAND, D.C. DEVIVO, E. BONILLA and S. DIMAURO: Mitochondrial myopathies and encephalomyopathies. In: F. Plum (Ed.), Advances in Contemporary Neurology. Philadelphia, Davis (1988) 95–133.

PELLOCK, J.M., M. BEHRENS, L. LEWIS, D. HOLUB, S. CARTER and L.P. ROWLAND: Kearns-Sayre syndrome and hypoparathyroidism. Ann. Neurol. 3 (1978) 455–458.

PETER, S.A.: Euthyroid Graves' disease: report of a case observed over a 12-year period. Am. J. Med. 80 (1986) 1197–1198.

PETTY, R.K., A.E. HARDING and J.A. MORGAN HUGHES: The clinical features of mitochondrial myopathy. Brain 109 (1986) 915–938.

PICCOLO, G. and G. MARTINO: Ocular palsy preceding chronic relapsing idiopathic polyneuropathy: poor response to plasma exchange. Ital. J. Neurol. Sci. 9 (1988) 383–384.

PIEROBON-BORMIOLI, S., M. ARMAIN, S.P. RINGEL, C. ANGELINI, L. VERGANI, R. BETTO and G. SALVIATI: Familial neuromuscular disease with tubular aggregates. Muscle Nerve 8 (1985) 291–298.

PONGRATZ, D., J. PERWEIN, G. HUBNER, C. KOPPEN-WALLNER, K. TOYKA and K.L. BIRNBERGER: [Muscle biopsy in progressive external ophthalmoplegia (author's transl.)]. Klin. Wochenschr. 57 (1979) 779–788.

POULTON, J., M.E. DEADMAN, S. RAMACHARAN and R.M. GARDINER: Germ-line deletions of mtDNA in mitochondrial myopathy. Am. J. Hum. Genet. 48 (1991) 649–653.

PROBST, A., W. TACKMANN, H.R. STOECKLI, F. JERUSALEM and J. ULRICH: Evidfence for a chronic axonal atrophy in oculopharyngeal 'muscular dystrophy'. Acta Neuropathol. 57 (1982) 209–216.

PRUMMEL, M.F., M.P. MOURITS, A. BERGHOUT, E.P. KRENNING, R. VAN DER GAAG, L. KOORNNEEF and W.M. WIERSINGA: Prednisone and cyclosporine in the treatment of severe Graves' ophthalmopathy. N. Engl. J. Med. 321 (1989) 1353–1359.

PURCELL, JR. J.J. and W.A. TAULBEE: Orbital myositis after upper respiratory tract infection. Arch. Ophthalmol. 99 (1981) 437–438.

RAPIN, F. and B. ECHENNE: [Tolosa-Hunt painful ophthalmoplegia: apropos of a case in a 10-year-old girl]. Arch. Fr. Pediatr. 44 (1987) 299–301.

REBEIZ, J.J., J.B. CAULFIELD and R.D. ADAMS: A pathological study of oculopharyngeal dystrophy. Second Congr. Neurogenet. Neuroophthal. (1967).

REICHMANN, H., F. DEGOUL, R. GOLD, B. MEURERS, U.P. KETELSEN, J. HARTMANN, C. MARSAC and P. LESTIENNE: Histological, enzymatic and mitochondrial DNA studies in patients with Kearns-Sayre syndrome and chronic progressive external ophthalmoplegia. Eur. Neurol. 31 (1991) 108–113.

REIFLER, D.M., J.N. HOLTZMAN and D.M. RINGEL: Sphenoid ridge meningioma masquerading as Graves' orbitopathy. Arch. Ophthalmol. 104 (1986) 1591.

REIFLER, D.M., D. LEDER and T. REXFORD: Orbital hemorrhage and eyelid ecchymosis in acute orbital myositis. Am. J. Ophthalmol. 107 (1989) 111–113.

RINGEL, S.P., W.B. WILSON and M.T. BARDEN: Extraocular muscle biopsy in chronic progressive external ophthalmoplegia. Ann. Neurol. 6 (1979) 326–339.

RIVNER, M.H., M. SHAMSNIA, T.R. SWIFT, J. TREFZ, R.A. ROESEL, A.L. CARTER, W. YANAMURA and F.A. HOMMES: Kearns-Sayre syndrome and complex II deficiency. Neurology 39 (1989) 693–696.

ROBERTS, A.H. and J. BAMFORTH: The pharynx and esophagus in ocular muscular dystrophy. Neurology 18 (1968) 645–652.

ROBINSON, B.H. and W.G. SHERWOOD: Lactic acidaemia. J. Inherit. Metab. Dis. 7 Suppl. 1 (1984) 69–73.

ROSEN, C.E. and R.M. BURDE: Pathophysiology and etiology of Graves' ophthalmology. In: S.A. Falk (Ed.), Tyroid Disease: Endocrinology, Surgery, Nuclear Medicine, and Radiotherapy. New York, Raven Press (1990) 255–263.

ROSENBERG, R.N., D.L. SCHOTLAND, R.E. LOVELACE and L.P. ROWLAND: Progressive ophthalmoplegia: report of cases. Arch. Neurol. 19 (1968) 362–376.

ROSS, A., D. LIPSCHUTZ, J. AUSTIN and J. SMITH JR.: External ophthalmoplegia and complete heart block. N. Engl. J. Med. 280 (1970) 313–316.

ROTHFUS, W.E. and H.D. CURTIN: Extraocular muscle enlargement: a CT review. Radiology 151 (1984) 677–681.

ROUSSELLE, C., C. CONFAVREUX, T. MOREAU, C. THIVOLET, J. MOINDROT, J. ORGIAZZI and G. AIMARD: [Pseudomyasthenic pre-Basedow ophthalmoplegia]. Rev. Neurol. 146 (1990) 308–310.

ROWLAND, L.P.: Progressive external ophthalmoplegia. In: P.J. Vinken and G.W. Bruyn (Eds.), Handbook of Clinical Neurology, Vol. 22. Amsterdam, North-Holland (1975) 177–202.

ROWLAND, L.P.: Ophthalmoplegia plus or Kearns-Sayre syndrome? [letter]. Arch. Neurol. 37 (1980) 256.

ROWLAND, L.P.: Molecular genetics, pseudogenetics, and clinical neurology. The Robert Wartenberg Lecture. Neurology 33 (1983) 1179–1195.

ROWLAND, L.P.: The Kearns-Sayre syndrome; does it exist? In: G. Serratrice, J.F. Pellissier, C. Desnuelle and J. Pouget (Eds.), Myélopathies, Neuropathies, et Myopathies: Acquisitions Récentes. Paris, Expansion Scientifique Française (1988) 322–326.

ROWLAND, L.P. and R.A. BERENBERG: Diagnosis of Leigh's disease questioned, defended [letter]. Neurology 24 (1974) 598–599.

ROWLAND, L.P., P.F.A. HOEFER, H. ARANOW JR. and H.H. MERRITT: Fatalities in myasthenia gravis: 39 cases with 26 autopsies. Neurology 6 (1956) 307–326.

ROWLAND, L.P., P.F.A. HOEFER and H. ARANOW JR.: Myasthenic syndromes. Res. Publ. Assoc. Res. Nerv. Ment. Dis. 38 (1960) 548–600.

ROWLAND, L.P., R.P. LISAK, D.L. SCHOTLAND, P.V. DEJESUS and P. BERG: Myasthenic myopathy and thymoma. Neurology 23 (1973) 282–288.

ROWLAND, L.P., A.P. HAYS, S. DIMAURO and D.C. DEVIVO: Diverse clinical disorders associated with morphological abnormalities of mitochondria. In: G. Scalarto and C.G. Cerri (Eds.), Mitochondrial Pathology in Muscle Diseases. Padua, Italy, Piccin Medical Books (1983) 141–158.

ROWLAND, L.P., I. HAUSMANOWA-PETRUSEWICZ, B. BARDURSKA, D. WARBURTON, I. NIBROJ DOBOSZ, S. DIMAURO, M. PALLAI and W.G. JOHNSON: Kearns-Sayre syndrome in twins: lethal dominant mutation or acquired disease? Neurology 38 (1988) 1399–1402.

ROWLAND, L.P., D.M. BLAKE, M. HIRANO, S. DIMAURO, E.A. SCHON, A.P. HAYS and D.C. DEVIVO: Clinical syndromes associated with ragged red fibers. Rev. Neurol. (Paris) 147 (1991) 467–473.

SAHASHI, K., K. OHNO, M. TANAKA, T. IBI, T. YAMAMOTO, M. TASHIRO, W. SATO, A. TAKAHASHI and T. OZAWA: Cytoplasmic body and mitochondrial DNA deletion. J. Neurol. Sci. 99 (1990) 291–300.

SALVI, M., Z.G. ZHANG, D. HAEGERT, M. WOO, A. LIBERMAN, L. CADARSO and J.R. WALL: Patients with endocrine ophthalmopathy not associated with overt thyroid disease have multiple thyroid immunological abnormalities. J. Clin. Endocrinol. Metab. 70 (1990) 89–94.

SANDBERG, M.A., S.G. JACOBSON and E.L. BERSON: Foveal cone electroretinograms in retinitis pigmentosa and juvenile maular degeneration. Am. J. Ophthalmol. 88 (1979) 702–707.

SANDIFER, P.H.: Chronic progressive ophthalmoplegia of myopathic origin. J. Neurol. Neurosurg. Psychiatry 9 (1948) 81–83.

SANDYK, R. and M.J. BRENNAN: Isolated ocular myopathy and celiac disease in childhood. Neurology 33 (1983) 792.

SANGLA, S., T. DE BROUCKER, F. CHERON, J. CAMBIER and H. DEHEN: [Improvement of Joseph's disease with sulfamethazole-trimethoprim]. Rev. Neurol. 146 (1990) 213–214.

SATOYOSHI, E., K. MURAKAMI, H. KOWA, M. KINOSHITA and J. TORII: Distal involvement of the extremities in ocular myopathy. Am. J. Ophthalmol. 59 (1965) 668–673.

SCARLATO, G., N. BRESOLIN, I. MORONI, C. DORIGUZZI, E. CASTELLI, G. COMI, C. ANGELINI and A. CARENZI: Multicenter trial with ubidecarenone: treatment of 44 patients with mitochondrial myopathies. Rev. Neurol. (Paris) 147 (1991) 542–548.

SCHIMKE, R.N., W.A. HORTON, D.L. COLLINS and L. THEROU: A new X-linked syndrome comprising progressive basal ganglion dysfunction, mental and growth retardation, external ophthalmoplegia, postnatal microcephaly and deafness. Am. J. Med. Genet. 17 (1984) 323–332.

SCHMITT, H.P. and K.H. KRAUSE: An autopsy study of a familial oculopharyngeal muscular dystrophy (OPMD) with distal spread and neurogenic involvement. Muscle Nerve 4 (1981) 296–305.

SCHNECK, I., M. ADACHI, P. BRIET, A. WOLINTZ and B.W. VOLK: Ophthalmoplegia plus, with morpho-muscle tissue. J. Neurol. Sci. 19 (1973) 37–44.

SCHNITZLER, E.R. and W.C. ROBERTSON JR: Familial Kearns-Sayre syndrome. Neurology 29 (1979) 1172–1174.

SCHONDER, A.A., R.C. CLIFT, J.W. BROPHY and L.W. DANE: Bilateral recurrent orbital inflammation associated with retroperitoneal fibrosclerosis. Br. J. Ophthalmol. 69 (1985) 783–787.

SCHOTLAND, D.L. and L.P. ROWLAND: Muscular dystrophy: features of ocular myopathy and myotonic dystrophy. Arch. Neurol. 10 (1964) 433–445.

SCHWARTZ, J.F., L.P. ROWLAND, H. EDER, P.A. MARKS, E.F. OSSERMAN, E. HIRSCHBERG and H. ANDERSON: Bassen-Kornzweig syndrome: deficiency of serum B-lipoprotein. Arch. Neurol. 8 (1963) 438–454.

SCHWARTZKOPFF, B., H. FRENZEL, B. LOSSE, M. BORGGREFE, K.V. TOYKA, W. HAMMERSTEIN, R. SEITZ, M. DECKERT and G. BREITHARDT: [Heart involvement in progressive external ophthalmoplegia

(Kearns-Sayre syndrome): electrophysiologic, hemodynamic and morphologic findings]. Z. Kardiol. 75 (1986) 161–169.

SCHWARTZMAN, M.J., H. MITSUMOTO, R.W. SHIELDS JR., M.L. ESTES, D.M. MEISLER and G.S. KOSMORSKY: Neurogenic muscle weakness in chronic progressive external ophthalmoplegia (CPEO). Muscle Nerve 13 (1990) 1183–1184.

SCHWARZ, G.A. and C. LIU: Chronic progressive external ophthalmoplegia: a clinical and neuropathologic report. Arch. Neurol. Psychiatr. 71 (1954) 31–53.

SEIGEL, R.S., J.F. SEEGER, T.O. GABRIELSEN and R.J. ALLEN: Computed tomography in oculocraniosomatic disease (Kearns-Sayre syndrome). Radiology 130 (1979) 159–164.

SERGOTT, R.C. and J.S. GLASER: Graves' ophthalmopathy: a clinical and immunologic review. Surv. Ophthalmol. 26 (1981) 1–21.

SERRATRICE, G. and J.F. PELLISSIER: [Ocular myopathies. Nosological study of 49 cases]. Presse Med. 16 (1987) 1969–1974.

SERRATRICE, G., J.F. PELLISSIER, C. DESNUELLE and J. POUGET: [Mitochondrial myopathies and ocular myopathies: 62 cases]. Rev. Neurol. (Paris) 147 (1991) 474–475.

SERVIDEI, S., M. ZEVIANI, G. MANFREDI, E. RICCI, G. SILVESTRI, E. BERTINI, C. GELLERA, S. DI MAURO, S. DI DONATO and P. TONALI: Dominantly inherited mitochondrial myopathy with multiple deletions of mitochondrial DNA: clinical, morphologic, and biochemical studies. Neurology 41 (1991) 1053–1059.

SHANSKE, S., C.T. MORAES, A. LOMBES, A.F. MIRANDA, E. BOUILLA, P. LEWIS, M.A. WHELAN, C.A. ELLSWORTH and S. DIMAURO: Widespread tissue distribution of mitochondrial DNA deletions in Kearns-Sayre syndrome. Neurology 40 (1990) 24–28.

SHER, J.H., A.B. RIMALOVSKI, T.J. ATHANASSIADES and S.M. ARONSON: Familial centronuclear myopathy: a clinical and pathological study. Neurology 17 (1967) 727–742.

SHIMIZU, N., Y. TAKIYAMA, Y. MIZUNO, M. MIZUNO, K. SAITO and M. YOSHIDA: Characteristics of oculomotor disorders of a family with Joseph's disease. J. Neurol. 237 (1990) 393–398.

SHORE, J.W. and C.D. MCCORD JR.: Anatomic changes in involutional blepharoptosis. Am. J. Ophthalmol. 98 (1984) 21–27.

SHOUBRIDGE, E.A., G. KARPATI and K.E. HASTINGS: Deletion mutants are functionally dominant over wild-type mitochondrial genomes in skeletal muscle fiber segments in mitochondrial disease. Cell 62 (1990) 43–49.

SHY, G.M., D.H. SILBERBERG, S.H. APPEL, M.M. MISHKIN and E.H. GODFREY: A generalized disorder of nervous system, skeletal muscle and heart resembling Refsum's disease and Hurler's syndrome. I. Clinical, pathologic and biochemical characteristics. Am. J. Med. 42 (1967) 163–168.

SIMON, L.T., D.S. HOROUPIAN, L.J. DORFMAN, M. MARKS, M.K. HERRICK, P. WASSERSTEIN and M.E. SMITH: Poly-

neuropathy, ophthalmoplegia, leukoencephalopathy, and intestinal pseudo-obstruction: POLIP syndrome. Ann. Neurol. 28 (1990) 349–360.

SLATT, B.: Hereditary external ophthalmoplegia. Am. J. Ophthal. 59 (1965) 1035–1041.

SLAVIN, M.L. and J.S. GLASER: Idiopathic orbital myositis: report of six cases. Arch. Ophthalmol. 100 (1982) 1261–1265.

SLUGA, E. and K. MOSER: Myopathy with glycogen storage and giant mitochondria: ultrastructural and biochemical findings. In: J.N. Walton, N. Canal and G. Scarlato (Eds.), Muscle Diseases. Amsterdam, Excerpta Medica (1970) 116–119.

SOLOMON, D.H., I.J. CHOPRA, U. CHOPRA and F.J. SMITH: Identification of subgroups of euthyroid Graves's ophthalmopathy. N. Engl. J. Med. 296 (1977) 181–186.

SPECTOR, R.H. and J.A. CARLISLE: Minimal thyroid ophthalmopathy. Neurology 37 (1987) 1803–1808.

SPECTOR, R.H. and M.S. FIANDACA: The 'sinister' Tolosa–Hunt syndrome. Neurology 36 (1986) 198–203.

SPIRO, A.J., G.M. SHY and N.K. GONATAS: Myotubular myopathy: Persistence of fetal muscle in an adolescent boy. Arch. Neurol. 14 (1966) 1–14.

STAAL, A., J.D. MEERWALDT, K.J. VAN DONGEN, P.G. MULDER and H.F. BUSCH: Non-familial degenerative disease and atrophy of brainstem and cerebellum. Clinical and CT data in 47 patients. J. Neurol. Sci. 95 (1990) 259–269.

STANNARD, K. and D.J. SPALTON: Sarcoidosis with infiltration of the external ocular muscles. Br. J. Ophthalmol. 69 (1985) 562–566.

STEELE, J.C., J.C. RICHARDSON and J. OLSZEWSKI: Progressive supranuclear palsy. Arch. Neurol. 10 (1964) 333–359.

STEINDLER, P., A.P. TORMENE, G.F. MICAGLIO and A. GALAN: Correlation of ERG and pigment epithelium changes in external progressive ophthalmoplegia (EPO). Doc. Ophthalmol. 60 (1985) 421–426.

STEINER, I., A. STEINBERG, Z. ARGOV, J. FABER, A. FICH and A. GILAI: Familial progressive neuronal disease and chronic idiopathic intestinal pseudo-obstruction. Neurology 37 (1987) 1046–1050.

STEPHENS, J.M., M.L. HOOVER and J. DENST: On familial ataxia, neural amyotrophy, and their association with progressive external ophthalmoplegia. Brain 81 (1958) 556–566.

STOTT, D.J., A.R. MCLELLAN, J. FINLAYSON, P. CHU and W.D. ALEXANDER: Elderly patients with suppressed serum TSH but normal free thyroid hormone levels usually have mild thyroid overactivity and are at increased risk of developing overt hyperthyroidism. Q.J. Med. 78 (1991) 77–84.

SUGIE, H., R. HANSON, G. RASMUSSEN and M.A. VERITY: Congenital neuromuscular disease with type I fibre hypotrophy, ophthalmoplegia and myofibril degeneration. J. Neurol. Neurosurg. Psychiatry 45 (1982) 507–502.

SURKS, M.I., I.J. CHOPRA, C.N. MARIASH, J.T. NICOLOFF and D.H. SOLOMON: American Thyroid Association guidelines for use of laboratory tests in thyroid disorders. J. Am. Med. Assoc. 263 (1990) 1529–1532.

SWASH, M. and M.S. SCHWARTZ: Familial multicore disease with focal loss of cross-striations and ophthalmoplegia. J. Neurol. Sci. 52 (1981) 1–10.

TAILLEFER, R. and A.C. DURANCEAU: Manometric and radionuclide assessment of pharyngeal emptying before and after cricopharyngeal myotomy in patients with oculopharungeal muscular dystrophy. J. Thorac. Cardiovasc. Surg. 95 (1988) 868–875.

TAKEDA, S., E. OHAMA and F. IKUTA: Involvement of extraocular muscle in mitochondrial encephalomyopathy. Acta Neuropathol. 80 (1990) 118–122.

TAMAI, H., T. NAKAGAWA, N. OHSAKO, O. FUKINO, H. TAKAHASHI, F. MATSUZUKA, K. KUMA and S. NAGATAKI: Changes in thyroid functions in patients with euthyroid Graves' disease. J. Clin. Endocrinol. Metab. 50 (1980) 108–112.

TAYLOR, E.W.: Progressive vagus-glossopharyngeal paralysis with ptosis: a contribution to the group of family diseases. J. Nerv. Ment. Dis. 42 (1915) 129–139.

TEASDALL, R.D., M.M. SCHUSTER and F.B. WALSH: Sphincter involvement in ocular myopathy. Arch. Neurol. (Chic.) 10 (1964) 446–448.

TELERMAN-TOPPET, N., C. COERS and J.J. DESNEUX: [External progressive ophthalmoplegia associated with a motor-sensitive polyneuropathy in a patient with celiac disease]. Rev. Neurol. 121 (1969) 57–70.

TER BRUGGEN, J.P., L.A. BASTIAENSEN, C.C. TYSSEN and G. GIELEN: Disorders of eye movement in myotonic dystrophy. Brain 113 (1990) 463–473.

THIEL, J.H.: Chronic progressive ophthalmoplegia with extra-ocular extension (ocular myopathy). Folia Psychiatr. Neurol. Neurochir. 57 (1954) 554–563.

THOMAS, D.J., M.C. CHARLESWORTH, F. AFSHAR and D.J. GALTON: Computerised axial tomography and magnetic resonance scanning in the Tolosa-Hunt syndrome. Br. J. Ophthalmol. 72 (1988) 299–302.

TOLOSA, E.: Periarteritic lesions of the carotid siphon with clinical features of carotid infraclinoidal aneurysm. J. Neurol. Neurosurg. Psychiatry 17 (1954) 300–302.

TOME, F.M.S. and M. FARDEAU: Nuclear inclusions in oculopharyngeal dystrophy. Acta Neuropathol. 49 (1980) 85–87.

TOME, F.M.S. and M. FARDEAU: Ocular myopathies. In: A.G. Engel and B.Q. Banker (Eds.), Myology: Basic and Clinical. New York, McGraw-Hill (1986) 1327–1347.

TOME, F.M.S., V. ASKANAS, W.K. ENGEL, R.B. ALVAREZ and C.S. LEE: Nuclear inclusions in innervated cultured muscle fibers from patients with oculopharyngeal muscular dystrophy. Neurology 39 (1989) 926–932.

TRIDON, P., J.J. MARTIN, M. VIDAILHET, J. FLOQUET, M. PHILIPPART and N. NEIMANN: [Juvenile cerebral spongiosis]. Pediatrie 29 (1974) 235–247.

TROKEL, S.L. and S.K. HILAL: Recognition and differential diagnosis of enlarged extraocular muscles in computed tomography. Am. J. Ophthalmol. 87 (1979) 503–512.

TROKEL, S.L. and S.K. HILAL: Submillimeter resolution CT scanning of orbital diseases. Ophthalmology 87 (1980) 412–417.

TROKEL, S.L. and F.A. JAKOBIEC: Correlation of CT scanning and pathologic features of ophthalmic Graves' disease. Ophthalmology 88 (1981) 553–564.

TROUNCE, I., E. BYRNE, S. MARZUKI, X. DENNETT, H. SUDOYO, F. MASTAGLIA and S.F. BERKOVIC: Functional respiratory chain studies in subjects with chronic progressive external ophthalmoplegia and large heteroplasmic mitochondrial DNA deletions. J. Neurol. Sci. 102 (1991) 92–99.

TRUONG, D.D., A.E. HARDING, F. SCARAVILLI, S.J. SMITH, J.A. MORGAN HUGHES and C.S. MARSDEN: Movement disorders in mitochondrial myopathies: a study of nine cases with two autopsy studies. Mov. Disord. 5 (1990) 109–117.

TULINIUS, M.H., E. HOLME, B. KRISTIANSSON, N.G. LARSSON and A. OLDFORS: Mitochondrial encephalomyopathies in childhood. I. Biochemical and morphologic investigations. J. Pediatr. 119 (1991) 242–250.

UPPAL, S.C.: Kearns' syndrome, a new form of cardiomyopathy. Br. Heart J. 35 (1973) 766–769.

VICTOR, M., R. HAYES and R.D. ADAMS: Oculopharyngeal muscular dystrophy: a familial disease of late life characterized by dysphagia and progressive ptosis of the eyelids. N. Engl. J. Med. 267 (1962) 1267–1272.

VITA, G., R. DATTOLA, M. SANTORO and C. MESSINA: Familial oculopharyngeal muscular dystrophy with distal spread. J. Neurol. 230 (1983) 57–64.

VOLPE, N.J. and J.W. SHORE: Orbital myositis associated with herpes zoster. Arch. Ophthalmol. 109 (1991) 471–472.

VON GRAEFE, A.: Verhandlungen ärztlicher Gesellschaften. Berl. Klin. Wochenschr. 5 (1868) 125–127.

VON NOORDEN, G.K.: Congenital hereditary ptosis with inferior rectus fibrosis: report of two cases. Arch. Ophthalmol. 83 (1970) 378–380.

WALSH, F.B.: Progressive external ophthalmoplegia (PEO). Trans. Pa. Acad. Ophthalmol. Otolaryngol. 22 (1969) 88–96.

WEINSTEIN, G.S., S.C. DRESNER, T.L. SLAMOVITS and J.S. KENNERDELL: Acute and subacute orbital myositis. Am. J. Ophthalmol. 96 (1983) 209–217.

WEITZNER, S.: Changes in the pharyngeal and esophageal musculature in oculopharyngeal muscular dystrophy: report of 2 cases. Am. J. Dig. Dis. 14 (1969) 805–810.

WERNER, S.C.: The severe eye changes of Graves' disease. J. Am. Med. Ass. 177 (1961) 551–555.

WIERSINGA, W.M., T. SMIT, A.L. SCHUSTER UITTENHOEVE, R. VAN DER GAAG and L. KOORNNEEF: Therapeutic outcome of prednisone medication and of orbital irradiation in patients with Graves' ophthalmopathy. Ophthalmologica 197 (1988) 75–84.

WILBRAND, H. and A. SAENGER: Die Ptosis bei der chronischen, progressiven aber isoliert bleibenden ophthalmoplegia exterior. In: Die Neurologie des Auges: Ein Handbuch für Nerven- und Augenärzte, Vol. 1. Wiesbaden, Bergmann (1900) 117–133.

WILSON, S.A.K.: Neurology, Vol. 2. Baltimore, Williams and Wilkins (1941) 1021–1023.

WOODS, B.T. and H.H. SCHAUMBURG: Nigro-spino-dentatal degeneration with nuclear ophthalmoplegia. A unique and partially treatable clinicopathological entity. J. Neurol. Sci. 17 (1972) 149–166.

YAMAMOTO, M. and I. NONAKA: Skeletal muscle pathology in chronic progressive external ophthalmoplegia with ragged-red fibers. Acta Neuropathol. 76 (1988) 558–563.

YODA, S., A. TERAUCHI, F. KITAHARA and T. AKABANE: Neurologic deterioration with progressive CT changes in a child with Kearns-Shy syndrome. Brain Dev. 6 (1984) 323–327.

ZEVIANI, M., C.T. MORAES, S. DIMAURO, H. NAKASE, E. BONILLA, E.A. SCHON and L.P. ROWLAND: Deletions of mitochondrial DNA in Kearns-Sayre syndrome. Neurology 38 (1988) 1339–1349.

ZEVIANI, M., S. SERVIDEI, C. GELLERA, E. BERTINI, S. DIMAURO and S. DIDONATO: An autosomal dominant disorder with mutliple deletions of mitochondrial DNA starting at the D-loop region. Nature (London) 339 (1989) 309–311.

ZEVIANI, M., N. BRESOLIN, C. GELLERA, A. BORDONI, M. PANNACCI, P. AMATI, M. MOGGIO, S. SERVIDEI, G. SCARLATO and S. DIDONATO: Nucleus-driven multiple large-scale deletions of the human mitochondrial genome: a new autosomal dominant disease. Am. J. Hum. Genet. 47 (1990a) 904–914.

ZEVIANI, M., C. GELLERA, M. PANNACCI, G. UZIEL, A. PRELLE, S. SERVIDEI and S. DIDONATO: Tissue distribution and transmission of mitochondrial DNA deletions in mitochondrial myopathies. Ann. Neurol. 28 (1990b) 94–97.

ZIERZ, S., G. JAHNS and F. JERUSALEM: Coenzyme Q in serum and muscle of 5 patients with Kearns-Sayre syndrome and 12 patients with ophthalmoplegia plus. J. Neurol. 236 (1989) 97–101.

ZINTZ, R.: Dystrophische Veränderungen in äusseren Augenmuskeln und Schultermuskeln bei der sog. progressiven Graefe'schen Ophthalmologie. In: E. Kuhn (Ed.), Muskeldystrophie, Myotonie, Myasthenie. Berlin, Springer (1966) 109–120.

ZINTZ, R. and W. VILLIGER: Elektronmikroskopische Befunde bei 3 Fällen von chronisch progressiver okulärer Muskeldystrophie. Ophthalmologica (Basel) 153 (1967) 439–459.

ZUPANC, M.L., C.T. MORAES, S. SHANSKE, C.B. LANGMAN, E. CIAFALONI and S. DIMAURO: Deletion of mitochondrial DNA in patients with combined features of Kearns-Sayre and MELAS syndromes. Ann. Neurol. 29 (1991) 680–683.

Handbook of Clinical Neurology, Vol. 18 (62): Myopathies
L.P. Rowland and S. DiMauro, editors
© 1992 Elsevier Science Publishers B.V. All rights reserved

CHAPTER 12

Congenital myopathies

HANS H. GOEBEL[1] and HANS G. LENARD[2]

[1]*Division of Neuropathology, University of Mainz and*
[2]*Department of Pediatrics, University of Düsseldorf, Germany*

The syndrome of the 'floppy infant' can be divided clinically into two groups. Hypotonia without marked weakness may be caused by a variety of pathological conditions ranging from cerebral defects to connective tissue abnormalities. In contrast, hypotonia plus limb weakness or paresis is encountered clinically in diseases of the motor unit, the neuromuscular disorders.

Until the mid-fifties, the spinal muscular atrophies were the only well-defined neuromuscular disorders of infancy and early childhood. Congenital muscular dystrophies had been described since Howard (1908), but are difficult to evaluate retrospectively because there was no morphologic documentation with modern techniques. Most patients were included in the ill-defined category of 'Oppenheim's amyotonia'.

This problem was clarified with the delineation of the 'new' myopathies (Dubowitz 1969), later more explicitly called 'congenital' myopathies; first among them was central core disease (Shy and Magee 1956). The number of congenital myopathies grew rapidly and now constitutes a firmly ensconced chapter in each textbook on neuromuscular disorders, nosologically corroborated by several review articles, including those which appeared in this Handbook (Bender 1979) (Table 1).

The concept of congenital myopathies developed – in parallel with the application of modern muscle pathology – slowly but imperturbably,

rather than by the brisk act of a literal definition and even though the term has been considered a misnomer (Heffner 1984).

The definitions are still imprecise, particularly in connection with the less well delineated and less frequently recognized forms that contrast with the classical group of central core disease, multicore disease, nemaline or rod myopathy and centronuclear/myotubular myopathy. However, there seems to be a consensus that congenital myopathies are a separate category of muscle disorders.

The definition of a congenital myopathy is based on both clinical features that are found in many different forms (Table 2) and on morphological criteria. Muscle biopsy is therefore mandatory for study by enzyme-histochemical, electron microscopic and, on occasion, morphometric methods. In the future, immunomocytochemical techniques and DNA hybridization are likely to be important.

A list of congenital myopathies would comprise, apart from the now more frequent classic forms, a varying number of rare conditions. The list would be distorted by the inclusions and exclusions by the individual composer, but might also grow in proportion to the number of consultant authors who would add their own claims. We suggest a necessarily preliminary classification (see Table 1). In addition to congenital myopathies marked by abnormal structures (often ultrastructural), we have recently introduced the concept of 'unstructured'

TABLE 1

Congenital myopathies.

Structured types	*References*
a. Abnormalities of sarcomeres	
1. Central core disease	Shy and Magee 1956[*]
Multicore disease	Engel et al. 1971[*]
Minicore disease	Currie et al. 1974
2. Myofibrillar lysis myopathy	Cancilla et al. 1971[*]
Selective myosin degeneration myopathy	Yarom and Shapira 1977[+]
Cap disease	Fidzianska et al. 1981[+]
3. Trilaminar fiber myopathy	Ringel et al. 1978[*+]
b. Z-disc abnormalities	
1. Nemaline/rod myopathy	Shy et al. 1963[*]
2. Mitochondria-jagged Z-line myopathy	Pellegrini et al. 1985[+]
3. Z-band plaque myopathy	Edström et al. 1990
4. Desmin-related forms	
Cytoplasmic body myopathy	Goebel et al. 1981
Spheroid body myopathy	Goebel et al. 1978
Mallory body myopathy	Fidzianska et al. 1983
Sarcoplasmic body myopathy	Edström et al. 1980
Granulo-filamentous body myopathy	Fardeau et al. 1978[+]
c. Nuclear abnormalities	
1. Myotubular myopathy	Spiro et al. 1966[*]
Centronuclear myopathy	Sher et al. 1967[*]
2. Nucleodegenerative myopathy	Schröder 1982b
a-c. Mixed myopathies	Dubowitz 1980
d. Abnormal inclusions	
1. Fingerprint body myopathy	Engel et al. 1972[*]
2. Zebra body myopathy	Lake and Wilson 1975[*]
3. Reducing body myopathy	Brooke and Neville 1972[*]
4. Tubulo-membranous inclusion myopathy	Fukuhara et al. 1981[+]
5. Cylindrical spirals myopathy	Bove et al. 1980
e. Abnormalities of organelles	
1. Sarcotubular myopathy	Jerusalem et al. 1973[*]
Tubular aggregate myopathy	Morgan-Hughes et al. 1970
Honeycomb myopathy	Carrier et al. 1976
2. Rimmed vacuole myopathy	Argov and Yarom 1984
Granulovacuolar lobular myopathy	Juguilon et al. 1982
f. Miscellaneous	
1. Abnormal myomuscular junction myopathy	Pierobon-Bormioli et al. 1980[+]
2. Minimal change myopathy	Dubowitz 1978
Unstructured types	
Congenital fiber type disproportion	Brooke 1973[*]
Type I fiber predominance	Brooke 1977
Uniform type I fiber myopathy	Oh and Danon 1983
Microfiber myopathy	Hanson et al. 1977[+]
Type II muscle fiber hypoplasia	Matsuoka et al. 1974[+]
Type I myofiber hypotrophy	Prince et al. 1972
Congenital myasthenic syndromes	Engel 1988
– Familial infantile myasthenia	
– Congenital endplate acetylcholinesterase deficiency	
– Slow channel syndrome	
– Congenital endplate AChR deficiency	
Partially characterized congenital myasthenic syndromes	
– AChR deficiency and increased affinity for d-tubocurarine	
– Decreased MEPP amplitude without AChR deficiency	
– Possible defect in ACh synthesis, mobilization or storage	
– Familial limb-girdle myasthenia	

[*] Congenital myopathy discussed by Bender (1979).

[+] Congenital myopathy of uncertain nosological connotation, i.e. only described in single individuals.

TABLE 2

Non-specific clinical criteria of congenital myopathies.

Early (childhood) onset
Slowly (or not) progressive course
Often hereditary
Generalized or proximal weakness
Generalized or proximal muscular atrophy
Muscular hypotonia
Contractures
Normal or mildly elevated serum CK
Normal or 'myopathic' EMG

congenital myopathies, that are characterized morphologically only by enzyme-histochemical and fiber diameter pathology (Lenard and Goebel 1980).

The nosological and terminological scope of congenital myopathies is probably similar to that of the muscular dystrophies, which also comprises an established core group of Duchenne, Becker, oculopharyngeal, and facioscapulohumeral muscular dystrophies as well as a variable number of fringe conditions.

We will not repeat material already presented by Bender (1979), but will supply brief definitions of previously canvassed congenital myopathies. We will add recent data, including immunological studies, and deal with congenital myopathies that were not mentioned earlier.

Clinical findings

Clinically, the patient with a congenital myopathy (see Table 2), is usually either a floppy infant or a hypotonic child with delayed or deficient motor development, with proximal or generalized limb weakness and wasting, and no or slow progression of the disease. The genetic nature of the disease may be demonstrable by family history or by examination of family members. However, even in familial cases, there may be late onset of clinical symptoms and of primarily distal weakness and atrophy (Goebel 1986).

Common abnormalities in these diseases include normal or mildly abnormal, chiefly 'myopathic' electromyographic (EMG) findings and normal or slightly elevated creatine kinase (CK) levels in serum. Several clinical features are specific but not pathognomonic, of some congenital myopathies,

including weakness of facial or ocular muscles, respiratory failure, skeletal deformities, or cardiac disorders (Table 3). More specific clinical findings, however, seem to be confined to only a few congenital myopathies (Table 4).

STRUCTURED CONGENITAL MYOPATHIES

Abnormalities of sarcomeres

Core diseases. The core diseases comprise central core disease (Figs. 1 and 2) and multicore or minicore disease (Fig. 3). The discovery of central core disease (Shy and Magee 1956) established the concept of congenital myopathies. It is probably also the most frequently encountered.

A core is a circumscribed lesion of myofibrils, located either in the center of the muscle fiber (central core), or at the periphery. Within the core, the sarcomeres may still be in register (structured cores) or disrupted (unstructured cores). Cores may be seen one to a myofiber or several (multi- or minicores). It is largely devoid of mitochondria. By enzyme-histochemical and ultrastructural criteria, several forms of cores are distinguished (Table 5).

Central core disease. The flood of papers dealing with clinical and morphological aspects of human central core disease (CCD) has abated in recent years because little has been added to what was known 10 years ago.

The clinical varieties include mild, moderate, and even severe forms. Affected individuals are often susceptible to malignant hyperthermia (MH) (Shuaib et al. 1987). The gene locus for MH has been localized to the chromosome 12q 12-13.2, but linkage studies of three families with CCD located their CCD gene to chromosome 19q (Kausch et al 1990). A non-CCD congenital myopathy was associated with MH in six American-Indian children (Stewart et al. 1988).

Clues to the morphological dynamics and pathogenesis of cores have come from experimental data. Bender (1979) described induction of cores by tenotomy, triethyltin sulfate and emetine intoxication. These cores were seen mostly in type I fibers. In human neuropathy after accidental organophosphate intoxication, core and targetoid lesions were seen in both type I and type II fibers,

TABLE 3

Specific, but not pathognomonic, clinical features of congenital myopathies.

	Facial involvement	Extraocular muscle weakness (ptosis and/or external ophthalmoplegia	Respiratory insufficiency	Skeletal deformities (face, trunk, limbs)	Cardiac involvement
Central core disease	x		x	x	
Multicore disease		x	x		x
Minicore disease	x	x			
Selective myosin degeneration myopathy				x	
Cap disease	x			x	
Trilaminar fiber myopathy	x		x		
Nemaline myopathy	x	x	x	x	x
Mitochondria-jagged Z-line myopathy	x				
Z-line disorganization myopathy			x		
Cytoplasmic body myopathy	x		x		x
Mallory body myopathy	x		x		
Sarcoplasmic body myopathy			x		x
Granulo-filamentous myopathy			x		x
Myotubular/centronuclear myopathy		x	x	x	
Reducing body myopathy			x		
Tubulo-membranous myopathy	x			x	
Tubular aggregate myopathy	x				x
Myomuscular junction myopathy	x	x			
Congenital fiber type disproportion	x	x	x	x	
Uniform type I fiber myopathy	x			x	
Congenital myasthenic syndromes	x	x	x		

more in type I (Fukuhara et al. 1977). The beige mouse spontaneously forms central core-like lesions in type I and type II fibers, and may be an animal model of CCD (Kirkeby 1981). Formation of cores by tenotomy seems to depend more on innervation of the myofibers than on fiber type, because cores were not formed when neurotomy (Otte et al. 1980) or spinal cord lesions (De Reuck et al. 1982) were performed simultaneously with tenotomy. When the relative amount of type II fibers in the soleus muscle was increased by experimentally-induced hyperthyroidism, core degeneration after tenotomy was also seen in those fibers (Hall-Craggs et al. 1983). Tenotomy-induced cores are not permanent, even when the tendon lesion is not repaired and the muscle remains in a permanent shortened position. Complete recovery was achieved not only by regeneration and reorientation of myofilaments and sarcomeres, but also by amplified reappearance of ribosomes and rough endoplasmic reticulum (Baker and Hall-Craggs 1980).

These experimental lesions were unstructured cores; they were lengthy central stripes of disrupted amitochondrial sarcomeres, often with a peripheral rim and therefore indistinguishable

TABLE 4

Special clinical features in congenital myopathies.

Myalgia	Spheroid body myopathy
	Cylindrical spirals myopathy
	Tubular aggregate myopathy
	Honeycomb myopathy
Cramps	Cylindrical spirals myopathy
	Tubular aggregate myopathy
Dysphagia	Spheroid body myopathy
	Cytoplasmic body myopathy
	Granulo-filamentous myopathy
Rigidity	Trilaminar fiber myopathy
Hypertrophy of calves	Cytoplasmic body myopathy
	Abnormal myomuscular junction myopathy
	Congenital fiber type disproportion
Myotonic reactions	Cylindrical spirals myopathy
	Granulovacuolar lobular myopathy
Mental retardation	Nucleodegenerative myopathy
	Fingerprint body myopathy
	Congenital fiber type disproportion
Myasthenic features	Tubular aggregate myopathy
	Congenital myasthenic syndrome
Temporomandibular ankylosis	Nemaline myopathy
Tremor	Trilaminar fiber myopathy
	Spheroid body myopathy
Malignant hyperthermia	Central core disease

from targets. Similarly, core fibers developed after immobilization in a shortened position of the soleus muscle, and such unstructured cores were morphologically identical to those seen after tenotomy and identical to target fibers (Bruce-Gregorios and Chou 1984). Mitochondria within the core lesions seem to degenerate as the cores develop, indicating that loss of mitochondria does not precede core formation under experimental conditions.

Thus, experimental data indicate that both mechanical (tenotomy, immobilization, or local tetanus) and neural factors are responsible for core formation. They also provide evidence that there are no essential differences (Mattle and Jerusalem 1981) between central cores, reversed cores (Radu et al. 1977) and locular cores (Mattle and Jerusalem 1981) (see Table 5), target and targetoid structures and multicores (Bruce-Gregorios and Chou 1984). The Z-band is frequently altered in these experimental core lesions, so that nemaline rods and cytoplasmic bodies (inclusions thought to be derived from Z-bands) may be seen with core lesions.

However, none of the factors thought to be re-

sponsible for experimental core formation has been unequivocally documented in patients with

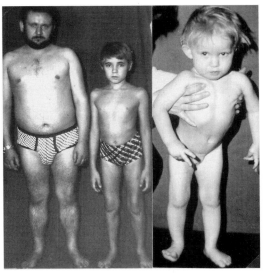

Fig. 1. Central core disease: a father and his two children are affected. Reprinted with permission by the editor, M. Adachi, M.D., from M. Adachi and J.H. Sher (Eds.), Neuromuscular Disease, Igaku-Shoin, New York, Tokyo (1990).

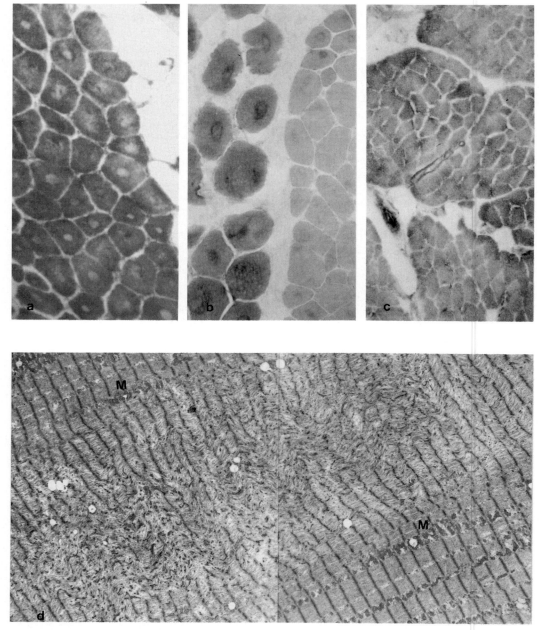

Fig. 2. Central core disease: (a) Cores largely marked by a central substrate-free area, some of them surrounded by a dark rim. NADH-tetrazolium reductase preparation, × 184. (b) A non-specific esterase preparation also demonstrates cores, though more enzymatically active than surrounding non-core areas, × 128. (c) Only type I fiber uniformity (as seen in a) but no cores are present in an affected patient, whose father and sister had central core disease. NADH preparation, × 136. (d) A core is marked by sarcomeres in disarray with marginal sarcomeres out of register compared to surrounding regular myofibrils which display rows of mitochondria (M), absent in core region, demonstrating aspects of both 'unstructured' and 'structured' core, ×3,870.

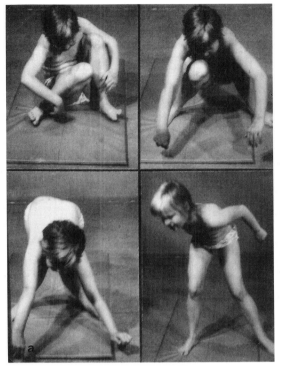
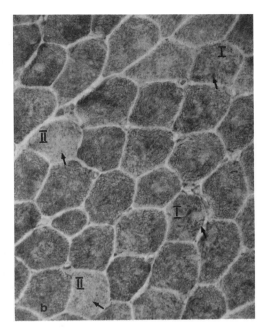

Fig. 3. Multicore disease: (a) A young girl demonstrates muscle weakness (by courtesy of Dr. Doose, University of Kiel, Germany). (b) Numerous multicores, i.e. substrate-free areas (arrows) are present in type I (dark) and II (light) myofibers. NADH preparation, × 320.

either central core or multicore disease. It can be stated only that the morphological features of ex-

perimentally-produced cores conform to the spontaneous cores in human conditions. Human cen-

TABLE 5

Morphological definition of cores.

Central cores (Greenfield et al. 1958)	
Structured cores	NADH-TR negative, ATPase positive, absence of mitochondria, contraction, but preservation of sarcomeres in register
Unstructured cores	NADH-TR negative, ATPase negative, absence of mitochondria excessive Z-band streaming with disruption of sarcomeric register
Reversed cores (Radu et al. 1977)	NADH-TR negative, rich in mitochondria
Locular cores (Mattle and Jerusalem 1981)	NADH-TR positive network from marginal zone results in locular appearance of central cores
Atypical cores (Kar et al. 1980)	small, NADH-TR positive, ATPase negative, myofibrillar lysis, absence of mitochondria, increased glycogen ribosomes, sarcotubular profiles
Multicores (Engel et al. 1979)	NADH-TR negative, absence of mitochondria, disruption of sarcomeres by Z-band streaming
Minicores (Currie et al. 1974)	
Miniature cores (Bethlem et al. 1978)	
Multifocal loss of cross striation (Engel 1967)	

tral cores show abnormalities of the actin and myosin components, and also of the network of intermediate filaments that are disorganized in the central core regions (Thornell et al. 1983). Abnormalities of sarcoplasmic reticulum and T-tubules may be seen within the cores and, to a lesser extent, outside cores (Hayashi et al. 1989).

Core myofibers have also been produced locally by applying tetanus toxin (Chou et al. 1981), establishing a reproducible sequence of morphological events within soleus type I myofibers. First, there is disruption of the Z-disc with multicore formation, followed by the appearance of central unstructured cores with loss of mitochondira and atrophy of muscle fibers. Finally, a stage of recovery is marked by formation of structured cores due to regeneration of myofilaments that originate from peripheral myofibrils (Chou 1985). The frequently observed increased numbers of central cores with advancing age of patients may be due to a failure of recovery from transient core formation. Structured and unstructured cores within the same muscle (Goebel 1986) may be different stages of the same basic process. Mutually exclusive occurrence of only structured (Frank et al. 1980) or unstructured cores (Cohen et al. 1978; Kar et al. 1980; Byrne et al. 1982) is difficult to reconcile with the experimental data and cannot be explained by insufficient sampling of muscle specimens, because both types of cores have been encountered in several biopsies of one patient (De Giacomo et al. 1970) or in several patients in the same family (Byrne et al. 1982). CCD has been considered a neurogenic disorder because of the morphological similarities of target or targetoid fibers found in neurogenic atrophy, and central cores (Mattle and Jerusalem 1981). Also, both structures are encountered mostly in type I fibers.

Furthermore, type I fiber predominance of CCD is thought to be due to abnormalities of innervation (Engel 1977). Other evidence of a neurogenic origin include the following: a double-peak histographic appearance of one fiber type (Pages and Pages 1981); late components of motor unit potentials with the EMG, attributed to collateral innervation of split myofibers (Lopez-Terradas and Conde Lopez 1979); and an increased terminal innervation ratio, likewise attributed to collateral innervation (Coërs et al. 1976).

Type I fiber predominance or uniformity in CCD may precede the formation of cores, as suggested by study of affected family members at different ages. The occurrence of cores in addition to type I fiber predominance was considered evidence of clinical deterioration by Pou-Serradell et al. (1980). The association of cores and type II fiber predominance has been observed in younger patients in a family where older patients showed type I fiber predominance, suggesting conversion of fiber type with longer duration of the disease (Fukunaga et al. 1980). Cohen et al. (1978) observed CCD in only one of identical twins, suggesting that the disorders may be not only genetic but also influenced by non-specific maturational arrest.

Evolution of multicore pathology into central core and rod lesions was described in a 10-year-old boy (Vallat et al. 1982), as seen in two biopsies obtained 4 years apart. These observations suggest that persistence of morphological features in the investigation of congenital myopathies may be more apparent than real because, especially in children, muscle biopsy is one of the less frequently repeated or performed diagnostic tests. For instance, in one study of 37 clinically affected members in five generations, only two had muscle biopsies (Byrne et al. 1982). Finally, long-term endurance training may improve clinical weakness in CCD patients (Hagberg et al. 1980).

Multicore/minicore myopathy. Ten years after the original description of CCD, the 'multifocal degeneration of muscle fibers' (Engel and Gomez 1966) was considered a separate congenital myopathy, 'multicore disease' (Engel et al. 1971), with morphological features (Fig. 3, Table 6) similar to lesions observed in other patients with congenital neuromuscular conditons, i.e. 'focal loss of cross striation' (Engel 1967) and 'target-like fibers' (Schotland 1969).

Other children, including siblings, with apparently the same muscle pathology, have been described as having minicore disease (Currie et al. 1974; Lake et al. 1977; Gullotta et al. 1982) or 'myopathy with multiple minicores' (Ricoy et al. 1980). Today, the terms multicore and minicore are used interchangeably (Bethlem et al. 1978; Penegyres and Kakulas 1991).

The clinical symptoms of patients with multi-

TABLE 6

Multicores in neuromuscular diseases.

Congenital myopathies	Non-congenital myopathies
Central core disease	Ischemia
Nemaline myopathy	Vacuolar myopathies
Centronuclear myopathy	Atypical type II glycogenosis
	Type III glycogenosis
	Sjøgren syndrome
	Steroid myopathy
	Osteomalacic myopathy
	Denervation
	Emetine myopathy
	Chloroquine myopathy
	Duchenne muscular dystrophy

core disease conform to the general clinical pattern of congenital myopathies. Weakness tends to affect the arms more than the legs. Mild facial weakness, external ophthalmoplegia (Swash and Schwartz 1981), scoliosis and respiratory failure (Fitzsimons and Tyler 1980), or cardiomyopathy –even fatal (Shuaib et al. 1988) – are occasionally additional findings. The prognosis is favorable for the majority of patients, but clinical deterioration may occur (Penegyres and Kakulas 1991). Pellissier et al. (1979) found multicores in a 4-year-old girl with type III glycogenosis whose brother had the same metabolic disorder but apparently no multicores. Koch et al. (1985) recorded post-anesthetic fever of unexplained cause in a $2\frac{1}{2}$-year-old boy who died 26 hours after the hyperthermic event; that may have been a forme fruste of CCD, because both central cores and multicores were seen in a mother and her son who were also susceptible to MH (Frank et al. 1980).

Beside the non-specific enzyme-histochemical features of congenital myopathies, i.e. type I fiber predominance and type I fiber atrophy (Lake et al. 1977; Pellissier et al. 1979), multi- or minicores characterize the morphological picture. These small lesions are seen within muscle fibers where the regular myofibrillar pattern of sarcomeres is altered due to loss of cross striation, and may affect several continuous sarcomeres in a pattern that resembles 'Z-band streaming'. Monosarcomeric streaming of the Z-disc is probably the original lesion, which then forms multicentric mini- or

multi-cores by spreading into adjacent sarcomeres. 'Structured' multi- or minicores have not been described unequivocally.

In addition, there may be loss of mitochondria, of glycogen, and of elements of the sarcotubular system. The lack of mitochondria gave rise to one of the pathogenetic explanations, i.e. primary loss of mitochondria resulting in local disruption of sarcomeric integrity (Engel et al. 1971), a theory supported by quantitative analysis that revealed areas devoid of mitochondria, but without concomitant disruption of sarcomeres.

Abnormally structured mitochondria have been observed in multicore or minicore disease (Fitzsimons and Tyer 1980; Gullotta et al. 1982), but the pathogenetic significance of this association remains obscure, especially because normal mitochondria were still present in some minicores (Gullotta et al. 1982), suggesting a defect in myosin and actin (Lake et al. 1977) that can be recognized as substrate-free areas in stains for adenosine triphosphatase (ATPase) (Engel et al. 1971). Similar areas of focal deficiency of ATPase have been seen in a myopathy with selective myosin degeneration (Yarom and Shapira 1977) and also in three infants where absence of heavy chain myosin was demonstrated by monoclonal antibodies (Fardeau et al. 1985). Z-band streaming is a frequent ultrastructural feature of many neuromuscular conditions; it is therefore not surprising to observe its larger variety, minicores or multicores, as a nonspecific phenomenon in other neuromuscular disorders. Minicores or multicores may not be present in each biopsy specimen of an individual patient (Penegyres and Kakulas 1991).

Myofibrillar lysis myopathy. In this disorder and the following congenital myopathies, the conditions may result from failure of protein synthesis.

Myofibrillar lysis was described in two siblings (Cancilla et al. 1971) who had the non-specific clinical findings of a congenital myopathy. Muscle fibers (Fig. 4), particularly those of type I, showed large areas that were devoid of sarcomeres, myofilaments and organelles, but were replete with a finely granular amorphous material of non-glycogen character. These morphological features were attributed to lysis of myofibrils (Cancilla et al. 1971). Later myomorphological studies of the

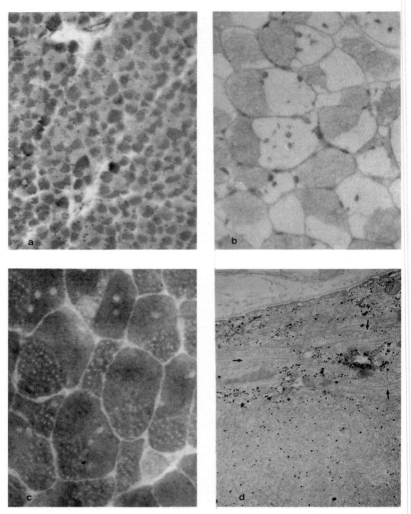

Fig. 4. Myofibrillar lysis myopathy in siblings: (a) Myofibers are separated into dark myofibrillar areas and light myofibril-free sectors, younger brother. Modified trichrome stain, ×113. (b) The light clear areas contain numerous large nuclei but are devoid of glycogen, older sister. PAS, ×300. (c) The ATPase preparation reveals homogeneous activity in myofibril-free sectors (enzyme-free spots represent floating nuclei) separated from myofibrillar areas marked by ice crystal artefacts, older sister. Alkaline ATPase, ×240. (d) Subsarcolemmal remnants of myofibrils (arrows) are sharply separated from finely granular area free of myofilaments, ×20,000.

same patients showed that these areas were devoid of oxidative (NADH, MAG) enzyme activities (Goebel 1986, 1990), phosphorylase, and acid phosphatase, while alkaline ATPase activity was present. Ultrastructurally, the areas were devoid of organelles, including autophagic vacuoles or lysosomes (Goebel 1986, 1990). We interpreted the lesions as defects in protein synthesis that resulted in focal, often largely sectorial absence of regular sarcomeres. The disorder ran a protracted course and

the siblings are now in wheelchairs, severely handicapped with respiratory problems (DeMyer, personal communication 1988).

Apparently unaware of the original report, Sahgal and Sahgal (1977) described an adult woman with a myopathy that had been stationary from early childhood. Her deltoid muscle contained subsarcolemmal areas – albeit defined as bodies rich in sulphur – that were ultrastructurally and histochemically similar to the intramyofibrillar lesions seen in the original patients of Cancilla et al. (1971). The term 'myofibrillar lysis' has also been used, in a different connotation, to describe minicores associated with type I fiber muscle hypotrophy and central nuclei (Radu et al. 1974). We consider this an inappropriate term for minicores or as used by Cancilla et al. (1971).

Yarom and Shapira (1977) described a 1-year-old child with 'myosin degeneration in a congenital myopathy'. There were focal areas within myofibrils in which the sarcomeres lacked myosin. Absence of myosin breakdown or increased lysosomal activity suggested that the condition was caused by a focal defect in myosin synthesis, comparable to the inhibition of myosin production by 5-bromodeoxyuridine in cultured chicken cardiac muscle fibers (Chacko et al. 1975).

'Cap' disease. Similar findings, i.e. large though subsarcolemmal areas of sarcomere-like structures without ATPase activity and myosin filaments termed 'caps', were described in a different stable congenital myopathy, seen in a 7-year-old boy (Fidzianska et al. 1981). The caps, by light microscopy, resembled sarcoplasmic masses, and were thought to indicate an abnormality in the fusion of immature muscle fibers; in that process, large areas devoid of myofibrils may also be seen transiently. However, in 'cap' disease the abnormal sarcomere-like inclusions were located in an abnormal position and were of an abnormal fine structure, with no phagocytosis or abnormal numbers of lysosomes, suggesting a 'defect in formation of myofibrils' rather than excessive breakdown.

A mildly progressive myopathy in a 17-year-old boy also showed similar subsarcolemmal 'cap' lesions, called 'subsarcolemmal-segmental myofibrillolysis'. The lesions were seen in two biopsies, obtained three years apart (Schröder 1982a).

Trilaminar muscle fiber myopathy. This congenital myopathy was described in an infant girl (Ringel et al. 1978) who had a peculiar set of symptoms — marked rigidity, reduced spontaneous movements, abnormally hard muscles on palpation and elevated CK values. The biceps muscle, biopsied at 7 weeks of age, disclosed numerous 'trilaminar fibers', consisting of an outer zone resembling sarcoplasmic masses, a middle zone of myofibrils separated from each other by extrajunctional acetylcholine receptors and an inner zone of glycogen, mitochondria and a few myofilaments. Another young girl (Schröder and Schönberger 1980) also had numerous 'trilaminar' myofibers as the morphological hallmark of a congenital myopathy with a much more malignant course. She died at age 10 months. Her symptoms differed from those of the first girl in that rigidity was absent, but there was a tremor, muscles were hypotonic and soft on palpation, and CK values were normal. Two older brothers had died of a clinically similar disorder, suggesting a familial disease but that was not ascertained by biopsy.

This trilaminar muscle fiber myopathy differs from 'double ring myopathy' (Coulter et al. 1990), in which glycogen accumulation and disrupted myofibrillar material more closely resembled caps or sarcoplasmic masses.

Z-disc abnormalities

Nemaline (rod) myopathy. In 1963, seven years after the concept of congenital myopathies was initiated by the description of CCD, another congenital myopathy was described independently as nemaline or rod myopathy (Shy et al. 1963) and as a congenital myopathy marked by 'myogranules' (Conen et al. 1963).[*]

These two reports were followed by a steady stream of case reports until nemaline or rod body myopathy was nosologically firmly established as a slowly or non-progressive disorder or one that might be rapidly fatal (Schmalbruch et al. 1987). The severe form was often marked by severe neonatal hypotonia (Adelaida Martinez and Lake 1987).

These patients, in addition to proximal or generalized limb weakness and wasting, often had skeletal deformities and facial weakness. A cardiomyopathy was associated with nemaline myopathy

---

[*] In this context, it is of historical interest to record that in 1958 Dr. R.D.K. Reye of Sydney, Australia, discovered rods in a paraffin-embedded, PTAH-stained muscle specimen of a 3-year-old boy with muscle weakness. He called them 'rod-like fragments' but others thought they were artifacts (Adams et al. 1965) (Fig. 5).

in one child (Van Antwerpen et al. 1988) and two adults, following the skeletal muscle symptoms (Stoessl et al. 1985) or preceding them (Meier et al. 1984). Repeated EMG recordings may parallel and document progression of the rod myopathy (Wallgren-Pettersson 1989; Wallgren-Pettersson et al. 1989).

The morphological pattern of rod myopathy (Fig. 5) is frequently marked by type I fiber predominance, type I fiber atrophy or irregular distribution, and by numerous aggregates of nemaline or rod bodies that are made up of regular giant lattices of Z-disc, occasionally even within nuclei (Paulus et al. 1988). Studies at different ages revealed progression of the rod myopathy and increasing lysosomal activity (Nonaka et al. 1989). Numerical increase and centralization of rods, with more variation and atrophy of myofibers gave further evidence of progression (Wallgren-Pettersson et al. 1988). Fiber type transformation (from type II to type I myofibers) was also more evident with increasing age (Miike et al. 1986).

Similar inclusions have also been seen in myocardial fibers in two adult sisters with nemaline myopathy and cardiomyopathy (Meier et al. 1984) The lesions were absent in cardiac myofibers of another patient with rods and cardiac myopathy (Stoessl et al. 1985).

Nemaline myopathy often runs a non-progressive or slightly progressive clinical course, but respiratory failure often leads to an untimely death in infancy (Matsuo et al. 1982; Norton et al. 1983) or later in life (Simpson and Hewlett 1982; Dodson et al. 1983). Autopsy studies of fatal nemaline myopathy have shown widespread formation of rods in numerous muscles, including the diaphragm (Dodson et al. 1983; Norton et al. 1983; Meier et al. 1984). Rarely, the diagnosis of nemaline myopathy was established only at autopsy (Simpson and Hewlett 1982).

Ophthalmoplegia has been seen with nemaline myopathy (Fukunaga et al. 1980). In their patients, abnormal mitochondria with crystalline inclusions were seen by electron microscopy. A few other combined instances of rods and abnormally structured mitochondria have been recorded (Kornfeld 1980; Tanaka et al. 1984), but the light microscopic equivalent of 'ragged red fibers' was not seen. Since a defect in mitochondrial metabolism has not

yet been documented, the implications of abnormally structured mitochondria in nemaline myopathy remain obscure. Equally unexplained is the occurrence of nemaline myopathy in some individuals and of a mitochondrial myopathy in other members of the same family (Shapira et al. 1981).

Nemaline bodies and abnormally structured mitochondria have also been seen in cricopharyngeal muscle specimens from patients with autosomal-dominant cricopharyngeal dysphagia (Hanna and Henderson 1980). The changes in cricopharyngeal muscle fibers were thought to be directly responsible for the functional impairment, establishing a close link between rod-related myopathology and loss of function. These patients had no clinical evidence of a generalized rod myopathy, which, however, could not be ruled out without biopsy because there were rods in muscle fibers of clinically unaffected relatives of other patients with rod myopathy (Kondo and Yuasa 1980). Similarly, temporomandibular ankylosis was the prominent symptom in one patient with atypical nemaline myopathy (Powers et al. 1980); myofibers biopsied from various head and limb muscles demonstrated ubiquitous presence of rods. Another link between myopathology and impaired muscle function may have been present in young infants who died of nemaline myopathy with respiratory failure and whose diaphragm myofibers were studded with rods (Tsujihata et a. 1983).

Based on studies of three families, there may be two genetic types of congenital rod myopathy, one autosomal-dominant and the other autosomal-recessive (Arts et al. 1978). An extensive review of 50 reported pedigrees of nemaline myopathy families, including asymptomatic family members whose biopsied muscles demonstrated nemaline bodies, suggested that inheritance is probably autosomal-dominant with reduced penetrance (Kondo and Yuasa 1980).

In Finland, however, autosomal-recessive transmission was suggested (Wallgren-Pettersson et al. 1990a). Morphologic signs of immaturity have been shown repeatedly in rod myopathy, i.e. an increased number of satellite cells (Nonaka et al. 1983a; Tsujihata et a. 1983) as well as small aggregates of tiny muscle fibers surrounded by a common basement membrane and frequent type IIC fibers (Nonaka et al. 1983a). This aspect of muscle

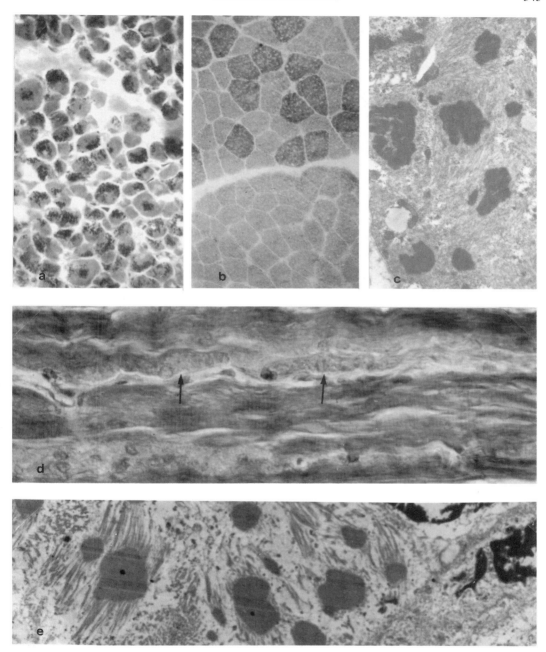

Fig. 5. Nemaline myopathy: (a) Almost every muscle fiber contains aggregates of dark rods (nemaline bodies). Modified trichrome stain, × 150. (b) Uneven type I fiber (light) predominance, not associated with rod formation. MAG (menadione-linked-alpha-glycerophosphate dehydrogenase) preparation, ×280. (c) Rods represent large areas of Z-disc material, ×23,000. (d) Rods in subsarcolemmal areas (arrows) of myofibers, observed by Dr. Reye of Sydney, Australia. Phosphotungstic acid stain, ×880 (by courtesy of Dr. Bale, Royal Alexandra Hospital for Children, Sydney, Australia). (e) Numerous rods and remnants of sarcomeres from the same muscle specimen as depicted in d, × 8,400 (by courtesy of Dr. Kan, Royal Alexandra Hospital for Children, Sydney, Australia).

immaturity appears largely confined to young infants.

Nemaline bodies are apparently derived from Z-discs, because they consist of actin (Yamaguchi et al. 1982), tropomyosin and alpha-actinin (Shimizu et al. 1990) and because the fine structural lattice is similar to that of the regular Z-disc. As alpha-actinin is increased in nemaline myopathy muscles (Stuhlfauth et al. 1983; Wallgren-Pettersson et al. 1990b), but normal in non-muscle cells (Jennekens et al. 1983) nemaline myopathy does not seem to be a multisystem disease. Unrelated to rods, aberrant development of fast-type heavy-chain myosin has been recorded (Shimizu et al. 1990) and persistence of fetal myosin in small myofibers has suggested delayed maturation (Sewry et al. 1990).

Anti-desmin antibodies mark areas of Z-band streaming, indicating increased presence of desmin in these areas, but anti-desmin antibodies do not react with nemaline bodies (Thornell et al. 1983). Optical diffraction studies have confirmed the similarity of nemaline bodies and Z-discs (Goldstein et al. 1980), but they also suggested an additional stabilizing protein (not actin) which is present in both nemaline bodies and Z-discs (Yamaguchi et al. 1975). Cryotechniques gave further evidence that nemaline bodies are composed of Z-disc material (Thornell 1978).

Apart from nemaline myopathy, rods have been encountered as scattered, numerically rare features in other neuromuscular conditions, including human-immunodeficiency-virus-related myopathy (Cabello et al. 1990).

Mitochondria-jagged Z-line myopathy. Another aspect of Z-band pathology associated with an abnormal mitochondrial network and type I fiber uniformity was described as 'jagged Z-lines' in a 9-year-old boy (Pellegrine et al. 1985). This may be another example, like the decrease or absence of mitochondria in core lesions, of abnormalities in spatial density and arrangement of mitochondria, although structural abnormalities of mitochondria are not part of the pattern.

Z-band plaque myopathy. A third example of Z-disc pathology was encountered in an autosomal-dominant myopathy. The clinical pattern was characterized by late onset, but early involvement of respiratory muscles (Edström et al. 1990). Muscle fibers contained peculiar F-actin-positive hypereosinophilic plaques, which ultrastructurally corresponded to Z-line-related electron-dense material, suggesting defective myofibrillogenesis.

The cytoplasmic-spheroid body myopathies

Cytoplasmic body myopathy. Cytoplasmic bodies were first described as a non-specific feature in several neuromuscular disorders (Engel 1962; Goebel 1986) and as another Z-band abnormality (MacDonald and Engel 1969). Neuromuscular disorders primarily characterized by cytoplasmic bodies – once also called 'myofibrillar aggregates' (Kinoshita et al. 1975b) – have only recently been described, but the number of recorded patients now warrants consideration of cytoplasmic body myopathy as a nosological entity. The lesions are not uniform in appearance, but include congenital (benign and severe), juvenile, and adult forms (Mizuno et al. 1989), as well as sporadic (Jerusalem et al. 1979; Goebel et al. 1981; Mizuno et al. 1989) and familial cases (Clark et al. 1978; Patel et al. 1983; Dickoff 1988). Apart from muscle weakness, respiratory insufficiency is frequent (Dickoff 1988; Mizuno et al. 1989) and life-threatening (Dickoff 1988; Spargo et al. 1988).

Myopathic changes in myofibers and pronounced predominance of type I fibers with numerous cytoplasmic bodies (Figs. 6a–c) were seen in a 15-year-old girl with a life-long history of limb weakness and wasting (Goebel et al. 1981). The cytoplasmic bodies were immunohistochemically shown to be composed of desmin intermediate filaments (Osborn and Goebel 1983). Thereafter, the presence of desmin has been documented repeatedly (Dickoff 1988; Sawicka et al. 1990), so often that desmin-related myopathies have now been fairly well delineated (Pellissier et al. 1989). Although desmin has now been firmly established as a component of cytoplasmic bodies, similar structures may be devoid of desmin or may contain actin (Schröder et al. 1990), alpha-actinin and cathepsins B, H and L (Fukuhara et al. 1990). These differences suggest heterogeneity rather than homogeneity of cytoplasmic bodies or of the cytoplasmic-spheroid complex – to be discussed below.

Ingestion of organophosphate (Fukuhara et al. 1977) and ipecac syrup (Mateer et al. 1985; Halbig et al. 1988) may result in intramuscular formation of cytoplasmic bodies. Together with desmin-prominent lesions, they have been induced experimentally by emetine injections in rats (Hopf and Goebel 1989). Abnormal desmin aggregation has also been documented in cardiac muscle fibers (Stoeckel et al. 1981; Voit et al. 1988), and cardiomyopathy has been recorded as a feature of cytoplasmic body myopathy (Spargo et al. 1988).

Spheroid body myopathy. In an autosomal-dominant slowly progressive neuromuscular disorder (Goebel et al. 1978), similar but usually larger and more numerous inclusions (Figs. 6d–e) were termed 'spheroid' bodies. Filamentous and granular components were less clearly demarcated from each other than in cytoplasmic bodies which themselves were not identified in this myopathy. The slowly progressive syndrome extended over more than fifty years. In a biopsy obtained 17 years after the original study from another affected relative, identical spheroid body complexes were surrounded immunohistochemically by desmin (by courtesy of J. Muller M.D. Indianapolis/USA).

This large kinship with 'spheroid body myopathy' has been explored by linking other patients from Oregon, independently communicated earlier (Clark et al. 1978), with one Indiana family (D.J. Dickoff, personal communication, 1988). Other genetically independent patients have had the same spheroid body myopathy (Dickoff 1988).

Granulo-filamentous myopathy. An autosomal-dominant familial congenital myopathy of late onset and associated with cardiomyopathy, was marked morphologically by numerous granulofilamentous non-membrane-bound inclusions. The changes were located in the periphery of the muscle fibers, often close to Z-disc, and predominantly in type I fibers which outnumbered type II fibers (Fardeau et al. 1978) Determination of phosphorylated desmin within the granulofilamentous inclusions in biopsies of four individuals from two families identified the inclusions as similar to the electron-dense component of leptomeres or zebra bodies (Rappaport et al. 1988). This myopathy is now considered a member of the cytoplasmic-spheroid body complex group (Chapon et al. 1989; Pellissier et al. 1989). Findings in two unrelated adults with slowly progressive muscle weakness included cytoplasmic bodies and larger inclusions of the 'spheroid' type (Fig. 6f), which also contained desmin by immunofluorescence (Fig. 6g), possibly a morphological transition between originally separate cytoplasmic bodies and the spheroid bodies (Halbig et al. 1991). Spheroid-cytoplasmic bodies have been induced experimentally in 'white' plantaris muscle by local tetanus (Chou and Mizuno 1984), and these bodies also contained desmin intermediate filaments (Chou 1985).

Mallory body-like inclusion myopathy. In three children from a large kinship (Goebel et al. 1980; Fidzianska et al. 1983) with a possibly autosomal-recessive neuromuscular condition, there were inclusions within muscle fibers that resembled hepatic Mallory bodies by electron microscopy and contained desmin by immunohistology. The children had a life-long history of muscle weakness and wasting. Two girls died suddenly of cardiorespiratory failure in early adolescence. Two boys were less severely affected. The older one, biopsied again after 11 years, still showed intramuscular inclusions.

The existence of Mallory bodies in muscle has been confirmed independently by Dubowitz (1985). The presence of desmin-intermediate filaments in intramuscular Mallory body-like inclusions is not surprising, because Mallory bodies in hepatic and renal epithelial cells contain cytokeratin, the epithelial type of intermediate filaments.

Sarcoplasmic body myopathy. The sarcoplasmic bodies of hereditary distal myopathy of late onset (Edström et al. 1980), together with filamentous bodies and intermediate filaments of the desmin – once called skeletin – type, were similar to cytoplasmic bodies. This is another familial form of cytoplasmic body myopathy.

Therefore, there is a group of neuromuscular conditions that affect either children or adults, with or without involvement of the heart, and in sporadic or familial form, which is marked by ultrastructural criteria of the spheroid-cytoplasmic complex (Halbig et al. 1991) and the presence of desmin-intermediate filaments or proteins. The group of conditions denoted by desmin pathology enlarges the nosological spectrum of neuromuscular disorders associated with a specific protein abnormality; the pathology of dystrophin in Duchenne and Becker muscular dystrophies is the most prominent example.

Abnormal inclusions

Fingerprint body myopathy (Engel et al. 1972). In addition to the previously recorded instances of fingerprint body myopathy (Bender 1979), and an earlier report of a long-standing benign myopathy with similar inclusions (Fardeau et al. 1976), there have been new reports. Curless et al. (1978) de-

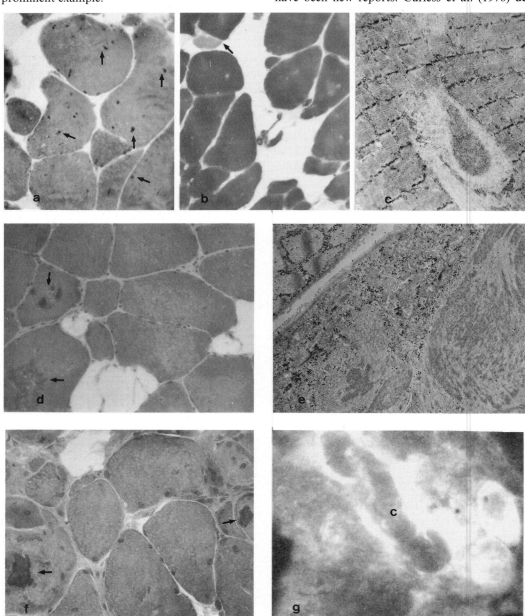

Fig. 6. (*For legend see next page*).

scribed twins, identifying this congenital myopathy as a familial neuromuscular disease. Mental retardation was prominent. Fingerprint bodies were sometimes associated with 'minicores' (Curless et al. 1978) or with peculiar 'circular structures' (Matsubara and Mair 1979). Fingerprint bodies have also been observed in diverse neuromuscular conditions (Goebel 1986), including myotonic dystrophy and oculopharyngeal muscular dystrophy.

Reducing body myopathy (Brooke and Neville 1972). Bender (1979) recorded four patients with reducing body myopathy, whose muscle fibers were morphologically marked by inclusions that, without enzyme activity, were positive in the menadione-linked alpha-glycerophosphate-dehydrogenase (MAG) preparation. The clinical course varied widely. Subsequently, five additional patients were recorded, including two sisters (Hübner

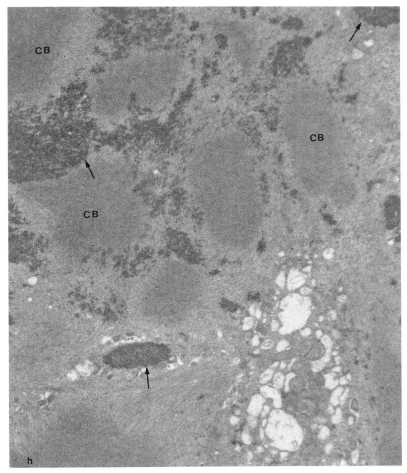

Fig. 6. Cytoplasmic spheroid body myopathy: Cytoplasmic body myopathy, 15-year-old girl with life-long muscle weakness: (a) Numerous small cytoplasmic inclusions (arrows). Modified trichrome stain, ×150. (b) There is only one (light) type II myofiber (arrow) among numerous large and atrophic type I (dark) myofibers. Acid ATPase stain, ×112. (c) A cytoplasmic body consists of a dense core and a halo of fine filaments. × 4,500. Spheroid body myopathy: (d) Several inclusions (arrows) in left half of photomicrograph indicate spheroid bodies. Modified trichrome stain, ×260. (e) A spheroid body consists of granular dense material randomly admixed with filamentous material, ×27,000. Spheroid-cytoplasmic body myopathy, marked by recently progressive muscle weakness in a 43-year-old man: (f) Large inclusions (arrows) within muscle fibers in a 'myopathic' muscle. Modified trichrome stain, ×225. (g) Spheroid-cytoplasmic body shows light halo of desmin around dense core (c) by immunofluorescence (by courtesy of Dr. Moll, University of Mainz, Germany), ×440. (h) A 'mixed' myopathy shows both cytoplasmic bodies (CB) and reducing bodies (arrows), ×12,900.

and Pongratz 1981; Oh et al. 1983; Carpenter et al. 1985), and one adult (Keith and Brownell 1990).

The clinical course was benign in all cases. The reducing bodies revealed two unidentifiable proteins and, in addition to the granular fine structure, there were 17 nm thick tubules, non-reactive with antibodies against tubulin or vimentin. They differed in arrangement and location from the tubulo-filamentous inclusions of inclusion body myositis (Carpenter et al. 1985). The possibility that these reducing bodies contained ribonucleic acid was not corroborated (Carpenter et al. 1985). The serum of this patient, a 7-year-old girl, did not react immunohistologically with her reducing bodies, but there were high antibody-titers against Coxsackie viruses B2, B4 and B5, a serological finding that had earlier been reported in another patient with reducing body myopathy (Dubowitz 1978). The reducing bodies also showed strong greenish autofluorescence, apparently different from the bright yellow autofluorescence of lipofuscin and ceroid-lipofuscin inclusions of the neuronal ceroid-lipofuscinoses (Goebel et al. 1975). Reducing bodies have not been encountered as a non-specific inclusion within muscle fibers of other neuromuscular disorders, but we recently (unpublished observations) encountered both reducing and cytoplasmic bodies in the same muscle fibers (Fig. 6h) of two separate muscle biopsies from an 11-year-old boy with a rigid spine syndrome; his maternal grandmother also had similar muscle fiber inclusions and a late-onset myopathy.

'Granular body myopathy' was another term for this condition (Hübner and Pongratz 1981); the origin and connotation of reducing bodies remain enigmatic.

Cylindric spirals myopathy. Cylindrical spirals are inclusion bodies of tubular profiles arranged in parallel circles, usually located in the subsarcolemmal region and thought to be derived from transverse (T) tubules. The inclusions were seen in a mother and her young son, both with symptoms of cramps, stiffness, posteffort muscle tightness and percussion myotonia (Bove et al. 1980). Autosomal dominance was suggested by late clinical onset in two related adults with similar clinical symptoms and cylindrical spirals (Taratuto et al. 1990).

Originally, cylindrical spirals were reported as findings in older patients afflicted with spinocerebellar degeneration, lymphoma (Carpenter et al. 1979), and later polyneuropathy (Gibbels et al. 1983). The origin and chemical composition of these spirals are not to be confused with concentric laminated bodies, a frequent non-specific finding in various neuromuscular conditions (Gambarelli et al. 1974).

Nuclear abnormalities

Myotubular/centronuclear myopathy. Myofiber nuclei are usually subsarcolemmal. Internal or even centrally-located nuclei are non-specific changes seen mostly in myopathies, but also in hereditary motor sensory neuropathies (Dubowitz and Brooke 1973). However, the central location of usually single large nuclei in round small muscle fibers is the hallmark of a congenital myopathy that is called myotubular or centronuclear myopathy. Symptoms may commence at birth (Collins et al. 1983) or may begin in adult years (Figs. 7–9). In adult-onset cases, centronuclear myopathy is largely sporadic (Goebel et al. 1984). Childhood forms may be autosomal-dominant or autosomal-recessive and there is an X-linked recessive, often malignant type (Fig. 7) (Barth et al. 1975) with a 90% fatality rate (De Angelis et al. 1991) that has been mapped to Xq28 (Lehesjoki et al. 1990). In one family, autosomal-dominant inheritance was suggested (Torres et al. 1985). Type I fiber hypotrophy without central nuclei – not an infrequent feature of childhood muscles (Iannaccone et al. 1987) – has been considered a separate nosological entity by some investigators, but not by others (Lo et al. 1990). From a review of 288 patients, De Angelis et al. (1991) concluded that centronuclear myopathy comprises two clinically and genetically distinct entities, an autosomal-dominant and an X-linked recessive (myotubular). The autosomal-recessive inheritance was difficult to unequivocally assess.

The original report (Spiro et al. 1966) propagated the term 'myotubular myopathy' because small round muscle fibers with central nuclei resembled myotubes. This was contested by Sher et al. (1967), who used the more descriptive designation 'centronuclear myopathy', a term preferred since then.

Histochemically, type I fiber predominance and often type I fiber hypotrophy were noted, leading to the consideration of 'type I fiber hypotrophy and central nuclei' (Engel et al. 1968) as a separate entity (Brooke 1977). Additional morphological features are numerous fat cells within muscle fascicles, and the occasional association with 'multicores' (Lee and Yip 1981; Fitzsimons and McLeod 1982; Goebel 1986) (Figs. 8 and 9), larger cores (Hülsmann et al. 1981), abnormally structured mitochondria (Canal et al. 1985), peculiar stacks of cylinders resembling proliferated junctional complexes (Ambler et al. 1984a,b), or fingerprint inclusions (Jadro-Santel et al. 1980).

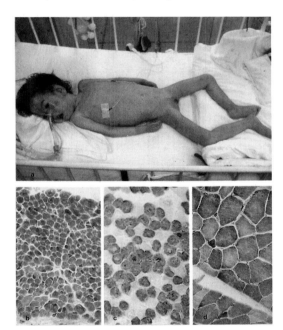

Fig. 7. Myotubular myopathy: (a) A male infant, whose brother died 4 days after birth of respiratory insufficiency, on artificial ventilation all of his life until death at age of four years (by courtesy of Dr. Boehnisch, Braunschweig, Germany). (b) His muscle showed large central nuclei in several muscle fibers. Modified trichrome stain, ×375. (c) Type I fiber predominance with central acid ATPase defects representing central nuclei, ×600. (d) His mother had an abnormal EMG and a 'myopathic' caliber spectrum by biopsy, with two atrophic dark type I myofibers depicted here. NADH preparation, ×110.

During the course of centronuclear myopathy, the hypotrophy of type I fibers with central nuclei may persist (Ambler et al. 1984b) or may recede (Ricoy and Cabello 1985). An increased number of central nuclei has also been observed in extraocular muscles (Bergen et al. 1980), which may account for the ptosis and external ophthalmoplegia that are prominent signs of this disorder.

While light microscopic features of the biopsied muscle in the original case of Spiro et al. (1966) suggested immaturity of the small muscle fibers, there was no electron microscopic evidence of ribosomal aggregates, which are typical of fetal muscle (Sher et al. 1967), militating against a fetal stage of these myofibers.

However, ultrastructural analysis has strengthened the concept that immature myotube-like muscle fibers are present in myotubular myopathy. Normal fetal muscle cells and small muscle fibers in X-linked recessive myotubular myopathy (Ambler et al. 1984b) were similar as regards myofibrils, nuclei, the arrangement of myofibers including satellite cells in clusters of immature cells surrounded by a common basement membrane, as well as alignment of triads and persistence of ribosomes. The exclusive presence of aggregates of proliferated sarcotubular profiles, not seen in normal fetal muscle fibers, were thought to indicate an innervational abnormality (Ambler et al. 1984a). Furthermore, energy disperse X-ray analysis of myotubular muscle fibers revealed a pattern of sodium, chloride and potassium that resembled that of human fetal muscle but not that of mature control muscle (Edström et al. 1982). These findings, seen in biopsies from both a father and his 14-year-old son, were thought to indicate persistence of fetal muscle features into adolescence and adulthood.

Although fetal muscle myosins as seen in developing and regenerating muscle fibers could not be ascertained in adult centronuclear myopathy (Fitzsimons and McLeod 1982), there was evidence of fetal vimentin and desmin (Sarnat 1990) and the prenatal myosin heavy chain isoform (Sawchak and Sher 1990) in a male-associated myotubular myopathy. Numerical increase of undifferentiated type IIC myofibers was considered additional evidence of delayed muscle fiber growth and differentiation (Sasaki et al. 1989). Matura-

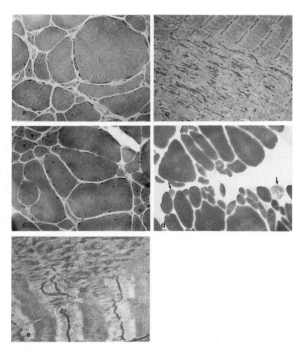

Fig. 8. Centronuclear myopathy associated with minicores, intrafamilial and persistent, male cousin to the patient whose muscle biopsy findings are depicted in Fig. 9II: (a) Numerous atrophic fibers, some with central nuclei, biopsy at age 21 years. Modified trichrome stain, ×250. (b) Cores are also a feature of this centronuclear myopathy, biopsied at the same age of 21 years, ×7,252. (c) Numerous atrophic fibers with central nuclei, biopsy taken at age of 28 years. Modified trichrome stain, ×180. (d) Pronounced type I fiber atrophy with only two light type II fibers (arrows). Central nuclei mark acid ATPase defects, biopsied at age of 28 years, ×110. (e) Cores mark persistence of this additional lesion 7 years after first biopsy, ×11,520.

tional arrest of myotubular myopathy myofibers has been related to impaired innervation because of abnormal EMG findings (Elder et al. 1983), abnormally structured neuromuscular junctions and nerves (Ambler et al. 1984b), degeneration and regeneration of sciatic axons (Sarnat et al. 1981), and segmental demyelination (Sugie et al. 1982a). Reduction of DNA by 50% in familial centronuclear myopathy was confined to central and subsarcolemmal nuclei of muscle fibers, but not encountered in nuclei of fibroblasts of the same biopsies. The mother of the three affected boys had normal diploid DNA contents of muscle fiber nuclei (Zimmermann and Weber 1979).

Type II fiber atrophy and myotube-like myofibers with central nuclei and a myopathic caliber spectrum were seen in several carriers of a large kinship with X-linked myotubular myopathy even when clinical, EMG and serological data were normal, indicating that muscle biopsy may be the most

sensitive way to detect carriers of X-linked myotubular myopathy (Barth et al. 1975; Oldfors et al. 1989).

Mild clinical symptoms, subtle EMG abnormalities, slightly elevated CK values (Zimmermann and Weber 1979; Dubois et al. 1982), non-specific myopathic features in biopsies, and scattered small fibers (Ambler et al. 1984a) (Fig. 7) have occasionally been reported in carriers. However, minimal abnormal findings do not distinguish heterozygote carriers and hemizygote patients with a mild autosomal-dominant variant of myotubular myopathy: in a mother with an 11- month-old girl who had myotubular myopathy, Kinoshita et al. (1975a) found only type I fiber atrophy. Peripheral nerve involvement has been documented mostly in adults with centronuclear myopathy (Sugie et al. 1982a; Torres et al. 1985), but occasionally also in the childhood form (Pavone et al. 1980). Central nuclei, usually found in type I fibers, were seen in

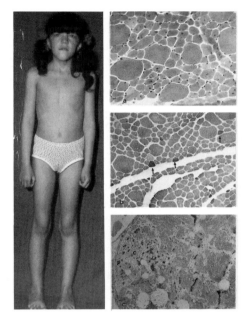

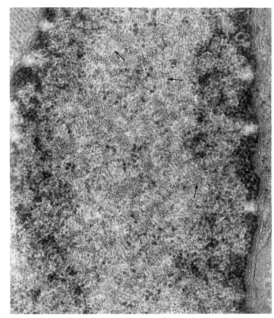

Fig. 9. Centronuclear myopathy. I. 8-year-old girl; II. Muscle biopsy findings from another girl, a 5-year-old female cousin of the patient whose muscle biopsy findings are depicted in Fig. 8: (a) Numerous atrophic fibers with central nuclei. Modified trichrome stain, ×190. (b) Pronounced type I fiber (light) predominance with only a few small scattered type II fibers (arrows), marked histochemically similar to male cousin's muscle. Alkaline ATPase, ×130. (c) Cores (c) are also a feature in this girl's centronuclear myopathy, further emphasizing genetic conformity to her male cousin's centronuclear myopathy. ×7,770. (III) Innumerable filaments of 8–10 nm size in a myofiber nucleus (arrows), oculopharyngeal muscular dystrophy, ×68,750.

type II myofibers of an adult patient (Sugie et al. 1982a) who also showed segmental demyelination in sural nerve. Supranuclear factors were also invoked as a cause of centronuclear myopathy (Serratrice et al. 1978; Torres et al. 1985).

Nucleodegenerative myopathy. A congenital muscular disease was described as 'nucleodegenerative myopathy' in a 10-year-old girl (Schröder 1982b). She had a non-progressive myopathy, cataracts, and mental retardation. Muscle fibers contained circular whorls of membranes, often around or close to myofiber nuclei and an occasional intranuclear inclusion. Sewry et al. (1988) recorded similar membranous structures associated with nuclei in three children with a myopathy in the Marinesco-Sjögren syndrome (MSS) and suggested from the clinical description that the patient of Schröder (1982b) also had MSS. The 'nucleodegenerative myopathy' – better designated a perinuclear degenerative myopathy – differed from 'granular nu-

clear inclusion body disease' in a 32-year-old woman with a chronically progressive motor disorder (Schröder et al. 1985). This patient had finely amorphous intranuclear inclusions in perivascular interstitial cells of anterior tibial muscle and sural nerve. There were also membrane-bound vacuoles located close to nuclei and more centrally inside myofibers.

Oculopharyngeal muscular dystrophy. Although this autosomal-dominant neuromuscular condition of late clinical onset is usually classified among the muscular dystrophies, the pathognomonic 8–10 nm filaments in myofiber nuclei (Fig. 9III) and familial occurrence render its classification with the structured congenital myopathies – or their late-onset forms – a reasonable alternative.

Mixed myopathies

Within the same muscle there may be more than

one structural lesion that is otherwise regarded as disease-specific. The concomitant presence of nemaline bodies and cores has been documented several times (Bethlem et al. 1978; Vallat et al. 1982; Seitz et al. 1984). Simultaneous occurrence of minicores and whorled myofibers prompted the term 'mixed myopathies' (Dubowitz 1980). Other structural abnormalities seen within the same muscle biopsy include: honeycomb structures and zebra bodies (Carrier et al. 1976); central nuclei and minicores (Lee and Yip 1981; Fitzsimons and McLeod 1982); cytoplasmic or filamentous bodies, fingerprint bodies, and minicores (Curless et al. 1978); caps and minicores (Fidzianska et al. 1981); cytoplasmic bodies and reducing bodies (Hübner and Pongratz 1981); a case of ours (Fig. 6h); filamentous bodies and tubular aggregates (Morgan-Hughes et al. 1970); fingerprint bodies, rods, and concentric laminated bodies (Fardeau et al. 1976); tubular aggregates and central cores (Castro et al. 1990). Some of the changes may be coincidental, others might be related. For instance, central cores and multi- or minicores, or rods and cores are both thought to involve the Z-disc. Similarly cytoplasmic bodies and minicore-like streaming of the Z-band have been encountered in human reversible ipecac myopathy (Mateer et al. 1985).

Vacuolar myopathies

Rimmed vacuole myopathy. Rimmed vacuoles within muscle fibers were originally described in patients with oculopharyngeal muscular dystrophy (Dubowitz and Brooke 1973), but were later found to be a non-specific finding also in other neuromuscular diseases (Fukuhara et al. 1980; Goebel et a. 1986). However, some patients show abundant rimmed vacuoles, and familial cases have been reported with late onset (Argov and Yarom 1984) or onset in childhood. Kalimo et al. (1988) described three brothers with slowly progressive proximal weakness, especially of the legs. Repeated biopsies suggested a numerical increase of rimmed vacuoles with age, a pattern of clinical and pathological development that resembles cylindrical spiral myopathy. Hence, 'rimmed vacuole' congenital myopathy may be added to the long list of these conditions.

Most familial cases of rimmed vacuole congeni-

tal myopathy showed a slowly or non-progressive course, but a fatal rimmed vacuole myopathy was described in a 5-year-old boy (Goebel et al. 1986) who had a life-long history of limb weakness; a quadriceps biopsy was marked by numerous rimmed vacuoles (Fig. 10). The rimmed or autophagic vacuole is often located close to the cell surface, demarcated by a fine granular rim in light microscopic sections. Mild acid phosphatase activity attests to its lysosomal nature which, by electron microscopy, is marked by a membrane-bound vacuole that may contain myelin figure-like lamellar material and other debris.

These pathological findings of abundant rimmed vacuoles resemble those seen in patients with lysosomal glycogen storage and normal acid maltase levels (Riggs et al. 1983; Byrne et al. 1986; Hart et al. 1987; Bru et al. 1988).

Granulovacuolar myopathy. 'Familial granulovacuolar lobular myopathy' (Juguilon et al. 1982), a myopathy with onset in late adolescence and associated with electrical myotonia, may be a variant of rimmed vacuole myopathy because about 30% of type I fibers were lobulated and there were numerous rimmed vacuoles. This phenomenon of 'lobulation' of myofibers is another non-specific feature, seen also in limb girdle-muscular dystrophy or facioscapulohumeral muscular dystrophy.

Tubulomembranous myopathy. Fukuhara et al. (1981) used this term to describe a sporadic mildly progressive myopathy in a 16-year-old boy, possibly another variant of congenital rimmed vacuole myopathy. The biopsy showed not only innumerable rimmed vacuoles, but also peculiar 'laminated tubulomembranous' structures and 'curvifilamentous material'.

Abnormalities of the sarcotubular system

Tubular aggregate myopathy. Tubular aggregates (Fig. 11a) were originally described (Engel 1964) as non-specific inclusions, chiefly in type II muscle fibers, reacting positively in nicotine adenine dinucleotide tetrazolium reductase (NADH), but negatively in MAG preparations. The inclusions were largely located in subsarcolemmal regions, but were also seen as smaller aggregates deep within the muscle fibers and consisted of agglomerates of empty or double ringed tubules that were thought to originate from sarcoplasmic reticulum.

Tubular aggregates were then encountered in patients with hypokalemic periodic paralysis (Meyers et al. 1972) and in chronic alcoholic myopathy (Del Villar Negro et al. 1982). Tubular aggregates have also been found in siblings (De Groot and Arts 1982) or as an autosomal-dominant myopathy (Rohkamm et al. 1983; Pierobon-Bormioli et al. 1985). The disease may start in adults (Rohkamm et al. 1983) or in childhood (Pierobon- Bormioli et al. 1985). Apart from the non-specific clinical features typical of congenital myopathies, patients with familial and sporadic tubular aggregate myopathy had experienced cramps or myalgia on exertion (Brumback et al.

1981). Originally, only type II muscle fibers were thought to be affected in sporadic tubular aggregate myopathy (Morgan-Hughes et al. 1970), but involvement of both type I and type II fibers was later documented (Rohkamm et al. 1983; Pierobon-Bormioli et al. 1985). The origin of tubular aggregates from sarcoplasmic reticulum, especially the terminal sacs, was demonstrated by electron microscopy (Engel et al. 1970) and by immunohistologic studies (Salviati et al. 1985) that showed an SR (sarcoplasmic reticulum) calcium protein and calsequestrin in tubular aggregates. Locally impaired calcium permeability may lead morphologically to compensatory hypertrophy of SR and formation of tubular aggregates, functionally to abnormal muscle contraction and relaxation, and clinical symptoms of cramps and myalgia.

Honeycomb myopathy. A possible familial X-linked recessive congenital myopathy of late onset (Carrier et al. 1976) was clinically marked by muscle weakness and fatiguability, myalgia on exertion, 'myopathic' EMG, and elevated CK values. Pathology included type I fiber atrophy, zebra bodies, numerous honeycomb structures (Fig. 11b), and other abnormalities of the T-tubules from which the honeycombs originated. A patient recorded by Niakan et al. (1985) had abundant honeycomb structures in muscle and clinical myalgia, which is a typical symptom of the myopathy with tubular aggregates.

Abnormal myomuscular junction myopathy. This apparently sporadic congenital myopathy was described in a 38-year-old man who, since childhood, had had ophthalmoplegia, proximal limb girdle and facial weakness, hypertrophy of calves, and wasting of neck, biceps, triceps, and interosseus muscles. CK values and EMG findings were normal (Pierobon-Bormioli et al. 1980). Triceps muscle disclosed abundant internally located nuclei, small and large diameter myofibers, and 'core-like lesions'.

Acetylcholinesterase activity was present in patches on longitudinal sections of type I and type II fibers and was arranged circularly around cross-sections. The circular arrangement corresponded, by electron microscopy, to numerous plasma membrane invaginations that simulated myomus-

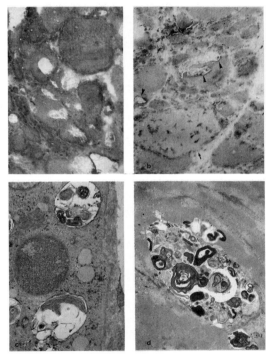

Fig. 10. Rimmed vacuole myopathy, 5-year-old boy: (a) Numerous vacuoles within myofibers, myopathic caliber spectrum. Modified trichrome stain, × 200. (b) Apart from a necrotic fiber (arrow) containing macrophages, the vacuoles (arrow heads) are acid phosphatase-positive, ×160. (c) Several subsarcolemmal membrane-bound vacuoles are filled with myelin-like debris, ×740. (d) A large membrane-bound autophagic vacuole is filled with myelin-like whorls, ×8,550.

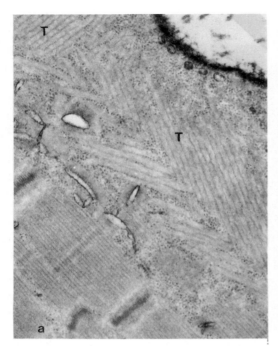 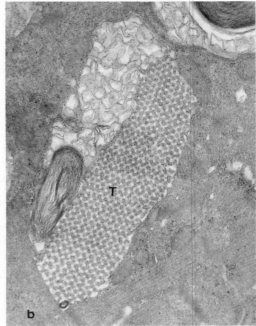

Fig. 11. Sarcotubular abnormalities: (a) Tubular (T) aggregates derived from the sarcoplasmic reticulum in subsarcolemmal position, ×41,250. (b) A honeycomb structure derived from transverse (T) tubules within a muscle fiber, ×42,000.

cular junctions and gave the muscle fibers a fragmented appearance. The condition was not familial and no other patients have been observed.

Other rare congenital myopathies. No new findings have been reported on sarcotubular myopathy (Jerusalem et al. 1973; Jerusalem, personal communication) or zebra body myopathy (Lake and Wilson 1975) since the previous review (Bender 1979), although no zebra bodies were detected in another muscle biopsy of the same patient 14 years later (Lake, written communication, November 1990).

UNSTRUCTURED CONGENITAL MYOPATHIES

Uniform type I fiber myopathy. Uniformity of one fiber type is unusual in human limb muscles. Physiologically, one fiber type – type IIC according to ATPase preparations – is seen in fetal muscle before the 18th week of gestation (Kumagai et al. 1984).

As a pathological sign, type I fiber uniformity (Fig. 12) has been observed in various neuromuscular disorders (Lambert 1974). Fiber type grouping subsequent to reinnervation of primarily denervated muscle fibers with regrowth of atrophic muscle fibers to normal size usually entails separate groups of type I and type II fibers. In a small specimen, only one fiber type, either of type I or type II, may represent giant motor units owing to reinnervation and retyping.

In some congenital myopathies with characteristic structural abnormalities, there may be uniformity of one fiber type, namely type I (Goebel 1986). Occasionally, in a well-defined structured congenital myopathy, such as CCD, other clinically affected, usually younger members of the same family may display uniformity of type I fibers, without the disease-specific structural abnormalities (see Fig. 2). The reverse developmental sequence was observed by Ricoy and Cabello (1985).

However, predominance of type I fibers without any differences in fiber size or additional structural abnormalities may be present in some children with congenital hypotonia (Brooke 1977; Dubowitz 1980). The extreme pattern of type I fiber predominance may actually amount to complete absence of type II fibers (Oh and Danon 1983).

Type I fiber uniformity may be the morphological sign of an underlying neuromuscular disease, because it is sometimes associated with proximal limb weakness, myopathic EMG, and variation in muscle fiber diameter. We found a favorable prognosis in a follow-up study of a woman now 26 years old, with non-progressive muscle weakness from early childhood (Fig. 12).

There may be more than the normal 3% of internal nuclei (Sarnat 1983) in such biopsied muscles, suggesting that this neuromuscular condition is a transitional stage between complete type I fiber uniformity and centronuclear myopathy (Oh and Danon 1983, patients 2 and 3). Minor structural defects such as moth-eaten lesions in the NADH-preparation or circumscribed areas of 'Z-disc streaming' (and a lack of mitochondria, i.e. possibly minicores) are non-specific findings that are overshadowed by the type I fiber uniformity, defined by more than 99% of type I fibers.

Uniformity of type IIC fibers has been recorded in a 7-year-old boy (Pearce and Harriman 1985). This finding was interpreted as either a congenital myopathy, indicating lack of maturation of myofibers, or a neurogenic neuromuscular disorder resulting in uniform conversion of myofibers to type IIC.

Congenital fiber type disproportion. Fiber type disproportion (FTD) represents, theoretically, any difference in size or distribution of type I and type II fibers. A number of such morphological patterns may occur, i.e. selective type I fiber atrophy, associated with or without hypertrophy of respective opposite fiber types as well as fiber type predominance or fiber type paucity. Sometimes, only type I fiber atrophy has been emphasized (Kinoshita et al. 1975c). The pattern of FTD is non-specific, encountered in a variety of neuromuscular conditions, such as in early myotonic dystrophy, in certain structured congenital myopathies (Goebel 1986), sometimes in the early, 'preclinical' phase of infantile spinal muscular atrophy (Dubowitz 1978), Krabbe's leukodystrophy

(Dehkharghini et al. 1981), Lowe syndrome (Kohyama et al. 1989), or the 'rigid spine syndrome' (Goebel et al. 1977). FTD should be considered as the diagnostic hallmark of a congenital myopathy only when there are additional clinical signs of a congenital myopathy.

The term 'congenital fiber type disproportion' (CFTD) has firmly been established in the terminology of neuromuscular conditions (Bender 1979), after it was delineated by Brooke (1973). It was originally defined by a difference in fiber size of more than 12% between type I and type II fibers, the type I fibers being smaller than the type II fibers (either the former ones being atrophic or the latter ones being hypertrophic or both), and by a disproportion in fiber type distribution, type I fibers outnumbering type II fibers. If CFTD is pronounced, it does not require additional morphometric confirmation. Brooke (1990) specified his definition of CFTD by giving a difference of at least 45% for fiber type size and of at least 65% for type I fiber predominance. Not infrequently, distribution of type I and type II fibers is uneven

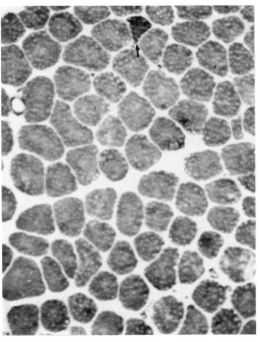

Fig. 12. Type I fiber uniformity with only a single (dark) type II fiber (arrow), 14-year-old girl with non-progressive muscle weakness. Alkaline ATPase, ×160.

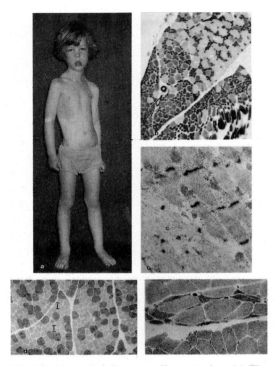

Fig. 13. Congenital fiber type disproportion: (a) The young boy shows flat fleet, slender build, and a 'myopathic' face. Reprinted with permission by the editor, M. Adachi, M.D., from M. Adachi and J.H. Sher (Eds.), Neuromuscular Disease, New York, Tokyo, Igaku-Shoin (1990). (b) Selective type I fiber atrophy and predominance and irregularly distributed subtypes of larger type II myofibers. Acid ATPase preparation, ×76.8. (c) A subsarcolemmal core (c) is also present, ×24,000. Childhood neurogenic atrophy: (d) Type I fiber atrophy and predominance occurred at 7 years of age. Alkaline ATPase, × 90 (by courtesy of Dr. Klein, Bremen, Germany). (e) Second biopsy taken 2 years later reveals groups of angular dark denervated fibers, indicating the spurious fiber type disproportion pattern of d, as an early morphological sign of denervation. NADH preparation, ×150.

within a given muscle. The distribution of individual subtypes of type II may vary considerably (Fig. 13) (Ter Laak et al. 1981), including the presence of type IIC fibers and lack of type IIB fibers.

Although patients with CFTD may occasionally be asymptomatic (Eisler and Wilson 1978), symptoms may vary (Clancy et al. 1980; Haltia et al. 1988). In the most severely affected patients, early death may follow respiratory insuffiency (Cavanagh et al. 1979), which may be a permanent threat

to life (Fukuda et al. 1983) and is thought to be correlated with more severe FTD (Mizuno and Komiya 1990). But there may be clinical improvement (Haltia et al. 1988).

External ophthalmoplegia or ptosis are rare clinical features (Owen et al. 1981). The disease may occur in families and has been documented in identical twins (Curless and Nelson 1977). Autosomal-dominant inheritance was suggested by Eisler and Wilson (1978).

In another family, autosomal-recessive inheritance was implied by multiple consanguineous marriages in a large kinship; a father and his two daughters were affected from childhood (Pelias and Thurmon 1979). There were no skeletal deformities and the patients had been floppy in infancy so the authors used the term 'congenital universal muscular hypoplasia', an inappropriate term that had been used earlier to describe insufficiently studied neuromuscular disorders (Krabbe 1947).

One 32-year-old man, whose daughter showed clinical and morphological signs of CFTD, had a neuromuscular problem from infancy and later had a 'rigid spine syndrome' marked by type I and type II fiber atrophy and type I fiber predominance in temporal muscle biopsy (Sulaiman et al. 1983).

Considering the limited spectrum of symptoms and pathology in congenital myopathies, precise separation and classification of some congenital neuromuscular conditions sometimes appear equivocal or artifactual. Others may be regarded nosologically as transitions between or variant forms of defined neuromuscular diseases. For instance, two unrelated children with ophthalmoplegia, non-progressive myopathy, and type I fiber hypotrophy resembling myotubes (Bender and Bender 1977; Sugie et al. 1982b) were thought to have a congenital myopathy that was placed nosologically between centronuclear myopathy and CFTD. In CFTD, when several muscles are studied, some may show centrally placed nuclei in type I fibers, suggesting centronuclear myopathy (Spiro et al. 1977).

The etiology of CFTD is not understood, although developmental defects in maturation have been suspected (Sarnat 1983; Argov et al. 1984). Two muscle biopsies taken from an adult woman 6 years apart, revealed persistence of the FTD; atrophy of type I fibers had decreased but did not completely disappear, and was interpreted as evidence of maturation in a primarily 'dysmaturative' myopathy (Ricoy and Cabello 1981). In a similar study, FTD persisted despite growth of muscle fibers over the years (Mizuno and Komiya 1990). It is not known whether developmental arrest of maturation is of neural origin or is due to other influences. Autopsy studies on patients with CFTD did not disclose anterior horn cell or peripheral nerve abnormalities (Spiro et al. 1977; Cavanagh et al. 1979), while numerical reduction of medially located anterior horn motor neurons was noted in an adult patient with 'rigid spine syndrome' whose daughter had CFTD (Sulaiman et al. 1983).

Muscle FTD in floppy infants may not always conform to the classical morphological pattern of CFTD but may show disproportion of muscle fibers in that type II fibers are smaller than type I fibers (Argov et al. 1984). Even a 'new myopathy' has been proposed, in which only type II muscle fibers appeared hypoplastic (Matsuoka et al. 1974), later supported by a similar report (Yoshioka et al. 1987). Type II fiber atrophy and paucity associated with numerous internal nuclei in type I and type II fibers were also found in another congenital myopathy of childhood (Scelsi et al. 1976).

Minimal change myopathy. This name refers to an ill-defined group of conditions with symptoms starting at birth, in infancy, or in early childhood (Dubowitz 1978); clinical and EMG findings do not point to any defined neuromuscular condition and biopsied muscles may show mild, completely non-specific findings. often only at the ultrastructural level (Ohtaki et al. 1990). Familial cases have also been recorded (Nonaka et al. 1983b). Owing to the subtlety of clinical and morphological findings in these patients, categorization as minimal change myopathy may be erroneous because other well-defined neuromuscular conditions may hide under this diagnosis until disclosed later – or never – by more refined techniques. The noncommittal term 'minimal change myopathy' may refer to several congenital myopathies (Nonaka et al. 1983b), but may imply both diagnostic incapability and a compulsion for diagnostic classification. Other benign myopathies commencing in early childhood

may also be grouped into this category (Mohire et al. 1988; Serratrice and Pellissier 1988; Mahon et al. 1989).

Congenital myasthenic syndromes

A group of congenital myopathies has emerged in which the primary apparently 'congenital' defect was localized in the neuromuscular junctions, resulting in the definition of 'congenital myasthenic syndromes' (see Table 1) (Engel 1988; Misulis and Fenichel 1989). These congenital myopathies are dealt with within the section on myasthenia.

CONCLUSION

After an early phase of pilot studies, congenital myopaties have now been defined as a separate group of neuromuscular conditions. The concept has reached a stage of terminological and nosological consolidation. In the last decade, numerous reports of congenital myopathies have enlarged the nosological spectrum, including follow-up studies. Late-onset or adult forms have led to criticism of the term 'congenital', but not necessarily to the nosological concept of congenital myopathies. However, late-onset forms or variants are well known and accepted in other nosological categories of – often hereditary – neurological disorders that usually affect children such as the lysosomal diseases.

The early enthusiastic emphasis on structural abnormalities has been queried, pointing to common denominators rather than special features of different types of congenital myopathies (Brooke et al. 1979; Fardeau et al. 1979; Fardeau 1987; Tomé and Fardeau 1988). The critics stressed histochemical abnormalities, especially type I fiber atrophy or hypotrophy and type I fiber predominance, changes that are common to several structurally different congenital myopathies. However, these histochemical deviations may only be a morphological framework within which the individual congenital myopathies can be delineated by individual structural hallmarks, comparable to the non-specific morphological patterns of individual muscular dystrophies or spinal muscular atrophies.

The frequently noted selective type I fiber atro-

TABLE 7

'Experimental' models of congenital myopathies.

Ipecac syrup ingestion	cytoplasmic bodies	man
Local tetanus	spheroid-cytoplasmic bodies	rat
Organophosphate	cytoplasmic bodies	man
Malignant hyperthermia-susceptible pigs	cytoplasmic bodies	pig
Organophosphate	core lesions	man
Local tetanus	core lesions and rods	rat
Tenotomy	core lesions and rods	rat
Bromodeoxyuridine	selective myosin absence	chick heart in tissue culture
Emetine	core like lesions	rat

phy and type I fiber predominance suggest neurogenic origin, although actual structural abnormalities have only rarely been documented in lower motor neurons (Dahl and Klutzow 1974), or peripheral nerves (Sarnat et al. 1981; Sugie et al. 1982a) including congenital fiber type disproportion in Krabbe's globoid cell leukodystrophy (Dehkharghani et al. 1981) or at the terminal innervation and neuromuscular junction (Coërs et al. 1976).

When available for study, these several levels of the peripheral nervous system have seemed to be normal more often than not (Coërs et al. 1976; Bender and Bender 1977). However, subtle neural factors have been claimed to play an etiological role in congenital myopathies; maturational arrest and maturational delay of muscle development before birth have been elaborated (Sarnat et al. 1981; Sarnat 1983; Ambler et al. 1984a,b; Nonaka et al. 1984). It is difficult to relate type I fiber atrophy and type I fiber predominance to any histochemical normal prenatal pattern. Uniformity of undifferentiated type IIC fibers in the first period of histochemical muscle fiber development is followed by gradual emergence of type I fibers and finally the differentiation of muscle fibers into type I, IIA, and IIB fibers (Sarnat 1983; Kumagai et al. 1984), but the histochemical abnormalities common to congenital myopathies can still be regarded as deviations from different prenatal physiological stages. In particular, the hypothesis of maturational arrest and delay fails to explain etiology, pathogenesis and thus nosological connotation of the structural characteristics. Some investigators consider these morphological features as of little importance. However, repeated observations in several muscles of the same patient and in muscles of different affected family members make it impossible to disregard these lesions. Conversely, the mounting evidence of biochemical maturational disturbances (Fidzianska et al. 1981, 1984; Edström et al. 1982; Fardeau et al. 1985) and ultrastructural features of fetal muscle (Edström et al. 1982; Nonaka et al. 1983a; Fidzianska et al. 1984) in several congenital myopathies cannot be ignored. Immunocytochemistry has only recently been applied to human muscle and may furnish new and important information. The pathology of dystrophin, in Duchenne and Becker muscular dystrophies, has provided a telling example that has revolutionized the interpretation of these two disorders.

After early periods of enthusiasm and then skepticism, the current phase of investigation of congenital myopathies has been marked by augmented attempts to explain the pathogenesis of structural particularities. Experimental cross-innervation of muscles has disclosed the dependency of muscle fibers on innervation, suggesting an explanation for fiber type grouping and perhaps fiber type predominance. Accidental and experimental observations (Table 7) have provided data on the pathogenesis and morphogenesis of congenital myopathies. Moreover, refined biochemical and immunological techniques have enriched our knowledge on structural proteins and their abnormalities in congenital myopathies.

New techniques are likely to include DNA analysis, immunology and biochemistry to identify fetal and structural parameters in human muscle tissues, incorporating aspects of normal fetal muscle development into nosological considerations of congenital myopathies. Experimental models will be explored. Finally tissue culture endeavors (Askanas et al. 1979) have furnished only scanty results but are also promising avenues to elucidate the congenital myopathies as a separate nosological category and as individual entities.

Acknowledgements

We are grateful to Lewis P. Rowland, M.D. and Thomas Voit, M.D. for critical advice, to Amanda Bastian, Marianne Messerschmidt and Astrid Wöber for editorial assistance, to Walter Meffert for photography, and to Margarete Schlie and Irene Warlo for innumerable light and electron microscopic preparations.

The 'Deutsche Gesellschaft zur Bekämpfung der Muskelkrankheiten e.V.', Freiburg Germany gave financial support.

REFERENCES

ADAMS, R.D., D. DENNY-BROWN and C.M. PEARSON: Diseases of Muscle. A Study in Pathology (2nd edit.) New York, Harper & Row (1965) 269–270.

ADELAIDA MARTINEZ, B. and B.D. LAKE: Childhood nemaline myopathy: a review of clinical presentation in relation to prognosis. Dev. Med. Child Neurol. 29 (1987) 815–820.

AMBLER, M.W., C. NEUAVE, B.G. TUTSCHAKA, S.M. PUESCHEL, J.M. ORSON and D.B. SINGER: X-linked recessive myotubular myopathy. I. Clinical and pathologic findings in a family. Hum. Pathol. 15 (1984a) 566–574.

AMBLER, M.W., C. NEAVE and D.B. SINGER: X-linked recessive myotubular myopathy. II. Muscle morphology and human myogenesis. Hum. Pathol. 15 (1984b) 1107–1120.

ARGOV. Z. and R. YAROM: 'Rimmed vacuole myopathy' sparing the quadriceps – a unique disorder in Iranian Jews. J. Neurol. Sci. 64 (1984) 33–43.

ARGOV. Z., D. GARDENER-MEDWIN, M.A. JOHNSON and F.L. MASTAGLIA: Patterns of muscle fiber-type disproportion in hypotonic infants. Arch. Neurol. 41 (1984) 53–57.

ARTS, W.F., J. BETHLEM, K.P. DINGEMANS and A.W. ERIKSSON: Investigations on the inheritance of nemaline myopathy. Arch. Neurol. 35 (1978) 72–77.

ASKANAS, V., W.K. ENGEL, N.B. REDDY, P.G. BARTH, J. BETHLEM, D.R. KRAUSS, M.E. HIBBERD, J.V. LAWRENCE and L.S. CARTER: X-linked recessive congenital muscle fiber hypotrophy with central nuclei. Arch. Neurol. 36 (1979) 604–609.

BAKER, J.H. and E.C.B. HALL-GRAGGS: Recovery from central core degeneration of the tenotomized rat soleus muscle. Muscle Nerve 3 (1980) 151–159.

BARTH, P.G., G.K. VAN WIJNGAARDEN and J. BETHLEM: X-linked myotubular myopathy with fatal neonatal asphyxia. Neurology 25 (1975) 531–536.

BENDER, A.N.: Congenital myopathies. In P.J. Vinken and G.W. Bruyn (Eds.), Handbook of Clinical Neurology, Vol. 41. Amsterdam North-Holland Publishing Co. (1979) 1–26.

BENDER, A.N. and M.B. BENDER: Muscle fiber hypotrophy with intact neuromuscular junctions: a study of a patient with congenital neuromuscular disease and ophthalmoplegia. Neurology 27 (1977) 206–212.

BERGEN, B.J., M.P. CARRY, W.B. WILSON, M.T. BARDEN and S.P. RINGEL: Centronuclear myopathy: extraocular- and limb-muscle findings in an adult. Muscle Nerve 3 (1980) 165–171.

BETHLEM, J., W.F. ARTZ and K.P. DINGEMANS: Common origin of rods, cores, miniature cores, and focal loss of cross-striations. Arch. Neurol. 35 (1978) 555–566.

BOVE, K.E., S.T. IANNACCONNE, P.K. HILTON and F. SAMAHA: Cylindrical spirals in a familial neuromuscular disorder. Ann. Neurol. 7 (1980) 550–556.

BROOKE, M.H.: Congenital fiber type dysproportion. In: B.A. Kakulas (Ed.), Clinical Studies in Myology, Part 2. Amsterdam, Excerpta Medica (1973) 147–159.

BROOKE, M.H.: A Clinician's View of Neuromuscular Diseases. Baltimore, Williams & Wilkins Co. (1977).

BROOKE, M.H.: Congenital fibre type disproportion. J. Neurol. Sci. 98 (1990) 100.

BROOKE, M.H. and H.E. NEVILLE: Reducing body myopathy. Neurology 22 (1972) 829–840.

BROOKE, M.H., J.E. CARROLL and S.P. RINGEL: Congenital hyotonia revisited. Muscle Nerve 2 (1979) 84–100.

BRU, P., J.F. PELLISSIER, J. GATAU-PELANCHON, G. FAUGERE, T. DE BARSY, S. LEVY and R. GÉRARD: Glycogénose lysosomiale cardio-musculaire de l'adulte sans déficit enzymatique connu. Arch. Mal. Coeur 1 (1988) 109–114.

BRUCE-GEGORIOS, J. and S.M. CHOU: Core myofibers and related alterations induced in rats' soleus muscle by immobilization in shortened position J. Neurol. Sci. 63 (1984) 267–275.

BRUMBACK, R.A., R.D. STATON and M.E. SUSAG: Exercise-induced pain, stiffness, and tubular aggregation in skeletal muscle. J. Neurol. Neurosurg. Psychiatry 44 (1981) 250–254.

BYRNE, E., P.C. BLUMBERGS and J.F. HALLPIKE: Central core disease: study of a family with five affected generations. J. Neurol. Sci. 53 (1982) 77–83.

BYRNE, E., X. DENNETT, B. CROTTY, I. TROUNCE, J.M. SANDS, R. HAWKINS, J. HAMMOND, S. ANDERSON, E.A. HAAN and A. POLLARD: Dominantly inherited cardioskeletal myopathy with lysosomal glycogen storage and normal acid maltase levels. Brain 109 (1986) 523–536.

CABELLO, A., P. MARTINEZ-MARTIN, E. GUTIÉRREZ-RIVAS and S. MADERO: Myopathy with nemaline structures associated with HIV infection. J. Neurol. 237 (1990) 64–66.

CANAL, N., G.C. COMI, M. COMALA, D. TESTA, M. MORA and F. CORNELIO: Centronuclear myopathy with unusual mitochondrial abnormalities. Clin. Neuropathol. 4 (1985) 23–27.

CANCILLA, P.A., K. KALYANARAMAN, M.A. VERITY, T. MUNSAT and C.M. PEARSON: Familial myopathy with probable lysis of myofibrils in type I fibers. Neurology 21 (1971) 579–585.

CARPENTER, S., G. KARPATI, Y. ROBITAILLE and C. MELMED: Cylindrical spirals in human skeletal muscle. Muscle Nerve 2 (1979) 282–287.

CARPENTER, S., G. KARPATI and P. HOLLAND: New observations in reducing body myopathy. Neurology 35 (1985) 818–827.

CARRIER, H., M. TOMMASI, N. KOPP and D. BOISSON: Prolifération du système tubulaire tansversal au cours d'une myopathie tardive et familiale. J. Neurol. Sci. 27 (1976) 499–512.

CASTRO, L., T. COELHO, A.V. PINHEIRO and A. GUIMARAES: Tubular aggregates in a case of central core disease. J. Neurol. Sci. 98 (1990) 337.

CAVANAGH, N.P.C., B.D. LAKE and P. MCMENIMAN: Congenital fibre type disproportion myopathy. Arch. Dis. Child. 54 (1979) 735–743.

CHACKO, S., A. KELLY, J. GARTRELL and C. HAN: Effect of 5-bromodeoxyuridine on myosin synthesis and myofibrillogenesis. J. Cell Biol. 67 (1975) 59a.

CHAPON, F., F. VILADER, M. FARDEAU, F. TOMÉ, N. DALUZEAU, CH. BERTHELIN, J. PH. THÉNINT and B. LECHEVALIER: Myopathie familiale avec inclusions de type 'corps cytoplasmiques' (ou 'sphéroides') révélée par une insuffisance respiratoire. Rev. Neurol. (Paris) 145 (1989) 460–465.

CHOU, S.M.: Congenital myopathies – a unifying concept. 'Neuromuscular Course' at the 61st Annual Meeting of the American Association of Neuropathologists, Boston, June 13–16 (1985).

CHOU, S.M. and Y. MIZUNO: Induction of spheroid cytoplasmic bodies in a rat muscle. Neurology 34 Suppl. 1 (1984) 82.

CHOU, S.M., T.M. CHOU and M. MORI: The 'core' myofibers induced by local tetanus and tenotomy in rats. J. Neuropathol. Exp. Neurol. 40 (1981) 300.

CLANCY, R.R., K.A. KELTS and J.W. OEHLERT: Clinical variability in congenital fiber type disproportion. J. Neurol. Sci. 46 (1980) 257–266.

CLARK, J.R., J. D'AGOSTINO, J. WILSON, R. BROOKS and G.C. COLE: Autosomal dominant myofibrillar inclusion body: clinical, histologic, histochemical, and ultrastructural characteristics. Neurology 28 (1978) 399.

COËRS, C., N. TELERMAN-TOPPET, J.M. GÉRARD, H. SZLIWOWSKI, J. BETHLEM and G.K. VAN WIJNGAARDEN: Changes in motor innervation and histochemical pattern of muscle fibers in some congenital myopathies. Neurology 26 (1976) 1046–1053.

COHEN, M.E., P.K. DUFFNER and R. HEFFNER: Central core disease in one of identical twins. J. Neurol. Neurosurg. Psychiatry 41 (1978) 659–663.

COLLINS, J.E., A. COLLINS, M.R. RADFORD and R.O. WELLER: Perinatal diagnosis of myotubular (centronuclear) myopathy: a case report. Clin. Neuropathol. 2 (1983) 79–82.

CONEN, P.E., E.G. MURPHY and W.L. DONOHUE: Light and electron microscopic studies of 'myogranules' in a child with hypotonia and muscle weakness. Can. Med. Assoc. J. 89 (1963) 983–986.

COULTER, C.L., W.A. MARKS, J.B. BODENSTEINER, R.W. LEECH and R.A. BRUMBACK: Double ring myopathy: a new storage myopathy presenting in adulthood Neurology 40 Suppl. 1 (1990) 205.

CURLESS, R.G. and M.B. NELSON: Congenital fiber type disproportion in identical twins. Ann. Neurol. 2 (1977) 455–459.

CURLESS, R.G., C.M. PAYNE and F.M. BRINNER: Fingerprint body myopathy: a report of twins. Dev. Med. Child Neurol. 20 (1978) 793–798.

CURRIE, S., M. NORONHA and D. HARRIMAN: 'Minicore' disease. IIIrd International Congress on Muscle Diseases. Amsterdam, Excerpta Medica (1974) 12.

DAHL, D.S. and F.W. KLUTZOW: Congenital rod disease: further evidence of innervational abnormalities as the basis of the clinicopathologic features. J. Neurol. Sci. 23 (1974) 371–385.

DE ANGELIS, M.S., L. PALMUCCI, M. LEONE and C. DORIGUZZI: Centronuclear myopathy: clinical, morphological and genetic characters. A review of 288 cases. J. Neurol. Sci 103 (1991) 2–9.

DE GIACOMO, P., L. MAZZARELLA and G.A. BUSCAINO: Central core disease; clinical, histochemical and electron microscopical observation on a case. In: Actualités de Pathologie Neuro-Musculaire, Advances in Neuromuscular Diseases (2eme Journées Internationales de Marseille) Paris, L'Expansion (1970) 325–331.

DE GROOT, J.G. and W.F. ARTS: Familial myopathy with tubular aggregates. J. Neurol. 227 (1982) 35–41.

DEHKHARGHANI, F., H.B. SARNAT, M.A. BREWSTER and S.I. ROTH: Congenital muscle fiber-type disproportion in Krabbe's leukodystrophy. Arch. Neurol. 38 (1981) 585–587.

DEL VILLAR NEGRO, A., J. MERINO ANGULO, J.M. RIVERA POMAR and C. AGUIRRE ERRASTI: Tubular aggregates in skeletal muscle of chronic alcoholic patients. Acta Neuropathol. (Berlin) 56 (1982) 250–254.

DE REUCK, J., W. DE COSTER, R. DE POTTER and H. VAN DER EECKEN: Inhibition of the target phenomenon in tenotomized rabbit muscle by medullary lesions. Acta Neuropathol. (Berlin) 56 (1982) 136–138.

DICKOFF, D.J.: Adult onset of inherited myopathies. Prog. Clin. Neurosci. 1 (1988) 65–80.

DODSON, R.F., G.O. CRISP JR., B. NICOTRA, L. MUNOZ and C.D. ALBRIGHT: Rod myopathy with extensive systemic and respiratory muscular involvement. Ultrastruct. Pathol. 5 (1983) 129–133.

DUBOIS, B., J.J. MARTIN, S. DEVELTER, M. GUÉRIN, G.A. LOEUILLE, P. DEBEUGNY and J.P. FARRIAUX: Myopathie centronucléaire: a propos d'une nouvelle observation. Ann. Pédiat. 29 (1982) 433–437.

DUBOWITZ, V: The 'new' myopathies. Neuropediatrics 1 (1969) 137–148.

DUBOWITZ, V: Muscle Disorders in Childhood. Philadelphia, Saunders (1978).

DUBOWITZ, V: The Floppy Infant. Philadelphia, Lippincott (1980).

DUBOWITZ, V: Muscle Biopsy. A Practical Approach. London Philadelphia Toronto, Baillière Tindall (1985).

DUBOWITZ, V. and M.H. BROOKE: Muscle Biopsy: A Modern Approach. Philadelphia, Saunders (1973).

EDSTRÖM, L., L.E. THORNELL and A. ERIKSSON: A new type of hereditary distal myopathy with characteristic sarcoplasmic bodies and intermediate (skeletin) filaments. J. Neurol. Sci. 47 (1980) 171–190.

EDSTRÖM, L., R. WROBLEWSKI and W.G.P. MAIR: Genuine myotubular myopathy. Muscle Nerve 5 (1982) 604–613.

EDSTRÖM, L., L.E. THORNELL, J. ALBO, S. LANDIN and M. SAMUELSSON: Myopathy with respiratory failure and typical myofibrillar lesions. J. Neurol. Sci. 96 (1990) 211–228.

EISLER, T. and J.H. WILSON: Muscle fiber-type disproportion: report of a family with symptomatic and asymptomatic members. Arch. Neurol. 35 (1978) 823–826.

ELDER, G.B., D. DEAN, A.J. MCCOMAS, B. PAES and D. DESA: Infantile centronuclear myopathy: evidence suggesting incomplete innervation. J. Neurol. Sci. 60 (1983) 79–88.

ENGEL, A.G.: Congenital myasthenic syndromes. J. Child Neurol. 3 (1988) 233–246.

ENGEL, A.G. and M.R. GOMEZ: Congenital myopathy associated with multifocal degeneration of muscle fibers. Trans. Am. Neurol. Assoc. 91 (1966) 222–223.

ENGEL, A.G., M.R. GOMEZ and R.V. GROOVER: Multicore disease: a recently recognized congenital myopathy associated with multifocal degeneration of muscle fibers. Mayo Clin. Proc. 46 (1971) 666–681.

ENGEL, A.G., C. ANGELINI and M.R. GOMEZ: Fingerprint body myopathy - a newly recognized congenital muscle disease. Mayo Clin. Proc. 47 (1972) 377–388.

ENGEL, W.K.: The essentiality of histo- and cytochemical studies of skeletal muscle in the investigation of neuromuscular disease. Neurology 12 (1962) 778–794.

ENGEL, W.K.: Mitochondrial aggregates in muscle disease. J. Histochem. Cytochem. 12 (1964) 46–48.

ENGEL, W.K.: Muscle biopsies in neuromuscular diseases. Ped. Clin. North Am. 14 (1967) 963–995.

ENGEL, W.K.: Central core disease and focal loss of cross-striations. In: E.S. Goldensohn and S.H. Appel (Eds.), Scientific Approaches to Clinical Neurology, Vol. 2. Philadelphia, Lea & Febiger (1977) 1555–1671.

ENGEL, W.K., G.N. GOLD and G. KARPATI: Type I fiber hypotrophy and central nuclei. Arch. Neurol. 18 (1968) 435–444.

ENGEL, W.K., D.W. BISHOP and G.G. CUNNINGHAM: Tubular aggregates in type II muscle fibers: ultrastructural and histochemical correlation. J. Ultrastruct. Res. 31 (1970) 507–525.

FARDEAU, M.: Some orthodox or non-orthodox considerations on congenital myopathies. In: R.J. Ellingson, N.M.F. Murray and A.M. Halliday (Eds.). The London Symposia (Suppl.) 39 (1987) 85–90.

FARDEAU, M., F.M.S. TOMÉ and S. DERAMBURE: Familial fingerprint body myopathy. Arch. Neurol. 33 (1976) 724–725.

FARDEAU, M., J. GODET-GUILLAIN, F.M.S. TOMÉ, H. COLLIN, S. GAUDEAU, C. BOFFETY and P. VERNANT: Une nouvelle affection musculaire familiale, définie par l'accumulation intra-sarco-plasmique d'un matériel granulo-filamentaire dense en microscopie électronique. Rev. Neurol. (Paris) 131 (1978) 411–425.

FARDEAU, M., J. GODET-GUILLAIN, F.M.S. TOMÉ, S. CARSON and R.G. WHALEN: Congenital neuromuscular disorders: a critical review. In: A.J. Aguayo and G. Karpati (Eds.) Current Topics In Nerve and Muscle Research. Amsterdam, Excerpta Medica (1979) 164–177.

FARDEAU, M., F.M.S. TOMÉ, M. CHEVALLAY and H. COLLIN: Difficult interpretation of muscle biopsies in puzzling cases of severe neonatal hypotonia. Tenth Oxford Symposium on Muscle Disease (1985).

FIDZIANSKA, A., B. BADURSKA, B. RYNIEWICZ and I. DEMBEK: 'Cap disease': new congenital myopathy. Neurology 31 (1981) 1113–1120.

FIDZIANSKA, A., H.H. GOEBEL, M. OSBORN, H.G. LENARD, G. OSSE and U. LANGENBECK: Mallory body-like inclusions in a hereditary congenital neuromuscular disease. Muscle Nerve 6 (1983) 195–200.

FIDZIANSKA, A., I. NIEBRÓJ-DOBOSZ, B. BADURSKA and B. RYNIEWICZ: Is central core disease with structural core a fetal defect? J. Neurol. 231 (1984) 212–219.

FITZSIMONS, R.B. and J.G. MCLEOD: Myopathy with pathological features of both centronuclear myopathy and multicore disease. J. Neurol. Sci. 57 (1982) 395–405.

FITZSIMONS, R.B. and H.D.D. TYER: A study of a myopathy presenting as idiopathic scoliosis: multicore disease or mitochondrial myopathy? J. Neurol. Sci. 46 (1980) 33–48.

FRANK, J.P., Y. HARATI, I.J. BUTLER, T.E. NELSON and C.I. SCOTT: Central core disease and malignant hyperthermia syndrome. Ann. Neurol. 7 (1980) 11–17.

FUKUDA, T., S. KOBAYASHI, Y. YAMAMOTO and S. KAMOSHITA: Congenital fiber type disproportion. No To Hattatsu 15 (1983) 425–431.

FUKUHARA, N., M. HOSHI and S. MORI: Core/targetoid fibres and multiple cytoplasmic bodies in organophosphate neuropathy. Acta Neuropathol. (Berlin) 40 (1977) 137–144.

FUKUHARA, N., T. KUMAMOTO and T. TSUBAKI: Rimmed vacuoles. Acta Neuropathol. (Berlin) 51 (1980) 229–235.

FUKUHARA, N., T. KUMAMOTO, H. HIRAHARA and T. TSUBAKI: A new myopathy with tubulomembranous inclusions. J. Neurol. Sci. 50 (1981) 95–107.

FUKUHARA, N., Y. INOUE, N. YOSHIMURA, H. SUGITA and K. LI: Histochemical and ultrastructural studies on cytoplasmic body myopathy. J. Neurol. Sci. 98 (1990) 101.

FUKUNAGA, H., M. OSAME and A. IGATA: A case of nemaline myopathy with ophthalmoplegia and mitochondrial abnormalities. J. Neurol. Sci. 46 (1980) 169–177.

GAMBARELLI, D., J. HASSOUN, J.F. PELLISSIER, M. BERARD and M. TOGA: Concentric laminated bodies in muscle pathology. Pathol. Eur. 9 (1974) 289–296.

GIBBELS, E., U. HENKE, H.J. SCHÄDLICH, W.F. HAUPT and W. FIEHN: Cylindrical spirals in skeletal muscle: a further observation with clinical, morphological, and biochemical analysis. Muscle Nerve 6 (1983) 646–655.

GOEBEL, H.H.: Neuropathological aspects of congenital myopathies. Prog. Neuropathol. 6 (1986) 231–262.

GOEBEL, H.H.: Congenital myopathies: In : M. Adachi and J.H. Sher (Eds.), Neuromuscular Disease. New York Tokyo, Igaku-Shoin (1990) 197–244.

GOEBEL, H.H., W. ZEMAN and H. PILZ: Significance of muscle biopsies in neuronal ceroid-lipofuscinoses. J. Neurol. Neurosurg. Psychiatry 38 (1975) 985–993.

GOEBEL, H.H., H.G. LENARD, W. GÖRKE and K. KUNZE: Fibre type disproportion in the rigid spine syndrome. Neuropediatrics 8 (1977) 467–477.

GOEBEL, H.H., J. MULLER, H.W. GILLEN and A.D. MERRITT: Autosomal dominant 'spheroid body myopathy'. Muscle Nerve 1 (1978) 14–26.

GOEBEL, H.H., H.G. LENARD, U. LANGENBECK and B. MEHL: A form of congenital muscular dystrophy. Brain Dev. 2 (1980) 387–400.

GOEBEL, H.H., H. SCHLOON and H.G. LENARD: Congenital myopathy with cytoplasmic bodies. Neuropediatrics 12 (1981) 166–180.

GOEBEL, H.H., H.M. MEINCK, M. REINECKE, K. SCHIMRIGK and U. MIELKE: Centronuclear myopathy with special consideration of the adult form. Eur. Neurol. 23 (1984) 425–434.

GOEBEL, H.H., S. VON LOH and J. GEHLER: Childhood neuromuscular disease with rimmed vacuoles. Eur. J. Pediatr. 144 (1986) 557–562.

GOLDSTEIN, M.A., M.H. STROMER, J.P. SCHROETER and R.L. SASS: Optical reconstruction of nemaline rods. Exp. Neurol. 70 (1980) 83–97.

GREENFIELD, J.G., T. CORNMAN and G.M. SHY: The prognostic value of the muscle biopsy in the 'floppy infant'. Brain 81 (1958) 461–484.

GULLOTTA, F., L. PAVONE, M. LA ROSA and A. GRASSO: Minicore myopathy. Klin. Wochenschr. 60 (1982) 1351–1355.

HAGBERG. J.M., J.E. CARROLL and M.H. BROOKE: Endurance exercise training in a patient with central core disease. Neurology 30 (1980) 1242–1244.

HALBIG, L., L. GUTMANN, H.H. GOEBEL, J.F. BRICK and S. SCHOCHET: Ultrastructural pathology in emetine-induced myopathy. Acta Neuropathol. (Berlin) 75 (1988) 577–582.

HALBIG, L., H.H. GOEBEL, H.C. HOPF and R. MOLL: Spheroid-cytoplasmic complexes in a congenital myopathy. Rev. Neurol. (Paris) 147 (1991) 300–307.

HALL-CRAGGS, E.C.B., S.R. MAX, M.M. WINES, T.M. MORELAND and J.R. HEBEL: Central core degeneration after tenotomy in soleus muscles of hyperthyroid rats. Exp. Neurol. 81 (1983) 722–732.

HALTIA, M., H. SOMER and S. REHUNEN: Congenital fibre type disproportion in an adult: a morphometric and microchemical study. Acta Neurol. Scand. 78 (1988) 65–71.

HANNA, W. and R.D. HENDERSON: Nemaline rods in cricopharyngeal dysphagia. Am. Soc. Clin. Pathol. 74 (1980) 186–191.

HANSON, P.A., A.F. MASTRIANNI and L. POST: Neonatal ophthalmoplegia with microfibers: a reversible myopathy? Neurology 27 (1977) 974–980.

HART, Z.H., S. SERVIDEI, P.L. PETERSON, C.H. CHANG and S. DIMAURO: Cardiomyopathy, mental retardation, and autophagic vacuolar myopathy. Neurology 37 (1987) 1065–1068.

HAYASHI, K., R.G. MILLER and A.K.W. BROWNELL: Central core disease: Ultrastructure of the sarcoplasmic reticulum and T-tubules. Muscle Nerve 12 (1989) 95–102.

HEFFNER, R.R., JR.: Muscle biopsy in the diagnosis of neuromuscular disease. Sem. Diagn. Pathol. 1 (1984) 114–151.

HOPF, N. and H.H. GOEBEL: Experimental emetine myopathy: enzyme histochemical, electron microscopic and immunomorphologic studies. Clin. Neuropathol. 8 (1989) 233.

HOWARD, R.: A case of congenital defect of the muscular system (dystrophia muscularis congenita) and its association with congenital talipes equinovarus. Proc. R. Soc. Med. 1 (1908) 157–166.

HÜBNER, G. and D. PONGRATZ: Granularkörpermyopathie (sog. Reducing Body Myopathy)? Beitrag zur Feinstruktur und Klassifizierung. Virchows Arch. A (Pathol. Anat.) 392 (1981) 97–104.

HÜLSMANN, N., F. GULLOTTA and H. CKUR: Cytopathology of an unusual case of centronuclear myopathy. J. Neurol. Sci. 50 (1981) 311–333.

IANNACCONE, S.T., K.E. BOVE, C.A. VOGLER and J.J. BUCHINO: Type I fiber size disproportion: morphometric data from 37 children with myopathic, neuropathic, or idiopathic hypotonia. Pediatr. Pathol. 7 (1987) 395–419.

JADRO-SANTEL, D., N. GRCEVIC, S. DOGAN, J. FRANJIC and H. BENC: Centronuclear myopathy with type I fibre hypotrophy and 'fingerprint' inclusions associated with Marfan's syndrome. J. Neurol. Sci. 45 (1980) 43–56.

JENNEKENS, F.G.I., J.J. ROORD, H. VELDMAN, J. WILLEMSE and B.M. JOKUSCH: Congenital nemaline myopathy. I. Defective organization of alpha-actinin is restricted to muscle. Muscle Nerve 6 (1983) 61–68.

JERUSALEM, F., A.G. ENGEL and M.R. GOMEZ: Sarcotubular myopathy; a newly recognized, benign, congenital, familial muscle disease. Neurology 23 (1973) 897–906.

JERUSALEM, F., H. LUDIN, A. BISCHOFF and G. HART-

MANN: Cytoplasmic body neuromyopathy presenting as respiratory failure and weight loss. J. Neurol. Sci. 41 (1979) 1–9.

JUGUILON, A., D. CHAD, W.G. BRADLEY, L. ADELMAN, J. KELEMEN, P. BOSCH and T.L. MUNSAT: Familial granulovacuolar lobular myopathy with electrical myotonia. J. Neurol. Sci. 56 (1982) 133–140.

KALIMO, H., M.L. SAVONTAUS, H. LANG, L. PALJÄRVI, V. SONNINEN, P.B. DEAN, K. KATEVUO and A. SALMINEN: X-linked myopathy with excessive autophagy: a new hereditary muscle disease. Ann. Neurol. 23 (1988) 258–265.

KAR, N.C., C.M. PEARSON and M.A. VERITY: Muscle fructose 1,6-diphosphatase deficiency associated with an atypical central core disease. J. Neurol. Sci. 48 (1980) 243–256.

KAUSCH, K., T. GRIMM, M. JANKA, F. LEHMANN-HORN, B. WIERINGA and C.R. MÜLLER: Evidence for linkage of the central core disease locus to chromosome 19q. J. Neurol. Sci. 98 (1990) 549.

KEITH, A. and W. BROWNELL: Reducing body myopathy in an adult. J. Neurol. Sci. 98 (1990) 336.

KINOSHITA, M., E. SATOYOSHI and N. MATSUO: 'Myotubular myopathy' and 'type I fiber atrophy' in a family. J. Neurol. Sci. 26 (1975a) 575–582.

KINOSHITA, M., E. SATOYOSHI and Y. SUZUKI: Atypical myopathy with myofibrillar aggregates. Arch. Neurol. 32 (1975b) 417–420.

KINOSHITA, M., E. SATOYOSHI and M. KUMAGAI: Familial type I fiber atrophy. J. Neurol. Sci. 25 (1975c) 11–17.

KIRKEBY, S.: Enzyme reactions in beige mouse muscle with central cores. Acta Neuropathol. (Berlin) 54 (1981) 325–327.

KOCH, B.M., T.E. BERTORINI, G.D. ENG and R. BOEHM: Severe multicore disease associated with reaction to anesthesia. Arch. Neurol. 42 (1985) 1204–1206.

KOHYAMA, J., F. NIIMURA, K. KAWASHIMA, Y. IWAKAWA and I. NONAKA: Congenital fiber type disproportion myopathy in Lowe syndrome. Pediatr. Neurol. 5 (1989) 373–376.

KONDO, K. and T. YUASA: Genetics of congenital nemaline myopathy. Muscle Nerve 3 (1980) 308–315.

KORNFELD, M.: Mixed nemaline-mitochondrial 'myopathy'. Acta Neuropathol. (Berlin) 51 (1980) 185–189.

KRABBE, K.H.: Kongenital generaliseret muskelaplasi. Nord. Med. 35 (1947) 1756.

KUMAGAI, T., S. HAKAMADA, K. HARA, T. TAKEUCHI, S. MIYAZAKI, K. WATANABE and K. KOMATSU: Development of human fetal muscles: a comparative histochemical analysis of the psoas and the quadriceps muscles. Neuropediatrics 15 (1984) 198–202.

LAKE, B.D. and J. WILSON: Zebra body myopathy. J. Neurol. Sci. 24 (1975) 437–446.

LAKE, B.D., N. CAVANAGH and J. WILSON: Myopathy with minicores in siblings. Neuropathol. Appl. Neurobiol. 3 (1977) 159–167.

LAMBERT, C.D.: Uniform muscle histochemistry: a clinical and biopsy study Trans. Am. Neurol. Assoc. 99 (1974) 120–125.

LEE, Y.S. and W.C.L. YIP: A fatal congenital myopathy with severe type I fibre atrophy, central nuclei and multicores. J. Neurol. Sci. 50 (1981) 277–290.

LEHESJOKI, A.E., E.M. SANKILA, J. MIAO, M. SOMER, R. SALONEN, J. RAPOLA and A. DE LA CHAPELLE: X-linked neonatal myotubular myopathy: one recombination detected with four polymorphic DNA markers from Xq28. J. Med. Genet. 27 (1990) 288–291.

LENARD, H.G. and H.H. GOEBEL: Congenital muscular dystrophies and unstructured congenital myopathies. Brain Dev. 2 (1980) 119–125.

LO, W.D., R.J. BAHRON, R.J. BOBULSKI, J. KEAN and J.R. MENDELL: Centronuclear myopathy and type-I hypotrophy without central nuclei: distinct nosologic entities? Arch. Neurol. 47 (1990) 273–276.

LOPEZ-TERRADAS, J.M. and M. CONDE LOPEZ: Late components of motor unit potentials in central core disease. J. Neurol. Neurosurg. Psychiatry 42 (1979) 461–464.

MACDONALD, R.D. and A.G. ENGEL: The cytoplasmic body: another structural anomaly of the Z disk. Acta Neuropathol. (Berlin) 14 (1969) 99–107.

MAHON, M., F. KRISTMUNDSDOTTIR, W.J.K. CUMMING and M.J. NORONHA: Sequential studies of a childhood myopathy: a clinical, histochemical and morphometric investigation, Neuropathol. Appl. Neurobiol. 15 (1989) 3–12.

MATEER, J.E., B.J. FARRELL, S.S.M. CHOU and L. GUTMANN: Reversible ipecac myopathy. Arch. Neurol. 42 (1985) 188–190.

MATSUBARA, S. and W.G.P. MAIR: Fingerprint inclusions and circular structures in the muscle. Acta Neuropathol. (Berlin) 47 (1979) 161–162.

MATSUO, T., T. TASHIRO, T. IKEDA, M. TSUJIHATA and C. SHIMOMURA: Fatal neonatal nemaline myopathy. Acta Pathol. Jpn. 32 (1982) 907–916.

MATSUOKA, Y., S.S. GUBBAY and B.A. KAKULAS: A new myopathy with type II muscle fibre hypoplasia. Proc. Aust. Assoc. Neurol. 11 (1974) 155–159.

MATTLE, H. and F. JERUSALEM: Central-Core-Myopathie: eine klinische und morphologische Studie zur diagnostischen Spezifiziät der zentralen Muskelfaserveränderungen. Schweiz. Med. Wochenschr. 111 (1981) 741–749.

MEIER, C., W. VOELLMY, M. GERTSCH, A. ZIMMERMANN and J. GEISSBÜHLER: Nemaline myopathy appearing in adults as cardiomyopathy: a clinicopathologic study. Arch. Neurol. 41 (1984) 443–445.

MEYERS, K.R., D.H. GILDEN, C.F. RINALDI and J.L. HANSEN: Periodic muscle weakness, normokalemia, and tubular aggregates. Neurology 22 (1972) 269–279.

MIIKE, T., Y. OHTANI, H. TAMARI, T. ISHITSU and Y. UNE: Muscle fiber type transformation in nemaline myopathy and congenital fiber type disproportion. Brain Dev. 8 (1986) 526–532.

MISULIS, K.E. and G.M. FENICHEL: Genetic forms of myasthenia gravis. Pediatr. Neurol. 5 (1989) 205–210.

MIZUNO, Y. and K. KOMIYA: A serial muscle biopsy study in a case of congenital fiber-type disproportion associated with progressive respiratory failure. Brain Dev. 12 (1990) 431–436.

MIZUNO, Y., Y. NAKAMURA and K. KOMIYA: The spectrum of cytoplasmic body myopathy: report of a congenital severe case. Brain Dev. 11 (1989) 20–25.

MOHIRE, M.D., R. TANDAN, T.J. FRIES, B.W. LITTLE, W.W. PENDLEBURY and W.G. BRADLEY: Early-onset benign autosomal dominant limb-girdle myopathy with contractures (Bethlem myopathy). Neurology 38 (1988) 573–580.

MORGAN-HUGHES, J.A., W.G.P. MAIR and P.T. LASCELLES: A disorder of skeletal muscle associated with tubular aggregates. Brain 93 (1970) 873–880.

NIAKAN, E., Y. HARATI and M.J. DANON: Tubular aggregates: their association with myalgia. J. Neurol. Neurosurg. Psychiatry 48 (1985) 882–886.

NONAKA, I., M. TOJO and H. SUGITA: Fetal muscle characteristics in nemaline myopathy. Neuropediatrics 14 (1983a) 47–52.

NONAKA, I., Y. NAKAMURA, M. TOJO, H. SUGITA, T. ISHIKAWA, A. AWAYA and N. SUGIYAMA: Congenital myopathy without specific features (minimal change myopathy). Neuropediatrics 14 (1983b) 237–241.

NONAKA, I., S. OKADA and Y. SAITO: Defects in muscle fiber maturation in congenital myopathies. In: G. Serratrice (Ed.), Neuromuscular Diseases. New York, Raven Press (1984) 207–212.

NONAKA, I., S. ISHIURA, K. ARAHATA, H. ISHIBASHI-UEDA, T. MARUYAMA and K.II: Progression in nemaline myopathy. Acta Neuropathol. (Berlin) 78 (1989) 484–491.

NORTON, P., P. ELLISON, A.R. SULAIMAN and J. HARB: Nemaline myopathy in the neonate. Neurology 33 (1983) 351–356.

OH, S.J. and M.J. DANON: Nonprogressive congenital neuromuscular disease with uniform type I fibre. Arch. Neurol. 40 (1983) 147–150.

OH, S.J., G.J. MEYERS, E.R. WILSON JR. and C.B. ALEXANDER: A benign form of reducing body myopathy. Muscle Nerve 6 (1983) 278–282.

OHTAKI, E., Y. YAMAGUCHI, Y. YAMASHITA, Y. MATSUISHI, K. TERASAWA, Y. KATAFUCHI and I. NONAKA: Complete external ophthalmoplegia in a patient with congenital myopathy without specific features (minimal change myopathy). Brain Dev. 12 (1990) 427–430.

OLDFORS A., M. KYLLERMAN, J. WAHLSTRÖM, C. DARNFORS and K.G. HENRIKSSON: X-linked myotubular myopathy: clinical and pathological findings in a family. Clin. Genet. 36 (1989) 5–14.

OSBORN, M. and H.H. GOEBEL: The cytoplasmic bodies in a congenital myopathy can be stained with antibodies to desmin, the muscle-specific intermediate filament protein. Acta Neuropathol. (Berlin) 62 (1983) 149–152.

OTTE, G., J. DE REUCK, W. DE COSTER and H. VANDER EECKEN: The target phenomenon in tenotomized rat gastrocnemius muscle: a comparative electrophysiological and morphological study. Acta Neurol. Belg. 80 (1980) 361–367.

OWEN, J.S., L.B. KLINE, S.J. OH, N.E. MILES and J.W. BENTON: Ophthalmoplegia and ptosis in congenital fiber type disproportion. J. Pediatr. Ophthalmol. Strabism 18 (1981) 55–60.

PAGES, A. and M. PAGES: La myopathie à axe central. Ann. Pathol. 1 (1981) 38–47.

PATEL, H., K. BERRY, P. MACLEOD and H.P. DUNN: Cytoplasmic body myopathy. J. Neurol. Sci. 60 (1983) 281–292.

PAULUS, W., J. PEIFFER, I. BECKER, W. ROGGENDORF and F. SCHUMM: Adult-onset rod disease with abundant intranuclear rods. J. Neurol. 235 (1988) 343–347.

PAVONE, L., F. MOLLICA, A. GRASSO and G. PERO: Familial centronuclear myopathy. Acta Neurol. Scand. 62 (1980) 41–54.

PEARCE, J.M.S. and D.G.F. HARRIMAN: Childhood neuromuscular disorder with type IIC fibres. Tenth Oxford Symposium on Muscle Disease (1985).

PELIAS, M.Z. and T.F. THURMON: Congenital universal muscular hypoplasia: evidence for autosomal recessive inheritance. Am. J. Hum. Genet. 31 (1979) 548–554.

PELLEGRINI, G., S. BARBIERI, M. MOGGIO, A. CHELDI, G. SCARLATO and C. MINNETI: A case of congenital neuromuscular disease with uniform type I fibers, abnormal mitochondrial network and jagged Z-line. Neuropediatrics 16 (1985) 162–166.

PELLISSIER, J.F., T. DE BARSY, M. FAUGÈRE and P. REBUFFEL: Type III glycogenosis with multicore structures. Muscle Nerve 2 (1979) 124–132.

PELLISSIER, J.F., J. POUGET, C. CHARPIN and D. FIGARELLA: Myopathy associated with desmin type intermediate filaments. J. Neurol. Sci. 89 (1989) 49–61.

PENEGYRES, P.K. and B.A. KAKULAS: The natural history of minicore-multicore myopathy. Muscle Nerve 14 (1991) 411–415.

PIEROBON-BORMIOLI, S., S. LÜCKE and C. ANGELINI: Abnormal myomuscular junctions and AChE in a congenital neuromuscular disease. Muscle Nerve 3 (1980) 240–247.

PIEROBON-BORMIOLI, S., M. ARMANI, S.P. RINGEL, C. ANGELINI, L. VERGANI, R. BETTO and G. SALVIATI: Familial neuromuscular disease with tubular aggregates. Muscle Nerve 8 (1985) 291–298.

POU-SERRADELL, A., M. AGUILAR, L. SOLER and I. FERRER: Myopathie congénitale bénigne avec prépondérance des fibres de type I et rares 'cores' chez la mère asymptomatique: Association avec des malformations de la ligne médiane. Rev. Neurol. (Paris) 12 (1980) 853–862.

POWERS, J.M., G.F. YOUNG, E.B. BASS and F.E. REED: Atypical nemaline myopathy with temporomandibular ankylosis. Neurology 30 (1980) 971–975.

PRINCE, A.D., W.K. ENGEL and J.R. WARMOLTS: type I myofiber smallness without central nuclei or myotonia. Neurology 22 (1972) 401.

RADU, H., V. IONESCU, A. RADU, V. PALER, A.M. ROSU and A. MARIAN: Hypotrophic type I muscle fibres with central nuclei, and central myofibrillar lysis preferentially involving type II fibres. Eur. Neurol. 11 (1974) 108–127.

RADU, H., A.M. ROSU-SERBU, V. IONESCU and A. RADU: Focal abnormalities in mitochondrial distribution in muscle: two atypical cases of so-called 'central core disease'. Acta Neuropathol. (Berlin) 39 (1977) 25–31.

RAPPAPORT, L., F. CONTARD, J.L. SAMUEL, C. DELCAYRE, F. MAROTTE, F. TOMÉ and M. FARDEAU: Storage of phosphorylated desmin in a familial myopathy. FEBS Lett. 231 (1988) 421–425.

RICOY, J.R. and A. CABELLO: Dysmaturative myopathy-evolution of the morphological picture in three cases. Acta Neuropathol. (Berlin) Suppl. VII (1981) 313–316.

RICOY, J.R. and A. CABELLO: Hypotrophy of type I fibres with central nuclei: recovery 4 years after diagnosis. J. Neurol. Neurosurg. Psychiatry 48 (1985) 167–171.

RICOY, J.R., A. CABELLO and G. GOIZUETA: Myopathy with multiple minicores: report of two siblings. J. Neurol. Sci. 48 (1980) 81–92.

RIGGS, J.E., S.S. SCHOCHET JR., L. GUTMANN, S. SHANSKE, W.A. NEAL and S. DIMAURO: Lysosomal glycogen storage disease without acid maltase deficiency. Neurology 33 (1983) 873–877.

RINGEL, S.P., H.E. NEVILLE, M.C. DUSTER and J.E. CARROLL: A new congenital neuromuscular disease with trilaminar muscle fibers. Neurology 28 (1978) 282–289.

ROHKAMM, R., K. BOXLER, K. RICKER and F. JERUSALEM: A dominantly inherited myopathy with excessive tubular aggregates. Neurology 33 (1983) 331–336.

SAHGAL, V. and S. SAHGAL: A new congenital myopathy: a morphological, cytochemical and histochemical study. Acta Neuropathol. (Berlin) 37 (1977) 225–230.

SALVIATI, G., S. PIEROBON-BORMIOLI, R. BETTO, E. DAMIANI, C. ANGELINI, S.P. RINGEL, S. SALVATORI and A. MARGRETH: Tubular aggregates: sarcoplasmic reticulum origin, calcium storage ability, and functional implications. Muscle Nerve 8 (1985) 299–306.

SARNAT, H.B.: Muscle Pathology and Histochemistry. Chicago, Am. Soc. Clin. Pathol. Press (1983).

SARNAT, H.B.: Myotubular myopathy: arrest of morphogenesis of myofibres associated with persistence of fetal vimentin and desmin. Can. J. Neurol. Sci. 17 (1990) 109–123.

SARNAT, H.B., S.I. ROTH and J.F. JIMINEZ: Neonatal myotubular myopathy: neuropathy and failure of postnatal maturation of fetal muscle. J. Can. Sci. Neurol. 8 (1981) 313–320.

SASAKI, T., K. SHIKURA, K. SUGAI, I. NONAKA and K. KUMAGAI: Muscle histochemistry in myotubular (centronuclear) myopathy. Brain Dev. 11 (1989) 26–32.

SAWCHAK, J.A. and J.H. SHER: Centronuclear myopathy heterogeneity: distinction of clinical types by myosin isoform patterns. Neurology 40 (Suppl. l) (1990) 204.

SAWICKA, E., A. FIDZIANSKA and H. STRUGALSKA: Cytoplasmic body myopathy. J. Neurol. Sci. 98 (1990) 335.

SCELSI, R., G. LANZI, L. NESPOLI and P. POGGI: Congenital non-progressive myopathy with type II fibre atrophy and internal nuclei. Eur. Neurol. 14 (1976) 285–293.

SCHMALBRUCH, H., Z. KAMIENIECKA and M. ARROE: Early fatal nemaline myopathy: case report and review. Dev. Med. Child Neurol. 29 (1987) 800–804.

SCHOTLAND, D.L.: An electron microscopic study of target fibers, target-like fibers and related abnormalities in human muscle. J. Neuropathol. Exp. Neurol. 28 (1969) 214–228.

SCHRÖDER, J.M.: Myopathie mit subsarkolemmal-segmentaler Myofibrillolyse. In: Pathologie der Muskulatur. Berlin Heidelberg New York, Springer-Verlag (1982a) 262.

SCHRÖDER, J.M.: Nukleodegenerative Myopathie. In: Pathologie der Muskulatur. Berlin Heidelberg New York, Springer-Verlag (1982b) 272.

SCHRÖDER, J.M. and W. SCHÖNBERGER: Neuromuskuläre Krankheit mit trilaminären Muskelfasern: der erste wahrscheinlich familiäre Fall. In: Fortschritte der Myologie, Band VI (1980) 54–62.

SCHRÖDER, J.M., K.G. KRÄMER and H.C. HOPF: Granular nuclear inclusion body disease: fine structure of tibial muscle and sural nerve. Muscle Nerve 8 (1985) 52–59.

SCHRÖDER, J.M., C. SOMMER and B. SCHMIDT: Desmin and actin associated with cytoplasmic bodies in skeletal muscle fibers: immunocytochemical and fine structural studies, with a note on unusual 18-to 20-nm filaments. Acta Neuropathol. (Berlin) 80 (1990) 406–414.

SEITZ, R.J., K.V. TOYKA and W. WECHSLER: Adult-onset mixed myopathy with nemaline rods, minicores, and central cores: a muscle disorder mimicking polymyositis. J. Neurol. 231 (1984) 103–108.

SERRATRICE, G. and J.F. PELLISSIER: Deux familles de myopathies bénignes prédominant sur les ceintures d'hérédité autosomique dominante. Rev. Neurol. (Paris) 144 (1988) 43–46.

SERRATRICE, G., J.F. PELLISSIER, M.C. FAUGÈRE and J.L. GASTAUT: Centronuclear myopathy: possible central nervous system origin. Muscle Nerve 1 (1978) 62– 69.

SEWRY, C.A., T. VOIT and V. DUBOWITZ: Myopathy with unique ultrastructural feature in Marinesco-Sjögren syndrome. Ann. Neurol. 24 (1988) 577–580.

SEWRY, C.A., E. RODILLO, J.Z. HECKMATT and V. DUBOWITZ: Immunocytochemical studies of congenital centronuclear and nemaline myopathies. J. Neurol. Sci. 98 (1990) 100.

SHAPIRA, Y.A., R. YAROM and A. BLANK: Nemaline myopathy and a mitochondrial neuromuscular disorder in one family. Neuropediatrics 12 (1981) 152–165.

SHER, J.H., A.B. RIMALOVSKI, T.J. ATHANASSIADES and S.M. ARONSON: Familial centronuclear myopathy: a clinical and pathological study. Neurology 17 (1967) 727–742.

SHIMIZU, T., K. HASHIMOTO, I. NONAKA and T. MANNEN:

Structural proteins in nemaline myopathy. J. Neurol. Sci. 98 (1990) 99.

SHUAIB, A., R.T. PAASUKE and A.K.W. BROWNELL: Central core disease: clinical features in 13 patients. Medicine 66 (1987) 389–396.

SHUAIB, A., J.M.E. MARTIN, L.B. MITCHELL and A.K.W. BROWNELL: Multicore myopathy: not always a benign entity. Can. J. Neurol. Sci. 15 (1988) 10–14.

SHY, G.M. and K.R. MAGEE: A new congenital non-progressive myopathy. Brain 79 (1956) 610–621.

SHY, G.M., W.K. ENGEL, J.E. SOMERS and T. WANKO: Nemaline myopathy: a new congenital myopathy. Brain 86 (1963) 793–810.

SIMPSON, R. and R. HEWLETT: Nemaline rod disease, with reference to the routine use of histochemical methods in autopsy investigations. Hum. Pathol. 13 (1982) 771–773.

SPARGO, E., B. DOSHI and H.L. WHITWELL: Fatal myopathy with cytoplasmic inclusions. Neuropathol. Appl. Neurobiol. 14 (1988) 516.

SPIRO, A.J., G.M. SHY and N.K. GONATAS: Myotubular myopathy. Arch. Neurol. 14 (1966) 1–14.

SPIRO, A.J., D.S. HOROUPIAN and D.R. SNYDER: Biopsy and autopsy studies of congenital muscle fiber type disproportion: a broadening concept. Neurology 27 (1977) 405.

STEWART, C.R., S.G. KAHLER and J.M. GILCHRIST: Congenital myopathy with cleft palate and increased susceptibility to malignant hyperthermia: King syndrome? Pediatr. Neurol. 4 (1988) 371–374.

STOECKEL, M.E., M. OSBORN, A. PORTE, A. SACREZ, A. BATZENSCHLAGER and K. WEBER: An unusual familial cardiomyopathy characterized by aberrant accumulation of desmin-type intermediate filaments. Virchows Arch. (Pathol. Anat.) 393 (1981) 53–60.

STOESSL, A.J., A.F. HAHN, D. MALOTT, D.T. JONES and M.D. SILVER: Nemaline myopathy with associated cardiomyopathy. Arch. Neurol. 42 (1985) 1084–1086.

STUHLFAUTH, I., F.G.I. JENNEKENS, J. WILLEMSE and B.M. JOKUSCH: Congenital nemaline myopathy. II. Quantitative changes in alpha-actinin and myosin in skeletal muscle. Muscle Nerve 6 (1983) 69–74.

SUGIE, J., R. HANSON, G. RASMUSSEN and M.A. VERITY: Adult onset type II fiber centronuclear neuromyopathy with segmental demyelination. Brain Dev. 4 (1982a) 7–12.

SUGIE, H., R. HANSON, G. RASMUSSEN and M.A. VERITY: Congenital neuromuscular disease with type I fibre hypotrophy, ophthalmoplegia and myofibril degeneration. J. Neurol. Neurosurg. Psychiatry 45 (1982b) 507–512.

SULAIMAN R.A., H.M. SWICK and D.S. KINDER: Congenital fibre type disproportion with unusual clinicopathologic manifestations. J. Neurol. Neurosurg. Psychiatry 46 (1983) 175–182.

SWASH, M. and M.S. SCHWARTZ: Familial multicore disease with focal loss of cross-striation and ophthalmoplegia. J. Neurol. Sci. 52 (1981) 1–10.

TANAKA, T., M. YOSHIMOTO, K. SHIRAI, S. KINOSHITA, M. WATANABE, T. YOSHIMURA and M. TSUJIHATA: A case of nemaline myopathy with abnormal mitochondria. No To Hattatsu 16 (1984) 476–480.

TARATUTO, A.L., M. MATTEUCCI and C. BARREIRO: Dominant autosomal neuromuscular disease with cylindrical spirals (CS). J. Neurol. Sci. 98 (1990) 336.

TER LAAK, H.J., H.H.J. JASPAR, F.J.M. GABREELS, T.J.M. BREUER, R.C.A. SENGERS, E.M.G. JOOSTEN, A.M. STADHOUDERS and A.A.W.M. GABREELS-FESTEN: Congenital fibre type disproportion: a morphometric study (4 cases). Clin. Neurol. Neurosurg. 83 (1981) 67–79.

THORNELL, L.-E.: Fine structure of nemaline rods in cryosections. J. Ultrastruct. Res. 63 (1978) 99–100.

THORNELL, L.-E., A. ERIKSSON and L. EDSTRÖM: Intermediate filaments in human myopathies. In: R.M. Dowben and J.W. Shay (Eds.), Cell and Muscle Motility. New York, Plenum (1983) 85–136.

TOMÉ, F.M.S. and M. FARDEAU: Congenital myopathies. Curr. Opin. Neurol. Neurosurg. 1 (1988) 782–787.

TORRES, C.F., R.C. GRIGGS and J.P. GOETZ: Severe neonatal centronuclear myopathy with autosomal dominant inheritance. Arch. Neurol. 42 (1985) 1011–1014.

TSUJIHATA, M., C. SHIMOMURA, T. YOSHIMURA, A. SATO, T. OGAWA, Y. TSUJI, S. NAGATAKI and T. MATSUO: Fatal neonatal nemaline myopathy: a case report. J. Neurol. Neurosurg. Psychiatry 46 (1983) 856–859.

VALLAT, J.M., L. DE LUMLEY, A. LOUBET, M.J. LEBOUTET, N. CORVISIER and R. UMDENSTOCK: Coexistence of minicores, cores and rods in the same muscle biopsy: a new example of mixed congenital myopathy. Acta Neuropathol. (Berlin) 58 (1982) 229–232.

VAN ANTWERPEN, C.L., S.M. GOSPE JR and M.P. DENTINGER: Nemaline myopathy associated with hypertrophic cardiomyopathy. Pediatr. Neurol. 4 (1988) 306–308.

VOIT, T., O. KROGMANN, H.G. LENARD, E. NEUEN-JACOB, W. WECHSLER, H.H. GOEBEL, G. RAHLF, A. LINDINGER and C. NIENABER: Emery-Dreifuss muscular dystrophy: disease spectrum and differential diagnosis. Neuropediatrics 19 (1988) 62–71.

WALLGREN-PETTERSSON, C.: Congenital nemaline myopathy: a clinical follow-up study of twelve patients. J. Neurol. Sci. 89 (1989) 1–14.

WALLGREN-PETTERSSON, C., J. RAPOLA and M. DONNER: Pathology of congenital nemaline myopathy: a follow-up study. J. Neurol. Sci. 83 (1988) 243–257.

WALLGREN-PETTERSSON, C., K. SAINIO and T. SALMI: Electromyography in congenital nemaline myopathy. Muscle Nerve 12 (1989) 587–593.

WALLGREN-PETTERSSON, C., H. KÄÄRIÄINEN, J. RAPOLA, T. SALMI, J. JÄÄSKELÄINEN and M. DONNER: Genetics of congenital nemaline myopathy: a study of 10 families. J. Med. Genet. 27 (1990a) 480–487.

WALLGREN-PETTERSSON, C., P. ARJOMAA and C. HOLMBERG: Alpha-actinin and myosin light chains in congenital nemaline myopathy. Pediatr. Neurol. 6 (1990b) 171–174.

YAMAGUCHI, M., R.G. CASSENS and D.S. DAHL: Congenital rod disease: some biochemical aspects of nemaline rods. Cytobiology 11 (1975) 335–345.

YAMAGUCHI, M., R.M. ROBSON, M.H. STROMER, D.S. DAHL and T. ODA: Nemaline myopathy rod bodies: structure and composition. J. Neurol. Sci. 56 (1982) 35–56.

YAROM, R. and Y. SHAPIRA: Myosin degeneration in a congenital myopathy. Arch. Neurol. 34 (1977) 114–115.

YOSHIOKA, M., S. KUROKI, K. OHKURA, Y. ITAGAKI and K. SAIDA: Congenital myopathy with type II muscle fiber hypoplasia. Neurology 37 (1987) 860–863.

ZIMMERMAN, P. and U. WEBER: Familial centronuclear myopathy: a haploid DNA disease? Acta Neuropathol. (Berlin) 46 (1979) 209–214.

Handbook of Clinical Neurology, Vol. 18 (62): Myopathies
L.P. Rowland and S. DiMauro, editors
© 1992 Elsevier Science Publishers B.V. All rights reserved

CHAPTER 13

Inflammatory myopathies

MARINOS C. DALAKAS

Neuromuscular Diseases Section, National Institute of Neurological Disorders and Stroke, National Institutes of Health, Bethesda, MD, USA

The inflammatory myopathies encompass a heterogeneous group of subacute or chronic acquired muscle diseases, characterized by proximal muscle weakness and inflammation in the muscle biopsy (Banker and Engel 1986; Dalakas 1988a, 1990a, 1991, 1992a,b; Karpati and Carpenter 1988; Engel and Esmslie-Smith 1989; Rowland 1989). Although these myopathies comprise a heterogeneous group of disorders, they have been viewed for years as being pathogenetically similar (Bohan and Peter 1975; Pearson and Bohan 1977; Ansell 1984), because of poor understanding of their pathogenetic mechanisms and lack of consensus regarding definition, classification, diagnosis and therapy. The participation of diagnosticians with diverse or opposing views such as neurologists, rheumatologists, internists, dermatologists or psychiatrists in the care of these patients, has further prolonged the uncertainties concerning diagnosis and management. Over the last 10 years, however, distinct clinical, laboratory, prognostic, therapeutic, demographic, histologic, and immunopathologic criteria have emerged that have unquestionably defined three major and distinct subsets of inflammatory myopathies: *polymyositis (PM), dermatomyositis (DM)* and *inclusion-body myositis (IBM)* (Karpati and Carpenter 1988; Engel and Esmslie-Smith 1989; Dalakas 1990a, 1991, 1992a). Each group retains its characteristic clinical, immunopathologic, and morphologic features re-gardless of whether it occurs separately or in connection with other systemic diseases.

The causes of PM, DM and IBM are still unknown. Their association, however, with other putative or definite autoimmune diseases or viruses, the evidence for a T-cell mediated myocytotoxicity or complement-mediated microangiopathy, the presence of a variety of autoantibodies, and their response to immunotherapies (Engel and Esmslie-Smith 1989; Dalakas 1990a, 1991) have all strengthened the hypothesis of an autoimmune role in their pathogenesis.

GENERAL CLINICAL FEATURES

Although unique features, which I will describe below, separate PM from DM or IBM, they all share in common a myopathy characterized by proximal and often symmetrical muscle weakness. In PM and DM the myopathy develops subacutely over weeks to months but rarely acutely, whereas in IBM it develops insidiously and slowly, over several months or years resembling a dystrophy. Tasks requiring the use of proximal muscles are compromised first, whereas fine motor movements that depend on the strength of distal muscles, are affected only late in DM and PM but fairly early in IBM. Falling is common in IBM owing to early involvement of the quadriceps muscle and buck-

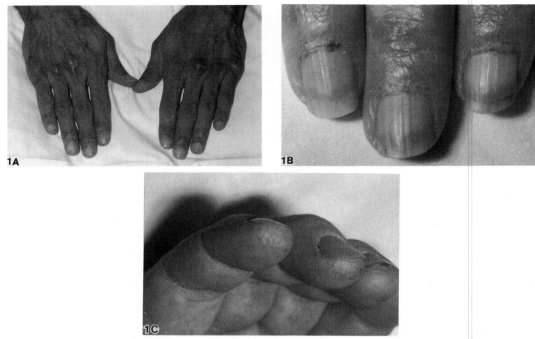

Fig. 1. (A) Characteristic scaly discoloration at the knuckles, sparing the phalanges, in a patient with DM. (B) Microinfarcts with dilatation of the capillaries at the base of the finger nails in a patient with DM. The shining appearance on the fingers is due to the characteristic erythematous lesions. (C) Cutaneous microinfarcts at the tips of the fingers (palmar surfaces) seen as irregular, 'dirty', black horizontal lines or dots resembling 'mechinic's hands', in a patient with adult dermatomyositis.

ling of the knees. Ocular muscles remain normal, even in advanced untreated cases, and if these muscles are affected the diagnosis of inflammatory myopathy should be in doubt. Facial muscles also remain normal although rare exceptions may exist. The pharyngeal and neck flexor muscles are often involved, causing varying degrees of dysphagia. In advanced cases and rarely in acute cases, respiratory muscles also may be affected. Severe weakness is almost always associated with muscular wasting. Sensation remains normal. The tendon reflexes are preserved, but may be absent in severely weakened or atrophied muscles. Myalgia and muscle tenderness may occur in some patients, usually early in the disease, and more often in DM than in PM. Unlike the progression of the weakness accompanying dystrophic processes, which is observable from one year to the next, the weakness in PM and DM may progress noticeably, often almost month by month. The exception is IBM, in

which painless weakness and atrophy develop and progress very slowly (over months or years), resulting in severe disease that resembles limb-girdle dystrophy.

DM affects both children and adults, and females more often than males, whereas PM is seen after the second decade of life and very rarely in childhood. IBM is more frequent in males (male: female ratio of 3:1) over the age of 50, and it is more common in whites than in blacks. A polymyositis-like disease with associated myoglobinuria (rhabdomyolysis) may rarely develop acutely after viral infections resulting in severe disease that often responds poorly to immunotherapies. It is my view that such postvirally induced rhabdomyolysis with necrotizing myopathy represents a distinct entity unrelated to the diseases described here. PM and DM may also begin during the last trimester of pregnancy, or within 2–3 months after delivery (Otero and Dalakas 1992).

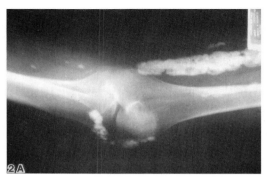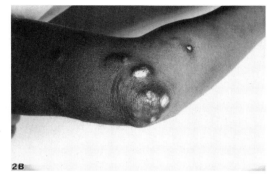

Fig. 2. (A) Subcutaneous calcifications with a floccular appearance in the thigh, leg and patella in a patient with dermatomyositis, as seen by X-ray film. (B) Calcium deposits extruding on to the skin in a 28-year-old patient with dermatomyositis.

DERMATOMYOSITIS

Dermatomyositis (DM) is a distinct clinical entity identified by a characteristic rash accompanying, or more often, preceding the muscle weakness. The skin manifestations include a heliotrope rash (blue-purple discoloration) on the upper eyelids with edema, a flat red rash on the face and upper trunk, and erythema of the knuckles with a raised violaceous scaly eruption (Gottron rash), that characteristically spares the phalanges. The initial erythematous lesions may result in scaling of the skin accompanied by pigmentation and depigmentation, giving at times a shiny appearance (Fig. 1A). The erythematous rash can also occur on other body surfaces, including the knees, elbows, malleoli, neck and anterior chest (often in a V sign), or back and shoulders (shawl sign), and may be exacerbated after exposure to the sun. The rash may be difficult to appreciate in dark-skinned people, and sometimes it is so faint, transient, or indiscernible that DM is diagnosed only in retrospect with the later discovery of subcutaneous calcifications (Rowland 1989). Dilated capillary loops with microinfarcts at the base of the fingernails are also characteristic of DM (Fig. 1B). The cuticles may be irregular, thickened and distorted, and the lateral and palmar areas of the fingers may become rough and cracked, with irregular, 'dirty' horizontal lines, resembling mechanic's hands (Fig. 1C). Subcutaneous calcification, although more often in childhood, occurs in some adults with DM and can be extensive (Fig. 2A). Sometimes calcium de-

posits extrude on the skin and cause ulcerations and infections (Fig. 2B). DM in children resembles the adult disease, except for more frequent extramuscular manifestations, as discussed later.

Occasionally patients with the classic skin lesions appear to have clinically normal strength even up to 3–5 years after onset. This form, often referred to as *amyopathic DM* or *DM sine myositis*, has an overall better prognosis because clinically the disease appears to be limited to the skin (Euwer and Southeimer 1991). The main neuromuscular symptoms of such patients are myalgia and fatigue. Although the strength appears normal, all these patients have a subclinical myopathy with histological and immunopathological features identical to those seen in the muscles of patients with the classic DM (Otero et al. 1992). Dermatomyositis is, therefore, a single disease with identical immunopathology affecting concurrently the skin and the muscle, but to varying degree among patients. DM usually occurs alone, but may be associated with systemic sclerosis, mixed connective tissue disease, other autoimmune conditions, or malignancies (Table 1). Fasciitis and skin changes similar to those found in DM were noticed in patients with the eosinophilia-myalgia syndrome associated with the ingestion of contaminated L-tryptophan (Hertzman et al. 1990). The contaminant found in the implicated lots, and less likely the L-tryptophan itself or its metabolites, were responsible for the changes noted in the skin, the fascia and the perimysium (Hertzman et al. 1990).

POLYMYOSITIS

In contrast with DM, which is a distinct entity because of the characteristic skin rash, polymyositis (PM) has no unique clinical features, and its diagnosis is one of exclusion. It is best defined as an inflammatory myopathy of subacute onset (weeks to months) and steady progression occurring in adults who do *not* have: (a) a rash, (b) involvement of eye and facial muscles, (c) a family history of a neuromuscular disease, (d) endocrinopathy, (e) a history of exposure to myotoxic drugs (such as cholesterol lowering agents) or toxins, and (f) neurogenic disease, dystrophy, biochemical muscle disease or IBM, as determined by muscle enzyme histochemistry and biochemistry. Unlike DM, in which the rash secures early recognition, the actual onset of PM cannot be easily determined, and the disease may exist for several months before the patient seeks medical advice. The reported seasonal onset of PM, determined on the basis of record review or the patient's own recollection of onset of symptoms (Leff et al. 1991), is unreliable because the disease may be lingering for months before it becomes noticeable by the patient. All claims of seasonal onset should therefore be viewed with skepticism.

Polymyositis appears to be a syndrome of diverse causes that may occur separately or in association with systemic autoimmune or connective tissue diseases and certain known viral or bacterial infections (Tables 1 and 2). It is very rare in childhood and if a diagnosis is made in patients below the age of 16 years, a careful review is needed to exclude another disease, especially certain dystrophies which may be associated with endomysial inflammation (Dalakas 1991). Other than D-penicillamine and zidovudine (AZT), which can cause a myopathy with endomysial inflammation like that seen in PM (Dalakas 1991, 1992b; Dalakas et al. 1990), myotoxic drugs, such as emetine, chloroquine, steroids, cimetidine, ipecac, and lovostatin, do not cause PM. Instead, they elicit a toxic *non primary inflammatory* myopathy that is histologically different from PM and does not require immunosuppressive therapy (Dalakas 1991).

TABLE 1

Inflammatory myopathies and associated conditions.

Associations	Dermatomyositis	Polymyositis	Inclusion-body myositis
Age of onset	Adults and children	Adults > 18 years	Adults > 50 years
Associated with connective tissue diseases	Yes, with scleroderma and mixed connective tissue disease	Yes	Yes, in up to 15% of cases
Overlaps with connective tissue diseases	Yes, with scleroderma and mixed connective tissue disease	No	No
Associated with systemic autoimmune diseases*	Infrequently	Frequently	Infrequently
Associated with malignancies	Probably	No	No
Associated with viruses	Unproved	With HIV, HTLV-1; possibly with other viral or postviral conditions	Unproved
Associated with parasites and bacteria	No	Yes[†]	No
Associated with drug-induced myotoxicity[‡]	Yes	Yes	No
Familial association	No	No	Yes, in some cases

*See Table 2.
[†]Parasitic (protozoa, cestodes, nematodes), tropical, and bacterial myositis, especially in patients with AIDS.
[‡]D-Penicillamine (for DM and PM), AZT (for PM), and contaminated L-tryptophan (for a DM-like illness). Other myotoxic drugs may cause myopathy but not *inflammatory* myopathy [adapted from Dalakas M.C. 1991].

Certain bacteria, such as *Borrelia burgdorferi* of Lyme disease and *Legionella pneumophila* of legionnaires' disease may infrequently be the cause of PM (Atlas et al. 1988; Warner 1991). Several animal parasites, such as protozoa (*Toxoplasma, Trypanosoma*), cestodes (cysticerci), and nematodes (trichinae), may produce a focal or diffuse inflammatory myopathy known as *parasitic* polymyositis. In the tropics, a suppurative myositis known as *tropical* polymyositis or *pyomyositis* may be produced by *Staphylococcus aureus, Yersinia, Streptococcus*, or other anaerobes (Banker 1986). Pyomyositis, previously rare in the West, can now be seen in patients with acquired immunodeficiency syndrome (AIDS), caused by *Staphylococcus aureus* or rarely by gram-negative bacteria (Gaut et al. 1988; Schwartzman et al. 1991; Widrow et al. 1991). Pyomyositis begins with low-grade fever, even without leukocytosis, and localized muscle pain and swelling. Imaging of the muscle with ultrasonography, magnetic resonance imaging (MRI) or computed tomography with contrast, reveals an enhancing lesion, often with a fluid density. The common colonization of HIV-positive patients by *S. aureus* and the reduced chemotaxis and bactericidal activity of their neutrophils against *S. aureus*, may be responsible for the bacterial infection of the muscle (Schwartzman et al. 1991).

INCLUSION-BODY MYOSITIS

Although inclusion-body myositis (IBM) is commonly suspected when a patient with presumed PM does not respond to therapy, involvement of distal muscles, especially foot extensors and finger flexors, in more than 50% of the cases, may be a clue to the early clinical diagnosis (Banker 1986; Lotz 1989; Dalakas 1988a, 1990, 1991, 1992a,b). Dysphagia can be prominent. The weakness and atrophy can be asymmetric with selective involvement of the quadriceps, iliopsoas, triceps and biceps muscles. The involvement of the deep flexors of the 3rd, 4th and 5th fingers due to selective weakness and atrophy of the flexor digitorum profundus muscle, is a clinically specific sign for IBM (Dalakas M.C., unpublished observations). Because of the distal weakness, and the early loss of the patellar reflex owing to severe weakness of the quadriceps muscle, a neurogenic disease is often suspected. The weakness and atrophy can also be asymmetrical and mimic lower motor neuron syndromes especially when the serum CK is not elevated. The diagnosis is always made by the characteristic findings on the muscle biopsy, as discussed later. IBM can be associated with systemic autoimmune or connective tissue diseases in up to 15% of the cases (Table 1). Familial cases, some of dominant inheritance (Neville et al. 1992), or others associated with leukoencephalopathy (Cole et al. 1988), may be found. Patients with IBM account for almost 50% of all the cases of 'polymyositis unresponsive to therapy' referred to our institution. It is my view that IBM is an overlooked and underdiagnosed disease, that occurs much more often than PM.

Table 2
Systemic autoimmune diseases most commonly associated with polymyositis*.

Crohn's disease	Psoriasis
Vasculitis	Hashimoto's disease
Sarcoidosis	Granulomatous diseases
Primary biliary cirrhosis	Agammaglobulinemia
Adult celiac disease	Monoclonal gammopathy
Chronic graft-versus-host disease	Hypereosinophilic syndrome
Discoid lupus	Lyme disease
Ankylosing spondylitis	Kawasaki disease
Behçet's disease	Autoimmune thrombocytopenia
Myasthenia gravis	Hyperagammaglobulinemic purpura
Acne fulminans	Hereditary complement deficiency
Dermatitis herpetiformis	IgA deficiency

*Banker, B.Q. and A.G. Engel 1986; Rowland, L.P. 1989; Carpenter, S. and G. Karpati 1984; Dalakas M.D. 1988a,b, 1992a.

Extramuscular manifestations

Aside from the primary disturbance of the skeletal muscles, the following extramuscular manifestations may be prominent to a varying degree in patients with PM, DM or IBM.

General symptoms. These include fever, malaise,

weight loss, arthralgia, and Raynaud's phenomenon especially when the inflammatory myopathy is associated with another connective tissue disorder. Such symptoms are more common in DM when it overlaps with mixed connective tissue disease or scleroderma.

Joint contractures. These occur mostly in DM and especially in children.

Gastrointestinal symptoms. At least 50% of the patients have dysphagia due to involvement of the oropharyngeal striated muscles and upper esophagus (Dalakas 1988a, 1991, 1992b; Plotz et al. 1989). Gastrointestinal ulcerations, seen more often in children with DM especially before the conventional use of immunosuppressive drugs, may result in melena or hematemesis due to vasculitis and infection. Ulceration may be rarely preceded by *pneumatosis intestinalis*, as the vasculitis allows a submucosal dissection of intraluminal bowel gas (Dalakas 1988a).

Cardiac disturbances. Cardiac abnormalities may be present in up to 40% of the patients and include atrioventricular conduction defects, tachyarrythmias, low ejection fraction and dilated cardiomyopathy (Dalakas 1991). Congestive heart failure and myocarditis, confirmed at autopsy, may be seen in up to 20% of the patients, either from the disease itself or from hypertension associated with long-term steroid use.

Pulmonary dysfunction. Pulmonary involvement, causing dyspnea, nonproductive cough, aspiration pneumonia and hypoxemia, may occur as the result of primary weakness of the thoracic muscles, drug-induced pneumonitis (e.g., from methotrexate), or interstitial lung disease (Dalakas 1991). Interstitial lung disease may precede the myopathy or occur early in the disease and develops in up to 10% of patients with PM or DM, half of whom have anti-Jo-1 antibodies (Plotz et al. 1989; Dalakas 1991), as discussed later.

Malignancies. It is still uncertain whether malignancies are more frequent in patients with PM and DM (Barnes 1976; Lakhanpal et al. 1986; Callen 1988; Richardson and Callen 1989). Methodologic deficiencies, such as the absence of diagnostic criteria for patient selection and the need for comparison of age-matched patients with other connective tissue diseases or autoimmune neuromuscular disorders treated with immunosuppressive drugs, have been predominantly responsible for the conflicting reports (Barnes 1976; Banker and Engel 1986; Lakhanpal et al. 1986; Dalakas 1988a, 1990a, 1991, 1992a,b; Callen 1988; Rowland 1989; Richardson and Callen 1989). These problems, along with a long-term preconceived bias that alerts the clinicians to a more vigorous search for malignancies in these patients, were the main reasons for the recently reported high incidence of malignancies in patients with PM and DM (Siguergeirsson et al. 1992). Based on more convincing data, however (Lakhanpal et al. 1986; Dalakas 1988a, 1990a, 1991, 1992a; Callen 1988; Richardson and Callen 1989), the incidence of malignancies appears to be increased only in patients with DM, but not PM or IBM. The extent of the search required for occult malignancy in adults with DM is still unsettled. Because often times the tumors are either uncovered at autopsy (in spite of a thorough search in life), or on the basis of abnormal findings on the medical history and physical examination (Lakhanpal et al. 1986; Callen 1988; Rowland 1989; Richardson and Callen 1989; Dalakas 1991), a blind, extensive and expensive radiologic search is not recommended. A complete *annual* physical examination, with pelvic and rectal examinations, urinalysis, complete blood-cell count, blood chemistry tests, chest X-ray film and a mammogram should suffice.

Overlap syndromes. This term is used loosely to define up to 20% of patients with PM or DM who may have features of another connective tissue disease, such as systemic lupus erythematosus (SLE), rheumatoid arthritis, Sjögren's syndrome, systemic sclerosis, or mixed connective tissue disease (Banker and Engel 1986; Dalakas 1988a, 1990a, 1991, 1992a,b; Rowland 1989; Plotz et al. 1989). However, 'overlap syndrome' indicates that the characteristics of two different disorders are common to both. In that sense, it is only DM, and not PM, that truly overlaps, and only with scleroderma and mixed connective tissue disease (Table 1). Specific signs of systemic sclerosis or mixed

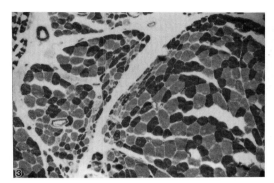

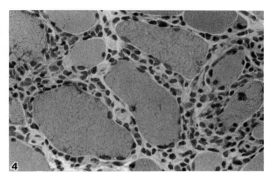

Fig. 3. Transverse frozen section of muscle biopsy specimen from a patient with adult dermatomyositis shows typical perifascicular atrophy (ATPase, ×160).

Fig. 4. Transverse section of a fresh-frozen muscle-biopsy specimen from a patient with polymyositis. Note the typical endomysial inflammation and lymphoid cells surrounding or beginning to invade healthy muscle fibers (Hematoxylin and eosin, ×430).

connective tissue disease, such as sclerotic thickening of the dermis, contractures, esophageal hypomotility, microangiopathy, and calcium deposits are present in DM but not PM, whereas signs of rheumatoid arthritis, SLE, or Sjögren's syndrome are very rare in DM. Patients with the overlap syndrome of DM/systemic sclerosis may have a specific antinuclear autoantibody, the anti-PM/Scl, directed against a nucleolar protein complex (Targoff 1990).

DIAGNOSIS

The clinically suspected diagnosis of PM, DM, or IBM is aided by evaluating the *serum muscle enzymes* and the *electromyography*, and it is confirmed with the *muscle biopsy*.

Muscle enzymes

The most sensitive enzyme is creatine kinase (CK), which in the presence of disease can be elevated as much as 50 times the normal level. Although CK usually parallels disease activity, it can be normal in active DM and, rarely, even in active PM. In IBM, CK is not usually elevated more than tenfold, and in some cases may be normal even from the beginning of the illness. CK may also be normal in patients with untreated, even active, adult or childhood DM and in some patients with PM or DM associated with a connective tissue disease, reflecting the concentration of the pathologic process in the intramuscular vessels and the peri-

mysium. Along with the CK, the serum SGOT, SGPT and LDH may be also elevated. The presence of high SGOT, SGPT, and LDH in a patient with early disease who has fatigue and minimal weakness often directs attention toward the diagnosis of liver disease and an unnecessary liver biopsy, if the CK level is not concurrently checked to exclude a myogenic origin of the 'liver enzyme' elevation (Dalakas 1991). If SGOT is higher than SGPT, a myogenic cause should be suspected; when SGPT is higher than SGOT, liver disease may be more likely.

Electromyography

Needle electromyography shows myopathic potentials characterized by short-duration, low-amplitude polyphasic units on voluntary activation, and increased spontaneous activity with fibrillations, complex repetitive discharges, and positive sharp waves (Barkhaus et al. 1990; Uncini et al. 1990; Dalakas 1991; Robinson 1991). This electromyographic pattern occurs in a variety of acute, toxic, and active myopathic processes and should not be considered diagnostic for the inflammatory myopathies (Dalakas 1988a, 1990a, 1991, 1992a). Mixed myopathic and neurogenic potentials (polyphasic units of short and long duration) are more often seen in IBM but they can be seen in both PM and DM as a consequence of muscle fiber regeneration and chronicity of the disease (Barkhaus et al.

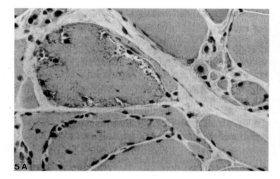
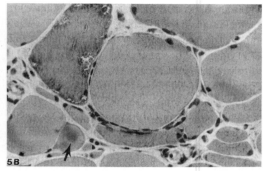
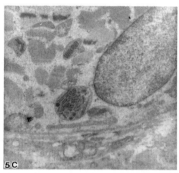

Fig. 5. Transverse frozen sections of a muscle biopsy from a patient with inclusion-body myositis shows the typical rimmed vacuoles (A,B). Small fibers in groups and one eosinophilic inclusion (arrow, 5B) are also noted. Electronmicrograph (5C) of a muscle biopsy from a patient with IBM demonstrates the typical filaments seen within the nucleus.

1990; Uncini et al. 1990; Robinson 1991). Electromyographic studies are generally useful for excluding neurogenic disorders and confirming either active or inactive myopathy (Dalakas 1988a, 1990a, 1991, 1992a). It is said that approximately 30% of patients with IBM have electromyographic signs of axonal neuropathy (Eisen et al. 1983). However, this study has not been confirmed with quantitative electromyography and age-matched nerve conduction studies. Such a study in IBM patients is underway in our institution to define if the sural nerves are abnormal and if the distally atrophic and weak muscles have myopathic or neurogenic features.

Muscle biopsy

The muscle biopsy is the definitive test not only for establishing the diagnosis of DM, PM, or IBM, but also for excluding other neuromuscular diseases. Although the presence of inflammation is the histological hallmark for these diseases, there

are additional unique histological features characteristic for each of DM, PM or IBM.

In DM the endomysial inflammation is predominantly perivascular or in the interfascicular septae and around rather than within the fascicles. The intramuscular blood vessels show endothelial hyperplasia with tubuloreticular profiles, fibrin thrombi, especially in children, and obliteration of capillaries (Carpenter and Karpati 1981; Banker and Engel 1986; Karpati 1988). The muscle fibers undergo necrosis, degeneration and phagocytosis often in groups involving a portion of a muscle fasciculus in a wedge-like shape, or at the periphery of the fascicle due to microinfarcts within the muscle. This results in perifascicular atrophy, characterized by two to ten layers of atrophic fibers at the periphery of the fascicles (Fig. 3). The presence of perifascicular atrophy is diagnostic of DM, *even in the absence of inflammation.*

In contrast, in PM the endomysial infiltrates are mostly within the fascicles (endomysial) surrounding individual, healthy, muscle fibers resulting in phagocytosis and necrosis (Fig. 4). There is no perivascular atrophy and the blood vessels are normal. When the disease is chronic, the connective tissue is increased.

IBM is characterized by endomysial inflammation, in a pattern similar to that described for PM. At times, however, the inflammation may be sparse or absent. The histological hallmark of IBM is the triad of: (a) basophilic granular inclusions distributed around the edge of slit-like vacuoles (rimmed vacuoles) (Fig. 5A,B); (b) small, round or angulated fibers often in small groups (Fig. 5A,B); and (c) eosinophilic cytoplasmic inclusions (Fig. 5B) (Karpati and Carpenter 1988; Lotz et al. 1989; Dalakas 1991). At high magnification, the vacuoles appear to contain granular material, which on electronmicroscopy, represent membranous whorls (Carpenter and Karpati 1981; Banker and Engel 1986; Karpati and Carpenter 1988; Lotz et al. 1989). Filamentous inclusions in the cytoplasm or myonuclei, prominent in the vicinity of the rimmed vacuoles, are thought to be pathognomonic of IBM (Fig. 5C). However, similar filamentous inclusions can be seen in other vacuolar myopathies such as oculopharyngeal muscular dystrophy or Welander's distal myopathy. The vacuolated muscle fibers contain strong ubiquitin immunoreactivity localized to the cytoplasmic tubulofilaments (Askanas et al. 1991). These vacuoles also contain Congo-red or crystal violet-positive amyloid deposits (Mendel et al. 1991) which immunoreact with β-amyloid protein (Askanas et al. 1992), the type of amyloid that has been sequenced from the amyloid fibrils of blood vessels of patients with Alzheimer's disease. The significance of these deposits is unclear. They most likely represent secondary products due to modification of normal proteins within an acidic degradative vacuolar

Table 3

Diagnostic criteria for inflammatory myopathies.

| Criteria | Polymyositis | | Dermatomyositis | | Inclusion-body myosysitis |
	Definite	Probable[§]	Definite	Mild or early	Definite
Muscle strength	Myopathic muscle weakness*	Myopathic muscle weakness*	Myopahtic muscle weakness*	Seemingly normal strength[†]	Myopathic muscle weakness with early involvement of distal muscles*
EMG[‡]	Myopathic	Myopathic	Myopathic	Myopathic or nonspecific	Myopathic with mixed potentials
Muscle enzymes	Elevated (up to 50-fold)	Elevated (up to 50-fold)	Elevated (up to 50-fold) or normal	Elevated (up to 10-fold) or normal	Elevated (up to 10-fold) or normal
Muscle biopsy[‡]	Diagnostic for inflammatory myopathy of the PM type	Nonspecific myopathy without signs of primary inflammation	Diagnostic for DM	Nonspecific or diagnostic for DM	Diagnostic for IBM
Rash or calcinosis	Absent	Absent	Present	Present	Absent

*Myopathic muscle weakness, affecting proximal muscles more than distal ones and sparing eye and facial muscles, is characterized by subacute onset (weeks to months) and rapid progression in patients who have no family history of neuromuscular disease, no endocrinopathy, no exposure to myotoxic drugs or toxins, or no biochemical muscle disease (excluded by muscle biopsy).
[†]Although strength is seemingly normal, patients often have new onset of easy fatigue, myalgia, and reduced endurance. Careful muscle testing may reveal mild muscle weakness.
[‡]See text for details.
[§]An adequate trial of prednisone or other immunosuppressive drugs is warranted in *probable* cases. If, in retrospect, the disease is unresponsive to therapy, repeat muscle biopsy should be considered to exclude other diseases or possible evolution to IBM, see text for details (adapted from Dalakas, M.C. 1991).

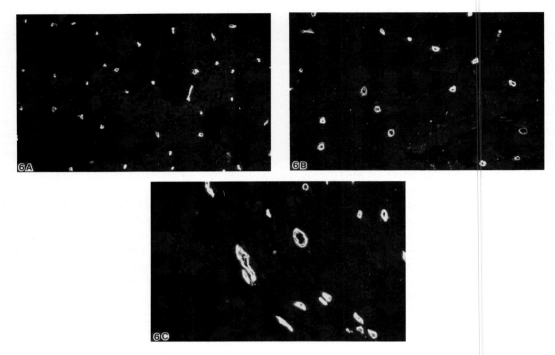

Fig. 6. Transverse frozen section of muscle biopsies from patients with dystrophy (A) and dermatomyositis (B,C) stained with ulex in a rhodamine-avidine immunofluorescence technique. Note marked reduction in the number of capillaries in the dermatomyositis patient. There is also an associated dilatation of the remaining capillaries in an effort to compensate for the ischemia that resulted from the loss or obstruction of the distal capillaries. In advanced cases, ony a few dilated capillaries have remained at the perifascicular region (C); the perifascicular atrophic fibers are seen in the background (C) (Same magnification, × 220).

compartment, rather than a primary phenomenon, as recently discussed (Dalakas 1992b).

Diagnostic criteria

Although the characteristic clinicopathological features of DM and IBM allow their diagnosis even in the absence of endomysial inflammation, in PM the presence of a primary intramuscular inflammatory response is an invariable feature of the disease and absence of inflammation should raise a critical concern about the diagnosis (Banker and Engel 1986; Dalakas 1991, 1992a,b). In the old diagnostic criteria introduced by Bohan and Peter 17 years ago (Bohan and Peter 1975), the main diagnostic features – proximal muscle weakness, myopathic findings on the electromyogram, elevated creatine kinase levels, and inflammation in the muscle biopsy – had equal diagnostic weight, and the diagnosis of 'PM' was acceptable even without the presence of specific-for-the disease muscle bi-

opsy findings. Consequently, IBM was overlooked, and various noninflammatory myopathies have been erroneously diagnosed as PM. This has prompted us to introduce the following new diagnostic criteria (Dalakas 1991) (Table 3).

The diagnosis of PM is *definite* when a patient has an acquired, subacute myopathy fulfilling the exclusion criteria described earlier, elevated CK, and a muscle biopsy with the histologic features of PM. The diagnosis is tentative, *probable* PM, if in such a patient the muscle biopsy shows nonspecific myopathic features without primary inflammation. A repeat muscle biopsy from another site may be more informative in these cases (Dalakas 1991). The diagnosis of DM is *definite* because of the characteristic rash even if inflammation is not present in the muscle biopsy. Cases of DM may be *mild or early* when a patient has the typical rash but seemingly normal strength, easy fatigue, and occasionally elevated CK. The diagnosis of IBM is *definite* when the characteristic histologic features

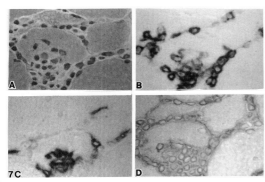

Fig. 7. Transverse sections of a fresh-frozen muscle-biopsy specimen from a patient with polymyositis. A single healthy, nonnecrotic muscle fiber, in serial sections, was stained sequentially with hematoxylin and eosin (A) and avidin-biotin immunoperoxidase with use of monoclonal antibodies to CD8 + cells (B), macrophages (C), and the MHC-I antigen (D). The predominant cells within the infiltrates that surround and invade the muscle fiber are CD8 + (suppressor-cytotoxic) T cells (B) and macrophages (C). The sarcolemma and the surrounding cells strongly express the MHC-I antigen, which is indicative of a cytotoxic process mediated by T cells and restricted to the MHC-I antigen (× 430).

are present in the muscle biopsy of a patient who has the appropriate clinical picture (Table 3).

Muscle imaging and inflammatory myopathies

In the last 2 years, the use of MRI, especially with a fat-suppressing imaging technique, has been advocated for the diagnosis and follow-up of patients with inflammatory myopathies (Keim et al. 1991; Frazer et al. 1991; Fujino et al. 1991; Fleckenstein et al. 1991; Hernandez et al. 1992). We feel however that this expensive procedure is unreliable for diagnosis and unnecessary to determine if a patient with PM or DM has improved. Furthermore, its accuracy and validity in differentiating PM or DM from a noninflammatory myopathy is questionable. The only clinically useful indication for muscle MRI is in the evaluation of HIV-associated focal muscle processes, and specifically pyomyositis (Fleckenstein et al. 1991; Dalakas 1992a). In pyomyositis, the characteristic radiologic sign is that of a muscle abscess with a rim of increased signal intensity corresponding to margins between drainable pus and edematous tissue. The subcutaneous tissues appear normal. Sparing of the subcu-

taneous tissues is felt to be a useful radiologic sign for distinguishing pyomyositis from diffuse soft-tissue swellings related to other conditions, such as lymphedema, venous thrombosis, cellulitis, or lymphoma, where both the muscle and the subcutaneous tissues are involved (Fleckenstein 1991).

PATHOGENESIS

Immunopathologic findings

In DM, the endomysial infiltrates have a higher percentage of B cells, a higher ratio of CD4 (helper cells) to CD8 (suppressor-cytotoxic T cells), close proximity of CD4 + cells to B cells and macrophages, and a relative absence of lymphocytic invasion of nonnecrotic muscle fibers, all of which suggest a mechanism primarily mediated by humoral processes (Banker and Engel 1986; Engel and Arahata 1986; Dalakas 1991). This immune process is directed against the intramuscular microvasculature and is mediated by the complement C5b-9 membranolytic attack complex (MAC), implying activation of the complement pathway by antibodies bound to microvascular components (Kissel et al. 1986; Emslie-Smith and Engel 1990; Dalakas 1991). On the basis of double immunolocalization of the capillaries, using the lectin *Ulex europaeus* as a specific endothelial marker and antibodies to complement C5b-9 membranolytic attack complex, it has been demonstrated that the deposit of complement on the capillaries, is the earliest and most specific lesion in DM and precedes inflammation or structural changes in the muscle fibers. This is followed by necrosis and marked reduction in the number of capillaries per each muscle fiber, as shown in Fig. 6, resulting in ischemia, muscle fiber destruction often resembling microinfarcts, and inflammation. Larger intramuscular blood vessels also may be affected in the same pattern, leading to actual infarcts. Residual perifascicular atrophy (Fig. 3) reflects the endofascicular hypoperfusion prominent distally. Compensatory dilatation of the remaining capillaries is easily demonstrated with ulex, as shown in Figure 6B,C.

In PM and IBM, there is evidence not of microangiopathy and muscle ischemia, as in DM, but of an antigen-directed cytotoxicity mediated by cytotoxic T cells (Banker and Engel 1986; Engel and Arahata 1986; Emslie-Smith et al. 1989; Dalakas 1991). This conclusion is supported by the presence of CD8+ cells, which along with macrophages initially surround healthy, nonnecrotic muscle fibers and eventually invade and destroy them (Fig. 7). The muscle fibers, either next to or remote from the areas of inflammation, strongly express the Class I major histocompatibility complex (MHC-I) antigen (Fig. 7), which is absent from the sarcolemma of normal muscle fibers (Karpati et al. 1988). Because cytotoxic T cells recognize antigenic targets in association with MHC-I antigen, these findings indicate that in PM and IBM the primary immunopathologic mechanism is mediated by cytotoxic T cells and is restricted to MHC-I antigen. The target antigen in the sarcolemma recognized by T cells is, however, still unknown. That these lymphocytes are cytotoxic and recognize muscle autoantigens has been supported by coculturing autologous myotubes with cells from T cell lines derived from the endomysial inflammatory cells (Hohlfeld and Engel 1991). Although most of the CD_8^+ T cells use the α/β T cell receptor for antigen recognition, CD_4^- and CD_8^- γ/δ cytotoxic T cells can also mediate muscle fiber injury (Hohlfeld et al. 1991). In vivo kinetics of indium-labeled autologous lymphocytes in PM patients shows increased uptake in the major muscle groups when compared with controls or with other non-muscle tissues such as liver, lung, or spleen (Plotz et al. 1989). The lymphocyte trafficking in the muscle is also proportional to the degree of inflammation in the concurrently done muscle biopsies (Dalakas M.C., Miller F., unpublished observations 1989). These in vivo data are consistent with previous in vitro studies and support the view that circulating sensitized lymphocytes are directed abnormally and specifically to muscle, in proportion to disease activity.

A necrotizing, probably inflammatory autoimmune myopathy was described in 3 patients by Emslie-Smith and Engel (Emslie-Smith and Engel 1991). The histologic hallmark of this condition was the presence of microangiopathy characterized by 'pipestem' vessels with deposits of C_{5b-9} membranolytic attack complex. A focal depletion of intramuscular capillaries was also present, but no other histologic or clinical signs suggestive of DM were found. All three patients were unresponsive to immunotherapies. This is a unique immune-mediated microangiopathy that differs from PM, DM, or IBM, and emphasizes the need to search for immune markers in the muscle biopsies of patients with unusual histologic features (Dalakas 1992b). We have recently seen a similar case in a 62-year-old female who also had concurrent Hashimoto's thyroiditis (Dalakas M.C., unpublished observations).

Role of autoantibodies

Various autoantibodies against *nuclear* and *cytoplasmic* antigens are found in up to 20% of patients with inflammatory myopathies (Mathews and Bernstein 1983; Bunn et al. 1986; Targoff 1989, 1990; Targoff and Arnett 1990). The antibodies against nuclear antigens include those against ribonucleoproteins, anti-Ro/SS-A, anti-Sm, or anti-La/SS-B which are not myositis-specific as they are more closely associated with the group of mixed connective tissue syndromes.

The antibodies to cytoplasmic antigens are directed against cytoplasmic ribonucleoproteins which are involved in translation and protein synthesis. They include antibodies against various *synthetases, translation factors*, and proteins of the *signal-recognition particles*. The most common are the anti-synthetases which are directed at the aminoacyl transfer RNA synthetase, a group of cytoplasmic enzymes that catalyze the attachment of a particular amino acid (histidine, alanine, threonine, isoleucine or glycine) to its cognate transfer RNA (Mathews and Bernstein 1983; Bunn et al. 1986; Targoff 1989, 1990; Plotz et al. 1989; Targoff and Arnett 1990; Dalakas 1991). The antibody directed against the histidyl-transfer RNA synthetase, called *anti-Jo-1*, accounts for 75% of all the anti-synthetases. At least 50% of PM or DM patients with anti-Jo-1 antibodies have also interstitial lung disease and often Raynaud's phenomenon or arthritis. Because some of these antibodies can inhibit the translation of messenger RNA or the transport of polypeptides across the endoplasmic reticulum, it has been suggested that they might be directed at antigens involved in protein synthesis (Mathews and Bernstein 1983; Targoff 1989; Targoff and Arnett 1990). Contrary to previous contentions, however (Plotz et. 1989; Leff et al. 1991), these antibodies are also non-muscle specific because they are directed against ubiquitous targets and represent epiphenomena of no pathogenic significance. Because they are almost always associated with interstitial lung disease and occur even in the absence of myositis, they are as likely to be specific markers for the interstitial lung disease as for the myopathy. At present, their only value is their tendency to identify serologically a small subset, up to 8%, of patients with PM or DM who also have interstitial lung disease. Contrary to other claims (Leff et al. 1991; Love et al. 1991), they have no value in defining distinct subsets or homogeneous groups of inflammatory myopathies.

Antiendothelial cell antibodies were found in the serum of 8 of 18 patients with DM, 6 of whom had interstitial lung disease (Cervera et al. 1991). Whether these antibodies are secondary, resulting from the MAC-mediated capillary destruction, or primary, initiating the immune-mediated microangiopathy, is presently unknown.

Viruses and inflammatory myopathies

A number of viruses, especially enteroviruses, have been linked to the pathogenesis of inflammatory myopathies, because many members of the picornaviridae family can cause myopathy in laboratory animals. The association of such viruses with the human disease, however, is circumstantial, unconvincing, and unconfirmed (Dalakas 1988a, 1990a, 1991, 1992a,b). A possible seasonal pattern in the onset of inflammatory myopathies has been liberally interpreted as an indirect sign of environmental (? viral) cause (Leff et al. 1991).

Picornaviruses. These viruses have been proposed as possible triggering agents for some PM or DM patients because of their structural homologies with the Jo-1 antigen (Mathews and Bernstein 1983; Bunn et al. 1986; Targoff 1989; Targoff and Arnett 1990). Jo-1, a histidyl-transfer RNA synthetase, is an enzyme that joins histidine to its cognate transfer RNA in the early stages of protein synthesis. This enzyme can interact with, and join, histidine not only with its native histidyl-transfer RNA but also with the genomic RNA of an animal picornavirus, the encephalomyocarditis virus,

whose tertiary structure has regions of homology with the histidyl-transfer RNA synthetase of *Escherichia coli* and muscle proteins (Mathews and Bernstein 1983; Bunn et al. 1986; Targoff 1989; Targoff and Arnett 1990). A theoretical interaction between this virus and the synthetase may result in a stable complex (such as that formed between the synthetase and its native transfer RNA substrate) that may be presented to the immune system as foreign, generating an autoimmune response. Alternatively, antibodies generated against a related picornavirus that a patient may be infected with could theoretically cross-react with the histidyl-transfer RNA synthetase (Jo-1) and also with muscle proteins in a phenomenon of molecular mimicry. Although there is no experimental data to support the theory that these reactions may result in myositis or that these autoantibodies are against muscle-specific antigens triggered by specific viruses, picornaviruses have been sought extensively in the muscles of both PM and DM patients. In spite of reports that Coxsackie or related enteroviral RNA have been found with in situ hybridization within the muscle fibers of PM and DM patients (Rosenberg et al. 1989; Yousel et al. 1990), our very sensitive technique using the polymerase chain reaction has not confirmed these reports (Leon-Monzon and Dalakas 1992; Leff et al. 1992).

Other viruses. The previously reported presence of mumps viral particles and antigens in the muscle biopsies of patients with IBM (Chou 1986), has not been confirmed by us or others (Nishino 1989; Leff et al. 1992). Our search for encephalomyocarditis viruses, adenoviruses, HIV and HTLV-I or II in the RNA extracted from muscle biopsies of 44 patients with PM, DM and IBM was also negative for any of these viruses (Leff et al. 1992).

Retroviruses. These are the most promising viral candidates in the cause of PM because they can trigger an inflammatory myopathy in both humans and primates. The retroviruses known to have an association with polymyositis are:

1. The *human immunodeficiency virus type 1 (HIV-1)*. In HIV-positive patients an inflammatory myopathy (HIV-PM) can occur either as an isolated clinical phenomenon, being the first clini-

cal indication of HIV infection, or concurrently with other manifestations of the acquired immune deficiency syndrome (AIDS) (Dalakas et al. 1986a; Dalakas and Pezeshkpour 1988; Dalakas and Illa 1991; Dalakas 1993). HIV seroconversion can also coincide with myoglobulinuria and acute myalgia, suggesting that myotropism for HIV may be symptomatic early in the infection.

Studies using in situ hybridization, immunocytochemistry, polymerase chain reaction, and electronmicroscopy failed to demonstrate viral antigens within the muscle fibers. Only occasional interstitial mononuclear cells in the proximity to the muscle fibers were positive for HIV antigens in some of the specimens (Dalakas and Illa 1991; Illa et al. 1991; Hantai et al. 1991; Dalakas 1993). Human myotubes in culture are also resistant to infection with intact HIV and to transfection with a naked HIV proviral-DNA construct, indicating that in HIV-PM there is no evidence of persistent infection of the muscle fiber with the virus and that viral replication does not take place within the human muscle (Lamperth et al. 1990; Leon-Monzon and Dalakas 1991).

In the muscle biopsy specimens of HIV-PM, the predominant cells are CD_8^+ cells and macrophages that invade or surround MHC-I-antigen-expressing non-necrotic muscle fibers (Dalakas et al. 1990; Illa et al. 1991) in a pattern analogous to that seen in polymyositis, as described above. This, along with a relative paucity of CD_4^+ cells suggests that the development of HIV-PM is independent of the CD_4^+ cell, and that a T-cell-mediated and MHC-I-restricted cytotoxic process is a common pathogenetic mechanism in both PM and HIV-PM.

These observations indicate that a systemic viral infection, such as HIV can trigger the immunopathologic mechanism that leads to development of PM even in the absence of the virus from the muscle fiber. The various cytokines or toxic lymphokines released by the HIV-infected endomysial inflammatory cells may contribute to this process by exposing new muscle antigens against which there is no self tolerance, generating a secondary autoimmune response (Dalakas 1992b, 1993). The resulting endomysial inflammation may subsequently become self-sustaining if it cannot be down-regulated by the host. Alternatively, activated T cells resulting from the persistent systemic

viral infection, might have recognized within the muscle putative sarcolemmal autoantigens leading to phagocytosis and muscle fiber necrosis. Molecular antigenic mimicry may also be a mechanism of self-sensitization because some antiribonucleoproteins react with polypeptides coded for the gag gene of the retroviruses (Rucheton et al. 1985).

2. The human T-cell lymphotrophic virus (HTLV-1). HTLV-1 does not only cause a myeloneuropathy – referred to as tropical spastic paraparesis (TSP) – but also PM, which may coexist with TSP or may be the only clinical manifestation of HTLV-1 infection (Morgan et al. 1989; Wiley et al. 1989). A T-cell mediated and MHC-1 restricted cytotoxic process triggered by the virus appears also to be the main mechanism of HTLV-PM (Dalakas et al. 1992; Dalakas 1993).

3. Human foamy retrovirus (HFR). HFR, although not pathogenic in humans, was found to be associated with a destructive myopathy in transgenic mice carrying the bel region of the retrovirus. The transgene was expressed in the striated muscle of these mice and viral RNA was present within the viable muscle fibers (Bothe 1991).

4. The Simian retrovirus type I (SRV-1) and the Simian immunodeficiency virus (SIV). These agents can cause immunodeficiency, Kaposi's sarcoma, and PM in infected monkeys (Dalakas et al. 1986b, 1987).

The association of PM with five different retroviruses resulting in disease identical to seronegative PM, without the presence of viral antigens within the muscle fibers, suggests that viruses can trigger the development of an inflammatory myopathy and provide a continuous stimulus for disease progression without necessarily causing a lytic, transient or persistent, viral infection of the muscle fiber.

AZT-myopathy versus HIV-myopathy. Some HIV-positive patients treated with azidothymidine (AZT), develop myopathy which is clinically indistinguishable to the HIV-associated primary inflammatory myopathy. The challenge of distinguishing one from the other has been resolved with the identification of unique histologic and biochemical features present only in the muscles of AZT-treated patients (Dalakas et al. 1990; Pezeshkpour et al. 1991; Mhiri et al. 1991; Arnaudo et al. 1991; Chalmers et al 1991; Jay et al. 1992). The unique histologic features of AZT-myopathy are 'ragged-red' fibers suggestive of mitochondrial abnormalities, subsarcolemmal or central accumulations of red granular material, or linear and circumferential 'red-rimmed cracks' demonstrable with the trichrome stain, pale granular degeneration, rods, varying degrees of endomysial inflammation, and increased neutral fat attributed to impaired mitochondrial control of fatty acid utilization. The combination of the above morphologic features within the muscle fiber characterize what we now call the 'AZT-fiber' (Dalakas 1992b, 1993). Electron microscopy has confirmed the presence of ultrastructurally abnormal mitochondria, present only in the AZT-treated patients and not in the HIV-myopathy patients (Pezeshkpour et al. 1991; Mhiri et al. 1991). Southern blots of the extracted DNA from the muscles of the AZT-myopathy patients has revealed an up to 80% reduction of mitochondrial DNA (Arnaudo et al. 1991). We have concluded that AZT, a DNA chain terminator, causes a *mitochondrial-DNA depleting myopathy* by inhiting the γ-DNA polymerase of the mitochondrial matrix resulting in termination of the muscle fiber mitochondrial DNA. Reduction of cytochrome C reductase and abnormal MRS spectroscopy suggestive of impaired oxidative metabolism, has further confirmed the abnormal function of the muscle mitochondria in the AZT-treated patients (Mhiri et al. 1991; Weissman et al. 1992; Soueidan et al. 1992).

TREATMENT

The evidence described above that immunopathologic mechanisms are primarily involved in PM, DM and IBM, justifies the need for treating these diseases with immunosuppressive therapies. All the treatment trials, however, have been *empirical* and large scale control therapeutic studies against the immunologically specific forms of childhood DM, adult PM or adult DM have not been conducted (Banker and Engel 1986; Dalakas 1988a,b 1989, 1990a,b, 1991, 1992a,b; Rowland 1989). As the specific target antigens are also unknown, these therapies are not selectively targeting the autoreactive T cells or the complement-mediated process on the intramuscular blood vessels but instead, they are inducing a non-selective immunosuppression or immunomodulation.

The goal of therapy in PM, DM and IBM is to improve the function in activities of-daily-living as the result of improvement in muscle strength. Although when the strength improves, the serum CK falls concurrently, the reverse is not always true because most of the immunosuppressive therapies can result in decrease of serum muscle enzymes without necessarily improving muscle strength. Unfortunately, this has been misinterpreted as 'chemical improvement', and has formed the basis for the common habit of 'chasing' or 'treating' the CK level instead of muscle weakness, a practice that has led to a prolonged use of unnecessary immunosuppressive drugs and erroneous assessment of their efficacy. The prudency of discontinuing these drugs if, after an adequate trial, they have only led to a reduction in CK and not to objective improvement in muscle strength has been stressed repeatedly (Dalakas 1988a, 1990a, 1991, 1992a,b). The following therapies are commonly applied to treat PM, DM or IBM.

Corticosteroids

Prednisone is the first in line drug of this *empirical* treatment. Because the effectiveness and relative safety of prednisone therapy will determine the future need for stronger immunosuppressive drugs, I prefer an aggressive approach of high-dose prednisone since early in the disease (Dalakas 1988a,b, 1989, 1990a,b, 1991, 1992a,b). A high dose of 80–100 mg/day as a *single* daily morning dose for an initial period of 3–4 weeks is preferable. Prednisone is then tapered over a 10 week period to 80–100 mg *single daily alternate-day* by gradually reducing an alternate 'off day' dose by 10 mg per week, or faster if necessitated by side effects, though this carries a greater risk of breakthrough of disease. If there is evidence of efficacy, and no serious side effects, the dosage is reduced gradually by 5–10 mg every 3–4 weeks until the lowest possible dose that controls the disease is reached. *In a patient responding to prednisone*, a 'maintenance' dose of 10–25 mg every other day has been empirically thought to be useful and safe in securing continuous improvement or prevent relapses. On the other hand, if by the time the dosage has been reduced to 80–100 mg every other day (approximately 14 weeks after initiating therapy), there is no objective benefit (defined as increased muscle strength, and not as lowering of the CK or a subjective feeling of increased energy), the patient may be considered unresponsive to prednisone and tapering is accelerated while the next-in-line immunosuppressive drug is started (Dalakas 1988a, 1990a, 1991, 1992a,b).

Steroid-myopathy versus disease activity. The long-term use of prednisone may cause worsening of muscle strength associated with a normal or unchanged CK level, which is referred to as 'steroid myopathy' (Engel and Dalakas 1982; Dalakas 1988a,b, 1989, 1990a,b, 1991, 1992a,b). The term steroid-myopathy is however a misnomer because in the absense of denervation steroids do not cause histological signs of myopathy but rather selective atrophy of type II muscle fibers. It may be difficult to distinguish between steroid-induced myopathy and increased weakness related to disease activity because the two can coexist, or the increased weakness may be caused by other factors, such as decreased mobility, infection, an associated systemic illness or no more response to steroids. The decision to adjust the prednisone dosage *in a patient who has previously responded* to treatment may be influenced by reviewing the past 1 to 2 months' history of the muscle strength, mobility, serum CK, medication changes, and associated medical conditions. For example, if in the past 2 months: (a) the level of steroid dosage has been unchanged or increased; (b) the CK level during the same period has been more or less stable; (c) new or increasing signs of steroid intoxication such as increased body weight, hypertension, striae, cushingoid features have developed; and (d) the patient's physical mobility is reduced, the most likely cause of increasing muscle weakness is *steroid myopathy* (atrophy of the type II muscle fibers). On the other hand, in a patient who for the past 1 to 2 months has increased CK levels, no development of new overt signs of steroid toxicity while the dosage of steroids is reduced or unchanged, and no evidence of a systemic illness or infection, an increasing muscle weakness is most likely due to exacerbation of the disease, which might again respond to higher doses of prednisone. When all these signs

are not clear, we arbitrarily lower or raise the prednisone dosage, waiting for the answer, which will be evident in about 2 to 8 weeks, according to the change in the patient's strength. A new muscle biopsy may not reveal the cause of increased weakness because active inflammatory disease can be present even when steroid intoxication was the cause of increasing weakness (Dalakas 1989, 1992b). Conversely, type II muscle fiber atrophy, the histological correlate of steroid myopathy, can coexist with active inflammation – even in a patient who has never received steroids – owing to disuse and immobilization of a limb from contractures, pain, or weakness. Electromyography is also of limited value because it is not informative in type II muscle fiber atrophy. However, it may occasionally be of some help if it shows fibrillations and positive sharp waves in many sites of proximal muscle groups, indicative of active myopathic disease. A clinical sign that I have found to be of some help in a few patients is the strength of neck flexor muscles which usually worsens with exacerbation of the disease but remains unchanged with steroid-induced muscle intoxication.

Side effects of corticosteroids. The most common side effects we have seen in patients with PM or DM treated for long periods with corticosteroids are (Dalakas 1990b, 1992a): (a) abnormality of fat distribution with generalized obesity – hence the need for strict caloric restriction; (b) lipolytic action resulting in hyperlipidemia, which can rarely cause fat emboli in the femoral head and aseptic necrosis of the hip. Epidural lipomatosis resulting in spinal cord compression is rare but should be suspected if a patient on long-term steroid therapy is developing back pain and signs of myelopathy; (c) diabetes – hence the reason for low-carbohydrate diet; (d) retarded growth in children with DM which may be minimized with the alternate-day-program; (e) menstrual irregularities; (f) edema and hypertension – hence the need for low-salt diet from the beginning of therapy; (g) osteoporosis especially in women – hence the suggestion of co-administering vitamin D or calcium supplements in women receiving long-term steroid therapy; (h) gastrointestinal complications including gastritis or stomach aches. Although not as common as believed, these can be helped with antacids or histamine blockers, as described above. In children with DM, who may have early signs of intestinal perforation related to vasculitis, steroids may potentially trigger gastrointestinal bleeding; (i) skin changes, including acne, ecchymosis, facial hirsutism and striae; (j) posterior subcapsular cataracts and, rarely, glaucoma necessitating the need for frequent eye examinations; and (k) central nervous system complications such as insomnia, irritability and exacerbation of the physiologic action tremor in the hands which are commonly seen with the high daily dose and always subside when the dose is reduced. Behavioral changes, such as psychosis and depression have not been frequent in our group of patients with PM or DM.

Among the interaction of prednisone with other drugs, it is worth remembering that anticonvulsants such as phenobarbital, carbamazepine, and phenytoin, which are inducers of the hepatic microsomal enzyme system, may alter the extent of prednisolone metabolism accelerating the elimination of prednisone and methylprednisolone (Dalakas 1990b). Somewhat higher than anticipated steroid doses may therefore be required in such patients to achieve therapeutic response.

'Prednisone failures' and non-steroidal immunosuppressive therapies

Although it is my view that almost all the patients with *bonafide* PM or DM respond to steroids to *some degree* and *for some period of time*, a number of them fail to respond or become steroid-resistant. Observations over the last few years, have strengthened our previous view that most of the patients with the initial diagnosis of PM who have not responded to any form of immunotherapy most likely have IBM or 'something else' (Dalakas 1992a). The decision to start an immunosuppressive drug in a patient with PM or DM is based on the following factors: (a) need for its 'steroid-sparing' effect, when in spite of steroid responsiveness the patient has developed significant complications; (b) attempts to lower a high-steroid dosage have repeatedly resulted in a new relapse; (c) adequate dose of prednisone for at least a 2–3 month period has been ineffective; and (d) rapidly progressive disease with evolving severe weakness and respiratory failure (Dalakas 1988b, 1989, 1990b,

1991). The preference for selecting the next in line immunosuppressive therapy is, however, empirical based on personal experience with each drug and its relative efficacy/safety ratio. Such therapies, routinely used in the treatment of patients with PM, DM or IBM (Dalakas 1991, 1992a) are:

Azathioprine, a derivative of 6-mercaptopurine, which is given orally. Although lower doses (1.5–2 mg/kg) are commonly used, for effective immunosuppression we prefer higher doses up to 3 mg/kg. This drug is my first preference because compared to the others, it is well tolerated, has fewer side effects and, empirically, it appears to be as effective as the other drugs whose value in PM and DM has also not been established. Because azathioprine is usually effective after 3–6 months of treatment, patience is required before its ineffectiveness is prematurely concluded. An elevation of liver enzymes, if slight, needs only observation. Most of the times, elevation of liver enzymes in these patients is related to a 'fatty liver' from the long-term steroid use rather than from the use of azathioprine. Azathioprine is metabolized by xanthine oxidase and if given concurrently with allopurinol, can be severely toxic to the liver or bone marrow; their combined use is not recommended.

Methotrexate, an antagonist of folate metabolism, is a drug heavily favored by rheumatologists, in spite of its disappointing efficacy in PM or DM. In our experience, methotrexate has been either ineffective or has shown only marginal and mostly subjective benefit at a cost of considerable toxicity to the patients. It can be given intravenously over 20–60 min at weekly doses of 0.4 mg/kg up to 0.8 mg/kg, or orally starting at 7.5 mg *weekly* for the first 3 weeks (given in a *total* of 3 doses, 2.5 mg every 12 hours), increasing it gradually by 2.5 mg/week up to a total of 25 mg weekly. A relevant side effect is *methotrexate-pneumonitis* which can be difficult to distinguish from the interstitial lung disease related to the primary disease, which is often associated with Jo-1 antibodies, as described above. Other side effects include stomatitis, gastrointestinal symptoms, leukopenia and hepatotoxicity.

Cyclophosphamide, an alkylating agent, given intravenously or orally (2.0–2.5 mg/kg, usually 50 mg p.o. 3 times/day) has been ineffective in our hands (Cronin et al. 1989) in spite of occasional promising results reported by others (Bombardieri et al. 1989). The drug may be helpful in a subset of patients who also have interstitial lung disease (Dalakas 1991). It is our belief that the risk/benefit ratio of this drug, only rarely justifies its use. Side effects include nausea, vomiting, alopecia, hemorrhagic cystitis, pulmonary fibrosis, bone marrow suppression, secondary malignancies or sterility. Monitoring of the complete blood count with care to maintain the WBC count above 3500 with adequate neutrophil count (no less than 1500–2000) and the lymphocyte count above 1000 are essential.

Cyclosporin has been used with limited success. Although the toxicity of the drug can now be monitored by measuring optimal trough serum levels (which vary between 100 to 250 ng/ml), its effectiveness is uncertain. A report that low doses of cyclosporin could be of benefit in children with DM or in adults with PM or DM needs confirmation (Heckmatt et al. 1989; Lueck et al. 1991). Based on the patients referred to us, the drug has been disappointing.

Plasmapheresis and *leukopheresis* are not helpful. In the first double-blind placebo controlled study that has been conducted in 39 PM and DM patients randomized to receive 12 treatments over a one month period of plasmapheresis, leukopheresis or sham apheresis, no differences were noted among the 3 groups (Miller et al. 1992). The results were convincing for demonstrating the ineffectiveness of this therapy in chronic disease.

Total lymphoid irradiation is an extreme remedy that can be considered for extreme situations. It has been helpful in 2 of our patients and had longlasting benefit (Dalakas and Engel 1988). The long-term side effects of this treatment, however, should be seriously considered before deciding on this experimental approach. Total lymphoid irradiation has been ineffective in IBM (Dalakas 1991; Kelly et al. 1986).

Intravenous immunoglobulin (IVIg), taken from human serum pools, is a new promising, but very expensive, therapy. In two uncontrolled studies, IVIg has been effective (Lang et al. 1991; Cherin et al. 1991). In one trial, 5 patients with juvenile DM improved during a 9 month course of IVIg, allowing steroids to be reduced or discontinued (Lang et al. 1991). In the other trial of 14 patients with

chronic PM and 6 patients with DM unresponsive to previous therapies (Cherin et al. 1991), IVIg, given in conjunction with prednisone, methotrexate, or plasmapheresis for a mean period of 4 months, resulted in marked clinical improvement in 15 of the 20 patients. Although its mode of action is unclear, IVIg appears to be a promising new agent in the management of PM and DM. We have preliminary evidence that IVIg may also be helpful to some patients with IBM (Soueidan and Dalakas 1992). It is safe and well tolerated, but prohibitively expensive, and a controlled study is required to secure its efficacy. Such a study, separate for PM, DM, and IBM, is now underway by us at the National Institutes of Health.

Cricopharyngeal myotomy can improve the dysphagia in patients with IBM (Verna et al. 1991; Darrow et al. 1992). When weakness of the pharyngeal musculature is advanced, the generated hypopharyngeal pressure may be insufficient to trigger reflex relaxation of the cricopharyngeal muscle. By eliminating the zone of elevated pressure between the pharynx and the esophagus, myotomy can produce a more efficient swallow, and may be considered in selected patients with IBM.

Calcinosis, seen in childhood or adult DM, does not usually resolve with immunotherapies. New calcium deposits may be prevented if the primary disease responds to the available therapies. Diphosphonates, aluminum hydroxide, probenecid, colchicine, low doses of warfarin, and surgical incision have all been tried without success.

Physical therapy to preserve existing muscle function and avoid disuse atrophy of the weak muscles or joint contractures should start early in the disease.

PRACTICAL DIAGNOSTIC AND THERAPEUTIC CONSIDERATIONS

Based on the information described above regarding the efficacy of these therapies and our experience with more than 300 patients referred to us, the following observations and practical tips may be useful in the diagnosis and management of patients with inflammatory myopathies.

a. Patients with *bona fide* PM and DM should

almost always respond to prednisone to a *certain degree* and for *some period* of time.

b. Patients with DM, as a group, respond better than patients with PM.

c. A patient with presumed PM who has not responded to any form of immunotherapy, most likely has IBM or another disease. In these cases, a repeat muscle biopsy and a more vigorous search for '*the other*' disease are recommended.

d. If prednisone or other immunosuppressive therapies have not helped, or have become ineffective in improving the patient's strength, they should be discontinued to avoid severe irreversible side effects for, contrary to common belief, their continuation does not 'maintain stability' or prevent further disease progression.

e. High-dose intravenous immunoglobulin (IVIg) is emerging as a new, promising and safe drug for the treatment of patients with *bona fide* PM or DM that have become resistant to therapies.

f. Patients with acute, fulminant course associated with rhabdomyolysis, may not respond to immunotherapies. Such therapies however should be aggressively tried before this conclusion is reached.

g. Inclusion-body myositis is generally resistant to all therapies. Some cases of probable PM, often referred to us as 'definite' PM, may also be resistant to all therapies, as found retrospectively. Whether such patients have an unidentified dystrophic process is unknown.

h. Patients with interstitial lung disease may have a high mortality. The treatment of choice in these patients is with cyclophosphamide.

i. When the treatment of PM is unsuccessful, the patient should be reevaluated and the muscle biopsy reexamined. A new biopsy might be considered to make sure that the diagnosis is correct or the disease has not evolved into IBM. Although it may be difficult to distinguish histologically an early case of IBM from classic PM, it is my belief that IBM develops de novo and it is not the evolution of a classic PM (Dalakas 1990a). Only high degree of clinical suspicion however may predict its early diagnosis. The disorders most commonly mistaken for PM are: IBM, sporadic limb-girdle dystrophy, metabolic myopathy (e.g. phosphorylase deficiency), endocrinopathy, and neurogenic muscular atrophies (Dalakas 1991).

REFERENCES

ANSELL, B.M. (Ed): Inflammatory diseases of muscle. Clin. Rheum. Dis. 10 (1984) 1–215.

ARNAUDO, E., M. DALAKAS, S. SHANSKE, C.T. MORAES, S. DIMAURO and E.A. SCHON: Depletion of muscle mitochondrial DNA in AIDS patients with zidovudine- induced myopathy. Lancet 337 (1991) 508–510.

ASKANAS, V., P. SERDAROGLU, W.K. ENGEL and R.B. ALVAREZ: Immunocytochemical localization of ubiquitin in muscle biopsies of patients with inclusion-body myositis and oculopharyngeal muscular dystrophy. Neurosci. Lett. 130 (1991) 73–76.

ASKANAS, V., W.K. ENGEL, R.B. ALVAREZ and G.G. GLENNER: β-Amyloid protein immunoreactivity in muscle of patients with inclusion-body myositis. Lancet 339 (1992) 560–561.

ATLAS, E., S.N. NOVAK, P.H. DURAY and A.C. STEERE: Lyme myositis: muscle invasion by *Borrelia burgdorferi*. Ann. Intern. Med. 109 (1988) 245–246.

BANKER, B.Q.: Parasitic myositis and other inflammatory myopathies. In: A.G. Engel and B.Q. Banker (Eds), Myology. New York, McGraw-Hill (1986) 1467–1524.

BANKER, B.Q. and A.G. ENGEL: The polymyositis and dermatomyositis syndrome. In: A.G. Engel and B.Q. Banker (Eds), Myology. New York, McGraw-Hill, (1986) 1385–1422.

BARKHAUS, P.E., S.D. NANDEDKAR and D.B. SANDERS: Quantitative EMG in inflammatory myopathy. Muscle Nerve 13 (1990) 247–253.

BARNES, B.E.: Dermatomyositis and malignancy: a review of the literature. Ann. Intern. Med. 84 (1976) 68–76.

BOHAN, A. and J.B. PETER: Polymyositis and dermatomyositis. N. Engl. J. Med. 292 (1975) 344–347, 403–407.

BOMBARDIERI, S., G.R.V. HUGHES, R. NERI, P. DEL BRAVO and L. DEL BONO: Cyclophosphamide in severe polymyositis. Lancet i (1989) 1138–1139.

BOTHE, K., A. AGUZZI, H. LASSMAN ET AL.: Progressive encephalopathy and myopathy in transgenic mice expressing human foamy virus gene. Science 253 (1991) 552–555.

BUNN, C.C., R.M. BERNSTEIN and M.B. MATHEWS: Autoantibodies against alanyl-tRNA synthetase and tRNAA 1a coexist and are associated with myositis. J. Exp. Med. 163 (1986) 1281–1291.

CALLEN, J.P.: Malignancy in polymyositis/dermatomyositis. Clin. Dermatol. 2 (1988) 55–63.

CARPENTER, S. and G. KARPATI: The major inflammatory myopathies of unknown causes. Pathol. Ann. 16 (1981) 205–237.

CARPENTER, S. and G. KARPATI: Pathology of skeletal muscle. New York, Churchill Livingston (1984) 516–592.

CARPENTER, S., G. KARPATI, L. HELLER ET AL.: Inclusion body myositis: a distinct variety of idiopathic inflammatory myopathy. Neurology 28 (1978) 8–17.

CERVERA, R., G. RAMIREZ, J. FERNANDEZ-SOLA, D.

D'CRUZ, J. CASADEMONT, J.M. GRAU, R.A. ASHERSON, M.A. KHAMASHTA, A. MARQUES AND G.R.V. HUGHS: Antibodies to endothelial cells in dermatomyositis: association with interstitial lung disease. Br. Med. J. 302 (1991) 880–881.

CHALMERS, A.C., C.M. GRECO and R.G. MILLER: Prognosis in AZT myopathy. Neurology (1991) 1181–1184.

CHERIN, P., S. HERSON, B. WECHSLER ET AL.: Efficacy of intravenous gammeglobulin therapy in chronic refractory polymyositis and dermatomyositis: an open study with 20 adult patients. Am. J. Med. 91 (1991) 162–168.

CHOU, S.M.: Inclusion body myositis: a chronic persistent mumps myositis. Hum. Pathol. 17 (1986) 765–777.

COLE, A.J., R. KUZNIECKY, G. KARPATI ET AL.: Familial myopathy with changes resembling inclusion body myositis and periventricular leucoencephalopathy. Brain 111 (1988) 1025–1037.

CRONIN, M.E., F.W. MILLER, J.E. HICKS, M. DALAKAS and P.H. PLOTZ: The failure of intravenous cyclophosphamide therapy in refractory idiopathic inflammatory myopathy. J. Rheumatol. 16 (1989) 1225–1228.

DALAKAS, M.C. (Ed.): Polymyositis and Dermatomyositis. Boston, Butterworths (1988a).

DALAKAS, M.C.: Treatment of polymyositis and dermatomyositis with corticosteroids: a first therapeutic approach. In: Dalakas M.C. (Ed.), Polymyositis and Dermatomyositis. Boston, Butterworths (1988b) 235–253.

DALAKAS, M.C.: Treatment of polymyositis and dermatomyositis. Curr. Opin. Rheumatol. 1 (1989) 443–449.

DALAKAS, M.C.: Inflammatory myopathies. Curr. Opin. Neurol. Neurosurg. 3 (1990a) 689–696.

DALAKAS, M.C.: Pharmacologic concerns of corticosteroids in the treatment of patients with immune-related neuromuscular diseases. Neurol. Clin. 8 (1990b) 93–118.

DALAKAS, M.C.: Polymyositis, dermatomyositis, and inclusion-body myositis. N. Engl. J. Med. 325 (1991) 1487–1498.

DALAKAS, M.C.: Inflammatory and toxic myopathies. Curr. Opin. Neurol. Neurosurg. (1992a) in press.

DALAKAS, M.C.: Inflammatory myopathies: pathogenesis and treatment. Neuropharmacology (1992b) in press.

DALAKAS, M.C.: Retroviral myopathies. In: Engel, A. (Ed.), Myology. New York, McGraw (1993) in press.

DALAKAS, M.C. and W.K. ENGEL: Total body irradiation in the treatment of intractable polymyositis/dermatomyositis. In: Dalakas, M.C. (Ed.), Polymyositis/Dermatomyositis. Boston, Butterworths (1988) 281–291.

DALAKAS, M.C. and I. ILLA: HIV-associated myopathies. In: Pizzo, A., C.M. Wilfelt (Eds), Pediatric AIDS. Baltimore, Williams & Wilkins (1991) 420–429.

DALAKAS, M.C. and G.H. PEZESHKPOUR: Neuromuscular diseases associated with human immunodeficiency virus infection. Ann. Neurol. 23 (S) (1988) 38–48.

DALAKAS, M.C., G.H. PEZESHKPOUR, M. GRAVELL and J.L. SEVER: Polymyositis in patients with AIDS. J. Am. Med. Ass. 256 (1986a) 2381–2383.

DALAKAS, M.C., W.T. LONDON, M. GRAVELL and J.L. SEVER: Polymyositis in an immunodeficiency disease in monkeys induced by a type D retrovirus. Neurology 36 (1986b) 569–572.

DALAKAS, M.C., M. GRAVELL, W.T. LONDON, G. CUNNINGHAM and J.L. SEVER: Morphological changes of an inflammatory myopathy in rhesus monkeys with simian acquired immunodeficiency syndrome (SAIDS). Proc. Soc. Exp. Biol. Med. 185 (1987) 368–376.

DALAKAS, M.C., I. ILLA, G.H. PEZESHKPOUR, J.P. LAUKAITIS, B. COHEN and J.L. GRIFFIN: Mitochondrial myopathy caused by long-term zidovudine therapy. N. Engl. J. Med. 322 (1990) 1098–1105.

DALAKAS, M.C., M. LEON-MONZON, I. ILLA, P. RODGERS-JOHNSON and O. MORGAN: Immunopathology of HTLV-I-associated-polymyositis (HTLV-PM): Studies in 6 patients. Neurology 42 (S) (1992) 301–302.

DARROW, D.H., H.T. HOFFMAN, G.J. BARNES and C.A. WILEY: Management of dysphagia in inclusion body myositis. Arch. Otolaryngol. Head Neck Surg. 118 (1992) 313–317.

EISEN, A., K. BERRY and G. GIBSON: Inclusion body myositis: myopathy or neuropathy? Neurology 33 (1983) 1109–1114.

EMSLIE-SMITH, A.M. and A.G. ENGEL: Microvascular changes in early and advanced dermatomyositis: A quantitative study. Ann. Neurol. 27 (1990) 343–356.

EMSLIE-SMITH, A.M. and A.G. ENGEL: Necrotizing myopathy with pipestem capillaries, microvascular deposition of the complement membrane attack complex (MAC) and minimal cellular infiltration. Neurology 41 (1991) 936–939.

EMSLIE-SMITH, A.M., K. ARAHATA and A.G. ENGEL: Major histocompatibility complex class-1 antigen expression, immunolocalization of interferon subtypes and T-cell-mediated cytotoxicity in myopathies. Hum. Pathol. 20 (1989) 224–231.

ENGEL, A.G. and K. ARAHATA: Mononuclear cells in myopathies: quantitation of functionally distinct subsets, recognition of antigen-specific cell-mediated cytotoxicity in some diseases, and implications for the pathogenesis of the different inflammatory myopathies. Hum. Pathol. 17 (1986) 704–721.

ENGEL, W.K. and M.C. DALAKAS: Treatment of neuromuscular diseases. In: Wiederhold, W.C. (Ed.), Therapy for Neurologic Diseases. New York, John Wiley (1982) pp 51–101.

ENGEL, A.G. and A.M. ESMSLIE-SMITH: Inflammatory myopathies. Curr. Opin. Neurol. Neurosurg. 2 (1989) 695–700.

EUWER, R.L. and R.D. SOUTHEIMER: Amyopathic dermatomyositis (dermatomyositis siné myositis). J. Am. Acad. Dermatol. 24 (1991) 959–966.

FLECKENSTEIN, J.L., D.K. BURNS, F.K. MURPHY, H.T. JAYSON and F.J. BONTE: Differential diagnosis of bacterial myositis in AIDS: evaluation with MR imaging. Radiology 179 (1991) 653–658.

FRAZER, D.D., J.A. FRANK, M. DALAKAS, F.W. MILLER, J.F. HICKS and P.H. PLOTZ: Magnetic resonance imaging in the idiopathic inflammatory myopathies. J. Rheumatol. 18 (1991) 1693–1700.

FUJINO, H., T. KOBAYASHI, I. GOTO and H. ONITSUKA: Diagnostic resonance imaging of the muscles in patients with polymyositis and dermatomyositis. Muscle Nerve 14 (1991) 716–720.

GAUT, P., P.K. WONG and R.D. MYER: Pyomyositis in a patient with the acquired immunodeficiency syndrome. Arch. Intern. Med. 148 (1988) 1608–1610.

HANTAI, D., J.G. FOURNIER, R. VAZEUX, H. COLLIN, M. BAUDRIMONT and M. FARDEAU: Skeletal muscle involvement in human immunodeficiency virus infection. Acta Neuropathol. 81 (1991) 496–502.

HECKMATT, J., N. HASSON, C. SAUNDERS ET AL.: Cyclosporin in juvenile dermatomyositis. Lancet i (1989) 1063–1066.

HERNANDEZ, R.J., D.R. KEIM, T.L. CHENEVERT, D.B. SULLIVAN and A.M. AISER: Fat-suppression MR imaging of myositis. Radiology 182 (1992) 217–219.

HERTZMAN, P.A., W.L. BLEVINS, J. MAYER, B. GREENFIELD, M. TING and G.J. GLEICH: Association of the eosinophilia-myalgia syndrome with the ingestion of tryptophan. N. Engl. J. Med. 322 (1990) 869–873.

HOHLFELD, R. and A.G. ENGEL: Coculture with autologous myotubes of cytotoxic T cells isolated from muscle in inflammatory myopathies. Ann. Neurol. 29 (1991) 498–507.

HOHLFELD, R., A.G. ENGEL, L.I. KUNIO and M.C. HARPER: Polymyositis mediated by T lymphocytes that express the gamma/delta receptor. N. Engl. J. Med. 324 (1991) 877–881.

ILLA, I., A. NATH and M.C. DALAKAS: Immunocytochemical and virological characteristics of HIV-associated inflammatory myopathies: Similarities with seronegative polymyositis. Ann. Neurol. 29 (1991) 474–481.

JAY, C., M. ROPKA, K. HENCH, C. GRADY and M. DALAKAS: Prospective study of myopathy during prolonged low-dose AZT: clinical correlates of AZT mitochondrial myopathy (AZT-MM) and HIV-associated inflammatory myopathy (HIV-IM). Neurology 42 (S) (1992) 146.

KARPATI, G. and S. CARPENTER: Idiopathic inflammatory myopathies. Curr. Opin. Neurol. Neurosurg. 1 (1988) 806–814.

KARPATI, G., Y. POULIOT and S. CARPENTER: Expression of immunoreactive major histocapability complex products in human skeletal muscles. Ann. Neurol. 23 (1988) 64–72.

KEIM, D.R., R.J. HERNANDEZ and D.B. SULLIVAN: Serial magnetic resonance imaging in juvenile dermato-

myositis. Arthritis Rheum. 34 (1991) 1580–1584.

KELLY, J.J. JR., H. MADOC-JONES, L.S. ADELMAN, P.L. ANDRES and T.L. MUNSAT: Total body irradiation not effective in inclusion body myositis. Neurology 36 (1986) 1264–1266.

KISSEL, J.T., J.R. MENDELL and K.W. RAMMOHAN: Microvascular deposition of complement membrane attack complex in dermatomyositis. N. Engl. J. Med. 314 (1986) 329–334.

LAKHANPAL, S., T.W. BUNCH, D.M. ILSTRUP and L.J. MELTON III: Polymyositis-dermatomyositis and malignant lesions: does an association exist? Mayo Clin. Proc. 61 (1986) 645–653.

LAMPERTH, L., I. ILLA and M. DALAKAS: In situ hybridization in muscle biopsies from patients with HIV-associated polymyositis (HIV-PM) using labelled HIV-RNA probes. Neurology 40 (S) (1990) 121.

LANG, B.A., R.M. LAXER, G. MURPHY, E.D. SILVERMAN and C.M. ROIFMAN: Treatment of dermatomyositis with intravenous immunoglobulin. Am. J. Med. 91 (1991) 169–172.

LEFF, R.L., L.A. LOVE, F.W. MILLER, S.J. GREENBERG, E.A. KLEIN, M.C. DALAKAS and P.H. PLOTZ: Viruses in the idiopathic inflammatory myopathies absence of candidate viral genomes in muscle. Lancet 339 (1992) 1192–1195.

LEFF, R.L., S.H. BURGESS, F.H. MILLER ET AL.: Seasonal onset of adult idiopathic inflammatory myopathy in patients with anti-Jo-1 and anti-signal recognition particle antibodies. Arthritis Rheum. 11 (1991) 1391–1396.

LEON-MONZON, M. and DALAKAS M.: Detection of HIV in muscle and nerve biopsies by DNA-amplification techniques. Neurology 41 (S) (1991) 376.

LEON-MONZON, M. and M.C. DALAKAS: Absence of persistent infection with enteroviruses in muscles of patients with inflammatory myopathies. Ann. Neurol. 32 (1992) 219–222.

LOVE, L.A., R.L. LEFF, D.D. FRAZER, I.N. TARGOFF, M. DALAKAS, P. PLOTZ and F.W. MILLER: A new approach to the classification of idiopathic inflammatory myopathy: myositis-specific autoantibodies define useful homogeneous patient groups. Medicine 70 (1991) 360–374.

LOTZ, B.P., A.G. ENGEL, H. NISHINO, J.C. STEVENS and W.J. LITCY: Inclusion body myositis. Brain 112 (1989) 727–742.

LUECK, C.J., P. TREND and M. SWASH: Cyclosporin in the management of polymyositis and dermatomyositis. J. Neurol. Neurosurg. Psychiatry 54 (1991) 1007–1008.

MATHEWS, M.B. and R.M. BERNSTEIN: Myositis autoantibody inhibits histidyl-tRNA synthetase: a model for autoimmunity. Nature (London) 304 (1983) 177–179.

MENDEL, J.R., Z. SAHENK, T. GALES and L. PAUL: Amyloid filaments in inclusion body myositis. Arch. Neurol. 48 (1991) 1228–1234.

MHIRI, C., M. BAUDRIMONT, G. BONNE, C. GENY, F. DE-

GOUL, C. MARSAC, E. ROULLET and R. GHERARDI: Zidovudine myopathy: a distinctive disorder associated with mitochondrial dysfunction. Ann. Neurol. 29 (1991) 606–614.

MILLER, F.W., S.F. LEITMAN, M.E. CRONIN ET AL.: A randomized double-blind controlled trial of plasma exchange and leukapheresis in patients with polymyositis/dermatomyositis. N. Eng. J. Med. 326 (1992) 1380–1384.

MORGAN, O. STC, P. RODGERS-JOHNSON, C. MORA and G. CHAR: HTLV-1 and polymyositis in Jamaica. Lancet ii (1989) 1184–1187.

NEVILLE, H.E., I.L. BAUMBACH, S.P. RINGEL, L.S. RUSSO, E. SUJANSKY and C.A. GARCIA: Familial inclusion body myositis: evidence for autosomal dominant inheritance. Neurology 42 (1992) 897–902.

NISHINO, H., A.G. ENGEL and B.K. RIMA: Inclusion body myositis: the mumps virus hypothesis. Ann. Neurol. 25 (1989) 260–264.

OTERO, C. and M.C. DALAKAS: Onset of polymyositis (PM) and dermatomyositis (DM) during pregnancy: Report of 4 cases. Ann. Neurol. 32 (1992) 251.

OTERO, C., I. ILLA and M.C. DALAKAS: Is there dermatomyositis (DM) without myositis? Neurology 42 (S) (1992) 388.

PEARSON, C.M. and A. BOHAN: The spectrum of polymyositis and dermatomyositis. Med. Clin. North. Am. 61 (1977) 439–457.

PEZESHKPOUR, G.H., I. ILLA and M.C. DALAKAS: Ultrastructural characteristics and DNA immunocytochemistry in HIV and AZT-associated myopathies. Hum. Pathol. 11 (1991) 1281–1288.

PLOTZ, P.H., M. DALAKAS, R.L. LEFF, L.A. LOVE, F.W. MILLER and M.E. CRONIN: Current concepts in the idiopathic inflammatory myopathies: polymyositis, dermatomyositis and related disorders. Ann. Intern. Med. 111 (1989) 143–157.

RICHARDSON, J.B. and J.P. CALLEN: Dermatomyositis and malignancy. Med. Clin. North. Am. 73 (1989) 1211–1220.

ROBINSON, L.R.: Polymyositis. Muscle Nerve 14 (1991) 310–315.

ROSENBERG, N.L., H.A. ROTBART, M.J. ABZUG, S.P. RINGEL and M.J. LEVIN: Evidence for a novel picornavirus in human dermatomyositis. Ann. Neurol. 26 (1989) 204–209.

ROWLAND, L.P.: Dermatomyositis and polymyositis. In: Rowland, L.P. (Ed.), Merritt's Textbook of Neurology. Philadelphia, Lea & Febiger (1989) 733–738.

RUCHETON, M., H. GRAAFLAND, H. FANTON, L. URSULE, P. FERRIER and C.J. LARSEN: Presence of circulating antibodies against *gag*-gene MuLV proteins in patients with autoimmune connective tissue disorders. Virology 144 (1985) 468–480.

SCHWARTZMAN, W.A., M.W. LAMPERTUS, C.A. KENNEDY and M.B. GOETZ: Staphylococcal pyomyositis in patients infected by the human immunodeficiency virus. Am. J. Med. 90 (1991) 595–600.

SIGUERGEIRSSON, B., B. LINDELOF, O. EDHAG and E. ALLANDER: Risk of cancer in patients with dermatomyositis and polymyositis. N. Engl. J. Med. 326 (1992) 363–367.

SOUEIDAN, S. and M.C. DALAKAS: High-dose intravenous immunoglobulin in the treatment of inclusion body myositis. Ann. Neurol. 32 (1992) 251.

SOUEIDAN, S., T. SINNWELL, C. JAY, J. FRANK, A. MCLAUGHLIN and M. DALAKAS: Impaired muscle energy metabolism in patients with AZT-myopathy: a blinded comparative study of exercise ^{31}P magnetic resonance spectroscopy (MRS) with muscle biopsy. Neurology 42 (S) (1992) 146.

TARGOFF, I.N.: Immunologic aspects of myositis. Curr. Opin. Rheumatol. 1 (1989) 432–442.

TARGOFF, I.N.: Immune mechanisms of myositis. Curr. Opin. Rheumatol. 2 (1990) 882–888.

TARGOFF, I.N. and F.C. ARNETT: Clinical manifestations in patients with antibody to PL-12 antigen (alanyl-tRNA synthetase). Am. J. Med. 88 (1990) 241–251.

UNCINI, A., D.J. LANGE, A.P. HAYES ET AL.: Long-duration polyphasic motor unit potentials in myopathies: a quantitative study with pathological correlation. Muscle Nerve 13 (1990) 263–267.

VERNA, A., W.G. BRADLEY, A.M. ADESINA, R. SOFFERMAN and W.W. PENBELBURY: Inclusion body myositis with cricopharyngeus muscle involvement and severe dysphagia. Muscle Nerve 14 (1991) 470–473.

WARNER, C.L., P.B. FAYAD and R.R. HEFFNER, JR.: Legionella myositis. Neurology 41 (1991) 750–752.

WIDROW, C.A., S.M. KELLIE, B.R. SALTZMAN and U. MATHUR-WAGH: Pyomyositis in patients with human immunodeficiency virus: an unusual form of disseminated bacterial infection. Am. J. Med. 91 (1991) 129–136.

WILEY, C.A., M. NERENBERG and D. CROSS. Soto-Aguilar MC: HTLV-1 Polymyositis in a patient also infected with the human immunodeficiency virus. N. Engl. J. Med. 320 (1989) 992–995.

WEISSMAN, J.D., I. CONSTANTINITIS, P. HUDGINS and D.C. WALLACE: ^{31}P magnetic resonance spectroscopy suggests impaired mitochondrial function in AZT-treated HIV-infected patients. Neurology 42 (1992) 519–623.

YOUSEF, G.E., D.A. ISENBERG and J.F. MOWBRAY: Detection of enterovirus specific RNA sequences in muscle biopsy specimens from patients with adult onset myositis. Ann. Rheum. Dis. 49 (1990) 310–315.

Handbook of Clinical Neurology, Vol. 18 (62): Myopathies
L.P. Rowland and S. DiMauro, editors

CHAPTER 14

Myasthenia gravis and myasthenic syndromes

ANDREW G. ENGEL

Department of Neurology and Neuromuscular Research Laboratory, Mayo Clinic, Rochester, MN, USA

Myasthenia gravis (MG) and the myasthenic syndromes are diseases affecting the neuromuscular junction (NMJ). The disorders are linked together in the sense that in each the safety margin of neuromuscular transmission is compromised by one or more specific mechanisms, and this expresses itself clinically in abnormal weakness and fatigability on exertion. Although uncommon, these diseases have been of unusual interest to students, clinicians, and investigators. For students they illustrate how the application of basic science principles can elucidate complicated disorders. For clinicians, they are a challenge in diagnosis and therapy. For investigators, they are experiments of nature that can illuminate one or another facet of synaptic transmission.

The tempo of investigation and the acquisition of new knowledge of end-plate disorders has accelerated during the past 2 decades. The reasons for this are easy to trace. Definitive study of the distribution of the acetylcholine receptor (AChR) at the end-plate and its role in MG was made possible by 2 initially unrelated discoveries in neurobiology; the electric organs of fish proved to be an excellent source of AChR, and the α-neurotoxins of elapid snakes were found to bind to AChR with high affinity. Hence, these toxins could be used to purify AChR by affinity chromatography, and AChR could be radiolabeled and quantitated using the toxins. Attempts to raise antibodies to AChR led

to the discovery of an autoimmune model of MG, and AChR liganded by radiolabeled toxin could be used to detect and quantitate antibodies directed against AChR in human MG. These developments, in turn, acted as a further stimulus for exploring the molecular architecture of AChR, the cloning of the genes encoding the AChR subunits, the analysis of the T and B cell responses to AChR in MG, and the investigation of AChR in the central and autonomic nervous system.

The recognition of the autoimmune origin of MG also implied that myasthenic disorders occurring in a genetic or congenital setting had a different etiology. As a result, a number of distinct myasthenic syndromes have been recognized and investigated by electrophysiological and ultrastructural methods. The newly recognized disorders are conditioned by divergent causes, such as a putative failure of acetylcholine (ACh) resynthesis, absence of acetylcholinesterase (AChE) from the end-plate, or abnormal gating properties of the AChR-associated ion channel. The genetic defects may either impair neuromuscular transmission directly, or result in secondary alterations at the end-plate that eventually compromise the safety margin of transmission.

Studies during the past decade have also elucidated the pathogenesis of the Lambert-Eaton myasthenic syndrome (LEMS). In the 1960s, microelectrode studies of LEMS revealed a defect in the calcium-mediated release of ACh from the nerve terminal. In the 1970s, data from several laboratories suggested that the membrane particles of the presynaptic active zones represented the voltage-sensitive calcium channels of the motor nerve terminal. In the 1980s, freeze-fracture studies of LEMS revealed clustering and depletion of the active zone particles and the physiologic and morphologic features of the disease were shown to be mediated by IgG. Therefore, LEMS is an autoimmune disease in which the voltage-sensitive calcium channel of the motor nerve terminal is the target of pathogenic autoantibodies.

Table 1 presents a classification of end-plate diseases. Only the acquired autoimmune and congenital disorders are considered in this Chapter. The classification is still tentative because future studies are likely to uncover additional disorders and may modify our concepts of the currently recognized diseases.

A clear grasp of the factors affecting the safety margin of neuromuscular transmission is central to understanding the various transmission defects. Therefore this topic will be considered next.

THE SAFETY MARGIN OF NEUROMUSCULAR TRANSMISSION

The factors affecting the safety margin of neuromuscular transmission can be readily deduced from the basic anatomy and physiology of the NMJ.

Basic features of end-plate organization

The end-plate consists of pre- and postsynaptic regions separated by a synaptic space (Fig. 1). Each presynaptic region is made up of a nerve terminal covered by a Schwann cell process. Each postsynaptic region is composed of junctional folds that overlie a specialized region of the muscle fiber, the junctional sarcoplasm. The synaptic space includes a single primary and multiple secondary clefts. ACh is stored in synaptic vesicles. The average synaptic vesicle contains about 6000–10,000 ACh molecules (Kuffler and Yoshikami 1975; Hartzell et al. 1976). The contents of the synaptic vesicle are discharged into the synaptic space by exocytosis (Heuser et al. 1979). A transmitter quantum is the

amount of ACh released from a single synaptic vesicle. The synaptic space is lined by basal lamina. Junctional AChE is associated with the basal lamina (Hall and Kelly 1971; Betz and Sakmann 1973; McMahan et al. 1978). AChE is distributed throughout the entire synaptic basal lamina at a density of about 2500 sites/μm^2 (Salpeter et al. 1978). AChR is concentrated on the terminal expansions of the junctional folds where its packing density is $10^4/\mu m^2$ (Matthews-Bellinger and Salpeter 1978; Heuser and Salpeter 1979; Land et al. 1980; Grohovaz et al. 1982; Hirokawa and Heuser 1982).

Events after the release of an ACh quantum

In the resting state there is a steady but random release of transmitter quanta from the nerve terminal associated with exocytosis of individual synaptic vesicles (Fatt and Katz 1952; Martin 1977). The focally released ACh molecules diffuse into the synaptic space, but most are not hydrolyzed ini-

TABLE 1

Classification of neuromuscular junction diseases.

Autoimmune
  Myasthenia gravis
  Lambert-Eaton syndrome
Congenital, well-characterized
  Familial infantile myasthenia (defect in ACh resynthesis/packaging)*
  Paucity of synaptic vesicles and reduced quantal release†
  End-plate AChE deficiency*
  Kinetic abnormality and deficiency of AChR
    Classical slow-channel syndrome‡
    AChR deficiency, long opentime, low conductance, ε subunit mutation†
    AChR deficiency short channel opentime†
  Kinetic abnormality of AChR
    High-conductance fast-channel syndrome*
    Abnormal ACh-AChR interaction†
Congenital, partially characterized
  AChR deficiency with paucity of secondary synaptic clefts*
  Other AChR deficiencies†
  Familial limb-girdle myasthenia*
Toxic
  Botulism
  Drug-induced

---

* Autosomal-recessive inheritance.
† Autosomal-recessive inheritance suspected.
‡ Autosomal-dominant inheritance.

tially by AChE due to saturation of AChE by the high concentration of ACh (Fertuck and Salpeter 1976; Salpeter 1983).

A high proportion of ACh reach the postsynaptic membrane and collide with AChRs. Some of the collisions result in the binding of ACh to specific sites on the 2 α subunits of AChR, and each AChR can bind 2 molecules of ACh. The binding of 2 ACh to an AChR molecule produces a conformational change, or isomerization, of AChR so that its cation-selective ion channel opens. After an exponentially distributed time interval, the ion channel closes and ACh dissociates from AChR (Anderson and Stevens 1973). The resultant depolarization, but not the current flow, is affected by the cable properties of the muscle fiber surface membrane. Once dissociated from AChR, ACh is rapidly hydrolyzed by AChE to choline and acetate. Choline is taken up by the nerve terminal by a high-affinity, sodium-dependent and hemicholinium-sensitive process (Beech et al. 1980). ACh is resynthesized by choline acetyltransferase and is then taken up by the synaptic vesicles (Parsons et al. 1982). Under normal conditions and with AChE fully active, ACh molecules in a single quantum exert their effect within a radius of 0.8 μm from the site of release (Hartzell et al. 1975). AChE limits the number of collisions of ACh with AChR and also the radius of spread of ACh. When AChE is inactive, the lateral spread of ACh is increased so that each ACh molecule can sequentially bind to multiple AChRs and open multiple ion channels before leaving the synaptic space by diffusion (Katz and Miledi 1973).

The depolarization and the concomitant current flow induced by a single quantum give rise to the miniature end-plate potential (MEPP) and miniature end-plate current (MEPC), respectively. The amplitude of the MEPP is a function of the number of ion channels opened by the quantum, the average depolarization per opening, and the input resistance of the muscle fiber (Martin 1976, 1977). The amplitude of the MEPC is a function of the number of ion channel openings and the mean current flow per opening. The number of channel openings is a function of the number of ACh in the quantum, the number of available AChR, and the geometry of the synaptic space which allows rapid diffusion of most ACh molecules from their site of

release to AChR. The duration of the MEPP depends on the average channel opentime (Katz and Miledi 1972; Anderson and Stevens 1973), the functional state of AChE (Katz and Miledi 1973), and the cable properties of the muscle fiber surface membrane (Fatt and Katz 1952). The duration of the MEPC is independent of the cable properties of the muscle fiber surface membrane, but otherwise is affected by the same factors that determine the amplitude and duration of the MEPP (Martin 1977).

Transmitter release by a single nerve impulse

Depolarization of the nerve terminal by nerve impulse is followed by an influx of calcium ions into the terminal via voltage-sensitive calcium channels, and it is this influx that mediates transmitter release (Katz and Miledi 1969; Miledi 1973; Llinas and Nicholson 1975; Charlton et al. 1982). In the terminal, calcium activates a calcium/calmodulin-dependent protein kinase which then phosphorylates synapsin I, a protein that anchors the synaptic vesicles to the cytoskeleton near their sites of release. Phosphorylated synapsin I can no longer anchor the synaptic vesicles which can therefore undergo exocytosis (Llinas et al. 1991). The calcium ingress also results in facilitation, so that the probability of release by a subsequent impulse is increased. The effects of calcium are antagonized by magnesium (Katz and Miledi 1967; Hubbard et al. 1968).

Nerve stimulation results in synaptic vesicle exocytosis adjacent to the active zones of the presynaptic membrane (Heuser et al. 1979). In the freeze-fractured presynaptic membrane, the active zones are represented by double parallel rows of large (10–12 nm) intramembrane particles (Rash and Ellisman 1974; Heuser et al. 1979; Ceccarelli and Hurlbut 1980). Studies of the squid giant synapse suggest that the active zone particles represent the voltage-sensitive calcium channels of the presynaptic membrane (Llinas et al. 1976; Pumplin et al. 1981). Light microscopic studies of the frog NMJ with fluorescent ω-conotoxin, which binds to neuronal voltage-sensitive calcium channels, are also consistent with this assumption (Robitaille et al. 1990).

The number of quanta released by nerve impulse

(m) depends on the probability of release (p) and on at least 1 additional factor designated as n, according to the formula $m = np$ (Del Castillo and Katz 1954). The factor n was originally defined as the number of quantal units capable of responding to the nerve impulse, but it more likely indicates the number of readily releasable quanta (Elmqvist and Quastel 1965a), the number of active release sites in the nerve terminal (Zucker 1973; Wernig 1975), or a combination of these variables.

Depolarization of the nerve terminal by a nerve impulse results in the nearly synchronous release of a relatively large number of quanta from the nerve terminal. The depolarization produced by the individual quanta summate, but not linearly, producing a larger depolarization, the end-plate potential (EPP). When the EPP exceeds a certain threshold, it triggers a muscle fiber action potential (Martin 1977). The EPP amplitude is affected by the same factors that affect the MEPP amplitude and, in addition, by the number of quanta released by nerve impulse. *The difference between the actual EPP amplitude and the EPP amplitude required to trigger the muscle fiber action potential represents the safety margin of neuromuscular transmission.*

Transmitter release by repetitive nerve stimulation

Repetitive stimulation under physiologic conditions results in a frequency-dependent depression of the EPP amplitude, and of the safety margin, to a certain plateau. The decline has been attributed to a decrease of m due to a decrease in the number of synaptic vesicles readily available for release, and the plateau to equilibrium between quantal release and mobilization (Elmqvist and Quastel 1965a). This might be an oversimplification, for a decrease in p (Christensen and Martin 1970) or in the number of ACh molecules per quantum might also occur during repetitive stimulation.

Under conditions of low quantal release (as in a high-magnesium or a low-calcium medium), repetitive nerve stimulation increases transmitter release (Martin 1977). At least 4 temporally distinct stimulation-dependent processes facilitate transmitter release by increasing p, n, or both (Magleby 1979). Although facilitation is best studied under conditions of low quantal release, it also occurs under physiological conditions and tends to antag-

onize the process of depression. The temporal profiles of the opposing processes are such that (a) a defect in neuromuscular transmission is most readily detected by a train of 5–10 stimuli delivered at a low (2–3 Hz) frequency and (b) tetanic stimulation results in a transient improvement and then a worsening of the defect.

In clinical electromyography (EMG) the transmission defect is demonstrated by repetitive supramaximal stimulation of a motor nerve and recording of the evoked compound muscle action potential by an electrode placed on the surface of the muscle. The amplitude of the evoked potential is proportionate to the number of muscle fibers activated by the nerve impulse. As neuromuscular transmission fails, the EPP becomes subthreshold for firing the muscle fiber action potential at an increasing number of junctions, and the amplitude of the compound muscle action potential decreases. This is referred to as the decremental EMG response.

The saturating disk model of neuromuscular transmission

The preceding sections reviewed the factors that can alter the safety margin of neuromuscular transmission. To understand how the safety margin is compromised, it is useful to consider the end-plate as a composite of units that respond to ACh quanta. Each unit, or disk, represents that region of the postsynaptic membrane on which a single quantum exerts its effect. Within the disk, effective interaction between the ACh quantum and the postsynaptic membrane is assured by (a) positioning of the presynaptic release sites in relation to the junctional folds, (b) the high density (10,000–20,000/μm^2) of ACh binding sites on the crests of the junctional folds, and (c) the lower density (2000–3000/μm^2) but uniform distribution of AChE throughout the synaptic basal lamina (Salpeter 1987).

After a quantum is released from the nerve terminal, ACh concentration is highest at the exocytotic site and gradually decreases as ACh diffuses into the synaptic space. By the time ACh has spread over about 0.2 μm of the junctional folds, its average concentration (about 3 mM) is still sufficiently high to swamp AChE. Further, by the

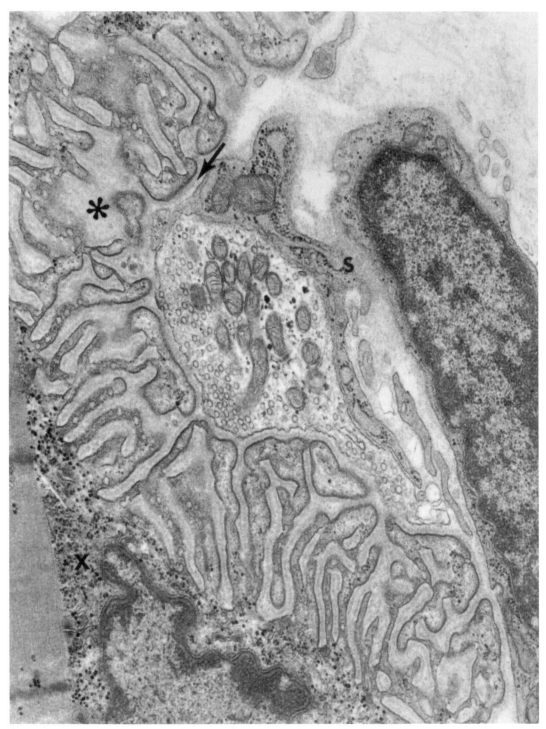

Fig. 1. Electron micrograph of normal human end-plate region. The right side of the nerve terminal is covered by Schwann cell (S); the left side of the nerve terminal faces the postsynaptic region of clefts and folds. Junctional sarcoplasm (x) contains glycogen granules, ribosomes, small tubular profiles and nucleus. The arrow and asterisk mark primary and secondary clefts, respectively. ×30,600. (Reproduced from Engel et al. (1975).)

time a quantal packet spreads over a distance of 0.3 μm along the top and down the side of the folds, all its ACh molecules will bind to AChR at a saturating concentration. The small size of the saturating disks assures that they do not overlap. Therefore only a small proportion of all available AChR is saturated with ACh when up to several hundred quanta are released by a nerve impulse. Also, since the presynaptic release sites are discrete and vary from impulse to impulse (Heuser et al. 1979; Heuser and Reese 1981), different sets of disks become saturated at finite intervals during normal activity. This prevents desensitization of AChR from continued exposure to ACh. The consequences of the model for some end-plate diseases are discussed below.

MYASTHENIA GRAVIS

Definition and basic concepts

Myasthenia gravis is an acquired autoimmune disorder of neuromuscular transmission associated with AChR deficiency at the NMJ. (Fig. 2). The principal symptoms are weakness and abnormal fatigability on exertion. The symptoms are improved by rest and anticholinesterase drugs. The disease involves the external ocular muscles selectively or is generalized in distribution. Anti-AChR antibodies are present in serum in 80–90% of the cases, and immunoglobulin and complement deposits are found at the NMJ in all cases.

Historical background

Early clinical observations. The earliest observations on myasthenia gravis are attributed to Thomas Willis. In 1672, he described patients with weakness of limb muscles which increased during the course of the day and progressive palsy of the tongue provoked by 'long, hasty or laborious speaking' (Guthrie 1903). In 1877, Wilks described a patient with possible MG; she had bulbar weakness and difficulty in walking and died of respiratory paralysis soon after the onset of her symptoms. Erb (1879) and Goldflam (1893) noted the fluctuating character of the symptoms, the selective involvement of the external ocular and other

cranial muscles, and the gradual worsening of the symptoms during the day. In 1895, Jolly made 3 notable contributions to MG. First, he gave the disease its name (myasthenia gravis pseudoparalytica); second, he described the myasthenic reaction as a progressive decline in the tetanic tension of indirectly stimulated muscles; and third, he suggested treatment of MG with physostigmine. By 1900, Campbell and Bramwell collected 60 cases of MG from the literature. They emphasized that the disease had a predilection for the most constantly used muscles, and that it fluctuated from day to day, week to week, or even disappeared for months or years and then reappeared. They inferred that MG was caused by a circulating toxin which acted selectively upon the lower motor neuron. In 1905, Buzzard, in an autopsy study, observed lymphorraphes in muscle and other organs and found lymphoid hyperplasia in the thymus gland. He attributed the disease to 'some toxic, possibly autotoxic, agent' and concluded that the best approach to the disease was through a study of the lymphatic system.

Therapeutic approaches to MG. In 1901, Laquer and Weigert observed an association between thymoma with myasthenia gravis, but it was only in 1936 that Blalock noted that thymectomy for a thymoma had a beneficial effect in a myasthenic patient. The usefulness of thymectomy in MG was further documented by Blalock in 1944. By 1960, it was known that about 15% of MG patients had thymoma and 65% had thymic hyperplasia (Viets and Schwab 1960). Although thymectomy was soon accepted as treatment for MG, the indications for and the long-term results of the procedure are still debated (McQuillen and Leone 1977).

In 1934, Mary Walker recognized the similarity of symptoms between MG and curare poisoning and suggested the use of physostigmine in MG. In the same year Dale and Feldberg (1934) established that ACh was the neurotransmitter at the NMJ. Neostigmine, which could be administered orally, was introduced in the following year (Pritchard 1935), while pyridostigmine was not employed until 2 decades later (Osserman 1955). Because of initial unfavorable experiences with corticotropin and cortisone in MG (Merritt 1952; Brunner et al. 1976), these medications fell in disfa-

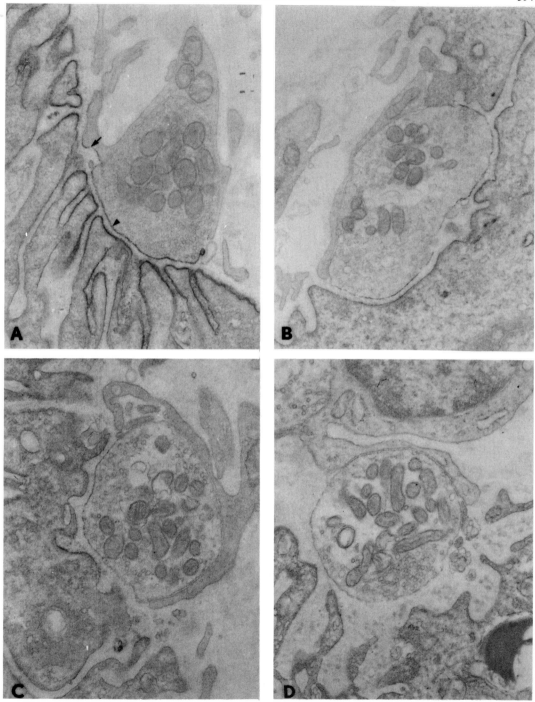

Fig. 2. Ultrastructural localization of AChR in control subject (A) and in moderately severe, generalized MG (B–D). (A) AChR is associated with the terminal expansions and deeper surfaces of the junctional folds; the presynaptic membrane (arrowhead) and Schwann cell membrane (arrow) facing the crests of the folds are lightly stained. The postsynaptic regions are simplified in B, C and D. In B and C, only segments of the simplified postsynaptic membrane react with AChR; no AChR is detected in D. ×21,000. (Reproduced from Engel et al. (1977c) by courtesy of the Editors of *Neurology*.)

vor until 1970 when the efficacy and relative safety of alternate-day prednisone therapy was recognized (Warmolts et al. 1970). Observations on the usefulness of other immunosuppressive drugs, such as azathioprine (Mertens et al. 1969), predated the recognition of the autoimmune etiology of MG. Since 1976, plasmapheresis had been used in the treatment of critically ill patients refractory to other forms of therapy (Pinching et al. 1976; Dau et al. 1977).

Electrophysiologic studies. In a classical EMG study of MG, Lindsley (1935) noted that the amplitudes of the motor unit potentials fluctuated abnormally during voluntary contraction but the rate and rhythm of firing of the potentials remained unaltered. From this, he inferred that (a) the motor nerves conducted impulses normally, but the impulses were blocked at the NMJ and (b) a variable number of fibers of a given motor unit were affected at any moment. Six years later, Harvey and Masland (1941) described the decremental response of the evoked compound muscle action potential in MG. This consisted of a rapid decline in the amplitude of the first few evoked muscle action potentials at low frequencies of stimulation. Even a single nerve stimulus was followed by a prolonged depression during which a second volley excited only a reduced number of muscle fibers. Quinine and ischemia worsened the defect.

During the next decade, in-vitro intracellular microelectrode studies established the quantal character of neuromuscular transmission. According to this, the MEPP represents the postsynaptic response to 1 quantum at rest, and the EPP the postsynaptic response to all quanta released from the nerve terminal by a single nerve impulse (Fatt and Katz 1952; Del Castillo and Katz 1954). In 1964, Elmqvist et al. used intracellular microelectrodes to analyze neuromuscular transmission in intercostal muscles in MG. They were able to show that in MG the amplitude of the MEPP was markedly decreased but the number of quanta liberated by nerve impulse was normal. Elmqvist et al. (1964) were unable to demonstrate a decreased postsynaptic sensitivity to ACh and concluded incorrectly that the defect in MG was presynaptic, consisting of an insufficient number of ACh molecules per quantum. It was not until 1976 that Rash

et al. showed that postsynaptic ACh sensitivity is reduced in MG, but by then there was overwhelming evidence from other sources that MG was caused by postsynaptic AChR deficiency.

The autoimmune etiology of MG. The concept that MG is an autoimmune disease was first proposed by Simpson (1960) and, in the same year, by Nastuk et al. (1960). Simpson's argument was based on the frequent thymus gland abnormalities in MG, the association of MG with other putative autoimmune diseases, the age and sex distribution of the patients, the fluctuating course of the diseases, and the occurrence of transient neonatal MG. Astutely, Simpson also suggested that anti-AChR antibodies competitively blocked neuromuscular transmission. Nastuk's argument was based mostly on the observation that serum complement levels fluctuated in the course of the disease. Also in 1960, Strauss et al. found anti-striated muscle antibodies in about one-third of MG sera, another hint that MG was an autoimmune disease. However, a relation of these antibodies to the neuromuscular transmission defect in MG has never been established.

In 1973, Patrick and Lindstrom observed myasthenic symptom in rabbits immunized with eel AChR and thus provided the first piece of direct evidence for the autoimmune etiology of MG. In the same year, Fambrough et al. (1973) demonstrated AChR deficiency at the NMJ in human MG. Between 1973 and 1976 experimental autoimmune myasthenia gravis (EAMG) was induced in several species and similarities of the experimental model to the essential features of the human disease were documented (Green et al. 1975; Lennon et al. 1975; Tarrab-Hazdai et al. 1975; Fulpius et al. 1976; Penn et al. 1976). By 1977, the autoimmune character of the human disease was firmly established by the demonstration of circulating anti-AChR antibodies in nearly 90% of the patients (Lindstrom et al. 1976c), by passive transfer with IgG of several features of the disease either from rat to rat (Lindstrom et al. 1976a) or from human to mouse (Toyka et al. 1977), by light microscopic and ultrastructural localization of IgG and complement at the NMJ in MG (Engel et al. 1977a), and by the beneficial effects of plasmapheresis (Pinching et al. 1976; Dau et al. 1977).

Clinical aspects

Epidemiology

Estimates of the annual incidence of MG per million population vary from a low of 2–5 (Kurtzke and Kurland 1977) to a high of 10.4 (Sorensen and Holm 1989). Estimates of prevalence vary from 25 to 125 per million (Kurtzke and Kurland 1977; Oosterhuis 1989; Sorensen and Holm 1989; Somnier et al. 1991). The age-adjusted death rate of MG was of the order of 1 per million population in 1977, but, because of continually improving methods of therapy, it may be lower now than stated in the literature. The ratio of female:male patients is 6:4 (Osserman and Genkins 1971). The disease may present at any age but the female incidence peaks in the third decade and the male incidence in the sixth or seventh decade of life (Osserman and Genkins 1971). The mean age at onset is 28 years in females and 42 years in males. Females are more common among the children and young adults and males are more common among patients older than 50 (Grob et al. 1981).

Although most cases of MG are sporadic, familial aggregations had been observed more commonly than could be explained by chance (Namba et al. 1971a,b; Osserman and Genkins 1971). Most familial cases present before the age of 20. In Osserman's series 5%, and in Namba's series 3.8% of the patients were related to each other. In light of current knowledge, an increased familial incidence of MG could be due to the inheritance of HLA haplotypes that predispose to sensitization to a given antigen. It is now also known that some patients with 'familial MG' suffer from genetically determined congenital myasthenic syndromes.

Characteristic clinical features

MG has the following characteristic clinical features (Campbell and Bramwell 1900; Wilson 1955; Osserman 1958; Simpson 1960; Viets and Schwab 1960; Simpson et al. 1966; Osserman and Genkins 1971; Gutmann 1978; Sethi 1987):
1. There is abnormal weakness and fatigability of some or all voluntary muscles.
2. The weakness increases with repeated or sustained exertion and in the course of the day, but is improved by rest; it is worsened by elevation of body temperature and is often improved by cold.

3. The symptoms respond to anticholinesterase drugs.
4. External ocular muscles are affected initially in approximately 50% and eventually in about 90% of the cases.
5. Other voluntary muscles innervated by cranial nerves (facial, masticatory, lingual, pharyngeal, and laryngeal muscles) and cervical, pectoral-girdle and hip flexor muscles are also frequently affected; proximal limb muscles are usually more severely affected than distal ones.
6. The tendon reflexes are brisk or normally active but may diminish if repeatedly elicited; reflexes may not be elicited from muscles that are too weak to be activated by voluntary effort.
7. There are no objective sensory deficits; subjective sensory symptoms are occasionally described.
8. The symptoms may fluctuate from day to day, week to week, or over longer periods of time.
9. Spontaneous remissions can occur and last for varying periods; long and complete spontaneous remissions are very rare; most of the significant spontaneous remissions occur during the first 3 years of the disease.

One can hardly improve the eloquent description of the symptoms of MG by Wilson (1955). Ptosis of the eyelid can be unilateral or bilateral. Weakness of other external ocular muscles also can be unilateral or bilateral but is more commonly bilateral than unilateral. The ocular palsies are typically asymmetrical and frequently affect muscles subserving convergence and vertical gaze. The facial lines may become ironed out so that the face seems smooth and expressionless. The mouth may be open with the lower lip slightly everted, or the jaw may drop and is supported by the patient's finger. The patient's smile resembles a snarl because the corners of the mouth are not drawn up and out while the levators expose the canines. Palatal weakness results in hypernasality of the voice and nasal regurgitation of liquids. Pharyngeal muscle weakness can result in blockage of the eustachian tube, tensor tympani paresis in low-frequency hearing loss, and stapedius paresis in hyperacusis (Simpson 1960). Laryngeal weakness may cause dysphonia and choking on food or secretions. Feeding may be associated with progressive difficulty in biting, chewing, swallowing, nasal regurgitation, choking, and breathlessness. While the patient is continuing to talk, the voice becomes first husky, then nasal, indistinct, soft and feebler, and finally anarthria and aphonia may occur. Weakness of the diaphragm, intercostal, and accessory respiratory muscles results in dyspnea either at rest or on mild or moderate exertion. Muscle wasting appears in about 14% of the cases and may involve the masseter, temporal, or facial muscles and, less frequently, shoulder or other muscles. The atrophic myasthenic tongue with a triple, shallow, longitudinal furrow was frequently observed by Wilson but infrequently by others.

In addition to exertion, extremes of temperature, viral or other infections, menses, and emotional upsets can worsen the symptoms. Exposure to bright light may increase the ocular signs. The effects of pregnancy on the disease may vary. According to Simpson et al. (1966) and Plauché (1979), during gestation MG worsens in 38–45% and improves in 24% of the patients; post partum, the disease worsens in 70% and improves in 30%. In patients progressing from mild to more severe disease, the weakness tends to spread from ocular to facial to lower bulbar muscles and then to truncal and limb muscles. However, this sequence may vary and different muscles may be affected either together or in succession.

The natural course of MG

A model clinical study by Grob et al. (1981) traced the course of MG in 1036 patients observed between 1940 and 1980. The initial symptoms were as follows: ptosis 25%; diplopia 25%; blurred vision 3%; leg weakness 13%; generalized fatigue 6%; dysphagia 6%; dysarthria 5%; weakness of the arms, face, and neck, in each 3%; and weakness of the torso or dyspnea 1%.

During the first month after the onset, the symptoms were purely ocular in 40%, generalized in 40%, involved only the extremities in 10%, and only the bulbar or oculobulbar muscles in 10%. Involvement of external ocular muscles typically began in the first year and at the end of the first year almost all patients had involvement of these muscles.

Although the disease was initially purely ocular in 40%, it remained confined to the ocular muscles in 16%. Eighty-seven percent of the generalizations occurred within 13 months after the onset. In pa-

tients with generalized disease, the interval from onset to the first episode of maximal weakness was less than 36 months in 83%, and more than 50% of deaths due to the disease took place in that period.

The fate of patients who had generalized disease was different between 1940 and 1960 as compared with 1960 and 1980. Significantly more patients improved (36% vs 20%) or were unchanged (42% vs 29%), and significantly fewer became worse (2% vs 7%) or died (12% vs 33%) in the 1960–1980 group. The improvement was primarily due to improved management of severely ill patients, and especially due to improved respiratory care during the first 3 years of the disease. The mortality rate of crisis fell from 30% between 1940 and 1960 to only 3% between 1970 and 1980. More than 90% of patients with even severe, generalized disease can now attain an improved or steady state. A more recent study of 1152 MG patients followed over a mean period of 4.9 years reports a 4% mortality; the principal risk factors were age over 40, short history of severe disease, and thymoma (Mantegazza et al. 1990).

Classifications of MG

The Osserman classifications of MG. The classification proposed by Osserman (1958) and subsequently modified by Osserman and Genkins (1971) divides MG into adult and pediatric types, accounting for 90% and 10% of all MG, respectively, and each type into several groups. The adult type is divided into the following groups; ocular (Group 1, 20%), mild generalized (Group 2A, 30%), moderately severe generalized (Group 2B, 20%), acute fulminating (Group 3, 11%), and late severe (Group 4, 9%). This classification remains useful for defining stages of the disease but there is no real reason to recognize separate childhood type of MG. The subdivision of pediatric MG into early and late-onset cases, as proposed by Bundey (1972), is of dubious merit as many early onset cases suffer from congenital myasthenic syndromes and not from autoimmune MG; both children and adults can be classified according to the severity of the disease; and in Caucasians early onset MG has the same course as adult onset MG (Batocchi et al. 1990). However, Chinese children tend to have mild MG (see below).

In Group 1, the clinical manifestations are limited to the external ocular muscles. However, careful examination often reveals slight weakness or abnormal fatigability in other muscles (Simpson 1958). Some patients in this group have EMG signs of the disease in limb muscles and carry an increased risk of developing more severe MG (Genkins et al. 1975). Approximately 40% of the patients in Group 1 develop more generalized disease, and this most commonly happens within 2 years after the onset of the initial symptoms. Conversely, many patients in other groups have previously belonged to Group 1.

In Group 2A, the mild generalized disease involves cranial, limb, and truncal muscles but tends to spare the respiratory muscles. The symptoms usually respond to anticholinesterase drugs and the mortality rate is relatively low.

In Group 2B, the moderately severe generalized disease is usually associated with significant diplopia, ptosis, dysarthria, dysphagia, feeding difficulty, limb weakness, and exercise intolerance.

In Group 3, the fulminating disease often begins abruptly and patients reach their worst state within 6 months. They show early respiratory muscle involvement, severe bulbar, limb and truncal muscle weakness, respond to anticholinesterase drugs poorly, and have frequent crises and a high mortality. Thymomas occur relatively frequently in this group.

Patients in Group 4 develop severe disease after having had milder disease for 2 or more years. The incidence of thymoma is higher in this group than in Groups 1 and 2 and the prognosis is relatively poor.

Other classifications and association with HLA haplotypes. Although the Osserman classification remains firmly entrenched, alternative classifications have been proposed. One classification (Compston et al. 1980) divides patients with MG according to age of onset and presence or absence of thymoma: (a) MG with thymoma: no sex or HLA-antigen association; relatively high AChR antibody titer; striated muscle antibodies present in 84%; low incidence of other organ-specific autoantibodies; (b) MG without thymoma, onset under age 40: female preponderance; association with HLA A1, B8, and DRw3 antigens; intermediate AChR antibody

titer; relatively high incidence or organ-specific autoantibodies; and (c) MG without thymoma, onset after age 40: male preponderance; increased association with HLA-A3, B7, and/or DRw2 antigens; low AChR antibody titer; anti-striated muscle antibodies in 47%, and relatively high incidence of other organ-specific autoantibodies.

The relationship of the HLA haplotype to the clinical phenotype indicated by the Compston classification applies only to Caucasians. Among Japanese patients early onset MG is associated with the DRw9, DRw13 and DQw3 haplotypes (Matsuki et al. 1990). In young females with thymic hyperplasia the B12 antigen is over-represented and in males over age 40 without thymoma the A10 antigen is over-represented (Yoshida et al. 1977). Another study of Japanese patients finds a strong association with DRw53 in females with early disease onset; further, 76% of the DRw53-positive patients, but only 19% of the controls, show a polymorphism of the DQB locus consisting of a 6.5- or a 6.2-kb *Bam*HI restriction fragment (Morita et al. 1991).

Chinese patients with onset under age 20 usually have ocular MG, a low antibody titer, and a strong association with HLA-DRw9; generalized MG occurs predominantly in patients with onset after age 20 (Hawkins et al. 1989).

Transient neonatal MG
The syndrome, first described by Strickroot et al. (1942), results from the transplacental transfer of circulating anti-AChR antibodies from the myasthenic mother to the fetus (Keesey et al. 1977). Transient neonatal MG develops in about 12% of infants born to myasthenic mothers (Namba et al. 1970). The symptoms usually appear within the first few hours after birth. Mean disease duration is 18 days, but 1 patient was ill for 47 days. The cardinal findings are feeding difficulty (87%), generalized weakness (69%), respiratory difficulty (65%), feeble cry (60%), facial paresis (54%), and ptosis (15%). A single report describes arthrogryposis associated with severe transient neonatal MG in 2 infants born to the same myasthenic mother (Moutard-Codou et al. 1987). There is no correlation between the severity of the disease in the mother and the infant (Namba et al. 1970). The reason for this, and why only some offspring of

MG mothers develop transient MG has been debated. Lefvert and Osterman (1983) found significantly different antibody specificities in affected infant-mother pairs and concluded that the affected infants produced antibodies independently because of adoptive transfer of immunocytes from the mother, or because fetal AChR damaged by maternal antibodies triggered a transient immune response in the fetus. According to Tzartos et al. (1990), however, the fine antigenic specificities of the anti-AChR antibodies of the infant-mother pairs are very similar, and according to Eymard et al. (1989) the maternal antibody titer correlates with the occurrence and severity of the disease in the newborn.

Association with other diseases

The association of MG with other autoimmune disorders was one of the tenets of Simpson's autoimmune hypothesis (1960). Of the 440 cases Simpson found 16 associated with rheumatoid arthritis, 4 with pernicious anemia, 1 with systemic lupus erythematosus, and 1 with sarcoid. Subsequent studies added Sjögren's syndrome, chronic ulcerative colitis, pure red cell anemia with thymoma, pemphigus, polymyositis, and Lambert-Eaton syndrome to the list (Downes et al. 1966; Wolf et al. 1966; Osserman et al. 1967; Penn et al. 1971; Taphoorn 1988; Oosterhuis 1989; Newsom-Davis et al. 1991). When the cases from the different series were combined, 3.6% had evidence of 1 of these disorders (Penn et al. 1971). Further, thyroid diseases, some of which are clearly autoimmune in origin, were observed in 13.1% of all MG patients (Osserman et al. 1967) and thyroiditis was found post-mortem in 19% of MG patients, but only in 0.9% of other patients (Becker et al. 1964). As discussed above, the susceptibility to develop MG and another autoimmune disease is likely to be related to HLA genes. Consistent with this is a report that among 44 MG patients 30% had a confirmed family history of autoimmune disease (Kerzin-Storrar et al. 1988). The association of hyperthyroidism with MG and whether the presence of 1 disease has a beneficial or adverse effect on the other had interested neurologists since the turn of the century, as reviewed by Engel (1961). Case studies (Millikan and Haines 1953) and the experimental induction of hyperthyroidism in MG patients (Engel 1961) showed that hyperthyroidism worsened the myasthenia. A plausible explanation for this effect was obtained by Hofmann and Denys (1972) who found a decrease in the amplitude of the MEPP in hyperthyroid animals.

Patients with hematologic disorders and especially those after bone marrow transplantation have a relatively high incidence of anti-AChR antibodies without clinical MG (Lefvert and Björkholm 1987). About one-third of bone marrow transplant recipients develop chronic graft-versus-host disease, but only 0.5% of the latter develop clinical MG (Bolger et al. 1986; Grau et al. 1990). Among the 3 patients observed by Grau, 2 carried the HLA B35 haplotype.

An association of MG with giant cell polymyositis, myocarditis, and thymoma has been repeatedly reported. Among 11 patients with MG and giant cell polymyositis, 8 had both thymoma and myocarditis (Pascuzzi et al. 1986). In another series of 13 patients with giant cell polymyositis, all had thymoma and myocarditis and 6 had MG (Namba et al. 1974).

The association of MG with thymoma or thymic hyperplasia will be considered in the sections on pathogenesis and treatment.

The pathogenesis of MG: clues from EAMG

EAMG has provided sharp insights into the pathogenesis of MG. Although the features of EAMG vary with species and strains within species, the source of AChR, the use of adjuvants, and the immunization schedule, the model has clearly established that sensitization to AChR produces pathogenic autoantibodies and that these cause AChR deficiency at the NMJ. Susceptibility to EAMG is related to the safety margin of neuromuscular transmission in the given species and genetic influences affecting immune responsiveness to AChR (Christadoss et al. 1981). Immunization with AChR can induce acute and then chronic, or only chronic EAMG. Use of pertussis vaccine plus Freund's complete adjuvant favors the induction of acute EAMG (Lennon et al. 1975), but even so the acute phase may not develop (Sahashi et al. 1978). Only chronic EAMG resembles human MG morphologically and electrophysiologically, but conceivably acute EAMG-like events might occur

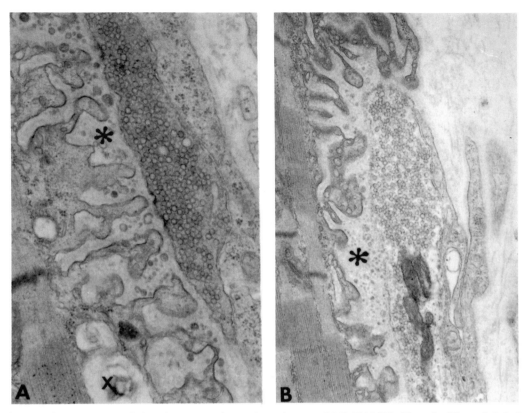

Fig. 3. Early acute (A) and chronic (B) experimental autoimmune MG (EAMG). Note shedding of globular fragments of the junctional folds (Asterisk in A and X in B). The shedding causes loss of AChR and widening of the synaptic space. (B) a small vacuole is present in the junctional sarcoplasm (X). (A) ×27,000, (B) ×18,800. (Reproduced from Engel et al. (1976b) by courtesy of the Editors of *Ann. N.Y. Acad. Sci.*)

at some human NMJ in some patients during the course of the disease (Pascuzzi and Campa 1988).

In the rat, a single immunization with Torpedo electric-organ AChR plus adjuvants induces acute EAMG accompanied by severe weakness after 7–11 days (Lennon et al. 1975; Lindstrom et al. 1976a). The initial ultrastructural change is disintegration of the terminal expansions of the junctional folds (Fig. 3A). Shortly after this, many degenerating postsynaptic regions split away from the underlying muscle fibers and are removed by macrophages (Engel et al. 1976a). This results in functional denervation of the affected muscle fibers; many NMJ become electrically silent and the quantal content of the EPP is reduced (Lambert et al. 1976). Some fibers undergo segmental necrosis which is centered on the NMJ. After day 11, the macrophages leave the NMJ, the nerve terminals

return to highly simplified postsynaptic regions, and the postsynaptic folds gradually regenerate (Engel et al. 1976a). Antibody against rat AChR is first detected in serum 3 days after immunization (Lindstrom et al. 1976b). The antibody titer is relatively low during the acute phase. It increases slowly until day 20 and then very rapidly until it exceeds the body content of AChR many times. The concentration of antibodies directed against electric-organ AChR is always higher than of antibodies directed against rat AChR.

Following the acute phase, animals regain their strength but again become weak with the onset of the chronic phase at about day 30 (Lennon et al. 1975). The junctional folds again degenerate and the postsynaptic regions become simplified (Fig. 3B) (Engel et al. 1976a). The ultrastructural changes now resemble those in human MG (Engel et al. 1976b). During chronic EAMG the muscle AChR content declines and 70% or more of what remains is complexed with antibody (Lindstrom et al. 1976b). AChR is decreased on the junctional folds but is detected on debris shed into the synaptic space by disintegrating folds (Engel et al. 1977b). IgG and C3 deposits are found on the terminal expansions of the remaining junctional folds, over patches of the simplified postsynaptic membrane, and on fragments of the folds shed into the synaptic space (Fig. 4) (Sahashi et al. 1978). Therefore, shedding of AChR-rich fragments of the junctional folds into the synaptic space represents a mechanism of AChR loss in chronic EAMG. That lytic complement components are important in causing AChR deficiency in EAMG is indicated by the fact that C5-deficient mice immunized with AChR produce anti-AChR antibodies but show no clinical signs of EAMG (Christadoss 1988). Accelerated internalization and destruction of AChR cross-linked with antibody (modulation) may also contribute to the AChR deficiency in chronic EAMG (Merlie et al. 1979; Fumagalli et al. 1982a,b). IgG from rats with chronic EAMG induces acute passive-transfer EAMG in recipient rats (Lindstrom et al. 1976a). The weakness begins within 24 hours after the transfer, peaks in 2–3 days, and improves in 5–10 days. As in acute EAMG, the NMJs are invaded by macrophages at the onset of weakness. Immunocytochemical studies reveal IgG, C3, C6, and C9 at the NMJ (Engel et al. 1979; Biesecker and Gomez 1989). Depletion of C3 with cobra venom factor (Lennon et al. 1978) or of C6 by anti-C6 Fab antibody fragments (Biesecker and Gomez 1989) prevents acute passive-transfer EAMG. The inhibition of EAMG by anti-C6 indicates that development of disease in the passive transfer model depends on the assembly of the lytic C5b-9 complement membrane attack complex (MAC) and on the MAC-mediated focal disruption of the postsynaptic membrane (Biesecker and Gomez 1989).

A still unresolved question is why IgG from rats with chronic EAMG should induce acute passive-transfer EAMG. One possible explanation is that the extremely rapid destruction of the junctional folds in EAMG requires a high density of antigen (i.e. AChR) for fixing IgG and complement and

that this condition is obtained only at previously undamaged NMJs (DuPont and Richman 1987).

The pathogenesis of MG: the afferent limb of the immune response

Viral or bacterial infections.   Unlike in EAMG, in human MG the event that triggers sensitization to AChR remains unknown. Clinical onset of MG after different viral infections has suggested that exposure to viral antigens may alter self-tolerance to AChR (Korn and Abramsky 1981). However, direct sensitization to a single epitope shared between a virus and AChR is unlikely because anti-AChR antibodies are polyclonal and different viral infections may precede the onset of MG. Further, in 37 patients with recent onset of MG, the incidence of antiviral antibody titers was not different from that of matched controls (Klavinskis et al. 1985). Attempts to isolate viruses from MG thymus glands have also failed (Klavinskis et al. 1986). Induction of MG by microbial antigens cross-reactive with AChR has been proposed because herpes simplex virus (Schwimbeck et al. 1989), and a number of bacteria, namely *Enterobacter cloacae*, *Serratia liquifaciens* (Dwyer et al. 1986), *Escherichia coli*, *Proteus vulgaris* and *Klebsiella pneumoniae* (Stefannson et al. 1986), share epitopes with AChR. However, the marked polyclonality of the anti-AChR antibodies in MG again suggests that this mechanism of sensitization is unlikely.

HLA association.   The association of different types of MG with specific HLA haplotypes (discussed above) implies that susceptibility to develop MG is genetically controlled. HLA class II molecules, expressed on the surface of antigen-presenting cells, bind peptide fragments of antigens and present them to helper T cells. Recognition of a peptide fragment in the context of the class II molecule by the T cell antigen receptor results in T cell activation, and activated antigen-specific T cells stimulate antibody production by B cells (Abbas et al. 1991). From this, one can infer that disease-associated HLA class II molecules are particularly prone to bind pathogenic peptides that will activate autoreactive T cells.

Penicillamine-induced MG. Penicillamine treatment induces MG in some patients. Further, the fine antigenic specificities of the anti-AChR antibodies are similar to those found in idiopathic MG (Tzartos et al. 1988b). This indicates that tolerance to self-AChR can be reversibly broken in susceptible individuals by a pharmacological agent (Masters et al. 1977). Penicillamine-dependent MG, however, provides no clue as to what triggers the autoimmune response in idiopathic MG. The mechanism of the drug effect is also controversial. In 1 study the drug stimulated the production of anti-AChR antibodies by sensitized lymphocytes in vitro (Fawcett et al. 1982), but this was not confirmed by another study (Scadding et al. 1983).

The role of the thymus gland

A series of observations implicate the thymus gland in the pathogenesis of MG.

1. Thymectomy has a beneficial effect in a high proportion of cases (Genkins et al. 1975).

2. Lymphofollicular hyperplasia of the thymic medulla occurs in 65% and thymoma is found in 15% of MG patients (Viets and Schwab 1960). In the hyperplastic gland germinal centers are surrounded by zones of T cells which are separated by a fenestrated basal lamina from bands of hypertrophic medullary epithelial cells (Bofill et al. 1985).

3. AChR is expressed on thymic myoid cells (Schluep et al. 1987; Kirchner et al. 1988). These cells, found in both normal and MG thymuses, lack HLA class II molecules and cannot present antigen to T cells, but they are intimately associated with class II-positive antigen-presenting interdigitating reticulum cells (Melms et al. 1988). Sparse and incompletely characterized cells found only in MG thymuses express both AChR and class II molecules (Schluep et al. 1987).

4. HLA class II-positive epithelial cells in thymuses and thymomas of MG patients express a 153-kilodalton (kD) protein that shares epitopes with the

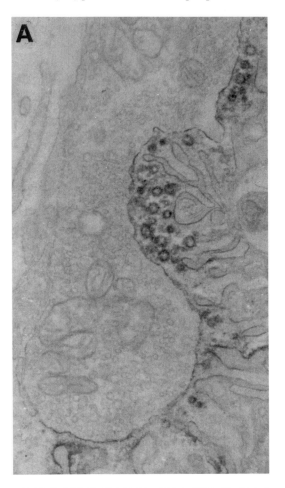

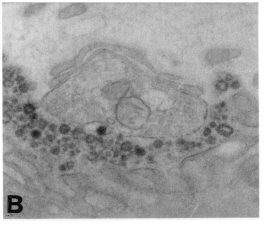

→
Fig. 4. Ultrastructural localization of IgG at the NMJ in chronic EAMG. The reaction product is on globular fragments of the junctional folds that had been shed into the synaptic space and on short segments of the postsynaptic membrane. The junctional folds are shortened and the synaptic space is widened by the destructive changes. (A) ×24,900, (B) ×24,400. (Reproduced from Sahashi et al. (1978) by courtesy of the Editors of the *J. Neuropathol. Exp. Neurol.*)

cytoplasmic domain of AChR (Kirchner et al. 1988; Marx et al. 1990) but has no ion channel properties (Siara et al. 1991).

5. The hyperplastic MG thymus is selectively enriched in AChR-specific T cells (Melms et al. 1988; Sommer et al. 1990).

6. The hyperplastic MG thymus contains an increased number of B cells (Lisak et al. 1976; Durelli et al. 1990), and cultured thymocytes of MG patients secrete anti-AChR antibodies of the same fine specificity as are found in the donors' sera (Scadding et al. 1981; Schluep et al. 1987).

7. Fragments of human MG thymus transplanted into SCID (severe combined immunodeficiency) mice secrete anti-AChR antibodies that bind to the NMJ of the recipients. Therefore, the MG thymus contains all cellular components for producing anti-AChR antibodies and can produce these antibodies in an immunocompromised host (Schönbeck et al. 1992).

On the basis of these observations the following intrathymic pathogenesis of MG has been postulated: (a) induction of AChR expression on selected thymic cells; (b) release of AChR from myoid cells, perhaps due to cell death, and uptake of myoid AChR by thymic antigen-presenting cells, followed by antigen presentation, or direct antigen presentation by HLA class II-positive epithelial cells of AChR epitopes derived from the 153-kD protein; (c) recognition of AChR fragments by specific autoreactive CD4+ T cells differentiating in the thymus; and (d) stimulation of antibody-secreting B cells by AChR-specific T cells in the thymus and, after their emigration from the thymus, also at the periphery (Wekerle and Ketelsen 1977; Wekerle et al. 1981; Melms et al. 1988; Marx et al. 1990).

Despite extensive evidence that the thymus participates in the induction and propagation of the immune response to AChR, the role of the gland in MG is not straightforward. The expression of AChR on thymic myoid cells is not sufficient to induce a pathologic immune response to AChR as such cells are even more abundant in the normal than the MG thymus (Schluep et al. 1987).

The normal thymus participates in inducing self-tolerance by deletion of self-reactive T cell clones or by conferring anergy on such clones. Either of these mechanisms could be abnormal in MG. It is now known, however, that clones of AChR-specific T cells occur not only in MG but also in the normal human immune repertoire (Zhang et al. 1990b; Salvetti et al. 1991; Sommer et al. 1991). *This implies that (a) clonal deletion of AChR-reactive T cells is not inevitable during normal ontogeny and (b) autoimmunity in MG could be due to a failure of a suppressor mechanism.* The inhibition of AChR-specific T cells could involve regulatory T cells or idiotypic–anti-idiotypic networks. Dwyer et al. (1986) and Eng and Lefvert (1988) proposed that a disturbance of idiotypic–anti-idiotypic networks could initiate the pathologic immune response to AChR, but no clear evidence for this has been found (Vincent 1988).

An immune response to AChR could also begin at extrathymic sites. Recent studies of rat Schwann cells in culture indicate that they can express HLA class II antigens, phagocytose AChR-enriched membrane fragments, and present AChR epitopes to AChR-specific T-cells (Zhang et al. 1990a). Thus, antigen presentation in MG could occur in muscle in the vicinity of the NMJ. There is no direct proof, however, that Schwann cells play such a role in MG.

Thymopoietin. An alternative theory of the pathogenesis of MG holds that the primary lesion is thymitis and that the AChR deficiency is a consequence of the release of the thymic hormone, thymopoietin (Venkatasubramanian et al. 1986). This notion is based on a series of observations by Goldstein and Schlesinger (1975). Thymopoietin is a polypeptide that induces the expression of thymic antigens on bone marrow cells in vitro. Goldstein has also suggested that the same polypeptide has a depressant effect on neuromuscular transmission. However, Goldstein's EMG studies could not be confirmed (Jones et al. 1971). Thymopoietin was found to bind with high affinity to the α-bungarotoxin-binding region of AChR (Venkatasubramanian et al. 1986). However, the depression of neuromuscular transmission by thymopoietin is only minimal, develops over several hours, and lasts for days. Rigorous proof that thymopoietin participates in the induction or perpetuation of autoimmune MG, or that it causes AChR deficiency at the NMJ, is lacking.

T and B cell responses to AChR epitopes

AChR is a T cell-dependent antigen (Lennon et al. 1976) and the production of anti-AChR antibodies requires both B cells and T helper cells (Hohlfeld et al. 1986). According to current concepts, the B cell response to a T cell-dependent antigen will include the following steps (Abbas et al. 1991).

1. B cells with distinct specificities recognize multiple, distinct, and conformation-dependent epitopes on the native antigen.

2. The antigen receptors on a given B cell are cross-linked by the native antigen; this stimulates endocytosis of the entire native antigen molecule and initiates B cell activation.

3. The endocytosed native antigen is processed and its degraded peptide fragments, consisting of 8–20 amino acid residues, are externalized on the surface of the B cell in the context of class II HLA molecules.

4. The antigen receptor of an antigen-specific CD4+ T helper cell binds to the class II molecule-peptide antigen on the surface of the B cell to form a trimolecular complex. This leads to activation of the antigen-specific T helper cell.

5. Activated T helper cells stimulate the antigen presenting B cells by lymphokine secretion and by direct physical contact.

6. The stimulated B cells undergo clonal expansion and many of the newly formed B cells differentiate into antibody-secreting plasma cells.

7. The antibody secreted by a given plasma cell clone is specific for the epitope that was originally recognized by the progenitor B cell.

8. Macrophages and some other class II-positive cells can also internalize native protein antigen, externalize peptide fragments of the processed antigen in the context of class II molecules, and present these to CD4+ T helper cells.

From this scheme it follows that (a) the T and B cell epitopes of AChR involved in the immune response in MG are not necessarily identical; (b) a single antigen-specific B cell clone can receive help from multiple antigen-specific T cell clones; and (c) a single antigen-specific T cell clone can provide help to multiple antigen-specific B cell clones.

Sensitization of T cells to AChR in MG was initially revealed by proliferative responses of peripheral blood lymphocytes on exposure to native AChR (Conti-Tronconi et al. 1977). After the amino acid sequence of the AChR subunits became known, synthetic AChR subunit peptides or fusion proteins comprising portions of subunits were tested for their ability to elicit proliferative responses in peripheral blood lymphocytes or T cell lines in MG and EAMG. The results of these studies can be briefly summarized.

1. A high proportion of identified T cell epitopes of AChR consists of sequences appearing in an amphipatic helix, as predicted by Delisi and Berzorfsky (1985), or contain a charged residue or glycine followed by 2 hydrophobic residues, as proposed by Rothbard and Taylor (1988).

2. Human rather than xenogeneic α-subunit sequences are required to detect relevant T cell reactivity in MG (Harcourt et al. 1988; Zhang et al. 1990b; Berrih-Aknin et al. 1991; Sommer et al. 1991).

3. The majority of epitopes stimulating T helper cells are found in the denatured α subunit of AChR (Hohlfeld et al. 1987), but with considerable cross-reactivity from 1 subunit to another (Krolick and Urso 1987; Tami et al. 1987). T cell lines also respond to a number of epitopes on the δ subunit (Protti et al. 1991b) and on the embryonic γ subunit (Protti et al. 1991a). The last finding would be consistent with intrathymic sensitization of AChR.

4. T cell reactivities to the human β and ϵ subunits of AChR have not been systematically investigated.

5. The B and T cell epitopes in a given patient are not identical (Hohlfeld et al. 1987).

6. The T cell response, like the B cell response, is polyclonal (Brocke et al. 1988; Harcourt et al. 1988; Hohlfeld et al. 1988b; Oshima et al. 1990; Zhang et al. 1990b; Berrih-Aknin et al. 1991; Salvetti et al. 1991; Sommer et al. 1991).

7. The same α-subunit sequences that stimulate T cells of MG patients also stimulate T cells of normal controls (Brocke et al. 1988; Harcourt et al. 1988; Zhang et al. 1990b; Berrih-Aknin et al. 1991; Salvetti et al. 1991; Sommer et al. 1991). Therefore, AChR-specific T cells are part of the normal immune repertoire, and the pathological immune response in MG could stem from a defect in immunoregulation.

8. Different T cell epitopes of AChR have been identified in the different studies and no single im-

munodominant T cell epitope has been identified. This may be related to the following circumstances. (a) Patients of different racial background and of different HLA haplotypes were used within or between the different studies (Brocke et al. 1988; Fuji and Lindstrom 1988; Oshima et al. 1990; Zhang et al. 1990b; Berrih-Aknin et al. 1991; Salvetti et al. 1991), but the T cell-epitope profile is affected by the HLA haplotype, and the T cell epitope-HLA haplotype relationship varies with racial background. (b) Only a few studies have examined the response to peptides that cover the entire extent or even the entire extracellular surface of an AChR subunit (Pachner et al. 1989; Oshima et al. 1990; Protti et al. 1990, 1991a,b). (c) Only a few patients or only a few subunit peptides were investigated in some studies.

9. Hypothetically, susceptibility to MG could also be influenced by genes encoding the T cell receptor (TCR): inherited polymorphisms or special somatic rearrangements in TCR genes could influence the ability of the TCR to recognize AChR fragments in the context of class II molecules during antigen presentation. However, there is as yet no evidence for inherited polymorphisms or a specific pattern of TCR variable gene usage in MG.

The pathogenesis of MG: the efferent limb of the immune response

The passive transfer of MG from man to mouse with IgG (Toyka et al. 1977), the beneficial effects of plasmapheresis (Pinching et al. 1976; Dau et al. 1977), and the deposition of IgG and complement at the NMJ in human MG (Engel et al. 1977b) indicate that the efferent limb of the immune response is humorally mediated.

The properties of the AChR antibodies
Competition studies employing monoclonal anti-AChR antibodies versus MG sera have established the polyclonality of anti-AChR antibodies and have revealed antibody specificities in MG patients. Anti-AChR antibodies can bind to a variety of sites on AChR (Tzartos and Lindstrom 1980), but most bind to epitopes in a circumscribed extracellular portion of the α subunit. This site, designated as the main immunogenic region (MIR), maps to residues 67–76 on the α subunit (Tzartos

et al. 1988a) and is distinct from the cholinergic binding site. Other MG antibodies are directed against sites on other AChR subunits, but only a few bind to sites on the α subunit outside the MIR (Tzartos et al. 1982, 1991). The antibodies against the MIR are pathologically significant because they account for most of the modulatory effects on AChR by MG sera (Conti-Tronconi et al. 1981b; Tzartos et al. 1985), are able to fix complement, and can passively transfer MG (Tzartos and Lindstrom 1980). The pattern of antibody specificities revealed by these studies does not correlate with the clinical state, duration of disease, or previous therapy (Tzartos et al. 1982).

Whiting et al. (1986), using 9 monoclonal anti-human AChR antibodies in competition with 36 different MG sera, found that the MG sera recognized 5 overlapping regions on AChR. This study disclosed some differences in antibody specificities in 3 forms of generalized MG (no thymoma, onset <40 years; no thymoma, onset >40 years; thymoma, all age groups). However, the residues on AChR recognized by the monoclonal antibodies were not identified, and it is not clear whether any regions recognized in these experiments were identical with the MIR.

Other AChR antibody characteristics. AChR antibodies are predominantly of the IgG1 and IgG2 subclasses and are relatively enriched in the κ-light-chain moiety (Vincent and Newsom-Davis 1982a). In the 3 forms of generalized MG, no differences in AChR antibodies were observed as regards the distribution of IgG light chains and subclasses, inhibition of α-bungarotoxin binding, and reactivity with human AChR from normal limb, denervated limb, or ocular muscle. By contrast, patients with ocular MG had a significantly greater proportion of κ-light-chain and IgG3-subclass antibodies, and their antibodies reacted somewhat better with AChR from ocular and normal than from denervated muscle. How these characteristics contribute to the clinical state is unclear, especially as ocular MG sera did not react better with AChR from ocular than from normal limb muscle (Vincent and Newsom-Davis 1982a). Some antibody characteristics may change with the duration of the disease. Thus, the percentage of κ-light-chain antibodies and anti-AChR avidity are lower in MG of recent onset than in generalized MG of more than 1 year's duration (Vincent and Newsom-Davis 1982b).

Pathogenic effects of the AChR antibodies

Two lines of evidence indicate the presence of AChR antibodies at the NMJ in MG. (a) A proportion of AChR isolated from myasthenic muscle is complexed with IgG (Lindstrom and Lambert 1978). (b) IgG is localized on the postsynaptic membrane of the NMJ of MG by immunoelectron microscopy (Fig. 5) (Engel et al. 1977b).

AChR antibodies could impair neuromuscular transmission in several ways: (a) by affecting the properties of the AChR ion channel; (b) by blocking the binding of ACh to AChR; (c) by causing complement-mediated destruction of the junctional folds; and (d) by accelerating the internalization and degradation of AChR (modulation of AChR by antibody). Only the last 2 mechanisms would be expected to decrease the density of AChR molecules at the NMJ.

Ion channel effects. An effect by AChR antibodies on AChR ion channel kinetics is unlikely. Measurements using noise analysis of AChR ion channel properties in MG (Cull-Candy et al. 1979), EAMG (Hohlfeld et al. 1981), and in-vitro systems (Dolly et al. 1988) have yielded normal results.

Blocking antibodies. Antibodies blocking the binding of α-bungarotoxin to AChR have been detected in 10–88% of MG sera (Mittag et al. 1981; Drachman et al. 1982; Pachner 1989). Although in most cases blocking antibodies account for only a minor fraction (Tzartos et al. 1982) or less than 1% (Vincent et al. 1983) of all AChR antibodies, their effect on neuromuscular transmission could be significant in some patients (Drachman et al. 1982; Gomez and Richman 1983; Pachner 1989). However, it is important to note that not all anti-AChR antibodies that block α-bungarotoxin binding in vitro affect neuromuscular transmission. For example, only 2 out of 10 antibodies that blocked toxin binding to AChR reduced the MEPP amplitude, and the effect was mild. Conversely, some anti-AChR antibodies can diminish the MEPP amplitude but do not block toxin binding (Dolly et al. 1988). When applied in vitro, MG sera typically

do not alter AChR function acutely (Green et al. 1975; Albuquerque et al. 1976) suggesting an absence of functionally significant blocking antibodies. However, in 1 exceptional case antibodies that recognized mainly the embryonic form of AChR did depress the MEPP amplitude at the human NMJ in vitro (Burges et al. 1990). In most cases of MG, the amplitude of the MEPP in MG correlates closely with the amount of AChR remaining at the NMJ (Engel et al. 1977c; Ito et al. 1978; Lindstrom and Lambert 1978). This again suggests that in most cases blocking antibodies do not significantly decrease the MEPP amplitude.

Complement effects. The pathogenesis of MG, like that of EAMG (discussed above), involves complement. Human anti-AChR antibodies readily fix complement in vitro and lyse myotubes in culture if complement is made available to the system (Ashizawa and Appel 1985; Childs et al. 1987). Complement fixation (Engel et al. 1977b), activation of the lytic phase of the complement reaction sequence (Sahashi et al. 1980) and MAC deposition (Fazekas et al. 1986; Engel and Arahata 1987) occur consistently at the myasthenic NMJ. The lytic complement components appear on segments of the postsynaptic membrane and on vesicular structures shed into the synaptic space (Fig. 6). The shedding of MAC-positive vesicles into the extracellular space by cells under complement attack is a general phenomenon for it also occurs when polymorphonuclear leukocytes (Campbell and Morgan 1985) and oligodendrocytes (Scolding et al. 1989) bind MAC. The shedding phenomenon may protect the cell from the influx of extracellular fluid and calcium through the MAC pores. At the MG NMJ, however, the continuous shedding of the terminal portions of the junctional folds causes AChR deficiency and progressive destruction of the folds (Fig. 6) (Sahashi et al. 1980; Engel et al. 1981). The same phenomenon disrupts the complex architecture of the folds, reduces the membrane surface available for the insertion of new AChR (Engel and Fumagalli 1982), and could even affect the metabolic stability of the junctional AChR (Fambrough et al. 1984). Destruction of the junctional folds occurs frequently in MG: an electron microscopic study of 255 NMJs in 42 cases of MG demonstrated such changes in 87% of the

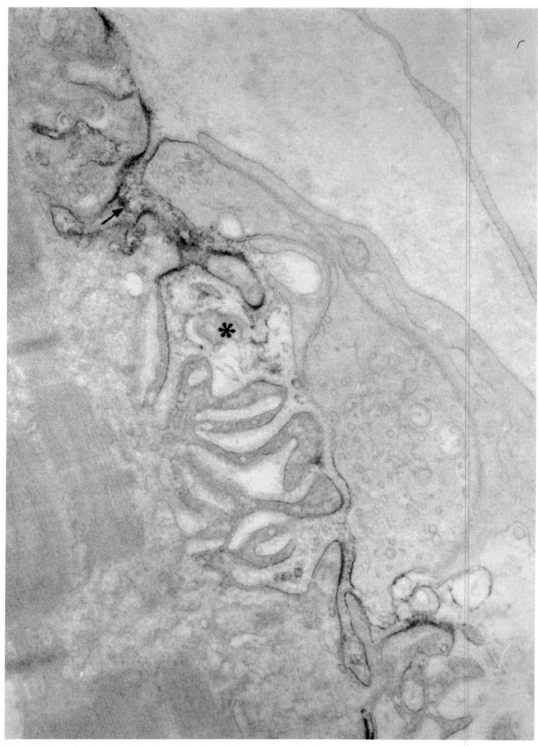

Fig. 5. Ultrastructural localization of IgG at a human myasthenic NMJ. IgG deposits appear on short segments of some junctional folds and on degenerate material in the synaptic space (arrow). Asterisk indicates degenerate material not reacting with IgG. Reciprocal staining of presynaptic membrane is due to diffusion artifact. ×33,900. (Reproduced from Engel et al. (1977b) by courtesy of the Editors of the *Mayo Clin. Proc.*)

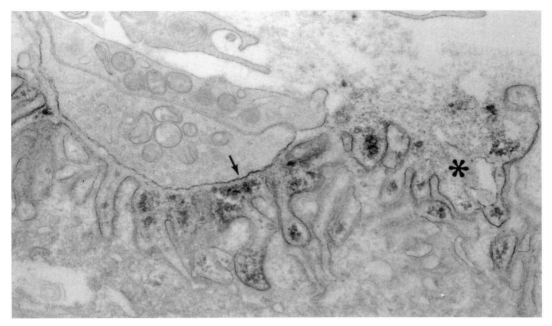

Fig. 6. Ultrastructural localization of the lytic C9 complement component at the myasthenic NMJ. C9 is detected on globular fragments of the junctional folds shed into the synaptic space and on short segments of the postsynaptic membrane. On the right, short junctional folds are not covered by the nerve terminal (asterisk). Presynaptic staining (arrow) is due to artifact. ×25,200. (Reproduced from Sahashi et al. (1980) by courtesy of the Editors of the *J. Neuropathol. Exp. Neurol.*)

junctions (Engel et al. 1981). Consistent with complement activation in MG, significantly elevated serum levels of SC5b-9 (MAC complexed to its serum inhibitor) were observed in 15 out of 16 cases of MG (Brey et al. 1991).

Another possible effect of complement fixation at the MG NMJ could be opsonization of the junctional folds for destruction by macrophages, as occurs in acute and in acute passive-transfer EAMG. Using anconeus muscle specimens from 8 patients who had MG for 1 month to 11 years, Maselli et al. (1991) observed inflammatory cells at the NMJ by light microscopy in 3 and by electron microscopy in 4 patients. The duration of the disease was not related to the presence of the inflammatory cells at the NMJ. However, cells at the NMJ cannot be identified as 'inflammatory' by light microscopy without the use of special markers; and none of the cells thought to be macrophages on electron microscopy showed signs of phagocytic activity. By contrast, electron microscopy analysis of 255 NMJs in intercostal muscle specimens of 42 MG patients revealed macrophage invasion of the

NMJ in only 1 patient who also had thymoma (Engel et al. 1981).

Antigenic modulation. Cross-linking of NMJ AChR by antibody accelerates AChR internalization and degradation by the mechanism of antigenic modulation (Heinemann et al. 1978; Stanley and Drachman 1978; Fumagalli et al. 1982b). AChR antibodies also increase the degradation of extrajunctional myotube AChR in vitro. Two studies correlated the modulation of myotube AChR with the clinical grade of MG and the antibody titer (Conti-Tronconi et al. 1981a; Drachman et al. 1982); subsequent studies, however, confirmed only the correlation with the total antibody titer (Appel et al. 1979; Hudgson et al. 1982; Elias and Appel 1983; Tzartos et al. 1986; Eymard et al. 1988). Since the severity of MG is not directly related to the antibody titer, the severity of MG may not be related to the presence of antibodies that cause modulation of AChR (Tzartos et al. 1986).

None of the above studies attempted to correlate clinical severity with destructive changes at the neuromuscular junction induced by complement. However, in a study of chronic EAMG in the rat, severe disease was consistently associated with severe structural injury to the junctional folds (Engel and Fumagalli 1982). Further, when AChR degradation rate was monitored by an external gamma counter in individual animals in vivo, no correlation was observed between the increase in the rate of AChR degradation rate and the severity of EAMG (Fumagalli et al. 1982a). In a study of EAMG in the mouse, the modulation of AChR was insufficient to account for the development of MG in the immunized mice (Berman and Heinemann 1984).

Neither complement attack nor accelerated internalization of AChR would cause AChR deficiency if the AChR loss were balanced by increased synthesis and increased membrane insertion of AChR. For example, marked acceleration of AChR internalization at the denervated NMJ is compensated by rapid AChR replacement (Levitt and Salpeter 1981). On the other hand, a marked increase in AChR synthesis, suggested by a 4- to 7-fold specific increase of α-subunit mRNA, does not prevent AChR deficiency in rats with EAMG (Asher et al. 1988).

It is likely that modulation and complement attack operate jointly in decreasing NMJ AChR. The complement attack has a dual effect: (a) it causes shedding of AChR into the synaptic space; and (b) by destroying segments of the junctional folds, it restricts the membrane surface available for the insertion of new AChR. Modulation of AChR also has a dual effect: (a) it accelerates the internalization of AChR; and (b) it helps the NMJ evade the complement attack by decreasing the density of AChR to which IgG and complement can bind, a paradigm also noted in other systems (Glennie and Stevenson 1982).

Diagnosis of MG

The diagnosis of MG is based on clinical history, physical findings, pharmacological tests, and special laboratory studies. The latter include EMG investigations (conventional needle EMG, study of the decremental response, and single fiber recordings), determination of the serum anti-AChR antibody titer, in-vitro microelectrode studies of neu-romuscular transmission (to measure the amplitude of the MEPP, the quantal content of the EPP and other parameters of transmission), and ultrastructural studies of the NMJ.

In patients who give a typical history of acquired weakness and fatigability increased by exertion and have generalized disease which also involves the external ocular muscles, a positive anticholinesterase drug test and a decremental EMG response are usually sufficient to confirm the diagnosis. However, there are patients with congenital symptoms, with no involvement of the external ocular muscles, with symptoms confined to the external ocular muscles which do not clearly fluctuate, or with atypical associated neurologic findings. Other patients give a history compatible with MG but show no weakness on examination. Such patients may or may not have a decremental EMG response or a positive anticholinesterase drug response and may or may not have MG. In these patients, determination of the anti-AChR antibody titer can be especially helpful. If this is negative, then more specialized electrophysiologic and morphologic studies are required to confirm or exclude the diagnosis of MG.

Anticholinesterase drug effects and tests

The decreased AChR density at the MG NMJ requires an increase in the diameter of the disk saturated with ACh for producing a normal quantal response. However, since AChE is not lost, an increased fraction of the ACh quantal packet binds to AChE before it can reach AChR, and the quantal response and the safety margin of neuromuscular transmission are reduced. AChE inhibitors are effective because they increase the number of ACh molecules available to bind to AChR and because they can now spread over a larger postsynaptic disk area to compensate for the reduced AChR density. However, an overdose of AChE inhibitors would cause overlapping of the saturated disks during physiological activity, and hence desensitization of AChR.

Edrophonium (Tensilon), an AChE inhibitor, acts within a few seconds and its effect lasts for a few minutes (Osserman and Teng 1956). For the test, 0.1–0.2 ml of a 10 mg/ml solution of the drug is injected intravenously over 15 seconds. If there is no response in 30 seconds, 0.8–0.9 ml of the drug is injected. Evaluation of the response requires objective assessment of changes in one or more clinical signs, such as the degree of ptosis, the range of ocular ductions, the force of the hand grip, etc. A placebo injection before the drug is given enhances the validity of the test. For infants, 0.05–0.1 ml of the drug is administered subcutaneously; for children, weighing up to 34 kg, the dose is 0.1 ml intravenously or 0.2 ml intramuscularly.

Edrophonium has also been used to distinguish between myasthenic weakness and weakness caused by overmedication with anticholinesterase drugs. In the latter instance, 0.1 ml of the drug increases rather than decreases weakness and may provoke fasciculations, flushing, tearing, abdominal cramps, nausea, vomiting, and diarrhea (Osserman and Teng 1956). However, this application of the test is limited by the fact that some muscles, or groups of muscles, may be overmedicated while others are still undermedicated.

Neostigmine methylsulfate, 0.5–1 mg administered subcutaneously acts up to 2 hours and allows a more detailed evaluation of changes in strength (Flacke 1973) or in the EMG decrement.

Anticholinesterase drug tests can also be positive in rare patients with both MG and the Lambert-Eaton syndrome (Oh and Cho 1990), as well as some congenital myasthenic syndromes (see below), and neurogenic atrophies e.g., amyotrophic lateral sclerosis, poliomyelitis, and peripheral neuropathies (Hamill and Walker 1935; Hodes 1948; Mulder et al. 1959). The positive tests in the neurogenic atrophies are attributed to a reduced quantal content of the EPP at reinnervated NMJs (Bennett and Florin 1974).

Provocative tests

Curare decreases the number of postsynaptic AChRs available to bind ACh and, therefore, the number of AChR ion channel openings that can contribute to the MEPP. Consequently, the amplitude of the MEPP and the EPP, and hence the safety margin of neuromuscular transmission, are decreased. Curare sensitivity is increased in MG because the number of AChRs available to bind ACh is already reduced. Increased curare sensitivity also occurs in other diseases in which the safety margin of neuromuscular transmission is reduced. Intravenously administered curare (1/20–1/50 of

the normal curarizing dose) has been used in the past to establish the diagnosis of MG in patients who gave a history compatible with MG but were not weak and did not have a decremental EMG response. The test is contraindicated in weak patients and in anyone with previous respiratory embarrassment (Osserman 1958). In this author's opinion, the original test should no longer be used.

Electromyographic studies

The number of quanta m released from the nerve terminal by nerve impulse is regulated by the processes of presynaptic depression and facilitation which, in turn, are modulated by the number and frequency of impulses (Martin 1977; Magleby 1979). Normally, the decreases in m are not sufficient to render the EPP subthreshold for triggering the muscle fiber action potential. In MG, however, the EPP amplitude is already reduced by the AChR deficiency. This plus the normally occurring decrease in m depress the EPP below the level required to activate the muscle fiber action potential. All clinical EMG phenomena in MG can be explained on this basis.

Needle electromyography. The fluctuations in the amplitude and duration of the motor unit potentials during voluntary activity (Lindsley 1935) are due to the occurrence of subthreshold EPPs and failure of transmission at a varying proportion of end-plates within individual motor units.

The decremental EMG response. At low frequencies of stimulation presynaptic depression transiently predominates over presynaptic facilitation and m decreases with the first 5–10 stimuli. Supramaximal stimuli are applied to a motor nerve and the evoked compound muscle action potential is recorded from the surface of the stimulated muscle. The decreases in m are not identical at all end-plates but, as m becomes subthreshold at an increasing number of end-plates, the amplitude of the compound muscle action potential declines.

For a few seconds after a 15–30-second period of maximal voluntary contraction, or after a short period of indirect tetanic stimulation of a muscle, facilitation predominates over depression, the safety margin of transmission is increased, and the decremental response is decreased. A few minutes later, a long-lasting process of post-tetanic depression predominates and the decremental response is enhanced. The usefulness of postexercise facilitation and depression in clinical EMG tests is limited because the duration of these phenomena varies from patient to patient (Özdemir and Young 1976).

The test for the decremental EMG response is most sensitive and reliable at 2–3-Hz stimulation. At this frequency, the maximal decrement usually occurs with the fifth evoked response; after this, the amplitude of the evoked action potential may either plateau or increase. In normal subjects, the fifth evoked response may decrease over the first, by not more than 7%. The test is not positive in all muscles of all patients (Desmedt and Borenstein 1970, 1976; Horowitz et al. 1976a; Özdemir and Young 1976). However, in 1 series 95% of the patients had a positive response in at least 1 muscle when 2 or more distal and 2 or more proximal muscles were examined; test on a single muscle was positive in only 50% of the patients (Özdemir and Young 1976). The sensitivity of the test can be increased by heating the limb (Desmedt and Borenstein 1976) or by conducting the test under ischemic conditions (Desmedt and Borenstein 1977; Vial et al. 1991).

Single-fiber electromyography (SFEMG). In SFEMG the action potentials generated by closely adjacent muscle fibers belonging to the same motor unit are recorded with a fine intramuscular electrode (Stalberg et al. 1976). When a motor unit is activated, the action potentials which invade the muscle fibers are not entirely synchronous. The mean interpotential interval between 2 fibers (or jitter) is normally less than 55 µs. In MG this interval often exceeds 100 µs and a certain proportion of the impulses fail to generate an action potential at 1 of the 2 fibers. The longer the interpotential interval, the more frequently does impulse blocking occur (Stalberg et al. 1976). The abnormal jitter in MG is related to the fact that the low-amplitude EPPs require an increased time to reach the threshold for activating the muscle fiber action potential. A progressive increase of the interpotential interval indicates a progressive decrease of the EPP amplitude in at least 1 fiber in the fiber pair, and if 1 of the 2 EPPs becomes subthreshold, 1 of the 2 action potentials disappears. SFEMG has also re-

vealed that in MG normal to highly abnormal endplates can exist within the same motor unit.

SFEMG can be combined with axonal microsimulation at various rates. As the stimulation frequency is increased from 0.5 to 20 Hz, the mean blocking frequency decreases slightly in MG and markedly in LEMS (Trontelj and Stalberg 1990).

In-vitro microelectrode studies of neuromuscular transmission
These studies require a strip of muscle intact from origin to insertion and specialized recording equipment. In MG the amplitude of the MEPP is abnormally small in rested muscle, but the number of quanta released by nerve impulse at 1-Hz stimulation is normal (Elmqvist et al. 1964). A low-amplitude MEPP in rested muscle also can occur in congenital myasthenic syndromes in which there is either AChR deficiency or an abnormal interaction of ACh with AChR (see section on congenital myasthenic syndromes). In autoimmune MG the reduction of the MEPP amplitude correlates linearly with electron microscopic (Engel et al. 1977b) and radiochemical (Ito et al. 1978; Lindstrom and Lambert 1978) estimates of AChR at the NMJ.

The AChR antibody test
The usefulness of the AChR antibody test has been repeatedly confirmed. The percentage of positive results is influenced by the source and method of preparation of AChR, the method of assay, the upper limit set for the titer in the control population, and the proportion of patients in remission or with ocular myasthenia included in the series. When human AChR labelled with ^{125}I-BGT and protected from proteolysis is used as the antigen in the Lindstrom immunoprecipitation assay (Lindstrom et al. 1976d), the test is positive in 84% of MG patients (pooled results on 1524 patients in 7 studies) (Tindall 1981; Garlepp et al. 1982; Raimond et al. 1982; Limburg et al. 1983; Vincent and Newsom-Davis 1985; Toyka and Heininger 1986; Howard et al. 1987). Similar results are obtained when the test utilizes AChR extracted from the human rhabdomyosarcoma cell line TE671 (Voltz et al. 1991). The percentage of positive tests may be even higher when both junctional extrajunctional AChR are used as antigens (Lefvert 1982; Vincent and Newsom-Davis 1985).

In 1 large and representative series the percentage of positive tests in different clinical forms of MG was as follows: remission 24%; ocular 50%; mild generalized 80%; moderately severe or acutely severe 100%; and chronic severe 89% (Tindall 1981). In ocular MG the antibody titer tends to be low. In generalized MG it tends to be high in the presence of thymoma; in the absence of thymoma, it tends to be intermediate or low depending on whether the onset is before or after the age of 40 (Compston et al. 1980; Limburg et al. 1983). Positive AChR antibody tests have been found in 5% of patients with LEMS and in a small proportion of patients with amyotrophic lateral sclerosis (Howard et al. 1987).

The serum antibody titer correlates poorly with the severity of MG when a group of patients is studied. In individual patients, a decrease of antibody titer of more than 50% sustained for more than 12 months is nearly always associated with sustained clinical improvement (Oosterhuis 1981; Seybold and Lindstrom 1981; Limburg et al. 1983). Changes in titer correlate strongly with long-term improvement induced by prednisone, azathioprine, or thymectomy, but not with response to anticholinesterase drugs or worsening due to infection or emotional stress (Oosterhuis et al. 1983; Kuks et al. 1991). The transient decrease in titer induced by plasmapheresis is associated with transient improvement (Newsom-Davis et al. 1979; Olarte et al. 1981).

Antibodies that block α-bungarotoxin binding to AChR are detected in 52% of the patients but only 1% of MG sera are positive for this type of antibody only (Howard et al. 1987). Antibodies that reduce the density of AChR on cultured human myotubes (modulating antibodies) are detected in 90% of MG sera. This test is useful when the usual immunoprecipitation assay gives negative results. Tests for blocking and modulating antibodies can be false positive with sera obtained from patients recently exposed to muscle relaxant drugs (Howard et al. 1987).

Striational and other antibodies. Striated muscle antibodies had been found in MG long before the existence of AChR antibodies was realized. The striational antibodies are heterogeneous: they can bind to either the A- or the I-band (Penn et al.

1986), and may recognize myosin, actin, α-actinin (Williams and Lennon 1986; Ohta et al. 1990), titin (Aarli et al. 1990), filamin, vinculin, and tropomyosin (Yamamoto et al. 1987).

The role of the striational antibodies in MG remains unknown, but their association with thymoma is clinically relevant (Limburg et al. 1983; Penn et al. 1986; Cikes et al. 1988). In a large series, striated muscle antibodies were present in 84% of patients with thymoma. In those without thymoma, they were found in 5% or 47%, depending on whether the age of onset was before or after the age of 40 (Limburg et al. 1983).

Antibodies to human neuroblastoma cells (Müller 1989) and to thymic epithelial cells (Safar et al. 1991) are detected in about 40% of MG sera. Patients with these antibodies usually also have thymic abnormalities.

Immunocytochemical studies

IgG and complement components C3, C9, and MAC can be localized at the MG NMJ in cryostat sections or resin sections of motor point muscle biopsy specimens (Engel et al. 1977b; Sahashi et al. 1980; Engel and Arahata 1987; Tsujihata et al. 1989). The immune complexes are present even when circulating antibodies are absent, and are also present at limb-muscle NMJs of patients who have only ocular signs of MG. C3 localization is technically the easiest and best suited for confirming the suspected diagnosis. Immune complexes have not been observed at the NMJ in other diseases, but it is probable that trace amounts of IgG bind to the presynaptic membrane in the Lambert-Eaton syndrome.

Differential diagnosis

The differential diagnosis of MG includes neurasthenia, oculopharyngeal muscular dystrophy, the mitochondrial myopathies with or without progressive external ophthalmoplegia (PEO), compressive lesions affecting cranial nerves, the Lambert-Eaton myasthenic syndrome, congenital myasthenic syndromes, and botulism.

Neurasthenia can be recognized by the lack of objective findings, by giving way on muscle testing on examination, and by the negative pharmacologic, EMG, and antibody tests. In oculopharyngeal muscular dystrophy the weakness does not fluctuate and diplopia is usually not a symptom. In the mitochondrial myopathies with or without PEO, fatigable weakness is often a symptom. PEO, if present, seldom causes diplopia; lactacidemia or multisystem features may be present and point to the correct diagnosis; the muscle biopsy may show ragged-red fibers, cytochrome c oxidase-negative fibers or both; and pharmacologic, EMG, and antibody tests are negative for MG. The clinical, electrophysiologic, and pharmacologic features of some of the myasthenic syndromes may resemble those of MG and will be discussed later in this Chapter. Compressive lesions affecting cranial nerves should be suspected if muscles of only 1 cranial nerve are involved, or if severe dysarthria, dysphagia, or both are present without other symptoms. Imaging studies usually establish the correct diagnosis (Moorthy et al. 1989; Straube and Witt 1990).

Botulism. Botulinum toxin selectively binds to the presynaptic membrane and is then translocated into the nerve terminal (Dolly et al. 1984). The type A and E botulinum toxins block the exocytotic release of ACh quanta by rendering a process required for release refractory to calcium ions. Effects of these toxins are antagonized by guanidine or the aminopyridines which enhance calcium entry into the nerve terminal, but even these medications do not shorten the course of the disease. The B, D and F botulinum toxins act similarly but also block a mechanism needed for the synchronization of evoked quantal release. The disease caused by these toxins responds poorly to guanidine or aminopyridines, as reviewed by Thesleff (1989).

Food botulism in adults follows the ingestion of food which had originally contained *Clostridium botulinum* spores that had germinated into bacilli that released toxin into the food. Food botulism in infants is caused by the ingestion of food contaminated with viable microorganisms that produce toxin in the intestine. The contaminated food usually remains unidentified (Arnon et al. 1977). Wound botulism occurs after open injuries, such as compound fractures and bullet or puncture wounds, that allow the growth of the anaerobic bacilli in devitalized tissue (Merson and Dowell 1973).

In infants, the syndrome is associated with constipation, hypotonia, multiple cranial nerve palsies, descending muscle weakness, and sudden apnea or progressive respiratory depression (Arnon et al. 1977). In adults, there may be a prodrome associated with nausea, vomiting, diarrhea and abdominal cramps, and a dry, painful throat. Blurred vision, loss of accommodation, and dilated pupils can occur at this stage. The external ocular and other cranial muscles are affected early and this can progress to generalized weakness that also involves the respiratory muscles. The tendon reflexes may or may not be preserved and there may or may not be a slight response to anticholinesterase drugs.

Clinical EMG studies show a decremental response at low (2–3-Hz), and an incremental response at high (40–50-Hz) frequencies of stimulation. However, the EMG abnormalities may not be present early in the disease, or can occur in 1 but not in another extremity (Cherington 1974). The diagnosis is confirmed by detection and identification of the toxin in the patient's serum and/or feces.

Therapy

Anticholinesterase drugs, alternate-day prednisone therapy, immunosuppressants other than prednisone, and plasmapheresis remain acceptable forms of therapy of MG. There is general agreement on four principles of therapy. (1) Anticholinesterase drugs are useful in all clinical forms of the disease. (2) These drugs are the mainstay of therapy in ocular MG (Oosterhuis 1982). (3) Plasmapheresis has only transient effects and does not confer greater long-term protection than immunosuppressants alone. Therefore its use should be limited to patients with severe or otherwise intractable disease (Newsom-Davis et al. 1979; Keesey et al. 1981; Olarte et al. 1981). (4) The presence of thymoma is an absolute indication for thymectomy. However, there are no universally accepted criteria (except for thymectomy in patients with thymoma) for the timing, sequence, or combination of thymectomy, prednisone, and other immunosuppressants in the management of different grades of generalized MG. Adequately controlled, large-scale, long-term, prospective clinical trials to establish such criteria have never been attempted. On the other hand, the relative risks and benefits of the various modalities of treatment are reasonably well-defined and allow rational management of most patients.

Anticholinesterase drugs
Pyridostigmine bromide (Mestinon) is now generally preferred to the shorter-acting neostigmine bromide (Prostigmine). Pyridostigmine is available in 60-mg tablets, in 180-mg 'timespan' tablets, and as a syrup. The drug acts within 60 minutes and its effects last 3–6 hours. It is less likely to cause muscarinic side effects and may be more effective in controlling bulbar weakness than neostigmine (Osserman 1955, 1967). The dose may range from 30 mg taken every 6 hours to 240 mg taken every 3–4 hours. The timespan preparation is used at bedtime for its effects last through the night. The syrup is useful in children and in patients requiring nasogastric feeding. Pyridostigmine bromide is also available in an intramuscularly injectable form in 2-ml ampules. Each ml contains 5 mg of the drug. Approximately 1/30th of the oral dose of the drug may be given parenterally for an equivalent effect. Intramuscular administration of the drug is useful postoperatively or when patients are unable to take medications orally.

Neostigmine bromide is available in 15-mg tablets and 1 tablet is equivalent to a 60 mg tablet of pyridostigmine bromide. Orally administered neostigmine acts within 30 minutes and its effects last for 2–3 hours. The dose may range from 7.5 mg taken every 4 hours to 60 mg taken every 2–3 hours. Higher doses of the drug cause muscarinic side effects, and especially abdominal cramps and diarrhea, which may require 0.5–1 mg atropine every 3–4 hours. Neostigmine methylsulfate is administered subcutaneously or intramuscularly and 1 mg of the parenterally administered drug is equivalent to 15 mg of the oral medication. Parenteral therapy is useful postoperatively (as after thymectomy) and when medications cannot be given orally.

Anticholinesterase drugs increase the radius of the postsynaptic AChR disk reached by ACh molecules in a quantum and allow single ACh molecules to bind more than once to AChR. However, overexposure of the remaining AChR to ACh can desensitize it, producing increasing weakness and, ultimately, a cholinergic crisis. Further, there is experimental evidence that chronic treatment with high doses of prostigmine causes degeneration of the junctional folds, a decrease in postsynaptic AChR, and a reduced MEPP amplitude (Engel et al. 1973; Chang et al. 1975). These changes are similar to those caused by MG itself. Because of their side effects, anticholinesterase drugs are used to decrease rather than to eliminate the symptoms of MG. Patients indoctrinated with this principle can regulate their own dosage on a demand basis. Patients requiring more than modest doses for partial relief of their symptoms are candidates for other modalities of treatment.

The management of crises. Progressive weakness when increasing amounts of anticholinesterase drugs are administered can be an early sign of a myasthenic or cholinergic crisis. A cholinergic crisis is usually accompanied by severe muscarinic side effects. However, cholinergic and myasthenic failure of transmission can exist simultaneously in different muscles, or even in different fibers of the same muscle. Therefore, critically ill patients not responding to relatively high doses of anticholinesterase drugs are best treated by withdrawal of all anticholinesterase drugs, ventilatory support, and intravenous feeding. Refractoriness to drug therapy usually disappears after a few days on this regimen.

Corticosteroid therapy
Alternate-day prednisone therapy is now considered to be a relatively safe and reliable form of treatment (W.K. Engel et al. 1974; Seybold and Drachman 1974; Brunner et al. 1976; Mann et al. 1976). It is indicated for disabling disease not responding adequately to moderate or large doses of anticholinesterase drugs. Ocular MG refractory to these drugs may respond well to alternate-day prednisone therapy (Fischer and Schwartzmann 1976), but prednisone should not be used when a lid-crutch or an eye-patch may also help. More compelling reasons to use alternate-day prednisone are to improve the condition of severely ill patients before thymectomy (Mann et al. 1976) or in the long-term management of patients after surgery for invasive thymoma (Goldman et al. 1975).

The medication is more effective in patients with disease onset after the age of 40 than in younger patients (Sghirlanzoni et al. 1984).

Various treatment regimens have been advocated, but no consensus has emerged on optimal therapy for patients with different grades of MG. One regimen advocates the use 100-mg prednisone on alternate days, and the dose is tapered after a period of 2–8 months, after improvement has occurred (W.K. Engel et al. 1974). Another regimen increases the dose gradually to reduce the risk of transient exacerbation (Miller et al. 1986) and to minimize the steroid side effects. An initial alternate-day dose of 25 mg is raised in 12.5 mg steps every 6 days until either 100 mg or maximum benefit is reached (Seybold and Drachman 1974). Complete remission attributed to alternate-day prednisone therapy has been reported in 37–69% and significant improvement in 20–37% of patients (Seybold and Drachman 1974, Mann et al. 1976; Cosi et al. 1991). In 1 study the average time for significant improvement was 4.9 months and the average dosage was 68 mg on alternate days (Tindall 1980). Significant clinical improvement induced by prednisone correlates with a decrease of antibody titer (Tindall 1980; Vincent and Newsom-Davis 1980; Seybold and Lindstrom 1981; Oosterhuis et al. 1983). The time required for depression of the antibody titer varies, but the most striking decreases are usually the most rapid and occur in less than 3 months (Seybold and Lindstrom 1981).

A total of 27–62% of those patients receiving long-term alternate-day prednisone therapy develop significant side effects (Brunner et al. 1976; Sghirlanzoni et al. 1984; Donaldson et al. 1990). These include gastrointestinal hemorrhage, aseptic necrosis of the femoral head, vertebral collapse secondary to osteoporosis, and posterior subcapsular cataracts. Cataracts, infection, and bone changes are especially frequent in patients older than 50. When the dose of the drug is reduced, about one-fifth of the patients develop short-lasting withdrawal symptoms consisting of arthralgias, myalgias, fever, and sterile joint effusions (Brunner et al. 1976). There is general agreement that patients receiving long-term prednisone therapy should be on a low-salt, high-protein diet, take antacids and supplementary potassium and calcium salts, and receive prophylactic antitubercu-

losis therapy if they have a positive tuberculin skin test.

Intravenous 'pulse' methylprednisolone (2 g per day at 5-day intervals ×3) has produced satisfactory improvement with minimal side effects in 12 out of 15 MG patients who had exacerbation of generalized MG. Improvement appeared within 2–3 days after each infusion. After improvement, oral prednisone therapy was used to maintain it (Arsura et al. 1985).

The mechanism of prednisone action in MG is not fully understood. That the drug modifies the immune response to AChR is evidenced by the decrease of the AChR antibody titer in patients responding to therapy (Seybold and Lindstrom 1981; Oosterhuis et al. 1983). Corticosteroids also stimulate AChR synthesis in cultured muscle cells (Braun et al. 1989; Kaplan et al. 1990) and perhaps increase AChR available for insertion into the NMJ.

Azathioprine and other immunosuppressants
Azathioprine therapy in doses of 2–3 mg/kg/day results in remission or significant improvement in 70–90% of the patients (Mertens et al. 1981; Mantegazza et al. 1988). The minimum time for improvement is 3 months. The percentage of positive responses is not significantly different in patients receiving azathioprine, azathioprine plus prednisone, or prednisone only (Mantegazza et al. 1988). About one-half of the patients relapse after treatment is stopped (Mertens et al. 1981). Among 271 MG patients treated with azathioprine, the side effects included hematologic reactions (18%), serious infections (7%), gastrointestinal irritation (8%) and signs of hepatotoxicity (6%) (pooled results of Mertens et al. 1981, Mantegazza et al. 1988 and Hohlfeld et al. 1988a). Cyclophosphamide therapy results in improvement in 84% of the patients (pooled results of Perez et al. 1981 and Niakan et al. 1986). This drug is seldom used on account of its potential toxicity (severe bone marrow depression, hemorrhagic cystitis, hair loss, anovulation, azoospermia) and because its therapeutic effect is not greater than that of azathioprine.

Cyclosporin has been used with good results in 20 MG patients not previously treated with thymectomy or immunosuppressants (Tindall et al. 1987). The drug also has produced significant improvement in 19 of 25 patients refractory to other forms of therapy (pooled results of Nyberg-Hansen and Gjerstad 1988 and Goulon et al. 1989). The usefulness of the medication is limited by its cost and by the high incidence of nephrotoxicity and hypertension during long-term therapy.

High-dose intravenous immune globulin (HDIVIG) therapy

More than 90 MG patients have been treated with HDIVIG since 1984 (Arsura 1989; Cosi et al. 1990). IVIG is infused at 400 mg/kg on each of 5 consecutive days, or at 1 g/kg on 2 consecutive days. Among 48 severely affected patients 70% responded favorably, the improvement occurring within 2–3 weeks from start of therapy. The mean duration of the response was 64 days in those patients also receiving corticosteroids, and 35 days in those who were not (Arsura 1989). HDIVIG is virtually riskless and is a valuable adjunct in the treatment of severely ill patients.

Thymectomy

Studies in the 1950s noted that the outcome of thymectomy was affected by sex, age, and thymic pathology (Schwab and Leland 1953; Eaton and Clagett 1955). The operation appeared to be especially effective in young females without thymoma and did not reduce the high mortality and morbidity for patients with thymoma. The operation was not recommended for patients after 10 years of illness and the benefits for male patients were regarded uncertain (Eaton and Clagett 1955). These criteria were subsequently revised by Simpson in 1958 who found that women with longer duration of symptoms also benefitted from thymectomy and that the trends for men were similar. Subsequent large studies confirmed Simpson's data. According to these, (a) the percentage of remissions after thymectomy increases steadily with time; (b) neither age, duration, nor sex represent contraindications for thymectomy; (c) females with shorter duration of the disease, hyperplastic glands, and high antibody titer improve more rapidly than males, but after 10 years the results tend to equalize; and (d) 7–10 years after thymectomy the remission rate is of the order of 40–60% in all categories of cases except for those with thymoma (Perlo et al. 1971;

Genkins et al. 1975; Horowitz et al. 1976b; Penn et al. 1981; Papatestas et al. 1987; Mulder et al. 1989).

Other studies emphasize that thymectomy is effective regardless of the age of the patient or the presence of thymic hyperplasia (Hankins et al. 1985; Evoli et al. 1988). Although thymectomy is effective in children with MG, it should be avoided during the first few years of life because of the risk of immunodeficiency (Brearly et al. 1987; Batocchi et al. 1990). Although thymectomy can be effective in ocular MG and may prevent generalization of the disease (Schumm et al. 1985), neurologists are reluctant to recommend it for ocular MG (Lanska 1990). However, as more than 50% of patients with ocular MG develop generalized disease (Grob et al. 1981), thymectomy can be justified as a preventive measure even in ocular MG. Old age is viewed as a relative contra-indication to thymectomy as older patients are more responsive to therapy (Sghirlanzoni et al. 1984), have a higher morbidity from the operation, and may not realize the full benefits of the procedure in their lifespan (Lanska 1990).

Transcervical thymectomy has been advocated because the procedure is relatively easy and the postoperative morbidity is minimal (Papatestas et al. 1987). This approach, however, may fail to remove the entire gland and will miss ectopic thymic tissue in the neck and mediastinum. Transsternal thymectomy combined with a search for thymic tissue in the cervical region is required to remove all thymic tissue predictably (Jaretzki et al. 1988).

Thymectomy also induces a short-lived but dramatic improvement in a very high proportion of patients in the immediate postoperative period. This usually occurs within 24 hours and disappears after a few days. During this time anticholinesterase drugs are discontinued or used only sparingly. The reason for this effect is not known.

As thymoma represents an absolute indication for thymectomy and greatly affects the prognosis, its detection is important. Up to 25% of the thymic tumors are not detected by anteroposterior and lateral chest roentgenograms. Radiogallium scans of the chest can reveal most of these (Swick et al. 1976), but computed tomography is now the procedure of choice for detecting thymic tumors (Batra et al. 1987; Mori et al. 1987).

Plasmapheresis

Plasmapheresis is a complicated and costly procedure. It should only be used as a lifesaving measure in severe generalized or fulminating MG not controlled by anticholinesterase drugs and refractory to all other forms of readily available therapy. There are no absolute rules as to the number of exchanges, the interval between the exchanges, or the amount of plasma exchanged per treatment (Am. Med. Assoc. 1986). Dau et al. (1977) have exchanged a volume of plasma equal to 5% of the body volume at each session and repeated this at 2–15-day intervals, depending on the patient's clinical status. Others exchanged 2 liters of plasma per session and repeated this treatment at daily intervals (Newsom-Davis et al. 1978). Objective improvement usually appears after a delay of a few days and can be correlated with a decrease of the AChR antibody titer. When treatment is stopped, the antibody titer increases again and the symptoms recur unless the patients are also receiving other immunosuppressant medications. It is now clear that plasmapheresis itself does not confer greater long-term protection than immunosuppressants alone (Hawkey et al. 1981; Keesey et al. 1981; Olarte et al. 1981).

A protein A-Sepharose resin removes anti-AChR antibodies from serum selectively, i.e., in excess over the total amount of IgG removed (Somnier and Langvad 1989). However, the resin may release traces of protein A into serum and is not yet approved for clinical use in humans.

Drugs with undesired effects

Aminoglycoside antibodies decrease the number of ACh quanta released by nerve impulse and reduce the safety margin of neuromuscular transmission (Elmqvist and Josefsson 1962; Wright and McQuillen 1971). Therefore, streptomycin, dihydrostreptomycin, polymyxin, colistin, neomycin, kanamycin, and gentamicin should be avoided or used with great caution in myasthenia gravis. Ampicillin (Argov et al. 1986), erythromycin (May and Calvert 1990), chlorpromazine, morphine, quinine, quinidine, procainamide (Osserman 1958; Kornfeld et al. 1976) and β-adrenergic blockers (Herishan and Rosenberg 1975; Confavreux et al. 1990) can also worsen the neuromuscular trans-mission defect and hence should be avoided or used with caution.

Experimental approaches to therapy

Based on current immunological dogma, a number of strategies to prevent or abolish the immune reaction to AChR have been be proposed and tested, as reviewed by Hohlfeld and Toyka (1985) and Steinman and Mantegazza (1990).

1. Anti-idiotypic antibodies could neutralize pathogenic anti-AChR antibodies. Immunization of rabbits with purified polyclonal anti-AChR antibodies has induced a polyclonal anti-idiotypic antibody response and partially or completely prevented the induction of EAMG by AChR (Fuchs et al. 1981). By contrast, immunization with a monoclonal anti-AChR antibody did not prevent sensitization to AChR (Lennon and Lambert 1981). This is not surprising because immunization with AChR induces a polyclonal antibody response. Anti-idiotypic antibodies could themselves stimulate the production of additional anti-AChR antibodies with different fine specificities. This suggests that anti-idiotypic antibody therapy will not be successful in clinical practice.

2. Peptides that bind to MG-associated HLA class II molecules but are not recognized by the T cell-antigen receptor could compete with peptide fragments of AChR in binding to class II molecules and thus block antigen presentation to T helper cells. This strategy has been successful in certain strains of mice with experimental allergic encephalitis, and has been shown to block T cell recognition of a segment of AChR, as reviewed by Steinman and Mantegazza (1990). For this approach to work in humans, different combinations of blocking peptides would have to be designed for patients with different MG-associated HLA haplotypes, and the approach would not apply to patients without such haplotypes.

3. Antibodies directed against class II molecules could also prevent antigen presentation to T helper cells. The immune response to AChR in certain strains of mice can be partially suppressed by anti-IA antibodies (Waldor et al. 1983; Christadoss and Dauphinee 1986; Todd et al. 1988). However, long-term treatment of humans with antibodies directed against monomorphic determinants of class II molecules could be immunosuppressive. Anti-

bodies raised against specific MG-associated class II molecules would be preferable but the same caveats mentioned with regard to blocking peptides again apply.

4. Theoretically, antibodies directed against clonotypic determinants of AChR-specific T cells could arrest the immune response to AChR. However, the diversity of T cell epitopes recognized in human MG (discussed above) is likely to limit this approach. Further, immunization of Lewis rats with AChR-specific polyclonal T cells potentiated rather than abolished the antibody response to AChR (Kahn et al. 1990).

5. Induction of AChR-specific suppressor cells could abolish the immune response to AChR. The induction of AChR-specific suppressor T cells has been accomplished in EAMG (Bogen et al. 1984; Pachner and Kantor 1984; McIntosh and Drachman 1986). This approach might succeed in humans provided the induced suppressor cell response is both AChR-specific and polyclonal.

Other experiments to suppress the immune reaction to AChR. Total lymphoid irradiation prevents the primary antibody response and abolishes the ongoing antibody response to AChR (de Silva et al. 1988), but does not facilitate the induction of specific tolerance to AChR (de Silva et al. 1990).

Theoretically, toxins targeted to components of the immune system could also alter the immune response to AChR. A hybrid molecule, that contains the toxic portion of diphtheria toxin and also binds to the interleukin-2 receptor on activated T cells, attenuates the immune response to AChR in mice in vivo (Balcer et al. 1991). Hybrid molecules of AChR coupled to ricin or daunomycin are toxic to AChR-specific B cells and attenuate the immune response to AChR in rats (Killen and Lindstrom 1984; Shelton et al. 1988). These approaches are of theoretical interest but none is presently applicable to human MG.

LAMBERT-EATON MYASTHENIC SYNDROME (LEMS)

Definitions and basic concepts

LEMS is an acquired autoimmune disease in which pathogenic autoantibodies deplete the voltage-sensitive calcium channels (VSCC) of the

motor nerve terminal (MNT) by antigenic modulation. The deficiency of VSCCs restricts the ingress of calcium into the MNT during activity. This reduces the probability of quantal release, the quantal content of the EPP, and the safety margin of neuromuscular transmission. A high proportion of LEMS patients have small-cell carcinoma of the lung, a neoplasm that expresses VSCCs. Sensitization to the VSCC of the carcinoma cell may result in the formation of antibodies that cross-react with the VSCC of the MNT.

Clinical features

The essential clinical features of LEMS were described by Lambert et al. in 1956 and by Eaton and Lambert in 1957. These papers were based on observations of 6 patients. Five of the 6 were men; 2 had small-cell and 1 had reticulum-cell carcinoma of the lung, and 2 others were suspected of having pulmonary malignancies. One patient also had cerebellar ataxia but no indication of carcinoma. They all had myasthenic symptoms but the distribution of the weakness, the EMG findings, and the response to anticholinesterase drugs were different from those observed in MG.

Subsequent investigations established that LEMS occurs more frequently in men than women; in 40 patients the male:female ratio was 4.7:1 (Elmqvist and Lambert 1968). Seventy percent of the males and 25% of the females have an associated malignancy, but malignancy is uncommon under age 40. Close to 80% of the tumors are small-cell carcinomas of the lung (Lambert and Lennon 1982; O'Neill et al. 1988). Approximately one-third of the patients have non-neoplastic LEMS, and in these the syndrome can present at any age. The youngest patient observed by the author was a 7-year-old boy whose symptoms appeared during the first year of life. Non-neoplastic LEMS can be associated with other autoimmune diseases, such as pernicious anemia, hypothyroidism, hyperthyroidism, Sjögren's syndrome, vitiligo, celiac disease, juvenile-onset diabetes mellitus, and MG (Gutmann et al. 1972; Lang et al. 1981; O'Neill et al. 1988; Newsom-Davis et al. 1991). Subacute cerebellar degeneration has been observed with either neoplastic (Satoyoshi et al. 1973) or non-neoplastic LEMS (Rooke et al. 1960).

The characteristic symptoms consist of weakness and increased fatigability of the truncal and mostly of the proximal limb muscles. Seventy percent of the patients have mild or transient ocular symptoms, and 80% have autonomic nervous system abnormalities. Decreased salivation, producing dryness of the mouth, is the commonest autonomic symptom; decreased lacrimation and sweating, orthostatism, impotence, and abnormally pupillary light reflexes can also occur. AChE inhibitors produce only slight or no improvement, but curare sensitivity is increased. Strength is reduced in rested extremity muscles but increases over a few seconds during the beginning of maximal voluntary contraction. This is best observed by asking the patient to squeeze the examiner's fingers or to continue to exert force with a proximal muscle such as the biceps or iliopsoas against resistance, or by repeated testing of the same muscle at short intervals. On continued maximal exertion, strength again decreases. The tendon reflexes are hypoactive or absent in nearly all patients, but may briefly return to normal immediately after exercise of those muscles activated by the reflex (Rooke et al. 1960; Lambert et al. 1961; Rubinstein et al. 1979; O'Neill et al. 1988).

Electrophysiologic aspects

Clinical EMG. Needle EMG shows low-amplitude motor unit potentials that fluctuate in amplitude and duration from moment to moment. With continued activity the potentials increase in amplitude and show less fluctuation. The improvement with continued activity is typical of LEMS (Lambert et al. 1961).

In most patients, the amplitude of the compound muscle action potential (CMAP) evoked by a single supramaximal nerve stimulus from rested muscle is markedly reduced. After a few seconds of maximal voluntary activity or tetanic stimulation, the CMAP amplitude increases. The increase can be several-fold, so that the CMAP attains normal amplitude. The postexercise facilitation is followed by subsequent depression. Indirect repetitive stimulation of rested muscle at 2–3 Hz further decreases the CMAP amplitude, but recovery sets in after the fourth to eighth stimulus. In most patients stimulation at 10 Hz or higher frequencies results in a progressive increase of the CMAP amplitude after the first stimulus. Similar stimulation in normal subjects has little effect on the CMAP amplitude (Lambert et al. 1961). In few patients the CMAP amplitude does not change on high frequency stimulation, or it decreases briefly before it increases (Oh 1989).

Single-fiber EMG shows increased jitter and frequent blocking of neuromuscular transmission even in minimally affected muscles. The abnormalities are worst on minimal activity and improve at higher rates of firing or on repetitive axonal stimulation. The opposite occurs in MG where blocking increases with increasing neural activity (Schwartz and Stalberg 1975; Trontelj and Stalberg 1990).

In-vitro electrophysiologic studies. In-vitro intracellular microelectrode studies indicate that the quantal content m of the EPP is abnormally low (Elmqvist and Lambert 1968; Lambert and Elmqvist 1971). This is not due to reduced synthesis or storage of ACh because the ACh content and choline acetyltransferase activity of LEMS muscle are normal (Molenaar et al. 1982). Due to the smallness of m, the EPP is often subthreshold for triggering the muscle fiber action potential. Repetitive stimulation, raising the external Ca^{2+} concentration, or adding guanidine or 3,4-diaminopyridine, which all increase the Ca^{2+} in the MNT, increase m by increasing p, the probability of quantal release, and improve the transmission defect in vitro. However, for a given increase in external $[Ca^{2+}]$, m increases less in LEMS than normal muscle (Cull-Candy et al. 1980). Depolarization of the nerve terminal by a high external K^+ concentration produces a subnormal increase in MEPP frequency, as is the case at the Mg^{2+}-blocked normal NMJ (Hubbard 1960). The MEPP amplitude is normal indicating that the amount of ACh per quantum and postsynaptic sensitivity to ACh are both normal.

The autoimmune etiology of LEMS

That LEMS is an autoimmune disease can be inferred from several observations. The non-neoplastic form of the disease responds to corticosteroids (Streib and Rothner 1981), has an increased association with other autoimmune disorders

(Gutmann et al. 1972; Lang et al. 1981), HLA-B8 and DRw3 antigens (Newsom-Davis et al. 1983), and organ-specific (thyroid and gastric) autoantibodies (Lennon et al. 1982; O'Neill et al. 1988). Plasmapheresis and immunosuppressive therapy improve both neoplastic and non-neoplastic LEMS (Newsom-Davis and Murray 1984). However, compelling evidence for the autoimmune etiology of both neoplastic and non-neoplastic LEMS came from passive transfer studies in which IgG from LEMS patients reproduced the electrophysiologic (Lang et al. 1981, 1983) and the morphologic (Fukunaga et al. 1983) features of the disease in mice.

The VSCCs of the MNT are the target of pathogenic LEMS antibodies. In 1982, Fukunaga et al. described striking freeze-fracture electron microscopy abnormalities of the LEMS MNT: the density of the active zones and active zone particles, and the number of particles per active zone were markedly reduced (Figs 7, 8). Further, many of the remaining active zone particles were aggregated into clusters (Engel et al. 1982a; Fukunaga et al. 1982). From these results and the evidence that the active zone particles represent VSCCs (See p. 393), the physiologic defect in LEMS could now be attributed to the loss of VSCCs from the MNT. By this time , Lang et al. (1981) had transferred the physiologic defect of LEMS to mice with IgG. Therefore, Fukunaga et al. could infer that the active zone particles (i.e., the VSCCs) where (a) targets of the LEMS autoantibodies, (b) aggregated because they were cross-linked by IgG, and (c) depleted by the mechanism of antigenic modulation (Fukunaga et al. 1982). Subsequent studies obtained adequate evidence for each of these postulates.

The mediation of the membrane lesions in LEMS by IgG was established by freeze-fracture studies on diaphragm muscles of mice treated with LEMS IgG: there was clustering and depletion of active zone particles in mice treated with pathogenic LEMS IgG (Fukunaga et al. 1983). That the active zone particles were close enough to be cross-linked by IgG was established by stereometric analysis of freeze-fracture replicas of the MNT, and aggregation of active zone particles was shown to be an early effect of LEMS IgG (Engel et al. 1987; Fukuoka et al. 1987a). Divalent IgG is required for antigenic modulation, and divalency of LEMS IgG was shown to be an essential requirement for the aggregation and depletion of the active zone particles (Nagel et al. 1988). Finally, immunoelectron microscopy studies demonstrated binding of LEMS IgG to NMJ active zones in mice injected with pathogenic LEMS IgG (Engel et al. 1987; Fukuoka et al. 1987b).

The principal idea that emerged from the morphologic studies was that LEMS IgG was directed against the VSCC of the MNT. Parallel electrophysiological studies were also consistent with this notion. In mice, LEMS IgG interfered with increase of the MEPP frequency induced by depolarization of the MNT (Lang et al. 1987). In other studies, LEMS IgG was shown to reduce functional VSCCs in small-cell carcinoma cells (De Aizapura et al. 1988; Lang et al. 1989), anterior pituitary cells (Login et al. 1987), adrenal chromaffin cells (Kim and Neher 1988), and neuroblastoma-glioma hybrid cells (Peers et al. 1987). Finally, IgG from some LEMS patients binds to ^{125}I-ω-conotoxin-labeled VSCCs extracted from human neuroblastoma (Sher et al. 1989; Leys et al. 1991) and small-cell carcinoma (Lennon and Lambert 1989) cells.

The observations on the adrenal chromaffin, anterior pituitary, and neuroblastoma-glioma hybrid cells suggest that LEMS IgG has a selective effect on L-type Ca^{2+} channels, whereas the binding of some LEMS IgG to ^{125}I-ω-conotoxin-labeled VSCCs implies that the autoantibodies recognize N-type and L-subtype of Ca^{2+} channels. These observations, however, do not mean that the MNT VSCC is either L or N in type, because LEMS IgG could readily recognize homologous and antigenically similar domains in different types of Ca^{2+} channels. More likely, the VSCC of the MNT belongs to a class of VSCC that is distinct from both N and L channels.

Diagnosis

The diagnosis of LEMS. This rests on clinical and EMG data. The combination of weakness and fatigability involving especially the proximal leg muscles, relative sparing of the cranial muscles, increase of strength at the beginning of voluntary activity, hyporeflexia, and dysautonomic features,

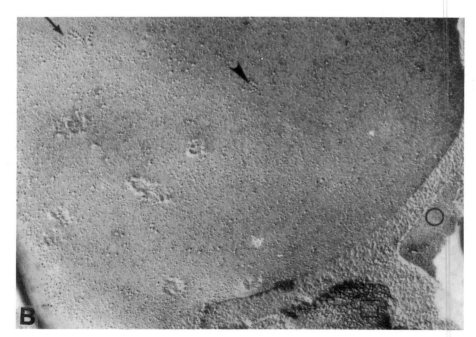

Fig. 7. Freeze-fracture electron micrographs of presynaptic membrane P faces of a normal (A) and a LEMS (B) motor nerve terminal. On the normal membrane, note the numerous double parallel rows of large membrane particles. These represent the voltage-sensitive calcium channels of the active zones. The LEMS membrane shows a marked depletion of active zones and active zone particles; only a short active zone (arrow) and single cluster of large membrane particles (arrowhead) can be detected. (A) ×121,000, (B) ×61,000. (Reproduced, respectively, from Engel (1986) and from Fukunaga et al. (1982) by courtesy of the Editors of *Muscle and Nerve*.)

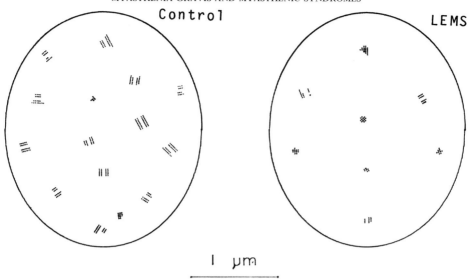

Fig. 8. Stereometric reconstruction of control and LEMS presynaptic membrane P faces. Control membrane contains 13 active zones and 2 clusters; LEMS membrane contains 3 active zones and 5 clusters. The active zones are smaller and the clusters larger in the LEMS than in the control membrane. (Reproduced from Fukunaga et al. (1982) by courtesy of the Editors of *Muscle and Nerve*.)

especially dry mouth, should point to the correct diagnosis. The salient EMG features useful for diagnosis were described above. An associated autoimmune disease or malignancy, and especially small-cell carcinoma of the lung, is a further diagnostic clue.

Surveillance for lung carcinoma. Any patients above the age of 40, and especially men, apparently free of malignancy at the time of diagnosis must be kept under surveillance for malignancy for at least 3 years. Although patients younger than 40 are unlikely to have carcinoma, periodic evaluation of these patients for malignancy is still advisable.

Serological tests for LEMS. The presence in some LEMS patients of antibodies that bind to ^{125}I-ω-conotoxin-labeled VSCCs extracted from neuroblastoma or small-cell carcinoma cells has formed the basis of serological tests for LEMS. However, with the ω-conotoxin-labeled neuroblastoma channels, significant antibody titers are detected in only 16% of all LEMS patients (Leys et al. 1991); and with ω-conotoxin-labeled small-cell carcinoma channels, significant titers are found in 76%

of neoplastic but only in 30% of non-neoplastic LEMS cases (Lennon and Lambert 1989). These findings suggest that ω-conotoxin is not a specific or reliable ligand for the sensitizing antigen in LEMS. Electrophysiologic studies, in fact, indicate that the Ca^{2+} channels of the mammalian MNT are relatively insensitive to ω-conotoxin (De Luca et al. 1991).

Differential diagnosis. This includes acquired autoimmune MG, Guillain-Barré syndrome, peripheral neuropathy, polymyositis, botulinum poisoning, magnesium intoxication, and, in the younger patients, a congenital myasthenic syndrome.

The distribution of the weakness, the edrophonium test, the EMG studies, and the test for circulating AChR antibodies distinguish acquired autoimmune MG from LEMS. Rare patients have both LEMS and MG by clinical or EMG criteria (Oh et al. 1974). In these, demonstration of both anti-AChR and anti-VSCC antibodies may provide evidence for both disorders (Newsom-Davis et al. 1991). The distribution of the weakness and the hyporeflexia may suggest the diagnosis of Guillain-Barré syndrome. However, LEMS has a more insidious onset, and the increase of strength during voluntary contraction, the normal spinal fluid examination, and the EMG findings should establish the diagnosis of LEMS. The hyporeflexia and the occasional peripheral paresthesia in LEMS may suggest a peripheral neuropathy. However, sensory and motor nerve conduction velocities are normal, fibrillation potentials are absent; and nerve stimulation studies reveal the typical features of LEMS.

Both botulinum poisoning (Cull-Candy et al. 1976) and magnesium intoxication (Del Castillo and Engbaek 1954) reduce the quantal content of the EPP and can closely mimic the clinical and EMG features of LEMS. However, the onset of these syndromes is more abrupt than that of LEMS. The diagnosis of botulinum poisoning was discussed above in connection with MG. Magnesium intoxication can arise from accidental poisoning or after abuse of magnesium-containing cathartics (Swift 1979). The history and determination of the serum magnesium level will establish the appropriate diagnosis.

The infrequent childhood cases of LEMS may resemble a congenital myasthenic syndrome. The combined use of clinical, EMG, morphologic, and in-vitro electrophysiologic criteria will readily distinguish between the different syndromes. The findings in the congenital syndromes are discussed below.

Therapy

In neoplastic LEMS the neuromuscular transmission defect may improve when chemotherapy and/or radiotherapy induces tumor regression (Lambert and Rooke 1965; Clamon et al. 1984; Chalk et al. 1990). However, the response is transient and LEMS recurs when the tumor recurs.

Anticholinesterase drugs, guanidine, 4-aminopyridine, 3,4-diaminopyridine, and immunosuppressive measures have been used in the treatment of neoplastic and non-neoplastic LEMS. Anticholinesterase drugs have only a slight effect in LEMS (Lambert et al. 1961; Kamenskaya et al. 1975) and are used only when other measures fail or are contraindicated.

In 1939, Minot et al. reported that guanidine, at a dose of 15 mg/kg/day, together with small doses of prostigmine, had a beneficial effect in MG.

Guanidine increases the quantal content of the EPP (Otsuka and Endo 1960), is most effective under conditions of low-quantal release (Kamenskaya et al. 1975), and has a beneficial clinical effect in LEMS (Lambert and Rooke 1965). However, the drug can also cause ataxia, paresthesias, gastrointestinal distress, confusion, dry skin, atrial fibrillation, hypotension, bone marrow depression, and renal failure due to tubular necrosis or interstitial nephritis (Minot et al. 1939; Norris 1974; Cherington 1976).

4-Aminopyridine and 3,4-diaminopyridine also facilitate quantal release by nerve impulse. These drugs prolong the duration of the presynaptic action potential by blocking the outward potassium current (Maeno 1980; Saint 1989). This increases the calcium entry into the nerve terminal which, in turn, increases quantal release (Katz and Miledi 1969). 4-Diaminopyridine is effective in LEMS (Lundh et al. 1977) but therapeutic doses frequently cause grand mal seizures, paresthesias, unsteadiness, and diaphoresis (Murray and Newsom-Davis 1981). 3,4-Diaminopyridine is less toxic and in doses up to 100 mg per day is effective in relieving both the motor and autonomic symptoms of LEMS (Lundh et al. 1984; McEvoy et al. 1989). Only 1 of the 12 patients studied by McEvoy et al. (1989) had seizures during therapy; other side effects were minimal and dose-related (McEvoy et al. 1989). The drug is still not approved for clinical use in the United States.

Plasmapheresis can be effective in LEMS (Newsom-Davis and Murray 1984), but it is expensive and repeated treatments are required. High-dose intravenous immunoglobulin therapy can also induce a temporary remission (Bird 1991). Either azathioprine or corticosteroids can improve LEMS. The 2 drugs tend to act synergistically with each other, but not with plasma exchange (Newsom-Davis and Murray 1984). Presently, optimal treatment of non-neoplastic LEMS consists of modest doses of alternate-day prednisone and 1.5–2 mg/kg/day of azathioprine. After clinical improvement, which is usually gradual, the prednisone dose can be gradually reduced over many months to establish the minimal maintenance requirement. In neoplastic LEMS, antitumor therapy may be combined with alternate-day prednisone treatment. Either form of LEMS should be treatable with 3,4-diaminopyridine when it becomes available in clinical practice.

CONGENITAL MYASTHENIC SYNDROMES

Table 1 lists the currently recognized congenital myasthenic syndromes. In familial infantile myasthenia and in the syndrome with a paucity of synaptic vesicles, the primary abnormality is presynaptic. In NMJ AChE deficiency, the absence of the enzyme from the synaptic cleft can induce presynaptic and also postsynaptic alterations which separately or together compromise the safety margin of transmission. In all other syndromes investigated in detail to date, there was a kinetic abnormality of AChR with or without a concomitant deficiency of AChR. Unraveling the pathogenesis of these syndromes requires special studies (Table 2).

TABLE 2

Investigation of myasthenic syndromes.

Clinical
  History, examination, response to AChE inhibitor
  EMG: conventional, stimulation studies, SFEMG
  Serologic tests (AChR antibodies, tests of botulism)
Morphologic studies
  Routine histochemical studies
  Cytochemical and immunocytochemical localizations of AChE, AChR, AChR subunits, IgG, C3, C9, MAC at the NMJ
  Estimate of NMJ size and shape on AChE-reacted teased muscle fibers
  Quantitative EM, immuno-EM, freeze-fracture EM
^{125}*I-bungarotoxin binding sites/NMJ*
In-vitro electrophysiology studies
  Conventional microelectrode studies: MEPP, MEPC, evoked quantal release ($m, n. p$)
  Noise analysis: channel kinetics
  Single-channel patch-clamp recordings

AChE, acetylcholinesterase; AChR, acetylcholine receptor; EM, electron microscopy; EMG, electromyography; MAC, C5b-9 complement membrane attack complex; MEPP, miniature end-plate potential; MEPC, miniature end-plate current; m, number of ACh quanta released by nerve impulse; n, number of readily releasable ACh quanta; p, probability of quantal release; NMJ, neuromuscular junction; SFEMG, single-fiber electromyography.

Familial infantile myasthenia (FIM)

The clinical features of FIM were recognized more than 3 decades ago (Greer and Schotland 1960; Robertson et al. 1980), but it was not differentiated from MG until the autoimmune origin of MG was established and electrophysiologic and morphologic differences were demonstrated between MG and the congenital syndrome (Hart et al. 1979; Engel and Lambert 1987; Mora et al. 1987).

Clinical aspects

The disease presents in early infancy or childhood. The typical history is one of fluctuating ptosis from birth; poor suck and cry, feeding difficulty, and secondary respiratory infections during infancy; and episodic exacerbation precipitated by fever, excitement, or vomiting throughout infancy and childhood. During exacerbation all symptoms worsen; apnea from respiratory muscle weakness may cause sudden death or anoxic brain injury. In childhood and between exacerbations, patients may appear normal or have only minimal weakness, but weakness can be induced by exercise. With increasing age, the exacerbations become less frequent. After age 10, some patients only complain of easy fatigability on sustained exertion; others have mild to moderate weakness of cranial, limb, and respiratory muscles even at rest, resembling patients with mild to moderately severe autoimmune myasthenia gravis. The tendon reflexes remain normally active. There is no loss of muscle bulk, and a permanent myopathy does not occur (Greer and Schotland 1960; Conomy et al. 1975; Robertson et al. 1980; Gieron and Korthals 1985; Mora et al. 1987).

Electrophysiological features

A decremental response at 2-Hz stimulation and SFEMG abnormalities similar to those observed in MG are found only in those muscles that are weak when tested. Weakness and EMG abnormalities can be induced in some, but not all, muscles either by exercise or by repetitive stimulation at 10 Hz for a few minutes (Hart et al. 1979; Mora et al. 1987). The EMG decrement, *when present*, can be corrected by edrophonium (Robertson et al. 1980). In vitro studies by Lambert on external intercostal muscles have elucidated the electrophysiologic basis of the disorder (Hart et al. 1979; Mora et al. 1987). Stimulation of small muscle bundles at 10 Hz in vitro results in an abnormal decrease of the amplitude of the CMAP and of the EPP. In contrast to autoimmune MG, the MEPP amplitude is normal in rested muscle but decreases abnormally after 10-Hz stimulation for 5 minutes. The safety margin of neuromuscular transmission is compromised during sustained activity by an abnormal decrease of the EPP, which is due to an abnormal decrease of the MEPP.

The electrophysiologic findings are like those noted in normal muscle poisoned by hemicholinium, an inhibitor of choline uptake by the nerve terminal (Fig. 9) (Elmqvist and Qvastel 1965b; Jones and Kwanbunbumpen 1970a,b; Wolters et al. 1974), and the decrease of the MEPP amplitude on prolonged stimulation from an initially normal to an abnormally low level suggests a progressive decrease in the ACh content of the synaptic vesicles. Alternatively, physiologic amounts of ACh released during stimulation induce abnormal desensitization of postsynaptic AChR. However, if desensitization was occurring then cholinesterase

inhibitors should worsen the defect, but in fact they improve it. On this basis, FIM could be caused by a defect in (a) the facilitated uptake of choline by the nerve terminal, (b) ACh resynthesis by choline acetyltransferase, or (c) the transport of ACh molecules into the synaptic vesicles.

Morphological observations

Muscle biopsy specimens show no histochemical abnormality. AChE-reacted sections demonstrate no abnormality of the NMJ. There are no immune complexes (IgG or complement) at the NMJ. The nerve terminals show normal immunoreactivity for choline acetyltransferase (Engel 1986). This, however, does not exclude the possibility of a mutation affecting the kinetic properties of the enzyme.

On electron microscopy, the NMJ appears normal on simple inspection. There is neither ultrastructural nor radioimmunochemical evidence for NMJ AChR deficiency (Hart et al. 1979; Engel et al. 1987).

Mora et al. (1987) evaluated the effect of stimulation on synaptic vesicle size in 3 FIM patients

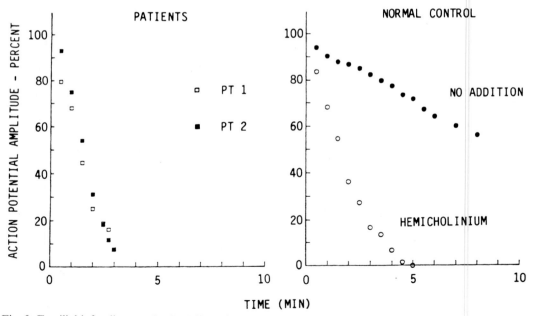

Fig. 9. Familial infantile myasthenia. Effect of 10-Hz stimulation on the amplitude of the evoked compound muscle action potential in external intercostal muscle strips in vitro in 2 patients (left panel) and in a normal subject (right panel). In the presence of 1 mg/dl of hemicholinium-3, the evoked action potential in normal muscle (open circles in right panel) declines as rapidly as the evoked action potential in FIM muscle in the absence of hemicholinium. (Reproduced from Mora et al. (1987) by courtesy of the Editors of *Neurology*.)

and 3 controls. Intercostal muscle strips were stimulated at 10 Hz for 10 minutes. In the FIM patients, but not in the controls, neuromuscular transmission failed during stimulation. In both patients and controls, stimulation reduced the density (number per unit area) of synaptic vesicles in the nerve terminal to the same extent. However, and unexpectedly, the size of the synaptic vesicles was significantly smaller in rested FIM than control muscles. Further, after stimulation, when the MEPP amplitude was markedly reduced in FIM, the FIM vesicles increased or did not change size. By contrast, after stimulation the control synaptic vesicles decreased or did not change in size.

There is no simple explanation for the lack of correlation between synaptic vesicle size and the MEPP amplitude in FIM. ACh is taken up by the synaptic vesicle by an ACh translocase that exchanges protons in the vesicles for cytosolic ACh (Whittaker 1984; Südhof and Jahn 1991). The proton accumulation in the vesicles, in turn, depends on an ATPase-driven proton pump. The synaptic vesicles contain not only ACh and protons, but also a relatively high concentration of ATP, GTP, Ca^{2+}, and Mg^{2+}, and a proteoglycan (Wagner et al. 1978; Whittaker 1984). These observations imply that (a) the number of ACh molecules in the vesicles is not the *only* determinant of vesicle volume and (b) a defect in any of the mechanisms that regulate the concentrations of any of the osmotically active substances in the synaptic vesicles could affect the vesicle volume. Thus, the fact that in FIM the synaptic vesicles are smaller than normal in rested muscle and increase paradoxically in size after stimulation suggests a defect in synaptic vesicle metabolism, but the character of the defect remains undefined.

Treatment

The muscle weakness, when present, responds well to small or modest doses of anticholinesterase drugs. Some patients are asymptomatic or have only minimal weakness except during crises, and require anticholinesterase drugs on an emergency basis only. Parents of affected children must be indoctrinated to anticipate sudden worsening of the weakness and possible apnea with febrile illnesses, excitement, or overexertion. The parents also must be familiar with the use of a hand-assisted ventilatory device, and should be able to administer appropriate doses of neostigmine intramuscularly during crises. Patients with a febrile illness and a previous history of crisis should be hospitalized for close observation and for ventilatory support as needed.

Paucity of synaptic vesicles and reduced quantal release

In this recently observed syndrome the safety margin of neuromuscular transmission is compromised by a deficiency of the number of synaptic vesicles in the nerve terminal (Engel et al. 1990; Walls et al. 1990). The patient, a 23-year-old woman, had fatigable weakness of bulbar and limb muscles since infancy. The symptoms responded to anticholinesterase drugs. Tests for anti-AChR antibodies were negative. A decremental EMG response was present at 2-Hz stimulation. In-vitro microelectrode studies revealed that the quantal content of the EPP (m) was markedly reduced and this was due to a decrease in the number of readily releasable quanta (n); the probability of quantal release (p) was normal. The amplitude and the decay time constant of the MEPP were normal. The presence of normally functioning voltage-sensitive calcium channels on the motor nerve terminal was indicated by 2 observations: (1) increased $[Ca^{2+}]$ in the bath increased m normally; and (2) increased $[K^+]$ in the bath increased the MEPP frequency normally. NMJ AChR content, estimated from the number of ^{125}I-α-bungarotoxin binding sites per NMJ, was normal. Quantitative ultrastructural studies of unstimulated NMJs demonstrated an approximately 80% decrease in synaptic vesicle density (no./μm^2), which was comparable to the decrease in n. Nerve terminal size, presynaptic membrane length, and the postsynaptic region were normal by ultrastructural criteria.

Synaptic vesicles are transported by fast axonal flow from the perikaryon of the anterior horn cell to the motor nerve terminal (Booj 1986; Llinas et al. 1989), where they undergo exocytosis and recycling. The paucity of synaptic vesicles in this syndrome could be due to impaired axonal transport of preformed vesicles to the nerve terminal, or to impaired vesicle recycling in the nerve terminal after activity. The fact that the synaptic vesicles were depleted even in unstimulated nerve terminals suggests that the defect involves the axonal transport mechanism.

Congenital end-plate AChE deficiency

Clinical aspects

This disorder was first recognized by Engel et al. (1977a) in a boy with lifelong symptoms. Since then, 4 other patients have been observed by Walls et al. (1989), Engel (1990) and Hutchinson et al. (1992), and a sixth patient was reported by Jennekens et al. (1992). An autosomal-recessive inheritance is likely because 2 patients were sisters, none of the parents were affected, and the disease occurs in either sex. Weakness and abnormal fatigability are present since birth or early childhood. Poor suck, cry, and episodes of respiratory distress can occur in infancy. Motor milestones are delayed. The weakness affects the facial, cervical, axial, and limb muscles. Ophthalmoparesis was detected in 3 of the 6 patients. The axial muscles are severely involved, so that on standing the patient may show increasing lordosis and scoliosis after a few seconds. Fixed scoliosis appears in the older patients. The oldest patient observed to date, a 41-year-old man, also had selectively severe weakness and atrophy of his dorsal forearm and intrinsic hand muscles (Hutchinson et al. 1992). The tendon reflexes were hypoactive in the first patient but not in the others. The 5 patients observed by the author all had abnormally slow pupillary light responses, but no other autonomic deficits. All symptoms are refractory to or worsened by anticholinesterase drugs.

Electrophysiological features

The EMG shows a decremental response at 2-Hz stimulation in all muscles. In 4 of the 6 patients there was a repetitive CMAP response to a single nerve stimulus. The repetitive response disappears at stimulation frequencies greater than 0.2 Hz, or with mild activity. Therefore, it can be overlooked unless a well-rested muscle is tested by single shocks.

In-vitro microelectrode studies revealed MEPPs of normal amplitude in 4 patients and of reduced amplitude in 2. However, in the absence of AChE from the NMJ, a MEPP of higher than normal

amplitude would be expected. The decay time constant of the MEPPs or MEPCs was prolonged in 3 of the 5 patients in whom this determination was obtained. Consistent with the absence of AChE from the NMJ, the addition of neostigmine to the bath had no additional effect on either the amplitude or the decay time constant of the potentials (Engel et al. 1977a; Walls et al. 1989; Hutchinson et al. 1992). The quantal content of the EPP was markedly decreased in 4 of the 5 patients in whom it was determined; the decrease was due to a reduction in the number of immediately releasable quanta; the probability of quantal release was normal or higher than normal (Engel et al. 1977a; Walls et al. 1989; Hutchinson et al. 1992). Analysis of the ACh-induced current noise, done in 2 patients, yielded normal values for single channel opentime and conductance (Hutchinson et al. 1992).

Morphological and biochemical observations

Conventional histologic studies of muscle are normal. The basic abnormality is a deficiency of AChE at the NMJ: no enzyme activity can be demonstrated by light microscopic cytochemistry; electron cytochemical studies show no AChE (Fig. 10) (Engel et al. 1977a), or only a trace of AChE (Jennekens et al. 1992) in the synaptic cleft. No immunoreactivity for AChE is detected at the NMJ by polyclonal or several monoclonal AChE antibodies (Engel 1986, 1990).

Quantitative electron microscopy reveals a statistically significant decrease in nerve terminal size (Fig. 11). At many NMJs, Schwann cell processes extend into the primary synaptic cleft and partially or even completely cover the presynaptic membrane; this reduces the surface available for ACh release from the nerve terminal. At some NMJs there is degeneration of the junctional folds and also of the underlying muscle fiber regions.

The distribution of AChR on the junctional folds is normal (Fig. 11) or reduced. The total number of AChRs at the NMJ, determined in 3 patients with ^{125}I-labeled α-bungarotoxin, was reduced in 2, falling 23% and 29% below the lower range of normal.

The total muscle AChE content is reduced. This is due to absence of the enzyme from the basal lamina of the NMJ. The kinetic properties of the

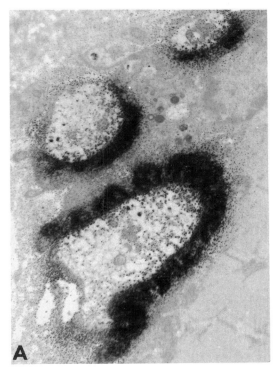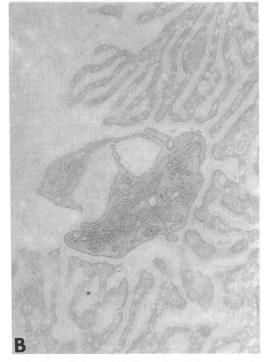

Fig. 10. Electron cytochemical localization of AChE in NMJ AChE deficiency. (A) Control end-plate is greatly over-reacted after incubation for 30 minutes at room temperature. (B) Patient's end-plate shows no reaction after incubation for 1 hour at room temperature. (A) ×8900, (B) ×20,500. (Reproduced from Engel et al. (1977a) by courtesy of the Editors of *Ann. Neurol.*)

residual extrajunctional AChE are normal. The activity and kinetic properties of erythrocyte AChE are also normal.

Pathogenetic mechanisms

Because AChE is absent from the NMJ, ACh-AChR interactions are terminated not by AChE but by diffusion of ACh from the synaptic space. Before leaving the synaptic space, ACh can bind to more than 1 AChR and this prolongs the decay of the MEPP and EPP (Katz and Miledi 1973). If the amplitude of the prolonged EPP is still high enough to evoke a muscle fiber action potential when the fiber recovers from the refractory period of the first action potential, then the prolonged EPP can evoke 1 or more additional muscle fiber action potentials.

The reduced number of readily releasable quanta (n) is adequately accounted for by the smallness of the nerve terminal and the reduced presynaptic membrane surface available for ACh release. However, the smallness of the nerve terminals and the decrease in n are not as constant as the AChE deficiency. For example, in all patients some nerve terminals were of normal size, and in 1 patient the quantal content of the EPP was normal, but there was AChE deficiency at all NMJs in all cases. This indicates that the AChE deficiency, and not the smallness of the nerve terminal, is the primary defect.

AChR is lost from NMJs at which there is focal degeneration of the junctional folds. The degeneration of the folds, in turn, can be accounted for by the ACh excess (Engel et al. 1973; Salpeter et al. 1979). However, at most NMJs the ACh excess is mild because ACh release is restricted by the smallness of n. The safety margin of neuromuscular transmission can be compromised by the smallness of n, by AChR deficiency, and possibly by desensitization or prolonged depolarization of the postsynaptic membrane by unhydrolyzed AChE.

The molecular forms of AChE show a tissue-

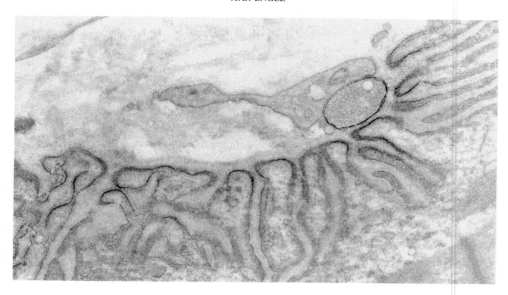

Fig. 11. AChR localization with peroxidase-labeled α-bungarotoxin in a patient with NMJ AChE deficiency. At this end-plate region there is a normal distribution of AChR on the terminal expansions of the junctional folds. The nerve terminal is abnormally small and covers only a small fraction of the postsynaptic region. ×23,400. (Reproduced from Engel et al. (1977a) by courtesy of the Editors of *Ann. Neurol.*)

specific distribution; and each form is anchored to the cell membrane by a post-translational modification. Symmetric and asymmetric forms of the enzyme exist. The symmetric forms are composed of 1, 2 or 4 catalytic subunits. The catalytic subunit is encoded by 3 invariant exons plus another exon (exon 3A); and the alternative use of exon 3A accounts for structural divergence in the carboxyl termini of the catalytic subunits in different tissues (Li et al. 1991).

NMJ AChE is asymmetric, consisting of 12 catalytic subunits anchored by 3 collagen-like tail subunits to a heparan sulfate proteoglycan receptor in the synaptic basal lamina (Massoulié and Bon 1982; Inestrosa and Perelman 1989). The tail portion of NMJ AChE is encoded by a gene different from that coding for the catalytic subunit (Krejci et al. 1991). Li et al. (1991) have now sequenced the entire gene encoding the catalytic subunit in humans, and they have also sequenced exon 3A in an AChE-deficient patient (Hutchinson et al. 1992). In the patient, that part of exon 3A which encodes the domain to which the tail subunit binds was normal. This observation, the presence of residual and kinetically normal AChE in muscle, and

the absence of AChE from the basal lamina of the NMJ suggest that NMJ AChE deficiency is caused by a defect in the tail subunit, in the basal lamina receptor for the tail subunit, or in a factor that promotes the assembly of the catalytic and tail subunits.

Slow-channel syndrome

Clinical aspects

This syndrome was described by Engel et al. (1982b). An additional report has been published by Oosterhuis et al. (1987). The disease is transmitted by an autosomal-dominant gene with high penetrance and variable expressivity. Sporadic cases also occur. The age at onset, the initial and eventual pattern of muscle involvement, the rate of progression, and the degree of weakness and fatigability vary from case to case. The disease may be symptomatic in infancy, childhood, or adult life. It progresses gradually or in an intermittent manner, remaining quiescent for years or decades between periods of worsening. Typically, there is selectively severe involvement of cervical, scapular, and finger extensor muscles (Figs 12, 13); mild to moderate weakness of the eyelid elevators and limitation

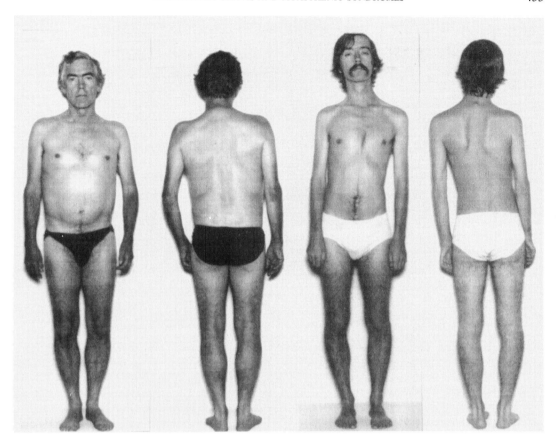

Fig. 12. Slow-channel syndrome: a 54-year-old man with his 29-year-old son. The father shows decreased bulk of the shoulder and forearm muscles and slight, asymmetric ptosis. The son has scoliosis, moderate to marked atrophy of cervical, shoulder, arm, forearm, and torso muscles, and pronounced ptosis. (Reproduced from Engel et al. (1982b) by courtesy of the Editors of *Ann. Neurol.*)

of ocular movements with only occasional double vision; and variable involvement of masticatory, facial and other upper extremity, respiratory, and trunk muscles. The lower limbs tend to be spared or can be less severely affected than the upper ones. The clinically affected muscles are weak, atrophic, and fatigue abnormally. The weakness and fatigability can fluctuate, but not as rapidly as in autoimmune MG. The tendon reflexes are usually normal, but can be reduced in severely affected limbs. The AChR antibody test is negative. AChE inhibitors are ineffective. Oosterhuis et al. (1987) found that the short-term use of flunarazine, a calcium channel blocker, was ineffective in 1 patient.

Electrophysiological features

As in some patients with NMJ AChE deficiency, single nerve stimuli evoke repetitive CMAPs (Fig. 14). In the slow-channel syndrome all patients observed to date displayed this phenomenon. The consecutive potentials occur at 5–10 msec intervals, each smaller than the preceding one, and disappear after a brief voluntary contraction. Two nerve volleys 1–3 msec apart increase rather than inhibit the amplitude of the second action potential, indicating that it is not produced by an axon reflex.

A decremental EMG response at 2–3-Hz stimulation is present, but only in clinically affected

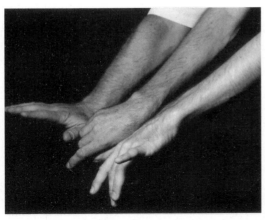

Fig. 13. Slow-channel syndrome. Patient attempting to extend wrists and fingers as shown by examiner (with sleeve). Note atrophy of patient's forearm muscles. (Reproduced from Engel et al. (1982b) by courtesy of the Editors of *Ann. Neurol.*)

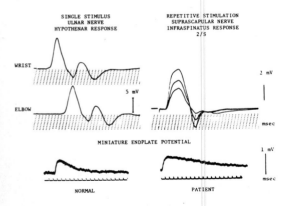

Fig. 14. Slow-channel syndrome. Upper left panel: Stimulus-linked repetitive compound muscle action potential in hypothenar muscles. Upper right panel: Decremental response of infraspinatus muscle action potential during 2-Hz stimulation of suprascapular nerve. Lower panels: The duration and decay time of the MEPP are longer in patient than in normal control. (Reproduced from Engel et al. (1982b) by courtesy of the Editors of *Ann. Neurol.*)

muscles. The motor unit potentials fluctuate in shape and amplitude during voluntary activity.

In-vitro microelectrode studies indicate that the decay of the MEPP (Fig. 14) and EPP is markedly prolonged and the duration of the potentials is further increased by prostigmine. The amplitude of the MEPP is significantly reduced in some but not all muscles, and the decrease is greater in the more severely affected muscles. The quantal content of the EPP is in the normal range. The prolonged duration of the MEPP and EPP is not due to an alteration in the cable properties of the muscle membrane because (a) the duration of the extracellularly recorded MEPP, which is independent of the cable properties of the membrane and reflects the duration of the miniature end-plate current (Del Castillo and Katz 1956), is also markedly prolonged (Engel et al. 1982b); and (b) in 1 patient the time constant of the ACh-induced voltage noise at the NMJ was abnormally prolonged (Oosterhuis et al. 1987); (c) patch-clamp analysis of single AChR ion channels in the author's laboratory has revealed a marked prolongation of the duration of channel openings.

Morphological and biochemical observations
Light microscopic histochemical studies show type 1 fiber predominance, isolated or small groups of atrophic fibers of either histochemical type, tubular aggregates, and vacuoles in fiber regions near NMJs. Other biopsy specimens show abnormal variation in fiber size, variable fiber splitting, and, in some instances, mild to moderate increase of endomysial or perimysial connective tissue.

AChE activity is present at all NMJs. In the more severely affected muscles, the configuration of the NMJs is often abnormal, with multiple, small, discrete regions distributed over an extended length of the muscle fiber. Focal calcium deposits were demonstrated at the NMJ by Engel et al. (1982b) in 1 out of 4 cases in which this test was performed.

On electron microscopy, the junctional folds at many NMJs contain myriad pinocytotic vesicles and labyrinthine membranous networks. In the more severely affected muscles, the junctional folds are frequently degenerating, causing widening of the synaptic space and accumulation of electron-dense debris (Fig. 15). Some of the highly abnormal postsynaptic regions are denuded of their nerve terminals. Unmyelinated nerve sprouts appear near some NMJs. The intramuscular nerves are normal. AChR is reduced at those NMJs that show degeneration of the junctional folds. Degenerative changes also occur in the junctional sarco-

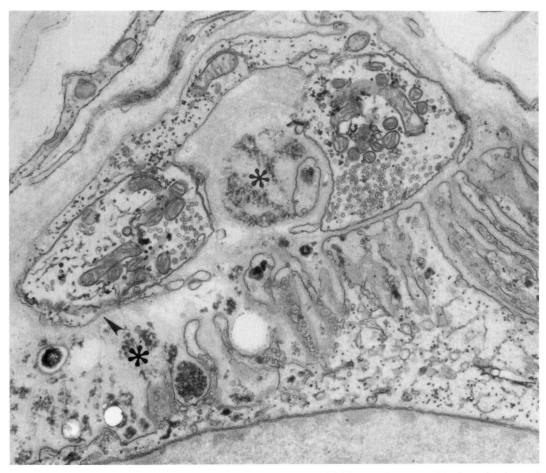

Fig. 15. Intercostal NMJ in the slow-channel syndrome. Junctional folds have degenerated in the region imaged at left. Highly electron-dense debris marks position of pre-existing folds (asterisk). Presynaptic membrane facing degenerated folds is partially covered by Schwann cell (arrowhead). ×20,100. (Reproduced from Engel et al. (1982b) by courtesy of the Editors of *Ann. Neurol.*)

plasm in nearby fiber regions. These involve the nuclei, myofibrils, mitochondria, and sarcoplasmic reticulum. The vacuoles near the NMJs contain cytoplasmic degradation products and are limited by membranes of transverse tubular origin or by proliferating transverse tubular system networks.

Morphometric reconstruction of the NMJ shows a 29–43% decrease of nerve terminal size and a 25–37% increase in synaptic vesicle density. The postsynaptic membrane length and density are significantly reduced due to degeneration of the junctional folds.

Although AChE inhibitors further increase the duration of the MEPP in vitro, and although NMJ AChE is preserved by cytochemical criteria, a partial deficiency or kinetic abnormality of AChE could still account for the prolonged MEPP. This possibility was excluded by demonstrating that the activity and K_m of AChE were normal in muscle in this syndrome (Engel et al. 1982b).

Pathogenetic mechanisms
The prolonged EPP can be attributed to the prolonged MEPP. Since AChE is intact by physiological, histochemical, and biochemical criteria, and since the prolonged MEPP cannot be attributed to abnormal cable properties of the muscle fiber plasma membrane, the prolonged MEPP is caused

A.G. ENGEL

by a prolonged opentime of the AChR ion channel.

As in congenital NMJ AChE deficiency, the repetitive muscle action potential can be explained by the prolonged EPP. The fact that the repetitive muscle action potential response is present in all muscles indicates that the AChR channel abnormality is ubiquitous and represents the primary disturbance. Thus, the weakness, wasting, and fatigability which appear in selected muscles at variable intervals are secondary phenomena.

The prolonged end-plate currents result in an abnormally increased cation flux into the junctional folds and nearby muscle fiber regions. Because a fraction of the current is carried by calcium (Takeuchi 1963; Evans 1974), transient or permanent calcium excess is likely to occur in the junctional folds and nearby fiber regions. The deleterious effects of the focal calcium excess, which include activation of intracellular proteases and stimulation of membrane-bound phospholipases (Ebashi and Sugita 1979; Jackson et al. 1984), can readily explain the focal degeneration of the junctional folds, the loss of AChR from the folds, and the end-plate myopathy. Further, some of the findings are similar to those noted in mouse muscle exposed to carbachol, a cholinergic agonist, and the carbachol-induced changes can be prevented by exclusion of calcium from the extracellular fluid (Leonard and Salpeter 1982). It is also possible that AChR synthesis is inhibited, as in cultured muscle exposed to cholinergic agonists (Gardner and Fambrough 1979). The decrease of AChR at the NMJ explains the reduced MEPP amplitude and the impaired safety margin of neuromuscular transmission.

The AChR ion channel is also slow in noninnervated muscle fibers and at newly formed end-plates (Schuetze 1987). Thus, hypothetically, the slow-channel syndrome could be due to a developmental failure of the slow to fast conversion of the ion channel. It is now known that this conversion depends on the replacement of the γ by an ϵ subunit in AChR (Gu and Hall 1988). Therefore, expression of the γ instead of the ϵ subunit in the slow-channel syndrome AChR would be evidence for abnormal developmental regulation. However, there is immunoreactivity for the ϵ but not for the γ subunit at slow-channel syndrome NMJs (Engel

1989). Therefore, the most plausible explanation is a mutation in an AChR subunit that hinders the closure of the AChR ion channel.

Congenital syndromes attributed to mutations in the ϵ subunit of AChR

Recently, Engel et al. (1992) observed molecular abnormalities of the AChR ϵ subunit in 2 patients. These were 2 unrelated women, 16 and 49 years of age (Patients 1 and 2), with generalized myasthenic symptoms since infancy, a decremental EMG response, no circulating AChR antibodies, no immune complexes at the NMJ, and a negative family history.

The EMG showed no repetitive CMAP response to single nerve stimuli. NMJ AChE activity was intact. Patient 1 had a severe and Patient 2 a mild end-plate myopathy. The quantal content of the EPP was normal or slightly increased. The number of AChR per NMJ was reduced. MEPPs and MEPCs were abnormally small and prolonged, and in Patient 2 their decay was biexponential. Analysis of ACh-induced current noise revealed that the AChR ion channel conductance was decreased by about 20% in Patient 1 and by 35% in Patient 2; the mean channel opentime was increased 1.8-fold in Patient 1; Patient 2 had biexponential channel kinetics with a shorter and a much longer than normal channel opentime. No immunoreactivity was found for the fetal γ subunit of AChR in either patient. Normal immunoreactivity was present for the α, β, and δ subunits in both patients. Analysis with ϵ-subunit specific antibodies revealed different ablations of internal ϵ epitopes in the 2 patients, but immunoreactivity for the carboxy terminus of the ϵ subunit was preserved in both patients.

In each patient, a mutation in the ϵ subunit is the likely cause of the kinetic ion channel abnormality, and this abnormality is the likely cause of the end-plate myopathy. The AChR deficiency cold be secondary to the end-plate myopathy or to reduced expression or increased lability of the mutant protein. The patients differ from those with the slow-channel syndrome (described above) because (a) the distribution of the weakness is more diffuse; (b) single-nerve stimuli do not evoke repetitive CMAPs; (c) identical immunocytochemical studies show no loss of ε subunit epitopes in the slow-channel syndrome; and (d) the slow-channel syndrome is an autosomal-dominant disorder. A more precise identification of the ε subunit mutations in the 2 patients will become feasible through molecular genetic analysis.

AChR deficiency and short channel opentime

A girl observed at the age of 27 months had bouts of respiratory insufficiency, facial diplegia, ophthalmoparesis, and ate poorly since birth. The symptoms responded to anticholinesterase drugs. A decremental EMG response was present in the facial muscles. AChR antibody tests were negative and no immune complexes were detected at the NMJ. The quantal content of the EPP was normal. MEPP and MEPC amplitudes were abnormally small, but were increased by neostigmine. Analysis of ACh-induced current noise revealed normal AChR channel conductance but the channel opentime was 30% shorter than in 9 normal controls ($p<0.001$). The number of AChR/NMJ determined with ^{125}I-α-bungarotoxin was severely reduced relative to the size of the NMJ. On electron microscopy, the density of AChR on the junctional folds, detected with peroxidase-labelled α-bungarotoxin, was markedly reduced but the junctional folds were well-preserved. It is highly likely that a mutation in an AChR subunit accounts for both the kinetic abnormality and the deficiency of AChR in this patient (Engel et al. 1990; Nagel et al. 1990).

Congenital myasthenic syndrome attributed to abnormal ACh-AChR interaction

In this disorder, the safety margin of neuromuscular transmission is compromised by a small MEPP, but there is no NMJ AChR deficiency. Detailed electrophysiological and morphological studies were recently carried out in a patient with this syndrome (Engel et al. 1990; Uchitel et al. 1990). This patient, a 21-year-old woman, had severe generalized myasthenic symptoms since birth that responded poorly to AChE inhibitors. A decremental EMG was present at 2-Hz stimulation. AChR antibody tests were negative and no immune complexes were detected at the NMJ. The number of AChR per NMJ, determined with ^{125}I-α-bungaro-

toxin, and the distribution of AChR on the junctional folds, determined with peroxidase-labeled α-bungarotoxin, were normal. Detailed quantitative analysis of the fine structure of the NMJ, including the density and diameter of the synaptic vesicles and the distribution of AChR on the junctional folds, gave normal results.

The MEPP and MEPC amplitudes were very small. The quantal content of the EPP potential was normal. Analysis of ACh-induced current noise revealed normal single-channel conductance but the noise power spectrum contained 2 components of a different time course. The small MEPP and MEPC without NMJ AChR deficiency could be due to a decreased ACh content of the synaptic vesicles or an abnormality in ACh-AChR interaction. The fact that synaptic vesicle size was not reduced suggests that synaptic vesicle ACh content was not reduced (Jones and Kwanbunbumpen 1970a; Nagel and Engel 1989), and the abnormal AChR channel kinetics point to a defect in AChR.

Biexponential channel kinetics could result from (a) dual population of AChR, as at the immature NMJ (Fischbach and Shuetze 1980); or (b) blocking of the AChR ion channel (Ruff 1977; Ogden and Colquhoun 1985; Adams 1987), a change in a rate constant for agonist dissociation (Colquhoun and Hawkes 1977), or another alteration in ACh-AChR interaction so that the closure of the diliganded channel is no longer rate-limiting (Colquhoun and Hawkes 1977; Adams 1987). A dual population of AChR with different rate constants of closure is unlikely because both types of AChR would have to have a low conductance to explain the small MEPP and MEPC, but the effective conductance of the channels that did open was normal (Engel et al. 1990; Uchitel et al. 1990). On the other hand, an abnormal interaction of ACh with AChR can account for both the small MEPC and the biexponential AChR ion channel kinetics. Single channel recordings will be required to further characterize the AChR abnormality in this syndrome.

Other less completely studied patients may have had instances of the same disorder. In a patient studied by Vincent et al. (1981) (Case 1), the MEPP amplitude was low, while the AChR content of the NMJ was normal. This could be due to decreased conductance of the AChR ion channel, decreased

number of ACh molecules per quantum, or reduced affinity of AChR for ACh. Specific studies
to distinguish between these possibilities were not
carried out. In a patient investigated by Lecky et
al. (1986) (Case 1), a reduced MEPP amplitude
was associated with normal α-bungarotoxin binding to the NMJ, end-plate structure, and D-
tubocurarine affinity. Limited observations on a
single NMJ suggested that the effectiveness of
ACh to open ion channels was reduced.

Congenital myasthenic syndrome with high conductance and fast closure of the AChR ion channel

This disorder was recognized by Engel et al.
(1990). The propositus, a 12-year-old girl, had mild
fatigable weakness of ocular, cervical, and selected
limb muscles since infancy; she responded only
slightly to pyridostigmine. A decremental EMG
response was present in the hand muscles after exercise and SFEMG was abnormal in the facial
muscles. Her 3-year-old sister had apneic episodes
and fluctuating ptosis since age 3 months. AChR
antibody tests were negative.

The quantal content of the EPP was normal.
MEPP and MEPC were abnormally large and
their decay time constants were abnormally short.
The MEPC alterations suggested abnormal AChR
channel kinetics. This was confirmed by analysis of
the ACh-induced current noise which revealed a
1.7-fold increase in the conductance and a 30% decrease in the opentime of the AChR ion channel.
No AChR deficiency was found at intercostal
NMJs. On electron microscopy, most NMJs
showed no abnormality, but a few were degenerating or simplified.

The ion channel abnormality in this syndrome
may stem from a mutant AChR subunit with an
increased negative charge near the external AChR
vestibule. Site-directed mutagenesis studies indicate that single amino acid substitutions in any
subunit that increase negative charges close to the
external ion channel vestibule enhance cation flux
through the channel (Imoto et al. 1988; Leonard et
al. 1988; Dani 1989). A similar mutation could account for the high channel conductance in this syndrome. The fact that the high channel conductance
is associated with a shorter than normal channel

opentime indicates that the mutation has a dual
effect on the kinetic properties of the ion channel.

The manner in which the physiologic defect produces clinical symptoms is unclear, but it may be
due to the development of an end-plate myopathy
and AChR deficiency in more severely affected
muscles, as in the slow-channel syndrome.

Partially characterized congenital syndromes

AChR deficiencies
In 3 patients described by Vincent et al. in 1981
(Cases 2, 4, 5), NMJ AChE was preserved. NMJ
AChR, estimated from the number of α-bungarotoxin binding sites, was reduced. The kinetic properties of AChR and NMJ ultrastructure were not
investigated. In a patient investigated by Lecky et
al. in 1986 (Case 2), the NMJ appeared elongated
on light microscopy, but NMJ ultrastructure was
normal. The number of NMJ α-bungarotoxin
binding sites was markedly reduced. No electrophysiological studies were done in this case. In another patient studied by Vincent et al. in 1981
(Case 3), the amount of α-bungarotoxin bound to
the NMJ was reduced only after prolonged wash-
ing. This may indicate a reduced affinity of AChR
for α-bungarotoxin and, by inference, for ACh. A
patient studied by Morgan-Hughes et al. (1981)
showed a slight decrease in NMJ AChR and a 10-
fold increase in the affinity of AChR for d-tubocurarine. The manner in which the latter abnormality
contributed to the defect of neuromuscular transmission was not defined.

AChR deficiency and paucity of secondary synaptic clefts
In 4 patients, NMJ AChR deficiency and small
MEPP were associated with poorly developed
junctional folds and a paucity of secondary synaptic clefts (Smit et al. 1984, 1988; Wokke et al. 1989).
In 2 of these patients, the disorder presented at
birth with arthogryposis, respiratory distress, feeding difficulty, and mild ptosis. Subsequently, they
had delayed motor development, cyanotic episodes, reduced exercise tolerance, and exacerbations with fever (Smit et al. 1984, 1988). A milder
form of the syndrome was recognized in 2 adult
siblings, a 64-year-old man and his 58-year-old sister (Wokke et al. 1989). Both had ptosis since early

childhood, exercise intolerance, and gradual worsening of symptoms in adult life. This disorder is associated with a decremental EMG response at low frequencies of stimulation and is responsive to AChE inhibitors. The kinetic properties of AChR were not investigated in any of the 4 patients.

Familial limb-girdle myasthenia

This is an autosomal recessive syndrome (McQuillen 1966; Johns et al. 1971). Weakness of limb-girdle muscles and easy fatigability appear during childhood or in the teens. Ocular and other cranial muscles are not affected. The symptoms respond to anticholinesterase drugs but not to prednisone. EMG studies show a decremental response. Joint contractures, cardiac repolarization defect, type 1 fiber atrophy, and abnormal electrical irritability of the muscle fibers were noted in 1 family (Dobkin and Verity 1978). Histochemical studies demonstrated tubular aggregates in the muscle fibers, but the position of the aggregates relative to the NMJs was not established. Detailed morphologic studies of the NMJ, in-vitro electrophysiologic studies of neuromuscular transmission, and measurements of NMJ AChR are not available in this syndrome.

Acknowledgements

Work conducted in the author's laboratory was supported in part by Research Grant NS-6277 from the National Institutes of Health, US Public Health Service, and by a Grant from the Muscular Dystrophy Association.

REFERENCES

AARLI, J.A., K. STEFANSSON, L.S.G. MARTON and R.L. WOLLMANN: Patients with myasthenia gravis and thymoma have in their sera IgG autoantibodies against titin. Clin. Exp. Immunol. 82 (1990) 284–288.

ABBAS, A.K., A.H. LICHTMAN and J.S. POBER: Cellular and Molecular Immunology. Philadelphia, PA, W.B. Saunders (1991).

ADAMS, P.R.: Transmitter action at endplate membrane. In: M.M. Salpeter (Ed.), The Vertebrate Neuromuscular Junction. New York, Alan Liss (1987) 317–359.

ALBUQUERQUE, E.X., F.J. LEBEDA, S.H. APPEL, R. ALMON, F.C. KAUFFMAN, R.F. MAYER, T. NARAHASHI and J.Z. YEH: Effects of normal and myasthenic serum factors on innervated and chronically denervated mammalian muscles. Ann. N.Y. Acad. Sci. 274 (1976) 475–492.

AMERICAN MEDICAL ASSOCIATION: Consensus Conference: The utility of therapeutic plasmapheresis. J. Am. Med. Assoc. 256 (1986) 1333–1337.

ANDERSON, C.R. and C.F. STEVENS: Voltage clamp analysis of acetylcholine produced end-plate current fluctuations at frog neuromuscular junction. J. Physiol. (London) 235 (1973) 655–691.

APPEL, S.H., S.B. ELIAS and P. CHAUVIN: The role of acetylcholine receptor antibodies in myasthenia gravis. Fed. Proc. 38 (1979) 2381–2385.

ARGOV, Z., T. BRENNER and O. ABRAMSKY: Ampicillin may aggravate clinical and experimental myasthenia gravis. Arch. Neurol. 43 (1986) 255–256.

ARNON, S.S., T.F. MIDURA, S.A. CLAY, R.M. WOOD and J. CHIN: Infant botulism. Epidemiological, clinical and laboratory aspects. J. Am. Med. Assoc. 237 (1977) 1946–1951.

ARSURA, E.: Experience with intravenous immunoglobulin in myasthenia gravis. Clin. Immunol. Immunopathol. 53 (1989) S170–S179.

ARSURA, E., N.G. BRUNNER, T. NAMBA and D. GROB: High-dose intravenous methylprednisolone in myasthenia gravis. Arch. Neurol. 42 (1985) 1149–1153.

ASHER, O., D. NEUMANN and S. FUCHS: Increased levels of acetylcholine receptor α-subunit mRNA in experimental autoimmune myasthenia gravis. FEBS Lett. 233 (1988) 277–281.

ASHIZAWA, T. and S.H. APPEL: Complement-dependent lysis of cultured rat myotubes by myasthenic immunoglobulins. Neurology 35 (1985) 1748–1753.

BALCER, L.J., K.R. MCINTOSH, J.C. NICHOLS and D.B. DRACHMAN: Suppression of immune response to acetylcholine receptor by interleukin 2-fusion toxin: in vivo and in-vitro studies. J. Neuroimmunol. 31 (1991) 115–122.

BATOCCHI, A.P., A. EVOLI, M.T. PALMISANI, M. LO MONACO, M. BARTOCCIONI and P. TONALI: Early-onset myasthenia gravis: clinical characteristics and response to therapy. Eur. J. Pediatr. 150 (1990) 66–68.

BATRA, P., C. HERRMANN and D. MULDER: Mediastinal imaging in myasthenia gravis. Correlation of chest radiography, CT, MR, and surgical findings. Am. J. Roentgenol. 148 (1987) 515–519.

BECKER, K.L., J.H. TITUS, W.M. MCCONAHEY and L.B. WOLLMAN: Morphologic evidence of thyroiditis in myasthenia gravis. J. Am. Med. Assoc. 187 (1964) 994–996.

BEECH, R.L., K. VACA and G. PILAR: Ionic and metabolic requirements for high-affinity choline uptake and acetylcholine synthesis in nerve terminals of a neuromuscular junction. J. Neurochem. 34 (1980) 1387–1398.

BENNETT, M.R. and T. FLORIN: A statistical analysis of the release of acetylcholine at newly formed synapses in striated muscle. J. Physiol. (London) 238 (1974) 93–107.

BERMAN, P.W. and S.P. HEINEMANN: Antigenic modulation of junctional acetylcholine receptor is not sufficient to account for the development of myasthenia gravis in receptor immunized mice. J. Immunol. 132 (1984) 711–714.

BERRIH-AKNIN, S., S. COHEN-KAMINSKY, V. LEPAGE, D. NEUMANN, J.-F. BACH and S. FUCHS: T-cell antigenic sites involved in myasthenia gravis: correlations with antibody titre and disease severity. J. Autoimmun. 4 (1991) 137–153.

BETZ, W. and B. SAKMANN: Effects of proteolytic enzymes on function and structure of frog neuromuscular junctions. J. Physiol. (London) 230 (1973) 673–688.

BIESECKER, G. and C.M. GOMEZ: Inhibition of acute passive transfer experimental autoimmune myasthenia gravis with Fab antibody to complement C6. J. Immunol. 142 (1989) 2654–2659.

BIRD, S.J.: Clinical and electrophysiologic improvement in the Lambert-Eaton syndrome with intravenous immunoglobulin therapy. Muscle Nerve 14 (1991) 913–914.

BLALOCK, A.: Thymectomy in the treatment of myasthenia gravis. Report of 20 cases. J. Thorac. Surg. 13 (1944) 316–339.

BLALOCK, A., M.F. MASON, H.F. MORGAN and S.S. RIVEN: Myasthenia gravis and tumors of the thymic region. Report of a case in which the tumor was removed. Ann. Surg. 110 (1939) 544–561.

BOFILL, M., G. JANOSSY, N. WILCOX, M. CHILOSI, L.K. TREJDOSIEWICZ and J. NEWSOM-DAVIS: Microenvironments in the normal thymus and the thymus in myasthenia gravis. Am. J. Pathol. 119 (1985) 462–473.

BOGEN, S., E. MOZES and S. FUCHS: Induction of acetylcholine receptor-specific suppression. An in vitro model of antigen-specific immunosuppression in myasthenia gravis. J. Exp. Med. 159 (1984) 292–304.

BOLGER, G.B., K.M. SULLIVAN, A.M. SPENCE, F.R. APPELBAUM and R. JOHNSTON: Myasthenia gravis after allogeneic bone marrow transplantation: relationship to chronic graft-versus-host disease. Neurology 36 (1986) 1087–1091.

BOOJ, S.: Axonal transport of synapsin I and cholinergic synaptic vesicle-like material. Further immunohistochemical evidence for transport of axonal cholinergic transmitter vesicles in motor neurons. Acta Physiol. Scand. 128 (1986) 155–165.

BRAUN, S., C. TRANCHANT, J.-T. VILQUIN, P. LABOURET, J.-M. WARTER and P. POIDRON: Stimulating effects of prednisolone on acetylcholine expression and myogenesis in primary culture of newborn rat muscle cells. J. Neurol. Sci. 92 (1989) 119–131.

BREARLY, S., T.A. GENTLE, M.I. BAYNHAM, K.D. ROBERTS, L.D. ABRAMS and R.A. THOMPSON: Immunodeficiency following neonatal thymectomy in man. Clin. Exp. Immunol. 70 (1987) 322–327.

BREY, R., R. MCATEE and R. BAROHN: Activated complement in human myasthenia gravis. Ann. Neurol. 30 (1991) 263.

BROCKE, S., C. BRAUTBAR, L. STEINMANN, O. ABRAMSKY, J. ROTHBARD, D. NEUMANN, S. FUCHS and E. MOZES: In vitro proliferative responses and antibody titers specific to human acetylcholine receptor synthetic peptides in patients with myasthenia gravis and relation to HLA class II genes. J. Clin. Invest. 82 (1988) 1894–1900.

BRUNNER, N.G., C.L. BERGER, T. NAMBA and D. GROB: Corticotropin and corticosteroids in generalized myasthenia gravis: Comparative studies and role in management. Ann. N.Y. Acad. Sci. 274 (1976) 577–595.

BUNDEY, S.: A genetic study of infantile and juvenile myasthenia gravis. J. Neurol. Neurosurg. Psychiat. 35 (1972) 41–51.

BURGES, J., D.W. WRAY, S. PIZZIGHELLA, Z. HALL and A. VINCENT: A myasthenia gravis plasma immunoglobulin reduces miniature endplate potentials at human endplates in vitro. Muscle Nerve 13 (1990) 407–413.

BUZZARD, E.F.: The clinical history and post-mortem examination of five cases of myasthenia gravis. Brain 28 (1905) 438–483.

CAMPBELL, A.K. and B.P. MORGAN: Monoclonal antibodies demonstrate protection of polymorophonuclear leukocytes against complement attack. Nature (London) 317 (1985) 164–166.

CAMPBELL, H. and E. BRAMWELL: Myasthenia gravis. Brain 23 (1900) 277–336.

CECCARELLI, B. and W.P. HURLBUT: Vesicle hypothesis of the release of quanta of acetylcholine. Physiol. Rev. 60 (1980) 396–441.

CHALK, C.H., N.M.F. MURRAY, J. NEWSOM-DAVIS, J.H. O'NEILL and S.G. SPIRO: Response of the Lambert-Eaton myasthenic syndrome to treatment of associated small-cell lung carcinoma. Neurology 40 (1990) 1552–1556.

CHANG, C.C., S.-T. CHUANG and M.C. HUANG: Effects of chronic treatment with various neuromuscular blocking agents on the number and distribution of acetylcholine receptors in the rat diaphragm. J. Physiol. (London) 250 (1975) 161–173.

CHARLTON, M.P., S.J. SMITH and R. ZUKER: Role of presynaptic calcium ions and channels in synaptic facilitation and depression at the squid giant synapse. J. Physiol. (London) 323 (1982) 173–193.

CHERINGTON, M.: Botulism. Ten year experience. Arch. Neurol. 30 (1974) 432–437.

CHERINGTON, M.: Guanidine and germine in Lambert-Eaton syndrome. Neurology 26 (1976) 944–946.

CHILDS, L.A., R. HARRISON and G.G. LUNT: Complement-mediated muscle damage produced by myasthenic sera. Ann. N.Y. Acad. Sci. 505 (1987) 180–193.

CHRISTADOSS, P.: C5 gene influences the development of murine myasthenia gravis. J. Immunol 140 (1988) 2589–2592.

CHRISTADOSS, P. and M.J. DAUPHINEE: Immunotherapy for myasthenia gravis: a murine model. J. Immunol. 136 (1986) 2437–2440.

CHRISTADOSS, P., V.A. LENNON, C.J. KRCO, E.H. LAMBERT and C.S. DAVID: Genetic control of autoimmunity to acetylcholine receptors: role of Ia molecules. Ann. N.Y. Acad. Sci. 377 (1981) 258–277.

CHRISTENSEN, B.N. and A.R. MARTIN: Estimates of

probability of transmitter release at the mammalian neuromuscular junction. J. Physiol. (London) 210 (1970) 933–945.

CIKES, N., M.Y. MOMOI, C.L. WILLIAMS, H.C. HOAGLAND, S. WHITTINGHAM and V.A. LENNON: Striational autoantibodies: quantitative detection by enzyme immunoassay in myasthenia gravis, thymoma, and recipients of D-penicillamine or allogeneic bone marrow. Mayo Clin. Proc. 63 (1988) 474–481.

CITTERIO, C.V., A. LOMBARDI, P.G. ROMANI and A. ARBETTA: Effectiveness of steroid treatment in myasthenia gravis: a retrospective study. Acta Neurol. Scand. 84 (1991) 33–39.

CLAMON, G.H., W.K. EVANS, F.A. SHEPHERD and J.G. HUMPHREY: Myasthenic syndrome and small cell cancer of the lung. Variable response to anti-neoplastic therapy. Arch. Intern. Med. 144 (1984) 999–1000.

COLQUHOUN, D. and A.G. HAWKES: Relaxation and fluctuations of membrane currents that flow through drug-operated ion channels. Proc. R. Soc. Lond. B. 199 (1977) 231–262.

COMPSTON, D.A.S., A. VINCENT, J. NEWSOM-DAVIS and J.R. BATCHELOR: Clinical, pathological, HLA antigen and immunological evidence for disease heterogeneity in myasthenia gravis. Brain 103 (1980) 579–601.

CONFAVREUX, C., C. NADINE and G. AIMARD: Fulminant myasthenia gravis soon after initiation of acebutolol therapy. Eur. Neurol. 30 (1990) 279–281.

CONOMY, J.P., M. LEVISOHN and A. FANAROFF: Familial infantile myasthenia gravis: a cause of sudden death in young children. J. Pediatr. 87 (1975) 428–429.

CONTI-TRONCONI, B.M., F. DI PADOVA, M. MORGUTI, A. MISSIROLI and L. FRATTOLA: Stimulation of lymphocytes by cholinergic receptor in myasthenia gravis. J. Neuropathol. Exp. Neurol. 36 (1977) 157–168.

CONTI-TRONCONI, B.M., A. BRIGONZI, G. FUMAGALLI, M. SHER, V. COSI, G. PICCOLO and F. CLEMENTI: Antibody-induced degradation of acetylcholine receptor in myasthenia gravis: clinical correlates and pathogenetic significance. Neurology 31 (1981a) 1440–1444.

CONTI-TRONCONI, B.M., S.J. TZARTOS and J. LINDSTROM: Monoclonal antibodies as probes of acetylcholine receptor structure. II. Binding to native receptor. Biochemistry 20 (1981b) 2181–2191.

COSI, V., M. LOMBARDI, G. PICCOLO and A. ERBETTA: Treatment of myasthenia gravis with high-dose intravenous immunoglobulin. Acta Neurol. Scand. 84 (1991) 81–84.

CULL-CANDY, S.G., H. LUND and S. THESLEFF: Effects of botulinum toxin on neuromuscular transmission in the rat. J. Physiol. (London) 260 (1976) 177–203.

CULL-CANDY, S.G., R. MILEDI and A. TRAUTMANN: Acetylcholine-induced channels and transmitter release at human end-plate. Nature (London) 271 (1978) 74–75.

CULL-CANDY, S.G., R. MILEDI and A. TRAUTMANN: On the release of transmitter at normal, myasthenia

gravis and myasthenic syndrome affected human end-plates. J. Physiol. (London) 299 (1980) 621–638.

DALE, H.H. and W. FELDBERG: Chemical transmission at motor nerve endings in voluntary muscle? J. Physiol. (London) 81 (1934) 39P–40P.

DANI, J.A.: Site-directed mutagenesis and single-channel currents define the ionic channel of the nicotinic acetylcholine receptor. Trends Neurosci. 12 (1989) 125–128.

DAU, P.C., J.M. LINDSTROM, C.K. CASSEL, E.H. DENYS, E.E. SHEV and L.E. SPITLER: Plasmapheresis and immunosuppressive drug therapy in myasthenia gravis. N. Engl. J. Med. 297 (1977) 1134–1140.

DE AIZAPURA, H.J., E.H. LAMBERT, G.E. GRIESMANN, B.M. OLIVERA and V.A. LENNON: Antagonism of voltage-gated calcium channels in small cell carcinomas of patients with and without Lambert-Eaton myasthenic syndrome by autoantibodies, ω-conotoxin and adenosine. Cancer Res. 48 (1988) 4719–4724.

DE LUCA, A., M.J. RAND, J.J. REID and D.F. STORY: Differential sensitivities of avian and mammalian neuromuscular junctions to inhibition of cholinergic transmission by ω-conotoxin GVIA. Toxicon 29 (1991) 311–320.

DE SILVA, S., J.E. BLUM, K. MCINTOSH, S. ORDER and D.B. DRACHMAN: Treatment of experimental myasthenia gravis with total lymphoid irradiation. Clin. Immunol. Immunopathol. 48 (1988) 31–41.

DE SILVA, S., K. MCINTOSH, J.E. BLUM, S. ORDER, D. MELLITS and D.B. DRACHMAN: Total lymphoid irradiation and antigen-specific tolerance: future therapy for experimental myasthenia gravis? J. Neuroimmunol. 29 (1990) 93–103.

DEL CASTILLO, J. and L. ENGBAEK: The nature of neuromuscular block caused by magnesium. J. Physiol. (London) 124 (1954) 370–384.

DEL CASTILLO, J. and B. KATZ: Quantal components of the end-plate potential. J. Physiol. (London) 124 (1954) 560–573.

DEL CASTILLO, J. and B. KATZ: Localization of active spots within the neuromuscular junction of the frog. J. Physiol. (London) 132 (1956) 160–649.

DELISI, C. and J.A. BERZOFSKY: T-cell antigenic sites tend to be amphipatic structures. Proc. Natl Acad. Sci. USA 82 (1985) 7048–7052.

DESMEDT, J.E. and S. BORENSTEIN: The testing of neuromuscular transmission. In: P.J. Vinken and G.W. Bruyn (Eds.), Handbook of Clinical Neurology, Vol. 7. Amsterdam, North-Holland Publ. Co. (1970) 104–115.

DESMEDT, J.E. and S. BORENSTEIN: Diagnosis of myasthenia gravis by nerve stimulation. Ann. N.Y. Acad. Sci. 274 (1976) 174–188.

DESMEDT, J.E. and S. BORENSTEIN: Double-step nerve stimulation for myasthenic block: sensitization of postactivation exhaustion by ischemia. Ann. Neurol. 1 (1977) 55–64.

DOBKIN, B.H. and M.A. VERITY: Familial neuromuscular disease with type 1 fiber hypoplasia, tubular aggregates, cardiomyopathy and myasthenic features. Neurology 28 (1978) 1135–1140.

DOLLY, J.O., J. BLACK, R.S. WILLIAMS and J. MELLING: Acceptors for botulinum neurotoxin reside on motor nerve terminals and mediate internalization. Nature (London) 307 (1984) 457–460.

DOLLY, J.O., M. GWILT, G. LACEY, J. NEWSOM-DAVIS, A. VINCENT, P. WHITING and D.W. WRAY: Action of antibodies directed against the acetylcholine receptor on channel function at mouse and rat motor endplates. J. Physiol. (London) 399 (1988) 577–589.

DONALDSON, D.H., M. ANSHER, S. HORAN, R.B. RUTHERFORD and S.P. RINGEL: The relationship of age to outcome in myasthenia gravis. Neurology 40 (1990) 786–790.

DOWNES, J.M., B.M. GREENWOOD and S.H. WRAY: Autoimmune aspects of myasthenia gravis. Q. J. Med. 25 (1966) 85–105.

DRACHMAN, D.B., R.N. ADAMS, L.F. JOSIFEK and S.G. SEL: Functional activities of autoantibodies to acetylcholine receptors and the clinical severity of 30 myasthenia gravis. N. Engl. J. Med. 307 (1982) 769–775.

DUPONT, B.L. and D.P. RICHMAN: Complement activation by anti-acetylcholine receptor monoclonal antibody in vivo correlates with potency of EAMG response in vivo. Ann. N.Y. Acad. Sci. 505 (1987) 725–727.

DURELLI, L., U. MASAZZO, G. POCCARDI, M.F. FERRIO, R. CAVALLO, C. MAGGI, C. CASCADIO, M. DISUMMA and L. BERGAMINI: Increased thymocyte differentiation in myasthenia gravis: a dust color immunofluorescence phenotypic analysis. Ann. Neurol. 27 (1990) 174–180.

DWYER, D.S., M. VAKIL and J.F. KEARNEY: Idiotypic network connectivity and a possible cause of myasthenia gravis. J. Exp. Med. 164 (1986) 1310–1318.

EATON, L.M. and O.T. CLAGETT: The value of thymectomy. Am. J. Med. 19 (1955) 703–717.

EATON, L.M. and E.H. LAMBERT: Electromyography and electrical stimulation of nerves in diseases of the motor unit: observations on a myasthenic syndrome associated with malignant tumors. J. Am. Med. Assoc. 163 (1957) 1117–1124.

EBASHI, S. and H. SUGITA: The role of calcium in physiological and pathological processes of skeletal muscle. Excerpta Medica ICS 455 (1979) 73–84.

ELIAS, S.B. and S.H. APPEL: Autoantibodies to acetylcholine receptors in myasthenia gravis. N. Engl. Med. 308 (1983) 402.

ELMQVIST, D. and J.O. JOSEFSSON: The nature of the neuromuscular block produced by neomycin. Acta Physiol. Scand. 54 (1962) 105–110.

ELMQVIST, D. and E.H. LAMBERT: Detailed analysis of neuromuscular transmission in a patient with the myasthenic syndrome sometimes associated with bronchogenic carcinoma. Mayo Clin. Proc. 43 (1968) 689–713.

ELMQVIST, D. and D.M.J. QUASTEL: A quantitative study of end-plate potentials in isolated human muscle. J. Physiol. (London) 178 (1965a) 505–529.

ELMQVIST, D. and D.M.J. QUASTEL: Presynaptic action of hemicholinium at the neuromuscular junction. J. Physiol. (London) 177 (1965b) 463–482.

ELMQVIST, D., W.W. HOFMANN, J. KUGELBERG and D.M.J. QUASTEL: An electrophysiological investigation of neuromuscular transmission in myasthenia gravis. J. Physiol. (London) 174 (1964) 417–434.

ENG, H. and A.K. LEFVERT: Isolation of an antiidiotypic antibody with acetylcholine-receptor-like binding properties from myasthenia gravis patients. Ann. Inst. Pasteur/Immunol. 139 (1988) 569–580.

ENGEL, A.G.: Thyroid function and myasthenia gravis. Arch. Neurol. (Chicago) 4 (1961) 95–106.

ENGEL, A.G.: Myasthenic syndromes. In: A.G. Engel and B.Q. Banker (Eds.), Myology. New York, McGraw-Hill (1986) 1955–1990.

ENGEL, A.G.: Changes in end-plate structure in neuromuscular transmission disorders. In: L.C. Selli, R. Libelius and S. Thesleff (Eds.), Neuromuscular Junction. New York, Elsevier (1989) 415–428.

ENGEL, A.G.: Congenital disorders of neuromuscular transmission. Sem. Neurol. 10 (1990) 12–26.

ENGEL, A.G. and K. ARAHATA: The membrane attack complex of complement at the end-plate in myasthenia gravis. Ann. N.Y. Acad. Sci. 505 (1987) 333–345.

ENGEL, A.G. and G. FUMAGALLI: Mechanisms of acetylcholine loss from the neuromuscular junction. In: D. Everest and J. Whelan (Eds.), Receptors, antibodies and disease. Ciba Foundation Symposium 90. London, Pitman Medical (1982) 197–224.

ENGEL, A.G. and E.H. LAMBERT: Congenital myasthenic syndromes. Electroencephalogr. Clin. Neurophysiol., Suppl. 39 (1987) 91–102.

ENGEL, A.G. and T. SANTA: Histometric analysis of the ultrastructure of the neuromuscular junction in myasthenia gravis and in the myasthenic syndrome. Ann. N.Y. Acad. Sci. 183 (1971) 46–63.

ENGEL, A.G., E.H. LAMBERT and T. SANTA: Study of long-term anticholinesterase therapy. Effects on neuromuscular transmission and motor end-plate fine structure. Neurology 23 (1973) 1273–1281.

ENGEL, A.G., M. TSUJIHATA and F. JERUSALEM: Quantitative assessment of motor end-plate ultrastructure in normal and diseased muscle. In: P.J. Dyck, P.K. Thomas and E.H. Lambert (Eds.), Peripheral Neuropathy, Vol. 2. Philadelphia, Pa., W.B. Saunders (1975) 1404–1415.

ENGEL, A.G., M. TSUJIHATA, E.H. LAMBERT, J.M. LINDSTROM and V.A. LENNON: Experimental autoimmune myasthenia gravis: a sequential and quantitative study of the neuromuscular junction ultrastructure and electrophysiological correlations. J. Neuropathol. Exp. Neurol. 35 (1976a) 569–587.

ENGEL, A.G., M. TSUJIHATA, J.M. LINDSTROM and V.A. LENNON: The motor end-plate in myasthenia gravis and in experimental autoimmune myasthenia gravis. A quantitative ultrastructural study. Ann. N.Y. Acad. Sci. 274 (1976b) 60–79.

ENGEL, A.G., E.H. LAMBERT and M.R. GOMEZ: A new myasthenic syndrome with end-plate acetylcholinesterase deficiency, small nerve terminals and reduced acetylcholine release. Ann. Neurol. 1 (1977a) 315–330.

ENGEL, A.G., E.H. LAMBERT and F.M. HOWARD: Immune complexes (IgG and C3) at the motor end-plate in myasthenia gravis. Ultrastructural and light microscopic localization and electrophysiological correlations. Mayo Clin. Proc. 52 (1977b) 267–280.

ENGEL, A.G., J.M. LINDSTROM, E.H. LAMBERT and V.A. LENNON: Ultrastructural localization of the acetylcholine receptor in myasthenia gravis and in its experimental autoimmune model. Neurology 27 (1977c) 307–315.

ENGEL, A.G., H. SAKAKIBARA, K. SAHASHI, J.M. LINDSTROM, E.H. LAMBERT and V.A. LENNON: Passively transferred experimental autoimmune myasthenia gravis. Sequential and quantative study of the end-plate fine structure and ultrastructural localization of immune complexes (IgG and C3) and of the acetylcholine receptor. Neurology 29 (1979) 179–188.

ENGEL, A.G., K. SAHASHI and G. FUMAGALLI: The immunopathology of acquired myasthenia gravis. Ann. N.Y. Acad. Sci. 377 (1981) 158–174.

ENGEL, A.G., H. FUKUNAGA and M. OSAME: Stereometric estimation of the area of the freeze-fractured membrane. Muscle Nerve 5 (1982a) 682–685.

ENGEL, A.G., E.H. LAMBERT, D.M. MULDER, C.F. TORRES, K. SAHASHI, T.E. BERTORINI and J.N. WHITAKER: A newly recognized congenital myasthenic syndrome attributed to a prolonged open time of the acetylcholine-induced ion channel. Ann. Neurol. 11 (1982b) 553–569.

ENGEL, A.G., T. FUKUOKA, B. LANG, J. NEWSOM-DAVIS, A. VINCENT and D.W. WRAY: Lambert-Eaton myasthenic syndrome IgG: early morphologic effects and immunolocalization at the motor end-plate. Ann. N.Y. Acad. Sci. 505 (1987) 333–345.

ENGEL, A.G., T.J. WALLS, A. NAGEL and O. UCHITEL: Newly recognized congenital myasthenic syndromes. I. Congenital paucity of synaptic vesicles and reduced quantal release. II. High-conductance fast-channel syndrome. III. Abnormal acetylcholine receptor (AChR) interaction with acetylcholine. IV. AChR deficiency and short channel open-time. Prog. Brain Res. 84 (1990) 125–137.

ENGEL, A.G., D. HUTCHINSON, S. NAKANO, C.M. HARPER, V. TRASTEK, L. MURPHY, J. FISCHER, Y. GU and Z.W. HALL: Congenital myasthemic syndrome attributed to a mutation of the epsilon subunit of the acetylcholine receptor. Neurology 42 (1992) 307.

ENGEL, W.K., B.W. FESTOFF, B.M. PATTEN, M.L. SWERDLOW, H.H. NEWBALL and M.D. THOMPSON: Myasthenia gravis. Ann. Intern. Med. 81 (1974) 225–246.

ERB, W.: Zur Casuistik der bulbären Lähmungen. Arch. Psychiatr. Nervenkr. 9 (1879) 325–350.

EVANS, R.H.: The entry of labelled calcium into the innervated region of the mouse diaphragm muscle. J. Physiol. (London) 240 (1974) 517–533.

EVOLI, A., A.P. BATOCCHI, C. PROVENZANO and P. TONALI: Thymectomy in the treatment of myasthenia gravis: report of 247 patients. J. Neurol. 235 (1988) 272–276.

EYMARD, B., S. DE LA PORTE, C. PANNIER, S. BERRIH-AKNIN, E. MOREL, M. FARDEAU, J.F. BACH and J. KOENIG: Effect of myasthenic patient sera on the number and distribution of acetylcholine receptors in muscle and nerve-muscle cultures from rat. Correlations with clinical state. J. Neurol. Sci. 86 (1988) 41–59.

EYMARD, B., E. MOREL, O. DULAC, M.L. MOUTARD-CODOU, E. JEANNOT, J.P. HARPEY, P. RONDOT and J.F. BACH: Myasténie et grossesse: une étude clinique et immunologique de 42 cas (21 myasthénies néonatales). Rev. Neurol. 145 (1989) 696–701.

FAMBROUGH, D.M.: Turnover of acetylcholine receptors: a brief overview and some cautions concerning significance in myasthenia gravis. In: G. Serratrice (Ed.), Neuromuscular Diseases. New York, Raven (1984) 465–470.

FAMBROUGH, D.M., D.B. DRACHMAN and S. SATYAMURTI: Neuromuscular junction in myasthenia gravis: decreased acetylcholine receptors. Science 182 (1973) 293–295.

FATT, P. and B. KATZ: Spontaneous subthreshold activity at motor nerve endings. J. Physiol. (London) 117 (1952) 109–128.

FAWCETT, P.R.W., S.M. MCLACHLAN, L.V.B. NICHOLSON, Z. ARGOV and F. MASTAGLIA: D-penicillamine associated myasthenia gravis: immunological and electrophysiological studies. Muscle Nerve 5 (1982) 328–334.

FAZEKAS, A., S. KOMOLY, B. BOZSIK and A. SZOBOR: Myasthenia gravis: demonstration of membrane attack complex in muscle end-plates. Clin. Neuropathol. 5 (1986) 78–83.

FERTUCK, H.C. and M.M. SALPETER: Quantitation of junctional and extrajunctional acetylcholine receptors by electron microscope autoradiography after ^{125}I-α-bungarotoxin binding at mouse neuromuscular junctions. J. Cell. Biol. 69 (1976) 144–158.

FISHBACH, G.D. and S.M. SCHUETZE: A post-natal decrease in acetylcholine channel open time at rat end-plates. J. Physiol. (London) 303 (1980) 125–137.

FISCHER, K.C. and R.J. SCHWARTZMAN: Oral corticosteroids in the treatment of ocular myasthenia gravis. Ann. N.Y. Acad. Sci. 274 (1976) 652–658.

FLACKE, W.: Treatment of myasthenia gravis. N. Engl. J. Med. 288 (1973) 27–31.

FUCHS, S., D. BARTFELD, D. MOCHLY-ROSEN, M. SOUROUJON and C. FEINGOLD: Acetylcholine receptor: molecular dissection and monoclonal antibodies in the study of experimental myasthenia. Ann. N.Y. Acad. Sci. 377 (1981) 110–124.

FUJI, Y. and J. LINDSTROM: Specificity of the T cell immune response to acetylcholine receptor in experimental autoimmune myasthenia gravis. Response to subunits and synthetic peptides. J. Immunol. 141 (1988) 1830–1837.

FUKUNAGA, H., A.G. ENGEL, M. OSAME and E.H. LAMBERT: Paucity and disorganization of presynaptic membrane active zones in the Lambert-Eaton my-asthenic syndrome. Muscle Nerve 5 (1982) 686–697.

FUKUNAGA, H., A.G. ENGEL, B. LANG, B. NEWSOM-DAVIS and A. VINCENT: Passive transfer of Lambert-Eaton myasthenic syndrome IgG from man to mouse depletes the presynaptic membrane active zones. Proc. Natl. Acad. Sci. USA 80 (1983) 7636–7640.

FUKUOKA, T., A.G. ENGEL, B. LANG, J. NEWSOM-DAVIS, C. PRIOR and D.W. WRAY: Lambert-Eaton myasthenic syndrome. I. Early morphologic effects of IgG on the presynaptic membrane active zones. Ann. Neurol. 22 (1987a) 193–199.

FUKUOKA, T., A.G. ENGEL, B. LANG, J. NEWSOM-DAVIS and A. VINCENT: Lambert-Eaton myasthenic syndrome. II. Immunoelectron microscopy localization of IgG at the mouse motor end-plate. Ann. Neurol. 22 (1987b) 200–211.

FULPIUS, B.W., A.D. ZURN, D.A. GRANATO and R.M. LEDER: Acetylcholine receptor and myasthenia gravis. Ann. N.Y. Acad. Sci. 274 (1976) 116–129.

FUMAGALLI, G., A.G. ENGEL and J. LINDSTROM: Estimation of acetylcholine receptor degradation rate by external gamma counting in vivo. Mayo Clin. Proc. 57 (1982a) 758–764.

FUMAGALLI, G., A.G. ENGEL and J. LINDSTROM: Ultrastructural aspects of acetylcholine receptor turnover at the normal end-plate and in autoimmune myasthenia gravis. J. Neuropathol. Exp. Neurol. 41 (1982b) 567–579.

GARDNER, J.M. and D.M. FAMBROUGH: Acetylcholine receptor degradation measured by density labeling: effects of cholinergic ligands and evidence against recycling. Cell 16 (1979) 661–674.

GARLEPP, M.J., P.H. KAY and R.L. DAWKINS: The diagnostic significance of autoantibodies to the acetylcholine receptor. J. Neurol. Sci. 3 (1982) 337–350.

GENKINS, G., A.E. PAPATESTAS, S.H. HOROWITZ and P. KORNFELD: Studies in myasthenia gravis: early thymectomy. Electrophysiologic and pathologic correlations. Am. J. Med. 58 (1975) 517–524.

GIERON, M.A. and J.K. KORTHALS: Familial infantile myasthenia gravis. Report of three cases with follow-up until adult life. Arch. Neurol. 42 (1985) 143–144.

GLENNIE, M.J. and G.T. STEVENSON: Univalent antibodies kill tumor cells in vitro and in vivo. Nature (London) 295 (1982) 712–714.

GOLDFLAM, S.: Über einen scheinbar heilbaren bulbärparalitischen Symptomenkomplex mit Beteiligung der Extremitäten. Dtsch. Z. Nervenheilkd. 4 (1893) 312–352.

GOLDMAN, A.J., C. HERRMAN JR., J.C. KEESEY, D.G. MULDER and W.J. BROWN: Myasthenia gravis and invasive thymoma: a 20 year experience. Neurology 25 (1975) 1021–1025.

GOLDSTEIN, G. and D.H. SCHLESINGER: Thymopoietin and myasthenia gravis: neostigmine-responsive neuromuscular block produced in mice by a synthetic peptide fragment of thymopoietin. Lancet ii (1975) 256–259.

GOMEZ, C.M. and D.P. RICHMAN: Anti-acetylcholine receptor antibodies directed against the α-bungarotoxin binding site induce a unique form of experimental myasthenia. Proc. Natl Acad. Sci. USA 80 (1983) 4089–4093.

GOULON, M., D. ELKHARRAT and P. GAJDOS Traitement de la myasthénie grave par la ciclosporine. Etude ouverte de 12 mois. Presse Med. (1989) 341–346.

GRAU, J.M., J. CASADEMONT, R. MONFORTE, P. MARIN, A. GRANENA and C. ROZMAN: Myasthenia gravis after allogeneic bone marrow transplantation: report of a new case and pathogenetic considerations. Bone Marrow Transplant. 5 (1990) 435–437.

GREEN, D.P.L., R. MILEDI and A. VINCENT: Neuromuscular transmission after immunization against acetylcholine receptors. Proc. R. Soc. London Ser. B. 189 (1975) 57–68.

GREER, M. and M. SCHOTLAND: Myasthenia gravis in the newborn. Pediatrics 26 (1960) 101–108.

GROB, D., N.G. BRUNNER and T. NAMBA: The natural course of myasthenia gravis and effect of therapeutic measures. Ann. N.Y. Acad. Sci 377 (1981) 652–669.

GROHOVAZ, F., A.R. LIMBRICK and R. MILEDI: Acetylcholine receptors at the rat neuromuscular junction as revealed by deep etching. Proc. R. Soc. London Ser. B 215 (1982) 147–154.

GU, Y. and Z.W. HALL: Immunological evidence for a change in subunits of the acetylcholine receptor in developing and denervated rat muscle. Neuron 1 (1988) 117–125.

GUTHRIE, L.B.: Myasthenia gravis in the 17th century. Lancet i (1903) 330.

GUTMANN, L.: Heat exacerbation of myasthenia gravis. Neurology 28 (1978) 398.

GUTMANN, L., T.W. CROSBY, M. TAKAMORI and J.D. MARTIN: The Eaton-Lambert syndrome and autoimmune disorders. Am. J. Med. 53 (1972) 354–356.

HALL, Z.W. and R.B. KELLY: Enzymatic detachment of endplate acetylcholinesterase from muscle. Nature (London) New Biol. 232 (1971) 62–63.

HAMILL, P. and M.B. WALKER: Action of 'Prostigmin' (Roche) in neuro-muscular disorders. J. Physiol. (London) 84 (1935) 36P.

HANKINS, J.R., R.F. MAYER, J.R. SATTERFIELD, S.Z. TURNEY, S. ATTAR, A.J. SEQUEIRA, B.W. THOMPSON and J.S. MCLAUGHIN: Thymectomy for myasthenia gravis: 14-year experience. Ann. Surg. 201 (1985) 618–625.

HARCOURT, G.C., N. SOMMER, J. ROTHBARD, H.N.A. WILCOX and J. NEWSOM-DAVIS: A juxtamembrane epitope on the human acetylcholine receptor recognized by T cells in myasthenia gravis. J. Clin. Invest. 82 (1988) 1295–1300.

HART, Z., K. SAHASHI, E.H. LAMBERT, A.G. ENGEL and J. LINDSTROM: A congenital, familial, myasthenic syndrome caused by a presynaptic defect of transmitter resynthesis of mobilization. Neurology 29 (1979) 559.

HARTZELL, H.C., S.W. KUFFLER and D. YOSHIKAMI: Postsynaptic potentiation: interaction between quanta of acetylcholine at the skeletal neuromuscular synapse. J. Physiol. (London) 251 (1975) 427–463.

HARTZELL, H.C., S.W. KUFFLER and D. YOSHIKAMI: The number of acetylcholine molecules in a quantum and the interaction between quanta at the subsynaptic membrane of the skeletal neuromuscular synapse. Symp. Quant. Biol. 40 (1976) 175–186.

HARVEY, A.M. and R.L. MASLAND: The electromyogram in myasthenia gravis. Bull. Johns Hopkins Hosp. 69 (1941) 1–13.

HAWKEY, C.J., J. NEWSOM-DAVIS and A. VINCENT: Plasma exchange and immunosuppressive drug treatment in myasthenia gravis: no evidence for synergy. J. Neurol. Neurosurg. Psychiatr. 44 (1981) 469–475.

HAWKINS, B.R., Y.L. YU, V. WONG, E. WOO, M.S.M. IP and R.L. DAWKINS: Possible evidence for a variant of myasthenia gravis based on HLA and acetylcholine receptor antibody in Chinese patients. Q. J. Med. 70 (1989) 235–241.

HEINEMANN, S., J. MERLIE and J. LINDSTROM: Modulation of acetylcholine receptor in rat diaphragm by anti-receptor sera. Nature (London) 274 (1978) 65–68.

HERISHAN, Y. and P. ROSENBERG: β-Blockers and myasthenia gravis. Ann. Intern. Med. 83 (1975) 834–835.

HEUSER, J.E. and T.S. REESE: Structural changes after transmitter release at the frog neuromuscular junction. J. Cell Biol. 88 (1981) 564–580.

HEUSER, J.E. and S.R. SALPETER: Organization of acetylcholine receptors in quick-frozen, frozen, deep-etched, and rotary-replicated Torpedo postsynaptic membrane. J. Cell Biol. 82 (1979) 150–173.

HEUSER, J.E., T.S. REESE, M.J. DENNIS, Y. JAN, L. JAN and L. EVANS: Synaptic vesicle exocytosis captured by quick freezing and correlated with quantal transmitter release. J. Cell Biol. 81 (1979) 275–300.

HIROKAWA, N. and J.E. HEUSER: Internal and external differentiation of the postsynaptic membrane at the neuromuscular junction. J. Neurocytol. 11 (1982) 487–510.

HODES, R.: Electromyographic study of defects of neuromuscular transmission in human poliomyelitis. Arch. Neurol. Psychiatr. 60 (1948) 457–473.

HOFMANN, W.W. and E.H. DENYS: Effects of thyroid hormone at the neuromuscular junction. Am. J. Physiol. 223 (1972) 283–287.

HOHLFELD, R. and K.V. TOYKA: Strategies for the modulation of neuroimmunological diseases at the level of autoreactive T-lymphocytes. J. Neuroimmunol. 9 (1985) 193–204.

HOHLFELD, R., R. STERZ, I. KALIES, K. PEPER and H. WEKERLE: Neuromuscular transmission in experimental autoimmune myasthenia gravis (EAMG). Quantitative ionophoresis and current fluctuation analysis at the normal and myasthenic rat endplates. Pfluegers Arch. 390 (1981) 156–160.

HOHLFELD, R., I. KALIES, B. KOHLEISEN, K. HEININGER, B. CONTI-TRONCONI and K. TOYKA: Myasthenia gravis: stimulation of antireceptor autoantibodies by autoreactive T cell lines. Neurology 36 (1986) 618–621.

HOHLFELD, R., K.V. TOYKA, S.J. TZARTOS, W. CARSON and B. CONTI-TRONCONI: Human T-helper lymphocytes in myasthenia gravis recognize nicotinic receptor α subunit. Proc. Natl Acad. Sci. USA 84 (1987) 5379–5383.

HOHLFELD, R., M. MICHELS, K. HEININGER, U. BESINGER and K. TOYKA: Azathioprine toxicity during long-term immunosuppression of generalized myasthenia gravis. Neurology 38 (1988a) 258–261.

HOHLFELD, R., K.V. TOYKA, L.L. MINER, S.L. WALGRAVE and B. CONTI-TRONCONI: Antipathic segment of the nicotinic receptor alpha subunit contains epitopes recognized by T lymphocytes in myasthenia gravis. J. Clin. Invest. 81 (1988b) 657–660.

HOROWITZ, S.H., G. GENKINS, P. KORNFELD and A.E. PAPATESTAS: Electrophysiological diagnosis of myasthenia gravis and the regional curare test. Neurology 26 (1976a) 410–417.

HOROWITZ, S.H., G. GENKINS, A.E. PAPATESTAS and P. KORNFELD: Electrophysiological evaluations of thymectomy in myasthenia gravis. Preliminary findings. Neurology 26 (1976b) 615–619.

HOWARD, F.M., V.A. LENNON, J. FINLEY, J. MATSUMOTO and L.R. ELVEBACK: Clinical correlations of antibodies that bind, block, or modulate human acetylcholine receptors in myasthenia gravis. Ann. N.Y. Acad. Sci. 505 (1987) 526–538.

HUBBARD, J.I.: The effect of calcium and magnesium on the spontaneous release of transmitter from the mammalian motor nerve endings. J. Physiol. (London) 159 (1960) 507–517.

HUBBARD, J.I., S.F. JONES and E.M. LANDAU: On the mechanism by which calcium and magnesium affect the release of transmitter by nerve impulses. J. Physiol. (London) 196 (1968) 75–87.

HUDGSON, P., M.W. MCADAMS, M.A. PEICAK-VANCE, A.M. EDWARDS and A.D. ROSES: Effects of sera from myasthenia gravis patients on acetylcholine receptors in myotube cultures. J. Neurol. Sci. 59 (1982) 37–45.

HUTCHINSON, D., S. NAKANO, T. WALLS, P. TAYLOR, S. CAMP, C.M. HARPER, R.V. GROOVER, M.R. GOMEZ, D.G. JAMIESON and A.G. ENGEL: Endplate acetylcholinesterase deficiency. Neurology 42 (1992) 307.

IMOTO, K., C. BUSCH, B. SAKMANN, M. MISHINA, T. KONNO, J. NAKAI, H. BUJO, Y. MORI, K. FUKUDA and S. NUMA: Rings of negatively charged amino acids determine the acetylcholine receptor channel conductance. Nature (London) 335 (1988) 645–648.

INESTROSA, N.C. and A. PERELMAN: Distribution and anchoring of molecular forms of acetylcholinesterase. Trends Pharmacol. Sci. 10 (1989) 325–329.

ITO, Y., R. MILEDI, P.C. MOLENAAR, A. VINCENT, R.L. POLAK, M. VAN GELDER and J. NEWSOME-DAVIS: Acetylcholine receptors and end-plate physiology in myasthenia gravis. Brain 101 (1978) 345–368.

JACKSON, M.J., D.A. JONES and R.H.T. EDWARDS: Experimental skeletal muscle damage: the role of calcium activated degenerative processes. Eur. J. Clin. Invest. 14 (1984) 369–375.

JARETZKI, A. III, A.S. PENN, D.S. YOUNGER, M. WOLFF, M.R. OLARTE, R.E. LOVELACE and L.P. ROWLAND: 'Maximal' thymectomy for myasthenia gravis. J. Thorac. Cardiovasc. Surg. 95 (1988) 747–757.

JENNEKENS, F.G.I., L.F.G.M. HESSELMANS, H. VELDMAN, E.N.H. JANSEN, F. SPAANS and P.C. MOLENAAR: Deficiency of acetylcholine receptors in a case of end-plate acetylcholinesterase deficiency: a histochemical investigation. Muscle Nerve 15 (1992) 63–72.

JOHNS, T.R., J.F. CAMPA, W.J. CROWLEY and J.Q. MILLER: Familial myasthenia with tubular aggregates. Neurology 21 (1971) 449.

JOLLY, F.: Über Myasthenia gravis pseudoparalytica. Berl. Klin. Wochenschr. 32 (1895) 1–7.

JONES, S.F. and S. KWANBUNBUMPEN: The effects of nerve stimulation and hemicholinium on synaptic vesicles at the mammalian neuromuscular junction. J. Physiol. (London) 207 (1970a) 31–50.

JONES, S.F. and S. KWANBUNBUMPEN: Some effects of nerve stimulation and hemicholinium on quantal transmitter release at the mammalian neuromuscular junction. J. Physiol. (London) 207 (1970b) 51–61.

JONES, S.F., J.L. BRENNAN and J.G. MCLEOD: An investigation of experimental myasthenia gravis. J. Neurol. Neurosurg. Psychiatr. 34 (1971) 339–403.

KAHN, C.R., K.R. MCINTOSH and D.B. DRACHMAN: T-cell vaccination in experimental myasthenia gravis: a two-edged sword. J. Autoimmun. 3 (1990) 659–669.

KAMENSKAYA, M.A., D. ELMQVIST and S. THESLEFF: Guanidine and neuromuscular transmission. II. Effect on transmitter release in response to repetitive nerve stimulation. Arch. Neurol. 32 (1975) 510–518.

KAPLAN, I., B.T. BLAKELY, G.K. PAVLATH, M. TRAVIS and H. BLAU: Steroids induce acetylcholine receptors on cultured human muscle: implications for myasthenia gravis. Proc. Natl. Acad. Sci. 87 (1990) 8100–8104.

KATZ, B. and R. MILEDI: The release of acetylcholine from nerve endings by graded electrical pulses. Proc. R. Soc. London Ser. B. 167 (1967) 23–38.

KATZ, B. and R. MILEDI: Tetrodotoxin-resistant electrical activity in presynaptic terminals. J. Physiol. (London) 203 (1969) 459–487.

KATZ, B. and R. MILEDI: The statistical nature of the acetylcholine potential and its molecular components. J. Physiol. (London) 224 (1972) 665–697.

KATZ, B. and R. MILEDI: The binding of acetylcholine to receptors and its removal from the synaptic cleft. J. Physiol. (London) 231 (1973) 549–574.

KEESEY, J., J. LINDSTROM, H. COKELEY and C. HERRMANN JR: Anti-acetylcholine receptor antibody in neonatal myasthenia gravis. N. Engl. J. Med. 296 (1977) 55.

KEESEY, J., D. BUFFKIN, D. KEBO, W. HO and C. HERRMANN: Plasma exchange alone as therapy for myasthenia gravis. Ann. N.Y. Acad. Sci. 377 (1981) 729–743.

KERZIN-STORRAR, L., R.A. METCALFE, P.A. DYER, G. KOWALSKA, I. FERGUSON and R. HARRIS: Genetic factors in myasthenia gravis. A family study. Neurology 38 (1988) 38–42.

KILLEN, J.A. and J. LINDSTROM: Specific killing of lymphocytes that cause experimental autoimmune myasthenia gravis by ricin toxin-acetylcholine receptor conjugates. J. Immunol. 133 (1984) 2549–2553.

KIM, Y.I. and E. NEHER: IgG from patients with Lambert-Eaton syndrome blocks voltage-dependent calcium channels. Science 239 (1988) 405–408.

KIRCHNER, T., S. TZARTOS, F. HOPPE, B. SCHALKE, H. WEKERLE and H.K. MÜLLER-HERMELINK: Pathogenesis of myasthenia gravis. Acetylcholine receptor-related antigenic determinants in tumor-free thymuses and thymic epithelial tumors. Am. J. Pathol. 130 (1988) 268–280.

KLAVINSKIS, L.S., H.N.A. WILLCOX, J.S. OXFORD and J. NEWSOM-DAVIS: Antivirus antibodies in myasthenia gravis. Neurology 35 (1985) 1381–1384.

KLAVINSKIS, L.S., H.N.A. WILCOX, J.E. RICHMOND and J. NEWSOM-DAVIS: Attempted isolation of viruses from myasthenia gravis thymus. J. Neuroimmunol. 11 (1986) 287–299.

KORN, I.L. and O. ABRAMSKY: Myasthenia gravis following viral infection. Eur. Neurol. 20 (1981) 435–439.

KORNFELD, P., S.H. HOROWITZ, G. GENKINS and A.E. PAPATESTAS: Myasthenia gravis unmasked by antiarrhythmic agents. Mt Sinai J. Med. N.Y. 43 (1976) 10–14.

KREJCI, E., F. COUSSEN, N. DUVAL, J.-M. CHATEL, C. LEGAY, M. PUYPE, J. VANDEKERCKHOVE, J. CARTAUD, S. BON and J. MASSOULIÉ: Primary structure of a collagenic tail peptide of *Torpedo* acetylcholinesterase: co-expression with catalytic subunit induces the production of collagen-tailed forms in transfected cells. Eur. Mol. Biol. Organ. J. 10 (1991) 1285–1293.

KROLICK, K.A. and E.A. URSO: Analysis of helper-T-cell function by acetylcholine receptor-reactive cell lines of defined AChR-subunit specificity. Cell. Immunol. 105 (1987) 75–85.

KUFFLER, S.W. and D. YOSHIKAMI: The number of transmitter molecules in the quantum: an estimate from iontophoretic application of acetylcholine at the neuromuscular synapse. J. Physiol. (London) 251 (1975) 465–482.

KUKS, J.B.M., H.J.G.H. OOSTERHUIS, P.C. LIMBURG and T.H. THE: Anti-acetylcholine receptor antibodies decrease after thymectomy in patients with myasthenia gravis. Clinical correlations. J. Autoimmun. 4 (1991) 197–211.

KURTZKE, J.F. and L.T. KURLAND: Epidemiology of neurologic disease. In: A.B. Baker and L.H. Baker

(Eds.), Clinical Neurology, Vol. 3, Ch. 48. Hagerstown, Md., Harper and Row (1977).

LAMBERT, E.H. and D. ELMQVIST: Quantal components of the end-plate potential in the myasthenic syndrome. Ann. N.Y. Acad. Sci. 183 (1971) 183–199.

LAMBERT, E.H. and E.D. ROOKE: Myasthenic state and lung cancer. In: W.R. Brain and F.H. Norris (Eds.), The Remote Effects of Cancer on the Nervous System. New York, Grune and Stratton (1965) 67–80.

LAMBERT, E.H. and V.A. LENNON: Neuromuscular transmission in nude mice bearing oat-cell tumors from Lambert-Eaton myasthenic syndrome. Muscle Nerve 5 (1982) S39–S45.

LAMBERT E.H., L.M. EATON and E.D. ROOKE: Defect of neuromuscular transmission associated with malignant neoplasm. Am. J. Physiol. 187 (1956) 612–613.

LAMBERT, E.H., E.D. ROOKE, L.M. EATON and C.H. HODGSON: Myasthenic syndrome occasionally associated with bronchial neoplasm: neurophysiologic studies. In: H.R. Viets (Ed.), Myasthenia gravis. Springfield, IL, Charles C. Thomas (1961) 362–410.

LAMBERT, E.H., J.M. LINDSTROM and V.A. LENNON: Endplate potentials in experimental autoimmune myasthenia gravis. Ann. N.Y. Acad. Sci. 274 (1976) 300–318.

LAND, B.R., E.E. SALPETER and M.M. SALPETER: Acetylcholine receptor site density affects the rising phase of miniature endplate currents. Proc. Natl Acad. Sci. USA 77 (1980) 3736–3740.

LANG, B., J. NEWSOM-DAVIS, D.W. WRAY, A. VINCENT and N. MURRAY: Autoimmune aetiology for myasthenic (Lambert-Eaton) syndrome. Lancet ii (1981) 224–226.

LANG, B., J. NEWSOM-DAVIS, C. PRIOR and D.W. WRAY: Antibodies to motor nerve terminals: an electrophysiological study of a human myasthenic syndrome transferred to mouse. J. Physiol. (London) 344 (1983) 335–345.

LANG, B., J. NEWSOM-DAVIS, C. PEERS and D.W. WRAY: The effect of myasthenic syndrome antibody on presynaptic calcium channels in mouse. J. Physiol. (London) 390 (1987) 257–270.

LANG, B., A. VINCENT, N.M.F. MURRAY and J. NEWSOM-DAVIS: Lambert-Eaton myasthenic syndrome: immunoglobulin G inhibition of Ca^{2+} flux in tumor cells correlates with disease severity. Ann. Neurol. 25 (1989) 265–271.

LANSKA, D.J.: Indications for thymectomy in myasthenia gravis. Neurology 40 (1990) 1828–1829.

LAQUER, L. and W.C. WEIGERT: Beiträge zur Lehre von der Erb'schen Krankheit. 1. Über die Erb'sche Krankheit (Myasthenia gravis) (Laquer). 2. Pathologisch-anatomischer Beitrag zur Erb'schen Krankheit (Myasthenia gravis). (Weigert). Neurol Zbl. 20 (1901) 594–601.

LECKY, B.R.F., J.A. MORGAN-HUGHES, N.M.F. MURRAY D.N. LANDON and D.W. WRAY: Congenital myasthenia. Further evidence of disease heterogeneity. Muscle Nerve 9 (1986) 233–242.

LEFVERT, A.K.: Differences in the interaction of acetylcholine receptor antibodies with receptor from normal, denervated and myasthenic human muscle. J. Neurol. Neurosurg. Psychiatr. 45 (1982) 70–73.

LEFVERT, A.K. and M. BJÖRKHOLM: Antibodies against the acetylcholine receptor in hematologic disorders: implications for the development of myasthenia gravis after bone marrow grafting. N. Engl. J. Med. 317 (1987) 170.

LEFVERT, A.K. and P.O. OSTERMAN: Newborn infants to myasthenic mothers: a clinical study and an investigation of acetylcholine receptor antibodies in 17 children. Neurology 33 (1983) 133–138.

LENNON, V.A. and E.H. LAMBERT: Monoclonal antibodies to acetylcholine receptors: evidence for a dominant idiotype and requirement of complement for pathogenicity. Ann. N.Y. Acad. Sci. 377 (1981) 77–96.

LENNON, V.A. and E.H. LAMBERT: Antibodies bind solubilized calcium channel-ω-conotoxin complexes from small cell lung carcinoma: a diagnostic aid for Lambert-Eaton myasthenic syndrome. Mayo Clin. Proc. 64 (1989) 1498–1504.

LENNON, V.A., J.M. LINDSTROM and M.E. SEYBOLD: Experimental autoimmune myasthenia gravis in rats and guinea pigs. J. Exp. Med. 141 (1975) 1365–1375.

LENNON, V.A., J.M. LINDSTROM and M.E. SEYBOLD: Experimental autoimmune myasthenia gravis: cellular and humoral immune responses. Ann. N.Y. Acad. Sci. 274 (1976) 283–299.

LENNON, V.A., M.E. SEYBOLD, J.M. LINDSTROM, C. COCHRANE and R. ULEVITCH: Role of complement in the pathogenesis of experimental autoimmune myasthenia gravis. J. Exp. Med. 147 (1978) 973–983.

LENNON, V.A., E.H. LAMBERT, S. WHITTINGHAM and V. FAIRBANKS: Autoimmunity in the Lambert-Eaton myasthenic syndrome. Muscle Nerve 5 (1982) S21–S25.

LEONARD, J.P. and M.M. SALPETER: Calcium-mediated myopathy at neuromuscular junctions of normal and dystrophic muscle. Exp. Neurol. 46 (1982) 121–138.

LEONARD, R.J., C.J. LABARCA, P. CHARNET, N. DAVIDSON and H.A. LESTER: Evidence that the M2 membrane spanning region lines the ion channel pore of the nicotinic receptor. Science 242 (1988) 1578–1581.

LEVITT, T.A. and M.M. SALPETER: Denervated endplates have a dual population of junctional acetylcholine receptors. Nature (London) 291 (1981) 239–241.

LEYS, K., B. LANG, I. JOHNSTON and J. NEWSOM-DAVIS: Calcium channel autoantibodies in the Lambert-Eaton myasthenic syndrome. Ann. Neurol. 29 (1991) 307–414.

LI, Y., S. CAMP, T.L. RACHINSKY, D. GETMAN and P. TAYLOR: Gene structure of mammalian acetylcholinesterase. Alternative exons dictate tissue specific expression. J. Biol. Chem. 266 (1991) 23083–23090.

LIMBURG, P.C., T.H. THE, E. HUMMEL-TAPPEL and

H.J.G.H. OOSTERHUIS: Anti-acetylcholine receptor antibodies in myasthenia gravis. Part 1. Relation to clinical parameters in 250 patients. J. Neurol. Sci. 58 (1983) 357–370.

LINDSLEY, D.B.: Myographic and electromyographic studies of myasthenia gravis. Brain 58 (1935) 470–482.

LINDSTROM, J.M. and E.H. LAMBERT: Content of acetylcholine receptor and antibodies bound to receptor in myasthenia gravis, experimental autoimmune myasthenia gravis, and Eaton-Lambert syndrome. Neurology 28 (1978) 130–138.

LINDSTROM, J.M., A.G. ENGEL, M.E. SEYBOLD, V.A. LENNON and E.H. LAMBERT: Pathological mechanisms in experimental autoimmune myasthenia gravis. II. Passive transfer of experimental autoimmune myasthenia gravis in rats with anti-acetylcholine receptor antibodies. J. Exp. Med. 144 (1976a) 739–753.

LINDSTROM, J.M., ,B.L. EINARSON, V.A. LENNON and M.E. SEYBOLD: Pathological mechanisms in experimental autoimmune myasthenia gravis. I. Immunogenicity of syngeneic muscle acetylcholine receptor and quantitative extraction of receptor and antibody-receptor complexes from muscles of rats with experimental autoimmune myasthenia gravis. J. Exp. Med. 144 (1976b) 726–738.

LINDSTROM, J.M., M.E. SEYBOLD, V.A. LENNON, S. WHITTINGHAM and D.D. DUANE: Antibody to acetylcholine receptor in myasthenia gravis. Neurology 26 (1976c) 1054–1059.

LISAK, R.P., N.I. ABDOU, B. ZWEIMAN, C. ZMIJEWSKI and A.S. PENN: Aspects of lymphocyte function in myasthenia gravis. Ann. N.Y. Acad. Sci. 274 (1976) 402–410.

LLINAS, R. and C. NICHOLSON: Calcium in depolarization secretion coupling: an aequorin study in squid giant synapse. Proc. Natl Acad. Sci. USA 72 (1975) 187–190.

LLINAS, R., I.Z. STEINBERG and K. WALTON: Presynaptic calcium currents and their relation to synaptic transmission: voltage clamp study in squid giant synapse and theoretical model for the calcium gate. Proc. Natl Acad. Sci. USA 73 (1976) 2918–2922.

LLINAS, R., M. SUGIMORI, J.W. LIN, P.L. LEOPOLD and S.T. BRADY: ATP-dependent directional movement of rat synaptic vesicles injected into the presynaptic terminal of the squid giant synapse. Proc. Natl Acad. Sci. USA 86 (1989) 5656–5660.

LLINAS, R., J.A. GRUNER, M. SUGIMORI, T.L. MCGUINESS and P. GREENGARD: Regulation by synapsin I and Ca^{2+}-calmodulin-dependent protein kinase II of transmitter release in squid giant synapse. J. Physiol. (London) 436 (1991) 257–282.

LOGIN, I.S., Y.I. KIM, A.M. JUDD, B.L. SPANGELO and R.M. MACLEOD: Immunoglobulins of Lambert-Eaton myasthenic syndrome inhibit rat pituitary hormone release. Ann. Neurol. 22 (1987) 610–614.

LUNDH, H., O. NILSSON and I. ROSEN: 4-Aminopyridine: a new drug in the treatment of Eaton-Lambert syndrome. J. Neurol. Neurosurg. Psychiatr. 40 (1977) 1109–1112.

LUNDH, H., O. NILSSON and I. ROSEN: Treatment of Lambert-Eaton syndrome: 3,4-diaminopyridine and pyridostigmine. Neurology 34 (1984) 1324–1330.

MAENO, T.: Kinetic analysis of a large facilitatory action of 4-aminopyridine on the motor nerve terminal of the neuromuscular junction. Proc. Jpn Acad. 56 (1980) 241–245.

MAGLEBY, K.L.: Facilitation, augmentation, and potentiation of transmitter release. Prog. Brain Res. 49 (1979) 175–182.

MANN, J.D., T.R. JOHNS and J.F. CAMPA: Long-term administration of corticosteroids in myasthenia gravis. Neurology 26 (1976) 729–740.

MANTEGAZZA, R., C. ANTOZZI, A. SGHIRLANZONI and F. CORNELIO: Azathioprine as a single drug or in combination with steroids in the treatment of myasthenia gravis. J. Neurol. 235 (1988) 449–453.

MANTEGAZZA, R., E. BEGHI, D. PAREYSON, C. ANTOZZI, D. PELUCHETTI, A. SGHIRLANZONI, V. COSI, M. LOMBARDI, G. PICCOLO, P. TONALI, A. EVOLI, E. RICCI, A.P. BATOCCHI, C. ANGELINI, G.F. MICAGLIO, G. MARCONI, R. TAIUTI, L. BERGAMINI, L. DURELLI and F. CORNELIO: A multicentre follow-up study of 1152 patients with myasthenia gravis in Italy. J. Neurol. 237 (1990) 339–344.

MARTIN, A.R.: Current concepts of pre- and post-junctional mechanisms in neuromuscular transmission. Ann. N.Y. Acad. Sci. 274 (1976) 3–5.

MARTIN, A.R.: Junctional transmission. II. Presynaptic mechanisms. In: Handbook of Neurophysiology, Section 1. The Nervous System. Vol. 1. Cellular Biology of Neurons. Bethesda, MD, American Physiological Society (1977) 329–355.

MARX, A., R. O'CONNOR, K.I. GEUDER, F. HOPPE, B. SCHALKE, S. TZARTOS, I. KALIES, T. KIRCHNER and H.K. MÜLLER: Characterization of a protein with an acetylcholine receptor epitoe from myasthenia gravis-associated thymomas. Lab. Invest. 62 (1990) 279–286.

MASSELI, R.A. and B.C. SOLIVEN: Analysis of the organophosphate-induced electromyographic response to repetitive nerve stimulation: paradoxical response to edrophonium and D-tubocurarine. Muscle Nerve 14 (1991) 1182–1188.

MASSELI, R.A., D.P. RICHMAN and R.L. WALLMANN: Inflammation at the neuromuscular junction in myasthenia gravis. Neurology 41 (1991) 1497–1504.

MASSOULIÉ, J. and S. BON: The molecular forms of cholinesterase and acetylcholinesterase in vertebrates. Annu. Rev. Neurosci. 5 (1982) 57–106.

MASTERS, C.L., R.L. DAWKINS, P.J. ZILKO, J.A. SIMPSON, R.J. LEEDMAN and J.M. LINDSTROM: Penicillamine-associated myasthenia gravis, antiacetylcholine receptor and antistriational antibodies. Am. J. Med. 63 (1977) 689–694.

MATSUKI, K., T. JUJI, K. TOKUNAGA, M. TAKAMIZAWA, H. MAEDA, M. SODA, Y. NOMURA and M. SEGAWA: HLA antigens in Japanese patients with myasthenia gravis. J. Clin. Invest. 86 (1990) 392–399.

MATTHEWS-BELLINGER, J. and M.M. SALPETER: Distribution of acetylcholine receptors at frog neuromuscular junctions with a discussion of some physiological implications. J. Physiol. (London) 279 (1978) 197–213.

MAY, E.F. and P.C. CALVERT: Aggravation of myasthenia gravis by erythromycin. Ann. Neurol. 28 (1990) 577–579.

MCEVOY, K.M., A.J. WINDEBANK, J.R. DAUBE and P.A. LOW: 3,4-Diaminopyridine in the treatment of Lambert-Eaton myasthenic syndrome. N. Engl. J. Med. 321 (1989) 1567–1571.

MCINTOSH, K.R. and D.B. DRACHMAN: Induction of suppressor cells specific for AChR in experimental autoimmune myasthenia gravis. Science 232 (1986) 401–403.

MCMAHAN, U.J., J.S. SANES and L.M. MARSHALL: Cholinesterase is associated with the basal lamina at the neuromuscular junction. Nature (London) 271 (1978) 172–174.

MCQUILLEN, M.P.: Familial limb-girdle myasthenia. Brain 89 (1966) 121–132.

MCQUILLEN, M.P. and M.G. LEONE: A treatment carol: thymectomy revisited. Neurology 27 (1977) 1144–1146.

MELMS, A., B.C.G. SCHALKE, TH. KIRCHNER, H.K. MÜLLER-ERMELINK and H. WEKERLE: Thymus in myasthenia gravis. Isolation of T-lymphocyte lines specific for the nicotinic acetylcholine receptor from thymuses of myasthenic patients. J. Clin. Invest. 81 (1988) 902–908.

MELMS, A., S. CHRESTEL, B.C.G. SCHALKE, H. WEKERLE, A. MAURON, M. BALLIVET and T. BARKAS: Autoimmune T lymphocytes in myasthenia gravis. Determination of target epitopes using T lines and recombinant products of the mouse nicotinic acetylcholine receptor gene. J. Clin. Invest. 83 (1989) 785–790.

MERLIE, J.P., S. HEINEMANN, B. EINARSON and J. LINDSTROM: Degradation of acetylcholine receptors in diaphragms of rats with experimental autoimmune myasthenia gravis. J. Biol. Chem. 254 (1979) 6328–6332.

MERRITT, H.H.: Corticotropin and cortisone in diseases of nervous system. Yale J. Biol. Med. 24 (1952) 466–473.

MERSON, M.H. and V.R. DOWELL: Epidemiologic, clinical and laboratory aspects of wound botulism. N. Engl. J. Med. 289 (1973) 1005–1010.

MERTENS, H.G., F. BALZEREIT and M. LEIPERT: The treatment of severe myasthenia gravis with immunosuppressive agents. Eur. Neurol. 2 (1969) 321–339.

MERTENS, H.G., H. HERTEL, P. REUTHER and K. RICKER: Effect of immunosuppressive drugs (Azathioprine). Ann. N.Y. Acad. Sci. 377 (1981) 691–699.

MILEDI, R.: Transmitter release by injection of calcium ions into nerve terminals. Proc. R. Soc. London Ser. B 183 (1973) 421–425.

MILLER, R.G., H.S. MILNER-BROWN and A. MIRKA:

Prednisone-induced worsening of neuromuscular function in myasthenia gravis. Neurology 36 (1986) 729–732.

MILLIKAN, C.H. and S.F. HAINES: The thyroid gland in relation to neuromuscular diseases. Arch. Intern. Med. 92 (1953) 5–39.

MINOT, A.S., K. DODD and S.S. RIVEN: Use of guanidine hydrochloride in the treatment of myasthenia gravis. J. Am. Med. Assoc. 113 (1939) 553–559.

MITTAG, T., T. MASSA and P. KORNFELD: Multiple forms of antiacetylcholine receptor antibody in myasthenia gravis. Muscle Nerve 4 (1981) 16–25.

MOLENAAR, P.C., J. NEWSOM-DAVIS, R.L. POLAK and A. VINCENT: Eaton-Lambert syndrome: Acetylcholine and choline acetyltransferase in skeletal muscle. Neurology 32 (1982) 1062–1065.

MOORTHY, G., M.M. BEHRENS, D.B. DRACHMAN, T.H. KIRKHAM, D.L. KNOX, N.R. MILLER, T.L. SLAMOVITZ and S.J. ZINREICH: Ocular pseudomyasthenia or ocular myasthenia 'plus': a warning to clinicians. Neurology 39 (1989) 1150–1154.

MORA, M., E.H. LAMBERT and A.G. ENGEL: Synaptic vesicle abnormality in familial infantile myasthenia. Neurology 37 (1987) 206–214.

MORGAN-HUGHES, J.A., B.R.F. LECKY, D.N. LANDON and N.M.F. MURAY: Alterations in the number and affinity of junctional acetylcholine receptors in a myopathy with tubular aggregates. A newly recognized receptor defect. Brain 4 (1981) 279–295.

MORI, K., K. EGUCHI, H. MORIYAMA, N. MIYAZAWA and T. KODAMA: Computed tomography of anterior mediastinal tumors. Differentiation between thymoma and germ cell tumor. Acta Radiol. 28 (1987) 395–398.

MORITA, K., J. MORIUCHI, H. INOKO, K. TSUJI and S. ARIMORI: HLA class II antigen and DNA restriction fragment length polymorphism in myasthenia gravis in Japan. Ann. Neurol. 29 (1991) 168–174.

MOUTARD-CODOU, M.L., M.M. DELLEUR, O. DULAC, E. MOREL, M. VOYER and E. DE GAMARRA: Myasthénie néo-natale sévère avec arthrogrypose. Presse Med. 16 (1987) 615–618.

MULDER, D.G., M. GRAVES and C. HERRMANN: Thymectomy for myasthenia gravis. Recent observations and comparisons with past experience. Ann. Thorac. Surg. 48 (1989) 551–555.

MULDER, D.W., E.H. LAMBERT and L.M. EATON: Myasthenic syndrome in patients with amyotrophic lateral sclerosis. Neurology 9 (1959) 627–631.

MÜLLER, K.M.I.: Antineuroblastoma antibodies in myasthenia gravis: clinical and immunological correlations. J. Neurol. Sci. 93 (1989) 263–275.

MURRAY, N.M.F. and J. NEWSOM-DAVIS: Treatment with oral 4-aminopyridine in disorders of neuromuscular transmission. Neurology 31 (1981) 265–271.

NAGEL, A. and A.G. ENGEL: Vesamicol decreases both quantal and synaptic vesicle size at the neuromuscular junction. Soc. Neurosci. (Abstr.) 15 (1989) 351–367.

NAGEL, A., A.G. ENGEL, B. LANG, J. NEWSOM-DAVIS and T. FUKUOKA: Lambert-Eaton syndrome IgG depletes presynaptic membrane active zone particles by antigenic modulation. Ann. Neurol. 24 (1988) 552–558.

NAGEL, A., A.G. ENGEL, T.J. WALLS, M.C. HARPER and H. WAISBURG: Congenital myasthenic syndrome with end-plate acetylcholine receptor deficiency and short channel opentime. Neurology 40 (1990) 277–278.

NAMBA, T., S.B. BROWN and D. GROB: Neonatal myasthenia gravis: report of two cases and a review of the literature. Pediatrics 45 (1970) 488–504.

NAMBA, T., N.G. BRUNNER, S.B. BROWN, M. MUGURAMA and D. GROB: Familial myasthenia gravis: report of 27 patients in 12 families and review of 164 patients 73 families. Arch. Neurol. (Chicago) 25 (1971a) 49–60.

NAMBA, T., M.S. SHAPIRO, N.G. BRUNNER and D. GROB: Myasthenia gravis occurring in twins. J. Neurol. Neurosurg. Psychiatr. 34 (1971b) 531–534.

NAMBA, T., N.G. BRUNNER and D. GROB: Idiopathic giant cell polymyositis. Arch. Neurol. (Chicago) 31 (1974) 27–30.

NASTUK, W.L., O. PLESCIA and K.E. OSSERMAN: Changes in serum complement activity in patients with myasthenia gravis. Proc. Soc. Exp. Biol. Med. 105 (1960) 177–184.

NEWSOM-DAVIS, J. and N.M.F. MURRAY: Plasma exchange and immunosuppressive drug treatment in the Lambert-Eaton myasthenic syndrome. Neurology 34 (1984) 480–485.

NEWSOM-DAVIS, J.W., A.J. PINCHING, A. VINCENT and S.G. WILSON: Function of circulating antibody to acetylcholine receptor in myasthenia gravis: investigation by plasma exchange. Neurology 28 (1978) 266–272.

NEWSOM-DAVIS, J.W., S.G. WILSON, A. VINCENT and C.D. WARD: Long term effects of repeated plasma exchange in myasthenia gravis. Lancet i (1979) 464–468.

NEWSOM-DAVIS, J.W., N.M.F. MURRAY, N. WILLCOX, C. LANG and K. WALSH: Lambert-Eaton myasthenic syndrome (LEMS): immunogenetic characteristics and response to immunosuppressive drug treatment (abstract). Neurology 33 (Suppl. 2) (1983) 156.

NEWSOM-DAVIS, J., K. LEYS, A. VINCENT, I. FERGUSON, G. MODI and K. MILLS: Immunological evidence for the co-existence of the Lambert-Eaton myasthenic syndrome and myasthenia gravis in two patients. J. Neurol. Neurosurg. Psychiatr. 54 (1991) 452–453.

NIAKAN, E., Y. HARATI and L.A. ROLAK: Immunosuppressive drug therapy in myasthenia gravis. Arch. Neurol. (Chicago) 43 (1986) 155–156.

NORRIS, F.H., P.R. CALANCHINI, R.J. FALLAH, R.P.T. PANCHARIK and B. JEWETT: The administration of guanidine in amyotrophic lateral sclerosis. Neurology 24 (1974) 721–728.

NYBERG-HANSEN, R. and L. GJERSTAD: Myasthenia gravis treated with cyclosporin. Acta Neurol. Scand. 77 (1988) 307–313.

OGDEN, D.C. and D. COLQUHOUN: Ion channel block by acetylcholine, carbachol and suberidylcholine at the frog neuromuscular junction. Proc. R. Soc. London Ser. B 225 (1985) 329–355.

OH, S.J.: The Eaton-Lambert syndrome in ocular myasthenia gravis. Arch. Neurol. 31 (1974) 183–186.

OH, S.J.: Diverse electrophysiological spectrum of the Lambert-Eaton myasthenic syndrome. Muscle Nerve 12 (1989) 464–469.

OH, S.J. and H.K. CHO: Edrophonium responsiveness not necessarily diagnostic of myasthenia gravis. Muscle Nerve 13 (1990) 187–191.

OHTA, M., K. OHTA, N. ITOH, M. KUROBE, K. HAYASHI and H. NISHITANI: Anti-skeletal muscle antibodies in the sera from myasthenic patients with thymoma: identification of anti-myosin, actomyosin, actin, and α-actinin antibodies by a solid-phase radioimmunoassay and Western blotting analysis. Clin. Chim. Acta 187 (1990) 255–264.

OLARTE, M.R., R.S. SCHOENFELDT, A.S. PENN, R.E. LOVELACE and L.P. ROWLAND: Effect of plasmapheresis in myasthenia gravis, 1978-1980. Ann. N.Y. Acad. Sci. 377 (1981) 725–728.

O'NEILL, J.H., N.M.F. MURRAY and J. NEWSOM-DAVIS: The Lambert-Eaton myasthenic syndrome. A review of 50 cases. Brain (1988) 577–596.

OOSTERHUIS, H.J.G.H.: Observations on the natural history of myasthenia gravis and the effect of thymectomy. Ann. N.Y. Acad. Sci. 377 (1981) 678–690.

OOSTERHUIS, H.J.G.H.: The ocular signs and symptoms of myasthenia gravis. Doc. Ophthalmol. 52 (1982) 363–378.

OOSTERHUIS, H.J.G.H.: The natural course of myasthenia gravis. J. Neurol. Neurosurg. Psychiatr. 52 (1989) 1121–1127.

OOSTERHUIS, H.J.G.H., P.C. LIMBURG, E. HUMMEL-TAPPEL and T.H. THE: Anti-acetylcholine receptor antibodies in myasthenia gravis. J. Neurol. Sci. 58 (1983) 371–385.

OOSTERHUIS, H.J.G.H., J. NEWSOM-DAVIS, J.H.J. WOKKE, P.C. MOLENAAR, T.V. WEERDEN, B.S. OEN, F.G.I. JENNEKENS, H. VELDMAN, A. VINCENT, D.W. WRAY, C. PRIOR and N.M.F. MURRAY: The slow channel syndrome. Two new cases. Brain 110 (1987) 1061–1079.

OSHIMA, M., T. ASHIZAWA, M.S. POLLACK and M.Z. ATASSI: Autoimmune T cell recognition of human acetylcholine receptor: the sites of T cell recognition in myasthenia gravis on the extracellular part of the α subunit. Eur. J. Immunol. 20 (1990) 2563–2569.

OSSERMAN, K.E.: Progress report on Mestinon bromide (pyridostigmine bromide). Am. J. Med. 19 (1955) 737–739.

OSSERMAN, K.E.: Myasthenia Gravis. New York, N.Y., Grune and Stratton (1958).

OSSERMAN, K.E.: Ocular myasthenia gravis. Invest. Ophthalmol. 6 (1967) 277–287.

OSSERMAN, K.E. and G. GENKINS: Studies in myasthenia gravis: review of a twenty-year experience in over 1200 patients. Mt Sinai J. Med. N.Y. 38 (1971) 497–537.

OSSERMAN, K.E. and P. TENG: Studies in myasthenia gravis: a rapid diagnostic test. Further progress with edrophonium (Tensilon) chloride. J. Am. Med. Assoc. 160 (1956) 153–155.

OSSERMAN, K.E., P. TSAIRIS and L.B. WEINER: Myasthenia gravis and thyroid disease: clinical and immunological correlation. J. Mt Sinai Hosp. N.Y. 34 (1967) 469–481.

OTSUKA, M. and M. ENDO: The effect of guanidine on neuromuscular transmission. J. Pharmacol. Exp. Ther. 128 (1960) 273–282.

ÖZDEMIR, C. and R.R. YOUNG: The results to be expected from electrical testing in the diagnosis of myasthenia gravis. Ann. N.Y. Acad. Sci. 274 (1976) 103–222.

PACHNER, A.R.: Anti-acetylcholine receptor antibodies block bungarotoxin binding to native human acetylcholine receptor on the surface of TE671 cells. Neurology 39 (1989) 1057–1061.

PACHNER, A.R. and F.S. KANTOR: Antigen specific suppression of experimental myasthenia by injection of suppressor lymphoblasts (abstract). Fed. Proc. 41 (1982) 954.

PACHNER, A.R., F.S. KANTOR, B. MULAC-JERICEVIC and M.Z. ATASSI: An immunodominant site of acetylcholine receptor in experimental myasthenia mapped with T lymphocyte clones and synthetic peptides. Immunol. Lett. 20 (1989) 199–204.

PAPATESTAS, A.E., G. GENKINS, P. KORNFELD, J.B. EISENKRAFT, R.P. FAGERSTROM, J. POZNER and A.H. AUFSES: Effects of thymectomy in myasthenia gravis. Ann. Surg. 206 (1987) 79–88.

PARSONS, S.M., R.S. CARPENTER, R. KOENIGSBERGER and J.E. ROTHLEIN: Transport in the cholinergic synaptic vesicle. Fed. Proc. 41 (1982) 2765–2768.

PATRICK, J. and J.M. LINDSTROM: Autoimmune response to acetylcholine receptor. Science 180 (1973) 871–872.

PASCUZZI, R.M. and J.F. CAMPA: Lymphorrhage localized to the muscle end-plate in myasthenia gravis. Arch. Pathol. Lab. Med. 112 (1988) 934–937.

PASCUZZI, R.M., K.L. ROOS and L.H. PHILLIPS: Granulomatous inflammatory myopathy associated with myasthenia gravis. A case report and review of the literature. Arch. Neurol. (Chicago) 43 (1986) 621–623.

PEERS, C., B. LANG, J. NEWSOM-DAVIS and D.W. WRAY: Selective action of Lambert-Eaton myasthenic syndrome antibodies on calcium channels in a rodent neuroblastoma × glioma hybrid cell line. J. Physiol. (London) 421 (1987) 293–308.

PENN, A.S., D.L. SCHOTLAND and L.P. ROWLAND: Immunology of muscle disease. Res. Publ. Ass. Res. Nerv. Ment. Dis. 49 (1971) 215–240.

PENN, A.S., H.W. CHANG, R.E. LOVELACE, W. NIEMI and A. MIRANDA: Antibodies to acetylcholine receptors in rabbits. Immunological and electrophysiological studies. Ann. N.Y. Acad. Sci. 274 (1976) 354–376.

PENN, A.S., A. JARETZKI III, M. WOLFF, H.W. CHANG and V. TENNYSON: Thymic abnormalities: antigen or antibody? Response to thymectomy in myasthenia gravis. Ann. N.Y. Acad. Sci. 377 (1981) 789–791.

PENN, A.S., D.L. SCHOTLAND and S. LAMME: Antimuscle and antiacetylcholine receptor antibodies in myasthenia gravis. Muscle Nerve 9 (1986) 407–415.

PEREZ, M.C., W.L. BUO and C. MARCADO-DONGUILAN: Stable remissions in myasthenia gravis. Neurology 31 (1981) 32–37.

PERLO, V.P., B. ARNASON, D. POSKANZER, B. CASTLEMAN, R.S. SCHWAB, K.E. OSSERMAN, A.E. PAPATESTAS, L. ALPERT and A. KARK: The role of thymectomy in the treatment of myasthenia gravis. Ann. N.Y. Acad. Sci. 183 (1971) 308–315.

PINCHING, A.J., D.K. PETERS and J. NEWSOM-DAVIS: Remission of myasthenia gravis following plasma exchange. Lancet ii (1976) 1373–1376.

PLAUCHÉ, W.C.: Myasthenia gravis in pregnancy: an update. Am. J. Obstetr. Gynecol. 135 (1979) 691–697.

PRITCHARD, E.A.B.: The use of 'Prostigmin' in the treatment of myasthenia gravis. Lancet i (1935) 432–434.

PROTTI, M.P., A.A. MANFREDI, C. STRAUB, X. WU, J.F. HOWARD and B.M. CONTI-TRONCONI: Use of synthetic peptides to establish anti-human acetylcholine receptor (CD4+ cell lines from myasthenia gravis patients. J. Immunol. 144 (1990) 1711–1720.

PROTTI, M.P., A.A. MANFREDI, J.F. HOWARD and B.M. CONTI-TRONCONI: T cells in myasthenia gravis specific for embryonic acetylcholine receptor. Neurology 41 (1991a) 1809–1814.

PROTTI, M.P., A.A. MANFREDI, X.D. WU, L. MOIOLA, J.F. HOWARD and B.M. CONTI-TRONCONI: Myasthenia gravis: T epitopes on the δ subunit of the human muscle acetylcholine receptor. J. Immunol. 146 (1991b) 2253–2261.

PUMPLIN, D.W., T.S. REESE and R. LLINAS: Are the presynaptic membrane particles calcium channels? Proc. Natl Acad. Sci. USA 78 (1981) 7210–7213.

RAIMOND, F., B. VERNET-DER-GARABEDIAN and E. MOREL: A lipid associated acetylcholine receptor as an antigen in diagnosis of myasthenia gravis. Clin. Exp. Immunol. 47 (1982) 345–350.

RASH, J.E. and M.M. ELLISMAN: Studies of excitable membranes. I. Macromolecular specializations of the neuromuscular junction and the nonjunctional sarcolemma. J. Cell Biol. 63 (1974) 567–586.

RASH, J.E., E.X. ALBUQUERQUE, C.S. HUDSON, R.F. MAYER and J.R. SATTERFIELD: Studies of human myasthenia gravis: electrophysiological and ultrastructural evidence compatible with antibody attachment to the acetylcholine receptor complex. Proc. Natl. Acad. Sci. (USA) 73 (1976) 4584–4588.

ROBERTSON, W.C., R.W.M. CHUN and S.E. KORNGUTH: Familial infantile myasthenia. Arch. Neurol. (Chicago) 37 (1980) 117–119.

ROBITAILLE, R., E.M. ADLER and M.P. CHARLTON: Strategic location of calcium channels at transmitter release sites of frog neuromuscular synapses. Neuron 5 (1990) 773–779.

ROOKE, E.D., L.M. EATON, E.H. LAMBERT and C.H. HODGSON: Myasthenia and malignant intrathoracic tumor. Med. Clin. North Am. 44 (1960) 977–988.

ROTHBARD, J.B. and W.R. TAYLOR: A sequence pattern common to T cell epitopes. Eur. Mol. Biol. Organ. J. 7 (1988) 93–100.

RUBENSTEIN, A.E., S.H. HOROWITZ and A.N. BENDER: Cholinergic dysautonomia and Lambert-Eaton syndrome. Neurology 29 (1979) 720–723.

RUFF, R.L.: A quantitative analysis of local anesthetic alteration of miniature end-plate currents and end-plate current fluctuations. J. Physiol. (London) 264 (1977) 89–124.

SAFAR, D., C. AIMÉ, S. COHEN-KAMINSKY and S. BERRIH-AKNIN: Antibodies to thymic epithelial cells in myasthenia gravis. J. Neuroimmunol. 35 (1991) 101–110.

SAHASHI, K., A.G. ENGEL, J.M. LINDSTROM, E.H. LAMBERT and V.A. LENNON: Ultrastructural localization of immune complexes (IgG and C3) at the end-plate in experimental autoimmune myasthenia gravis. J. Neuropathol. Exp. Neurol. 37 (1978) 212–223.

SAHASHI, K., A.G. ENGEL, E.H. LAMBERT and F.M. HOWARD: Ultrastructural localization of the terminal and lytic ninth complement component (C9) at the motor end-plate in myasthenia gravis. J. Neuropathol. Exp. Neurol. 39 (1980) 160–172.

SAINT, D.A.: The effects of 4-aminopyridine and tetraethylammonium on the kinetics of transmitter release at the mammalian neuromuscular synapse. Can. J. Physiol. Pharmacol. 67 (1989) 1045–1050.

SALPETER, M.M.: Molecular organization of the neuromuscular synapse. In: E.X. Albuquerque and A.T. Eldefrawi (Eds.), Myasthenia Gravis. New York, Chapman and Hall (1983) 105–129.

SALPETER, M.M.: Vertebrate neuromuscular junctions: general morphology, molecular organization, and functional consequences. In: M.M. Salpeter (Ed.), The Vertebrate Neuromuscular Junction. New York, Alan Liss (1987) 1–54.

SALPETER, M.M., A.W. ROGERS, H. KASPRZAK and F.A MCHENRY: Acetylcholinesterase in the fast extraocular muscle of the mouse by light and electron microscopy autoradiography. J. Cell Biol. 78 (1978) 274–285.

SALPETER, M.M., H. KASPRZAK, H. FENG and H. FERTUCK: End-plates after esterase inactivation in vivo: correlation between esterase concentration, functional response and fine structure. J. Neurocytol. 8 (1979) 95–115.

SALVETTI, M., S. JUNG, S.-F. CHANG, H. WILL, B.C.G. SCHALKE and H. WEKERLE: Acetylcholine receptor-specific T-lymphocyte clones in the normal human immune repertoire: target epitopes, HLA restriction, and membrane phenotypes. Ann. Neurol. 29 (1991) 508–516.

SATOYOSHI, E., H. KOWA and N. FUKUNAGA: Subacute cerebellar degeneration in Eaton-Lambert syndrome with bronchogenic carcinoma. Neurology 23 (1973) 764–768.

SCADDING, G.K., A. VINCENT, J. NEWSOM-DAVIS and K. HENRY: Acetylcholine receptor antibody synthesis by thymic lymphocytes: correlation with thymic histology. Neurology 31 (1981) 935–943.

SCADDING, G.K., L. CALDER and J. NEWSOM-DAVIS: The in-vitro effects of penicillamine upon anti-AChR production by thymic and peripheral blood lymphocytes from patients with myasthenia gravis. Muscle Nerve 6 (1983) 656–660.

SCHLUEP, M., N. WILCOX, A. VINCENT, G.K. DHOOT and J. NEWSOM-DAVIS: Acetylcholine receptors in human thymic myoid cells in situ: an immunological study. Ann. Neurol. 22 (1987) 212–222.

SCHÖNBECK, S., F. PADBERG, R. HOHLFELD and H. WEKERLE: Transplantation of thymic autoimmune microenvironment to severe combined immunodeficiency mice: a new model of myasthenia gravis. J. Clin. Invest. 90 (1992) 245–250.

SCHUETZE, S.M.: Developmental regulation of acetylcholine receptors. Annu. Rev. Neurosci. 10 (1987) 403–457.

SCHUMM, F., H. WIETHÖLTER, A. FATEH-MOGHADAM and J. DICHGANS: Thymectomy in myasthenia with pure ocular symptoms. J. Neurol. Neurosurg. Psychiatr. 48 (1985) 332–337.

SCHWAB, R.S. and C. LELAND: Sex and age in myasthenia gravis as critical factors in incidence and remissions. J. Am. Med. Assoc. 153 (1953) 1270–1273.

SCWARTZ, M.S. and E. STALBERG: Myasthenic syndrome studied with single fiber electromyography. Arch. Neurol. 32 (1975) 815–817.

SCHWIMBECK, P., T. DRYBERG, D. DRACHMAN and M.B.A. OLDSTONE: Molecular mimicry and myasthenia gravis: an autoantigenic site of the AChR that has biologic activity and reacts immunochemically with herpes simplex virus. J. Clin. Invest. 84 (1989) 1174–1180.

SCOLDING, N.J., B.P. MORGAN, W.A.J. HOUSTON, C. LININGTON, A.K. CAMPBELL and D.A.S. COMPSTON: Vesicular removal by oligodendrocytes of membrane attack complexes formed by activated complement. Nature (London) 339 (1989) 620–622.

SETHI, K.D., M.H. RIVNER and T.R. SWIFT: Ice pack test for myasthenia gravis. Neurology 37 (1987) 1383–1385.

SEYBOLD, M.E. and D.B. DRACHMAN: Gradually increasing doses of prednisone in myasthenia gravis: reducing the hazards of treatment. N. Engl. J. Med. 290 (1974) 81–84.

SEYBOLD, M.E. and J. LINDSTROM: Patterns of acetylcholine receptor antibody fluctuation in myasthenia gravis. Ann. N.Y. Acad. Sci. 377 (1981) 292–306.

SGHIRLANZONI, A., D. PELUCHETTI, R. MANTEGAZZA, F. FIACCHINO and F. CORNELIO: Myasthenia gravis: prolonged treatment with steroids. Neurology 34 (1984) 170–174.

SHELTON, D., Y. FUJII, W. KNOGGE and J. LINDSTROM: Specific suppression of the antibody response to acetylcholine receptor in vitro and in vivo by daunomycin-acetylcholine receptor conjugates. Ann. N.Y. Acad. Sci. 540 (1988) 530–532.

SHER, E., C. GOTTI, N. CANAL, C. SCOPETTA, G. PICCOLO, A. EVOLI and F. CLEMENTI: Specificity of calcium channel autoantibodies in Lambert-Eaton myasthenic syndrome. Lancet ii (1989) 640–643.

SIARA, J., R. RÜDEL and A. MARX: Absence of acetylcholine-induced current in epithelial cells from thymus glands and thymomas of myasthenia gravis patients. Neurology 41 (1991) 128–131.

SIMPSON, J.A.: An evaluation of thymectomy in myasthenia gravis. Brain 81 (1958) 112–144.

SIMPSON, J.A.: Myasthenia gravis: a new hypothesis. Scott. Med. J. 5 (1960) 419–436.

SIMPSON, J.F., M.R. WESTBERY and K.R. MAGEE: Myasthenia gravis. An analysis of 295 cases. Acta Neurol. Scand. Suppl. 23 (1966) 1–27.

SMIT, L.M.E., G. HAGEMAN, H. VELDMAN, P.C. MOLENAAR, B.S. OEN and F.G.I. JENNEKENS: A myasthenic syndrome with congenital paucity of secondary synaptic clefts: CPSC syndrome. Muscle Nerve 11 (1988) 337–348.

SMIT, L.M.E., F.G.I. JENNEKENS, H. VELDMAN and P.G. BARTH: Paucity of secondary synaptic clefts in a case of congenital myasthenia with multiple contractures: ultrastructural morphology of a developmental disorder. J. Neurol. Neurosurg. Psychiat. 47 (1984) 1091–1097.

SOMMER, N., N. WILCOX, G.C. HARCOURT and J. NEWSOM-DAVIS: Myasthenic thymus and thymoma are selectively enriched in acetylcholine receptor-reactive T cells. Ann. Neurol. 28 (1990) 312–319.

SOMMER, N., G.C. HARCOURT, N. WILLCOX, D. BEESON and J. NEWSOM-DAVIS: Acetylcholine receptor-reactive T lymphocytes from healthy subjects and myasthenia gravis patients. Neurology 41 (1991) 1270–1276.

SOMNIER, F.E. and E. LANGVAD: Plasma exchange with selective immunoadsorption of anti-acetylcholine receptor antibodies. J. Neuroimmunol. 22 (1989) 123–127.

SOMNIER, F.E., N. KEIDING and O.B. PAULSON: Epidemiology of myasthenia gravis in Denmark. A longitudinal and comprehensive population survey. Arch. Neurol. (Chicago) 48 (1991) 733–739.

SORENSEN, T.T. and E.-B. HOLM: Myasthenia gravis in the county of Viborg, Denmark. Eur. Neurol. 29 (1989) 177–179.

STÅLBERG, E.J., V. TRONTELJ and M.S. SCHWARTZ: Single-muscle-fiber recording of the jitter phenomenon in patients with myasthenia gravis and in members of their families. Ann. N.Y. Acad. Sci. 274 (1976) 189–202.

STANLEY, E.F. and D.B. DRACHMAN: Effect of myasthenic immunoglobulin in acetylcholine receptors of intact mammalian neuromuscular junctions. Science 200 (1978) 1285–1287.

STEFANSSON, K., B.S. DIEPERINK, D.P. RICHMAN, C.M. GOMEZ and L.S. CHORTON: Sharing of antigenic determinants between the nicotinic AChR and proteins in E. coli, proteus vulgaris and K. pneumoniae. N. Engl. J. Med. 312 (1986) 221–225.

STEINMAN, L. and R. MANTEGAZZA: Prospects for specific immunotherapy in myasthenia gravis. FASEB J. 4 (1990) 2726–2731.

STRAUBE, A. and T.N. WITT: Oculo-bulbar myasthenic symptoms as the sole sign of tumor involving or compressing the brain stem. J. Neurol. 237 (1990) 369–371.

STRAUSS, A.J.L., B.C. SEEGAL, K.C. HSU, P.M. BURKHOLDER, W.L. NASTUK and K.E. OSSERMAN: Immunofluorescence demonstration of a muscle binding, complement fixing serum globulin fraction in myasthenia gravis. Proc. Soc. Exp. Biol. Med. 105 (1960) 184–191.

STREIB, E.W. and A.D. ROTHNER: Eaton-Lambert myasthenic syndrome: long-term treatment of three patients with prednisone. Ann. Neurol. 10 (1981) 448–453.

STRICKROOT, F.L., B.L. SCHAEFFER and H.L. BERGO: Myasthenia gravis occurring in an infant born of a myasthenic mother. J. Am. Med. Assoc. 120 (1942) 1207–1209.

SÜDHOF, T.C. and R. JAHN: Proteins of synaptic vesicles involved in exocytosis and membrane recycling. Neuron 6 (1991) 665–677.

SWICK, H.M., D.F. PRESTON and M.P. MCQUILLEN: Gallium scans in myasthenia gravis. Ann. N.Y. Acad. Sci. 274 (1976) 536–554.

SWIFT, T.R.: Weakness from magnesium-containing cathartics: electrophysiologic studies. Muscle Nerve 2 (1979) 295–298.

TAKEUCHI, N.: Effects of calcium on the conductance change of the end-plate membrane during the action of the transmitter. J. Physiol. (London) 167 (1963) 141–155.

TAMI, J.A., O.E. URSO and K.A. KROLICK: T cell hybridomas reactive with the acetylcholine receptor and its subunits. J. Immunol. 138 (1987) 732–738.

TAPHOORN, M.J.B., H. VAN DUIJN and E.C.H. WOLTERS: A neuromuscular transmission disorder: combined myasthenia gravis and Lambert-Eaton syndrome in one patient. J. Neurol. Neurosurg. Psychiatr. 51 (1988) 880–882.

TARRAB-HAZDAI, R., A. AHARONOV, I. SILMAN, S. FUCHS and O. ABRAMSKY: Experimental autoimmune myasthenia induced in monkeys by purified acetylcholine receptor. Nature (London) 256 (1975) 128–130.

THESLEFF, S.: Botulinal neurotoxins as tools in studies of synaptic mechanisms. Q. J. Exp. Physiol. 74 (1989) 1003–1017.

TINDALL, R.S.A.: Humoral immunity in myasthenia gravis. Effects of steroids and thymectomy. Neurology 30 (1980) 554–557.

TINDALL, R.S.A.: Humoral immunity in myasthenia gravis: biochemical characterization of acquired antireceptor antibodies and clinical correlations. Ann. Neurol. 10 (1981) 437–447.

TINDALL, R.S., J.A. ROLLINS, J.T. PHILLIPS, R.G. GREENLEE, L. WELLS and G. BELENDIUK: Preliminary results of a double-blind, randomized placebo trial of cyclosporine in myasthenia gravis. N. Engl. J. Med. 316 (1987) 719–724.

TODD, J.A., H. ACHA-ORBEA, J.I. BELL, N. CHAO, Z. FRONEK, C.O. JACOB, M. MCDERMOTT, A.A. SINHA, L. TIMMERMAN, L. STEINMAN and H.O. MCDEVITT: Molecular basis for MHC class II-associated autoimmunity. Science 240 (1988) 1003–1009.

TOYKA, K.V. and K. HEININGER: Acetylcholin-Rezeptor-Antikörper in der Diagnostic der Myasthenia gravis. Dtsch. Med. Wochenschr. 111 (1986) 1435–1439.

TOYKA, K.V., D.B. DRACHMAN, D.E. GRIFFIN, A. PESTRONK, J.A. WINKELSTEIN, K.H. FISCHBECK JR and I. KAO: Myasthenia gravis: study of humoral immune mechanisms by passive transfer to mice. N. Engl. J. Med. 296 (1977) 125–131.

TRONTELJ, J.V. and E. STALBERG: Single motor end-plates in myasthenia gravis and LEMS at different firing rates. Muscle Nerve 14 (1990) 226–232.

TSUJIHATA, M., T. YOSHIMURA, A. SATOH, I. KINOSHITA, H. MATSUO, M. MORI and S. NAGATAKI: Diagnostic significance of IgG, C3, and C9 at the limb muscle motor end-plate in minimal myasthenia gravis. Neurology 39 (1989) 1359–1363.

TZARTOS, S.J. and J. LINDSTROM: Monoclonal antibodies used to probe acetylcholine receptor structure: localization of the main immunogenic region and detection of similarities between subunits. Proc. Natl Acad. Sci. USA 77 (1980) 755–759.

TZARTOS, S.J., M.E. SEYBOLD and J. LINDSTROM: Specificities of antibodies to acetylcholine receptors in sera from myasthenia gravis patients measured by monoclonal antibodies. Proc. Natl Acad. Sci. USA 70 (1982) 188–192.

TZARTOS, S.J., D. SOPHIANOS and A. EFTHIMIADIS: Role of the main immunogenic region of acetylcholine receptor in myasthenia gravis. An Fab monoclonal antibody protects against antigenic modulation by human sera. J. Immunol. 134 (1985) 2343–2349.

TZARTOS, S.J., D. SOPHIANOS, K. ZIMMERMAN and A. STARZINSKI-POWITZ: Antigenic modulation of human myotube acetylcholine receptor by myasthenic sera. Serum titer determines receptor internalization rate. J. Immunol. 136 (1986) 3231–3238.

TZARTOS, S.J., A. KOKLA, S.L. WALGRAVE and B.M. CONTI-TRONCONI: Localization of the main immunogenic region of human acetylcholine receptor to residues 67-76 of the α-subunit. Proc. Natl Acad. Sci. USA 85 (1988a) 2899–2903.

TZARTOS, S.J., E. MOREL, A. EFTHIMIADIS, A.F. BUSTARRET, J.D. D'ANGLEJAN, A.A. DROSOS and H.A. MOUTSOPOULOS: Fine antigenic specificities of antibodies in sera from patients with D-penicillamine-induced myasthenia gravis. Clin. Exp. Immunol. 73 (1988b) 80–86.

TZARTOS, S.J., A. EFTHIMIADIS, E. MOREL, B. EYMARD and J.-F. BACH: Neonatal myasthenia gravis: antigenic specifities of antibodies in sera from mothers and their infants. Clin. Exp. Immunol. 80 (1990) 376–380.

TZARTOS, S.J., T. BARKAS, M.T. CUNG, A. KORDOSSI, H. LOUTRARI, M. MARRAUD, I. PAPADOULI, C. SAKAREL-LOS, D. SOPHIANOS and V. TSIKARIS: The main immunogenic region of the acetylcholine receptor. Structure and role in myasthenia gravis. Autoimmunity 8 (1991) 259–270.

UCHITEL, O., A.G. ENGEL, T.J. WALLS, A. NAGEL, V. BRIL and V.F. TRASTEK: Congenital myasthenic syndrome attributed to abnormal acetylcholine-acetylcholine receptor interaction. Neurology 40 (1990) 278.

VENKATASUBRAMANIAN, K., T. AUDHYA and G. GOLDSTEIN: Binding of thymopoetin to the acetylcholine receptor. Proc. Natl Acad. Sci. USA 83 (1986) 3171–3174.

VIAL, C., N. CHARLES, G. CHAUPLANNAZ and B. BADY: Myasthenia gravis in childhood and infancy. Usefulness of electrophysiologic studies. Arch. Neurol. 48 (1991) 847–849.

VIETS, H.R. and R.S. SCHWAB: Thymectomy for Myasthenia Gravis. Springfield, IL, Charles C. Thomas (1960).

VINCENT, A.: Are spontaneous anti-idiotypic antibodies against anti-acetylcholine receptor antibodies present in myasthenia gravis? J. Autoimmun. 1 (1988) 131–142.

VINCENT, A. and J. NEWSOM-DAVIS: Anti-acetylcholine receptor antibodies. J. Neurol. Neurosurg. Psychiatr. 43 (1980) 590–600.

VINCENT, A. and J. NEWSOM-DAVIS: Acetylcholine receptor antibody characteristics in myasthenia gravis. I. Patients with generalized myasthenia or disease restricted to ocular muscles. Clin. Exp. Immunol. 49 (1982a) 257–265.

VINCENT, A. and J. NEWSOM-DAVIS: Acetylcholine receptor antibody characteristics in myasthenia gravis. II. Patients with penicillamine-induced myasthenia or idiopathic myasthenia of recent onset. Clin. Exp. Immunol. 49 (1982b) 266–272.

VINCENT, A. and J. NEWSOM-DAVIS: Acetylcholine receptor antibody as a diagnostic test for myasthenia gravis: results in 153 validated cases and 2697 diagnostic assays. J. Neurol. Neurosurg. Psychiatr. 48 (1985) 1246–1252.

VINCENT, A., S.G. CULL-CANDY, J. NEWSOM-DAVIS, A. TRAUTMANN, P.C. MOLENAAR and R.L. POLAK: Congenital myasthenia: end-plate acetylcholine receptors and electrophysiology in five cases. Muscle Nerve 4 (1981) 306–318.

VINCENT, A., J. NEWSOM-DAVIS, P. NEWTON and N. BECK: Acetylcholine receptor antibody and clinical response to thymectomy in myasthenia gravis. Neurology 33 (1983) 1276–1282.

VOLTZ, R., R. HOHLFELD, A. FATEH-MOGHADAM, TH.N. WITT, M. WICK, C. REIMERS, B. SIEGELE and H. WEKERLE: Myasthenia gravis: measurement of AChR autoantibodies using cell line TE671. Neurology 41 (1991) 1836–1838.

WAGNER, J.A., S.C. CARLSON and R.B. KELLY: Chemical and physical characterization of cholinergic synaptic vesicles. Biochemistry 17 (1978) 1199–1206.

WALDOR, M.K., S. SRIRAM, H.O. MCDEVITT and L. STEINMAN: In vivo therapy with monoclonal anti-I-A antibody suppresses immune responses to acetylcholine receptor. Proc. Natl Acad. Sci. USA 80 (1983) 2713–2717.

WALKER, M.B.: Treatment of myasthenia gravis with physostigmine. Lancet i (1934) 1200–1201.

WALLS, T.J., A.G. ENGEL, C.M. HARPER, R.V. GROOVER and H.A. PETERSON: Congenital neuromuscular junction acetylcholinesterase deficiency. Ann. Neurol. 26 (1989) 147.

WALLS, T.J., A.G. ENGEL, A. NAGEL, M.C. HARPER and V.F. TRASTEK: Congenital myasthenic syndrome with paucity of synaptic vesicles and reduced quantal release. Neurology 40 (1990) 278.

WARMOLTS J.R., W.K. ENGEL and J.N. WHITAKER: Alternate-day prednisone in myasthenia gravis. Lancet ii (1970) 1198–1199.

WEKERLE, H. and U.-P. KETELSEN: Hypothesis: intrathymic pathogenesis and dual genetic control of myasthenia gravis. Lancet i (1977) 678–680.

WEKERLE, H.R., R. HOHLFELD, U.-P. KETELSEN, J.R. KALDEN and I. KALIES: Thymic myogenesis, T lymphocytes and the pathogenesis of myasthenia gravis. Ann. N.Y. Acad. Sci. 377 (1981) 455–475.

WERNIG, A.: Estimates of statistical release parameters from crayfish and frog neuromuscular junctions. J. Physiol. (London) 244 (1975) 207–221.

WHITING, P.J., A. VINCENT and J. NEWSOM-DAVIS: Myasthenia gravis: monoclonal antihuman acetylcholine receptor antibodies used to analyze antibody specificities and responses to treatment. Neurology 36 (1986) 612–617.

WHITTAKER, V.B.: The structure and function of cholinergic synaptic vesicles. Biochem. Soc. Trans. 12 (1984) 561–578.

WILKS, S.: On cerebritis, hysteria and bulbar paralysis, as illustrative of arrest of function of the cerebrospinal centers. Guy's Hosp. Rep. 22 (1877) 7–55.

WILLIAMS, C.L. and V.A. LENNON: Thymic B lymphocyte clones from patients with myasthenia gravis secrete monoclonal striational autoantibodies reacting with myosin, α-actinin, or actin. J. Exp. Med. 164 (1986) 1043–1059.

WILSON, K.S.A.: Neurology, Vol. 3. London, Butterworth, 2nd Ed. (1955) 1730.

WOKKE, J.H.J., F.G.I. JENNEKENS, P.C. MOLENAAR, C.J.M. VAN DEN ORD, B.S. OEN and H.F.M. BUSCH: Congenital paucity of secondary synaptic clefts (CPSC) syndrome in 2 adult sibs. Neurology 39 (1989) 648–654.

WOLF, S.M., L.P. ROWLAND, D.L. SCHOTLAND, A.S. MCKINNEY, P.F.A. HOEFFER and H. ARANOW JR.: Myasthenia as an autoimmune disease: clinical aspects. Ann. N.Y. Acad. Sci. 135 (1966) 517–535.

WOLTERS, CH.M.J., R.S. LEEUWIN and K.V. VAN WIJNGAARDEN: The effect of prednisolone on the rat phrenic nerve-diaphragm preparation treated with hemicholinium. Eur. J. Pharmacol. 29 (1974) 165–167.

WRIGHT, E.A. and M.P. MCQUILLEN: Antibiotic-induced neuromuscular blockade. Ann. N.Y. Acad. Sci. 183 (1971) 358–368.

YAMAMOTO, T., T. SATO and H. SUGITA: Antifilamin, antivinculin, and antitropomyosin antibodies in myasthenia gravis. Neurology 37 (1987) 1329–1333.

YOSHIDA, T., M. TSUCHIYA, A. ONO, H. YOSHIMATSU, E. SATOYOSHI and K. TSUJI: HLA antigens and myasthenia gravis in Japan. J. Neurol. Sci. 32 (1977) 195–201.

ZHANG, Y., S. PORTER and H. WEKERLE: Schwann cells in myasthenia gravis. Preferential uptake of soluble and membrane-bound AChR by normal and immortalized Schwann cells, and immunogenic presentation to AChR-specific T line lymphocytes. Am. J. Pathol. 136 (1990a) 111–122.

ZHANG, Y., M. SCHLUEP, S. FRUTIGER, G.J. HUGHES, M. JEANNET, A. STECK and T. BARKAS: Immunological heterogeneity of autoreactive T lymphocytes against the nicotinic acetylcholine receptor in myasthenic patients. Eur. J. Immunol. 20 (1990b) 2577–2583.

ZUCKER, R.S.: Changes in the statistics of transmitter release during facilitation. J. Physiol. (London) 229 (1973) 787–810.

Handbook of Clinical Neurology, Vol. 18 (62): Myopathies
L.P. Rowland and S. DiMauro, editors

CHAPTER 15

Periodic paralysis

J.J. SCHIPPERHEYN, A.R. WINTZEN and O.J.S. BURUMA

Departments of Cardiology and Neurology, University Hospital, Leiden, The Netherlands

Familial periodic paralysis is the name of a group of relatively rare, hereditary muscle diseases, all characterized by transient attacks of muscle weakness of varying duration and severity, but without impairment of consciousness, sensation or coordination. The attacks are often accompanied by changes of the serum potassium concentration. The different forms of periodic paralysis are distinguished on the basis of the serum potassium changes during the attacks, and are called hypo-, hyper- and normokalemic periodic paralysis. Separate forms exist that occur only in patients with thyrotoxicosis. Permanent muscle weakness persisting in the periods between the attacks is a common feature. Myotonic symptoms occur occasionally, especially in the hyperkalemic type, and may dominate the clinical picture. The hyperkalemic type, which is also called adynamia episodica hereditaria (Gamstorp), and paramyotonia congenita (Eulenburg) are different phenotypes of the same genetic disorder (De Silva et al. 1990). In a few families, both hypokalemic and hyperkalemic attacks have affected the same patient (Layzer et al. 1967).

Attacks of muscle weakness, but as a rule not complete paralysis, may occur in other diseases. These attacks follow severe potassium depletion in renal disease, primary or secondary hyperaldosteronism, and hypothalamic tumors, and attacks have been observed with hypermagnesemia

(Emser 1982). These attacks should be distinguished from familial periodic paralysis.

For an earlier review on periodic paralysis the reader is referred to Vol. 41 of the first series of this Handbook (Buruma and Schipperheyn 1979).

Hypokalemic periodic paralysis

The attacks. Severe attacks usually begin during the night. The patient wakes up early in the morning with limb weakness, or, in a full-blown attack, completely helpless and unable to raise the head, move the limbs, or even change position. Rotation of the head is usually possible and, in severe attacks, some functions are spared, e.g. deglutition, mastication, phonation and eye movements. Although respiration is usually not impaired, the accessory respiratory and intercostal muscles may be paralysed; the diaphragm, however, is never involved. Coughing and clearing of the throat may be difficult, but ventilation remains adequate (Putman 1900). Rare instances of ptosis and diplopia have been reported (Singer and Goodbody 1901). Putman (1900) and Taylor (1898) noted weakness of facial muscles during an attack.

In the hypokalemic form, limb weakness is usually symmetrical, but sometimes it involves arms or legs only partially, or only one limb is paralysed. The paralysis may be unilateral (Burr 1893; Mankowski 1929). The weakness is always of the flaccid type, with hypotonia and loss of tendon reflexes. There is no disturbance of the sphincters of rectum and bladder. Shy et al. (1961) found that urinary production is reduced during the attacks, probably because of intracellular accumulation of water in skeletal muscle (Wexberg 1917; Kastan 1921; Neel 1928). For prodromata of the attacks see Buruma and Schipperheyn (1979).

Abortive attacks. Episodes of muscle weakness may be brief. If so, they usually occur more often than the severe ones. In most of these abortive attacks, paresis of the legs predominates, affecting the extensor muscles more than the flexors; the paresis may be restricted to a single muscle group. Occasionally, an abortive attack is not more than a vague feeling of stiffness or tiredness in the legs, which may persist for days or weeks. This is uncommon, however, because the duration usually correlates with the severity of the attack, and varies from hours to as long as 8 days (Maclachlan 1932). The duration seldom exceeds 72 hours. The duration of incomplete attacks can be shortened by mild exercise. As a rule, the muscles that were last to become paralysed recover first. The duration of the recovery phase is roughly equal to the duration of the development of paralysis.

Attack rate. The frequency of attacks varies greatly from patient to patient, and in the same patient at different times. In some, one single attack occurs in an entire life, while in others the attacks are numerous, even daily (Zabriskie and Frantz 1932). The age at onset may range, even in a single family, from 4 to 35 years. Attacks usually begin late in the first decade or early in puberty. The rate usually diminishes after age 30 and attacks are rare after age 50.

Provocative factors. Rest after exertion is the best recognized precipitating factor. The more strenuous and unusual the activity and the more strict the period of rest, the more easily the attacks develop. Mild exercise may even have a preventive effect (Westphal 1885). Carbohydrate-rich meals are another attack-provoking factor. Less consistently reported precipitating factors include exposure to cold, emotional stress, intake of alcohol, trauma, or infection. Bender (1935) mentioned an influence

of the menstrual cycle and Sarova-Pinhas et al. (1981) also described a 17-year-old woman who had typical attacks with each menstrual cycle. Although her family history was negative and the laboratory studies were said to have been normal, the serum potassium levels ranged from 3.2 to 4.4 mmol/l and acetazolamide prevented the attacks. Among our patients, the attacks sometimes increased during pregnancy.

The administration of carbohydrates can be used to provoke attacks experimentally. Glucose administration and, in later years, glucose and insulin have been used. The procedure is not always effective and prolonged hyperglycemia seems to be essential (Shinosalki 1925) to induce an attack.

Tarsanen et al. (1983) mentioned that the time-course of the exercise induced rise of the serum potassium concentration in hypokalemic periodic paralysis patients is unusually flat and slow, which might be a diagnostic test. It has the advantage of not inducing a paralytic attack.

Interictal features, myotonia and permanent weakness. Characteristically, there are no symptoms between attacks, but there may be permanent limb weakness or myotonia.

Myotonia is found mainly in the hyperkalemic form of periodic paralysis, but may be seen in the hypokalemic form (Resnick and Engel 1967). In Resnick and Engel's report, three sporadic cases had myotonic lid-lag in the attack-free intervals, as also observed by Odor et al. (1967) and Griggs et al. (1970).

Persistent weakness probably develops in all affected persons (Buruma et al. 1985; Links et al. 1990). The prevalence of interictal weakness is hard to determine, however, because mild weakness may be missed unless specifically sought. In the younger patients, it is sometimes hard to tell whether there is permanent muscular weakness, or a prolonged state of abortive attacks. Permanent weakness occurs usually later in life (Holtzapple 1905; De Fine Olivarius and Christensen 1965; Gruner 1966), but Dyken et al. (1969) reported permanent muscular weakness at age 9 and 12 years, and Griggs et al. (1970) described an 8-year-old patient. Dyken et al. (1969) noted a strong male preponderance. The weakness starts and predominates in proximal muscles, involving the legs and pelvic girdle more than the arms, shoulder girdle and neck. Early muscular involvement can be detected by computed tomography before weakness is clinically apparent. The same applies to vacuolar changes in muscle biopsies (Links et al. 1990). Tendon reflexes are usually spared until late, with a close correlation to the severity of permanent weakness. Hypertrophy of the calves has been noted (Bernhardt 1895; Serko 1919), but is more likely in the hyperkalemic form. Proximal wasting rather than hypertrophy is more common when there is permanent muscular weakness. Permanent weakness is always slowly progressive. In advanced stages it can be disabling, confining the patient to a wheel-chair, or even to bed.

Neurophysiologic studies. Troni et al. (1983) demonstrated slowing of conduction in muscle fibers between attacks in hypokalemic periodic paralysis. During the attack the conduction slowed even more. Zwarts et al. (1988) and Links et al. (1990) found that asymptomatic heterozygotes could be detected by impaired muscle fiber conduction.

Life expectancy. Paralytic attacks may be fatal (Schachnowitz cited by Talbott 1941; Holtzapple 1905; Byrne 1916; Schmidt 1919; Serko 1919; Shinosaki 1925; Mankowski 1929; Maclachlan 1932; Zabriskie and Frantz 1932). In 1941, Talbott estimated a 10% mortality rate, but that high figure was a manifestation of the rudimentary care available in those days. The cause of death was usually not made clear and the available information is often too scanty to allow conclusions.

Inability to clear the trachea (Shinosaki 1925) or aspiration pneumonia was often the immediate cause of death. Although complete respiratory paralysis is probably never seen in periodic paralysis, it has been reported as a cause of death. Grotemeyer and Jörg (1979) described a patient with respiratory failure who had to be intubated, but they did not state how insufficient the respiration actually was. In recent years, fatal paralytic attacks are virtually unknown. The only recent autopsy study in a fatal case of periodic paralysis was made by Ionasescu et al. (1974). Death was ascribed to respiratory paralysis followed by cardiac arrest; at autopsy, there was severe bilateral pneumonia with signs of anoxic encephalopathy.

Cardiac abnormalities. Several authors have reported cardiac abnormalities during paralytic attacks (Goldflam 1891; Oppenheim 1891; Hirsch 1894; Singer and Goodbody 1901; Fuchs 1905; Holtzapple 1905; Schlesinger 1905; Wexberg 1917; Janota and Weber 1928), but all reported findings seem to be explained by hypokalemia alone. The prevalence of arrhythmias is inversely related to the serum potassium concentration (Sagild 1959). There is no evidence of acutely diminished contractility of the heart muscle during paralytic attacks. The acute dilation of the heart during attacks noted on chest X-ray pictures already at the beginning of this century by Wexberg (1917) does not necessarily indicate cardiac failure. Bradycardia, volume shift caused by adynamia and flattening of the body due to loss of muscle tone may all increase the apparent size of the heart. More convincing evidence of acute cardiac failure, such as low blood pressure or high filling pressures of the right and the left heart, has never been presented. One single case of Kramer et al. (1979) is a rare exception: a 19-year-old patient developed signs of left ventricular failure during an attack with high serum concentrations of creatine kinase and lactate dehydrogenase isoenzymes of myocardial origin. The myocardium was probably damaged, but the damage may have been caused by ischemia due to the severe bradycardia, vasospasm, or coronary thrombosis; a direct effect of severe hypokalemia on the integrity of myocardial membranes cannot be excluded.

In a normal person, lowering of the serum potassium concentration hyperpolarizes skeletal and myocardial muscle cells. The affected skeletal muscle fibers of patients with hypokalemic periodic paralysis do not hyperpolarize if serum potassium falls, but they lose excitability. The cardiac muscle cells remain fully excitable, however, and the electrocardiographic changes make it likely that myocardial cells are somewhat hyperpolarized by the hypokalemia. Moderate hyperpolarization increases size and duration of ventricular transmembrane action potentials. As a result, the R-wave is expected to become tall and slender and the T-wave to change into the diphasic or inverted forms described long ago (Zabriskie and Frantz 1932; Stewart et al. 1940). Hyperpolarization may cause intraventricular conduction disturbances with widening of the QRS-complex as found by Stewart et al. (1940), as well as bradycardia caused by sino-atrial conduction blockade and different types of supraventricular arrhythmias as reported in the past (Goldflam 1891; Schlesinger 1905; Wexberg 1917). Ventricular arrhythmias were described in two patients (Levitt et al. 1972; Stubbs 1976) but there is doubt that these cases were hypokalemic. In fact, it is surprising that ventricular arrhythmias are so rare during the attacks, despite the frequent occurrence of serum potassium concentrations below 3.0 mmol/l.

Between attacks, the heart is normal in size. Ten patients with familial hypokalemic periodic paralysis and permanent muscular weakness were examined by means of echocardiography and the size of the cardiac chambers was normal. Thus, no evidence of permanent weakness of the cardiac muscle was found in these patients (Schipperheyn et al. 1978). In one member of a hypokalemic periodic paralysis family with permanent weakness, biopsy of the left ventricular wall was taken when the patient underwent left heart catheterization for possible coronary artery disease. The biopsy showed an unusual amount of intermyofibrillary glycogen (Buruma et al. 1981) but no loss of muscle fibers. The deposition of glycogen in the cardiac muscle resembled that found in skeletal muscle biopsies from the same patient. Although this was only a single case, it seems warranted to state that the cardiac muscle may show minor abnormalities resembling those in skeletal muscle, but clinically significant cardiomyopathy does not develop. In an autopsy study of a 70-year-old man the myocardium appeared normal under the light microscope (Links et al. 1990). For reasons incompletely understood, the occasionally severe hypokalemia causes only relatively benign arrhythmias.

Anesthesia and surgery. Anesthesia or the stress of operation may induce an attack in the postoperative phase. Reviews of the anesthetic and postoperative management of patients with hypokalemic periodic paralysis once focused on the possible influence of anesthetics and muscle relaxants. A clear-cut effect of specific agents or types of anesthesia on postoperative attacks of paralysis has not been established. Reports now stress the importance of avoiding intravenous administration of glucose or drugs that may lead to hypokalemia. Periods of severe hypotension may induce hypokalemia because of sympathetic activation (Feurstein 1980; Boulton and Hardisty 1982; Fozard 1983; Melnick et al. 1983; Behne and André 1984).

Mode of inheritance. Hypokalemic periodic paralysis is transmitted as an autosomal-dominant condition. The report of Khan (1935), who suggests an X-linked inheritance, must be incorrect and could be explained by variable expression of the gene, specially in females. This was demonstrated in the reports of Biemond and Polak Daniels (1934) and Olivier et al. (1944).

In women, the disease tends to be less severe than in men, and some women who inherit the gene do not have overt attacks, or may have only one attack in a lifetime. These women consider themselves free of the disease. It is a common observation in inherited neurological disorders of late onset that patients with minor symptoms deny being affected. A relative high prevalence of apparently unaffected females led to the impression that the gene is often non-penetrant in women, which explained the phenomenon of skipping of generations (Goldflam 1891; Schmidt 1919). In a family described by Woratz (1965), the gene seemed to be completely non-penetrant in the women. In most families, however, a regular autosomal-dominant mode of inheritance is the rule.

Partial non-penetrance in women explains the male preponderance observed by Talbott (1941) with a ratio of 3:1, or even 4:1 as observed by Cerny and Katzenstein-Sutro (1952). The high ratio possibly results from sporadic cases. In the sporadic cases the disease is also less severe in females than in males, and the female sporadic cases with only minor symptoms are more likely to pass unrecognized than the mildly affected females from well-studied families.

In a linkage study in a rather large family with hypokalemic periodic paralysis and permanent muscular weakness, close linkage was excluded with 28 genetic marker systems (Buruma et al. 1985). Fontaine et al. (1990) reported one recombinant in a family with hypokalemic periodic paralysis in a linkage study of the sodium channel α-subunit gene.

Sporadic cases. Sporadic cases of hypokalemic periodic paralysis are not rare. It is difficult to believe that all sporadic cases are due to incomplete penetrance and insufficient information about the family, or to an incorrect diagnosis of muscle weakness secondary to hypokalemia. Gaupp and Kalden (1942) presented evidence that, in at least one of their cases, no other members of the family were affected. The occurrence of sporadic cases does not necessarily mean that there is an acquired form of hypokalemic periodic paralysis, distinct from the familial disease. Sporadic cases may be new mutations, but it might take a long time before an attack in children of that patient provides proof. Ropers and Szliwowski (1979) described two daughters of an apparently unaffected father, in a family otherwise free of hypokalemic periodic paralysis.

The relatively high incidence of sporadic cases is still not understood. However, the sporadic cases do not differ clinically from familial cases, and the muscle biopsy changes are histologically indistinguishable. It is therefore believed that many sporadic and familial cases are the same disease, and that occasional non-penetrance explains why so many cases are sporadic.

For an overview on the geographical and racial distribution of hypokalemic periodic paralysis the reader is referred to Buruma and Schipperheyn (1979). Two sporadic cases have been reported (Sadeh et al. 1980; Toglia et al. 1982) in which hypokalemic periodic paralysis was associated with multiple sclerosis.

Thyrotoxic periodic paralysis

Attacks. Thyrotoxic periodic paralysis resembles the hypokalemic form of the disease, but it is seen only in thyrotoxic patients. The attacks disappear as soon as the patient becomes euthyroid (Conway et al. 1974). As in hypokalemic periodic paralysis, attacks can be provoked by administration of glucose and insulin, and the attacks normally occur after a period of rest and a carbohydrate-rich meal. The prevalence of periodic paralysis among thyrotoxic patients is estimated at 0.1 to 0.2% in whites (McFadzean and Yeung 1969), and about 1 to 2% in Asian men (Kelley et al. 1989).

Inheritance. Although thyrotoxic periodic paralysis seems to be a sporadic disease, the incidence of periodic paralysis among members of a family with thyrotoxicosis is higher than in a general thyrotoxic population (McFadzean and Yeung 1969). Together with the racial difference, this indicates that thyrotoxic periodic paralysis is actually an inherited disease that is unmasked by thyrotoxicosis. Assuming that thyrotoxicosis is truly sporadic, or, if it is familial, that the mode of inheritance is unrelated to that of the muscle disease, the prevalence of the genetic defect is high.

Hyperkalemic periodic paralysis

The attacks. In hyperkalemic periodic paralysis, the attacks usually start with a sense of heaviness or stiffness in the muscles. Paresthesias in the extremities or in the lips and face are common. In children, especially babies, an early symptom of an attack is staring of the eyes (McArdle 1962), which is caused by myotonic lagging of the upper eyelid when a downward gaze follows an upward gaze. More severe attacks occur when the patient rests after exercise, but in contrast to the hypokalemic form, the hyperkalemic attacks occur mostly during the day (Sagild 1959). When attacks occur during the night, they start in the first hours and not early in the morning, as in the hypokalemic form. Nocturnal attacks are usually more prolonged and more severe.

Paresis tends to begin in the thighs and calves. Sometimes the weakness is noted first in the small of the back (McArdle 1962). Weakness then progresses to the arms and the neck. Facial muscles may be affected subsequently and, in severe attacks, there may be difficulty in coughing and swallowing. Serious involvement of respiratory muscles, however, is rare (Gamstorp 1956). Complete helplessness is uncommon, and in contrast to hypokalemic attacks, there is often residual mobility in the affected muscles (Sagild 1959). Tendon reflexes, especially in the legs, are reduced during attacks and may be lost in severe attacks. Recovery is rapid, from a few minutes to one or two hours. In 75% of the patients, the duration of attacks never exceeds one hour (Gamstorp 1956). As in hypokalemic periodic paralysis, the sphincters are spared, but weakness of the abdominal muscles may interfere with voiding and defecation.

Duration and frequency of attacks. In children the attacks tend to be shorter; the duration increases in puberty. Although attacks are usually brief, they may last up to 12 hours or more (Pearson 1964). In a family described by Van der Meulen (1961), seven of sixteen patients had attacks lasting more than 24 hours.

As in hypokalemic periodic paralysis, the muscles first affected are the last to recover (Sagild 1959). Also, as in the hypokalemic form, patients can often 'walk off' attacks. Rest also elicits attacks and exercise promotes recovery. If exercise is started when the incipient sensations of weakness are experienced, it may prevent an attack.

In some patients, 'walking off' is ineffective (McArdle 1962), and results in pain, tenderness and hardening of the calf muscles persisting for hours or days. As a method to postpone attacks, patients found regular exercise undesirable, because it deprived them of the prolonged attack-free periods that usually followed severe attacks.

The attack rate varies from patient to patient, and in the same patient may vary over the years from several times a day to one a year (Gamstorp 1956). Most patients, however, have about one attack a week. Slight weakness may persist after an attack, usually not enough to interfere with daily life, but sufficient to keep them from heavy labor. This weakness may persist for several days, and is accompanied by a dull pain in the affected muscles.

The first attack usually occurs in the first decade, even in infancy (Carson and Pearson 1964), but may be delayed until after age 20, and one patient started at age 48 (Feneck and Soler 1968). As in the hypokalemic form, the frequency of attacks tends to decrease with advancing age (Gamstorp 1956; Sagild 1959).

Provocative factors. Attacks are provoked by some factors that differ from those of hypokalemic attacks. A long period of rest, fasting, and exposure to cold are the best known. Inactivity after strenuous or prolonged exercise is probably the most important. Subramony and Wee (1986) found that immobility of the muscle is more important than the preceding exercise. The attack rate may increase in winter, probably due to the promoting effect of exposure to cold (Sagild 1959). Unlike the immediate effect in the hypokalemic form, hyperkalemic attacks develop many hours after a meal, when the patient feels hungry again. Also in contrast to the hypokalemic type, the ingestion of sugar and other carbohydrates, especially bread, may prevent and reduce attacks, rather than provoke them (Gamstorp 1956). Administration of potassium salt is not beneficial as in the hypokalemic type, but is detrimental. The intake of potassium-rich beer seemed to cause attacks in two of Bradley's (1969) patients. Mental stress may provoke attacks (McArdle 1962; Bradley et al. 1990). Pregnancy may increase both rate and severity of the attacks.

Experimentally, attacks are elicited by oral administration of potassium salts. They are invariably induced by the ingestion of 0.5 mg of potassium chloride per kg body weight (Gamstorp 1956). Attempts to induce attacks by means of local cooling (icewater) were unsuccessful (Sagild 1959; Layzer et al. 1967), although cooling seems to promote the attacks. The administration of sodium bicarbonate orally may provoke attacks within 30 minutes (Bradley 1969).

The interictal period and associated features. Permanent muscular weakness was seen in practically all members of the family described by McArdle (1962), but was rare in other families (Gamstorp 1956; Bradley 1969). In a Danish family described by Sagild (1959), none of the patients had persistent weakness. Bradley et al. (1990) thought that permanent weakness is as frequent as in the hypokalemic form. Permanent weakness may be confined to hip flexors and triceps, but abdominal muscles and neck flexors may be affected. In contrast to muscular wasting seen in the hypokalemic form, some hyperkalemic patients show calf hypertrophy, which may be noted without permanent weakness (Bradley 1969). Muscular wasting, however, may also be seen (Saunders et al. 1968).

Myotonic signs. Apart from a myotonic lid-lag, all eye muscles may show symptoms of myotonia, causing transient blurring of vision and transient diplopia. Chvostek's sign has been observed during attacks in 50% of patients (Gamstorp 1956;

Kaplan et al. 1957; Van der Meulen et al. 1961). Percussion myotonia of the tongue has also been observed (Bradley 1969). In the family described by Gamstorp, however, myotonia was not observed, but it was seen in some families by McArdle (1962) and Carson and Pearson (1964). Sometimes myotonia was seen only on electromyography. Myotonia, especially lingual myotonia, may be the only manifestation of disease (especially in women) in otherwise asymptomatic members of a family (Layzer et al. 1967; Bradley 1969; Lisak et al. 1972). Three types of affected individuals may occur in the families with myotonia (Layzer et al. 1967): patients with EMG myotonia only, patients with clinical myotonia and no paralytic attacks, and patients with both myotonia and attacks of weakness. Patients with myotonia only did not have persistent weakness either.

Genetic aspects. Hyperkalemic periodic paralysis was first recognized as a separate disease in 1952 (Gamstorp 1956). However, in retrospect, several families described in earlier publications could have had the hyperkalemic form. For instance, an English family described by Buzzard (1901), and one described by Schoenthal (1934) in the United States were probably hyperkalemic. Genetic aspects, occurrence and distribution of hyperkalemic periodic paralysis were studied by Hellweg-Larsen et al. (1955) and Gamstorp (1956) in Scandinavia. Gamstorp called the hyperkalemic form 'adynamia episodica', seen in an uninterrupted transmission through five and possibly seven generations in two families. She concluded that the mode of inheritance was autosomal-dominant. Because one asymptomatic individual seemed to have transmitted the disease, she stated that there was 'almost complete' penetrance. Bradley et al. (1990) studied four unrelated families and confirmed the autosomal-dominant type of inheritance, with slight interfamilial differences in clinical manifestation. They suggested that the differences might be due to minor variations in the mutation of ion channel proteins.

Dyken and Timmons (1963) suggested an autosomal-recessive inheritance in an aberrant sporadic case with hypocalcemia as well as hyperkalemia because there was also a family history of painful muscle cramps in both maternal and paternal lines. A similar mode of inheritance was suggested by Riggs et al. (1981), who studied a patient with typical hyperkalemic periodic paralysis with no other affected family members. Potassium provocation tests had no effect in members of the immediate family.

True sporadic cases are rare (Benedek and Von Angyal 1942; Dyken and Timmons 1963; Herman and McDowell 1963; Feneck and Soler 1968; Riggs et al. 1981) and may all be new mutations of the autosomal-dominant disease. Fontaine et al. (1990) studied one large family. Using portions of the adult muscle sodium channel α-subunit gene, they mapped it to chromosome 17, and demonstrated tight linkage with hyperkalemic periodic paralysis.

Normokalemic periodic paralysis

In 1951 Tyler et al. described a kindred in which 33 members had attacks very similar to those of hypokalemic periodic paralysis, but the serum potassium level did not fall. Similar cases were noted by Talbott (1941) and Watson (1946). Poskanzer and Kerr (1961) described another such family. Attacks began at the age of 10, just as in the family described by Tyler et al. (1951). Both families showed autosomal-dominant inheritance. The attacks lasted much longer than hypokalemic attacks. Complete paralysis might last a week, with recovery over one or two additional weeks. The frequency of attacks ranged from one in two months to a few attacks in a lifetime. Bladder and bowel functions were never impaired. In contrast to the other forms, however, one person had loss of sensation in at least three attacks.

Variant types of periodic paralysis

In 1972 Meyers et al. reported on a mother and son, both showing attacks of weakness and occasional paralysis of certain groups of muscles accompanied by stiffness and muscle ache. In both patients, the age at onset was in the second decade and the frequency of attacks varied in both from once a month to once a year. The duration also varied, usually a few hours, but one severe attack lasted three weeks. No permanent muscular weakness was observed in these patients initially. Spontaneous attacks invariably began with rest after exercise. Occasionally, the weakness was confined to muscles used actively before the attack. Between attacks, serum potassium levels were normal, and attacks were not induced by hypokalemia or hyperkalemia. Most attacks were hypokalemic but some were hyperkalemic. Layzer et al. (1967) also compiled reports of patients who had documented attacks with hypokalemia or hyperkalemia at different times. In hyperkalemic attacks, the serum potassium level rose to 6.1 mmol/l; during hypokalemic attacks it dropped, but never below 3.3 mmol/l. Other laboratory tests were all normal except for a consistent elevation of creatine kinase (CK) in the boy only. Muscle biopsies were obtained during an asymptomatic period. The biopsy of the boy showed vacuole-like structures in type II fibers. These structures reacted for NADH and, on ultrastructural analysis, they appeared to be tubular aggregates. In the mother's biopsy, light microscopic examination revealed no significant pathological changes, but histochemically there was atrophy of some type II fibers. On ultrastructural analysis, the sarcoplasmic reticulum was slightly dilated. Later, the son developed slight permanent weakness (Chesson et al. 1979) as attacks of hyperkalemic and hypokalemic paralysis continued. The attacks came on spontaneously but they were also provoked by administration of potassium and also by giving insulin. Similar observations were made by Pearson (1964), Layzer et al. (1967) and by Lander et al. (1985). The kindred described by Lander also showed high CK-levels.

Klein et al. (1963) claimed the discovery of what they considered to be another variant type of periodic paralysis. They studied two girls, one with a family history, the other a sporadic case. The type of attacks, age at onset, duration, and precipitating factors were similar to those of hypokalemic periodic paralysis. The serum potassium levels, however, did not fall. Administration of potassium chloride had a beneficial effect and sodium chloride was deleterious. However, administration of glucose had a protective rather than a provocative effect. The patients showed premature ectopic heartbeats of varying origin and frequency, with occasional periods of bigeminy and non-sustained ventricular tachycardia.

Similar cardiac arrhythmias were described by Lisak et al. (1972) in a family with an autosomal-dominant pattern of lingual myotonia. One girl, with myotonia, had attacks of weakness that started at age 18 months. They usually lasted two days and occurred about once a month. The attacks were never pronounced and did not confine her to bed. They were provoked by administration of potassium chloride but not by glucose and insulin. Serum potassium levels were normal during attacks. From age 15 she had frequent, self-limiting attacks of tachycardia comparable to those described by Klein et al. (1963). Another case of periodic paralysis with frequent premature ventricular contractions and subnormal or normal serum potassium levels during attacks was described by Yoshimura et al. (1983). Attacks were provoked by potassium chloride and not by glucose and insulin. Gould et al. (1985) described a mother and son with bidirectional ventricular tachycardia, short stature, microcephaly and clinodactyly. Attacks were induced by potassium chloride. The mother had lingual myotonia, but no attacks of weakness. Cardiac dysrhythmia in the boy was associated with hypokalemia and a nearly fatal arrest occurred when he ate a large quantity of sweet chocolates.

Levitt et al. (1972) and Stubbs (1976) described patients with periodic paralysis with minimal changes of serum potassium levels during attacks and with pronounced cardiac dysrhythmia. A case described by Anderson et al. (1971) was associated with multiple congenital anomalies but was not sufficiently documented to classify the attacks.

Apart from the two cases of Klein et al. (1963), these cases of variant periodic paralysis were similar to the hyperkalemic form. However, the changes of the serum potassium level were small and there was an association with cardiac dysrhythmia. Administration of potassium salts provoked the attacks but contrary to experience with the hyperkalemic form, attacks were also induced by glucose and insulin. Lisak et al. (1972) suggested that this variant form should be considered as a separate hyperkalemic entity.

PATHOLOGY

Vacuolar changes are the most remarkable feature

of the muscle biopsies from patients with periodic paralysis, and were first noted by in 1895 by Goldflam. They appeared to be empty, or to contain unidentifiable material. Schmidt (1919) showed that the contents had an affinity for glycogen stains, and Pearson (1973) demonstrated that pretreatment with amylase prevented the staining. The vacuoles contain glycogen or a closely related substance.

Engel (1970) noted other abnormalities, including abnormal variation of fiber diameter, excess of centrally located nuclei, and excessive fiber splitting. Intravacuolar macrophages and regenerating fibers were rare, and some non-vacuolated fibers showed regional rarefaction. Vacuoles were usually empty, but some contained granular material and some had compartments. Engel confirmed the presence of myriads of vesicles, or vacuoles, and irregularly curved lamellae of various shapes in small abnormal regions of the fibers. Acid phosphatase-active regions were seen in some fibers, adjacent to well-formed vacuoles, or within the center or at one pole of a larger vacuole. Some smaller fibers showed diffuse acid phosphatase activity and some of these fibers appeared to originate from larger fibers by splitting.

PAS-positive material occasionally filled the vacuoles entirely; more often it was localized within one of the vacuolar compartments or in small subsarcolemmal or deeper intermyofibrillar spaces. Engel's ultrastructural studies distinguished stages in the evolution of the vacuoles. Non-vacuolated fiber regions were usually normal, but occasionally, fibers in apparently normal regions showed proliferation of the T-tubules, as well as dilatation of components of the sarcoplasmic reticulum, considered to be early stages. Later, the proliferating T-system seemed to act as a membrane source and to form a membrane-bound space. The membrane-bound spaces contained remnants of evolving vacuoles. Eventually the trapped components disappeared, and the vacuole had 'matured'.

When horseradish peroxidase was added to the extracellular space, the marker soon appeared inside the vacuoles, implying free communication between the lumen and the extracellular space, which was confirmed ultrastructurally by Buruma (1978).

Some authors were convinced that the vacuoles contained only remnants of vesicles arising from the sarcoplasmic reticulum (Shy et al. 1961; Howes et al. 1966; Ionasescu et al. 1971). Others thought the membranes lining the vacuoles were related to the T-system, and that the contents did not originate from the sarcoplasmic reticulum (Schutta and Armitage 1969). Engel suggested that both the sarcoplasmic reticulum and the T-system contributed to the vacuolar contents, but the observations of Gérard et al. (1978) did not support that view.

Tubular aggregates have been noted in both hypokalemic (Gruner 1966; Odor et al. 1967; Engel et al. 1970) and hyperkalemic periodic paralysis. The aggregates seem to be proliferations of the lateral sacs of the longitudinal components of the sarcoplasmic reticulum, and are seen beneath the sarcolemma. The tubular aggregates seem to occur only in type II fibers and do not contain mitochondria (Engel et al. 1970). Pellissier et al. (1980) and Faugère et al. (1981) stressed that tubular aggregates are found in later stages of the disease when the vacuoles are less common. Tubular aggregates are also found in myotonic dystrophy, mitochondrial myopathy (Spiro et al. 1970) and perhaps may be induced by diazepam (Engel 1970).

Weller and McArdle (1971) described intracellular basophilic granular degeneration, which they attributed to deposition of calcium. Calcium deposits are observed with some frequency. Porte et al. (1979) observed calcium accumulation in the T-system instead of in the terminal cisternae of the sarcoplasmic reticulum during a paralytic attack. They suggested that a defect of uptake and storage of calcium ions in the sarcoplasmic reticulum plays a pathogenic role in hypokalemic periodic paralysis, but the calcium deposition could be secondary to altered sodium transport and a high intracellular concentration of sodium. Hartlage and Soudmand (1981) noted internalized capillaries in 20% of the type I fibers in a patient with familial hypokalemic paralysis. Others observed this phenomenon in Duchenne muscular dystrophy (Hastings et al. 1980).

The concentration of mitochondria and degradation of mitochondria into myelin bodies (Buruma 1978) and changes in the cristae led Guarino et al. (1984) to believe that hypokalemic periodic paralysis is primarily a mitochondrial cytopathy.

The histopathology indicates that periodic paralysis is primarily a muscle disease. The abnormalities indicate irreversible permanent changes (Engel et al. 1970; Pellissier 1980; Martin et al. 1984). With only a few rare exceptions (MacDonald et al. 1969), the presence of vacuoles and other abnormalities seems to be the anatomical counterpart of permanent muscular weakness. Bekény (1961) discussed the relationship between structural changes, and the attacks and the permanent weakness. The first possibility, he suggested, is that the structural changes cause the attacks as well as the permanent weakness. The second option is that the attacks occur independently from a mild form of progressive muscular dystrophy. The third possibility is that permanent weakness results from damage induced by the paralytic attacks. Bekény favored the third possibility because biopsies from patients with hypokalemic periodic paralysis without permanent weakness are often completely normal (Resnick and Engel 1967; MacDonald et al. 1969; Engel 1970).

In the hyperkalemic form, the structural changes are usually mild, consisting of increased centrally localized nuclei, occasional isolated muscle fiber degeneration, and increased glycogen content in some fibers, especially beneath the sarcolemma (Bradley 1969). Large variation in fiber size may occur (MacDonald et al. 1969). Small vacuoles have been seen in the perinuclear area of some muscle fibers (Bradley 1969). The vacuoles are smaller and less numerous than in the hypokalemic form. Many contain PAS-positive material and they are usually membrane-bound.

Vacuolization in hyperkalemic paralysis was first reported by Bekény et al. (1961). In a few earlier studies (Gamstorp 1956; Sagild 1959; Klein et al. 1960), no such abnormalities were noticed, but vacuolar changes were confirmed by Pearson (1964). The results of histochemical examinations were normal, except for accumulation of the glycogen-like material, occasional excess of sarcoplasmic fat in small fibers, or an abnormal staining for mitochondrial enzymes (Bradley 1969).

Ultrastructural analysis in hyperkalemic paralysis revealed vacuoles in some fibers, which were not seen with the light microscope (MacDonald et al. 1968). In the larger vacuoles, finely dispersed granular material and scattered glycogen granules confirmed the histochemical observations (MacDonald et al. 1968). Moderately dilated sarcoplasmic reticulum vesicles, very similar to those noted in hypokalemic paralysis, were also observed in the hyperkalemic form (MacDonald et al. 1968). Danowski et al. (1975) reported dilation of the sarcoplasmic reticulum in apparently unaffected siblings in a family with hyperkalemic paralysis. Tubular aggregates have been reported on several occasions (Engel et al. 1970). The muscle pathology of thyrotoxic paralysis shows the same features as found in the hypokalemic form (Engel 1961; Brody and Dudley 1969). Except for quantitative differences, the three forms of periodic paralysis are histologically very similar to each other.

It is generally accepted that the vacuolar changes found in all types of periodic paralysis are associated with permanent weakness and not with a propensity to paralytic attacks (Engel 1970; Links et al. 1990). The histopathology of permanent muscular weakness bears no resemblance to that of progressive muscular dystrophy. The development of permanent muscular weakness does not appear to depend on the number or severity of the attacks. Danowski et al. (1975) have reported ultrastructurally a moderate dilation of the sarcoplasmic reticulum in an unaffected sibling of a family with hyperkalemic periodic paralysis. Such dilations are commonly found in the hypo-, hyper- and normokalemic form (Shy et al. 1961; Howes et al. 1966; MacDonald et al. 1969). Buruma and Bots (1978) described a girl aged 9 years from a family with hypokalemic periodic paralysis, whose muscle biopsy showed classical abnormalities although she had experienced only two mild attacks. Her brother's biopsy had an increased glycogen content and atrophy of muscle fibers. He had slight proximal weakness but had not had a single attack. These observations underline that the myopathy may develop independently from the attacks.

PATHOPHYSIOLOGY

The electrical properties of muscle cells from patients with hypokalemic or hyperkalemic periodic paralysis are determined by the extracellular potassium concentration (Lehmann-Horn et al. 1983; Rüdel et al. 1984). In hypokalemic paralysis, the fibers become completely inexcitable during an attack. The membrane potential remains normal, despite the low external potassium concentration, and the membrane resistance is lowered. The only plausible explanation for a normal membrane potential at a low external potassium concentration is a disturbance of the inactivation of the sodium channel; some Na-channels remain open and fail to inactivate, while others remain inactivated and fail to reactivate.

Fontaine et al. (1990) found that the genetic defect of hyperkalemic periodic paralysis is linked to the sodium channel α-subunit gene. The defect maps near the human growth hormone locus (GHl) on chromosome 17. The gene for hypokalemic periodic paralysis may also map close to this locus. It seems likely that all forms of periodic paralysis are caused by similar mutations. Like the hemoglobin molecule, the α-subunit appears to occur in a number of genetically determined isoforms of which only a few cause a manifest disease.

Localization close to the GHl-locus would explain the variant forms of periodic paralysis that are combined with disturbances of growth (Cousot 1887). Fontaine's work proves that the disease is primarily a disorder of transmembrane sodium ion transport, and that cellular metabolism is only secondarily involved.

The sodium channel of muscle cells comprises α- and β-subunits; the α-subunit carries the sodium current. The structure of the α-subunit contains four domains, each comprising six transmembrane segments that are connected by intracellular and extracellular segments (Catterall 1988). Some of the connecting segments are involved in the inactivation process. In hyperkalemic paralysis, the intracellular connecting segment between the second and third domain seems to be defective. To explain all the abnormalities found in the different forms of periodic paralysis, one has to assume that the structure of at least one of the connecting segments is abnormal and that, as a result, the inactivation of the sodium channel is defective. Inactivation of the sodium channel is normally only weakly voltage-sensitive, and it derives its apparently strong voltage dependence from the coupling to voltage-dependent activation. In periodic paralysis, the inactivation depends on the potassium concentration immediately outside the membrane

(Lehmann-Horn et al. 1983; Rüdel et al. 1984). In the hypokalemic form, inactivation and reactivation are delayed or even completely inhibited by a low extracellular potassium concentration. If the extracellular concentration of potassium is sufficiently lowered, some sodium channels fail to inactivate and remain open while an even larger number fail to reactivate and remain closed and unexcitable. In the hyperkalemic form, the same happens at high extracellular concentrations of potassium (Rudel et al. 1989).

Sodium conductance is abnormally high at rest, even in the interictal state. Between attacks, the probability that a sodium channel is activated may not be decreased, but due to the inherited abnormality of the channel the probability that inactivation occurs in an activated channel is reduced even at a normal potassium concentration outside the fiber. As a result, the sodium influx at rest is high and the intracellular sodium concentration is constantly slightly elevated. The resting activity of the sodium-potassium pump is stimulated to compensate for the high passive influx of sodium ions. The balance is unstable because any increase of pump activity would reduce the pericellular potassium concentration, which, in turn, would increase sodium influx and enhance Na/K-pump activity even further. In that way, any accidental increase of pump activity would induce a vicious development that would end in complete paralysis. The same might happen if the serum potassium concentration were lowered.

The paralytic state is maintained as long as the potassium concentration directly outside the fiber remains sufficiently low. Potassium ions from the blood eventually correct the problem and the time needed to do so depends on the circulation inside the muscle. In our opinion, differences in blood supply of the muscle provide the clue for the absence of paralysis in constantly working muscles, including heart muscle. The high blood flow in working muscle serves to stabilize the potassium concentration in the extracellular space at a sufficiently high level to prevent complete inactivation. In resting muscle, blood flow is at such a low level that the pericellular potassium concentration can easily drop to an extremely low value. Changes in local blood flow may also explain why attacks can be aborted by exercising the muscle.

Provoking factors induce attacks by stimulating the sodium-potassium pump (rest after exercise, administration of glucose and insulin, sympathetic stimulation), by directly lowering the extracellular potassium concentration or, possibly, by reducing the enhancing effect on inactivation of the phosphorylation of intracellular segments of the sodium channel (Catterall 1988).

The high steady-state activity of the Na/K-pump raises glucose utilization at rest and increases uptake of insulin (Hofmann et al. 1983; Johnsen and Beck-Nielsen 1979). It seems plausible that the continuously high insulin uptake causes excessive storage of glycogen, which eventually leads to the formation of vacuoles as discussed above. Moreover, the high sodium content of the cell reduces the capacity of the cell to remove calcium ions by exchange against sodium; intracellular accumulation of calcium eventually leads to mitochondrial damage and cell death and should no longer be considered an unrelated finding. Calcium accumulation is not necessarily related to the number of attacks (Bekény 1961).

Thyrotoxicosis affects the properties of the sodium channel even in normals. Sodium conductance at rest is enhanced and the activity of the sodium-potassium pump is increased. These effects are abolished by β-adrenergic blockade, which indicates that they may be mediated by an increased expression of mRNA for the formation of adrenergic receptors. In 0.1 to 0.2% of white people, thyrotoxicosis discloses an abnormal sensitivity of sodium channel inactivation and reactivation to low potassium outside the fiber. In Asian men the prevalence of this abnormality is about ten times as high (Engel 1961). The abnormality in Asian men differs from that in European whites (Marx et al. 1989). Apparently, the peculiar sodium channel of hypokalemic paralysis is not the only genetically determined isoform. The structure of the α-subunit of the adult sodium channel of muscle seems to vary in sensitivity to changes of the pericellular potassium concentration. If a rat is severely depleted of potassium, even the normal sodium channel becomes permanently inactivated and the muscles are paralysed, if the animals are given glucose and insulin (Offerijns et al. 1958). In humans at least one variety of the sodium channel becomes

inactivated by moderately low extracellular potassium only in case of thyrotoxicosis. In the thyrotoxic state, the properties of affected muscle fibers resemble those of familial hypokalemic periodic paralysis. The uptake of glucose and insulin is raised (Marx et al. 1989) in the interictal state and, probably due to some down-regulation of insulin receptors, the patients develop insulin resistance and hyperinsulinemia (Lee et al. 1991). It is probably important clinically, because in thyrotoxic periodic paralysis an abnormal sodium channel with an unusually high sodium uptake at rest induces hyperinsulinemia. In unexplained hyperinsulinemia with a tendency to retention of sodium, some glucose intolerance and hypertension may be caused by a high sodium uptake at rest balanced by a high Na/K-pump activity resulting from a genetically determined abnormality of the sodium channel.

In the hyperkalemic form, sodium influx at rest is also elevated and balanced by a high activity of the Na/K-pump, as in the hypokalemic form. In case of hyperkalemic paralysis, however, a rise instead of a drop of the pericellular potassium concentration blocks inactivation. A rise of outside potassium may result from potassium loss from the muscle during exercise or from temporary inhibition of the Na/K-pump by exposure to cold. A high extracellular potassium concentration would block inactivation and reactivation of the sodium channel but would also depolarize the fiber. Depolarization activates sodium channels, in part counteracting the blockade of reactivation by the potassium ion concentration. Therefore, it will be unpredictable whether the fiber becomes myotonic or paralysed if the extracellular potassium concentration is raised. In our opinion, hyperkalemic periodic paralysis (adynamia episodica) and paramyotonia congenita (Eulenburg) are therefore genetically the same disease (De Silva 1990).

In cases of hyperkalemic paralysis, the influx of sodium is high in the interictal state but it drops during the attacks. While the patient is paralysed, the intracellular sodium concentration is maintained at a low level. As a result, pump activity is reduced, and the high extracellular potassium concentration is maintained for an extended time. The muscle is paralysed and stays that way until the excess of potassium ions around the fibers is removed by the circulation.

As in the hypokalemic form, the muscles with a high blood flow at rest remain free and it is possible to walk off an attack. It is now clear why walking-off has disadvantages in the hyperkalemic form. Although it may stop the paralysis by removing excess potassium, it promotes additional potassium loss from the muscle and that may induce another attack later. Provoking factors in hyperkalemic paralysis act by reducing pump activity, by inducing loss of potassium ions from the muscle, or possibly by reducing phosphorylation of the sodium channel by suppression of sympathetic activation (Bendheim et al. 1985).

TREATMENT

Hypokalemic periodic paralysis. Muscle fibers in hypokalemic periodic paralysis are characterized by a high sodium permeability at rest. No known drug can correct the insufficient inactivation of the sodium channel. Tetrodotoxin or other sodium channel blockers inhibit activation of the channel, but do not correct the abnormal sodium conductance in vitro (Rüdel et al. 1984); these agents bind to the sodium channel only in its active state (Hille 1977; Hondeghem and Katzung 1977) and not to the inactivated channels.

Inhibition of sympathetic activity was thought to have a beneficial effect because an increased sympathetic drive stimulates Na/K-pump activity, which was thought to be instrumental in initiating attacks. The effect of β-adrenergic blockade was studied by Johnsen (1977) and Buruma (1978). In both studies, β-blockade, after an initial protective effect, induced severe attacks after 7–11 days. Sympathetic activity via a cAMP-dependent phosphorylation of the sodium channel enhances inactivation (Catterall 1988), so β-blockade should impair inactivation. With the wisdom of hindsight, we now think that β-blockade ultimately reduced sodium channel phosphorylation to spoil the initially beneficial effect of pump inhibition.

As the attacks are attributed to a rapid fall of the extracellular potassium concentration, it seems advantageous to keep serum potassium levels high and to give extra potassium salts as soon as an attack begins. The beneficial effect of potassium salts was not fully appreciated until Herrington (1937) reported successful treatment of two cases of familial periodic paralysis with potassium citrate. They gave it on the appearance of prodromata usually in a dose of 5 g orally repeated every hour until muscle strength returned to normal. The beneficial effect of potassium salts was confirmed by Aitken et al. (1937). After that, administration of potassium salts became the therapy of choice, also prophylactically, but the irritating effect of potassium salts on the stomach and esophagus was a problem. Talbott (1941) recommended a 25% potassium chloride solution in water because it was less irritating. Later, mixtures were prepared containing different organic water-soluble potassium salts, with the advantage of being tasteless (Sagild 1959). These solutions could also be administered through a gastric tube, or injected intravenously if necessary. When potassium salts are given prophylactically, the daily dose should be taken before going to bed, to prevent the development of paralysis in the morning.

Conn et al. (1957) reported that urine aldosterone excretion increases before attacks. This is probably due to activation of the renin-angiotensin system by the shift of sodium ions and water into the muscle cells. They inferred that dietary sodium restriction might be of value prophylactically. Rowley and Kliman (1960) found retention of sodium and water during attacks, but in contrast to Conn et al. (1957), they found normal urinary aldosterone excretion before the attack. Gérard et al. (1978) found high aldosterone activity during an attack induced by insulin and glucose. Sodium loading reduced muscle strength, and sodium depletion enhanced it. The authors concluded that a low sodium diet is a more rational prophylaxis than the administration of potassium salts. Earlier, however, a beneficial effect of sodium restriction was not found by De Graeff and Lameyer (1965), who provoked attacks at whatever sodium load they had chosen, and even severe restriction of sodium intake to less than 10 meq per day did not reduce the attack rate. Conn's observation prompted others to study the effect of aldosterone antagonists. Rowley and Kliman (1960) used spironolactone combined with sodium depletion, and found loss of sodium and retention of potassium. In subsequent therapeutic studies, spironolactone had a beneficial effect on attack rate (Poskanzer and Kerr 1961; De Graeff and Lameyer 1965).

De Graeff and Lameyer (1965) discovered that, in one patient with hypokalemic periodic paralysis, attacks stopped completely one week after they administered glucocorticosteroids. The attacks reappeared when the medication was discontinued because of side effects. During treatment the patients were on a sodium-restricted diet, to prevent urinary loss of potassium. The beneficial effect might have been due to the inhibitory effect of glucocorticosteroids on glucose utilization. If so, it would help only to prevent attacks. In balance studies, however, De Graeff and Lameyer (1965) observed an increase of sodium chloride excretion, implying that muscle uptake of sodium was reduced. This suggests an effect of the corticosteroids on sodium conductance between attacks and on the properties of the sodium channel. The effect is interesting from a theoretical point of view, but corticosteroids cannot be given as a long-term prophylactic therapy because of side effects.

McArdle (1962) reported one beneficial effect of acetazolamide in patients with hyperkalemic periodic paralysis. In 1968, Resnick et al. described one preventive effect of acetazolamide in a case of what was thought to be hyperkalemic. Acetazolamide reduced the frequency and severity of attacks, and when acetazolamide was discontinued because a subsequent work-up showed the patient to have the hypokalemic form, a severe attack occurred four days later. In this and one other patient, Resnick et al. (1968) deemed acetazolamide more effective in preventing attacks than spironolactone and potassium salts. Griggs et al. (1970) treated 12 patients with hypokalemic periodic paralysis in a placebo-controlled trial with acetazolamide and observed a dramatic improvement in ten. Four of these 12 patients were sporadic; the others were familial. In both groups one failure of the therapy was observed. Ten of the 12 patients had permanent weakness and in 7 out of the 10, strength during the attack-free intervals also responded favorably to the acetazolamide treatment. This has been confirmed by Dalakas and Engel (1983), Thompson and Hutchinson (1984), and Links et al. (1988). However, Buruma (1978) found that, after a longer follow-up, the ir-

reversible myopathy component had progressed irrevocably. The initial beneficial effect is probably due to functional improvement of the muscle fibers, but the myopathic process is probably not slowed down.

Resnick et al. (1968) argued that the preventive effect of acetazolamide was not due to sodium depletion and potassium retention, but to the metabolic acidosis caused by the inhibition of carbonic anhydrase. They based that conclusion on the observation that another diuretic with an even stronger natriuretic and even a potassium-saving effect, triamterene, failed to prevent attacks in one patient in whom acetazolamide was successful. The hypothesis that the effect of acetazolamide is based on the metabolic acidosis was tested by giving ammonium chloride instead of acetazolamide to two patients (Jarrell et al. 1976). With a regimen of 12 g per day, metabolic acidosis was induced in both patients. Balance studies demonstrated that the response of the serum potassium level to the administration of glucose and insulin was blunted by the acidosis. In one patient, the insulin-glucose test invariably resulted in a severe attack. While using ammonium chloride, however, only mild proximal weakness was observed.

Goulon et al. (1978) observed that the prophylactic effect of acetazolamide against glucose-insulin provocation was opposed by an injection of sodium bicarbonate. Vroom et al. (1975) demonstrated that acetazolamide prevents the strong potassium influx in muscle cells during induced attacks. This observation is compatible with a direct inhibitory effect of pH on Na/K-pump activity. When metabolic alkalosis was induced by bicarbonate administration, there were paralytic attacks that were then abolished with intravenous acetazolamide (Viskoper et al. 1973a,b).

Although it seems established that acetazolamide acts through induction of tissue acidosis, some observations point in another direction. Metabolic acidosis might prevent attacks by inhibiting the sodium-potassium pump, but this cannot be the only mechanism by which acetazolamide prevents attacks, because the drug works also in hyperkalemic patients, where inhibition of the pump has an adverse effect. A direct effect of acidosis or of the drug itself on sodium permeability, by an enhancement of Na-channel inactivation, is a plausible explanation.

Dalakas and Engel (1983) gave dichlorophenamide, a potent carboanhydrase inhibitor, in a single-blind, placebo-controlled trial to three patients who no longer responded to acetazolamide. Dichlorophenamide dramatically reduced the number and severity of attacks and also improved interictal muscle strength. However, the drug produced only a light metabolic acidosis, compared to the more profound acidosis of acetazolamide. Moreover, in one patient they administered ammonium chloride, and strength decreased rather than increased. It appears, therefore, that acidosis is not the only possible mode of therapeutic action of carbonic anhydrase inhibitors, at least not in all patients.

Another possible mode of action was suggested by Vroom et al. (1975). They noted that acetazolamide reduces blood glucose levels during glucose-induced attacks. Acetazolamide restores glucose levels back to normal more quickly than in untreated controls. Riggs et al. (1984) measured the effect of acetazolamide on arteriovenous glucose and potassium differences across the forearm in healthy persons after oral glucose loading. Acetazolamide increased glucose uptake, in agreement with the reduction of blood glucose levels shown by Vroom et al. (1975). This is unexpected, because metabolic acidosis normally decreases the uptake of glucose, and it reduces tissue sensitivity to insulin. This led Johnsen (1977) to suggest that acetazolamide acts indirectly by blunting blood glucose levels. Riggs et al. (1984) also found that, in healthy persons, the uptake of potassium after glucose loading was reduced by acetazolamide, suggesting that acetazolamide has a direct inhibitory effect on resting sodium conductance or enhances inactivation even in healthy people.

Side effects of acetazolamide are said to be minimal; transient or mild paresthesias, fatigue, mild polyuria and polydipsia, or occasional painful muscle cramps. In our own experience with 12 patients, paresthesias were a problem in only one (Buruma 1978). Dangerous complications such as agranulocytosis, thrombocytopenia and crystalluria have been mentioned, but seem to be rare. In a follow-up period of more than five years, acetazolamide continued to be effective in controlling the attacks (Buruma 1977). Others have noted loss of responsiveness to acetazolamide (Dalakas and Engel 1983), failure to respond in the first place (Griggs et al. 1970, 1983; Roeltgen 1979) and even exacerbation of attacks (Torres et al. 1981). These failures of therapy remain unexplained. The patients who failed to respond usually developed the same degree of tissue acidosis, and the kaliuretic effect of acetazolamide was not especially pronounced in these patients. However, the patients of Torres et al. (1981) responded well to subsequent triamterene therapy.

Grafe et al. (1990) described a beneficial effect of cromakalim, a drug that increases the potassium conductance of the muscle membrane. In line with the hypothesis on the induction of attacks by rapid reduction of the extracellular potassium concentration, the increased opening of potassium channels is expected to prevent the attack to develop because it balances the potassium concentration immediately outside the muscle cell.

Hypokalemic periodic paralysis should be treated primarily by avoiding all situations that promote development of attacks. A sudden drop of the serum potassium concentration is unlikely to occur, but alkalosis, either metabolic or respiratory, may develop more easily and should be avoided. Stimulation of the sodium-potassium pump is the most likely trigger of attacks. Sympathetic activity may induce an attack, and especially the activation of the pump by glucose and insulin after carbohydrate-rich meals should be avoided.

Acetazolamide reduces the risk of developing attacks either by the inhibitory effect of tissue acidosis on the Na/K-pump or by a direct effect of the drug on sodium permeability. It may slow down the deterioration of muscle cells and the development of permanent weakness, although it cannot prevent this. Similar preventive and also long-term beneficial effects are expected from the new K-channel openers like cromakalim. Derivatives of cromakalim are presently being developed for their vasodilating properties. A related drug, nicorandil, with additional nitrate-like properties may also be of value.

Administration of potassium salts serves primarily to terminate an attack. Prophylactic treatment with potassium salts has largely been abandoned. Incidental use of β-blockers may be of value to prevent an attack after unusually severe exercise or stress. The continued use of β-blockers should be avoided. In conclusion, drug treatment of hypokalemic periodic paralysis is best started with acetazolamide. If this fails, triamterene or dichlorophenamide is worth a try.

Hyperkalemic periodic paralysis. The development of attacks in hyperkalemic paralysis is triggered by a rise of the potassium concentration outside the muscle fibers. The most likely source of potassium ions is the muscle itself. Potassium ions escape from the muscle during exercise, and simultaneous inhibition of the Na/K-pump by cold, fasting, or β-blockade will promote the development of an attack. Patients should refrain from a high intake of potassium salts and from vigorous exercise; frequent meals with sufficient quantities of carbohydrates are recommended.

A low serum potassium lowers the risk of developing attacks. Treatment with thiazides (McArdle 1962; Bradley 1969) is therefore an effective measure, because it induces hypokalemia. It even protects in some instances (Carson and Pearson 1964) against the adverse effect of administration of potassium salts. Restriction of sodium intake may have a sufficient effect on the potassium balance (Brillman and Pincus 1973) to have a beneficial effect of its own.

Acute attacks of weakness may be shortened and muscle strength may improve when calcium gluconate is administered intravenously (Gamstorp 1956; Van der Meulen et al. 1961).

In an attempt to treat hyperkalemic periodic paralysis by inducing kaliuresis, McArdle (1962) used acetazolamide. With a dosage of 125–375 mg/day, he obtained most satisfactory results. Acetazolamide promotes the elimination of potassium as well as sodium, bicarbonate and water by inhibiting carbonic anhydrase in cells of the renal tubules. The therapeutic effect was ascribed to the kaliuresis until, in 1968, Resnick et al. found it beneficial in the hypokalemic form as well. We think that acetazolamide acts by reducing sodium permeability, and that the effect is due to a direct action of the drug or is mediated by acidosis. If acetazolamide acted only by a direct or indirect effect on the Na/K-pump, a beneficial effect in both hypokalemic and hyperkalemic paralysis would be

impossible; the hypokalemic form requires inhibition and the hyperkalemic a potentiation of the pump to prevent attacks. A similar reasoning holds true for an effect mediated by potassium permeability or extracellular potassium concentrations.

Metabolic alkalosis, produced by administration of bicarbonate, provokes attacks in both hypokalemic and hyperkalemic paralysis (Viskoper et al. 1973a,b). This also supports our view of an effect of pH on sodium conductance.

To analyse the efficacy of acetazolamide in hyperkalemic paralysis, the acute effects of the drug were studied by Riggs et al. (1980) in two patients and 13 controls. Acetazolamide lowered mean venous potassium concentrations in all subjects without preventing large plasma potassium fluctuations. The authors concluded that the effectiveness of acetazolamide is probably due to the kaliopenic effect, but that view makes it difficult to understand how the drug could be equally effective in hypokalemic paralysis (Hoskins 1977).

Wang and Clausen (1976, 1978), Dahl-Jörgensen and Michalsen (1979) and Clausen et al. (1980) found that daily inhalation of salbutamol is a simple and effective treatment for hyperkalemic paralysis. Salbutamol is an adrenergic drug that favors intracellular accumulation of potassium by stimulating the sodium-potassium pump. Other β-adrenergic agonists, such as metaproterenol, were also effective both in treating and preventing attacks (Bendheim et al. 1985). That salbutamol acts through activation of the pump is supported by the observation that the drug makes the hypokalemic form worse. Streib (1985) suggested that salbutamol may aggravate myotonic symptoms, perhaps because sympathetic stimulation enhances not only the inactivation but also the activation of the sodium channel.

Hopf et al. (1981) observed disappearance of weakness and paresthesia after intravenous treatment with 500 mg xylocaine, but the effect of long-term treatment is unknown. This effect is, according to our views, due to blockade of persistently activated Na-channels. Similar beneficial effects have been found in vitro with the experimental Na-channel blocker tetrodotoxin (Lehmann-Horn et al. 1983) and with tocainide (Ricker et al. 1980, 1983; Streib 1987). The dose-effect relation of the

Na-channel blockers probably makes the long-term use of these agents impractical because of side-effects.

REFERENCES

AITKEN, R.S., E.N. ALLOT, L.I.M. CASTELDEN and M. WALKER: Observations on a case of familial periodic paralysis. Clin. Sci. 3 (1937) 47–57.

ANDERSEN, E.D., P.A. KRASILNIKOFF and H. OVERVAD: Intermittent muscular weakness, extrasystoles, and multiple developmental anomalies. Acta Paediatr. Scand. 60 (1971) 559–564.

BEHNE, M. and A. ANDRÉ: Über einen postoperativen Anfall von familiärer periodischer hypokaliämischer Lähmung. Anästh. Intensivher. Notfallmed. 19 (1984) 75–77.

BEKÉNY, G.: Über irreversibele Muskeländerungen bei der paroxysmaler Lähmung, auf Grund bioptischer Muskeluntersuchung. Dtsch. Z. Nervenheilkd. 182 (1961) 119–154.

BEKÉNY, G, T. HASZNOS and F. SOLTI: Über die hyperkalämische Form der paroxysmalen Lähmung: zur Frage der Adynamia episodica hereditaria. Dtsch. Z. Nervenheilkd. 182 (1961) 92–118.

BENDER, J.: Family periodic paralysis in a girl aged seventeen. Arch. Neurol. Psychiatry 35 (1935) 131–135.

BENDHEIM, P.E., E.O. REALE and B.O. BERG: Adrenergic treatment of hyperkalemic periodic paralysis. Neurology 35 (1985) 746–749.

BENEDEK, L. and L. VON ANGYAL: Beiträge zur Pathogenese der paroxysmalen Lähmung. Z. Ges. Neurol. Psychiatr. 174 (1942) 213–228.

BERNHARDT, M.: Notiz über die familiäre Form der Dystrophia musculorum progressiva und deren Combination mit periodisch auftredender paroxysmaler Lähmung. Dtsch. Z. Nervenheilkd. 8 (1895) 111–118.

BIEMOND, A. and A. POLAK DANIELS: Familial periodic paralysis and its transition into spiral muscular atrophy. Brain 57 (1934) 91–108.

BOULTON, A.J.M. and C.A. HARDISTY: Hypokalaemic periodic paralysis precipitated by diuretic therapy and minor surgery. Postgrad. Med. J. 58 (1982) 106–107.

BRADLEY, W.G.: Adynamia episodica hereditaria. Brain 92 (1969) 345–347.

BRADLEY, W.G., R. TAYLOR, D.R. RICE, I. HAUSMANOVA-PETRAZEWICZ, L.S. ADELMAN, M. JENKISON, H. JEDRZEJOWSKA, H. DRAC and W. PENDLEBURY: Progressive myopathy in hyperkalemic periodic paralysis. Arch. Neurol. 47 (1990) 1013–1017.

BRILLMAN, J. and J.H. PINCUSS: Myotonic periodic paralysis improved by negative sodium balance. Arch. Neurol. 29 (1973) 67–69.

BRODY, I.A. and A.W. DUDLEY: Thyrotoxic hypokalemic periodic paralysis. Arch. Neurol. (Chic.) 21 (1969) 1–63.

BURR, C.W.: Periodic paralysis, with the report of a case. Univ. Med. Mag. Philadelphia 5 (1893) 836–840.

BURUMA, O.J.S.: Acetazolamide treatment in hypokalaemic periodic paralysis. Abstracts, 11th World Congress of Neurology. International Congress Series, Excerpta Medica 427 (1977) 195–196.

BURUMA, O.J.S.: Familial hypokalaemic periodic paralysis. Leiden, thesis (1978) 122–124.

BURUMA, O.J.S. and G.TH.A.M. BOTS: Myopathy in familial hypokalaemic periodic paralysis independent of paralytic attacks. Acta Neurol. Scand. 57 (1978) 171–179.

BURUMA, O.J.S. and J.J. SCHIPPERHEYN: Periodic paralysis. In: P.J. Vinken and G.W. Bruyn (Eds.), Handbook of Clinical Neurology, Vol. 41. Amsterdam, North-Holland Publishing Co (1979) 147–174.

BURUMA, O.J.S., J.J. SCHIPPERHEYN and G.TH.A.M. BOTS: Heart muscle disease in familial hypokalaemic periodic paralysis. Acta Neurol. Scand. 64 (1981) 12–21.

BURUMA, O.J.S., G.TH.A.M. BOTS and L.N. WENT: Familial hypokalemic periodic paralysis. Arch. Neurol. 42 (1985) 28–31.

BUZZARD, E.F.: Three cases of family periodic paralysis, with a consideration of the pathology of the disease. Lancet (1901) 1564–1567.

BYRNE, J.: Case of family periodic paralysis: death occurring in attack. J. Nerv. Ment. Dis. 43 (1916) 159–166.

CARSON, M.J. and C.M. PEARSON: Familial hyperkalemic periodic paralysis with myotonic features. J. Pediatr. 64 (1964) 853–865.

CATTERALL, W.A.: Structure and function of voltage-sensitive ion channels. Science 242 (1988) 50–61.

CERNY, A. and E. KATZENSTEIN-SUTRO: Die paroxysmale Lähmung. Schweiz. Arch. Neurol. Psychiatr. 70 (1952) 259–338.

CHESSON, A.L., S.S. SCHOCHET and B.H. PETERS: Biphasic periodic paralysis. Arch. Neurol. 36 (1979) 700–704.

CLAUSEN, T., P. WANG, H. ORSKOV and O. KRISTENSEN: Hyperkalemic periodic paralysis: relationships between changes in plasma water, electrolytes, insulin and catecholamines during attacks. Scand. J. Clin. Lab. Invest. 40 (1980) 211–220.

CONN, J.W., S.S. FAJANS, L.H. LOUIS, D.H.P. STREETEN and R.D. JOHNSON: Intermittent aldosteronism in periodic paralysis. Lancet i (1957) 802–805.

CONWAY, M.J., J.A. SEIBEL and R.P. EATON: Thyrotoxicosis and periodic paralysis: improvement with beta-blockade. Ann. Intern. Med. 81 (1974) 332–336.

COUSOT, G.: Paralysie périodique. Rev. Méd. (Paris) 7 (1887) 190–203.

DAHL-JORGENSEN, K. and H. MICHALSEN: Adynamia episodica hereditaria. Acta Paediatr. Scand. 68 (1979) 538–585.

DALAKAS, M.C. and W.K. ENGEL: Treatment of 'permanent' muscle weakness in familial hypokalemic periodic paralysis. Muscle Nerve 6 (1983) 182–186.

DANOWSKI, T.S., E.R. FISCHER, C. VIDALON, J.W. VESTER, R. THOMPSON, S. NOLAN, T. STEPHAN and J.H. SUNDER: Clinical and ultrastructural observations in a kindred with normo-kalaemic periodic paralysis. J. Med. Genet. 12 (1975) 20–28.

DE GRAEFF, J. and L.D.F. LAMEYER: Periodic paralysis. Am. J. Med. 39 (1965) 70–80.

DE FINE OLIVARIUS, B. and E. CHRISTENSEN: Histopathological muscular changes in familial, periodic paralysis. Acta Neurol. Scand. 41 (1965) 1–18.

DE SILVA, S.M., R.W. KUNCL, J.W. GRIFFIN, D.R. CORNBLATH and S. CHAVOUS-TIC: Paramyotonia congenita or hyperkalemic periodic paralysis? Clinical and electrophysiological features of each entity in one family. Muscle Nerve 13 (1990) 21–26.

DYKEN, H., W. ZEMAN and T. RUSCHE: Hypokalemic periodic paralysis: children with permanent myopathic weakness. Neurology 7 (1969) 691–699.

DYKEN, M. and G.D. TIMMONS: Hyperkalemic periodic paralysis with hypocalcemic episodes. Arch. Neurol. 9 (1963) 508–517.

EMSER, W.: Hypermagnesemic periodic paralysis: treatment with digitalis and lithium carbonate. Arch. Neurol. 39 (1982) 727–730.

ENGEL, A.G.: Thyroid function and periodic paralysis. Am. J. Med. 30 (1961) 327–333.

ENGEL, A.G.: Evolution and content of vacuoles in primary hypokalemic periodic paralysis. Mayo Clin. Proc. 45 (1970) 774–814.

ENGEL, W.K., D.W. BISHOP and G.G. CUNNINGHAM: Tubular aggregates in type II muscle fibers: Ultrastructural and histochemical correlation. J. Ultrastruct. Res. 31 (1970) 507–525.

FAUGERE, M.C., J.F. PELLISSIER and M. TOGA: Subsequent morphological changes in periodic paralysis: a study of seven cases. Acta Neuropathol. 7 (1981) 301–304.

FENECK, F.F. and N.G. SOLER: Hyperkalaemic periodic paralysis starting at age 48. Br. Med. J. 2 (1968) 472–473.

FONTAINE, B., T.S. KHURANA, E.P HOFFMAN, G.A BRUNS, J.L HAINES, J.A. TROFATTER, M.P. HANSON, J. RICH, H. MCFARLANE, D. MCKENNA YASEK, D. ROMANO, J.F. GUSELLA and R.H. BROWN: Hyperkalemic periodic paralysis and the adult muscle sodium channel α-subunit gene. Science 250 (1990) 1000–1002.

FOZARD, J.R.: Anaesthesia and familial hypokalaemic periodic paralysis. Anaesthesia 38 (1983) 293–294.

FUCHS, A.: Periodische Extremitätenlähmung. Wien. Klin. Wochenschr. 18 (1905) 21.

FUERSTEIN, V.: Report on familial paroxysmal hypokalaemic paralysis in relation to postoperative intensive care. Anaesthesist 29 (1980) 632–634.

GAMSTORP, I.: Adynamia episodica hereditaria. Acta Paediatr. Scand. 45, Suppl. 108 (1956) 1–126.

GAUPP, R. and O. KALDEN: Erblichkeitsuntersuchungen bei paroxysmaler Lähmung. Z. Gesamte Neurol. Psychiatr. 174 (1942) 194–212.

GÉRARD, J.M., P. KHOUBESSERIAN, N. TELERMAN-TOPPET, TH. DE BARSY and C. COERS: Paralysie périodique familiale avec hypokaliémie, hyperaldostéronisme et vacuolisation extra-cellulaire. Rev. Neurol. (Paris) 134 (1978) 761–772.

GOLDFLAM, S.: Über eine eigentümliche Form von periodischer, familiärer wahrscheinlich autointoxicatorischer Paralyse. Z. Klin. Med 19 (1891) 240–285.

GOLDFLAM, S.: Weitere mitteilung über die paroxysmaler familiäre Lähmung. Dtsch. Z. Nervenheilkd. 7 (1895) 1–31.

GOULD, R.J., C.N. STEEG, A.B. EASTWOOD, A.S PENN, L.P. ROWLAND and D.C. DE VIVO: Potentially fatal cardiac dysrhythmia and hyperkalemic periodic paralysis. Neurology 35 (1985) 1208–1212.

GOULON, M., J.C. RAPHAEL and N. SIMON: Paralysie periodique familiale avec hypokaliémie. Rev. Neurol. 11 (1978) 655–672.

GRAFE, P., S. QUASTHOFF, M. STRUPP and F. LEHMANN-HORN: Enhancement of K^+ conductance improves in vitro the contraction force of skeletal muscle in hypokalemic periodic paralysis. Muscle Nerve 13 (1990) 451–457.

GRIGGS, R.C., W.K. ENGEL and J.S. RESNICK: Acetazolamide treatment of hypokalemic periodic paralysis. Ann. Intern. Med. 73 (1970) 39–48.

GRIGGS, R.C., J.S. RESNICK and W.K. ENGEL: Intravenous treatment of hypokalemic periodic paralysis. Arch. Neurol. 40 (1983) 539–540.

GROTEMEYER, K.H. and J. JÖRG: Neue Aspekte zur paroxysmalen familiären hypokaliämischen Lähmung mit Ateminsuffizienz. Nervenarzt 50 (1979) 649– 652.

GRUNER, J.E.: Anomalies du reticulum sarcoplasmique et prolifération des tubules dans le muscle de la paralysie périodique familiale. C.R. Soc. Biol. (Paris) 160 (1966) 93–195.

GUARINO, M., A. TARATETA and A. DI STEFANO: Mitochondrial cytopathy in familial periodic hypokalaemic paralysis. Lancet ii (1984) 49.

HARTLAGE, P.L. and R. SOUDMAND: Internalized capillaries in hypokalemic periodic paralysis. Arch. Neurol. 38 (1981) 602.

HASTINGS, B.A., D.R. GROOTHUIS and N.A. VICK: Dominantly inherited pseudohypertrophic muscular dystrophy with internalized capillaries. Arch. Neurol. 37 (1980) 709–714.

HELLWEG-LARSEN, H.F., M. HAUGE and U. SAGILD: Hereditary transient muscular paralysis in Denmark, genetic aspects of family periodic paralysis and family periodic adynamia. Acta Genet. 5 (1955) 263–280.

HERMAN, R.H. and M.K. MCDOWELL: Hyperkalemic paralysis (adynamia episodica hereditaria): report of four cases and clinical studies. Am. J. Med. 35 (1963) 749–767.

HERRINGTON, M.S.: Successful treatment of two cases of familial periodic paralysis with potassium citrate. J. Am. Med. Assoc. 108 (1937) 1339.

HILLE, B.: Local anesthetics: hydrophylic and hydrophobic pathways for the drug-receptor reaction. J. Gen. Physiol. 69 (1977) 497–515.

HIRSCH, K.: Über einen Fall von periodischer familiärer Paralyse. Dtsch. Med. Wochenschr. 32 (1894) 646–649.

HOFMANN, W.F., B.T. ADORNATO and H. REICH: The relationship of insulin receptors to hypokalemic periodic paralysis. Muscle Nerve 6 (1983) 48–51.

HOLTZAPPLE, G.E.: Periodic paralysis. J. Am. Med. Assoc. 45 (1905) 1224–1231.

HONDEGHEM, L.M. and B.G. KATZUNG: Time- and voltage-dependent interaction of antiarrhythmic drugs with cardiac sodium channels. Biochim. Biophys. Acta 472 (1977) 373–398.

HOPF, H.C.H., K. VON BARDELEBEN and K. LOWITZSCH: Stopping a paralytic attack in hyperkalemic paralysis by lidocaine derivatives. Abstract 12th World Congress of Neurology. International Congress Series, Excerpta Medica Amsterdam-Oxford-Princeton (1981) 548.

HOSKINS, B.: Studies on the mechanism of action of acetazolamide in the prophylaxis of hyperkalemic periodic paralysis. Life Sci. 20 (1977) 343–350.

HOWES, E.L., H B. PRICE, C.M. PEARSON and J.M. BLUMBERG: Hypokalemic periodic paralysis: electronmicroscopic changes in the sarcoplasm. Neurology (Minneap.) 16 (1966) 242–256.

IONASESCU, V., H. RADU and P. NICOLESCU: Ultrastructural changes in hypokalaemic periodic paralysis. Rev. Roum. Neurol. 8 (1971) 419–425.

IONASESCU, V., S.S. SCHOCHET JR., J.M. POWERS, K. KOOB and T.N. CONWAY: Hypokalemic periodic paralysis, low activity of sarcoplasmic reticulum and muscle ribosomes during an induced attack. J. Neurol. Sci. 21 (1974) 419–429.

JANOTA, D. and K. WEBER: Die paroxysmale Lähmung, Abh. Neurol. Psychiatr. Psychol. Ihren Grenzgeb. 46 (1928).

JARRELL, M.A., M. GREER and T.H. MAREN: The effect of acidosis in hypokalemic periodic paralysis. Arch. Neurol 33 (1976) 791–793.

JOHNSEN, T.: Endogenous insulin fluctuations during glucose-induced paralysis in patients with familial periodic hypokalemia. Metabolism 26 (1977) 1185–1190.

JOHNSEN, T. and H. BECK-NIELSEN: Insulin receptors, insulin secretion, and glucose disappearance rate in patients with periodic hypokalaemic paralysis. Acta Endocrinol. 90 (1979) 272–282.

KAPLAN, M., P. STRAUS, R. GRUMBACH and B. AYMARD: L'adynamie épisodique héréditaire. Presse Méd. 65 (1957) 1305–1308.

KASTAN, M.: Über einige allgemein als familiär bekannte Nervenkrankheiten. Arch. Psychiatr. Nervenkr. 63 (1921) 361–389.

KELLEY, D.E., H. GHARIB, F.P. KENNEDY, R.J. DUDA JR and P.G. MCMANIS: Thyrotoxic periodic paralysis. Report of 10 cases and review of electromyographic findings. Arch. Intern. Med. 149 (1989) 2597–2600.

KHAN, M.Y.: Familial periodic paralysis. Indian Med. Gaz. 70 (1935) 28–29.

KLEIN, R., T. EGAN and P. USHER: Changes in sodium, potassium and water in hyperkalemic familial periodic paralysis. Metabolism 9 (1960) 1005–1024.

KLEIN, R., R. GANELIN, J.F. MARKS, P. USHER and C. RICHARDS: Periodic paralysis with cardiac arrhythmia. J. Pediatr. 62 (1963) 371–385.

KRAMER, L.D., J.P. COLE, J.C. MESSENGER and M.H. ELLESTAD: Cardiac dysfunction in a patient with familial hypokalemic periodic paralysis. Chest 75 (1979) 189–192.

LANDER, C.M., M. MURPHY and J.L. ALLSOP: Familial periodic paralysis: a kindred with hypokalaemic, normokalaemic and hyperkalaemic attacks. Clin. Exp. Neurol. Proc. Aust. Assoc. Neurol. 21(1985) 334.

LAYZER, R.B., R.E. LOVELACE and L.P. ROWLAND: Hyperkalemic periodic paralysis. Arch. Neurol. 16 (1967) 455–472.

LEE, K-O., E.A. TAYLOR, V.M.S. OH, J-S. CHEAH and S-E. AW: Hyperinsulinaemia and thyrotoxic hypokalaemic paralysis. Lancet 337 (1991) 1063–1064.

LEHMANN-HORN, F., R. RUDEL, K. RICKER, H. LORKOVIC, R. DENGLER and H.C. HOPF: Two cases of adynamia episodica hereditaria: in vitro investigation of muscle cell membrane and contraction parameters. Muscle Nerve 6 (1983) 113–121.

LEVITT, L.P., L.I. ROSE and D.M. DAWSON: Hypokalemic periodic paralysis with arrhythmia. N. Engl. J. Med. 286 (1972) 253–254.

LISAK, R.P., J. LEBEAU, S.H TUCKER and L.P ROWLAND: Hyperkalemic periodic paralysis and cardiac arrhythmia. Neurology 22 (1972) 810–815.

LINKS, T.P., M.J. ZWARTS and J.G.H. OOSTERHUIS: Improvement of muscle strength in familial hypokalaemic paralysis with acetazolamide. J. Neurol. Neurosurg. Psychiatry 51 (1988) 1142–1145.

LINKS, T.P., M.J. ZWARTS, J.T. WILMINK, W.M. MOLENAAR and H.J.G.H. OOSTERHUIS: Permanent muscle weakness in familial hypokalaemic periodic paralysis. Brain 113 (1990) 1873–1889.

MACDONALD, R.D., N.B. REWCASTLE and J.G. HUMPHREY: The myopathy of hyperkalaemic periodic paralysis. Arch. Neurol. 19 (1968) 274–283.

MACDONALD, R.D., N.B. REWCASTLE and J.G. HUMPHREY: Myopathy of hypokalaemic periodic paralysis. Arch. Neurol. 20 (1969) 565–585.

MACLACHLAN, T.K.: Familial periodic paralysis: a description of six cases occurring in three generations of one family. Brain 55 (1932) 47–76.

MANKOWSKY, B.N.: Über die paroxysmale Lähmung. Arch. Psychiatr. Nervenkr. 87 (1929) 280–326.

MARTIN, J.J., C. CEUTERICK, R. MERCELIS and D. AMROM: Familial periodic paralysis with hypokalaemia: study of a muscle biopsy in the myopathic stage of the disorder. Acta Neurol. Belg. 84 (1984) 233–242.

MARX, A., J.P. RUPPERSBERG, C. PIETRZYK and R. RUDEL: Thyrotoxic periodic paralysis and the sodium/potassium pump. Muscle Nerve 12 (1989) 810–815.

MCARDLE, B.: Adynamia episodica heriditaria and its treatment. Brain 85 (1962) 121–148.

MCFADZEAN, A.J.S. and R. YEUNG: Familial occurrence of thyrotoxic periodic paralysis. Br. Med. J. 1 (1967) 760.

MELNICK, B., J. CHANG, C.E. LARSON and R.C. BEDGER: Hypokalemic familial periodic paralysis. Anesthesiology 58 (1983) 263–265.

MEYERS, K.R., D.H. GILDEN, C.F. RINALDI and J.L. HANSEN: Periodic muscle weakness, normokalemia, and tubular aggregates. Neurology 22 (1972) 269–279.

NEEL, A.V.: Zwei von einander unabhängige Fälle von Myoplegia paroxysmatica (periodica familiaris). Z. Neurol. Psychol. 118 (1928) 269.

ODOR, D.L., A.N. PATEL and L.A. PEARCE: Familial hypokalemic periodic paralysis with permanent myopathy. J. Neuropathol. Exp. Neurol. 26 (1967) 98–114.

OFFERIJNS, F.G.J., D. WESTERINK and A.F. WILLEBRANDS: The relation of potassium deficiency to muscular paralysis by insulin. J. Physiol. (London) 141 (1958) 377–384.

OLIVIER, C.P., M. ZIEGLER and I. MCQUARRIE: Hereditary periodic paralysis in a family showing varied manifestations. Am. J. Dis. Child. 68 (1944) 308–320.

OPPENHEIM, H.: Neue Mitteilungen über den von Professor Westphal beschriebenen Fall von periodischer Lähmung aller vier Extremitäten. Charité Am. 16 (1891) 350–372.

PEARSON, C.M.: The periodic paralyses: differential feature and pathological observation in permanent myopathic weakness. Brain 87 (1964) 341–353.

PEARSON, C.M.: The pathologic features of the periodic paralysis. The striated muscle. Baltimore, Williams and Williams (1973) 427–441.

PELLISSIER, J.F., M.C. FAUGERE, D. GAMBARELLI and M. TOGA: Les paralysies périodiques. Sem. Höp. Paris 33 (1980) 1393–1404.

PORTE, A., M.E. STOECKEL, J. Y. LEDEAUT, S . GUEZ, C. HINDELANG-GERTNER and G. MACK: Modifications in the sarcoplasmic reticulum and subcellular calcium distribution in skeletal muscle in a case of Westphal's disease (hypokalemic periodic paralysis). Virchows Arch. Path. Anat. Histol. 383 (1979) 345–350.

POSKANZER, D.C. and D.N.S. KERR: Periodic paralysis with response to spironolactone. Lancet ii (1961) 511–513.

PUTMAN, J.J.: A case of family periodic paralysis. Am. J. Med. Sci. 119 (1900) 160–170.

RESNICK, J.S. and W.K. ENGEL: Myotonic lid lag in hypokalaemic periodic paralysis. J. Neurol. Neurosurg. Psychiatry 30 (1967) 47–51.

RESNICK, J.S., W.K. ENGEL, R.C. GRIGGS and A.C. STAM: Acetazolamide prophylaxis in hypokalemic periodic paralysis. N. Engl. J. Med. 278 (1968) 582–586.

RICKER, K., A. HAASS, R. RUDEL, R. BÖHLEN and H.G. MERTENS: Successful treatment of paramyotonica congenita (Eulenburg): muscle stiffness and weakness prevented by tocainide. J. Neurol. Neurosurg. Psychiatry 43 (1980) 268–271.

RICKER, K., R. BÖHLEN and R. ROHKAMM: Different effectiveness of tocainide and hydrochlorothiazide in paramyotonia congenita with hypokalemic episodic paralysis. Neurology (Cleveland) 33 (1983) 1615–1618.

RIGGS, J.E., R.C. GRIGGS, R.T. MOXLEY and E.D. LEWIS: Mechanism of acetazolamide action in hyperkalemic periodic paralysis. Neurology (NY) 30 (1980) 380.

RIGGS, J.E., R.T. MOXLEY, R.C. GRIGGS and F.A. HORNER: Hyperkalemic periodic paralysis: an apparent sporadic case. Neurology (NY) 31(1981) 1157–1159.

RIGGS, J.E., R.C. GRIGGS and R.T. MOXLEY: Dissociation of glucose and potassium arterial-venous differences across the forearm by acetazolamide. Arch. Neurol. 41 (1984) 35–38.

ROELTGEN, D.: Hypokalemic periodic paralysis. Arch. Neurol. 36 (1979) 453.

ROPERS, H.H. and H.B. SZLIWOWSKI: Periodic hypokalemic paralysis transmitted by an unaffected male with negative family history: a delayed mutation? Hum. Genet. 48 (1979) 113–116.

ROWLEY, P.T. and B. KLIMAN: The effect of sodium loading and depletion on muscular strength and aldosterone secretion in familial periodic paralysis. Am. J. Med. 28 (1960) 328–376.

RÜDEL, R., F. LEHMANN-HORN, K. RICKER and G. KUTHER: Hypokalemic periodic paralysis: in vitro investigation of muscle fiber membrane parameters. Muscle Nerve 7 (1984) 110–120.

RÜDEL, R., J.P. RUPPERSBERG and W. SPITTELMEISTER: Abnormalities of the fast sodium current in myotonic dystrophy, recessive generalized myotonia, and adynamia episodica. Muscle Nerve 12 (1989) 281–287.

SADEH, M., A. OHRY, I. SAROVA-PINHAS and J. BRAHAM: Hypokalemic periodic paralysis associated with multiple sclerosis. Eur. Neurol. 19 (1980) 252–253.

SAGILD, U.: Hereditary transient paralysis: with special reference to the metabolism of potassium. Thesis. Copenhagen, Munksgaard (1959).

SAROVA-PINHAS, I., J. BRAHAM and A. SHALEV: Premenstrual periodic paralysis. J. Neurol. Neurosurg. Psychiatry 44 (1981) 1162–1164.

SAUNDERS, M., B.E.H. ASHWORTH, A.E.H. EMERY and J.E.G. BENEDIKZ: Familial myotonic periodic paralysis with muscle wasting. Brain 91 (1968) 295–304.

SCHIPPERHEYN, J.J., O.J.S. BURUMA and P.J. VOOGD: Hypokalaemic periodic paralysis and cardiomyopathy. Acta Neurol. Scand. 57 (1978) 374–378.

SCHLESINGER, H.: Über die periodisch auftretende (paroxysmale) Lähmung. Wien. Klin. Wochenschr. 18 (1905) 323–327.

SCHMIDT, A.K.E.: Die paroxysmale Lähmung. Monogr. Gesamte Neurol. Psychiatry. H18 (1919) 2–56.

SCHOENTHAL, L.: Family periodic paralysis. Am. J. Dis. Child. 48 (1934) 799–813.

SCHUTTA, H.S. and J.L. ARMITAGE: The sarcoplasmic

reticulum in thyrotoxic hypokalemic periodic paralysis. Metabolism 18 (1969) 81–83.

SERKO, A.: Ein Fall von familiärer periodischer Lähmung. Wien. Klin. Wochenschr. 32 (1919) 1138–1140.

SHINOSAKI, T.: Klinische Studien über die periodische Extremitätenlähmung. Z. Neurol. Psychol. 100 (1925) 564.

SHY, G.M., T. WANKO, P.T. ROWLEY and A.G. ENGEL: Studies in familial periodic paralysis. Exp. Neurol. 3 (1961) 53–121.

SINGER, H.D. and F.N. GOODBODY: A case of family periodic paralysis with critical digest of the literature. Brain 3 (1901) 257–285.

SPIRO, A.J., J.W. PRINEAS and C.L. MOORE: A new mitochondrial myopathy in a patient with salt craving. Arch. Neurol. (Chic.) 22 (1970) 259–269.

STEWART, H.J., J.J. SMITH and A.T. MILHORAT: Electrocardiographic and serum potassium changes in familial periodic paralysis. Am. J. Med. Sci. 199 (1940) 789–795.

STREIB, E.W.: Beta-adrenergic treatment of hyperkalemic periodic paralysis. Neurology 35 (1985) 1805.

STREIB, E.W.: Paramyotonia congenita: successful treatment with tocainide: clinical and electrophysiologic findings in seven patients. Muscle Nerve 10 (1987) 155–162.

STUBBS, W.A.: Bidirectional ventricular tachycardia in familial hypokalaemic periodic paralysis. Proc. R. Soc. Med. 69 (1976) 223–224.

SUBRAMONY, S.H. and A.S. WEE: Exercise and rest in hyperkalemic periodic paralysis. Neurology 36 (1986) 173–177.

TALBOTT, J.H.: Periodic paralysis: a clinical syndrome. Medicine 20 (1941) 85–143.

TARSANEN, L.T., I.M. KANTOLA and M.E. HUIKKO: Serum potassium exercise test in the diagnosis of familial periodic paralysis. Acta. Neurol. Scand. 68 (1983) 30–33.

TAYLOR, E.W.: Family periodic paralysis. J. Nerv. Ment. Dis. 9 (1898) 719–746.

THOMPSOM, A.J. and M. HUTCHINSON: Myopathy in hypokalaemic periodic paralysis: reversal with acetazolamide. Ir. Med. J. 77 (1984) 171–172.

TOGLIA, J.U., S. MANDEL and G. KOSMORSKY: Multiple sclerosis and hypokalemic periodic paralysis in the same patient. Arch. Neurol. 39 (1982) 530–531.

TORRES, C.F., R.C. GRIGGS, R.T. MOXLEY and A.N. BENDER: Hypokalemic periodic paralysis exacerbated by acetazolamide. Neurology 31 (1981) 1423–1428.

TRONI, W., C. DORIGUZZI and T. MORGINI: Interictal conduction slowing in muscle fibers in hypokalemic periodic paralysis. Neurology (Cleveland) 33 (1983) 1522–1555.

TYLER, F.H., F.E. STEPHENS, F.D. GUNN and G.T. PERKOFF: Studies in disorders of muscle. VII. Clinical manifestations and inheritance of a type of periodic paralysis without hypopotassemia. J. Clin. Invest. 30 (1951) 492–502.

VAN DER MEULEN, J.P., G.J GILBERT and C.A. KANE: Familial hyperkalemic paralysis with myotonia. N. Engl. J. Med. 264 (1961) 1–6.

VISKOPER, R.J., A. LICHT, J. FIDEL and J. CHACO: Acetazolamide treatment in hypokalemic periodic paralysis. Am. J. Med. Sci. 266 (1973) 119–123.

VISKOPER, R.J., J. FIDEL, T. HORN, D. TZIVONI and J. CHACO: On the beneficial action of acetazolamide in hypokalemic periodic paralysis. Am. J. Med. Sci. 266 (1973) 125–129.

VROOM, F.W., M.A. JARRELL and T.H. MAREN: Acetazolamide treatment of hypokalemic periodic paralysis. Arch. Neurol. 32 (1975) 385–392.

WANG, P. and T. CLAUSEN: Treatment of atacks in hyperkalaemic familial periodic paralysis by inhalation of salbutamol. Lancet 31 (1976) 221–223.

WANG, P., T. CLAUSEN and H. ORSKOV: Salbutamol inhalations suppress attacks of hyperkalemia in familial periodic paralysis. Monogr. Hum. Genet. 10 (1978) 62–65.

WATSON, C.W.: Familial periodic paralysis. Yale J. Biol. Med. 19 (1946) 127–135.

WELLER, R.O. and B. MCARDLE: Calcification within fibers in periodic paralysis. Brain 94 (1971) 263–272.

WESTPHAL, G.: Über einen merkwürdigen Fall von periodischer Lähmung aller 4 Extremitäten mit gleichzeitigem Erlöschen der elektrischen Erregbarkeit während der Lähmung. Berl. Klin. Wochenschr. 22 (1885) 489–491.

WEXBERG, E.: Eine neue Familie mit periodischer Lähmung. Jahrb. Psychiatr. Neurol, 37 (1917) 108–131.

WORATZ, G.: Hypokalämische paroxysmale Lähmung. Münch. Med. Wochenschr. 46 (1965) 1–15.

YOSHIMURA, T., M. KANEUJI, T. OKUNO, M. YOSHIOKA, T. UEDA, H. MIKAWA, T. KOWATA and T. KAMIYA: Periodic paralysis with cardiac arrhythmia. Eur. J. Pediatr. 140 (1983) 338–343.

ZABRISKIE, E.G. and A.M. FRANTZ: Familial periodic paralysis. Bull. Neurol. Inst. N.Y. 2 (1932) 57–74.

ZWARTS, M.J., T.W. VAN WEERDEN, T.P. LINKS, H.T.M. HAENEN and H.J.G.H. OOSTERHUIS: The muscle fiber conduction velocity and power spectra in familial hypokalemic paralysis. Muscle Nerve 11 (1988) 166–173.

Handbook of Clinical Neurology, Vol. 18 (62): Myopathies
L.P. Rowland and S. DiMauro, editors

CHAPTER 16

Metabolic myopathies

SALVATORE DIMAURO[3], PAOLA TONIN[1] and SERENELLA SERVIDEI[2]

[1]*Department of Neurology, Università di Verona and*
[2]*Department of Neurology, Università Cattolica, Roma, Italy and*
[3]*H. Houston Merritt Clinical Research Center for Muscular Dystrophy
and Related Diseases, College of Physicians and Surgeons of Columbia
University, New York, NY, USA*

GLYCOGEN STORAGE DISEASES

Historical considerations. The first cases of glycogen storage disease were reported by Parnas and Wagner in 1922 and by Van Creveld in 1928. Von Gierke in 1929 described postmortem studies of two children with glycogen accumulation in liver and kidney. Involvement of skeletal and cardiac muscle was first reported independently by Pompe and by Putschar in 1932. In 1950 Di Sant'Agnese et al. proposed a clinical classification of the glycogenoses and, in the same period, major contributions to our knowledge of glycogen metabolism were made by Carl and Gerti Cori and their collaborators, who also recognized the defects of branching and debranching enzymes (Cori 1954; Illingworth and Cori 1952; Illingworth-Brown and Brown 1966).

In 1951, McArdle elegantly demonstrated a block in muscle glycogenolysis in a patient with myophosphorylase deficiency, although the enzyme defect was documented only eight years later (Schmidt and Mahler 1959; Mommaerts et al. 1959). In 1963, Hers discovered acid maltase deficiency in Pompe's disease and formulated the important concept of inborn lysosomal diseases (Hers 1965). The reports of phosphofructokinase deficiency by Tarui et al. (1965) and by Layzer et

al. (1967) separated that syndrome from McArdle's disease, despite the similarity of the clinical picture. Three additional defects of glycolysis were documented in the early 1980s involving phosphoglycerate kinase (DiMauro et al. 1981a; Rosa et al. 1982), phosphoglycerate mutase (DiMauro et al. 1981b), and lactate dehydrogenase (Kanno et al. 1980).

Although no new enzyme defects have been reported in the last decade, many examples of genetic heterogeneity have been described, manifesting as clinical or biochemical variants of known diseases. Molecular genetic analysis of the glycogenoses is making great strides and is revealing the molecular basis for the different variants. The pathogenetic relationship between biochemical error and clinical syndrome (usually chronic weakness or exercise intolerance, with or without myoglobinuria) remains unclear, although standardized exercise tests combined with ^{31}P-NMR spectroscopy are providing clues to the mechanism of normal fatigue and to the pathogenesis of the premature fatigue seen in patients with glycogenoses.

We shall describe the myopathic syndromes in the sequence they were numbered (Fig. 1). Glycogenoses type I (glucose-6-phosphatase deficiency) and type VI (hepatic phosphorylase deficiency) do not cause neuromuscular disorders.

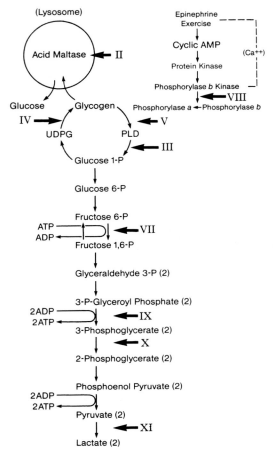

Fig. 1. Scheme of glycogen metabolism and glycolysis. Roman numbers refer to glycogen storage diseases resulting from defects of the following enzymes: II, acid maltase; III, debrancher; IV, brancher; V, muscle phosphorylase; VII, phosphofructokinase (PFK); VIII, phosphorylase b kinase; IX, phosphoglycerate kinase (PGK); X, phosphoglycerate mutase (PGAM); XI, lactate dehydrogenase (LDH). (Reproduced from DiMauro et al. (1986), with permission of the publisher of *Am. J. Med. Genet.*)

Acid maltase deficiency

Clinical manifestations. Acid maltase deficiency (AMD; glycogenosis type II) causes two major syndromes, a severe and generalized disease of infancy, first described by Pompe (1932) in Holland and Putschar (1932) in Germany, or a more benign neuromuscular disorder manifesting in childhood or adult life.

Infantile AMD (Pompe's disease) causes diffuse hypotonia and weakness in the first weeks or months of life, although muscle bulk may be increased and macroglossia is common. There is massive cardiomegaly and modest hepatomegaly. Weakness of respiratory muscles increases the risk of pulmonary infection, and death, due to cardiac or pulmonary failure, occurs invariably before 2 years of age and usually within the first year.

In *childhood AMD*, symptoms are usually manifested by delay of independent walking after 18 months. The disorder is milder than the infantile form, with slowly progressive weakness of limb-girdle and trunk muscles. In boys, the clinical picture (toe-walking, waddling gait, excessive lumbar lordosis, calf hypertrophy) may resemble Duchenne muscular dystrophy. Respiratory muscles are involved early, and death from ventilatory insufficiency usually occurs in the second or third decade. In addition to the milder manifestations, lack of cardiomegaly is the main feature distinguishing childhood from infantile AMD.

Adult AMD is characterized by a slowly progressive myopathy starting in the third or fourth decade, occasionally even later. The initial diagnosis in most cases is 'limb-girdle dystrophy' or 'polymyositis'. Ventilatory insufficiency, often an early symptom and out of proportion to the moderate limb-weakness, is a useful diagnostic clue. Visceral organs are not affected.

Laboratory investigations. Serum CK activity is variably but consistently increased in all forms of AMD. The forearm ischemic exercise causes a normal rise of venous lactate, indicating that phosphorolytic glycogen breakdown and glycolysis are normal in muscle. Electromyography (EMG) shows myopathic features associated with fibrillation potentials, positive waves, bizarre, high-frequency discharges, and myotonic discharges. In adults, these abnormalities may be more apparent in paraspinal muscles. In infantile AMD, chest radiogram shows severe cardiomegaly, and electrocardiography (ECG) shows characteristic, though not specific, changes: short P-R interval, giant QRS complexes, and signs of left ventricular of biventricular hypertrophy. In adult patients, pulmonary function tests show markedly reduced vital capacity, maximal breathing capacity, maximal expiratory and inspiratory static pressure, and early diaphragmatic fatigue.

Pathology. Muscle biopsy shows a vacuolar myopathy in all three forms of AMD. In the infantile form, all muscles and all fibers contain many vacuoles which often coalesce, producing a 'lacework' appearance. In the later-onset forms, and especially in adult AMD, the vacuoles are less numerous, and biopsies from clinically unaffected muscles may appear normal. Most vacuoles contain PAS-positive, diastase-digestible material and stain intensely for acid phosphatase. In infantile AMD and, to a much lesser extent, in the later-onset forms, there is also accumulation of a different material that has the staining properties of an acid mucopolysaccharide; the precise chemical nature has not been clarified. Electron microscopy confirms the presence of excess glycogen in two compartments: within lysosomal vacuoles limited by a single membrane, and free in the cytoplasm.

In infantile AMD, there is accumulation of intralysosomal and free glycogen in all tissues, especially severe in the heart. In the central nervous system (CNS), anterior horn cells of the spinal cord and neurons of brainstem nuclei are more severely affected than neurons of the cerebral cortex. This may contribute to the severe weakness (and occasional tongue fasciculations) of these infants, and may explain why seizures or mental retardation are not seen in their short life. Peripheral nerve biopsies show glycogen accumulation in Schwann cells.

Postmortem studies of childhood AMD showed glycogen storage in heart and brain of two patients in whom myopathy was associated with cardiopathy, mental retardation, or both. Vacuolar changes were seen in the heart of one patient with adult AMD.

Inheritance. The three forms of AMD are allelic disorders transmitted as autosomal recessive traits, consistent with the assignment of the gene encoding acid maltase (AM) to chromosome 17. Allelic diversity with various combinations of homo- and hetero-allelic mutant genotypes has been suggested as the basis for the clinical and biochemical heterogeneity of AMD (Hoefsloot et al. 1990). Prenatal diagnosis can be established reliably by measuring AM activity in cultured amniocytes.

Biochemistry and molecular genetics. AM is a lysosomal hydrolase with both α-1,4- and α-1,6-glucosidase activities and, therefore, capable of digesting glycogen completely to glucose (Illingworth-Brown and Brown 1965; Illingworth-Brown et al. 1970; Jeffrey et al. 1970). There are no tissue-specific isozymes of AM, and a defect of this enzyme should result in a generalized, progressive accumulation of glycogen within lysosomes. As suggested by morphological observations, glycogen concentration is markedly increased in all tissues of infantile AMD patients, especially in heart and muscle, while in later-onset forms of AMD, glycogen content in muscle is generally lower and may be normal in patients with adult-onset. It is not clear why free glycogen is so abundant in skeletal and cardiac muscle in infantile AMD; if glycogen is released from ruptured lysosomes, it ought to be digested by the glycogenolytic enzymes of the cytoplasm. Nor is it clear why, in childhood and adult AMD, symptoms are confined to muscle although the enzyme defect is, in fact, generalized. This striking difference in clinical expression appears to be due to the presence of a small but critical amount of residual AM activity in the later-onset forms but not in infantile AMD (Mehler and DiMauro 1977; Reuser et al. 1987; Van der Ploeg et al. 1988).

Although the presence or absence of residual AM activity seems to be the crucial discriminant between infantile and later-onset forms of AMD, the underlying biochemical causes of the enzyme defect are diverse. Like other lysosomal enzymes, AM is synthesized as a larger precursor (110 kDa) which, while being transported to the lysosomal system, undergoes glycosylation, phosphorylation of high-mannose oligosaccharide chains, and proteolytic trimming down to the 76 kDa molecular mass of the mature enzyme (Hasilik and Neufeld 1982a,b). Experiments in fibroblasts have revealed a variety of biochemical defects between and within the two main clinical variants. In patients with infantile AMD, molecular defects included: (1) complete lack of enzyme synthesis; (2) altered processing of a normal precursor leading to undetectable mature enzyme; (3) lack of processing of a smaller-than-normal precursor: (4) lack of activity of the mature enzyme (Beratis et al. 1978; Hasilik and Neufeld 1982b; Reuser et al. 1985; Van der Ploeg et al. 1989). In patients with childhood or adult AMD, the precursor was normal or reduced in amount; no defects of glycosylation were detected, but phosphorylation was defective in two cases (Reuser et al. 1985). On the basis of clinical manifestations and abnormal features of mutant enzymes, Reuser et al. (1987) counted at least 10 different variants of AMD.

The remarkable clinical and biochemical heterogeneity of AMD is now being explored at the molecular level, using the full-length cDNA encoding human AM (Hoefsloot et al. 1988; Martiniuk et al. 1990a), and an equally impressive genetic heterogeneity is emerging. In a study of 14 patients, Martiniuk et al. (1990b) found that mRNA was lacking in 5 of 10 infantile AMD cases and was present in all four adult cases, although it was shorter than normal in two. mRNA was also lacking in two other unrelated patients with infantile AMD, while the message was present and apparently normal in two sisters with abnormally small AM precursor (Van der Ploeg 1989).

Therapy. Unsuccessful therapeutic attempts in infants with Pompe's disease have included lysosome-labilizing agents, activators of glycogenolysis, and enzyme replacement using AM alone, 'packaged' within liposomes, or bound to albumin or to low-density lipoproteins. A more promising approach to enzyme replacement may be offered by the use of an AM precursor containing phosphorylated, N-linked, high-mannose carbohydrate chains, which was efficiently taken up by cultured muscle cells from patients (Van der Ploeg et al. 1988). In mice, intravenous injection of the precursor isolated from bovine testis showed that the enzyme was taken up by the heart and muscle, but not by brain (Van der Ploeg et al. 1991).

In patients with childhood or adult AMD, respiratory insufficiency may require permanent of intermittent mechanically assisted ventilation. Controversial results were obtained with high-protein diet, which seemed to improve strength and respiratory function in some (Slonim et al. 1983; Isaacs et al. 1989; Margolis and Hill 1986; Umpleby et al. 1987), but not all patients (Umpleby et al. 1989).

Attempts to increase residual AM activity in muscle by administration of thyroid hormone were unsuccessful in one study (Isaacs et al. 1986) and in our own experience (Braun et al. 1982).

Debrancher deficiency

Clinical manifestations. Debrancher deficiency (glycogenosis type III; Cori-Forbes disease) usually causes a benign hepatopathy of infancy or childhood, characterized by hepatomegaly, growth retardation, and fasting hypoglycemia, sometimes accompanied by seizures. Although the enzyme defect involves both liver and muscle in most cases, clinical myopathy is not common and often manifests in adult life, long after liver symptoms have remitted (Brunberg et al. 1971; DiMauro et al. 1979; Cornelio et al. 1984; Smit et al. 1990). Weakness usually starts in the third or fourth decade, and is often accompanied by wasting of distal leg muscles and intrinsic hand muscles, which may lead to the erroneous diagnosis of motor neuron disease or peripheral neuropathy. The course is slowly progressive and the myopathy is rarely incapacitating. Myoglobinuria was reported in one patient (Brown 1986). Peripheral neuropathy may contribute to the distal weakness and wasting of adult patients, but sensory abnormalities have been described only in one case (Moses et al. 1986).

Cardiac involvement is demonstrable by laboratory tests in virtually all patients with myopathy, but clinical cardiopathy is infrequent (DiMauro et al. 1979; Moses et al. 1989).

Laboratory investigations. Because the liver is affected by the metabolic block, administration of epinephrine or glucagon in the fasting state does not increase blood glucose. The forearm ischemic exercise causes little or no rise of venous lactate. Serum CK is variably, sometimes markedly, increased in patients with myopathy. EMG shows myopathic features associated with fibrillations, positive sharp waves, and myotonic discharges. Nerve conduction velocity may be decreased, confirming the frequent peripheral nerve involvement. ECG and echocardiography show left ventricular or biventricular hypertrophy in most patients with myopathy (DiMauro et al. 1979; Cornelio et al. 1984; Moses et al. 1989).

Inheritance. The inheritance is autosomal-recessive, with a predominance of men among patients with myopathy. Prenatal diagnosis can be based on immunoblot analysis of cultured amniocytes or on the demonstration of abnormal polysaccharide after exposure of amniocytes to glucose-free medium (Yang et al. 1990).

Pathology. Muscle biopsy shows severe vacuolar myopathy with accumulation of PAS-positive material under the sarcolemma and between myofibrils. Ultrastructurally, the vacuoles correspond to large pools of free glycogen (DiMauro et al. 1979). Excessive glycogen was also documented in skin biopsies (Sancho et al. 1990), cultured muscle fibers (Miranda et al. 1981), intramuscular nerves (Powell et al. 1985), and in both Schwann cells and axons of sural nerve biopsies (Moses et al. 1986; Ugawa et al. 1986).

Biochemistry and molecular genetics. The debranching enzyme is a single 160 kDa polypeptide with two distinct and independent catalytic functions, oligo-1,4-1,4-glucantransferase and amylo-1,6-glucosidase. After phosphorylase has shortened the peripheral chains of glycogen to about 4 glucosyl units (a form of glycogen called phosphorylase-limit-dextrin, PLD), the transferase activity of the debrancher enzyme removes a maltotriosyl unit, leaving behind a single glucosyl unit in an α-1,6-glucosidic link. This glucosyl unit is hydrolyzed by the amylo-1,6-glucosidic activity of the debranching enzyme.

Using a combination of enzymatic and immunological assays, Chen et al. (1987) and Ding et al. (1990) classified glycogenosis type III into three subgroups: (1) IIIA is characterized by lack of both transferase and α-1,6-glucosidase activities and of immunologically cross-reacting material (CRM) in both liver and muscle; most patients belong to this group. (2) IIIB is also characterized by lack of both enzymatic functions and of CRM in liver, but spares muscle and heart. (3) IIIC is limited to a few patients who lack only the transferase activity in both liver and muscle; predictably, CRM is present in these cases. While allelic defects of the gene encoding the single debrancher protein can easily explain the biochemical phenotypes of groups IIIA and IIIC, it is more difficult to envision a molecular defect that causes lack of CRM in some tissues but not in others, as in group IIIB.

In patients with myopathy, the concentration of glycogen in muscle is markedly increased and, as expected, the stored polysaccharide has abnormally short peripheral chains, compatible with PLD. Anaerobic glycolysis of muscle extracts in vitro causes no production of lactate with endogenous glycogen, while a small amount of lactate is formed with exogenous glycogen, probably due to the action of phosphorylase on the peripheral chains of the normal polysaccharide (Cornelio et al. 1984; Servidei and DiMauro 1989).

The pathogenesis of weakness is not understood. It cannot be attributed solely to the disruption of muscle architecture by glycogen accumulation because similar changes can be seen in patients without overt weakness. A defect of energy production could play a role, but other blocks of glycogen breakdown of glycolysis cause exercise intolerance, cramps, and myoglobinuria rather than weakness, and the difference in clinical phenotype is puzzling.

A full-length cDNA for human debrancher enzyme has been obtained (Yang et al. 1992), and the gene has been assigned to chromosome 1 (Yang-Feng et al. 1992). Northern analysis of mRNA isolated from lymphoblastoid cells in four patients provided evidence of genetic heterogeneity (Ding et al. 1989).

Therapy. The therapy is limited to protecting affected children from fasting hypoglycemia through frequent feeding and nocturnal gastric infusions of glucose and uncooked cornstarches. High-protein diet was beneficial in a child with weakness (Slonim et al. 1982) but remains to be confirmed in more patients.

Brancher deficiency

Clinical manifestations. Brancher deficiency (glycogenosis type IV; Andersen's disease) is a rapidly progressive disease of early childhood characterized by hepatosplenomegaly, cirrhosis, and chronic hepatic failure, causing death before 4 years of age. Muscle involvement is usually over-shadowed by liver disease, but hypotonia, wasting, and contractures were prominent in three children (Fernandes and Huijing 1968; Zellweger et al. 1972). Cardiomyopathy dominated the clinical picture in three unrelated older children with less severe liver disease (Farrans et al. 1966; Servidei et al. 1987a; Tonin et al., unpublished): the diagnosis was suggested by the presence of PAS-positive, diastase-resistant material in muscle and endomyocardial biopsies.

Laboratory tests. In patients with liver disease, serum CK is only slightly and inconsistently increased. In one patient with cardiomyopathy, chest radiograms showed cardiomegaly and M-mode echocardiography showed left ventricular dilatation with impaired shortening (Servidei et al. 1987a). In patients with adult polyglucosan body disease (APBD), electrodiagnostic tests show evidence of axonal sensorimotor neuropathy, and cystometry suggests neurogenic bladder.

Inheritance. The inheritance is autosomal-recessive. Prenatal diagnosis can be established through enzyme analysis in cultured amniocytes or in cultures obtained from chorionic villi (Brown and Brown 1989).

Pathology. The hallmark of the disease is the presence of basophilic, intensely PAS-positive polysaccharide which is partially resistant to diastase digestion and which, ultrastructurally, consists of filamentous and finely granular material. The abnormal polysaccharide has been demonstrated in skin, liver, muscle, heart, and CNS, but the amount varies markedly in different tissues from the same patient.

Biochemistry. The branching enzyme catalyzes the last step in glycogen synthesis by adding short glucosyl chains (consisting of about 7 glucosyl units) to linear peripheral chains of glycogen in α-1,6-glucosidic bonds. The newly added 'twigs' are elongated by glycogen synthetase. The enzyme purified from rabbit muscle is a monomeric protein of 77,000 Da (Caudwell and Cohen 1980).

Although brancher activity has to be measured indirectly because no specific substrate is availa-ble, there is good evidence that the enzyme defect is generalized, at least in affected children. Brancher activity was lacking in muscle, liver, heart, and brain from a girl who died of cardiopathy (Servidei et al. 1987a).

The polysaccharide that accumulates in branching enzyme deficiency has longer-than-normal peripheral chains and fewer branching points. thus resembling amylopectin. However, the coexistence of normal glycogen with the polyglucosan and the presence of some branching in the abnormal polysaccharide have not been explained.

Therapy. Liver transplantation was beneficial in 10 children (Shelby et al. 1991), but this approach does not prevent appearance of the disease in other organs; two years after liver transplantation, one child developed intractable cardiopathy (Sokal et al. 1992).

Adult polyglucosan body disease. Branching enzyme deficiency has also been identified in leukocytes from two Israeli patients with APBD (Lossos et al. 1991). We found the same abnormality in leukocytes and peripheral nerve from two Ashkenazi Jewish patients with APBD, but not in a French Canadian patient with the same syndrome (Bruno et al. 1992). APBD is characterized by late onset (fifth or sixth decade), upper and lower motor neuron signs, glove-stocking sensory loss, sphincter problems, neurogenic bladder and, in about half of the patients, dementia (Cafferty et al. 1991). In sural nerve biopsies, polyglucosan bodies are seen in the axoplasm of myelinated fibers and, less frequently, in unmyelinated fibers and Schwann cells (Cafferty et al. 1991).

In Ashkenazi Jewish patients with APBD, branching activity is decreased in leukocytes, peripheral nerve, and, presumably, brain, but is normal in muscle (Cafferty et al. 1991; Lossos et al. 1991; Bruno et al. 1992). The molecular basis for the differences in organs affected and clinical course of 'typical' glycogenosis type IV and APBD remains to be explained. Undoubtedly, diverse biochemical defects can cause accumulation of polyglucosan bodies, because branching enzyme activity was normal in leukocytes and peripheral nerve from a French Canadian patient with APBD (Bruno et al. 1992) and in muscle and brain from a patient with Lafora's disease (Gambetti et al. 1971; Ponzetto-Zimmerman and Gold 1982).

Myophosphorylase deficiency

Clinical manifestations. Muscle phosphorylase deficiency (glycogenosis type V; McArdle's disease) is characterized by exercise intolerance with premature fatigue, myalgia, and cramps in exercising muscles, relieved by rest (DiMauro and Bresolin 1986a). Symptoms are more likely to occur with intense isometric exercise, such as lifting heavy weights or pushing a stalled car, or with less intense but sustained dynamic exercise, such as walking uphill or climbing stairs. Moderate exercise, such as walking on level ground, is usually well tolerated. Most patients experience a 'second wind', that is, if they slow down or pause briefly at the first appearance of symptoms, they can resume exercising at the original pace without difficulty. About half of the patients have acute muscle necrosis and myoglobinuria after exercise, and approximately half of them develop renal failure. Fixed weakness is seen in one third of cases and is more common in older patients. Although onset of exercise intolerance is in childhood, cramps and myoglobinuria develop later and the diagnosis is rarely established in children.

Atypical clinical manifestations include: (1) some patients have only 'tiredness' or poor stamina, without cramps or myoglobinuria – these symptoms may be considered psychogenic; (2) progressive limb weakness may start in the sixth or seventh decade, without prior history of cramps or myoglobinuria; (3) conversely, there may be severe generalized weakness at or soon after birth, with respiratory insufficiency and death in infancy (DiMauro and Hartlage 1978; Milstein et al. 1989); (4) there may be delayed psychomotor development and mild proximal weakness (Cornelio et al. 1983).

Laboratory tests. Resting serum CK activity is variably increased in over 90% of patients. Between attacks of myoglobinuria, electromyograms are either normal or compatible with mild myopathy. During spontaneous or ischemic exercise-induced contractures, shortened muscles are electrically silent. Motor and sensory conduction velocities are normal. The forearm ischemic exercise tests, described in detail by DiMauro and Bresolin (1986a), causes virtually no rise of venous lactate. The lack of acidification of the myoplasm after exercise is also detectable by ^{31}P-NMR spectroscopy (Argov and Bank 1991).

Inheritance. The inheritance is autosomal-recessive, and the gene for myophosphorylase has been localized on chromosome 11 (Lebo et al. 1984). Reports of autosomal dominant inheritance are probably explained by the presence in the same family of homozygotes and manifesting heterozygotes in whom the residual enzyme activity falls below a critical threshold needed to prevent muscle dysfunction (Schmidt et al. 1987; Papadimitriou et al. 1990). Although the relatively benign nature of McArdle's disease makes prenatal diagnosis less urgent than in other glycogeneses, a few restriction fragment length polymorphisms (RFLPs) close to the gene on chromosome 11q13 are informative in 75% of patients at risk (Lebo et al. 1990).

Pathology. Muscle biopsy shows subsarcolemmal and intermyofibrillar vacuoles filled with glycogen, but glycogen accumulation can be mild enough to escape detection. Ultrastructurally, there are deposits of normal-looking, free glycogen particles. The histochemical reaction for phosphorylase shows no staining of muscle fibers, while smooth muscle of intramuscular vessels stains normally. However, false positive staining of regenerating muscle fibers, which express an immature isozyme, may occur, especially soon after an episode of myoglobinuria (Mitsumoto 1979).

Biochemistry and molecular genetics. Phosphorylase catalyzes the phosphorylytic stepwise removal of α-1,4-glucosyl residues from the outer branches of glycogen with liberation of glucose-1-P. This action proceeds until the peripheral chains have been shortened to about 4 glucosyl units, and the resulting PLD can be acted upon by the debranching enzyme. Muscle phosphorylase is a dimer of two identical subunits, of 97,000 Da each.

Studies of anaerobic glycolysis in vitro show that muscle extracts from McArdle patients produce no lactate with endogenous or added glycogen, while lactate production is normal with glucose-1-P or other hexose-phosphate glycolytic intermediates (DiMauro and Bresolin 1986a). Glycogen concentration in muscle is increased two- or three-fold in most patients, but can be normal. The accumulated glycogen has normal structure.

Most patients lack immunologically detectable enzyme protein in muscle (Servidei et al. 1988a; McConchie et al. 1991). In contrast to this remarkable biochemical homogeneity, molecular genetic analysis has revealed striking genetic heterogeneity. In Northern analysis of muscle from 21 patients, mRNA was absent in 12 biopsies, decreased in 4, normal in 4, and truncated in one (Gautron et al. 1987; Servidei et al. 1988a; McConchie et al. 1991).

Normal mature muscle has a single phosphorylase isozyme. Heart and brain share this isozyme with muscle, but also contain a distinct isozyme (the 'brain' isozyme), as well as a third isoform, that is a hybrid of the muscle and brain isozymes (Bresolin et al. 1983a). The partial defect of phosphorylase in the heart and brain of McArdle patients is compensated functionally by the quantitatively predominant muscle isozyme.

Liver, like muscle, has a single isozyme which is different from both muscle and brain phosphorylase and under separate genetic control. Liver phosphorylase deficiency (glycogenosis type VI, Hers' disease) causes hepatomegaly and hypoglycemia in childhood (Fernandes 1990).

The apparently paradoxical presence of phosphorylase activity in muscle cultures or regenerating muscle from patients with McArdle's disease (Roelofs et al. 1972) has been attributed to the transient expression of a 'fetal' isozyme (DiMauro et al. 1978) which appears to be identical to brain phosphorylase (Crerar et al. 1988).

The genes for the three phosphorylase isozymes have been cloned, sequenced, and assigned to the following chromosomes: 11q13 for the muscle (M) isoform; 14 for the liver (L) isoform; and 10 and 20 for the brain (B) isoform.

Physiopathology. The observation that isometric exercise and intense dynamic exercise are more likely to cause symptoms in McArdle patients agrees with present knowledge that energy for these types of exercise derives mainly from anaerobic or aerobic glycolysis (Lewis and Haller 1991).

The block of glycogen utilization leads to a shortage of pyruvate and, therefore, of acetyl-CoA, the main anaplerotic substrate of the Krebs cycle, and to a decreased mitochondrial energy output (Haller et al. 1985). In agreement with this concept that oxidative phosphorylation is limited by substrate availability, Haller et al. (1985) found that oxygen extraction and maximal oxygen uptake were decreased in McArdle patients, but were at least partially restored by intravenous glucose infusion.

The 'second wind' results from an adaptation to the limited substrate availability of McArdle muscle and is mediated by several mechanisms, including increased mobilization and delivery of free fatty acids, increased blood flow to exercising muscles, and increased glucose utilization (Braakhekke et al. 1986; Mineo et al. 1990).

The pathogenesis of premature fatigue is not understood, but increasing evidence suggests that it may be due to impaired excitation-contraction coupling mediated by excessive accumulation of ADP (Lewis and Haller 1991; Ruff and Weissman 1992). Depletion of ATP was considered the logical cause of contracture and myoglobinuria, but repeated attempts have failed to document ATP depletion during premature fatigue or during ischemic exercise-induced contracture, both using biochemical measurements in flash-frozen biopsies (Rowland et al. 1965) and ^{31}P-NMR spectroscopy (Argov and Bank 1991).

Therapy. Improvement of muscle endurance and strength by high-protein diet was reported in a patient with fixed weakness (Slonim and Goans 1985) and documented in another patient through formal exercise testing and ^{31}P-NMR spectroscopy (Jensen et al. 1990).

Phosphorylase b kinase deficiency

Clinical manifestations. Four main clinical syndromes are associated with phosphorylase b kinase (PBK) deficiency (glycogenosis type VIII) (Van der Berg and Berger 1990) : (1) liver disease, usually a benign condition of infancy or childhood characterized by hepatomegaly, growth retardation, delayed motor development, and fasting hypoglycemia, usually inherited as an X-linked trait;

(2) liver and muscle disease, with a static myopathy, apparently transmitted as an autosomal-recessive trait; (3) myopathy alone, with apparently autosomal-recessive transmission; and (4) fatal infantile cardiomyopathy (Mizuta et al. 1984; Servidei et al. 1988b).

Myopathy has been reported in 8 patients, one a hypotonic infant (Ohtani et al. 1982). Of the other 7 patients, six complained of exercise intolerance, with stiffness and weakness of exercising muscles (Iwamasa et al. 1983; Abarbanel et al. 1986; Servidei et al. 1987b; Carrier et al. 1990; Clemens et al. 1990), while one had progressive limb weakness starting at age 46 (Clemens et al. 1990, case 1). Three of the six patients with exercise intolerance also had episodes of pigmenturia after intense exercise (Abarbanel et al. 1986; Carrier et al. 1990; Clemens et al. 1990).

Laboratory tests. In patients with myopathy, serum CK is variably but consistently increased. Forearm ischemic exercise caused no rise of venous lactate in one patient (Abarbanel et al. 1986), but the response was normal in five other patients so tested. EMG showed non-specific myopathic features.

Pathology. Muscle biopsy shows subsarcolemmal accumulations of glycogen predominantly in type IIb fibers. Ultrastructurally, the glycogen appears normal and is free in the cytoplasm. Histochemical stain for phosphorylase is normal.

Biochemistry and molecular genetics. PBK is composed of four different subunits, α, β, γ and δ, forming a hexadecameric enzyme $(\alpha\beta\gamma\delta)_4$. The γ subunit is catalytic, activity is regulated by the degree of phosphorylation of the α and β subunits, and calcium sensitivity is conferred by the δ subunit, which is a calmodulin. Muscle PBK can be activated either through a cascade of reactions involving epinephrine, G proteins, adenylate cyclase, protein kinase, and resulting in the phosphorylation of the α and β subunits of PBK, or, more directly, through binding of calcium to the δ subunit of PBK. In patients with muscle PBK deficiency, the normal or only partially decreased rise of venous lactate after ischemic exercise is probably due to direct activation of PBK by cytoplasmic calcium released from the sarcoplasmic reticulum during contraction.

The involvement of different tissues in the various clinical forms of PBK deficiency is difficult to explain and will require a combination of immunoblot and molecular genetic analysis of tissues from patients with different phenotypes. This task is facilitated by availability of cDNAs for the different subunits. The genes encoding subunits α and β have been assigned to chromosome Xq12-q13 and 16q12-q13 (Francke et al. 1989; Davidson et al. 1992), and the gene for the γ subunit to chromosome 7 (Jones et al. 1990).

Phosphofructokinase deficiency

Clinical manifestations. In its typical presentation, muscle phosphofructokinase (PFK) deficiency (glycogenosis type VII; Tarui's disease) is indistinguishable from myophosphorylase deficiency: intolerance to vigorous exercise, often accompanied by cramps and relieved by rest (Rowland et al. 1986). Myoglobinuria seems to be less frequent than in McArdle's disease and the 'second wind' phenomenon is described by some but not all patients. When present, jaundice caused by mild hemolysis and gouty arthritis are useful clues to the differential diagnosis.

Clinical variants include: (1) hemolytic anemia without myopathy; (2) fixed weakness, often of late onset (Serratrice et al. 1969; Hays et al. 1981; Danon et al. 1988); and (3) severe and often fatal myopathy of infancy, sometimes accompanied by brain involvement (Guibaud et al. 1978; Danon et al. 1981a; Servidei et al. 1986).

Laboratory tests. Serum CK activity is consistently increased. Forearm ischemic exercise causes no rise of venous lactate. Because of the hemolytic trait, there is moderate reticulocytosis and serum bilirubin is often increased. Uric acid is elevated in most patients. EMG may be normal or it may show myopathic and 'irritative' features.

Results of ^{31}P-NMR spectroscopy differ from those in McArdle patients because glycolytic intermediates accumulate in muscle during exercise as phosphorylated monoesters (PME) and are detected as a discrete peak (Argov and Bank 1991).

Inheritance. The inheritance is autosomal-recessive with a predominance of affected men. All patients reported in the United States have been of Ashkenazi Jewish descent.

Pathology. Muscle biopsy shows subsarcolemmal and intermyofibrillar accumulation of normal-looking glycogen. However, a distinctive feature is the additional presence of an abnormal polysaccharide in some, especially older patients (Agamanolis et al. 1980; Hays et al. 1981; Danon et al. 1988). The abnormal polysaccharide stains intensely with PAS but is resistant to diastase digestion. Ultrastructurally, it appears composed of finely granular and filamentous material, similar to that described in branching enzyme deficiency, in APBD, and in Lafora disease. A specific histochemical stain for PFK is available (Bonilla and Schotland 1970).

Biochemistry and molecular genetics. PFK is a tetrameric enzyme under the control of three structural loci encoding muscle (M), liver (L), and platelet (P) subunits (Vora 1982a). Mature muscle expresses only the M subunit and contains the homotetramer M4. Erythrocytes express both the M and L subunits and contain five isozymes, the homotetramers M4 and L4 plus the three hybrid forms. In PFK deficiency, genetic defects of the M subunit cause total lack of PFK activity in muscle and partial enzyme deficiency in erythrocytes, where the residual homotetramer L4 accounts for about 50% of total normal activity. Because the M subunit is a major component of heart and brain PFK, the lack of clinical cardiopathy or encephalopathy in patients with muscle PFK deficiency is puzzling.

Immunological studies of muscle in one patient (Layzer et al. 1967) and of erythrocytes in another (Vora 1981) showed lack of CRM, but CRM was present in muscle from three other patients (Agamanolis et al. 1980; Hays et al. 1981), suggesting genetic heterogeneity. A distinct mutation rendered the M subunit unstable or incompetent to tetramerize in one of the patients with late-onset myopathy (Vora et al. 1987; Danon et al. 1988).

Glycogen concentration in muscle is only moderately increased. The presence of some glycogen with longer than normal peripheral chains has been attributed to the accumulation of glucose-6-P, a physiological activator of glycogen synthetase, possibly resulting in an inbalance between synthetase and branching enzyme activities (Hays et al. 1981).

Anaerobic glycolysis in vitro shows no lactate production by PFK-deficient muscle extracts with glycogen, glucose-1-P, glucose-6-P, and fructose-6-P, but normal formation of lactate with fructose-1,6-P as substrate (Layzer et al. 1967; Servidei and DiMauro 1989).

The genes encoding subunits M, P, and L have been localized to chromosomes 1, 10 and 21 (Vora 1982b; Vora et al. 1983; Van Keuren et al. 1986). The full-length cDNA for the M subunit has been isolated and sequenced (Nakajima et al. 1988) and the genetic lesion in the family described by Tarui et al. in 1965 has been elucidated; a G-to-T transversion activates a cryptic 5'-splice site upstream, which results in a 75-base in-frame deletion in the coding region (Nakajima et al. 1991).

Pathophysiology. Many of the considerations discussed for McArdle's disease apply to the pathophysiology of muscle PFK deficiency. However, because PFK-deficient muscle can utilize neither glycogen nor glucose, its maximal oxygen uptake ought to be directly dependent on the availability of free fatty acids (FFA), a hypothesis elegantly proven in four patients with large experimentally induced variations of plasma FFA (Haller and Lewis 1991). The exercise intolerance of patients with PFK deficiency is worsened by high-carbohydrate meals, because glucose lowers blood levels of FFA and ketones, alternative fuels responsible for the 'second wind' phenomenon. This negative effect of glucose has been aptly dubbed the 'out-of-wind' phenomenon (Haller and Lewis 1991).

Although it is more common in patients with PFK deficiency, hyperuricemia is also seen in patients with phosphorylase or debrancher deficiency, and has been ascribed to excessive exercise-induced degradation of muscle purine nucleotides when there is a block of muscle glycogenolysis or glycolysis; the increase of blood inosine and hypoxanthine is followed by a liver-mediated increase of uric acid ('myogenic hyperuricemia', Mineo et al. 1987)

Defects of terminal glycolysis

Phosphoglycerate kinase (PGK) deficiency. PGK deficiency (glycogenosis type IX) may be clinically silent or it may cause a syndrome characterized by hemolytic anemia, seizures, and mental retardation (DiMauro and Bresolin 1986b). Myopathy alone was seen in three males (DiMauro et al. 1982, 1983a; Rosa et al. 1982; Tonin et al. 1989), and was characterized by cramps, intolerance to intense exercise, and myoglobinuria. A fourth patient, an 11-year-old boy, had both recurrent myoglobinuria and hemolytic anemia with mental retardation (Sugie et al. 1989).

Resting serum CK activity is inconsistently increased, and electromyogram is normal. Forearm ischemic exercise caused no rise of venous lactate in two patients (DiMauro et al. 1983a; Tonin et al. 1989), and an inadequate rise in one (Rosa et al. 1982). Muscle biopsy showed non-specific changes; mild, diffuse increase in PAS stain was seen in one patient (DiMauro et al. 1983a).

Glycogen concentration in muscle was normal in all patients. Anaerobic glycolysis in vitro showed markedly decreased, but not absent, lactate formation with glycogen and with all four hexose phosphate glycolytic intermediates, suggesting a block below the PFK reaction. Biochemical analysis showed an isolated defect of PGK. In studies of physical and kinetic characteristics, the mutant enzymes differed in each of the four patients with myopathy; these PGK mutants have been labeled Creteil (Rosa et al. 1982), New Jersey (DiMauro et al. 1983a), Alberta (Tonin at al. 1989), and Hamamatsu (Sugie et al. 1989).

Inheritance is X-linked recessive. Because PGK is a single polypeptide in all tissues except spermatogenic cells, the defect is expressed in fibroblasts and prenatal diagnosis is possible. A full-length cDNA for the X-linked gene has been obtained (Michelson et al. 1983). Using this cDNA as a probe, we have documented genetic heterogeneity. Northern analysis of muscle mRNA from two patients showed a normal message in one but an abnormally large mRNA in the other (Shanske et al. 1991).

^{31}P-NMR spectroscopy in the patients with PGK-Creteil showed changes similar to those observed in patients with PFK deficiency: during exercise, there was a marked increase of PME, a correspondingly smaller increase of Pi, and, due to the partial nature of the metabolic block, a slight decrease of intramuscular pH (Duboc et al. 1987).

Phosphoglycerate mutase (PGAM) deficiency. PGAM deficiency (glycogenosis type X) has been described in four patients, two men and two women (DiMauro et al. 1981b, 1982; Bresolin et al. 1983b; Kissel et al. 1985; Vita et al. 1990); all had intolerance to strenuous exercise, cramps, and recurrent myoglobinuria.

Forearm ischemic exercise causes abnormally low, though not absent, rise of venous lactate. Muscle biopsy may be normal or show diffuse or patchy glycogen storage.

Glycogen concentration in muscle is normal or mildly increased. Anaerobic glycolysis in vitro results in decreased but not absent lactate production with glycogen and all hexose-phosphate glycolytic intermediates. Enzymatic analysis shows an isolated defect of PGAM, with residual activities in muscle varying between 4 and 6% of the normal mean.

PGAM is a dimeric enzyme containing various proportions of a muscle (M) subunit and a brain (B) subunit. About 95% of total PGAM activity in normal mature human muscle is represented by the MM homodimer. The small residual activity found in muscle biopsies of patients consisted of the BB isozyme, suggesting a genetic defect of the M subunit. The predominance of the BB isozyme in nonmuscle tissues explains why symptoms are confined to skeletal muscle. In heart, where the MM and MB isoforms together account for about 50% of total activity, the residual BB homodimer must be sufficient to prevent the clinical expression of cardiopathy.

Inheritance is autosomal-recessive. Family history was non-contributory in all patients, but partial PGAM deficiency was documented in muscle biopsies from one set of parents (Bresolin et al. 1983b). A full-length cDNA (Shanske et al. 1987) and the genomic clone containing the entire gene for PGAM-M (Tsujino et al. 1989) have been isolated and sequenced, and the gene has been assigned to chromosome 7 (Edwards et al. 1989). Southern blot analysis of genomic DNA from one patient with PGAM-M deficiency showed no difference in the number or intensity of hybridizing bands, ruling out major deletions, insertions, or rearrangements as the underlying defect (Shanske et al. 1987).

Results of ^{31}P-NMR spectroscopy studies in one patient were virtually identical to those obtained in the patient with PGK deficiency: exercise caused accumulation of PME and mild intracellular acidosis (Argov et al. 1987).

Lactate dehydrogenase (LDH) deficiency. LDH deficiency (glycogenosis type XI) was documented in two young men with exercise intolerance and recurrent myoglobinuria (Kanno et al. 1980; Bryan et al. 1990), and in an asymptomatic woman identified through blood screening (Maekawa et al. 1984). In all three patients, forearm ischemic exercise caused inadequate rise of venous lactate but excessive increase of pyruvate.

LDH is a tetrameric enzyme composed of two subunits, one (M) predominant in skeletal muscle, the other (H) predominant in heart muscle. Random tetramerization results in the formation of five isozymes, the two homotetramers M4 and H4, and three heterotetramers. Tissues from patients contained only the H4 isozyme, suggesting a defect of LDH-M. The small residual LDH activity (about 5% of normal) present in muscle of patients corresponds to the proportion of LDH-H4 present in normal muscle.

Inheritance is autosomal-recessive. The gene encoding LDH-M has been localized to chromosome 11 (Boone et al. 1972).

Other glycogenoses

Cardiomyopathy, mental retardation, and autophagic vacuolar myopathy. Eight male patients had a familial disorder characterized by non-obstructive hypertrophic cardiomyopathy, proximal weakness, and mental retardation, with onset in childhood or adolescence (Danon et al. 1981b; Riggs et al. 1983; Bergia et al. 1986; Byrne et al. 1986; Hart et al. 1987; Dworzak et al. 1991). Cardiomyopathy dominated the clinical picture and caused death in the second or third decade in all except one patient who received a cardiac transplant at age 24. Muscle biopsy showed autophagic vacuoles and accumulation of intralysosomal and

free glycogen. Although these morphologic changes are reminiscent of AMD, acid maltase activity was normal in all patients. Family history was positive in most cases, but women were less severely affected: the mean age at death was 35 years in women and 16 years in men (Hart et al. 1987). This pattern suggests X-linked dominant inheritance, but autosomal-dominant transmission cannot be excluded.

This syndrome seems to be due to a lysosomal disorder, but glycogen metabolism may not be primarily affected, because the autophagic vacuoles contain various cytoplasmic degradation products beside glycogen. The biochemical cause is unknown.

Cerebral glycogenosis. The hallmark of this disorder is the accumulation in the brain of glycogen β-particles, often forming 'rosettes', which are normally seen only in liver. Three patients have been reported, with severe encephalopathy since birth: two died before 1 year of age, the third at age 20 (Resibois-Gregoire and Dourov 1966; Kornfeld and LeBaron 1984; Towfighi et al. 1989). Clinically and pathologically the disease was confined to the brain, but glycogen concentration was markedly increased in the heart of one patient (Towfighi et al. 1989). Selective involvement of brain and heart could be explained by a defect of the brain isozyme of phosphorylase, which might not be compensated by the smaller amount of muscle isozyme also expressed in these tissues (see above). This hypothesis, however, awaits biochemical confirmation.

DISORDERS OF LIPID METABOLISM

Historical considerations. Muscle diseases due to altered lipid metabolism were first suggested in some patients by excessive accumulation of lipid droplets in muscle biopsies (Bradley et al. 1969, 1972; W.K. Engel et al. 1970; A.G. Engel and Siekert 1972; Johnson et al. 1973), or by the results of metabolic studies in vivo (W.K. Engel et al. 1970). However, biochemical errors of muscle lipid metabolism were first described in 1973, when A.G. Engel and Angelini documented carnitine deficiency in a young woman with weakness, and DiMauro and Melis-DiMauro identified carnitine

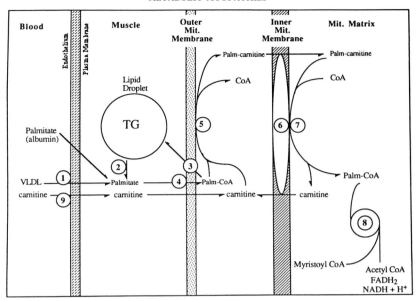

Fig. 2. Schematic representation of long-chain fatty acid metabolism in muscle. Blood-borne substrates are represented by fatty acids bound to albumin and by triglycerides in the form of very-low-density lipoproteins (VLDL). Endogenous lipid stores are triglycerides (TG) in lipid droplets. Enzymes or enzyme complexes are indicated by circled numbers adjacent to or within the membrane to which they are bound. Mit., mitochondrial; 1, lipoprotein lipase; 2, intracellular neutral tri-, di-, and mono-glyceride lipase; 3, triglyceride synthetic pathway; 4, palmitoylCoA synthetase; 5, CPT I; 6, carnitine-acylcarnitine translocase; 7, CPT II; 8, beta-oxidation pathway; 9, active transport system for carnitine. (Modified from DiMauro and Papadimitriou (1986), with permission.)

palmitoyltransferase (CPT) deficiency as the cause of recurrent myoglobinuria in two brothers. Two years later, systemic carnitine deficiency was distinguished from the myopathic form (Karpati et al. 1975) and, in 1981, a hepatic form of CPT deficiency was identified in an infant with Reye-like syndrome (Bougneres et al. 1981). During the 1980s, systemic studies of β-oxidation led to the recognition of multiple defects along this pathway (Stanley et al. 1983; Turnbull et al. 1984; Hale et al. 1985; Amendt et al. 1987; Rocchiccioli et al. 1990; Roe et al. 1990; Wanders et al. 1990) and showed that defects of β-oxidation were the most common causes of systemic carnitine deficiency. On the other hand, Treem et al. (1988) have documented a defect of carnitine transport in a form of primary systemic carnitine deficiency, often associated with cardiomyopathy (Tein et al. 1990).

In recent years, the genes encoding many of the enzymes of lipid metabolism have been isolated and sequenced, and the stage is set for the definition of the lipid disorders of muscle at the molecular level.

Biochemistry of fatty acid oxidation (Fig. 2).
Long-chain fatty acids (LCFA), the 'currency' of muscle lipid metabolism, can derive from exogenous or intramuscular sources. Exogenous, blood-borne sources are LCFA bound to albumin, plasma triglycerides in the form of very low-density lipoproteins (VLDL), and chylomicrons. Fatty acids are released from VLDL by a lipoprotein lipase located on the endothelial surface of capillaries. The mechanism of transport of LCFA across the plasma membrane is not known, although a plasma membrane fatty acid-binding protein has been described in liver (Stremmel et al. 1985), and a cytosolic fatty acid-binding protein is found in heart and muscle (Ockner et al. 1972). Intramuscular sources of LCFA are triglycerides stored as lipid droplets in close proximity to mitochondria. Fatty acids are liberated from these intracellular triglycerides through the action of a poorly studied neutral triglyceride lipase (distinct from the lysosomal acid triglyceride lipase).

In the myoplasm, LCFA are activated to acyl-CoAs at the expense of ATP by a LCFAcyl-CoA synthetase located on the outer mitochondrial membrane. LCFAcyl residues are then transferred from coenzyme A to carnitine (gamma-trimethyl-amino-beta-hydroxybutyrate) by carnitine palmitoyltransferase I (CPT I), also located on the outer mitochondrial membrane (Murthy and Pande 1987). The acylcarnitines cross the inner mitochondrial membrane in exchange for carnitine, a reaction catalyzed by carnitine:acylcarnitine translocase (Pande 1975). Inside the mitochondrion, a second carnitine palmitoyltransferase (CPT II), bound to the inner face of the inner mitochondrial membrane, reconverts LCFAcyl-carnitines to fatty acyl-CoAs, which undergo beta-oxidation. It is still controversial whether CPT I and CPT II are the same or distinct proteins. The recent cloning of the genes encoding human (Finocchiaro et al. 1991) and rat (Woeltje et al. 1990) liver CPT II will help resolve this question.

The first reaction of beta-oxidation is the dehydrogenation of acyl-CoAs to 2-trans-enoyl-CoAs catalyzed by three acyl-CoA dehydrogenases with different but overlapping specificities for short-chain (SCAD), medium-chain (MCAD), and long-chain (LCAD) fatty acids. These enzymes are homotetramers and each subunit contains a flavin adenine dinucleotide (FAD). Electrons are transferred to a second flavoprotein, electron transfer flavoprotein (ETF), which donates them to an iron-sulphur (FeS) protein, ETF-ubiquinone oxidoreductase. This enzyme funnels electrons derived from beta-oxidation (but also from the oxidation of branched-chain aminoacids and of glutaryl-CoA) into the respiratory chain.

In the second reaction of beta-oxidation, 2-trans-enoyl-CoAs are hydrated to L-3-hydroxy-acyl-CoAs by two enoyl-CoA hydratases, a short-chain enzyme (better known as crotonase), and a long-chain enzyme. The third reaction of beta-oxidation is also catalyzed by two enzymes, a short-chain 3-hydroxyacyl-CoA dehydrogenase (SCHAD) and a long-chain 3-hydroxyacyl-CoA dehydrogenase (LCHAD). The last reaction of beta-oxidation is catalyzed by two distinct 3-oxoacyl-CoA thiolases: an acetoacetyl-CoA thiolase, which is only active with acetoacetyl-CoA, and a 'general' 3-oxoacyl-CoA thiolase, which generates acetyl- CoA and an acyl-CoA chain-shortened by two carbon atoms.

Carnitine deficiency

Carnitine (beta-hydroxy-gamma-N-trimethylamino-butyrate) has a crucial role not only in the transport of LCFA into mitochondria but also in the buffering of the acyl-CoA/CoASH ratio, in the scavenging of potentially toxic acyl-CoA groups, in peroxisomal fatty acid oxidation, and in the oxidation of branched-chain aminoacids (Hoppel 1992). About 75% of the carnitine comes from dietary sources (especially red meat and dairy products), and the remaining 25% is synthesized in liver and kidney from its immediate precursor, gamma-butyrobetaine. The highest concentrations of total (i.e. free plus esterified) carnitine are found in tissues with high fatty acid oxidative metabolism, skeletal and cardiac muscle, and about 95% of total body carnitine is stored in muscle. Because carnitine in tissues is 20–60 times more concentrated than in plasma, active transport mechanisms must exist, especially in tissues that have high concentrations of carnitine but are not capable of synthesizing it, such as muscle and heart. Receptors with high, low, or intermediate affinity for carnitine have been identified in various tissues (DiDonato et al. 1992). The kidney has a crucial role in the regulation of plasma and tissue carnitine because active re-absorption in the proximal renal tubule minimizes loss of carnitine in the urine.

Myopathic carnitine deficiency. This is characterized clinically by slowly progressive axial and proximal limb weakness often fluctuating in severity and, pathologically, by severe lipid storage myopathy, which is best revealed by the oil-red-O (ORO) stain for neutral lipids. Onset is usually in childhood or young adult life. Non-muscle tissues are clinically spared (DiDonato et al. 1992), and there is no abnormal excretion of organic acids in the urine.

Biochemically, there is severe deficiency of muscle carnitine contrasting with normal levels of plasma carnitine. As expected in a primary defect, the defective LCFA oxidation by muscle homogenate is corrected by the addition of exogenous carnitine (Engel and Angelini 1973; Angelini et al. 1992). Lack of this corrective effect of carnitine in vitro in some patients with the clinical features of myopathic carnitine deficiency (Willner et al. 1979) suggests that secondary forms may also exist. A primary and selective defect of carnitine transport into skeletal muscle has been postulated to explain the myopathic form of carnitine deficiency, a hypothesis supported by kinetic compartmental analysis in one patient (Rebouche and Engel 1984).

Oral supplementation of L-carnitine (2–6 g/day) is usually beneficial, and repeat muscle biopsies may show less severe lipid storage, although the concentration of carnitine may not have changed (Engel 1986). However, the beneficial effect of L-carnitine may not be sustained, even when the doses are increased. In these cases, administration of L-acetylcarnitine (Angelini et al. 1992) or corticosteroids (Engel 1986) may be effective.

Systemic carnitine deficiency. This disorder was first described by Karpati et al. (1975) in an 11-year-old boy with recurrent episodes of hepatic encephalopathy from age 3 years and limb weakness starting at age 10. Many other patients were reported with similar Reye-like episodes, lipid storage myopathy, and low levels of carnitine in muscle, blood, and other tissues (hence the label 'systemic'). A block in the synthetic pathway could explain the generalized nature of the defect, but was not identified (Engel 1986). It soon became apparent that most cases of systemic carnitine deficiency are secondary to other inborn errors of metabolism, most commonly MCAD deficiency but also other defects of beta-oxidation, defects of branched-chain aminoacid metabolism, and defects of the respiratory chain (Engel 1986; DiDonato et al. 1992). In these disorders, there is a tendency for various acyl-CoAs to accumulate. These potentially toxic compounds are esterified to acylcarnitines, which are excreted in the urine, resulting in a net loss of carnitine. Treatment with oral L-carnitine has been effective in many of these patients, especially during the metabolic attacks (Engel 1986; DiDonato et al. 1992).

Primary systemic carnitine deficiency. A special form of systemic carnitine deficiency, and the only one so far that can be considered primary, is associated with childhood cardiomyopathy, which is often familial and is invariably fatal if untreated (Stanley et al. 1991; Tein and DiMauro 1992). There is also weakness and hypotonia, sometimes

accompanied by hypoketotic hypoglycemic encephalopathy, failure to thrive, and anemia. Carnitine concentration is very low in plasma (< 10 μM; normal 35–70 μM) and in muscle (less than 5% of the normal mean), and there is severe renal carnitine leak. There is no urinary excretion of abnormal organic acids. The electrocardiogram characteristically shows enlargement or peaking of the T-waves, together with signs of left or biventricular hypertrophy and conduction disturbances. Muscle biopsy shows lipid storage.

There is good evidence that this condition is due to a defect of the specific high-affinity, low-concentration, carrier-mediated carnitine uptake mechanism (Eriksson et al. 1988; Treem et al. 1988; Tein et al. 1990; Stanley et al. 1991). Although the defect has been documented only in cultured fibroblasts, the same uptake system is probably shared by muscle, heart, and kidney, thus explaining the lipid storage myopathy, the cardiomyopathy, and the renal loss of carnitine. Oral L-carnitine supplementation causes dramatic improvement in cardiac function (with normalization of cardiac size and electrocardiogram within a month), strength, and growth (Tein and DiMauro 1992). The disorder is inherited as an autosomal-recessive trait. This has been suggested by pedigree analysis and has been confirmed by the observation that asymptomatic parents have intermediate levels of plasma and carnitine concentrations and a partial defect of carnitine uptake in cultured fibroblasts (Tein et al. 1990; Tein and DiMauro 1992). Given the frequent history of unexplained sibling death in these families and the curative effect of L-carnitine, early identification of affected children is of the utmost importance.

Secondary systemic carnitine deficiency. Most cases of systemic carnitine deficiency are due to defects of beta-oxidation and are described below. In rare instances, however, such as in alcoholics with cirrhosis and malnutrition, in patients with kwashiorkor, or, more commonly, in premature infants receiving total parenteral nutrition, the combination of impaired hepatic synthesis and poor dietary intake may lower carnitine levels (Engel 1986). Systemic deficiency may also result from excessive carnitine loss, as in renal Fanconi syndrome, or in patients with chronic renal failure treated by hemodialysis. An iatrogenic cause of carnitine deficiency is valproate therapy; valproyl-CoA and other short-chain acyl-CoAs derived from its metabolism are probably esterified with carnitine and excreted, with a net loss of free carnitine (DiDonato et al. 1992).

Carnitine palmitoyltransferase (CPT) deficiency

Myopathic CPT deficiency. In a review of 77 adult patients with myoglobinuria (Tonin et al. 1990), CPT deficiency was the most common biochemical cause (17 of 36 patients with identified enzymopathies), followed by myophosphorylase deficiency (10 of 36). The clinical course tends to be stereotyped, as shown by a review of 39 cases (DiMauro and Papadimitriou 1986). The patients, usually young men, are normal individuals who, after prolonged though not necessarily intense exercise, develop myalgia, 'tightness', and weakness of exercising muscles, followed by myoglobinuria. The other major precipitating factor is prolonged fasting, either alone or in association with exercise. Minor causative factors include cold exposure, lack of sleep, and intercurrent infection. Renal failure occurs in about one fourth of cases. Unlike patients with glycogen diseases, who are usually 'warned' of impending myoglobinuria by painful cramps, patients with CPT deficiency have no such warning signs and therefore tend to have multiple episodes. In addition, while only exercising muscles are damaged in the glycogenoses, all muscles can be affected in CPT deficiency, especially after prolonged fasting, and some patients present with acute respiratory insufficiency.

Forearm ischemic exercise causes a normal rise of venous lactate. Prolonged (24–72 hours) fasting (to be performed only under strict medical supervision because of the risk of myoglobinuria) causes a sharp rise of serum CK activity and a blunted or delayed ketogenic response in about half of the patients.

Muscle biopsy may be completely normal or may show some degree of lipid storage, especially when taken soon after an episode of myoglobinuria. The accumulation of lipid droplets is usually far less marked than in patients with carnitine deficiency.

The enzyme defect has been documented in non-muscle tissues, such as liver, leukocytes, and cultured fibroblasts, suggesting that there are no tissue-specific isozymes, but also raising the question of why only muscle is clinically affected. One explanation could be that the enzyme is present but abnormally inhibited by increasing concentrations of substrate (Zierz and Engel 1985); this would make muscle especially vulnerable because of its critical dependence on LCFA metabolism during prolonged exercise and during fasting. It would also explain why symptoms are intermittent and why so little fat accumulates in muscle. Another question is whether both CPT I and CPT II are affected by the genetic defect (as would be expected if they were the same protein), or whether one enzyme is specifically involved (if CPT I and CPT II are distinct isozymes under separate genetic control). Immunological studies using antibodies that appear to react specifically to one or the other enzyme have suggested that the myopathic form of CPT deficiency is due to a defect of CPT II (Singh et al. 1988; Demaugre et al. 1988). Availability of a full-length cDNA for human CPT (Finocchiaro et al. 1991) will help clarify the molecular basis of CPT deficiency.

There is no specific therapy, but a high-carbohydrate, low-fat diet seems to reduce the number of myoglobinuria attacks. Patients should be warned about the risks of prolonged exercise and skipping meals.

Hepatic CPT deficiency. This causes severe infantile hypoglycemic hypoketotic encephalopathy (Bougneres et al. 1981; Tein et al. 1989). Biochemical studies in cultured fibroblasts suggested that CPT I was specifically affected in this form.

Defects of beta-oxidation

Myopathy is rarely the predominant feature in this group of disorders, which are characterized by liver dysfunction, with non-ketotic hypoglycemia, metabolic encephalopathy, and dicarboxylic aciduria. Therefore, we will only briefly describe each disease, with particular attention to muscle involvement.

Long-chain acyl-CoA dehydrogenase (LCAD) deficiency usually causes a fatal infantile syndrome characterized by failure to thrive, hepatomegaly, cardiomegaly, recurrent metabolic encephalopathy, and hypotonia. There is dicarboxylic aciduria. Total and free carnitine are decreased while long-chain acyl-carnitines are increased in blood, liver, and muscle. Inheritance is autosomal-recessive. Carnitine administration seems to improve cardiac function and reduce the frequency of encephalopathic attacks. A few patients survive the metabolic crises of infancy and may later develop chronic myopathy or myalgia and myoglobinuria under stress, similar to the myopathic form of CPT deficiency (Stanley 1987).

Medium-chain acyl-CoA dehydrogenase (MCAD) deficiency is one of the most common inborn errors of metabolism, affecting 1 in 5,000 to 1 in 10,000 live births (Roe and Coates 1989). It affects infants or young children causing recurrent episodes of hypoketotic hypoglycemic encephalopathy, and hepatomegaly with fatty degeneration, a presentation closely resembling Reye's syndrome. MCAD deficiency has also been associated to some cases of sudden infant death syndrome. Total carnitine is markedly decreased in blood, and the pattern of fatty acid excretion in the urine offers useful diagnostic clues (Rinaldo et al. 1988). Inheritance is autosomal-recessive.

Short-chain acyl-CoA dehydrogenase (SCAD) deficiency has been described in three infants with failure to thrive, developmental delay, vomiting, and non-ketotic hypoglycemia (DiDonato et al. 1992). One of the three also had severe weakness with lipid storage myopathy and decreased muscle carnitine (Coates et al. 1988). A lipid storage myopathy of adult onset was attributed to SCAD deficiency in a 53-year-old woman with muscle carnitine deficiency (Turnbull et al. 1984). Because she produced ketone bodies normally in response to fasting, and SCAD activity was normal in cultured fibroblasts, the biochemical defect seemed to be confined to skeletal muscle. However, since there is no evidence that there are tissue-specific isozymes of SCAD, this patient may have had riboflavin-responsive multiple acyl-CoA dehydrogenase (MAD) deficiency, in which the defect of SCAD predominated (Coates et al. 1988; see below).

Long-chain 3-hydroxyacyl-CoA dehydrogenase (LCHAD) deficiency has been described in five children (Glasgow et al. 1983; Bertini et al. 1990; Hale et al. 1989; Rocchiccioli et al. 1990; Wanders et al. 1990). All had recurrent hypoketotic hypoglycemic encephalopathy, but four of them also had marked weakness and cardiomyopathy, similar to patients with LCAD deficiency. Muscle biopsy in three patients with myopathy showed carnitine deficiency and lipid storage.

Short-chain 3-hydroxyacyl-CoA dehydrogenase (SCHAD) deficiency was documented in two patients. One had Reye-like episodes and died at 11 months (Hale et al. 1989). The other was a 16-year-old girl with hypoketotic hypoglycemic encephalopathy, cardiomyopathy, and recurrent myoglobinuria (Tein et al. 1991).

Multiple acyl-CoA dehydrogenase (MAD) deficiency (glutaric aciduria type II) is due to partial or complete defects of the electron-transferring flavoproteins, ETF or ETF-dehydrogenase and is, therefore, a biochemically heterogeneous entity, characterized by the urinary excretion of multiple organic acids due to the dysfunction of all six acyl-CoA dehydrogenases that funnel their reducing equivalents into ETF and ETF-dehydrogenase. However, irrespective of the specific biochemical defect, two main forms of glutaric aciduria II (GA II) have been reported: a severe, invariably fatal, neonatal form with lethargy, hypotonia, nonketotic hypoglycemia, and acidosis (GA II-A); and a later-onset, milder form with recurrent coma, vomiting, and hypoglycemia, often triggered by fasting (GA II-B). In addition, several children or young adults with GA II have had prominent myopathy, with or without episodic encephalopathy (Dusheiko et al. 1979; DiDonato et al. 1986; Turnbull et al. 1988; Bell et al. 1990). The myopathy is chronic, with myalgia, exercise intolerance, and proximal weakness. Muscle biopsy shows lipid storage and decreased levels of free carnitine. A specific defect of ETF dehydrogenase has been documented in two patients (DiDonato et al. 1986; Bell et al. 1990). Administration of carnitine or riboflavin can be of some benefit.

A distinct form of GA II is characterized by the dramatic response to riboflavin administration (*ri-*

boflavin-responsive GA II): clinical manifestations are otherwise similar to those of patients with GA II-A and GA II-B. Some patients have severe lipid storage myopathy, which also improves markedly with riboflavin (Carroll et al. 1981; De Visser et al. 1986; DiDonato et al. 1989). In one patient, a 12-year-old girl, riboflavin administration not only increased strength and eliminated the organic aciduria, but also restored to normal the activity of SCAD in muscle, which had been especially defective, and caused the 'reappearance' of SCAD protein in Western blots (DiDonato et al. 1989).

Secondary defects of beta-oxidation may accompany blocks of the respiratory chain, as documented by Watmough et al. (1990) in muscle from a patient with complex I deficiency.

Triglyceride storage disease

Five patients had a disorder characterized by congenital ichthyosis, myopathy, and lipid storage in multiple tissues, including cultured fibroblasts and muscle (Chanarin et al. 1975; Miranda et al. 1979; Angelini et al. 1980; Radom et al. 1987; DiDonato et al. 1988). The stored lipid was identified as triglyceride by thin-layer chromatography and blood and muscle concentrations of carnitine were normal, as were the activities of CPT and acid lipase in muscle (Miranda et al. 1979). Impaired oxidation of fatty acids was reported in one case (Angelini et al. 1980), but studies in cultured fibroblasts from two other patients showed normal oxidation of fatty acids and revealed a specific impairment in the degradation of endogenously synthesized triglycerides (DiDonato et al. 1988). The enzyme defect may involve the intracellular neutral triglyceride lipase, but this remains to be proven.

MITOCHONDRIAL DISEASES

Historical considerations. The concept of mitochondrial disease was introduced in 1962, when Luft and coworkers described a young woman with hypermetabolism of non-thyroid origin, morphologically abnormal muscle mitochondria, and 'loose coupling' of oxidation and phosphorylation in isolated muscle mitochondria (Luft et al. 1962). After this elegant biochemical study, the attention

of clinical scientists was mainly directed to morphology for about a decade, when systematic ultrastructural studies of muscle biopsies were pioneered by Shy and Gonatas (Shy and Gonatas 1964; Shy et al. 1966), and a modification of the Gomori trichrome histochemical stain, introduced by W.K. Engel and Cunningham (1963), allowed recognition of fibers with pathologic accumulations of mitochondria by light microscopy ('ragged-red fibers', RRF).

In the 1970s, systematic biochemical analysis of muscle biopsies from patients with mitochondrial myopathies or multisystem disorders – the term 'mitochondrial encephalomyopathies' was introduced by Shapira et al. (1977) to stress the special vulnerability of muscle and brain to defects of oxidative metabolism – led to the description of many specific biochemical defects, such as pyruvate dehydrogenase complex (PDHC) deficiency (Blass et al. 1970), CPT deficiency (DiMauro and Melis-DiMauro 1973), carnitine deficiency (Engel and Angelini 1973; Karpati et al. 1975), and defects of individual complexes of the respiratory chain (Spiro et al. 1970; Willems et al. 1977).

In the 1980s, progress in the field of mitochondrial encephalomyopathies has been extraordinary, leading to a rational biochemical classification (DiMauro et al. 1985, 1987; Morgan Hughes 1986). More important, attention has been directed to mitochondrial DNA (mtDNA) and maternal inheritance. Molecular genetic analysis of patients' tissues has led, first, to the description of large-scale deletions of mtDNA (Holt et al. 1988) and, soon thereafter, to the identification of a point mutation in patients with Leber's hereditary optic neuropathy (Wallace et al. 1988). In less than two years, several more point mutations have been associated with mitochondrial diseases, the deletions have been fairly well characterized, and two disorders apparently due to faulty communication between nuclear and mitochondrial genomes have been discovered (Shoffner and Wallace 1990; Moraes et al. 1991a; Zeviani and DiDonato 1991).

Morphologic considerations. Ultrastructural alterations of mitochondria in skeletal muscle include: (1) abnormal proliferation of normal-looking mitochondria ('pleoconial myopathy'; Shy et al. 1966); (2) enlarged mitochondria with disoriented cristae ('megaconial myopathy', Shy and Gonatas 1964; Shy et al. 1966); (3) intramitochondrial 'paracrystalline' inclusions or osmiophilic inclusions. In the light microscope, besides the modified Gomori trichrome, histochemical stains for mitochondrial enzymes, such as succinate dehydrogenase (SDH), NADH-tetrazolium reductase (NADH-TR), and cytochrome *c* oxidase (COX) are useful to confirm excessive accumulations of mitochondria and to demonstrate specific enzyme defects.

While morphologic criteria offer useful diagnostic clues, they have at least two important limitations: (1) RRF or, more commonly, ultrastructural alterations are also seen in some patients with disorders of non-mitochondrial etiology; (2) conversely, and more important, RRF are *not* seen in some primary mitochondrial diseases, especially those due to defects outside the respiratory chain, such as CPT, PDHC, beta-oxidation, or fumarase deficiencies. Even some defects of the respiratory chain, such as COX deficiency in Leigh's syndrome, may not be accompanied by RRF.

In mitochondrial encephalomyopathies, morphologic abnormalities of mitochondria are more difficult to document in non-muscle tissues and are then often more difficult to interpret because of postmortem artifacts. An apparent increase in the number of mitochondria in extraocular or cardiac muscle from patients may be deceiving because these tissues are normally rich in mitochondria. Intense SDH staining in the walls of intramuscular blood vessels has been suggested as a useful, but not entirely specific, clue to the diagnosis of the MELAS syndrome (Hasegawa et al. 1991; see below).

Clinical considerations. Mitochondrial diseases are clinically heterogeneous. Pure myopathies vary in age at onset from birth to adult years, course (rapid or slowly progressive, static, or even reversible), and distribution of weakness (generalized with respiratory insufficiency, proximal more than distal, facioscapulohumeral, or orbicularis and extraocular muscles with ptosis and ophthalmoplegia).

Mitochondrial encephalomyopathies are especially difficult to classify because of the many overlapping symptoms and signs. A distinct neuropa-

thology has helped define Leigh's syndrome; irrespective of biochemical etiology this is a devastating encephalopathy of infancy or childhood characterized by psychomotor regression and brainstem involvement, with optic atrophy, nystagmus, and respiratory abnormalities. More controversial is the identification of Alpers syndrome, in which earlier onset, more severe involvement of the cortical gray matter with seizures, and liver dysfunction have been considered distinctive features from Leigh's syndrome (Egger et al. 1987).

A controversy between 'lumpers' and 'splitters' has surrounded the classification of the three syndromes illustrated in Table 1. According to the splitters, Kearns-Sayre syndrome (KSS) is characterized by progressive external ophthalmoplegia (PEO), pigmentary retinopathy, heart block, and cerebrospinal (CSF) protein above 100 mg/dl. The syndrome of myoclonus epilepsy with RRF (MERRF) is denoted by myoclonus, epilepsy, ataxia, and myopathic limb weakness. Mitochondrial encephalomyopathy with lactic acidosis and stroke-like episodes (MELAS) is evidenced by migraine headache and vomiting, and acute, often reversible, stroke-like manifestations such as cortical blindness, hemianopia, and hemiparesis. We now know that all three syndromes are due to mutations of mtDNA. The fact that three *distinct* mutations underlie the three syndromes has provided the splitters a molecular genetic basis for their classification. On the other hand, better understanding of the biochemical consequences of these genetic defects is necessary to understand why clinical expression, in fact, differs in the three syndromes.

Skeletal muscle is clinically spared in Leber's hereditary optic neuropathy (LHON), which is characterized by acute or subacute bilateral visual loss, usually in young adult men, and often accompanied by peripapillary microangiopathy and cardiovascular abnormalities, such as the pre-excitation syndrome (Nikoskelainen et al. 1987).

In the X-linked recessive Menkes' disease (trichopoliodystrophy), mitochondrial dysfunction is secondary to defective intestinal absorption of copper; the clinical picture is characterized by developmental regression, seizures, hair abnormalities, fragile bones, tortuous arteries, hypopigmentation, and temperature instability.

Biochemical considerations. A schematic overview of mitochondrial metabolism (Fig. 3) allows recognition of five main steps: (1) cytoplasmic metabolites have to be carried across the inner mitochondrial membrane through active transport systems, or translocases; (2) in the matrix, metabolites are further oxidized, pyruvate through the action of PDHC, fatty acids through the beta-oxidation pathway; (3) acetyl-CoA, the common product of intramitochondrial metabolism, is oxidized through the reactions of the Krebs cycle; (4) the reducing equivalents derived from acetyl-CoA oxidation flow down the respiratory chain through a series of reactions in which the final hydrogen acceptor is molecular oxygen and the final product is water. (5) The energy liberated in this series of oxidation/reduction events is harnessed to pump protons from one side of the inner membrane to the other, and the resulting electrochemical proton gradient is used to synthesize ATP at three sites of the respiratory chain (oxidation/phosphorylation coupling).

Mitochondrial diseases can also be conveniently classified into five groups according to the area of mitochondrial metabolism specifically affected: (1) defects of transport; (2) defects of substrate utilization; (3) defects of the Krebs cycle; (4) defects of the respiratory chain; and (5) defects of oxidation/phosphorylation coupling (Table 2).

Genetic considerations. What makes mitochondrial diseases uniquely interesting from a genetic point of view is that mitochondria are the only subcellular organelles endowed with their own DNA (mtDNA) and capable of synthesizing a small but vital set of proteins. Human mtDNA, a small (16.5 kb), circular, double-stranded molecule, has been sequenced in its entirety (Anderson et al. 1981). It encodes 13 structural proteins, all of them subunits of respiratory chain complexes, and also two rRNAs and 22 tRNAs needed for translation (Fig. 4).

Several unique features of mtDNA are important to understand mitochondrial diseases: (1) the genetic code differs from that of nuclear DNA (nDNA); (2) genetic information is tightly packed

TABLE 1

Clinical and laboratory features in three syndromes with mitochondrial encephalomyopathy[a].

Features	KSS	MERRF	MELAS
Ophthalmoplegia	+	–	–
Retinal degeneration	+	–	–
Heart block	+	–	–
CSF protein >100 mg/dl	+	–	–
Myoclonus	–	+	–
Ataxia	+	+	–
Weakness	+	+	+
Seizures	–	+	+
Dementia	+	+	+
Short stature	+	+	+
Episodic vomiting	–	–	+
Cortical blindness	–	–	+
Hemiparesis, hemianopia	–	–	+
Sensorineural hearing loss	+	+	+
Lactic acidosis	+	+	+
Positive family history	–	+	+
Ragged-red fibers	+	+	+
Spongy degeneration	+	+	+

[a]KSS: Kearns-Sayre syndrome; MERRF: myoclonus epilepsy with ragged-red fibers; MELAS: mitochondrial encephalopathy, myopathy, lactic acidosis, and stroke-like episodes; CSF: cerebrospinal fluid; boxes highlight positive and negative differential features (see text).

TABLE 2

Biochemical classification of the 'mitochondrial myopathies'.

1. Defects of transport
   (a) CPT deficiency
   (b) Carnitine deficiency
2. Defects of substrate utilization
   (a) Pyruvate carboxylase deficiency
   (b) Pyruvate dehydrogenase complex deficiency
   (c) Defects of β-oxidation
3. Defects of the Krebs cycle
   (a) Fumarase deficiency
   (b) α-Ketoglutarate dehydrogenase (dihydrolipoyl dehydrogenase) deficiency
4. Defects of oxidation-phosphorylation coupling
   (a) Luft's syndrome (loose coupling of muscle mitochondria)
5. Defects of the respiratory chain
   (a) Complex I deficiency
   (b) Complex II deficiency
   (c) Complex III deficiency
   (d) Complex IV deficiency
   (e) Complex V deficiency
   (f) Combined defects of respiratory chain components

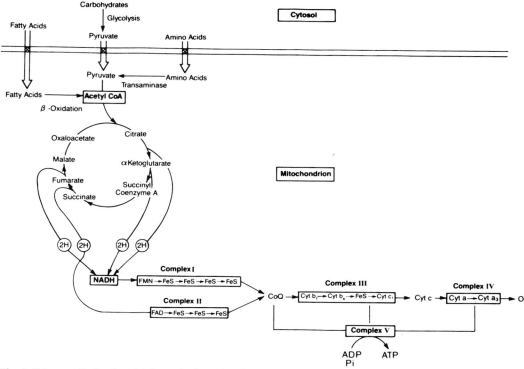

Fig. 3. Scheme of mitochondrial metabolism showing substrate transport, the Krebs cycle, and the respiratory chain. NADH, nicotinamide adenine dinucleotide, reduced; FMN, flavin mononucleotide; FAD, flavin adenine dinucleotide; FeS, non-heme iron-sulfur protein; CoQ, coenzyme Q (ubiquinone); Pi, inorganic phosphate. (Reproduced from DiMauro and DeVivo (1989), with persmission of Raven Press.)

because there are virtually no introns; (3) spontaneous mutations are more frequent than in nDNA; (4) repair systems are less efficient than in nDNA; (5) there are hundreds or thousands of copies of mtDNA per cell, as opposed to two alleles for each nDNA gene; (6) mtDNA is transmitted by maternal inheritance.

In the formation of the zygote (Fig. 5), mtDNA is contributed exclusively by the oocyte; a mutation affecting some mtDNAs in the ovum or in the zygote will be passed on randomly to subsequent generations of cells, some of which may receive few or no mutant genomes (normal or wild-type homoplasmy), while others may receive predominantly or exclusively mutant genomes (mutant homoplasmy), and still others may receive a mixed population of mutant and wild-type mtDNAs (heteroplasmy). If the mtDNA mutation illustrated in Fig. 5 impairs oxidation/phosphorylation and the three

cells A, B, and C represent stem-cells of different tissues, we can see how different tissues can be either spared or variously affected by an mtDNA mutation.

The concepts of maternal inheritance and heteroplasmy have important implications in human pathology: (1) inheritance of mtDNA-related diseases is maternal as in X-linked traits, but differs in that both sexes are equally affected; (2) because each cell contains multiple copies of mtDNA, the phenotypic expression of a mtDNA mutation will depend on the relative proportions of mutant and wild-type genomes, and a minimum critical number of mutant genomes will be necessary for expression – this is the 'threshold effect'; (3) at cell division, this proportion may shift in daughter cells ('mitotic segregation'), and the phenotype may change accordingly; (4) subsequent generations are affected by a pathologic mutation as in au-

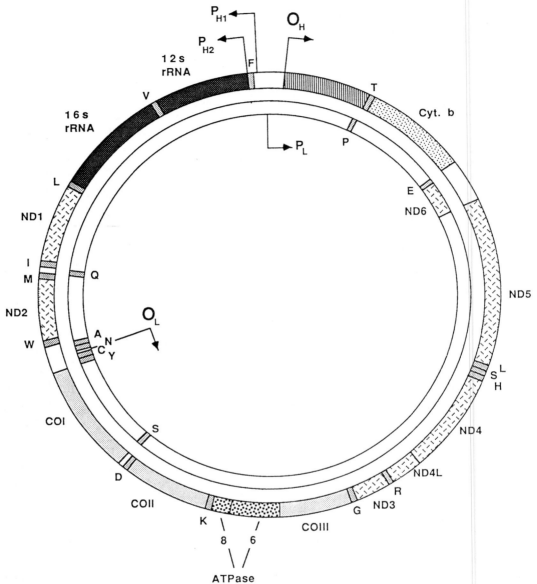

Fig. 4. Map of the human mitochondrial DNA. Genes are represented by differently shaded areas. Individual patterns identify genes encoding the rRNAs, the tRNAs, or subunits of the same respiratory complex. The inner circle represents the light strand, and the outer circle the heavy strand. Abbreviations: O_H, O_L, origin of replication of the heavy (H) or light (L) strand (arrows indicate the direction of duplication or transcription); P_{H1}, P_{H2}, P_L, promoters of transcription for the heavy and light strand; Cyt.b, cytochrome b; ND1 to ND6, subunits of NADH dehydrogenase complex (complex I); COI to COIII, subunits of cytochrome c oxidase (complex IV). Single capital letters indicate tRNA genes and identify the corresponding aminoacids by conventional single-letter code. (Modified from Zeviani et al. (1989b) with permission of the publisher of *Neurol. Clinics.*)

tosomal-dominant transmission, but the number of affected individuals in each generation ought to

be higher than in autosomal-dominant diseases. In fact, all children of an affected mother would be

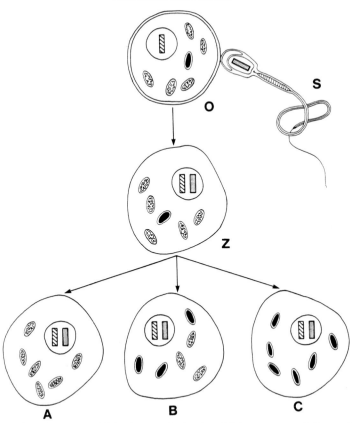

Fig. 5. Cartoon illustrating the maternal inheritance of mitochondrial genomes and the random distribution of mutant and wild-type genomes in daughter cells of the zygote. For simplicity, the relative sizes of the oocyte and the sperm have not been respected, and it has been assumed that individual mitochondria contain either a single mitochondrial genome or uniform populations of mutant (filled mitochondria) or wild-type (open mitochondria) genomes. Abbreviation: O, oocyte; S, sperm; Z, zygote; A, B, C, daughter cells representing stem cells of different tissues. (Reproduced from DiMauro et al. (1990b) with permission of the publisher of *Neurol. Clinics.*)

clinically affected, were it not for the 'threshold effect'. The 'threshold effect' is a relative concept, because the critical number of mutant mitochondrial genomes needed to cause tissue dysfunction may vary in different tissues depending on their vulnerability to impairment of oxidative phosphorylation.

Although functionally important, mtDNA-encoded proteins represent only about 10% of all mitochondrial proteins. Most mitochondrial proteins are encoded by nDNA, synthesized in the cytoplasm, and imported into mitochondria. The transport of proteins from the cytoplasm into mitochondria and the targeting of imported proteins to intramitochondrial compartments require a series of post-translational events and a complex

translocation machinery (Schatz 1991). Briefly, mitochondrial peptides are synthesized as larger precursors that contain, in addition to the mature protein sequence, amino-terminal 'leader peptides', which function as 'address signals' by recognizing specific receptors located at points of contact between inner and outer mitochondrial membranes. Interaction of peptide and receptors somehow allows the translocation of precursor peptides across the mitochondrial membranes, an energy-requiring process. Within the matrix, the leader peptides are cleaved by mitochondrial peptidases, and the mature proteins are assembled at their final intramitochondrial location.

The dual genetic control of mitochondrial proteins and the complexity of post-translational

TABLE 3

Mitochondrial diseases.

Site of defect	Heredity	Clinical features	Biochemistry
Nuclear DNA (nDNA)			
Tissue-specific gene	Mendelian	Tissue-specific syndrome	Tissue-specific monoenzymopathy
Non-tissue-specific gene	Mendelian	Multisystemic disorder	Generalized monoenzymopathy
Mitochondrial DNA (mtDNA)			Generalized monoenzymopathy (structural genes)
Point mutations	Maternal	Multisystemic, heterogeneous	Generalized multienzymopathy (tRNA genes)
Deletions or duplications	Sporadic	PEO; KSS; Pearson	Generalized (±) multienzymopathy
nDNA/mtDNA communication			
Multiple mtDNA deletions	Mendelian (AD)	PEO ± other features	Generalized multienzymopathy
mtDNA depletion	Mendelian (AR)	Myopathy ± nephropathy; hepatopathy; encephalopathy	Tissue-specific multienzymopathy

events required for the transport and correct assembly of proteins synthesized in the cytoplasm explain why inherited mitochondrial diseases can be due to diverse genetic errors. Advances in understanding the molecular basis of mitochondrial diseases allow us to propose a genetic classification that takes into account defects of mtDNA, defects of nDNA, and defects of communication between the two genomes (Table 3).

Diseases due to defects of nuclear DNA

In describing these disorders, we will consider first the five biochemical groups listed in Table 2, then the defects of mitochondrial protein transport.

1. Defects of substrate transport. These affect mainly lipid metabolism (e.g. *CPT deficiency* or *carnitine deficiency*) and have been described above.

2. Defects of substrate oxidation. Defects in the *beta-oxidation pathway* have been described above.

Defects of the *pyruvate dehydrogenase complex (PDHC)* can affect each of the three catalytic components, E_1 (pyruvate decarboxylase), E_2 (dihydro-

lipoyl dehydrogenase), or E_3 (dihydrolipoyl transacetylase), as well as either of the two regulatory components, PDH-kinase, which inactivates the enzyme, and PDH-phosphatase, which activates it. Because muscle involvement in these disorders is limited and overshadowed by brain disease, we will only summarize their features.

There are three main variants of E_1 *(pyruvate decarboxylase) deficiency*: (1) a severe and invariably fatal *neonatal form*, with failure to thrive, hypotonia, episodic apnea and lethargy, seizures, severe lactic acidosis, and dysmorphic features reminiscent of the fetal alcohol syndrome; (2) an *infantile form* clinically and neuropathologically conforming to *Leigh syndrome*, with symptoms appearing before 6 months of age and consisting of psychomotor regression, hypotonia, seizures, episodes of apnea and lethargy, ataxia, ophthalmoplegia, optic atrophy, and mild to moderate lactic acidosis; and (3) a *benign form*, characterized by intermittent ataxia or exercise intolerance, apparently responsive to thiamine administration.

E_2 *(dihydrolipoyl dehydrogenase) deficiency* has been documented in a single child with psychomotor delay, microcephaly, hyperammonemia, and lactic acidosis.

E_3 (dihydrolipoyl transacetylase) deficiency has been described only in three patients with devastating infantile encephalopathy.

PDH-phosphatase deficiency has been described in four patients, three of whom had the clinical and neuropathological features of *Leigh syndrome*.

Defects of pyruvate carboxylase (PC), the first enzyme in gluconeogenesis, cause two main syndromes, a severe neonatal disorder with hepatomegaly, metabolic acidosis, and death before 3 months of age, and a milder syndrome starting in the first months of life and characterized by delayed psychomotor development, metabolic acidosis, and death in childhood.

3. Defects of the Krebs cycle.

These defects have been documented at three steps, involving α-ketoglutarate dehydrogenase, fumarase, and aconitase.

α-ketoglutarate dehydrogenase deficiency is accompanied by deficiencies of PDH-E_3 (see above) and branched-chain ketoacid dehydrogenase because the three enzymes share the same subunit.

Fumarase deficiency, first described by Zinn et al. (1986), is a devastating encephalomyopathy of infancy characterized by poor feeding, lethargy, persistent vomiting, and, neurologically, by microcephaly, lethargy alternating with irritability, impaired vision, hypotonia, and hyporeflexia. Death usually occurs before 1 year of age. Muscle biopsy, reported only in one case, showed no RRF, but increased numbers of mitochondria were seen by electron microscopy. Biochemical studies show markedly decreased fumarase activity in all tissues, including fibroblasts. The defect involves both the cytoplasmic and the mitochondrial isozymes, which are encoded by the same gene. The partial enzyme defect observed in fibroblasts from asymptomatic parents suggests autosomal recessive transmission (Gellera et al. 1990).

Aconitase deficiency has been reported in a patient with exercise intolerance and myoglobinuria, who also had complex II deficiency (Haller et al. 1991; see below).

4. Defects of the respiratory chain.

These may result from genetic errors in either the nuclear or the mitochondrial genome. Here, we will consider those disorders that are presumably due to defects of nDNA.

Complex I deficiency. Complex I (NADH-CoQ reductase) contains at least 25 different peptides (7 of which are encoded by mtDNA), and several non-protein components, such as flavin mononucleotide (FMN), 8 non-heme iron-sulfur clusters, and phospholipid (Hatefi 1985). Defects of complex I have been documented by polarographic studies of freshly isolated mitochondria (showing impaired respiration with NADH-generating substrates, but normal respiration with $FADH_2$-generating substrates), or by measurement of partial enzyme reactions, such as rotenone-sensitive NADH-CoQ reductase, NADH-cytochrome *c* reductase, or NADH-dehydrogenase. To identify the defective subunits, immunoblots of mitochondrial preparations have been performed in several patients using antibodies against the holoenzyme or against individual subunits.

From the clinical point of view, Complex I deficiency can be divided into three broad groups: (1) *Fatal infantile multisystem disorders*, characterized by severe congenital lactic acidosis, psychomotor delay, diffuse hypotonia and weakness, cardiopathy, and cardiorespiratory failure causing death in infancy (Moreadith et al. 1984; Robinson et al. 1986; Hoppel et al. 1987). The enzyme defect was documented in multiple organs. Because the molecular defect is not known in any of these cases, there is only circumstantial evidence that they are due to nDNA defects. This is based on lack of maternal inheritance, lack of symptoms in the parents, parental consanguinity in one case (Robinson et al. 1986, case 2), presence of both affected and unaffected siblings, and, in one patient (Robinson et al. 1986, case 1), lack of a 20-kDA nDNA-encoded subunit of complex I (Slipetz et al. 1991b). Therapeutic trials with thiamine, biotin, carnitine, and ketogenic diet have been unsuccessful. (2) *Myopathy*, starting in childhood or adult life with exercise intolerance followed by fixed weakness (Morgan-Hughes et al. 1988). There is usually lactic acidosis at rest which is exaggerated by exercise. The tissue-specific nature of this disorder suggests that one or more nDNA-encoded subunits of complex I may exist as tissue-specific isoforms, a concept reinforced by the demonstration that complex I activity was normal in the liver of a patient with myopathy (Watmough et al. 1989). However, immunoblot analysis in a few patients has shown a generalized decrease of all bands rather than a selective defect of any one band (Morgan-Hughes et al. 1988; Bet et al. 1990). In agreement with the postulated nuclear nature of these disorders, family history is usually negative or compatible with mendelian inheritance. Therapy with riboflavin was beneficial to one patient (Arts et al. 1983). (3) *Mitochondrial encephalomyopathy* (excluding MELAS), with onset in childhood or adult life, and variable combinations of symptoms and signs, including ophthalmoplegia, seizures, dementia, ataxia, neurosensory hearing loss, pigmentary retinopathy, sensory neuropathy, and involuntary movements (Morgan-Hughes et al. 1988). This heterogeneous group is likely to include some patients with nDNA and others with mtDNA defects because most cases were reported before systematic analysis of mtDNA was conducted.

Complex II deficiency. This diagnosis was suggested by more or less severe defects of succinate-cytochrome c reductase activity in muscle from five patients with encephalomyopathy (DiMauro 1993). More convincing biochemical evidence, together with complete lack of SDH stain in muscle biopsies, was offered in two patients with myopathy. One was a 22-year-old man with exercise intolerance and recurrent myoglobinuria, who also had a defect of aconitase, a Krebs cycle enzyme (Haller et al. 1991). The other patient was a 14-year-old girl with weakness and exercise intolerance but no myoglobinuria (Garavaglia et al. 1990).

Coenzyme Q_{10} (CoQ_{10}) deficiency was documented in two sisters with exercise intolerance, recurrent myoglobinuria, and slowly progressive weakness of axial and proximal limb muscles (Ogasahara et al. 1989). Brain involvement was manifested by seizures in one sister, cerebellar syndrome in the other, and learning disability in both. There was lactic acidosis and increased serum CK. Muscle biopsies showed increased numbers of lipid droplets and mitochondria in type I fibers. Cytochrome spectra showed normal levels of all cytochromes, but CoQ_{10} concentration was only about 5% of normal in muscle from both patients. Because CoQ_{10} concentration was normal in fibroblasts, it was concluded that the primary defect probably involved a tissue-specific isozyme in the

CoQ_{10} synthetic pathway of muscle and brain. Replacement therapy ameliorated both muscle and brain symptoms.

Complex III deficiency. Complex III is composed of 11 subunits, including two high-molecular weight core proteins, the apoprotein of cytochrome b (encoded by mtDNA), the apoprotein of cytochrome c_1, and a non-heme iron-sulfur protein known as Rieske protein (Hatefi 1985). Polarographic analysis of isolated mitochondria show impaired utilization of both NAD-linked and FAD-linked substrates, and spectra of reduced-minus-oxidized cytochromes show lack of reducible cytochrome b in some but not all patients. Clinical manifestations can be generalized (encephalomyopathies) or confined to individual tissues, such as muscle or heart.

Among the *encephalomyopathies*, a fatal infantile form has been described in a child with severe lactic acidosis and hypotonia manifesting a few hours after birth, accompanied by generalized aminoaciduria and, terminally, by dystonia, seizures, and coma (Birch-Machin et al. 1989). Muscle biopsy was histochemically normal, but biochemical analysis showed an isolated defect of complex III, and immunoblot showed a severe defect of the Rieske protein. *Encephalomyopathies of later onset* (childhood to adult life) were accompanied by various combinations of weakness, short stature, dementia, ataxia, sensorineural deafness, pigmentary retinopathy, sensory neuropathy, and pyramidal signs (Morgan-Hughes et al. 1985; Kennaway 1988). *Myopathy*, with exercise intolerance followed by fixed weakness, is accompanied by low levels of cytochrome b in most (Morgan-Hughes et al. 1977; Hayes et al. 1984; Kennaway et al. 1984) but not all patients (Reichmann et al. 1986). Immunoblot analysis in one patient showed decreased amounts of cytochrome b, core proteins 1 and 2, the Rieske protein, and peptide VI, while the concentration of cytochrome c_1 was normal (Darley-Usmar et al. 1983). The existence of tissue-specific subunits is suggested by the normal activity of complex III in fibroblasts and lymphoid cells from a patient with pure myopathy (Darley-Usmar et al. 1986). Complex III deficiency has also been described in one patient with autosomal dominant facioscapulohumeral (FSH) dystrophy (Slipetz et al. 1991a). Administration of menadione (vitamin

K$_3$) and ascorbate (vitamin C) markedly improved the clinical condition of one patient with myopathy (Eleff et al. 1984; Argov et al. 1986), but was ineffective in other patients (Reichmann et al. 1986; Birch-Machin et al. 1989). *Cardiopathy* was associated with marked decrease of complex III activity and cytochrome *b* content in the myocardium (but not in liver or muscle) in a patient with a rare and invariably fatal disease known as 'histiocytoid cardiomyopathy of infancy' (Papadimitriou et al. 1984).

Complex IV deficiency Complex IV (cytochrome *c* oxidase, COX) is composed of 13 polypeptides: the three largest subunits (I, II, and III) are encoded by mtDNA, while the 10 smaller ones (IV, V, VIa, VIb, VIc, VIIa, VIIb, VIIc, VIII, according to Kadenbach et al. [1983]) are encoded by nDNA and synthesized in the cytoplasm. The smaller subunits may modulate COX activity, optimizing it to the metabolic requirements of different tissues. This concept is supported by biochemical and molecular genetic evidence that two of these subunits (VIa and VIIa) are, in fact, tissue-specific (DiMauro et al. 1990).

There are two major clinical presentations, one characterized by myopathy, the other by encephalopathy. Two forms of *myopathy* have been described, both presenting soon after birth with severe diffuse weakness, respiratory insufficiency, and lactic acidosis, but with very different outcome. *Fatal infantile myopathy* causes death before 1 year of age. Although heart, liver, and brain are clinically spared, many patients have renal dysfunction with DeToni-Fanconi-Debré syndrome. Pedigree analysis in informative families suggests autosomal-recessive transmission. Immunotitration and immunocytochemistry showed decreased amount of CRM in muscle of patients, but electrophoretic analysis of the mutant enzyme after immunoprecipitation failed to show a defect of any specific subunit (Bresolin et al. 1985). However, using antibodies against specific subunits, Tritschler et al. (1991) provided immunocytochemical evidence of a selective defect of subunit VIIa, which would agree with the postulated nuclear and tissue-specific nature of this disorder. Patients with *benign infantile myopathy* also have severe diffuse weakness in the neonatal period, but improve spontaneously and are usually normal by 2 or 3 years of age (DiMauro et al. 1983b, 1990). Immunocytochemistry is useful in differentiating the malignant from the benign form of infantile myopathy because muscle biopsies from children with benign myopathy lack both subunit VIIa and subunit II (Tritschler et al. 1991). The spontaneous recovery in children with benign infantile myopathy corresponds to a gradual return of COX activity in muscle, which can be documented both histochemically and biochemically (DiMauro et al. 1983b). It is still unclear whether the benign form is due to a defect in the nuclear or in the mitochondrial genome (DiMauro et al. 1990).

Among the *encephalomyopathies*, the most important is *Leigh syndrome*: about 40 patients have been reported and this appears to be the most common known biochemical cause of Leigh syndrome (DiMauro et al. 1990; Van Coster et al. 1991). Patients with Leigh syndrome have a generalized but partial defect of COX, and residual activities vary in different tissues but tend to be similar in the same tissue from different patients. Because the enzyme defect is expressed in fibroblasts in most (but not all) patients, prenatal diagnosis ought to be possible in families with one affected child, but has been performed only in one instance (Ruitenbeek et al. 1988). In immunological studies, the amount of CRM was variably decreased in different tissues, and no alteration of the subunit pattern was detected in mitochondria isolated from brain or fibroblasts (DiMauro et al. 1990). Northern analysis in different tissues using cDNA probes for eight nDNA-encoded COX subunits showed messages of normal size and abundance (Lombes et al. 1991). Thus, Leigh syndrome seems to be due to a mutation in a nuclear regulatory gene that controls assembly or stability of complex IV, rather than to the mutation of a gene that encodes a specific COX subunit.

MNGIE syndrome (myo-neuro-gastrointestinal disorder and encephalopathy), first described by Bardosi et al. (1987), is characterized by PEO, limb weakness, peripheral neuropathy, gastroenteropathy with chronic diarrhea and intestinal pseudo-obstruction, leukodystrophy, lactic acidosis, and RRF. Seven new patients with similar clinical picture have been reported (Blake et al. 1990; Simon et al. 1990); four of them were classified under the acronym POLIP (polyneuropathy, ophthalmoplegia, leukoencephalopathy, intestinal pseudo-obstruction). Partial COX deficiency was found in muscle and liver of the original patient (Bardosi et al. 1987) and in muscle biopsies from two other patients (Blake et al. 1990). Southern analysis of muscle mtDNA in two patients failed to show any large deletion (Blake et al. 1990).

Complex V deficiency. Complex V (mitochondrial ATPase) consists of two parts, a membrane portion Fo and a catalytic portion F1, joined by a stalk. Of the 12–14 subunits, two (subunits 6 and 8) are encoded by mtDNA. Indirect demonstration of ATPase deficiency was offered in two clinically different patients, a 37-year-old woman with congenital, slowly progressive myopathy, RRF, and abundant paracrystalline inclusions (Schotland et al. 1976), and a 17-year-old boy with a multisystem disorder characterized by weakness, ataxia, retinopathy, dementia, and peripheral neuropathy (Clark et al. 1983).

Combined defects of the respiratory chain have been reported in several patients with myopathy or, more often, encephalomyopathy. Although usually associated with genetic errors of mtDNA, such as deletions of point mutations in tRNA genes, combined defects can be due to nDNA defects. Potential mechanisms include: (1) mutations of regulatory genes controlling multiple complexes; (2) mutations of subunits shared by two or more complexes; (3) mutations affecting the physical microenvironment of the complexes, such as the phospholipid composition of the inner membrane; (4) defects of mitochondrial protein transport affecting subunits of multiple complexes. A mutation in the mature isoform of an nDNA-encoded, tissue-specific, and developmentally regulated subunit common to complexes I and IV was postulated in an infant with growth retardation, weakness, cardiomyopathy, hepatic insufficiency, severe lactic acidosis, RRF, and combined deficiencies of complex I and IV in muscle, heart, and liver (Zheng et al. 1989). Conversely, a mutation affecting the fetal isoform of the same subunit was suggested to explain the spontaneously reversible myopathy of another child with a combined defect of complex I and IV (Roodhooft et al. 1986). A combined defect of complexes III and IV was reported in an infant who died at 5 months of age with generalized weakness, lactic acidosis, and cardiorespiratory insufficiency (Takamiya et al. 1986).

5. Defects of oxidation/phosphorylation coupling. Non-thyroidal hypermetabolism has been reported only in two cases, a Swedish woman (Luft et al. 1962) and a Jordanian woman (Haydar et al. 1971; DiMauro et al. 1976), both sporadic and with no clinically affected relatives. Onset was in adolescence with hyperthermia, heat intolerance, profuse perspiration, polyphagia, polydipsia, and resting tachycardia. There was exercise intolerance but weakness was mild. Although basal metabolic rate was greatly increased, thyroid function was normal. Muscle biopsy showed many RRF and abundant capillaries, and electron microscopy showed enlarged mitochondria, many of which contained osmiophilic inclusions. Both women died in middle-age, but autopsies were not performed. Polarographic studies of isolated muscle mitochondria showed maximal respiratory rate even in the absence of ADP, indicating loss of respiratory control. Respiration proceeded maximally independent of phosphorylation (loose coupling) and energy was wasted as heat. A vicious cycling of calcium uptake and release was suggested as the cause of loose coupling in the second patient (DiMauro et al. 1976). In the second case, mtDNA contained no large-scale deletions (Moraes et al. 1989).

Defects of mitochondrial protein transport. Direct demonstration of errors in protein transport have been provided in non-neurological disorders (Ledley et al. 1990; Purdue et al. 1990), but convincing, if indirect, evidence of this pathogenetic mechanism was offered in a 14-year-old girl with congenital myopathy (Schapira et al. 1990). Muscle biopsy showed increased number of lipid droplets, scattered COX-negative fibers, complete lack of SDH reaction by histochemistry, but no RRF. Biochemical analysis showed combined defects of the respiratory chain, with a specific defect of the 27.7 kDa iron-sulfur protein of SDH and the Rieske protein of complex III. Because the Rieske protein was present both in muscle homogenate and in the cytosol, but not in isolated mitochondria, it was proposed that the primary defect involved mitochondrial protein transport.

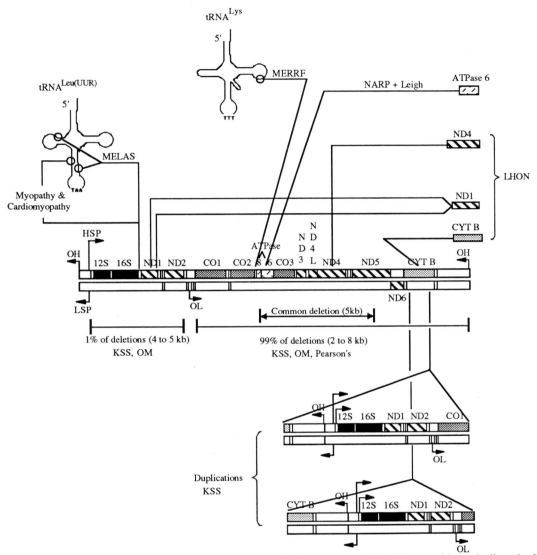

Fig. 6. Schematic summary of qualitative mutations of mtDNA in mitochondrial diseases. Above the linearized mtDNA map are depicted the point mutations associated with mitochondrial encephalomyopathy, lactic acidosis, and stroke-like episodes (MELAS), myopathy and cardiomyopathy, myoclonic epilepsy and ragged-red fibers (MERRF), neurogenic atrophy, ataxia, and retinitis pigmentosa (NARP), and Leber's hereditary optic neuropathy (LHON). Below the mtDNA map are mutations (deletions and duplications) associated with Kearns-Sayre syndrome (KSS), ocular myopathy with ragged-red fibers (OM), and Pearson's syndrome. For the nomenclature of the mtDNA map, see Fig. 4. (Modified from Moraes et al. (1991a), with permission of Mosby Year Book.)

Diseases due to defects of mitochondrial DNA (Fig. 6)

1. Deletions and duplications. Here, we are considering only patients harboring each a *single deletion* of mtDNA, identical in all tissues of any one patient. The number of normal and deleted genomes, however, varies in different tissues (heteroplasmy). Three major syndromes are associated with deletions of mtDNA: the Kearns-Sayre syndrome

(KSS); sporadic PEO with RRF; and Pearson's syndrome. Intermediate cases, such as patients with 'incomplete' KSS, or evolving cases, such as patients with Pearson's syndrome developing KSS, have been reported, stressing the pathogenetic homogeneity of this group of disorders.

Kearns-Sayre syndrome is defined by the invariant triad of (1) onset before age 20; (2) PEO; and (3) pigmentary retinopathy, plus at least one of the following: heart block, cerebellar syndrome, or CSF protein above 100 mg/dl (Rowland et al. 1991). Other common but non-specific features include dementia, sensorineural deafness, and endocrine abnormalities (short stature, diabetes, hypoparathyroidism). Muscle biopsy shows RRF and variable numbers of COX-negative fibers. Prognosis is poor, even after placement of a pacemaker, with a relentlessly downhill neurological disorder and death usually in the third or fourth decade. Numerous series from various parts of the world have shown that almost all patients with typical KSS had mtDNA deletions (DiMauro 1993). However, a few KSS patients – 3 of 15 in our series of 1989 (Moraes et al. 1989) – had normal mtDNA by Southern analysis. Because no deletions have been found in muscle mtDNA from mothers of KSS patients (Moraes et al. 1989; Zeviani et al. 1990b) or children of affected mothers (Larsson et al. 1992), the deletions seem to be new mutations, probably arising in the zygote and affecting somatic rather than germ-line cells. Duplications of mtDNA were described in two patients with KSS (Poulton et al. 1989a,b).

Sporadic PEO with RRF is a relatively benign condition characterized by ophthalmoplegia, ptosis, and proximal limb weakness. Onset is usually in adolescence or young adult age, and the course is slowly progressive. Muscle biopsy shows RRF and COX-negative fibers. About 50% of all patients with PEO have mtDNA deletions (Moraes et al. 1989). The negative family history helps distinguish this from other forms of PEO, such as late-onset autosomal-dominant oculopharyngeal muscular dystrophy and autosomal-dominant PEO with multiple rather than single deletions (see below). As women with this condition reproduce normally and have unaffected children, the mutation must affect only somatic cells or, if it affects germ-line cells, must be eliminated in the course of oogenesis.

Pearson's marrow/pancreas syndrome is a disease of infancy characterized by refractory sideroblastic anemia, vacuolization of marrow precursors, and exocrine pancreatic dysfunction. Death usually occurs in early childhood due to sepsis after bone-marrow failure (Rotig et al. 1990). The virtual identity of this condition with KSS has been demonstrated by the few patients with Pearson's syndrome who survive into adolescence only to develop symptoms and signs of KSS (Larsson et al. 1990; McShane et al. 1991).

Deletions of mtDNA: molecular biology. The deletions range in size from approximately 2.0 to 8.5 kb and are most often confined to an 11-kb region of mtDNA (Fig. 6). Irrespective of the associated clinical picture, 30% to 40% of all deletions are identical and span 4,977 bp from the ATPase gene to the ND5 gene. This 'common deletion' is flanked by a perfect 13-bp direct repeat (Schon et al. 1989). About 70% of all deletions are flanked by perfect direct repeats (class I), while the others (class II) are not (Mita et al. 1990), implying different pathogenetic mechanisms for the two types of deletions.

Analysis of mtDNA heteroplasmy has shown abundant deletions in most postmortem tissues from patients with KSS (Moraes et al. 1989; Ponzetto et al. 1990; Shanske et al. 1990; Zeviani et al. 1990b). The presence of small amounts of deleted mitochondrial genomes has been documented by PCR polymerase chain reaction in non-muscle tissues from patients with PEO (Zeviani et al. 1990b). These data and studies of patients with Pearson's syndrome evolving into KSS strongly suggest that the clinical phenotype is determined by the distribution and relative abundance of mtDNA deletions in different tissues.

Studies combining in situ hybridization, histochemistry, and immunocytochemistry in muscle biopsies (Mita et al. 1989; Shoubridge et al. 1990), and Northern analysis and mitochondrial protein synthesis in clonal fibroblast and myoblast cultures (Nakase et al. 1990) have led to the following conclusions: (1) focal COX deficiencies in muscle fibers correlate with local abundance of deleted mtDNAs; (2) COX subunits encoded by mtDNA (even by genes *not* encompassed by the deletion) were lacking in COX-deficient fibers; (3) abnormal fusion genes spanning the deletion are transcribed but are apparently not translated, because the corresponding abnormal proteins are not seen.

2. Point mutations. In three years (1988–1991), five human maternally inherited diseases were linked to point mutations of mtDNA: Leber's hereditary optic neuroretinopathy (LHON; Wallace et al. 1988a); MERRF (Shoffner et al. 1990); ATPase 6 mutation syndrome (Holt et al. 1990); MELAS (Goto et al. 1990; Kobayashi et al. 1990); and a fatal infantile myopathy and cardiopathy (Zeviani et al. 1991b) (Fig. 6).

LHON is characterized by acute or subacute loss of vision due to severe bilateral optic atrophy, with onset usually between 18 and 30 years, and marked predominance in men (Newman and Wallace 1990). There is circumpapillary teleangiectatic microangiopathy and pseudoedema of the optic disk. Associated features may include hyperreflexia, cerebellar ataxia, peripheral neuropathy, or cardiac conduction anomalies. Muscle is clinically spared, but there are morphological changes of mitochondria and decreased activity of complex I has been documented (Larsson et al. 1991). Three distinct point mutations of mtDNA have been reported: a G-to-A transition at nt 11778 in the gene encoding subunit 4 of complex I (ND4); and two mutations in the ND1 gene, one at nt 3460, the other at nt 4160.

MERRF is characterized by myoclonus, seizures, mitochondrial myopathy, and cerebellar ataxia. Additional, less common, signs include dementia, hearing loss, optic atrophy, peripheral neuropathy, and spasticity. Evidence of maternal inheritance is another important diagnostic feature. However, as is typical of mtDNA mutations, the disease may be fully expressed only in few members of a given family, while other maternal relatives may show only some symptoms and signs (oligosymptomatic patients) or be totally asymptomatic (Wallace et al.1988b). Therefore, in the absence of at least one 'typical' patient, the diagnosis of MERRF (or MELAS, see below) can be missed and maternal inheritance may be overlooked. A highly specific, though not exclusive, point mutation at nt 8344 in the tRNA[Lys] gene makes it possible to identify patients regardless of clinical severity (Shoffner et al. 1990). Muscle biopsy shows RRF as implied by the acronym, but one patient with otherwise typical symptoms had RRF in only one of two biopsies (Berkovic et al. 1989), and another patient with the MERRF mutation did not have RRF (Hammans et al. 1991). Histochemistry shows numerous COX-negative fibers, and immunocytochemistry shows two populations of mitochondria in individual muscle fibers, a population with normal COX activity and immunoreactivity for subunit II, and a population with decreased COX activity and decreased COX II immunoreactivity (Lombes et al. 1989). Neuropathology shows neuronal loss and gliosis affecting especially the dentate nucleus and the inferior olives. Biochemical studies of muscle have given inconsistent results, with defects of complex III, complexes II and IV, complexes I and IV, complexes I, III, and IV, or complex IV alone (DiMauro 1993). Chomyn et al. (1991) used human cell lines completely devoid of mtDNA (rho° cells) and repopulated them with exogenous mitochondria containing wild-type or mutant mitochondrial genomes; they found a severe defect of protein synthesis in the myoblasts from a patient with MERRF and in the rho° cells repopulated with mitochondria from the patient's myoblasts. The point mutation was present in all tissues of a patient studied postmortem (Zeviani et al. 1991a) and was detected in blood cells from 5 of 5 patients (Hammans et al. 1991), suggesting that the diagnosis can be established in easily accessible tissues. There is no specific therapy, but CoQ_{10} administration seemed to benefit a mother and daughter with the MERRF mutation (Wallace et al. 1991).

The *MELAS* syndrome is characterized primarily by the stroke-like episodes that begin before age 40. In the absence of strokes, the syndrome would be difficult to diagnose clinically. As with MERRF, the identification of a highly specific, though not exclusive, point mutation at nt 3243 in the tRNA[Leu(UUR)] gene has provided an indispensable diagnostic tool. When we defined MELAS on the basis of: (1) stroke-like episodes (with CT or MRI evidence of focal brain lesions); (2) lactic acidosis, RRF, or both; and (3) at least two of the following: focal or generalized seizures, dementia, recurrent headache and vomiting, we found that

21 of 23 patients with MELAS had the mutation (Ciafaloni et al. 1992). We also found the mutation in 11 of 11 oligosymptomatic and 12 of 14 asymptomatic maternal relatives. Onset was before age 15 in 62% of our MELAS patients. Hemianopia or cortical blindness was the most common manifestation, and the first stroke-like episode occurred before age 40 in all but 4 patients. Seizures were present in all cases and lactic acidosis in all but one (Ciafaloni et al. 1992). CSF protein was increased in about half of the patients in our series, but exceeded 100 mg/dl only in one case. Basal ganglia calcification was seen in 6 of 22 patients, and cardiopathy in four. Although muscle biopsy usually shows RRF, these were lacking in three patients of ours (Ciafaloni et al. 1992) and in two of Hammans et al. (1991). Abnormal accumulations of mitochondria have been described in smooth muscle cells of intramuscular vessels, whose profiles stand out with the SDH stain (Sakuta and Nonaka 1989; Hasegawa et al. 1991). Biochemical studies of muscle had shown complex I deficiency in many patients (Kobayashi et al. 1987; Nishizawa et al. 1987; Ichiki et al. 1988, 1989; Koga et al. 1988). We found multiple defects of the respiratory chain affecting complexes I, III, and especially IV (Ciafaloni et al. 1992). The point mutation was found in all tissues in two postmortem studies (Ciafaloni et al. 1991, 1992). The number of mutant genomes is lower in blood than in muscle and may not be detectable in some patients. Thus, screening of blood cells for the MELAS mutation is useful but not absolutely reliable.

Five patients with typical MELAS in Goto's series (Goto et al. 1990) and two in ours (Ciafaloni et al. 1992) did not have the mutation at nt 3243, suggesting genetic heterogeneity. At least one other mutation, also affecting the tRNA$^{Leu(UUR)}$ gene but at nt 3271, has been found (Goto et al. 1991). In patients with the full syndrome, the prognosis is dismal. Therapeutic trials have included corticosteroids and CoQ$_{10}$, but results have been inconsistent.

The *ATPase 6 mutation syndrome* is a maternally inherited multisystem disorder that was first described by Holt et al. (1990), and is characterized by retinitis pigmentosa, dementia, seizures, ataxia, proximal limb weakness, and sensory neuropathy. There were no RRF in the muscle biopsy. A heteroplasmic point mutation was identified at nt 8993 within the ATPase 6 gene. In a new family with the same mutation, an infant presented with the clinical and neuropathological features of Leigh's syndrome (Tatuch et al. 1992), adding to the genetic and biochemical heterogeneity of this syndrome, in which inheritance may be autosomal-recessive (as in COX deficiency), X-linked (as in some cases of PDHC deficiency), or maternal.

Myopathy and cardiopathy. A third heteroplasmic point mutation in the tRNA$^{Leu(UUR)}$ gene, at nt 3260, was identified by Zeviani et al. (1991b) in a family with maternally inherited myopathy, cardiopathy, or a combination of myopathy and cardiopathy. Serum CK activity and lactate levels were increased, and muscle biopsies showed RRF. Biochemical analysis showed combined defects of complex I and IV. It is not clear how different mutations in the same tRNA gene can cause clinical phenotypes as strikingly different as MELAS on the one hand and myopathy with cardiopathy on the other.

3. Defects of communication between nuclear and mitochondrial genome. Multiple mtDNA deletions was first described by Zeviani et al. (1989) in an Italian family with autosomal-dominant transmission of PEO, exercise intolerance, weakness of proximal limb and respiratory muscles, cataracts, and early death. Southern analysis showed several populations of partially deleted mtDNA instead of the single population seen in KSS and related diseases. Laboratory examination showed lactic acidosis and RRF. Several other families have been described: all had autosomal-dominantly inherited PEO, but associated symptoms and signs varied, including hearing loss, tremor, ataxia, peripheral neuropathy (Zeviani et al. 1990a), ataxia, mental retardation, hypoparathyroidism (Cormier et al. 1991), nystagmus, and abnormal EEG (Iannaccone et al. 1974).

Clinical and family histories were different in two sets of Japanese brothers. One set of brothers had ptosis but no ophthalmoplegia, myopathy, optic atrophy, and peripheral neuropathy (Yuzaki et al. 1989). The other two brothers had no involvement of extraocular muscles but suffered from recurrent myoglobinuria triggered by exercise, fasting, or alcohol intake (Ohno et al. 1991).

Transmission appeared to be autosomal-recessive, as parents were clinically unaffected in both families.

Depletion of mtDNA is the first hereditary human disorder characterized by a quantitative rather than qualitative abnormality of mtDNA (Moraes et al. 1991b). *Severe* depletion (2% to 17% residual mtDNA) was observed initially in four infants, two cousins, one of whom had died of a mitochondrial myopathy and the other of a mitochondrial hepatopathy (Boustany et al. 1983), and two unrelated children with myopathy and renal dysfunction. All patients with myopathy developed symptoms in the first weeks of life and died between 3 and 9 months of age. They had severe lactic acidosis and abundant RRF. The patient with liver disease died at 9 months of age of hepatic failure and postmortem examination of the liver showed abnormal mitochondria. Family history suggested autosomal-recessive inheritance; all parents were asymptomatic and there was no evidence of maternal transmission. In each case, mtDNA depletion was observed only in affected (muscle, liver, kidney), but not in unaffected tissues (Moraes et al. 1991b). *Partial* mtDNA depletion (8% to 36% residual mtDNA) was observed in four children with myopathy; onset was later, between 5 and 12 months, and the course was slower, but two children died at 11 months and 3 years, and one was tetraplegic and ventilator-dependent at 3.5 years (Tritschler et al. 1992). Blood lactates were normal. Muscle biopsies showed RRF, but a biopsy obtained early in the course of the disease had only non-specific myopathic features. Two of the four patients were siblings and all parents were asymptomatic.

In both severe and partial mtDNA depletion, biochemical studies showed combined defects of respiratory chain complexes, and the severity of the defects agreed with the degree of mtDNA depletion. The decrease of mtDNA was documented by densitometry of Southern blots (using nuclear 18S ribosomal DNA as normalizing control), and confirmed by immunocytochemistry with anti-DNA antibodies, and by in situ hybridization (Moraes et al. 1991b; Tritschler et al. 1992). As predicted by the scarcity of mtDNA and suggested by the biochemical results, immunocytochemistry using subunit-specific antibodies showed that mtDNA- encoded subunits were absent or markedly reduced, while nDNA-encoded subunits were present. The apparently mendelian inheritance of mtDNA depletion suggests that the genetic defect may involve a nuclear gene controlling mtDNA replication, but the tissue-specific expression of the depletion is difficult to explain (Moraes et al. 1991b).

A *secondary*, iatrogenic form of mtDNA depletion (Arnaudo et al. 1991) is probably the cause of the mitochondrial myopathy described in patients with AIDS after prolonged treatment with zidovudine (AZT) (Dalakas et al. 1990; Mhiri et al. 1991). This was suggested by Southern analysis showing up to 78% depletion of muscle mtDNA in treated but not in untreated AIDS patients, and by the finding that the depletion was reversible after discontinuation of AZT. The effect of the drug is probably due to the fact that AZT is incorporated into mtDNA by the gamma polymerase, and this inhibits mtDNA replication (Simpson et al. 1989).

MYOADENYLATE DEAMINASE DEFICIENCY

Clinical presentation. First described in 1978 by Fishbein and coworkers in five patients with exercise-related myalgia and cramps, myoadenylate deaminase (mAMPD) deficiency seems to be a common metabolic defect, detected in 1% to 3% of all muscle biopsies (Sabina 1993). However, the pathogenetic significance of this enzyme defect has been questioned for two main reasons: (1) in patients with otherwise unexplained myopathy, symptoms are generally mild, ill-defined, and often subjective (the 'aches, pains, and cramps' syndrome); and (2) over 50% of the patients had either well-defined myopathies, such as Duchenne or Becker muscular dystrophy or McArdle's disease, or other neuromuscular disorders, such as spinal muscular atrophy or amyotrophic lateral sclerosis. Fishbein (1985) distinguished two forms of mAMPD deficiency: (1) primary (hereditary) mAMPD deficiency, characterized clinically by myopathy alone, with exercise intolerance, myalgia, and cramps, and biochemically by negligible (less than 1% of normal) residual activity and lack of CRM in muscle; (2) secondary (acquired) mAMPD deficiency, associated with well-defined myopathies or neuromuscular disorders; muscle biopsies had higher residual activity and detectable CRM. Although this distinction lacks solid biochemical or molecular genetic criteria, nonetheless our experience based on biochemical analyses of hundreds of biopsies suggests that mAMPD deficiency is often the only enzyme defect found in muscle from patients with exercise-related cramps and pains and normal morphology. We also found an apparently isolated defect of mAMPD in three patients with recurrent myoglobinuria (Tonin et al. 1990). Although the cause of myoglobinuria in these patients may be some other coincidental and undisclosed metabolic defect, we cannot exclude a role of mAMPD deficiency. The considerations that follow regard patients with the 'primary' form of mAMPD deficiency.

Laboratory tests. Resting serum CK activity is usually normal, but variably increased values may follow activity. The forearm ischemic exercise produces a normal rise of venous lactate but no rise in the products of the mAMPD reaction, ammonia and inosine monophosphate (IMP, detected in blood as its diffusible catabolites, inosine and hypoxanthine). EMG may be normal or may show non-specific myopathic features.

Pathology. Muscle biopsy is usually normal. There is a simple histochemical stain for mAMPD, based on the pH change that accompanies the liberation of ammonia (Fishbein et al. 1980).

Biochemistry and molecular genetics. There are three tissue-specific subunits, M (muscle), L (liver), and E (erythrocyte), under separate genetic control, that form homo- and hetero-tetramers. Adult human muscle contains almost exclusively the M homotetramer. AMPD converts AMP to IMP and ammonia and is one of the three enzymes of the purine nucleotide cycle. Five functions have been attributed to mAMPD: (1) to stabilize the energy potential when ATP depletion displaces the myokinase reaction ($2ADP \leftrightarrow ATP+AMP$) toward the formation of ATP, with accumulation of AMP; by removing AMP, the mAMPD reaction reduces the adenylate pool but increases the energy state; (2) to activate glycolysis through the action of IMP on phosphorylase and of ammonia on PFK; (3) to buffer the lactate produced during exercise; (4) to provide Krebs cycle intermediates, such as fumarate, that is formed in the anabolic phase of the purine nucleotide cycle; (5) to 'trap' and keep within the cell nucleotides, such as the relatively non-diffusible IMP. However, the functional significance of each of these roles during exercise in vivo has been questioned (Sabina 1993). The physiological consequences of mAMPD deficiency in exercise studies of patients have also been contradictory (Sabina 1993).

The cDNA and the gene for mAMPD have been cloned and sequenced; the gene has been localized to the short arm of chromosome 1 (Sabina et al. 1990). Southern analysis in 10 patients failed to show any major alterations, and Northern analysis of muscle biopsies from nine patients showed normal amounts of normal-size mRNA, suggesting point mutations as the underlying molecular defects (Sabina et al. 1992).

Note added in proof

A homozygous point mutation in the gene for CPTII (a C-to-T transition at nt 1992, resulting in the substitution of an arginine with a cysteine at position 631) has been described in a child with hypoketotic hypoglycemia and cardiomyopathy (Taroni et al. 1992). The same mutation was present in three patients with the typical myopatic form of CPTII deficiency, who were compound heterozygotes for this and another, as yet unknown, mutation.

Acknowledgements

Part of the work described here was supported by Center Grant NS 11766 from the National Institute for Neurological Diseases and Stroke, by a grant from the Muscular Dystrophy Association, and by a donation from Libero and Graziella Danesi (Milan, Italy). We thank Dr. Lewis P. Rowland for his support and for reviewing the manuscript.

REFERENCES

ABARBANEL, J.M., N. BASHAN, R. POTASHNIK, A. OSIMANI, S.W. MOSES and Y. HERISHANU: Adult muscle phosphorylase b kinase deficiency. Neurology 36 (1986) 560–562.

AGAMANOLIS, D.P., A.D. ASKARI, S. DI MAURO, A.P. HAYS, K. KUMAR, M. LIPTON and A. RAYNOR: Muscle phosphofructokinase deficiency: two cases with unusual polysaccharide accumulation and immunologically active enzyme protein. Muscle Nerve 3 (1980) 456–467.

AMENDT, B.A., C. GREEN, L. SWEETMAN, J. CLOHERTY, V. SHIH, A. MOON, L. TEEL and W.J. RHEAD: Short-chain acyl-CoA dehydrogenase deficiency: clinical and biochemical studies in two patients. J. Clin. Invest. 79 (1987) 1303–1309.

ANDERSON, S., A.T. BANKIER, B.G. BARREL, M.H.L. DE BRUIJN, A.R. COULSON, J. DROUIN, I.C. EPERON, D.P. NIERLICH, B.A. ROE, F. SANGER, P.H. SCHREIER, A.J.H. SMITH, R. STADEN and I.G. YOUNG: Sequence and organization of the human mitochondrial genome. Nature (London) 290 (1981) 457–465.

ANGELINI, C., M. PHILIPPART, C. BORRONE, N. BRESOLIN, M. CANTINI and S. LUCKE: Multisystem triglyceride storage disorder with impaired long-chain fatty acid oxidation. Ann. Neurol. 7 (1980) 5–10.

ANGELINI, C., A. MARTINUZZI and L. VERGANI: Treatment with L-carnitine of the infantile and adult form of primary carnitine deficiency. In: R. Ferrari, S. DiMauro and G. Sherwood (Eds.), L-Carnitine and Its Role in Medicine: From Function to Therapy. London, Academic Press (1992) 139–153.

ARGOV, Z. and W.J. BANK: Phosphorus magnetic resonance spectroscopy (^{31}P MRS) in neuromuscular disorders. Ann. Neurol. 30 (1991) 90–97.

ARGOV, Z., W.J. BANK, J. MARIS, S. ELEFF, N.G. KENNAWAY, R.E. OLSON and B. CHANCE: Treatment of mitochondrial myopathy due to complex III deficiency with vitamins K_3 and C: a ^{31}P-NMR follow-up study. Ann. Neurol. 19 (1986) 598–602.

ARGOV, Z., W.J. BANK, B. BODEN, Y-I RO and B. CHANCE: Phosphorus magnetic resonance spectroscopy of partially blocked muscle glycolysis. Arch. Neurol. 44 (1987) 614–617.

ARNAUDO, E., M. DALAKAS, S. SHANSKE, C.T. MORAES, S. DIMAURO and E.A. SCHON: Depletion of muscle mitochondrial DNA in AIDS patients with zidovudine-induced myopathy. Lancet 337 (1991) 508–510.

ARTS, W.F.M., H.R. SCHOLTE, J.M. BOGAARD, K.F. KERREBIJN and I.E.M. LUYT-HOUWEN: NADH-CoQ reductase deficient myopathy: successful treatment with riboflavin. Lancet ii (1983) 581–582.

BARDOSI, A., W. CREUTZFELDT, S. DI MAURO, K. FELGENHAUER, R.L. FRIEDE, H.H. GOEBEL, A. KOHLSCHUTTER, G. MEYER, G. RAHLF, S. SERVIDEI, G. VAN LESSEN and T. WETTERLING: Myo-, Neuro-, Gastro-Intestinal Encephalopathy (MNGIE syndrome) due to the partial deficiency of cytochrome c oxidase. Acta Neuropathol. 74 (1987) 248–258.

BELL, R.B., A.K.W. BROWNELL, C.R. ROE, A.G. ENGEL, S.I. GOODMAN, F.E. FRERMAN, D.W. SECCOMBE and F.F. SNYDER: Electron transfer flavoprotein: ubiquinone oxidoreductase (ETF:QO) deficiency in an adult. Neurology 40 (1990) 1779–1782.

BERATIS, N.G., G.U. LABADIE and K. HIRSCHHORN: Characterization of the molecular defect in infantile and adult acid alpha-glucosidase deficiency fibroblasts. J. Clin. Invest. 62 (1978) 1264–1274.

BERGIA, B., H.D. SYBERS and I.J. BUTLER: Familial lethal cardiomyopathy with mental retardation and scap-

uloperoneal muscular dystrophy. J. Neurol. Neurosurg. Psychiatry 49 (1986) 1423–1426.

BERKOVIC, S.F., S. CARPENTER, A. EVANS, G. KARPATI, E.A. SHOUBRIDGE, F. ANDERMANN, E. MEYER, J.L. TYLER, M. DIKSIC, D. ARNOLD, L.S. WOLFE, E. ANDERMANN and A.M. HAKIM: Myoclonus epilepsy and ragged-red fibers (MERRF). 1. A clinical, pathological, biochemical, magnetic resonance spectrographic and positron emission tomographic study. Brain 112 (1989) 1231–1260.

BERTINI, E., M. SABATELLI, B. GARAVAGLIA, A.B. BURLINA, M. RIMOLDI, E. RICCI, C. DIONISI-VICI, A. BARTULI, G. SABETTA and S. DI DONATO: Myopathy and sensorimotor polyneuropathy in long-chain 3-OH acylCoA dehydrogenase deficiency. J. Neurol. Sci. 98 (1990) 273.

BET, L., N. BRESOLIN, M. MOGGIO, G. MEOLA, A. PRELLE, A.H. SCHAPIRA, T. BINZONI, A. CHOMYN, F. FORTUNATO, P. CERRETELLI and G. SCARLATO: A case of mitochondrial myopathy, lactic acidosis and complex I deficiency. J Neurol. 237 (1990) 399–404.

BIRCH-MACHIN, M.A., I.M. SHEPHERD, N.J. WATMOUGH, H.S.A. SHERRATT, K. BARTLETT, V.M. DARLEY-USMAR, D.W.A. MILLIGAN, R.J. WELCH, A. AYNSLEY-GREEN and D.M. TURNBULL: Fatal lactic acidosis in infancy with a defect of complex III of the respiratory chain. Pediatr. Res. 25 (1989) 553–559.

BLAKE, D., A. LOMBES, C. MINETTI, E. BONILLA, A.P. HAYS, R.E. LOVELACE, I. BUTLER and S. DIMAURO: MNGIE syndrome: report of 2 new patients. Neurology 40 (1990) 294.

BLASS, J.P., J. AVIGAN and B.W. UHLENDORF: A defect in pyruvate decarboxylase in a child with intermittent movement disorder. J. Clin. Invest. 49 (1970) 423–432.

BONILLA, E. and D.L. SCHOTLAND: Histochemical diagnosis of muscle phosphofructokinase deficiency. Arch. Neurol. 22 (1970) 8–12.

BOONE, C.M., T.R. CHEN and F.H. RUDDLE: Assignment of three human genes to chromosomes (LDH-A to 11, TK to 17, and IDH to 20) and evidence for translocation between human and mouse chromosomes in somatic cell hybrids. Proc. Natl. Acad. Sci. (USA) 69 (1972) 510–514.

BOUGNERES, P.F., J-M SAUDUBRAY, C. MARSAC, O. BERNARD, M. ODIEVRE and J. GIRAND: Fasting hypoglycemia resulting from hepatic carnitine palmitoyltransferase deficiency. J. Pediatr. 98 (1981) 742–746.

BOUSTANY, R.M., J.R. APRILLE, J. HALPERIN, H. LEVY and G.R. DE LONG: Mitochondrial cytochrome deficiency presenting as a myopathy with hypotonia, external ophthalmoplegia, and lactic acidosis in an infant and as fatal hepatopathy in a second cousin. Ann. Neurol. 14 (1983) 462–470.

BRAAKHEKKE, J.P., M.I. DE BRUIN, D.F. STEGMAN, R.A. WEVERS, R.A. BINKHORST and E.M.G. JOOSTEN: The second wind phenomenon in McArdle's disease. Brain 109 (1986) 1087–1101.

BRADLEY, W.G., P. HUDGSON, D. GARDNER-MEDWIN and J.N. WALTON: Myopathy associated with abnormal lipid metabolism in skeletal muscle. Lancet i (1969) 495–498.

BRADLEY, W.G., M. JENKISON, D.C. PARK, P. HUDGSON, D. GARDNER-MEDWIN, R.J.T. PENNINGTON and J.N. WALTON: A myopathy associated with lipid storage. J. Neurol. Sci. 16 (1972) 137–154.

BRAUN, N., W. MARINO, T. JACOBS, R. SCHOENFELDT, A. TAHMOUSH, S. SHANSKE and S. DIMAURO: Therapeutic trial of thyroid hormone in patients with acid maltase deficient myopathy: a preliminary report (abstract). 5th International Congress on Neuromuscular Diseases, Marseilles (1982).

BRESOLIN, N., A.F. MIRANDA, M.P. JACOBSON, J.H. LEE, T. CAPILUPI and S. DIMAURO: Phosphorylase isoenzymes of human brain. Neurochem. Path. 1 (1983a) 171–178.

BRESOLIN, N., Y. RO, M. REYES, A.F. MIRANDA and S. DIMAURO: Muscle phosphoglycerate mutase (PGAM) deficiency: a second case. Neurology 33 (1983b) 1049–1053.

BRESOLIN, N., M. ZEVIANI, E. BONILLA, R.H. MILLER, R.W. LEECH, S. SHANSKE, M. NAKAGAWA and S. DIMAURO: Fatal infantile cytochrome c oxidase deficiency: decrease of immunologically detectable enzyme in muscle. Neurology 35 (1985) 828–833.

BROWN, B.I.: Debranching and branching enzyme deficiencies. In: A.G. Engel and B.Q. Banker (Eds.), Myology. New York, McGraw-Hill (1986) 1653–1661.

BROWN, B.I. and D.H. BROWN: Branching enzyme activity of cultured amniocytes and chorionic villi; prenatal testing for type IV glycogen storage disease. Am. J. Hum. Genet. 44 (1989) 378–381.

BRUNBERG, J.A., W.F. MCCORMICK and S.S. SCHOCHET: Type III glycogenosis: an adult with diffuse weakness and muscle wasting. Arch. Neurol. 25 (1971) 171–178.

BRUNO, C., S. SERVIDEI, S. SHANSKE, G. KARPATI, S. CARPENTER, D. MCKEE, R. BAROHN, M. HIRANO, Z. RIFAI, and S. DIMAURO: Glycogen branching enzyme in adult polyglucosan body disease. Ann. Neurol. (1992) in press.

BRYAN, W., S.F. LEWIS, L. BERTOCCI, M. GUNDER, K. AYYAD, P. GUSTAFSON and R.G. HALLER: Muscle lactate dehydrogenase derficiency: a disorder of anaerobic glycogenolysis associated with exertional myoglobinuria. Neurology 40 (1990) 203.

BYRNE, E., X. DENNETT, B. CROTTY, I. TROUNCE, J.M. SANDS, R. HAWKINS, J. HAMMOND, S. ANDERSON, E.A. HAAN and A. POLLARD: Dominantly inherited cardioskeletal myopathy with lysosomal glycogen storage and normal acid maltase levels. Brain 109 (1986) 523–536.

CAFFERTY, M.S., R.E. LOVELACE, A.P. HAYS, S. SERVIDEI, S. DIMAURO and L.P. ROWLAND: Polyglucosan body disease. Muscle Nerve 14 (1991) 102–107.

CARRIER, H., I. MAIRE, C. VIAL, G. RAMBAUD, F. FLOCARD and A. FLECHAIRE: Myopathic evolution of an exertional muscle pain syndrome with phosphorylase b kinase deficiency. Acta Neuropathol. 81 (1990) 84–88.

CARROLL, J.E., J.B. SHUMATE, M.H. BROOKE and J.M. HAGBERG: Riboflavin-responsive lipid myopathy and carnitine deficiency. Neurology 31 (1981) 1557–1559.

CAUDWELL, F.B. and P. COHEN: Purification and subunit structure of glycogen-branching enzyme from rabbit skeletal muscle. Eur. J. Biochem. 109 (1980) 391–394.

CHANARIN, L., A. PATEL, G. SLAVIN, E.J. WILLS, T.M. ANDREWS and G. STEWART: Neutral-lipid storage disease: a new disorder of lipid metabolism. Br. Med. J. 1 (1975) 553–555.

CHEN, Y-T., J-K., HE, J-H. DING and B.I. BROWN: Glycogen debranching enzyme: purification, antibody characterization, and immunoblot analyses of type III glycogen storage disease. Am. J. Hum. Genet. 41 (1987) 1002–1015.

CHOMYN, A., G. MEOLA, N. BRESOLIN, S.T. LAI, G. SCARLATO and G. ATTARDI: In vitro genetic transfer of protein synthesis and respiration defects to mitochondrial DNA-less cells with myopathy-patient mitochondria. Mol. Cell Biol. 11 (1991) 2236–2244.

CIAFALONI, E., E. RICCI, S. SERVIDEI, S. SHANSKE, G. SILVESTRI, G. MANFREDI, E.A. SCHON and S. DIMAURO: Wide-spread tissue distribution of a tRNA$^{Leu(UUR)}$ mutation in the mitochondrial DNA of a patient with MELAS syndrome. Neurology 41 (1991) 1663–1665.

CIAFALONI, E., E. RICCI, S. SHANSKE, C.T. MORAES, G. SILVESTRI, M. HIRANO, S. SIMONETTI, C. ANGELINI, A. DONATI, C. GARCIA, A. MARTINUZZI, R. MOSEWICH, S. SERVIDEI, E. ZAMMARCHI, E. BONILLA, D.C. DE VIVO, L.P. ROWLAND, E.A. SCHON and S. DIMAURO: MELAS: Clinical features, biochemistry, and molecular genetics. Ann. Neurol. 31 (1992) 391–398.

CLARK, J.B., D.J. HAYES, E. BYRNE and J.A. MORGAN-HUGHES: Mitochondrial myopathies: defects in mitochondrial metabolism in human skeletal muscle. Biochem. Soc. Trans. 11 (1983) 626–627.

CLEMENS, P.R., M. YAMAMOTO and A.G. ENGEL: Adult phosphorylase b kinase deficiency. Ann. Neurol. 28 (1990) 529–538.

COATES, P.A., D.E. HALE, G. FINOCCHIARO, K. TANAKA and S.C. WINTER: Genetic deficiency of short-chain acyl-CoA dehydrogenase in cultured fibroblasts from a patient with muscle carnitine deficiency and severe skeletal muscle weakness. J. Clin. Invest. 81 (1988) 171–175.

CORI, G.T.: Glycogen structure and enzyme deficiencies in glycogen storage disease. Harvey Lect. 48 (1954) 145–171.

CORMIER, V., A. ROTIG, M. TARDIEU M. COLONNA, J-M SAUDUBRAY and A. MUNNICH: Autosomal dominant deletions of the mitochondrial genome in a case of progressive encephalomyopathy. Am. J. Hum. Genet. 48 (1991) 643–648.

CORNELIO, F., N. BRESOLIN, S. DIMAURO, M. MORA and M.R. BALESTRINI: Congenital myopathy due to phosphorylase deficiency. Neurology 33 (1983) 1383–1385.

CORNELIO, F., N. BRESOLIN, P.A. SINGER, S. DIMAURO and

L.P. ROWLAND: The clinical varieties of neuromuscular disease in debrancher deficiency. Arch. Neurol. 41 (1984) 1027–1932.

CRERAR, M.M., J.W. HUDSON, K.E. MATTHEWS, E.S. DAVID and G.B. GOLDING: Studies on the expression and evolution of the glycogen phosphorylase gene family in the rat. Genome 30 (1988) 582–590.

DALAKAS, M., I. ILLA and G.H. PEZESHKPOUR: Mitochondrial myopathy caused by long-term AZT (zidovudine) therapy: management and differences from HIV-associated myopathy. N. Engl. J. Med. 322 (1990) 1098–1105.

DANON, M.J., S. CARPENTER, J.R. MANALIGOD and L.H. SCHLISELFELD: Fatal infantile glycogen storage disease: deficiency of phosphofructokinase and phosphorylase b kinase. Neurology 31 (1981a) 1303–1307.

DANON, M.J., S.J. OH, S. DIMAURO, J.R. MANALIGOD, A. EASTWOOD, S. NAIDU and L.H. SCHLISELFELD: Lysosomal glycogen storage disease with normal acid maltase. Neurology 31 (1981b) 51–57.

DANON, M.J., S. SERVIDEI, S. DIMAURO and S. VORA: Late-onset muscle phosphofructokinase deficiency. Neurology 38 (1988) 956–960.

DARLEY-USMAR, V.M., N.G. KENNAWAY, N.R. BUIST and R.A. CAPALDI: Deficiency in ubiquinone cytochrome c reductase in a patient with mitochondrial myopathy and lactic acidosis. Proc. Natl. Acad. Sci. (USA) 80 (1983) 5103–5106.

DARLEY-USMAR, M.M., M. WATANABE, Y. UCHIYAMA, I. KONDO, N.G. KENNAWAY, L. GRONKE and H. HAMAGUCHI: Mitochondrial myopathy: tissue-specific expression of a defect in ubiquinol-cytochrome c reductase. Clin. Chim. Acta 158 (1986) 253–261.

DAVIDSON, J.J., T. OZCELIK, C. HAMACHER, P.J. WILLEMS, U. FRANCKE and M.W. KILIMAN: cDNA cloning of a liver isoform of the phosphorylase kinase alpha subunit and mapping of the gene to Xp 22.2–p22.1, the region of human X-linked liver glycogenosis. Proc. Natl. Acad. Sci. USA 89 (1992) 2096–2106.

DEMAUGRE, F., J.P. BONNEFONT, G. MITCHELL, N. NGUYEN-HOANG, A. PELET, M. RIMOLDI, S. DIDONATO and J-M SAUDUBRAY: Hepatic and muscular presentations of carnitine palmitoyltransferase deficiency: two distinct entities. Pediatr. Res. 24 (1988) 308–311.

DE VISSER, M., H.R. SCHOLTE, R.B.H. SCHUTGENS, P.A. BOLHUIS, I.E.M. LUYT-HOUWEN, M.H.M. VAANDRAGER-VERDUIN, H.A. VEDER and P.L. OEY: Riboflavin-responsive lipid-storage myopathy and glutaric aciduria type II of early adult onset. Neurology 36 (1986) 367–372.

DIDONATO, S., F.E. FRERMAN, M. RIMOLDI, P. RINALDO, F. TARONI and U.N. WIESMANN: Systemic carnitine deficiency due to lack of electron transfer flavoprotein: ubiquinone oxidoreductase. Neurology 36 (1986) 957–963.

DIDONATO, S., B. GARAVAGLIA, P. STRISCIUGLIO, C. BORRONE and G. ANDRIA: Multisystem triglyceride storage disease is due to a specific defect in the degradation of endocellularly synthesized triglycerides. Neurology 38 (1988) 1107–1110.

DIDONATO, S., C. GELLERA, D. PELUCHETTI, G. UZIEL, A. ANTONELLI, G. LUS and M. RIMOLDI: Normalization of short-chain acylcoenzyme A dehydrogenase after riboflavin treatment in a girl with multiple acylcoenzyme A dehydrogenase-deficient myopathy. Ann. Neurol. 25 (1989) 479–484.

DIDONATO, S., B. GARAVAGLIA, M. RIMOLDI and F. CARRARA: Clinical and biomedical phenotypes of carnitine deficiencies. In: R. Ferrari, S. DiMauro and G. Sherwood (Eds.), L.-Carnitine and Its Role in Medicine: From Function to Therapy. London, Academic Press (1992) 81–98.

DIMAURO, S.: Mitochondrial encephalomyopathies. In: R.N. Rosenberg (Ed.), Molecular and Genetic Basis of Neurological Disease. Stoneham, CT Butterworth (1993) 665–694.

DIMAURO, S. and N. BRESOLIN: Phosphorylase deficiency. In: A.G. Engel and B.Q. Banker (Eds.), Myology. New York, McGraw-Hill (1986a) 1585–1601.

DIMAURO, S. and N. BRESOLIN: Newly recognized defects of terminal glycolysis. In: A.G. Engel and B.Q. Banker (Eds.), Myology. New York, McGraw-Hill (1986b) 1619–1628.

DIMAURO, S. and P. HARTLAGE: Fatal infantile form of muscle phosphorylase deficiency. Neurology 28 (1978) 1124–1129.

DIMAURO, S. and P. MELIS-DIMAURO: Muscle carnitine palmityltransferase deficiency and myoglobinuria. Science 182 (1973) 929–931.

DIMAURO, S. and A. PAPADIMITRIOU: Carnitine palmitoyltransferase (CPT) deficiency. In: A.G. Engel and B.Q. Banker (Eds.), Myology. New York, McGraw-Hill (1986) 1697–1708.

DIMAURO, S. and D.C. DEVIVO: Diseases of carbohydrate, fatty acid, and mitochondrial metabolism. In: G.J. Siegel, B.W. Agranoff, R.W. Albers and P.B. Molinoff (Eds.), Basic Neurochemistry. New York, Raven Press (1989) 647–670.

DIMAURO, S., E. BONILLA, C.P. LEE, D.L. SCHOTLAND, A. SCARPA, H. CONN and B. CHANCE: Luft's disease: further biochemical and ultrastructural studies of skeletal muscle in the second case. J. Neurol. Sci. 27 (1976) 217–232.

DIMAURO, S., S. ARNOLD, A.F. MIRANDA and L.P. ROWLAND: McArdle disease: the mystery of reappearing phosphorylase activity in muscle culture: a fetal isoenzyme. Ann. Neurol. 3 (1978) 60–66.

DIMAURO, S., G.B. HARTWIG, A.P. HAYS, A.B. EASTWOOD, R. FRANCO, M. OLARTE, M. CHANG, A.D. ROSES, M. FETELL, R.S. SCHOENFELDT and L.Z. STERN: Debrancher deficiency: neuromuscular disorder in five adults. Ann. Neurol. 5 (1979) 422–436.

DIMAURO, S., M. DALAKAS and A.F. MIRANDA: Phosphoglycerate kinase deficiency: a new cause of recurrent myoglobinuria. Trans. Am. Neurol. Assoc. 106 (1981a) 202–205.

DIMAURO, S., A.F. MIRANDA, S. KHAN, K. GITLIN and R.

FRIEDMAN: Human muscle phosphoglycerate mutase deficiency: a newly discovered metabolic myopathy. Science 212 (1981b) 1277–1279.

DIMAURO, S., A.F. MIRANDA, M. OLARTE, R. FRIEDMAN and A.P. HAYS: Muscle phosphoglycerate mutase deficiency. Neurology 32 (1982) 548–591.

DIMAURO, S., M. DALAKAS and A.F. MIRANDA: Phosphoglycerate kinase deficiency: another cause of recurrent myoglobinuria. Ann. Neurol. 13 (1983a) 11–19.

DIMAURO, S., J.F. NICHOLSON, A.P. HAYS, A.B. EASTWOOD, A. PAPADIMITRIOU, R. KOENIGSBERGER and D.C. DE VIVO: Benign infantile mitochondrial myopathy due to reversible cytochrome c oxidase deficiency. Ann. Neurol. 14 (1983b) 226–234.

DIMAURO, S., E. BONILLA, M. ZEVIANI, M. NAKAGAWA and D.C. DE VIVO: Mitochondrial myopathies. Ann. Neurol. 17 (1985) 521–538.

DIMAURO, S., A.F. MIRANDA, S. SAKODA, E.A. SCHON, S. SERVIDEI, S. SHANSKE and M. ZEVIANI: Metabolic myopathies. Am. J. Med. Genet. 25 (1986) 635–651.

DIMAURO, S., E. BONILLA, M. ZEVIANI, S. SERVIDEI, D.C. DE VIVO and E.A. SCHON: Mitochondrial myopathies. J. Inherit. Metab. Dis. 10 (1987) 113–128.

DIMAURO, S., A. LOMBES, H. NAKASE, S. MITA, G.M. FABRIZI, H.-J. TRITSCHLER, E. BONILLA, A.F. MIRANDA, D.C. DE VIVO and E.A. SCHON: Cytochrome c oxidase deficiency. Pediatr. Res. 28 (1990a) 526–541.

DIMAURO, S., E. BONILLA, A. LOMBES, S. SHANSKE, C. MINETTI and C.T. MORAES: Mitochondrial encephalomyopathies. Neurol. Clinics 8 (1990b) 483–506.

DING, J-H., D.A. HARRIS, B-Z. YANG and Y-T. CHEN: Cloning of cDNA for human glycogen debrancher, the enzyme deficient in type III glycogen storage disease (GSD III). Pediatr. Res. 25 (1989) 140A.

DING, J-H., TH. DE BARSY, B.I. BROWN, R.A. COLEMAN and Y-T. CHEN: Immunoblot analyses of glycogen debranching enzyme in different subtypes of glycogen storage disease type III. J. Pediatr. 116 (1990) 95–100.

DI SANT'AGNESE, P.A., D.H. ANDERSEN, H.H. MASON and W.A. BAUMAN: Glycogen storage disease of the heart. I. Report of two cases in siblings with chemical and pathological studies. Pediatrics 6 (1950) 402–424.

DUBOC, D., P. JEHENSON, S.T. DINH, C. MARSAC, A. SYROTAZ and M. FARDEAU: Phosphorus NMR spectroscopy study of muscular enzyme deficiencies involving glycogenolysis and glycolysis. Neurology 37 (1987) 663–671.

DUSHEIKO, G., M.G. KEW, B.I. JOFFE, J.R. LEWIN, S. MANTAGOS and K. TANAKA: Recurrent hypoglycemia associated with glutaric aciduria type II in an adult. N. Engl. J. Med. 301 (1979) 1405–1409.

DWORZAK, F., F. CASAZZA, L. MORANDI, M. MORA, F. CARRARA and F. CORNELIO: Heart transplant in a patient with lysosomal glycogen storage disease with normal acid maltase. Neurology 41 (1991) 421.

EDWARDS. Y.H., S. SAKODA, E.A. SCHON and S. POVEY: The gene for human muscle-specific phosphoglycerate mutase, PGAMM, mapped to chromosome 7

by polymerase chain reaction. Genomics 5 (1989) 948–951.

EGGER, J., B.N. HARDING, S.G. BOYD, J. WILSON and M. ERDOHAZI: Progressive neuronal degeneration of childhood (PNDC) with liver disease. Clin. Pediatr. 26 (1987) 167–173.

ELEFF, S., N.G. KENNAWAY, N.R. BUIST, V.M. DARLEY-USMAR, R.A. CAPALDI, W.J. BANK and B. CHANCE: ^{31}P NMR study of improvement in oxidative phosphorylation by vitamins K_3 and C in a patient with a defect in electron transport at complex III in skeletal muscle. Proc. Natl. Acad. Sci (USA) 81 (1984) 3529–3533.

ENGEL, A.G.: Carnitine deficiency syndromes and lipid storage myopathies. In: A.G. Engel and B.Q. Banker (Eds.), Myology. New York, McGraw-Hill (1986) 1663–1696.

ENGEL, A.G. and C. ANGELINI: Carnitine deficiency of human skeletal muscle with associated lipid storage myopathy: a new syndrome. Science 179 (1973) 899–902.

ENGEL, A.G. and R.G. SIEKERT: Lipid storage myopathy responsive to prednisone. Arch. Neurol. 27 (1972) 174–181.

ENGEL, W.K. and G.C. CUNNINGHAM: Rapid examination of muscle tissue: an improved trichrome method for fresh-frozen biopsy sections. Neurology 13 (1963) 919–923.

ENGEL, W.K., N.K. VICK, J. GLUECK and R.I. LEVY: A skeletal muscle disorder associated with intermittent symptoms and a possible defect in lipid metabolism. N. Engl. J. Med. 282 (1970) 697–704.

ERIKSSON, B.O., B. GUSTAFSON, S. LINDSTEDT and I. NORDIN: Hereditary defect in carnitine membrane transport is expressed in skin fibroblasts. Eur. J. Pediatr. 147 (1988) 662–663.

FARRANS, V.J., R.G. HIBBS, J.J. WALSH and G.E. BURCH: Cardiomyopathy, cirrhosis of the liver and deposits of a fibrillar polysaccharide. Am. J. Cardiol. 17 (1966) 457–469.

FERNANDES, J.: The glycogen storage diseases. In: J. Fernandes, J-M. Saudubray and K. Tada (Eds.), Inborn Metabolic Diseases. Berlin, Spring-Verlag (1990) 69–88.

FERNANDES, J. and F. HUIJING: Branching enzyme deficiency glycogenosis: studies in therapy. Arch. Dis. Child. 43 (1968) 347–352.

FINOCCHIARO, G., F. TARONI, M. ROCCHI, A. LIRAS MARTIN, I. COLOMBO, G. TORRI TARELLI and S. DIDONATO: cDNA cloning, sequence analysis, and chromosomal localization of the gene for human carnitine palmitoyltransferase. Proc. Natl. Acad. Sci. 88 (1991) 661–665.

FISHBEIN, W.N.: Myoadenylate deaminase deficiency: inherited and acquired forms. Biochem. Med. 33 (1985) 158–169.

FISHBEIN, W.N., V.W. ARMBRUSTMACHER and J.L. GRIFFIN: Myoadenylate deaminase deficiency: a new disease of muscle. Science 200 (1978) 545–548.

FISHBEIN, W.N., J.L. GRIFFIN and V.W. ARMBRUSTMACHER: Stain for skeletal muscle adenylate deaminase: an effective tetrazolium stain for frozen biopsy specimens. Arch. Path. Lab. Med. 104 (1980) 462–466.

FRANCKE, U., B.T. DARRAS, N.F. ZANDER and M.W. KILIMANN: Assignment of human genes for phosphorylase kinase subunits alpha (PHKA) to Xq12-q13 and beta (PHKB) to 16q12-q13. Am. J. Hum. Genet. 45 (1989) 276–282.

GAMBETTI, P.L., S. DIMAURO, L. HIRT and R.P. BLUME: Myoclonic epilepsy with Lafora bodies. Arch. Neurol. 25 (1971) 483–493.

GARAVAGLIA, B., C. ANTOZZI, F. GIROTTI, D. PELUCHETTI, M. RIMOLDI, M. ZEVIANI and S. DIDONATO: A mitochondrial myopathy with complex II deficiency. Neurology 40 (1990) 294.

GAUTRON, S., D. DAEGELEN, F. MENNECIER, D. DUBOCQ, A. KAHN and J-C. DREYFUS: Molecular mechanisms of McArdle's disease (muscle glycogen phosphorylase deficiency). J. Clin. Invest. 79 (1987) 275–281.

GELLERA, C., G. UZIEL, M. RIMOLDI, M. ZEVIANI, A. LAVERDA, F. CARRARA and S. DIDONATO: Fumarase deficiency is an autosomal recessive encephalopathy affecting both the mitochondrial and the cytosolic enzymes. Neurology 40 (1990) 495–499.

GLASGOW, A.M., A.G. ENGEL, D.M. BIER, L.W. PERRY, M. DICKIE, J. TODARO, B.I. BROWN and M.F. UTTER: Hypoglycemia, hepatic dysfunction, muscle weakness, cardiomyopathy, free carnitine deficiency and long-chain acylcarnitine excess responsive to medium-chain triglyceride diet. Pediatr. Res. 17 (1983) 319–326.

GOTO, Y-I., I. NONAKA and S. HORAI: A mutation in the tRNA$^{Leu(UUR)}$ gene associated with the MELAS subgroup of mitochondrial encephalomyopathies. Nature (London) 348 (1990) 651–653.

GOTO, Y-I., I. NONAKA and S. HORAI: A new mtDNA mutation associated with mitochondrial myopathy, encephalopathy, lactic acidosis and stroke-like episodes (MELAS). Biochim. Biophys. Acta 1097 (1991) 238–240.

GUIBAUD, R., H. CARRIER, M. MATHIEU, C. DORCHE, B. PAREHOUX, M. BETHENOD and F. LARBRE: Observation familiale de dystrophie musculaire congénitale par déficit en phosphofructokinase. Arch. Fr. Pediatr. 35 (1978) 1105–1115.

HALE, D.E. and C. THORPE: Short-chain 3-OH acyl-CoA dehydrogenase deficiency. Pediatr. Res. 25 (1989) 199A.

HALE, D.E., M.L. BATSHAW, P.M. COATES, F.E. FRERMAN, S.I. GOODMAN, I. SINGH and C.A. STANLEY: Long-chain acyl-coA dehydrogenase deficiency: an inherited cause of nonketotic hypoglycemia. Pediatr. Res. 19 (1985) 666–671.

HALE, D.E., C. THORPE, K. BRAAT, J.H. WRIGHT, P.M. COATES and A.M. GLASGOW: Long-chain 3-OH acyl-CoA dehydrogenase deficiency. Pediatr. Res. 25 (1989) 199A.

HALLER, R.G. and S.F. LEWIS: Glucose-induced exertional fatigue in muscle phosphofructokinase deficiency. N. Engl. J. Med. 324 (1991) 364–369.

HALLER, R.G., S.F. LEWIS, J.D. COOK and C.G.

BLOMQVIST: Myophosphorylase deficiency impairs muscle oxidative metabolism. Ann. Neurol. 17 (1985) 196–199.

HALLER, R.G., K.G. HENRIKSSON, L. JORFELDT, E. HULTMAN, R. WIBON, K. SAHLIN, N-H ARESKOG, M. GUNDER, K. AYYAD, C.G. BLOMQVIST, R.E. HALL, P. THUILLIER, N.G. KENNAWAY and S.F. LEWIS: Deficiency of skeletal muscle succinate dehydrogenase and aconitase: pathophysiology of exercise in a novel human muscle oxidative defect. J. Clin. Invest. 88 (1991) 1197–1206.

HAMMANS, S.R., M.G. SWEENEY, M. BROCKINGTON, J.A. MORGAN-HUGHES and A.E. HARDING: Mitochondrial encephalomyopathies: molecular genetic diagnosis from blood samples. Lancet 337 (1991) 1311–1313.

HART, Z.H., S. SERVIDEI, P.L. PETERSON, C.H. CHANG and S. DIMAURO: Cardiopathy, mental retardation, and autophagic vacuolar myopathy. Neurology 37 (1987) 1065–1068.

HASEGAWA, H., T. MATSUOKA, Y. GOTO and I. NONAKA: Strongly succinate dehydrogenase-reactive blood vessels in muscles from patients with mitochondrial myopathy, encephalopathy, lactic acidosis, and stroke-like episodes. Ann. Neurol. 29 (1991) 601–605.

HASILIK, A. and E.F. NEUFELD: Biosynthesis of lysosomal enzymes in fibroblasts: synthesis as precursors of higher molecular weight. J. Biol. Chem. 255 (1982a) 4937–4945.

HASILIK, A. and E.F. NEUFELD: Biosynthesis of lysosomal enzymes in fibroblasts: phosphorylation and mannose residues. J. Biol. Chem. 255 (1982b) 4946–4950.

HATEFI, Y.: The mitochondrial electron transport and oxidative phosphorylation system. Annu. Rev. Biochem. 54 (1985) 1015–1069.

HAYDAR, N.A., H.L. CONN, A. AFIFI, N. WAKID, S. BALLAS and K. FAWAZ: Severe hypermetabolism with primary abnormality of skeletal muscle mitochondria. Ann. Intern. Med. 74 (1971) 548–558.

HAYES, D.J., B.R.F. LECKY, D.N. LANDON, J.A. MORGAN-HUGHES and J.B. CLARK: A new mitochondrial myopathy: biochemical studies revealing a deficiency in the cytochrome bc$_1$ complex (complex III) of the respiratory chain. Brain 107 (1984) 1165–1177.

HAYS, A.P., M. HALLETT, J. DELFS, J. MORRIS, A. SOTREL, M.M. SHEVCHUK and S. DIMAURO: Muscle phosphofructokinase deficiency: abnormal polysaccharide in a case of late-onset myopathy. Neurology 31 (1981) 1077–1086.

HERS, H.G.: α-Glucosidase deficiency in generalized glycogen storage disease (Pompe's disease). Biochem. J. 86 (1963) 11–16.

HERS, H.G.: Inborn lysosomal diseases. Gastroenterology 48 (1965) 625–633.

HOEFSLOOT, L.H., M. HOOGEVEEN-WESTERVELD, M. KROOS, J. VAN BEEUMEN and A.J.J. REUSER: Primary structure and processing of lysosomal alpha-glucosidase: homology with the intestinal sucrase-isomaltase complex. EMBO J. 7 (1988) 1697–1704.

HOEFSLOOT, L.H., A.T. VAN DER PLOEG, M.A. KROOS, M.

HOOGEVEEN-WESTERVELD, B.A. OOSTRA and A.J.J. REUSER: Adult and infantile glycogenosis type II in one family, explained by allelic diversity. Am. J. Hum. Genet. 46 (1990) 45–52.

HOLT, I.J., A.E. HARDING and J.A. MORGAN-HUGHES: Deletions of muscle mitochondrial DNA in patients with mitochondrial myopathies. Nature (London) 331 (1988) 717–719.

HOLT, I.J., A.E. HARDING, R.K.H. PETTY and J.A. MORGAN-HUGHES: A new mitochondrial disease associated with mitochondrial DNA heteroplasmy. Am. J. Hum. Genet. 46 (1990) 428–433.

HOPPEL, C.: The physiological role of carnitine. In: R. Ferrari, S. DiMauro and G. Sherwood (Eds.), L-Carnitine and Its Role in Medicine: From Function to Therapy. London, Academic Press (1992) 5–19.

HOPPEL, C.L., D.S. KERR, B. DAHMS and U. ROESSMAN: Deficiency of the reduced nicotinamide adenine nucleotide dehydrogenase component of complex I of the respiratory chain. J. Clin. Invest. 80 (1987) 71–77.

IANNACCONE, S.T., R.C. GRIGGS, W.R. MARKESBERY and R.J. JOYNT: Familial progressive external ophthalmoplegia and ragged-red fibers. Neurology 24 (1974) 1033–1038.

ICHIKI, T., M. TANAKA, M. NISHIKIMI, H. SUZUKI, T. OZAWA, M. KOBAYASHI and Y. WADA: Deficiency of subunits of complex I and mitochondrial encephalomyopathy. Ann. Neurol. 23 (1988) 287–294.

ICHIKI, T., M. TANAKA, M. KOBAYASHI, N. SUGIYAMA, H. SUZUKI, M. NISHIKIMI, T. OHNISHI, I. NONAKA, Y. WADA and T. OZAWA: Disproportionate deficiency of iron-sulfur clusters and subunits of complex I in mitochondrial encephalomyopathy. Pediatr. Res. 25 (1989) 194–201.

ILLINGWORTH, B. and G.T. CORI: Structure of glycogen and amylopectins. III. Normal and abnormal human glycogen. J. Biol. Chem. 199 (1952) 653–660.

ILLINGWORTH-BROWN, B. and D.H. BROWN: The subcellular distribution of enzymes in type II glycogenosis and the occurrence of an oligo-1,4-glucan glucohydrolase in human tissues. Biochim. Biophys. Acta 110 (1965) 124–133.

ILLINGWORTH-BROWN, B. and D.H. BROWN: Lack of an α-1,4-glucan 6 glucosyl transferase in a case of type IV glycogenosis. Proc. Natl. Acad. Sci. USA 56 (1966) 725–729.

ILLINGWORTH-BROWN, B., D. BROWN and P.L. JEFFREY: Simultaneous absence of alpha-1,4-glucosidase and alpha-1,6-glucosidase activities (pH 4.0) in tissues of children with type II glycogen storage disease. Biochemistry 9 (1970) 1423–1428.

ISAACS, H., N. SAVAGE, M. BADENHORST and T. WHISTLER: Acid maltase deficiency: a case study and review of the pathophysiological changes and proposed therapeutic measures J. Neurol. Neurosurg. Psychiatry 49 (1986) 1011–1018.

IWAMASA, T., S. FUKUDA, S. TOKUMITSU, N. NINOMIYA, J. MATSUDA and M. OSAME: Myopathy due to glycogen

storage disease. Exp. Mol. Pathol. 38 (1983) 405–420.

JEFFREY, P.L., D.H. BROWN and B. ILLINGWORTH-BROWN: Studies of lysosomal α-glucosidase. II. Kinetics of action of the rat liver enzyme. Biochemistry 9 (1970) 1416–1422.

JENSEN, K.E., J. JAKOBSEN, C. THOMSEN and O. HENRIKSEN: Improved energy kinetics following high protein diet in McArdle's syndrome. A ^{31}P magnetic resonance spectroscopy study. Acta Neurol. Scand. 81 (1990) 499–503.

JOHNSON, M.A., J.J. FULTHORPE and P. HUDGSON: Lipid storage myopaty: a recognizable clinicopathological entity? Acta Neuropathol. 24 (1973) 97–106.

JONES, T.A., E.F. DACRUZ E SILVA, N.K. SPURR, D. SHEER, and P.I.W. COHEN: Localization of the gene encoding the catalytic gamma subunit of phosphorylase kinase to human chromosome bands 7p12–q21. Biochim. Biophys. Acta. 1048 (1990) 24–29.

KADENBACH, B., J. JARAUSCH, R. HARTMAN and P. MERLE: Separation of mammalian cytochrome c oxidase into 13 polypeptides by a sodium dodecylsulfate-gel electrophoretic procedure. Anal. Biochem. 129 (1983) 517–521.

KANNO, T., K. SUDO, I. TAKEUCHI, S. KANADA, N. HONDA, Y. NISHIMURA and K. OYAMA: Hereditary deficiency of lactate dehydrogenase M-subunit. Clin. Chim. Acta 108 (1980) 267–276.

KARPATI, G., S. CARPENTER, A.G. ENGEL, G. WATTERS, J. ALLEN, S. ROTHMAN, G. KLASSEN and O.A. MAMER: The syndrome of systemic carnitine deficiency. Neurology 25 (1975) 16–24.

KENNAWAY, N.G.: Defects in the cytochrome bc_1 complex in mitochondrial diseases. J. Bioenerg. Biomembr. 20 (1988) 325–352.

KENNAWAY, N.G., N.R. BUIST, V.M. DARLEY-USMAR, A. PAPADIMITRIOU, S. DIMAURO, R.I. KELLEY, R.A. CAPALDI, N.K. BLANK and A. D'AGOSTINO: Lactic acidosis and mitochondrial myopathy associated with deficiency of several components of complex III of the respiratory chain. Pediatr. Res. 18 (1984) 991–999.

KISSEL, J.T., W. BEAM, N. BRESOLIN, G. GIBBONS, S. DIMAURO and J.R. MENDELL: The physiological assessment of a newly described myopathy, phosphoglycerate mutase deficiency, through incremental exercise testing. Neurology 35 (1985) 828–833.

KOBAYASHI, M., H. MORISHITA, N. SUGIYAMA, K. YOKOCHI, M. NAKANO, Y. WADA, Y. HOTTA, A. TERAUCHI and I. NONAKA: Two cases of NADH-coenzyme Q reductase deficiency: relationship to MELAS syndrome. J. Pediatr. 110 (1987) 223–227.

KOBAYASHI, Y., M.Y. MOMOI, K. TOMINAGA, T. MOMOI, K. NIHEI, M. YANAGISAWA, Y. KAGAWA and S. OHTA: A point mutation in the mitochondrial tRNA$^{Leu(UUR)}$ gene in MELAS. Biochem. Biophys. Res. Comm. 173 (1990) 816–822.

KOGA, Y., I. NONAKA, M. KOBAYASHI, M. TOIYO and K. NIHEI: Findings in muscle in complex I (NADH Coenzyme Q reductase) deficiency. Ann. Neurol. 24 (1988) 749–756.

KORNFELD, M. and M. LE BARON: Glycogenosis type VIII. J. Neuropathol. Exp. Neurol. 43 (1984) 568–579.

LARSSON, N-G., E. HOLME and B. KRISTANSSON: Progressive increase of the mutated mitochondrial DNA fraction in Kearns-Sayre syndrome. Pediatr. Res. 28 (1990) 131–136.

LARSSON, N-G., O. ANDERSEN, E. HOLME, A. OLDFORS and J. WAHLSTROM: Leber's hereditary optic neuropathy and complex I deficiency in muscle. Ann. Neurol. 30 (1991) 701–708.

LARSSON, N-G., H.G. EIKEN, H. BOMAN, E. HOLME, A. OLD-FORS and M.H. TULINIUS: Lack of transmission of mtDNA from a woman with Kearns-Sayre syndrome to her child. Am. J. Hum. Genet. 50 (1992) 360–363.

LAYZER, R.B., L.P. ROWLAND and H.M. RANNEY: Muscle phosphofructokinase deficiency. Arch. Neurol. 17 (1967) 512–523.

LEBO, R.V., F. GORIN, R.J. FLETTERICK, F-T. KAO, M-C., CHEUNG, B.D. BRUCE and Y.W. KAN: High-resolution chromosome sorting and DNA spot-blot analysis assign McArdle's syndrome to chromosome 11. Science 225 (1984) 57–59.

LEBO, R.V., L.A. ANDERSON, S. DIMAURO, E. LYNCH, P. HWANG and R. FLETTERICK: Rare McArdle disease locus polymorphic site on 11q13 contains CpG sequences. Hum. Genet. 86 (1990) 17–24.

LEDLEY, F.D., R. JANSEN, S-U. NHAM, W.A. FENTON and L.E. ROSENBERG: Mutation eliminating mitochondrial leader sequence of methylmalonyl-CoA mutase causes mut° methylmalonic acidemia. Proc. Natl. Acad. Sci. (USA) 87 (1990) 3147–3150.

LEWIS, S.F. and R.G. HALLER: Fatigue in skeletal muscle disorders. In: G. Atlan, L. Beliveau and P. Bouissou (Eds.), Muscle Fatigue: Biochemical and Physiological Aspects. Paris, Masson (1991) 119–134.

LOMBES, A., J.R. MENDELL, H. NAKASE, R.J. BAROHN, E. BONILLA, M. ZEVIANI, A.J. YATES J. OMERZA, T.L. GALES, K. NAKAHARA, R. RIZZUTO, W.K. ENGEL and S. DIMAURO: Myoclonic epilepsy and ragged-red fibers (MERRF) with cytochrome oxidase deficiency: neuropathology, biochemistry, and molecular genetics. Ann. Neurol. 26 (1989) 20–33.

LOMBES, A., H. NAKASE, H-J. TRITSCHLER, B. KADEN-BACH, E. BONILLA, D.C. DE VIVO, E.A. SCHON and S. DIMAURO: Biochemical and molecular analysis of cytochrome c oxidase deficiency in Leigh syndrome. Neurology 41 (1991) 491–498.

LOSSOS, A., V. BARASH, D. SOFFER, Z. ARGOV, M. GOMORI, Z. BEN-NARIAH, O. ABRAMSKY and I. STEINER: Hereditary branching enzyme dysfunction in adult polyglucosan body disease: a possible metabolic cause in two patients. Ann. Neurol. 30 (1991) 655–662.

LUFT, R., D. IKKOS, G. PALMIERI, L. ERNSTER and B. AFZELIUS: A case of severe hypermetabolism of nonthyroid origin with a defect in the maintenance of mitochondrial respiratory control: a correlated clinical, biochemical, and morphological study. J. Clin. Invest. 41 (1962) 1776–1804.

MAEKAWA, M., S. KANDA, K. SUDO and T. KANNO: Estimation of the gene frequency of lactate dehydrogenase subunit deficiencies. Am. J. Hum. Genet. 36 (1984) 1204–1214.

MARGOLIS, M.L. and A.R. HILL: Acid maltase deficiency in an adult. Am. Rev. Respir. Dis. 134 (1986) 328–331.

MARTINIUK, F., M. MEHLER, S. TZALL, G. MEREDITH and R. HIRSCHHORN: Sequence of the cDNA and 5′-flanking region for human acid alpha-glucosidase, detection of an intron in the 5′untranslated leader sequence, definition of 18-bp polymorphisms, and differences with previous cDNA and amino acid sequence. DNA Cell Biol. 9 (1990a) 85–94.

MARTINIUK, F., M. MEHLER, S. TZALL, G. MEREDITH and R. HIRSCHHORN: Extensive genetic heterogeneity in patients with alpha glucosidase deficiency as detected by abnormalities of DNA and mRNA. Am. J. Hum. Genet. 47 (1990b) 73–78.

MCARDLE, B.: Myopathy due to a defect in muscle glycogen breakdown. Clin. Sci. 10 (1951) 13–33.

MCCONCHIE, S.M., J. COAKLEY, R.H.T. EDWARDS and R.J. BEYNON: Molecular heterogeneity in McArdle's disease. Biochim. Biophys. Acta 1096 (1991) 26–32.

MCSHANE, M.A., S.R. HAMMANS, M. SWEENEY, I.J. HOLT, T.J. BEATTIE, E.M. BRETT and A.E. HARDING: Pearson syndrome and mitochondrial encephalopathy in a patient with a deletion of mtDNA. Am. J. Hum. Genet. 48 (1991) 39–42.

MEHLER, M. and S. DIMAURO: Residual acid maltase activity in late-onset acid maltase deficiency. Neurology 27 (1977) 178–184.

MHIRI, C., M. BAUDRIMONT, G. BONNE, C. GENY, F. DE-GOUL, C. MARSAC, E. ROULLET and R. GHERARDI: Zidovudine myopathy: a distinctive disorder associated with mitochondrial dysfunction. Ann. Neurol. 29 (1991) 606–614.

MICHELSON, A.M., A.F. MARKHAM and S.H. ORKIN: Isolation and DNA sequence of a full-length cDNA clone for human X chromosome-encoded phosphoglycerate kinase. Proc. Natl. Acad. Sci. (USA) 80 (1983) 472–476.

MILSTEIN, J.M., T.M. HERRON and J.E. HAAS: Fatal infantile muscle phosphorylase deficiency. J. Child Neurol. 4 (1989) 186–188.

MINEO, I., N. KONO, N. HARA, T. SHIMIZU, Y. YAMADA M. KAWACHI, H. KIYOKAWA, Y.L. WANG and S. TARUI: Myogenic hyperuricemia. N. Engl. J. Med. 317 (1987) 75–80.

MINEO, I., N. KONO, Y. YAMADA, N. HARA, H. KIYOKAWA, T. HAMAGUCHI, M. KAWACHI, T. YA-MASAKI, H. NAKAJIMA, M. KUWAJIMA and S. TARUI: Glucose infusion abolishes the excessive ATP degradation in working muscles of a patient with McArdle's disease. Muscle Nerve 13 (1990) 618–620.

MIRANDA, A.F., S. DIMAURO, A.B. EASTWOOD, A.P. HAYS, W.G. JOHNSON, M. OLARTE, R. WHITLOCK, R. MAYEUX and L.P. ROWLAND: Lipid storage, ichthyosis and steatorrhea. Muscle Nerve 2 (1979) 1–13.

MIRANDA, A.F., S. DIMAURO, A. ANTLER, L.Z. STERN and L.P. ROWLAND: Glycogen debrancher deficiency is reproduced in muscle culture. Ann. Neurol. 9 (1981) 283–289.

MITA, S., B. SCHMIDT, E.A. SCHON, S. DIMAURO and E. BONILLA: Detection of deleted mitochondrial genomes in cytochrome *c* oxidase-deficient muscle fibers of a patient with Kearns-Sayre syndrome. Proc. Natl. Acad. Sci. (USA) 86 (1989) 9509–9513.

MITA, S., R. RIZZUTO, C.T. MORAES, S. SHANSKE, E. AR-NAUDO, G.M. FABRIZI, Y. KOGA, S. DIMAURO and E.A. SCHON: Recombination via flanking direct repeats is a major cause of large-scale deletions of human mitochondrial DNA. Nucleic Acids Res. 18 (1990) 561–567.

MITSUMOTO, H.: McArdle disease: phosphorylase activity in regenerating muscle fibers. Neurology 29 (1979) 258–262.

MIZUTA, K., E. HASHIMOTO, A. TSUTOU, Y. EISHI, T. TAKEMURA, K. NARISAWA and H. YAMAMURA: A new type of glycogen storage disease caused by a deficiency of cardiac phosphorylase kinase. Biochim. Biophys. Res. Commun. 119 (1984) 582–587.

MOMMAERTS, W.F.H.M., B. ILLINGWORTH, C.M. PEARSON, R.J. GUILLORY and K. SERAYDARIAN: A functional disorder of muscle associated with the absence of phosphorylase. Proc. Natl. Acad. Sci. (USA) 45 (1959) 791–797.

MORAES, C.T., S. DIMAURO, M. ZEVIANI, A. LOMBES, S. SHANSKE, A.F. MIRANDA, H. NAKASE, E. BONILLA, L.C. WERNECK, S. SERVIDEI, I. NONAKA, Y. KOGA, A.J. SPIRO, A.K.W. BROWNELL, B. SCHMIDT, D.L. SCHOT-LAND, M. ZUPANC, D.C. DE VIVO, E.A. SCHON and L.P. ROWLAND: Mitochondrial DNA deletions in progressive external ophthalmoplegia and Kearns-Sayre syndrome. N. Engl. J. Med. 320 (1989) 1293–1299.

MORAES, C.T., E.A. SCHON and S. DIMAURO: Mitochondrial diseases: toward a rational classification. In: S.H. Appel (Ed.), Current Neurology, Vol. 11. St. Louis, MO, Mosby Year Book (1991a) 83–119.

MORAES, C.T., S. SHANSKE, H-J. TRITSCHLER, J.R. APRILLE, F. ANDREETTA, E. BONILLA, E.A. SCHON and S. DIMAURO: mtDNA depletion with variable tissue expression: a novel genetic abnormality in mitochondrial diseases. Am. J. Hum. Genet. 48 (1991b) 492–501.

MOREADITH, R.W., M.L. BATSHAW, T. OHNISHI, D. KERR, B. KNOX, D. JACKSON, R. HRUBAN, J. OLSON, B. REYNAFARJE and A.L. LEHNINGER: Deficiency of the iron-sulfur clusters of mitochondrial reduced nicotinamide-adenine dinucleotide-ubiquinone oxidoreductase (complex I) in an infant with congenital lactic acidosis. J. Clin. Invest. 74 (1984) 685–697.

MORGAN-HUGHES, J.A.: The mitochondrial myopathies. In: A.G. Engel and B.Q. Banker (Eds.), Myology. New York, McGraw-Hill (1986) 1709–1743.

MORGAN-HUGHES, J.A., P. DARVENIZA, S.N. KAHN, D.N.

LANDON, R.M. SHERRATT, J.M. LAND and J.B. CLARK: A mitochondrial myopathy characterized by a deficiency in reducible cytochrome b. Brain 100 (1977) 617–640.

MORGAN-HUGHES, J.A., D.J. HAYES, M. COOPER and J.B. CLARK: Mitochondrial myopathies: deficiencies localized to complex I and complex III of the mitochondrial respiratory chain. Biochem. Soc. Trans. 13 (1985) 648–650.

MORGAN-HUGHES, J.A., A.H.V. SCHAPIRA, J.M. COOPER and J.B. CLARK: Molecular defects of NADH-ubiquinone oxidoreductase (complex I) in mitochondrial diseases. J. Bioenerg. Biomembr. 20 (1988) 365–382.

MOSES, S.W., N. GADOTH, E. BEN-DAVID, A. SLONIM and K.L. WANDERMAN: Neuromuscular involvement in glycogen storage disease type III. Acta Pediatr. Scand. 5 (1986) 764–766.

MOSES, S.W., K.L. WANDERMAN, A. MYROZ and M. FRYDMAN: Cardiac involvement in glycogen storage disease type III. Eur. J. Pediat. 148 (1989) 764–766.

MURTHY, M.R. and S.V. PANDE: Malonyl-CoA binding site and overt carnitine palmitoyltransferase activity reside on opposite sides of the outer mitochondrial membrane. Proc. Natl. Acad. Sci. (USA) 84 (1987) 378–382.

NAKAJIMA, H., T. NOGUCHI, T. YAMASAKI, N. KONO, T. TANAKA and S. TARUI: Cloning of human muscle phosphofructokinase cDNA. FEBS Lett. 223 (1988) 113–116.

NAKAJIMA, H., N. KONO, T. YAMASAKI, K. HOTTA, M. KA-WACHI, M. KUWAJIMA, T. NOGUCHI, T. TANAKA and S. TARUI: Genetic defect in muscle phosphofructokinase deficiency. J. Biol. Chem. 265 (1991) 9292–9295.

NAKASE, H., C.T. MORAES, R. RIZZUTO, A. LOMBES, S. DIMAURO and E.A. SCHON: Transcription and translation of deleted mitochondrial genomes in Kearns-Sayre syndrome: implications for pathogenesis. Am. J. Hum. Genet. 46 (1990) 418–427.

NEWMAN, N.J. and D.C. WALLACE: Mitochondria and Leber's hereditary optic neuropathy. Am. J. Ophthalmol. 109 (1990) 726–730.

NIKOSKELAINEN, E.K., M.L. SAVONTAUS, O.P. LUANNE, M. KATILA and K.U. NUMMELINKU: Leber's hereditary optic neuroretinopathy, a maternally inherited disease. Arch. Ophthalmol. 105 (1987) 665-671.

NISHIZAWA, M., K. TANAKA, K. SHINOZAWA, T. KUWA-BARA, T. ATSUMI, T. MIYATAKE and E. OHAMA: A mitochondrial encephalomyopathy with cardiomyopathy: a case revealing a defect of complex I of the respiratory chain. J. Neurol. Sci. 78 (1987) 189–201.

OCKNER, R.K., J.A. MANNING, R.B. POPENHAUSEN and W.K.L. HO: A binding protein for fatty acids in cytosol of intestinal mucosa, liver, myocardium and other tissues. Science 177 (1972) 56–58.

OGASAHARA, S., A.G. ENGEL, D. FRENS and D. MACK: Muscle coenzyme Q deficiency in familial mitochondrial encephalomyopathy. Proc. Natl. Acad. Sci. (USA) 86 (1989) 2379–2382.

OHNO, K., M. TANAKA, K. SAHASHI, T. IBI, W. SATO, T. YAMAMOTO, A. TAKAHASHI and T. OZAWA: Mitochondrial DNA deletions in inherited recurrent myoglobinuria. Ann. Neurol. 29 (1991) 364–369.

OHTANI, Y., I. MASUDA, T. IWAMASA, H. TAMARI, Y. ORIGUCHI and T. MIIKE: Infantile glycogen storage myopathy in a girl with phosphorylase kinase deficiency. Neurology 32 (1982) 833–838.

PANDE, S.V.: A mitochondrial carnitine acylcarnitine translocase system. Proc. Natl. Acad. Sci. (USA) 72 (1975) 883–887.

PAPADIMITRIOU, A., H.B. NEUSTEIN, S. DIMAURO, R. STANTON and N. BRESOLIN: Histiocytoid cardiomyopathy of infancy: deficiency of reducible cytochrome b in heart mitochondria. Pediatr. Res. 18 (1984) 1023–1028

PAPADIMITRIOU, A., P. MANTA, R. DIVARI, A. KARABETSOS, E. PAPADIMITRIOU and N. BRESOLIN: McArdle's disease: two clinical expressions in the same pedigree. J. Neurol. 237 (1990) 267–270.

PARNAS, J.K. and R. WAGNER: Beobachtungen über Zuckerneubildung. Biochem. Z. 127 (1922) 55–65.

POMPE, J.C.: Over idiopatische hypertrophie van het hart. Ned. Tijdschr. Geneeskd. 76 (1932) 304–311.

PONZETTO, C., N. BRESOLIN, A. BORDONI, M. MOGGIO, G. MEOLA, L. BET, A. PRELLE and G. SCARLATO: Kearns-Sayre syndrome: different amounts of deleted mitochondrial DNA are present in several autoptic tissues. J. Neurol. Sci. 96 (1990) 207–210.

PONZETTO-ZIMMERMAN, C. and A.M. GOLD: Glycogen branching enzyme in Lafora myoclonus epilepsy. Biochem. Med. 28 (1982) 83–93.

POULTON, J., M.E. DEADMAN and R.M. GARDINER: Duplications of mitochondrial DNA in mitochondrial myopathy. Lancet i (1989a) 236–240.

POULTON, J., M.E. DEADMAN and R.M. GARDINER: Tandem direct duplications of mitochondrial DNA in mitochondrial myopathy: analysis of nucleotide sequence and tissue distribution. Nucleic Acids Res. 17 (1989b) 10223–10229.

POWELL, H.C., R. HAAS, C.H. HALL, J.A. WOLFF, W. NYHAN and B.I. BROWN: Peripheral nerve in type III glycogenosis: selective involvement of unmyelinated fiber Schwann cells. Muscle Nerve 8 (1985) 667–671.

PURDUE, P.E., Y. TAKADA and C.J. DANPURE: Identification of mutations associated with peroxisome-to-mitochondrion mistargeting of alanine/glyoxylate aminotransferase in primary hyperoxaluria type 1. J. Cell Biol. 111 (1990) 2341–2351.

PUTSCHAR, W.: Über angeborene Glykogenspeicherkrankheit des Herzens: 'Thesaurismosis glycogenica'. Beitr. Pathol. Anat. 90 (1932) 222–223.

RADOM, J., R. SALVAYRE, A. NEGRE, A. MARET and L. DOUSTE-BLAZY: Metabolism of neutral lipids in cultured fibroblasts from multisystem (or type 3) lipid storage myopathy. Eur. J. Clin. Invest. 164 (1987) 703–708.

REBOUCHE, C.J. and A.G. ENGEL: Kinetic compartmental analysis of carnitine metabolism in the human carnitine deficiency syndromes: evidence for alterations in tissue carnitine transport. J. Clin. Invest. 73 (1984) 857–867.

REICHMANN, H., R. ROHKAMM, M. ZEVIANI, S. SERVIDEI, K. RICKER and S. DIMAURO: Mitochondrial myopathy due to complex III deficiency with normal reducible cytochrome b concentration. Arch. Neurol. 43 (1986) 957–961.

RESIBOIS-GREGOIRE, A. and N. DOUROV: Electron microscopic study of a case of cerebral glycogenosis. Acta Neuropathol. 6 (1966) 70–79.

REUSER, A.J.J., M. KROOS, R. WILLEMSEN, D. SWALLOW, J.M. TAGER and H. GALJAARD: Clinical diversity in glycogenosis type II. J. Clin. Invest. 79 (1987) 1689–1699.

REUSER, A.J.J., M. KROOS, R.P.J.O. ELFERINK and J.M. TAGER: Defect in synthesis, phosphorylation, and maturation of acid alpha-glucosidase in glycogenosis type II. J. Biol. Chem. 260 (1985) 8336–8341.

REUSER, A.J.J., M. KROOS, R. WILLEMSEN, D. SWALLOW, J.M. TAGER and H. GALJAARD: Clinical diversity in glycogenosis type II. J. Clin. Invest. 79 (1987) 1689–1699.

RIGGS, J.E., S.S. SCHOCHET, L. GUTMAN, S. SHANSKE, W.A. NEAL and S. DIMAURO: Lysosomal glycogen storage disease without acid maltase deficiency. Neurology 33 (1983) 873–877.

RINALDO, P., J.J. O'SHEA, P.M. COATES, D.E. HALE, C.A. STANLEY and K. TANAKA: Medium-chain acyl-CoA dehydrogenase deficiency: diagnosis by stable isotope dilution measurement of urinary n-hexanoylglycine and 3-phenylpropionylglycine. N. Engl. J. Med. 319 (1988) 1308–1319.

ROBINSON, B.H., J. WARD, P. GOODYER and A. BEAUDET: Respiratory chain defects in the mitochondria of cultured skin fibroblasts from three patients with lactacidemia. J. Clin. Invest. 77 (1986) 1422–1427.

ROCCHICCIOLI, F., R.J.A. WANDERS, P. AUBOURG, C. VIANEY-LIAUD, L. IJLST, M. FABRE, N. CARTIER and P-F. BOUGNERES: Deficiency of long-chain 3-hydroxyacyl-CoA dehydrogenase: a cause of lethal myopathy and cardiomyopathy in early childhood. Pediatr. Res. 28 (1990) 657–662.

ROE, C.R. and P.M. COATES: Acyl-CoA dehydrogenase deficiencies. In: C.R. Scriver, A.L. Beaudet, W.S. Sly and D. Valle (Eds.), The Metabolic Basis of Inherited Diseases. New York, McGraw-Hill (1989) 889–914.

ROE, C.R., D.S. MILLINGTON, D.L. NORWOOD, N. KODO, H. SPRECHER, B.S. MOHAMMED, M. NADA, H. SCHULZ and R. MCVIE: 2,4-Dienoyl-CoA reductase deficiency: a possible new disorder of fatty acid oxidation. J. Clin. Invest. 85 (1990) 1703–1707.

ROELOFS, R.I., W.K. ENGEL and P.B. CHAUVIN: Histochemical phosphorylase activity in regenerating muscle fibers from myophosphorylase-deficient patients. Science 177 (1972) 795–797.

ROODHOOFT, A.M., K.J. VAN ACKER, J.J. MARTIN, C. CEUTERICK, H.R. SCHOLTE and I.E.M. LUYT-HOUWEN: Benign mitochondrial myopathy with deficiency of NADH-CoQ reductase and cytchrome c oxidase. Neuropediatrics 17 (1986) 221–226.

ROSA, R., C. GEORGE, M. FARDEAU, M.C. CALVIN, M. RAPIN and J. ROSA: A new case of phosphoglycerate kinase deficiency: PGK Creteil associated with rhabdomyolysis and lacking hemolytic anemia. Blood 60 (1982) 84–91.

ROTIG, A., V. CORMIER, S. BLANCHE, J-P., BONNEFONT, F. LEDEIST, N. ROMERO, J. SCHMITZ, P. RUSTIN, A. FISCHER, J-M. SAUDUBRAY and A. MUNNICH: Pearson's marrow-pancreas syndrome: a multisystem mitochondrial disorder in infancy. J. Clin. Invest. 86 (1990) 1601–1608.

ROWLAND, L.P., S. ARAKI and P. CARMEL: Contracture in McArdle's disease. Arch. Neurol. 13 (1965) 541–544.

ROWLAND, L.P., S. DIMAURO and R.B. LAYZER: Phosphofructokinase deficiency. In: A.G. Engel and B.Q. Banker (Eds.), Myology. New York, McGraw-Hill (1986) 1603–1617.

ROWLAND, L.P., D.M. BLAKE, M. HIRANO, S. DIMAURO, E.A. SCHON, A.P. HAYS and D.C. DE VIVO: Clinical syndromes associated with ragged red fibers. Rev. Neurol. 147 (1991) 467–473.

RUFF, R.L. and J. WEISSMAN: Iodoacetate-induced contracture in rat skeletal muscle: possible role of ADP. J. Physiol. (1992) in press.

RUITENBEEK, W., R. SENGERS M. ALBANI, F. TRIJBELS, A. JANSSEN, O. VAN DIGGELEN and J. BAKKEREN: Prenatal diagnosis of cytochrome c oxidase deficiency by biopsy of chorionic villi. N. Engl. J. Med. 319 (1988) 1095.

SABINA, R.L.: Myoadenylate deaminase deficiency. In: R.N. Rosenberg (Ed.), The Molecular and Genetic Basis of Neurological Disease. Stoneham, CT, Butterworth (1993) in press.

SABINA, R.L., T. MORISAKI, P. CLARKE, R. EDDY, T.B. MORTON and E.W. HOLMES: Characterization of the human and rat myoadenylate deaminase genes. J. Biol. Chem. 265 (1990) 9423–9433.

SABINA, R.L., W.N. FISHBEIN, G. PEZESHKPOUR, P.R.H. CLARKE and E.W. HOLMES: Molecular analysis of the myoadenylate deaminase deficiencies. Neurology 42 (1992) 170–179.

SAKUTA, R. and I. NONAKA: Vascular involvement in mitochondrial myopathy. Ann. Neurol. 25 (1989) 594–601.

SANCHO, S., C. NAVARRO, J.M. FERNANDEZ, C. DOMINGUEZ, A. ORTEGA, M. ROIG and C. CERVERA: Skin biopsy findings in glycogenosis III: clinical, biochemical, and electrophysiological correlations. Ann. Neurol. 27 (1990) 480–486.

SCHAPIRA, A.H.V., J.M. COOPER, J.A. MORGAN-HUGHES, D.N. LANDON and J.B. CLARK: Mitochondrial myopathy with a defect of mitochondrial protein transport. N. Engl. J. Med. 323 (1990) 37–42.

SCHATZ, G.: The mitochondrial protein import machinery. In: T. Sato and S. DiMauro (Eds.), Mitochondrial encephalomyopathies. New York, Raven (1991) 57–74.

SCHMID, R. and R. MAHLER: Chronic progressive myopathy with myoglobinuria: demonstration of a glycogenolytic defect in the muscle. J. Clin. Invest. 38 (1959) 2044–2058.

SCHMIDT, B., S. SERVIDEI, A.A. GABBAI, A.C. SILVA, A. DE SOUSA BULLE DE OLIVEIRA and S. DIMAURO: McArdle's disease in two generations: autosomal recessive transmission with manifesting heterozygote. Neurology 37 (1987) 1558–1561.

SCHON, E.A., R. RIZZUTO, C.T. MORAES, H. NAKASE, M. ZEVIANI and S. DIMAURO: A direct repeat is a hotspot for large-scale deletion of human mitochondrial DNA. Science 244 (1989) 346–349.

SCHOTLAND, D.L. S. DIMAURO, E. BONILLA, A. SCARPA and C.P. LEE: Neuromuscular disorder associated with a defect in mitochondrial energy supply. Arch. Neurol. 33 (1976) 475–479.

SELBY, R., T.E. STARZL, E. YUNIS, B.I. BROWN, R.S. KENDALL and A. TZAKIS: Liver transplantation for type IV glycogen storage disease. N. Engl. J. Med. 324 (1991) 39–42.

SERRATRICE, G., A. MONGES, H. ROUX, R. AQUARON and D. GAMBARELLI: Forme myopathique du déficit en phosphofructokinase. Rev. Neurol. 120 (1969) 271–277.

SERVIDEI, S. and S. DIMAURO: Disorders of glycogen metabolism of muscle. Neurol. Clin. 7 (1989) 159–178.

SERVIDEI, S., E. BONILLA, R.G. DIEDRICH, M. KORNFELD, J.D. OATES, M. DAVIDSON, S. VORA and S. DIMAURO: Fatal infantile form of phosphofructokinase deficiency. Neurology 36 (1986) 1465–1470.

SERVIDEI, S., R.E. RIEPE, C. LANGSTON, L.Y. TANI, J.T. BRICKER, N. CRISP-LINDGREN H. TRAVERS, D. ARMSTRONG and S. DIMAURO: Severe cardiomyopathy in branching enzyme deficiency. J. Pediatr. 111 (1987a) 51–56.

SERVIDEI, S., L.A. METLAY, F.A. BOOTH, K.R. RAO and S. DIMAURO: Clinical and biochemical heterogeneity of phosphorylase b kinase deficiency. Neurology 37 (1987b) 139.

SERVIDEI, S., S. SHANSKE, M. ZEVIANI, R. LEBO, R. FLETTERICK and S. DIMAURO: McArdle disease: biochemical and molecular genetic studies. Ann. Neurol. 24 (1988a) 774–781.

SERVIDEI, S., L.A. METLAY, J. CHODOSH and S. DIMAURO: Fatal infantile cardiopathy caused by phosphorylase b kinase deficiency. J. Pediatr. 113 (1988b) 82–85.

SHANSKE, S., S. SAKODA, M.A. HERMODSON, S. DIMAURO and E.A. SCHON: Isolation of a cDNA encoding the muscle-specific subunit of human phosphoglycerate mutase. J. Biol. Chem. 262 (1987) 14612–14617.

SHANSKE, S., C.T. MORAES, A. LOMBES, A.F. MIRANDA, E. BONILLA, P. LEWIS, M.A. WHELAN, C.A. ELLSWORTH and S. DIMAURO: Wide-spread tissue distribution of mitochondrial DNA deletions in Kearns-Sayre syndrome. Neurology 40 (1990) 24–28.

SHANSKE, S., P. TONIN, A.F. MIRANDA, A.K. BROWNELL and S. DIMAURO: Phosphoglycerate kinase (PGK) deficiency: biochemical and molecular genetic studies. Neurology 41 (1991) 179.

SHAPIRA, Y., S. HAREL and A. RUSSEL: Mitochondrial encephalomyopathies: a group of neuromuscular disorders with defects in oxidative metabolism. Isr. J. Neurol. Sci. 13 (1977) 161–164.

SHOFFNER, J.M. and D.C. WALLACE: Oxidative phosphorylation diseases: disorders of two genomes. Adv. Hum. Genet. 19 (1990) 267–330.

SHOFFNER, J.M., M.T. LOTT, A.M.S. LEZZA, P. SEIBEL, S.W. BALLINGER and D.C. WALLACE: Myoclonic epilepsy and ragged-red fibers (MERRF) is associated with a mitochondrial DNA tRNA^{Lys} mutation. Cell 61 (1990) 931–937.

SHOUBRIDGE, E.A., G. KARPATI and K.E.M. HASTINGS: Deletion mutants are functionally dominant over wild-type mitochondrial genomes in skeletal muscle fiber segments in mitochondrial diseases. Cell 62 (1990) 43–49.

SHY, G.M. and N.K. GONATAS: Human myopathy with giant abnormal mitochondria. Science 145 (1964) 493–496.

SHY, G.M., N.K. GONATAS and M. PEREZ: Two childhood myopathies with abnormal mitochondria: I. Megaconial myopathy. II. Pleoconial myopathy. Brain 89 (1966) 133–158.

SIMON, L.T., D.S. HOROUPIAN, L.J. DORFMAN, M. MARKS, M.K. HERRICK, P. WASSERSTEIN and M.E. SMITH: Polyneuropathy, ophthalmoplegia, leukoencephalopathy, and intestinal pseudo-obstruction: POLIP syndrome. Ann. Neurol. 28 (1990) 349–360.

SIMPSON, M.V., C.D. CHIN, S.A. KEILBAUGH, T-S. LIN and W.H. PRUSOFF: Studies on the inhibition of mitochondrial DNA replication by 3′-acido-3′-deoxythymidine and other dideoxynucleoside analogs which inhibit HIV-1 replication. Biochem. Pharmacol. 38 (1989) 1033–1036.

SINGH, R., I.M. SHEPHERD, J.P. DERRICK, R.R. RAMSEY, H.S.A. SHERRATT and D.M. TURNBULL: A case of carnitine palmitoyltransferase II deficiency in human skeletal muscle. FEBS Lett. 241 (1988) 126–130.

SLIPETZ, D.M., J.R. APRILLE, P.R. GOODYER and R. ROZEN: Deficiency of complex III of the mitochondrial respiratory chain in a patient with facioscapulohumeral disease. Am. J. Hum. Genet. 48 (1991a) 502–510.

SLIPETZ, D.M., P.R. GOODYER and R. ROZEN: Congenital deficiency of a 20-kDa subunit of mitochondrial complex I in fibroblasts. Am. J. Hum. Genet. 48 (1991b) 1121–1126.

SLONIM, A.E. and P.J. GOANS: Myopathy in McArdle's syndrome: improvement with a high-protein diet. N. Engl. J. Med. 312 (1985) 355–359.

SLONIM, A.E., C. WEISBERG, P. BENKE, O.B. EVANS and I.M. BURR: Reversal of debrancher deficiency myopathy by the use of high-protein nutrition. Ann. Neurol. 11 (1982) 420–422.

SLONIM, A.E., R.A. COLEMAN, M.A. MCELLIGOT, J. NAJJAR, K. HIRSCHHORN, G.U. LA BADIE, R. MRAK, O.B. EVANS, E. SHIPP and R. PRESSON: Improvement of muscle function in acid maltase deficiency by high-protein therapy. Neurology 33 (1983) 34–38.

SMIT, G.P.A., J. FERNANDES, J.V. LEONARD, E.E. MAT-

THEWS, S.W. MOSES, M. ODIEVRE and K. ULLRICH: The long-term outcome of patients with glycogen storage diseases. J. Inherit. Metab. Dis. 13 (1990) 411–418.

SOKAL, E.M., F. VAN HOOF and D. ALBERTI: Progressive cardiac failure following successful orthotopic liver transplantation for type IV glycogenosis. Eur. J. Pediatr. 15 (1992) 200–203.

SPIRO, A.J., C.L. MOORE, J.W. PRINEAS, P.M. STRASBERG and I. RAPIN: A cytochrome-related inherited disorder of the nervous system and muscle. Arch. Neurol. 23 (1970) 103–112.

STANLEY, C.A.: New genetic defects in mitochondrial fatty acid oxidation and carnitine deficiency. Adv. Pediatr. 34 (1987) 59–88.

STANLEY, C.A., D.E. HALE, P.M. COATES, C.L. HALL, B.E. CORKEY, W. YANG, R.I. KELLY, E.L. GONZALES, J.R. WILLIAMSON and L. BAKER: Medium-chain acyl-CoA dehydrogenase deficiency in children with nonketotic hypoglycemia and low carnitine levels. Pediatr. Res. 17 (1983) 877–884.

STANLEY, C.A., S. DE LEEUW, P.A. COATES, C. VIANEY-LIAUD, P. DIVRY, J-P. BONNEFONT, J-M. SAUDUBRAY, M. HAYMOND, F.K. TREFZ, G.N. BRENINGSTALL, R.S. WAPPNER, D.J. BYRD, C. SANSARICQ, I. TEIN, W. GROVER, D. VALLE, S.L. RUTLEDGE and W.R. TREEM: Chronic cardiomyopathy and weakness or acute coma in children with a defect in carnitine uptake. Ann. Neurol. 30 (1991) 709–716.

STREMMEL, W., G. STROHMEYER, G. BORCHARD, F. KOCHWA and P.D. BERK: Isolation and partial characterization of a fatty acid binding protein in rat liver plasma membranes. Proc. Natl. Acad. Sci. (USA) 83 (1985) 4–8.

SUGIE, H., Y. SUGIE, M. NISHIDA, I. MASATAKA, S. TSURUI, M. SUZUKI, R. MIYAMOTO and Y. IGARASHI: Recurrent myoglobinuria in a child with mental retardation: phosphoglycerate kinase deficiency. J. Child Neurol. 4 (1989) 95–99.

TAKAMIYA, S., W. YANAMURA, R.A. CAPALDI, N.G. KENNAWAY, R. BART, R.C.A. SENGERS, J.M.F. TRIJBELS and W. RUITENBEEK: Mitochondrial myopathies involving the respiratory chain: a biochemical analysis. Ann. N.Y. Acad. Sci. 488 (1986) 33–43.

TARONI, F., E. VERDERIO, S. FIORUCCI, P. CAVADINI, G. FINOCCHIARO, G. UZIEL, E. LAMANTEA, C. GELLERA and S. DIDONATO: Molecular characterization of inherited carnitine palmitoyltransferase II deficiency. Proc. Natl. Acad. Sci. USA 89 (1992) 8429–8433.

TARUI, S., G. OKUNO, Y. IKUA, T. TANAKA, M. SUDA and M. NISHIKAWA: Phosphofructokinase deficiency in skeletal muscle: a new type of glycogenosis. Biochem. Biophys. Res. Commun. 19 (1965) 517–523.

TATUCH, Y., J. CHRISTDOULOU, A. FEIGENBAUM, J.T.R. CLARKE, J. WHERRET, C. SMITH, N. RUDD, R. PETROVA-BENEDICT and B.H. ROBINSON: Heteroplasmic mtDNA mutation (T->G) at 8993 can cause Leigh's disease when the percentage of abnormal ntDNA is high. Am. J. Hum. Genet. 50 (1992) 852–858.

TEIN, I. and S. DIMAURO: Primary systemic carnitine

deficiency manifested by carnitine-responsive cardiomyopathy. In: R. Ferrari, S. DiMauro and G. Sherwood (Eds.), L-Carnitine and Its Role in Medicine: From Function to Therapy. London, Academic Press (1992) 155–184.

TEIN, I., F. DEMAUGRE, J-P. BONNEFONT and J-M. SAUDUBRAY: Normal muscle CPT1 and CPT2 activities in hepatic presentation patients with CPT1 deficiency in fibroblasts. J. Neurol. Sci. 92 (1989) 229–245.

TEIN, I., D.C. DE VIVO, F. BIERMAN, P. PULVER, L.J. DE MEIRLEIR, L. CIVITANOVIC-SOJAT, R.A. PAGON, E. BERTINI, C. DIONISI-VICI, S. SERVIDEI and S. DIMAURO: Impaired skin fibroblast carnitine uptake in primary systemic carnitine deficiency manifested by childhood carnitine-responsive cardiomyopathy. Pediatr. Res. 28 (1990) 247–255.

TEIN, I., D.C. DE VIVO, D.E. HALE, J.T.R. CLARKE, H. ZINMAN, R. LAXER, A. SHORE and S. DIMAURO: Short-chain L-3-hydroxyacyl-CoA dehydrogenase deficiency in muscle: a new cause for recurrent myoglobinuria and encephalopathy. Ann. Neurol. 30 (1991) 415–419.

TONIN, P., S. SHANSKE, A.K. BROWNELL, J.P. WYSE and S. DIMAURO: Phosphoglycerate kinase deficiency: a third case with recurrent myoglobinuria. Neurology 39 (1989) 359–360.

TONIN, P., P. LEWIS, S. SERVIDEI and S. DIMAURO: Metabolic causes of myoglobinuria. Ann. Neurol. 27 (1990) 181–185.

TOWFIGHI, J., B.S. YOSS, W.W. WASIEWSKI, R.C. VANNUCCI, M.S. BENTZ and A. MAMOURIAN: Cerebral glycogenosis, alpha particle type: morphological and biochemical observations in an infant. Hum. Pathol. 29 (1989) 1210–1215.

TREEM, W.R., C.A. STANLEY, D.N. FINEGOLD, D.E. HALE and P.M. COATES: Primary carnitine deficiency due to a failure of carnitine transport in kidney, muscle and fibroblasts. N. Engl. J. Med. 319 (1988) 1331–1336.

TRITSCHLER, H-J., E. BONILLA, A. LOMBES, F. ANDREETTA, S. SERVIDEI, B. SCHNEYDER, A.F. MIRANDA, E.A. SCHON, B. KADENBACH and S. DIMAURO: Differential diagnosis of fatal and benign cytochrome c oxidase-deficient myopathy of infancy: an immunohistochemical approach. Neurology 41 (1991) 300–305.

TRITSCHLER, H-J., F. ANDREETTA, C.T. MORAES, E. BONILLA, E. ARNAUDO, M.J. DANON, B.M. ZELAYA, E. VAMOS, N. TELERMAN-TOPPET, S. SHANSKE, B. KADENBACH, S. DIMAURO and E.A. SCHON: Mitochondrial myopathy of childhood associated with depletion of mitochondrial DNA. Neurology 42 (1992) 209–217.

TSUJINO, S., S. SAKODA, R. MIZUNO, T. KOBAYASHI, T. SUZUKI, S. KISHIMOTO, S. SHANSKE, S. DIMAURO and E.A. SCHON: Structure of the gene encoding the muscle-specific subunit of human phosphoglycerate mutase. J. Biol. Chem. 264 (1989) 15334–15337.

TURNBULL, D.M., K. BARTLETT, D.L. STEVENS, K.G.M.M.

ALBERTI, G.J. GIBSON, M.A. JOHNSON, A.J. MCCULLOCH and H.S.A. SHERRATT: Short-chain acyl-CoA dehydrogenase deficiency associated with a lipid-storage myopathy and secondary carnitine deficiency. N. Engl. J. Med. 311 (1984) 1232–1236.

TURNBULL, D.M., I.M. SHEPHERD, B. ASHWORTH, K. BARTLETT, M.A. JOHNSON, M.J. CULLEN, S. JACKSON and H.S.A. SHERRATT: Lipid storage myopathy associated with low acyl-CoA dehydrogenase activities. Brain 111 (1988) 815–828.

UGAWA, Y., K. INOUE, T. TAKEMURA and T. IWAMASA: Accumulation of glycogen in sural nerve axons in adult-onset type III glycogenosis. Ann. Neurol. 19 (1986) 294–297.

UMPLEBY, A.M., C.M. WILES, P.ST.J. TREND, I.N. SCOBIE, A.F. MCLEOD, G.T. SPENCER and P.H. SONKSEN: Protein turnover in acid maltase deficiency before and after treatment with a high-protein diet. J. Neurol. Neurosurg. Psychiatry 50 (1987) 587–592.

UMPLEBY, A.M., P.ST.J. TREND, D. CHUBB, J.V. CONAGLEN, C.D. WILLIAMS, R. HESP, I.N. SCOBIE, C.M. WILES, G. SPENCER and P.H. SONKSEN: The effect of a high-protein diet on leucine and alanine turnover in acid maltase deficiency. J. Neurol. Neurosurg. Psychiatry 52 (1989) 954–961.

VAN COSTER, R., A. LOMBES, D.C. DE VIVO, T.L. CHI, W.E. DODSON, S. ROTHMAN, E.J. ORRECCHIO, W. GROVER, G.T. BERRY, J.F. SCHWARTZ, A. HABIB and S. DIMAURO: Cytochrome c oxidase-associated Leigh syndrome: phenotypic features and pathogenetic speculations. J. Neurol. Sci. 104 (1991) 97–111.

VAN CREVELD, S.: Over een bijzondere stoornis in de koolhydraatstofwisseling in de kinderleeftijd. Ned. Tijdschr. Geneeskd. 15 (1928) 349–359.

VAN DER BERG, I.E.T. and R. BERGER: Phosphorylase b kinase deficiency in man: a review. J. Inherit. Metab. Dis. 13 (1990) 442–451.

VAN DER PLOEG, A.T., P.A. BOLHUIS, R.A. WOLTERMAN, J.W. VISSER, C.M.B. LOONEN, H.F.M. BISCH and A.J.J REUSER: Prospect for enzyme therapy in glycogenosis type II variants: a study on cultured muscle cells. J. Neurol. 235 (1988) 392–396.

VAN DER PLOEG, A.T., L.H. HOEFSLOOT, M. HOOGEVEEN-WESTERVELD, E.M. PETERSEN and A.J.J. REUSER: Glycogenosis type II: Protein and DNA analysis in five South African families from various ethnic origins. Am. J. Hum. Genet. 44 (1989) 787–793.

VAN DER PLOEG, A.T., M.A. KROOS, R. WILLEMSEN, N.H.C. BRONS and A.J.J. REUSER: Intravenous administration of phosphorylated acid alpha-glucosidase leads to uptake of enzyme in heart and skeletal muscle of mice. J. Clin. Invest. 87 (1991) 513–518.

VAN KEUREN, M., H. DRABKIN, I. HART, D. HARKER, D. PATTERSON and S. VORA: Regional assignment of human liver-type phosphofructokinase to chromosome 21q22.3 by using somatic cell hybrids and a monoclonal anti-L antibody. Hum. Genet. 74 (1986) 34–40.

VITA, G., A. TOSCANO, N. BRESOLIN, G. MEOLA, B. BARBIROLI, A. BARADELLO and C. MESSINA: Muscle phosphoglycerate mutase (PGAM) deficiency in the first Caucasian patient. Neurology 40 (1990) 297.

VON GIERKE, E.: Hepato-nephromegalia glykogenica. Beitr. Pathol. Anat. 82 (1929) 497–513.

VORA, S.: Isozymes of human phosphofructokinase in blood cells and cultured cell lines: molecular and genetic evidence for a trigenic system. Blood 57 (1981) 724–731.

VORA, S.: Isozymes of phosphofructokinase. Curr. Topics Biol. Med. Res. 6 (1982a) 119–167.

VORA, S., S. DURHAM, B. DE MARTINVILLE, D.L. GEORGE and U. FRANCKE: Assignment of the human gene for muscle-type phosphofructokinase (PFKM) to chromosome 1 (region cen->q32) using somatic cell hybrids and monoclonal anti-M antibody. Som. Cell Genet. 8 (1982b) 95–104.

VORA, S., A.F. MIRANDA, E. HERNANDEZ and U. FRANCKE: Regional assignment of the human gene for platelet-type phosphofructokinase (PFKP) to chromosome 10p: a novel use of polyspecific rodent antisera to localize human enzyme genes. Hum. Genet. 63 (1983) 374–379.

VORA, S., S. DIMAURO, D. SPEAR, D. HARKER and M.J. DANON: Characterization of the enzymatic defect in late-onset muscle phosphofructokinase deficiency. J. Clin. Invest. 80 (1987) 1479–1485.

WALLACE, D.C., G. SINGH, M.T. LOTT, J.A. HODGE, T.G. SCHURR, A.M.S. LEZZA, L.J. ELSAS and E.K. NIKOSKE-LAINEN: Mitochondrial DNA mutation associated with Leber's hereditary optic neuropathy. Science 242 (1988a) 1427–1430.

WALLACE, D.C., X. ZHENG, M.T. LOTT, J.M. SHOFFNER, J.A. HODGE, R.I. KELLEY, C.M. EPSTEIN and L.C. HOPKINS: Familial mitochondrial encephalomyopathy (MERRF): genetic, pathophysiological and biochemical characterization of a mitochondrial DNA disease. Cell 56 (1988b) 601–610.

WALLACE, D.C., J.M. SHOFFNER and M.T. LOTT: Myoclonic epilepsy and ragged-red fiber disease (MERRF): a mitochondrial tRNA^{Lys} mutation responsive to Coenzyme Q$_{10}$ (CoQ) therapy. Neurology 41 (1991) 280.

WANDERS, R.J.A., L. IJLST, A.H. VAN GENNIP, C. JAKOBS, J.P. DE JAGER, L. DORLAND, F.J. VAN SPRANG and M. DURAN: Long-chain 3-hydroxyacyl-CoA dehydrogenase deficiency: identification of a new inborn error of mitochondrial fatty acid beta-oxidation. J. Inherit. Metab. Dis. 13 (1990) 311–314.

WATMOUGH, N.J., M.A. BIRCH-MACHIN, L.A. BINDOFF, A. AYNSLEY-GREEN, K. SIMPSON, C.I. RAGAN, H.S.A. SHERRATT and D.M. TURNBULL: Tissue specific defect of complex I of the mitochondrial respiratory chain. Biochem. Biophys. Res. Commun. 160 (1989) 623–627.

WATMOUGH, N.J., L.A. BINDOFF, M.A. BIRCH-MACHIN, S. JACKSON, K. BARTLETT, C.I. RAGAN, J. POULTON, R.M. GARDINER, H.S.A. SHERRATT and D.M. TURNBULL: Impaired mitochondrial beta-oxidation in a patient with an abnormality of the respiratory chain. J. Clin. Invest. 85 (1990) 177–184.

WILLEMS, J.L., L.A.H. MONNENS, J.M.F. TRIJBELS, J.H. VEER-KAMP, A.E. MAYER, K. VAN DAM and U. VAN HAELST: Leigh's encephalomyelopathy in a patient with cytochrome c oxidase deficiency in muscle tissue. Pediatrics 60 (1977) 850–857.

WILLNER, J.H., S. DIMAURO, A. EASTWOOD, A. HAYS, F. ROOHI and R.E. LOVELACE: Muscle carnitine deficiency: genetic heterogeneity. J. Neurol. Sci. 41 (1979) 235–246.

WOELTJE, K.F., V. ESSER, B.C. WEIS, A. SEN, W.F. COX, M.J. MCPHAUL, C.A. SLAUGHTER, D.W. FOSTER and J.D. MCGARRY: Cloning, sequencing, and expression of a cDNA encoding rat liver mitochondrial carnitine palmitoyltransferase II. J. Biol. Chem. 265 (1990) 10720–10725.

YANG, B-Z., J-H. DING, B.I. BROWN and Y-T. CHEN: Definitive prenatal diagnosis for type III glycogen storage disease. Am. J. Hum. Genet. 47 (1990) 735–739.

YANG, B-Z., J.H. DING, J.J. ENGHILD, Y. BAO and Y.-T. CHEN: Molecular cloning and nucleotide sequence of cDNA encoding human muscle glycogen debranching enzyme. J. Biol. Chem. 267 (1992) 9294–9299.

YANG-FENG, T.L., K. ZHENG, J. YU, B.Z. YANG, Y.-T. CHEN and F.T. KAO: Assignment of the human glycogen debrancher gene to chromosome Ip21. Genomics (1992) in press.

YUZAKI, M., N. OHKOSHI, I. KANAZAWA, Y. KAGAWA and S. OHTA: Multiple deletions in mitochondrial DNA at direct repeats of non-D-loop regions in cases of familial mitochondrial myopathy. Biochem. Biophys. Res. Commun. 164 (1989) 1352–1357.

ZELLWEGER, H., S. MUELLER, V. IONASESCU, S.S. SCHOCHET and W.F. MCCORMICK: Glycogenosis type IV: a new cause of infantile hypotonia. J. Pediatr. 80 (1972) 842–844.

ZEVIANI, M. and S. DIDONATO: Neuromuscular disorders due to mutations of the mitochondrial genome. Neuromusc. Disord. 1 (1991) 165–172.

ZEVIANI, M., S. SERVIDEI, C. GELLERA, E. BERTINI, S. DIMAURO and S. DIDONATO: An autosomal dominant disorder with multiple deletions of mitochondrial DNA starting at the D-loop. Nature (London) 339 (1989a) 309–311.

ZEVIANI, M., E. BONILLA, D.C. DE VIVO and S. DIMAURO: Mitochondrial diseases. Neurol. Clinics 7 (1989b) 123–156.

ZEVIANI, M., N. BRESOLIN, C. GELLERA, A. BORDONI, M. PANNACCI, P. AMATI, M. MOGGIO, S. SERVIDEI, G. SCARLATO and S. DIDONATO: Nucleus-driven multiple large-scale deletions of the human mitochondrial genome: a new autosomal dominant disease. Am. J. Hum. Genet. 47 (1990a) 904–914.

ZEVIANI, M., C. GELLERA, M. PANNACCI, G. UZIEL, A. PRELLE, S. SERVIDEI and S. DIDONATO: Tissue distribution and transmission of mitochondrial DNA deletions in mitochondrial myopathies. Ann. Neurol. 28 (1990b) 94–97.

ZEVIANI, M., P. AMATI, N. BRESOLIN, C. ANTOZZI, G. PICCOLO, A. TOSCANO and S. DIDONATO: Rapid detection of the A-to-G$^{(8344)}$ mutation of mtDNA in Italian families with myoclonus epilepsy and ragged-red fibers (MERRF). Am. J. Hum. Genet. 48 (1991a) 203–211.

ZEVIANI, M., C. GELLERA, C. ANTOZZI, M. RIMOLDI, L. MORANDI, F. VILLANI, V. TIRANTI and S. DIDONATO: Maternally inherited myopathy and cardiomyopathy: association with mutation in mitochondrial DNA tRNA$^{Leu(UUR)}$. Lancet 338 (1991b) 143–147.

ZHENG, X., J.M. SHOFFNER, M.A. LOTT, A.S. VOLJAVEC, N.S. KRAWIECKI, K. WINN and D.C. WALLACE: Evidence in a lethal infantile mitochondrial disease for a nuclear mutation affecting respiratory complexes I and IV. Neurology 39 (1989) 1203–1209.

ZIERZ, S. and A.G. ENGEL: Regulatory properties of a mutant carnitine palmitoyltransferase in human skeletal muscle. Eur. J. Biochem. 149 (1985) 207–214.

ZINN, A.B., D. KERR and C.L. HOPPEL: Fumarase deficiency: a new cause of mitochondrial encephalomyopathy. N. Engl. J. Med. 315 (1986) 469–475.

Handbook of Clinical Neurology, Vol. 18 (62): Myopathies
L.P. Rowland and S. DiMauro, editors

CHAPTER 17

The endocrine myopathies

JOHN T. KISSEL and JERRY R. MENDELL

Division of Neuromuscular Disease, Department of Neurology, Ohio State University, Columbus, OH, USA

Endocrine myopathies have overlapping features despite the unique aspects of individual hormonal disorders. Most endocrine myopathies are manifest by weakness, predominantly affecting the proximal limb muscles with prominent fatigue and occasional muscle cramps or pain. The symptoms and signs now tend to be mild, although as originally described, many of the endocrine myopathies were quite debilitating. This shift in clinical manifestations coincided with the advent of sophisticated laboratory methods that facilitate early diagnosis.

However, knowledge of pathophysiologic mechanisms is disappointingly incomplete for all the endocrine myopathies and specific abnormalities in muscle biochemistry, structure, or function cannot usually be demonstrated. Nevertheless, diagnosis is important because treatment of the underlying endocrine disturbance almost always reverses the muscle weakness.

Use of the term 'endocrine myopathy' is an important semantic issue. Most endocrine disorders cause non-specific muscle pathology, usually consisting of varying degrees of atrophy, particularly of the type 2 fibers (Fig. 1). Therefore, some myologists insist that the term myopathy is inappropriate. Nevertheless, terms like 'thyrotoxic myopathy' and 'steroid myopathy' are ingrained in the medical literature and we prefer to follow this precedent. While the shortcomings are apparent, the alternatives seem no better; use of terms like thyroid or parathyroid atrophy would only add to the confusion.

THYROID DISORDERS

Abnormalities of the thyroid gland are associated with an array of muscle disorders (Table 1), in part reflecting the protean effects of the two metabolically active thyroid hormones, L-thyroxine (T4) and 3,5,3′-triiodo-L-thyronine (T3). Both hormones regulate a wide range of metabolic processes that affect skeletal muscle metabolism and function. Specifically, thyroid hormones enhance

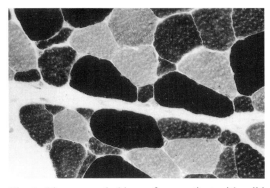

Fig. 1. Biceps muscle biopsy from patient with mild weakness on chronic high-dose corticosteroid therapy. There is severe atrophy of the type 2B fibers. Myofibrillar ATPase, pH 4.6 (× 200).

or accelerate all phases of lipid synthesis and deg-radation, stimulate protein synthesis and enzyme production, and regulate many aspects of carbohy-drate absorption and metabolism. Thyroid hormones also stimulate calorogenesis in muscle, increase muscle demand for vitamins, and enhance muscle sensitivity to circulating catecholamines (Ingbar 1985; Lavin 1989).

Disorders associated with thyrotoxicosis

The term thyrotoxicosis refers to the complex of clinical and biochemical features resulting from ex-cess circulating thyroid hormone. Thyrotoxicosis is not a specific disease, but rather has many causes (Table 2). The term hyperthyroidism, although often mistakenly used interchangeably with thyro-toxicosis, more correctly refers to conditions aris-ing from true hyperfunction of the thyroid gland (Table 2). Thyrotoxicosis of any cause can be asso-ciated with thyrotoxic myopathy. Thyroid-associ-ated ophthalmopathy, however, is usually seen in the setting of Graves' disease, an autoimmune thy-roid disease of uncertain etiology (Utiger 1991). Graves' disease is the most common cause of hy-perthyroidism in the United States (Amino et al. 1980; Ingbar 1985).

Thyrotoxic myopathy. Graves (1835) and Von Basedow (1840) both recognized the presence of muscle weakness in hyperthyroid patients, but these observations were generally ignored until they were re-emphasized by Ramsay (1965, 1966, 1974). In 5% of thyrotoxic patients muscle weak-ness is the dominant symptom, but more than 75% have weakness on clinical examination (Havard et al. 1963; Satoyoshi et al. 1963a; Ramsay 1974; Puvanendran et al. 1979a). Although thyrotoxico-sis is more common in women, the myopathy is

equally distributed between the sexes, suggesting greater susceptibility in men (Ramsay 1974).

Muscle symptoms usually begin several months after the onset of thyrotoxicosis (McEachern and Ross 1942; Ramsay 1974). Proximal limb weak-ness is characteristic, usually associated with mild wasting. In some patients, the arms are more af-fected than the legs. In fact, prominent shoulder girdle wasting with scapular winging (Fig. 2) is a distinctive feature that was first recognized by Ramsay (1966). Distal weakness is seen in about 20% and may be an isolated manifestation (Puva-nendran et al. 1979a). Rarely, limb weakness may be accompanied by dysphagia, dysphonia, aspira-tion or respiratory compromise (Gaan 1967; Joasoo et al. 1970; Kammer and Hamilton 1974; Weinstein et al. 1975; Marks et al. 1980; Sweatman and Chambers 1985). However, these cases have not been clearly differentiated from myasthenia gravis (Puvanendran et al. 1979b), although some authors use the term 'acute thyrotoxic myopathy' (Kammer and Hamilton 1974; Stern and Fagan 1979; Marks et al. 1980). Muscle stretch reflexes are usually normal or hyperactive with shortened relaxation times (Lambert 1951). Cramps, myal-gia, and fasciculations or myokymia are occasion-ally encountered, possibly as a result of superim-posed peripheral nerve or anterior horn cell dys-function (Havard et al. 1963; McComas et al. 1974; Feibel and Campa 1976). In fact, several patients have been reported with hyperthyroidism and clin-

TABLE 2

Differential diagnosis of thyrotoxicosis.

A. Disorders of excess hormone production by thy-roid gland (hyperthyroidism)
   1. Graves' disease
   2. Toxic multinodular goiter
   3. Toxic adenoma
   4. Excess thyroid-stimulating hormone secretion by pituitary
   5. Trophoblastic tumor with TSH secretion
   6. Iodide-induced hyperthyroidism

B. Disorders without excess hormone production by thyroid gland
   1. Excess thyroid hormone ingestion
   2. Ectopic thyroid tissue (e.g. metastatic thyroid carcinoma)
   3. Subacute thyroiditis
   4. Chronic thyroiditis with transient thyrotoxico-sis

TABLE 1

Myopathic syndromes associated with thyroid dis-ease.

---

1. Thyrotoxic myopathy
2. Thyroid-associated ophthalmopathy
3. Thyrotoxic periodic paralysis
4. Hypothyroid myopathy
5. Myasthenia gravis

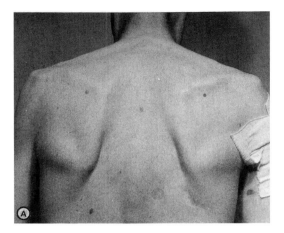

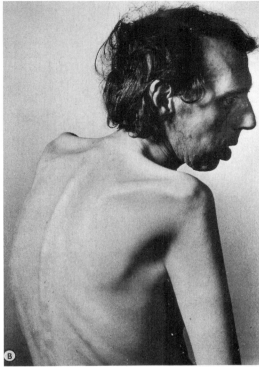

Fig. 2. Patients (A and B) with thyrotoxic myopathy show marked wasting of shoulder girdle muscles and scapular winging. (Photographs courtesy of Dr. I. Ramsay.)

ical syndromes that closely resemble amyotrophic lateral sclerosis, including the presence of upper motor neuron signs (Rowland 1982; Pou Serradell et al. 1990).

Serum creatine kinase (CK) levels and other sarcoplasmic enzymes (aldolase and aspartate aminotransferase) are usually normal or even low in thyrotoxic myopathy (Graig and Smith 1965; Docherty et al. 1984), probably related to increased clearance from blood (Karlsberg and Roberts 1978). Serum CK activity may be extremely high, with rhabdomyolysis and myoglobinuria, in patients with thyroid storm (Bennett and Huston 1984).

Electromyographic (EMG) abnormalities are found in most patients, particularly in proximal limb muscles (Ramsay 1965; Ramsay 1974; Puvanendran et al. 1979a). The myopathic pattern includes excessively recruited motor unit action potentials of decreased duration and increased polyphasia. Fasciculations, fibrillations and positive sharp waves are unusual but may be found in distal muscles (Ludin et al. 1969; Ramsay 1974). The electrodiagnostic features correlate with the degree of weakness and wasting but not with the severity of thyrotoxicosis (Havard et al. 1963; Buchthal 1970).

The frequency of histologic abnormalities on muscle biopsy increases with the duration of thyrotoxic symptoms but does not correlate with degree of thyrotoxicosis (Satoyoshi et al. 1963a; Ramsay 1974). Most patients show non-specific changes consisting of atrophy (usually of both fiber types), slight fatty infiltration and connective tissue proliferation (Gruener et al. 1975; Wiles et al. 1979; Korenyi-Both et al. 1981). Non-specific ultrastructural changes include mitochondrial enlargement and hyperplasia with localized areas of mitochondrial loss. In addition, focal myofibrillar degeneration, dilatation of the transverse tubular system, subsarcolemmal glycogen deposits and papillary sarcolemmal projections have been noted (Engel 1966a, 1972).

Although numerous observations have been

made on the effects of thyroid hormone excess on skeletal muscle structure and function, the specific pathogenesis of thyrotoxic myopathy is unclear. Evidence does not support the early hypothesis that uncoupling of oxidative phosphorylation is the primary defect; mitochondrial respiratory chain enzyme activity is normal in thyrotoxicosis (Gustafsson et al. 1965; Stocker et al. 1968; Argov et al. 1988). Rather, the clinical findings in thyrotoxic myopathy appear to arise from a combination of factors resulting from the effects of thyroid hormones on muscle biochemistry, metabolism and physiology.

Fundamentally, excess thyroid hormone results in accelerated muscle protein catabolism (Goldberg et al. 1977; Rodier et al. 1984), an effect that could be mediated by enhanced lysosomal protease activity (Demartino and Goldberg 1978; Brown and Millward 1980; Decker and Wildenthal 1981). In addition, thyroid hormone disrupts the anabolic effects of insulin on amino acid and protein metabolism and therefore leads to decreased muscle protein synthesis (Kaciuba-Uscilko and Brzezinska 1979; Kingston 1983; Morrison et al. 1988). This relative insulin resistance also disrupts carbohydrate metabolism, resulting in glycogen depletion and decreased glucose uptake in thyrotoxic skeletal muscle (Okajima and Ui 1979; Ruff 1986).

The muscle membrane and neuromuscular junction are also affected by excess thyroid hormone (Puvanendran et al. 1979b). Experimentally, hyperthyroidism causes a reduction in the amplitude of the miniature end-plate potentials and muscle membrane potentials (Hofmann and Denys 1972). In clinical studies, Gruener et al. (1975) demonstrated partial depolarization of thyrotoxic muscle and resulting reduced membrane excitability to repetitive stimulation.

An end-result of the many biochemical and electrophysiologic derangements is reduced efficiency of contraction of thyrotoxic muscle; ATP expenditure for a given amount of work is increased twofold while work output per muscle area is significantly decreased in thyrotoxicosis. Both abnormalities are reversed by restoration of the euthyroid state (Wiles et al. 1979; Zürcher et al. 1989). These findings are compatible with the clinical observation that weakness in thyrotoxic patients is usually more severe than the degree of muscle wasting.

Successful treatment of thyrotoxicosis results in improvement of the myopathy, usually over a period of months (Engel 1966b; Ramsay 1966; Buchthal 1970; Ramsay 1974). Strength improves before reversal of muscle wasting. Propranolol, known to ameliorate many of the systemic signs and symptoms of hyperthyroidism, improves muscle strength in proximal limb myopathy (Pimstone et al. 1968) and the oropharyngeal muscles (Kammer and Hamilton 1974; Weinstein et al. 1975).

Thyroid-associated ophthalmopathy. The exophthalmos and ocular muscle dysfunction seen with thyroid disease have been discussed under a variety of names including: thyroid-associated ophthalmopathy (TAO), thyrotoxic ocular myopathy, Graves' ophthalmopathy, exophthalmic ophthalmoplegia, endocrine exophthalmos or ophthalmopathy, infiltrative ophthalmopathy, Graves' ophthalmoplegia, dysthyroid orbitopathy and malignant exophthalmos. (This condition is also described in Chapter 10.) Although TAO accompanies Graves' hyperthyroidism in almost all cases, the condition can also occur with Hashimoto's thyroiditis (5% of cases). TAO in euthyroid individuals is called 'euthyroid Graves' disease' (Salvi et al. 1990); most patients in this group later develop clinical or immunologic evidence of thyroid disease (Agapitos and Hart 1987; Salvi et al. 1990).

The cardinal manifestations of Graves' disease (goiter, ophthalmopathy and infiltrative dermopathy) may follow independent courses. The ophthalmopathy is extremely variable, but usually has an insidious onset and course; however, there is an acute fulminating form. Involvement of both eyes occurs in most cases but asymmetry is common (Hall et al. 1970; Wiersinga et al. 1989). Although ocular manifestations usually begin within months after the onset of thyrotoxicosis, ocular symptoms may appear months or years after successful treatment of hyperthyroidism (Havard 1979).

The American Thyroid Association has classified Graves' ophthalmopathy according to severity, a schema sometimes referred to as the 'NO-SPECS' system (an acronym derived from the first letter of the cardinal manifestation in each grade, Table 3) (Werner 1977). Lid-lag and stare (Class 1)

TABLE 3

Detailed classification of ocular changes in Graves' disease (NOSPECS system).

(After Werner 1977; Wiersinga, et al. 1989)

Grade	Suggestions for grading
Class 0	No physical signs or symptoms
Class 1	Only signs (limited to upper lid retraction, stare and lid-lag)
Class 2	Soft tissue involvement with symptoms and signs
0	Absent
a	Minimal
b	Moderate
c	Marked
Class 3	Proptosis 3 mm or more in excess of upper normal limits
0	Absent
a	3–4 mm increase over upper normal (23–24 mm)
b	5–7 mm (25–27 mm)
c	8 mm or more increase ($\geq$ 28 mm)
Class 4	Extraocular muscle involvement (usually with diplopia)
0	Absent
a	Limitation of motion at extremes of gaze
b	Evident restriction of motion
c	Fixation of a globe or globes
Class 5	Corneal involvement (primarily due to lagophthalmos)
0	Absent
a	Stippling of cornea
b	Ulceration
c	Clouding, necrosis, perforation
Class 6	Sight loss (due to optic nerve involvement)
0	Absent
a	Disc pallor or choking, or visual field defect; vision 20/20 — 20/60
b	Same, but vision 20/70 —20/200
c	Blindness, i.e. failure to perceive light vision < 20/200

have been attributed to increased sensitivity of the sympathetic nervous system to catecholamines (Wall et al. 1991). In more advanced cases, inflammation and fibrosis of the levator palpebrae superioris and Müller's muscles also contribute to lid-lag (Hodes et al. 1979). Ocular irritation, conjunctival injection and edema, lid fullness and exophthalmos are the early manifestations of true ophthalmopathy (Class 2 and 3). When ocular muscles are affected, diplopia ensues (Class 4). The inferior rectus and inferior oblique muscles are involved first, disrupting vertical before horizontal gaze (Hodes et al. 1979; Wiersinga et al. 1989). As

the disease advances, vision may be impaired by exposure-induced corneal ulceration and keratitis (Class 5) or the optic nerve may be compromised by retro-orbital congestion and vascular compression (Class 6). Visual loss may be irreversible.

The diagnosis of TAO is apparent in 90% of patients harboring the classic triad of Graves' disease (ophthalmopathy, goiter and dermopathy). Difficulties arise in patients with typical TAO in the absence of hyperthyroidism (Amino et al. 1980; Wall et al. 1981; Salvi et al. 1990). Nevertheless, euthyroid Graves' disease is defined by abnormalities in the T3 suppression test, thyrotropin-releasing hormone stimulation test or thyroid immunologic testing (Salvi et al .1990); about half eventually become clinically and biochemically hypothyroid (Tamai et al. 1980; Agapitos and Hart 1987). Orbital ultrasound, computed tomography (CT) and magnetic resonance imaging (MRI) demonstrate the characteristic eye muscle enlargement and increased retro-orbital tissue (Yamamoto et al. 1983; Forbes et al. 1986; Nadalo et al. 1991).

The pathogenesis of the ophthalmopathy is poorly understood but thought to have an autoimmune basis (Alper 1990; Volpe 1990; Wall et al. 1991). One theory suggests that ocular problems result from cell-mediated immune mechanisms directed against an unidentified retro-orbital muscle antigen (Hiromatsu et al. 1987). Other evidence suggests that exophthalmos can be accounted for by an 'exophthalmos-producing immunoglobulin' that acts in conjunction with a part of the thyroid-stimulating hormone (TSH) molecule to stimulate TSH receptors in retro-orbital fat cells (Havard 1979). Thyroglobulin-antithyroglobulin immune complexes with high affinity for extraocular muscles have been demonstrated in the orbit and may reach this location through retrograde lymphatic flow from the thyroid gland (Konishi et al. 1974). According to this view, the immune complexes might initiate retro-orbital inflammation mediated through cytotoxic lymphocytes and killer cells.

The relationship of hyperthyroidism to ophthalmopathy remains unclear. However, TAO patients may have circulating antibodies that react with antigens in ocular muscles as well as thyroid cells (Salvi et al. 1988; McKenzie and Zakarija 1989).

Graves' ophthalmopathy must be treated independently from the hyperthyroidism (Dresner and

Kennerdell 1985; Bahn and Gorman 1987) because successful management of hyperthyroidism improves the ocular disease in less than 5% of patients (Jones et al. 1969). A variety of immunosuppressive drugs have been used to treat TAO, but controlled trials have not been performed. The drugs used include corticosteroids, azathioprine, cyclophosphamide, or cyclosporine (Wall et al. 1981; Bahn and Gorman 1987). Plasma exchange (Dandona et al. 1979) and supervoltage radiation (Teng et al. 1980; Yamamoto et al. 1983; Dresner and Kennerdell 1985) have also been used. Surgical decompression may be necessary for threatened visual loss due to excessive intra-orbital pressure or exposure keratitis due to proptosis. Muscle resection can be considered to correct diplopia due to fibrosis of the muscles (Dyer 1984).

Thyrotoxic periodic paralysis. Periodic paralysis is a recognized complication of thyrotoxicosis, occurring predominantly in Asians. Approximately 75% of reported patients were of Japanese or Chinese descent; in some series, up to 13% of Asian thyrotoxic men have had periodic paralysis (Satoyoshi et al. 1963b; Ramsay 1982). Susceptibility to thyrotoxic periodic paralysis may be an autosomal-dominant trait in Asians (Kufs et al. 1989) and occurs in association with certain human leukocyte antigen (HLA) haplotypes including A2, Bw22, AW 19, B17, and DRw8 (Yeo et al. 1978; Tamai et al. 1987). Men have thyrotoxic periodic paralysis 6–20 times more frequently than women despite the higher incidence of hyperthyroidism in women (Satoyoshi et al. 1963b; Engel 1972; Tamai et al. 1987; Kelley et al. 1989).

Clinical features of thyrotoxic periodic paralysis are identical with familial hypokalemic periodic paralysis (Engel 1961a, 1966b). Attacks may be precipitated by exercise, exposure to cold, ingestion of alcohol, or a high salt or high carbohydrate diet. Paralysis commonly affects the proximal limb and trunk muscles and usually spares the oropharynx and diaphragm (Engel 1972). Weakness typically lasts for hours but may persist for up to a week (Kendall-Taylor and Turnbull 1983). The frequency of attacks varies; in some individuals attacks occur several times a week.

The similarities between thyrotoxic periodic paralysis and the familial hypokalemic form extend to laboratory findings (Kelley et al. 1989; Oh et al. 1990). During an attack, serum potassium levels usually fall (McFadzean and Yeung 1967). Between episodes, potassium levels are normal. For diagnostic purposes, attacks may be induced by infusion of insulin and glucose.

Correction of thyrotoxicosis prevents further attacks in almost all patients. Acute episodes of paralysis respond well to oral administration of potassium. Propranolol is useful in eliminating or reducing attacks (Conway et al. 1974; Yeung and Tse 1974; Kufs et al. 1989). Other forms of periodic paralysis are discussed in Chapter 12.

Hypothyroid myopathy

Muscle symptoms are common in hypothyroid patients. Although there have been few studies of unselected hypothyroid patients, clinical weakness occurs in 30–40% of patients irrespective of the etiology of hypothyroidism (Table 4) (Nickel et al. 1961; Collins et al. 1964; Ramsay 1974; Rao et al. 1980). The severity of the muscle symptoms generally correlates with the degree and chronicity of hypothyroidism (Khaleeli et al. 1983a). Muscle cramps, pain and stiffness may be precipitated by exercise or cold weather (Wilson and Walton 1959; Golding 1970). The 'myotonoid' or 'pseudomyotonic' features of slowed muscle contraction and relaxation occur in 25% of patients (Ramsay 1974).

Weakness, especially of proximal limb muscles, develops insidiously over months. Most patients have normal muscle bulk but wasting has been observed, especially in chronic hypothyroidism (Ramsay 1974; Rao et al. 1980; Khaleeli et al. 1983a). Muscular hypertrophy, often mentioned in early descriptions, is now rare. The Achilles tendon reflex time is prolonged in most hypothyroid patients (Lambert et al. 1951; Ringqvist 1970; Khaleeli and Edwards 1984) but the diagnostic utility of this sign is limited because it has been found in other diseases and can also be caused by pharmacologic agents (Swanson et al. 1981; Layzer 1985). Myoedema or the 'mounding phenomenon' corresponds to the electrically silent, painless, muscle contracture produced by direct percussion. This phenomenon is seen in about one-third of hypothyroid patients but can also be evoked in cachectic people and sometimes in normal subjects (Swanson et al. 1981; Mizusawa et al. 1983).

Hypothyroidism may be associated with distinct clinical features. The Kocher-Debré-Sémélaigne syndrome occurs in hypothyroid children, especially boys. Weakness, slowness of movement, and striking muscular hypertrophy cause an 'infant Hercules' appearance (Najjar 1974). In adults, the combination of muscle enlargement, stiffness, and 'myotonoid' features is called 'Hoffmann's syndrome' (Klein et al. 1981). The cause of muscle enlargement in hypothyroidism is not known; there are no distinctive light or ultrastructural features (Afifi et al. 1974; Ruff 1986).

Serum CK activity is typically elevated in hypothyroid myopathy (Graig and Smith 1965; Khaleeli et al. 1983a; Docherty et al. 1984; Khaleeli and Edwards 1984), probably due to decreased clearance from the blood (Karlsberg and Roberts 1978; Docherty et al. 1984). Hypothyroid patients often have high CK activity without clinical manifestations of myopathy. Aspartate aminotransferase, aldolase, lactic dehydrogenase, other sarcoplasmic enzymes, and myoglobulin level are also increased (Docherty et al. 1984).

A typical myopathic pattern is seen in electrophysiologic studies with a decrease in the amplitude and duration of motor unit potentials as well as an increase in the number of polyphasic motor unit potentials (Ramsay 1974; Rao et al. 1980; Khaleeli et al. 1983a). Fasciculations, fibrillations and positive sharp waves are uncommon (Scarpalezos et al. 1973; Ramsay 1974). Myotonia may

TABLE 4

Differential diagnosis of hypothyroidism.

A. Associated with thyroid atrophy or loss
  1. Primary hypothyroidism
  2. Postradiation or postsurgical hypothyroidism
  3. Congenital thyroid dysplasia
B. Associated with goiter
  1. Hashimoto's thyroiditis
  2. Heritable biosynthetic defects
  3. Iodine deficiency
  4. Antithyroid agents
  5. Peripheral resistance to thyroid hormone
C. Associated with suprathyroid dysfunction
  1. Postpartum pituitary necrosis (Sheehan's syndrome)
  2. Tumor infiltration of pituitary or hypothalamus

be seen (Venables et al. 1978). A decrementing response to repetitive nerve stimulation that corrects with edrophonium has also been reported with hypothyroidism even when there is no clinical evidence of myasthenia gravis (Norris 1966; Norris and Panner 1966; Takamori et al. 1972). This neuromuscular transmission defect corrects with treatment of hypothyroidism. The presence of a hypothyroid polyneuropathy may confuse the electrodiagnostic picture (Dyck and Lambert 1970; Scarpalezos et al. 1973; Meier and Bischoff 1977; Rao et al. 1980).

Muscle biopsy changes of type 2 atrophy usually parallel the degree of clinical wasting and weakness. Type 1 fiber hypertrophy has been described, particularly in women (Wiles et al. 1979; Khaleeli et al. 1983b; Klein and Levey 1984). Isolated necrotic fibers may occasionally be seen (Ramsay 1974). In severely affected muscles, intracellular glycogen deposits may be found in scattered muscle fibers (McKeran et al. 1980), and there may be single or multiple large vacuoles (Carpenter and Karpati 1984). Non-specific ultrastructural features include myofibrillar degeneration, Z-disc streaming, lipofuscin accumulation, and mitochondrial alterations (Norris and Panner 1966; Khaleeli et al. 1983b; Ruff 1986). No distinctive light microscopic or ultrastructural features accompany the muscle enlargement of hypothyroidism (Afifi et al. 1974; Ruff 1986).

The weakness of hypothyroidism is attributed to a complex set of effects of thyroid hormone deficiency on skeletal muscle structure and function. Impaired muscle energy metabolism appears to be the primary factor. Mitochondrial enzyme activity and oxidation capacity are diminished (Janssen et al. 1978), findings supported by magnetic resonance spectroscopy (Argov et al. 1988). Hypothyroidism also impairs the generation of free fatty acids from lipids and reduces muscle glycogenolysis, resulting in decreased energy sources during activity (McDaniel et al. 1977; Wiles et al. 1979; Baldwin et al. 1980a,b). Skeletal muscle protein turnover is relatively unchanged in hypothyroid patients, even though total body protein synthesis and breakdown are reduced (Morrison et al. 1988).

Most of the physiologic alterations of hypothyroidism, including myoedema, can be attributed to reduced myosin ATPase activity (Ianuzzo et al.

1977; Wiles et al. 1979; Nwoye et al. 1982) and decreased calcium uptake by the sarcoplasmic reticulum (Peter et al. 1970). These alterations include decreased velocity of muscle shortening, reduced tension generation and prolonged relaxation phase (Gold et al. 1970; Takamori et al. 1971; Wiles et al. 1979; Everts et al. 1981). Myoedema results when a percussion-induced liberation of calcium from the sarcoplasmic reticulum causes a muscle contraction that persists because of delay in calcium re-uptake (Fanburg 1968; Nwoye et al. 1982; Mizusawa et al. 1983). The multiple biochemical and physiologic derangements induced by hypothyroidism result, as in thyrotoxicosis, in reduced muscle efficiency measured as work output per muscle area (Zürcher et al. 1989).

Successful treatment of hypothyroidism improves most of the muscular abnormalities. Serum CK activity and Achilles reflex time return to normal shortly after thyroid replacement. Muscle hypertrophy, if present, takes months to recede (Najjar 1974; Ramsay 1974; Klein and Levey 1984). Weakness improves slowly and can persist up to one year after a euthyroid state is achieved (Khaleeli and Edwards 1984). The biopsy changes also improve, although slowly, and some degree of type 2 atrophy may persist (McKeran et al. 1980; Khaleeli et al. 1983b).

Thyroid disease and myasthenia gravis

The prevalence of myasthenia gravis in thyrotoxic patients is about 30 times greater than in the general population (Simpson 1968; Ramsay 1974). Conversely, the prevalence of thyrotoxicosis in a myasthenic population is about 5%, significantly greater than the 0.4–1.0% prevalence of thyrotoxicosis in the general population (Ramsay 1974; Ingbar 1985). Women with myasthenia develop thyrotoxicosis 5 times more commonly than men (Ramsay 1974). Although Graves' disease is the most common thyroid disease seen with myasthenia gravis, other causes of thyrotoxicosis (see Table 2) have been associated with myasthenia. Thyrotoxicosis usually precedes the onset of myasthenia, but in 20–30% of cases the myasthenia gravis appears first (Millikan and Haines 1953; Rowland et al. 1966; Kissel et al. 1970; Ramsay 1974).

Hypothyroidism may also occur with myasthe-

nia gravis (Sahay et al. 1965; Simpson et al. 1966; Osserman et al. 1967; Kiessling et al. 1981), most often in association with Hashimoto's thyroiditis (lymphadenoid goiter) (see Table 4). Sometimes, proof of Hashimoto's disease is found only at postmortem examination (Becker et al. 1964; Ramsay 1974).

The clinical picture and diagnosis do not differ from myasthenia gravis without thyroid disease. Recognition of both disorders is important because the aberrations of thyroid hormone aggravate the myasthenic symptoms.

There seems to be a genetic predisposition for the autoimmune origin of both myasthenia gravis and thyroid disease. Forty percent of patients with combined myasthenia gravis and thyroid disease have demonstrable circulating thyroid antibodies; even myasthenic patients without symptoms of hyperthyroidism have significant levels of antithyroid antibodies (Ramsay 1974; Kiessling et al. 1981). The HLA haplotypes B8 and DR3 are risk factors for both Graves' disease and myasthenia gravis in white individuals (Behan 1980; Ingbar 1985).

Despite the simultaneous occurrence of thyroid disease and myasthenia gravis, each requires independent treatment. Both hyper- and hypothyroidism adversely affect the neuromuscular junction and aggravate myasthenia gravis (Engel 1961b; Drachman 1962; Norris 1966; Gaelen and Levitan 1968; Hofmann and Denys 1972; Takamori et al. 1972).

ADRENAL DISORDERS

Separate zones of the adrenal cortex synthesize specific hormones, each associated with a different clinical disorder. The zona fasciculata secretes glucocorticoids, of which cortisol (hydrocortisone) is the most important, predominantly regulating intermediary metabolism. The physiologic actions of glucocorticoids are primarily anti-insulin and include the regulation of protein, carbohydrate, lipid and nucleic acid metabolism. Since the actions are mainly catabolic, conditions associated with glucocorticoid excess cause a myopathy. In contrast, glucocorticoid deficiency rarely causes muscle disease (Layzer 1985; Mor et al. 1987).

The major mineralocorticoid, aldosterone, is secreted from the zona glomerulosa. Aldosterone regulates extracellular fluid volume and profoundly affects potassium homeostasis. Primary aldosteronism due to a secreting adenoma causes hypokalemia and muscle weakness.

A third group of hormones, the adrenal androgens, derive primarily from the zona reticularis. Excess production of adrenal androgens causes virilism, a condition not usually associated with neuromuscular manifestations. This condition will therefore not be discussed further in this chapter, even though the use of anabolic steroids to improve athletic performance has been a subject of expanding interest. After androgen therapy in hypogonadal men, enlargement of muscle mass parallels an increase in diameter of muscle fibers (Chauhan et al. 1986). However, controlled trials of the effects of anabolic steroids on muscle bulk, strength or athletic performance in healthy adults have shown inconsistent and often conflicting results (Haupt and Rovere 1984; Lamb 1984; Griffin and Wilson 1985; Wilson 1988; Rogozkin 1991).

Cushing's syndrome and steroid myopathy

The widespread clinical use of glucocorticoids accounts for steroid myopathy as the most commonly diagnosed endocrine muscle disease (Ramsay 1982). Cushing first called attention to the association of excess glucocorticoid levels and muscle weakness in patients with pituitary adenomas (Cushing 1932), observations later expanded by others (Plotz et al. 1952; Dubois 1958; Freyberg et al. 1958; Harman 1959; Kendall and Hart 1959; MacLean and Schurr 1959; Perkoff et al. 1959; Williams 1959). Dubois (1958) reported muscle weakness as the most serious side effect of a 'new synthetic fluorinated, unsaturated, anti-inflammatory hormone, triamcinolone'.

Steroid excess produces varying degrees of proximal limb weakness especially in the legs, sparing facial and oropharyngeal muscles. Muscle wasting may be striking (Fig. 3) (Askari et al. 1976; Khaleeli et al. 1983c; Bowyer et al. 1985). A cushingoid appearance invariably precedes or accompanies clinical signs of myopathy.

Duration of treatment and dosage do not necessarily predict weakness, but rather an individual sensitivity to the hormone; women seem more susceptible to steroid-induced weakness (Bunch et al. 1980). Steroid myopathy occurs in every age group and with virtually every synthetic glucocorticoid analogue; muscle damage may appear sooner with the 9α-fluorinated glucocorticoids (Golding et al. 1961; Byers et al. 1962; Pearce 1963; Fauldi et al. 1964, 1966; Coomes 1965; Afifi et al. 1968; Pleasure et al. 1970; Askari et al. 1976). Patients on daily prednisone doses exceeding 30 mg are more vulnerable to myopathy than those who take lower doses or alternate-day regimens (Bowyer et al. 1985).

In common with other endocrine disorders, muscle atrophy predominantly affects type 2 fibers, especially type 2B (Pleasure et al. 1970; Askari et al. 1976; Bunch et al. 1980; Khaleeli et al. 1983c) (see Fig. 1). Early clinical reports strongly emphasizing the presence of necrotic and regenerating fibers as the major histologic feature were probably in error (Müller and Kugelberg 1959; Perkoff et al. 1959; Byers et al. 1962). Ultrastructural changes include a biphasic response of mitochondrial proliferation and aggregation, followed by degeneration and mitochondrial loss (Pearce 1963; Engel 1966a; Tice and Engel 1967; Afifi and Bergman 1969; Goldberg and Goodman 1969; Pleasure et al. 1970). There may be irregularities of Z lines, myofibrillar disarray and myofilament loss (Tice and Engel 1967; Afifi and Bergman 1969; Khaleeli et al. 1983c).

Serum CK activity is normal in steroid myopathy, but creatinuria is expected (Golding et al. 1961; Perkoff et al. 1959; Askari et al. 1976). A selective increase of lactic dehydrogenase has not been confirmed (Kanayama et al. 1981). Meager EMG findings are the rule, although a myopathic pattern may be seen (Müller and Kugelberg 1959; Golding et al. 1961; Prabhu and Oester 1971; Khaleeli et al. 1983c).

Diagnosis of steroid myopathy is controversial and challenging, especially because the underlying disease can usually account for the muscle weakness. A constellation of features indicating steroids as the offending agent includes: cushingoid appearance, normal serum CK activity, and minimal findings by EMG, particularly a lack of membrane irritability. Creatinuria is useful only if a previous baseline had been established, because creatinine excretion does not correlate with steroid dose or severity of weakness (Bowyer et al. 1985). Muscle biopsy may be misleading since type 2 atrophy can frequently occur with any underlying systemic disease.

The pathogenesis of steroid-induced muscle weakness has been studied extensively. Muscle atrophy seems to be due mainly to decreased muscle protein synthesis (Goldberg and Goldspink 1975; Shoji 1989; Pacy and Halliday 1989) and at high-dose therapy, a concomitant increased protein catabolism occurs, especially with high-dose therapy (Goldberg 1967; Shoji and Pennington 1977). Protein catabolism relates to increased skeletal muscle protease activity (Mayer et al. 1976; Clark and Vignos 1981). Catabolic effects of glucocorticoids are more pronounced in less active muscles, which might account for the selective atrophy of fast-twitch or type 2 motor units with relative sparing of type 1 or slow-twitch motor units (Goldberg and Goodman 1969). These findings are also consistent with observations that steroid atrophy is lessened by exercise (Gardiner et al. 1980; Horber et al. 1985, 1987) and that the number of glucocorticoid receptors increases in immobilized muscle (DuBois and Almon 1980). Glucocorticoids also interfere with both muscle carbohydrate metabolism and oxidative capacity, although the significance of these changes is uncertain because ATP levels in muscle do not change with steroid treatment (Afifi and Bergman 1969; Peter et al. 1970; Peters et al. 1970; Vignos and Greene 1973; Koski et al. 1974; Ruff 1986).

The discrepancy between the severity of muscle weakness and the paucity of histologic findings has led to speculation that physiologic changes account for the diminished muscle strength. Despite extensive investigation, however, there is no convincing evidence that glucocorticoids alter muscle membrane excitability, calcium uptake or release by sarcoplasmic reticulum, or excitation-contraction coupling (Peter et al. 1970; Vignos and Greene 1973; Shoji et al. 1976; Ruff et al. 1982; Laszewski and Ruff 1985).

Treatment of steroid myopathy requires reduction in dose and conversion to an alternate-day regimen. Unsuccessful treatment programs have included administration of phenytoin (Bowyer et al. 1985) or anabolic steroids (Coomes 1965) Horber et al. (1985, 1987) found that isokinetic ex-

ercises reversed muscle wasting and improved strength. Exercise can also prevent type 2 atrophy from other causes (Aniansson et al. 1980).

Myopathy with aldosterone excess

In primary hyperaldosteronism or Conn's syndrome, neuromuscular complications are due to potassium depletion. The usual clinical picture is one of persistent muscle weakness, although it is sometimes episodic (Conn et al. 1964). Long-standing hyperaldosteronism may lead to proximal limb weakness and wasting that return to normal when K^+ levels are repleted (Sambrook et al. 1972). There may be combinations of episodic weakness simulating hypokalemic periodic paralysis and fixed weakness (Gallai 1977; Atsumi et al. 1979).

CK levels may be elevated and a myopathic pattern may be seen in the EMG. Muscle biopsy may demonstrate necrotic and regenerating fibers, vacuoles, and tubular aggregates (Palmucci et al. 1985). These changes with hyperaldosteronism are attributed to hypokalemia, not excess aldosterone.

Adrenal insufficiency/corticosteroid withdrawal

Muscle fatigue and subjective weakness commonly occur with adrenal insufficiency (Addison's disease). Rare myopathic syndromes with objective weakness have associated electrolyte disturbances or concurrent endocrinopathies (Pollen and Williams 1960; Mor et al. 1987). Severe weakness should prompt evaluation of other etiologies.

Fatigue, subjective weakness, and myalgia are characteristic of withdrawal from corticosteroid therapy, particularly if the dosage of steroids was high, the duration of therapy long, and the speed of taper rapid. Arthralgia, weight loss, skin changes, and occasionally fever can also occur as part of a steroid withdrawal syndrome, and may be confused with an exacerbation of the underlying disease (Axelrod 1989; Kissel and Rammohan 1991). A reduction in the speed of the corticosteroid withdrawal often eliminates these symptoms.

Myopathy with pheochromocytoma

The adrenal medulla secretes epinephrine and

norepinephrine, hormones that bind to alpha or beta receptors on the surface of target cells and control a wide range of metabolic functions that involve essentially every organ in the body, especially the cardiovascular system (Goldfien 1991). Catecholamine excess most commonly occurs with pheochromocytoma, a tumor of the chromaffin cells arising in the adrenal gland in 85% of cases. Skeletal muscle complications of pheochromocytoma are rare, although proximal limb weakness and marked elevation in CK level have been seen (Oristrell-Salva and Mirada-Canals 1984; Bhatnagar et al. 1986). The pathogenesis of the muscle damage may relate to the marked vasoconstriction and increased force of muscle contraction with high levels of circulating catecholamines.

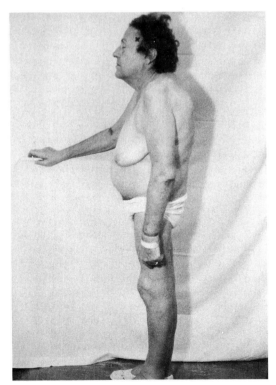

Fig. 3. Patient with Cushing's syndrome and myopathy demonstrating significant proximal muscle atrophy. (Photograph courtesy of Dr. E. Mazzaferri.)

PITUITARY DISORDERS

Myopathy and Nelson's syndrome

The effects of high serum adrenocorticotropin (ACTH) levels on muscle are poorly defined; a direct effect on muscle in the absence of excess glucocorticoids is unproven. The myopathy of Nelson's syndrome, a condition that follows total bilateral adrenalectomy (for many years the standard treatment for Cushing's disease), raised the possibility of an ACTH-induced myopathy. However, patients with Nelson's syndrome were routinely treated with glucocorticoids, making it impossible to differentiate the effects of ACTH and corticosteroids. As originally described, weakness and easy fatiguability were accompanied by generalized hyperpigmentation (Prineas et al. 1968). A myopathic pattern may be seen on EMG with increased insertional activity, positive waves and pseudomyotonic discharges (Prineas et al. 1968). Muscle biopsy showed increased lipid in both type 1 and type 2 fibers, especially in a subsarcolemmal location. Serum CK levels were normal.

Nelson's syndrome has not been reported since 1968 because most adrenal hyperplasia results from pituitary hypersecretion of ACTH from a microadenoma that is now removed by transsphenoidal resection (Tyrell et al. 1978; Layzer 1985).

Myopathy with acromegaly and pituitary gigantism

The association of acromegaly with muscle weakness was mentioned in the first description of the disease by Marie (1886). There are many different causes of weakness in these patients, including peripheral neuropathy (Schiller and Kolb 1954; Skanse 1961; Low et al. 1974), compression of nerve roots in the vertebral foramina (Duchesneau 1892) and spinal cord compression by soft tissue enlargement of the vertebral column (Horenstein et al. 1971). Joint involvement causes arthralgia and, in some patients, crippling degenerative joint disease and scoliosis (Daughaday 1985). With these manifestations, it may be difficult to identify a myopathy, which is said to occur in as many as 50% of patients (Pickett et al. 1975).

The muscle weakness in acromegaly develops insidiously and is a late manifestation of the endocrinopathy. The symptoms are mild and affect proximal limb muscles without wasting (Pickett et al. 1975). Serum CK activity may be normal or mildly elevated and most patients with proximal muscle weakness have EMG changes consisting of small motor unit potentials of diminished amplitude and duration (Lundberg et al. 1970; Mastaglia et al. 1970; Stern et al. 1974; Pickett et al. 1975). Muscle biopsy findings have been inconsistent. Lundberg et al. (1970) and Pickett et al. (1975) found no structural muscle changes, while others (Mastaglia et al. 1970) observed segmental necrosis. Ultrastructural abnormalities include myofibrillar loss (Stern et al. 1974; Pickett et al. 1975), increased glycogen (Mastaglia et al. 1970; Mastaglia 1973; Stern et al. 1974) and increased numbers of satellite cells (Mastaglia 1973). Stern et al. (1974) observed muscle hypertrophy confined to type 2B fibers. Others found hypertrophy and atrophy of all three fiber types, 1, 2A and 2B (Mastaglia et al. 1970; Mastaglia 1973; Nagulesparen et al. 1976).

A myopathy associated with pituitary gigantism was described in only one well documented case (Lewis 1972). A 26-year-old man had symptoms of growth hormone excess beginning at age 5. He differed from acromegalics in that diffuse proximal and distal muscle wasting was present. There was a myopathic EMG pattern and non-specific histologic changes included variability in fiber size, with occasional necrotic, regenerating, and ring fibers. Features of denervation atrophy were also observed and a sural nerve biopsy showed loss of myelinated fibers.

The cause of the muscle changes associated with growth hormone excess have not been elucidated. Attempts to correlate the myopathy with growth hormone levels have not proved useful (Pickett et al. 1975; Nagulesparen et al. 1976). In fact, the duration of acromegaly rather than serum growth hormone levels seems to be a better predictor of myopathy (Pickett et al. 1975; Nagulesparen et al. 1976). Experimental growth hormone administration in rats can lead to muscle hypertrophy (Plattener and Reed 1939; Bigland and Jehring 1952) and increased protein synthesis (Prysor-Jones and Jenking 1980) not accompanied by improved muscle strength (Bigland and Jehring 1952). Prysor-Jones and Jenkins (1980) injected cultured pitui-

tary tumor subcutaneously into rats and produced selective type I fiber hypertrophy similar to the human biopsy findings of Nagulesparen et al. (1976). A series of biochemical abnormalities have also been identified, which could provide some insight into the myopathy. The respiratory quotient of resting forearm muscle in acromegalics is lower than normal (Rabinowitz and Zierler 1963). This is accompanied by an increase in basal oxygen uptake by 50% and a subnormal uptake of glucose in response to intra-arterial administration of insulin. These findings indicate that muscle may use lipid preferentially under the influence of growth hormone (Layzer 1985), a suggestion supported by experimental growth hormone administration which increases free fatty acid oxidation in muscle (Winckler et al. 1964).

PARATHYROID DISORDERS

Myopathy with hyperparathyroidism and alterations in bone metabolism

Muscle weakness is an integral part of primary and secondary hyperparathyroidism and disorders of bone mineralization. Various causes of abnormalities in calcium and phosphorus homeostasis and bone metabolism associated with myopathies are listed and the complexity of interactions is illustrated in Fig. 4. Of these interrelated conditions, a myopathy most commonly occurs with osteomalacia (Layzer 1985), an abnormality characterized by the accumulation of unmineralized bone matrix. Muscle weakness also accompanies rickets (Yalowitz et al. 1984), a condition similar to osteomalacia except that defective mineralization affects the growing skeleton, including the bone and the cartilaginous matrix of the growth plate (Glorieux 1991). Persistently elevated levels of parathyroid hormone produce resorption of bone mineral and matrix with replacement by fibrous tissue, a condition called osteitis fibrosa or, if severe enough, osteitis fibrosa cystica. This occurs in primary hyperparathyroidism due to an autonomously functioning parathyroid gland, either an adenoma or hyperplasia. In secondary hyperparathyroidism there is resistance to the metabolic actions of parathyroid hormone, leading to mild degrees of hypocalcemia, at least in the ionized

Disorders of bone metabolism and calcium and phosphorus homeostasis associated with myopathies

I. Vitamin D: Deficiency and abnormal metabolism
  Nutritional deficiency   Skaria et al. 1975; Irani 1976
  Malabsorption   Smith and Stern 1967, 1969; Morgan et al. 1970;
                 Mallette et al. 1975; Palmucci et al. 1982
  Anticonvulsants   Marsden et al. 1973; Mallette et al. 1975
  Chronic renal failure   Floyd et al. 1984; Henderson et al. 1974
  Tumor-induced osteomalacia   Drezner and Feinglos 1977; Parker et al. 1981

II. Increased urinary calcium loss
  Renal tubular acidosis   Vicale 1949; Smith and Stern 1967; Mallette et al.
                 1975

III. Phosphate depletion
  Nutritional   Isenberg et al. 1982
  Antacid abuse   Insogna et al. 1980; Ravid and Robson 1976
  Adult-onset vitamin D resistant hypophosphatemia   Dent and Stamp 1971
  Tumor-induced osteomalacia   Pollack et al. 1973

IV. Parathyroid hormone excess
  Primary hyperparathyroidism   Patten et al. 1974
  Secondary hyperparathyroidism   Mallette et al. 1975

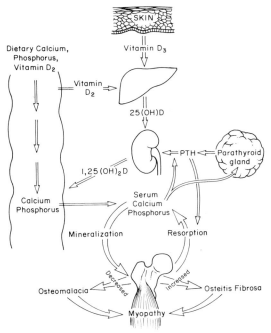

Fig. 4. Conditions associated with myopathies are listed in relation to specific abnormalities of calcium and phosphorus homeostasis. The diagram illustrates intestinal, hepatic, renal and skeletal interactions regulating calcium and phosphorus homeostasis leading to poor bone mineralization (osteomalacia) or accelerated bone resorption (osteitis fibrosa). Interstitial calcium absorption takes place under the influence of vitamin D derived from skin (vitamin D₃, cholecalciferol) and diet (vitamin D₂, ergocalciferol). 25-Hydroxyvitamin D (25(OH)D) is made in the liver and conversion to the most potent metabolite, 1,25-dihydroxyvitamin D (1,25(OH)D) takes place in the kidney. Parathyroid hormone (PTH) regulates blood calcium levels by promoting bone resorption, augmenting renal calcium absorption and phosphorus excretion and enhancing 1,25(OH)$_2$D conversion. Serum phosphorus levels are mostly dependent on dietary uptake, intestinal absorption and renal excretion.

fraction, and hypophosphatemia, which in turn may cause osteomalacia.

Muscle weakness had long been a recognized manifestation of hyperparathyroidism and osteomalacia (Hirschberg 1889; Von Recklinghausen 1891), but modern physicians were reintroduced to this condition by Vicale (1949). He described three patients with more severe clinical findings than are likely to be encountered today. One patient had renal tubular acidosis, secondary hyperparathyroidism, and osteomalacia and two others had primary hyperparathyroidism. All three had severe proximal limb weakness with a waddling gait. These patients had extensive bone disease, with prominent bone tenderness and a gait restricted by pain. Muscle wasting was conspicuous and possibly worsened by immobility. Many subsequent reports confirmed Vicale's observations in patients with different causes of hyperparathyroidism and osteomalacia. The overlapping clinical neuromuscular features, irrespective of the cause of calcium and phosphorus imbalance, make discussion of each individual condition superfluous; citations to specific examples are provided in Fig. 4.

The incidence of muscle symptoms in these disorders was prospectively assessed by Smith and Stern (1969) in 55 consecutive patients who were seen in 18 months. One of 41 patients with primary hyperparathyroidism and eight of 11 with osteomalacia had proximal limb weakness. In another study, Lafferty (1981) identified muscle weakness in seven of 100 patients with primary hyperparathyroidism. These figures imply that conditions producing osteomalacia are more often complicated by muscle weakness than is primary hyperparathyroidism. Because better diagnostic methods are now available for diagnosing hyperparathyroidism (Heath et al. 1980), neuromuscular symptoms are usually mild (Patten et al. 1974; Mallette et al. 1975; Turken et al. 1989). Weakness of the legs, muscle wasting and brisk tendon reflexes are the main features with less distortion of gait because bone disease is usually mild. Abnormal tongue movements resemble fasciculations in either primary (Patten et al. 1974) or secondary (Mallette et al. 1975) hyperparathyroidism.

Serum muscle enzyme levels are normal in these conditions (Patten et al. 1974; Mallette et al. 1975; Schott and Wills 1975). EMG reveals a myopathic pattern without positive waves or fibrillations (Bischoff and Esslen 1965; Prineas et al. 1965; Smith and Stern 1967; Frame et al. 1968; Patten et al. 1974; Mallette et al. 1975; Schott and Wills 1975; Skaria et al. 1975; Irani 1976). Muscle biopsies show varying degrees of atrophy (Smith and Stern 1967; Frame et al. 1968; Prineas et al. 1968; Dastur et al. 1975; Skaria et al. 1975; Frame 1976) affecting particularly the type 2 fibers (Patten et al. 1974; Mallette et al. 1975).

The muscle weakness of primary and secondary hyperparathyroidism has been attributed to an underlying neuropathy rather than myopathy (Aurbach et al. 1973; Patten et al. 1974; Mallette et al. 1975) because of normal serum CK levels, type 2 muscle fiber atrophy, and an EMG pattern with motor unit potentials suggesting a neuropathic process. This was emphatically refuted by Layzer (1985). Although the controversy cannot be entirely resolved, hyperparathyroidism certainly does not produce a neuropathy in any conventional sense. The constellation of signs, symptoms and laboratory data are shared by virtually every other endocrine disorder, none of which are deemed neurogenic.

The pathogenesis of muscle weakness in hyperparathyroidism and osteomalacia has not been elucidated. Multiple factors have been considered because the causes of osteomalacia and hyperparathyroidism are so heterogeneous (see Fig. 4). Parathyroid hormone (PTH) excess itself may be important because in animals, PTH enhances muscle proteolysis (Garber 1983) and impairs energy production, transfer, and utilization in skeletal muscle (Baczynski et al. 1985). PTH might also decrease the calcium sensitivity of myofibrillar proteins through a cAMP-dependent mechanism (Ruff 1986).

However, the variability of PTH levels in the many disorders associated with parathyroid abnormalities (see Fig. 4) emphasize the lack of a single explanation. Serum calcium and phosphorus levels also show no correlation with clinical manifestations in any of these syndromes (Smith and Stern 1969; Patten et al.1974; Frame 1976).

Also, vitamin D has direct effects on muscle independent of calcium and phosphorus. Vitamin D increases muscle ATP concentration and accelerates amino acid incorporation into muscle protein

(Birge and Haddad 1975). In addition, skeletal muscle appears to have a receptor for binding 25-hydroxyvitamin D (Haddad and Birge 1971). The uptake of calcium by sarcoplasmic reticulum (Curry et al. 1974; Pointon et al. 1979) is influenced by vitamin D, and the concentration of troponin C, the calcium binding component of the troponin complex, decreases in vitamin D-deficient animals (Pointon et al. 1979). Effects on muscle contractility have also been demonstrated in vitamin D-deficient rats (Rodman and Baker 1978) and chicks (Pleasure et al. 1979).

If treatment of the underlying disorder responsible for hyperparathyroidism or osteomalacia is successful, muscle strength will improve. This is best illustrated in primary hyperparathyroidism where removal of the adenoma or portions of the hyperplastic gland results in complete resolution of weakness (Frame et al. 1968; Patten et al. 1974).

Myopathy and chronic renal failure

Myopathy may be a manifestation of chronic renal failure, separate and distinct from the better known uremic polyneuropathy (Floyd et al. 1974; Henderson et al. 1974). Abnormalities of calcium and phosphorus homeostasis and bone metabolism in chronic renal failure result from a reduction in 1,25-dihydroxyvitamin D, leading to decreased intestinal absorption of calcium. Hypocalcemia, further accentuated by hyperphosphatemia due to decreased renal phosphate clearance, leads to secondary hyperparathyroidism. In turn, a complex bone disease (renal osteodystrophy) includes features of osteomalacia because of reduced calcium availability and osteitis fibrosa from the excess of parathyroid hormone. An additional component of renal osteodystrophy is osteosclerosis, due to bone remodeling and redistribution.

The clinical picture of the myopathy of chronic renal failure is identical to that of primary hyperparathyroidism and osteomalacia. There is proximal limb weakness with bone pain. Serum CK levels are normal and muscle biopsy demonstrates type 2 muscle fiber atrophy (Floyd et al. 1974; Lazaro and Kirshner 1980). Some patients respond to vitamin D in high doses (Floyd et al. 1974; Henderson et al. 1974), but others, especially those on chronic hemodialysis, are vitamin D-resistant (Floyd et al. 1974; Alvarez-Ude et al. 1978; Hodsman et al. 1981). Lack of response to vitamin D has been attributed to a contaminant in the dialysate that originates in the local water supply and can be removed by purification of the water. Aluminum has been blamed because it causes vitamin D-resistant osteomalacia when given to rats (Ellis et al. 1979).

A separate, rare, sometimes fatal complication of chronic renal failure is gangrenous calcification (Richardson et al. 1969; Goodhue et al. 1972); widespread arterial calcification affects the media and external elastic lamina with intimal proliferation in small subcutaneous and intramuscular arteries that lead to ischemia. Skin necrosis may be extensive and there may be a painful myopathy affecting the proximal muscles with myoglobinuria (Richardson et al. 1969).

Myopathy with hypoparathyroidism

Overt myopathy due to hypocalcemia is rarely seen (Frame 1976) in contrast to the usual neuromuscular manifestations of paresthesias and localized or generalized tetany. Serum CK levels may be increased with only minor histologic changes and no clinical weakness (Hower et al. 1972; Wolf et al. 1972; Snowdon et al. 1976; Walters 1979; Shane et al. 1980; Kruse et al. 1982). The described abnormalities may be examples of muscle damage that follow tetany rather than a direct effect of hypocalcemia. Rarely there is proximal weakness as well as elevated serum CK (Kruse et al. 1982). In one patient, symptoms occurred one year after subtotal thyroidectomy and were reversed by treatment of hypocalcemia (Wolf et al. 1972). One patient with pseudohypoparathyroidism had no loss of strength but there was histochemical evidence of phosphorylase, a deficiency without alterations in muscle histology (Cape 1969); serum CK activity was elevated.

Hypoparathyroidism has been seen with the Kearns-Sayre syndrome (Marks et al. 1976; Toppet al. 1977; Horwitz and Roessmann 1978; Pellock et al. 1978) and in one patient with ataxia, retinal degeneration, neuromyopathy and mental subnormality (Gomez et al. 1972).

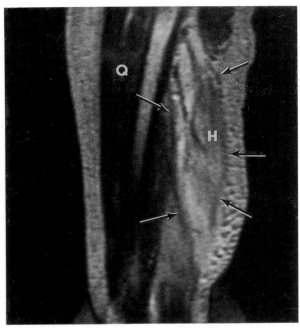

Fig. 5. Sagittal view of MRI scan of the left thigh in a 25-year-old female with type 1 diabetes mellitus. The patient presented with pain, focal swelling and tenderness in the posterior aspect of the left thigh. The T2-weighted scan reveals an area of swelling and diffuse high signal (arrows) in the hamstring muscle group (H). The bony structures and other soft tissue elements, including the quadriceps muscles (Q), are normal.

PANCREATIC DISORDERS

Neuromuscular complications of diabetes mellitus are most often related to neuropathy with cranial and peripheral nerve palsies, or distal sensorimotor polyneuropathy. 'Diabetic amyotrophy' is sometimes thought to be a primary muscle complication, but there is overwhelming evidence that this is a type of neuropathy, at the level of the proximal major nerve trunks and lumbosacral plexus (Raff et al. 1968; Asbury 1977; Thomas and Eliasson 1984; Barohn et al. 1991). More appropriate terms include diabetic proximal neuropathy, lumbosacral plexopathy, or the eponymic designation, Bruns-Garland syndrome (Barohn et al. 1991).

Thigh infarction in diabetes mellitus

The most notable myopathy of diabetes mellitus is ischemic infarction of thigh muscles. This syndrome has been reported in poorly controlled diabetes, usually with other end-organ complications of diabetes (polyneuropathy, nephropathy, and retinopathy) (Chester and Banker 1986; Barohn and Kissel 1992). There is acute onset of pain, tenderness and edema of one thigh with a palpable mass that most frequently involves the vastus lateralis, thigh adductors, and biceps femoris. CK levels have been normal in most cases and EMG demonstrates brief polyphasic motor units and fibrillation potentials in the involved muscles (Barohn and Kissel 1992). CT (Levinsohn and Bryan 1979), technetium 99 radionuclide scans (Reich et al. 1985) and MRI (Reich et al. 1985; Barohn and Kissel 1992) demonstrate focal abnormalities in the region of the mass and may preclude the need for muscle biopsy (Fig. 5). Pathologic specimens reveal true muscle infarction, with large areas of muscle necrosis, edema and inflammatory cell infiltration. Connective tissue and fat infiltration follow the acute phase. Small and medium-sized arteries have thickened media and arterioles and are frequently occluded by calcium, fibrin or lipid (Banker and Chester 1973).

Although the infarction itself tends to resolve spontaneously over several weeks, the condition may recur in the same or contralateral thigh (Barohn and Kissel 1992). Treatment consists of supportive care directed to relief of pain and immobilization of the limb. Surgical exploration of the swollen thigh should be avoided (Banker and Chester 1973), but muscle biopsy may be necessary for diagnosis if there are unusual clinical or radiologic features (Barohn and Kissel 1992).

Acknowledgments

The authors would like to thank Jennifer Omerza for technical assistance and Nancy Hodges and Bobbie Swank for preparing the manuscript. Dr. Jack George, Chief of the Division of Endocrinology at Ohio State University, reviewed the manuscript and provided useful discussion.

REFERENCES

AFIFI, A.K. and R.A. BERGMAN: Steroid myopathy: a study of the evolution of the muscle lesion in rabbits. Johns Hopkins Med. J. 124 (1969) 66–86.

AFIFI, A.K., R.A. BERGMAN and J.C. HARVEY: Steroid myopathy: clinical, histologic and cytologic observations. Johns Hopkins Med. J. 123 (1968) 158–174.

AFIFI, A.K., S.S. NAJJAR, J. MIRE-SALMAN and R.A. BERGMAN: The myopathy of the Kocher-Debré-Sémélaigne syndrome. J. Neurol. Sci. 22 (1974) 445–470.

AGAPITOS, P.J. and I.R. HART: Long-term follow-up of ophthalmic Graves' disease. Can. Med. Assoc. J. 136 (1987) 369–372.

ALPER, M.: Endocrine ophthalmopathy. In: K.L. Becker (Ed.), Principles and Practice of Endocrinology and Metabolism. Philadelphia, J.B. Lippincott (1990) 347–359.

ALVAREZ-UDE, F., T.G. FEEST, M.K. WARD, A.M. PIERIDES, H.A. ELLIS, K.M. PEART, W. SIMPSON, D. WEIGHTMAN and D.N.S. KERR: Hemodialysis bone disease: correlation between clinical, histologic, and other findings. Kidney Int. 14 (1978) 68–73.

AMINO, N., T. YUASA, Y. YABA, K. MIYAI and Y. KUMAHARA: Exophthalmos in autoimmune thyroid disease. J. Clin. Endocrinol. Metab. 51 (1980) 1232–1234.

ANIANSSON, A., G. GRIMBY, A. RUNDGREN, A. SVANBORG and J. ORLANDER: Physical training in old men. Age Aging 9 (1980) 186–187.

ARGOV, Z., P.F. RENSHAW, B. BODEN, A. WINOKUR and W.J. BANK: Effects of thyroid hormones on skeletal muscle bioenergetics. J. Clin. Invest. 18 (1988) 1695–1701.

ASBURY, A.K.: Proximal diabetic neuropathy. Ann. Neurol. 2 (1977) 179–180.

ASKARI, A., P.J. VIGNOS and R.W. MOSKOWITZ: Steroid myopathy in connective tissue disease. Am. J. Med. 61 (1976) 485–492.

ATSUMI, T., S. ISHIKAWA, T. MIYATAKE and M. YOSHIDA: Myopathy and primary aldosteronism: electromicroscopic study. Neurology 29 (1979) 1348–1353.

AURBACH, G.D., L.E. MALLETTE, B.M. PATTEN, D.A. HEATH, J.L. DOPPMAN and J.B. BILEZIHIAN: Hyperparathyroidism: recent studies. NIH Conference. Ann. Intern. Med. 79 (1973) 566–581.

AXELROD, L.: Glucocorticoids. In: W.N. Kelly, E.D. Harris, S. Ruddy and C.B. Sledy (Eds.), Textbook of Rheumatology. New York, W.B. Saunders (1989) 845–861.

BACZYNSKI, R., S.G. MASSRY, M. MAGOTT, S. EL-BELBESSI, R. KOHAN and N. BRANTBAR: Effect of parathyroid hormone on energy metabolism of skeletal muscle. Kidney Int. 28 (1985) 722–727.

BAHN, R.S. and G.A. GORMAN: Choice of therapy and criteria for assessing treatment outcome in thyroid associated ophthalmopathy. Endocrinol. Metab. Clin. North Am. 16 (1987) 391–407.

BALDWIN, K.M., S.B. ERNST, R.E. HERRICK, A.M. HOOKER and W.J. MULLIN: Exercise capacity and cardiac function in trained and untrained thyroid-deficient rats. J. Appl. Physiol. 49 (1980a) 1022–1026.

BALDWIN, K.M., A.M. HOOKER, R.E. HERRICK and L.F. SCHRADER: Respiratory capacity and glycogen depletion in thyroid-deficient muscle. J. Appl. Physiol. 49 (1980b) 102–106.

BANKER, B.Q. and C.S. CHESTER: Infarction of thigh muscle in the diabetic patient. Neurology 23 (1973) 667–677.

BAROHN, R.J. and J.T. KISSEL: Painful thigh mass in a young woman: diabetic muscle infarction. Muscle Nerve 15 (1992) 850–855.

BAROHN, R.J., Z. SAHENK, J.R. WARMOLTS and J.R. MENDELL: The Bruns-Garland syndrome ('diabetic amyotrophy'): revisited 100 years later. Arch. Neurol. 48 (1991) 1130–1135.

BECKER, K.L., J.H. TITUS, W.M. MCCONAHEY and L.B. WOOLNER: Morphologic evidence of thyroiditis in myasthenia gravis. J. Am. Med. Assoc. 187 (1964) 994–1000.

BEHAN, P.O.: Immune disease and HLA associations with myasthenia gravis. J. Neurol. Neurosurg. Psychiatry 43 (1980) 611–621.

BENNETT, W.R. and D.P. HUSTON: Rhabdomyolysis in thyroid storm. Am. J. Med. 77 (1984) 733–735.

BHATNAGAR, D., P. CAREY and A. POLLARD: Focal myositis and elevated creatine kinase levels in a patient with phaeochromocytoma. Postgrad. Med. J. 62 (1986) 197–198.

BIGLAND, B. and B. JEHRING: Muscle performance in rats, normal and treated with growth hormone. J. Physiol. 116 (1952) 129–136.

BIRGE, S.J. and J.G. HADDAD: 25-Hydroxycholecalciferol stimulation of muscle metabolism. J. Clin. Invest. 56 (1975) 1100–1107.

BISCHOFF, A. and E. ESSLEN: Myopathy with primary hyperparathyroidism. Neurology (Minneap.) 15 (1965) 64–68.

BOWYER, S.L., M.P. LAMOTHE and J.R. HOLLISTER: Steroid myopathy: incidence and detection in a population with asthma. J. Allergy Clin. Immunol. 76 (1985) 234–242.

BROWN, J.G. and D.J. MILLWARD: The influence of thyroid status on skeletal muscle protein metabolism. Biochem. Soc. Trans. 8 (1980) 366–367.

BUCHTHAL, F.: Electrophysiological abnormalities in metabolic myopathies and neuropathies. Acta Neurol. Scand. 46 (Suppl.) (1970) 129–176.

BUNCH, T.W., J.W. WORTHINGTON, J.J. COMBS, D.M. ILSTRUP and A.G. ENGEL: Azathioprine with prednisone for polymyositis: a controlled clinical trial. Ann. Intern. Med. 92 (1980) 365–369.

BYERS, R.L., A.B. BERGMAN and M.C. JOSEPH: Steroid myopathy: report of 5 cases occurring during treatment of rheumatic fever. Pediatrics 29 (1962) 26–36.

CAPE, C.A.: Phosphorylase a deficiency in pseudo-hypo-parathyroidism. Neurology (Minneap.) 19 (1969) 167–172.

CARPENTER, S. and G. KARPATI: Pathology of Skeletal Muscle. New York, Churchill Livingstone (1984) 515–516.

CHAUHAN, A.K., B.C. KATIYAR, S. MISRA, A.K. THACKER and N.K. SINGH: Muscle dysfunction in male hypogonadism. Acta Neurol. Scand. 73 (1986) 466–471.

CHESTER, C. and B. BANKER: Focal infarctions of muscle in diabetes. Diabetic Care 9 (1986) 623–630.

CLARK, A.F. and P.J. VIGNOS: The role of proteases in experimental glucocorticoid myopathy. Muscle Nerve 4 (1981) 219–222.

COLLINS, J.A., F.E. ZIMMER, W.J. JOHNSON and R.H. KOUGH: The many faces of hypothyroidism. Postgrad. Med. J. 36 (1964) 371–384.

CONN, J.W., R.F. KNOPF and R.M. NESBIT: Clinical characteristics of primary aldosteronism from an analysis of 145 cases. Am. J. Surg. 107 (1964) 159–172.

CONWAY, M.J., J.A. SEIBEL and R.P. EATON: Thyrotoxicosis and periodic paralysis: improvement with beta blockage. Ann. Intern. Med. 81 (1974) 332–336.

COOMES, E.M.: Corticosteroid myopathy. Ann. Rheum. Dis. 24 (1965) 465–472.

CURRY, O.B., J.F. BASTEN, M.J.O. FRANCIS and R. SMITH: Calcium uptake by sarcoplasmic reticulum of muscle from vitamin D-deficient in rabbits. Nature (London) 249 (1974) 83–84.

CUSHING, H.: The basophil adenomas of the pituitary body and their clinical manifestations (pituitary basophilism). Bull. Johns Hopkins Hosp. (1932) 137–195.

DANDONA, P., N.J. MARSHALL, S.P. BIDEY, A. NATHAN and C.W. HAVARD: Successful treatment of exophthalmos and pretibial myxoedema with plasmapheresis. Br. Med. J. 1 (1979) 374–376.

DASTUR, D.K., B.M. GAGRAT, N.H. WADIA, M.M. DESAI and E.P. BHARUCHA: Nature of muscular change in osteomalacia: light- and electron microscope observations. J. Pathol. 117 (1975) 211–228.

DAUGHADAY, W.H.: The anterior pituitary. In: J.D. Wilson and D.W. Foster (Eds.), Williams Text-

book of Endocrinology, 7th edit. Philadelphia, W.B. Saunders Company (1985) 568–613.

DECKER, R.S. and K. WILDENTHAL: Lysosomal alterations in heart, skeletal muscle, and liver of hyperthyroid rabbits. Lab. Invest. 44 (1981) 455–465.

DEMARTINO, G.N. and A.L. GOLDBERG: Thyroid hormones control lysosomal enzyme activities in liver and skeletal muscle. Proc. Natl. Acad. Sci. (USA) 75 (1978) 1369–1373.

DENT, C.E. and T.C.B. STAMP: Hypophosphataemic osteomalacia presenting in adults. Q. J. Med. 40 (1971) 303–329.

DOCHERTY, I., J.S. HARROP, K.R. HINE, M.R. HOPTON, H.L. MATTHEWS and C.J. TAYLOR: Myoglobin concentration, creatine kinase activity, and creatine kinase B subunit concentrations in serum during thyroid disease. Clin. Chem. 30 (1984) 42–45.

DRACHMAN, D.B.: Myasthenia gravis and the thyroid gland. N. Engl. J. Med. 266 (1962) 330–333.

DRESNER, S.C. and J.S. KENNERDELL: Dysthyroid orbitopathy. Neurology 35 (1985) 1628–1634.

DREZNER, M.K. and M.N. FEINGLOS: Osteomalacia due to 1,25-dihydroxycholecalciferol deficiency: association with a giant cell tumor of bone. J. Clin. Invest. 60 (1977) 1046–1053.

DUBOIS, D.C. and R.R. ALMON: Disuse atrophy of skeletal muscle is associated with an increase in number of glucocorticoid receptors. Endocrinology 107 (1980) 1649–1651.

DUBOIS, E.L.: Triamcinolone in the treatment of systemic lupus erythematosus. J. Am. Med. Assoc. 167 (1958) 1590–1599.

DUCHESNEAU, G.: Contribution à l'étude anatomique et clinique de l'acromégalie et en particulier d'une forme amyotrophique de cette maladie. Paris, Ballière et Fils (1892).

DYCK, P.J. and E.H. LAMBERT: Polyneuropathy associated with hypothyroidism. J. Neuropathol. Exp. Neurol. 29 (1970) 631–658.

DYER, J.A.: Ocular muscle surgery. In: C.A. Gorman (Ed.), The Eye and Orbit in Thyroid Disease. New York, Raven Press (1984) 253–261.

ELLIS, H.A., J.H. MCCARTHY and J. HERRINGTON: Bone aluminum in hemodialyzed patients and in rats injected with aluminum chloride: relationship to impaired bone mineralization. J. Clin. Pathol. 32 (1979) 832–844.

ENGEL, A.G.: Thyroid function and periodic paralysis. Am. J. Med. 30 (1961a) 327–333.

ENGEL, A.G.: Thyroid function and myasthenia gravis. Arch. Neurol. 4 (1961b) 95–106.

ENGEL, A.G.: Electron microscopic observations in thyrotoxic and corticosteroid-induced myopathies. Mayo Clin. Proc. 41 (1966a) 758–796.

ENGEL, A.G.: Treatment of metabolic and endocrine myopathies. Mod. Treat. 3 (1966b) 313–325.

ENGEL, A.G.: Neuromuscular manifestations of Graves' disease. Mayo Clin. Proc. 47 (1972) 919–925.

EVERTS, M.E., C. VAN HARDEVELD, H.E.D.J. TER KEURS

and A.A.H. KASSENAAR: Force development and metabolism in skeletal muscle of euthyroid and hypothyroid rats. Acta Endocrinol. 97 (1981) 221–225.

FANBURG, B.L.: Calcium transport by skeletal muscle sarcoplasmic reticulum in the hypothyroid rat. J. Clin. Invest. 47 (1968) 2499–2506.

FAULDI, G., L.C. MILLS and Z.W. CHAYES: Effect of steroids on muscle. Acta Endocrinol. 45 (1964) 68–78.

FAULDI, G., J. GOTLIEB and J. MEYERS: Factors influencing the development of steroid-induced myopathies. Ann. N.Y. Acad. Sci. 138 (1966) 61–72.

FEIBEL, J.H. and J.F. CAMPA: Thyrotoxic neuropathy (Basedow's paraplegia). J. Neurol. Neurosurg. Psychiatry 39 (1976) 491–497.

FLOYD, M., D.R. AYYAR, D.D. BARWICK, P. HUDGSON and D. WEIGHTMAN: Myopathy in chronic renal failure. Q. J. Med. 53 (1974) 509–524.

FORBES, G., C.A. GORMAN and M.D. BRENNAN: Ophthalmopathy of Graves' disease: computerized volume measurements of the orbital fat and muscle. AJNR 7 (1986) 651–656.

FRAME, B.: Neuromuscular manifestations of parathyroid disease. In: P.J. Vinken, G.W. Bruyn and H.L. Klawans (Eds.), Handbook of Clinical Neurology, Vol. 27: Metabolic and Deficiency Diseases of the Nervous System, Part I. Amsterdam, North-Holland Publishing Company (1976) 283–320.

FRAME, B., E.G. HEINZE, M.A. BLOCK and G.A. MANSON: Myopathy in primary hyperparathyroidism: observations in three patients. Ann. Intern. Med. 68 (1968) 1022–1027.

FREYBERG, R.H., C.A. BERNSTSEN and L. HELLMAN: Further experience with triamcinolone in treatment of patients with rheumatoid arthritis. Arthritis Rheum. I (1958) 218–229.

GAAN, D.: Chronic thyrotoxic myopathy with involvement of respiratory and bulbar muscles. Br. Med. J. 3 (1967) 415–416.

GAELEN, L.H. and S. LEVITAN: Myasthenia gravis and thyroid function. Arch. Neurol. 18 (1968) 107–110.

GALLAI, M.: Myopathy with hyperaldosteronism. J. Neurol. Sci. 32 (1977) 337–345.

GARBER, A.J.: Effects of parathyroid hormone on skeletal muscle protein and amino acid metabolism in the rat. J. Clin. Invest. 71 (1983) 1806–1821.

GARDINER, P.F., B. HIBL, D.R. SIMPSON, R. ROY and V.R. EDGERTON: Effects of a mild weight-lifting program on the progress of glucocorticoid-induced atrophy in rat hindlimb muscles. Pflügers Arch. 385 (1980) 147–153.

GLORIEUX, F.H.: Rickets, the continuing challenge. N. Engl. J. Med. 325 (1991) 1875–1877.

GOLD, H.K., J.F. SPANN and E. BRAUNWALD: Effect of alterations in the thyroid state on the intrinsic contractile properties of isolated rat skeletal muscle. J. Clin. Invest. 49 (1970) 849–854.

GOLDBERG, A.L.: Work-induced growth of skeletal muscle in normal and hypophysectomized rats. Am. J. Physiol. 213 (1967) 1193–1198.

GOLDBERG, A.L. and D. GOLDSPINK: Influence of food deprivation and adrenal steroids on DNA synthesis in various mammalian tissues. Am. J. Physiol. 228 (1975) 310–317.

GOLDBERG, A.L. and H.M. GOODMAN: Relationship between cortisone and muscle work in determining muscle size. J. Physiol. (London) 200 (1969) 667–675.

GOLDBERG, A.L., G.E. GRIFFIN and J.F. DICE: Regulation of protein turnover in normal and dystrophic muscle. In: L.P. Rowland (Ed.), Pathogenesis of Human Muscular Dystrophies. Amsterdam, Excerpta Medica (1977) 376–385.

GOLDFIEN, A.: Adrenal medulla. In: F.S. Greenspan (Ed.), Basic and Clinical Endocrinology, 3rd edit. Norwalk, Appleton and Lange (1991) 380–399.

GOLDING, D.N.: Hypothyroidism presenting with musculoskeletal symptoms. Ann. Rheum. Dis. 29 (1970) 10–14.

GOLDING, D.N., S.M. MURRAY, G.W. PEARCE and M. THOMPSON: Corticosteroid myopathy. Ann. Phys. Med. 6 (1961) 171–177.

GOMEZ, M.R., A.G. ENGEL and P.J. DYCK: Progressive ataxia, retinal degeneration, neuropathy, and mental subnormality in a patient with true hypoparathyroidism, dwarfism, malabsorption, and cholelithiasis. Neurology (Minneap.) 22 (1972) 849–855.

GOODHUE, W.W., J.N. DAVIS and R.S. PORRO: Ischemic myopathy in uremic hyperparathyroidism. J. Am. Med. Assoc. 221 (1972) 911–912.

GRAIG, F.A. and J.C. SMITH: Serum creatine phosphokinase activity in altered thyroid states. J. Clin. Endocrinol. 25 (1965) 723–731.

GRAVES, R.J.: Clinical lectures. Lond. Med. Surg. J. 7 (1835) 516–529.

GRIFFIN, J.E. and J.D. WILSON: Disorders of the testes and male reproductive tract. In: J.D. Wilson and D.W. Foster (Eds.), Williams Textbook of Endocrinology, 7th edit. Philadelphia, W.B. Saunders Company (1985) 259–311.

GRUENER, R.G., L.Z. STERN, C. PAYNE and L. HANNAPEL: Hyperthyroid myopathy: intracellular electrophysiological measurements in biopsied human intercostal muscle. J. Neurol. Sci. 24 (1975) 339–349.

GUSTAFSSON, R., J.R. TATA, O. LINDBERG and L. ERNSTER: The relationship between the structure and activity of rat skeletal muscle mitochondria after thyroidectomy and thyroid hormone treatment. J. Cell. Biol. 26 (1965) 555–578.

HADDAD, J.G. and S.J. BIRGE: 25-Hydroxycholecalciferol: specific binding by ricketic tissue extracts. Biochem. Biophys. Res. Commun. 45 (1971) 829–834.

HALL, R., K. KIRKHAM, D. DONIACH and D.E. KABIR: Ophthalmic Graves' disease: diagnosis and pathogenesis. Lancet i (1970) 375–378.

HARMAN, J.B.: Muscular wasting and corticosteroid therapy. Lancet i (1959) 887.

HAUPT, H.A. and G.D. ROVERE: Anabolic steroids: a review of the literature. Am. J. Sports Med. 12 (1984) 469–484.

HAVARD, C.W.H.: Progress in endocrine exophthalmos. Br. Med. J. 1 (1979) 1001–1004.

HAVARD, C.W.H., E.D.R. CAMPBELL, H.B. ROSS and A.W. SPENCE: Electromyographic and histological findings in the muscle of patients with thyrotoxicosis. Q. J. Med. 32 (1963) 145–163.

HEATH, H., S.F. HODGSON and M.A. KENNEDY: Primary hyperparathyroidism: incidence, morbidity, and potential economic impact in a community. N. Engl. J. Med. 302 (1980) 189–193.

HENDERSON, R.G., R.G.G. RUSSELL, J.G.G. LEDINGHAM, R. SMITH, D.O. OLIVER, R.J. WATON, D.G. SMALL, C. PRESTON and G.T. WARNER: Effects of 1,25-dihydroxycholecalciferol on calcium absorption, muscle weakness, and bone disease in chronic renal failure. Lancet i (1974) 379–384.

HIROMATSU, Y., P.W. WANG, L. WOSU, J. HOW and J.R. WALL: Mechanisms of immune damage in Graves ophthalmopathy. Hormone Res. 62 (1987) 198–207.

HIRSCHBERG, K.: Zur Kenntnisse der Osseomalacie und Ostitis malacissans. Beitr. Pathol. 6 (1889) 513–524.

HODES, B.L., L. FRAZEE and S. SZMYD: Thyroid orbitopathy: an update. Ophthalmic Surg. 10 (1979) 25–33.

HODSMAN, A.B., D.J. SHERRARD, E.G.C. WONG, A.S. BRICKMAN, D.B.N. LEE, A.C. ALFREY, F.R. SINGER, A.W. NORMAN and J.W. COBURN: Vitamin D-resistant osteomalacia in hemodialysis patients lacking secondary hyperparathyroidism. Ann. Intern. Med. 94 (1981) 629–637.

HOFMANN, W.W. and E.H. DENYS: Effects of thyroid hormone at the neuromuscular junction. Am. J. Physiol. 223 (1972) 283–287.

HORBER, F.F., J.R. SCHEIDEGGER, B.E. GRÜNIG and F.J. FREY: Evidence that prednisone-induced myopathy is reversed by physical training. J. Clin. Endocrinol. Metab. 61 (1985) 83–88.

HORBER, F.F., H. HOOPELER, J.R. SCHEIDEGGER, B.E. GRÜNIG, H. HOWALD and F.J. FREY: Impact of physical training on the ultrastructure of midthigh muscle in normal subjects and in patients treated with glucocorticoids. J. Clin. Invest. 79 (1987) 1181–1190.

HORENSTEIN, S., G. HAMBROOK and E. EYERMAN: Spinal cord compression by vertebral acromegaly. Trans. Am. Neurol. Assoc. 96 (1971) 254–266.

HORWITZ, S.J. and U. ROESSMANN: Kearns-Sayre syndrome with hypoparathyroidism. Ann. Neurol. 3 (1978) 513–518.

HOWER, J., H. STRUCK, W. TACKMAN and H. STOLECKE: CPK activity in hypoparathyroidism. N. Engl. J. Med. 287 (1972) 1098.

IANUZZO, D., P. PATEL, V. CHEN, P. O'BRIEN and C. WILLIAMS: Thyroidal trophic influence on skeletal muscle myosin. Nature (London) 270 (1977) 74–76.

INGBAR, S.H.: The thyroid gland. In: J.D. Wilson and D.W. Foster (Eds.), Williams Textbook of Endocrinology, 7th edit. Philadelphia, W.B. Saunders Company (1985) 682–815.

INSOGNA, K.L., D.R. BORDLEY, J.F. CARO and D.H. LOCKWOOD: Osteomalacia and weakness from excessive antacid indigestion. J. Am. Med. Assoc. 244 (1980) 2544–2546.

IRANI, P.F.: Electromyography in nutritional osteomalacic myopathy. J. Neurol. Neurosurg. Psychiatry 39 (1976) 686–693.

ISENBERG, D.A., D. NEWHAM, R.H.T. EDWARDS, C.M. WILES and A. YOUNG: Muscle strength and pre-osteomalacia in vegetarian Asian women. Lancet i (1982) 52.

JANSSEN, J.W., C. VAN HARDEVELD and A.A.H. KASSENAAR: Evidence for a different response of red and white skeletal muscle of the rat in different thyroid states. Acta Endocrinol. 87 (1978) 768–775.

JOASOO, A., I.P.C. MURRAY and A.W. STEINBECK: Involvement of bulbar muscles in thyrotoxic myopathy. Aust. Ann. Med. 4 (1970) 338–340.

JONES, D.I.R., D.S. MUNRO and G.M. WILSON: Observations on the course of exophthalmos after ^{131}I therapy. Proc. R. Soc. Med. 62 (1969) 15–18.

KACIUBA-USCILKO, H. and Z. BRZEZINSKA: Effect of thyroxine treatment on metabolic responses to a single insulin injection. Horm. Metab. Res. 11 (1979) 561–566.

KAMMER, G.M. and C.R. HAMILTON: Acute bulbar muscle dysfunction and hyperthyroidism: a study of four cases and review of the literature. Am. J. Med. 56 (1974) 464–470.

KANAYAMA, Y., K. SHIOTA, T. HORIGUCHI, N. KATO, A. OHE and T. INOUE: Correlation between steroid myopathy and serum lactic dehydrogenase in systemic lupus erythematosus. Arch. Intern. Med. 141 (1981) 1176–1179.

KARLSBERG, R.P. and R. ROBERTS: Effect of altered thyroid function on plasma creatine kinase clearance in the dog. Am. J. Physiol. 235 (1978) E614–E618.

KELLEY, D.E., H. GHARIB, F.P. KENNEDY, R.J. DUDA and P.G. MCMANIS: Thyrotoxic periodic paralysis: report of 10 cases and review of electromyographic findings. Arch. Intern. Med. 149 (1989) 2597–2600.

KENDALL, P.H. and M.F. HART: Side effects following triamcinolone. Br. Med. J. 1 (1959) 682–685.

KENDALL-TAYLOR, P. and D.M. TURNBULL: Endocrine myopathies. Br. Med. J. 287 (1983) 705–708.

KHALEELI, A.A. and R.H.T. EDWARDS: Effect of treatment on skeletal muscle dysfunction in hypothyroidism. Clin. Sci. 66 (1984) 63–68.

KHALEELI, A.A., D.G. GRIFFITH and R.H.T. EDWARDS: The clinical presentation of hypothyroid myopathy and its relationship to abnormalities in structure and function of skeletal muscle. Clin. Endocrinol. 19 (1983a) 365–376.

KHALEELI, A.A., K. GOHIL, G. MCPHAIL, J.M. ROUND and R.H.T. EDWARDS: Muscle morphology and metabolism in hypothyroid myopathy: effects of treatment. J. Clin. Pathol. 36 (1983b) 519–526.

KHALEELI, A.A., R.H.T. EDWARDS, K. GOHIL, G. MCPHAIL, M.J. RENNIE, J. ROUND and E.J. ROSS: Corticosteroid myopathy: a clinical and pathological study. Clin. Endocrinol. 18 (1983c) 155–166.

KIESSLING, W.R., K.W. PFLUGHAUPT, K. RICKER, I. HAUBITZ and H.G. MERTENS: Thyroid function and circulating antithyroid antibodies in myasthenia gravis. Neurology 31 (1981) 771–774.

KINGSTON, W.J.: Endocrine myopathies. Semin. Neurol. 3 (1983) 258–264.

KISSEL, J.T. and K.W. RAMMOHAN: Pathogenesis and therapy of nervous system vasculitis. Clin. Neuropharm. 14 (1991) 28–48.

KISSEL, P., J. SCHMITT, M. DUC and M.L. DUC: Myasthenia and thyrotoxicosis. In: J.W. Walton, N. Canal and G. Scarlato (Eds.), Proceedings of an International Congress on Muscle Diseases, Milan 1969. Amsterdam, Excerpta Medica (1970) 464–481.

KLEIN, I. and G.S. LEVEY: Unusual manifestations of hypothyroidism. Arch. Intern. Med. 144 (1984) 123–128.

KLEIN, I., M. PARKER, R. SHEBERT, D.R. AYYAR and G.S. LEVEY: Hypothyroidism presenting as muscle stiffness and pseudohypertrophy: Hoffmann's syndrome. Am. J. Med. 70 (1981) 891–894.

KONISHI, J., M.M. HERMAN and J.P. KRISS: Binding of thyroglobulin and thyroglobulin-antithyroglobulin immune complex to extraocular muscle membrane. Endocrinology 95 (1974) 434–446.

KORENYI-BOTH, A., I. KORENYI-BOTH and B.C. KAYES: Thyrotoxic myopathy: pathomorphological observations of human material and experimentally induced thyrotoxicosis in rats. Acta Neuropathol. (Berlin) 53 (1981) 237–248.

KOSKI, C.L., D.H. RIFENBERICK and S.R. MAX: Oxidative metabolism of skeletal muscle in steroid atrophy. Arch. Neurol. (Chicago) 31 (1974) 407–410.

KRUSE, K., W. SCHEUNEMANN, W. BAIER and J. SCHAUB: Hypocalcemic myopathy in idiopathic hypoparathyroidism. Eur. J. Pediatr. 138 (1982) 280–282.

KUFS, W.M., M. MCBILES and T. JURNEY: Familial thyrotoxic periodic paralysis. West. J. Med. 150 (1989) 461–463.

LAFFERTY, F.W.: Primary hyperparathyroidism: changing clinical spectrum, prevalence of hypertension, and discriminant analysis of laboratory tests. Arch. Intern. Med. 141 (1981) 1761–1766.

LAMB, D.R.: Anabolic steroids: how well do they work and how dangerous are they? Am. J. Sports Med. 12 (1984) 31–38.

LAMBERT, E.H., L.O. UNDERDAHL, S. BECKETT and L.O. MEDEROS: A study of the ankle jerk in myxedema. J. Clin. Endocrinol. Metab. 11 (1951) 1186–1205.

LASZEWSKI, B. and R.L. RUFF: The effects of glucocorticoid treatment and excitation-contraction coupling. Am. J. Physiol. 248 (1985) E363–E369.

LAVIN, T.N.: Mechanisms of action of thyroid hormone. In: L.V. DeGroot (Ed.), Endocrinology, 2nd edit. Philadelphia, W.B. Saunders Company (1989) 562–573.

LAYZER, R.B.: Neuromuscular Manifestations of Systemic disease. Philadelphia, F.A. Davis Company (1985).

LAZARO, R.P. and H.S. KIRSHNER: Proximal muscle weakness in uremia: case reports and review of the literature. Arch. Neurol. 37 (1980) 555–558.

LEVINSOHN, E.M. and P.J. BRYAN: Computed tomography in unilateral extremity swelling of unusual cause. J. Comput. Assist. Tomogr. 3 (1979) 67–70.

LEWIS, P.D.: Neuromuscular involvement in pituitary giantism. Br. Med. J. 2 (1972) 499–500.

LOW, P.A., J.G. MCLEOD, J.R. TURTLE, P. DONNELLY and R.G. WRIGHT: Peripheral neuropathy in acromegaly. Brain 97 (1974) 139–152.

LUDIN, H.P., H. SPIESS and M.P. KOENIG: Neuromuscular dysfunction associated with thyrotoxicosis. Eur. Neurol. 2 (1969) 269–278.

LUNDBERG, P.O., P.O. OSTERMAN and E. STALBERG: Neuromuscular signs and symptoms in acromegaly. In: J.N. Walton, N. Canal and G. Scarlato (Eds.), Proceedings of an International Congress on Muscle Disease, Milan 1969. Amsterdam, Excerpta Medica (1970) 531–534.

MACLEAN, K. and P.H. SCHURR: Reversible amyotrophy complicating treatment with fluorocortisone. Lancet i (1959) 701–703.

MALLETTE, L.E., B.M. PATTEN and W.K. ENGEL: Neuromuscular disease in secondary hyperparathyroidism. Ann. Intern. Med. 82 (1975) 474–483.

MARIE, P.: Sur deux cas d'acromégalie: hypertrophie singulière, non congénitale, des extrémités supérieures, inférieures et céphalique. Rev. Med. (Paris) 6 (1886) 297–333.

MARKS, H., J. HEADINGTON and R.J. ALLEN: Hypoparathyroidism associated with oculosomatic syndrome. Presented at the 5th Annual Meeting, Child Neurology Society (1976).

MARKS, P., J. ANDERSON and R. VINCENT: Thyrotoxic myopathy presenting with dysphagia. Postgrad. Med. J. 56 (1980) 669–670.

MARSDEN, C.D., E.H. REYNOLDS, V. PARSONS, R. HARRIS and L. DUCHEN: Myopathy associated with anticonvulsant osteomalacia. Br. Med. J. 4 (1973) 526–527.

MASTAGLIA, F.L.: Pathological changes in skeletal muscle in acromegaly. Acta Neuropathol. 24 (1973) 273–286.

MASTAGLIA, F.L., D.B. BARWICK and R. HALL: Myopathy in acromegaly. Lancet ii (1970) 907–909.

MAYER, M., E. SHAFRIR, N. KAISER, R.J. MILHOLLAND and F. ROSEN: Interaction of glucocorticoid hormones with rat skeletal muscle: catabolic effects and hormone binding. Metabolism 25 (1976) 157–167.

MCCOMAS, A.J., R.E.P. SICA, A.R. MCNABB, W.M. GOLDBERG and A.R.M. UPTON: Evidence for reversible motoneurone dysfunction in thyrotoxicosis. J. Neurol. Neurosurg. Psychiatry 37 (1974) 548–558.

MCDANIEL, H.G., C.S. PITTMAN, S.J. OH and S. DIMAURO: Carbohydrate metabolism in hypothyroid myopathy. Metabolism 26 (1977) 867–873.

MCEACHERN, D. and W.D. ROSS: Chronic thyrotoxic myopathy: a report of three cases with a review of previously reported cases. Brain 65 (1942) 181–192.

MCFADZEAN, A.J.S. and R. YEUNG: Periodic paralysis complicating thyrotoxicosis in Chinese. Br. Med. J. 1 (1967) 451–455.

MCKENZIE, J.M. and M. ZAKARIJA: Hyperthyroidism. In: L.J. DeGroot (Ed.), Endocrinology. Philadelphia, W.B. Saunders Company (1989) 646–682.

MCKERAN, R.O., G. SLAVIN, P. WARD, E. PAUL and W.G.P. MAIR: Hypothyroid myopathy: a clinical and pathological study. Pathology 132 (1980) 35–54.

MEIER, C. and A. BISCHOFF: Polyneuropathy in hypothyroidism: clinical and nerve biopsy study of 4 cases. J. Neurol. 215 (1977) 103–114.

MILLIKAN, C.H. and S.F. HAINES: The thyroid gland in relation to neuromuscular disease. Arch. Intern. Med. 92 (1953) 5–39.

MIZUSAWA, H., A. TAKAGI, H. SUGITA and Y. TOYOKURA: Mounding phenomenon: an experimental study in vitro. Neurology 33 (1983) 90–93.

MOR, F., P. GREEN and A. WYSENBEEK: Myopathy in Addison's disease. Ann. Rheum. Dis. 46 (1987) 81–83.

MORGAN, D.B., G. HUNT and C.R. PATERSON: The osteomalacia syndrome after stomach operations. Q. J. Med. 39 (1970) 395–410.

MORRISON, W.L., J.N.A. GIBSON, R.T. JUNG and M.J. RENNIE: Skeletal muscle and whole body protein turnover in thyroid disease. Eur. J. Clin. Invest. 18 (1988) 62–68.

MÜLLER, R. and E. KUGELBERG: Myopathy in Cushing's syndrome. J. Neurol. Neurosurg. Psychiatry 22 (1959) 314–319.

NADALO, L.A., J. EASTERBROOK, C. MCARDLE, D.B. MENDELSOHN and T.H. PONDER: The neuroradiology of visual disturbances. Neurol. Clin. 9 (1991) 1–33.

NAGULESPAREN, M., R. TRICKEY, M.J. DAVIES and J.S. JENKINS: Muscle changes in acromegaly. Br. Med. J. 2 (1976) 914–915.

NAJJAR, S.S.: Muscular hypertrophy in hypothyroid children: the Kocher-Debré-Sémélaigne syndrome: a review of 23 cases. J. Pediatr. 85 (1974) 236–239.

NICKEL, S.N., B. FRAME, J. BEBIN, W.W. TOURTELLOTTE, J.A. PARKER and B.R. HUGHES: Myxedema neuropathy and myopathy: a clinical and pathologic study. Neurology (Minneap.) 11 (1961) 125–137.

NORRIS, F.H.: Neuromuscular transmission in thyroid disease. Ann. Intern. Med. 64 (1966) 81–86.

NORRIS, F.H. and B.J. PANNER: Hypothyroid myopathy: clinical, electromyographical and ultrastructural observations. Arch. Neurol. (Chicago) 14 (1966) 574–589.

NWOYE, L., W.F.H.M. MOMMAERTS, D.R. SIMPSON, K. SERAYDARIAN and M. MARUSICH: Evidence for a direct action of thyroid hormone in specifying muscle properties. Am. J. Physiol. 242 (1982) R401–R408.

OH, V.M.S., E.A. TAYLOR, S.H. YEO and K.O. LEE: Cation transport across lymphocyte plasma membranes in euthyroid and thyrotoxic men with and without hypokalemic periodic paralysis. Clin. Sci. 78 (1990) 199–206.

OKAJIMA, F. and M. UI: Metabolism of glucose in hyper- and hypo-thyroid rats in vivo. Biochem. J. 182 (1979) 565–575.

ORISTRELL-SALVA, J. and A. MIRADA-CANALS: Phaeo-chromocytoma and rhabdomyolysis. B. Med. J. 288 (1984) 1198.

OSSERMAN, K.E., P. TSAIRIS and L.B. WEINER: Myasthenia gravis and thyroid disease: clinical and immunological correlation. Mt. Sinai J. Med. (New York) 34 (1967) 469–483.

PACY, P.J. and D. HALLIDAY: Muscle protein synthesis in steroid-induced proximal myopathy: a case report. Muscle Nerve 12 (1989) 378–381.

PALMUCCI, L., A. BERTOLOTTO, C. DORIGUZZI, T. MONGINI and R. CODA: Osteomalacic myopathy in a case of diffuse nodular lipomatosis of the small bowel. Acta Neurol. Belg. 82 (1982) 65–71.

PALMUCCI, L., C. DORIGUZZI and A.P. ANZIL: Myopathy with tubular aggregates in a patient adrenalectomized for Cushing's syndrome. J. Neurol. 232 (1985) 374–377.

PARKER, M.S., I. KLEIN, M.R. HAUSSLER and D.H. MINTZ: Tumor-induced osteomalacia: evidence of a surgically correctable alteration of vitamin D metabolism. J. Am. Med. Assoc. 245 (1981) 492–493.

PATTEN, B.M., J.P. BILEZIKIAN, L.E. MALLETTE, A. PRINCE, W.K. ENGEL and G.D. AURBACH: Neuromuscular disease in primary hyperparathyroidism. Ann. Intern. Med. 80 (1974) 182–193.

PEARCE, G.W.: Electron microscopy in the study of muscular dystrophy. In: G.H. Bourne and M.N. Golarz (Eds.), Muscular Dystrophy in Man and Animals. New York, Hafner Publishing Company (1963) 159–175.

PELLOCK, J.M., M. BEHRENS, I. LEWIS, D. HOLUB, S. CARTER and L.P. ROWLAND: Kearns-Sayre syndrome and hypoparathyroidism. Ann. Neurol. 3 (1978) 455–458.

PERKOFF, G.T., R. SILBER, F.H. TYLER, G.E. CARTWRIGHT and M.M. WINTROBE: Studies in disorder of muscle. Part XII: Myopathy due to the administration of therapeutic amounts of 17-hydroxycorticosteroids. Am. J. Med. 26 (1959) 891–898.

PETER, J.B., D.A. VERHAAG and M. WORSFOLD: Studies of steroid myopathy: examination of the possible effect of triamcinolone on mitochondria and sarcotubular vesicles of rat skeletal muscle. Biochem. Pharmacol. 19 (1970) 1627–1636.

PETERS, R.F., M.C. RICHARDSON, M. SMALL and A.M. WHITE: Some biochemical effects of triamcinolone acetomide on rat liver and muscle. Biochem. J. 116 (1970) 349–355.

PICKETT, J.B., R.B. LAYZER, S.R. LEVIN, V. SCHNEIDER, M.J. CAMPBELL and A.J. SUMNER: Neuromuscular complications of acromegaly. Neurology (Minneap.) 25 (1975) 638–645.

PIMSTONE, N., N. MARINE and B. PIMSTONE: Beta-adrenergic blockade in thyrotoxic myopathy. Lancet ii (1968) 1219–1220.

PLATTENER, E.B. and C.I. REED: Study of muscular efficiency in rats injected with anterior pituitary growth factor. Endocrinology 25 (1939) 401–404.

PLEASURE, D.E., G.O. WALSH and W.K. ENGEL: Atrophy of skeletal muscle in patients with Cushing's syndrome. Arch. Neurol. (Chicago) 22 (1970) 118–125.

PLEASURE, D., B. WYSZYNSKI, D. SUMNER, D.L. SCHOTLAND, B. FELDMANN, N. NUGENT, K. HITZ and D.B.P. GOODMAN: Skeletal muscle calcium metabolism and contractile force in vitamin D-deficient chicks. J. Clin. Invest. 64 (1979) 1157–1167.

PLOTZ, C.M., A.I. KNOWLTON and C. RAGAN: The natural history of Cushing's syndrome. Am. J. Med. 13 (1952) 597–614.

POINTON, J.J., M.J.O. FRANCIS and R. SMITH: Effect of vitamin D deficiency on sarcoplasmic reticulum and troponin C concentration of rabbit skeletal muscle. Clin. Sci. 57 (1979) 257–263.

POLLACK, J.A., A.L. SCHILLER and J.D. CRAWFORD: Rickets and myopathy cured by removal of nonossifying fibroma of bone. Pediatrics 52 (1973) 364–371.

POLLEN, R.H. and R.H. WILLIAMS: Hyperkalemic neuromyopathy in Addison's disease. N. Engl. J. Med. 263 (1960) 273–278.

POU SERRADELL, A., J. ROQUER GONZALEZ, J.M. COROMINAS TORRES, J. LLORTA TRULL, J. OLIVA BIELSA and A. UGARTE ELOLA: [Amyotrophic lateral sclerosis syndrome and hyperthyroidism. Cure with antithyroid drugs]. Rev. Neurol. (Paris) 146 (1990) 219–220.

PRABHU, V.G. and Y.T. OESTER: Electromyographic studies of skeletal muscle of rats given cortisone. Arch. Neurol. (Chicago) 24 (1971) 253–258.

PRINEAS, J.W., A.S. MASON and R.A. HENSON: Myopathy in metabolic bone disease. Br. Med. J. 1 (1965) 1034–1036.

PRINEAS, J.W., R. HALL, D.D. BARWICK and A.J. WATSON: Myopathy associated with pigmentation following adrenalectomy for Cushing's syndrome. Q. J. Med. 37 (1968) 63–77.

PRYSOR-JONES, R.A. and J.S. JENKINS: Effect of excessive secretion of growth hormone on tissues of the rat, with particular reference to the heart and skeletal muscle. J. Endocrinol. 85 (1980) 75–82.

PUVANENDRAN, K., J.S. CHEAH, N. NAGANATHAN and P.K. WONG: Thyrotoxic myopathy: a clinical and quantitative analytic electromyographic study. J. Neurol. Sci. 42 (1979a) 441–451.

PUVANENDRAN, K., J.S. CHEAH, N. NAGANATHAN, P.P.B. YEO and P.K. WONG: Neuromuscular transmission in thyrotoxicosis. J. Neurol. Sci. 43 (1979b) 47–57.

RABINOWITZ, D. and K.L. ZIERLER: Differentiation of active from inactive acromegaly by studies of forearm metabolism and response to intra-arterial insulin. Bull. Johns Hopkins Hosp. 113 (1963) 211–224.

RAFF, M.C., V. SANGALANG and A.K. ASBURY: Ischemic mononeuropathy multiplex associated with diabetes mellitus. Arch. Neurol. 18 (1968) 487–499.

RAMSAY, I.D.: Electromyography in thyrotoxicosis. Q. J. Med. 34 (1965) 255–267.

RAMSAY, I.D.: Muscle dysfunction in hyperthyroidism. Lancet ii (1966) 931–934.

RAMSAY, I.D.: Thyroid disease and muscle dysfunction. Year Book Medical Publishers, Chicago (1974).

RAMSAY, I.D.: Endocrine myopathies. Practitioner 226 (1982) 1075–1080.

RAO, S.N., B.C. KATTYAR, K.R.P. NAIR and S. MISRA: Neuromuscular status in hypothyroidism. Acta Neurol. Scand. 61 (1980) 167–177.

RAVID, M. and M. ROBSON: Proximal myopathy caused by iatrogenic phosphate depletion. J. Am. Med. Assoc. 236 (1976) 1380–1381.

REICH, S., S.N. WIENER, S. CHESTER and R. RUFF: Clinical and radiologic features of spontaneous muscle infarction in the diabetic. Clin. Nucl. Med. 10 (1985) 876–879.

RICHARDSON, J.A., G. HERRON, R. REITZ and R. LAYZER: Ischemic ulcerations of the skin and necrosis of muscle in azotemic hyperparathyroidism. Ann. Intern. Med. 71 (1969) 129–138.

RINGQVIST, I.: Achilles reflex time as a measure of thyroid function. Acta Med. Scand. 188 (1970) 231–239.

RODIER, M., J.L. RICHARD, J. BRINGER, G. CAVALIE, H. BELLET and J. MIROUZE: Thyroid status and muscle protein breakdown as assessed by urinary 3-methylhistidine excretion: study in thyrotoxic patients before and after treatment. Metabolism 33 (1984) 97–100.

RODMAN, J.S. and T. BAKER: Changes in the kinetics of muscle contraction in vitamin D-depleted rats. Kidney Int. 13 (1978) 189–193.

ROGOZKIN, V.A.: Metabolism of Anabolic Androgenic Steroids. Boca Raton, CRC Press (1991) 123–142.

ROWLAND, L.P.: Diverse forms of motor neuron diseases. In: L.P. Rowland (Ed.), Human Motor Neuron Diseases. New York, Raven Press (1982) 1–11.

ROWLAND, L.P., H. ARANOW JR. and P.F.A. HOEFER: Endocrine aspects of myasthenia gravis. In: E. Kuhn (Ed.), Progressive Muskel Dystrophie Mytonie Myasthenie. Berlin, Springer (1966) 416–426.

RUFF, R.L.: Endocrine myopathies (hyper- and hypofunction of adrenal, thyroid, pituitary, and parathyroid glands and iatrogenic steroid myopathy). In: A.G. Engel and B.Q. Banker (Eds.), Myology. New York, McGraw-Hill (1986) 1871–1906.

RUFF, R.L., W. STUHMER and W. ALMERS: Effect of glucocorticoid treatment on the excitability of rat skeletal muscle. Pflugers Arch. 395 (1982) 132–137.

SAHAY, B.M., L.M. BLENDIS and R. GREENE: Relation between myasthenia gravis and thyroid disease. Br. Med. J. 1 (1965) 762–765.

SALVI, M., H. FUKAYAWA, N. BERNARD, Y. HIRAMATSU, J. HOW and J.R. WALL: Role of autoantibodies in the pathogenesis and association of endocrine autoimmune disorders. Endocrine Rev. 9 (1988) 450–466.

SALVI, M., Z.G. ZHANG, D. HALGERT, M. WOO, A. LIBERMAN, L. CADARSO and J.R. WALL: Patients with endocrine ophthalmopathy not associated with overt thyroid disease have multiple thyroid immunologic abnormalities. J. Clin. Endocrinol. Metab. 70 (1990) 89–94.

SAMBROOK, M.A., J.R. HERON and G.M. ABER: Myopathy in association with primary hyperaldosteronism. J. Neurol. Neurosurg. Psychiatry 35 (1972) 202–207.

SATOYOSHI, E., E. MURAKAMI, H. KOWA, M. KINOSHITA, K. NOGUCHI, S. HOSHINA, Y. NISHIYAMA and K. ITO:

Myopathy in thyrotoxicosis with special emphasis on an effect of potassium ingestion on serum and urinary creatine. Neurology (Minneap.) 13 (1963a) 645–658.

SATOYOSHI, E., K. MURAKAMI, H., KOWA, M. KINOSHITA and Y. NISHIYAMA: Periodic paralysis in hyperthyroidism. Neurology (Minneap.) 13 (1963b) 746–752.

SCARPALEZOS, S., C. LYGIDAKIS, C. PAPAGEORGIOU, S. MALIARA and A.S. KOUKOULOMMATI: Neural and muscular manifestations of hypothyroidism. Arch. Neurol. 29 (1973) 140–144.

SCHILLER, F. and F.O. KOLB: Carpal tunnel syndrome in acromegaly. Neurology (Minneap.) 4 (1954) 271–282.

SCHOTT, G.D. and M.R. WILLS: Myopathy in hypophosphataemic osteomalacia presenting in adult life. J. Neurol. Neurosurg. Psychiatry 38 (1975) 297–304.

SHANE, E., K.A. MCCLANE, M.R. OLARTE and J.P. BILEZIKIAN: Hypoparathyroidism and elevated serum enzymes. Neurology 30 (1980) 192–195.

SHOJI, S.: Myofibrillar protein catabolism in rat steroid myopathy measured by 3-methylhistidine excretion in the urine. J. Neurol. Sci. 93 (1989) 333–340.

SHOJI, S. and R. PENNINGTON: The effect of cortisone on protein breakdown and synthesis in rat skeletal muscle. Mol. Cell Endocrinol. 6 (1977) 159–169.

SHOJI, S., A. TAKAGI, H. SUGITA and Y. TOYOKURA: Dysfunction of sarcoplasmic reticulum in rabbit and human steroid myopathy. Exp. Neurol. 51 (1976) 304–309.

SIMPSON, J.A.: The correlations between myasthenia gravis and disorders of the thyroid gland. In: Research in Muscular Dystrophy, Proceedings IVth Symposium, edited by members of the Research Committee of the Muscular Dystrophy Group. London, Pitman Medical (1968) 31–44.

SIMPSON, J.F., M.R. WESTERBERG and K.R. MAGEE: Myasthenia gravis: analysis of 295 cases. Acta Neurol. Scand. 42 (Suppl.) (1966) 1–27.

SKANSE, B.: Carpal tunnel syndrome in myxedema and acromegaly. Acta Chir. Scand. 121 (1961) 476–480.

SKARIA, J., B.C. KATTYAR, T.P. SRIVASTAVA and B. DUBE: Myopathy and neuropathy associated with osteomalacia. Acta Neurol. Scand. 51 (19750 37–58.

SMITH, R. and G. STERN: Myopathy, osteomalacia and hyperparathyroidism. Brain 90 (1967) 593–602.

SMITH, R. and G. STERN: Muscular weakness in osteomalacia and hyperparathyroidism. J. Neurol. Sci. 8 (1969) 511–520.

SNOWDON, J.A., A.C. MACFIE and J.B. PEARCE: Hypocalcaemic myopathy with paranoid psychosis. J. Neurol. Neurosurg. Psychiatry 39 (1976) 48–52.

STERN, L.Z. and J.M. FAGAN: The endocrine myopathies. In: P.J. Vinken and G.W. Bruyn (Eds.), Handbook of clinical Neurology, Vol. 41. Amsterdam, North-Holland Publishing Company (1979) 235–258.

STERN, L.Z., C.M. PAYNE and L.K. HANNAPEL:

Acromegaly: histochemical and electron microscopic changes in deltoid and intercostal muscle. Neurology (Minneap.) 24 (1974) 589–593.

STOCKER, W.W., F.J. SAMAHA and L.J. DEGROOT: Coupled oxidative phosphorylation in muscle of thyrotoxic patients. Am. J. Med. 44 (1968) 900–909.

SWANSON, J.W., J.J. KELLY and W.M. MCCONAHEY: Neurologic aspects of thyroid dysfunction. Mayo Clin. Proc. 56 (1981) 504–512.

SWEATMAN, M.C.M. and L. CHAMBERS: Disordered oesophageal motility in thyrotoxic myopathy. Postgrad. Med. J. 61 (1985) 619–620.

TAKAMORI, M., L. GUTMANN and S.R. SHANE: Contractile properties of human skeletal muscle: normal and thyroid disease. Arch. Neurol. 25 (1971) 535–546.

TAKAMORI, M., L. GUTMANN, T.W. CROSBY and J.D. MARTIN: Myasthenic syndromes in hypothyroidism: electrophysiological study of neuromuscular transmission and muscle contraction in two patients. Arch. Neurol. 26 (1972) 326–335.

TAMAI, H., T. NAKAGAWA, N. OHSAKO, O. FUKIN, H. TAKAHASHI, F. MATSUZUKA, K. KUMA and S. NAGATAKI: Changes in thyroid functions in patients with euthyroid Graves' disease. J. Clin. Endocrinol. Metab. 50 (1980) 108–112.

TAMAI, H., K. TANAKA, G. KOMAKI, S. MATSUBAYASHI, Y. HIROTA, K. MORI, K. KUMA, L.F. KUMAGAI and S. NAGATAKI: HLA and thyrotoxic periodic paralysis in Japanese patients. J. Clin. Endocrinol. Metab. 64 (1987) 1085–1078.

TENG, C.S., A.L. CROMBIE, R. HALL and W.M. ROSS: An evaluation of supervoltage orbital irradiation for Graves' ophthalmopathy. Clin. Endocrinol. 13 (1980) 545–551.

THOMAS, P.K. and S.G. ELIASSON: Diabetic neuropathy. In: P.J. Dyck, P.K. Thomas, E.H. Lambert and R. Bunge (Eds.), Peripheral Neuropathy. Philadelphia, W.B. Saunders Company (1984) 1773–1810.

TICE, L.W. and A.G. ENGEL: The effects of glucocorticoids on red and white muscles in the rat. Am. J. Pathol. 50 (1967) 311–333.

TOPPET, M., N. TELLERMAN-TOPPET, H.B. SZLIWOWSKI, M. VAINSEL and C. COERS: Oculocraniosomatic neuromuscular disease with hypoparathyroidism. Am. J. Dis. Child. 131 (1977) 437–441.

TURKEN, S.A., M. CAFFERTY, S.J. SILVERBERG, L. DE LA CRUZ, C. CIMINO, D.J. LANGE, R.E. LOVELACE and J.P. BILEZIKIAN: Neuromuscular involvement in mild, asymptomatic, primary hyperparathyroidism. Am. J. Med. 87 (1989) 553–557.

TYRELL, J.B., R.M. BROOKS, P.A. FITZGERALD, P.B. COFOID, P.H. FORSHAM and C.B. WILSON: Cushing's disease: selective trans-sphenoidal resection of pituitary microadenomas. N. Engl. J. Med. 298 (1978) 753–758.

UTIGER, R.D.: The pathogenesis of autoimmune thyroid disease. N. Engl. J. Med 325 (1991) 278–279.

VENABLES, G.S., D. BATES and D.A. SHAW: Hypothyroidism with true myotonia. J. Neurol. Neurosurg. Psychiatry 41 (1978) 1013–1015.

VICALE, C.T.: The diagnostic features of a muscular syndrome resulting from hyperparathyroidism, osteomalacia owing to renal tubular acidosis, and perhaps to related disorders of calcium metabolism. Trans. Am. Neurol. Assoc. 74 (1949) 143–147.

VIGNOS, P.J. and R. GREENE: Oxidative respiration of skeletal muscle in experimental corticosteroid myopathy. J. Lab. Clin. Med. 81 (1973) 365–378.

VOLPE, R.: Autoimmunity in endocrine disease. In: K.L. Becker (Ed.), Principles and Practice of Endocrinology and Metabolism. Philadelphia, J.B. Lippincott (1990) 1560–1570.

VON BASEDOW, C.A.: Exophthalmos durch Hypertrophie des Zellgewebes in der Augenhöhle. Wochenschr. Ges. Heilkd. 6 (1840) 197–202.

VON RECKLINGHAUSEN, F.: Die fibröse oder deformierende Ostitis, die osteomalacie und osteoplastische Carcinose in ihren gegenseitigen Beziehungen. Festschrift Rudolph Virchow zu Seinem 71. Geburtstag. Berlin, G. Reimer (1891) 1–89.

WALL, JR., J. HENDERSON, C.R. STRAKOSCH and D.M. JOYNER: Graves' ophthalmology. Can. Med. Assoc. J. 124 (1981) 855–866.

WALL, J.R., M. SALVI, N.F. BERNARD, A. BOUCHER and D. HAEGERT: Thyroid associated ophthalmopathy — a model for the association of organ-specific autoimmune disorders. Immunol. Today 12 (1991) 150–153.

WALTERS, R.O.: Idiopathic hypoparathyroidism and extrapyramidal and myopathic features. Arch. Dis. Child. 54 (1979) 236–238.

WEINSTEIN, R., R. SCHWARTZMAN and G.S. LEVEY: Propranolol reversal of bulbar dysfunction and proximal myopathy in hyperthyroidism. Ann. Intern. Med. 82 (1975) 540–541.

WERNER, S.C.: Modification of the classification of the eye changes of Graves' disease. recommendations of the ad hoc committee of the American Thyroid Association. J. Clin. Endocrinol. 44 (1977) 203–204 (letter).

WIERSINGA, W.M., T. SMITT, R. VAN DER GAAG, M. MOURITS and L. KOORNNEEF: Clinical presentation of Graves' ophthalmopathy. Ophthalmic Res. 21 (1989) 73–82.

WILES, C.M., A. YOUNG, D.A. JONES and R.H.T. EDWARDS: Muscle relaxation rate, fibre-type composition and energy turnover in hyper- and hypothyroid patients. Clin. Sci. 57 (1979) 375–384.

WILLIAMS, R.S.: Triamcinolone myopathy. Lancet i (1959) 698–701.

WILSON, J.: Androgen abuse by athletes. Endocr. Rev. 9 (1988) 181–199.

WILSON, J. and J.N. WALTON: Some muscular manifestations of hypothyroidism. J. Neurol. Neurosurg. Psychiatry 22 (1959) 320–324.

WINCKLER, B., R. STEELE, N. ALTSZULLER, R.C. DE BODO and C. BJERKNES: Effect of growth hormone on free fatty acid metabolism. Am. J. Physiol. 206 (1964) 174–178.

WOLF, S.M., W. LUSK and L. WEISBERG: Hypocalcemia myopathy. Bull. Los Angeles Neurol. Soc. 37 (1972) 167–177.

YALOWITZ, D.L., A.S. BRETT and J.M. EARLL: Far-advanced primary hyperparathyroidism in an 18 year-old young man. Am. J. Med. 77 (1984) 545–548.

YAMAMOTO, K., K. SAITO, T. TOKAI: Diagnosis of exophthalmos using orbital ultrasonography and treatment of malignant exophthalmos with steroid therapy, orbital radiation therapy, and plasmapheresis. Prog. Clin. Biol. Res. 116 (1983) 189–205.

YEO, P.P.B., S.H. CHAN, K.F. LUI, G.B. WEE, P. LIM and J.S. CHEAH: HLA and thyrotoxic periodic paralysis. Br. Med. J. 2 (1978) 930.

YEUNG, R.T.T. and T.T. TSE: Thyrotoxic periodic paralysis: effect of propranolol. Am. J. Med. 57 (1974) 584–590.

ZÜRCHER, R.M., F.F. HARBER, B.E. GRUNIG and F.J. FREY: Effect of thyroid dysfunction on thigh muscle efficiency. J. Clin. Endocrinol. Metab. 69 (1989) 1082–1086.

Handbook of Clinical Neurology, Vol. 18 (62): Myopathies
L.P. Rowland and S. DiMauro, editors

CHAPTER 18

Myoglobinuria

INGRID TEIN[1], SALVATORE DIMAURO[2] and LEWIS P. ROWLAND[3]

[1]*Division of Neurology, The Hospital for Sick Children, Toronto, Ontario, Canada,*
[2]*H. Houston Merritt Clinical Research Center for Muscular Dystrophy and Related Diseases, College of Physicians and Surgeons of Columbia University, New York, NY, and*
[3]*Neurological Institute, Columbia-Presbyterian Medical Center, New York, NY, USA*

Myoglobinuria is a form of pigmenturia, and dark urine has a peculiar relationship to some of the darker attributes of modern life: war, drug abuse, cruel physical abuse, and misuse of therapeutic drugs are all common causes. The syndrome has also been sensitive to technological advances; methods to detect myoglobin in body fluids have become more sensitive and clinical correlations have changed.

DEFINITION

Myoglobinuria literally means a state in which myoglobin is present in urine. However, myoglobin is normally present in urine in amounts too small to be detected by the naked eye; the medical definition refers to a pathological state, a syndrome of acute muscle necrosis that is manifest clinically by limb weakness, myalgia, swelling, and gross pigmenturia or heme-positive urine without hematuria. Under some circumstances, especially when there is heme-induced nephropathy with anuria, there may be no pigmenturia, but the syndrome can then be recognized by the combination of renal failure and high serum levels of sarcoplasmic enzymes.

The word 'rhabdomyolysis' has formally replaced 'myoglobinuria' in the Index Medicus. However, it seems reasonable to maintain the original term to describe the clinical syndrome and associated biochemical abnormalities. The nature of the debate arises in the history of the disorder.

HISTORY

The first description is attributed to Meyer-Betz (1910), but the syndrome gained prominence in World War II when Bywaters et al. (1941; Bywaters and Popjack 1942) recognized the crush syndrome victims of fallen debris created by the bombing of London.

By 1964 there had been only 118 case reports of idiopathic myoglobinuria, and 103 of these came after 1951 (Rowland et al. 1964). By that time, however, the first heritable cause had been identified with the description of phosphorylase deficiency (Schmid and Mahler 1959). Rowland et al. (1964) were impressed by the diversity of the apparent causes they encountered in 12 patients. Within a few years, they saw 38 more cases (Rowland and Penn 1972). From that time on, the syndrome was transformed from one considered rare to one that causes 5 to 25% of all cases of acute renal failure (Rowland 1984; Thomas and Ibels 1985; Penn 1986).

NOMENCLATURE

Bowden et al. (1956) introduced the term 'rhabdomyolysis' to emphasize the primary role of muscle necrosis in causing the syndrome and also because, during an attack, myoglobin is not the only constituent of muscle to appear in the plasma and urine. However, 'myoglobinuria' was still a popular word. The 2 terms were used interchangeably until Gabow et al. (1982) used rhabdomyolysis to designate any condition in which the serum creatine kinase (CK) level was more than 5 times normal (in the absence of brain or heart disease). The new term seemed reasonable because myoglobin can be found in the urine by modern sensitive methods even without the clinical syndrome, and myoglobin may not be demonstrable in urine during a severe attack if there is a state of anuria.

There is now no sense in prolonging the lexical debate; rhabdomyolysis has won by official (if arbitrary) sanction. Nevertheless, continued use of 'myoglobinuria' seems justified for several reasons (Rowland 1984). 1. The clinical syndrome has been recognized by that name for 82 years. 2. The imagined precision of 'rhabdomyolysis' is flawed by ambiguity because there is no generalized destruction of muscle (as in the parallel state of hemolysis) and because there is no histological hallmark of rhabdomyolysis, which cannot be distinguished anatomically from necrosis. The word has hardly ever been used as a pathological term (Zimmer et al. 1991). Without any prompting from the editor, Karpati and Carpenter (see Chapter 1, this Volume) did not use the word 'rhabdomyolysis' once in their detailed review of muscle pathology. 3. The defining CK level proposed by Gabow et al. (1982) would, in our hospital, include anyone with a CK value of 250 units, in contrast to levels between 5000 and 10,000 that are encountered in Duchenne dystrophy or polymyositis; values of 50,000 or more are seen in attacks of myoglobinuria. 4. The definition of Gabow et al. (1982) would not distinguish between the chronically high values in patients with Duchenne dystrophy and the attacks of myoglobinuria they sometimes experience with general anesthesia. Similarly, high values of CK are sometimes encountered in people who have no other evidence of myopathy, a condition therefore called 'asymptomatic hyperCKemia'. To lump both chronic states and the acute changes under the same term would be like making no distinction between angina and myocardial infarction or between transient ischemic attack and cerebral infarction (Rowland 1984). 5. The main hazard of an attack is renal failure, and the main nephrotoxin still seems to be myoglobin, influenced undoubtedly by other biochemical and metabolic abnormalities that occur during an attack. 6. We are concerned with a clinical disorder of diverse cause, not a biochemical definition. In contrast, some authors have used 'rhabdomyolysis' to describe the rise in serum enzymes of syndromes that have never been associated with clinical myoglobinuria (Fernandez-Sola et al. 1987; De Smet et al. 1991).

DIAGNOSIS OF THE CLINICAL SYNDROME

If the patient is alert, there is myalgia or limb weakness, and the urine is darker than normal (Table 1). The color is usually brownish, rather than red. The urine gives positive chemical tests for both albumin and heme (a concentration of at least 4 µg/ml), but there are few or no red blood cells; casts are often present and may include the pigment, which is now identified by immunochemical methods. Serum content of CK and other sarcoplasmic enzymes is usually more than 100 times the upper limit of normal. Hyperuricemia, hyperphosphatemia, and hypo- or hypercalcemia are inconstant features; if there is renal failure, serum levels of potassium and calcium may rise.

If the patient is comatose, or if the presenting disorder is one of acute renal failure, there may be no muscle symptoms or signs. The diagnosis can then be made without even examining the urine if 2 conditions are met: (a) the serum content of sarcoplasmic enzymes is 100 times normal; and (b) there is renal failure. Most cases of this nature arise from prolonged coma or extreme muscular exertion, but there are other causes.

IDENTIFICATION OF MYOGLOBIN IN BODY FLUIDS

When myoglobin is present in sufficient amount, the urine appears darker than normal. The color is brown, not red, because the myoglobin is rapidly oxidized. In the US, the color is often described as resembling coca-cola; in France, that of a dark burgundy wine. Gross coca-cola coloured urine reflects an excretion of more than 250 µg/ml of myoglobin (Penn 1986).

The pigment can be identified provisionally

from the clinical circumstances. Acute intermittent porphyria may be associated with an acute peripheral neuropathy but not the acute myopathy seen with myoglobinuria. Tests for porphobilinogen are positive in porphyria.

The only other conditions involved in the differential diagnosis of pigmenturia are unusual drug excretions, which are usually evident and do not give reactions for heme. The major differential is hemoglobinuria, which involves the distinction between hematuria and excretion of hemoglobin after hemolysis (Hamilton et al. 1989). In hematuria, the urine contains red blood cells and the serum is clear. In hemolysis the serum is pink, because hemoglobin is bound to haptoglobin and the threshold for excretion is higher than that for myoglobin. Myoglobin, on the other hand, is not bound in the serum and the renal threshold is 0.3–2.0 µg/ml. The serum is clear. Most important, the clinical syndromes of hemolysis and myoglobinuria differ, and the serum levels of sarcoplasmic enzymes do not rise in hemolysis.

Proof that the urine pigment is myoglobin depends on some specific tests (Table 2). In the past, myoglobin was identified by solubility characteristics, spectroscopy, or electrophoresis. All of these methods have been replaced by immunochemical

methods. Antibodies raised against purified myoglobin do not react with albumin, hemoglobin, or other muscle or serum proteins. Radioimmunoassays are sufficiently sensitive to measure the low levels found in normal serum or urine (Cloonan et al. 1979; Roxin et al. 1979; Adams 1980). Normal serum contains 3–80 ng/ml, and amounts up to 12 ng/ml have been found in urine from about 66% of normal people (Penn 1986).

MYOGLOBIN LEVELS IN OTHER DISEASES

The use of these methods, introduced in the 1970s, transformed concepts of the diagnostic significance of the presence of myoglobin in urine. 'Myoglobinuria' was no longer restricted to the gross amounts identified by earlier methods of detection, those encountered in patients with the acute syndrome. With immunochemical detection, abnormally elevated levels were found in serum and urine whenever there was a rise in sarcoplasmic enzyme levels. Myoglobin is then merely another muscle constituent that appears in serum. Under these circumstances, serum levels of 300–1900 ng/ml (up to 40 times normal) were found in boys with Duchenne muscular dystrophy (Kagen et al. 1975; Ando et al. 1978; Askmark et al. 1981; Kiessling et al. 1981; Nicholson 1981). Kiessling and Beckmann (1981) found a correlation coefficient of 0.17, between serum myoglobin levels and serum CK in Duchenne dystrophy, but others found no relationship (Kagen et al. 1975; Hische and Van

TABLE 1

Myoglobinuria: definition of clinical disorder.

If patient is alert:
  Myalgia or limb weakness
  Pigmenturia
    Test in urine is positive but there are few red blood cells in urine.
    Identification as myoglobin by immunochemical method.
    Serum creatine kinase (CK) and other sarcoplasmic enzyme levels usually >100 times the upper normal limit during acute attack.
    Inconstant features: increased serum uric acid level, increased PO_4, increased or decreased Ca^{2+} level; if renal failure, serum K^+ and Ca^{2+} levels increase.

If patient is comatose or in acute renal failure, there may be no muscular symptoms or signs, but:
  Serum sarcoplasmic enzyme levels are 100 times normal.
  There is biochemical evidence of renal failure.

Taken from Rowland (1984) and Tein et al. (1990b).

TABLE 2

Sensitivity of assays for myoglobin.

	µg/ml
Urine	
Radial immunodiffusion	5.0
Peroxidase-sensitive chromogens (orthotolidine, o-dianisidine, benzidine)	0.5
Hemagglutination inhibition	0.3
Complement fixation	0.15
Serum	
Hemagglutination inhibition	0.005
Radioimmunoassay	0.005
ELISA	0.005

Taken from Penn (1986) with permission of McGraw-Hill, Inc.

Der Helm 1979). Abnormal levels were also found in carriers of the gene (Adornato et al. 1978; Miyoshi et al. 1978).

Similarly, abnormal serum myoglobin levels have been found in Becker dystrophy, dermato-myositis, polymyositis, facioscapulohumeral dystrophy, myotonic dystrophy, and motor neuron diseases (Miyoshi et al. 1978; Kiessling et al. 1981). Serum levels rise (500–5500 ng/ml) in the first 4–12 hours after myocardial infarction, even before CK levels increase (Kagen et al. 1975; Reichlin et al. 1978). Levels up to 100 ng/ml may be seen in asymptomatic people after exercise.

CONCOMITANTS AND CONSEQUENCES OF MYOGLOBINURIA: TREATMENT

It has been calculated that clinical myoglobinuria results from destruction of at least 200 grams of skeletal muscle. It is presumed that all of the constituents of the necrotic muscle are discharged into the blood. The circumstances that cause the catastrophe are also likely to cause dehydration, shock, or acidosis and these conditions all increase the risk of heme-induced nephropathy (Braun et al. 1970; Knochel 1976, 1982; Koffler et al. 1976). Renal failure was seen in 33% of the cases of Gabow et al. (1982), a figure that may have been inordinately high because renal failure often calls attention to myoglobinuria.

The combination of increased entry and decreased exit of muscle constituents leads to raised plasma levels of sarcoplasmic enzymes, amino acids (especially taurine), creatine, creatinine, potassium, phosphate, and urate. These changes may adversely affect kidneys and heart, exaggerating the damage (Warren et al. 1975; Knochel 1976). Hyperphosphatemia may lead to secondary hypocalcemia, and later there may be hypercalcemia because of increased secretion of parathormone and vitamin D, with deposition of calcium in injured muscle followed by release. It is therefore urgent to prevent renal failure.

Once the syndrome is identified, administration of mannitol or other osmotic diuretics may prevent renal failure (Eneas et al. 1979). Alkalinization of the urine and volume replacement are also important (Better and Stein 1990). Once established, however, renal failure may have to be treated by dialysis. Myoglobin itself is not dialyzable (Hart et al. 1982) and plasmapheresis has been used (Kuroda et al. 1981). Even with modern intensive care, some patients die because of the nature of the precipitating event or complicating medical conditions (Gabow et al. 1982). Fasciotomy has been used to relieve muscle swelling in compartment syndromes (Owen et al. 1979), but this should be approached cautiously (Odeh 1991).

Among survivors there is no documented case of permanent renal disease (McCarron et al. 1980). Nerve injuries, however, may result from compression or traction in cases of coma-crush myoglobinuria (Penn et al. 1972; Chaikin 1980; Akmal and Massry 1983).

PATHOGENESIS

The etiology of myoglobinuria comprises diverse situations, but the common features are few. Either there is a direct injury to the surface membranes, or there is a failure of energy supply to maintain the integrity of surface membranes. Direct injury may be traumatic, as in the crush syndrome, or it may be the result of myotoxins such as ethanol or any of a myriad of drugs. These distinctions provide the basis for a functional classification of the syndrome (Rowland et al. 1964; Rowland 1984; Penn 1986) (Table 3).

The complex nature of individual attacks is illustrated by drug-induced coma. Most of the damage is due to the resulting crush injury, which includes direct trauma and ischemia from compression of vessels. However, the sedative drug may also depress respiration in muscle, brain, and other organs. Cerebral depression may lead to alveolar hypoventilation, hypoxia, and respiratory acidosis. Shock and metabolic acidosis may lead to heme-induced renal failure to further aggravate the already complex metabolic disorder. The myoglobinuria of drug-induced coma is therefore related to the offending drug, trauma, and several other factors.

MUSCLE PATHOLOGY

There have been few detailed studies of muscle pathology because the findings depend on the time the biopsy is taken and the severity of the attack, as

TABLE 3

The diverse causes of myoglobinuria: sporadic (acquired) disorders.

Exertion
Military training, wrestling, squat jump or situp exercises, long distance running, skiing
Anterior tibial syndrome/other compartmental syndromes
Convulsions (status epilepticus)
Agitated delirium, delirium tremens, restraints
High-voltage electric shock, lightning stroke
Status asthmaticus
Prolonged myoclonus or acute dystonia

Crush
Compression by fallen weights
Compression by body in prolonged coma

Ischemic
Occlusion of major artery
Ischemia in compression or anterior tibial syndromes
Coagulopathy in sickle cell disease or syndromes with disseminated intravascular coagulation
Ligation of vena cava

Metabolic depression or distortion
Carbon monoxide, barbiturates, narcotics
Diabetic ketoacidosis
Nonketotic hyperglycemia hyperosmolar states
Renal tubular acidosis
Hypernatremia
Hyponatremia
Hypokalemia (see Table 15)
Hypophosphatemia
    Intravenous fluid therapy for acute and chronic alcohol abuse
    Diabetic ketoacidosis
    Parathyroidectomy

Drugs and toxins (see Table 15)

Abnormalities of body temperature
Hypothermia
    Exposure to cold
    Hypothyroidism
Fever
    Tetanus toxin
    Thyroid vaccine
    Heat injury: heat cramps, heat exhaustion, heat stroke
    Malignant hyperthermia
    Malignant neuroleptic syndrome

Infections
Viral: influenza A, influenza B, herpes simplex, infectious mononucleosis, Coxsackie, HZV, HIV
Bacterial: typhoid fever, *E. coli* sepsis, *Shigella*, *Staphylococcus*
Other organisms: *Mycoplasma*
Toxic shock syndrome

Progressive muscle disease
Polymyositis and dermatomyositis

Modified from Rowland (1984) by courtesy of the Editors of the *Can. J. Neurol. Sci.*

well as the site of biopsy. Moreover, the findings are confined to typical changes of necrosis and regeneration that are described so well by Karpati and Carpenter (see Chapter 1, this Volume). Early, the picture is dominated by signs of degeneration, which are followed by evidence of regeneration (Rowland et al. 1964). Inflammatory cells are not prominent but may be seen (Rowland et al. 1964).

TABLE 4

Heritable causes of myoglobinuria.

Biochemical abnormality known
Glycolysis/glycogenolysis
 * Phosphorylase (McArdle 1951)
   Phosphofructokinase (Tarui et al. 1965; Layzer et al. 1967)
 * Phosphoglycerate kinase (DiMauro et al. 1981a)
 * Phosphoglycerate mutase (DiMauro et al. 1981b)
 * Lactate dehydrogenase (Kanno et al. 1980)
   Phosphorylase 'b' kinase (Abarbanel et al. 1986)
Fatty acid oxidation
 * Carnitine palmitoyltransferase (DiMauro and DiMauro 1973)
   Long-chain acyl-CoA dehydrogenase (Roe 1986)
 * Short-chain L-3-hydroxyacyl-CoA dehydrogenase (Tein et al. 1990a)
Pentose phosphate pathway
 * G6PDH (Bresolin et al. 1989)
Purine nucleotide cycle
   Myoadenylate deaminase (Hyser et al. 1989)
Respiratory chain
 * Complex II and aconitase (Haller et al. 1991)
   Coenzyme Q deficiency (Ogasahara et al. 1989)

Biochemical abnormality incompletely characterized
 * Impaired long-chain fatty acid oxidation (Engel et al. 1970)
 * Impaired function of the sarcoplasmic reticulum (?) in familial malignant hyperthermia (predisposition in central core disease, Duchenne muscular dystrophy, Becker muscular dystrophy, myotonic dystrophy, myotonia congenita, Schwartz-Jampel syndrome, King-Denborough syndrome)
 * Abnormal composition of the sarcolemma in Duchenne and Becker muscular dystrophy (Bonilla et al. 1989; Hoffman et al. 1989; Medori et al. 1989)

Biochemical abnormality unknown
 * Familial recurrent myoglobinuria
 * Repeated attacks in sporadic cases

Modified from Tein et al. (1990b) by courtesy of the Editors of *Adv. Pediatr.* (Mosby-Year Book, Inc.)
* Etiologies that have been documented to cause recurrent myoglobinuria beginning in childhood.

Histochemical stains and immunocytochemical reactions can be used for rapid diagnosis of conditions caused by lack of phosphorylase or phosphofructokinase. Also, observation of ragged red fibers would lead to suspicion of a mitochondrial myopathy.

CLASSIFICATION BASED ON ETIOLOGY AND PATHOGENESIS

A functional classification of myoglobinuria can begin with a division into heritable and sporadic forms, but there may be more than 1 cause in an attack. For instance, heritable myoglobinuria may be precipitated by exercise, fasting, infection, fever, or cold exposure. Some sporadic exercise-induced attacks are due to an identified enzyme disorder and some of the others may be due to currently unidentified enzyme aberrations.

Twenty years ago, there were only 2 identified heritable causes of myoglobinuria, lack of phosphorylase or phosphofructokinase (Rowland and Penn 1972). A decade ago, 6 had been identified (Rowland 1984); now there are 13 (Table 4). All are autosomal-recessive conditions except for X-linked phosphoglycerate kinase and glucose-6-phosphate dehydrogenase (G6PDH) deficiencies. G6PDH (Bresolin et al. 1989) and myoadenylate deaminase (Hyser et al. 1989; Tonin et al. 1990) deficiencies, however, have not been linked conclusively to a logical theory of causation because both enzymes are often totally absent in asymptomatic people.

In all of the other conditions, there is a disorder of energy metabolism and it has been suspected that maintenance of muscle surface membranes fails when intracellular levels of ATP fall below some critical value. Substrates for glycolysis or fatty acid oxidation are used to replenish high-energy phosphate compounds. Phosphocreatine and phosphoenolpyruvate, in turn, are used to regenerate ATP. The state of activity of muscle modifies the dependence on glycolysis or fatty acids and vulnerability to different metabolic blocks. Resting muscle depends mostly on fatty acid oxidation (Felig and Wahren 1975). At rest, glucose utilization accounts for 10–15% of total oxygen consumption (Wahren 1977). Both slow and fast twitch fibers have similar levels of glycogen content at rest (Essen 1977). In working muscle, energy is derived from either the combination of triglycerides and stored glycogen or the combination of glucose and free fatty acids, a choice that depends on the type, intensity, and duration of exercise (Gollnick et al. 1974; Essen 1978). High-energy phosphates are used first to regenerate ATP, followed by use of muscle glycogen for the first 5–10 minutes of moderate exercise.

The lactate produced is an indication of glycogen breakdown and rises sharply in the first 10 minutes. The blood lactate level then drops as muscle triglycerides and blood-borne fuels are used (Felig and Wahren 1975; Lithell et al. 1979). After 90 minutes, the major fuels are glucose and

free fatty acids. During 1–4 hours of mild to moderate prolonged exercise, free fatty acid uptake by muscle increases by about 70%. After 4 hours, free fatty acids are used twice as much as carbohydrate sources.

Inherited diseases

Frequency of different causes of heritable myoglobinuria

Tonin et al. (1990) analyzed 77 patients, ages 15–65 years, whose muscle biopsies had been studied biochemically at Columbia Presbyterian Medical Center (CPMC) between 1985 and 1988 (Tables 5, 6). Men were affected 3.5 times more often than

TABLE 5

Clinical and laboratory features of 77 adults with myoglobinuria.

Enzyme defect	CPT[a]	Ph[a]	PhK[a]	PGK[a]	MAD[a]	None[a]
No. of episodes						
1	3/17	2/10	0/4	0/1	0/3	10/41
1	14/17	8/10	4/4	1/1	3/3	31/41
Precipitating factors						
Exercise	15/17	10/10	3/4	1/1	2/3	32/37
Fasting	5/17	0/10	0/4	0/1	1/3	7/37
Infections	5/17	0/10	1/4	0/1	1/3	8/37
Interictal CK						
Increased	1/16	7/10	1/3	0/1	1/2	15/29
Normal	14/16	1/10	2/3	1/1	1/2	14/29
Unknown	1/16	2/10	0/3	0/1	0/2	0/29
Ischemic exercise[b]						
Normal	14/16	0/8	3/4	0/1	3/3	25/36
Abnormal	0/16	6/8	0/4	1/1	0/3	4/36
Not done	2/16	2/8	1/4	0/1	0/3	7/36
Muscle biopsy						
Inc. glycogen[c]	0/17	3/9	0/4	0/1	0/3	2/36
Inc. lipid[c]	3/17	0/9	0/4	0/1	0/3	2/36
Nonspecific changes	2/17	1/9	2/4	1/1	1/3	6/36
Histochemical defects[d]		4/9			2/3	9/36
Normal	12/17	1/9	2/4	0/1	0/3	15/36

[a]The denominator reflects the number of patients for whom information was available; thus, the denominator may change in different rows of the same column.
[b]Refers to the response of serum lactate.
[c]Refers to morphological observations.
[d]Could be documented only for Ph and MAD.
CPT, carnitine palmitoyltransferase; Ph, phosphorylase; PhK, phosphorylase kinase; PGK, phosphoglycerate kinase; MAD, myoadeylate deaminase.
Taken from Tonin et al. (1990) by courtesy of the Editors of *Ann. Neurol.* (Little, Brown and Company).

women. By selection, none had been exposed to alcohol or other drugs. Myoglobinuria had been documented in 44 patients and suspected in 33. Assays were carried out for 8 enzymes: phosphorylase (PPL), PPL b kinase, phosphofructokinase (PFK), phosphoglycerate kinase (PGK), phosphoglycerate mutase (PGAM), lactate dehydrogenase (LDH), carnitine palmitoyltransferase (CPT), and myoadenylate deaminase (MAD).

The enzyme abnormality was identified in 36 patients (47%). The distribution of abnormalities was as follows: CPT deficiency, 17 patients; PPL, 10; PPL b kinase, 4; MAD, 3; PGK, 1; and combined CPT and MAD, 1. The number attributed to lack of PPL b kinase was higher than expected.

Biopsy samples showed glycogen storage in only 3 of 9 patients with PPL deficiency and values were normal in those lacking PPL b kinase or PGK. Venous lactate responses were abnormal in all patients lacking PPL or PGK; responses were normal in all the others so studied, including PPL b kinase deficiency. Between attacks serum CK levels were high in patients with glycogen diseases but not in those with CPT deficiency.

Differential diagnosis

There are few clinical clues to the differential diagnosis of recurrent myoglobinuria; it is difficult to predict a patient's biochemical abnormality on clinical grounds, except that hemolytic anemia ac-companies a decrease in PFK activity. In the series of Tonin et al. (1990), the precipitating event was almost always exercise; this was true in patients with or without an identified enzyme disorder. Patients lacking CPT may be more likely to have an attack if they miss a meal before vigorous exercise, as dramatically indicated in the first cases described by DiMauro and DiMauro (1973). This was also seen by Tonin et al. (1990), who noted that some attacks were linked to the lack of appetite in patients with fever.

Tonin et al. (1990) also found enzyme abnormalities in 30 patients who had symptoms of exercise intolerance without recognized bouts of myoglobinuria. A total of 14 patients (58%) lacked PPL, 3 PPL b kinase, 3 PFK, 1 PGAM, and 9 MAD.

Myoglobinuria in children

The original patient of Meyer-Betz (1910) was a 12-year-old boy who had 3 attacks of pigmenturia, but there have been few analyses of the syndrome in children (Savage et al. 1971; Chamberlain 1991). Tein et al (1990b) identified 60 reports between 1910 and 1989 of patients with at least 2 episodes of myoglobinuria before age 16 years with no obvious cause. They divided the cases (as per Korein et al. 1959; Savage et al. 1971) into 2 groups, one related to exertion, the other to infection or fever and peripheral leukocytosis (Table 7).

In group 1 (exertional), attacks were precipi-

TABLE 6

Biochemical defects in 77 adults with myoglobinuria.

Enzyme defect	No. of patients		Sex (M:F)	Age (mean) (years)
	Myoglobinuria documented	Myoglobinuria suspected		
Detected	23	13	26:10	
CPT	13	4	13:4	15–41(26)
Phosphorylase	3	7	5:5	16–63(36)
Phosphorylase kinase	3	1	4:0	18–22(21)
PGK	1	0	1:0	38
MAD	2	1	3:0	21–35(27)
CPT + MAD	1	0	0:1	53
None detected	21	20	34:7	15–65(33)
Total	44	33	60:17	15–65(31)

CPT, carnitine palmitoyltransferase; PGK, phosphoglycerate kinase; MAD, myoadenylate deaminase.
Taken from Tonin et al. (1990) by courtesy of the Editors of *Ann. Neurol.* (Little, Brown and Company).

TABLE 7

Literature review of idiopathic cases of recurrent childhood myoglobinuria (1910-1989).

Feature	Group I (Exertional)[a]	Group II (Toxic)[b]
No. of patients	26	14
Male:female	21:5	7:7
Average age of onset (years)	10.8 ± 4.0	4.3 ± 4.0
Positive family history	2/25	4/13
Precipitating factors		
Infection	3/13	10/14*
Fever	1/15	14/14*
Exercise	26/26*	2/14
Fasting	3/3	3/3
Ictal features		
Cramps/stiffness	16/16	6/6
Bulbar signs	0/1	2/2
Encephalopathy	NR	1/3
Respiratory distress	5/16	2/12
Cardiac	1/8	2/2
Renal failure	8/13	6/13
Death	3	4
Interictal features		
Fixed weakness	3/8	1/1
Muscle atrophy	3/24	2/13
Muscle hypertrophy	9/25	
Laboratory features		
Increased lactate	9/9	NR
Abnormal EMG	7/7	1/2
Abnormal ECG	1/8	2/2
Muscle biopsy; nec/reg	8/19	7/8

[a]Group I references: Bywaters and Dible 1943; Louw and Nielson 1944; Hed 1947; Kreutzer et al. 1948; Acheson and McAlpine 1953; Spaet et al. 1954; Hed 1955; Reiner et al. 1956; Javid et al. 1959; Segar 1959; Lyons 1963; Larsson et al. 1964; Rowland et al. 1964; Tavill et al. 1964; Hinz et al. 1965; Ford 1966; Goldberg and Chakrabarti 1966; Ozsoylu and Agkun 1966; Engel et al. 1970.

[b]Group II references: Debre et al. 1934; Buchanan and Steiner 1951; Berenbaum et al. 1955; Bowden et al. 1956; Watson and Ainbender 1959; Haase and Engel 1960; Wheby and Miller 1960; Bailie 1964; Bacon 1967; Favara et al. 1967; Casenave et al. 1972. * major feature; NR, not recorded; EMG, electromyogram; ECG, electrocardiogram; nec/reg, necrosis/regeneration.

Taken from Tein et al. (1990b) by courtesy of the Editors of *Adv. Pediatr.* (Mosby-Year Book, Inc.).

tated by exertion in 46 patients. The enzyme abnormality was identified in 18; 26 were idiopathic and 2 were partially characterized. Boys outnumbered girls by 4:1. Some had evidence of fixed myopathy with relatives who had limb weakness (Louw and Nielsen 1944; Wissler 1948; Acheson and McAlpine 1953), suggesting that the myoglobinuria was related to the underlying myopathy. The identified enzyme abnormalities were as follows: CPT, 11 patients; PPL, 2; PGK, 1; PGAM, 3; and LDH, 1. Two were suspected of having a defect in long-chain fatty acid oxidation (Engel et al. 1970).

In group 2 (*toxic*), infection or fever was associated with attacks in 14 patients. No enzyme abnormality was identified in this group, but as in the exertional group, about a third had similarly affected relatives. Compared to the exertional group, the toxic patients tended to have fewer but more severe attacks, with renal failure in 46%, muscle wasting in 15%, and a higher mortality rate of 28% (however, intensive care has vastly improved since some of the fatal cases were first recognized). Boys and girls were affected equally, and the age of onset tended to be younger in this group (4.3 ± 4.0 years). The exertional group was characterized by later age of onset (10.8 ± 4.0 years for idiopathic cases, 14.2 ± 1.6 years for glycolytic cases, and 10.4 ± 4.4 years for CPT deficiency). The exertional patients had more frequent but less severe recurrences over a longer time (up to years) with renal failure in 43%, muscle wasting in 8%, and a lower mortality rate of 6.5%.

The cases of childhood myoglobinuria associated with glycolytic defects or CPT deficiency are summarized in Table 8. In both groups, family history was positive in about one-third of cases, there were no fatalities or syndromes of permanent limb weakness, and renal failure was an important complication.

Tein et al. (1990b) also analyzed the features of 40 children whose muscle biopsies had been studied in DiMauro's laboratory (CPMC series) (Tables 9, 10). Ten cases were exertional, 23 toxic, and 7 uncertain. There was some clinical overlap; 30% of the 'exertional' cases also had attacks precipitated by infection, and 17% of the 'toxic' cases had exertional attacks. The causes of infection included *Shigella* and other diarrheas, streptococcal

TABLE 8

Literature review of group I exertional recurrent childhood myoglobinuria with identified defects (1910–1989).

Feature	Glycolytic defects[a]	CPT deficiency[b]
No. of cases	7	11
Male:female	6:1	10:1
Average age of onset (years)	14.2 ± 1.6	10.4 ± 4.4
Positive family history	2/6	3/9
Associated exertion	7/7*	10/11*
Associated fasting	NR	4/5
Associated cold	1/1	2/2
Ictal features		
Stiffness/cramping	5/5*	9/10*
Renal failure	2/6	2/9
Death	0	0
Interictal features		
Fixed weakness	0/5*	0/9*
Muscle atrophy	0/5*	0/8*
Second wind	2/2	NR
Laboratory features		
Increased CK interictally	4/4*	2/5
Increased serum triglycerides	NR	1/9
Increased serum cholesterol	0/1	2/9
Abnormal fasting ketogenesis	NR	7/7*
Abnormal forearm ischemic lactate test	7/7*	0/10
Abnormal EMG ictally	0/4 timing	2/2
Abnormal EMG interictally	not specified	3/7
Abnormal EKG	1/2	1/6
Muscle biopsy		
Normal		5/11
Increased lipid	1/6	4/11
Increased glycogen	3/6	0/11
Decreased carnitine	NR	0/9

[a]Glycolytic case references: PGAM (DiMauro et al. 1982; Bresolin et al. 1983; Kissel et al. 1985); PGK (DiMauro et al. 1983); LDH (Kanno et al. 1980); phosphorylase (Tobin and Coleman 1965; Sengers et al. 1980).
[b]CPT case references: Bank et al. 1975; Cumming et al. 1976; Herman and Nadler 1977; Carroll et al. 1978; Reza et al. 1978; Layzer et al. 1980; Angelini et al. 1981; Trevisan et al. 1984, 1987.
* major feature; NR, not recorded; EMG, electromyogram; EKG, electrocardiogram, CK, creatine kinase; CPT, carnitine palmitoyltransferase.
Modified from Tein et al. (1990b) by courtesy of the Editors of *Adv. Pediatr.* (Mosby-Year Book, Inc.).

TABLE 9

CPMC series: idiopathic childhood myoglobinuria (1980–1989).

Feature	Group I (Exertional)	Group II (Toxic/Infectious)	Group III
No. of cases	10	23	7
Male:female	8:2	9:14	3:4
Average age at diagnosis (years)	6.4 ± 3.6	3.9 ± 4.0	3.2 ± 3.5
Positive family history	3/10	5/16	2/3
Recurrence	10/10	17/23	3/7
No. of recurrences	7.8 ± 7.0	2.9 ± 3.0	1.5
Precipitating factors			
Exercise	10*	4	
Infection	3	23*	
Fasting	0	5	
Cold	2	1	
No apparent cause	1	2	
Ictal features			
Weakness	5/5	18/21	1/1
Muscle cramps/pain	9/9	12/14*	1/1
Bulbar signs	0/3	8/18	
Encephalopathy	0/4	4/19	1/3
Seizures	0/3	2/9	
Respiratory weakness	0/3	8/11	1/1
Cardiac abnormalities	0/1	6/13	1/1
Renal failure	0/3	2/20	
Death	0/7	5/15	
Interictal features			
Fixed weakness	4/10	5/16	
Fixed atrophy	0/3	2/6	
Bulbar signs	0/7	1/5	
Failure to thrive	0/5	2/15	
Abnormal development	2/6	4/17	
Dysmorphism	0/4	2/9	
Cardiac abnormality	1/2	1/6	

* Major features.
Modified from Tein et al. (1990b) by courtesy of the Editors of *Adv. Pediatr.* (Mosby-Year Book, Inc.).

pharyngitis, *Escherichia coli* urinary infection, pneumonia, Epstein-Barr virus infection, herpetic stomatitis, and varicella. This diversity suggests that fever and associated metabolic abnormalities were responsible for the myoglobinuria. There was

TABLE 10

CPMC series: idiopathic childhood myoglobinuria (1980–1989) — laboratory features.

	Group I (Exertional) (10 cases)	Group II (Toxic) (23 cases)	Group III (7 cases)
Biochemistry			
Myoglobinuria	6/9	18/22	5/5
Abnormal forearm ischemic lactate test	0/4	0/2	1/1
Acidosis	0/1	5/9	
Increased triglycerides	0/3	1/6	0/1
Increased cholesterol	0/3	1/7	0/1
Abnormal fasting ketogenesis		2/2	
Electrophysiology			
ECG abnormalities	1/4	4/11	
Myopathic EMG	1/4 (mild)	2/9 (mild)	1/2 (NS)
	1/4 (moderate)	1/9 (moderate)	
Abnormal NCS	1/3 (axonal)	1/8 (demyelinating)	
Abnormal EEG	0/1	3/5	
Muscle biopsy	1/7 (mildly myop)	4/19 (mildly myop)	
	1/7 (severely myop)	1/19 ($\uparrow$ lipid)	1/4 lipid storage
	3/7 (mild NS)	5/19 (mild NS)	1/4 (mild NS)
	1/7 (mild $\uparrow$ lipid/glyc/ abn mit)	2/19 (chronic denervation)	1/4 ($\uparrow$ lipid/glyc/abn mit)
	3/5 (neg/reg)	8/17 (nec/reg)	1/3 (nec/reg)

NCS, nerve conduction studies; EMG, electromyogram; ECG, electrocardiogram; EEG, electroencephalogram; NS, nonspecific; myop, myopathic; glyc, glycogen; abn mit, abnormal mitochondria; nec/reg, necrosis/regeneration.
Modified from Tein et al. (1990b) by courtesy of the Editors of *Adv. Pediatr.* (Mosby-Year Book, Inc.).

an unexplained male predominance in the exertional group and slight female predominance in the toxic group. Fixed limb weakness was reported in 40% of the exertional group and 31% of the toxic group and family history was positive in 30% of both groups.

Signs of encephalopathy and cranial nerve abnormalities were seen in a few patients in the toxic group, but in none in the exertional group. Four of the 8 children with bulbar signs, e.g. dysphagia and dysarthria died, in contrast to 1 fatality among 13 children lacking these signs. Two children with encephalopathy were diagnosed as having a 'Reye-like' syndrome and 1 was queried to have 'Leigh's syndrome. Respiratory failure with ventilator dependence was seen in 5 toxic patients, but in none of the exertional cases. Only 2 toxic patients had renal failure. Five children in the toxic group died (33%) during the acute episode, while no deaths occurred in 7 children in the exertional group.

In the series of Tein et al. (1990b), 5 of 40 children had CPT deficiency. Several had partial deficiencies (in the range of heterozygotes) of enzymes involved in mitochondrial β-oxidation; the significance of these partial deficiencies is uncertain but might make a child more susceptible to external challenges. One 13-year-old girl in the toxic group had recurrent myoglobinuria with hypoketotic hypoglycemic encephalopathy, and finally developed a dilatative and hypertrophic cardiomyopathy which was attributed to a deficiency of muscle short-chain L-3-hydroxyacyl-CoA dehydrogenase (SCHAD) activity (Tein et al. 1991).

Compared to adult-onset cases (Table 11), fewer children have been diagnosed biochemically (Tein et al. 1990b). CPT deficiency was the most common identified cause in all patients, including exertional episodes in children or adults and toxic attacks in children. There were no glycogen disorders in children with toxic attacks. Exertional attacks occurred commonly in children and adults, but toxic attacks were much more common in children and were more likely to be fatal. The childhood exertional group and adult-onset group had a marked male predominance contrasting with a slight female predominance in the toxic childhood form. Moreover, the presence of additional clinical features including ictal bulbar signs (8/18), encephalopathy (4/19), seizures (2/7), developmental delay (4/17), and dysmorphic features (2/9) in the toxic childhood group suggest the presence of more generalized diseases and different biochemical etiologies than in adults.

Four of 5 children with CPT deficiency had toxic attacks, which were seen in only 3 of 39 adult-onset patients (Table 12). Other toxic childhood cases tended to have a peak age at diagnosis of 1–2 years, but childhood-onset CPT deficiency was seen throughout the first decade. The slight female predominance in childhood CPT deficiency and the marked male predominance in adults is a further differentiating feature.

Chamberlain (1991) recorded cases in children due to *Neisseria* sepsis, dialysis disequilibrium, and diabetic ketoacidosis.

Enzyme abnormality known

Glycogen disorders. 1. General considerations.
Of the 6 recognized defects of glycogen metabolism or glycolysis, the most common is phosphorylase deficiency, first described by McArdle in 1951. DiMauro and Bresolin (1986) reviewed 112 cases proven biochemically or histochemically. Males out-numbered females 79:33. The cases included 76 from the literature, 7 studied at CPMC, and 29 studied at other institutions but analyzed biochemically by DiMauro et al. Myoglobinuria was documented in 56 of the 112 cases; 32 had no more than 2 attacks. Although children might have exercise intolerance or painful contractures, only 1 had an attack of myoglobinuria before adolescence (Sengers et al. 1980).

Phosphofructokinase deficiency seems less common than McArdle's disease. Rowland et al. (1986) found only 29 reported cases, dating back to the first descriptions by Tarui at al. (1965) and Layzer et al. (1967). They divided the cases into 3 groups. In group 1, there were 20 patients and the muscle (M) subunit of PFK was affected. These patients had myoglobinuria (8 cases) or myopathy (3 cases), some with high serum levels of sarcoplasmic enzymes between attacks. They also had evidence of hemolysis. Eighteen of the 20 patients were men, and all but 1 were adults at time of diagnosis, although all recalled exercise intolerance in childhood. Eight of the 20 patients recalled episodes of probable myoglobinuria in childhood (Tarui et al. 1965; Layzer et al. 1967; Waterbury and Frankel 1972; Tobin et al. 1973; Oda et al. 1977; Noble et al. 1980; Tani et al. 1983a,b; Vora et al. 1983), although the pigment was identified in only 2 (Agamanolis et al. 1980; Vora et al. 1980, 1983). In group 2, the liver (L) subunit was affected or the M-subunit may have been unstable. These 6 patients had hemolysis with no evidence of myopathy. In group 3, there were infants who had fixed limb weakness and no hemolysis.

TABLE 11

Comparison of childhood versus adult onset myoglobinuria.

Feature	Childhood form[a]		Adult form[b]
	Group I (Exertional)	Group II (Toxic)	
No. of cases	56	37	77
Male:female	45:11	16:21	60:17
Age at diagnosis	Span childhood, small peak at 7 years	Peak 1–2 years	Late adolescence & early adulthood
Leading precipitant	Exercise	Infection	Exercise
Mortality	3 (5%)	9 (24%)	
Etiology	12 CPT	4 CPT	17 CPT
	7 Glycolytic[c]	1 SCHAD*	15 Glycolytic[d]
			3 MAD
			1 MAD + CPT
Diagnosed %	34	13	46

[a]Combined CPMC and literature (1910–1989) (Tein et al. 1990b).
[b]CPMC series (1985–1988) (Tonin et al. 1990).
[c]2 phosphorylase, 1 phosphoglycerate kinase, 3 phosphoglycerate mutase, 1 lactate dehydrogenase.
[d]10 phosphorylase, 4 phosphorylase kinase, 1 phosphoglycerate kinase.
*Defect subsequently identified in 1990.
CPT, carnitine palmitoyltransferase; SCHAD, short-chain L-3-hydroxy-acyl-CoA dehydrogenase; MAD, myoadenylate deaminase.
Modified from Tein et al. (1990b) by courtesy of the Editors of *Adv. Pediatr.* (Mosby-Year Book, Inc.).

The other glycogen and glycolytic disorders are much less common. There have been only 5 cases of PGAM deficiency in adults (DiMauro and Bresolin 1986; Meola et al. 1989) and 2 of PGK deficiency (Rosa et al. 1982; Tonin et al. 1989). PPL b kinase deficiency was identified in a single patient (Abarbanel et al. 1986). These disorders are discussed in more detail by DiMauro (see Chapter 16, this Volume).

2. Pathophysiology. In glycolytic disorders, impaired utilization of glycogen could lead to a net decrease of ATP content. The initial stages of intense exercise, would be the time of greatest vulnerability because carbohydrate is then the major source of energy. Sometimes, patients note that they have to rest soon after beginning exercise, but then they can go on for a longer time; this 'second-wind' phenomenon has been attributed to a metabolic switch from carbohydrate to fatty acid utilization (Felig and Wahren 1975). A drop in ATP content could first cause the electrically silent shortening of muscle (contracture) that is seen in PPL-deficiency, and even more severe depletion could cause myoglobinuria (Rowland 1984). However, this has not been proven.

In rats, Brumback et al. (1983) found that iodoacetate impairs replenishment of ATP and induces both contracture and myoglobinuria. In PPL-deficient humans, however, muscle was biopsied before and after ischemic exercise-induced contracture, but there was no change in total ATP content (Rowland et al. 1965). Failure to demonstrate the expected change in ATP could have been due to regeneration of ATP during the few seconds between excision and freezing of the tissue. However, more rapid freezing in later studies gave similar results (Edwards et al. 1980) and so did a study by nuclear resonance spectroscopy, in which freez-

TABLE 12

CPT deficiency 'muscular' form: comparison of childhood and adult onset cases.

	Childhood cases (CPMC Series)		Adult series[a]	
Feature				
No. of cases	5		39	
Female: Male	3:2		1:5.5	
	Incidence	%	Incidence	%
Age at diagnosis (15 yrs)	5/5	100	4/39	10
Recurrence +	5/5	100	31/39	82
Family history	2/2		16/39	41
Precipitating factors				
Exercise	2/3	66	35/39	90
Infection	4/4	100	3/39	8
Ictal features				
Cramps/pain	5/5	100		
Bulbar signs	1/5	20		
Renal failure	0/4	0	9/39	23
Death	0/4	0		
Interictal				
Fixed weakness	0/3	0	3/39	8
Laboratory features				
ECG abnormal	1/4	25		
EMG myopathic	2/3	66		
Muscle biposy myopathic	1/4	25		
Myoglobinuria	5/5	100	38/39	97

[a]This series includes 4 children; DiMauro and Papadimitriou (1986).
ECG, electrocardiogram; EMG, electromyogram; CPT, carnitine palmitoyltransferase.
Modified from Tein et al. (1990b) by courtesy of the Editors of *Adv. Pediatr.* (Mosby-Year Book, Inc.)

ing was not needed (Ross et al. 1981). These methods would not detect a change in ATP restricted to the sarcoplasmic reticulum, although this kind of focal depletion of ATP is not known in any other condition. Brandt et al. (1977) found that excised normal muscle might shorten vigorously if only 20% of all fibers were contracted. It is therefore still unproven that ATP depletion causes contracture or myoglobinuria, but no better hypothesis has emerged.

Disorders of lipid metabolism. 1. Pathogenesis. In fatty acid oxidation, 3 defects have been identified as causing recurrent myoglobinuria. Lack of CPT in skeletal muscle (CPT$_2$) was the first of these to be described (DiMauro and DiMauro 1973) and is probably the most common form of inherited myoglobinuria (Tonin et al. 1990; Tein et al. 1990b). CPT is the pivotal enzyme that controls transport of long-chain fatty acids (LCFAs) across the mitochondrial membrane where they then undergo β-oxidation. Lack of long-chain acyl-CoA dehydrogenase (LCAD) is another cause of myoglobinuria and was first described by Roe in 1987 (cited by Stanley 1987). This enzyme catalyzes the first step of intramitochondrial β-oxidation. Short-chain L-3-hydroxyacyl-CoA dehydrogenase (SCHAD) catalyzes the third step of intramito-chondrial β-oxidation and lack of this enzyme may also cause myoglobinuria (Tein et al. 1990a, 1991).

In disorders of fatty acid oxidation, patients are prone to attacks of myoglobinuria during prolonged mild or moderate exercise or during fasting, when muscle glycogen stores and glucose have been depleted and fatty acids are the key substrate. Again, a drop in ATP content might lead to myoglobinuria. In addition, increased levels of LCFAs might arise from the metabolic block and, because they have detergent-like properties, might be toxic to muscle membranes. In experimental ischemia of rat heart, Subramanian et al. (1987) found that myocardial damage was potentiated by reperfusion with an excess of free fatty acids, including LCFAs such as palmitate. Recovery was enhanced by treatment with L-propionyl carnitine or propionyl carnitine taurine amide, which might have reduced formation of free radicals. Mak et al. (1986) found that peroxidation injury was exaggerated in canine heart sarcolemma that had been pretreated

with palmitoyl-CoA or palmitoylcarnitine. Long-chain acyl-CoAs may deform mitochondrial or cellular membranes to potentiate the toxic effects of hydroperoxides, high cellular content of calcium, or high temperatures (Siliprandi et al. 1987).

Increased content of short- or medium-chain fatty acids and particularly their dicarboxylic metabolites from compensatory ω-oxidation may inhibit gluconeogenesis, β-oxidation, or the citric acid cycle (Tonsgard and Getz 1985; Tonsgard 1986; Corkey et al. 1988), leading to a decrease in muscle ATP content.

Other risk factors arise in disorders of fatty acid oxidation. During a fast, the block in hepatic fatty acid oxidation impedes ketogenesis and further exacerbates the already diminished energy substrates, including glycogen and blood glucose. In cold exposure, ketogenesis is stimulated in normal individuals (Johnson et al. 1961), but this response is limited if there is a defect of an essential enzyme of fatty acid metabolism. Moreover, shivering is an involuntary form of muscle activity and also depends on LCFA oxidation (Bell and Thompson 1979).

In addition to the 'myoglobinuric' or myopathic form of CPT$_2$ deficiency, there is a rare 'hepatic' or encephalopathic form of CPT$_1$ deficiency. This form is manifest before the age of 1 year by hypoglycemic hypoketotic encephalopathy and seizures, which are precipitated by fasting or infection, in which there are no muscular manifestations (Bougneres et al. 1981; Bonnefont et al. 1985; Demaugre et al. 1988; Tein et al. 1989). In the 'hepatic' form, 'total' CPT deficiency was demonstrated in liver in 1 girl in whom it was measured (Bougneres et al. 1981) and CPT$_1$ deficiency was shown in fibroblasts (Demaugre et al. 1988, 1990) with normal CPT$_1$ and CPT$_2$ activities in muscle (Tein et al. 1989).

In the 'myopathic' form, CPT$_2$ deficiency has been documented in fibroblasts (Demaugre et al. 1988, 1990) and in muscle (Tein et al. 1989). Demaugre et al. (1991) recently described a unique infantile form of CPT$_2$ deficiency ('overlap' case) in a 3-month-old boy with fasting hypoketotic hypoglycemic encephalopathy, elevated serum CK, and cardiac arrhythmias and cardiomegaly. Usually, 'muscular' CPT$_2$ deficiency is characterized by later onset of recurrent myoglobinuria with delayed ketogenesis and otherwise normal interictal muscle function, with no cardiac symptoms or overt liver disease. CPT_2 activity was decreased by 90% in the fibroblasts of the severe infantile CPT_2 deficiency in contrast to the 75% decrease in 2 patients with 'conventional' muscular CPT_2 deficiency (Demaugre et al. 1991).

  2. Clinical considerations. DiMauro and Papadimitriou (1986) reviewed 39 patients with CPT deficiency including 6 patients at CPMC, 7 studied elsewhere clinically and biochemically at CPMC, and 26 from the literature. Thirty-three were men; 4 had been diagnosed before age 15. All patients had had attacks of myoglobinuria, but exercise-induced myalgia may be the only clinical manifestation (Angelini et al. 1981; Desnuelle et al. 1990). Differences between childhood and adult-onset CPT deficiency are discussed above.

Mitochondrial diseases. Haller et al. (1991) described a combined complex II-aconitase deficiency in the muscle of a 22-year-old man with recurrent myoglobinuria. The clinical picture in this Swedish man was strikingly similar to the syndrome reported by Larsson et al. (1964). Larsson described a large Swedish kindred which, in addition to myoglobinuria, was characterized by exercise intolerance, dyspnea, palpitations, and excessive exertional rise in lactate and pyruvate content. Ogasahara et al. (1989) described coenzyme Q_{10} deficiency in 2 sisters with encephalomyopathy and recurrent myoglobinuria. Ohno et al. (1991) described 2 brothers with recurrent exertional myoglobinuria who had multiple deletions of mitochondrial DNA and no other recognized disorder.

Incompletely characterized syndromes
There are 2 disorders in this category. Engel et al. (1970) described twin girls with recurrent myoglobinuria; their clinical and biochemical features suggested an abnormality of LCFA oxidation. CPT deficiency was excluded by biochemical assay (DiMauro and Papadimitriou 1986). As discussed later, malignant hyperthermia is another form of myoglobinuria that may be inherited (Gronert 1986; Wood et al. 1982).

  Lack of dystrophin has been found in some families with an X-linked pattern of recurrent myoglobinuria as the only manifestation of myopathy;

presumably this syndrome is a variant of Becker dystrophy (Fischbeck et al. 1988). Also, some patients with typical manifestations of Becker dystrophy have had exertional myoglobinuria (Bonilla et al. 1989; Hoffman et al. 1989; Medori et al. 1989). DiMauro (personal communication) has suggested that only patients with mild myopathy have myoglobinuria because more severely affected patients cannot exercise. Patients with Duchenne dystrophy are at increased risk of myoglobinuria after general anesthesia, a distinct syndrome described below with malignant hyperthermia.

Biochemical abnormality undefined
The largest single group of patients with myoglobinuria comprises those with no identifiable cause after all enzymatic tests for known etiologies have been performed. This includes patients with recurrent attacks, and some with 1 or more attacks who have similarly affected relatives. It also includes adult- (Tonin et al. 1990) and childhood-onset (Tein et al. 1990b) cases.

Malignant hyperthermia: a special problem

Definition
Malignant hyperthermia (MH) is a syndrome defined clinically (Britt et al. 1969; Gronert 1980). After induction of general anesthesia by halothane or other volatile hydrocarbons and almost always also after exposure to a depolarizing muscle relaxant, there is a sharp rise in body temperature with stiffness of masseter and limb muscles. A metabolic catastrophe includes metabolic acidosis, tachycardia and arrhythmia, and myoglobinuria. More than half of the reported cases were fatal before the introduction of dantrolene prophylaxis and therapy.

  In the future, it may be possible to define the condition by demonstrating an abnormality of the calcium-release channel in the sarcoplasmic reticulum (SR) or the ryanodine receptor (MacLennan et al. 1990; McCarthy et al 1990a,b), or it may be possible to define the condition by identifying a mutant DNA. The common physiological abnormality seems to be entry of excessive amounts of calcium into skeletal muscle.

Epidemiology
In children, the incidence of MH in North America

and Europe is about 1 in every 15,000 exposures to general anesthesia. In adults the range is estimated between 1:50,000 and 1:150,000 (Britt and Kalow 1970). The higher incidence in children may correspond to the more frequent use of succinylcholine as well as the inhalational agent (Schwartz et al. 1984).

Inheritance

From the time the first case was described (Denborough and Lovell 1960), the condition has been considered heritable, but there has also been evidence of clinical heterogeneity. Both views have been proven correct by molecular genetics. There is an autosomal-dominant condition that maps to chromosome 19q12-13.2 (MacLennan et al. 1990; McCarthy et al. 1990a,b), but there is also evidence that not all families with well-characterized MH map to the same locus (Levitt et al. 1991). The story can be divided into 2 eras. The first extends from the first description of the syndrome to 1990, when the second period began with the localization of the gene.

Clinical evidence. It has been said that 50% of cases are inherited in an autosomal-dominant pattern (Britt et al. 1969; McPherson and Taylor 1982). However, among 89 patients reviewed by Britt and Kalow (1970), about one-third had relatives who were also affected with MH. Among 18 families reviewed by King et al. (1972), only 1 propositus had an affected relative. Moreover, there have been few families in which affected members had the full clinical syndrome; in these reports more than 1 person was affected, but the number was rarely sufficient to suggest a particular mode of transmission (Rowland 1984). In 4 families, the pattern was thought to imply autosomal-dominant inheritance (Britt et all. 1976a). In 1 remarkable case (Denborough et al. 1988), a girl survived an attack of MH. Her father worked for a company that made fire extinguishers and had an attack himself after unusually strong exposure to a gas in the tanks; the offending gas resembled halothane in structure.

Defining MH susceptibility. Genetic analysis has been impeded because it has proven difficult to identify MH susceptibility. Few victims of overt

attacks have similarly affected relatives, and investigators have therefore sought some other marker to identify 'affected' individuals. For a while, the marker was a serum CK level above normal (Kelstrup et al. 1973; Plotz 1980). Then, physiologic tests were used.

Markers other than DNA are of uncertain validity. Normal CK values have been found in at least 20% of those who survive overt attacks, so that there are false negatives with this test (Rowland 1984). There are undoubtedly false positives, too, because asymptomatic populations have not been studied simultaneously with the MH families. Although no one would expose a presumably susceptible person to general anesthesia for diagnostic purposes, many people with high CK values have had anesthesia for surgery without difficulty (Owen and Kerry 1974).

The most popular physiological test is to excise muscle strips by biopsy, then measure the tension generated by strips of muscle on exposure to halothane, caffeine, or both; survivors of overt clinical attacks are said to be abnormally sensitive to these tests. It has been thought that any survivor would have an abnormal response in 1 of these 3 tests and that there are no false negatives (Nelson and Flewellen 1983; Allen et al. 1990). However, Ørding et al. (1987) found that 88% of 40 survivors were sensitive to both halothane and caffeine, while 12% were sensitive to 1 but not both agents. It seems likely that there are other false negatives; people who have survived but are not sensitive to either precipitant in these tests. It also seems likely that there are false positives because clinical penetrance is about 0.33; only a third of subjects so identified actually have an MH attack on exposure to general anesthesia (Kalow 1987). Also, in 1 study, 15% of the population was deemed 'MH-susceptible' (Nelson et al. 1983). Denborough et al. (1984) suggested that MH susceptibility might underlie attacks induced by exercise. Nevertheless, as defined by these tests, MH susceptibility is inherited in an autosomal-dominant pattern in almost all families studied by the combined test (Healy et al. 1991).

Other physiologic tests of muscle function have not survived (Britt et al. 1977c; Eng et al. 1984; Lennmarken et al. 1987). Quinlan et al. (1989) have evaluated twitch properties of ankle dorsiflexors without excising the muscle. Biochemical tests on muscle have also been evaluated (Willner et al. 1979, 1980, 1981) but have not proven reliable, including depletion of ATP (Harrison et al. 1969; Britt et al. 1976b; Gronert 1980), glycolytic metabolites (Verburg et al. 1986), activities of PPL a and b (Ono et al. 1977a,b; Willner et al. 1980; Traynor et al. 1983; Ellis et al. 1984), adenylate cyclase (Ono et al. 1977a,b; Willner et al. 1981; Stanec et al. 1984), adenylate kinase (Schmitt et al. 1974; Cerri et al. 1981; Marjanen and Denborough 1982), and adenylate deaminase (Fishbein et al. 1985).

Numerous fruitless attempts have been made to find an abnormality in serum (Whittaker et al. 1977; Whittaker and Britten 1981), platelets (Zsigmond et al. 1978; Rosenberg et al. 1981; Sullivan et al. 1982; Gerrard et al. 1983; Giger and Kaplan 1983; Solomons and Masson 1984; Lee et al. 1985; Britt and Scott 1986), erythrocyte fragility (Kelstrup et al. 1973; Zsigmond et al. 1978), or HLA typing (Kikuchi et al. 1982; Lutsky et al. 1982).

Molecular genetics. In families with a clear autosomal-dominant pattern, the human gene for susceptibility has been mapped to chromosome 19q12-13.2 (MacLennan et al. 1990; McCarthy et al. 1990a,b; Healy et al. 1991). The clue for this observation came from MH analysis in pigs. Investigators found that the disease mapped with the gene for the ryanodine receptor, the calcium release channel of SR. This locus was tightly linked to the gene for glucose phosphate isomerase (GPI), which is part of a linkage group that has been highly conserved across species. Human GPI had been mapped to 19q12-13.2 and that gave the clue to the successful mapping of human MH. When family patterns are informative, susceptibility can be diagnosed with an accuracy of 99.7% with linked DNA markers (Healy et al. 1991). In an informative family, the DNA studies will supplant contracture tests (Healy et al. 1991), but sporadic cases cannot yet be identified reliably by this method.

At present, MH can be defined as an autosomal-dominant disease due to an abnormality of the ryanodine receptor, but the molecular abnormality of the presumed gene product must still be proven.

DNA studies have also indicated that the clinical syndrome is genetically diverse, because not all families map to the ryanodine receptor locus (Levitt et al. 1991; MacKenzie et al 1991). This kind of heterogeneity had been suspected on clinical grounds.

Clinical MH is a syndrome of diverse etiology
Before 1990, the problems of identifying a marker of susceptibility to MH led to the suspicion that it may be a syndrome with more than 1 etiology, probably induced by more than 1 genetic fault, and perhaps not always genetic. Several lines of evidence were deduced.

1. The syndrome is hard to define clinically. Problems arise if more than 1 of the major characteristics is missing (Kripke et al. 1983; Flewellen and Nelson 1982). It has been stated that rigidity is not seen in 20% of otherwise acceptable cases (Britt and Kalow 1970). Even fever may be lacking (Bernhardt and Horder 1978). Sometimes, a child has masseter rigidity at the start of anesthesia and there is no fever because cooling measures are instituted promptly (Carballo 1975; Caseby 1975; Schmitt et al. 1975; Davis 1977; Inoue et al. 1977; Jago and Payne 1977; Donlon et al. 1978; Dodd et al. 1981). Under these circumstances it would be difficult to distinguish an attack of MH that has been aborted by treatment from some other kind of reaction to succinylcholine. Also, full attacks of MH may start with masseter spasm, but masseter spasm may occur alone or with myoglobinuria but no other manifestations of MH; not all of these patients have 'positive' tests for MH susceptibility (Flewellen and Nelson 1982).

There may be a syndrome of suxamethonium-induced muscle rigidity that differs from MH and can be identified by the following features (Ellis and Halsall 1984; Rosenberg and Fletcher 1986): suxamethonium used for intubation; induction of anesthesia with halothane or ultrashort-acting barbiturates; patients usually children between ages 4 and 15 years; no fever or arrhythmia; muscle rigidity for 2–3 minutes; and no family history of MH. It is said that contracture tests for MH are abnormal in about half of those with masseter rigidity (Ellis and Halsall 1984; Flewellen and Nelson 1984; Rosenberg and Fletcher 1986).

There are other problems of clinical definition. For instance, serum levels of sarcoplasmic enzymes and myoglobin commonly rise after administration of succinylcholine (Lewandowski 1981), but what level defines an attack of MH? How much tachycardia or arrhythmia suffices? Is cardiac arrest a manifestation of MH if there is no preceding tachycardia? How much of a rise in temperature is needed to qualify? Are all cases of anesthesia-induced myoglobinuria (Bernhardt and Horder 1978; Chaboche et al. 1982) manifestations of some form of MH? Is it possible to state that a person is not susceptible to MH? One man had an attack after 12 uneventful operations with general anesthesia (Puschel et al. 1978).

2. When so many cases are sporadic, multifactorial inheritance seems likely (Ellis et al. 1978) or the condition may be heterogeneous, including several different disorders. Willner et al. (1984) found that only 2 of 50 survivors of MH attacks had a relative who also had an attack. Halsall et al. (1979) encountered no MH when general anesthesia was given to 321 relatives of MH patients.

3. Among the genetic causes of MH are identifiable disorders, especially central core disease (Frank et al. 1980), which is an autosomal-dominant disorder. Shuaib et al. (1987) found that the 2 conditions were linked in 11 patients and Harriman (1986) reported abnormal in-vitro contracture tests in almost all patients with central core disease. It therefore seems advisable to consider all patients with central core disease to be MH-susceptible. The gene for central core disease has been mapped to the same position as MH (Kausch et al. 1990), but it is uncertain whether they are fully allelic disorders, and it is not known why the clinical manifestations of myopathy appear in some but not all people with a mutation at this site.

MH has also appeared in people with other conditions, suggesting a possible linkage, but not so often to be definite. Among these are the King syndrome (a combination of skeletal abnormalities and static myopathy) (King and Denborough 1973; Kaplan et al. 1977; McPherson and Taylor 1981), Schwartz-Jampel syndrome (Seay and Ziter 1978), Fukuyama congenital muscular dystrophy (Nakazato et al. 1983), Becker dystrophy (Heimann-Patterson et al. 1986), periodic paralysis (Brownell 1988), myotonia congenita (Gordon et al. 1986), and a condition called the sarcoplasmic reticulum adenosine triphosphatase deficiency

syndrome (Karpati et al. 1986). One boy with proven deficiency of PPL gave MH-like responses to tension tests; his father had an attack of MH (Isaacs et al. 1989). Lehmann-Horn and Iaizzo (1990) concluded that myotonic muscle may give falsely abnormal tension tests because of the electrical after-activity that characterizes the muscle.

There has been a difference of opinion about the nature of the susceptibility of boys with Duchenne dystrophy to attacks of myoglobinuria (Brownell et al. 1983, 1985) induced by general anesthesia. On the one hand, Rowland (1983) and McKishnie et al. (1983) noted that myoglobinuria is the essential feature of both Duchenne syndrome and MH. Also, the heart is affected in the myoglobinuric attacks of both conditions and there may be fever. In contrast, Karpati and Watters (1980) analyzed data from 20 Duchenne patients who had anesthetic reactions and they noted clinical differences in the 2 conditions (Table 13). Willner et al. (1984) called attention to the same differences. Gronert et al. (1992) found that patients with either Duchenne or Becker muscular dystrophy may not give abnormal responses to tension tests and they believe there are 3 different kinds of these reactions in Duchenne patients: clinically typical MH; 'rhabdomyolysis' (not further defined); and cardiac arrest without preceding clinical abnormality (Bush

TABLE 13

Classic malignant hyperthermia and Duchenne MD anesthetic reaction

Feature	DMD anesthetic reaction	Malignant hyperthermia
Heart rate	Decreased often proceeding to cardiac arrest	Increased
Myoglobinuria	Common	Less common now
Temperature	Normal or slightly increased	Very elevated
Rigidity	Uncommon	Common
Response to dantrolene	No significant alteration of course in 3 boys so treated	+

DMD = Duchenne muscular dystrophy.

and Dubowitz 1991). Because the affected gene products differ in MH and Duchenne muscular dystrophy, the clinically similar reactions to anesthesia in these conditions (and perhaps others) must arise from different kinds of pathogenesis. In both, however, the final triggering agent is presumably the same; excessive ingress of calcium into muscle is induced by exposure to halothane and a depolarizing agent.

4. As described later, there are similarities between MH and other heat-related causes of myoglobinuria, such as the neuroleptic malignant syndrome, heat exhaustion, and heat stroke (Table 14). These conditions, however, are not inherited and are thought to differ from MH. They are discussed under Sporadic causes.

These arguments about the genetics of MH have practical implications because it seems unlikely that there is a single cause of 'MH myopathy' or that any single test will identify all people at risk, unless 1 person in a family has had an attack, and that family is also informative for currently available probes. All first-degree relatives of a survivor of MH should probably be deemed at risk and wear a warning bracelet so that anesthesiologists could be forewarned in case of emergency (Rowland 1984). This may not be feasible, however, as we try to educate people about the problem. Even the name of the condition is a handicap. 'Hyperthermia' is not now constant because treatment may commence early, and 'malignant' raises questions about cancer. An eponym, the Denborough-Britt-Kalow syndrome for example, would have advantages.

Pathophysiology of MH

Kalow et al. (1979) attributed MH to 3 factors: inherited susceptibility; internal and external factors that establish or protect from vulnerability; and the triggering agent. For 20 years, all clues have pointed to an abnormality of the calcium-sequestering capacity of the SR. That view has been validated by DNA analysis implicating the ryanodine receptor of the SR.

Normally, the SR binds calcium during relaxation and releases the cation during contraction (Sandow 1970). In MH, caffeine induces tension that is greater than normal (Britt et al. 1975a,b; Bennet et al. 1977). It is thought that the SR in affected people cannot accumulate and retain calcium properly (Britt et al. 1975a,b; Bennett et al. 1977) because the excessive contractions can be blocked by dantrolene (Ellis and Carpenter 1972; Britt et al. 1975a,b; Anderson and Jones 1976; Britt et al. 1984) or procainamide (Inesi and Watanabe 1967; Johnson and Inesi 1969; Beldaus et al. 1971; Thorpe and Seeman 1973). Several possibilities might account for the excessive uptake of calcium, including the following: excessive release of calcium from SR or decreased re-uptake (Gronert et al. 1979); decreased re-uptake of calcium by mitochondria (Britt 1971, 1973; Britt et al. 1975a, b; Cheah and Cheah 1978); increased sensitivity of the SR to calcium-induced calcium release (Endo 1982; Nelson 1982); excessively fragile sarcolemma allowing excessive calcium influx (Heffron and Gronert 1977); or increased adrenergic innervation with indirect effects (Williams 1976; Williams et al. 1975, 1976, 1978). The most favored theory now is an abnormality of the calcium-induced calcium release channel, the ryanodine receptor.

Treatment of MH

Most anesthesiologists are now alert to the problem of MH and recognize the warning signs. Administration of the inhalant is stopped, cooling measures are instituted, and dantrolene is given (Snyder et al. 1967; Denborough et al. 1970; Digby et al. 1971; Ryan 1973, 1979; Lietman et al. 1974; Owen and Kerry 1974; Britt et al. 1977a,b; Mazzia and Simon 1977; Gronert 1980; Kolb et al. 1982; Weinreich and Silvay 1982). In 1 multicenter study, all patients with established MH survived when dantrolene was given within 24 hours of onset (Kolb et al. 1982). The recommended loading dose is 2.5 mg/kg body weight, with a maximum of 10 mg/kg given over 15 minutes. That dosage can be repeated every 15 minutes if symptoms recur. When temperature and heart rate decrease and the muscles relax, a maintenance infusion is given at a rate of 1–2 mg/kg every 3–4 hours until all evidence of the crisis has passed (Tomarken and Britt 1987).

Other aspects of treatment include cooling and attention to the metabolic acidosis, hyperkalemia, arrhythmia, disseminated intravascular coagulation, and myoglobinuria. Other aspects of treatment of an attack of myoglobinuria are described below in the section on crush syndrome.

If a person has already had an episode of MH, or is otherwise considered at risk, dantrolene can be given prophylactically before elective operations.

Most drugs can be given safely to people at risk for MH. Tomarken and Britt (1987) advised caution, however, in the use of cardiac glycosides, calcium salts, sympathomimetics, and quinidine.

Sporadic (acquired) causes of myoglobinuria

The causes of acquired myoglobinuria are numerous and diverse, and undoubtedly overlap with genetic causes, which make a person more susceptible to physical or toxic stress on muscles. Also, recurrent myoglobinuria starts with a single attack. Therefore, enzymatic abnormalities may be found in patients who have had a single attack induced by exertion or fever. The causes of an attack may be multiple. For instance, drug-induced myoglobinuria may be due to direct effects on muscle membranes, depression of oxidative metabolism, agitation against restraint, and fever. In some exertional cases it is not possible to determine whether fever caused or resulted from myoglobinuria. Nevertheless, it is possible to separate the numerous causes of myoglobinuria for this discussion.

Exercise

Genetic susceptibility is not held responsible for most exercise-induced attacks of myoglobinuria. Furthermore, it seems likely that anyone, stressed sufficiently, might have an attack (Gitlin and Demos 1974). How this comes about is not known. Muscle may be injured by trauma, and running downhill is more traumatic than level running (Friden et al. 1981; Hikida et al. 1983; Schwane et al. 1983). Other factors include ischemia of contracted muscle, especially during prolonged isometric contraction. Also, there are metabolic alterations brought on by prolonged running, distorted heat regulation or blood flow, or by hypokalemic effects on the circulation (Knochel 1978). Unsuspected viral infections may complicate the more apparent exercise. Patients with sickle cell disease may be at special risk (Helzlsouer et al. 1983). Involuntary movement disorders may be among the

causes, but those patients are likely to receive neuroleptic drugs (Lazarus and Toglia 1985; Jankovic 1986).

Salt depletion has been held responsible for heat exhaustion in those who work hard at high ambient temperatures (Rowland and Penn 1972, 1974) and these workers are advised to take supplemental salt. However, marathon runners are advised to drink water without salt and rarely have myoglobinuria (Williams et al. 1981; Rowland 1984).

Many episodes of exertional myoglobinuria affected physically untrained military recruits or athletes. Training protects against exercise-induced muscle injury (Maxwell and Bloor 1981; Ross et al. 1983; Penn 1986) but it is not clear how this comes about; fiber types tend to become more oxidative than glycolytic, and mitochondria increase in number. Professional distance runners have muscle levels of succinate dehydrogenase that are 3–4 times

TABLE 14

Heat, fever, and myoglobinuria.

	Exercise-induced myoglobinuria	MH	MNS	Heat exhaustion/heat stroke
Myoglobinuria	+	+	+	+
Provoking factor	Exercise	Halothane	Neuroleptics	Exercise/exposure
Tachycardia	+	+	+	+
Acidosis	+	+	+	+
DIC	+	+	+	+
Muscle rigidity or spasm	0	+	+	0
Onset duration	Minutes	Minutes	Days	Minutes
Familial attacks	Rare*	Rare	None	None

*Heritable biochemical abnormality may be identified.

MH, malignant hyperthermia; MNS, malignant neuroleptic syndrome; DIC, disseminated intravascular coagulation.

Taken from Rowland (1984) by courtesy of the Editors of *Can. J. Neurol. Sci.*.

higher than sedentary controls (Essen 1977; Fink et al. 1977; Saltin et al. 1977). Cardiovascular training increases athletic endurance. In some military cases and many marathon cases of myoglobinuria, the subjects had been well-trained and were physically fit. Sometimes, however, a well-trained soldier may undertake a new and vigorous exercise (Hurley 1989). Among experienced runners, the competitors set unrealistic goals for speed or distance or were exposed to high ambient temperatures (England et al. 1982). The role of heat exhaustion is considered later. Nevertheless, even well-conditioned body builders may suffer an attack (Doriguzzie et al. 1990).

Exertion is involved in other forms of myoglobinuria, including wrestling and skiing, status epilepticus, status asthmaticus (Chugh et al. 1978), and struggling against restraints (Rowland 1984). Exertion could also be a factor in myoglobinuria that accompanies the convulsions of lightning or electric shock or after an overdose of drugs (Rowland 1984), including amphetamines (Kendrik et al. 1977), phencyclidine (Cogen et al. 1978; Barton et al. 1980; Patel and Connor 1985), phenmetrazine, phentermine (Kendrick et al. 1977; Black and Murphy 1984; Chavanet et al. 1984; Penn 1986), strychnine (Boyd et al. 1983), loxapine (Tam et al. 1980), water hemlock (Carlton et al 1979), or doxylamine (Koppel et al. 1987a,b). Attacks have been caused by accidental ingestion of paraphenylenediamine (Baud et al. 1983). Theophylline overdose causes agitation, seizures, and myoglobinuria (MacDonald et al. 1985; Modi et al. 1985; Rumpf et al. 1985; Wight et al. 1987). Drug-induced tardive dyskinesia (Lazarus and Toglia 1985) or dystonia (Jankovic and Penn 1982), decerebrate posturing (Briggs and Smith 1986), and prolonged myoclonus in viral encephalomyelitis (Langston et al. 1977) have also been implicated. Sometimes, drug-induced agitation leads to restraints that may aggravate the muscle damage (Rowland and Penn 1972; Goode and Meltzer 1976; Goode et al. 1977; Mercieca and Brown 1984). Also drugs and electricity may have more direct effects on muscle (Undterdorfer and Lederer 1975).

Poels et al. (1991) found that 5 of 6 patients with unexplained and recurrent myoglobinuria gave tension tests that were characteristic of MH. In 1 patient with recurrent myoglobinuria of unknown cause but induced by exercise, prophylactic therapy with dantrolene seemed to arrest the attacks (Haverkort-Poels et al. 1987).

Heat

In addition to MH, heat plays a role in 3 other syndromes: neuroleptic malignant syndrome, heat stroke, and heat exhaustion.

The clinical manifestations of MH and the 'neuroleptic malignant syndrome' (NMS) are almost identical (Eiser et al. 1982; Smego and Duralk 1982), including fever, acidosis, rigidity, myoglobinuria, and relief by dantrolene (Kolb et al. 1982; Day et al. 1983; Goulon et al. 1983). The main difference is that MH is an acute disorder with a defined onset, exposure to anesthesia, and a brief course that is measured in hours, in contrast to the gradual onset of NMS with exposure to 1 or more neuroleptic drugs and a duration of days or weeks (Table 14). The offending drugs include tricyclic antidepressants, monoamine oxidase inhibitors, phenetylzine, lithium, tetrabenazine, levodopa, and dothiepin (Heyland and Sauve 1991).

Nevertheless, MH and NMS have been considered different conditions because no one has yet had both conditions, which would be expected if susceptibility to both were induced by the same gene. Also, at least 1 survivor of NMS later had general anesthesia safely (Lotstra et al. 1983) and, with few exceptions (Caroff et al. 1983), tests for MH susceptibility have been negative in survivors of NMS (Burke et al. 1981; Tollefson 1982; Krivosic-Horber et al. 1987). There is evidence of a cerebral disorder in NMS (Henderson and Wooten 1981) and bromocriptine therapy has had apparent benefit (Dhib-Jahlbut et al. 1983; Granato et al. 1983; Mueller et al. 1983).

Heat stroke is another disorder with similar manifestations (Table 14) (Jardon 1982); the diagnostic criteria include metabolic encephalopathy with depressed consciousness, hemoconcentration, lack of sweating (anhydrosis), and fever (Clowes and O'Donnell 1974; Knochel 1974; O'Donnell 1975; Beard et al. 1979; Sprung et al. 1980). Myoglobinuria is common in heat stroke, especially when it is incurred by physical labor in hot climates (Malamud et al. 1946; Berry and King 1963; Vertel and Knochel 1967; Demos and Gitlin

1974; Demos et al. 1974; Cook et al. 1990). Mala-mud et al. (1946) described autopsies of 125 soldiers who died of heat stroke; most had served less than 2 months before the attack and 19 had pigmented casts in renal tubules. Because of these similarities to MH, dantrolene has been used to treat heat syndromes (Lydiatt and Hill 1981).

Toxic shock syndrome is still another disorder that incorporates fever and myoglobinuria, with additional essential elements of hypotension, rash, thrombocytopenia, encephalopathy, hepatic failure, and renal failure (Clayton et al. 1982). It is difficult to classify some cases (Mason and Thomas 1976). Disseminated intravenous coagulopathy is seen in MH and all other heat-related syndromes. Only in MH is genetics thought to play a role.

Cold
Myoglobinuria may also follow prolonged exposure to cold, sometimes complicated by the ischemia of frostbite (Raifman et al. 1978).

Crush and ischemia
Crush and ischemia involve common mechanisms. Crush directly traumatizes muscle and also compresses blood vessels, adding ischemic insult. If major arteries are occluded, the muscles may become necrotic and edematous, further compressing small vessels (Olivero and Ayus 1978; Rowland 1984). Swelling within a fascial compartment may aggravate the condition (Mubarak and Owen 1975; Owen et al 1979). Reperfusion may play a role in the systemic effects, and prevention of peroxidation in the damaged tissue might protect against renal failure (Odeh 1991).

The conditions that lead to crush injury may also cause hypotensive shock and disseminated intravascular clotting (DIC); Bywaters and Stead (1945) noted muscle infarcts in some of the original cases. DIC may also play a role in heat syndromes and the hemolytic-uremic syndrome (Andreoli and Bergstein 1982). ε-Aminocaproic acid has been thought to cause myoglobinuria by inducing clotting in small vessels (Britt et al. 1980; Kennard et al. 1980; Van Renterghem et al. 1984; Kane et al. 1988), but that is unproven. Ischemia may have been important in the exotic case of myoglobinuria

induced by intravenous injection of peanut oil (Lynn 1975) and in 2 patients with myoglobinuria after treatment of gastrointestinal hemorrhage with vasopressin (Affarah et al. 1984).

When comatose patients lie immobile for many hours, there may be crush injuries of limbs or torso (Gordon and Newman 1953; Penn et al. 1972). Similar problems may arise when people are held in an unusual position under anesthesia for prolonged surgery (Larcan et al. 1980). Owen et al. (1979) measured intramuscular pressure in volunteers who were placed in typical positions of people who had taken an overdosage of drugs. Pressures were 100–225 mmHg on hard surfaces in contrast to 20 mmHg on soft surfaces. Edema and prolonged immobility are also likely to lead to nerve injuries by compression, traction, and ischemia (Akmal and Massry 1983).

Sad to relate, civilians still suffer crush syndrome due to war bombardment (Better 1989). Treatment, as reviewed by Better and Stein (1990), must be prompt; if intravenous volume replacement is delayed for more than 6 hours, acute renal failure is likely. Forced solute-alkaline diuresis may protect against renal failure. They believe that fasciotomy should be reserved for patients who clearly have had occlusion of major arteries. In catastrophes, they recommend searching for survivors for several days; 1 woman was saved after 5 days under the rubble of an earthquake (Better and Stein 1990).

Metabolic depression
Cerebral depressant drugs may also affect muscle directly or secondarily if cardiorespiratory controls are lost (Table 15). Sometimes this results from attempted suicide (Hojgaard et al. 1988). Myoglobinuria has accompanied the encephalopathies of hypothyroidism (Halverson et al. 1979), hypothermia (Raifman et al. 1978; Rosenthal et al. 1981), diabetic ketoacidosis (Buckingham et al. 1981), and nonketotic hyperglycemia (Rumpf et al. 1981). Hypothyroidism may affect muscle directly or through hypothermia. Fever alone may cause myoglobinuria (Berg and Frenkel 1958) or may play a role when there is sepsis (Heinrich et al. 1980; Kalish et al. 1982; Nahas et al. 1983). Sepsis may also be important in newborns (Gilboa and Swanson 1976; Haftel et al. 1978).

Altered salt and water metabolism

Chronic hypokalemia can cause a vacuolar myopathy that may be expressed by increased serum enzyme levels alone, by fixed limb weakness, or by acute myoglobinuria (Knochel and Schlein 1972; Dominic et al. 1978; Nadel et al. 1979; Altenwerth 1982) (Table 15). Hypokalemic myoglobinuria may be induced by diuretic therapy with carbenoxolone (Mohamed et al. 1966; Barnes and Leonard 1971; Mitchell 1971), thiazides, licorice abuse (exposure to the precursor of carbenoxolone) (Gross et al. 1966; Tourtelotte and Hirst 1970), renal tubu-

TABLE 15

Drugs, toxins, metabolic disorders and myoglobinuria.

Mechanism	Drugs	Toxins or metabolic disorders
Metabolic depression	Barbiturates Other sedatives Carbon monoxide Ethanol Fenfluramine Glutethimide	Hypothyroidism Diabetic acidosis Nonketotic hyperglycemia Hyperosmolarity Hypernatremia
Hypokalemia	Amphotericin Carbenoxolone Glycyrrhizate Thiazides Other kaluretics Laxative abuse	Diarrhea Hyperaldosteronism Renal tubular acidosis
Hypophosphatemia		Parenteral fluids for acute alcoholism Diabetic ketoacidosis Parathyroidectomy
Ischemia	ε-Aminocaproic acid Cocaine Vasopressin	Total parenteral nutrition Air embolism
Direct membrane effect (?)	Clofibrate Bezafibrate Toluene	Venoms: hornet, mulga snake, tiger snake, sea snake
Agitation	Amphetamines Lithium Loxapine Paracetamol Phencyclidine Salicylates Strychnine Water hemlock Mercuric chloride Cocaine Theophylline	Electroconvulsive therapy Tetanus Myoclonus (prolonged) Dystonia
Extremes of body temperature Hypothermia Fever		Hypothyroidism Tetanus toxin Typhoid vaccine

Modified from Rowland (1984) by courtesy of the Editors of *Can. J. Neurol. Sci.*.

lar acidosis (Campion et al. 1972), chronic alcoholism (Martin et al. 1971), amphotericin B (Drutz et al. 1970), primary aldosteronism (Crawhall et al. 1976; Dominic et al. 1978), regional enteritis (Heitzman et al. 1962), parenteral hyperalimentation (Nadel et al. 1979), or laxative abuse. The alkalosis of hypokalemia may protect against heme-induced nephropathy, because renal failure has been relatively uncommon in these cases (Penn 1986). Also, myoglobinuria has not been recorded in familial hypokalemic periodic paralysis, so that the insult must be caused by the chronic condition, not by intermittently low levels.

Distorted sodium metabolism, either hyper- or hyponatremia, may also lead to myoglobinuria. Hypernatremic myoglobinuria may result from excessive administration of salt with replacement fluids for children with gastroenteritis (Penn 1980). Myoglobinuria has been seen in hyponatremia with water intoxication (Browne 1979) and nonketotic hyperosmolar coma (Grossman et al. 1974); these metabolic disorders were complex. Complex abnormalities may also be responsible for the myoglobinuria seen in pancreatitis (Nankivell and Gillies 1991) and in the bulimic vomiting of patients with anorexia nervosa (Abe et al. 1990); hypokalemia may also be induced by laxative abuse in anorexia nervosa (Dive et al. 1991).

Hypophosphatemia may be partly responsible for alcoholic myoglobinuria (Knochel et al. 1975). Knochel et al. (1978) found increased serum CK levels and muscle necrosis in dogs rendered deficient in phosphorus and calories; the changes were prevented by adding elemental phosphorus to the diet. Myoglobinuria was seen in 1 patient with hypophosphatemia after parathyroidectomy (Woolridge et al. 1978).

All of these metabolic disorders are complex, involving more than a single salt. In a retrospective analysis of 133 patients with diabetic ketoacidosis, 12 had heme-positive urine and 1 had a serum CK level of 1200 IU (Buckingham at al. 1981); concomitant abnormalities probably included hyperosmolality, hypernatremia, or hypophosphatemia.

Alcohol

Chronic alcoholism may cause persistent myopathic limb weakness; acute attacks of myoglobinuria may be superimposed on this chronic myopa-

thy (Fahlgren et al. 1957). Some authors have reported myoglobinuria particularly in binge drinkers (Perkoff et al. 1966; Lafair and Myerson 1968; Walsh and Conomy 1977). In addition to the metabolic effects, there are other factors: coma and crush mechanisms; the exercise of agitation, convulsions, or delirium tremens; and metabolic depression from barbiturates (Fahlgren et al. 1957).

Poor nutrition has been held responsible on the basis of experimental studies (Haller and Drachman 1980; Haller and Knochel 1984); exposure to ethanol in fasted rats was more likely to cause myoglobinuria than in fed rats. Fed rats also had less histological evidence of muscle necrosis (Haller et al. 1984). However, alcohol may also be a myotoxin (Rubenstein and Wainapel 1977; Rubin 1979). Other factors include hypokalemia (Martin et al. 1971) and hypophosphatemia (Knochel et al. 1975). In any series of myoglobinuria, alcoholism is usually a prominent cause (Pariente et al. 1983; Saltissi et al. 1984).

Drugs and myotoxins

Few agents seem to affect muscle membranes directly. Among the best examples are drugs used to lower serum cholesterol levels and which may alter the lipid composition of muscle membranes. The drugs comprise 2 different classes. One includes clofibrate and bezafibrate, which are branched chain fatty acid esters that inhibit hepatic synthesis of triglycerides (Cotton 1972) and reduce serum levels of lipoproteins and cholesterol (Langer and Levy 1968). Muscle toxicity has been reported in at least 44 patients, sometimes a week after starting therapy (Rimon et al. 1984). With chronic use, some patients have cramps, myoglobinuria, or myotonic symptoms (Table 15) (Langer and Levy 1968; Smals et al. 1977; Schneider et al. 1980; Bock 1981; Heidemann et al. 1981).

The other cholesterol-lowering drug that causes myoglobinuria is lovastatin, an inhibitor of 3-hydroxy-3-methyl glutaryl CoA reductase, the rate-limiting enzyme in cholesterol biosynthesis (The Lovastatin Study Group II 1986; Havel et al. 1987). Myoglobinuria has occurred in at least 15 people taking this drug. Again, the mechanism is uncertain because so few have been affected and some were taking other drugs at the same time, e.g. cyclosporine or gemfibrozil (Corpier et al. 1988;

East et al. 1988; Norman et al. 1988; Pierce et al. 1990), nicotinic acid (Reaven and Witzium 1988), or erythromycin (Ayanian et al. 1988). Nicotinic acid itself has been incriminated (Litin and Anderson 1989).

Other direct actions may occur after bites by snakes (Furtado and Lester 1968; Rowlands et al. 1969) or insects (Ownby and Odell 1983), but the mechanisms are not known. Some of these toxins are phospholipases (Fohlman and Eaker 1977; Mebs and Samejima 1980). Others may affect SR membranes (Hoh et al. 1966; Shilkin et al. 1972; Ownby et al. 1976; Ownby and Odell 1983).

Myoglobinuria may follow use of heroin, with or without coma (Richter et al. 1971; Penn et al. 1972; Katherein et al. 1983; Aeschlimann et al. 1984; Arcangeli et al. 1984; Rohling et al. 1984; De Gans et al. 1985); it is not clear whether the narcotic is responsible or some adulterant (D'Agostino and Arnett 1979). Chronic use of methadone may also lead to myoglobinuria (Nanji and Filipenko 1983; Weston et al. 1986), perhaps due to direct muscle injury (Koppel 1989). Cocaine has also been incriminated as causing myoglobinuria (Schwartz and McAfee 1987; Krohn et al. 1988; Reinhart and Stricker 1988; Zamora-Quezada et al. 1988); vasoconstriction (Reinhart and Striker 1988) and NMS have also been suggested causes (Kosten and Kleber 1987; Merigan and Roberts 1987). Antihistamines may also have direct effects on muscle to cause myoglobinuria (Koppel et al. 1987a,b; Mendoza et al. 1987). Terbutaline may also directly affect muscle when it is given for tocolysis (stopping premature labor) (Ditzian-Kadanoff et al. 1988).

As indicated in other sections of this Chapter, drug-induced myoglobinuria may be mediated by excessive muscle activity, ischemia, hypokalemia or coma. We have tried to list most known cases of drug-induced myoglobinuria in Table 15, and Koppel (1989) provided a comprehensive review of drug-induced myoglobinuria. Among those incriminated are isoniazid (Cronkright and Szymaniak 1989), oxprenolol, perhaps through hypotension (Schofield et al. 1985), and theophylline (MacDonald et al. 1985; Modi et al. 1985; Rumpf et al. 1985; Wight et al. 1987). Some causes of myoglobinuria are still mysterious, including the epidemic called Haff disease, which was attributed to an un-

known pollutant in Baltic waters (Berlin 1948), and also outbreaks among people eating quail (Rosner 1970; Billis et al. 1971; Carlton et al. 1979).

Infection

There have been many reports of myoglobinuria during infections. The incriminated viruses include influenza (Gamboa et al. 1979; Zamkoff and Rosen 1979; Canaud et al. 1983), herpes simplex (Schlesinger et al. 1978), Epstein-Barr virus with mononucleosis (Kantor et al. 1978), Coxsackie (Dunnet et al. 1981) and ECHO viruses (Bardelas et al. 1977; Wilfert et al. 1977; Jehn and Fink 1980).

Bacteria have also been implicated. The toxic shock syndrome, attributed to staphylococcal toxin was defined by Todd et al. (1978) as including hypotension, fever, rash, and involvement of 4 organ systems: liver; kidney; brain; and muscle. Myopathy has not been emphasized but it may be a major manifestation (McKenna et al. 1980; Saul et al. 1980; Shands et al. 1980) and may play a role in renal failure. Myoglobinuria may also be associated with staphylococcal infiltration or abscesses of muscle (Armstrong 1978; Adamski et al. 1980).

Other bacterial infections have been associated with myoglobinuria, including Legionnaire disease, typhoid, *E. coli*, *Mycoplasma*, leptospirosis, *Shigella*, *Salmonella*, *Clostridium*, tularemia, and pneumococcus (Bowden et al. 1956; Rowland and Penn 1972; Rothstein and Kenny 1979; Heinrich et al. 1980; Ho and Scully 1980; Posner et al. 1980; Rowland 1984; Kaiser et al. 1985; Hroncich and Rudinger 1989). One case was associated with *Candida fungemia*, but the patient was also receiving ε-aminocaproic acid and other drugs (Hinds and Rowe 1988).

In all of these infections the possible mechanisms include direct effects on muscle (Armstrong et al. 1978; Huber and Job 1983), fever, or seizures.

Progressive muscle disease and myoglobinuria

There have been extraordinarily few cases of myoglobinuria in patients with chronic myopathies. Walton and Adams (1954) included bouts of myoglobinuria as evidence of acute polymyositis. There have been only a few cases of acute myoglobinuria documented in people with polymyositis (Gunther 1924; Paul 1924; Kagen 1971; Skrabal et al. 1972; Korz and Volz-Boers 1974; Sloan et al.

1978). In some, limb weakness was severe and there was respiratory failure in a few (Wynne et al. 1977). However, these cases mostly differed from ordinary polymyositis because, if they recovered in weeks or months, there was no permanent weakness.

Some of these could have been due to viral infection. Some were diagnosed as polymyositis because there were inflammatory cells in muscle biopsy, but that histological change can be seen in myoglobinuria of any cause, including marathon running (Hikida et al. 1983). 'Polymyositis' seems to be an appropriate designation for these cases, but the acute disorder with myoglobinuria differs from the subacute or chronic myopathy that usually goes by the same name.

In a few cases there was a chronic myopathy. The patient of Gamboa et al. (1979) had a gait disorder for months before the attack of myoglobinuria that was attributed to influenza virus. Another patient (Kreitzer et al. 1978) had limb weakness for a year before an attack of myoglobinuria, and one had fever, arthralgia, and limb weakness for a month before pigmenturia (Pirovino et al. 1979). Samuels et al. (1989) described a similar case. Two patients had dermatomyositis with myoglobinuria (Kessler et al. 1972; Marks et al. 1976), neither straightforward.

These cases suggest overlap with the ill-defined syndrome of polymyositis, but there has been no well-documented case of progressive muscular dystrophy with myoglobinuria, except for those associated with anesthesia (Duchenne dystrophy) and exertion (Becker dystrophy). One infant had an attack in the neonatal period and later proved to have Duchenne dystrophy (Breningstall et al. 1988); the original attack could have been an intercurrent infection.

AN OVERALL VIEW

Myoglobinuria was once considered a rare event. However, it now seems to be common and it is recognized readily by physicians in emergency rooms. This change can be attributed to several developments. The syndrome is recognized more frequently because the essentials are taught to house officers and medical students. This, in turn, results from the increased attention of nephrolo-gists to acute renal failure, and of neurologists to different forms of muscle disease. Technological advances have made it possible to identify 13 different biochemical causes of the syndrome, and technological advances in intensive care medicine have reduced the mortality rate. Studying the syndrome has increased knowledge of human genetics as well as the normal physiology of muscle and kidney. Sports medicine, occupational medicine, pharmacology, toxicology, and virology have also played roles. Myoglobinuria is a prime example of the value of rare diseases in promoting better care for patients and advancing the knowledge of human biology.

REFERENCES

ABARBANEL, J.M., N. BASHAN, R. POTASHNIK, A. OSIMANI, S.W. MOSES and Y. HERISHANU: Adult muscle phosphorylase 'b' kinase deficiency. Neurology 36 (1986) 560–562.

ABE, K., T. MEZAKI, N. HIRONO, F. UDAKA and M. KAMEYAMA: A case of anorexia nervosa with acute renal failure resulting from rhabdomyolysis. Acta Neurol. Scand. 81 (1990) 82–83.

ACHESON, D. and D. MCALPINE: Muscular dystrophy associated with paroxysmal myoglobinuria and excessive excretion of ketosteroids. Lancet ii (1953) 372–375.

ADAMS, E.C.: Differentiation of myoglobin and hemoglobin in biological fluids. Ann. Clin. Lab. Sci. 10 (1980) 493–499.

ADAMSKI, G.B., E.H. GARIN, W.E. BALLINGER and S.T. SHULMAN: Generalized nonsuppurative myositis with staphylococcal septicemia. J. Pediatr. 96 (1980) 694–697.

ADORNATO, B.T., L.J. KAGAN and W.K. ENGEL: Myoglobinaemia in Duchenne muscular dystrophy patients and carriers: a new adjunct to carrier detection. Lancet ii (1978) 499–501.

AESCHLIMANN, A., T. MALL, P. SANDOZ and A. PROBST: Verlauf und Komplikationen der Rhabdomyolyse nach Heroin-Intoxikation. Schweiz. Med. Wochenschr. 114 (1984) 1236–1240.

AFFARAH, H.B., R.L. MARS, A. SOMEREN, H.W. SMITH III and S.B. HEYMSFIELD: Myoglobinuria and acute renal failure associated with intravenous vasopressin infusion. South. Med. J. 77 (1984) 918–921.

AGAMANOLIS, D.P., A.D. ASKARI, S. DIMAURO, A. HAYS, K. KUMAR, M. LIPTON and A. RAYNOR: Muscle phosphofructokinase deficiency: two cases with unusual polysaccharide accumulation and immunologically active enzyme protein. Muscle Nerve 3 (1980) 456–467.

AKMAL, M. and S.G. MASSRY: Peripheral nerve damage in patients with nontraumatic rhabdomyolysis. Arch. Intern. Med. 143 (1983) 835–836.

ALLEN, G.C., H. ROSENBERG and J.E. FLETCHER: Safety of general anesthesia in patients previously tested negative for malignant hyperthermia susceptibility. Anesthesiology 72 (1990) 619–622.

ALTENWERTH, F.J.: Rhabdomyolyse durch Hypokaliamie? Dsche Med. Wochenschr. 107 (1982) 196–197.

ANDERSON, I.L. and E.W. JONES: Porcine malignant hyperthermia: effect on dantrolene sodium on in-vitro halothane-induced contraction of susceptible muscle. Anesthesiology 44 (1976) 57–61.

ANDO, T., T. SHIMIZU, T. KATO, M. OHSAWA and Y. FUKUYAMA: Myoglobinemia in children with progressive muscular dystrophy. Brain Dev. 10 (1978) 25–27.

ANDREOLI, S.P. and J.M. BERGSTEIN: Acute rhabdomyolysis associated with hemolytic-uremic syndrome. J. Pediatr. 103 (1983) 78–80.

ANGELINI, C., L. FREDDO, P. BATTISTELLA, N. BRESOLIN, S. PIEROBON-BORMIOLI, M. ARMANI and L. VERGANI: Carnitine palmitoyltransferase deficiency: clinical variability, carrier detection, and autosomal-recessive inheritance. Neurology 31 (1981) 883–886.

ARCANGELI, A., F. CAVALIERE, P. CARDUCCI, R. PROIETTI and S.I. MAGALINI: Acute rhabdomyolysis during heroin abuse. Ital. J. Med. 2 (1984) 68–70.

ARMSTRONG, C.L., A.F. MIRANDA, K.C. HSU and E.T GAMBOA: Susceptibility of human skeletal muscle culture to influenza virus infection. Cytopathology and immunofluorescence. J. Neurol. Sci. 35 (1978) 43–57.

ARMSTRONG, J.H.: Tropical pyomyositis and myoglobinuria. Arch. Intern. Med. 138 (1978) 1145–1146.

ASKMARK, H., P.O. OSTERMAN, L.E. ROXIN and P. VENGE: Radioimmunoassay of serum myoglobin in neuromuscular disease. J. Neurol. Neurosurg. Psychiatr. 44 (1981) 68–72.

AYANIAN, J.Z., C.S. FUCHS and R.M. STONE: Lovastatin and rhabdomyolysis. Ann. Intern. Med. (1988) 682.

BACON, A.E. JR.: Acute idiopatic rhabdomyolysis with myoglobinuria. Del. Med. J. 39 (1967) 302–305.

BAILIE, M.D.: Primary paroxysmal myoglobinuria. N. Engl. J. Med. 271 (1964) 186–189.

BANK, W.J., S. DIMAURO, E. BONILLA, D.M. CAPUZZI and L.P. ROWLAND: A disorder of muscle lipid metabolism and myoglobinuria. Absence of carnitine palmityl transferase. N. Engl. J. Med. 292 (1975) 443–449.

BARDELAS, J.A., J.A. WINKELSTEIN, D.S.Y. SETO, T. TSCI and A.D. ROGOL: Fatal ECHO 24 infection in a patient with hypogammaglobulinemia: relationship to dermatomyositis-like syndrome. J. Pediatr. 90 (1977) 396–399.

BARNES, P.C. and J.H.C. LEONARD: Hypokalemic myopathy and myoglobinuria due to carbenoxolone sodium. Postgrad. Med. J. 47 (1971) 813–814.

BARTON, C.H., M.L. STERLING and N.D. VAZIRI: Rhabdomyolysis and acute renal failure associated with phencyclidine intoxication. Arch. Intern. Med. (1980) 568–569.

BAUD, F., C. BISMUTH, M. GALLIOT, R. GARNIER and A.

PERALMA: Rhabdomyolysis in paraphenylenediamine intoxication. Lancet ii (1983) 514.

BEARD, M.E., J.W. HAMER, G. HAMILTON and A.H. MASLOWSKI: Jogger's heat stroke. N. Z. Med. J. 631 (1979) 159–161.

BELDAUS, J., V. SMALL, D.A. COOPER and B.A. BRITT: Post operative malignant hyperthermia: a case report. Can. Anaesth. Soc. J. 18 (1971) 202–212.

BELL, A.W. and G.E. THOMPSON: Free fatty acid oxidation in bovine muscle in vivo: effects of cold exposure and feeding. Am. J. Physiol. 237 (1979) E309–315.

BENNETT, C., P.A. CAIN, F.R. ELLIS, C.F. LOUIS and M. STANTON: Calcium and magnesium contents of malignant hyperpyrexia-susceptible human muscle. Br. J. Anaesth. 49 (1977) 979–982.

BERENBAUM, M.C., C.A. BIRCH and J.D. MORELAND: Paroxysmal myoglobinuria. Lancet i (1955) 892–896.

BERG, P. and E.P. FRENKEL: Myoglobinuria after spontaneous and induced fever: report of a case. Ann. Intern. Med. 48 (1958) 380–389.

BERLIN, R.: Haff disease in Sweden. Acta Med. Scand. 129 (1948) 560–572.

BERNHARDT, D. and M.H. HORDER: Anestheia–induced myoglobinuria without hyperpyrexia: an abortive form of malignant hyperthermia? In: A. Aldrete and B.A. Britt (Eds), Proceedings, 2nd International Symposium on Malignant Hyperthermia. New York, Grune and Stratton (1978) 419–425.

BERRY, M.E. and B.A. KING: Heatstroke. S. Afr. Med. J. 36 (1963) 455–461.

BETTER, O.S.: Traumatic rhabdomyolysis ('crush syndrome') updated 1989. Isr. J. Med. Sci. 25 (1989) 69–72.

BETTER, O.S. and J.H. STEIN: Early management of shock and prophylaxis of acute renal failure in traumatic rhabdomyolysis. N. Engl. J. Med. 322 (1990) 825–829.

BILLIS, A.G., S. KASTANAKIS, H. GIAMARELLOU and G.K. DAIKAS: Acute renal failure after a meal of quail. Lancet ii (1971) 702.

BLACK, W.D. and W.M. MURPHY: Nontraumatic rhabdomyolysis and acute renal failure associated with oral phenmetrazine hydrochloride. J. Tenn. Med. Assoc. 77 (1984) 80–81.

BOCK, K.D.: Acute muscular syndrome after bezafibrate. Klin. Wochenschr. 59 (1981) 1321.

BONILLA, E., H.W. CHANG, A.F. MIRANDA, R. MEDORI and S. DIMAURO: Duchenne muscular dystrophy: the disease, the protein. In: G. Benzi (Ed.), Advances in Myochemistry II. London, Paris, John Libbey Eurotext (1989) 173–184.

BONNEFONT, J.P., H. OGIER, G. MITCHELL, F. DEMAUGRE, A. PELET, J.M. SAUDUBRAY and J. FREZAL: Heterogeneitie des deficits en palmitoyl carnitine transferase. Arch. Fr. Pediatr. 42 (1985) 613–617.

BOUGNERES, P.F., J.M. SAUDUBRAY, C. MARSAC, O. BERNARD, M. ODIEVRE and J. GIRARD: Fasting hypoglycemia resulting from hepatic carnitine palmitoyltransferase deficiency. J. Pediatr. 98 (1981) 742–746.

BOWDEN, D.H., D. FRASER, S.H. JACKSON and N.F. WALKER: Acute recurrent rhabdomyolysis (paroxysmal myohemoglobinuria). A report of three cases and a review of literature. Medicine (Baltimore) 35 (1956) 335–353.

BOYD, R.E., P.T. BRENNAN, J.F. DENG, F. ROCHESTER and D.A. SPYKER: Strychnine poisoning. Recovery from profound lactic acidosis, hyperthermia, rhabdomyolysis. Am. J. Med. 74 (1983) 507–512.

BRANDT, N.J., F. BUCHTHAL, F. EBBESEN, Z. KARMIENIECKA and C. KRARUP: Post-tetanic mechanical tension and evoked action potentials in McArdle's disease. J. Neurol. Neurosurg. Psychiatr. 40 (1977) 920–925.

BRAUN, S.R., R.F. WEISS, A.L. KELLER, J.R. CICCONE and H.G. PREUSS: Evaluation of the renal toxicity of heme proteins and their derivatives: a role in the genesis of acute tubular necrosis. J. Exp. Med. 131 (1970) 443–460.

BRENINGSTALL, G.N., W.D. GROVER, S. BARBERA and H.G. MARKS: Neonatal rhabdomyolysis as a presentation of muscular dystrophy. Neurology 38 (1988) 1271–1272.

BRESOLIN, N., Y.I. RO, M. REYES, A.F. MIRANDA and S. DIMAURO: Muscle phosphoglycerate mutase (PGAM) deficiency: a second case. Neurology 33 (1983) 1049–1053.

BRESOLIN, N., L. BET, M. MOGGIO, G. MEOLA, F. FORTUNATO, G. COMI, L. ADOBATTI, L. GEREMIA, S. PITTALIS and G. SCARLATO: Muscle glucose-6-phosphate dehydrogenase deficiency. J. Neurol. 236 (1989) 193–198.

BRIGGS, T.B. and R.R. SMITH: Exertional rhabdomyolysis associated with decerebrate posturing. Neurosurgery 19 (1986) 297–299.

BRITT, B.A.: Malignant hyperthermia: an investigation of three patients. Ann. R. Coll. Surg. Engl. 48 (1971) 73–75.

BRITT, B.A.: Malignant hyperthermia: an investigation of five patients. Can. Anaesth. Soc. J. 20 (1973) 431–467.

BRITT, B.A. and W. KALOW: Malignant hyperthermia: a statistical review. Can. Anaesth. Soc. J. 17 (1970) 293–315.

BRITT, B.A. and E. SCOTT: Failure of the platelet-halothane nucleotide depletion test as a diagnostic or screening test for malignant hyperthermia. Anesth. Analg. (Cleveland) 65 (1986) 171–175.

BRITT, B.A., W.G. LOCHER and W. KALOW: Hereditary aspects of malignant hyperthermia. Can. Anaesth. Soc. J. 16 (1969) 89–98.

BRITT, B.A., W. KALOW, A. GORDON, J. HUMPHREY and N.B. REWCASTLE: Malignant hyperthermia: an investigation of five patients. Can. Anaesth. Soc. J. 20 (1973) 431–467.

BRITT, B.A., L. ENDRENYI, R.L. BARCLAY and D.L. CADMAN: Total calcium content of skeletal muscle isolated from humans and pigs susceptible to malignant hyperthermia. Br. J. Anaesth. 47 (1975a) 647–653.

BRITT, B.A., L. ENDRENYI, D.L. CADMAN, H. MANFAN and H.Y.-K. FUNG: Porcine malignant hyperthermia: effects of halothane on mitochondrial respiration and calcium accumulation. Anesthesiology 42 (1975b) 292–300.

BRITT, B.A., L. ENDRENYI and P.L. PETERS: Screening of malignant hyperthermia susceptible families by creatine phosphokinase measurement and other clinical investigations. Can. Anaesth. Soc. J. 23 (1976a) 263–284.

BRITT, B.A., L. ENDRENYI, W. KALOW and P.L. PETERS: The adenosine triphosphate (ATP) depletion test: comparison with the caffeine contracture test as a method of diagnosing malignant hyperthermia susceptibility. Can. Anaesth. Soc. J. 23 (1976b) 624–635.

BRITT, B.A., F.H.F. KWONG and L. ENDRENYI: The clinical and laboratory features of malignant hyperthermia management: a review. In: E.O. Henschell (Ed.), Malignant Hyperthermia: Current Concepts. New York, Appleton-Century-Crofts (1977a) 9–46.

BRITT, B.A., F.H.F. KWONG and L. ENDRENYI: Endrenyl: management of malignant hyperthermia susceptible patients. A review. In: E.O. Henschell (Ed.), Malignant Hyperthermia: Current Concepts. New York, Appleton-Century-Crofts (1977b) 63–77.

BRITT, B.A., A.J. MCCOMAS, L. ENDRENYI and W. KALOW: Motor unit counting and the caffeine contracture test in malignant hyperthermia. Anesthesiology 47 (1977c) 490–497.

BRITT, B.A., E. SCOTT, W. FRODIS, M.J. CLEMENTS and L. ENDRENYI: Dantrolene: in vitro studies in malignant hyperthermia susceptible (MHS) and normal skeletal muscle. Can. Anaesth. Soc. J. 31 (1984) 130–154.

BRITT, C.W. JR., R.R. LIGHT, B.H. PETERS and S.S. SCHOCHET JR: Rhabdomyolysis during treatment with epsilon-aminocaproic acid. Arch. Neurol. 37 (1980) 187–188.

BROWNE, P.M.: Rhabdomyolysis and myoglobinuria associated with acute water intoxication. West. J. Med. 130 (1979) 459–461.

BROWNELL, A.K.W.: Malignant hyperthermia: relationship to other diseases. Br. J. Anaesth. 60 (1988) 303–308.

BROWNELL, A.K.W., R.T. PASSUKE, A. ELASH, S.B. FOWLOW, C.G. SEAGRAM, R.J. DIEWOLD and C. FRIESEN: Malignant hyperthermia in Duchenne muscular dystrophy. Anesthesiology 58 (1983) 180–182.

BROWNELL, A.K.W., S.B. FOWLOW and R.T. PAASUKE: Coexistence of malignant hyperthermia and Duchenne muscular dystrophy in the same pedigree. Neurology 35, Suppl. 1 (1985) 195.

BRUMBACK, R.A., J.W. GERST and H.R. KNULL: High energy phosphate depletion in a model of defective muscle glycolysis. Muscle Nerve 6 (1983) 52–55.

BUCHANAN, D. and P.E. STEINER: Myoglobinuria with paralysis (Meyer-Betz disease). Arch. Neurol. Psychiatr. 66 (1951) 107.

BUCKINGHAM, B.A., T.F. ROE and J.W. YOON: Rhabdomyolysis in diabetic ketoacidosis. Am. J. Dis. Child. 135 (1981) 352–354.

BURKE, R.E., S. FAHN, R. MAYEUX, H. WEINBERG, K. LOUIS and J.H. WILLNER: Neuroleptic malignant syndrome caused by dopamine-depleting drugs in a patient with Huntington disease. Neurology 31 (1981) 1022–1026.

BUSH, A. and V. DUBOWITZ: Fatal rhabdomyolysis complicating general anaesthesia in a child with Becker muscular dystrophy. Neuromusc. Disord. 1 (1991) 201–204.

BYWATERS, E.G.L. and J.H. DIBLE: Acute paralytic myohaemoglobinuria in man. J. Pathol. Bacteriol. 55 (1943) 7–15.

BYWATERS, E.G.L. and G. POPJACK: Experimental crushing injury: peripheral circulatory collapse and other effects of muscle necrosis in rabbit. Surg. Gynecol. Obstet. 75 (1942) 612–627.

BYWATERS, E.G.L. and J.K. STEAD: Thrombosis of the femoral artery with myohaemoglobinuria and low serum potassium concentration. Clin. Sci. 5 (1945) 195–204.

BYWATERS, E.G.L., G.E. DELORY, C. RIMMINGTON and J. SMILES: Myohemoglobin in urine of air raid casualties with crushing injury. Biochem. J. 35 (1941) 1164–1168.

CAMPION, D.S., J.M. ARIAS and W. CARTER: Rhabdomyolysis and myoglobinuria. Association with hypokalemia of renal tubular acidosis. J. Am. Med. Assoc. 220 (1972) 967–969.

CANAUD, B., C. LEGENDRE, J.J. BERAUD and J. MIROUZE: Rhabdomyolyse aigue au cours d'une grippe 'A' maligne. Presse Med. 12 (1983) 299.

CARBALLO, A.S.: Aborted malignant hyperthermia: case report. Can. Anaesth. Soc. J. 22 (1975) 227–229.

CARLTON, B.E., E. TUFTS and D.E. GIRARD: Water Hemlock poisoning complicated by rhabdomyolysis and renal failure. Clin. Toxicol. 14 (1979) 87–92.

CAROFF, S., H. ROSENBERG and J.C. GERBER: Neuroleptic malignant syndrome and malignant hyperthermia. Lancet i (1983) 244.

CARROLL, J.E., M.H. BROOKE, D.C. DE VIVO, K.K. KAISER and J.M. HAGBERG: Biochemical and physiologic consequences of carnitine palmitoyltransferase deficiency. Muscle Nerve 1 (1978) 103–110.

CASEBY, N.G.: Muscle hypertonus after intravenous suxamethonium: a clinical problem. Br. J. Anaesth. 47 (1975) 1101–1106.

CASENAVE, C., J.P. LAMAGNERE, J. LAUGIER, P. JOBARD, H. MOURAY and B. GRENIER: Myoglobinurie paroxystique. Etude biochimique et étude morphologique au microscope électronique. Arch. Fr. Pediatr. 29 (1972) 539–549.

CERRI, C.G., J.H. WILLNER, B.A. BRITT and D.S. WOOD: Adenylate kinase deficiency and malignant hyperthermia. Hum. Genet. 57 (1981) 325–326.

CHABOCHE, C., Y. NORDMANN, J.L. FONTAINE and C. LEJEUNE: Myoglobinurie postanesthesique. Arch. Fr. Pediatr. 39 (1982) 169–171.

CHAIKIN, H.L.: Rhabdomyolysis secondary to drug overdose and prolonged coma. South. Med. J. 73 (1980) 990–994.

CHAMBERLAIN, M.C.: Rhabdomyolysis in children: a 3-year retrospective study. Pediatr. Neurol. 7 (1991) 226–228.

CHAVANET, P., H. PORTIER and M. DUMAS: Malignant hyperpyrexia and fulminating rhabdomyolysis after amphetamine ingestion. Sem. Hop. Paris 60 (1984) 2919–2920.

CHEAH, K.S. and A.M. CHEAH: Calcium movements in skeletal muscle mitochondria of malignant hyperthermic pigs. Fed. Eur. Biochem. Soc. 95 (1978) 307–310.

CHUGH, K.S., P.C. SINGHAL and G.K. KHATRI: Rhabdomyolysis and renal failure following status asthmaticus. Chest 73 (1978) 879–880.

CLAYTON, A.J., J.E. PEACOCKE and P.E. EWAN: Toxic shock syndrome in Canada. Can. Med. J. 126 (1982) 776–779.

CLOONAN, M.J., G.A. BISHOP, P.O. WILTON-SMITH, I.W. CARTER, R.M. ALLAN and D.E.L. WILCKEN: An enzyme-immunoassay for myoglobin in normal serum and urine. Pathology 2 (1979) 689–699.

CLOWES, G.H.A. JR. and T.F. O'DONNELL: Heat stroke. N. Engl. J. Med. 291 (1974) 364–367.

COGEN, F.C., G. RIGG, J.L. SIMMONS and E.F. DOMINO: Phencyclidine-associated acute rhabdomyolysis. Ann. Intern. Med. 88 (1978) 210–212.

COOK, J.A., J.M. HILL and J.H. TURNEY: Rhabdomyolysis and a 'greenhouse effect'. Lancet ii (1990) 1136–1137.

CORKEY, B.E., D.E. HALE, M.C. GLENNON, R.I. KELLEY, P.M. COATES, L. KILPATRICK and C.A. STANLEY: Relationship between unusual hepatic acyl coenzyme A profiles and the pathogenesis of Reye syndrome. J. Clin. Invest. 82 (1988) 782–788.

CORPIER, C.L., P.H. JONES, W.N. SUKI, E.D. LEDERER, M.A. QUINONES, S.W. SCHMIDT and J.B. YOUNG: Rhabdomyolysis and renal injury with lovastatin use. Report of two cases in cardiac transplant recipients. J. Am. Med. Assoc. 260 (1988) 239–241.

COTTON, R.S.: The action of Atromid-S on lipids and other plasma factors which may be involved in arterial thrombosis. Acta Cardiol. Suppl. 15 (1972) 163–171.

CRAWHALL, J.C., G. TULIS and D. ROY: Elevation of serum creatine kinase in severe hypokalemic hyperaldosteronism. Clin. Biochem. 9 (1976) 237–240.

CRONKRIGHT, P.J. and G. SZYMANIAK: Isoniazid and rhabdomyolysis. Ann. Intern. Med. 110 (1989) 945.

CUMMING, W.J.K., M. HARDY, P. HUDGSON and J. WALLS: Carnitine-palmitoyltransferase deficiency. J. Neurol. Sci. 30 (1976) 247–258.

D'AGOSTINO, R.S. and E.N. ARNETT: Acute myoglobinuria and heroin snorting. J. Am. Med. Assoc. 70 (1979) 277.

DAVIS, G.C.: Prolonged muscle rigidity following administration of succinylcholine. South. Med. J. 70 (1977) 1139–1140.

DAY, D.C., S.W. MORRIS, R.M. STEWART, B.J. FENTON and

F.A. GAFFNEY: Neuropleptic malignant syndrome: response to dantrolene sodium. Ann. Intern. Med. 98 (1983) 183–184.

DE GANS, J., J. STAM and G.K. VAN WIJNGAARDEN: Rhabdomyolysis and concomitant neurological lesions after intravenous heroin abuse. J. Neurol. Neurosurg. Psychiatr. 48 (1985) 1057–1059.

DE SMET, Y., M. JAMINET, U. JAEGER, J. JACOB, H. NEURAY, G. HAUS, D. LEDESCH-CAMUS and R. MEYERS: Myopathie aigue cortisonique de l'asthmatique. Rev. Neurol. (Paris) 147 (1991) 682–685.

DEBRE, R., C. GERNEZ and G. SEE: Crises myopathiques paroxystiques avec hémoglobinurie. Bull. Mem. Soc. Méd. Hôp. Paris 50 (1934) 1640–1649.

DEMAUGRE, F., J.P. BONNEFONT, G. MITCHELL, N. NGUYEN-HOANG, A. PELET, M. RIMOLDI, S. DI DONATO and J.M. SAUDUBRAY: Hepatic and muscular presentations of carnitine palmitoyltransferase deficiency: two distinct entities. Pediatr. Res. 24 (1988) 308–311.

DEMAUGRE, F., J.P. BONNEFONT, C. CEPANEC, J. SCHOLTE, J.M. SAUDUBRAY and J.P. LEROUX: Immunoquantitative analysis of human carnitine pamitoyltransferase I and II defects. Pediatr. Res. 27(5) (1990) 497–500.

DEMAUGRE, F., J.P. BONNEFONT, M. COLONNA, C. CEPANEC, J.P. LEROUX and J.M. SAUDUBRAY: Infantile form of carnitine palmitoyltransferase II deficiency with hepatomuscular symptoms and sudden death. J. Clin. Invest. 87 (1991) 859–864.

DEMOS, M.A. and E.L. GITLIN: Acute exertional rhabdomyolysis. Arch. Intern. Med. 133 (1974) 233–239.

DEMOS, M.A., E.L. GITLIN and L.J. KAGEN: Exercise myoglobinemia and acute exertional rhabdomyolysis. Arch. Intern. Med. 134 (1974) 669–673.

DENBOROUGH, M.A. and R.R.H. LOVELL: Anesthetic deaths in a family. Lancet ii (1960) 45.

DENBOROUGH, M.A., P. EBELING, J.O. KING and P. ZAPF: Myopathy and malignant hyperpyrexia. Lancet i (1970) 1138–1140.

DENBOROUGH, M.A., S.P. COLLINS and K.C. HOPKINSON: Rhabdomyolysis and malignant hyperpyrexia. Br. Med. J. 288 (1984) 1878.

DENBOROUGH, M.A., K.C. HOPKINSON and D.G. BANNEY: Firefighting and malignant hyperthermia. Br. Med. J. 296 (1988) 1442–1443.

DESNUELLE, C., J.F. PELLISSIER, T. DE BARSY and G. SERRATRICE: Intolérance à l'exercice par déficit en carnitine palmitoyl-transférase. (Exercise intolerance with carnitine palmitoyl transferase deficiency). Rev. Neurol. 146 (1990) 231–234.

DHIB-JAHLBUT, S., R. HESSELBROCK, T. BROTT and D. SILBERGELD: Treatment of the neuroleptic malignant syndrome with bromocriptine. J. Am. Med. Assoc. 250 (1983) 484–485.

DIGBY, L.M., G.B. LEWIS JR., E.G. SCOTT and W. HERBERT: Successful treatment of malignant hyperthermia. Anesth. Analg. (Cleveland) 50 (1971) 39–42.

DIMAURO, S. and N. BRESOLIN: Phosphorylase defi-

ciency. In: A.G. Engel and B.Q. Banker (Eds), Myology. New York, McGraw-Hill Book Co. (1986) 1585–1601.

DIMAURO, S. and P.M.M. DIMAURO: Muscle carnitine palmitoyltransferase deficiency and myoglobinuria. Science 182 (1973) 929–931.

DIMAURO, S. and A. PAPADIMITRIOU: Carnitine palmitoyltransferase deficiency. In: A.G. Engel and B.Q. Banker (Eds), Myology. New York, McGraw-Hill Book Co. (1986) 1697–1708.

DIMAURO, S., M. DALAKAS and A.F. MIRANDA: Phosphoglycerate kinase (PGK) deficiency: a new cause of recurrent myoglobinuria. Ann. Neurol. 10 (1981a) 90.

DIMAURO, S., A.F. MIRANDA, S. KHAN, K. GITLIN and R. FRIEDMAN: Human muscle phosphoglycerate mutase deficiency: a newly discovered metabolic myopathy. Science 212 (1981b) 1277–1279.

DIMAURO, S., A.F. MIRANDA, M. OLARTE, R. FRIEDMAN and A.P. HAYS: Muscle phosphoglycerate mutase deficiency. Neurology 32 (1982) 584–591.

DIMAURO, S., M. DALAKAS and A.F. MIRANDA: Phosphoglycerate kinase deficiency: cause of recurrent myoglobinuria. Ann. Neurol. 13 (1983) 11–19.

DITZIAN-KADANOFF, R., J.D. REINHARD, C. THOMAS and A.S. SEGAL: Polymyositis with rhabdomyolysis in pregnancy: a report and review of the literature. J. Rheumatol. 15 (1988) 513–514.

DIVE, A., J. DONCKIER, D. LEJEUNE and M. BUYSSCHAERT: Hypokalemic rhabdomyolysis in anorexia nervosa. Acta Neurol. Scand. 83 (1991) 419.

DODD, M.J., P. PHATTIYAKUL and S. SILPASUVAN: Suspected malignant hyperthermia in a strabismus patient. Arch. Ophthalmol. 99 (1981) 1247–1250.

DOMINIC, J.A., M. KOCH, G.P. GUTHRIE and J.H. GALLA: Primary aldosteronism presenting as myoglobinuric acute renal failure. Arch. Intern. Med. 138 (1978) 1433–1434.

DONLON, J.V., P. MEWFIELD, F. SRETER and J.F. RYAN: Implications of masseter spasm after succinylcholine. Anaesthesiology 49 (1978) 298–301.

DORIGUZZIE, C., L. PALMUCCI and T. MONGINI: Body building and rhabdomyolysis. J. Neurol. Neurosurg. Psychiatr. 53 (1990) 806–807.

DRUTZ, D.J., J.H. FAN, T.Y. TAI, J.T. CHENG and W.C. HSIEH: Hypokalemic rhabdomyolysis and myoglobinuria following amphotericin B therapy. J. Am. Med. Assoc. 211 (1970) 824–826.

DUNNET, J., J.Y. PATON and C.E. ROBERTSON: Acute renal failure and Coxsackie viral infection. Clin. Nephrol. 16 (1981) 262–263.

EAST, C., P.A. ALIZIVATOS, S.M. GRUNDY, P.H. JONES and J.A. FARMER: Rhabdomyolysis in patients receiving lovastatin after cardiac transplantation. N. Engl. J. Med. 318 (1988) 47–48.

EDWARDS, R.H.T., A. YOUNG and M. WILES: Needle biopsy of skeletal muscle in the diagnosis of myopathy and the clinical study of muscle functions and repair. N. Engl. J. Med. 302 (1980) 261–271.

EISER, A.R., M.S. NEFF and R.F. SLIFKIN: Acute myoglobinuric renal failure. A consequence of the neuroleptic malignant syndrome. Arch. Intern. Med. 142 (1982) 601–603.

ELLIS, F.R. and P.J. HALSALL: Suxamethonium spasm: a differential diagnostic conundrum. Br. J. Anaesth. 56 (1984) 381–383.

ELLIS, F.R., P.J. HALSALL, P. ALLAM and E. HAY: A biochemical abnormality found in muscle from unstressed malignant-hyperpyrexia-susceptible humans. Biochem. Soc. Trans. 12 (1984) 357–358.

ELLIS, K.O. and J.F. CARPENTER: Studies on the mechanism of action of dantrolene sodium, a skeletal muscle relaxant. Naunyn–Schmiedeberg's Arch. Pharmacol. 275 (1972) 83.

ELLIS, R.F., P.A. CAIN and D.G.F. HARRIMAN: Multifactorial inheritance of malignant hyperthermia susceptibility. In: J.A. Aldrete and B.A. Britt (Eds), Proceedings, 2nd International Symposium on Malignant Hyperthermia. New York, Grune and Stratton (1978) 329–338.

ENDO, M.: Role of calcium in control of muscle contraction and muscle metabolism. Presented at the Third International Workshop on Malignant Hyperthermia, Banff, Alberta, Canada (October 1982).

ENEAS, J.F., P.Y. SCHOENFELD and M.H. HUMPHREYS: The effect of infusion of mannitol-sodium bicarbonate on the clinical course of myoglobinuria. Arch. Intern. Med. 139 (1979) 801–805.

ENG, G.D., M.J. BECKER and S.M. MULDOON: Electrodiagnostic tests in the detection of malignant hyperthermia. Muscle Nerve 7 (1984) 618–625.

ENGEL, W.K., N.A. VICK, C.J. GLUECK and R.I. LEVY: A skeletal-muscle disorder associated with intermittent symptoms and a possible defect of lipid metabolism. N. Engl. J. Med. 282 (1970) 697–704.

ENGLAND, A.C. III, D.W. FRASER, A.W. HIGHTOWER, R. TIRINNANZI, D.J. GREENBERG, K.E. POWELL, C.M. SLOVIS and R.A. VARSHA: Preventing severe heat injury in runners: suggestions from the 1979 Peachtree Road Experience. Ann. Intern. Med. 97 (1982) 196–201.

ESSEN, B.: Intramuscular substrate utilization during prolonged exercise. Ann. NY. Acad. Sci. 301 (1977) 30–44.

ESSEN, B.: Glycogen depletion of different types in human skeletal muscle during intermittent and continuous exercise. Acta Physiol. Scand. 103 (1978) 446–455.

FAHLGREN, H., R. HED and C. LUNDMARK: Myonecrosis and myoglobinuria in alcohol and barbiturate intoxication. Acta Med. Scand. 158 (1957) 405–412.

FAVARA, B.E., G.F. VAWTER, R. WAGNER, S. KEVY and E.G. PORTER: Familial paroxysmal rhabdomyolysis in children: a myoglobinuric syndrome. Am. J. Med. 42 (1967) 196–207.

FELIG, P. and J. WAHREN: Fuel homeostasis in exercise. N. Eng. J. Med. 293 (1975) 1078–1084.

FERNANDEZ-SOLA, J., A. CASES, R. MONFORTE, J.C. PEDRO-BOTET, R. ESTRUCH, J.M. GRAU and A. URBANO-MARQUEZ: A possible pathogenic mechanism for rhabdomyolysis associated with multiple myeloma. Acta Haematol. 77 (1987) 231–233.

FINK, W.J., D.L. COSTILL and M.L. POLLOCK: Submaximal and maximal working capacity of elite distance runners. Part II. Muscle fiber composition and enzyme activities. Ann. N.Y. Acad. Sci. 301 (1977) 323–327.

FISCHBECK, K., J. KAMHOLZ and Y.J. SHI: X–linked myoglobinuria. Neurology 38, Suppl. 1 (1988) 174.

FISHBEIN, W.N., S.M. MULDOON, P.A. DEUSTER and V.W. ARMBRUSTMACHER: Myoadenylate deaminase deficiency and malignant hyperthermia susceptibility: is there a relationship? Biochem. Med. 34 (1985) 344–354.

FLEWELLEN, E.H. and T.E. NELSON: Masseter spasm induced by succinylcholine in children: contracture testing for malignant hyperthermia: report of six cases. Can. Anaesth. Soc. 29 (1982) 42–49.

FLEWELLEN, E.H. and T.E. NELSON: Halothane-succinylcholine induced masseter spasm: indicative of malignant hyperthermia susceptibility? Anesth. Analg. (Cleveland) 63 (1984) 693–697.

FOHLMAN, J. and D. EAKER: A lethal myotoxic phospholipase A from sea snake. Toxicon 15 (1977) 385–392.

FORD, F.R.: Diseases of the Nervous System in Infancy, Childhood and Adolescence. Springfield IL, Charles C. Thomas (1966).

FRANK, J.P., Y. HARATI, I.J. BUTLER, T.E. NELSON and C.I. SCOTT: Central core disease and malignant hyperthermia. Ann. Neurol. 7 (1980) 11–17.

FRIDEN, J.M., M. SJOSTROM and B. EKBLOM: A morphological study of delayed muscle soreness. Experientia 37 (1981) 506–507.

FURTADO, M.A. and I.A. LESTER: Myoglobinuria following snakebite. Med. J. Aust. 1 (1968) 674–676.

GABOW, P.A., W.D. KAEHNY and S.P. KELLEHER: The spectrum of rhabdomyolysis. Medicine 61 (1982) 141–152.

GAMBOA, E.T., A.B. EASTWOOD, A.P. HAYS, J. MAXWELL and A.S. PENN: Isolation of influenza virus from muscle in myoglobinuric polymyositis. Neurology 29 (1979) 1323–1335.

GERRARD, J.M., P.G. DUNCAN, S.A. KOSHYK, S.M. GLOVER and J.M. MCCREA: Halothane stimulates the aggregation of platelets of both normal individuals and those susceptible to malignant hyperthermia. Br. J. Anaesth. 55 (1983) 1249–1257.

GIGER, U. and R.F. KAPLAN: Halothane-induced ATP-depletion in platelets from patients susceptible to malignant hyperthermia and from controls. Anesthesiology 58 (1983) 347–352.

GILBOA, N. and J.R. SWANSON: Serum creatine phosphokinase in normal newborns. Arch. Dis. Child. 51 (1976) 283–285.

GITLIN, E.L. and M.A. DEMOS: Acute exertional rhabdomyolysis: a syndrome of increasing importance to the military physician. Mil. Med. 139 (1974) 33–36.

GOLDBERG, W.M. and S.B. CHAKRABARTI: Idiopathic paroxysmal myoglobinuria with transient renal failure. Can. Med. Assoc. J. 94 (1966) 681–683.

GOLLNICK, P.D., K. PIEHL and B. SALTIN: Selective glycogen depletion pattern in human muscle fibres after exercise of varying intensity and at varying pedalling rates. J. Physiol. 241 (1974) 45–57.

GOODE, D.J. and H.Y. MELTZER: Effects of isometric exercise on serum creatine phosphokinase activity. Arch. Gen. Psychiatr. 33 (1976) 120–121.

GOODE, D.J., D.H. WEINBERG, T.A. MAZURA, G. CURTIS, R.J. MURETTI and H.Y. MELTZER: Effect of limb restraints on serum creatine phosphokinase activity in normal volunteers. Biol. Psychiatr. 12 (1977) 743–755.

GORDON, A.S., B.A. BRITT and K.P.H. PRITZKER: Myotonia and malignant hyperpyrexia. Muscle Nerve 5, Suppl. (1986) 221.

GORDON, B.S. and W. NEWMAN: Lower nephron syndrome following prolonged knee-chest position. J. Bone Jt Surg. 35 (1953) 764–768.

GOULON, M., P. DE ROHAN-CHABOT, D. ELKHARRAT, P. GAJDOS, C. BISMUTH and F. CONSO: Beneficial effects of dantrolene in the treatment of neuroleptic malignant syndrome: a report of two cases. Neurology 33 (1983) 516–518.

GRANATO, J.E., B.J. STERN, A. RINGEL, A.H. KARIM, A. KRUMHOLZ, J. COLE and S. ADLER: Neuroleptic malignant syndrome: successful treatment with dantrolene and bromocriptine. Ann. Neurol. 14 (1983) 89–90.

GRONERT, G.A.: Malignant hyperthermia. Anesthesiology 53 (1980) 395–423.

GRONERT, G.A.: Malignant hyperthermia. In: A.G. Engel and B.Q. Banker (Eds), Myology. New York, McGraw-Hill Book Co. (1986) 1763–1784.

GRONERT, G.A., J.J. HEFFRON and S.R. TAYLOR: Skeletal muscle sarcoplasmic reticulum in porcine malignant hyperthermia. Eur. J. Pharmacol. 58 (1979) 179–187.

GRONERT, G.A., W. FOWLER, G.H. CARDINET III, A. GRIX JR., W.G. ELLIS and M.Z. SCHWARTZ: Absence of malignant hyperthermia contractures in Becker-Duchenne dystrophy at age 2. Muscle Nerve 15 (1992) 52–56.

GROSS, E.G., J.D. DEXTER and R.G. ROTH: Hypokalemic myopathy with myoglobinuria associated with licorice ingestion. N. Engl. J. Med. 274 (1966) 602–606.

GROSSMAN, R.A., R.W. HAMILTON, B.M. MORSE, A.S. PENN and M. GOLDBERG: Nontraumatic rhabdomyolysis and acute renal failure. N. Engl. J. Med. 291 (1974) 807–811.

GUNTHER, H.: Kasuistische Mitteilung uber Myositis myoglobinurica. Virchow's Arch. A 251 (1924) 141–149.

HAASE, G.R. and A.G. ENGEL: Paroxysmal recurrent rhabdomyolysis. Arch. Neurol. 2 (1960) 410–419.

HAFTEL, A.J., J. EICHNER and M.L. WILSON: Myoglobinuric renal failure in a newborn infant. J. Pediatr. 93 (1978) 1015–1016.

HALLER, R.G. and D.B. DRACHMAN: Alcoholic rhabdomyolosis: an experimental model in the rat. Science 208 (1980) 412–415.

HALLER, R.G. and J.P. KNOCHEL: Skeletal muscle disease in alcoholism. Med. Clin. North Am. 68 (1984) 91–103.

HALLER, R.G., N.W. CARTER, E. FERGUSON and J.P. KNOCHEL: Serum and muscle potassium in experimental alcoholic myopathy. Neurology 34 (1984) 529–532.

HALLER, R.G., K.G. HENRIKSSON, L. JORFELDT, E. HULTMAN, R. WIBOM, K. SAHLIN, N.-H. ARESKOG, M. GUNDER, K. AYYAD, C.G. BLOMQVIST, R.E. HALL, P. THUILLIER, N.G. KENNAWAY and S.F. LEWIS: Deficiency of skeletal muscle succinate dehydrogenase and aconitase. Pathophysiology of exercise in a novel human muscle oxidative defect. J. Clin. Invest. 88 (1991) 1197–1206.

HALSALL, P.J., P.A. CAIN and F.R. ELLIS: Retrospective analysis of anesthetics received by patients before susceptibility to malignant hyperpyrexia was recognized. Br. J. Anaesth. 51 (1979) 949–954.

HALVERSON, P.B., F. KOZIN, L.M. RYAN and A.R. SULAIMAN: Rhabdomyolysis and renal failure in hypothyroidism. Ann. Intern. Med. 91 (1979) 57–58.

HAMILTON, R.W., M.B. HOPKINS and Z.K. SHIHABI: Myoglobinuria, hemogloburia, and acute renal failure. Clin. Chem. 35 (1989) 1713–1720.

HARRIMAN, D.G.F.: The differentiation of normal muscle and of malignant hyperthermic myopathy, and the association of myopathies with the malignant hyperthermia trait. Muscle Nerve 5, Suppl. (1986) 222.

HARRISON, G.G., S.J. SAUNDERS, J.F. BIEBUYCK, R. HICKMAN, D.M. DENT, V. WEAVER and J. TERBLANCHE: Anaesthetic-induced malignant hyperpyrexia and a method for its prediction. Br. J. Anaesth. 41 (1969) 844–855.

HART, P.M., D.A. FEINFELD, A.M. BRISCOE, H.M. NURSE, J.L. HOTCHKISS and G.E. THOMSON: The effect of renal failure and hemodialysis on serum and urine myoglobin. Clin. Nephrol. 18 (1982) 141–143.

HAVEL, R.J., D.B. HUNNINGHAKE, D.R. ILLINGWORTH, R.S. LEES, E.A. STEIN, J.A. TOBERT, S.R. BACON, J.A. BOLOGNESE, P.H. FROST and G.E. LAMKIN: Lovastatin (Mevinolin) in the treatment of heterozygous familial hypercholesterolemia. A multicenter study. Ann. Intern. Med. 107 (1987) 609–615.

HAVERKORT-POELS, P.J., E.M. JOOSTEN and W. RUITENBEEK: Prevention of recurrent exertional rhabdomyolysis by dantrolene sodium. Muscle Nerve 10 (1987) 45–46.

HEALY, J.M.S., J.J.A. HEFFRON, M. LEHANE, D.G. BRADLEY, K. JOHNSON and T.V. MCCARTHY: Diagnosis of susceptibility to malignant hyperthermia with flanking DNA markers. Br. Med. J. 303 (1991) 1225–1228.

HED, R.: En familjar form av paroxysmal myoglobinurie. Nord. Med. 35 (1947) 1586.

HED, R.: Myoglobinuria in man: with special reference to a familial form. Acta Med. Scand. 151, Suppl. 303 (1955) 1.

HEFFRON, J.J.A. and G.A. GRONERT: Calcium binding

and respiratory activities of skeletal muscle mitochondria of pigs susceptible to the malignant hyperthermia syndrome. Ir. J. Med. Sci. 146 (1977) 87.

HEIDEMANN, H., K.D. BOCK and E. KREUZFELDER: Rhabdomyolysis with acute renal failure due to bezafibrate. Klin. Wochenschr. 59 (1981) 413–414.

HEIMANN-PATTERSON, T.D., H.N. NATTER, H.R. ROSENBERG, J.E. FLETCHER and A.J. TAHMOUSH: Malignant hyperthermia susceptibility in X-linked muscle dystrophies. Pediatr. Neurol. 2 (1986) 356–358.

HEINRICH, W.L., D. PROPHET and J.P. KNOCHEL: Rhabdomyolysis associated with Escherichia coli septicemia. South. Med. J. 73 (1980) 936–937.

HEITZMAN, E.J., J.F. PATTERSON and M.M. STANLEY: Myoglobinuria and hypokalemia in regional enteritis. Arch. Intern. Med. 110 (1962) 117–124.

HELZLSOUER, K.J., F.G. KAYDEN and A.D. ROGOL: Severe metabolic complications in a cross-country runner with sickle cell trait. J. Am. Med. Assoc. 24 (1983) 777–779.

HENDERSON, V.W. and G.F. WOOTEN: Neuroleptic malignant syndrome: a pathogenetic role for dopamine receptor blockade? Neurology 31 (1981) 132–137.

HERMAN, J. and H.L. NADLER: Recurrent myoglobinuria and muscle carnitine palmitoyltransferase deficiency. J. Pediatr. 91 (1977) 247–250.

HEYLAND, D. and M. SAUVE: Neuroleptic malignant syndrome without the use of neuroleptics. Can. Med. Assoc. J. 145 (1991) 817–819.

HIKIDA, R.S., R.S. STARON, F.C. HAGERMAN, W.M. SHERMAN and D.L. COSTILL: Muscle fiber necrosis associated with human marathon runners. J. Neurol. Sci. 59 (1983) 185–203.

HINDS, S.W. JR. and J.M. ROWE: Extreme rhabdomyolysis in a patient with acute leukemia: association with Candida kruseii fungemia. N.Y. State J. Med. 88 (1988) 599–600.

HINZ, C.F. JR., W.R. DRUKER and J. LARNER: Idiopathic myoglobinuria; metabolic and enzymatic studies on three patients. Am. J. Med. 39 (1965) 49–57.

HISCHE, E.A.H. and H.J. VAN DER HELM: The significance of the estimation of serum myoglobin in neuromuscular diseases. J. Neurol. Sci. 43 (1979) 243–251.

HO, K.J. and K.T. SKULLY: Acute rhabdomyolysis and renal failure in Weil's disease. Ala. J. Med. Sci. 17 (1980) 133–137.

HOFFMAN, E.P., L.M. KUNKEL, C ANGELINI, A. CLARKE, M. JOHNSON and J.B. HARRIS: Improved diagnosis of Becker muscular dystrophy by dystrophin testing. Neurology 39 (1989) 1011–1017.

HOH, T.K., C.L. SOONG and C.T. CHENG: Fatal hemolysis from wasp and hornet sting. Singapore Med. J. 7 (1966) 122–126.

HOJGAARD, A.D., P.T. ANDERSON and J. MOLLER-PETERSEN: Rhabdomyolysis and acute renal failure following an overdose of doxepine and nitrazepam. Acta Med. Scand. 23 (1988) 79–82.

HRONCICH, M.E. and A.N. RUDINGER: Rhabdomyolysis

with pneumococcal pneumonia: a report of two cases. Am. J. Med. 86 (1989) 467–468.

HUBER, S.A. and L.P. JOB: Cellular immune mechanisms in Coxsackievirus Group B, Type 3 induced myocarditis in Balb/C mice. In: J.J. Spitzer (Ed.), Myocardial Injury. New York, Plenum (1983) 491–508.

HURLEY, J.K.: Severe rhabdomyolysis in well conditioned athletes. Mil. Med. 154 (1989) 244–245.

HYSER, C.L., P.R.H. CLARKE, S. DIMAURO and J.R. MENDELL: Myoadenylate deaminase deficiency and exertional myoglobinuria. Neurology 39, Suppl. 1 (1989) 335 (abstract).

INESI, G. and S. WATANABE: Temperature dependence of ATP hydrolysis and calcium uptake by fragmented sarcoplasmic membranes. Arch. Biochem. Biophys. 121 (1967) 665–671.

INOUE, R., M. KAWAMATA, Y. YAMAMURA and M. FUJITA: Generalized muscular rigidity associated with increased serum enzymes and post-operative muscular weakness induced by general anesthesia without hyperthermia. Hiroshima J. Anesth. 13 (1977) 232–235.

ISAACS, H., M.E. BADENHORST and C. DU SAUTOY: Myophosphorylase B deficiency and malignant hyperthermia. Muscle Nerve 12 (1989) 203–205.

JAGO, R.H. and M.J. PAYNE: Malignant hyperpyrexia: the difficulty of diagnosis. Anesthesiology 30 (1977) 30–75.

JANKOVIC, J.: Myoglobinuric renal failure in Huntington's chorea. Neurology 36 (1986) 138–139.

JANKOVIC, J. and A.S. PENN: Severe dystonia and myoglobinuria. Neurology 32 (1982) 1195–1197.

JARDON, O.M.: Physiologic stress, heat stroke, malignant hyperthermia: perspective. Mil. Med. 147 (1982) 8–14.

JAVID, J., H.I. HOROWITZ, A.R. SANDERS and T.H. SPAET: Idiopathic paroxsmal myoglobinuria. Arch. Intern. Med. 104 (1959) 628.

JEHN, U.W. and M.K. FINK: Myositis, myoglobinuria associated with enterovirus Echo 9 infection. Arch. Neurol. 37 (1980) 457–458.

JOHNSON, P.N. and G. INESI: The effect of methylxanthines and local anaesthetics on fragmented sarcoplasmic reticulum. J. Pharmacol. Exp. Ther. 169 (1969) 308–314.

JOHNSON, R.E., R. POSSMORE and F. SARGENT II: Multiple factors in experimental human ketosis. Arch. Intern. Med. 107 (1961) 432.

KAGEN, L.J.: Myoglobinemia and myoglobinuria in patients with myositis. Arthritis Rheum. 14 (1971) 457–464.

KAGEN, L.J., S. SCHEIDT, L. ROBERTS, A. PORTER and H. PAUL: Myoglobinemia following acute myocardial infarction Am. J. Med. 58 (1975) 177–182.

KAISER, A.B., D. RIEVES, A.H. PRICE, M.R. GELFAND, R.E. PARISH, M.D. DECKER and M.E. EVANS: Tularemia and rhabdomyolysis. J. Am. Med. Assoc. 253 (1985) 241–243.

KALISH, S.B., M.S. TALLMAN, F.V. COOK and E.A.

BLUMEN: Polymicrobial septicemia associated with rhabdomyolysis, myoglobinuria and acute renal failure. Arch. Intern. Med. 142 (1982) 133–134.

KALOW, W.: Inheritance of malignant hyperthermia. In: B.A. Britt (Ed.), Malignant Hyperthermia. Boston, Amsterdam, Martinus-Nijhoff (1987) 155–179.

KALOW, W., B.A. BRITT and F.Y. CHAN: Epidemiology and inheritance of malignant hyperthermia. In: B.A. Britt (Ed.), Malignant Hyperthermia. Int. Anaesthesiol. Clin. 17 (1979) 119–139.

KANE, M.J., L.R. SILVERMAN, J.H. RAND, P.A. PACIUCCI and J.F. HOLLAND: Myonecrosis as a complication of the use of epsilon amino-caproic acid: a case report and review of literature. Am. J. Med. 85 (1988) 861–863.

KANNO, T., K. SUDO, I. TAKEUCHI, S. KANDA, N. HONDA, Y. NICHIMURA and K. OYAMA: Hereditary deficiency of lactate dehydrogenase M-subunit. Clin. Chim. Acta 108 (1980) 267–276.

KANTOR, R.J., C.W. NORDEN and T.P. WEIN: Infectious mononucleosis associated with rhabdomyolysis and renal failure. South. Med. J. 71 (1978) 346–348.

KAPLAN, A.M., P.S. BEREGSON, S.A. GREGG and R.G. CURLESS: Malignant hyperthermia associated with myopathy and normal muscle enzymes. J. Pediatr. 91 (1977) 431–434.

KARPATI, G. and G.V. WATTERS: Adverse anesthetic reactions in Duchenne dystrophy. In: C. Angelini, G.A. Danieli and D. Fontanari (Eds), Muscular Dystrophy Research Advances and New Trends. International Congress Series No. 527. Amsterdam, Excerpta Medica (1980) 206–217.

KARPATI, G., J. CHARUK, S. CARPENTER, C. JABLECKI and P. HOLLAND: Myopathy caused by a deficiency of Ca2+-adenosine triphosphatase in sarcoplasmic reticulum (Brody's disease). Ann. Neurol. 20 (1986) 38–49.

KATHEREIN, H., W. KIRCHMAIR, P.V. KONIG and P. VON DITTRICH: Rhabdomyolyse mit akutem Nierenversagen nach Heroin-Intoxikation. Dtsch. Med. Wochenschr. 108 (1983) 464–467.

KAUSCH, K., T. GRIMM, M. JANKA, F. LEHMANN-HORN, B. WIERINGA and C.R. MULLER: Evidence for linkage of the central core disease locus to chromosome 19q. J. Neurol. Sci. 98 (1990) 549.

KELSTRUP, J., J. HAASE, J. JØRNI, E. RESKE-NIELSEN and H.K. HANEL: Malignant hyperthermia in a family. Acta Anaesthesiol. Scand. 17 (1973) 283–284.

KENDRIK, W.C., A.R. HULL and J.P. KNOCHEL: Rhabdomyolysis and shock after intravenous amphetamine administration. Ann. Intern. Med. 86 (1977) 381–387.

KENNARD, C., M. SWASH and R.A. HENSON: Myopathy due to epsilon amino-caproic acid. Muscle Nerve 3 (1980) 202–206.

KESSLER, E., I. WEINBERGER and J.B. ROSENFELD: Myoglobinuric acute renal failure in a case of dermatomyositis. Isr. J. Med. Sci. 8 (1972) 978–983.

KIESSLING, W.R. and R. BECKMANN: Duchenne muscular dystrophy: does serum myoglobin correlate with serum creatine kinase? Muscle Nerve 4 (1981) 257.

KIESSLING, W.R., K. RICKER, K.W. PFLUGHAUPT, H.G. MERTENS and I. HAUBITZ: Serum myoglobin in primary and secondary skeletal muscle disorders. J. Neurol. 224 (1981) 229–233.

KIKUCHI, H., FUKUDA, M. MORIA and K. DOHI: Human leucocyte antigen (HLA) family study and MH. Hiroshima J. Anaesth. 19 (supplement) (1983) 41–45.

KING, J.O. and M.A. DENBOROUGH: Anesthetic-induced malignant hyperpyrexia in children. J. Pediatr. 83 (1973) 37–40.

KING, J.O., M.A. DENBOROUGH and P.W. ZAPF: Inheritance of malignant hyperpyrexia. Lancet i (1972) 365–370.

KISSEL, J.T., W. BREAM, N. BRESOLIN, G. GIBBONS, S. DIMAURO and J. MENDELL: Physiologic assessment of phosphoglycerate mutase deficiency: incremental exercise tests. Neurology 35 (1985) 828–833.

KNOCHEL, J.P.: Environmental heat illness. An eclectic review. Arch. Intern. Med. 133 (1974) 841–864.

KNOCHEL, J.P.: Renal injury in muscle disease. In: W.N. Suki and G. Eknoyan (Eds), The Kidney in Systemic Disease. New York, John Wiley and Sons Inc. (1976) 129–140.

KNOCHEL, J.P.: Rhabdomyolysis and effects of potassium deficiency on muscle structure and function. Cardiovasc. Med. 3 (1978) 247–261.

KNOCHEL, J.P.: Rhabdomyolysis and myoglobinuria. Annu. Rev. Med. 33 (1982) 435–443.

KNOCHEL, J.P. and E.M. SCHLEIN: On the mechanism of rhabdomyolysis in potassium depletion. J. Clin. Invest. 51 (1972) 1750–1758.

KNOCHEL, J.P., G.L. BILBREY, T.J. FULLER and N.W. CARTER: The muscle cell in chronic alcoholism: the possible role of phosphate depletion in alcoholic myopathy. Ann. N.Y. Acad. Sci. 252 (1975) 274–286.

KNOCHEL, J.P., C. BARCENAS, J.R. COTTON JR., T.J. FULLER, R.G. HALLER and N.W. CARTER: Hypophosphatemia and rhabdomyolysis. J. Clin. Invest. 62 (1978) 1240–1246.

KOFFLER, A., R.M. FRIEDLER and S.G. MASSRY: Acute renal failure due to nontraumatic rhabdomyolysis. Ann. Intern. Med. 85 (1976) 23–28.

KOLB, M.E., M.L. HORNE and R. MARTZ: Dantrolene in human malignant hyperthermia. Anesthesiology 56 (1982) 254–262.

KOPPEL, C.: Clinical features, pathogenesis, and management of drug-induced rhabdomyolysis. Med. Toxicol. Adverse Drug Exper. 4 (1989) 108–126.

KOPPEL, C., K. IBE and U. OBERDISSE: Rhabdomyolysis in doxylamine overdose. Lancet i (1987a) 442–443.

KOPPEL, C., J. TENCZER and K. IBE: Poisoning with over-the-counter doxylamine preparations: an evaluation of 109 cases. Hum. Toxicol. 6 (1987b) 355–359.

KOREIN, J., D.R. CODDON and F.H. MOWREY: The clinical syndrome of paroxysmal paralytic myoglobinuria. Report of 2 cases and an analytical review of the literature. Neurology 9 (1959) 767–785.

KORZ, R. and U. VOLZ-BOERS: Akutes Nierenversagen

nach massiver Myolyse bei pseudomyopathischer Polymyositis. Dtsch. Med. Wochenschr. 99 (1974) 1084–1087.

KOSTEN, T.R. and H.D. KLEBER: Sudden death in cocaine abusers: relation to neuroleptic malignant syndrome. Lancet i (1987) 1198–1199.

KREITZER, S.M., M. EHRENPREIS, E. MIGUEL and J. PETRASEK: Acute myoglobinuric renal failure in polymyositis. N.Y. State J. Med. 78 (1978) 295–297.

KREUTZER, F.L., L. STRAIT and W.J. KERR: Spontaneous myohaemoglobinuria in man. Arch. Intern. Med. 81 (1948) 249.

KRIPKE, B.J., T.J.J. BLANCK, D.A. SIZEMORE, F.L. COMUNALE, J. CHRISTIANSEN and R. GRUENER: Association of post-anesthetic hyperthermia with abnormal muscle characteristics: a case report. Can. Anaesth. Soc. J. 30 (1983) 290–294.

KRISOVIC-HORBER, R., P. ADNET, E. GUEVART, D. THEUNYNCK and P. LESTAVEL: Neuroleptic malignant syndrome and malignant hyperthermia. Br. J. Anesth. 59 (1987) 1554–1556.

KROHN, K.D., S. SLOWMAN-KOVACS and S.B. LEAPMANN: Cocaine and rhabdomyolysis. Ann. Intern. Med. 108 (1988) 639–640.

KURODA, M., K. KATSUKI, H. UEHARA, T. KITA, S. ASAKA, R. MIYAZAKI, T. AKIYAMA, Y. TOFUKU and R. TAKEDA: Successful treatment of fulminating complications associated with extensive rhabdomyolysis by plasma exchange. Artif. Organs 5 (1981) 372–378.

LAFAIR, J.S. and R.M. MYERSON: Alcoholic myopathy. With special reference to the significance of creatine phosphokinase. Arch. Intern. Med. 122 (1968) 417–422.

LANGER, T. and R. LEVY: Acute muscular syndrome associated with administration of clofibrate. N. Engl. J. Med. 279 (1968) 856–858.

LANGSTON, J.W., D.R. RICCI and C. PORTLOCK: Nonhypoxemic hazards of prolonged myoclonus. Neurology 27 (1977) 542–545.

LARCAN, A., H. LAMBERT, M.C. LABPREVOTE-HEULLY and M. VERDAGUER: Rhabdomyolyses post-opératoires de cause posturale. A propos de deux cas. Ann. Chir. (Paris) 34 (1980) 769–774.

LARSSON, L.E., H. LINDERHOLM, R. MULLER, T. RINGQVIST and R. SORNAS: Hereditary metabolic myopathy with paroxysmal myoglobinuria due to abnormal glycolysis. J. Neurol. Neurosurg. Psychiatr. 27 (1964) 361–380.

LAYZER, R.B., L.P. ROWLAND and H.M. RANNEY: Muscle phosphofructokinase deficiency. Arch. Neurol. 17 (1967) 512–523.

LAYZER, R.B., R.J. HAVEL and M.B. MCILROY: Partial deficiency of carnitine palmitoyltransferase: physiologic and biochemical consequences. Neurology 30 (1980) 627–633.

LAZARUS, A.L. and J.U. TOGLIA: Fatal myoglobinuric renal failure in a patient with tardive dyskinesia. Neurology 35 (1985) 1055–1057.

LEE, M.B., M.G. ADRAGNA and L. EDWARDS: The use of a platelet nucleotide assay as a possible diagnostic test for malignant hyperthermia. Anesthesiology 63 (1985) 311–315.

LEHMANN-HORN, F. and P.A. IAIZZO: Are myotonias and periodic paralyses associated with susceptibility to malignant hyperthermia? Br. J. Anaesth. 65 (1990) 692–697.

LENNMARKEN, C., H. RUTBERG and K.G. HENRIKSSON: Abnormal relaxation rates in subjects susceptible to malignant hyperthermia. Acta Neurol. Scand. 75 (1987) 81–83.

LEVITT, R.C., N. NOURI, A.E. JEDLICKA, V.A. MCKUSICK, A.R. MARKS, J.G. SHUTACK, J.E. FLETCHER, H. ROSENBERG and D.A. MEYERS: Evidence for genetic heterogeneity in malignant hyperthermia susceptibility. Genomics 11 (1991) 543–547.

LEWANDOWSKI, K.B.: Rhabdomyolysis, myoglobinuria and hyperpyrexia caused by suxamethonium in a child with increased serum CK concentrations. Br. J. Anaesth. 53 (1981) 981–984.

LIETMAN, P.S., R.H.A. HASLAM and J.R. WALCHER: Pharmacology of dantrolene sodium in children. Arch. Phys. Med. Rehabil. 55 (1974) 388–392.

LITHELL, H., J. ORLANDER, R. SCHELE, B. SJODIN and J. KARLSSON: Changes in lipoprotein-lipase activity and lipid stores in human skeletal muscle with prolonged heavy exercise. Acta Physiol. Scand. 107 (1979) 257–261.

LITIN, S.C. and C.F. ANDERSON: Nicotinic acid-associated myopathy: a report of three cases. Am. J. Med. 86 (1989) 481–483.

LOTSTRA, F., P. LINKOWSKI and J. MENDLEWICZ: General anesthesia after neuroleptic malignant syndrome. Biol. Psychiatr. 246 (1983) 243–247.

LOUW, A. and H.E. NIELSON: Paroxysmal paralytic hemoglobinuria. Acta Med. Scand. 117 (1944) 424–436.

LUTSKY, I., J. WITKOWSKI and E.O. HENSCHEL: HLA typing in a family prone to malignant hyperthermia. Anesthesiology 56 (1982) 224–226.

LYDIATT, J.S. and G.E. HILL: Treatment of heat stroke with dantrolene. J. Am. Med. Assoc. 246 (1981) 41–42.

LYNN, K.L.: Acute rhabdomyolysis and acute renal failure after intravenous self administration of peanut oil. Br. Med. J. 2 (1975) 385–386.

LYONS, R.H.: Myoglobinuria. N.Y. State J. Med. 63 (1963) 2512–2517.

MACDONALD, J.B., H.M. JONES and R.A. COWAN: Rhabdomyolysis and acute renal failure after theophylline overdose. Lancet i (1985) 932–933.

MACKENZIE, A.E., G. ALLEN, D. LAHEY, M.L. CROSSAN, K. NOLAN, G. METTLER, R.G. WORTON, D.H. MACLENNAN and R. KORNELUK: A comparison of the caffeine halothane muscle contracture test with the molecular genetic diagnosis of malignant hyperthermia. Anesthesiology 75 (1991) 4–8.

MACLENNAN, D.H., C. DUFF, F. ZORZATO, J. FUJII, M. PHILLIPS, R.G. KORNELUK, W. FRODIS, B.A. BRITT and R.G. WORTON: Ryanodine receptor gene: is a candidate gene for predisposition to malignant hyperthermia. Nature (London) 343 (1990) 559–561.

MAK, I.T., J.H. KRAMER and W.B. WEGLICKI: Potentiation of free radial-induced lipid peroxidative injury to sarcolemmal membranes by lipid amphiphiles. J. Biol. Chem. 26 (1986) 1153–1157.

MALAMUD, N., W, HAYMAKER and R.P. CUSTER: Heatstroke: a clinico-pathologic study of 125 fatal cases. Mil. Surg. 99 (1946) 397–449.

MARJANEN, L.A. and M.A. DENBOROUGH: Adenylate kinase and malignant hyperpyrexia. Br. J. Anaesth. 54 (1982) 949–952.

MARKS, S.H., D.J. MCSHANE and D.M. MITCHELL: Dermatomyositis following rhabdomyolysis. J. Rheumatol. 3 (1976) 224–226.

MARTIN, J.B., J.W. CRAIG, R.E. ECKEL and J. MUNGER: Hypokalemic myopathy in chronic alcoholism. Neurology 21 (1971) 1160–1168.

MASON, J. and E. THOMAS: Rhabdomyolysis from heat hyperpyrexia. Severe hypocalcemia and hypophosphatemia as complicating factors. J. Am. Med. Assoc. 235 (1976) 633–634.

MAXWELL, J.H. and C.M. BLOOR: Effects of conditioning on exertional rhabdomyolysis and serum creatine kinase after severe exercise. Enzyme 26 (1981) 177–181.

MAZZIA, V.D.B. and A. SIMON: Medicolegal implications of malignant hyperthermia. In: J.A. Aldrete and B.A. Britt (Eds), Second International Symposium on Malignant Hyperthermia. New York, Grune and Stratton (1977) 545–551.

MCARDLE, B.: Myopathy due to defect in muscle glycogen breakdown. Clin. Sci 10 (1951) 13–35.

MCCARRON, D.A., K.A. ROYER, D.C. HOUGHTON and W.M. BENNETT: Chronic tubulointerstitial nephritis caused by recurrent myoglobinuria. Arch. Intern. Med. 140 (1980) 1106–1107.

MCCARTHY, T.V., S.J.M. HEALY, J.J.A. HEFFRON, M. LEHANE, T. DEUFEL, F. LEHMANN-HORN, M. FARRAL and K. JOHNSON: Localization of the malignant hyperthermia susceptibility locus to human chromosome 19q 12-13.2 Nature (London) 343 (1990a) 562–564.

MCCARTHY, T.V., J.M. HEALY, M. LEHANE and J.J. HEFFRON: Recent developments in the molecular genetics of malignant hyperthermia: implications for future diagnosis at the DNA level. Acta Anesthesiol. Belg. 41 (1990b) 107–112.

MCKENNA, U.G., J.A. MEADOWS III, N.S. BREWSER, W.R. WILSON and J. PERRAULT: Toxic shock syndrome, a newly recognized disease entity. Roport of II cases. Mayo Clin. Proc. 55 (1980) 663–672.

MCKISHNIE, J.D., J.M. MUIR and D.P. GIRVAN: Anaesthesia-induced rhabdomyolysis. Can. Anaesth. Soc. J. 30 (1983) 295–298.

MCPHERSON, E.W. and C.A. TAYLOR JR.: The King syndrome: malignant hyperthermia, myopathy and multiple anomalies. Am. J. Med. Genet. 8 (1981) 159–165.

MCPHERSON, E.W. and C.A. TAYLOR JR.: The genetics of malignant hyperthermia: evidence for heterogeneity. Am. J. Med. Genet. 11 (1982) 273–285.

MEBS, D. and Y. SAMEJIMA: Myotoxic phospholipases from snake venom, Pseudechis colletti, producing myoglobinuria in mice. Experientia 36 (1980) 868–869.

MEDORI, R., M.H. BROOKE and R. WATERSTON: Two dissimilar brothers with Becker's dystrophy have an identical genetic defect. Neurology 39 (1989) 1493–1496.

MENDOZA, F.S., J.O. ATIBA, A.M. KRENSKY and L.M. SCANNELL: Rhabdomyolysis complicating doxylamine overdose. Clin. Pediatr. 26 (1987) 595–597.

MEOLA, G., A. TOSCANO, M. VELICOGNA, S. SERTORELLI, N. BRESOLIN, M. MOGGIO, S. JANN, G. VITA, F. FORTUNATO and G. SCARLATO: Muscle phosphoglycerate mutase (PGAM) deficiency in aneural and innervated cultures. Neurology 39, Suppl. 1 (1989) 233 (Abstract).

MERCIECA, J. and E.A. BROWN: Acute renal failure due to rhabdomyolysis associated with use of a straitjacket in lysergide intoxication. Br. Med. J. 288 (1984) 1949–1950.

MERIGAN, K.S. and J.R. ROBERTS: Cocaine intoxication: hyperpyrexia, rhabdomyolysis and acute renal failure. J. Toxicol. Clin. Toxicol. 25 (1987) 135–148.

MEYER-BETZ, F.: Beobachtungen an einem eigenartigen mit Muskellähmungen verbundenen Fall von Hämoglobinuria. Dtsch. Arch. Klin. Med. 101 (1910) 85.

MITCHELL, A.B.S.: Duogastrone-induced hypokalaemic nephropathy and myopathy with myoglobinuria. Postgrad. Med. J. 47 (1971) 807–813.

MIYOSHI, K., S. SAITO, H. KAWAI, A. KONDO, M. IWASA, T. HAYASHI and M. YAGITA: Radioimmunoassay for human myoglobin: methods and results in patients with skeletal muscle or myocardial disorders. J. Lab. Clin. Med. 92 (1978) 341–352.

MODI, K.B., E.H. HORN and S.M. BRYSON: Theophylline poisoning and rhabdomyolysis. Lancet ii (1985) 160–161.

MOHAMED, S.D., R.S. CHAPMAN and J. CROOKS: Hypokalemia, flaccid quadruparesis, and myoglobinuria with carbenoxolone (biogastrone). Br. Med. J. 1 (1966) 1581–1582.

MUBARAK, S. and C.A. OWEN: Compartmental syndrome and its relation to the crush syndrome: a spectrum of disease. A review of 11 cases of prolonged limb compression. Clin. Orthop. 113 (1975) 81–89.

MUELLER, P.S., J. VESTER and J. FERMAGLICH: Neuroleptic malignant syndrome. Successful treatment with bromocriptine. J. Am. Med. Assoc. 249 (1983) 386–388.

NADEL, S.M., J.W. JACKSON and D.W. PLOTH: Hypokalemic rhabdomyolysis and acute renal failure. Occurrence following total parenteral nutrition. J. Am. Med. Assoc. 241 (1979) 2294–2296.

NAHAS, A.M.E., K. FARRINGTON, S. QUYYUMI, J.F. MOORHEAD and P. SWEENEY: Rhabdomyolysis and systemic infection. Br. Med. J. 1 (1983) 349–350.

NAKAZATO, A., H. SHIME, K. MOROOKA, K. NONAKA and

A. TAKOGI: Anesthesia-induced rhabdomyolysis in a patient with Fukuyama-type congenital muscular dystrophy. Brain Dev. 5 (1983) 243.

NANJI, A.A. and J.D. FILIPENKO: Rhabdomyolysis and acute myoglobinuric renal failure associated with methadone intoxication. J. Toxicol. Clin Toxicol. 20 (1983) 353–360.

NANKIVELL, B.J. and A.H.B. GILLIES: Acute pancreatitis and rhabdomyolysis: a new association. Aust. N.Z. J. Med. 21 (1991) 414–417.

NELSON, T.E.: Diagnostic tests for malignant hyperthermia. Presented, 3rd International Workshop on Malignant Hyperthermia, Banff, Alberta, Canada (October 1982). N.Z. Med. J. 89 (1982) 159–161.

NELSON, T.E. and E.H. FLEWELLEN: The malignant hyperthermia syndrome. N. Engl. J. Med. 209 (1983) 416–418.

NELSON, T.E., E.H. FLEWELLEN and D.F. GLOYNA: Spectrum of susceptibility to malignant hyperthermia: diagnostic dilemma. Anesth. Analg. (Cleveland) 62 (1983) 545–552.

NICHOLSON, L.V.B.: Serum myoglobin in muscular dystrophy and carrier detection. J. Neurol. Sci. 51 (1981) 411–426.

NOBLE, N.A., L.H. KUWASHIMA and K.R. TANAKA: Evidence for an unstable muscle subunit in erythrocyte phosphofructokinase deficiency. Clin. Res. 28 (1980) 548a.

NORMAN, D.J., D.R. ILLINGWORTH, J. MUNSON and J. HOSENPUD: Myolysis and acute renal failure in a heart- transplant recipient receiving lovastatin. N. Engl. J. Med. 318 (1988) 46–47.

ODA, S., T. ODA and K.R. TANAKA: Erythrocyte phosphofructokinase (PFK) deficiency: characterization and metabolic studies. Clin. Res. 25 (1977) 344A (Abstract).

ODEH, M.: The role of reperfusion-induced injury in the pathogenesis of the crush syndrome. N. Engl. J. Med. 324 (1991) 1417–1422.

O'DONNELL, T.F. JR.: Acute heat stroke. Epidemiologic, biochemical, renal, and coagulation studies. J. Am. Med. Assoc. 234 (1975) 824–828.

OGASAHARA, S., A.G. ENGEL, D. FRENS and D. MACK: Muscle coenzyme Q deficiency in familial mitochondrial encephalomyopathy. Proc. Natl Acad. Sci. USA 86 (1989) 2379–2382.

OHNO, K., M. TANAKA, K. SAHASHI, T. IBI, W. SATO, T. YAMAMOTO, A. TAKAHASHI and T. OZAWA: Mitochondrial DNA deletions in inherited recurrent myoglobinuria. Ann. Neurol. 29 (1991) 364–369.

OLIVERO, J. and J.C. AYUS: Rhabdomyolysis and acute myoglobinuric renal failure. Complications of inferior vena cava ligation. Arch. Intern. Med. 138 (1978) 1548–1549.

ONO, K., D.G. TOPEL and T.G. ALTHEN: Adenylate cyclase and cyclic 3′,5′-nucleotide phosphodiesterase activities in muscles from stress-susceptible and control pigs. J. Food Sci. 42 (1977a) 111–112.

ONO, K., D.G. TOPEL, L.L. CHRISTIAN and T.G. ALTHEN: Relationship of cyclic-AMP and phosphorylase A

in stress-susceptible and control pigs. J. Food Sci. 42 (1977b) 108–110.

ØRDING, H.: The European MH group protocol for in vitro diagnosis of susceptibility to MH and preliminary results. In: B.A. Britt (Ed.), Malignant Hyperthermia. Boston, Amsterdam, Martinus-Nijhoff (1987) 267–277.

OWEN, C.A., S.J. MUBARAK, A.R. HARGENS, L. RUTHERFORD, L.P. GARETTO and W.H. AKESON: Intramuscular pressures with limb compression. Clarification of the pathogenesis of the drug-induced muscle–compartment syndrome. N. Engl. J. Med. 300 (1979) 1169–1172.

OWEN, G. and R.J. KERRY: Anesthesia during raised creatine phosphokinase activity. Br. Med. J. 4 (1974) 75–76.

OWNBY, C.L. and G.V. ODELL: Pathogenesis of skeletal muscle necrosis induced by tarantula venom. Exp. Mol. Pathol. 38 (1983) 283–296.

OWNBY, C.L., D. CAMERON and A.T. TU: Isolation of myotoxic component from rattlesnake (Crotalus viridis) venom. Electron microscopic analysis of muscle damage. Am. J. Pathol. 85 (1976) 149–156.

OZSOYLU, S. and S. AGKUN: Idiopathic paroxysmal myoglobinuria. Turk. J. Pediatr. 8 (1966) 99–108.

PARIENTE, E.Z., O. NOUEL, J. BERNUAU, F. FRAISSE, C. DEGOTT and B. RUEFF: Rhabdomyolyse aigue chez des malades alcoholiques. Presse Med. 12 (1983) 339–343.

PATEL, R. and G. CONNOR: A review of thirty cases of rhabdomyolysis-associated acute renal failure among phencyclidine users. J. Toxicol. Clin. Toxicol. 23 (1985) 547–556.

PAUL, F.: Parolytic hemoglobinuria. Wien. Z. Inn. Med. Ihre Grenzgeb. 7 (1924) 531–554.

PENN, A.S.: Myoglobin and myoglobinuria. In: P.J. Vinken and G.W. Bruyn (Eds), Handbook of Clinical Neurology, Vol. 41. Amsterdam, North-Holland Publishing Co. (1980) 259–285.

PENN, A.S.: Myoglobinuria. In: A.G. Engel en B.Q. Banker (Eds), Myology. New York, McGraw-Hill Book Co. (1986) 1785–1805.

PENN, A.S., L.P. ROWLAND and D.W. FRASER: Drugs, coma and myoglobinuria. Arch. Neurol. 26 (1972) 336–343.

PERKOFF, G.T., P. HARDY and E. VELEZ-GARCIA: Reversible acute muscular syndrome in chronic alcoholism. N. Engl. J. Med. 274 (1966) 1277–1285.

PIERCE, L.R., D.K. WYSOWSKI and T.P. GROSS: Myopathy and rhabdomyolysis associated with lovastatin-gemfibrozil combination therapy. J. Am. Med. Assoc. 264 (1990) 71–75.

PIROVINO, M., M.S. NEFF and E. SHARON: Myoglobinuria and acute renal failure with acute polymyositis. N.Y. State J. Med. 79 (1979) 764–767.

PLOTZ, J.: Maligne Hyperthermie. II. Befragung und Untersuchung von Familienmitgliedern einer Erkrankten. Anaesthetist 29 (1980) 94–98.

POELS, P.J.E., E.M.G. JOOSTEN, R.C.A. SENGERS, A.M. STADHOUDERS, J.H. VEERKAMP and A.A.G.M. BENDERS: In vitro contraction test for malignant hyperth-

ermia in patients with unexplained recurrent rhabdomyolysis. J. Neurol. Sci. 105 (1991) 67–72.

POSNER, M.R., M.A. CAUDILL, R. BRASS and E. ELLIS: Legionnaires' disease associated with rhabdomyolysis
and myoglobinuria. Arch. Intern. Med. 140 (1980)
848–850.

PUSCHEL, K., I. SCHUBERT-THIELE, L. HIRTH, H.G. BENK-
MANN and B. BRINKMANN: Maligne Hyperthermie in
der 13. Vollnarkose. Anaesthetist 27 (1978) 448–
491.

QUINLAN, J.G., P.A. IAIZZO, E.H. LAMBERT and G.A.
GRONERT: Ankle dorsiflexor twitch properties in malignant hyperthermia. Muscle Nerve 12 (1989) 119–
125.

RAIFMAN, M.A., M. BERANT and M. LENARSKY: Cold
weather and rhabdomyolysis. J. Pediatr. 93 (1978)
970–971.

REAVEN, P. and J.L. WITZTUM: Lovastatin, nicotinic
acid, and rhabdomyolysis. Ann. Intern. Med. 109
(1988) 597–598.

REICHLIN, M., J.P. VISCO and F.J. KLOCKE: Radioimmunoassay for human myoglobin: experiences in patients with coronary heart disease. Circulation 57
(1978) 52–56.

REINER, L., N. KONIKOFF, M.D. ALTSCHULE, G.J. DAMMIN
and J.P. MERRILL: Idiopathic paroxysmal myoglobinuria. Arch. Intern. Med. 97 (1956) 537–550.

REINHART, W.H. and H. STRICKER: Rhabdomyolysis
after intravenous cocaine. Am. J. Med. 85 (1988)
597.

REZA, M.J., N.C. KAR, C.M. PEARSON and R.A. KARK: Recurrent myoglobinuria due to muscle carnitine
palmitoyltransferase deficiency. Ann. Intern. Med.
88 (1978) 610–615.

RICHTER, R.W., Y.B. CHALLENOR, J. PEARSON, L.J.
KLAGEN and L.L. HAMILTON: Acute myoglobinuria
associated with heroin addiction. J. Am. Med.
Assoc. 216 (1971) 1172–1176.

RIMON, D., R. LUDATSCHER and L. COHEN: Clofibrate-
induced muscular syndrome. Isr. J. Med. Sci. 20
(1984) 1082–1086.

ROHLING, T., E. HEEKING and T. WEIBRAUCH: Heroin-
induzierte Rhabdomyolyse mit akutem Nierenversagen. Med. Klin. 79 (1984) 616–618.

ROSA, R., C. GEORGE, M. FARDEAU, M.C. CALVIN, M.
RAPIN and J. ROSA: A new cause of phosphoglycerate
kinase deficiency: PGK Creteil associated with
rhabdomyolysis and lacking hemolytic anemia.
Blood 60 (1982) 84–91.

ROSENBERG, H. and J.E. FLETCHER: Masseter muscle rigidity and malignant hyperthermia susceptibility.
Anesth. Analg. (Cleveland) 65 (1986) 161–164.

ROSENBERG, H., C.A. FISHER, S.B. REED and V.P. ADDON-
IZIO: Platelet aggregation in patients susceptible to
malignant hyperthermia. Anesthesiology 55 (1981)
621–624.

ROSENTHAL, L., R. KLOIBER, R. GAGNON, B. DAMTEW
and J. LOUGH: Frostbite with rhabdomyolysis and
renal failure: radionuclide study. Am. J. Roentgenol. 137 (1981) 387–390.

ROSNER, F.: Biblical quail incident. J. Am. Med.
Assoc. 211 (1970) 1544.

ROSS, B.D., G.K. RADDA, D.G. GADIAN, G. ROCKER, M.
ESIRI and J. FALCONER-SMITH: Examination of a case
of suspected McArdle's syndrome by 31P nuclear
magnetic resonance. N. Engl. J. Med. 304 (1981)
1338–1342.

ROSS, J.H., E.C. ATTWOOD, G.E. ATKIN and R.N. VILLAR:
A study on the effects of severe repetitive exercise
on serum myoglobin, creatine kinase, transami-
nases and lactate dehydrogenate. Q. J. Med. 52
(1983) 268–279.

ROTHSTEIN, T.L. and G.E. KENNY: Cranial neuropathy,
myeloradiculopathy, and myositis: complications
of Mycoplasma pneumoniae infection. Arch. Neurol. 36 (1979) 476–477.

ROWLAND, L.P.: Molecular genetics, pseudogenetics
and clinical neurology. Neurology 33 (1983) 1179–
1195.

ROWLAND, L.P.: Myoglobinuria, 1984. Can. J. Neurol.
Sci. 11 (1984) 1–13.

ROWLAND, L.P. and A.S. PENN: Myoglobinuria. Med.
Clin. North Am. 56 (1972) 1233–1256.

ROWLAND, L.P. and A.S. PENN: Heat-related muscle
cramps. Arch. Intern. Med. 134 (1974) 133–134.

ROWLAND, L.P., S. FAHN, E. HIRSCHBERG and D.H. HAR-
TER: Myoglobinuria. Arch. Neurol. 10 (1964) 537–
562.

ROWLAND L.P., S. ARAKI and P. CARMEL: Contracture
in McArdle's disease. Stability of ATP during contracture in ATP-deficient human muscle. Arch.
Neurol. 24 (1965) 629–644.

ROWLAND, L.P., S. DIMAURO and R.B. LAYZER:
Phosphofructokinase deficiency. In: A.G. Engel
and B.Q. Banker (Eds), Myology. New York,
McGraw-Hill Book Co. (1986) 1603–1617.

ROWLANDS, J.B., F.L. MASTAGLIA, B.A. KAKULAS and D.
HAINSWORTH: Clinical and pathological aspects of a
fatal case of mulga (Pseudechis australis) snakebite.
Med. J. Aust. 1 (1969) 226–230.

ROXIN, L.E., P. VENGE, G. FRIMAN and R. HALLGREN:
Radioimmunoassays of human myoglobin in
serum and urine. Scand. J. Clin Lab. Invest. 39
(1979) 37–46.

RUBENSTEIN, A.E. and S.F. WAINAPEL: Acute hypokalemic myopathy in alcoholism. Arch. Neurol. 34
(1977) 553–555.

RUBIN, E.: Alcoholic myopathy in heart and skeletal
muscle. N. Engl. J. Med. 301 (1979) 28–33.

RUMPF, K.W., H. KAISER, H.J. GRONE, V.E. TRAPP, H.M.
MEINCK, H.H. GOEBEL, E. KUNZE, H. KREUZER and F.
SCHELER: Myoglobinurisches Nierenversagen bei
hyperosmolaren diabetischem Koma. Dtsch. Med.
Wochenschr. 106 (1981) 708–711.

RUMPF, K.W., H. WAGNER, C.P. CRIEE, H. SCHWARCK, H.
KLEIN and F. SCHELER: Rhabdomyolysis after theophylline overdose. Lancet i (1985) 1451–1452.

RYAN, J.F.: The early treatment of malignant hyperthermia. In: R.A. Gordon, B.A. Britt and W. Kalow
(Eds), International Symposium on Malignant Hy-

perthermia. Springfield, IL, Charles C. Thomas (1973) 430–440.

RYAN, J.F.: Treatment of acute hyperthermia crises. Int. Anaesthesiol. Clin. 17 (1979) 153–168.

SALTIN, B., J. HENRIKSSON, E. NYGAARD, P. ANDERSEN and E. JANSSON: Fiber types and metabolic potentials of skeletal muscles in sedentary man and endurance runners. Ann. N.Y. Acad. Sci. 301 (1977) 3–29.

SALTISSI, D., P.S. PARFREY, J.R. CURTIS, P.E. GOWER, M.E. PHILLIPS, D.F. WOODROW, B. VALKOVA, G.D. PERKIN and K.D. SETHI: Rhabdomyolysis and acute renal failure in chronic alcoholics with myopathy, unrelated to acute alcohol ingestion. Clin. Nephrol. 21 (1984) 294–300.

SAMUELS, A.J., S.N. BERNEY, C.D. TOURTELLOTTE and R. ARTYMYSHYN: Coexistence of adult onset Still's disease and polymyositis with rhabdomyolysis successfully treated with methotrexate and corticosteroids. J. Rheumatol. 16 (1989) 685–687.

SANDOW, A.: Skeletal muscle. Annu. Rev. Physiol. 32 (1970) 87–138.

SAUL, R.A., M. VERNON, C. ROE and S.G. OSOFSKY: Rhabdomyolysis in a patient with nonoliguric renal failure: similarities to the toxic-shock syndrome. South. Med. J. 73 (1980) 261–263.

SAVAGE, D.C.L., M. FORBES and G.W. PEARCE: Idiopathic rhabdomyolysis. Arch. Dis. Child. 46 (1971) 594–607.

SCHLESINGER, J.J., D. GANDARA and K.G. BENSCH: Myoglobinuria associated with herpes-group viral infections. Arch. Intern. Med. 138 (1978) 422–424.

SCHMID, R. and R. MAHLER: Chronic progressive myopathy with myoglobinuria: demonstration of a glycogenolytic defect in the muscle. J. Clin. Invest. 38 (1959) 2044–2058.

SCHMID, R., P.W. ROBBINS and R.R. TRAUT: Glycogen synthesis in muscle lacking phosphorylase. Proc. Natl Acad. Sci. USA 45 (1959) 1236.

SCHMITT, H.P., J.H. SIMMENDINGER, H. WAGNER, B. VOLK, C.M. BUSING and M. STENZEL: Severe morphological changes in skeletal muscles of a 5-month-old infant dying from an anesthetic complication with hyperthermia? Neuropaediatrie 6 (1975) 102–111.

SCHMITT, J., K. SCHMIDT and H. RITTER: Hereditary malignant hyperpyrexia associated with muscle adenylate kinase deficiency. Humangenetik 24 (1974) 253–257.

SCHNEIDER, J., G. MUHLFELLNER and H. KAFFARNIK: Creatine-kinase in hyperlipoproteinemic patients treated with clofibrate. Artery 8 (1980) 164–170.

SCHOFIELD, P.M., S.V. BEATH, T.G. MANT and R. BHAMRA: Recovery after severe oxprenolol overdose complicated by rhabdomyolysis. Hum. Toxicol. 4 (1985) 57–60.

SCHWANE, J.A., S.R. JOHNSON, C.B. VANDENAKKER and R.B. ARMSTRONG: Delayed-onset muscular soreness and plasma CPK and LDH activities after downhill running. Med. Sci. Sports Exercise 15 (1983) 51–56.

SCHWARTZ, J.G. and R.D. MCAFEE: Cocaine and rhabdomyolysis. J. Fam. Pract. 24 (1987) 209.

SCHWARTZ, L., M.A. ROCKOFF and B.V. KOKA: Masseter spasm with anaesthesia: incidence and implications. Anesthesiology 61 (1984) 772–775.

SEAY, A.R. and F.A. ZITER: Malignant hyperpyrexia in a patient with Schwartz-Jampel syndrome. J. Pediatr. 93 (1978) 83–84.

SEGAR, W.E.: Idiopathic paroxysmal myoglobinuria. Pediatrics 23 (1959) 12–17.

SENGERS, C.A., A.M. STADHOUDERS, H.H.J. JASPAR, K.J. LAMMERS, J.M. TRIJBELS and S.L. NOTERMANS: Muscle phosphorylase deficiency in childhood. Eur. J. Pediatr. 134 (1980) 161–165.

SHANDS, K.N., B.B. DAN, P. SCHMID, D. BLUM, R.J. GUIDOTTI, N.T. HARGRETT, R.L. ANDERSON, D.L. HILL, C.V. BROOME, J.D. BOND and D.W. FRASER: Toxic shock syndrome in menstruating women: association with tampon use and Staphylococcus aureus and clinical features in 52 cases. N. Engl. J. Med. 303 (1980) 1436–1442.

SHILKIN, K.B., B.T.M. CHEN and O.T. KHOO: Rhabdomyolysis caused by hornet venom. Br. Med. J. (1972) 156–157.

SHUAIB, A., R.T. PAASUKE and A.K.W. BROWNELL: Central core disease. A reappraisal of its clinical features and new management recommendations. Medicine (Baltimore) 66 (1987) 389–396.

SILIPRANDI, N., F. DILISA and A. PIVETTA: Transport and function of L-Carnitine and L-propionyl carnitine. Relevance to some cardiomyopathies and cardiac ischemia. Z. Kardiol. 5, Suppl. (1987) 34–40.

SKRABAL, F., D. BALOGH and P. DITTRICH: Akutes Nierenversagen bei Polymyositis. Wien Klin. Wochenschr. 84 (1972) 109.

SLOAN, M.F., A.J. FRANKS, K.A. EXLEY and A.M. DAVISON: Acute renal failure due to polymyositis. Br. Med. J. 1 (1978) 1457.

SMALS, A.G.H., L.V.A.M. BEEX and P.W.C. KLOPPENBORG: Clofibrate-induced muscle damage with myoglobinuria and cardiomyopathy. N. Engl. J. Med. 296 (1977) 942.

SMEGO, R.A. JR. and D.T. DURALK: The neuroleptic malignant syndrome. Arch. Intern. Med. 142 (1982) 1183–1185.

SNYDER, H.R. JR., C.S. DAVIS, R.K. BICKERTON and R.P. HALLIDAY: 1-[(5-arylfurfurylidene amino] hydantoins. A new class of muscle relaxants. J. Med. Chem. 10 (1967) 807–810.

SOLOMONS, C.C. and N.C. MASSON: Platelet model for halothane-reduced effects on nucleotide metabolism applied to malignant hyperthermia. Acta Anaesthesiol. Scand. 28 (1984) 185–190.

SPAET, T.H., M.C. ROSENTHAL and W. DAMESHEK: Idiopathic myoglobinuria in man. Blood 9 (1954) 881–896.

SPRUNG, C.L., C.J. PORTOCARRERO, A.V. FERNAINE and P.F. WEINBERG: The metabolic and respiratory alterations of heat stroke. Arch. Intern. Med. 140 (1980) 665–669.

STANEC, A. and G. STEFANO: Cyclic AMP in normal

and malignant hyperpyrexia susceptible individuals following exercise. Br. J. Anaesth. 56 (1984) 1243–1246.

STANLEY, C.A.: New genetic defects in mitochondrial fatty acid oxidation and carnitine deficiency. Adv. Pediatr. 34 (1987) 59–88.

SUBRAMANIAN, R., S. PLEHN, J. NOONAN, M. SCHMIDT and A.L. SHUG: Free radical-mediated damage during myocardial ischemia and reperfusion and protection by carnitine esters. Z. Kardiol. 76 Suppl. 5 (1987) 41–45.

SULLIVAN, J.S., N.G. ARDLIE and M.A. DENBOROUGH: Platelet function in malignant hyperpyrexia. Br. J. Anaesth. 54 (1982) 900–901.

TAM, C.W., B.R. OLIN III and A.E. RUIZ: Loxapine-associated rhabdomyolysis and acute renal failure. Arch. Intern. Med. 140 (1980) 975–976.

TANI, K., H. FUJII, S. MIWA, F. IMANAKA, A. KURAMOTO and H. ISHIKAWA: Phosphofructokinase deficiency associated with congenital nonspherocytic hemolytic anemia and mild myopathy: biochemical and morphological studies on the muscle. Tohoku J. Exp. Med. 141 (1983a) 287–293.

TANI, K., H. FUJII, S. TAKEGAWA, S. MIWA, W. KOYAMA, M. KANAYAMA, A. IMANAKA, F. IMANAKA and A. KURAMOTO: Two cases of phosphofructokinase deficiency associated with congenital hemolytic anemia found in Japan. Am. J. Hematol. 14 (1983b) 165–174.

TARUI, S., G. OKUNO, Y. IKURA, T. TANAKA, M. SUDA and M. NISHIKAWA: Phosphofructokinase deficiency in skeletal muscle: a new type of glycogenosis. Biochem. Biophys. Res. Commun. 19 (1965) 517–523.

TAVILL, A.S., J.M. EVANSON, S.B. BAKER, C. DE and V. HEWITT: Idiopathic paroxysmal myoglobinuria with acute renal failure and hypercalcemia. N. Engl. J. Med. 271 (1964) 283–287.

TEIN, I., F. DEMAUGRE, J.P. BONNEFONT and J.M. SAUDUBRAY: Normal muscle CPT1 and CPT2 activities in hepatic presentation patients with CPT1 deficiency in fibroblasts: tissue-specific isoforms of CPT1. J. Neurol. Sci. 92 (1989) 229–245.

TEIN, I., D.C. DE VIVO, J.T.R. CLARKE, H. ZINMAN, R. LAXER and S. DIMAURO: Short-chain L-3-hydroxyacyl-CoA dehydrogenase deficiency. A new cause for recurrent myoglobinuria and encephalopathy. Ann. Neurol. Abstr 28(3) (1990a) 437.

TEIN, I., S. DIMAURO and D.C. DE VIVO: Recurrent childhood myoglobinuria. Adv. Pediatr. 37 (1990b) 77–117.

TEIN, I., D.C. DE VIVO, D.E. HALE, J.T.R. CLARKE, H. ZINMAN, R. LAXER, A. SHORE and S. DIMAURO: Short-chain L-3-hydroxyacyl-CoA dehydrogenase deficiency: a new cause for recurrent myoglobinuria and encephalopathy. Ann. Neurol. 30 (1991) 415–419.

THE LOVASTATIN STUDY GROUP II: Therapeutic response to lovastatin (mevinolin) in nonfamilial hypercholesterolemia. J. Am. Med. Assoc. 256 (1986) 2829–2834.

THOMAS, M.A.B. and L.S. IBELS: Rhabdomyolysis and acute renal failure. Aust. N.Z. J. Med. 15 (1985) 623–628.

THORPE W. and P. SEEMAN: Drug-induced contracture of muscle. In: R.A. Gordon, B.A. Britt and W. Kalow (Eds), International Symposium on Malignant Hyperthermia. Springfield, IL, Charles C. Thomas (1973) 152–162.

TOBIN, R.B. and W.A. COLEMAN: A family study of phosphorylase deficiency in muscle. Ann. Intern. Med. 62 (1965) 313–327.

TOBIN, W.E., F. JUIJING, R.S. PORRO and R.T. SALZMAN: Muscle phosphofructokinase deficiency. Arch. Neurol. 28 (1973) 128–130.

TODD, J., M. FISHAUT, F. KAPRAL and T. WELCH: Toxic-shock syndrome associated with phage-group I staphylococci. Lancet ii (1978) 1116–1118.

TOLLEFSON, G.: A case of neuroleptic malignant syndrome: in vitro muscle comparison with malignant hyperthermia. J. Clin. Psychopharmacol. 2 (1982) 266–270.

TOMARKIN, J.L. and B.A. BRITT: Malignant hyperthermia. Ann. Emerg. Med. 16 (1987) 1253–1265.

TONIN, P., S. SHANSKE, A.K. BROWNELL, J.P. WYSE and S. DIMAURO: Phosphoglycerate kinase (PGK) deficiency: a third case with recurrent myoglobinuria. Neurology 39, Suppl. 1 (1989) 359–360 (Abstract).

TONIN, P., P. LEWIS, S. SERVIDEI and S. DIMAURO: Metabolic causes of myoglobinuria. Ann. Neurol. 27 (1990) 181–185.

TONSGARD, J.H.: Serum dicarboxylic acids in patients with Reye syndrome. J. Pediatr. 109 (1986) 440–445.

TONSGARD, J.H. and G.S. GETZ: Effect of Reye's syndrome serum on isolated chinchilla liver mitochondria. J. Clin. Invest. 76 (1985) 816–825.

TOURTELOTTE, C.R. and A.E. HIRST: Hypokalemia, muscle weakness, and myoglobinuria due to licorice ingestion. Calif. Med. 113 (1970) 51–53.

TRAYNOR, C.A., R.A. VAN DYKE and G.A. GRONERT: Phosphorylase ratio and susceptibility to malignant hyperthermia. Anesth. Analg. (Cleveland) 62 (1983) 324–326.

TREVISAN, C.P., C. ANGELINI, L. FREDDO, G. ISAYA and A. MARTINUZZI: Myoglobinuria and carnitine palmitoyltransferase (CPT) deficiency: studies with malonyl-CoA suggest absence of only CPT-II. Neurology 34 (1984) 353–356.

TREVISAN, C.P., G. ISAYA and C. ANGELINI: Exercise-induced myoglobinuria: defective activity of inner carnitine palmitoyltransferase in muscle mitochondria of two patients. Neurology 37 (1987) 1184–1188.

UNTERDORF, H. and B. LEDERER: Elektrotraumatischer Muskelnekrosen nach Unfallen mit niedergespannten Wechselstrom. Wochenschr. Unfallheilkd. 78 (1975) 333–338.

VAN RENTERGHEM, D., J. DE REUCK, K. SCHELSTRAETE, W. ELINCK and M. VAN DER STRAETEN: Epsilon amino caproic acid myopathy: additional features. Clin. Neurol. Neurosurg. 86 (1984) 153–157.

VERBURG, M.P., B.A. BRITT, F.T.J.J. OERLEMANS, B.

SCOTT, J. VAN EGMOND and C.H.M.M. DE BRUIJN: Comparison of metabolites in skeletal muscle biopsies from normal humans and those susceptible to malignant hyperthermia. Anesthesiology 65 (1986) 654–657.

VERTEL, R.M. and J.P. KNOCHEL: Acute renal failure due to heat injury. An analysis of ten cases associated with a high incidence of myoglobinuria. Am. J. Med. 43 (1967) 435–451.

VORA, S., L. CORASH, W.K. ENGEL, S. DURHAM, C. SEAMAN and S. PIOMELLI: The molecular mechanism of the inherited phosphofructokinase deficiency associated with hemolysis and myopathy. Blood 55 (1980) 629–635.

VORA, S., M. DAVIDSON, C. SEAMAN, A.F. MIRANDA, N.A. NOBLE, K.R. TANAKA, E.P. FRENKEL and S. DIMAURO: Heterogeneity of the molecular lesions in inherited phosphofructokinase deficiency. J. Clin. Invest. 72 (1983) 1995–2006.

WAHREN, J.: Glucose turnover during exercise in man. Ann. N.Y. Acad. Sci. 301 (1977) 45–55.

WALSH, J.C. and A.B. CONOMY: The effect of ethyl alcohol on striated muscle: some clinical and pathological observations. Aust. N.Z. J. Med. 7 (1977) 485–490.

WALTON, J.N. and R.D. ADAMS: Polymyositis. Baltimore, Williams and Wilkins (1954).

WARREN, D.J., A.C. LEITCH and R.J.E. LEGGETT: Hyperuricaemic acute renal failure after epileptic seizures. Lancet ii (1975) 385–387.

WATERBURY, L. and E.P. FRANKEL: Hereditary nonspherocytic hemolysis with erythrocyte phosphofructokinase deficiency. Blood 39 (1972) 415–425.

WEINREICH, A.I. and G. SILVAY: A hyperthermia emergency cart. Anaesthesiol. Rev. 9 (1982) 31–32.

WESTON, M.D., N.P. HIRSCH and J.A. JONES: Narcotic overdose and acute rhabdomyolysis. Anaesthesiology 41 (1986) 1269.

WHEBY, M.S. and H.S. MILLER: Idiopathic paroxysmal myoglobinuria. Am. J. Med. 29 (1960) 599.

WHITTAKER, M. and J.J. BRITTEN: Malignant hyperthermia and the fluoride-resistant gene. Br. J. Anaesth. 53 (1981) 241–244.

WHITTAKER, M., R. SPENCER and J. SEARLE: Plasma cholinesterase and malignant hyperthermia (letter). Br. J. Anaesth. 49 (1977) 393.

WIGHT, J.P., J. LAURENCE, S. HOLT and A.R. FORREST: Rhabdomyolysis with hyperkalaemia after aminophylline overdose. Med. Sci. Law 27 (1987) 103–105.

WILFERT, C.M., R.H. BUCKLEY, T. MOHANAKUMAR, J.F. GRIFFITH, S.L. KATZ, J.K. WHISNANT, P.A. EGGLESTON, M. MOORE, E. TREADWELL, M.N. OXMAN and F.S. ROSEN: Persistent and fatal central-nervous-system ECHO-virus infections in patients with agammaglobulinemia. N. Engl. J. Med. 296 (1977) 1485–1489.

WILLIAMS, C.H.: Some observations on the etiology of the fulminant hyperthermia-stress syndrome. Perspect. Biol. Med. 20 (1976) 120–130.

WILLIAMS, C.H., C. HOUCHINS and M.D. SHANKLIN: Energy metabolism in pigs susceptible to the fulminant hyperthermia stress syndrome. Br. Med. J. 3 (1975) 411–413.

WILLIAMS, C.H., D.H. STUBBS, M.D. SHANKLIN and H.B. HENDRICK: Energy metabolism and hemodynamics in fulminant hyperthermic stress syndrome swine. Proceedings of the International Pig Veterinary Society Congress. Ames, IA (1976).

WILLIAMS, C.H., G.P. HOECH and J.T. ROBERTS JR.: Experimental malignant hyperthermia. Anesthesiology 49 (1978) 58–59.

WILLIAMS, R.S., D.D. SCHOCKEN, M. MOREY and F.P. KOISCH: Medical aspects of competitive distance running: guidelines for community physicians. Postgrad. Med. 70 (1981) 41–44, 47–48, 51.

WILLNER, J.H.: Malignant hyperthermia. Pediatr. Ann. 13 (1984) 128–134.

WILLNER, J.H., C.G. CERRI and D.S. WOOD: Malignant hyperthermia: abnormal cyclic AMP metabolism in skeletal muscle. Neurology 29 (1979) 557.

WILLNER, J.H., D.S. WOOD, C.C. CERRI and B. BRITT: Increased myophosphorylase A in malignant hyperthermia. N. Engl. J. Med. 303 (1980) 138–140.

WILLNER, J.H., C.G. CERRI and D.S. WOOD: High skeletal muscle adenylate cyclase in malignant hyperthermia. J. Clin. Invest. 68 (1981) 1119–1124.

WILLNER, J., M. NAKAGAWA and D. WOOD: Drug-induced fiber necrosis in Duchenne dystrophy. Ital. J. Neurol. Sci. 5, Suppl. 1 (1984) 117–121.

WISSLER, H.: Paroxysmal myoglobinurie. Helv. Paediatr. Acta 3 (1948) 334–337.

WOOD, D.S., J.H. WILLNER and G. SALVIATI: Malignant hyperthermia: the pathogenesis of abnormal caffeine contracture. In: D.L. Schotland (Ed.), Diseases of the Motor Unit. New York, Wiley (1982) 597–609.

WOOLRIDGE, T.D., J.D. BOWER and N.C. NELSON: Rhabdomyolysis. A complication of parathyroidectomy and calcium supplementation. J. Am. Med. Assoc. 239 (1978) 643.

WYNNE, J.W., J.B. GOSLEN and W.E. BALLINGER JR.: Rhabdomyolysis with cardiac and respiratory involvement. South. Med. J. 70 (1977) 1125–1127.

ZAMKOFF, K. and N. ROSEN: Influenza and myoglobinuria in brothers. Neurology 29 (1979) 340–345.

ZAMORA-QUEZADA, J.C., H. DINERMAN, M.J. STADECKER and J.J. KELLY: Muscle and skin infarction after free-basing cocaine (crack). Ann. Intern. Med. 108 (1988) 564–566.

ZIMMER, C., H. ALTENKRICH, S. DORFMÜLLER-KÜCHIN, D. PONGRATZ, I. PAETZKE and G. GOSZTOKRYI: Type 2a fibre rhabdomyolysis in myoadenylate deaminase deficiency. J. Neurol. 238 (1991) 31–33.

ZSIGMOND, E.K., J. PENNER and S.P. KOTHARY: Normal erythrocyte fragility and abnormal platelet aggregation in MH families: a pilot study. In: J.A. Aldrete and B.A. Britt (Eds), Second International Symposium on Malignant Hyperthermia. New York, Grune & Stratton (1978) 213–219.

Handbook of Clinical Neurology, Vol. 18 (62): Myopathies
L.P. Rowland and S. DiMauro, editors

CHAPTER 19

Toxic myopathies

F.L. MASTAGLIA

*Department of Medicine, University of Western Australia and Department of Neurology, Queen Elizabeth II
Medical Centre, Perth, Australia*

There has been increasing awareness of the adverse effects of therapeutic agents and exogenous toxins on the structure and function of skeletal muscle. It has been recognized that drugs used therapeutically in various branches of medicine, as well as alcohol and other drugs of addiction, may produce muscular symptoms, either through a direct effect on the skeletal muscles or by interfering with neuromuscular transmission or peripheral nerve function. The resulting clinical syndrome varies from one characterized by mild muscle pain and cramping to profound myalgia, paralysis, and myoglobinuria. Similarly, a variety of chemicals, biological toxins, and venoms are myotoxic or neurotoxic.

The possibility of a toxic myopathy should be considered in drug addicts or any patient who develops muscular symptoms while on drug therapy, particularly with newly introduced agents. It is important to recognize the nature of these disorders because symptoms are usually reversible if the offending agent is withdrawn, whereas failure to do so often leads to increasing symptoms and incapacity. It is particularly important to consider the possibility of these adverse effects in patients with a pre-existing neuromuscular disorder, as they may be more sensitive to and less able to cope with these effects.

Experimental studies have elucidated the mechanism of action of many drugs and toxins on skeletal muscle. These studies have also provided important insights into the pathological and physiological reactions of skeletal muscle in disease. Moreover, several experimental models of naturally occurring myopathies have emerged from these investigations.

BASIC MECHANISMS

The mechanisms of action of drugs and toxins on skeletal muscle are diverse. Some have a direct toxic effect, either locally following intramuscular injection or more diffusely after systemic administration or absorption of the agent. Some drugs are not intrinsically myotoxic but damage muscle by an immunological process that is initiated by the drug, by causing hypokalemia or ischemia, or, with a narcotic drug overdose, by causing coma, with muscle compression during periods of prolonged unconsciousness and immobility ('crush syndrome') (Penn et al. 1972; Mastaglia 1982). For drugs such as tetrabenazine, phencyclidine, and cholinesterase inhibitors, muscle fiber necrosis is due to excessive neural driving or to the accumulation of acetylcholine at the neuromuscular junction experimentally, thus can be prevented by prior denervation (Wecker et al. 1978; Max et al. 1986).

Morphological and biochemical studies in animals and humans have provided some understanding of the basic cellular mechanisms of action of drugs and toxins, but this is still incomplete. Some drugs, venoms, and other chemicals cause muscle fiber necrosis by acting directly on the muscle plasma membrane; the outer boundary of the cell is exposed to the full extracellular concentration of the toxin and is therefore the most vulnerable component of the cell (Pritchard 1979). Derangement of ionic permeability of the plasma membrane may increase entry of calcium ions into the cell, leading to myofibrillar contracture and initiating a chain of events that culminates in cell death (Mastaglia 1982; Steer et al. 1986). Mobilization of calcium from intracellular stores, such as the sarcoplasmic reticulum or mitochondria, may also increase intracellular calcium levels (Steer et al. 1986). The muscle fiber necrosis caused by cholinesterase inhibitors is due to calcium influx into the muscle fiber at the end-plate region, where the degenerative changes commence; the process can be prevented in vitro by removing calcium from the incubating medium (Leonard and Salpeter 1979). Increased myoplasmic calcium levels are also thought to cause the myofibrillar contracture and muscle necrosis of malignant hyperpyrexia, when susceptible individuals are exposed to specific anesthetic agents and other drugs (Britt 1979).

A change in the electrical properties and excitability of the muscle fiber plasma membrane is the basis for the myotonia that may be induced by drugs or chemicals. A similar mechanism is adduced to explain the muscular weakness of patients who become severely hypokalemic after treatment with diuretics or other drugs. Changes in the electrical properties of the cell membrane may also underlie the myalgia and cramps induced by drugs (see below).

Some agents interfere with aerobic or anaerobic pathways of energy generation in muscle and have been used to induce experimental models of human metabolic myopathies. Uncoupling of oxidative phosphorylation with 2,4-dinitrophenol produces a myopathy comparable to human mitochondrial myopathy (Melmed et al. 1975; Sahgal et al. 1979; Hayes et al. 1985). Administration of brominated vegetable oil leads to impaired β-oxidation of medium- and short-chain fatty acids and a lipid storage myopathy (Brownell and Engel 1978). Iodoacetate blocks the glycolytic enzyme glyceraldehyde-3-phosphate dehydrogenase and

results in a condition that resembles the human disorders of muscle glycolysis and glycogenolysis (Brumback et a. 1983).

Some drugs interfere with protein synthesis and degradation in muscle. This occurs particularly with the natural and synthetic glucocorticoids which inhibit protein synthesis and also enhance protein degradation (Max et al. 1986). The local anesthetic agent, bupivacaine, also inhibits muscle protein synthesis and increases protein degradation (Steer and Mastaglia 1986).

Chloroquine and other amphiphilic cationic drugs cause a myopathy characterized by autophagic degeneration and phospholipid accumulation in muscle. These drugs are soluble in both water and lipid, and being poorly ionized at the pH of plasma, they readily penetrate the cell membrane in the lipid phase to become adsorbed to intracellular membranes and to form inert intralysosomal drug-phospholipid complexes that accumulate in the form of membranous and crystalloid structures within autophagic vacuoles (Lullmann et al. 1978). Some drugs of this type may become integrated into the plasma membrane, leading to conformational changes in the membrane and altered membrane function (Drenckhahn and Lullmann-Rauch 1979).

Drugs and toxins may cause muscle weakness or paralysis by interfering with neuromuscular transmission. These drugs have been discussed in detail elsewhere (Argov and Mastaglia 1979b; Mastaglia and Argov 1981). Those most frequently implicated are the aminoglycoside antibiotics, which act both pre- and post-synaptically. D-Penicillamine leads to the formation of antibodies to the acetylcholine receptor, resulting in a myasthenic syndrome that is clinically indistinguishable from myasthenia gravis but which is usually reversible when the drug is withdrawn.

DRUG-INDUCED DISORDERS

Myalgia, cramps, fasciculations and myokymia

Muscle pain, stiffness, and cramps may be caused by drugs (Table 1) and usually subside rapidly once the drug is withdrawn. In some patients transient elevations of serum creatine kinase (CK) activity may accompany these symptoms, indicating a direct effect of the drug on muscle (Spaulding 1979; Pomara et al. 1984). The mechanisms of action of most of these agents on muscle have not been investigated. Similar symptoms may occur in patients with drug-induced myotonia and may also herald the onset of a more severe necrotizing myopathy (see below).

Muscle pain and fasciculations are commonly seen after administration of suxamethonium for muscle relaxation during anesthesia and may be prevented by the administration of d-tubocurarine, diazepam, or calcium gluconate (Eisenberg et al. 1979; Shrivastava et al. 1983). Diffuse fasciculations and myokymia have also been reported in patients treated with D-penicillamine or gold (Reeback et al. 1979; Mitsumoto et al. 1982; Pinals 1983).

Myotonic disorders

Some drugs induce or exacerbate myotonia in humans or experimental animals (Kwiecinski 1981). The drug which has been most fully investigated is the now discarded hypocholesterolemic agent, 20,25-diazacholesterol, which caused a reversible form of myotonia (Somers and Winer 1966). The drug interferes with cholesterol biosynthesis by blocking the sterol Δ^{24}-reductase, leading to the accumulation of desmosterol in the serum as well as the sarcolemma and sarcoplasmic reticulum (Peter and Fiehn 1973). The activity of (Na^+-K^+)-ATPase in the sarcolemma increases and the calcium concentrating capacity of the sarcoplasmic reticulum

TABLE 1

Some of the drugs reported to cause myalgia or muscle cramps.

Suxamethonium	Isoetherine
Danazol	Zimeldine
Clofibrate	Labetalol
Clofibride	Pindolol
Salbutamol	Cimetidine
Lithium	D-penicillamine
Captopril	Gold
Diuretics	Enalapril
Colchicine	Rifampicin
Procainamide	L-tryptophan
Metolazone	Mercaptopropionyl glycine
Cytotoxics	Ethchlorvynol
Zidovudine	Nifedipine

is reduced (Chalikian and Barchi 1982; Mastaglia 1982). Myotonia develops particularly in fast-twitch muscle fibers and requires an intact muscle innervation (Caccia et al. 1975). Other Δ^{24}-reductase-blocking agents, such as triparanol and zuclomiphene, also induce myotonia experimentally. The pathogenesis of myotonia is not fully understood and there have been conflicting reports about possible reduction of sarcolemmal chloride conductance (Furman and Barchi 1981; D'Alonzo and McArdle 1982).

Clofibrate, which inhibits the biosynthesis of cholesterol by blocking production of mevalonate, also induces myotonia experimentally in animals but not in humans (Eberstein et al. 1978; Kwiecinski 1978). Myotonia may be caused by the main metabolite of clofibrate, chlorophenoxyisobutyric acid (Dromgoole et al. 1975), which belongs to the group of monocarboxylic aromatic acids that induce myotonia in animals (Bryant and Morales-Aguilera 1971).

Other drugs may exacerbate or unmask previously undetected myotonia (Mastaglia 1982). These include the depolarizing muscle relaxants (e.g. suxamethonium), which can markedly exacerbate myotonia during general anesthesia. Non-depolarizing relaxants do not have this effect and are therefore preferable for use in patients with known myotonia. The β_2-adrenergic blockers, propranolol and pindolol, and the β_2-adrenergic agonists, fenoterol and ritrodrine, also exacerbate myotonia (Sholl et al. 1985). Diuretics, including frusemide (furosemide), ethacrynic acid, mersalyl, and acetazolamide, induce myotonia in animal muscles and should be used with caution in individuals with hereditary forms of myotonia (Bretag et al. 1980a).

A wide range of chemical agents produce myotonia or repetitive firing in motor nerves and skeletal muscle fibers experimentally (Kwiecinski 1981). The most fully investigated are the monocarboxylic acids, which act directly on skeletal muscle to induce myotonia by blocking chloride conductance (Bryant 1973). Chemical agents that induce experimental myotonia include 2,4-dichlorphenoxyacetate, anthracene-9-carboxylic acid, phenanthrene-9-carboxylic acid, polychlorbenzoic acids, and para-substituted mercuribenzoates (Bryant 1973; Iyer et al. 1977; Danon et al. 1978; Eberstein and Goodgold 1979; Bretag et al.

1980a,b; Roed 1982). Colchicine, guanidine, and amines may induce myotonia, but the mechanisms have not been investigated (Bryant 1973). Other agents such as veratrine compounds, germine esters, DDT and pyrethrins, desoxycorticosterone acetate, licorice, tetraethylammonium, zinc, and uranyl also cause repetitive firing in muscle and peripheral nerves (Bryant 1973).

Necrotizing myopathies

Drugs may cause a diffuse myopathy that evolves for days or weeks. Muscle pain and tenderness may be seen in the more rapidly developing cases. Weakness involves the proximal limb and axial muscles, but is at times generalized; the tendon reflexes are usually preserved, unless the myopathy is severe or associated with a peripheral neuropathy. Serum CK activity is elevated, sometimes markedly, and myoglobinuria may follow (Tein et al., Chapter 18, this Volume).

ε-Aminocaproic acid. Myopathy is a recognized but uncommon complication of treatment with this antifibrinolytic agent that is used in patients with subarachnoid hemorrhage or hereditary angioneurotic edema (Lane et al. 1979; Brown et al. 1982). The myopathy usually develops after 4–6 weeks of continuous treatment with doses of over 18 g daily and has varied in severity from a mild self-limiting condition to severe life-threatening myoglobinuria with renal failure (Britt et al. 1980). Histological changes in biopsies from proximal limb muscles consist of disseminated fiber necrosis and regeneration (Fig. 1A). Selective involvement of type I fibers has been noted in some cases (Britt et al. 1980).

The occurrence of myopathy appears to be both time- and dose-dependent. However, many patients treated with large doses of the drug do not develop muscle symptoms, arguing against a simple dose-related toxic effect; other factors are probably involved. ε-Aminocaproic acid is an analogue of lysine and may become incorporated into cellular membranes in place of lysine, leading to altered membrane function. An ischemic basis for the muscle damage has been suggested by the finding of capillary occlusions and fibrinogen deposition (Mastaglia 1982).

Hypocholesterolemic drugs. A rapidly developing myopathy characterized by severe muscle pain, cramps, tenderness, weakness, and markedly elevated serum CK levels occurs in some patients with hyperlipidemia treated with clofibrate or one of its congeners (Mastaglia 1982; Haubenstock et al. 1984; Rimon et al. 1984; Rumpf et al. 1984). Elevated serum CK levels indicate a subclinical myopathy in 8–16% of patients treated with clofibrate, bezafibrate, etofibrate, or beclofibrate (Afifi et al. 1984). Symptomatic myopathy is uncommon when clofibrate is administered in conventional doses, but is more likely to occur in patients with the nephrotic syndrome, renal failure, and possibly hypothyroidism, situations in which serum levels of the active metabolite, chlorophenoxyisobutyric acid, are increased. Symptoms develop abruptly, usually within 3 weeks of commencing treatment. Withdrawal of the drug or dose reduction is followed by gradual recovery; readministration may lead to recurrence of symptoms. Some patients also have a mild peripheral neuropathy and experimental studies have shown that the drug is both myotoxic and neurotoxic (Teravainen et al. 1977; Afifi et al. 1984).

In the rat, clofibrate leads to the development of myotonia which has been presumed to be due to the inhibitory effects on cholesterol biosynthesis, altering the composition and function of the muscle cell membrane (Kwiecinski 1978). In addition, and probably of greater relevance to the pathogenesis of human myopathy, is the finding of reduced carnitine palmitoyltransferase activity and impaired fatty acid and glucose oxidation in muscles of clofibrate-treated animals (Paul and Adibi 1979).

A painful myopathy, at times with muscle necrosis and myoglobinuria, has also been reported in patients treated with the newer hypocholesterolemic agents, lovastatin (Walravens et al. 1989; Marais and Larson 1990) and gemfibrozil (Magarian et al. 1991).

Cardiac glycosides. In Australia a proximal myopathy has been seen in opiate addicts who consume large quantities of the opiate-based cough suppressant, Linctus Codeine (Australian Pharmaceutical Formulary) (Kennedy 1981; Kilpatrick et al. 1982; Seow 1984). One of the components of

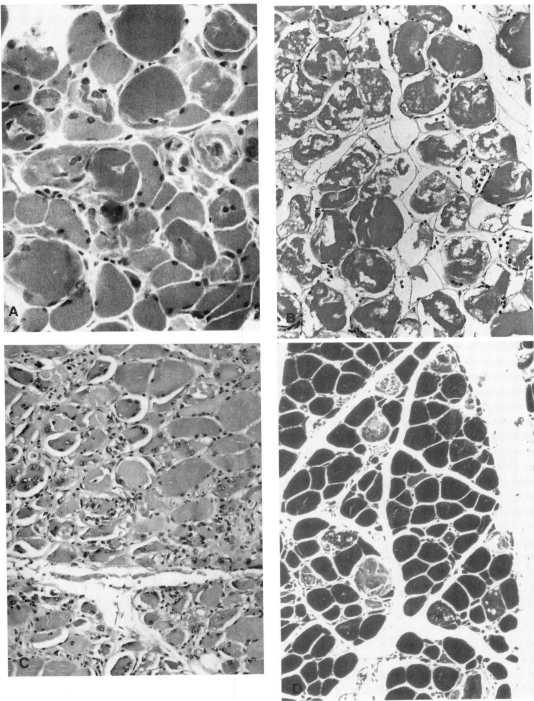

Fig. 1. (A) Necrotizing myopathy due to ε-aminocaproic acid. H & E (×320). (Reproduced from Argov and Mastaglia (1988) with permission of Churchill Livingstone.) (B) Massive acute rhabdomyolysis due to envenomation by *Pseudechis australis*. H & E (×160). (C) Acute corticosteroid myopathy complicating high-dose dexamethasone therapy. H & E (×160). (Courtesy of Dr. V. Ojeda.) (D) Acute alcoholic myopathy. PTAH (×160). (Courtesy of Dr. V. Ojeda.)

Linctus codeine is Squill (an extract of the bulb of *Urginea maritama*), which contains the cardiac glycosides, scillarin A and B (Kennedy 1981). Muscle pain and tenderness were a feature in 1 case; myasthenic features were seen in another. Serum CK levels were elevated, up to 25-fold in 1 case, and muscle biopsy showed a necrotizing myopathy. Another feature was the presence of electrocardiographic abnormalities of cardiac glycoside toxicity.

The mechanism of the myopathy has not been investigated. Cardiac glycosides have an inhibitory effect on the cell membrane Na^+-K^+ pump; both the skeletal myopathy and the cardiac abnormalities in these cases may therefore be due to the effects of scillarin on muscle cell membranes.

Emetine. Reversible generalized muscle weakness was recognized as a common side effect of this ipecac alkaloid in treating patients with amebiasis (Klatskin and Friedman 1948). However, it was not clear whether the weakness was due to a toxic effect on skeletal muscle, neuromuscular transmission, or peripheral nerve. Severe myopathy with weakness of neck, oropharynx, proximal limb, and trunk muscles has been seen in patients taking emetine for amebiasis or alcohol aversion therapy, or using ipecac syrup as an emetic agent, sometimes in bulimia (Fewings et al. 1973; Brotman et al. 1981; Bennett et al. 1982; Friedman 1984; Sugie et al. 1984; Mateer et al. 1985; Palmer and Guay 1985). Serum CK activity is elevated up to 14-fold in some cases, but has been normal in others. Electromyography (EMG) shows a myopathic pattern with short-duration motor unit potentials and fibrillations in some cases. Histological changes in biopsied muscles include scattered muscle fiber necrosis and regeneration with type II fiber atrophy and, in some cases, focal core-targetoid areas with loss of enzyme activity and myofibrillar disorganization, particularly in type I fibers (Bennett et al. 1982; Sugie et al. 1984; Palmer and Guay 1985).

The drug may also have cardiotoxic effects as shown by electrocardiographic changes, abnormalities of left ventricular function, and cardiac failure in a number of cases (Fewings et al. 1973; Brotman et al. 1981; Palmer and Guay 1985). Gradual recovery over weeks or months was the

rule in most cases after withdrawal of the drug, but the condition may be fatal (Fewings et al. 1973).

Experimentally the drug has a pure myotoxic action leading to mitochondrial and myofibrillar changes followed by muscle fiber necrosis and regeneration, with no damage to intramuscular nerves or motor end-plates (Duane and Engel 1970; Bradley et al. 1976; Bindoff and Cullen 1978). However, emetine alters chloride and potassium conductance and action potential generation in the peroneal nerve of rats (Conte-Camerino et al. 1982). Emetine inhibits protein synthesis and mitochondrial respiration and, in high concentrations, disrupts cell membranes (Duane and Engel 1970).

Other drugs. Necrotizing myopathy has been reported in heroin (Richter et al. 1971) and phencyclidine addicts (Cogen et al. 1978). A necrotizing myopathy with prominent mitochondrial changes occurs in patients with AIDS who are treated with zidovudine (Pinching et al. 1989; Panegyres et al. 1990; Chalmers et al. 1991). Muscle fiber necrosis may also occur in the myopathies induced by a number of other drugs including vincristine, colchicine, and plasmocid and in severe drug-induced hypokalemia, but it is usually inconspicuous. In addition, a variety of drugs may induce localized muscle necrosis following administration by intramuscular injection. These include paraldehyde, pentazocine, chlorpromazine, and a number of antibiotics. Other drugs which have been shown to cause muscle necrosis experimentally include corticosteroids, disulfiram, imipramine, pargyline, tetrabenazine, serotonin, marcaine, clonidine, DMSO, p-phenylenediamine, diphenylhydantoin, cholinesterase inhibitors, and azathioprine.

Acute muscle necrosis (rhabdomyolysis)

This is the most serious and acute form of toxic myopathy encountered in clinical practice. It may follow general anesthesia, alcohol intoxication, prolonged drug-induced coma, self-administration of heroin, cocaine, and other narcotic drugs, or intoxication with one of a number of other drugs (Tein et al., Chapter 18, this Volume; Rowland et al. 1964; Penn et al. 1972; Gabow et al. 1982; Briner et al. 1986; Grob 1990). The same syndrome may follow envenomation or other forms of poisoning (Table 2) (Gabow et al. 1982; Tein et al., Chapter 18, this Volume).

The condition is characterized by widespread muscle pain, tenderness, and areflexic weakness evolving within 24–48 hours. Marked swelling of limb muscle groups may lead to limb ischemia and peripheral nerve entrapment (*compartment syn-*

TABLE 2

Drugs and toxins that have been implicated in causing acute myoglobinuria.

Drug-induced coma, seizures, dyskinesia (9,11,17,30,38)	*Other drugs* Amphetamines (17)
Barbiturates (30)	Phenmetrazine (6)
Heroin (28, 31, 43)	Phencyclidine (14)
Methadone (27, 40)	Glutethimide (30)
Cocaine (41)	Chlorpromazine (25)
Phenylpropanolamine (8, 20)	Diazepam (9)
Morphine (7)	Rohypnol (9)
Dihydrocodeine (7)	Lithium (39)
LSD (17)	Amoxapine (1, 22)
Salicylates (17)	Lovastatin (42)
Clofibrate/bezafibrate/fenofibrate (10, 17)	ε-aminocaproic acid (10, 17)
Phenelzine/phenformin/fenfluramine (29)	Isoniazid (40)
Meprobamate (40)	Loxapine (37)
Antihistamines/paracetamol (20, 38)	Theophylline (26)
Oxprenolol (33)	Pentamidine (34)
Ethanol (19)	Vasopressin (2)
	Gemfibrozil (42)
Postanesthetic	
Suxamethonium	*Toxins*
Malignant hyperpyrexia	Ethanol (17, 19)
	Isopropyl alcohol (17)
Neuroleptic malignant syndrome (15, 23)	Carbon monoxide (17)
Haloperidol	Mercuric chloride (12)
Stelazine	Ethylene glycol (17)
Fluphenazine	Copper sulfate (13)
Other neuroleptics	Zinc phosphide (13)
	Strychnine (40)
Hypokalemia (17)	Metaldehyde (40)
Diuretics	Chloralose (40)
Carbenoxolone	Paraphenylenediamine (3)
Amphotericin B	Toluene (paint sniffing) (17)
Licorice	Gasoline (sniffing) (24)
	Lindane/benzene (21)
	Snake venom
	Hornet or wasp venom (35)
	Brown spider venom (17)
	Haff disease (4)
	Quail ingestion (?hemlock) (5)

1. Abreo et al. (1982); 2. Affarah et al. (1984); 3. Baud et al. (1983); 4. Berlin (1948); 5. Billis et al. (1971); 6. Black et al. (1984); 7. Blain et al. (1985); 8. Blewitt and Siegel (1983); 9. Briner et al. (1986); 10. Britt et al. (1980); 11. Chiaken (1980); 12. Chugh et al. (1978); 13. Chugh et al. (1979); 14. Cogen et al. (1978); 15. Eiser et al. (1982); 16. Frendin and Swainson (1985); 17. Gabow et al. (1982); 18. Gabry et al. (1982); 19. Haller and Knochel (1984); 20. Jaeger et al. (1984); 22. Jennings et al. (1983); 23. Kleinknecht et al. (1982); 24. Kovanen et al. (1983); 25. Lazarus and Toglia (1985); 26. Kodi et al. (1985); 27. Nanji and Filipenko (1983); 28. Nicholls et al. (1982); 29. Palmucci et al. (1978); 30. Penn et al. (1972); 31. Richter et al. (1971); 32. Rumpf et al. (1984); 33. Schofield et al. (1985); 34. Senskovic et al. (1985); 35. Shilkin et al. (1972); 36. Swenson et al. (1982); 37. Tam et al. (1980); 38. Thomas and Ibels (1985); 39. Unger et al. (1982); 40. Wattel et al. (1978); 41. Roth et al. (1988); 42. Marais and Larson (1990); 43. Gibb et al. (1985).

dromes), sometimes treated by fasciotomy. Myoglobinuria is usually an early feature, with brownish discoloration of the urine and a blood-positive orthotolidine (Hematest) dipstick reaction in the absence of hematuria or hemoglobinuria. False-negative tests for urine myoglobin may be found with this procedure, but myoglobinemia and myoglobinuria can be confirmed by the more sensitive technique of radioimmunoassay (Thomas and Ibels 1985). Serum CK activity is markedly elevated.

Acute oliguric renal failure may follow severe myoglobinuria and responds to dialysis (Hampel et al. 1983). Hyperkalemia, hyperphosphatemia, and hypocalcemia are common in these patients. Hypocalcemia is attributed to deposition of calcium salts in damaged muscles, while hypercalcemia may appear in the recovery phase (Llach et al. 1981; Gabow et al. 1982; Knochel 1982; Thomas and Ibels 1985). The myocardium is also affected in some cases of heroin overdose (Scherrer et al. 1985). Radionuclide scanning with ^{99m}Tc-diphosphonate may be useful in determining the extent of muscle damage. EMG reveals florid myopathic motor unit changes with spontaneous discharges.

Muscle biopsy shows widespread myofiber necrosis and mild reactive inflammatory changes (Fig. 1B). Regenerative changes are often also a feature but may be absent in patients who die at an early stage. Although the prognosis for recovery is generally good, some patients with severe renal failure die as a result of multiple organ failure and other complications (Briner et al. 1986).

Although a variety of drugs have been implicated in causing acute myoglobinuria (Table 2), few have been shown to have a direct myotoxic action. In most instances, other factors are probably responsible for the muscle damage. In many cases, alcohol abuse or drug overdose leads to prolonged periods of unconsciousness and immobility, muscle compression and ischemia, as well as the effects of hypoxia and hypotension (Penn et al. 1972). In another group of cases with drug-induced seizures (Tam et al. 1980; Jennings et al. 1983; Modi et al. 1985), dyskinesias (Lazarus and Toglia 1985), or acute dystonic reactions in patients with phencyclidine intoxication (Cogen et al. 1978), the common factor seems to be muscular hyperactivity.

However, acute myoglobinuria can occur without other contributory factors in some cases of alcohol or narcotic abuse, suggesting a direct toxic effect of the drug (Richter et al. 1971). In the case of alcohol, other factors include starvation, refeeding, hypokalemia, and hypophosphatemia (see below). The action of the narcotic drugs has been attributed to changes in cell membrane, calcium ion fluxes, and cellular energy production (Nicholls et al. 1982). Use of illicit drugs raises the possibility of myotoxic contaminants.

Some cases of postanesthetic myoglobinuria were attributed to suxamethonium, but were probably incomplete forms of malignant hyperpyrexia. However, there are patients, particularly children, who do not have malignant hyperpyrexia or a muscle enzyme defect, but do have myoglobinuria after the use of suxamethonium for anesthesia (Gibbs 1978; Chaboche et al. 1982; Blumberg and Marti 1984). Myoglobinuria without other features of malignant hyperpyrexia may also follow anesthesia in Duchenne muscular dystrophy (Karpati and Watters 1980; Willner et al. 1984).

Mitochondrial myopathy

Dalakas et al. (1990) described a distinctive myopathy characterized by numerous *ragged red* fibers containing paracrystalline inclusions in AIDS patients who had been on long-term treatment with the nucleoside analogue zidovudine (AZT), which inhibits mitochondrial DNA replication (Arnaudo et al. 1991). Panegyres et al. (1990) described similar cases in which electron microscopy showed striking mitochondrial enlargement, vacuolation, and abnormal cristae (Fig. 2). The myopathy is characterized clinically by myalgia, proximal or generalized muscle weakness, and elevated serum CK levels; it usually improves when the drug is withdrawn (Chalmers et al. 1991). Some patients also have an HIV-related inflammatory myopathy. This syndrome has been attributed to depletion of mitochondrial DNA (Arnaudo et al. 1991).

Mitochondrial myopathy has been induced in experimental animals by chemicals that block aerobic metabolic pathways, including 2,4-dinitrophenol, β-guanidine propionic acid, and diphenyleneiodium (Cooper et al. 1988; Gori et al. 1988).

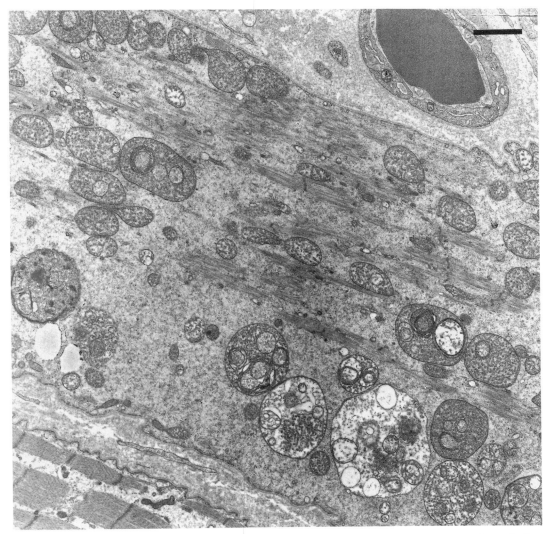

Fig. 2. Mitochondrial myopathy due to zidovudine. Electron micrograph of a grossly disorganized muscle fiber containing enlarged mitochondria with abnormal cristae and inclusions. Bar: 1 μm. (Courtesy of Professor B.A. Kakulas).

Mitochondrial abnormalities and reduced cytochrome c oxidase activity were also prominent features of a myopathy induced by germanium (Higuchi et al. 1989).

Hypokalemic myopathy

Hypokalemia of sufficient severity to cause muscular weakness and hypotonia may develop in patients treated with thiazide diuretics, chlorthal- idone, amphotericin B, carbenoxolone, or fluoro-prednisolone-containing nasal sprays or in patients with purgative abuse (Mastaglia 1982; Vita et al. 1986). Hypokalemic myopathy has also been reported in people who consume large quantities of licorice and licorice extracts that are found in traditional Chinese drugs. The problem may also arise in individuals who use large quantities of snuff or chewing tobacco (Cumming et al. 1980; Valeriano et al. 1983; Piette et al. 1984; Mori et al.

1985). The common ingredient and the one that causes hypokalemia is glycyrrhizic acid, a potent mineralocorticoid analogue (Valeriano et al. 1983).

The weakness is usually generalized and may be profound, with hypotonia and loss of tendon reflexes, sometimes resembling the Guillain-Barré syndrome. Although the condition is usually painless, myalgia and muscle tenderness may be seen in rapidly evolving cases. Sometimes the weakness is episodic, resembling familial hypokalemic periodic paralysis (Basser 1979). The serum CK level is usually markedly elevated and overt myoglobinuria occurs in some cases. Histological changes in muscle tend to be relatively inconspicuous, with scattered fibers being swollen and vacuolated; in severe cases, there is myofiber necrosis and regeneration. Complete recovery is the rule after potassium replacement.

Less commonly, profound muscle weakness due to hyperkalemia is seen in patients treated with potassium-retaining diuretics (Udezue and Harrold 1980).

Inflammatory myopathies

Drugs have been associated with the development of an inflammatory myopathy (Mastaglia and Argov 1982), most frequently with D-penicillamine treatment of patients with rheumatoid arthritis, progressive systemic sclerosis, or Wilson's disease. The average dose of penicillamine used in reported cases was 600 mg daily and the average duration of treatment before the onset of myopathic symptoms was 12 months (Takahashi et al. 1986). In some cases, the myopathy appeared after only a few weeks of treatment with doses as low as 50–100 mg daily. The myopathy is usually indistinguishable clinically and pathologically from other forms of polymyositis, except that prompt improvement has followed cessation of penicillamine therapy; some patients were treated with corticosteroids or plasmapheresis, and some cases were fatal (Mastaglia 1982).

The incidence of inflammatory myopathy seems to be higher in patients with rheumatoid arthritis who are treated with penicillamine than in those who do not take the drug, suggesting that penicillamine was causally involved in the myopathy (Takahashi et al. 1986). Sometimes the myopathy

reappeared after a second course of treatment with penicillamine (Takahashi et al. 1986). Some patients failed to develop myositis during a second course of the drug, and others have continued to take the drug in reduced dosage (Halla et al. 1984). Inflammatory myopathy has been attributed to altered immunoregulatory mechanisms caused by the drug (Mastaglia and Argov 1982).

An interstitial form of eosinophilic myositis and fasciitis has been ascribed to certain preparations that contain L-tryptophan (the *eosinophilia-myalgia syndrome*) (Eidson et al. 1990; Hertzman et al. 1990; Medsger 1990; Silver et al. 1990; Van Garsse and Boeykens 1990). Over 1500 cases have occurred in the United States (Kaufman 1990). The syndrome is characterized by severe myalgia, muscle tenderness, and hyperesthesia with edema and induration of the skin of the extremities, resembling scleroderma, and a peripheral blood eosinophilia (Varga et al. 1990). Some patients have had polyneuropathy or other systemic disorders (Kaufman 1990). The bulk source of the tryptophan preparation in the American cases was traced to one manufacturer, and the syndrome has been attributed to a chemical contaminant of the preparation (Belongia et al. 1990; Center for Disease Control 1990).

An interstitial form of myositis characterized by muscle pain, stiffness, and mild weakness may develop in patients treated with other drugs, including procainamide, hydralazine, phenytoin, mesantoin, and levodopa. Dermatomyositis has followed intramuscular injections of penicillin (Mastaglia and Argov 1982; Harney and Glasberg 1983; Lewis et al. 1986).

β-Adrenoreceptor blockers

Patients treated with β-blockers commonly complain of muscle fatigue and reduced exercise tolerance. Physiological studies of normal subjects and hypertensive patients on long-term treatment indicate that these symptoms probably result from the combined effects of a reduction in cardiac output and the effects of the drug on muscle metabolism during exercise. Subjects taking β-blockers show a greater than normal depletion of muscle ATP and creatine phosphate levels during exercise, probably due to a reduction in the supply of free fatty acid substrates, rather than to an effect on glycogen utilization (Frisk- Holmberg et al. 1979; Kaiser et al. 1985). Endurance exercise capacity is reduced to a greater extent with nonselective β-blockers, such as propranolol, and more so in subjects with higher proportions of slow-twitch (type I) fibers in leg muscles (Karlsson 1983).

Although β-blockers are not inherently myotoxic, one patient had severe painless proximal myopathy with elevated serum CK levels during treatment with propranolol and sotalol; such drugs should therefore be used with caution in patients with pre-existing muscular symptoms (Forfar et al. 1979). Muscle pain and raised serum CK levels were reported in a hypertensive patient treated with labetalol (Teicher et al. 1981). The mechanism of this myopathy remains uncertain, although in both cases there was prompt improvement after withdrawal of the drugs, suggesting a causal relationship.

β-Adrenergic blockers may also interfere with neuromuscular transmission and there have been occasional reports of propranolol, oxprenolol, and practolol unmasking myasthenia gravis or inducing a myasthenic syndrome de novo (Argov and Mastaglia 1979b). Propranolol and pindolol may also exacerbate myotonia (Blessing and Walsh 1977; Richter et al. 1978).

Corticosteroid myopathy

Proximal myopathy is a common complication of prolonged corticosteroid therapy. Symptomatic myopathy is most likely to develop in patients treated with 9-α-fluorinated steroids, such as triamcinolone, betamethasone, and dexamethasone (Dropcho and Soong 1991), but may occur with any of the other corticosteroids when they are administered for prolonged periods, although there is considerable interindividual variability in the dose of drug and duration of treatment that leads to myopathy. Myopathy is most likely to occur in patients taking daily doses of prednisone over 40 mg, but may occur with even lower doses (over 10 mg prednisone daily or its equivalent) if taken for prolonged periods (Askari et al. 1976; Bowyer et al. 1985). Bowyer et al. (1985) found that myopathy did not develop in patients on an alternate-day prednisone regimen. Quantitative studies of muscle function in patients on long-term daily therapy frequently show reductions in muscle performance (Khaleeli et al. 1983; Rothstein et al. 1983) and EMG studies show a high incidence of subclinical myopathy (Coomes 1965a; Yates 1970). Among patients with brain tumors who were on daily dexamethasone therapy, the risk of steroid myopathy was lower in those who were also taking phenytoin, which increases the hepatic metabolism of dexamethasone (Dropcho and Soong 1991).

Muscle weakness develops insidiously, tending to involve the quadriceps and other pelvic girdle muscles initially. It is usually painless, although myalgia has been reported after commencement of high-dose corticosteroid therapy (Thompson et al. 1981). Cranial muscles are usually spared, but dysphonia may occur in patients treated with inhaled corticosteroids and has been attributed to a reversible form of corticosteroid myopathy of the laryngeal muscles (Williams et al. 1983). Bowyer et al. (1985) have also drawn attention to the development of diaphragmatic weakness in asthmatic patients on corticosteroids.

Serum levels of CK and other enzymes are normal in steroid myopathy; if CK is elevated some other type of myopathy should be considered. Urinary creatine and 3-methylhistidine excretion are increased, but are not helpful diagnostically (Askari et al. 1976; Khaleeli et al. 1983). EMG shows typical myopathic changes in proximal limb muscles with reduction in motor unit duration and amplitude without spontaneous muscle fiber potentials.

The characteristic muscle biopsy finding is that of selective atrophy of type II fibers (Fig. 3A), particularly of the type IIb subgroup, which is also most severely affected in experimental steroid myopathy (Braunstein and Girolami 1981; Livingstone et al. 1981; Hudgson and Hall 1982). Muscle fiber necrosis, regeneration, and other degenerative changes are not a feature, but have been found in the severe generalized form of myopathy that occurs in asthmatic patients who were treated with high intravenous doses of hydrocortisone (MacFarlane and Rosenthal 1977; Van Maarle and Woods 1980; Mastaglia 1982) and in some patients treated with high doses of dexamethasone for cerebral edema (Ojeda 1982) (Fig. 1C). The histological changes in these cases were similar to those found in experimental cortisone myopathy in rabbits (Afifi and Bergman 1969).

The biochemical and physiological effects of corticosteroids on muscle have been extensively investigated in animals. There are changes in oxidative metabolism (Koski et al. 1974), glycogen metabolism (Shoji et al. 1974), lipid content (Wakata et al. 1983), calcium uptake by the sarcoplasmic reticulum (Shoji et al. 1976), myofibrillar ATPase activity (Clark and Vignos 1979), muscle contractile proteins (Heiner et al. 1980), protein synthesis and degradation (Shoji and Pennington 1977a; Santidrian et al. 1981; Clark et al. 1986), membrane excitability (Gruener and Stern 1972), and muscle contractile properties (Grossie and Albuquerque 1978; Gardiner and Edgerton 1979). The basic cellular action of corticosteroids seems to be an inhibition of messenger RNA synthesis which, in turn, influences the translation and synthesis of muscle-specific proteins (Rannels et al. 1978; Karpati 1984).

The factors that contribute to the greater susceptibility of fast-twitch glycolytic (type IIB) fibers are uncertain. Comparing glucocorticoid receptors, there was a higher concentration of cytosol-binding sites in the slow-contracting soleus than in the fast-contracting extensor digitorum longus (Shoji and Pennington 1977b; DuBois and Almon 1984). There was a significant increase in the cytosol receptor concentration with disuse and after denervation. Therefore endogenous glucocorticoids may play a part in causing the muscle fiber atrophy that is seen in these situations (DuBois and Almon 1980, 1981). This would also account for the observation that the degree of atrophy induced by dexamethasone in denervated muscles is greater than that expected from denervation or dexamethasone alone (Livingstone et al. 1981). Moreover, relative disuse of muscles due to physical inactivity may render them more susceptible to the effects of corticosteroids. Observations in the rat suggest that the particular susceptibility of fast-twitch glycolytic fibers may be linked to a particularly severe reduction of myophosphorylase activity in these fibers and their comparative inability to utilize alternative energy sources, especially substrates derived from free fatty acids (Livingstone et al. 1981).

Corticosteroid myopathy is usually reversible if the drug can be withdrawn or the dose reduced or, to some extent, if an alternate-day regimen can be implemented, or if prednisone can be substituted. Anabolic steroids and B group vitamins prevent the development of corticosteroid myopathy in the rat (Sakai et al. 1978), but are not useful in humans (Coomes 1965b). Phenytoin, vitamin E, and potassium supplementation are also ineffective. In rats and humans, glucocorticoid-induced muscle atrophy and weakness can be partially prevented or reversed by a regular program of physical training (Hickson and Davis 1981; Horber et al. 1985).

Autophagic myopathies

Drugs with amphiphilic cationic properties interfere with lysosomal digestion and lead to autophagic degeneration and accumulation of phospholipids in muscle and other tissues experimentally in animals (Drenckhahn and Lullmann-Rauch 1979; Lullmann-Rauch 1979). Three of these drugs, chloroquine, amiodarone, and perhexiline, as well as vincristine and colchicine, are also known to cause human myopathy or neuromyopathy.

Chloroquine. This antimalarial and antirheumatic drug may cause myopathy or neuromyopathy after treatment with doses of 250–750 mg daily for periods of several weeks to 4 years (Whisnant et al. 1963; Eadie and Ferrier 1966; Mastaglia et al. 1977). There is insidious development of painless weakness, particularly of proximal muscle groups, often associated with wasting that may be severe in advanced cases. Loss of tendon reflexes, mild sensory changes, and abnormal nerve conduction studies point to an associated peripheral neuropathy. Diplopia has been present in some cases. Serum enzyme levels are normal or slightly ele-

→

Fig. 3. (A) Type II fiber atrophy in a case of corticosteroid myopathy. Myosin ATPase, pH 9.4 (×205). (B) Vacuolar change and type II fiber atrophy in a case of chloroquine neuromyopathy. Myosin ATPase, pH 7.2 (×205). (C, D) Colchicine neuromyopathy showing scattered necrotic and regenerating fibers (C) and fibers with central core-like areas (D). (C) H & E (×205); (D) Myosin ATPase, pH 4.6 (×160). (Reproduced from Kakulas and Mastaglia (1992) with permission of Churchill Livingstone.)

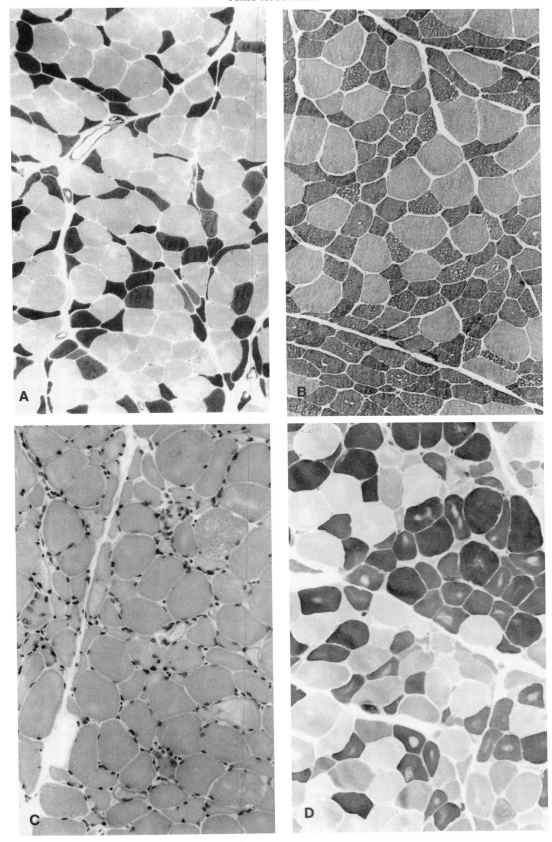

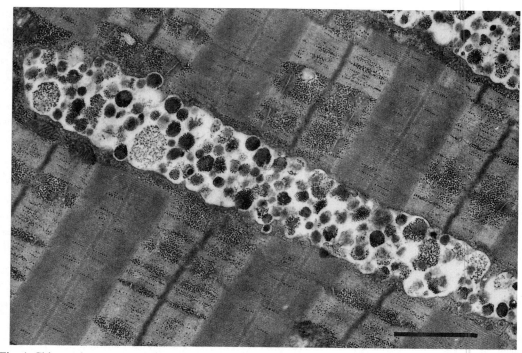

Fig. 4. Chloroquine myopathy. Electron micrograph showing autophagic vacuoles in muscle fibers. Bar = 1.0 μm. (Reproduced from Kakulas and Mastaglia (1992) with permission of Churchill Livingstone.)

vated, and EMG may show spontaneous muscle fiber potentials in addition to the typical motor unit changes of myopathy (Eadie and Ferrier 1966; Mastaglia et al. 1977). Cardiomyopathy was prominent in some reported cases (Hughes et al. 1971; Estes et al. 1987). The myopathy is reversible once the drug is withdrawn, but full recovery may take many months.

Histologically, the myopathy is characterized by vacuolar changes, usually in both major fiber types (Fig. 3B). Electron microscopy shows that the vacuolation is due to autophagic degeneration (Fig. 4) with associated exocytosis and accumulation of lamellated membrane bodies (*myeloid bodies*) in muscle fibers and interstitial cells, as well as curvilinear bodies (Mastaglia et al. 1977; Neville et al. 1979). Quantitative studies of muscle lipids indicated abnormal accumulation of phospholipids and, to a lesser extent, of neutral lipids (Mastaglia et al. 1977). Experimentally, there is swelling of the sarcoplasmic reticulum in the early stages (Schmalbruch 1980; Trout et al. 1981), a marked increase

in lysosomal enzyme activity, particularly of acid proteases, and an increase in the desmin content of muscle fibers (Stauber et al. 1981; Sano 1985).

Perhexiline. This drug may lead to the development of a demyelinating sensorimotor peripheral neuropathy in a small proportion of treated patients (Argov and Mastaglia 1979a). In some patients there are additional features of myopathy with pain, tenderness, and weakness of proximal as well as distal muscle groups. Proximal myopathy may occur without peripheral neuropathy (Tomlinson and Rosenthal 1977). Ultrastructural studies in humans and mice with an experimentally-induced form of perhexiline neuromyopathy have shown numerous membranous and granular inclusions of probable lysosomal origin in muscle fibers as well as endothelial cells and Schwann cells (Fardeau et al. 1979).

Amiodarone. Demyelinating sensorimotor peripheral neuropathy has appeared in patients treated with this antiarrhythmic drug (Argov and Mastaglia 1988). In some cases there were additional myopathic features in proximal limb muscles with fiber vacuolation, autophagic degeneration with membrane-bound dense bodies (Meier et al. 1979), and myofiber necrosis (Clouston and Donnelly 1989).

Vincristine. The alkaloid vincristine, which interferes with RNA and protein synthesis and with the polymerization of tubulin into microtubules, commonly causes an axonal peripheral neuropathy. In some patients there is also a proximal myopathy (Bradley et al. 1970). Electron microscopic studies in humans and experimental animals have shown that the drug has a profound effect on membrane systems leading to the formation of complex *spheromembranous bodies*, thought to be derived from the sarcoplasmic reticulum, and autophagic degeneration of muscle fibers (Slotwiner et al. 1966; Anderson et al. 1967; Bradley 1970).

Colchicine. Like vincristine, this drug prevents the polymerization of tubulin into microtubules and may cause an axonal neuropathy or myopathy in humans or experimental animals (Kontos 1962; Riggs et al. 1986; Kuncl et al. 1987). Severe neuromyopathy may follow prolonged administration of high doses of the drug. It may also occur in patients taking conventional doses for the treatment of gout, particularly in the presence of renal insufficiency. Serum CK levels are usually elevated 10–20-fold. Prompt recovery occurs on withdrawal of the drug.

Characteristic histological findings in biopsies from proximal limb muscles comprise excessive variation in fiber size, the presence of small vacuoles in muscle fibers, or of central areas of altered staining on hematoxylin and eosin preparations, and loss of enzyme activity resembling cores or core-targetoids in histochemical preparations (Fig. 3C,D). Some vacuoles contain granular hematoxyphilic material. The major findings on electron microscopy are the autophagic vacuoles and *spheromembranous bodies* as in experimental colchicine myopathy. Muscle fiber necrosis and regeneration rarely occur. Denervation changes may also be found in distal limb muscles.

Other drugs. Other drugs rarely implicated in causing myopathy include rifampicin (Jenkins and Emerson 1981), mercaptoproprionyl glycine (Hales et al. 1982), cimetidine (Treves et al. 1985), tetracycline (Sinclair and Philips 1982), adenine arabinoside (Mak et al. 1990), and ethchlorvynol (Placidyl), which has been associated with the presence of tubular aggregates in muscle fibers (Petajan et al. 1986).

Alcoholic myopathy

There is ample clinical and experimental evidence that ethanol is myotoxic. Acute, subacute, and chronic forms of myopathy are well-documented in alcoholics (Klinkerfuss et al. 1967; Perkoff et al. 1967; Martin et al. 1982; Haller and Knochel 1984).

Acute alcoholic myopathy. This is a condition of variable severity ranging from transient asymptomatic elevation of serum CK activity and myoglobin levels to overt myoglobinuria and renal failure. It has been reported predominantly in male alcoholics following binge-drinking. The condition is probably more frequent than is generally appreciated judging from the finding of significant and at times marked elevations of serum CK activity in a high proportion of alcoholics admitted to hospital in an intoxicated state or with alcohol withdrawal (Haller and Knochel 1984).

Since the original description of acute myopathy by Hed et al. in 1962, ethanol has been recognized as one of the commonest causes in most series of acute myoglobinuria (Gabow et al. 1982; Haller and Knochel 1984). The onset in such cases is usually abrupt with widespread muscle pain, swelling, weakness, myoglobinuria, and acute renal failure. The onset may be less acute and the condition less severe with only proximal limb weakness. In others, there is focal weakness of the muscles of one or both calves and the clinical picture may resemble that of venous thrombophlebitis (Hed et al. 1962; Walsh and Conomy 1977). Attacks of acute myopathy may occur on a number of occasions following alcoholic binges. The prognosis for recovery is usually good with abstinence from drinking, but full recovery may take several months.

The myopathological changes consist of scattered muscle fibers necrosis, which is particularly widespread in the severe form, together with evidence of muscle fiber regeneration (Fig. 1D). Other changes include a mild mononuclear cellular infiltration in some cases, and patchy loss of oxidative enzyme activity, especially in type I fibers (Kahn and Meyer 1970; Martinez et al. 1973).

Chronic alcoholic myopathy. Ekbom et al. (1964) first drew attention to a more slowly evolving proximal myopathy in chronic alcoholics. The clinical picture is that of painless progressive weakness with wasting of pelvic and shoulder girdle muscles. Affected individuals frequently show other features of chronic alcoholism, such as peripheral neuropathy, Wernicke encephalopathy, hepatic cirrhosis, and nutritional deficiency. The serum CK activity is usually normal. EMG shows myopathic potentials or mixed myopathic and neuropathic changes in proximal limb muscles.

The typical histological finding in biopsies of proximal limb muscles, such as the quadriceps femoris, is type II fiber atrophy (Martin et al. 1985). In addition, there is accumulation of triglyceride in muscle fibers (Sunnasy et al. 1983). Type II fiber atrophy is accompanied by a reduction in glycolytic and glycogenolytic enzyme activity (Langohr et al. 1983; Martin et al. 1984). This is presumably the basis for the finding of reduced lactic acid production after ischemic exercise in alcoholics (Perkoff et al. 1966). Type II fiber atrophy is also a frequent finding in biopsies from the quadriceps femoris in alcoholics without symptoms of muscle weakness suggesting that, as in the case of acute alcoholic myopathy, chronic alcoholic myopathy is frequently subclinical (Martin et al. 1982). In some cases, particularly those with associated peripheral neuropathy, histological changes of denervation atrophy are found in the proximal leg muscles suggesting that the atrophy and weakness is of neurogenic origin (Faris and Reyes 1971; Urbano-Marquez et al. 1985). Tubular aggregates have been found in type II fibers in biopsies from some chronic alcoholics (Chui et al. 1975).

The proximal myopathy may improve gradually with abstinence and type II fiber atrophy is also reversible (Slavin et al. 1983).

Pathogenesis. There is a good deal of evidence that at least the acute necrotizing form of alcoholic myopathy is due to a direct toxic effect of ethanol itself or its metabolite, acetaldehyde. Experimental observations in humans and in rats have shown acute elevations in serum CK activity of skeletal muscle origin after administration of ethanol, with a direct relationship between the serum enzyme and blood alcohol levels (Haller and Drachman 1980; Lane and Radoff 1981; Schubert et al. 1981; Spargo 1984; Haller 1985). In addition, ultrastructural changes were found in muscle fibers in human volunteers after regular ingestion of large quantities of ethanol for 1 month (Song and Rubin 1972). In-vitro studies indicate that ethanol alters the configuration, fluidity, and (Na^+,K^+)-ATPase activity of cell membranes and inhibits calcium uptake by the sarcoplasmic reticulum (Haller and Knochel 1984). Ethanol also causes marked inhibition of oxidation of palmitic acid and glucose-6-phosphate, two of the major substrates for energy production in skeletal muscle (Anderson and Torrance 1984).

Other contributory factors may be involved in some cases of alcoholic myopathy. Food deprivation, a common accompaniment of binge drinking,

TABLE 3

Snake venoms associated with rhabdomyolysis and their myotoxic components.

Elapid	
Notechis scutatus scutatus	Notexin
(Tiger snake)	Notechis II-5
Oxyuranus scutellatus	Taipoxin
(Taipan)	
Pseudechis australis	Mulgatoxin a
(Mulga)	
Enhydrina schistosa	Phospholipase A
(Seasnake)	
Dendroaspis jamesoni	
(Jameson's mamba)	
Crotalid	
Crotalus viridis viridis	Myotoxin a
(Prairie rattlesnake)	
Crotalus durissus terrificus	Crotamine
(South American rattlesnake)	Crotoxin
Crotalus atrox	
(Western diamondback rattle-	
snake)	
Bothrops asper	
Bothrops nummifer	
(Jumping viper of Costa Rica)	

retards the metabolism of ethanol, allowing the development of high blood levels that may be toxic to skeletal muscle. In the rat model of experimental alcoholic myopathy, muscle necrosis may be triggered by a period of food deprivation which leads to markedly elevated blood ethanol levels (Haller 1985). Hypokalemia is present in some cases of alcohol withdrawal and may be severe enough to cause a hypokalemic myopathy in some cases (Rubenstein and Wainapel 1977). Phosphate depletion, which may develop with chronic alcohol ingestion, may also play a role in acute myopathy of some chronically malnourished alcoholics.

Focal myopathy

Localized areas of muscle damage follow intramuscular injections as a result of needle insertion (*needle myopathy*) and local effects of the injected agent (Mastaglia 1982). Diazepam, digoxin, and lignocaine (lidocaine) cause more extensive muscle necrosis when injected into animals and lead to elevated serum CK levels (Steiness et al. 1977; Yagiela et al. 1981). Other drugs that have a local myotoxic effect include chloroquine, opiates, paraldehyde, cephalothin, and chlorpromazine, which may cause severe tissue damage and abscess formation (Mastaglia and Argov 1981; Saito et al. 1982).

Repeated intramuscular injections may lead to marked fibrosis and fibrous contractures. This has been seen after prolonged courses of antibiotic injections into the quadriceps and deltoid muscles in children and in drug addicts who may develop multiple limb muscle contractures following repeated self-injection of pethidine or pentazocine (Mastaglia et a. 1971; Hoefnagel et al. 1978; Rousseau et al. 1979; Mariani et al. 1981; Adams et al. 1983; Robertson and Dimon 1983; Choucair and Ziter 1984).

MYOPATHIES DUE TO ENVENOMATION

Snake venoms

The venoms of a number of Crotaline and Elapine snakes and seasnakes have myotoxic properties and may cause myoglobinuria in addition to postsynaptic neuromuscular blockade (Table 3) (Meldrum 1965; Tu 1977; Sunderland 1983). Myolysis restricted to the site of the bite may be caused by many venoms, but those shown in Table 3 have been associated with more widespread muscle necrosis and myoglobinuria (Fig. 1B). The myotoxic components of a number of these venoms have been isolated and found to be polypeptides and their mechanisms of action have been studied in the experimental animal. Those most fully investigated are the venoms of the Australian tiger snake (*Notechis scutatus scutatus*) (Harris et al. 1975; Sutherland and Coulter 1977; Pluskal et al. 1978; Ng and Howard 1980); the taipan (*Oxyuranus scutellatus*) (Harris and Maltin 1982); the mulga (*Pseudechis australis*) (Rowlands et al. 1969; Papadimitriou and Mastaglia 1973; Leonard et al. 1979); the seasnake *Enhydrina schistosa* (Reid 1961; Fohlman and Eaker 1977); the coral snake (*Micrurus nigrocinctus*) (Gutierrez et al. 1986); the rattlesnakes of North and South America (the prairie rattlesnake, *Crotalus viridis viridis*, the Western diamondback rattlesnake *Crotalus atrox*, and the South American rattlesnake *Crotalus durissus terrificus*) (Cameron and Tu 1978; Ownby et al. 1979; Huang and Perez 1982; Azevedo-Marques et al. 1985); and the Costa Rican vipers (*Bothrops nummifer* and *Bothrops asper*) (Gutierrez et al. 1984, 1989).

Spider venoms

The venoms of the Arkansas and Honduran tarantulas (*Dugesiella hentzi* and *Aphonophelma spp.*) are intensely myotoxic and cardiotoxic. Both the crude venom and purified necrotoxin (MW 6700 daltons) cause rapid irreversible injury to the muscle fiber plasma membrane leading to necrosis and marked accumulation of calcium and phosphate in muscle fibers (Ownby and Odel 1983). The brown spider has also been reported to cause rhabdomyolysis (Gabow et al. 1982). The venom of the black widow spider (*Lactrodectus*) acts at the neuromuscular junction leading to depletion of ACh from presynaptic terminals and permanent neuromuscular block (Gorio and Mauro 1979; Duchen 1981).

Wasp venoms

There have been reports of severe widespread muscle necrosis and renal failure following envenomation by the wasp *Vespa cincta* and the hornet *Vespa affinis* (Shilkin et al. 1972; Sitprija and Boonpucknavig 1972). The venom of *Vespa affinis* is known to contain polypeptides and phospholipases, but the myotoxic components have not been identified (Shilkin et al. 1972). Ishay et al. (1975) demonstrated specific involvement of the transverse tubular system by the venom of the oriental hornet (*Vespa orientalis*).

HAFF DISEASE

This condition occurred in epidemic form in East Prussia, Russia, and Sweden between 1923 and 1943 and has not been reported from other parts of the world (Berlin 1948). It is estimated that over 1000 cases occurred in the 2 major epidemics in the Koenigsberg Haff in East Prussia. In each epidemic, the people most affected were those who had eaten fish from nearby waters or who had fished these waters. Low-grade muscle discomfort for a few days was followed by the sudden onset of severe widespread myalgia and tenderness, particularly in the calves, back, and neck, with dark-brown discoloration of the urine (presumed to be myoglobinuria), with leukocytosis and creatinuria. Rapid recovery over a period of 24–72 hours was the rule. The condition was presumed to be toxic in origin, but the nature of the toxic agent has never been identified in spite of extensive investigation.

QUAIL MYOPATHY

This condition, which has been reported from Algiers and Greece, particularly on the island of Lesbos, is characterized by the onset of severe myalgia, dark red pigmenturia and acute renal failure after ingestion of quail (Billis et al. 1971). The myopathological changes and mechanism of the myopathy have not been investigated. The toxic agent may be derived from the seeds of the plant *Conium maculatum* (hemlock) on which the quail feed. On the other hand, repeated attacks in some individuals led to the suggestion that there might be an underlying enzymic deficiency that predisposes individuals to myoglobinuria.

MICROBIAL TOXINS

Clostridial toxins

Clostridial toxins may have profound effects on the neuromuscular system. *C. welchii* and *C. perfringens*, which cause gas gangrene, produce a number of toxins. The one thought to be primarily responsible for muscle damage is the α-toxin (lecithinase C), which has been shown experimentally to cause focal lysis of the muscle fiber plasma membrane and necrosis (Strunk et al. 1967).

Botulinum toxin causes neuromuscular block by preventing ACh release from motor nerve terminals. Experimental studies have shown degenerative changes in muscle fibers and motor end-plates with prominent sprouting of nerve terminals after intramuscular injection of the toxin (Duchen 1971a,b).

In addition to its central action on inhibitory spinal cord synapses, tetanus toxin also acts on motor nerve terminals interfering with transmitter release and causing prolonged weakness or paralysis (Duchen 1973). Degenerative changes were found in muscle fibers but not in motor nerve terminals or end-plates in biopsies from patients with tetanus (Agostini and Noetzel 1970). In an experimental study in the mouse, intramuscular injection of tetanus toxin caused sprouting of motor nerve terminals in slow-twitch but not in fast-twitch muscle fibers (Duchen 1973).

Monensin toxicity

Monensin, a polyether antibiotic produced by *Streptomyces cinnamonensis*, is a Na^+-selective carboxylic ionophore that can produce cardiac and skeletal muscle necrosis when given therapeutically to animals. Selective involvement of type 1 fibers has been found in experimental studies (Van Vleet and Ferrans 1984).

ORGANOPHOSPHATES

Myopathy may appear in individuals exposed to organophosphate insecticides, but this is much less frequent than the polyneuropathy that follows acute or chronic exposure to these agents (Wadia et al. 1974; Ahlgren et al. 1979; Vasilescu et al. 1984). Experimental studies in the rat have shown that organophosphates, which are irreversible cholinesterase inhibitors, produce a progressive dose-related necrosis of muscle fibers that begins in the motor end-plate region. The myopathy seems to be due to increased neurotransmitter release and can be prevented by prior denervation or administration of pyridine-2-aldoxime methiodide (Wecker et al. 1978).

GASOLINE SNIFFING

Marked serum CK elevations or an acute myopathy with myoglobinuria has been seen in gasoline sniffers (Kovanen et al. 1983; Fortenberry 1985). In 2 cases the myopathy was associated with signs of encephalopathy; in 1 case, myopathy was the predominant manifestation (Kovanen et al. 1983). It is not known which of the various organic solvents and other components of gasoline is myotoxic.

SOLVENTS

Clinical myoglobinuria has been reported after intoxication with the organic solvent, toluene, which is used in paint sprays, lacquer thinners, and household glues (Streicher et al. 1981). Hypokalemic periodic paralysis due to renal tubular acidosis has followed chronic toluene exposure (Bennett and Forman 1980). Serum CK activity was elevated in subjects with occupational exposure to solvents. Further studies are required to determine the frequency of neuromuscular effects in these workers (Pedersen et al. 1980).

METAL FUME FEVER

A case of probable myopathy with associated myocardial involvement was seen in a welder who had recurrent episodes of 'metal fume fever' and myalgia, limb weakness, ECG abnormalities, and serum CK elevation. Inhalation of zinc oxide fumes was thought to have been responsible (Shusterman and Neal 1986).

Acknowledgements

The author acknowledges with thanks the expert secretarial assistance of Mrs. S. Moncrieff. Certain portions of this Chapter have also been included in a review on Drug-induced, Toxic and Nutritional Myopathies by B.A. Kakulas and F.L. Mastaglia in *Skeletal Muscle Pathology*, F.L. Mastaglia and J.N. Walton (Eds.), Churchill Livingstone, Edinburgh (in press).

REFERENCES

ABREO, K., W.D. SHELP, A. KOSSEFF and S. THOMAS: Amoxapine-associated rhabdomyolysis and acute renal failure: case report. J. Clin. Psychiatr. 43 (1982) 426–427.

ADAMS, E.M., H.M. HOROWITZ and W.R. SUNDSTROM: Fibrous myopathy in association with pentazocine. Arch. Intern. Med. 143 (1983) 2203–2204.

AFFARAH, H.B., R.L. MARS, A. SOMEREN, H.W. SMITH and S.B. HEYMSFIELD: Myoglobinuria and acute renal failure associated with intravenous vasopressin infusion. South. Med. J. 77 (1984) 918–921.

AFIFI, A.K. and R.A. BERGMAN: Steroid myopathy. A study of the evolution of the muscle lesion in rabbits. Johns Hopkins Med. J. 124 (1969) 66–86.

AFIFI, A.K., G.A. HAJJ, S. SAAD, A. TEKIAN, R.A. BERGMAN, N.B. BAHUTH and N. ABOURIZK: Clofibrate-induced myotoxicity in rats. Temporal profile of myopathology. Eur. Neurol. 23 (1984) 182–197.

AGOSTINI, B. and H. NOETZEL: Morphological study of muscle fibres and motor end-plates in tetanus. In: J.N. Walton, N. Canal and G. Scarlato (Eds.), Muscle Diseases. Amsterdam, Excerpta Medica (1970) 123–127.

AHLGREN, J.D., H.J. MANZ and J.C. HARVEY: Myopathy of chronic organo-phosphate poisoning: a clinical entity. South. Med. J. 72 (1979) 555–559.

ANDERSON, P.J., S.K. SONG and P. SLOTWINER: The fine structure of spheromembranous degeneration of skeletal muscle induced by vincristine. J. Neuropathol. Exp. Neurol. 25 (1967) 15.

ANDERSON, T.L. and C.A. TORRANCE: Metabolic mechanisms of acute alcoholic myopathy. Neurology 34 (1984) 81.

ARGOV, Z. and F.L. MASTAGLIA: Drug-induced peripheral neuropathies. Br. Med. J. 1 (1979a) 663–666.

ARGOV, Z. and F.L. MASTAGLIA: Disorders of neuromuscular transmission caused by drugs. N Engl. J. Med. 301 (1979b) 409–413.

ARGOV, Z. and F.L. MASTAGLIA: Drug-induced neuromuscular disorders in man. In: J.N. Walton (Ed.), Disorders of Voluntary Muscle. Edinburgh, Churchill Livingstone, 4th Ed. (1988) 981–1014.

ARNAUDO, E., M. DALAKAS, S. SHANSKE, C.T. MORRAES, S. DI MAURO and E.A. SCHON: Depletion of muscle mitochondrial DNA in AIDS patients with zidovudine-induced myopathy. Lancet 337 (1991) 508–510.

ASKARI, A., P.J. VIGNOS and R.W. MOSKOWITZ: Steroid

myopathy in connective tissue disease. Am. J. Med. 61 (1976) 485–492.

AZEVEDO-MARQUES, M.M., P. CUPO, T.M. COIMBRA, S.E. HERING, M.A. ROSSI and C.J. LAURE: Myonecrosis, myoglobinuria and acute renal failure induced by South American rattlesnake (*Crotalus durissus terrificus*) envenomation in Brazil. Toxicon 23 (1985) 631–636.

BASSER, L.S.: Purgatives and periodic paralysis. Med. J. Aust. 1 (1979) 47–49.

BAUD, F., C. BISMUTH, M. GALLIOT, R. GARNIER and A. PERALMA: Rhabdomyolysis in paraphenylenediamine intoxication. Lancet ii (1983) 514.

BELONGIA, E.A., C.W. HEDBERG, G.J. GLEICH, K.E. WHITE, A.N. MAYENO, D.A. LOEGERING, S.L. DUNNETTE, P.L. PIRIE, K.L. MACDONALD and M.T. OSTERHOLM: An investigation of the cause of the eosinophilia-myalgia syndrome associated with tryptophan use. N. Engl. J. Med. 323 (1990) 357.

BENNETT, H.S., A.J. SPIRO, M.A. POLLACK and P. ZUCKER: Ipecac-induced myopathy simulating dermatomyositis. Neurology 32 (1982) 91–94.

BENNETT, R.H. and H.R. FORMAN: Hypokalemic periodic paralysis in chronic toluene exposure. Arch. Neurol. 37 (1980) 673.

BERLIN, R.: Haff disease in Sweden. Acta Med. Scand. 129 (1948) 560–572.

BILLIS, A.G., S. KASYANAKIS, H. GIAMARELLOU and G.K. DAIKOS: Acute renal failure after a meal of quail. Lancet i (1971) 702.

BINDOFF, L. and M.J. CULLEN: Experimental (-)emetine myopathy. Ultrastructural and morphometric observations. J. Neurol. Sci. 39 (1978) 1–15.

BLACK, W.D. and W.M. MURPHY: Nontraumatic rhabdomyolysis and acute renal failure associated with oral phenmetrazine hydrochloride. J. Tenn. Med. Assoc. 77 (1984) 80–81.

BLESSING, W. and J.C. WALSH: Myotonia precipitated by propranolol therapy. Lancet i (1977) 73–74.

BLEWITT, G.A. and E.B. SIEGEL: Renal failure, rhabdomyolysis and phenylpropanolamine. J. Am. Med. Assoc. 249 (1983) 3017–3018.

BLUMBERG, A. and H.R. MARTI: Akute Rhabdomyolyse nach Succinylcholin. Schweiz. Med. Wochenschr. 114 (1984) 1068–1071.

BOWYER, S.L., M.P. LAMONTHE and J.R. HOLLISTER: Steroid myopathy: incidence and detection in a population with asthma. J. Allergy Clin. Immunol. 76 (1985) 234–242.

BRADLEY, W.G.: The neuromyopathy of vincristine in the guinea pig. An electrophysiological and pathological study. J. Neurol. Sci. 10 (1970) 133–162.

BRADLEY, W.G., L.P. LASSMAN, G.W. PEARCE and J.N. WALTON: The neuromyopathy of vincristine in man. Clinical, electrophysiological and pathological studies. J. Neurol. Sci. 10 (1970) 107–131.

BRADLEY, W.G., J.D. FEWINGS, J.B. HARRIS and M.A. JOHNSON: Emetine myopathy in the rat. Br. J. Pharmacol. 57 (1976) 29–41.

BRAUNSTEIN, P.W. and U. DE GIROLAMI: Experimental

corticosteroid myopathy. Acta Neuropathol. 55 (1981) 167–172.

BRETAG, A.H., S.R. DAWE and A.G. MOSKWA: Chemically induced myotonia in amphibia. Nature (London) 286 (1980a) 625–626.

BRETAG, A.H., S.R. DAWE, D.I.B. KERR and A.G. MOSKWA: Myotonia as a side effect of diuretic action. Br. J. Pharmacol. 71 (1980b) 467–471.

BRINER, V., A. COLOMBI, W. BRUNNER and B. TRUNIGER: Die akute Rhabdomyolyse. Schweiz. Med. Wochenschr. 116 (1986) 198–208.

BRITT, B.A.: Etiology and pathophysiology of malignant hyperthermia. Fed. Proc. 38 (1979) 44–48.

BRITT, C.W., R.R. LIGHT, B.H. PETERS and S.S. SCHOCHET: Rhabdomyolysis during treatment with epsilon-aminocaproic acid. Arch. Neurol. 37 (1980) 187–188.

BROTMAN, M.C., N. FORBATH, P.E. GARFINKEL and J.G. HUMPHREY: Myopathy due to ipecac syrup poisoning in a ptient with anorexia nervosa. Can. Med. Assoc. J. 125 (1981) 453–454.

BROWN, J.A., R.L. WOLLMANN and S. MULLAN: Myopathy induced by epsilon aminocaproic acid. J. Neurosurg. 57 (1982) 130–134.

BROWNELL, A.K.W. and A.G. ENGEL: Experimental lipid storage myopathy. A quantitative ultrastructural and biochemical study. J. Neurol. Sci. 35 (1978) 31–41.

BRUMBACK, R.A., J.W. GERST and H.R. KNULL: High energy phosphate depletion in a model of defective muscle glycolysis. Muscle Nerve 6 (1983) 52–55.

BRYANT, S.: The electrophysiology of myotonia, with a review of congenital myotonia of goats. In: J.E. Desmedt (Ed.), New Developments in Electromyography and Clinical Neurophysiology, Vol. 1. Basel, Karger (1973) 420–450.

BRYANT, S. and A. MORALES-AGUILERA: Chloride conductance in normal and myotonic muscle fibres and the action of monocarboxylic aromatic acids. J. Physiol. 219 (1971) 367–383.

CACCIA, M.R., A. BOIARDI, L. ANDREUSI and F. CORNELIO: Nerve supply and experimental myotonia in rats. J. Neurol. Sci. 24 (1975) 145–150.

CAMERON, D.L. and A.T. TU: Chemical and functional homology of myotoxin A from prairie rattlesnake venom and crotamine from South American rattlesnake venom. Biochim. Biophys. Acta 532 (1978) 147–154.

CENTERS FOR DISEASE CONTROL: Analysis of L-tryptophan for the etiology of eosinophilia-myalgia syndrome. MMWR 39 (1990) 589–591.

CHABOCHE, C., Y. NORDMANN, J.L. FONTAINE and C. LEJEUNE: Myoglobinuria postanesthesique. Arch. Fr. Pediatr. 39 (1982) 169–171.

CHAIKIN, H.L.: Rhabdomyolysis after drug overdose with coma. South. Med. J. 73 (1980) 990–994.

CHALIKIAN, D.M. and R.L. BARCHI: Membrane desmosterol and the kinetics of the sarcolemmal Na^+, K^+-ATPase in myotonia induced by 20,25-diazacholesterol. Exp. Neurol. 77 (1982) 578–589.

CHALMERS, A.C., C.M. GRECO and R.G. MILLER: Prognosis in AZT myopathy. Neurology 41 (1991) 1181–1184.

CHOUCAIR, A.K. and F.A. ZITER: Pentazocine abuse masquerading as familial myopathy. Neurology 34 (1984) 524–527.

CHUGH, K.S., P.C. SINGHAL and H.S. UBEROI: Rhabdomyolysis and renal failure in acute mercuric chloride poisoning. Med. J. Aust. 2 (1978) 125–126.

CHUGH, K.S., I.V.S. NATH, H.S. UBROI, P.C. SINGHAL, S.K. PAREEK and A.K. SARKAR: Acute renal failure due to non-traumatic rhabdomyolysis. Postgrad. Med. J. 55 (1979) 386–392.

CHUI, L.A., H. NEVSTEIN and T.L. MUNSAT: Tubular aggregates in subclinical alcoholic myopathy. Neurology 25 (1975) 405–412.

CLARK, A.F. and P.J. VIGNOS: Experimental corticosteroid myopathy: effect on myofibrillar ATPase activity and protein degradation. Muscle Nerve 2 (1979) 265–273.

CLARK, A.F., G.N. DEMARTINO and K. WILDENTHAL: Effects of glucocorticoid treatment on cardiac protein synthesis and degradation. Am. J. Physiol. 250 (1986) C821–C827.

CLOUSTON, P.D. and P.E. DONNELLY: Acute necrotising myopathy associated with amiodarone therapy. Aust. N. Z. J. Med. 19 (1989) 483.

COGEN, F.C., F. RIGG, J.L. SIMMONS and E.F. DOMINO: Phencyclidine-associated acute rhabdomyolysis. Ann. Intern. Med. 88 (1978) 210–212.

CONTE-CAMERINO, D., S.H. BRYANT and D. MITOLO-CHIEPPA: Electrical properties of rat extensor digitorum longus muscle after chronic application of emetine to the motor nerve. Exp. Neurol. 77 (1982) 1–11.

COOMES, E.N.: Corticosteroid myopathy. Ann. Rheum. Dis. 24 (1965a) 465.

COOMES, E.N.: The rate of recovery of reversible myopathies and the effects of anabolic agents. Neurology 15 (1965b) 523.

COOPER, J.M., R.K. PETTY, D.J. HAYES, R.A.J. CHALLIS, M.J. BROSNAN, E.A. SHOUBRIDGE, G.K. RADDA, J.A. MORGAN-HUGHES and J.B. CLARK: An animal model of mitochondrial myopathy: a biochemical and physiological investigation of rats treated in vivo with the NAD-Coq reductase inhibitor diphenyleneiodonium. J. Neurol. Sci. 83 (1988) 335–347.

CUMMING, A.M.M., K. BODDY, J.J. BROWN, R. FRASER, A.F. LEVER, P.L. PADFIELD and J.I.S. ROBERTSON: Severe hypokalaemia with paralysis induced by small doses of liquorice. Postgrad. Med. J. 56 (1980) 526–569.

D'ALONZO, A.J. and J.J. MCARDLE: An evaluation of fast- and slow-twitch muscle from rats treated with 20,25-diazacholesterol. Exp. Neurol. 78 (1982) 46–66.

DALAKAS, M.C., I. ILLA, G.H. PEZESHKPOUR, J.P. LLAUKAITIS, B. COHEN and J.L. GRIFFIN: Mitochondrial myopathy caused by long-term zidovudine therapy. N. Engl. J. Med. 322 (1990) 1098–1105.

DANON, J.M., G. KARPATI and S. CARPENTER: Subacute skeletal myopathy induced by 2,4-dichlorophenoxyacetate in rats and guinea pigs. Muscle Nerve 1 (1978) 89–102.

DRENCKHAHN, D. and R. LULLMANN-RAUCH: Experimental myopathy induced by amphiphilic cationic compounds including several psychotropic drugs. Neuroscience 4 (1979) 549–562.

DROMGOOLE, S., D. CAMPION and J. PETER: Myotonia induced by clofibrate and sodium chlorophenoxyisobutyrate. Biochem. Med. 14 (1975) 238–240.

DROPCHO, E.J. and S. SOONG: Steroid-induced weakness in patients with primary brain tumors. Neurology 41 (1991) 1235–1239.

DUANE, D.D. and A.G. ENGEL: Emetine myopathy. Neurology 20 (1970) 733–739.

DUBOIS, D.C. and R.R. ALMON: Disuse atrophy of skeletal muscle is associated with an increase in number of glucocorticoid receptors. Endocrinology 107 (1980) 1649–1651.

DUBOIS, D.C. and R.R. ALMON: A possible role for glucocorticoids in denervation atrophy. Muscle Nerve 4 (1981) 370–373.

DUBOIS, D.C. and R.R. ALMON: Glucocorticoid sites in skeletal muscle: adrenalectomy, maturation, fiber type, and sex. Am. J. Physiol. 247 (1984) E118–E125.

DUCHEN, L.W.: An electronmicroscopic study of the changes induced by botulinum toxin in the motor end-plates of slow and fast skeletal muscle fibres of the mouse. J. Neurol. Sci. 14 (1971a) 47–60.

DUCHEN, L.W.: Changes in the electron microscopic structure of slow and fast skeletal muscle fibres of the mouse after the local injection of botulinum toxin. J. Neurol. Sci. 14 (1971b) 61–74.

DUCHEN, L.W.: The effects of tetanus toxin on the motor end-plates of the mouse. An electron microscopic study. J. Neurol. Sci. 19 (1973) 153–167.

DUCHEN, L.W.: The neuromuscular junction of the mouse after black widow spider venom. J. Physiol. (London) 316 (1981) 279–291.

EADIE, M.J. and T.M. FERRIER: Chloroquine myopathy. J. Neurol. Neurosurg. Psychiatr. 29 (1966) 331–337.

EBERSTEIN, A. and J. GOODGOLD: Experimental myotonia induced in denervated muscles by 2,4-D. Muscle Nerve 2 (1979) 364–368.

EBERSTEIN, A., J. GOODGOLD and R. JOHNSTON: Clofibrate-induced myotonia in the rat. Experientia 34 (1978) 1607.

EIDSON, M., R.M. PHILEN, C.M. SEWELL, R. VOORHEES and E.M. KILBOURNE: L-Triptophan and eosinophilia-myalgia syndrome in New Mexico. Lancet 335 (1990) 645.

EISENBERG, M., S. HALSLEY and R.L. KATZ: Effects of diazepam on succinylcholine-induced myalgia, potassium increase, creatine phosphokinase elevation, and relaxation. Anest. Analg. (Cleveland) 58 (1979) 314–317.

EISER, A.R., M.S. NEFF and R.F. SLIFKIN: Acute myoglobinuric renal failure. A consequence of the neuroleptic malignant syndrome. Arch. Intern. Med. 142 (1982) 601–603.

EKBOM, K., R. HED, L. KIRSTEIN and K.-E. ASTROM: Muscular affections in chronic alcoholism. Arch. Neurol. 10 (1964) 449–458.

ESTES, M.L. D. EWING-WILSON, S.M. CHOU, H. MITSUMOTO, M. HANSON, E. SHIREY and N.B. RATLIFF: Chloroquine neuromyotoxocity. Am. J. Med. 82 (1987) 447.

FARDEAU, M., F.M.S. TOME and P. SIMON: Muscle and nerve changes induced by perhexiline maleate in man and mice. Muscle Nerve 2 (1979) 24–36.

FARIS, A.A. and M.G. REYES: Reappraisal of alcoholic myopathy. J. Neurol. Neurosurg. Psychiatr. 34 (1971) 86–92.

FEWINGS, J.D., R.J. BURNS and B.A. KAKULAS: A case of acute emetine myopathy. In: B.A. Kakulas (Ed.), Clinical Studies in Myology. Amsterdam, Excerpta Medica (1973) 594–598.

FOHLMAN, J. and D. EAKER: Isolation and characterization of a lethal myotoxic phospholipase A from the venom of the common sea snake Enhydrina schistosa causing myoglobinuria in mice. Toxicon 15 (1977) 385–393.

FORFAR, J.C., G.J. BROWN and R.E. CULL: Proximal myopathy during beta blockade. Br. Med. J. 279 (1979) 1331–1332.

FORTENBERRY, J.D.: Gasoline sniffing. Am. J. Med. 79 (1985) 740–744.

FRENDIN, T.J. and C.P. SWAINSON: Acute renal failure secondary to non-traumatic rhabdomyolysis following amoxapine overdose. N. Z. Med. J. 98 (1985) 690–691.

FRIEDMAN, E.J.: Death from ipecac intoxication in a patient with anorexia nervosa. Am. J. Psychiatr. 141 (1984) 702–703.

FRISK-HOLMBERG, M., L. JORFELDT, A. JUHLIN-DANNFELT and J. KARLSSON: Metabolic changes in muscle on long-term alprenolol therapy. Clin. Pharmacol. Ther. 26 (1979) 566–571.

FURMAN, R.E. and R.L. BARCHI: 20,25-Diazacholesterol myotonia: an electrophysiological study. Ann. Neurol. 10 (1981) 251–260.

GABOW, P.A., W.D. KAEHNY and S.P. KELLEHER: The spectrum of rhabdomyolysis. Medicine 61 (1982) 141–152.

GABRY, A.L., J.L. POURRIAT, P. HOANG THE DAN and M. CUPA: Acute rhabdomyolysis associated with heroin addiction. Ann. Fr. Anesth. Reanim. 1 (1982) 179–181.

GARDINER, P.F. and V.R. EDGERTON: Contractile responses of rat fast-twitch and slow-twitch muscles to glucocorticoid treatment. Muscle Nerve 2 (1979) 274–281.

GIBB, W.R.G. and I.C. SHAW: Myoglobinuria due to heroin abuse. J. R. Soc. Med. 78 (1985) 862–863.

GIBBS, J.M.: A case of rhabdomyolysis associated with suxamethonim. Anesth. Intens. Care 6 (1978) 141–142.

GORI, Z., V. DE TATA, M. POLLERA and E. BERGAMINI: Mitochondrial myopathy in rats fed with a diet containing beta-guanidine propionic acid, an inhibitor of creatine entry in muscle cells. Br. J. Exp. Pathol. 69 (1988) 639–650.

GORIO, A. and A. MAURO: Reversibility and mode of action of black widow spider venom on the vertebrate neuromuscular junction. J. Gen. Physiol. 73 (1979) 245–263.

GROB, D.: Rhabdomyolysis and drug-related myopathies. Curr. Opin. Rheumatol. 2 (1990) 908–915.

GROSSIE, J. and E.X. ALBUQUERQUE: Extensor muscle responses to triamcinolone. Exp. Neurol. 58 (1978) 435–445.

GRUENER, R. and L.Z. STERN: Corticosteroids. Effects on muscle membrane excitability. Arch. Neurol. 26 (1972) 131–181.

GUTIERREZ, J.M., C.L. OWNBY and G.V. ODELL: Isolation of a myotoxin from Bothrops asper venom: partial characterisation and action on skeletal muscle. Toxicon 22 (1984) 115.

GUTIERREZ, J.M., O. ARRJOYO, F. CHAVES, B. LOMONTE and L. CERDAS: Pathogenesis of myonecrotis induced by coral snake (Micrurus nigrocinctus) venom in mice. Br. J. Exp. Pathol. 67 (1986) 1.

GUTIERREZ, J.M., F. CHAVES, J.A. GENSE, B. LOMONTE and Z. COMACHO: Myonecrosis induced in mice by a basic myotoxin isolated from the venom of the snake Bothrops nummifer (jumping viper) from Costa Rica. Toxicon 27 (1989) 785.

HALES, D.S.M., R. SCOTT and H.J.E. LEWI: Myopathy due to mercaptoproprionyl glycine. Br. Med. J. 285 (1982) 939.

HALLA, T., S. FALLAHI and W.J. KOOPMAN: Penicillamine-induced myositis. Observations and unique features in two patients and review of the literature. Am. J. Med. 77 (1984) 719–722.

HALLER, R.G.: Experimental acute alcoholic myopathy: a histochemical study. Muscle Nerve 8 (1985) 195–203.

HALLER, R.G. and D.B. DRACHMAN: Alcoholic rhabdomyolysis. An experimental model in the rat. Science 208 (1980) 412–415.

HALLER, R.G. and J.P. KNOCHEL: Skeletal muscle disease in alcoholism. Med. Clin. North Am. 68 (1984) 91–103.

HAMPEL, G., H.H HORSTKOTTE and K.W. RUMPF: Myoglobinuric renal failure due to drug-induced rhabdomyolysis. Hum. Toxicol. 2 (1983) 197–203.

HARNEY, J. and M.R. GLASBERG: Myopathy and hypersensitivity in phenytoin. Neurology 33 (1983) 790–791.

HARRIS, J.B. and C.A. MALTIN: Myotoxic activity of the crude venom and the principal neurotoxin, taipoxin, of the Australian taipan, Oxyuranus scutellatus. Br. J. Pharmacol. 76 (1982) 61–75.

HARRIS, J.B., M.A. JOHNSON and E. KARLSSON: Pathological responses of rat skeletal muscle to a single subcutaneous injection of a toxin isolated from the venom of the Australian tiger snake, Notechis scutatus scutatus. Clin. Exp. Pharmacol. Physiol. 2 (1975) 383–404.

HAUBENSTOCK, A., P. SCHMIDT, J. ZAZGORNIK, P.

BALCKE, H. KOPSA and K. SERTL: Bezafibrat-induzierte Rhabdomyolyse bei einem Patienten mit eingeschrankter Nierenfunktion. Dtsch. Med. Wochenschr. 109 (1984) 157–158.

HAYES, D.J., E. BYRNE, E.A. SHOUBRIDGE, J.A. MORGAN-HUGHES and J.B. CLARK: Experimentally induced defects of mitochondrial metabolism in rat skeletal muscle. Biochem. J. 229 (1985) 109–117.

HED, R., C. LUNDMARK, H. FAHLGREN and S. ORELL: Acute muscular syndrome in chronic alcoholism. Acta Med. Scand. 171 (1962) 585–599.

HEINER, L., O. TAKACS and F. GUBA: Changes in the contractile proteins of skeletal muscles induced by steroid myopathy. Acta Physiol. Acad. Sci. Hung. 55 (1980) 51–55.

HERTZMAN, P.A., W.L. BLEVINS, J. MAYER, B. GREENFIELD, M. TING and G.J. GLEICH: Association of the eosinophilia-myalgia syndrome with the ingestion of tryptophan. N. Engl. J. Med. 322 (1990) 869.

HICKSON, R.C. and J.R. DAVIS: Partial prevention of glucocorticoid-induced muscle atrophy by endurance training. Am. J. Physiol. 241 (1981) E226–E232.

HIGUCHI, I., S. IZUMO, M. KURIYAMA, M. SUCHARA, M. NAKAGAWA, H. FUKUNAGA, M. OSAME, S. OHTSUBO and K. MIGABA: Germanium myopathy. Clinical and experimental pathological studies. Acta Neuropath. (Berlin) 79 (1989) 300–304.

HOEFNAGEL, D., E.O. JALBERT, D.G. PUBLOW and A.J. RICHSMEIER: Progressive fibrosis of the deltoid muscles. J. Pediatr. 92 (1978) 79–81.

HORBER, F.F., J.R. SCHEIDEGGER, B.E. GRUNIG and F.J. FREY: Thigh muscle mass and function in patients treated with glucocorticoids. Eur. J. Clin. Invest. 15 (1985) 302–307.

HUANG, S.Y. and J.C. PEREZ: A comparative electron microscopic study of myonecrosis induced by *Crotalus atrox* (Western diamondback rattlesnake) in gray woodrats and mice. Toxicon 20 (1982) 443–449.

HUDGSON, P. and R. HALL: Endocrine myopathies. In: F.L. Mastaglia and J.N. Walton (Eds.), Skeletal Muscle Pathology. Edinburgh, Churchill Livingstone (1982) 393–408.

HUGHES, J.T., M. ESIRI, J.M. OXBURY and C.W.M. WHITTY: Chloroquine myopathy. Q. J. Med. 40 (1971) 85–93.

IYER, V., N.A. RANISH and G.M. FENICHEL: The effect of denervation on subsequent in vitro induction of myotonia. Neurology 27 (1977) 669–671.

JAEGER, U., A. PRODCZECK, A. HAUBENSTOCK, K. PIRICH, A. DONNER and K. HRUBY: Acute oral poisoning with lindane-solvent mixtures. Vet. Med. Toxicol. 26 (1984) 11–14.

JENKINS, P. and P.A. EMERSON: Myopathy induced by rifampicin. Br. Med. J. 283 (1981) 105–106.

JENNINGS, A.E., A.S. LEVEY and J.T. HARRINGTON: Amoxapine-associated acute renal failure. Arch. Intern. Med. 143 (1983) 1525–1527.

KAHN, L.B. and J.S. MEYER: Acute myopathy in chronic alcoholism: a study of 22 autopsy cases with ultrastructural observations. Am. J. Clin. Pathol. 53 (1970) 516–530.

KAISER, P., P.A. TESCH, A. THORSSON and J. KARLSSON: Skeletal muscle glycolysis during submaximal exercise following acute-adrenergic blockade in man. Acta Physiol. Scand. 123 (1985) 285–291.

KAKULAS, B.A. and F.L. MASTAGLIA: Drug induced, toxic and nutritional myopathies. In: F.L. Mastaglia and J.N. Walton (Eds.), Skeletal Muscle Pathology. Edinburgh, Churchill Livingstone (1992) 511–540.

KARLSSON, J.: Muscle fibre composition, short term 1-+ 2- and 1-blockade and endurance exercise performance in healthy young men. Drugs 25 Suppl. 2 (1983) 241–246.

KARPATI, G.: Denervation and disuse atrophy of skeletal muscles: involvement of endogenous glucocorticoid hormones? Trends Neurosci. 7 (1984) 61–62.

KARPATI, G. and G.V. WATTERS: Adverse anaesthetic reactions in Duchenne dystrophy. In: C. Angelini, G.A. Danieli and D. Fontanari (Eds.), Muscular Dystrophy Research: Advances and New Trends. International Congress Series No. 527. Amsterdam, Excerpta Medica (1980) 206–217.

KAUFMAN, L.E.: Neuromuscular manifestations of the L-tryptophan-associated eosinophilia-myalgia syndrome. Curr. Opin. Rheumatol. 2 (1990) 896–900.

KENNEDY, M.: Cardiac glycoside toxicity. An unusual manifestation of drug addiction. Med. J. Aust. 1 (1981) 686–689.

KHALEELI, A.A., R.H.T. EDWARDS, K. GOHIL, G. MCPHAIL, M.J. RENNIE, J. ROUND and E.J. ROSS: Corticosteroid myopathy: a clinical and pathological study. Clin. Endocrinol. 18 (1983) 155–160.

KILPATRICK, C., W. BRAUND and R. BURNS: Myopathy with myasthenic features possibly induced by codeine linctus. Med. J. Aust. 2 (1982) 410.

KLATSKIN, G. and H. FRIEDMAN: Emetine toxicity in man: studies on the nature of early toxic manifestations, their relation to the dose level, and their significance in determining safe dosage. Ann. Intern. Med. 28 (1948) 892–915.

KLEINKNECHT, D., A. PARENT, P. BLOT, G. BOCHEREAU, P.Y. LASLLEMENT and J.L. POURRIAT: Rhabdomyolysis with acute renal failure and malignant neuroleptic syndrome. Ann. Med. Intern. (Paris) 133 (1982) 549–552.

KLINKERFUSS, G., V. BLEISCH, M.M. DIOSO and G.T. PERKOFF: A spectrum of myopathy associated with alcoholism. II. Light and electron microscopic observations. Ann. Intern. Med. 67 (1967) 493–510.

KNOCHEL, J.P.: Rhabdomyolysis and myoglobinuria. Annu. Rev. Med. 33 (1982) 435–443.

KONTOS, H.A.: Myopathy associated with chronic colchicine toxicity. N. Engl. J. Med. 266 (1962) 38–39.

KOSKI, C.L., D.H. RIFENBERICK and S.R. MAX: Oxidative metabolism of skeletal muscle in steroid atrophy. Arch. Neurol. 31 (1974) 407–410.

KOVANEN, J., H. SOMER and P. SCHRODER: Acute myopathy associated with gasoline sniffing. Neurology 33 (1983) 629–631.

KUNCL, R.W., G. DUNCAN, D. WATSON, K. ALDERSON, M. ROGAWSKI and M. PEPER: Colchicine myopathy and neuropathy. N. Engl. J. Med. 316(25) (1987) 1562.

KWIECINSKI, H.: Myotonia induced with clofibrate in rats. J. Neurol. 219 (1978) 107–116.

KWIECINSKI, H.: Myotonia induced by chemical agents. CRC Crit. Rev. Toxicol. 8 (1981) 279–310.

LANE, R.J.M. and F.M. RADOFF: Alcohol and serum creatine kinase levels. Ann. Neurol. 10 (1981) 581–582.

LANE, R.J.M., A.M. MARTIN, NJ. MCLELLAND and F.L. MASTAGLIA: Epsilon aminocaproic acid (EACA) myopathy. Postgrad. Med. J. 55 (1979) 282–285.

LANGOHR, H.D., H. WIETHOLTER and J. PFEIFFER: Muscle wasting in chronic alcoholics: comparative histochemical and biochemical studies. J. Neurol. Neurosurg. Psychiatr. 46 (1983) 248–254.

LAZARUS, A.L. and J.U. TOGLIA: Fatal myoglobinuric renal failure in a patient with tardive dyskinesia. Neurology 35 (1985) 1055–1057.

LEONARD, J.P. and M.M. SALPETER: Agonist-induced myopathy at the neuromuscular junction is mediated by calcium. J. Cell Biol. 82 (1979) 811–819.

LEONARD, T.M., M.E.H. HOWDEN and I. SPENCE: A lethal myotoxin isolated from the venom of the Australian king brown snake (Pseudechis australis). Toxicon 17 (1979) 549–555.

LEWIS, C.A., N. BOHEIMER, P. ROSE and G. JACKSON: Myopathy after short term administration of procainamide. Br. Med. J. 292 (1986) 593–594.

LIVINGSTONE, I. M.A. JOHNSON and F.L. MASTAGLIA: Effects of dexamethasone on fibre subtypes in rat muscle. Neuropathol. Appl. Neurobiol. 7 (1981) 381–398.

LLACH, F., A.J. FELSENFELD and M.R. HAUSSLER: The pathophysiology of altered calcium metabolism in rhabdomyolysis-induced acute renal failure. N. Engl. J. Med. 305 (1981) 117–123.

LULLMANN, H., R. LULLMANN-RAUCH and O. WASSERMANN: Lipidosis induced by amphiphilic cationic drugs. Biochem. Pharmacol. 27 (1978) 1103–1108.

LULLMANN-RAUCH, R.: Drug-induced lysosomal storage disorders. In: J.T. Dingle, P.J. Jacques and I.H. Shaw (Eds.). Lysosomes in Applied and Therapeutics, Chapter 3. Amsterdam, North-Holland Publishing Co. (1979) 40–130.

MACFARLANE, I.A. and F.D. ROSENTHAL: Severe myopathy after status asthmaticus. Lancet i (1977) 615.

MAGARIAN, G.J., L.M. LUCAS and C. COLLEY: Gemfibrozil-induced myopathy. Arch. Intern. Med. 151 (1991) 1873–1874.

MAK, K.H., S.H. WAN, M.L. BOEY and Y.S. LEE: Myocardial and skeletal muscle injuries following adenine arabinoside therapy. Aust. N. Z. J. Med. 20 (1990) 811–813.

MARAIS, G.E. and K.K. LARSON: Rhabdomyolysis and acute renal failure induced by combination lovastatin and gemfibrozil therapy. Ann. Intern. Med. 112(3) (1990) 228.

MARIANI, C., G. MEOLA, P.L. MERONI, C. GUAITA and G. SCARLATO: Pentazocine-induced neuromuscular syndromes: clinical, immunological and histopathological studies in two cases. Acta Neuropathol. (Berlin), Suppl. 7 (1981) 246–248.

MARTIN, F.C., G. SLAVIN and A.J. LEVI: Alcoholic muscle disease. Br. Med. Bull. 38 (1982) 53–56.

MARTIN, F.C., A.J. LEVI, G. SLAVIN and T.J. PETERS: Glycogen content and activities of key glycolytic enzymes in muscle biopsies from control subjects and patients with chronic alcoholic skeletal myopathy. Clin. Sci. 66 (1984) 69–78.

MARTIN, F.C., G. SLAVIN, J. LEVI and T.J. PETERS: Alcoholic skeletal myopathy, a clinical and pathological study. Q. J. Med. 55 (1985) 233–251.

MARTINEZ, A.J., H. HOOSHMAND and A.A. FARIS: Acute alcoholic myopathy. J. Neurol. Sci. 20 (1973) 245–252.

MASTAGLIA, F.L.: Adverse effects of drugs on muscle. Drugs 24 (1982) 304–321.

MASTAGLIA, F.L. and Z. ARGOV: Drug induced neuromuscular disorders in man. In: J.N. Walton (Ed.), Disorders of Voluntary Muscle. Edinburgh, Churchill Livingstone (1981) 873–906.

MASTAGLIA, F.L. and Z. ARGOV: Immunologically mediated drug-induced neuromuscular disorders. In: P. Dukor, P. Kallos, H.D. Schlumberger and G.B. West (Eds.), Pseudo-allergic Reactions. Involvement of Drugs and Chemicals, Vol. 3. Basel, Karger (1982) 62–86.

MASTAGLIA, F.L., D. GARDNER-MEDWIN and P. HUDGSON: Muscle fibrosis and contractures in a pethidine addict. Br. Med. J. 4 (1971) 532–533.

MASTAGLIA, F.L., J.M. PAPADIMITRIOU, R.L. DAWKINS and B. BEVERIDGE: Vacuolar myopathy associated with chloroquine, lupus erythematosus and thymoma. J. Neurol. Sci. 34 (1977) 315–328.

MATEER, J.E., B.J. FARRELL, S.S.M. CHOU and L. GUTMANN: Reversible ipecac myopathy. Arch. Neurol. 42 (1985) 188–190.

MAX, S.R., M. KONAGAYA and Y. KONAGAYA: Drug-induced myopathies: examples of cellular mechanisms. Muscle Nerve 9 (1986) 33.

MEDSER, T.A.: Tryptophan-induced eosinophilia-myalgia syndrome. N. Engl. J. Med. 322 (13) (1990) 926–928.

MEIER, C., B. KAUER, U. MULLER and H.P. LUDIN: Neuromyopathy during chronic amiodarone treatment. A case report. J. Neurol. 220 (1979) 231–239.

MELDRUM, B.S.: The actions of snake venoms on nerve and muscle. The pharmacology of phosphilipase A and of polypeptide toxins. Pharmacol. Rev. 17 (1965) 393–445.

MELMED, C., G. KARPATI and S. CARPENTER: Experimental mitochondrial myopathy produced by in vivo uncoupling of oxidative phosphorylation. J. Neurol. Sci. 26 (1975) 305–318.

MITSUMOTO, H., A.J. WILBOURN and S.H. SUBRAMONY: Generalized myokymia and gold therapy. Arch. Neurol. 39 (1982) 449–450.

MODI, K.B., E.H. HORN and S.M. BRYSON: Rhabdomyolysis in theophylline poisoning. Lancet ii (1985) 161.

MORI, M., A. SATOH, M. TSUJIHATA, H. IWANGAGA and S. NAGATAKI: Myotonic discharges in a case of licorice-induced hypokalemic myopathy. Clin. Neurol. 25 (1985) 560–564.

NANJI, A.A. and J.D. FILIPENKO: Rhabdomyolysis and acute myoglobinuric renal failure associated with methadone intoxication. J. Toxicol. Clin. Toxicol. 20 (1983) 353–360.

NEVILLE, H.E., C.A. MAUNDER-SEWRY, J. MCDOUGALL and V. DUBOWITZ: Chloroquine-induced cytosomes with curvilinear profiles in muscle. Muscle Nerve 2 (1979) 376–381.

NG, R.H. and B.D. HOWARD: Mitochondria and sarcoplasmic reticulum as model targets for neurotoxic and myotoxic phospholipases A2. Proc. Natl. Acad. Sci. USA 77 (1980) 1346–1350.

NICHOLLS, K., J.F. NIALL and J.E. MORAN: Rhabdomyolysis and renal failure. Complications of narcotic abuse. Med. J. Aust. 2 (1982) 387–389.

OJEDA, V.J.: Necrotizing myopathy associated with steroid therapy. Report of two cases. Pathology 14 (1982) 435–438.

OWNBY, C.L. and G.V. ODELL: Pathogenesis of skeletal muscle necrosis induced by tarantula venom. Exp. Mol. Pathol. 38 (1983) 283–296.

OWNBY, C.L., W.M. WOODS and G.V. ODELL: Antiserum to myotoxin from prairie rattlesnake (Crotalus viridis viridis) venom. Toxicon 17 (1979) 373–380.

PALMER, E.P. and A.T. GUAY: Reversible myopathy secondary to abuse of ipecac in patients with major eating disorders. N. Engl. J. Med. 313 (1985) 1457–1459.

PALMUCCI, L., A. BERTOLOTTO and D. SCHIFFER: Acute muscle necrosis after chronic overdosage of phenformin and fenfluramine. Muscle Nerve 1 (1978) 245–247.

PANEGYRES, P.K., J.M. PAPADIMITRIOU, P.N. HOLLINGSWORTH, J.A. ARMSTRONG and B.A. KAKULAS: Vesicular changes in the myopathies of AIDS ultrastructural observations and their relationship to zidovudine treatment. J. Neurol. Neurosurg. Psychiatr. 53 (1990) 649–655.

PAPADIMITRIOU, J.M. and F.L. MASTAGLIA: Myopathy induced by mulga snake venom: a model for the study of muscle degeneration and regeneration. In: Basic Research in Myology, Proceedings, Second International Congress on Muscle Diseases, Perth, Australia, 22–26 November 1971. International Congress Series No. 294. Amsterdam, Excerpta Medica (1973) 426–437.

PAUL, H.S. and S.A. ADIBI: Paradoxical effects of clofibrate on liver and muscle metabolism in rats. J. Clin. Invest. 64 (1979) 405–412.

PEDERSON, L.M., E. NYGAARD, O.S. NIELSEN and B. SALTIN: Solvent-induced occupational myopathy. J. Occup. Med. 22 (1980) 603–606.

PENN, A.S., L.P. ROWLAND and D.W. FRASER: Drugs, coma, and myoglobinuria. Arch. Neurol. 26 (1972) 336–344.

PERKOFF, G.T., P. HARDY and E. VELEZ-GARCIA: Reversible acute muscular syndrome in chronic alcoholism. N. Engl. J. Med. 274 (1966) 1277–1285.

PERKOFF, G.T., M.M. DIOSO, V. BLEISCH and G. KLINKERFUSS: A spectrum of myopathy associated with alcoholism. I. Clinical and laboratory features. Ann. Intern. Med. 67 (1967) 481–492.

PETAJAN, J.H., J. TOWNSEND and K.M. CURREY: Ethchlorvynol (Placidyl) may produce tubular aggregates in skeletal muscle. Electroencephalogr. Clin. Neurophysiol. 64 (1986) 54P.

PETER, J.B. and W. FIEHN: Diazacholesterol myotonia: accumulation of desmosterol and increased adenosine triphosphatase activity of sarcolemma. Science 179 (1973) 910–912.

PIETTE, A-M., D. BAUER and A. CHAPMAN: Severe hypokalemia with rhabdomyolysis secondary to absorption of an alcohol-free liquorice beverage. Ann. Med. Intern. 135 (1984) 296–298.

PINALS, R.S.: Diffuse fasciculations induced by D-penicillamine. J. Rheumatol. 10 (1983) 809–810.

PINCHING, A.J., M. HERBERT, B. PEDDLE, D. ROBINSON, K. JANES, D. GOR, D. JEFFRIES, C. STONEHAM, D. MITCHELL, A.E. KOCSIS, J. MANN, S.M. FORSTER and J.R.W. HARRIS: Clinical experience with zidovudine for patients with acquired immune deficiency syndrome and acquired immune deficiency syndrome-related complex. J. Infect. 18, Suppl. 1 (1989) 33.

PLUSKAL, M.G., J.B. HARRIS, R.J. PENNINGTON and D. EAKER: Some biochemical responses of rat skeletal muscle to a single subcutaneous injection of a toxin (notexin) isolated from the venom of the Australian tiger snake Notechis scutatus scutatus. Clin. Exp. Pharmacol. 5 (1978) 131–141.

POMARA, N., K.L. COFFMAN, D.F. BUSCH and G.S. LAFAYETTE: Myalgia and elevation in muscle creatine phosphokinase during zimelidine treatment. J. Clin. Psychopharmacol. 4 (1984) 220–222.

PRITCHARD, J.B.: Toxic substances and cell membrane function. Fed. Proc. 38 (1979) 2220–2225.

RANNELLS, S.R., D.E. RANNELS, A.E. PEGG and L.S. JEFFERSON: Glucocorticoid effects on peptide-chain initiation in skeletal muscle and heart. Am. J. Physiol. 235 (1978) E134–E139.

REEBACK, J., S. BENTON and M. SWASH: Penicillamine-induced neuromyotonia. Br. Med. J. 1 (1979) 1464–1465.

REID, H.A.: Myoglobinuria and sea-snake bite poisoning. Br. Med. J. 1 (1961) 1284–1286.

RICHTER, K., A. HAASS and F. GLOTZNER: Fenoterol precipitating myotonia in a minimally affected case of recessive myotonia. J. Neurol. 219 (1978) 279–282.

RICHTER, R.W., Y.B. CHALLENOR, J. PEARSON, L.J. KAGEN, L.L. HAMILTON and W.H. RAMSEY: Acute myoglobinuria associated with heroin addiction. J. Am. Med. Assoc. 216 (1971) 1172–1176.

RIGGS, J.E., S.S. SCHOCHET JR., L. GUTMANN, T.W. CROSBY and A.G. DIBARTOLOMEO: Chronic human colchicine neuropaty and myopathy. Arch. Neurol. 43 (1986) 521–523.

RIMON, D., R. LUDATSCHER and L. COHEN: Clofibrate-induced muscular syndrome. Case report with ultrastructural findings and review of the literature. Isr. J. Med. Sci. 20 (1984) 1082–1086.

ROBERTSON, J.R. and J.H. DIMON: Myofibrosis and joint contractures caused by injections of pentazocine. J. Bone J. Surg. Am. 65 (1983) 1007–1009.

ROED, A.: Myotonia in the rat diaphragm preparation caused by the sulfhydryl inhibiting para-substituted mercuribenzoates. Acta Physiol. Scand. 115 (1982) 31–38.

ROTH, D., F.J. ALARCON, J.A. FERNANDEZ, R.A. PRESTON and J.J. BOURGOIGNIE: Acute rhabdomyolysis associated with cocaine intoxication. N. Engl. J. Med. 319 (1988) 673–677.

ROTHSTEIN, J.M., A. DELITTO, D.R. SINCAORE and S.J. ROSE: Muscle function in rheumatic disease patients treated with corticosteroids. Muscle Nerve 6 (1983) 128–135.

ROUSSEAU, J-J., M. REZNIK, G.N. LEJEUNE and G. FRANCK: Sciatic nerve entrapment of pentazocine-induced muscle fibrosis. A case report. Arch. Neurol. 36 (1979) 723–724.

ROWLAND, L.P., S. FAHN, E. HIRSCHBERG and D.H. HARTER: Myoglobinuria. Arch. Neurol. 10 (1964) 537–562.

ROWLANDS, J.B., F.L. MASTAGLIA, B.A. KAKULAS and D. HAINSWORTH: Clinical and pathological aspects of a fatal case of mulga (Pseudechis australis) snakebite. Med. J. Aust. 1 (1969) 226–230.

RUBENSTEIN, A.E. and S.F. WAINAPEL: Acute hypokalemic myopathy in alcoholism. A clinical entity. Arch. Neurol. 34 (1977) 553–555.

RUMPF, K.W., M. BARTH, M. BLECH, H. KAISER, I. KOOP, R. ARNOLD and F. SCHELER: Bezafibrate-induced myolysis and myoglobinuria in patients with impaired renal function. Klin. Wochenschr. 62 (1984) 346–348.

SAHGAL, V., V. SUBRAMANI, R. HUGHES, A. SHAH and H. SINGH: On the pathogenesis of mitochondrial myopathies. Acta Neuropathol. (Berlin) 46 (1979) 177–183.

SAITO, K., M. KAKEI, S. UCHIMURA, T. KASHIMA and H. TANAKA: Toxic effects of chlorpromazine on red and white muscles in rats: an ultrastructural study. Toxicol. Appl. Pharmacol. 65 (1982) 347–353.

SAKAI, Y., K. KOBAYASHI and N. IWATA: Effects of an anabolic steroid and vitamin B complex upon myopathy induced by corticosteroids. Eur. J. Pharmacol. 52 (1978) 353–359.

SANO, M.: Biochemical studies in experimental chloroquine myopathy. Clin. Neurol. 25 (1985) 627–635.

SANTIDRIAN, S., M. MOREYRA, M.N. MUNRO and V.R. YOUNG: Effect of corticosterone and its route of administration on muscle protein breakdown, measured in vivo by urinary excretion of N'-methylhistidine in rats: response to different levels of dietary protein and energy. Metab. Clin. Exp. 30 (1981) 798–804.

SCHERRER, P., A. DELALOYE-BISCHOF, G. TURINI and C. PERRET: Myocardial involvement in nontraumatic rhabdomyolysis following an opiate overdose. Schweiz. Med. Wochenschr. 115 (1985) 1166–1170.

SCHMALBRUCH, H.: The early changes in experimental myopathy induced by chloroquine and chlorphentermine. J. Neuropathol. Exp. Neurol. 39 (1980) 65–81.

SCHOFIELD, P.M., S.V. BEATH, T.G.K. MANT and R. BHAMRA: Recovery after severe oxprenolol overdose complicated by rhabdomyolysis. Hum. Toxicol. 4 (1985) 57–60.

SCHUBERT, D.S.P., K. BROCCO, F. MILLER and M. PATTERSON: Brief and mild alcohol intake can increase serum creatine phosphokinase. Ann. Neurol. 9 (1981) 200–201.

SENSAKOVIC, J.W., M. SUAREZ, G. PEREZ, E.S. JOHNSON and L.G. SMITH: Pentamidine treatment of Pneumocystis carinii pneumonia in the acquired immunodeficiency syndrome. Association with acute renal failure and myoglobinuria. Arch. Intern. Med. 145 (1985) 2247.

SEOW, S.S.W.: Abuse of APF linctus codeine and cardiac glycoside oxicity. Med. J. Aust. 140 (1984) 54.

SHILKIN, K.B., B.T.M. CHEN and O.T. KHOO: Rhabdomyolysis caused by hornet venom. Br. Med. J. 1 (1972) 156–157.

SHOJI, S. and R.J.T. PENNINGTON: The effect of cortisone on protein breakdown and synthesis in rat skeletal muscle. Mol. Cell. Endocrinol. 6 (1977a) 159–169.

SHOJI, S. and R.J.T. PENNINGTON: Binding of dexamethasone and cortisol to cytosol receptors in rat extensor digitorum longus and soleus muscles. Exp. Neurol. 57 (1977b) 342–348.

SHOJI, S., A. TAKAGI, H. SUGITA and Y. TOYOKURA: Muscle metabolism in steroid-induced myopathy in rabbits. Exp. Neurol. 45 (1974) 1–7.

SHOJI, S., A. TAKAGI, H. SUGITA and Y. TOYOKURA: Dysfunction of sarcoplasmic reticulum in rabbit and human steroid myopathy. Exp. Neurol. 51 (1976) 304–309.

SHOLL, J.S., M.J. HUGHEY and R.A. HIRSCHMANN: Myotonic muscular dystrophy associated with ritodrine tocolysis. Am. J. Obstet. Gynecol. 151 (1985) 83–86.

SHRIVASTAVA, O.P., S. CHATTERJI, S. KACHHAWA and S.R. DAGA: Calcium gluconate pretreatment for prevention of succinylcholine-induced myalgia. Anesth. Analg. (Cleveland) 62 (1983) 59–62.

SHUSTERMAN, D. and E. NEAL: Skeletal muscle and myocardial injury associated with metal fume fever. J. Fam. Prac. 23 (1986) 159–160.

SILVER, R.M., M.P. HEYES, J.C. MAIZE, B. QUEARRY, M. VIONNET-FUASSET and E.M. STERNBERG: Scleroderma, fascitis and eosinophilia associated with the ingestion of tryptophan. N. Engl. J. Med. 322 (1990) 874–881.

SINCLAIR, D. and C. PHILIPS: Transient myopathy apparently due to tetracycline. N. Engl. J. Med. 307 (1982) 821–822.

SITPRIJA, V. and V. BOONPUCKNAVIG: Renal failure and myonecrosis following wasp-stings. Lancet i (1972) 749–750.

SLAVIN, G., F. MARTIN, P. WARD, J. LEVI and T.J. PETERS:

Chronic alcohol excess is associated with selective but reversible injury to type 2B muscle fibres. J. Clin. Pathol. 36 (1983) 772–777.

SLOTWINER, P., S.K. SONG and P.J. ANDERSON: Spheromembranous degeneration of muscle induced by vincristine. Arch. Neurol. 15 (1966) 172–176.

SOMERS, J.E. and N. WINER: Reversible myopathy and myotonia following administration of a hypocholesterolemic agent. Neurology 16 (1966) 761–765.

SONG, S.K. and E.R. RUBIN: Ethanol produces muscle damage in human volunteers. Science 175 (1972) 327–328.

SPARGO, E.: The acute effects of alcohol on plasma creatine kinase (CK) activity in the rat. J. Neurol. Sci. 63 (1984) 307–316.

SPAULDING, W.B.: Myalgia and elevated creatine phosphokinase with danazol in hereditary angioedema. Ann. Intern. Med. 90 (1979) 654.

STAUBER, W.T., A.M. HEDGE, J.J. TROUT and B.A. SCHOTTELIUS: Inhibition of lysosomal function in red and white skeletal muscles by chloroquine. Exp. Neurol. 71 (1981) 295–306.

STEER, J.H., and F.L. MASTAGLIA: Protein degradation in bupivacaine-treated muscles. The role of extracellular calcium. J. Neurol. Sci. 75 (1986) 343–351.

STEER, J.H., F.L. MASTAGLIA, J.M. PAPADIMITRIOU and I. VAN BRUGGEN: Bupivacaine-induced muscle injury. The role of extracellular calcium. J. Neurol. Sci. 73 (1986) 205–217.

STEINNESS, E., F. RASMUSSEN, O.S VENDSEN and P. NIELSEN: A comparative study of serum creatine phosphokinase (CPK) activity in rabbits, pigs and humans after intramuscular injection of local damaging drugs. Acta Pharmacol. Toxicol. 42 (1977) 357–364.

STREICHER, H.Z., P.A. GABOW, A.H. MOSS, D. KONO and W.D. KAEHNY: Syndromes of toluene sniffing in adults. Ann. Intern. Med. 94 (1981) 758–762.

STRUNK, S., C.W. SMITH and J.M. BLUMBERG: Ultrastructural studies on the lesion produced in skeletal muscle fibers by crude type A Clostridium perfringens toxin and its purified alpha fraction. Am. J. Pathol. 50 (1967) 89–107.

SUGIE, H., R. RUSSIN and M.A. VERITY: Emetine myopathy: two case reports with pathobiochemical analysis. Muscle Nerve 7 (1984) 54–59.

SUNDERLAND, S.K.: Australian Animal Toxins. Melbourne, Oxford University Press (1983).

SUNNASY, D., S.R. CAIRNS, F. MARTIN, G. SLAVIN and T.J. PETERS: Chronic alcoholic skeletal muscle myopathy: a clinical histological and biochemical assessment of muscle lipid. J. Clin. Pathol. 36 (1983) 778–784.

SUTHERLAND, S.K. and A.R. COULTER: Three instructive cases of tiger snake (Notechis scutatus) envenomation — and how a radioimmunoassay proved the diagnosis. Med. J. Aust. 2 (1977) 177–180.

SWENSON, R.D., T.A. GOLPER and W.M. BENNETT: Acute renal failure and rhabdomyolysis after ingestion of phenylpropanolamine-containing diet pills. J. Am. Med. Assoc. 248 (1982) 1216.

TAKAHASHI, K., T. OGITA, H. OKUDAIRA, S. YOSHINOYA, H. YOSHIZAWA and T. MIYAMOTO: D-Penicillamine-induced polymyositis in patients with rheumatoid arthritis. Arthritis Rheum. 29 (1986) 560–564.

TAM, C.W., B.Y. OLIN and A.E. RUIZ: Loxapine-associated rhabdomyolysis and acute renal failure. Arch. Intern. Med. 140 (1980) 975–976.

TEICHER, A., T. ROSENTHAL, E. KISSIN and I. SAROVA: Labetalol-induced toxic myopathy. Br. Med. J. 282 (1981) 1824–1825.

TERAVAINEN, H., A. LARSEN and M. HILLBOM: Clofibrate-induced myopathy in the rat. Acta Neuropathol. (Berlin) 39 (1977) 135–138.

THOMAS, M.A.B. and L.S. IBELS: Rhabdomyolysis and acute renal failure. Aust. N. Z. J. Med. 15 (1985) 623–628.

THOMPSON, M.D., L.K. JIM, R.R. BENTZ and D.K. ALEXANDER: Acute prednisone-induced disabling myalgias in scleroderma. Arthritis Rheum. 24 (1981) 639–640.

TOMLINSON, L.W. and F.D. ROSENTHAL: Proximal myopathy after perhexiline maleate treatment. Br. Med. J. 1 (1977) 1319–1320.

TREVES, R., M. ARNAUD, F. TABARAUD, F. BURKI, J.M. VALLAT and R. DESPROGES-GOTTERON: Myopathie au cours d'un traitement a la cimetidine. Rev. Rhum. Mal. Osteo-Articulaires 52 (1985) 133–134.

TROUT, J.J., W.T. STAUBER and B.A. SCHOTTELIUS: Chloroquine-induced alterations in phasic muscles. II. Sarcoplasmic reticulum. Exp. Mol. Pathol. 34 (1981) 237–243.

TU, A.T.: Venoms: Chemistry and Molecular Biology. New York, Wiley (1977) 459–526.

UDEZUE, E.O. and B.P. HARROLD: Hyperkalaemic paralysis due to spironolactone. Postgrad. Med. J. 56 (1980) 254–255.

UNGER, J., G. DECAUX and M. L'HERMITE: Rhabdomyolysis, acute renal failure, endocrine alterations and neurological sequelae in a case of lithium self poisoning. Acta Clin. Belg. 37 (1982) 216–223.

URBANO-MARQUEZ, A., R. ESTRUCH, J.M. GRAU, J. FERNANDEZ-HUERTA and M. SALA: On alcoholic myopathy. Ann. Neurol. 17 (1985) 418.

VALERIANO, J., P. TUCKER and J. KATTAH: An unusual cause of hypokalemic muscle weakness. Neurology 33 (1983) 1242–1243.

VAN GARSSE, L.G.M.M. and P.P.H. BOEYKENS: Two patients with eosinophilia myalgia syndrome associated with tryptophan. Br. Med. J. 301 (1990) 21.

VAN MARLE, W. and K.L. WOODS: Acute hydrocortisone myopathy. Br. Med. J. 281 (1980) 271–272.

VAN VLEET, J.R. and V.J. FERRANS: Ultrastructural alterations in skeletal muscle of pigs with acute monensin myotoxicosis. Am. J. Pathol. 114 (1984) 461–471.

VARGA, J., J. PELTONEN, J. UITTO and S. JIMENEZ: Development of diffuse fasciitis with eosinophilia during l-tryptophan treatment: demonstration of elevated type I collagen gene expression in affected tissues. Ann. Intern. Med. 112 (1990) 344.

VASILESCU, C., M. ALEXIANU and A. DAN: Delayed neuropathy after organophosphorus insecticide (Dipterex) poisoning: a clinical, electrophysiological and nerve biopsy study. J. Neurol. Neurosurg. Psychiatr. 47 (1984) 543–548.

VITA, G., S. BARTOLONE, M. SANTORO, A. TOSCANO, G. CAROZZA, P. GIRLANDA and N. FRISINA: Hypokalemic myopathy induced by fluroprednizolone-containing nasal spray. Acta Neurol. (Napoli) 8 (1986) 108–109.

WADIA, R.S., C. SADAGOPAN, R.B. AMIN and H.V. SANDESAI: Neurological manifestations of organophosphorous insecticide poisoning. J. Neurol. Neurosurg. Psychiatr. 27 (1974) 841–847.

WAKATA, N., Y. KAWAMURA, Y. ARAKI, N. YAMADA and N. KINOSHITA: Study on steroid-induced muscular change. Clin. Neurol. 23 (1983) 430–435.

WALRAVENS, P.A., C. GREENE and F.E. FRERMAN: Lovastatin, isoprenes and myopathy. Lancet ii (1989) 1097.

WALSH, J.C. and A.B. CONOMY: The effect of ethyl alcohol on striated muscle: some clinical and pathological observations. Aust. N. Z. J. Med. 7 (1977) 485–490.

WATTEL, F., C. CHOPIN, A. DUROCHER and B. BERZIN: Rhabdomyolyses au cours des intoxications aigues. Nouv. Presse Med. 7 (1978) 2553–2560.

WECKER, L., M.B. LASKOWSKI and W.-D. DETTBARN: Neuromuscular dysfunction induced by acetylcholinesterase inhibition. Fed. Proc. 37 (1978) 2818–2822.

WHISNANT, J.P., R.E. ESPINOSA, R.R. KIERLAND and E.H. LAMBERT: Chloroquine neuromyopathy. Staff Meet. Mayo Clin. 38 (1963) 501–513.

WILLIAMS, A.J., M.S. BAGHAT, D.E. STABLEFORTH, R.M. LAYTON, P.M. SHENOI and C. SKINNER: Dysphonia caused by inhaled steroids: recognition of a characteristic laryngeal abormality. Thorax 38 (1983) 813–821.

WILLNER, J., M. NAKAGAWA and D. WOOD: Drug-induced fiber necrosis in Duchenne dystrophy. Ital. J. Neurol. 5 (1984) 117–121.

YAGIELA, J.A. P.W. BENOIT, R.D. BUONCRISTIANI, M.P. PETERS and N.F. FORT: Comparison of myotoxic effects of lidocaine with epinephrine in rats and humans. Anesth. Analg. (Cleveland) 60 (1981) 471–480.

YATES, D.A.H.: Steroid myopathy. In: J.N. Walton, N. Canal and G. Scarlato (Eds.), Muscle Diseases. International Congress Series 199. Amsterdam, Excerpta Medica (1970) 482–488.

Index

Prepared by W. van Ockenburg

Entries in this index refer both to the present Volume and to Volumes 1–43 in the preceding series of the Handbook of Clinical Neurology

A band
  muscle fiber, 62/26
Abetalipoproteinemia, *see* Bassen-Kornzweig
  syndrome
Abiotrophia retinae pigmentosa, *see* Primary
  pigmentary retinal degeneration *and* Secondary
  pigmentary retinal degeneration
Abnormal acetylcholine-acetylcholine receptor
  interaction syndrome
  congenital myasthenic syndrome, 62/437
  EMG, 62/437
  neuromuscular junction disease, 62/392
Abnormal myomuscular junction myopathy
  congenital myopathy, 62/332
  core like lesion, 62/353
  facial weakness, 62/334
  symptom, 62/353
Abortion
  myotonic dystrophy, 62/228
Acanthocyte
  McLeod myopathy, 62/127
Acetazolamide
  hyperkalemic periodic paralysis, 62/470, 472
Acetylcholine receptor
  muscle tissue culture, 40/186, 62/97
  myasthenia gravis, 40/129, 41/106, 114, 125, 62/391
  myasthenic syndrome, 62/391
Acetylcholine receptor antibody
  myasthenia gravis, 41/114, 120, 129, 368, 62/408
Acetylcholine receptor epitope
  myasthenia gravis, 62/407
ε-Acetylcholine receptor subunit mutation syndrome
  congenital, *see* Congenital ε-acetylcholine receptor
  subunit mutation syndrome
Acetylcholinesterase
  muscle tissue culture, 62/97
Acid maltase deficiency, 41/175-181

adult, *see* Adult acid maltase deficiency
allelic diversity, 62/481
biochemistry, 41/178, 62/481
chromosome 17, 62/481
clinical features, 40/280, 317, 41/59, 175, 62/480
creatine kinase, 62/480
ECG, 62/480
EMG, 62/480
genetic heterogeneity, 62/482
α-1,4-glucosidase, 62/481
α-1,6-glucosidase, 62/481
glycogen storage disease, 40/53, 99, 153, 190, 255,
  280, 317, 438, 62/480
infantile, *see* Infantile acid maltase deficiency
inheritance, 62/481
ischemic exercise, 62/480
late infantile, *see* Late infantile acid maltase
  deficiency
metabolic myopathy, 62/480
molecular genetics, 62/481
mRNA, 62/482
muscle biopsy, 62/481
muscle fiber feature, 62/17
muscle tissue culture, 40/190, 62/99
pathology, 62/481
prenatal diagnosis, 43/178, 62/481
rimmed vacuole distal myopathy, 62/481
treatment, 41/180, 62/482
Acidosis
  lactic, *see* Lactic acidosis
  metabolic, *see* Metabolic acidosis
Aconitase deficiency, *see* Aconitate hydratase
  deficiency
Aconitate hydratase deficiency
  exercise intolerance, 62/503
  mitochondrial disease, 62/503
  myoglobinuria, 62/503

Acquired immune deficiency syndrome
   tropical polymyositis, 62/373
Acquired myoglobinuria
   amphetamine, 62/573
   bezafibrate, 62/576
   Candida, 62/577
   chronic myopathy, 62/577
   clofibrate, 62/576
   Clostridium, 62/577
   cocaine, 62/577
   cold, 62/574
   Coxsackie virus, 62/577
   crush, 62/574
   cyclosporine, 62/576
   diamorphine, 62/577
   doxylamine, 62/573
   drug, 62/575
   drug overdose, 62/573
   ECHO virus, 62/577
   Epstein-Barr virus, 62/577
   erythromycin, 62/577
   Escherichia coli, 62/577
   exercise, 62/572
   gemfibrozil, 62/576
   Haff disease, 62/577
   heat stroke, 62/572
   herpes simplex, 62/577
   infection, 62/577
   influenza, 62/577
   ischemia, 62/574
   isoniazid, 62/577
   legionellosis, 62/577
   leptospirosis, 62/577
   loxapine, 62/573
   marathon runner, 62/573
   metabolic depression, 62/574
   methadone, 62/577
   mevinolin, 62/576
   mycoplasma, 62/577
   neuroleptic malignant syndrome, 62/572
   osmolarity derangement, 62/575
   oxprenolol, 62/577
   phencyclidine, 62/573
   phenmetrazine, 62/573
   phentermine, 62/573
   phenylenediamine, 62/573
   polymyositis, 62/577
   Salmonella, 62/577
   salt depletion, 62/572
   Shigella, 62/577
   sickle cell anemia, 62/572
   snake venom, 62/577
   Staphylococcus, 62/577
   status asthmaticus, 62/573
   status epilepticus, 62/573
   Streptococcus pneumoniae, 62/577
   strychnine, 62/573
   terbutaline, 62/577
   theophylline, 62/573, 577
   toxic agent, 62/575
   toxic shock syndrome, 62/574, 577
   toxin, 62/576
   trauma, 62/572
   tularemia, 62/577
   typhoid fever, 62/577
   viral infection, 62/572
   vitamin PP, 62/577
Acromegalic myopathy
   creatine kinase, 62/538
   EMG, 62/538
   muscle biopsy, 62/538
   muscle ultrastructure, 62/538
   pituitary gigantism myopathy, 62/537
   proximal limb muscle, 62/538
ACTH induced myopathy
   EMG, 62/537
   fatigue, 62/537
   hyperpigmentation, 62/537
   muscle biopsy, 62/537
   muscle weakness, 62/537
   Nelson syndrome, 62/537
α-Actinin
   dystrophin, 62/126
   muscle fiber, 62/26, 35
Action potential
   compound muscle, *see* Compound muscle action
     potential
Acute anterior poliomyelitis, *see* Poliomyelitis
Acute childhood encephalopathy, *see* Reye syndrome
Acute inflammatory demyelinating neuropathy, *see*
    Guillain-Barré syndrome
Acute inflammatory demyelinating polyneuropathy,
    see Guillain-Barré syndrome
Acute polymyositis
   EMG, 62/71
Acute polyradiculoneuritis, *see* Guillain-Barré
    syndrome
Acute serous encephalitis, *see* Reye syndrome
Acute toxic encephalitis, *see* Reye syndrome
Adenosine monophosphate deaminase
   myoglobinuria, 62/558
Adenosine monophosphate deaminase deficiency
   biochemistry, 62/512
   chromosome 1p, 62/512
   creatine kinase, 62/512
   differential diagnosis, 62/511
   histopathology, 62/512
   ischemic exercise, 62/512
   limb girdle syndrome, 62/190
   muscle cramp, 62/511
   myalgia, 62/511
   myoglobinuria, 62/512
   primary type, 62/511
   secondary type, 62/511
Adenosine triphosphatase 6 mutation syndrome
   ataxia, 62/510
   dementia, 62/510
   epilepsy, 62/510
   metabolic encephalopathy, 62/510
   metabolic myopathy, 62/510
   mitochondrial disease, 62/510
   mitochondrial DNA point mutation, 62/509

multiple neuronal system degeneration, 62/510
muscle weakness, 62/510
proximal limb weakness, 62/510
secondary pigmentary retinal degeneration, 62/510
sensory neuropathy, 62/510
subacute necrotizing encephalomyelopathy, 62/510
Adenosine triphosphatase
    multicore myopathy, 62/339
    muscle fiber type, 62/4
Adenylate deaminase
    muscle, see Adenosine monophosphate deaminase
Adrenal gland
    Conn syndrome, see Hyperaldosteronism
Adrenal insufficiency myopathy
    arthralgia, 62/536
    corticosteroid withdrawal myopathy, 62/536
    fatigue, 62/536
    fever, 62/536
    myalgia, 62/536
    weight loss, 62/536
Adult acid maltase deficiency
    limb girdle syndrome, 62/190
    myopathy, 27/229, 62/480
    respiratory dysfunction, 62/480
    tardive myopathy, 62/480
Adult celiac disease, see Celiac disease
Adult dominant myotubular myopathy
    rimmed muscle fiber vacuole content, 62/18
    rimmed muscle fiber vacuole type, 62/18
Adult polyglucosan body disease
    axonal sensorimotor neuropathy, 62/484
    brancher deficiency, 62/484
    dementia, 62/484
    lower motoneuron sign, 62/484
    metabolic encephalopathy, 62/484
    neurogenic bladder, 62/484
    sensory loss, 62/484
    upper motoneuron sign, 62/484
Adverse drug reaction
    myasthenia gravis, 62/420
Adynamia episodica, see Hyperkalemic periodic
    paralysis
Agammaglobulinemia
    polymyositis, 62/373
AIDS, see Acquired immune deficiency syndrome
Alcohol
    fetal intoxication, see Fetal alcohol syndrome
    toxic myopathy, 62/601
Alcohol syndrome
    fetal, see Fetal alcohol syndrome
Alcoholic myopathy, see Alcoholic polyneuropathy
Alcoholic polyneuropathy
    acute type, 62/609
    chronic type, 62/610
    histopathology, 62/609
    pathogenesis, 62/610
    symptom, 62/609
Aldolase, see Fructose bisphosphate aldolase
Allelic diversity
    acid maltase deficiency, 62/481
Allelic expansion

myotonic dystrophy, 62/212
Alopecia
    myotonic dystrophy, 62/238
Alpers disease
    mitochondrial disease, 62/497
ALS, see Amyotrophic lateral sclerosis
Alveolar hypoventilation
    myotonic dystrophy, 40/517, 62/238
Amethopterin, see Methotrexate
Aminoacid sequence
    dystrophin, 62/125
Aminoaciduria
    complex III deficiency, 62/504
ε-Aminocaproic acid
    toxic myopathy, 62/598, 601
4-Aminopyridine
    Eaton-Lambert myasthenic syndrome, 41/360,
        62/426
Amiodarone
    toxic myopathy, 62/606
Amiodarone myopathy
    sensorimotor polyneuropathy, 62/608
Amoxapine
    toxic myopathy, 62/601
Amphetamine
    acquired myoglobinuria, 62/573
    toxic myopathy, 62/601
Amphotericin B
    toxic myopathy, 62/601
Amylo-1,6-glucosidase
    debrancher deficiency, 62/483
Amylo-1,6-glucosidase deficiency, see Debrancher
    deficiency
Amyloid deposit
    dermatomyositis, 62/377
    inclusion body myositis, 62/377
    polymyositis, 62/377
Amyloid β-protein
    dermatomyositis, 62/377
    inclusion body myositis, 62/377
    polymyositis, 62/377
Amyotrophic lateral sclerosis, 22/281-330
    see also Motoneuron disease
    edrophonium test, 62/413
    polyneuropathy, 62/413
    progressive external ophthalmoplegia
        classification, 62/290
    thyrotoxic myopathy, 62/529
Amyotrophy
    scapuloperoneal, see Scapuloperoneal spinal
        muscular atrophy
Andersen disease, see Brancher deficiency
Anemia, 38/15-28
    Fanconi, see Fanconi syndrome
    hemolytic, see Hemolytic anemia
    myasthenia gravis, 62/402
    sickle cell, see Sickle cell anemia
    sideroblastic, see Sideroblastic anemia
Anesthesia
    general, see General anesthesia
    hypokalemic periodic paralysis, 62/460

Anesthesia complication
    myotonic dystrophy, 62/244
Animal model
    hereditary myotonic syndrome, 62/272
    myotonic syndrome, 62/263
Ankylosing spondylitis
    polymyositis, 62/373
Ankylosis
    temporomandibular, see Temporomandibular
        ankylosis
Ankyrin
    dystrophin, 62/126
Anoxic seizure, see Syncope
Anterior poliomyelitis, see Poliomyelitis
Anthracene-9-carboxylic acid intoxication, see 9-
        Anthroic acid intoxication
9-Anthroic acid intoxication
    drug induced myotonia, 62/271
    drug induced myotonic syndrome, 62/271
Anticholinesterase
    myasthenia gravis, 62/412
Antigen
    HLA, see HLA antigen
Antihistaminic agent
    toxic myopathy, 62/601
Apamin
    myotonic dystrophy, 62/222, 225
Aphthous uveitis
    relapsing, see Behçet syndrome
Apnea
    episodic, see Episodic apnea
    familial infantile myasthenia, 62/427
Areflexia
    see also Hyporeflexia
    Emery-Dreifuss muscular dystrophy, 62/149
    inclusion body myositis, 62/373
    scapuloperoneal spinal muscular atrophy, 62/170
Argyrophilic dystrophy
    subcortical, see Progressive supranuclear palsy
Armadillo syndrome, see Isaacs syndrome
Aromatic carboxylic acid intoxication
    myotonic syndrome, 62/264
Arrhythmia
    malignant hyperthermia, 62/567
Arthralgia
    adrenal insufficiency myopathy, 62/536
    corticosteroid withdrawal myopathy, 62/536
    dermatomyositis, 62/374
    inclusion body myositis, 62/374
    polymyositis, 62/374
Arthritis
    rheumatoid, see Rheumatoid arthritis
Arthrogryposis
    congenital acetylcholine receptor deficiency/
        synaptic cleft syndrome, 62/438
    muscle fibrosis, 62/173
Aspartate aminotransferase
    hypothyroid myopathy, 62/533
    thyrotoxic myopathy, 62/529
Asphyxia
    X-linked myotubular myopathy, 62/156

Aspiration pneumonia
    dermatomyositis, 62/374
    inclusion body myositis, 62/374
    polymyositis, 62/374
Ataxia, 1/309-348
    adenosine triphosphatase 6 mutation syndrome,
        62/510
    complex I deficiency, 62/504
    complex III deficiency, 62/504
    complex V deficiency, 62/506
    Friedreich, see Friedreich ataxia
    hereditary, see Hereditary ataxia
    Kearns-Sayre-Daroff-Shy syndrome, 62/508
    Leber hereditary optic neuropathy, 62/509
    MERRF syndrome, 62/509
    multiple mitochondrial DNA deletion, 62/510
    periodic, see Periodic ataxia
    spastic, see Spastic ataxia
    ubidecarenone, 62/504
Atrial arrhythmia, see Emery-Dreifuss muscular
        dystrophy
Atrial paralysis, see Emery-Dreifuss muscular
        dystrophy
Atromid-S, see Clofibrate
Atrophy
    distal spinal muscular, see Spinal muscular atrophy
    infantile spinal muscular, see Infantile spinal
        muscular atrophy
    multiple system, see Multiple neuronal system
        degeneration
    muscle fiber, 40/11, 142, 200, 62/13
    muscular, see Muscular atrophy
    myopathy, 62/13
    neuromuscular disease, 62/13
    olivopontocerebellar, see Olivopontocerebellar
        atrophy
    optic, see Optic atrophy
    optic nerve, see Optic atrophy
    spinal muscular, see Spinal muscular atrophy
Attack
    hyperkalemic periodic paralysis, 62/461
    hypokalemic periodic paralysis, 62/457
    thyrotoxic periodic paralysis, 62/461
    transient ischemic, see Transient ischemic attack
Autoallergy, see Autoimmune disease
Autoimmune antibody
    dermatomyositis, 62/380
    inclusion body myositis, 62/380
    polymyositis, 62/380
Autoimmune disease
    myositis, 62/369
Autoimmune thyroiditis, see Hashimoto thyroiditis
Autoimmunity, see Autoimmune disease
Autonomous micturition, see Neurogenic bladder
Autophagic vacuole
    muscle fiber, 62/23
Autosomal recessive generalized myotonia
    age at onset, 62/266
    chloride conductance, 62/267
    fall, 62/266
    muscle hypertrophy, 62/267

muscle stretch reflex, 62/266
muscle weakness, 62/266
muscular atrophy, 62/267
myotonia distribution, 62/266
myotonic dystrophy, 62/240
myotonic syndrome, 62/263
prevalence, 62/267
startle, 62/266
Axial tomography
computerized, *see* Computer assisted tomography
Axonal neuropathy
dermatomyositis, 62/376
inclusion body myositis, 62/376
polymyositis, 62/376
Axonal polyneuropathy
creatine kinase, 62/609
muscle biopsy, 62/609
spheromembranous body, 62/609
Azacosterol
drug induced myotonia, 62/271
Azathioprine
dermatomyositis, 62/385
Eaton-Lambert myasthenic syndrome, 62/426
inclusion body myositis, 62/385
myasthenia gravis, 41/96, 133, 136, 62/418
polymyositis, 41/73, 62/385
Azidothymidine, *see* Zidovudine
AZT, *see* Zidovudine

B-cell, *see* B-lymphocyte
B-lymphocyte
myositis, 62/42
Bacterial myositis
polymyositis, 62/372
Barbiturate
toxic myopathy, 62/601
Barth syndrome type I, *see* X-linked myotubular
myopathy
Barth syndrome type II, *see* X-linked neutropenic
cardioskeletal myopathy
Basal ganglion calcification, *see* Striatopallidodentate
calcification
Basement membrane, *see* Sarcolemma
Basilar arterial intermittent insufficiency, *see*
Transient ischemic attack
Basophilic granular inclusion
dermatomyositis, 62/377
inclusion body myositis, 62/377
polymyositis, 62/377
Bassen-Kornzweig syndrome, 13/413-430, 29/401-424
progressive external ophthalmoplegia, 22/191, 204,
207, 62/304
progressive external ophthalmoplegia
classification, 62/290
Batten disease
see also Neuronal ceroid lipofuscinosis
muscle fiber feature, 62/17
Batten-Mayou disease, *see* Batten disease
Batten-Mayou-Spielmeyer-Vogt disease, *see* Batten
disease
Batten-Vogt disease, *see* Batten disease

Bechterew disease, *see* Ankylosing spondylitis
Becker mild X-linked dystrophy, *see* Becker muscular
dystrophy
Becker muscular dystrophy
age at onset, 62/135
clinical features, 40/306, 312, 317, 321, 323, 337,
342, 388, 442, 62/135
differential diagnosis, 40/438, 62/135
drug trial, 62/136
Duchenne muscular dystrophy, 62/123
history, 40/434, 62/134
incidence, 40/387, 62/135
incomplete dystrophin deficiency, 62/135
malignant hyperthermia, 62/570
muscle biopsy, 62/123
myoglobinuria, 62/556, 558
prevalence, 62/135
related phenotype, 62/135
treatment, 62/136
X-linked muscular dystrophy, 40/387-389, 62/117
Bee venom toxin, *see* Apamin
Behçet syndrome, 34/475-506
polymyositis, 62/373
Benzine, *see* Gasoline
Beta adrenergic receptor blocking agent
toxic myopathy, 62/604
Beta oxidation
metabolic myopathy, 62/492
Bethlem-Van Wijngaarden syndrome
see also Limb girdle syndrome
autosomal dominant, 62/181
early onset, 62/181
Emery-Dreifuss muscular dystrophy, 62/155
features, 62/184
muscle contraction, 62/155, 181
Biliary cirrhosis
primary, *see* Primary biliary cirrhosis
Biopsies
muscle, *see* Muscle biopsy
nerve, *see* Nerve biopsy
Bladder
neurogenic, *see* Neurogenic bladder
Blindness
cerebral, *see* Cortical blindness
cortical, *see* Cortical blindness
Blood clotting factor VIII
Emery-Dreifuss muscular dystrophy, 62/157
Blurred vision, *see* Visual impairment
Body
inclusion, *see* Inclusion body
zebra, *see* Zebra body
Body length
short, *see* Short stature
Body temperature, *see* Thermoregulation
Bone marrow transplantation
myasthenia gravis, 62/402
Bone pain
chronic renal failure myopathy, 62/541
Borrelia burgdorferi
see also Lyme disease
polymyositis, 62/373

Botulism
   Eaton-Lambert myasthenic syndrome, 62/425
   myasthenia gravis, 62/415
   toxic myopathy, 62/612
Boulimia, *see* Bulimia
Brain
   dystrophin, 62/125
   glycogen storage disease, 62/490
Brain failure, *see* Dementia
Brain infarction
   see also Stroke
   MELAS syndrome, 62/509
Brain stone, *see* Striatopallidodentate calcification
Brain venous infarction, *see* Brain infarction
Brancher deficiency
   adult polyglucosan body disease, 62/484
   biochemistry, 62/484
   cardiomyopathy, 62/484
   clinical features, 27/232, 41/184, 62/483
   creatine kinase, 62/484
   glycogen storage disease, 62/483
   hepatosplenomegaly, 62/483
   histopathology, 62/484
   infantile hypotonia, 62/484
   inheritance, 62/484
   liver cirrhosis, 62/483
   metabolic myopathy, 62/483
   muscle contraction, 62/484
   muscle fiber feature, 62/17
   treatment, 27/232, 41/185, 62/484
Branching enzyme deficiency, *see* Brancher deficiency
Breast milk
   intoxication, *see* Subacute necrotizing
      encephalomyelopathy
Brown spider venom
   toxic myopathy, 62/601, 611
Brueghel syndrome, *see* Dystonia musculorum
      deformans
Bruns-Garland syndrome, *see* Diabetic proximal
      neuropathy
Bulimia
   oxidation-phosphorylation coupling defect, 62/506
α-Bungarotoxin
   myasthenia gravis, 41/113, 125, 62/409

Cachectin
   muscle tissue culture, 62/92
Calcification
   basal ganglion, *see* Striatopallidodentate
      calcification
   corticostriatopallidodentate, *see*
      Striatopallidodentate calcification
   familial basal ganglion, *see* Striatopallidodentate
      calcification
   muscle fiber, 62/18
   striatopallidocorticodentate, *see*
      Striatopallidodentate calcification
   toxic striatopallidodentate, *see*
      Striatopallidodentate calcification
Calcinosis
   brain, *see* Striatopallidodentate calcification

   cerebral, *see* Striatopallidodentate calcification
   dermatomyositis, 62/386
   inclusion body myositis, 62/386
   polymyositis, 62/386
Calcinosis nucleorum cerebri, *see*
      Striatopallidodentate calcification
Calcium channel, 28/528-532
   Eaton-Lambert myasthenic syndrome, 62/421, 423
   malignant hyperthermia, 62/567, 569
   muscle tissue culture, 62/91
Calcium channel blocking agent
   myotonia, 62/276
Calf hypertrophy
   infantile distal myopathy, 62/201
   late infantile acid maltase deficiency, 62/480
Candida
   acquired myoglobinuria, 62/577
Cane fever, *see* Leptospirosis
Canicola fever, *see* Leptospirosis
Cap disease
   congenital myopathy, 62/332
   facial weakness, 62/334
   inclusion body, 62/341
   sarcoplasmic mass, 62/341
Caps
   muscle fiber, 62/44
Captopril
   toxic myopathy, 62/597
Carbenoxolone
   toxic myopathy, 62/601
Carbon monoxide
   toxic myopathy, 62/601
Carcinoma
   see also Malignancy
   limb girdle syndrome, 62/188
Cardiac abnormality, *see* Heart disease
Cardiac conduction
   Leber hereditary optic neuropathy, 62/509
   myotonic dystrophy, 62/217
Cardiac glycoside
   toxic myopathy, 62/598
Cardiac involvement
   childhood myoglobinuria, 62/561
   congenital myopathy, 62/334
   debrancher deficiency, 62/482
   Duchenne muscular dystrophy, 62/119
   limb girdle syndrome, 62/185
Cardiomegaly
   infantile acid maltase deficiency, 27/195, 227,
      62/480
   long chain acyl coenzyme A dehydrogenase
      deficiency, 62/495
Cardiomyopathy
   see also Heart disease
   brancher deficiency, 62/484
   carnitine deficiency, 43/175, 62/493
   chloroquine myopathy, 62/608
   combined complex I-V deficiency, 62/506
   dermatomyositis, 62/374
   Emery-Dreifuss muscular dystrophy, 62/145, 152
   glycogen storage disease, 62/490

inclusion body myositis, 62/374
incomplete dystrophin deficiency, 62/135
infantile histiocytoid, *see* Infantile histiocytoid cardiomyopathy
Kearns-Sayre-Daroff-Shy syndrome, 22/211, 43/142, 62/310
lipid metabolic disorder, 62/493
long chain 3-hydroxyacyl coenzyme A dehydrogenase deficiency, 62/495
nemaline myopathy, 62/341
phosphorylase kinase deficiency, 62/487
polymyositis, 62/374
primary systemic carnitine deficiency, 62/493
progressive external ophthalmoplegia, 22/192, 62/310
short chain 3-hydroxyacyl coenzyme A dehydrogenase deficiency, 62/495
Cardiopathy
complex I deficiency, 62/503
complex III deficiency, 62/505
MELAS syndrome, 62/510
Cardiorespiratory failure
complex I deficiency, 62/503
Mallory body myopathy, 62/345
Carnitine deficiency
cardiomyopathy, 43/175, 62/493
ECG, 62/493
Fanconi syndrome, 62/493
hemodialysis, 62/494
hepatic encephalopathy, 62/493
hypoglycemic encephalopathy, 62/493
infantile hypotonia, 62/493
inheritance, 62/493
kwashiorkor, 62/493
lipid metabolic disorder, 41/196-201, 62/492
lipid storage myopathy, 40/95, 258, 280, 308, 317, 422, 43/175, 184, 62/493
metabolic myopathy, 62/492
primary systemic carnitine deficiency, 62/493
Reye like episode, 62/493
secondary systemic carnitine deficiency, 62/493
systemic carnitine deficiency, 62/493
valproic acid, 62/494
Carnitine deficiency myopathy
age at onset, 62/492
fluctuating weakness, 62/492
Carnitine palmitoyltransferase
childhood myoglobinuria, 62/563
myoglobinuria, 62/558
Carnitine palmitoyltransferase deficiency
carnitine palmitoyltransferase isoenzyme, 62/494
cold exposure, 62/494
exercise, 62/494
fasting, 62/494
ischemic exercise, 62/494
lipid metabolic disorder, 41/204-207, 62/494
liver, 62/494
metabolic myopathy, 62/494
muscle biopsy, 62/494
muscle cramp, 62/494
myalgia, 43/176, 62/494
myoglobinuria, 41/204, 265, 428, 43/176, 62/494
myopathy, 62/494
precipitating factor, 62/494
renal failure, 62/494
sex ratio, 62/494
Cataract
multiple mitochondrial DNA deletion, 62/510
myotonic dystrophy, 40/512, 41/486, 43/153, 62/217, 233
progressive external ophthalmoplegia, 62/306
Cave foot, *see* Pes cavus
CD8+ cytotoxic T-lymphocyte
myositis, 62/42
Celiac disease
Eaton-Lambert myasthenic syndrome, 62/421
Kearns-Sayre-Daroff-Shy syndrome, 62/313
polymyositis, 62/373
progressive external ophthalmoplegia, 62/313
Celiac sprue, *see* Celiac disease
Cells
B, *see* B-lymphocyte
inflammatory, *see* Inflammatory cell
Central core disease, *see* Central core myopathy
Central core myopathy
chromosome 19q, 62/333
clinical variety, 62/333
congenital myopathy, 41/3-8, 62/332
core morphology, 62/337
core myofiber, 62/338
definition, 62/333
experimental core lesion, 62/334
facial weakness, 62/334
facioscapulohumeral syndrome, 62/168
histopathology, 62/336
malignant hyperthermia, 41/3, 43/81, 62/333, 570
multicore lesion, 62/338
multicore pathology, 62/338
muscle biopsy, 62/338
muscle fiber type I predominance, 41/4, 62/338
myoglobinuria, 62/558
organophosphate intoxication, 62/333
split myofiber, 62/338
tenotomy, 62/334
tetanus toxin, 62/338
Central nervous system, *see* CNS
Centronuclear myopathy, *see* Myotubular myopathy
Cerebellar syndrome, 2/392-426
Kearns-Sayre-Daroff-Shy syndrome, 62/308
progressive external ophthalmoplegia, 62/308
Cerebellolental degeneration, *see* Marinesco-Sjögren syndrome
Cerebral blindness, *see* Cortical blindness
Cerebral infarction, *see* Brain infarction
Cerebromuscular dystrophy, *see* Fukuyama syndrome
Cerebrospinal fluid, *see* CSF
Channel
calcium, *see* Calcium channel
potassium, *see* Potassium channel
sodium, *see* Sodium channel
Charcot disease, *see* Amyotrophic lateral sclerosis

Charcot-Marie disease, *see* Charcot-Marie-Tooth disease
Charcot-Marie-Tooth disease, 21/271-309
   see also Hereditary motor and sensory neuropathy type I, Hereditary motor and sensory neuropathy type II *and* Hypertrophic interstitial neuropathy
   myotonic dystrophy, 62/234
   X-linked neuromuscular dystrophy, 62/117
Chickenpox
   childhood myoglobinuria, 62/562
   congenital myopathy, 62/562
Childhood encephalopathy
   acute, *see* Reye syndrome
Childhood myoglobinuria
   age at onset, 62/561
   attack feature, 62/561
   bulbar symptom, 62/561
   cardiac involvement, 62/561
   carnitine palmitoyltransferase, 62/563
   chickenpox, 62/562
   death, 62/561
   dysarthria, 62/564
   dysmorphic feature, 62/564
   dysphagia, 62/564
   ECG, 62/561
   EMG, 62/561
   encephalopathy, 62/561
   epilepsy, 62/564
   Epstein-Barr virus, 62/562
   Escherichia coli, 62/562
   exertional type, 62/561
   familial history, 62/561
   fatty acid oxidation, 62/566
   glycogen storage disease, 62/564
   2 groups, 62/560
   herpetic stomatitis, 62/562
   hypoglycemic encephalopathy, 62/563, 566
   idiopathic myoglobinuria, 62/562
   incomplete syndrome, 62/567
   infantile, 62/560
   infection, 62/561
   lipid metabolic disorder, 62/566
   malignant hyperthermia, 62/567
   metabolic encephalopathy, 62/561-563
   mitochondrial disease, 62/567
   muscle biopsy, 62/561
   muscle cramp, 62/561
   muscle hypertrophy, 62/561
   muscular atrophy, 62/561
   myophosphorylase deficiency, 62/564
   Neisseria, 62/564
   pathophysiology, 62/565
   permanent weakness, 62/561
   phosphofructokinase deficiency, 62/564
   phosphoglycerate kinase deficiency, 62/565
   phosphoglycerate mutase deficiency, 62/565
   phosphorylase kinase deficiency, 62/565
   precipitating factor, 62/561
   psychomotor retardation, 62/564
   renal failure, 62/561
   respiratory distress, 62/561
   Reye like syndrome, 62/563
   sex ratio, 62/561
   Shigella, 62/561
   short chain 3-hydroxyacyl coenzyme A dehydrogenase, 62/563
   streptococcal pharyngitis, 62/561
   subacute necrotizing encephalomyelopathy, 62/563
   toxic type, 62/561
Chloride channel
   myotonic dystrophy, 62/224
Chloroquine
   nonprimary inflammatory myopathy, 62/372
   toxic myopathy, 62/606
Chloroquine myopathy
   cardiomyopathy, 62/608
   diplopia, 62/606
   EMG, 62/608
   histopathology, 62/608
   myeloid body, 62/608
   neuropathy, 62/606
   symptom, 62/606
Chlorpromazine
   toxic myopathy, 62/601
Chondrodysplastic myotonia, *see* Chondrodystrophic myotonia
Chondrodystrophic myotonia
   clinical symptom, 62/269
   hyporeflexia, 62/269
   malignant hyperthermia, 62/570
   mental deficiency, 62/269
   muscle biopsy, 62/269
   muscle hypertrophy, 40/284, 62/269
   myoglobinuria, 62/558
   myotonic syndrome, 62/262
Chorea
   complex I deficiency, 62/504
   Schimke syndrome, 62/304
Choreatic movement, *see* Chorea
Choreic dyskinesia, *see* Chorea
Chromakalim, *see* Cromakalim
Chronic brain syndrome, *see* Dementia
Chronic progressive external ophthalmoplegia
   EMG, 62/70
Chronic renal failure myopathy
   bone pain, 62/541
   clinical features, 62/541
   creatine kinase, 62/541
   fatal gangrenous calcification, 62/541
   proximal limb weakness, 62/541
Chronic ulcerative colitis
   myasthenia gravis, 62/402
Chvostek sign
   hyperkalemic periodic paralysis, 62/462
   myotonic hyperkalemic periodic paralysis, 62/462
Ciclosporin
   dermatomyositis, 62/386
   inclusion body myositis, 62/386
   polymyositis, 62/386
Cimetidine
   nonprimary inflammatory myopathy, 62/372

toxic myopathy, 62/597, 609
Cirrhosis
    liver, *see* Liver cirrhosis *and* Liver disease
    primary biliary, *see* Primary biliary cirrhosis
Classification
    congenital myopathy, 40/286, 62/331
    distal myopathy, 62/198
    myoglobinuria, 62/558
    myotonia, 41/165, 62/263
    myotonic dystrophy, 40/282, 537, 62/216
    myotonic syndrome, 40/537, 62/263
    neuromuscular junction disease, 62/392
    progressive external ophthalmoplegia, *see*
        Progressive external ophthalmoplegia
        classification
    thyroid associated ophthalmopathy, 62/530
Classification history
    limb girdle syndrome, 62/179
Claw foot, *see* Pes cavus
Clofibrate
    acquired myoglobinuria, 62/576
    toxic myopathy, 62/597, 601
Clofibrate intoxication
    myotonic syndrome, 62/264
Clofibride
    toxic myopathy, 62/597
Clomipramine
    myotonia, 62/276
Clostridium
    acquired myoglobinuria, 62/577
Clostridium botulinum, *see* Botulism
Clostridium perfringens, *see* Clostridium welchii
Clostridium welchii
    toxic myopathy, 62/612
CNS, 1/45-74
    Duchenne muscular dystrophy, 62/121
    myotonic dystrophy, 40/517, 62/235
Coats disease
    facioscapulohumeral muscular dystrophy, 62/167
    facioscapulohumeral syndrome, 62/167
    hearing loss, 62/167
Cocaine
    acquired myoglobinuria, 62/577
    toxic myopathy, 62/601
Coenzyme Q10, *see* Ubidecarenone
Colchicine
    toxic myopathy, 62/597, 600, 606
Colchicine myopathy
    creatine kinase, 62/609
    muscle biopsy, 62/609
    spheromembranous body, 62/609
Cold
    acquired myoglobinuria, 62/574
Collagen vascular disease
    see also Inflammatory myopathy
    inclusion body myositis, 62/373
    mixed, *see* Mixed connective tissue disease
Coma, 3/62-77
    complex III deficiency, 62/504
    glutaric aciduria type II, 62/495
    multiple acyl coenzyme A dehydrogenase

deficiency, 62/495
Combined complex I-V deficiency
    cardiomyopathy, 62/506
    encephalomyopathy, 62/506
    lactic acidosis, 62/506
    myopathy, 62/506
    ragged red fiber, 62/506
Complement mediated microangiopathy
    myositis, 62/369
Complex I deficiency
    ataxia, 62/504
    cardiopathy, 62/503
    cardiorespiratory failure, 62/503
    chorea, 62/504
    dementia, 62/504
    epilepsy, 62/504
    fatal infantile multisystem disorder, 62/503
    hearing loss, 62/504
    infantile hypotonia, 62/503
    lactic acidosis, 62/503
    metabolic myopathy, 62/503
    mitochondrial disease, 62/503
    mitochondrial encephalomyopathy, 62/504
    multiple neuronal system degeneration, 62/503
    myopathy, 62/503
    ophthalmoplegia, 62/504
    pigmentary retinopathy, 62/504
    psychomotor retardation, 62/503
    reduced nicotinamide adenine dinucleotide
        dehydrogenase, 62/503
    sensory neuropathy, 62/504
Complex II deficiency
    encephalomyopathy, 62/504
    exercise intolerance, 62/504
    metabolic myopathy, 62/504
    mitochondrial disease, 62/504
    myoglobinuria, 62/504
    myopathy, 62/504
    succinate cytochrome-c reductase, 62/504
Complex III deficiency
    aminoaciduria, 62/504
    ataxia, 62/504
    cardiopathy, 62/505
    coma, 62/504
    dementia, 62/504
    dystonia, 62/504
    encephalomyopathy, 62/504
    epilepsy, 62/504
    exercise intolerance, 62/504
    infantile histiocytoid cardiomyopathy, 62/505
    infantile hypotonia, 62/504
    lactic acidosis, 62/504
    metabolic myopathy, 62/504
    mitochondrial disease, 62/504
    multiple neuronal system degeneration, 62/504
    muscle weakness, 62/504
    myopathy, 62/504
    pigmentary retinopathy, 62/504
    pyramidal syndrome, 62/504
    sensorineural deafness, 62/504
    sensory neuropathy, 62/504

short stature, 62/504
Complex IV deficiency
    benign infantile myopathy, 62/505
    cytochrome-c oxidase, 62/505
    DeToni-Fanconi-Debré syndrome, 62/505
    encephalopathy, 62/505
    fatal infantile myopathy, 62/505
    lactic acidosis, 62/505
    metabolic myopathy, 62/505
    mitochondrial disease, 62/505
    myopathy, 62/505
    respiratory insufficiency, 62/505
    subacute necrotizing encephalomyelopathy, 62/505
Complex V deficiency
    ataxia, 62/506
    dementia, 62/506
    metabolic encephalopathy, 62/506
    metabolic myopathy, 62/506
    mitochondrial disease, 62/506
    multiple neuronal system degeneration, 62/506
    myopathy, 62/506
    neuropathy, 62/506
    paracrystalline inclusion, 62/506
    ragged red fiber, 62/506
    retinopathy, 62/506
Compound muscle action potential
    familial infantile myasthenia, 62/428
Computer assisted tomography
    Kearns-Sayre-Daroff-Shy syndrome, 62/306, 311
    orbital myositis, 62/299, 305
    progressive external ophthalmoplegia, 62/295, 306,
        311
    thyroid associated ophthalmopathy, 62/531
Computer model
    myotonia, 62/277
Congenital acetylcholine receptor deficiency/short
        channel opentime syndrome
    congenital myasthenic syndrome, 62/437
    EMG, 62/437
    facial diplegia, 62/437
    neuromuscular junction disease, 62/392
    ophthalmoplegia, 62/437
Congenital acetylcholine receptor deficiency/synaptic
        cleft syndrome
    arthrogryposis, 62/438
    congenital myasthenic syndrome, 62/438
    cyanotic episode, 62/438
    exacerbation, 62/438
    exercise intolerance, 62/438
    neuromuscular junction disease, 62/392
    ptosis, 62/438
    respiratory distress, 62/438
Congenital ε-acetylcholine receptor subunit mutation
        syndrome
    congenital myasthenic syndrome, 62/436
    EMG, 62/436
    neuromuscular junction disease, 62/392
    symptom, 62/436
Congenital end plate acetylcholine receptor
        deficiency
    congenital myopathy, 62/332

Congenital end plate acetylcholinesterase deficiency
    congenital myasthenic syndrome, 62/430
    congenital myopathy, 62/332
    EMG, 62/430
    muscle biopsy, 62/430
    neuromuscular junction disease, 62/392
    pathogenesis, 62/431
    symptom, 62/430
Congenital facial diplegia, see Möbius syndrome
Congenital familial limb girdle myasthenia
    congenital myasthenic syndrome, 62/439
    EMG, 62/439
    joint contracture, 62/439
    neuromuscular junction disease, 62/392
    symptom, 62/439
Congenital fiber type disproportion
    congenital myopathy, 41/14-16, 62/332
    facial weakness, 62/334
    fatality rate, 62/356
    genetics, 41/431, 62/356
    globoid cell leukodystrophy, 62/355
    infantile spinal muscular atrophy, 62/355
    Lowe syndrome, 62/355
    respiratory dysfunction, 62/356
    rigid spine syndrome, 62/355
    twin, 62/356
Congenital high conductance fast channel syndrome
    congenital myasthenic syndrome, 62/438
    EMG, 62/438
    episodic apnea, 62/438
    neuromuscular junction disease, 62/392
    ptosis, 62/438
    symptom, 62/438
Congenital ichthyosis
    triglyceride storage disease, 62/496
Congenital myasthenic syndrome
    abnormal acetylcholine-acetylcholine receptor
        interaction syndrome, 62/437
    congenital acetylcholine receptor deficiency/short
        channel opentime syndrome, 62/437
    congenital acetylcholine receptor deficiency/
        synaptic cleft syndrome, 62/438
    congenital ε-acetylcholine receptor subunit
        mutation syndrome, 62/436
    congenital end plate acetylcholinesterase
        deficiency, 62/430
    congenital familial limb girdle myasthenia, 62/439
    congenital high conductance fast channel
        syndrome, 62/438
    congenital myopathy, 62/332
    congenital slow channel syndrome, 62/432
    congenital synaptic vesical paucity syndrome,
        62/429
    Eaton-Lambert myasthenic syndrome, 62/425
    edrophonium test, 62/413
    facial weakness, 62/334
    familial infantile myasthenia, 62/427
    myasthenia gravis, 62/415
    polyneuropathy, 62/413
Congenital myopathy, 41/1-22
    see also Infantile hypotonia

abnormal myomuscular junction myopathy, 62/332
aspecific clinical criteria, 62/333
cap disease, 62/332
cardiac involvement, 62/334
central core myopathy, 41/3-8, 62/332
chickenpox, 62/562
classification, 40/286, 62/331
concept history, 62/331
congenital end plate acetylcholine receptor
    deficiency, 62/332
congenital end plate acetylcholinesterase
    deficiency, 62/332
congenital fiber type disproportion, 41/14-16,
    62/332
congenital myasthenic syndrome, 62/332
creatine kinase, 62/333, 350
cytoplasmic inclusion body myopathy, 62/332
definition, 62/331
dysphagia, 62/335
EMG, 62/69, 333, 350
experimental model, 62/358
external ophthalmoplegia, 62/333
extraocular muscle weakness, 62/334
facial weakness, 62/334
familial infantile myasthenia, 62/332
fingerprint body myopathy, 41/20, 62/332
granulofilamentous myopathy, 62/332
granulovacuolar lobular myopathy, 62/332
hereditary cylindric spirals myopathy, 62/332
honeycomb myopathy, 62/332
hypotrophy, 62/349
malignant hyperthermia, 62/335
Mallory body myopathy, 62/332
mental deficiency, 62/335
microfiber myopathy, 62/332
minimal change myopathy, 62/332
mitochondria jagged Z line myopathy, 62/332
multicore myopathy, 41/16, 62/332
muscle cramp, 62/335
muscle fiber type I, 40/146, 62/349
muscle fiber type I hypotrophy, 62/332
muscle fiber type I predominance, 62/332
muscle fiber type II hypoplasia, 62/332
muscle hypertrophy, 62/335
myalgia, 62/335
myasthenia, 62/335
myofibrillar lysis myopathy, 62/332
myotonia, 62/335
myotubular myopathy, 41/11-14, 62/332
nemaline myopathy, 41/8-11, 62/332
nucleodegenerative myopathy, 62/332
progressive external ophthalmoplegia, 62/297, 333
ptosis, 62/297, 333, 349
reducing body myopathy, 41/17-20, 62/332
respiration, 62/334
respiratory dysfunction, 62/334, 349
rigidity, 62/335
rimmed vacuole distal myopathy, 62/332
sarcoplasmic body myopathy, 62/332
sarcotubular myopathy, 41/21, 62/332
selective myosin degeneration myopathy, 62/332

skeletal deformity, 62/334
slow channel syndrome, 62/332
spheroid body myopathy, 62/332
structured type, 62/332
temporomandibular ankylosis, 62/335
tremor, 62/335
trilaminar myopathy, 41/22, 62/332
tubular aggregate myopathy, 62/332
tubulomembranous inclusion myopathy, 62/332
uniform muscle fiber type I myopathy, 62/332
unstructured type, 62/332
X-linked myotubular myopathy, 62/117
X-linked neutropenic cardioskeletal myopathy,
    62/117
Z band plaque myopathy, 62/332
zebra body myopathy, 41/21, 62/332
Congenital ptosis
    ophthalmoplegia, 62/297
    progressive external ophthalmoplegia, 62/297
Congenital slow channel syndrome
    chemistry, 62/434
    congenital myasthenic syndrome, 62/432
    EMG, 62/432
    muscle biopsy, 62/434
    neuromuscular junction disease, 62/392
    pathogenesis, 62/435
    ptosis, 62/432
    symptom, 62/432
Congenital synaptic vesical paucity syndrome
    congenital myasthenic syndrome, 62/429
    EMG, 62/429
    miniature motor end plate potential, 62/429
    neuromuscular junction disease, 62/392
Congestive heart failure
    dermatomyositis, 62/374
    inclusion body myositis, 62/374
    myotonic dystrophy, 62/227
    polymyositis, 62/374
Conn syndrome, see Hyperaldosteronism
Connectin
    limb girdle syndrome, 62/187
Connective tissue
    muscle, 62/40
Connective tissue disease, see Collagen vascular
    disease
Continuous muscle fiber activity syndrome, see Isaacs
    syndrome
Contraction
    muscular, see Muscle contraction
Contracture
    see also Muscle cramp
    Duchenne muscular dystrophy, 40/337, 361,
        41/465, 43/106, 62/119, 133
Copper sulfate
    toxic myopathy, 62/601
Core disease
    central, see Central core myopathy
Cori disease
    type II, see Infantile acid maltase deficiency
    type III, see Debrancher deficiency
    type IV, see Brancher deficiency

type V, *see* Myophosphorylase deficiency
Cori-Forbes disease, *see* Debrancher deficiency
Cortical blindness
    MELAS syndrome, 62/510
Corticoid, *see* Corticosteroids
Corticosteroid withdrawal myopathy
    adrenal insufficiency myopathy, 62/536
    arthralgia, 62/536
    fatigue, 62/536
    fever, 62/536
    myalgia, 62/536
    weight loss, 62/536
Corticosteroids
    see also Glucocorticoids
    dermatomyositis, 62/383
    Duchenne muscular dystrophy, 40/378, 62/129
    inclusion body myositis, 62/383
    myasthenia gravis, 41/93, 132, 62/417
    polymyositis, 62/383
    steroid myopathy, 41/250, 62/384
    toxic myopathy, 62/605
Corticostriatopallidodentate calcification, *see*
    Striatopallidodentate calcification
Coxsackie virus
    acquired myoglobinuria, 62/577
CPT deficiency, *see* Carnitine palmitoyltransferase
    deficiency
Cramp
    muscle, *see* Muscle cramp
Cranial neuropathy
    Kearns-Sayre-Daroff-Shy syndrome, 62/313
    progressive external ophthalmoplegia, 62/313
Creatine kinase
    acid maltase deficiency, 62/480
    acromegalic myopathy, 62/538
    adenosine monophosphate deaminase deficiency,
        62/512
    axonal polyneuropathy, 62/609
    brancher deficiency, 62/484
    chronic renal failure myopathy, 62/541
    colchicine myopathy, 62/609
    congenital myopathy, 62/333, 350
    debrancher deficiency, 62/482
    dermatomyositis, 62/375
    Duchenne muscular dystrophy, 40/281, 376,
        43/106, 62/118, 124
    facioscapulohumeral muscular dystrophy, 40/420,
        62/166
    fatal infantile myopathy/cardiopathy, 62/510
    hyperaldosteronism myopathy, 62/536
    hypoparathyroid myopathy, 62/541
    hypothyroid myopathy, 62/533
    inclusion body myositis, 62/373, 375
    limb girdle syndrome, 40/449, 62/185
    muscle tissue culture, 62/87
    myophosphorylase deficiency, 62/485
    pheochromocytoma myopathy, 62/537
    phosphofructokinase deficiency, 62/487
    phosphorylase kinase deficiency, 62/487
    polymyositis, 41/60, 62/375
    quadriceps myopathy, 62/174

rimmed vacuole distal myopathy, 62/204
steroid myopathy, 62/535, 605
thyrotoxic myopathy, 62/529
toxic myopathy, 62/596
ubidecarenone, 62/504
Creatine kinase MM
    muscle tissue culture, 62/89
Creatine phosphokinase, *see* Creatine kinase
Creatinuria
    steroid myopathy, 62/535
Cricopharyngeal dysphagia
    nemaline myopathy, 62/342
Cricopharyngeal myotomy
    dermatomyositis, 62/386
    dysphagia, 62/386
    inclusion body myositis, 62/386
    polymyositis, 62/386
Crohn disease
    polymyositis, 62/373
Cromakalim
    myotonia, 62/276
Crotamine
    toxic myopathy, 62/610
Crotoxin
    toxic myopathy, 62/610
CSF
    Kearns-Sayre-Daroff-Shy syndrome, 38/225, 228,
        43/142, 62/311
    progressive external ophthalmoplegia, 62/311
CSF protein
    Kearns-Sayre-Daroff-Shy syndrome, 62/508
CT scanning, *see* Computer assisted tomography
Curare
    myasthenia gravis, 62/413
Curschmann-Steinert disease, *see* Myotonic
    dystrophy
Cushing syndrome
    see also Nelson syndrome
    steroid myopathy, 40/162, 41/250-253, 62/535
Cyclophosphamide
    dermatomyositis, 62/385
    inclusion body myositis, 62/385
    myasthenia gravis, 41/96, 136, 62/418
    polymyositis, 62/385
Cyclosporin A, *see* Ciclosporin
Cylindric spirals myopathy
    hereditary, *see* Hereditary cylindric spirals
        myopathy
Cysticercus
    inflammatory myopathy, 62/373
    polymyositis, 62/373
Cytochrome aa$_3$, *see* Cytochrome-c oxidase
Cytochrome-c oxidase
    complex IV deficiency, 62/505
    Kearns-Sayre-Daroff-Shy syndrome, 62/314
    mitochondrial disease, 62/497
    progressive external ophthalmoplegia, 62/293, 314
Cytochrome-c oxidase deficiency
    combined complex deficiency, 62/509
    mitochondrial DNA deletion, 62/508
    POLIP syndrome, 62/506

Cytochrome oxidase, *see* Cytochrome-c oxidase
Cytoplasmic inclusion body myopathy
    age, 62/344
    congenital myopathy, 62/332
    desmin intermediate filament, 62/344
    emetine intoxication, 62/345
    experimental injury, 62/345
    facial weakness, 62/334
    ipecac intoxication, 62/345
    organophosphate intoxication, 62/345
Cytoskeleton
    muscle fiber, 62/19
    muscle fiber type, 62/3
Cytosol
    muscle fiber type, 62/4
Cytotoxic agent
    toxic myopathy, 62/597

Danazol
    toxic myopathy, 62/597
Davidenkow syndrome, *see* Scapuloperoneal spinal
    muscular atrophy
Dawidenkow syndrome, *see* Scapuloperoneal spinal
    muscular atrophy
Deafness
    see also Hearing loss *and* Sensorineural deafness
    Kearns-Sayre-Daroff-Shy syndrome, 62/313
    progressive external ophthalmoplegia, 22/193,
        62/313
    sensorineural, *see* Sensorineural deafness
Death
    childhood myoglobinuria, 62/561
Debrancher deficiency
    amylo-1,6-glucosidase, 62/483
    biochemistry, 62/483
    cardiac involvement, 62/482
    clinical course, 62/482
    creatine kinase, 62/482
    ECG, 62/482
    echoCG, 62/483
    EMG, 62/482
    epilepsy, 62/482
    glycogen storage disease, 27/222, 40/256, 41/181,
        62/482
    hepatomegaly, 62/482
    histopathology, 62/483
    inheritance, 62/483
    liver disease, 62/482
    metabolic myopathy, 62/482
    molecular genetics, 62/483
    muscle biopsy, 62/483
    muscle fiber feature, 62/17
    muscle tissue culture, 62/99
    muscular atrophy, 62/482
    oligo-1,4→1,4-glucantransferase, 62/483
    phosphorylase limit dextrin, 62/483
    rimmed vacuole distal myopathy, 62/483
    type IIIA, 62/483
    type IIIB, 62/483
    type IIIC, 62/483
Deflazacort

Duchenne muscular dystrophy, 62/131
Degenerations
    heterogeneous system, *see* Progressive
        supranuclear palsy
    multiple neuronal system, *see* Multiple neuronal
        system degeneration
    multisystem, *see* Multiple neuronal system
        degeneration
    primary pigmentary retinal, *see* Primary
        pigmentary retinal degeneration
    secondary pigmentary retinal, *see* Secondary
        pigmentary retinal degeneration
    subacute cerebellar, *see* Subacute cerebellar
        degeneration
    tapetoretinal, *see* Secondary pigmentary retinal
        degeneration
Dejerine-Sottas disease, *see* Hypertrophic interstitial
    neuropathy
Delta lesion
    see also Sarcolemma
    muscle fiber, 62/15
Dementia
    adenosine triphosphatase 6 mutation syndrome,
        62/510
    adult polyglucosan body disease, 62/484
    complex I deficiency, 62/504
    complex III deficiency, 62/504
    complex V deficiency, 62/506
    Kearns-Sayre-Daroff-Shy syndrome, 62/308, 508
    MELAS syndrome, 62/509
    MERRF syndrome, 62/509
    progressive external ophthalmoplegia, 62/308
Demyelinating neuropathy
    acute inflammatory, *see* Guillain-Barré syndrome
    Guillain-Barré, *see* Guillain-Barré syndrome
Demyelination
    myotonic dystrophy, 62/237
    scapuloperoneal spinal muscular atrophy, 62/170
Denervation
    see also Neuropathy
    motor unit, 62/62
    muscle contraction, 62/53
Denervation atrophy
    muscle fiber, 62/13
    myopathy, 62/13
    neuromuscular disease, 62/13
    pituitary gigantism myopathy, 62/538
Depakene, *see* Valproic acid
Depression
    mental, *see* Mental depression
Dermatitis herpetiformis
    polymyositis, 62/373
Dermatomyositis
    see also Inflammatory myopathy *and* Polymyositis
    amyloid deposit, 62/377
    amyloid β-protein, 62/377
    amyopathic dermatomyositis, 62/371
    anti-Jo-1 antibody, 62/374
    arthralgia, 62/374
    aspiration pneumonia, 62/374
    autoimmune antibody, 62/380

axonal neuropathy, 62/376
azathioprine, 62/385
basophilic granular inclusion, 62/377
calcinosis, 62/386
cardiac disorder, 62/374
cardiomyopathy, 62/374
ciclosporin, 62/386
congestive heart failure, 62/374
corticosteroids, 62/383
creatine kinase, 62/375
cricopharyngeal myotomy, 62/386
cyclophosphamide, 62/385
diagnosis, 62/375
diagnostic criteria, 62/378
dyspnea, 62/374
electronmicroscopy, 62/377
EMG, 1/642, 62/375
eosinophilia myalgia syndrome, 62/371
eosinophilic cytoplasmic inclusion, 62/377
fatigue, 62/371
fever, 62/373
gastrointestinal symptom, 62/374
Gottron rash, 62/371
heliotrope rash, 41/57, 62/371
hematemesis, 62/374
hypoxemia, 62/374
immunopathology, 62/379
immunosuppressive therapy, 62/385
intravenous Ig, 62/386
joint contracture, 62/374
lactate dehydrogenase, 62/375
leukopheresis, 62/386
lymphoid irradiation, 62/386
malaise, 62/373
malignancy, 62/371, 374
melena, 62/374
membranolytic attack complex, 62/379
methotrexate, 62/385
mixed connective tissue disease, 62/371, 374
muscle biopsy, 62/375
muscle weakness, 62/371
myalgia, 62/371
myocarditis, 62/374
myoglobinuria, 62/556
nuclear magnetic resonance, 62/379
overlap syndrome, 62/374
pathogenesis, 62/379
physiotherapy, 62/386
picornavirus, 62/381
plasmapheresis, 62/386
pneumatosis intestinalis, 62/374
pneumonitis, 62/374
practical guideline, 62/386
primary inflammatory myositis, 62/369
pulmonary dysfunction, 62/374
Raynaud phenomenon, 62/374
rheumatoid arthritis, 62/374
rimmed muscle fiber vacuole, 62/376
rimmed muscle fiber vacuole content, 62/18
rimmed muscle fiber vacuole type, 62/18
scleroderma, 62/374

serum alanine aminotransferase, 62/375
serum aspartate aminotransferase, 62/375
serum muscle enzyme, 62/375
shawl sign, 62/371
Sjögren syndrome, 62/374
steroid myopathy, 62/384
systemic lupus erythematosus, 62/374
systemic sclerosis, 62/371, 374
tachyarrhythmia, 62/374
treatment, 62/383
tubulofilament, 62/377
viral myositis, 62/381
weight loss, 62/374
Desmin intermediate filament
    cytoplasmic inclusion body myopathy, 62/344
    sarcoplasmic body myopathy, 62/345
Diabetes mellitus, 42/543-545
    juvenile, see Juvenile diabetes mellitus
    Kearns-Sayre-Daroff-Shy syndrome, 62/309, 508
    progressive external ophthalmoplegia, 62/309
Diabetic amyotrophy, see Diabetic proximal
    neuropathy
Diabetic myopathy, see Diabetic proximal
    neuropathy
Diabetic proximal neuropathy
    pathogenesis, 62/542
Diabetic thigh muscle infarction
    clinical features, 62/542
    histopathology, 62/542
    recurrence, 62/542
Diagnoses
    prenatal, see Prenatal diagnosis
Dialysis
    see also Hemodialysis
    renal, see Renal dialysis
3,4-Diaminopyridine
    Eaton-Lambert myasthenic syndrome, 62/426
Diamorphine
    acquired myoglobinuria, 62/577
    toxic myopathy, 62/601
Diamorphine addiction
    toxic myopathy, 62/600
Diarrhea
    MNGIE syndrome, 62/505
20,25-Diazacholesterol, see Azacosterol
Diazepam
    toxic myopathy, 62/601
Dicarboxylic aciduria
    lipid metabolic disorder, 62/494
    long chain acyl coenzyme A dehydrogenase
        deficiency, 62/495
2,4-Dichlorophenoxyacetic acid intoxication
    drug induced myotonia, 62/271
    drug induced myotonic syndrome, 62/271
Diffuse cerebral sclerosis, see Diffuse sclerosis and
    Leukodystrophy
Diffuse cerebral sclerosis (Krabbe), see Globoid cell
    leukodystrophy
Diffuse lewy body disease
    progressive external ophthalmoplegia, 62/304
Diffuse sclerosis

see also Leukodystrophy
  globoid cell type, *see* Globoid cell leukodystrophy
  Krabbe type, *see* Globoid cell leukodystrophy
Dihydrocodeine
  toxic myopathy, 62/601
Dilantin, *see* Phenytoin
Diltiazem
  myotonia, 62/276
DiMauro disease, *see* Carnitine palmitoyltransferase
  deficiency
Diphenylhydantoin, *see* Phenytoin
Diplegia
  congenital facial, *see* Möbius syndrome
  facial, *see* Facial diplegia
Diplococcus pneumoniae, *see* Streptococcus
  pneumoniae
Diplopia
  chloroquine myopathy, 62/606
  myotonic hyperkalemic periodic paralysis, 62/462
  thyroid associated ophthalmopathy, 62/531
Distal muscular dystrophy
  distal myopathy, 62/199, 201
Distal myopathy
  classification, 62/198
  diagnosis, 62/197
  diagnostic criteria, 62/198
  differential diagnosis, 62/198
  distal muscular dystrophy, 62/199, 201
  heterogeneity, 62/197
  infantile, *see* Infantile distal myopathy
  juvenile, 62/199, 201
  myotonic dystrophy, 62/240
  rimmed muscle fiber vacuole, 40/475, 62/204
  rimmed vacuole distal myopathy, 62/199
  treatment, 62/206
  Wohlfart-Kugelberg-Welander disease, 62/199
Diuretic agent
  toxic myopathy, 62/597, 601
Dive bomber discharge, *see* Myotonic discharge
Dominant inheritance
  inclusion body myositis, 62/373
  myotonia fluctuans, 62/266
  myotonic syndrome, 62/263
Doxylamine
  acquired myoglobinuria, 62/573
Drug addiction, 37/365-394
  diamorphine, *see* Diamorphine addiction
  opiate, *see* Opiate addiction
  phencyclidine, *see* Phencyclidine addiction
Drug induced myotonia
  9-anthroic acid intoxication, 62/271
  azacosterol, 62/271
  2,4-dichlorophenoxyacetic acid intoxication,
    62/271
Drug induced myotonic syndrome
  9-anthroic acid intoxication, 62/271
  2,4-dichlorophenoxyacetic acid intoxication,
    62/271
Drug trial
  Becker muscular dystrophy, 62/136
Duchenne muscular dystrophy

age at onset, 62/118
atypical form, 62/122
Becker muscular dystrophy, 62/123
biochemistry, 40/376-382, 62/124
cardiac involvement, 62/119
chromosome Xp21, 62/125
chronic ventilatory failure, 62/134
CNS, 62/121
contracture, 40/337, 361, 41/465, 43/106, 62/119,
  133
corticosteroids, 40/378, 62/129
course, 62/119
creatine kinase, 40/281, 376, 43/106, 62/118, 124
cyclosporine, 62/131
deflazacort, 62/131
differential diagnosis, 41/176, 62/122
dystrophin, 62/121, 125
fructose bisphosphate aldolase, 62/118
functional grading, 62/120
gastrointestinal features, 62/134
gastrointestinal manifestation, 62/120
genetic diagnosis, 62/128
genetics, 40/352-357, 41/487, 489, 62/125
genotype, 62/126
glycerol kinase deficiency, 62/127
Gowers sign, 40/359, 62/117
heart failure, 62/134
histology, 62/123
histopathology, 62/118
history, 40/350, 434, 62/117
incidence, 62/118
intestinal pseudo-obstruction, 62/120
malignant hyperthermia, 62/570
McLeod myopathy, 62/127
mdx mouse model, 62/127
mental deficiency, 40/363, 41/479, 62/121
muscle biopsy, 62/123
muscle tissue culture, 40/191, 375, 62/100
myoblast transfer, 62/131
myoglobinuria, 62/555, 558
natural history, 62/119
phenotype, 62/126
physiotherapy, 62/132
prevalence, 43/106, 62/118
respiratory failure, 62/122, 133
scoliosis, 40/362, 62/119
surgical treatment, 62/132
symptom, 62/118
treatment, 62/129
treatment trial, 62/129
urinary carnosine, 62/124
urinary creatinine, 62/124
urinary 3-methylhistidine, 62/124
urinary putrescine, 62/124
urinary spermidine, 62/124
urinary spermine, 62/124
X-linked muscular dystrophy, 40/277, 349-391,
  41/408, 415, 62/117
Duhring-Brocq disease, *see* Dermatitis herpetiformis
Duhring disease, *see* Dermatitis herpetiformis
Dysarthria

childhood myoglobinuria, 62/564
Dysbasia lordotica progressiva (Oppenheim-
    Fraenkel), *see* Dystonia musculorum deformans
Dyskinesia
    choreic, *see* Chorea
Dysphagia
    childhood myoglobinuria, 62/564
    congenital myopathy, 62/335
    cricopharyngeal, *see* Cricopharyngeal dysphagia
    cricopharyngeal myotomy, 62/386
    inclusion body myositis, 62/206, 373
    thyrotoxic myopathy, 62/528
Dysphonia
    thyrotoxic myopathy, 62/528
Dyspnea
    dermatomyositis, 62/374
    inclusion body myositis, 62/374
    polymyositis, 62/374
Dysthyroid orbitopathy, *see* Thyroid associated
    ophthalmopathy
Dystonia
    see also Dystonia musculorum deformans
    complex III deficiency, 62/504
    hereditary torsion, *see* Dystonia musculorum
        deformans
    oculofacial, *see* Progressive supranuclear palsy
    oculofaciocervical, *see* Progressive supranuclear
        palsy
    progressive external ophthalmoplegia, 62/304
    torsion, *see* Dystonia musculorum deformans
Dystonia musculorum deformans, 6/517-541
    see also Dystonia
    progressive external ophthalmoplegia
        classification, 62/290
Dystonia musculorum deformans (Ziehen-
    Oppenheim), *see* Dystonia musculorum
    deformans
Dystrophia myotonica, *see* Myotonic dystrophy
Dystrophia retinae pigmentosa, *see* Primary
    pigmentary retinal degeneration *and* Secondary
    pigmentary retinal degeneration
Dystrophin
    α-actinin, 62/126
    aminoacid sequence, 62/125
    ankyrin, 62/126
    brain, 62/125
    Duchenne muscular dystrophy, 62/121, 125
    limb girdle syndrome, 62/187
    muscle, 62/125
    muscle fiber, 62/35
    spectrin, 62/126
Dystrophin deficiency
    incomplete, *see* Incomplete dystrophin deficiency
Dystrophy
    Becker muscular, *see* Becker muscular dystrophy
    distal muscular, *see* Distal muscular dystrophy
    Duchenne muscular, *see* Duchenne muscular
        dystrophy
    Emery-Dreifuss muscular, *see* Emery-Dreifuss
        muscular dystrophy
    facioscapulohumeral muscular, *see*

    Facioscapulohumeral muscular dystrophy
    limb girdle, *see* Limb girdle syndrome
    myotonic, *see* Myotonic dystrophy
    oculopharyngeal muscular, *see* Oculopharyngeal
        muscular dystrophy
    scapulohumeral muscular, *see* Scapulohumeral
        muscular dystrophy
    scapuloperoneal muscular, *see* Scapuloperoneal
        muscular dystrophy
    subcortical argyrophilic, *see* Progressive
        supranuclear palsy
    X-linked muscular, *see* X-linked muscular
        dystrophy
    X-linked neuromuscular, *see* X-linked
        neuromuscular dystrophy

E. coli, *see* Escherichia coli
Eaton-Lambert myasthenic syndrome
    see also Myasthenic syndrome
    acquired autoimmune myasthenia gravis, 62/425
    4-aminopyridine, 41/360, 62/426
    anticholinesterase agent, 62/426
    atypical case, 62/79
    autoimmune etiology, 62/422
    azathioprine, 62/426
    botulism, 62/425
    calcium channel, 62/421, 423
    celiac disease, 62/421
    clinical features, 41/101, 111, 129, 319, 349-352,
        62/421
    congenital myasthenic syndrome, 62/425
    definition, 62/421
    diagnosis, 62/423
    3,4-diaminopyridine, 62/426
    differential diagnosis, 41/290, 356-358, 62/425
    edrophonium test, 62/413
    EMG, 62/78, 422
    guanidine, 41/358, 62/426
    Guillain-Barré syndrome, 62/425
    hyperthyroidism, 62/421
    hyporeflexia, 62/426
    hypothyroidism, 62/421
    immunosuppression, 62/426
    juvenile diabetes mellitus, 62/421
    lung carcinoma, 62/425
    magnesium intoxication, 62/425
    myasthenia gravis, 62/391, 402, 415, 421
    myasthenic syndrome, 62/391
    neoplasm, 62/421
    neuromuscular junction disease, 62/392
    paresthesia, 62/426
    pernicious anemia, 62/421
    plasmapheresis, 62/426
    polymyositis, 62/425
    polyneuropathy, 62/413, 425
    presynaptic membrane, 62/425
    repetitive nerve stimulation, 62/79
    serologic test, 62/425
    single fiber electromyography, 62/79
    Sjögren syndrome, 62/421
    subacute cerebellar degeneration, 62/421

treatment, 41/358-361, 62/426
vitiligo, 62/421
Eaton-lambert syndrome, *see* Eaton-Lambert myasthenic syndrome
EB virus, *see* Epstein-Barr virus
ECG
  acid maltase deficiency, 62/480
  carnitine deficiency, 62/493
  childhood myoglobinuria, 62/561
  debrancher deficiency, 62/482
  Emery-Dreifuss muscular dystrophy, 62/153
  Kearns-Sayre-Daroff-Shy syndrome, 62/311
  myotonic dystrophy, 62/226
  progressive external ophthalmoplegia, 62/311
ECHO virus
  acquired myoglobinuria, 62/577
EchoCG
  debrancher deficiency, 62/483
  myotonic dystrophy, 62/227
Edrophonium test
  amyotrophic lateral sclerosis, 62/413
  congenital myasthenic syndrome, 62/413
  Eaton-Lambert myasthenic syndrome, 62/413
  myasthenia gravis, 40/298, 303, 321, 41/126, 240, 62/413
  poliomyelitis, 62/413
  progressive external ophthalmoplegia, 62/303
EEG
  multiple mitochondrial DNA deletion, 62/510
Elapid snake venom
  myasthenia gravis, 62/391
  myasthenic syndrome, 62/391
Electrocardiography, *see* ECG
Electroencephalography, *see* EEG
Electromyography, *see* EMG
Emery-Dreifuss muscular dystrophy
  areflexia, 62/149
  atrial flutter, 62/171
  autosomal dominant, 62/155
  Bethlem-Van Wijngaarden syndrome, 62/155
  blood clotting factor VIII, 62/157
  cardiomyopathy, 62/145, 152
  carrier, 62/154
  centronuclear, 62/156
  chromosome Xq28, 62/156
  differential diagnosis, 62/154
  ECG, 62/153
  EMG, 62/150
  facioscapulohumeral muscular dystrophy, 62/155
  genetics, 62/156
  heredofamilial myosclerosis, 62/173
  histopathology, 62/151
  history, 62/146
  limb girdle syndrome, 62/189
  mild form, 62/149
  muscle atrophy distribution, 62/148
  muscle contraction, 62/145
  muscle fiber type I atrophy, 62/151
  neutropenia, 62/156
  prognosis, 62/154
  rigid spine syndrome, 40/390, 62/155

scapuloperoneal syndrome, 40/392, 426, 62/171
scapuloperoneal weakness, 62/171
severe form, 62/150
single fiber electromyography, 62/150
symptom, 62/146
syncope, 62/153
transient ischemic attack, 62/154
winged scapula, 62/146
X-linked muscular dystrophy, 40/389, 62/117
X-linked recessive, 62/154
Emery-Dreifuss syndrome, *see* Emery-Dreifuss muscular dystrophy
Emetine
  nonprimary inflammatory myopathy, 62/372
  toxic myopathy, 62/600
Emetine intoxication
  cytoplasmic inclusion body myopathy, 62/345
EMG
  abnormal acetylcholine-acetylcholine receptor interaction syndrome, 62/437
  acid maltase deficiency, 62/480
  acromegalic myopathy, 62/538
  ACTH induced myopathy, 62/537
  acute polymyositis, 62/71
  childhood myoglobinuria, 62/561
  chloroquine myopathy, 62/608
  chronic progressive external ophthalmoplegia, 62/70
  congenital acetylcholine receptor deficiency/short channel opentime syndrome, 62/437
  congenital ε-acetylcholine receptor subunit mutation syndrome, 62/436
  congenital end plate acetylcholinesterase deficiency, 62/430
  congenital familial limb girdle myasthenia, 62/439
  congenital high conductance fast channel syndrome, 62/438
  congenital myopathy, 62/69, 333, 350
  congenital slow channel syndrome, 62/432
  congenital synaptic vesical paucity syndrome, 62/429
  debrancher deficiency, 62/482
  dermatomyositis, 1/642, 62/375
  Eaton-Lambert myasthenic syndrome, 62/78, 422
  Emery-Dreifuss muscular dystrophy, 62/150
  endocrine myopathy, 62/69
  facioscapulohumeral muscular dystrophy, 40/420, 62/166
  familial hypokalemic periodic paralysis, 62/69
  familial infantile myasthenia, 62/427
  hyperaldosteronism myopathy, 62/536
  hyperparathyroid myopathy, 62/540
  hypokalemic periodic paralysis, 28/583, 62/459
  hypothyroid myopathy, 62/533
  inclusion body myositis, 62/375
  limb girdle syndrome, 40/453, 62/186
  metabolic myopathy, 62/69
  motor unit, 19/275, 40/138, 62/49
  muscle biopsy, 62/124
  muscle contraction, 62/52
  myasthenia gravis, 1/639, 644, 62/398, 413

myasthenic syndrome, 62/78
myopathy, 62/49
myotonia congenita, 40/539, 62/70
myotonic dystrophy, 40/519, 62/70
myotubular myopathy, 62/69
Nelson syndrome, 62/537
phosphofructokinase deficiency, 41/191, 62/487
pituitary gigantism myopathy, 62/538
polymyositis, 1/642, 62/375
progressive external ophthalmoplegia, 22/181,
    62/295
quadriceps myopathy, 62/174
rigid spine syndrome, 62/173
rimmed vacuole distal myopathy, 62/204
scapulohumeral spinal muscular atrophy, 62/171
scapuloperoneal muscular dystrophy, 42/99, 62/169
scapuloperoneal spinal muscular atrophy, 62/170
steroid myopathy, 62/535
thyrotoxic myopathy, 62/529
EMI, see Computer assisted tomography
Emotion, 3/343-363
myotonia congenita, 62/264
Enalapril
toxic myopathy, 62/597
Encephalomyelopathy
subacute necrotizing, see Subacute necrotizing
    encephalomyelopathy
Encephalomyopathy
combined complex I-V deficiency, 62/506
complex II deficiency, 62/504
complex III deficiency, 62/504
mitochondrial, see Mitochondrial
    encephalomyopathy
α-oxoglutarate dehydrogenase deficiency, 62/503
Encephalopathy
see also Leukoencephalopathy
acute childhood, see Reye syndrome
acute toxic, see Reye syndrome
childhood myoglobinuria, 62/561
complex IV deficiency, 62/505
episodic, see Episodic encephalopathy
hepatic, see Hepatic encephalopathy
hypoglycemic, see Hypoglycemic encephalopathy
Leigh, see Subacute necrotizing
    encephalomyelopathy
metabolic, see Metabolic encephalopathy
End plate, see Motor end plate
Endocrine disorder
Kearns-Sayre-Daroff-Shy syndrome, 62/309
progressive external ophthalmoplegia, 62/309
Endocrine exophthalmos, see Thyroid associated
    ophthalmopathy
Endocrine myopathy, 41/235-253
EMG, 62/69
fatigue, 62/527
hyperthyroidism, 62/528
hypothyroid myopathy, 41/243-246, 298, 62/528
muscle cramp, 62/527
muscle fiber type II atrophy, 62/527
myasthenia gravis, 41/240, 62/528
thyroid associated ophthalmopathy, 62/528

thyrotoxic myopathy, 41/235-240, 62/528
thyrotoxic periodic paralysis, 41/165, 241-243,
    62/528
Entactin
muscle fiber, 62/31
Enteropathy
gluten induced, see Celiac disease
Enzyme histochemistry
multicore myopathy, 62/339
muscle tissue culture, 62/94
myofibrillar lysis myopathy, 62/339
tubular aggregate myopathy, 62/352
Eosinophilia myalgia syndrome
dermatomyositis, 62/371
toxic myopathy, 62/604
Eosinophilic cytoplasmic inclusion
dermatomyositis, 62/377
inclusion body myositis, 62/377
polymyositis, 62/377
Epidemiology
malignant hyperthermia, 62/567
myasthenia gravis, 41/101, 62/399
progressive external ophthalmoplegia, 62/291
Epidermal growth factor
muscle tissue culture, 62/88
Epilepsy
see also Status epilepticus
adenosine triphosphatase 6 mutation syndrome,
    62/510
anoxic, see Syncope
childhood myoglobinuria, 62/564
complex I deficiency, 62/504
complex III deficiency, 62/504
debrancher deficiency, 62/482
Kearns-Sayre-Daroff-Shy syndrome, 62/308
Lafora progressive myoclonus, see Lafora
    progressive myoclonus epilepsy
MELAS syndrome, 62/509
MERRF syndrome, 62/509
phosphoglycerate kinase deficiency, 62/489
progressive external ophthalmoplegia, 62/308
pyruvate decarboxylase deficiency, 62/502
state, see Status epilepticus
ubidecarenone, 62/504
Epileptic state, see Status epilepticus
Episodic apnea
congenital high conductance fast channel
    syndrome, 62/438
pyruvate decarboxylase deficiency, 62/502
Episodic ataxia, see Periodic ataxia
Episodic encephalopathy
glutaric aciduria type II, 62/495
multiple acyl coenzyme A dehydrogenase
    deficiency, 62/495
Epithelioma
Malherbe, see Pilomatricoma
myotonic dystrophy, 62/238
Epstein-Barr virus
acquired myoglobinuria, 62/577
childhood myoglobinuria, 62/562
Erythematosus

systemic lupus, *see* Systemic lupus erythematosus
Erythromycin
    acquired myoglobinuria, 62/577
Escherichia coli
    acquired myoglobinuria, 62/577
    childhood myoglobinuria, 62/562
Ethchlorvynol
    toxic myopathy, 62/597, 609
Ethylene glycol
    toxic myopathy, 62/601
Eulenberg paramyotonia congenita, *see*
        Paramyotonia congenita
Exacerbation
    congenital acetylcholine receptor deficiency/
        synaptic cleft syndrome, 62/438
    familial infantile myasthenia, 62/427
Exercise
    acquired myoglobinuria, 62/572
    carnitine palmitoyltransferase deficiency, 62/494
    ischemic, *see* Ischemic exercise
    myotonic dystrophy, 62/249
Exercise intolerance
    aconitate hydratase deficiency, 62/503
    complex II deficiency, 62/504
    complex III deficiency, 62/504
    congenital acetylcholine receptor deficiency/
        synaptic cleft syndrome, 62/438
    glutaric aciduria type II, 62/495
    lactate dehydrogenase deficiency, 62/490
    multiple acyl coenzyme A dehydrogenase
        deficiency, 62/495
    multiple mitochondrial DNA deletion, 62/510
    myophosphorylase deficiency, 62/485
    oxidation-phosphorylation coupling defect, 62/506
    phosphofructokinase deficiency, 62/487
    phosphoglycerate kinase deficiency, 62/489
    phosphoglycerate mutase deficiency, 62/489
    phosphorylase kinase deficiency, 62/487
    pyruvate decarboxylase deficiency, 62/502
    ragged red fiber, 62/292
    ubidecarenone, 62/504
Exercise myoglobinuria
    myophosphorylase deficiency, 62/485
Exertional myalgia
    incomplete dystrophin deficiency, 62/135
Exophthalmic ophthalmoplegia, *see* Thyroid
        associated ophthalmopathy
Expectancy
    life, *see* Life expectancy
Experimental injury
    cytoplasmic inclusion body myopathy, 62/345
Experimental model
    congenital myopathy, 62/358
Experimental myotonia
    toxic myopathy, 62/597
External ophthalmoplegia
    chronic progressive, *see* Chronic progressive
        external ophthalmoplegia
    congenital myopathy, 62/333
    progressive, *see* Progressive external
        ophthalmoplegia

Extraocular muscle weakness
    see also Progressive external ophthalmoplegia
    congenital myopathy, 62/334

Fabry disease type I
    muscle fiber feature, 62/17
Facial diplegia
    congenital, *see* Möbius syndrome
    congenital acetylcholine receptor deficiency/short
        channel opentime syndrome, 62/437
Facial nerve
    abnormality, *see* Möbius syndrome
Facial weakness
    abnormal myomuscular junction myopathy, 62/334
    cap disease, 62/334
    central core myopathy, 62/334
    congenital fiber type disproportion, 62/334
    congenital myasthenic syndrome, 62/334
    congenital myopathy, 62/334
    cytoplasmic inclusion body myopathy, 62/334
    Mallory body myopathy, 62/334
    mitochondria jagged Z line myopathy, 62/334
    multicore myopathy, 62/334
    nemaline myopathy, 62/334
    trilaminar myopathy, 62/334
    tubular aggregate myopathy, 62/334
    tubulomembranous inclusion myopathy, 62/334
    uniform muscle fiber type I myopathy, 62/334
Facilitating myasthenic syndrome, *see* Eaton-
        Lambert myasthenic syndrome
Facioscapulohumeral muscular dystrophy
    age at onset, 62/163
    asymptomatic case, 62/164
    clinical features, 41/485, 62/162
    Coats disease, 62/167
    creatine kinase, 40/420, 62/166
    Emery-Dreifuss muscular dystrophy, 62/155
    EMG, 40/420, 62/166
    genetics, 62/163
    histopathology, 62/166
    laboratory features, 62/166
    lactate dehydrogenase, 62/166
    Möbius syndrome, 40/419, 422, 62/162
    muscle computerized assisted tomography, 62/162
    myoglobinuria, 62/556
    myotonic dystrophy, 62/240
    myotubular myopathy, 62/168
    penetrance, 62/165
    prevalence, 43/98, 62/165
    retinopathy, 62/166
    scapulohumeral distribution, 62/170
    scapuloperoneal muscular dystrophy, 62/169
    serum alanine aminotransferase, 62/166
    treatment, 62/168
    winged scapula, 40/417, 62/162
Facioscapulohumeral syndrome, 40/415-423
    central core myopathy, 62/168
    Coats disease, 62/167
    differentiation, 62/161
    mitochondrial myopathy, 62/168
    muscle computerized assisted tomography, 62/162

myasthenia gravis, 62/168
nemaline myopathy, 62/168
polymyositis, 62/167
Fahr disease, *see* Striatopallidodentate calcification
Familial basal ganglion calcification, *see*
Striatopallidodentate calcification
Familial hypokalemic periodic paralysis
EMG, 62/69
Familial infantile myasthenia
age improvement, 62/427
apnea, 62/427
compound muscle action potential, 62/428
congenital myasthenic syndrome, 62/427
congenital myopathy, 62/332
EMG, 62/427
exacerbation, 62/427
muscle biopsy, 62/428
neuromuscular junction disease, 62/392
ptosis, 62/427
single fiber electromyography, 62/427
treatment, 62/429
Familial limb girdle myasthenia
congenital, *see* Congenital familial limb girdle
myasthenia
Familial pancytopenia, *see* Fanconi syndrome
Familial periodic paralysis
hyperkalemic periodic paralysis, 62/457
hypokalemic periodic paralysis, 62/457
normokalemic periodic paralysis, 62/457
Familial relapsing ophthalmoplegia
orbital myositis, 62/302
Familial spastic paraplegia, 22/421-430
extrapyramidal sign, *see* Ferguson-Critchley
syndrome
ophthalmoplegia, *see* Ferguson-Critchley
syndrome
progressive external ophthalmoplegia
classification, 62/290
Fanconi anemia, *see* Fanconi syndrome
Fanconi syndrome
carnitine deficiency, 62/493
lipid metabolic disorder, 62/493
secondary systemic carnitine deficiency, 62/493
Fasciculation
Isaacs syndrome, 62/273
motor unit, 62/50
myopathy, 40/330, 62/50
toxic myopathy, 62/596
Fasciculation discharge
features, 62/50
motor unit, 62/50
myopathy, 62/50
Fasting
carnitine palmitoyltransferase deficiency, 62/494
glutaric aciduria type II, 62/495
multiple acyl coenzyme A dehydrogenase
deficiency, 62/495
painful dominant myotonia, 62/266
Fatal infantile myopathy/cardiopathy
creatine kinase, 62/510
mitochondrial disease, 62/510

mitochondrial DNA point mutation, 62/509
ragged red fiber, 62/510
Fatality rate
congenital fiber type disproportion, 62/356
myotubular myopathy, 62/348
Fatigue
ACTH induced myopathy, 62/537
adrenal insufficiency myopathy, 62/536
corticosteroid withdrawal myopathy, 62/536
dermatomyositis, 62/371
endocrine myopathy, 62/527
myophosphorylase deficiency, 62/485
Nelson syndrome, 62/537
tremor, 6/818, 62/52
Fatty acid oxidation
childhood myoglobinuria, 62/566
lipid metabolic disorder, 62/491
metabolic myopathy, 62/491
myoglobinuria, 62/558
Fenfluramine
toxic myopathy, 62/601
Fenofibrate
toxic myopathy, 62/601
Fenoterol
myotonia, 62/276
Ferguson-Critchley syndrome, 22/433-442
progressive external ophthalmoplegia, 62/304
Ferritin H
muscle tissue culture, 62/92
Fetal alcohol syndrome
pyruvate decarboxylase deficiency, 62/502
Fetus
harlequin, *see* Congenital ichthyosis
Fever
adrenal insufficiency myopathy, 62/536
corticosteroid withdrawal myopathy, 62/536
dermatomyositis, 62/373
inclusion body myositis, 62/373
polymyositis, 62/373
typhoid, *see* Typhoid fever
Fiber electromyography
single, *see* Single fiber electromyography
Fiber type disproportion
congenital, *see* Congenital fiber type disproportion
Fiber vacuole
muscle, *see* Rimmed muscle fiber vacuole
Fibers
muscle, *see* Muscle fiber
ragged red, *see* Ragged red fiber
Fibrillation potential
features, 62/61
motor unit, 62/50, 61, 67
myopathy, 62/50, 61, 67
Fibroblast
muscle tissue culture, 62/87
Field fever, *see* Leptospirosis
Fingerprint body
muscle fiber, 40/39, 43, 62/44
Fingerprint body myopathy
congenital myopathy, 41/20, 62/332
differential diagnosis, 62/347

mental deficiency, 62/347
Floppy infant, *see* Infantile hypotonia
Flunitrazepam
   toxic myopathy, 62/601
Fluphenazine
   toxic myopathy, 62/601
Focal myopathy
   incomplete dystrophin deficiency, 62/135
   needle, 62/611
   toxic myopathy, 62/611
Forbes limit dextrinosis, *see* Debrancher deficiency
Foster Kennedy syndrome
   X-linked neuromuscular dystrophy, 62/117
Friedreich ataxia, 21/319-359
   progressive external ophthalmoplegia, 21/347,
      22/191, 62/304
Friedreich tabes, *see* Friedreich ataxia
Fructose bisphosphate aldolase
   Duchenne muscular dystrophy, 62/118
   hypothyroid myopathy, 62/533
   limb girdle syndrome, 62/185
   thyrotoxic myopathy, 62/529
Fukuyama cerebromuscular dystrophy, *see*
      Fukuyama syndrome
Fukuyama congenital muscular atrophy, *see*
      Fukuyama syndrome
Fukuyama syndrome
   malignant hyperthermia, 62/570
Fumarase deficiency
   mitochondrial disease, 62/497
   α-oxoglutarate dehydrogenase deficiency, 62/503

Galactocerebroside-β-galactosidase deficiency, *see*
      Globoid cell leukodystrophy
Galactosylceramide lipidosis, *see* Globoid cell
      leukodystrophy
Gammopathy
   monoclonal, *see* Monoclonal gammopathy
Gamstorp-Wohlfart syndrome, *see* Isaacs syndrome
Ganglioside sialidase deficiency, *see* Mucolipidosis
      type IV
Gasoline
   toxic myopathy, 62/601
Gasoline intoxication
   toxic myopathy, 62/613
Gee disease, *see* Celiac disease
Gee-Herter disease, *see* Celiac disease
Gemfibrozil
   acquired myoglobinuria, 62/576
   toxic myopathy, 62/601
Gene instability
   myotonic dystrophy, 62/213
Gene product
   myotonic dystrophy, 62/213
General anesthesia
   malignant hyperthermia, 62/567
Genetics
   see also Inheritance
   congenital fiber type disproportion, 41/431, 62/356
   Duchenne muscular dystrophy, 40/352-357, 41/487,
      489, 62/125

Emery-Dreifuss muscular dystrophy, 62/156
facioscapulohumeral muscular dystrophy, 62/163
hereditary cylindric spirals myopathy, 62/348
hyperkalemic periodic paralysis, 41/160, 425,
   62/463
Kearns-Sayre-Daroff-Shy syndrome, 62/312
limb girdle syndrome, 41/419, 486, 62/181
mitochondrial disease, 62/498
molecular, *see* Molecular genetics
myotonic dystrophy, 41/421, 486, 62/209
nemaline myopathy, 41/432, 62/342
oculopharyngeal muscular dystrophy, 62/296
progressive external ophthalmoplegia, 22/185,
   41/420, 62/295, 312
thyrotoxic periodic paralysis, 62/461
Genotype
   Duchenne muscular dystrophy, 62/126
Giant cell polymyositis
   myasthenia gravis, 62/402
Globoid cell leukodystrophy, 10/67-90, 42/489-491
   congenital fiber type disproportion, 62/355
α-1,4-Glucan-6-glucosyltransferase deficiency, *see*
   Brancher deficiency
Glucocorticoids
   see also Corticosteroids
   muscle tissue culture, 62/90
Glucocorticosteroids, *see* Glucocorticoids
Glucose-6-phosphate dehydrogenase
   myoglobinuria, 62/558
α-1,4-Glucosidase
   acid maltase deficiency, 62/481
α-1,6-Glucosidase
   acid maltase deficiency, 62/481
Glucosidase deficiency
   Isaacs syndrome, 62/263
α-1,4-Glucosidase deficiency, *see* Acid maltase
   deficiency
Glutamate oxaloacetate transaminase, *see* Aspartate
   aminotransferase
Glutaric aciduria type II
   see also Multiple acyl coenzyme A dehydrogenase
      deficiency
   coma, 62/495
   episodic encephalopathy, 62/495
   exercise intolerance, 62/495
   fasting, 62/495
   hypoglycemia, 62/495
   infantile hypotonia, 62/495
   lethargy, 62/495
   2 main forms, 62/495
   metabolic encephalopathy, 62/495
   metabolic myopathy, 62/495
   myalgia, 62/495
   recurrent coma, 62/495
Gluten induced enteropathy, *see* Celiac disease
Glutethimide
   toxic myopathy, 62/601
Glycerol kinase deficiency
   characteristics, 62/127
   Duchenne muscular dystrophy, 62/127
Glycogen

see also Progressive myoclonus epilepsy
  muscle fiber, 62/20
Glycogen debrancher enzyme deficiency, *see*
    Debrancher deficiency
Glycogen storage disease, 27/221-237
  acid maltase deficiency, 40/53, 99, 153, 190, 255,
      280, 317, 438, 62/480
  brain, 62/490
  brancher deficiency, 62/483
  cardiomyopathy, 62/490
  childhood myoglobinuria, 62/564
  debrancher deficiency, 27/222, 40/256, 41/181,
      62/482
  history, 62/479
  limb girdle syndrome, 62/190
  mental deficiency, 62/490
  metabolic myopathy, 41/175-194, 62/479
  myoglobinuria, 41/265, 62/558
  myophosphorylase deficiency, 40/4, 53, 189, 257,
      317, 327, 339, 342, 438, 62/485
  phosphofructokinase deficiency, 40/53, 190, 327,
      339, 342, 62/487
  phosphoglycerate kinase deficiency, 62/489
  phosphoglycerate mutase deficiency, 62/489
  phosphorylase kinase deficiency, 62/486
  progressive external ophthalmoplegia, 62/298
  progressive external ophthalmoplegia
      classification, 62/290
  rimmed muscle fiber vacuole content, 62/18
  rimmed muscle fiber vacuole type, 62/18
  rimmed vacuole distal myopathy, 62/490
  type II, *see* Acid maltase deficiency
  type III, *see* Debrancher deficiency
  type IV, *see* Brancher deficiency
  type V, *see* Myophosphorylase deficiency
  type VII, *see* Phosphofructokinase deficiency
  type VIII, *see* Phosphorylase kinase deficiency
  type IX, *see* Phosphoglycerate kinase deficiency
  type X, *see* Phosphoglycerate mutase deficiency
  type XI, *see* Lactate dehydrogenase deficiency
  variant type, 62/490
Glycogenosis, *see* Glycogen storage disease
Glycoside
  cardiac, *see* Cardiac glycoside
Glycyrrhiza
  toxic myopathy, 62/601, 603
Gold
  toxic myopathy, 62/597
Golgi system
  muscle fiber, 62/21
Gottron rash
  dermatomyositis, 62/371
Gougerot-Sjögren syndrome, *see* Sjögren syndrome
Gowers sign
  Duchenne muscular dystrophy, 40/359, 62/117
G6PD, *see* Glucose-6-phosphate dehydrogenase
Grande aphthose de Touraine, *see* Behçet syndrome
Granulofilamentous myopathy
  congenital myopathy, 62/332
  histopathology, 62/345
Granulomatous disease

polymyositis, 62/373
Granulovacuolar lobular myopathy
  congenital myopathy, 62/332
  rimmed muscle fiber vacuole variant, 62/352
Graves disease, *see* Hyperthyroidism
Graves ophthalmopathy, *see* Thyroid associated
    ophthalmopathy
Graves ophthalmoplegia, *see* Thyroid associated
    ophthalmopathy
Growth hormone
  myotonic dystrophy, 62/230, 249
Guanidine
  Eaton-Lambert myasthenic syndrome, 41/358,
      62/426
Guillain-Barré polyradiculoneuropathy, *see* Guillain-
    Barré syndrome
Guillain-Barré syndrome, 7/495-507
  Eaton-Lambert myasthenic syndrome, 62/425

Haff disease
  acquired myoglobinuria, 62/577
  toxic myopathy, 62/601, 612
Hair loss, *see* Alopecia
Haloperidol
  toxic myopathy, 62/601
Harlequin fetus, *see* Congenital ichthyosis
Harvest fever, *see* Leptospirosis
Hashimoto disease, *see* Hashimoto thyroiditis
Hashimoto thyroiditis
  myasthenia gravis, 62/534
  polymyositis, 62/373
  thyroid associated ophthalmopathy, 62/530
Hearing loss
  see also Deafness
  Coats disease, 62/167
  complex I deficiency, 62/504
  MERRF syndrome, 62/509
  multiple mitochondrial DNA deletion, 62/510
  myotonic dystrophy, 62/234
  Schimke syndrome, 62/304
Heart block
  Kearns-Sayre-Daroff-Shy syndrome, 38/224,
      62/508
Heart disease
  see also Cardiomyopathy
  hypokalemic periodic paralysis, 62/459
  myotonic dystrophy, 40/507, 522, 62/226
Heart failure
  Duchenne muscular dystrophy, 62/134
  infantile acid maltase deficiency, 62/480
Heat intolerance
  oxidation-phosphorylation coupling defect, 62/506
Heat stroke, 38/543-549
  see also Thermoregulation
  acquired myoglobinuria, 62/572
  malignant hyperthermia, 62/571, 573
  metabolic encephalopathy, 62/573
Heine-Medin disease, *see* Poliomyelitis
Heliotrope rash
  dermatomyositis, 41/57, 62/371
Hematemesis

dermatomyositis, 62/374
inclusion body myositis, 62/374
polymyositis, 62/374
Heme induced nephropathy
myoglobinuria, 62/556
Hemeralopic retinosis, *see* Primary pigmentary
retinal degeneration *and* Secondary pigmentary
retinal degeneration
Hemodialysis
see also Dialysis
carnitine deficiency, 62/494
secondary systemic carnitine deficiency, 62/494
Hemolytic anemia
phosphofructokinase deficiency, 43/183, 62/487
phosphoglycerate kinase deficiency, 62/489
Heparan sulfate
muscle fiber, 62/31
Heparitin sulfate, *see* Heparan sulfate
Hepatic carnitine palmitoyltransferase deficiency
hypoglycemic encephalopathy, 62/494
Hepatic cirrhosis, *see* Liver cirrhosis
Hepatic disease, *see* Liver disease
Hepatic encephalopathy
carnitine deficiency, 62/493
lipid metabolic disorder, 62/493
systemic carnitine deficiency, 62/493
Hepatomegaly
see also Hepatosplenomegaly
debrancher deficiency, 62/482
long chain acyl coenzyme A dehydrogenase
deficiency, 62/495
medium chain acyl coenzyme A dehydrogenase
deficiency, 62/495
Hepatopathy, *see* Liver disease
Hepatophosphorylase kinase deficiency, *see*
Phosphorylase kinase deficiency
Hepatosplenomegaly
see also Hepatomegaly
brancher deficiency, 62/483
Hereditary ataxia
progressive external ophthalmoplegia, 62/304
progressive external ophthalmoplegia
classification, 62/290
Hereditary cylindric spirals myopathy
congenital myopathy, 62/332
genetics, 62/348
muscle tissue culture, 62/108
percussion myotonia, 62/348
T tubule, 62/348
Hereditary motor and sensory neuropathy type I
see also Charcot-Marie-Tooth disease
scapuloperoneal spinal muscular atrophy, 62/170
Hereditary motor and sensory neuropathy type II
see also Charcot-Marie-Tooth disease
scapuloperoneal spinal muscular atrophy, 62/170
Hereditary motor and sensory neuropathy type III,
see Hypertrophic interstitial neuropathy
Hereditary myotonic syndrome
animal model, 62/272
Hereditary neurogenic muscular atrophy, *see*
Charcot-Marie-Tooth disease

Hereditary oligophrenic cerebellolental degeneration,
see Marinesco-Sjögren syndrome
Hereditary spastic paraplegia, *see* Familial spastic
paraplegia
Hereditary spinal ataxia, *see* Friedreich ataxia
Hereditary torsion dystonia, *see* Dystonia
musculorum deformans
Heredofamilial myosclerosis
Emery-Dreifuss muscular dystrophy, 62/173
genetic heterogeneity, 62/173
rigid spine syndrome, 62/173
Heroin, *see* Diamorphine
Herpes simplex, 34/145-156
acquired myoglobinuria, 62/577
Herpetic stomatitis
childhood myoglobinuria, 62/562
Heterogeneity
distal myopathy, 62/197
quadriceps myopathy, 62/174
Heterogeneous system degeneration, *see* Progressive
supranuclear palsy
High conductance fast channel syndrome
congenital, *see* Congenital high conductance fast
channel syndrome
Histiocytoid cardiomyopathy
infantile, *see* Infantile histiocytoid cardiomyopathy
HLA antigen
muscle tissue culture, 62/92
HLA haplotype
thyrotoxic periodic paralysis, 62/532
Hoffmann disease, *see* Hypertrophic interstitial
neuropathy
Hoffmann syndrome
see also Kocher-Debré-Sémélaigne syndrome
hypothyroid myopathy, 62/533
Hollow foot, *see* Pes cavus
Honeycomb myopathy
clinical features, 62/353
congenital myopathy, 62/332
zebra body, 62/353
Hormone
growth, *see* Growth hormone
parathyroid, *see* Parathyroid hormone
Hornet venom
toxic myopathy, 62/601, 612
HTLV-1, *see* Human T-lymphotropic virus type I
Human immunodeficiency virus type 1
polymyositis, 62/381
Human leukocyte antigen, *see* HLA antigen
Human T-lymphotropic virus type I
polymyositis, 62/382
Hyaline body, *see* Neurofibrillary tangle
Hyperagammaglobulinemic purpura
polymyositis, 62/373
Hyperaldosteronism
periodic paralysis, 62/457
Hyperaldosteronism myopathy
creatine kinase, 62/536
EMG, 62/536
hypokalemic periodic paralysis, 62/536
muscle biopsy, 62/536

periodic weakness, 62/536
potassium depletion, 62/536
Hypercalcemia
    myoglobinuria, 62/556
Hypereosinophilic syndrome
    polymyositis, 62/373
Hyperinsulinism
    myotonic dystrophy, 62/230
Hyperkalemic periodic paralysis, 28/592-597
    see also Paramyotonia congenita
    acetazolamide, 62/470, 472
    age, 62/462
    attack, 62/461
    attack duration, 62/462
    attack frequency, 62/462
    attack provocation, 62/462
    channel α-subunit gene, 62/463
    chromosome 17, 62/463
    Chvostek sign, 62/462
    familial periodic paralysis, 62/457
    genetics, 41/160, 425, 62/463
    myotonia, see Myotonic hyperkalemic periodic
        paralysis
    orciprenaline, 62/472
    permanent muscular weakness, 62/462
    salbutamol, 62/472
    symptom, 62/461
Hypermagnesemia
    periodic paralysis, 62/457
Hyperparathyroid myopathy
    EMG, 62/540
    incidence, 62/540
    muscular atrophy, 62/540
    neuropathy, 62/540
    osteomalacia, 62/538
    pathogenesis, 62/540
    proximal limb weakness, 62/540
    rickets, 62/538
    treatment, 62/541
    waddling gait, 62/540
Hyperphosphatemia
    myoglobinuria, 62/556
Hyperpigmentation
    ACTH induced myopathy, 62/537
    Nelson syndrome, 62/537
Hyperpyrexia, see Hyperthermia
Hyperreflexia
    Leber hereditary optic neuropathy, 62/509
Hypersomnia
    see also Lethargy and Sleep
    myotonic dystrophy, 40/517, 62/217, 235
Hypersomnolence, see Hypersomnia
Hyperthermia
    malignant, see Malignant hyperthermia
    oxidation-phosphorylation coupling defect, 62/506
Hyperthyroidism
    differential diagnosis, 40/438, 41/209, 335, 337,
        62/528
    Eaton-Lambert myasthenic syndrome, 62/421
    endocrine myopathy, 62/528
    myasthenia gravis, 41/101, 43/157, 62/402, 534

myopathy, see Thyrotoxic myopathy
    periodic paralysis, 27/268, 28/597, 40/285, 41/165,
        241, 62/457, 468
Hypertonia
    extrapyramidal, see Rigidity
    pyramidal, see Spasticity
Hypertrophic interstitial neuropathy, 21/145-165
    see also Charcot-Marie-Tooth disease
    scapuloperoneal spinal muscular atrophy, 62/170
Hypertrophic nerve
    scapuloperoneal spinal muscular atrophy, 62/170
Hypertrophic neuropathy, see Hypertrophic
        interstitial neuropathy
Hypertrophied muscle, see Muscle hypertrophy
Hypertrophy
    muscle, see Muscle hypertrophy
    muscle fiber, 62/15
Hypocalcemia
    hypoparathyroid myopathy, 62/541
    myoglobinuria, 62/556
Hypoglycemia
    glutaric aciduria type II, 62/495
    lipid metabolic disorder, 62/494
    multiple acyl coenzyme A dehydrogenase
        deficiency, 62/495
Hypoglycemic encephalopathy
    carnitine deficiency, 62/493
    childhood myoglobinuria, 62/563, 566
    hepatic carnitine palmitoyltransferase deficiency,
        62/494
    long chain 3-hydroxyacyl coenzyme A
        dehydrogenase deficiency, 62/495
    medium chain acyl coenzyme A dehydrogenase
        deficiency, 62/495
    primary systemic carnitine deficiency, 62/493
Hypokalemia
    toxic myopathy, 62/603
Hypokalemic periodic paralysis, 28/582-593, 41/149-
        158
    abortive attack, 62/458
    anesthesia, 62/460
    attack, 62/457
    attack provocation, 62/458
    attack rate, 62/458
    EMG, 28/583, 62/459
    familial, see Familial hypokalemic periodic
        paralysis
    familial periodic paralysis, 62/457
    female nonpenetrance, 62/460
    heart disease, 62/459
    hyperaldosteronism myopathy, 62/536
    inheritance mode, 62/460
    life expectancy, 62/459
    male preponderance, 62/460
    myotonia, 43/170, 62/458
    permanent weakness, 62/458
    sodium channel α-subunit gene, 62/460
    sporadic case, 62/461
    surgery, 62/460
    symptom, 62/457
    treatment, 28/591, 62/469

Hypoparathyroid myopathy
  creatine kinase, 62/541
  hypocalcemia, 62/541
  Kearns-Sayre-Daroff-Shy syndrome, 62/541
  paresthesia, 62/541
  tetany, 62/541
Hypoparathyroidism
  Kearns-Sayre-Daroff-Shy syndrome, 41/215,
    43/142, 62/309, 508
  multiple mitochondrial DNA deletion, 62/510
  progressive external ophthalmoplegia, 62/309
Hyporeflexia
  see also Areflexia
  chondrodystrophic myotonia, 62/269
  Eaton-Lambert myasthenic syndrome, 62/426
  α-oxoglutarate dehydrogenase deficiency, 62/503
Hypothalamic tumor
  periodic paralysis, 62/457
Hypothyroid myopathy
  aspartate aminotransferase, 62/533
  biochemistry, 62/533
  creatine kinase, 62/533
  EMG, 62/533
  endocrine myopathy, 41/243-246, 298, 62/528
  fructose bisphosphate aldolase, 62/533
  Hoffmann syndrome, 62/533
  hypothyroid polyneuropathy, 62/533
  Isaacs syndrome, 62/263
  Kocher-Debré-Sémélaigne syndrome, 62/533
  lactate dehydrogenase, 62/533
  muscle biopsy, 62/533
  muscle cramp, 41/243, 246, 62/532
  muscle stretch reflex time, 62/532
  muscle ultrastructure, 62/533
  myoedema, 62/532
  precipitating factor, 62/532
  pseudomyotonic, 62/532
  treatment, 62/534
Hypothyroid polyneuropathy
  hypothyroid myopathy, 62/533
Hypothyroidism
  differential diagnosis, 40/438, 62/533
  Eaton-Lambert myasthenic syndrome; 62/421
  myasthenia gravis, 62/534
Hypotonia
  infantile, see Infantile hypotonia
Hypoventilation
  alveolar, see Alveolar hypoventilation
  myotonic dystrophy, 62/235
Hypoxemia
  dermatomyositis, 62/374
  inclusion body myositis, 62/374
  myotonic dystrophy, 62/235
  polymyositis, 62/374

I band
  muscle fiber, 62/26
Iatrogenic disease
  mitochondrial DNA depletion, 62/511
  secondary systemic carnitine deficiency, 62/494
Ichthyosis

congenital, see Congenital ichthyosis
Idiopathic steatorrhea, see Celiac disease
IgA deficiency
  polymyositis, 62/373
Imipramine
  myotonia, 62/276
Immunocytochemistry
  muscle biopsy, 62/2
  myasthenia gravis, 62/415
Immunodeficiency syndrome
  acquired, see Acquired immune deficiency
    syndrome
Immunosuppression
  Eaton-Lambert myasthenic syndrome, 62/426
  thyroid associated ophthalmopathy, 62/532
Inclusion body
  cap disease, 62/341
  nemaline myopathy, 62/342
  Tomé-Fardeau body, 62/103
Inclusion body myositis
  age at onset, 62/204
  amyloid deposit, 62/377
  amyloid β-protein, 62/377
  anti-Jo-1 antibody, 62/374
  areflexia, 62/373
  arthralgia, 62/374
  aspiration pneumonia, 62/374
  autoimmune antibody, 62/380
  axonal neuropathy, 62/376
  azathioprine, 62/385
  basophilic granular inclusion, 62/377
  calcinosis, 62/386
  cardiac disorder, 62/374
  cardiomyopathy, 62/374
  ciclosporin, 62/386
  collagen vascular disease, 62/373
  congestive heart failure, 62/374
  corticosteroids, 62/383
  creatine kinase, 62/373, 375
  cricopharyngeal myotomy, 62/386
  cyclophosphamide, 62/385
  diagnosis, 62/375
  diagnostic criteria, 62/378
  distal muscle, 62/373
  dominant inheritance, 62/373
  dysphagia, 62/206, 373
  dyspnea, 62/374
  electronmicroscopy, 62/377
  EMG, 62/375
  eosinophilic cytoplasmic inclusion, 62/377
  extramuscular manifestation, 62/373
  fever, 62/373
  filamentous inclusion, 62/206
  gastrointestinal symptom, 62/374
  hematemesis, 62/374
  histopathology, 62/206
  hypoxemia, 62/374
  immunopathology, 62/379
  immunosuppressive therapy, 62/385
  intravenous Ig, 62/386
  joint contracture, 62/374

lactate dehydrogenase, 62/375
leukoencephalopathy, 62/373
leukopheresis, 62/386
limb girdle syndrome, 62/187
lymphoid irradiation, 62/386
malaise, 62/373
malignancy, 62/374
melena, 62/374
membranolytic attack complex, 62/379
methotrexate, 62/385
mixed connective tissue disease, 62/374
muscle biopsy, 62/373, 375
muscle tissue culture, 62/104
myocarditis, 62/374
nuclear magnetic resonance, 62/379
oculopharyngeal muscular dystrophy, 62/297
overlap syndrome, 62/374
pathogenesis, 62/379
physiotherapy, 62/386
picornavirus, 62/381
plasmapheresis, 62/386
pneumatosis intestinalis, 62/374
pneumonitis, 62/374
practical guideline, 62/386
primary inflammatory myositis, 62/369
pulmonary dysfunction, 62/374
Raynaud phenomenon, 62/374
rheumatoid arthritis, 62/374
rimmed muscle fiber vacuole, 62/376
rimmed muscle fiber vacuole content, 62/18
rimmed muscle fiber vacuole type, 62/18
scleroderma, 62/374
serum alanine aminotransferase, 62/375
serum aspartate aminotransferase, 62/375
serum muscle enzyme, 62/375
sex ratio, 62/204
Sjögren syndrome, 62/374
steroid myopathy, 62/384
systemic lupus erythematosus, 62/374
systemic sclerosis, 62/374
tachyarrhythmia, 62/374
treatment, 62/383
tubulofilament, 62/377
viral myositis, 62/381
weight loss, 62/374
Wohlfart-Kugelberg-Welander disease, 62/204
Incomplete dystrophin deficiency
    Becker muscular dystrophy, 62/135
    cardiomyopathy, 62/135
    exertional myalgia, 62/135
    focal myopathy, 62/135
    malignant hyperthermia, 62/135
    myoglobinuria, 62/135
    quadriceps myopathy, 62/135
Indonesian Weil disease, see Leptospirosis
Infantile acid maltase deficiency
    cardiomegaly, 27/195, 227, 62/480
    heart failure, 62/480
    infantile hypotonia, 27/195, 227, 62/480
    late, see Late infantile acid maltase deficiency
    limb girdle syndrome, 62/190

    macroglossia, 62/480
    muscle hypertrophy, 62/480
    pulmonary failure, 62/480
Infantile distal myopathy
    calf hypertrophy, 62/201
    clinical course, 62/201
    foot drop, 62/201
    pedigree, 62/201
    taxonomy, 62/199
Infantile histiocytoid cardiomyopathy
    complex III deficiency, 62/505
Infantile hypotonia
    see also Congenital myopathy
    brancher deficiency, 62/484
    carnitine deficiency, 62/493
    complex I deficiency, 62/503
    complex III deficiency, 62/504
    glutaric aciduria type II, 62/495
    infantile acid maltase deficiency, 27/195, 227,
        62/480
    long chain acyl coenzyme A dehydrogenase
        deficiency, 62/495
    multiple acyl coenzyme A dehydrogenase
        deficiency, 62/495
    α-oxoglutarate dehydrogenase deficiency, 62/503
    pyruvate decarboxylase deficiency, 62/502
Infantile myopathy/cardiopathy
    fatal, see Fatal infantile myopathy/cardiopathy
Infantile nemaline myopathy
    muscle tissue culture, 62/99, 107
Infantile paralysis, see Poliomyelitis
Infantile periarteritis nodosa, see Kawasaki syndrome
Infantile Pompe disease, see Infantile acid maltase
    deficiency
Infantile progressive poliodystrophy, see Alpers
    disease
Infantile spinal muscular atrophy, 22/81-100
    congenital fiber type disproportion, 62/355
    progressive external ophthalmoplegia
        classification, 62/290
Infarction
    brain, see Brain infarction
    brain venous, see Brain infarction
    diabetic thigh muscle, see Diabetic thigh muscle
        infarction
    myocardial, see Myocardial infarction
Infection
    acquired myoglobinuria, 62/577
    childhood myoglobinuria, 62/561
    lung, see Pneumonitis
    myoglobinuria, 41/275, 62/557
    viral, see Viral infection
Infiltrative ophthalmopathy, see Thyroid associated
    ophthalmopathy
Inflammatory cell
    muscle, 62/42
Inflammatory demyelinating neuropathy
    acute, see Guillain-Barré syndrome
Inflammatory myopathy, 41/51-89
    see also Collagen vascular disease,
        Dermatomyositis, Myositis and Polymyositis

associated condition, 62/372
  clinical features, 40/307, 313, 317, 323, 337, 339,
    342, 41/370-372, 62/369
  cysticercus, 62/373
  parasitic polymyositis, 62/373
  toxic myopathy, 62/604
  Toxoplasma gondii, 62/373
  Trichinella, 62/373
  Trypanosoma, 62/373
Inflammatory scapuloperoneal muscular dystrophy
  facioscapulohumeral muscular dystrophy type,
    62/170
Influenza
  acquired myoglobinuria, 62/577
INH, *see* Isoniazid
INH polyneuropathy, *see* Isoniazid
Inheritance
  see also Genetics
  acid maltase deficiency, 62/481
  brancher deficiency, 62/484
  carnitine deficiency, 62/493
  debrancher deficiency, 62/483
  dominant, *see* Dominant inheritance
  lactate dehydrogenase deficiency, 62/490
  malignant hyperthermia, 62/568
  medium chain acyl coenzyme A dehydrogenase
    deficiency, 62/495
  mitochondrial DNA depletion, 62/511
  multiple mitochondrial DNA deletion, 62/510
  myophosphorylase deficiency, 62/485
  normokalemic periodic paralysis, 62/463
  phosphofructokinase deficiency, 62/488
  phosphoglycerate kinase deficiency, 62/489
  phosphoglycerate mutase deficiency, 62/489
  thyrotoxic periodic paralysis, 62/532
Inheritance mode
  hypokalemic periodic paralysis, 62/460
Injuries
  experimental, *see* Experimental injury
Insertional activity
  features, 62/50
  motor unit, 62/50
  myopathy, 62/50
Insolation, *see* Heat stroke
Insulin
  muscle tissue culture, 62/88
Insulin growth factor 1
  muscle tissue culture, 62/90
Insulin resistance
  myotonic dystrophy, 62/230
γ-Interferon
  muscle tissue culture, 62/92
Intermittent ataxia, *see* Periodic ataxia
Interstitial neuropathy
  hypertrophic, *see* Hypertrophic interstitial
    neuropathy
Intestinal pseudo-obstruction
  Duchenne muscular dystrophy, 62/120
  Kearns-Sayre-Daroff-Shy syndrome, 62/313
  mitochondrial myopathy, 62/313
  MNGIE syndrome, 62/505

progressive external ophthalmoplegia, 62/313
Intoxication
  9-anthroic acid, *see* 9-Anthroic acid intoxication
  aromatic carboxylic acid, *see* Aromatic carboxylic
    acid intoxication
  breast milk, *see* Subacute necrotizing
    encephalomyelopathy
  clofibrate, *see* Clofibrate intoxication
  2,4-dichlorophenoxyacetic acid, *see* 2,4-
    Dichlorophenoxyacetic acid intoxication
  emetine, *see* Emetine intoxication
  gasoline, *see* Gasoline intoxication
  heroin, *see* Diamorphine
  ipecac, *see* Ipecac intoxication
  magnesium, *see* Magnesium intoxication
  metal fume, *see* Metal fume intoxication
  monensin, *see* Monensin intoxication
  mother milk, *see* Subacute necrotizing
    encephalomyelopathy
  organophosphate, *see* Organophosphate
    intoxication
  toluene, *see* Toluene intoxication
  triperinol, *see* Triperinol intoxication
  valproic acid, *see* Valproic acid intoxication
Intracerebral calcification, 42/534-536
  see also Striatopallidodentate calcification
  corticostriatopallidodentate, *see*
    Striatopallidodentate calcification
  Fahr disease, *see* Striatopallidodentate
    calcification
Intranuclear inclusion body
  nucleodegenerative myopathy, 62/351
Ipecac intoxication
  cytoplasmic inclusion body myopathy, 62/345
Isaacs-Mertens syndrome, *see* Isaacs syndrome
Isaacs syndrome
  cervical radiculopathy, 62/274
  characteristics, 62/273
  chronic muscle denervation, 62/263
  clinical symptom, 62/273
  fasciculation, 62/273
  glucosidase deficiency, 62/263
  hypothyroid myopathy, 62/263
  myokymic discharge, 62/263
  myotonia, 43/160, 62/263, 273
  myotonic syndrome, 62/263, 273
  neuromyotonic discharge, 62/263
  repetitive electrical activity, 62/263
Ischemia
  acquired myoglobinuria, 62/574
Ischemic attack
  transient, *see* Transient ischemic attack
Ischemic exercise
  acid maltase deficiency, 62/480
  adenosine monophosphate deaminase deficiency,
    62/512
  carnitine palmitoyltransferase deficiency, 62/494
  lactate dehydrogenase deficiency, 62/490
  myophosphorylase deficiency, 62/485
  phosphoglycerate mutase deficiency, 62/489
  phosphorylase kinase deficiency, 62/487

Isoetarine
    toxic myopathy, 62/597
Isoetharine, *see* Isoetarine
Isoniazid
    acquired myoglobinuria, 62/577
    toxic myopathy, 62/601
Isonicotinic acid hydrazide, *see* Isoniazid
Isopropyl alcohol, *see* 2-Propanol

Japanese autumn fever, *see* Leptospirosis
Joseph disease, *see* Joseph-Machado disease
Joseph-Machado disease, 42/261-263
    progressive external ophthalmoplegia
        classification, 62/290
Juvenile diabetes mellitus
    Eaton-Lambert myasthenic syndrome, 62/421
Juvenile periarteritis nodosa, *see* Kawasaki syndrome
Juvenile proximal spinal muscular atrophy, *see*
        Wohlfart-Kugelberg-Welander disease
Juvenile spinal muscular atrophy, *see* Wohlfart-
        Kugelberg-Welander disease

Kalemia
    hypo, *see* Hypokalemia
Kawasaki syndrome
    see also Polyarteritis nodosa
    polymyositis, 62/373
Kearns-Sayre-Daroff-Shy syndrome, 38/221-230
    see also Progressive external ophthalmoplegia
    ataxia, 62/508
    biochemistry, 62/311
    cardiac disorder, 62/309
    cardiomyopathy, 22/211, 43/142, 62/310
    celiac disease, 62/313
    cerebellar syndrome, 62/308
    clinical features, 38/222-227, 40/302, 62/307
    computer assisted tomography, 62/306, 311
    cranial neuropathy, 62/313
    CSF, 38/225, 228, 43/142, 62/311
    CSF protein, 62/508
    cytochrome-c oxidase, 62/314
    deafness, 62/313
    definition, 62/307
    dementia, 62/308, 508
    diabetes mellitus, 62/309, 508
    ECG, 62/311
    endocrine disorder, 62/309
    epilepsy, 62/308
    genetics, 62/312
    heart block, 38/224, 62/508
    histopathology, 62/310
    hypoparathyroid myopathy, 62/541
    hypoparathyroidism, 41/215, 43/142, 62/309, 508
    intestinal pseudo-obstruction, 62/313
    mental deficiency, 38/225, 43/142, 62/308
    MEPOP syndrome, 62/315
    metabolic myopathy, 62/508
    mitochondrial disease, 62/497
    mitochondrial DNA defect, 62/507
    mitochondrial encephalomyopathy, 62/498
    mitochondrial myopathy, 41/215, 62/313

    MNGIE syndrome, 62/313
    molecular genetics, 62/311
    muscle biopsy, 62/508
    muscle histopathology, 62/310
    neuropathy, 62/313
    onset before age 20, 62/508
    OSPOM syndrome, 62/313
    Pearson syndrome, 62/508
    pigmentary retinopathy, 62/308, 508
    POLIP syndrome, 62/313
    polyneuropathy, 62/309, 313
    progressive external ophthalmoplegia, 22/192, 211,
        62/287, 306, 508
    progressive external ophthalmoplegia
        classification, 62/290
    ragged red fiber, 62/314, 508
    renal dysfunction, 62/309
    retinopathy, 62/311
    sensorineural deafness, 43/142, 62/508
    short stature, 43/142, 62/508
    treatment, 38/229, 62/312
Kearns-Sayre-Daroff syndrome, *see* Kearns-Sayre-
        Daroff-Shy syndrome
Kearns-Shy syndrome, *see* Kearns-Sayre-Daroff-Shy
        syndrome
Kell blood group
    McLeod myopathy, 62/127
Kennedy syndrome, *see* Foster Kennedy syndrome
Kern calcinosis, *see* Striatopallidodentate
        calcification
α-Ketoglutarate dehydrogenase deficiency, *see* α-
        Oxoglutarate dehydrogenase deficiency
King-Denborough syndrome
    malignant hyperthermia, 62/570
    myoglobinuria, 62/558
Kinky hair disease, *see* Trichopoliodystrophy
Kinky hair syndrome, *see* Trichopoliodystrophy
Kocher-Debré-Sémélaigne syndrome
    see also Hoffmann syndrome
    hypothyroid myopathy, 62/533
Krabbe disease, *see* Globoid cell leukodystrophy
Krabbe leukodystrophy, *see* Globoid cell
        leukodystrophy
Krebs cycle deficiency, *see* Tricarboxylic acid cycle
        deficiency
Kufs disease, 42/465-467
    muscle fiber feature, 62/17
Kufs-Hallervorden disease, *see* Kufs disease
Kugelberg-Welander disease, *see* Wohlfart-
        Kugelberg-Welander disease
Kugelberg-Welander syndrome, *see* Wohlfart-
        Kugelberg-Welander disease
Kwashiorkor
    carnitine deficiency, 62/493
    lipid metabolic disorder, 62/493
    secondary systemic carnitine deficiency, 62/493

Labetalol
    toxic myopathy, 62/597
Lactate dehydrogenase
    dermatomyositis, 62/375

facioscapulohumeral muscular dystrophy, 62/166
hypothyroid myopathy, 62/533
inclusion body myositis, 62/375
limb girdle syndrome, 62/185
myoglobinuria, 62/558
polymyositis, 62/375
Lactate dehydrogenase deficiency
chromosome 11, 62/490
exercise intolerance, 62/490
inheritance, 62/490
ischemic exercise, 62/490
metabolic myopathy, 62/490
myoglobinuria, 62/490
Lactic acidosis, 42/585-589
combined complex I-V deficiency, 62/506
complex I deficiency, 62/503
complex III deficiency, 62/504
complex IV deficiency, 62/505
MELAS syndrome, 62/509
mitochondrial DNA depletion, 62/511
MNGIE syndrome, 62/505
multiple mitochondrial DNA deletion, 62/510
ubidecarenone, 62/504
Lafora body disease, see Lafora progressive
myoclonus epilepsy
Lafora disease, see Lafora progressive myoclonus
epilepsy
Lafora progressive myoclonus epilepsy, 27/171-180,
184-189
muscle fiber feature, 62/17
rimmed muscle fiber vacuole content, 62/18
rimmed muscle fiber vacuole type, 62/18
Lambert-Eaton syndrome, see Eaton-Lambert
myasthenic syndrome
Lamina basalis
muscle fiber, 62/31
Laminin
muscle fiber, 62/31
Landouzy-Dejerine disease, see Facioscapulohumeral
muscular dystrophy
Landouzy-Dejerine dystrophy, see
Facioscapulohumeral muscular dystrophy
Landouzy-Dejerine syndrome, see
Facioscapulohumeral muscular dystrophy
Landry-Guillain-Barré-Strohl disease, see Guillain-
Barré syndrome
Landry-Guillain-Barré syndrome, see Guillain-Barré
syndrome
Late adult muscular dystrophy (Nevin)
see also Limb girdle syndrome
autosomal dominant, 62/181
features, 62/184
late onset, 62/181
limb girdle syndrome, 62/179
Pelger-Huët abnormality, 62/184
Late infantile acid maltase deficiency
calf hypertrophy, 62/480
delayed walking, 62/480
lumbar lordosis, 62/480
respiratory dysfunction, 62/480
toe walker, 27/228, 62/480

waddling gait, 62/480
Lateral sclerosis
amyotrophic, see Amyotrophic lateral sclerosis
LDH, see Lactate dehydrogenase
Leber hereditary optic neuropathy
age at onset, 62/509
ataxia, 62/509
cardiac conduction, 62/509
clinical features, 62/509
hyperreflexia, 62/509
mitochondrial disease, 62/497
mitochondrial DNA point mutation, 62/509
neuropathy, 62/509
optic atrophy, 62/509
sex ratio, 62/509
Legal aspects, see Death
Legionella feelii, see Legionellosis
Legionella micdadei, see Legionellosis
Legionella pneumophila
polymyositis, 62/373
Legionellosis
acquired myoglobinuria, 62/577
polymyositis, 62/373
Legionnaires disease, see Legionellosis
Leigh disease, see Subacute necrotizing
encephalomyelopathy
Leigh necrotizing encephalomyelopathy, see
Subacute necrotizing encephalomyelopathy
Leigh syndrome, see Subacute necrotizing
encephalomyelopathy
Leptospirosis, 33/395-414
acquired myoglobinuria, 62/577
Lethargy
see also Hypersomnia and Sleep
glutaric aciduria type II, 62/495
multiple acyl coenzyme A dehydrogenase
deficiency, 62/495
α-oxoglutarate dehydrogenase deficiency, 62/503
pyruvate decarboxylase deficiency, 62/502
Leukodystrophy
see also Diffuse sclerosis
globoid cell, see Globoid cell leukodystrophy
MNGIE syndrome, 62/505
Leukoencephalopathy
see also Encephalopathy
inclusion body myositis, 62/373
Leukopheresis
dermatomyositis, 62/386
inclusion body myositis, 62/386
polymyositis, 62/386
Lewy body disease
diffuse, see Diffuse lewy body disease
Leyden-Möbius dystrophy, see Limb girdle syndrome
Leyden-Möbius pelvifemoral muscular dystrophy,
see Pelvifemoral muscular dystrophy (Leyden-
Möbius)
LGB syndrome, see Guillain-Barré syndrome
Lhermitte sign
glucocorticosteroids, see Glucocorticoids
Licorice, see Glycyrrhiza
Lid lag

myotonic hyperkalemic periodic paralysis, 62/462
thyroid associated ophthalmopathy, 62/530
Life expectancy
    hypokalemic periodic paralysis, 62/459
Limb girdle dystrophy, *see* Limb girdle syndrome
Limb girdle syndrome
    see also Bethlem-Van Wijngaarden syndrome, Late
        adult muscular dystrophy (Nevin), Muscular
        dystrophy, Pelvifemoral muscular dystrophy
        (Leyden-Möbius), Quadriceps myopathy *and*
        Scapulohumeral muscular dystrophy (Erb)
    adenosine monophosphate deaminase deficiency,
        62/190
    adult acid maltase deficiency, 62/190
    autosomal recessive, 62/179
    carcinoma, 62/188
    cardiac involvement, 62/185
    cervical neurofibroma, 62/188
    classification history, 62/179
    clinical features, 40/306, 312, 316, 322, 342, 449,
        62/182
    connectin, 62/187
    creatine kinase, 40/449, 62/185
    differential diagnosis, 40/437, 41/177, 62/187-190
    dystrophin, 62/187
    Emery-Dreifuss muscular dystrophy, 62/189
    EMG, 40/453, 62/186
    etiology, 62/190
    fructose bisphosphate aldolase, 62/185
    genetics, 41/419, 486, 62/181
    glycogen storage disease, 62/190
    incidence, 62/182
    inclusion body myositis, 62/187
    infantile acid maltase deficiency, 62/190
    infectious cause, 62/187
    lactate dehydrogenase, 62/185
    late adult muscular dystrophy (Nevin), 62/179
    muscle biopsy, 62/186
    muscle histopathology, 62/186
    muscle pseudohypertrophy, 62/182
    myopathy pattern, 62/182
    myophosphorylase deficiency, 62/190
    myotonic dystrophy, 62/188, 240
    non-REM, 62/185
    pelvifemoral muscular dystrophy (Leyden-
        Möbius), 62/179
    polymyositis, 40/457, 62/187
    prevalence, 43/104, 62/182
    pseudomyopathic spinal muscular atrophy, 62/188
    pulmonary involvement, 62/185
    pyruvate kinase, 62/185
    recent classification, 62/181
    REM, 62/185
    scapulohumeral muscular dystrophy (Erb), 62/179
    secondary type, 62/181
    single fiber electromyography, 62/186
    sporadic, 62/179
    titin, 62/187
    toxic cause, 62/187
    treatment, 62/190
    vitamin D deficiency, 40/438, 62/188

vitamin E deficiency, 62/188
Wohlfart-Kugelberg-Welander disease, 40/433,
    62/188
Limit dextrin type glycogen
    glycogen storage disease, *see* Debrancher
        deficiency
Limit dextrinosis, *see* Debrancher deficiency
Lindane
    toxic myopathy, 62/601
Lipid metabolic disorder
    beta oxidation defect, 62/494
    biochemistry, 62/491
    cardiomyopathy, 62/493
    carnitine deficiency, 41/196-201, 62/492
    carnitine palmitoyltransferase deficiency, 41/204-
        207, 62/494
    childhood myoglobinuria, 62/566
    dicarboxylic aciduria, 62/494
    Fanconi syndrome, 62/493
    fatty acid oxidation, 62/491
    hepatic encephalopathy, 62/493
    history, 62/490
    hypoglycemia, 62/494
    kwashiorkor, 62/493
    long chain acyl coenzyme A dehydrogenase
        deficiency, 62/494
    long chain 3-hydroxyacyl coenzyme A
        dehydrogenase deficiency, 62/495
    medium chain acyl coenzyme A dehydrogenase
        deficiency, 62/495
    metabolic encephalopathy, 62/494
    metabolic myopathy, 41/194-207, 62/490
    multiple acyl coenzyme A dehydrogenase
        deficiency, 62/495
    short chain acyl coenzyme A dehydrogenase
        deficiency, 62/495
    short chain 3-hydroxyacyl coenzyme A
        dehydrogenase deficiency, 62/495
    systemic types, 62/493
    triglyceride storage disease, 62/496
    valproic acid intoxication, 62/493
Lipid storage disease
    rimmed muscle fiber vacuole content, 62/18
    rimmed muscle fiber vacuole type, 62/18
Lipid storage myopathy
    carnitine deficiency, 40/95, 258, 280, 308, 317, 422,
        43/175, 184, 62/493
Lipidosis
    galactosylceramide, *see* Globoid cell
        leukodystrophy
Liquorice, *see* Glycyrrhiza
Lithium
    toxic myopathy, 62/597, 601
Liver cirrhosis
    brancher deficiency, 62/483
Liver disease
    debrancher deficiency, 62/482
    encephalopathy, *see* Hepatic encephalopathy
    mitochondrial DNA depletion, 62/511
    phosphorylase kinase deficiency, 62/486
Long chain acyl coenzyme A dehydrogenase

deficiency
    cardiomegaly, 62/495
    dicarboxylic aciduria, 62/495
    hepatomegaly, 62/495
    infantile hypotonia, 62/495
    lipid metabolic disorder, 62/494
    metabolic encephalopathy, 62/495
    metabolic myopathy, 62/495
    myalgia, 62/495
    myoglobinuria, 62/495
    myopathy, 62/495
Long chain 3-hydroxyacyl coenzyme A
    dehydrogenase deficiency
    cardiomyopathy, 62/495
    hypoglycemic encephalopathy, 62/495
    lipid metabolic disorder, 62/495
    metabolic encephalopathy, 62/495
    metabolic myopathy, 62/495
    muscle biopsy, 62/495
Lordosis
    see also Scoliosis
    lumbar, see Lumbar lordosis
Lovastatin, see Mevinolin
Lowe syndrome
    congenital fiber type disproportion, 62/355
Lower motoneuron sign
    adult polyglucosan body disease, 62/484
LSD, see Lysergide
Lumbar lordosis
    late infantile acid maltase deficiency, 62/480
Lung infection, see Pneumonitis
Lupus erythematosus
    systemic, see Systemic lupus erythematosus
Lyme disease
    see also Borrelia burgdorferi
    polymyositis, 62/373
Lymphocyte
    B, see B-lymphocyte
D-Lysergic acid diethylamide, see Lysergide
Lysergide
    toxic myopathy, 62/601
Lysosome
    muscle fiber, 40/36, 62/22
    muscle fiber type, 62/4

M protein
    muscle fiber, 62/26
Macroglossia
    infantile acid maltase deficiency, 62/480
Magnesemia
    hyper, see Hypermagnesemia
Magnesium intoxication
    Eaton-Lambert myasthenic syndrome, 62/425
Magnetic resonance
    nuclear, see Nuclear magnetic resonance
Magnetic resonance imaging, see Nuclear magnetic
    resonance
Malherbe epithelioma, see Pilomatricoma
Malignancy
    see also Carcinoma
    dermatomyositis, 62/371, 374

inclusion body myositis, 62/374
    polymyositis, 62/374
Malignant exophthalmos, see Thyroid associated
    ophthalmopathy
Malignant hyperpyrexia, see Malignant hyperthermia
Malignant hyperthermia, 38/550-556
    arrhythmia, 62/567
    Becker muscular dystrophy, 62/570
    calcium channel, 62/567, 569
    central core myopathy, 41/3, 43/81, 62/333, 570
    childhood myoglobinuria, 62/567
    chondrodystrophic myotonia, 62/570
    chromosome 12q12-13.2, 62/333
    chromosome 19q12-13.2, 62/568
    clinical course, 62/569
    congenital myopathy, 62/335
    definition, 62/567
    Duchenne muscular dystrophy, 62/570
    epidemiology, 62/567
    Fukuyama syndrome, 62/570
    general anesthesia, 62/567
    heat stroke, 62/571, 573
    incomplete dystrophin deficiency, 62/135
    inheritance, 62/568
    King-Denborough syndrome, 62/570
    metabolic acidosis, 62/567
    molecular genetics, 62/569
    myoglobinuria, 41/268, 62/567
    myotonia congenita, 62/570
    neuroleptic malignant syndrome, 62/571
    pathophysiology, 62/571
    periodic paralysis, 62/570
    ryanodine receptor, 62/567, 569
    sarcoplasmic reticulum adenosine triphosphate
      deficiency syndrome, 62/570
    succinylcholine, 40/546, 41/268, 270, 62/568
    susceptibility, 62/568
    suxamethonium induced muscle rigidity, 62/569
    tachycardia, 43/117, 62/567
    toxic myopathy, 62/601
    toxic shock syndrome, 62/574
    treatment, 38/555, 62/571
Mallory body myopathy
    cardiorespiratory failure, 62/345
    congenital myopathy, 62/332
    facial weakness, 62/334
Mannitol
    myoglobinuria, 62/556
Mannosidosis
    muscle fiber feature, 62/17
Marathon runner
    acquired myoglobinuria, 62/573
Marie-Charcot-Tooth disease, see Charcot-Marie-
    Tooth disease
Marinesco-Sjögren-Garland disease, see Marinesco-
    Sjögren syndrome
Marinesco-Sjögren syndrome, 21/555-560
    nucleodegenerative myopathy, 62/351
Marker
    muscle fiber type, 62/3
McArdle disease, see Myophosphorylase deficiency

McLeod myopathy
    acanthocyte, 62/127
    characteristics, 62/127
    Duchenne muscular dystrophy, 62/127
    Kell blood group, 62/127
Medicolegal aspects, *see* Death
Medium chain acyl coenzyme A dehydrogenase
        deficiency
    hepatomegaly, 62/495
    hypoglycemic encephalopathy, 62/495
    inheritance, 62/495
    lipid metabolic disorder, 62/495
    metabolic encephalopathy, 62/495
    metabolic myopathy, 62/495
    prevalence, 62/495
    Reye syndrome, 62/495
Megaconial myopathy, *see* Mitochondrial myopathy
Melancholy, *see* Mental depression
MELAS syndrome
    see also Progressive myoclonus epilepsy
    brain infarction, 62/509
    cardiopathy, 62/510
    cortical blindness, 62/510
    dementia, 62/509
    epilepsy, 62/509
    lactic acidosis, 62/509
    mitochondrial disease, 62/497
    mitochondrial DNA point mutation, 62/509
    mitochondrial encephalomyopathy, 62/498
    progressive external ophthalmoplegia
        classification, 62/290
    ragged red fiber, 62/292, 509
    striatopallidodentate calcification, 62/510
Melena
    dermatomyositis, 62/374
    inclusion body myositis, 62/374
    polymyositis, 62/374
Menkes disease, *see* Trichopoliodystrophy
Menkes steely hair disease, *see* Trichopoliodystrophy
Menkes syndrome, *see* Trichopoliodystrophy
Mental deficiency
    chondrodystrophic myotonia, 62/269
    congenital myopathy, 62/335
    Duchenne muscular dystrophy, 40/363, 41/479,
        62/121
    fingerprint body myopathy, 62/347
    glycogen storage disease, 62/490
    Kearns-Sayre-Daroff-Shy syndrome, 38/225,
        43/142, 62/308
    multiple mitochondrial DNA deletion, 62/510
    phosphoglycerate kinase deficiency, 62/489
    progressive external ophthalmoplegia, 62/308
    Schimke syndrome, 62/304
Mental depression
    myotonic dystrophy, 62/235
Mental retardation, *see* Mental deficiency
MEPOP syndrome
    Kearns-Sayre-Daroff-Shy syndrome, 62/315
    progressive external ophthalmoplegia, 62/315
Meprobamate
    toxic myopathy, 62/601

Mercaptopropionyl glycine, *see* Tiopronin
Mercuric chloride
    toxic myopathy, 62/601
Merosin
    muscle fiber, 62/31
MERRF syndrome
    see also Progressive myoclonus epilepsy
    ataxia, 62/509
    dementia, 62/509
    epilepsy, 62/509
    hearing loss, 62/509
    mitochondrial disease, 62/497
    mitochondrial DNA point mutation, 62/509
    mitochondrial encephalomyopathy, 62/498
    myoclonus, 62/509
    myopathy, 62/509
    neuropathy, 62/509
    optic atrophy, 62/509
    progressive external ophthalmoplegia
        classification, 62/290
    ragged red fiber, 62/292
    spasticity, 62/509
Mestinon, *see* Pyridostigmine bromide
Metabolic acidosis
    malignant hyperthermia, 62/567
Metabolic disorder
    myoglobinuria, 62/557
Metabolic encephalopathy
    adenosine triphosphatase 6 mutation syndrome,
        62/510
    adult polyglucosan body disease, 62/484
    beta oxidation defect, 62/494
    cerebral glycogenosis, 62/490
    childhood myoglobinuria, 62/561-563
    complex V deficiency, 62/506
    glutaric aciduria type II, 62/495
    glycogen storage disease variant, 62/490
    heat stroke, 62/573
    lipid metabolic disorder, 62/494
    long chain acyl coenzyme A dehydrogenase
        deficiency, 62/495
    long chain 3-hydroxyacyl coenzyme A
        dehydrogenase deficiency, 62/495
    medium chain acyl coenzyme A dehydrogenase
        deficiency, 62/495
    MNGIE syndrome, 62/505
    multiple acyl coenzyme A dehydrogenase
        deficiency, 62/495
    phosphoglycerate kinase deficiency, 62/489
    primary systemic carnitine deficiency, 62/493
    short chain 3-hydroxyacyl coenzyme A
        dehydrogenase deficiency, 62/495
    systemic carnitine deficiency, 62/493
    toxic shock syndrome, 62/574
Metabolic myopathy, 41/175-220
    acid maltase deficiency, 62/480
    adenosine triphosphatase 6 mutation syndrome,
        62/510
    beta oxidation, 62/492
    brancher deficiency, 62/483
    carnitine deficiency, 62/492

carnitine palmitoyltransferase deficiency, 62/494
complex I deficiency, 62/503
complex II deficiency, 62/504
complex III deficiency, 62/504
complex IV deficiency, 62/505
complex V deficiency, 62/506
debrancher deficiency, 62/482
EMG, 62/69
fatty acid oxidation, 62/491
glutaric aciduria type II, 62/495
glycogen storage disease, 41/175-194, 62/479
history, 62/490
Kearns-Sayre-Daroff-Shy syndrome, 62/508
lactate dehydrogenase deficiency, 62/490
lipid metabolic disorder, 41/194-207, 62/490
long chain acyl coenzyme A dehydrogenase, 62/492
long chain acyl coenzyme A dehydrogenase
    deficiency, 62/495
long chain fatty acid, 62/491
long chain 3-hydroxyacyl coenzyme A
    dehydrogenase deficiency, 62/495
medium chain acyl coenzyme A dehydrogenase,
    62/492
medium chain acyl coenzyme A dehydrogenase
    deficiency, 62/495
MNGIE syndrome, 62/505
multiple acyl coenzyme A dehydrogenase
    deficiency, 62/495
myophosphorylase deficiency, 62/485
oxidation-phosphorylation coupling defect, 62/506
phosphofructokinase deficiency, 62/487
phosphoglycerate kinase deficiency, 62/489
phosphoglycerate mutase deficiency, 62/489
phosphorylase kinase deficiency, 62/486
primary systemic carnitine deficiency, 62/493
short chain acyl coenzyme A dehydrogenase,
    62/492
short chain 3-hydroxyacyl coenzyme A
    dehydrogenase deficiency, 62/495
systemic carnitine deficiency, 62/493
Metal fume intoxication
    toxic myopathy, 62/613
Metaldehyde
    toxic myopathy, 62/601
Metaproterenol, see Orciprenaline
Methadone
    acquired myoglobinuria, 62/577
    toxic myopathy, 62/601
Methotrexate
    dermatomyositis, 62/385
    inclusion body myositis, 62/385
    polymyositis, 41/80, 62/385
Metolazone
    toxic myopathy, 62/597
Mevinolin
    acquired myoglobinuria, 62/576
    nonprimary inflammatory myopathy, 62/372
    toxic myopathy, 62/601
Mexiletine
    myotonia, 62/275
    myotonic dystrophy, 62/241

Microcephaly, 30/507-520
    α-oxoglutarate dehydrogenase deficiency, 62/503
Migraine
    ophthalmoplegic, see Ophthalmoplegic migraine
Milk intoxication, see Subacute necrotizing
    encephalomyelopathy
Mimicking disorder
    Parkinson disease, see Multiple neuronal system
        degeneration
Miniature motor end plate potential
    congenital synaptic vesical paucity syndrome,
        62/429
    features, 62/50
    motor unit, 62/50
    myasthenia gravis, 41/129, 62/414
    myopathy, 62/50
Minicore disease, see Multicore myopathy
Minicore myopathy, see Multicore myopathy
Minimal change myopathy
    congenital myopathy, 62/332
    symptom, 62/357
Mitochondria
    muscle fiber, 40/31, 90, 216, 62/23
    muscle fiber type, 62/3
Mitochondria jagged Z line myopathy
    case, 62/344
    congenital myopathy, 62/332
    facial weakness, 62/334
Mitochondrial disease
    see also Ragged red fiber
    aconitate hydratase deficiency, 62/503
    adenosine triphosphatase 6 mutation syndrome,
        62/510
    Alpers disease, 62/497
    biochemistry, 62/498
    childhood myoglobinuria, 62/567
    clinical features, 62/498
    complex I deficiency, 62/503
    complex II deficiency, 62/504
    complex III deficiency, 62/504
    complex IV deficiency, 62/505
    complex V deficiency, 62/506
    cytochrome-c oxidase, 62/497
    fatal infantile myopathy/cardiopathy, 62/510
    fumarase deficiency, 62/497
    genetics, 62/498
    history, 62/496
    Kearns-Sayre-Daroff-Shy syndrome, 62/497
    laboratory finding, 62/498
    Leber hereditary optic neuropathy, 62/497
    MELAS syndrome, 62/497
    MERRF syndrome, 62/497
    mitochondrial DNA, 62/500
    mitochondrial DNA defect, 62/506
    mitochondrial DNA deletion, 62/508
    mitochondrial DNA depletion, 62/511
    mitochondrial encephalomyopathy, 62/496, 498
    mitochondrial metabolism, 62/498
    mitochondrial morphology, 62/496
    mitochondrial myopathy, 62/496
    mitochondrial protein transport defect, 62/506

MNGIE syndrome, 62/505
nuclear DNA disease, 62/502
oxidation-phosphorylation coupling defect, 62/506
α-oxoglutarate dehydrogenase deficiency, 62/503
progressive external ophthalmoplegia, 62/497
pyruvate carboxylase deficiency, 62/503
pyruvate dehydrogenase complex deficiency,
  62/496, 502
pyruvate dehydrogenase phosphatase deficiency,
  62/503
ragged red fiber, 62/496
subacute necrotizing encephalomyelopathy,
  41/212, 62/497
succinate dehydrogenase, 62/497
tetrazolium reductase, 62/497
tricarboxylic acid cycle deficiency, 62/503
trichopoliodystrophy, 62/497
Mitochondrial DNA defect
  Kearns-Sayre-Daroff-Shy syndrome, 62/507
  mitochondrial disease, 62/506
  Pearson syndrome, 62/508
  sporadic progressive external ophthalmoplegia
    with ragged red fiber, 62/508
Mitochondrial DNA deletion
  cytochrome-c oxidase deficiency, 62/508
  mitochondrial disease, 62/508
  multiple, see Multiple mitochondrial DNA
    deletion
Mitochondrial DNA depletion
  iatrogenic disease, 62/511
  inheritance, 62/511
  lactic acidosis, 62/511
  liver disease, 62/511
  mitochondrial disease, 62/511
  myopathy, 62/511
  ragged red fiber, 62/511
  renal dysfunction, 62/511
  tetraplegia, 62/511
  zidovudine, 62/511
Mitochondrial DNA point mutation
  adenosine triphosphatase 6 mutation syndrome,
    62/509
  fatal infantile myopathy/cardiopathy, 62/509
  Leber hereditary optic neuropathy, 62/509
  MELAS syndrome, 62/509
  MERRF syndrome, 62/509
Mitochondrial encephalomyopathy
  complex I deficiency, 62/504
  Kearns-Sayre-Daroff-Shy syndrome, 62/498
  MELAS syndrome, 62/498
  MERRF syndrome, 62/498
  microscopic features, 62/25
  mitochondrial disease, 62/496, 498
Mitochondrial myopathy
  see also Progressive external ophthalmoplegia
  biochemical classification, 62/498
  facioscapulohumeral syndrome, 62/168
  intestinal pseudo-obstruction, 62/313
  Kearns-Sayre-Daroff-Shy syndrome, 41/215,
    62/313
  mitochondrial disease, 62/496

muscle tissue culture, 62/100
myasthenia gravis, 62/415
progressive external ophthalmoplegia, 62/289, 313
Mixed connective tissue disease
  dermatomyositis, 62/371, 374
  inclusion body myositis, 62/374
  polymyositis, 62/374
MNGIE syndrome
  diarrhea, 62/505
  intestinal pseudo-obstruction, 62/505
  Kearns-Sayre-Daroff-Shy syndrome, 62/313
  lactic acidosis, 62/505
  leukodystrophy, 62/505
  metabolic encephalopathy, 62/505
  metabolic myopathy, 62/505
  mitochondrial disease, 62/505
  multiple neuronal system degeneration, 62/505
  POLIP syndrome, 62/505
  polyneuropathy, 62/505
  progressive external ophthalmoplegia, 62/305, 313,
    505
  progressive external ophthalmoplegia
    classification, 62/290
  ragged red fiber, 62/292, 505
Möbius syndrome
  facioscapulohumeral muscular dystrophy, 40/419,
    422, 62/162
  progressive external ophthalmoplegia
    classification, 62/290
Moersch-Woltman syndrome, see Stiff-man
  syndrome
Molecular genetics
  acid maltase deficiency, 62/481
  debrancher deficiency, 62/483
  Kearns-Sayre-Daroff-Shy syndrome, 62/311
  malignant hyperthermia, 62/569
  myophosphorylase deficiency, 62/485
  phosphofructokinase deficiency, 62/488
  phosphorylase kinase deficiency, 62/487
  progressive external ophthalmoplegia, 62/311
Monensin intoxication
  toxic myopathy, 62/612
Monoclonal gammopathy
  polymyositis, 62/373
Mood, see Emotion
Morbus Weil disease, see Leptospirosis
Morphine
  toxic myopathy, 62/601
Mother milk intoxication, see Subacute necrotizing
  encephalomyelopathy
Motoneuron disease
  see also Amyotrophic lateral sclerosis
  myoglobinuria, 62/556
Motor conduction velocity
  scapuloperoneal spinal muscular atrophy, 62/170
Motor end plate
  see also Motor end plate potential
  muscle fiber, 62/28
  muscle tissue culture, 62/92, 97
Motor end plate noise
  features, 62/50

motor unit, 62/50
myopathy, 62/50
Motor end plate potential
   see also Motor end plate
   miniature, see Miniature motor end plate potential
Motor retardation
   psycho, see Psychomotor retardation
Motor unit
   axonal stimulation, 62/59
   complex repetitive discharge, 62/62, 67
   conventional EMG, 62/49
   denervation, 62/62
   EMG, 19/275, 40/138, 62/49
   EMG method, 62/49
   extradischarge, 62/50, 67
   fasciculation, 62/50
   fasciculation discharge, 62/50
   fatigue index, 62/54
   fibrillation potential, 62/50, 61, 67
   insertional activity, 62/50
   intramuscular stimulation, 62/58
   macroelectromyography, 62/55
   macromotor unit potential, 62/55
   miniature motor end plate potential, 62/50
   motor end plate noise, 62/50
   motor unit counting technique, 62/60
   motor unit potential, 62/49, 67
   muscle denervation, 62/62
   muscle fiber density, 62/54
   muscle fiber stimulation, 62/59
   muscle reinnervation, 62/62
   myogenic jitter, 62/59
   myotonic discharge, 62/50, 61, 67
   neuromuscular jitter, 62/54
   positive sharp wave, 62/50, 67
   propagation velocity, 62/53
   rate coding, 62/52
   recruitment, 62/52
   reinnervation, 62/62
   scanning electromyography, 62/57
   single fiber electromyography, 62/49, 53
   size, 62/55
   spontaneous activity, 62/50, 67
   tremor, 62/52
   velocity recovery function, 62/54
   voluntary contraction, 62/52
Motor unit potential
   area, 62/51
   complex motor unit potential, 62/51
   definition, 62/50
   duration, 62/50
   motor unit, 62/49, 67
   muscle denervation, 62/64
   muscle reinnervation, 62/64
   myopathy, 62/49, 67
   normal value, 62/49
   polyphasia, 62/51
   satellite potential, 62/51
   serrated motor unit potential, 62/51
   spike, 62/51
   turn, 62/51

Movement
   choreatic, see Chorea
MRI, see Nuclear magnetic resonance
mRNA
   acid maltase deficiency, 62/482
Mucocutaneous lymph node syndrome, see Kawasaki
   syndrome
Mucolipidosis type IV
   muscle fiber feature, 62/17
Mulgatoxin a
   toxic myopathy, 62/610
Multicore disease, see Multicore myopathy
Multicore myopathy
   adenosine triphosphatase, 62/339
   clinical symptom, 62/338
   congenital myopathy, 41/16, 62/332
   differential diagnosis, 62/338
   enzyme histochemistry, 62/339
   facial weakness, 62/334
   mitochondria loss, 62/339
Multiple acyl coenzyme A dehydrogenase deficiency
   see also Glutaric aciduria type II
   coma, 62/495
   episodic encephalopathy, 62/495
   exercise intolerance, 62/495
   fasting, 62/495
   hypoglycemia, 62/495
   infantile hypotonia, 62/495
   lethargy, 62/495
   lipid metabolic disorder, 62/495
   2 main forms, 62/495
   metabolic encephalopathy, 62/495
   metabolic myopathy, 62/495
   myalgia, 62/495
   recurrent coma, 62/495
Multiple mitochondrial DNA deletion
   ataxia, 62/510
   cataract, 62/510
   EEG, 62/510
   exercise intolerance, 62/510
   hearing loss, 62/510
   hypoparathyroidism, 62/510
   inheritance, 62/510
   lactic acidosis, 62/510
   mental deficiency, 62/510
   muscle weakness, 62/510
   myoglobinuria, 62/510
   neuropathy, 62/510
   nystagmus, 62/510
   optic atrophy, 62/510
   progressive external ophthalmoplegia, 62/510
   ptosis, 62/510
   ragged red fiber, 62/510
   tremor, 62/510
Multiple neuronal system degeneration
   adenosine triphosphatase 6 mutation syndrome,
      62/510
   complex I deficiency, 62/503
   complex III deficiency, 62/504
   complex V deficiency, 62/506
   MNGIE syndrome, 62/505

ubidecarenone, 62/504
Multiple system atrophy, *see* Multiple neuronal system degeneration
Multiple system degeneration, *see* Multiple neuronal system degeneration
Muscle
  connective tissue, 62/40
  dystrophin, 62/125
  inflammatory cell, 62/42
  intramuscular nerve, 62/44
  vascularization, 62/38
Muscle action potential
  compound, *see* Compound muscle action potential
Muscle atrophy, *see* Muscle fiber type II atrophy *and* Muscular atrophy
Muscle biopsy
  acid maltase deficiency, 62/481
  acromegalic myopathy, 62/538
  ACTH induced myopathy, 62/537
  axonal polyneuropathy, 62/609
  Becker muscular dystrophy, 62/123
  carnitine palmitoyltransferase deficiency, 62/494
  central core myopathy, 62/338
  childhood myoglobinuria, 62/561
  chondrodystrophic myotonia, 62/269
  colchicine myopathy, 62/609
  congenital end plate acetylcholinesterase deficiency, 62/430
  congenital slow channel syndrome, 62/434
  cytochemistry, 62/2
  debrancher deficiency, 62/483
  dermatomyositis, 62/375
  Duchenne muscular dystrophy, 62/123
  EMG, 62/124
  familial infantile myasthenia, 62/428
  hyperaldosteronism myopathy, 62/536
  hypothyroid myopathy, 62/533
  immunocytochemistry, 62/2
  inclusion body myositis, 62/373, 375
  interpretation, 40/54, 62/44
  Kearns-Sayre-Daroff-Shy syndrome, 62/508
  limb girdle syndrome, 62/186
  long chain 3-hydroxyacyl coenzyme A dehydrogenase deficiency, 62/495
  microscopy, 62/2
  muscle pathology, 62/1
  myophosphorylase deficiency, 41/187, 62/485
  myotonic dystrophy, 62/225
  myotonic syndrome, 62/274
  needle biopsy, 62/2
  Nelson syndrome, 62/537
  opaque fiber, 62/15
  α-oxoglutarate dehydrogenase deficiency, 62/503
  periodic paralysis, 62/465
  phosphofructokinase deficiency, 41/191, 62/488
  phosphoglycerate kinase deficiency, 62/489
  phosphoglycerate mutase deficiency, 62/489
  phosphorylase kinase deficiency, 62/487
  polymyositis, 62/375
  processing method, 62/1
  removal method, 62/1

scapulohumeral spinal muscular atrophy, 62/171
small caliber fiber, 62/14
split fiber, 62/16
sporadic progressive external ophthalmoplegia with ragged red fiber, 62/508
storage disease, 62/17
thyrotoxic myopathy, 62/529
ubidecarenone, 62/504
Wohlfart-Kugelberg-Welander disease, 62/200
Muscle cirrhosis, *see* Myosclerosis
Muscle contraction
  Bethlem-Van Wijngaarden syndrome, 62/155, 181
  brancher deficiency, 62/484
  denervation, 62/53
  Emery-Dreifuss muscular dystrophy, 62/145
  EMG, 62/52
  interference pattern, 62/53
  muscle fibrosis, 62/173
  myophosphorylase deficiency, 62/485
  myotonia, 62/262, 273
  myotonic syndrome, 62/262, 273
  recruitment pattern, 62/53
Muscle contracture, *see* Contracture
Muscle cramp
  see also Contracture
  adenosine monophosphate deaminase deficiency, 62/511
  carnitine palmitoyltransferase deficiency, 62/494
  childhood myoglobinuria, 62/561
  congenital myopathy, 62/335
  endocrine myopathy, 62/527
  hypothyroid myopathy, 41/243, 246, 62/532
  myophosphorylase deficiency, 62/485
  myotonia, 40/327, 533, 62/261, 273
  myotonic syndrome, 62/261, 273
  phosphofructokinase deficiency, 62/487
  phosphoglycerate kinase deficiency, 62/489
  phosphoglycerate mutase deficiency, 62/489
  thyrotoxic myopathy, 62/528
  toxic myopathy, 62/596
Muscle culture, *see* Muscle tissue culture
Muscle denervation
  macroelectromyography, 62/65
  motor unit, 62/62
  motor unit potential, 62/64
  scanning electromyography, 62/66
  single fiber, 62/62
Muscle disease, *see* Myopathy
Muscle fiber
  A band, 62/26
  α-actinin, 62/26, 35
  atrophy, 40/11, 142, 200, 62/13
  atrophy cause, 62/13
  autophagic vacuole, 62/23
  C protein, 62/26
  calcification, 62/18
  caps, 62/44
  change, 62/27
  collagen type IV, 62/31
  cytoskeleton, 62/19
  delta lesion, 62/15

denervation atrophy, 62/13
dense core tubule, 62/23
dystrophin, 62/35
dystrophin related protein, 62/35
entactin, 62/31
exocytosis, 62/18
extralysosomal storage, 62/16
fingerprint body, 40/39, 43, 62/44
forked fiber, 62/11
glycogen, 62/20
Golgi system, 62/21
heparan sulfate, 62/31
hypercontraction, 62/15
hypertrophy, 62/15
I band, 62/26
lamina basalis, 62/31
laminin, 62/31
lipid globule, 62/22
lysosomal storage, 62/16
lysosome, 40/36, 62/22
M protein, 62/26
merosin, 62/31
mitochondria, 40/31, 90, 216, 62/23
motor end plate, 62/28
myonucleus, 62/29
nebulin, 62/26
necrosis, *see* Muscle necrosis
neural cell adhesion molecule, 62/33
nuclear inclusion, 62/31
opaque fiber, 62/15
partial invasion, 62/9
phagocytosis, 62/6
plasma membrane, 62/32
ragged red fiber, 62/24
reducing body, 62/44
regeneration, 40/20, 89, 62/9
repair, 62/19
rimmed muscle fiber vacuole type, 62/18
ryanodine binding protein, 62/37
sarcolemma, 62/31
sarcoplasmic mass, 40/27, 244, 497, 62/44
sarcoplasmic reticulum, 62/37
segmental necrosis, 62/13
small caliber fiber, 62/13
spectrin, 62/35
split fiber, 62/16
T tubule, 62/23, 37
titin, 62/26
vacuole, *see* Rimmed muscle fiber vacuole
Z band, 62/7, 26
Muscle fiber band
  A, *see* A band
  I, *see* I band
  Z, *see* Z band
Muscle fiber type
  adenosine triphosphatase, 62/4
  criteria, 62/2
  cytoskeleton, 62/3
  cytosol, 62/4
  determinant molecule, 62/3
  display mode, 62/3

fast mature myosin, 62/4
  lysosome, 62/4
  marker, 62/3
  mitochondria, 62/3
  myopathy, 62/2
  neuromuscular disease, 62/2
  nucleus, 62/4
  organelle distribution, 62/3
  ribosome, 62/4
  sarcolemma, 62/3
  sarcoplasmic reticulum, 62/3
  T tubule, 62/3
  type grouping, 62/4
Muscle fiber type I
  congenital myopathy, 40/146, 62/349
  criteria, 62/4
  predilection, 62/5
Muscle fiber type I atrophy
  Emery-Dreifuss muscular dystrophy, 62/151
Muscle fiber type I hypertrophy, *see* Myotubular
    myopathy
Muscle fiber type I predominance
  central core myopathy, 41/4, 62/338
  congenital myopathy, 62/332
  uniform muscle fiber type I myopathy, 62/354
Muscle fiber type II atrophy
  see also Muscular atrophy
  endocrine myopathy, 62/527
  steroid myopathy, 62/605
Muscle fiber type II hypoplasia
  congenital myopathy, 62/332
Muscle fiber type IIA
  criteria, 62/4
  predilection, 62/5
Muscle fiber type IIB
  criteria, 62/4
  predilection, 62/5
Muscle fiber type IIC
  criteria, 62/4
Muscle fibrosis
  arthrogryposis, 62/173
  muscle contraction, 62/173
  rigid spine syndrome, 62/173
Muscle histopathology
  Kearns-Sayre-Daroff-Shy syndrome, 62/310
  limb girdle syndrome, 62/186
  myotonic dystrophy, 62/219
  progressive external ophthalmoplegia, 62/310
Muscle hypertrophy
  autosomal recessive generalized myotonia, 62/267
  childhood myoglobinuria, 62/561
  chondrodystrophic myotonia, 40/284, 62/269
  congenital myopathy, 62/335
  infantile acid maltase deficiency, 62/480
  myotonia congenita, 40/539, 41/298, 43/161, 62/266
Muscle necrosis
  histologic features, 62/6
  myoglobinuria, 62/554
  myophosphorylase deficiency, 62/485
Muscle pain, *see* Myalgia
Muscle phosphofructokinase deficiency, *see*

Phosphofructokinase deficiency
Muscle pseudohypertrophy
    limb girdle syndrome, 62/182
Muscle reinnervation
    macroelectromyography, 62/65
    motor unit, 62/62
    motor unit potential, 62/64
    scanning electromyography, 62/66
    single fiber, 62/62
Muscle relaxation
    myotonic dystrophy, 62/224
Muscle rigidity
    suxamethonium induced, see Suxamethonium
        induced muscle rigidity
Muscle stiffness
    painful dominant myotonia, 62/266
Muscle strength scoring
    myotonic dystrophy, 62/248
Muscle stretch reflex
    autosomal recessive generalized myotonia, 62/266
    thyrotoxic myopathy, 62/528
Muscle stretch reflex time
    hypothyroid myopathy, 62/532
Muscle tissue culture, 40/183-194
    acetylcholine receptor, 40/186, 62/97
    acetylcholine receptor clustering, 62/89
    acetylcholinesterase, 62/97
    acid maltase deficiency, 40/190, 62/99
    aneural culture, 62/86
    cachectin, 62/92
    calcium channel, 62/91
    clonal culture, 62/94
    cloned satellite cell, 62/88
    contractile activity, 62/96
    creatine kinase, 62/87
    creatine kinase MM, 62/89
    debrancher deficiency, 62/99
    Duchenne muscular dystrophy, 40/191, 375, 62/100
    electrical property, 62/96
    enzyme histochemistry, 62/94
    epidermal growth factor, 62/88
    ferritin H, 62/92
    fibroblast, 62/87
    fibroblast growth factor, 62/88
    genetical abnormal muscle, 62/85
    glucocorticoids, 62/90
    hereditary cylindric spirals myopathy, 62/108
    HLA antigen, 62/92
    inclusion body myositis, 62/104
    infantile nemaline myopathy, 62/99, 107
    innervated cultured, 62/92
    innervation method, 62/92
    insulin, 62/88
    insulin growth factor 1, 62/90
    γ-interferon, 62/92
    methodologic consideration, 62/86
    mitochondrial myopathy, 62/100
    monolayer culture, 62/93
    motor end plate, 62/92, 97
    muscle fiber phenotype, 62/96
    muscle phenotype, 62/88
        muscle specific isoenzyme, 62/94
        myoblast fusion, 62/91
        myogenesis, 62/88
        myopathy, 62/85
        myophosphorylase deficiency, 40/189, 62/109
        myotonic dystrophy, 62/105, 222
        myotubule, 62/93
        normal muscle, 62/85
        oculopharyngeal muscular dystrophy, 62/102
        phosphoglycerate mutase, 62/94
        phosphorylase, 62/94, 109
        phosphorylase deficiency, 62/87
        postsynaptic membrane organization, 62/98
        primary monolayer culture, 62/87
        ribosomal S6 protein, 62/90
        Tomé-Fardeau body, 62/103
        ultrastructure, 62/94
Muscle wasting, see Muscular atrophy
Muscle weakness
    ACTH induced myopathy, 62/537
    adenosine triphosphatase 6 mutation syndrome,
        62/510
    autosomal recessive generalized myotonia, 62/266
    complex III deficiency, 62/504
    dermatomyositis, 62/371
    extraocular, see Extraocular muscle weakness
    multiple mitochondrial DNA deletion, 62/510
    Nelson syndrome, 62/537
    proximal, see Proximal muscle weakness
    sporadic progressive external ophthalmoplegia
        with ragged red fiber, 62/508
    thyrotoxic myopathy, 62/528
    ubidecarenone, 62/504
Muscular atrophy
    see also Muscle fiber type II atrophy
    autosomal recessive generalized myotonia, 62/267
    childhood myoglobinuria, 62/561
    debrancher deficiency, 62/482
    distal spinal, see Spinal muscular atrophy
    hyperparathyroid myopathy, 62/540
    infantile spinal, see Infantile spinal muscular
        atrophy
    juvenile spinal, see Wohlfart-Kugelberg-Welander
        disease
    scapulohumeral spinal, see Scapulohumeral spinal
        muscular atrophy
    scapuloperoneal, see Scapuloperoneal spinal
        muscular atrophy
    spinal, see Spinal muscular atrophy
    steroid myopathy, 62/535
Muscular dystrophy
    see also Limb girdle syndrome
    Becker, see Becker muscular dystrophy
    distal, see Distal muscular dystrophy
    Duchenne, see Duchenne muscular dystrophy
    Emery-Dreifuss, see Emery-Dreifuss muscular
        dystrophy
    facioscapulohumeral, see Facioscapulohumeral
        muscular dystrophy
    inflammatory scapuloperoneal, see Inflammatory
        scapuloperoneal muscular dystrophy

late adult Nevin type, *see* Late adult muscular
   dystrophy (Nevin)
oculopharyngeal, *see* Oculopharyngeal muscular
   dystrophy
pelvifemoral Leyden-Möbius type, *see*
   Pelvifemoral muscular dystrophy (Leyden-
   Möbius)
peroneal, *see* Charcot-Marie-Tooth disease
pseudohypertrophic, *see* Duchenne muscular
   dystrophy
scapulohumeral, *see* Scapulohumeral muscular
   dystrophy
scapulohumeral Erb type, *see* Scapulohumeral
   muscular dystrophy (Erb)
scapuloperoneal, *see* Scapuloperoneal muscular
   dystrophy
X-linked, *see* X-linked muscular dystrophy
Myalgia
   adenosine monophosphate deaminase deficiency,
      62/511
   adrenal insufficiency myopathy, 62/536
   carnitine palmitoyltransferase deficiency, 43/176,
      62/494
   congenital myopathy, 62/335
   corticosteroid withdrawal myopathy, 62/536
   dermatomyositis, 62/371
   exertional, *see* Exertional myalgia
   glutaric aciduria type II, 62/495
   long chain acyl coenzyme A dehydrogenase
      deficiency, 62/495
   multiple acyl coenzyme A dehydrogenase
      deficiency, 62/495
   myophosphorylase deficiency, 62/485
   thyrotoxic myopathy, 62/528
   toxic myopathy, 62/596
Myasthenia
   congenital myopathy, 62/335
   familial infantile, *see* Familial infantile myasthenia
   ocular, *see* Ocular myasthenia gravis
Myasthenia gravis
   see also Myasthenic syndrome
   acetylcholine quantum release, 62/392
   acetylcholine receptor, 40/129, 41/106, 114, 125,
      62/391
   acetylcholine receptor antibody, 41/114, 120, 129,
      368, 62/408
   acetylcholine receptor antibody test, 62/414
   acetylcholine receptor epitope, 62/407
   adverse drug reaction, 62/420
   age at onset, 62/399
   anemia, 62/402
   anticholinesterase, 62/412
   anticholinesterase agent, 41/131, 62/417
   antigenic modulation, 62/411
   Asian race, 62/401
   atypical case, 62/79
   autoimmune etiology, 62/398
   autoimmune experiment, 62/402
   azathioprine, 41/96, 133, 136, 62/418
   B-lymphocyte response, 62/407
   bacterial infection, 62/404

blocking antibody, 62/409
bone marrow transplantation, 62/402
botulism, 62/415
α-bungarotoxin, 41/113, 125, 62/409
chronic ulcerative colitis, 62/402
clinical course, 62/400
clinical features, 40/297, 303, 306, 308, 317, 323,
   330, 335, 41/98, 288, 364-366, 62/399
complement effect, 62/409
Compston classification, 62/401
congenital myasthenic syndrome, 62/415
corticosteroids, 41/93, 132, 62/417
crisis, 62/417
curare, 62/413
cyclophosphamide, 41/96, 136, 62/418
cyclosporin, 62/418
decremental EMG response, 62/413
definition, 62/396
diagnosis, 62/412
differential diagnosis, 62/415
Eaton-Lambert myasthenic syndrome, 62/391, 402,
   415, 421
edrophonium, 62/412
edrophonium test, 40/298, 303, 321, 41/126, 240,
   62/413
elapid snake venom, 62/391
EMG, 1/639, 644, 62/398, 413
endocrine myopathy, 41/240, 62/528
epidemiology, 41/101, 62/399
experimental treatment, 62/420
facioscapulohumeral syndrome, 62/168
giant cell polymyositis, 62/402
Hashimoto thyroiditis, 62/534
high dose intravenous Ig, 62/419
history, 41/95, 62/396
HLA association, 62/404
HLA haplotype classification, 62/401
hyperthyroidism, 41/101, 43/157, 62/402, 534
hypothyroidism, 62/534
immunocytochemistry, 62/415
incidence, 62/399
ion channel effect, 62/409
lymphoid irradiation, 62/421
miniature motor end plate potential, 41/129, 62/414
mitochondrial myopathy, 62/415
neostigmine, 41/96, 131, 62/413, 417
neurasthenia, 62/415
neuroblastoma, 62/415
neuromuscular junction disease, 62/392
oculopharyngeal muscular dystrophy, 62/415
Osserman classification, 62/400
pathogenesis, 41/111-126, 367, 62/402
pemphigus, 62/402
penicillamine, 41/126, 62/405
plasmapheresis, 41/96, 136, 62/420
polymyositis, 43/157, 62/373, 402
polyneuropathy, 62/413
progressive external ophthalmoplegia, 22/181, 188,
   204, 62/415
progressive external ophthalmoplegia
   classification, 62/290

provocative test, 41/127, 62/413
pyridostigmine bromide, 41/131, 62/417
repetitive nerve stimulation, 62/79, 394
rheumatoid arthritis, 41/240, 62/402
saturating disc model, 62/394
sex ratio, 43/157, 62/399
single fiber electromyography, 41/128, 62/77, 79,
    414
single nerve impulse, 62/393
Sjögren syndrome, 39/430, 41/240, 62/402
striated antibody, 62/415
systemic lupus erythematosus, 62/402
T-lymphocyte response, 62/407
thymectomy, 41/134, 62/419
thymoma, 41/361, 43/157, 62/402
thymopoietin, 62/406
thymus gland, 62/405
transient neonatal, *see* Transient neonatal
    myasthenia gravis
transmitter release, 62/393
treatment, 41/80, 130-136, 367, 62/416
viral infection, 62/404
Myasthenia levis, *see* Ocular myasthenia gravis
Myasthenic myopathic syndrome, *see* Eaton-Lambert
    myasthenic syndrome
Myasthenic syndrome
    see also Eaton-Lambert myasthenic syndrome *and*
        Myasthenia gravis
    acetylcholine quantum release, 62/392
    acetylcholine receptor, 62/391
    congenital, *see* Congenital myasthenic syndrome
    Eaton-Lambert myasthenic syndrome, 62/391
    elapid snake venom, 62/391
    EMG, 62/78
    motor end plate organization, 62/392
    repetitive nerve stimulation, 62/394
    saturating disc model, 62/394
    single nerve impulse, 62/393
    transmitter release, 62/393
Mycoplasma
    acquired myoglobinuria, 62/577
Myelin
    demyelinating disease, *see* Demyelination
Myelin breakdown, *see* Demyelination
Myelin decomposition, *see* Demyelination
Myelin loss, *see* Demyelination
Myoadenylate deaminase, *see* Adenosine
    monophosphate deaminase
Myoadenylate deaminase deficiency, *see* Adenosine
    monophosphate deaminase deficiency
Myocardial infarction
    myoglobinuria, 62/556
Myoclonus, 38/575-588
    MERRF syndrome, 62/509
Myoedema
    hypothyroid myopathy, 62/532
Myofibril, *see* Muscle fiber
Myofibrillar lysis myopathy
    congenital myopathy, 62/332
    enzyme histochemistry, 62/339
    histopathology, 62/339

protein synthesis, 62/339
Myoglobinuria
    see also Rhabdomyolysis
    aconitate hydratase, 62/558
    aconitate hydratase deficiency, 62/503
    acquired, *see* Acquired myoglobinuria
    adenosine monophosphate deaminase, 62/558
    adenosine monophosphate deaminase deficiency,
        62/512
    assay sensitivity, 62/555
    asymptomatic creatine kinasemia, 62/554
    Becker muscular dystrophy, 62/556, 558
    biochemical defect, 62/560
    carnitine palmitoyltransferase, 62/558
    carnitine palmitoyltransferase deficiency, 41/204,
        265, 428, 43/176, 62/494
    cause, 62/557
    central core myopathy, 62/558
    child, *see* Childhood myoglobinuria
    childhood *vs* adult onset, 62/565
    chondrodystrophic myotonia, 62/558
    classification, 62/558
    clinical diagnosis, 62/554
    coma crush, 62/556
    complex II, 62/558
    complex II deficiency, 62/504
    concomitant, 62/556
    consequence, 62/556
    crush, 41/270, 62/557
    definition, 62/553
    dermatomyositis, 62/556
    diagnostic criteria, 62/555
    differential diagnosis, 40/338-340, 41/167, 62/560
    drug, 41/271, 62/557
    Duchenne muscular dystrophy, 62/555, 558
    exercise, *see* Exercise myoglobinuria
    exertion, 41/266, 62/557, 560
    facioscapulohumeral muscular dystrophy, 62/556
    familial malignant hyperthermia, 62/558
    familial recurrent, 62/557
    fatty acid oxidation, 62/558
    glucose-6-phosphate dehydrogenase, 62/558
    glycogen storage disease, 41/265, 62/558
    heme induced nephropathy, 62/556
    hereditary cause, 62/558
    hereditary disease, 62/559
    history, 62/553
    hypercalcemia, 62/556
    hyperphosphatemia, 62/556
    hypocalcemia, 62/556
    incomplete dystrophin deficiency, 62/135
    infection, 41/275, 62/557
    ischemic, 62/557
    King-Denborough syndrome, 62/558
    lactate dehydrogenase, 62/558
    lactate dehydrogenase deficiency, 62/490
    long chain acyl coenzyme A dehydrogenase, 62/558
    long chain acyl coenzyme A dehydrogenase
        deficiency, 62/495
    long chain fatty acid oxidation, 62/558
    malignant hyperthermia, 41/268, 62/567

mannitol, 62/556
metabolic disorder, 62/557
motoneuron disease, 62/556
multiple mitochondrial DNA deletion, 62/510
muscle fasciotomy, 62/556
muscle necrosis, 62/554
muscle pathology, 62/556
myocardial infarction, 62/556
myoglobin identification, 62/554
myophosphorylase deficiency, 27/233, 41/186, 265,
    43/181, 62/485, 558
myotonia congenita, 62/558
myotonic dystrophy, 62/556, 558
nomenclature, 62/553
parathyroid hormone, 62/556
pathogenesis, 62/556
pentose phosphate pathway, 62/558
phosphofructokinase, 62/558
phosphofructokinase deficiency, 27/235, 41/190,
    265, 43/183, 62/487
phosphoglycerate kinase, 62/558
phosphoglycerate kinase deficiency, 62/489
phosphoglycerate mutase, 62/558
phosphorylase kinase, 62/558
plasmapheresis, 62/556
polymyositis, 41/58, 60, 274, 62/556
postexercise, 62/556
precipitating factor, 62/559
progressive muscle disease, 62/557
recurrent sporadic, 62/558
renal dialysis, 62/556
rhabdomyolysis, 62/553
sex ratio, 62/559
short chain 3-hydroxyacyl coenzyme A
    dehydrogenase, 62/558
short chain 3-hydroxyacyl coenzyme A
    dehydrogenase deficiency, 62/495
symptom, 62/554
thermoregulation, 62/557
thyrotoxic myopathy, 62/529
toxic shock syndrome, 62/574
toxin, 41/271, 62/557
treatment, 62/556
ubidecarenone, 62/504
ubiquinone deficiency, 62/558
Myokymia
    myotonia, 43/160, 62/273
    myotonic syndrome, 62/273
    thyrotoxic myopathy, 62/528
    toxic myopathy, 62/596
Myokymic discharge
    definition, 62/263
    Isaacs syndrome, 62/263
    myotonia, 62/263
    myotonic syndrome, 62/263
Myoneural junction, see Motor end plate
Myopathy
    see also Neuromuscular disease
    abnormal myomuscular junction, see Abnormal
        myomuscular junction myopathy
    acromegalic, see Acromegalic myopathy

ACTH induced, see ACTH induced myopathy
adrenal insufficiency, see Adrenal insufficiency
    myopathy
adult acid maltase deficiency, 27/229, 62/480
adult dominant myotubular, see Adult dominant
    myotubular myopathy
alcoholic, see Alcoholic polyneuropathy
amiodarone, see Amiodarone myopathy
atrophy, 62/13
atrophy cause, 62/13
axonal stimulation, 62/59
carnitine deficiency, see Carnitine deficiency
    myopathy
carnitine palmitoyltransferase deficiency, 62/494
centronuclear, see Myotubular myopathy
chloroquine, see Chloroquine myopathy
chronic renal failure, see Chronic renal failure
    myopathy
colchicine, see Colchicine myopathy
combined complex I-V deficiency, 62/506
complex repetitive discharge, 62/62, 67
complex I deficiency, 62/503
complex II deficiency, 62/504
complex III deficiency, 62/504
complex IV deficiency, 62/505
complex V deficiency, 62/506
congenital, see Congenital myopathy
conventional EMG, 62/49
corticosteroid withdrawal, see Corticosteroid
    withdrawal myopathy
cricopharyngeal, see Cricopharyngeal myotomy
cytoplasmic inclusion body, see Cytoplasmic
    inclusion body myopathy
denervation atrophy, 62/13
diabetic, see Diabetic proximal neuropathy
diagnostic electromyography value, 62/71
diagnostic macroelectromyography value, 62/73
diagnostic scanning electromyography value, 62/74
diagnostic single fiber electromyography value,
    62/72
distal, see Distal myopathy
EMG, 62/49
EMG method, 62/49
endocrine, see Endocrine myopathy
extradischarge, 62/50, 67
fasciculation, 40/330, 62/50
fasciculation discharge, 62/50
fatigue index, 62/54
fibrillation potential, 62/50, 61, 67
fingerprint body, see Fingerprint body myopathy
forked fiber, 62/11
granulofilamentous, see Granulofilamentous
    myopathy
granulovacuolar lobular, see Granulovacuolar
    lobular myopathy
hereditary cylindric spirals, see Hereditary
    cylindric spirals myopathy
honeycomb, see Honeycomb myopathy
hyperaldosteronism, see Hyperaldosteronism
    myopathy
hyperparathyroid, see Hyperparathyroid

myopathy
hyperthyroid, *see* Thyrotoxic myopathy
hypoparathyroid, *see* Hypoparathyroid myopathy
hypothyroid, *see* Hypothyroid myopathy
infantile distal, *see* Infantile distal myopathy
infantile nemaline, *see* Infantile nemaline
    myopathy
inflammatory, *see* Inflammatory myopathy
insertional activity, 62/50
intramuscular stimulation, 62/58, 74
lipid storage, *see* Lipid storage myopathy
long chain acyl coenzyme A dehydrogenase
    deficiency, 62/495
macroelectromyography, 62/55
macromotor unit potential, 62/55
Mallory body, *see* Mallory body myopathy
McLeod, *see* McLeod myopathy
megaconial, *see* Mitochondrial myopathy
MERRF syndrome, 62/509
metabolic, *see* Metabolic myopathy
miniature motor end plate potential, 62/50
minimal change, *see* Minimal change myopathy
mitochondria jagged Z line, *see* Mitochondria
    jagged Z line myopathy
mitochondrial, *see* Mitochondrial myopathy
mitochondrial DNA depletion, 62/511
motor end plate noise, 62/50
motor unit counting, 62/76
motor unit counting technique, 62/60
motor unit potential, 62/49, 67
multicore, *see* Multicore myopathy
muscle fiber density, 62/54
muscle fiber hypercontraction, 62/15
muscle fiber hypertrophy, 62/15
muscle fiber stimulation, 62/59
muscle fiber type, 62/2
muscle pathology, 62/1
muscle tissue culture, 62/85
myasthenic, *see* Eaton-Lambert myasthenic
    syndrome
myofibrillar lysis, *see* Myofibrillar lysis myopathy
myogenic jitter, 62/59
myotonic discharge, 62/50, 61, 67
myotubular, *see* Myotubular myopathy
nemaline, *see* Nemaline myopathy
neuromuscular jitter, 62/54
nonprimary inflammatory, *see* Nonprimary
    inflammatory myopathy
nucleodegenerative, *see* Nucleodegenerative
    myopathy
ocular, *see* Ocular myopathy
partial invasion, 62/9
perhexiline, *see* Perhexiline myopathy
pheochromocytoma, *see* Pheochromocytoma
    myopathy
phosphorylase kinase deficiency, 62/487
pituitary gigantism, *see* Pituitary gigantism
    myopathy
pleoconial, *see* Mitochondrial myopathy
positive sharp wave, 62/50, 67
quadriceps, *see* Quadriceps myopathy

quail, *see* Quail myopathy
reducing body, *see* Reducing body myopathy
regeneration, 62/9
rimmed vacuole distal, *see* Rimmed vacuole distal
    myopathy
rod body, *see* Nemaline myopathy
sarcoplasmic body, *see* Sarcoplasmic body
    myopathy
sarcotubular, *see* Sarcotubular myopathy
scanning electromyography, 62/57
segmental necrosis, 62/13
single fiber electromyography, 62/49, 53
small caliber fiber, 62/13
spheroid body, *see* Spheroid body myopathy
split fiber, 62/16
spontaneous activity, 62/50, 67
steroid, *see* Steroid myopathy
storage disease, 62/16
tardive, *see* Tardive myopathy
thyrotoxic, *see* Thyrotoxic myopathy
toxic, *see* Toxic myopathy
trilaminar, *see* Trilaminar myopathy
tubular aggregate, *see* Tubular aggregate
    myopathy
tubulomembranous inclusion, *see*
    Tubulomembranous inclusion myopathy
uniform muscle fiber type I, *see* Uniform muscle
    fiber type I myopathy
velocity recovery function, 62/54
vincristine, *see* Vincristine myopathy
X-linked myotubular, *see* X-linked myotubular
    myopathy
X-linked neutropenic cardioskeletal, *see* X-linked
    neutropenic cardioskeletal myopathy
Z band, 62/7
Z band plaque, *see* Z band plaque myopathy
zebra body, *see* Zebra body myopathy
zidovudine, *see* Zidovudine myopathy
Myophosphorylase deficiency
biochemistry, 41/188, 62/485
childhood myoglobinuria, 62/564
chromosome 11q13, 62/485
clinical features, 27/233, 40/317, 41/185, 62/485
creatine kinase, 62/485
exercise intolerance, 62/485
exercise myoglobinuria, 62/485
fatigue, 62/485
glycogen storage disease, 40/4, 53, 189, 257, 317,
    327, 339, 342, 438, 62/485
histopathology, 41/187, 62/485
inheritance, 62/485
ischemic exercise, 62/485
limb girdle syndrome, 62/190
metabolic myopathy, 62/485
molecular genetics, 62/485
muscle biopsy, 41/187, 62/485
muscle contraction, 62/485
muscle cramp, 62/485
muscle fiber feature, 62/17
muscle necrosis, 62/485
muscle tissue culture, 40/189, 62/109

myalgia, 62/485
myoglobinuria, 27/233, 41/186, 265, 43/181, 62/485, 558
   phosphorylase isoenzyme, 62/486
   phosphorylase limit dextrin, 62/485
   physiopathology, 62/486
   renal failure, 27/233, 62/485
   rimmed vacuole distal myopathy, 62/485
   treatment, 41/190, 62/486
Myosclerosis
   hereditary, *see* Heredofamilial myosclerosis
   rigid spine syndrome, 62/173
Myositis
   see also Inflammatory myopathy
   age, 62/370
   autoimmune disease, 62/369
   B-lymphocyte, 62/42
   cause, 62/369
   CD8+ cytotoxic T-lymphocyte, 62/42
   clinical course, 62/369
   clinical features, 62/369
   complement mediated microangiopathy, 62/369
   fall, 62/369
   helper inducer lymphocyte, 62/42
   inclusion body, *see* Inclusion body myositis
   muscle involvement distribution, 62/370
   orbital, *see* Orbital myositis
   postpartum period, 62/370
   pregnancy, 62/370
   primary inflammatory, *see* Primary inflammatory myositis
   rhabdomyolysis, 62/370
   sex ratio, 62/370
   T-lymphocyte mediated myocytotoxicity, 62/369
   tropical, *see* Tropical polymyositis
   viral, *see* Viral myositis
   viral disease, 62/369
Myotonia
   ambient temperature, 62/262
   calcium channel blocking agent, 62/276
   chloride channel defect, 62/277
   chloride conductance, 40/553, 62/276
   chondrodystrophic, *see* Chondrodystrophic myotonia
   chromosome 17q23.1-25.3, 62/279
   classification, 41/165, 62/263
   clomipramine, 62/276
   computer model, 62/277
   congenital myopathy, 62/335
   cromakalim, 62/276
   definition, 62/261
   differential diagnosis, 62/273
   diltiazem, 62/276
   discharge duration, 62/262
   drug induced, *see* Drug induced myotonia
   fenoterol, 62/276
   hypokalemic periodic paralysis, 43/170, 62/458
   imipramine, 62/276
   Isaacs syndrome, 43/160, 62/263, 273
   mexiletine, 62/275
   muscle contraction, 62/262, 273

muscle cramp, 40/327, 533, 62/261, 273
   myokymia, 43/160, 62/273
   myokymic discharge, 62/263
   myotonia congenita, 62/262
   myotonic discharge, 62/262
   neuromyotonic discharge, 62/263
   nifedipine, 62/276
   orciprenaline, 62/276
   painful dominant, *see* Painful dominant myotonia
   paradoxical, *see* Paradoxical myotonia
   paramyotonia, 62/262
   paramyotonia congenita, 40/542, 62/224
   pathophysiology, 40/549-557, 62/276
   percussion, *see* Percussion myotonia
   phenytoin, 62/275
   pindolol, 62/276
   procainamide, 62/275
   pseudo, *see* Isaacs syndrome
   quinine, 40/548, 558, 41/297, 62/275
   sodium channel, 62/278
   stiff-man syndrome, 62/274
   T tubule system, 62/277
   taurine, 62/276
   tocainide, 62/275
   toxic myopathy, 62/597
   treatment, 62/274
   warm up phenomenon, 62/261
Myotonia atrophica, *see* Myotonic dystrophy
Myotonia congenita
   ambient temperature, 62/264
   clinical features, 40/324, 328, 539, 41/298, 62/264
   EMG, 40/539, 62/70
   emotion, 62/264
   incidence, 62/266
   malignant hyperthermia, 62/570
   muscle hypertrophy, 40/539, 41/298, 43/161, 62/266
   myoglobinuria, 62/558
   myotonia, 62/262
   myotonic dystrophy, 62/240
   myotonic syndrome, 62/262
   pregnancy, 62/264
   progressive external ophthalmoplegia, 62/305
   sex ratio, 62/266
Myotonia fluctuans
   dominant inheritance, 62/266
   myotonic syndrome, 62/263
   pain, 62/266
Myotonic discharge
   features, 62/61
   motor unit, 62/50, 61, 67
   myopathy, 62/50, 61, 67
   myotonia, 62/262
   myotonic syndrome, 62/262
Myotonic dystrophy, 40/485-524
   abortion, 62/228
   adult form, 62/216
   allelic expansion, 62/212
   alopecia, 62/238
   alveolar hypoventilation, 40/517, 62/238
   anesthesia complication, 62/244
   anticipation, 40/488, 62/214

apamin, 62/222, 225
autosomal recessive generalized myotonia, 62/240
cardiac conduction, 62/217
cataract, 40/512, 41/486, 43/153, 62/217, 233
Charcot-Marie-Tooth disease, 62/234
Charcot-Marie-Tooth like syndrome, 62/234
chloride channel, 62/224
chromosome 19q13.3, 62/210
classification, 40/282, 537, 62/216
CNS, 40/517, 62/235
cognitive impairment, 62/235
congenital form, 40/283, 490, 62/216, 218
congestive heart failure, 62/227
course, 62/217
demyelination, 62/237
differential diagnosis, 40/422, 472, 481, 520, 41/13,
    245, 62/239
distal myopathy, 62/240
drug treatment, 62/245
ECG, 62/226
echoCG, 62/227
EMG, 40/519, 62/70
epithelioma, 62/238
ethnic distribution, 62/214
exercise, 62/249
facioscapulohumeral muscular dystrophy, 62/240
gastrointestinal symptom, 62/233
gene instability, 62/213
gene lesion, 62/211
gene product, 62/213
genetics, 41/421, 486, 62/209
gonadal abnormality, 62/229
growth hormone, 62/230, 249
hearing loss, 62/234
heart disease, 40/507, 522, 62/226
human menopausal gonadotropin secretion, 62/229
hyperinsulinism, 62/230
hypersomnia, 40/517, 62/217, 235
hypoventilation, 62/235
hypoxemia, 62/235
immunologic change, 62/239
insulin resistance, 62/230
laboratory data, 62/247
limb girdle syndrome, 62/188, 240
luteinizing hormone secretion, 62/229
management, 62/240-244
maternal complication, 62/229
mental depression, 62/235
mexiletine, 62/241
muscle biopsy, 62/225
muscle histopathology, 62/219
muscle relaxation, 62/224
muscle strength scoring, 62/248
muscle tissue culture, 62/105, 222
myoglobinuria, 62/556, 558
myotonia congenita, 62/240
myotonia physiology, 62/224
myotubular myopathy, 62/240
nemaline myopathy, 62/240
neurofibrillary tangle, 62/237
neuropathy, 62/234

nuclear magnetic resonance, 62/237
obstetric complication, 62/229
ocular symptom, 62/233
oculopharyngeal muscular dystrophy, 62/240
orthosis, 62/241
personality change, 62/235
phenytoin, 40/522, 62/241
pilomatricoma, 43/153, 62/239
pituitary hormone release, 62/229
postoperative care, 62/244
potassium channel, 62/225
pregnancy, 62/228
prenatal diagnosis, 43/153, 62/215
prevalence, 43/153, 62/214
progressive external ophthalmoplegia, 22/187,
    62/302
progressive external ophthalmoplegia
    classification, 62/290
respiratory system, 62/238
sleep, 62/235
sleep apnea, 62/217, 235
sodium channel, 62/224
testosterone, 62/248
testosterone secretion, 62/229
thalamic inclusion, 62/237
thyrotropic hormone secretion, 62/230
warm up phenomenon, 62/226
white matter lesion, 62/237
Myotonic hyperkalemic periodic paralysis
    ambient temperature, 62/267
    attack symptom, 62/462
    chromosome 17, 62/224, 267
    Chvostek sign, 62/462
    diplopia, 62/462
    features, 62/267
    lid lag, 62/462
    serum potassium, 62/267
    sodium channel, 62/224
    tongue myotonia, 62/462
    visual impairment, 62/462
Myotonic muscular dystrophy, see Myotonic
    dystrophy
Myotonic syndrome, 40/533-563
    acquired type, 62/264
    ambient temperature, 62/262
    animal model, 62/263
    aromatic carboxylic acid intoxication, 62/264
    autosomal recessive generalized myotonia, 62/263
    azacosterol intoxication, 62/264
    chondrodystrophic myotonia, 62/262
    classification, 40/537, 62/263
    clofibrate intoxication, 62/264
    differential diagnosis, 40/545, 62/273
    discharge duration, 62/262
    dominant inheritance, 62/263
    drug induced, see Drug induced myotonic
        syndrome
    hereditary, see Hereditary myotonic syndrome
    Isaacs syndrome, 62/263, 273
    mechanism, 62/261
    muscle biopsy, 62/274

muscle contraction, 62/262, 273
muscle cramp, 62/261, 273
myokymia, 62/273
myokymic discharge, 62/263
myotonia congenita, 62/262
myotonia fluctuans, 62/263
myotonic discharge, 62/262
neuromyotonic discharge, 62/263
painful dominant myotonia, 62/263
paradoxical myotonia, 62/261
paramyotonia, 62/262
paramyotonia congenita, 40/542, 62/263
sarcolemma, 62/261
stiff-man syndrome, 62/274
triperinol intoxication, 62/264
warm up phenomenon, 62/261
Myotubular myopathy
  adult dominant, *see* Adult dominant myotubular
    myopathy
  age at onset, 62/348
  chromosome Xq28, 62/348
  clinical form, 62/30
  congenital myopathy, 41/11-14, 62/332
  EMG, 62/69
  facioscapulohumeral muscular dystrophy, 62/168
  fatality rate, 62/348
  histochemistry, 40/46, 62/349
  myotonic dystrophy, 62/240
  progressive external ophthalmoplegia, 22/179,
    41/11, 62/297
  progressive external ophthalmoplegia
    classification, 62/290
  ptosis, 43/113, 62/297
  transmission type, 62/348
Myotubule
  muscle tissue culture, 62/93

NADH-tetrazolium reductase, *see* Tetrazolium
    reductase
Nebulin
  muscle fiber, 62/26
Necrosis
  muscle, *see* Muscle necrosis
  neuromuscular disease, 62/6
Necrotizing encephalomyelopathy
  subacute, *see* Subacute necrotizing
    encephalomyelopathy
Neisseria
  childhood myoglobinuria, 62/564
Nelson syndrome
  see also Cushing syndrome
  ACTH induced myopathy, 62/537
  EMG, 62/537
  fatigue, 62/537
  hyperpigmentation, 62/537
  muscle biopsy, 62/537
  muscle weakness, 62/537
Nemaline myopathy
  antidesmin antibody, 62/344
  cardiomyopathy, 62/341
  clinical course, 62/342

congenital myopathy, 41/8-11, 62/332
  cricopharyngeal dysphagia, 62/342
  facial weakness, 62/334
  facioscapulohumeral syndrome, 62/168
  fatal form, 62/341
  genetics, 41/432, 62/342
  histopathology, 62/342
  inclusion body, 62/342
  myogranule, 62/341
  myotonic dystrophy, 62/240
  ophthalmoplegia, 62/342
  temporomandibular ankylosis, 62/342
Neoplasm
  Eaton-Lambert myasthenic syndrome, 62/421
Neostigmine
  myasthenia gravis, 41/96, 131, 62/413, 417
Nephropathy, *see* Renal disease
Nerve
  hypertrophic, *see* Hypertrophic nerve
Nerve atrophy
  optic, *see* Optic atrophy
Nerve biopsy
  scapuloperoneal spinal muscular atrophy, 62/170
Nerve fiber
  muscle, *see* Muscle fiber
Nervous system
  central, *see* CNS
Neural cell adhesion molecule
  muscle fiber, 62/33
Neurasthenia
  myasthenia gravis, 62/415
Neuro-Behçet syndrome, *see* Behçet syndrome
Neuroblastoma
  myasthenia gravis, 62/415
Neurofibrillary tangle, 22/366-415
  myotonic dystrophy, 62/237
Neurogenic bladder
  adult polyglucosan body disease, 62/484
Neurogenic muscular atrophy
  hereditary, *see* Charcot-Marie-Tooth disease
Neuroleptic malignant syndrome
  acquired myoglobinuria, 62/572
  malignant hyperthermia, 62/571
  symptom, 62/573
Neuromuscular disease
  see also Myopathy
  atrophy, 62/13
  atrophy cause, 62/13
  denervation atrophy, 62/13
  forked fiber, 62/11
  muscle fiber hypercontraction, 62/15
  muscle fiber hypertrophy, 62/15
  muscle fiber type, 62/2
  muscle pathology, 62/1
  necrosis, 62/6
  oculopharyngeal, *see* Oculopharyngeal muscular
    dystrophy
  partial invasion, 62/9
  phagocytosis, 62/6
  regeneration, 62/9
  segmental necrosis, 62/13

small caliber fiber, 62/13
Z band, 62/7
Neuromuscular junction, *see* Motor end plate
Neuromuscular junction disease
abnormal acetylcholine-acetylcholine receptor
interaction syndrome, 62/392
autoimmune, 62/392
classification, 62/392
congenital acetylcholine receptor deficiency/short
channel opentime syndrome, 62/392
congenital acetylcholine receptor deficiency/
synaptic cleft syndrome, 62/392
congenital ε-acetylcholine receptor subunit
mutation syndrome, 62/392
congenital end plate acetylcholinesterase
deficiency, 62/392
congenital familial limb girdle myasthenia, 62/392
congenital high conductance fast channel
syndrome, 62/392
congenital partial characterized, 62/392
congenital slow channel syndrome, 62/392
congenital synaptic vesical paucity syndrome,
62/392
Eaton-Lambert myasthenic syndrome, 62/392
familial infantile myasthenia, 62/392
myasthenia gravis, 62/392
Neuromyotonia, *see* Isaacs syndrome
Neuromyotonic discharge
definition, 62/263
Isaacs syndrome, 62/263
myotonia, 62/263
myotonic syndrome, 62/263
Neuronal ceroid lipofuscinosis
see also Batten disease *and* Progressive myoclonus
epilepsy
adult, *see* Kufs disease
Neuronal system degeneration
multiple, *see* Multiple neuronal system
degeneration
Neuropathology
progressive external ophthalmoplegia, 22/182,
62/289
Neuropathy
see also Denervation *and* Polyneuropathy
acute inflammatory demyelinating, *see* Guillain-
Barré syndrome
axonal, *see* Axonal neuropathy
chloroquine myopathy, 62/606
complex V deficiency, 62/506
cranial, *see* Cranial neuropathy
diabetic proximal, *see* Diabetic proximal
neuropathy
hyperparathyroid myopathy, 62/540
hypertrophic interstitial, *see* Hypertrophic
interstitial neuropathy
Kearns-Sayre-Daroff-Shy syndrome, 62/313
Leber hereditary optic neuropathy, 62/509
MERRF syndrome, 62/509
multiple mitochondrial DNA deletion, 62/510
myotonic dystrophy, 62/234
peripheral, *see* Polyneuropathy

poly, *see* Polyneuropathy
progressive external ophthalmoplegia, 43/136,
62/313
sensory, *see* Sensory neuropathy
thyrotoxic myopathy, 62/528
Neutropenia
Emery-Dreifuss muscular dystrophy, 62/156
Nevin late adult muscular dystrophy, *see* Late adult
muscular dystrophy (Nevin)
Niacin, *see* Vitamin PP
Nicotinamide, *see* Vitamin PP
Nicotinamide adenine dinucleotide tetrazolium
reductase, *see* Tetrazolium reductase
Nicotinic acid, *see* Vitamin PP
Nifedipine
myotonia, 62/276
toxic myopathy, 62/597
NMR, *see* Nuclear magnetic resonance
Nonprimary inflammatory myopathy
chloroquine, 62/372
cimetidine, 62/372
emetine, 62/372
ipecac, 62/372
mevinolin, 62/372
Nonsystematic system degeneration, *see* Progressive
supranuclear palsy
Nontropical sprue, *see* Celiac disease
Normokalemic periodic paralysis
age, 62/463
familial periodic paralysis, 62/457
inheritance, 62/463
symptom, 62/463
NOSPECS system
thyroid associated ophthalmopathy, 62/530
Notechis II-5
toxic myopathy, 62/610
Notexin
toxic myopathy, 62/610
Nuclear magnetic resonance
dermatomyositis, 62/379
inclusion body myositis, 62/379
myotonic dystrophy, 62/237
orbital myositis, 62/299
polymyositis, 62/379
thyroid associated ophthalmopathy, 62/531
tropical polymyositis, 62/373
Nucleodegenerative myopathy
congenital myopathy, 62/332
intranuclear inclusion body, 62/351
Marinesco-Sjögren syndrome, 62/351
membrane whorl, 62/351
Nucleus
muscle fiber type, 62/4
Nystagmus
multiple mitochondrial DNA deletion, 62/510

Obstructions
intestinal pseudo, *see* Intestinal pseudo-obstruction
Ocular myasthenia gravis
progressive external ophthalmoplegia, 62/302
Ocular myopathy

see also Progressive external ophthalmoplegia
progressive external ophthalmoplegia, 22/186, 41/215, 62/288, 291
progressive external ophthalmoplegia classification, 62/290
Oculocerebrorenal syndrome, *see* Lowe syndrome
Oculocraniosomatic neuromuscular disease, *see* Progressive external ophthalmoplegia
Oculofacial dystonia, *see* Progressive supranuclear palsy
Oculofacial palsy, *see* Möbius syndrome
Oculofaciocervical dystonia, *see* Progressive supranuclear palsy
Oculopharyngeal muscular dystrophy
    clinical features, 40/302, 62/296
    definition, 62/295
    8-10 nm filament, 62/351
    genetics, 62/296
    histopathology, 62/296
    inclusion body myositis, 62/297
    muscle tissue culture, 62/102
    myasthenia gravis, 62/415
    myotonic dystrophy, 62/240
    nuclear inclusion, 62/296
    progressive external ophthalmoplegia, 22/179, 187, 206, 40/302, 41/215, 62/295
    progressive external ophthalmoplegia classification, 62/290
    rimmed muscle fiber vacuole, 40/253, 62/297
    sporadic progressive external ophthalmoplegia with ragged red fiber, 62/508
    Tomé-Fardeau body, 62/62-103
Oligophrenia, *see* Mental deficiency
Oligophrenic cerebellolental degeneration hereditary, *see* Marinesco-Sjögren syndrome
Olivopontocerebellar atrophy
    progressive external ophthalmoplegia, 62/304
Olivopontocerebellar degeneration, *see* Olivopontocerebellar atrophy
Ophthalmoparesis, *see* Ophthalmoplegia
Ophthalmopathy, *see* Thyroid associated ophthalmopathy
Ophthalmoplegia
    see also Progressive external ophthalmoplegia
    chronic progressive external, *see* Chronic progressive external ophthalmoplegia
    complex I deficiency, 62/504
    congenital acetylcholine receptor deficiency/short channel opentime syndrome, 62/437
    congenital ptosis, 62/297
    external, *see* External ophthalmoplegia
    familial relapsing, *see* Familial relapsing ophthalmoplegia
    nemaline myopathy, 62/342
    nuclear progressive, *see* Progressive external ophthalmoplegia
    progressive external, *see* Progressive external ophthalmoplegia
    sporadic progressive external ophthalmoplegia with ragged red fiber, 62/508
Ophthalmoplegia plus syndrome, *see* Progressive external ophthalmoplegia
Ophthalmoplegic migraine
    orbital myositis, 62/301
Opiate addiction
    toxic myopathy, 62/598
Optic atrophy
    Leber hereditary optic neuropathy, 62/509
    MERRF syndrome, 62/509
    multiple mitochondrial DNA deletion, 62/510
    progressive external ophthalmoplegia, 13/37, 82, 62/305
    pyruvate decarboxylase deficiency, 62/502
Optic nerve atrophy, *see* Optic atrophy
Orbital myositis
    clinical features, 62/299
    computer assisted tomography, 62/299, 305
    definition, 62/298
    differential diagnosis, 62/298, 305
    familial relapsing ophthalmoplegia, 62/302
    histopathology, 62/299
    nuclear magnetic resonance, 62/299
    ophthalmoplegic migraine, 62/301
    progressive external ophthalmoplegia, 62/298
    progressive external ophthalmoplegia classification, 62/290
    T3, 62/300
    thyroid associated ophthalmopathy, 62/298
    thyrotropic hormone, 62/300
    Tolosa-Hunt syndrome, 62/301
Orciprenaline
    hyperkalemic periodic paralysis, 62/472
    myotonia, 62/276
Organophosphate intoxication, 37/541-559
    central core myopathy, 62/333
    cytoplasmic inclusion body myopathy, 62/345
    toxic myopathy, 62/612
OSPOM syndrome
    Kearns-Sayre-Daroff-Shy syndrome, 62/313
    progressive external ophthalmoplegia, 62/313
    progressive external ophthalmoplegia classification, 62/290
Osteomalacia
    hyperparathyroid myopathy, 62/538
Overeating, *see* Bulimia
Oxidation-phosphorylation coupling defect
    bulimia, 62/506
    exercise intolerance, 62/506
    heat intolerance, 62/506
    hyperthermia, 62/506
    metabolic myopathy, 62/506
    mitochondrial disease, 62/506
    polydipsia, 62/506
    sweating, 62/506
    tachycardia, 62/506
α-Oxoglutarate dehydrogenase deficiency
    encephalomyopathy, 62/503
    fumarase deficiency, 62/503
    hyporeflexia, 62/503
    infantile hypotonia, 62/503
    lethargy, 62/503
    microcephaly, 62/503

mitochondrial disease, 62/503
muscle biopsy, 62/503
pyruvate dehydrogenase-dihydrolipoyl
   dehydrogenase, 62/503
pyruvate dehydrogenase-dihydrolipoyl
   transacetylase, 62/502
vision, 62/503
vomiting, 62/503
Oxprenolol
acquired myoglobinuria, 62/577
toxic myopathy, 62/601

Pain, 1/114-142
bone, see Bone pain
muscular, see Myalgia
myotonia fluctuans, 62/266
Painful dominant myotonia
chromosome 17q, 62/266
fasting, 62/266
muscle stiffness, 62/266
myotonic syndrome, 62/263
oral potassium, 62/266
Pancytopenia
familial, see Fanconi syndrome
Paracetamol
toxic myopathy, 62/601
Paracrystalline inclusion
complex V deficiency, 62/506
Paradoxical myotonia
myotonic syndrome, 62/261
paramyotonia congenita, 62/268
Paralysis
familial periodic, see Familial periodic paralysis
hyperkalemic periodic, see Hyperkalemic periodic
   paralysis
hypokalemic periodic, see Hypokalemic periodic
   paralysis
infantile, see Poliomyelitis
normokalemic periodic, see Normokalemic
   periodic paralysis
periodic, see Periodic paralysis
progressive supranuclear, see Progressive
   supranuclear palsy
Paramyotonia
myotonia, 62/262
myotonic syndrome, 62/262
Paramyotonia congenita
see also Hyperkalemic periodic paralysis
age at onset, 62/267
ambient temperature, 62/267
autosomal dominant, 62/267
myotonia, 40/542, 62/224
myotonic syndrome, 40/542, 62/263
paradoxical myotonia, 62/268
sodium channel gene, 62/269
Paramyotonia congenita (Eulenberg), see
   Paramyotonia congenita
Paraparesis
spastic, see Spastic paraparesis
Paraplegia
familial spastic, see Familial spastic paraplegia

spastic, see Spastic paraplegia
Paraskepsia, see Dementia
Parathormone, see Parathyroid hormone
Parathyroid hormone
myoglobinuria, 62/556
Paresthesia
Eaton-Lambert myasthenic syndrome, 62/426
hypoparathyroid myopathy, 62/541
Parkinson disease, 6/173-207
mimicking disorder, see Multiple neuronal system
   degeneration
Parkinson plus syndrome, see Multiple neuronal
   system degeneration
Partial status epilepticus, see Status epilepticus
Pathophysiology
childhood myoglobinuria, 62/565
malignant hyperthermia, 62/571
myotonia, 40/549-557, 62/276
periodic paralysis, 41/167, 62/466
phosphofructokinase deficiency, 62/488
Peapicker disease, see Leptospirosis
Pearson syndrome
Kearns-Sayre-Daroff-Shy syndrome, 62/508
mitochondrial DNA defect, 62/508
pancreatic dysfunction, 62/508
sideroblastic anemia, 62/508
Pelger-Huët abnormality
late adult muscular dystrophy (Nevin), 62/184
Pelvifemoral muscular dystrophy (Leyden-Möbius)
see also Limb girdle syndrome
features, 62/183
limb girdle syndrome, 62/179
Pemphigus
myasthenia gravis, 62/402
Penetrance
facioscapulohumeral muscular dystrophy, 62/165
Penicillamine
myasthenia gravis, 41/126, 62/405
polymyositis, 62/372
toxic myopathy, 62/597
Pentamidine
toxic myopathy, 62/601
Pentose phosphate pathway
myoglobinuria, 62/558
Percussion myotonia
hereditary cylindric spirals myopathy, 62/348
Perhexiline
toxic myopathy, 62/606
Perhexiline myopathy
sensorimotor polyneuropathy, 62/608
ultrastructure, 62/608
Pericentronuclear myopathy, see Myotubular
   myopathy
Periodic ataxia
pyruvate decarboxylase deficiency, 62/502
Periodic paralysis, 28/581-598, 41/147-170
calcium deposition, 62/465
familial, see Familial periodic paralysis
familial hypokalemic, see Familial hypokalemic
   periodic paralysis
hyperaldosteronism, 62/457

hyperkalemic, *see* Hyperkalemic periodic paralysis
hypermagnesemia, 62/457
hyperthyroidism, 27/268, 28/597, 40/285, 41/165, 241, 62/457, 468
hypokalemic, *see* Hypokalemic periodic paralysis
hypothalamic tumor, 62/457
malignant hyperthermia, 62/570
muscle biopsy, 62/465
normokalemic, *see* Normokalemic periodic paralysis
pathology, 62/464
pathophysiology, 41/167, 62/466
precipitating factor, 62/468
renal disease, 62/457
rimmed muscle fiber vacuole, 40/29, 106, 219, 41/153, 160, 62/465
rimmed muscle fiber vacuole content, 62/18
rimmed muscle fiber vacuole type, 62/18
sodium conductance, 62/467
sodium-potassium pump, 62/468
T tubule proliferation, 62/465
thyrotoxic, *see* Thyrotoxic periodic paralysis
treatment, 62/469
tubular aggregate, 40/229, 41/154, 162, 62/465
variant type, 62/463
Peripheral neuropathy, *see* Polyneuropathy
Pernicious anemia
    Eaton-Lambert myasthenic syndrome, 62/421
Peroneal muscular atrophy, *see* Charcot-Marie-Tooth disease
Personality change
    myotonic dystrophy, 62/235
Perspiration, *see* Sweating
Pes arcuatus, *see* Pes cavus
Pes cavus
    scapuloperoneal spinal muscular atrophy, 62/170
Pes excavatus, *see* Pes cavus
Phagocytosis
    muscle fiber, 62/6
    neuromuscular disease, 62/6
Phencyclidine
    acquired myoglobinuria, 62/573
    toxic myopathy, 62/601
Phencyclidine addiction
    toxic myopathy, 62/600
Phenelzine
    toxic myopathy, 62/601
Phenformin
    toxic myopathy, 62/601
Phenmetrazine
    acquired myoglobinuria, 62/573
    toxic myopathy, 62/601
Phenomenon
    Raynaud, *see* Raynaud phenomenon
Phenotype
    Duchenne muscular dystrophy, 62/126
Phentermine
    acquired myoglobinuria, 62/573
Phenylenediamine
    acquired myoglobinuria, 62/573
    toxic myopathy, 62/601

Phenylpropanolamine
    toxic myopathy, 62/601
Phenytoin
    myotonia, 62/275
    myotonic dystrophy, 40/522, 62/241
Pheochromocytoma myopathy
    creatine kinase, 62/537
    proximal muscle weakness, 62/537
Phosphofructokinase
    myoglobinuria, 62/558
Phosphofructokinase deficiency, 41/190-192
    biochemistry, 41/191, 62/488
    childhood myoglobinuria, 62/564
    chromosome 1, 62/488
    chromosome 10, 62/488
    chromosome 21, 62/488
    clinical features, 27/235, 40/327, 339, 342, 41/190, 62/487
    creatine kinase, 62/487
    EMG, 41/191, 62/487
    exercise intolerance, 62/487
    glycogen storage disease, 40/53, 190, 327, 339, 342, 62/487
    hemolytic anemia, 43/183, 62/487
    histopathology, 41/191, 62/488
    inheritance, 62/488
    metabolic myopathy, 62/487
    molecular genetics, 62/488
    muscle biopsy, 41/191, 62/488
    muscle cramp, 62/487
    muscle fiber feature, 62/17
    myoglobinuria, 27/235, 41/190, 265, 43/183, 62/487
    pathophysiology, 62/488
    permanent weakness, 62/487
    phosphorus 31 nuclear magnetic resonance, 62/487
Phosphoglycerate kinase
    myoglobinuria, 62/558
Phosphoglycerate kinase deficiency
    childhood myoglobinuria, 62/565
    epilepsy, 62/489
    exercise intolerance, 62/489
    glycogen storage disease, 62/489
    hemolytic anemia, 62/489
    inheritance, 62/489
    mental deficiency, 62/489
    metabolic encephalopathy, 62/489
    metabolic myopathy, 62/489
    muscle biopsy, 62/489
    muscle cramp, 62/489
    myoglobinuria, 62/489
    phosphorus 31 nuclear magnetic resonance, 62/489
    X-linked recessive disorder, 62/489
Phosphoglycerate mutase
    muscle tissue culture, 62/94
    myoglobinuria, 62/558
Phosphoglycerate mutase deficiency
    childhood myoglobinuria, 62/565
    exercise intolerance, 62/489
    glycogen storage disease, 62/489
    inheritance, 62/489
    ischemic exercise, 62/489

metabolic myopathy, 62/489
muscle biopsy, 62/489
muscle cramp, 62/489
phosphorus 31 nuclear magnetic resonance, 62/490
Phospholipase A
  toxic myopathy, 62/610
Phosphorylase
  muscle tissue culture, 62/94, 109
Phosphorylase isoenzyme
  chromosome 10, 62/486
  chromosome 11q13, 62/485
  chromosome 14, 62/486
  chromosome 20, 62/486
  myophosphorylase deficiency, 62/486
Phosphorylase kinase
  myoglobinuria, 62/558
Phosphorylase kinase deficiency
  biochemistry, 62/487
  cardiomyopathy, 62/487
  childhood myoglobinuria, 62/565
  chromosome 7, 62/487
  chromosome 11, 62/487
  chromosome 16q12-q13, 62/487
  chromosome Xq12-q13, 62/487
  clinical features, 27/222, 62/486
  creatine kinase, 62/487
  exercise intolerance, 62/487
  fasting hypoglycemia, 62/486
  glycogen storage disease, 62/486
  hepatomyopathy, 62/487
  histopathology, 62/487
  ischemic exercise, 62/487
  liver disease, 62/486
  metabolic myopathy, 62/486
  molecular genetics, 62/487
  muscle biopsy, 62/487
  myopathy, 62/487
  pigmenturia, 62/487
Physiotherapy, 41/457-497
  dermatomyositis, 62/386
  Duchenne muscular dystrophy, 62/132
  inclusion body myositis, 62/386
  polymyositis, 62/386
Picornavirus
  dermatomyositis, 62/381
  inclusion body myositis, 62/381
  polymyositis, 62/381
Pigmentary retinopathy
  complex I deficiency, 62/504
  complex III deficiency, 62/504
  Kearns-Sayre-Daroff-Shy syndrome, 62/308, 508
  primary pigmentary retinal degeneration, 62/308
  progressive external ophthalmoplegia, 62/308
Pilomatricoma
  myotonic dystrophy, 43/153, 62/239
Pindolol
  myotonia, 62/276
  toxic myopathy, 62/597
Pitressin, see Vasopressin
Pittsburgh pneumonia agent, see Legionellosis
Pituitary gigantism myopathy

acromegalic myopathy, 62/537
denervation atrophy, 62/538
EMG, 62/538
growth hormone excess, 62/538
Plasma membrane
  muscle fiber, 62/32
Plasmapheresis
  dermatomyositis, 62/386
  Eaton-Lambert myasthenic syndrome, 62/426
  inclusion body myositis, 62/386
  myasthenia gravis, 41/96, 136, 62/420
  myoglobinuria, 62/556
  polymyositis, 62/386
Plasmocid
  toxic myopathy, 62/600
Pleoconial myopathy, see Mitochondrial myopathy
Pneumatosis intestinalis
  dermatomyositis, 62/374
  inclusion body myositis, 62/374
  polymyositis, 62/374
Pneumococcus, see Streptococcus pneumoniae
Pneumonia
  aspiration, see Aspiration pneumonia
Pneumonitis
  dermatomyositis, 62/374
  inclusion body myositis, 62/374
  polymyositis, 62/374
Poliodystrophy
  infantile progressive, see Alpers disease
  progressive, see Alpers disease
Poliomyelitis, 34/93-127
  edrophonium test, 62/413
  polyneuropathy, 62/413
Poliovirus infection, see Poliomyelitis
POLIP syndrome
  cytochrome-c oxidase deficiency, 62/506
  Kearns-Sayre-Daroff-Shy syndrome, 62/313
  MNGIE syndrome, 62/505
  progressive external ophthalmoplegia, 62/313
Polyarteritis nodosa, 39/295-309
  see also Kawasaki syndrome
  infantile, see Kawasaki syndrome
  juvenile, see Kawasaki syndrome
Polydipsia
  oxidation-phosphorylation coupling defect, 62/506
Polyglucosan body disease
  adult, see Adult polyglucosan body disease
Polymyositis
  see also Dermatomyositis and Inflammatory
    myopathy
  acne fulminans, 62/373
  acquired myoglobinuria, 62/577
  acute, see Acute polymyositis
  agammaglobulinemia, 62/373
  age, 62/372
  amyloid deposit, 62/377
  amyloid β-protein, 62/377
  ankylosing spondylitis, 62/373
  anti-Jo-1 antibody, 62/374
  arthralgia, 62/374
  aspiration pneumonia, 62/374

autoimmune antibody, 62/380
autoimmune thrombocytopenia, 62/373
axonal neuropathy, 62/376
azathioprine, 41/73, 62/385
bacterial myositis, 62/372
basophilic granular inclusion, 62/377
Behçet syndrome, 62/373
Borrelia burgdorferi, 62/373
calcinosis, 62/386
cardiac disorder, 62/374
cardiomyopathy, 62/374
celiac disease, 62/373
chronic graft vs host disease, 62/373
ciclosporin, 62/386
clinical course, 62/372
congestive heart failure, 62/374
corticosteroids, 62/383
course, 62/372
creatine kinase, 41/60, 62/375
cricopharyngeal myotomy, 62/386
Crohn disease, 62/373
cyclophosphamide, 62/385
cysticercus, 62/373
dermatitis herpetiformis, 62/373
diagnosis, 62/375
diagnostic criteria, 62/378
dyspnea, 62/374
Eaton-Lambert myasthenic syndrome, 62/425
electronmicroscopy, 62/377
EMG, 1/642, 62/375
eosinophilic cytoplasmic inclusion, 62/377
erythematosus discoid lupus, 62/373
exclusion criteria, 62/372
facioscapulohumeral syndrome, 62/167
fever, 62/373
gastrointestinal symptom, 62/374
giant cell, see Giant cell polymyositis
granulomatous disease, 62/373
Hashimoto thyroiditis, 62/373
hematemesis, 62/374
hereditary complement deficiency, 62/373
human foamy retrovirus, 62/382
human immunodeficiency virus type 1, 62/381
human T-lymphotropic virus type I, 62/382
hyperagammaglobulinemic purpura, 62/373
hypereosinophilic syndrome, 62/373
hypoxemia, 62/374
IgA deficiency, 62/373
immunopathology, 62/379
immunosuppressive therapy, 62/385
intravenous Ig, 62/386
joint contracture, 62/374
Kawasaki syndrome, 62/373
lactate dehydrogenase, 62/375
Legionella pneumophila, 62/373
legionellosis, 62/373
leukopheresis, 62/386
limb girdle syndrome, 40/457, 62/187
Lyme disease, 62/373
lymphoid irradiation, 62/386
malaise, 62/373

malignancy, 62/374
melena, 62/374
membranolytic attack complex, 62/379
methotrexate, 41/80, 62/385
mixed connective tissue disease, 62/374
monoclonal gammopathy, 62/373
muscle biopsy, 62/375
myasthenia gravis, 43/157, 62/373, 402
myocarditis, 62/374
myoglobinuria, 41/58, 60, 274, 62/556
nuclear magnetic resonance, 62/379
overlap syndrome, 62/374
pathogenesis, 41/74, 62/379
penicillamine, 62/372
physiotherapy, 62/386
picornavirus, 62/381
plasmapheresis, 62/386
pneumatosis intestinalis, 62/374
pneumonitis, 62/374
practical guideline, 62/386
primary biliary cirrhosis, 62/373
primary inflammatory myosis, 62/369
psoriasis, 62/373
pulmonary dysfunction, 62/374
ragged red fiber, 62/292
Raynaud phenomenon, 41/56, 58, 62/374
retrovirus, 62/381
rheumatoid arthritis, 38/484, 41/77, 62/374
rimmed muscle fiber vacuole, 62/204, 376
sarcoidosis, 62/373
scleroderma, 62/373
seasonal onset, 62/372
serum alanine aminotransferase, 62/375
serum aspartate aminotransferase, 62/375
serum muscle enzyme, 62/375
simian immunodeficiency virus, 62/382
simian retrovirus type I, 62/382
Sjögren syndrome, 8/17, 62/374
steroid myopathy, 62/384
systemic lupus erythematosus, 62/374
systemic sclerosis, 62/374
tachyarrhythmia, 62/374
Toxoplasma gondii, 62/373
treatment, 41/79, 62/383
Trichinella, 62/373
Trypanosoma, 62/373
tubulofilament, 62/377
vasculitis, 62/373
viral myositis, 62/381
weight loss, 62/374
zidovudine, 62/372, 382
Polyneuropathy
    see also Neuropathy
    alcohol, see Alcoholic polyneuropathy
    amyotrophic lateral sclerosis, 62/413
    axonal, see Axonal polyneuropathy
    congenital myasthenic syndrome, 62/413
    Eaton-Lambert myasthenic syndrome, 62/413, 425
    hypothyroid, see Hypothyroid polyneuropathy
    Kearns-Sayre-Daroff-Shy syndrome, 62/309, 313
    MNGIE syndrome, 62/505

myasthenia gravis, 62/413
poliomyelitis, 62/413
progressive external ophthalmoplegia, 22/207,
  62/309, 313
sensorimotor, *see* Sensorimotor polyneuropathy
Polyphagia, *see* Bulimia
Polyradiculoneuritis
  acute, *see* Guillain-Barré syndrome
Polyradiculoneuropathy
  Guillain-Barré, *see* Guillain-Barré syndrome
Polysaccharide storage disease
  muscle fiber feature, 62/17
Pompe disease, *see* Acid maltase deficiency
Pontiac fever, *see* Legionellosis
Potassium channel
  myotonic dystrophy, 62/225
Potentials
  compound muscle action, *see* Compound muscle
    action potential
  miniature motor end plate, *see* Miniature motor
    end plate potential
Prednisone, *see* Corticosteroids
Pregnancy
  myositis, 62/370
  myotonia congenita, 62/264
  myotonic dystrophy, 62/228
Prenatal diagnosis
  acid maltase deficiency, 43/178, 62/481
  myotonic dystrophy, 43/153, 62/215
Primary biliary cirrhosis
  polymyositis, 62/373
Primary inflammatory myositis
  dermatomyositis, 62/369
  inclusion body myositis, 62/369
  polymyositis, 62/369
Primary pigmentary retinal degeneration, 13/148-340
  see also Secondary pigmentary retinal degeneration
  pigmentary retinopathy, 62/308
Procainamide
  myotonia, 62/275
  toxic myopathy, 62/597
Prognosis
  Emery-Dreifuss muscular dystrophy, 62/154
Progressive external ophthalmoplegia, 13/309-314,
  22/177-196
  see also Extraocular muscle weakness, Kearns-
    Sayre-Daroff-Shy syndrome, Mitochondrial
    myopathy, Ocular myopathy *and*
    Ophthalmoplegia
  Bassen-Kornzweig syndrome, 22/191, 204, 207,
    62/304
  biochemistry, 62/311
  cardiac disorder, 62/309
  cardiomyopathy, 22/192, 62/310
  cataract, 62/306
  celiac disease, 62/313
  cerebellar syndrome, 62/308
  chronic, *see* Chronic progressive external
    ophthalmoplegia
  clinical features, 22/180, 209, 40/302, 342, 62/292,
    307

computer assisted tomography, 62/295, 306, 311
concept history, 62/287
congenital fibrosis, 62/297
congenital myasthenia gravis, 62/297
congenital myopathy, 62/297, 333
congenital ocular muscle fibrosis, 62/297
congenital ptosis, 62/297
cranial neuropathy, 62/313
criteria, 62/287
CSF, 62/311
curare sensitive ocular myopathy, 22/188, 62/302
cytochrome-c oxidase, 62/293, 314
deafness, 22/193, 62/313
definition, 62/287, 307
dementia, 62/308
diabetes mellitus, 62/309
diagnostic test, 62/294
diffuse lewy body disease, 62/304
dystonia, 62/304
ECG, 62/311
edrophonium test, 62/303
EMG, 22/181, 62/295
endocrine disorder, 62/309
epidemiology, 62/291
epilepsy, 62/308
Ferguson-Critchley syndrome, 62/304
Friedreich ataxia, 21/347, 22/191, 62/304
genetics, 22/185, 41/420, 62/295, 312
glycogen storage disease, 62/298
hereditary ataxia, 62/304
histopathology, 62/293, 310
hypoparathyroidism, 62/309
intestinal pseudo-obstruction, 62/313
Kearns-Sayre-Daroff-Shy syndrome, 22/192, 211,
  62/287, 306, 508
limb weakness, 62/291
mental deficiency, 62/308
MEPOP syndrome, 62/315
mitochondrial disease, 62/497
mitochondrial DNA, 62/294
mitochondrial myopathy, 62/289, 313
MNGIE syndrome, 62/305, 313, 505
molecular genetics, 62/311
motoneuron degeneration, 62/305
multiple mitochondrial DNA deletion, 62/510
muscle histopathology, 62/310
myasthenia gravis, 22/181, 188, 204, 62/415
myasthenic ocular myopathy, 62/302
myotonia congenita, 62/305
myotonic dystrophy, 22/187, 62/302
myotubular myopathy, 22/179, 41/11, 62/297
neurogenic ophthalmoplegia, 62/304
neuropathology, 22/182, 62/289
neuropathy, 43/136, 62/313
ocular myasthenia gravis, 62/302
ocular myopathy, 22/186, 41/215, 62/288, 291
oculopharyngeal muscular dystrophy, 22/179, 187,
  206, 40/302, 41/215, 62/295
olivopontocerebellar atrophy, 62/304
optic atrophy, 13/37, 82, 62/305
orbital myositis, 62/298

OSPOM syndrome, 62/313
pathogenesis, 62/294
pigmentary retinopathy, 62/308
POLIP syndrome, 62/313
polyneuropathy, 22/207, 62/309, 313
ptosis, 22/179, 62/297
ragged red fiber, 62/289, 291-293, 305, 314
renal dysfunction, 62/309
retinopathy, 62/305, 311
Schimke syndrome, 62/304
senile ptosis, 62/302
spastic ataxia, 62/304
spastic paraparesis, 62/305
spastic paraplegia, 13/307, 309, 62/304
treatment, 22/195, 62/312
Progressive external ophthalmoplegia classification
amyotrophic lateral sclerosis, 62/290
Bassen-Kornzweig syndrome, 62/290
congenital myopathic ptosis, 62/290
dystonia musculorum deformans, 62/290
familial spastic paraplegia, 62/290
glycogen storage disease, 62/290
hereditary ataxia, 62/290
infantile spinal muscular atrophy, 62/290
Joseph-Machado disease, 62/290
Kearns-Sayre-Daroff-Shy syndrome, 62/290
MELAS syndrome, 62/290
MERRF syndrome, 62/290
MNGIE syndrome, 62/290
Möbius syndrome, 62/290
myasthenia gravis, 62/290
myotonic dystrophy, 62/290
myotubular myopathy, 62/290
ocular myopathy, 62/290
oculopharyngeal muscular dystrophy, 62/290
oculopharyngeal myopathy, 62/290
ophthalmoplegia variant, 62/290
orbital myositis, 62/290
OSPOM syndrome, 62/290
practical guideline, 62/289
progressive supranuclear palsy, 62/290
senile ptosis, 62/290
thyroid associated ophthalmopathy, 62/290
Wohlfart-Kugelberg-Welander disease, 62/290
Progressive myoclonus epilepsy
see also Glycogen, MELAS syndrome, MERRF
syndrome and Neuronal ceroid lipofuscinosis
Lafora, see Lafora progressive myoclonus epilepsy
Progressive nuclear external ophthalmoplegia, see
Progressive external ophthalmoplegia
Progressive poliodystrophy, see Alpers disease
Progressive spinal muscular atrophy, 42/86-88
proximal, see Wohlfart-Kugelberg-Welander
disease
pseudomyopathic, see Wohlfart-Kugelberg-
Welander disease
Progressive supranuclear palsy, 22/217-227
progressive external ophthalmoplegia
classification, 62/290
Progressive torsion spasm in child (Flatau-Sterling),
see Dystonia musculorum deformans

2-Propanol
toxic myopathy, 62/601
Protein
CSF, see CSF protein
M, see M protein
ribosomal S6, see Ribosomal S6 protein
Protein synthesis
myofibrillar lysis myopathy, 62/339
Proximal muscle weakness
pheochromocytoma myopathy, 62/537
Proximal neuropathy
diabetic, see Diabetic proximal neuropathy
Proximal pseudomyopathic spinal muscular atrophy,
see Wohlfart-Kugelberg-Welander disease
Pseudohypertrophic muscular dystrophy, see
Duchenne muscular dystrophy
Pseudomyasthenia, see Eaton-Lambert myasthenic
syndrome
Pseudomyopathic familial spinal muscular atrophy,
see Wohlfart-Kugelberg-Welander disease
Pseudomyotonia, see Isaacs syndrome
Pseudo-obstruction
intestinal, see Intestinal pseudo-obstruction
Pseudoretinitis pigmentosa, see Secondary
pigmentary retinal degeneration
Psoriasis
polymyositis, 62/373
Psychomotor retardation
childhood myoglobinuria, 62/564
complex I deficiency, 62/503
pyruvate decarboxylase deficiency, 41/211, 62/502
short chain acyl coenzyme A dehydrogenase
deficiency, 62/495
Ptosis
congenital, see Congenital ptosis
congenital acetylcholine receptor deficiency/
synaptic cleft syndrome, 62/438
congenital high conductance fast channel
syndrome, 62/438
congenital myopathy, 62/297, 333, 349
congenital slow channel syndrome, 62/432
familial infantile myasthenia, 62/427
multiple mitochondrial DNA deletion, 62/510
myotubular myopathy, 43/113, 62/297
progressive external ophthalmoplegia, 22/179,
62/297
senile, see Senile ptosis
spastic ataxia, 62/304
sporadic progressive external ophthalmoplegia
with ragged red fiber, 62/508
Pyomyositis, see Tropical polymyositis
Pyramidal sign
hypertonia, see Spasticity
Pyramidal syndrome
complex III deficiency, 62/504
Pyridostigmine bromide
myasthenia gravis, 41/131, 62/417
Pyruvate carboxylase deficiency
mitochondrial disease, 62/503
Pyruvate decarboxylase deficiency
epilepsy, 62/502

episodic apnea, 62/502
exercise intolerance, 62/502
fetal alcohol syndrome, 62/502
infantile hypotonia, 62/502
lethargy, 62/502
optic atrophy, 62/502
periodic ataxia, 62/502
psychomotor retardation, 41/211, 62/502
severe lactic acidosis, 62/502
subacute necrotizing encephalomyelopathy, 62/502
Pyruvate dehydrogenase complex deficiency
mitochondrial disease, 62/496, 502
Pyruvate dehydrogenase phosphatase deficiency
mitochondrial disease, 62/503
Pyruvate kinase
limb girdle syndrome, 62/185

Quadriceps myopathy
see also Limb girdle syndrome
characteristics, 62/174
creatine kinase, 62/174
EMG, 62/174
features, 62/184
heterogeneity, 62/174
incomplete dystrophin deficiency, 62/135
Quadriplegia, see Tetraplegia
Quail myopathy
toxic myopathy, 62/612
Quantal squander syndrome, see Isaacs syndrome
Quinine
myotonia, 40/548, 558, 41/297, 62/275

Ragged red fiber
see also Mitochondrial disease
combined complex I-V deficiency, 62/506
complex V deficiency, 62/506
exercise intolerance, 62/292
fatal infantile myopathy/cardiopathy, 62/510
human menopausal gonadotropin myopathy,
62/292
infantile cytochrome-c oxidase deficiency, 62/292
Kearns-Sayre-Daroff-Shy syndrome, 62/314, 508
MELAS syndrome, 62/292, 509
MERRF syndrome, 62/292
mitochondrial disease, 62/496
mitochondrial DNA depletion, 62/511
MNGIE syndrome, 62/292, 505
multiple mitochondrial DNA deletion, 62/510
muscle fiber, 62/24
polymyositis, 62/292
progressive external ophthalmoplegia, 62/289, 291-
293, 305, 314
spinal muscular atrophy, 62/292
sporadic progressive external ophthalmoplegia, see
Sporadic progressive external ophthalmoplegia
with ragged red fiber
zidovudine myopathy, 62/383
Rapid eye movement, see REM
Raynaud phenomenon
dermatomyositis, 62/374
inclusion body myositis, 62/374

polymyositis, 41/56, 58, 62/374
Receptor
acetylcholine, see Acetylcholine receptor
Receptor antibody
acetylcholine, see Acetylcholine receptor antibody
Recessive generalized myotonia
autosomal, see Autosomal recessive generalized
myotonia
Reducing body myopathy
clinical course, 62/347
congenital myopathy, 41/17-20, 62/332
Coxsackie virus B2, 62/348
Coxsackie virus B4, 62/348
Coxsackie virus B5, 62/348
rigid spine syndrome, 62/348
Reflex, 1/61-72, 237-255
muscle stretch, see Muscle stretch reflex
tendon, see Muscle stretch reflex
Regeneration
muscle fiber, 40/20, 89, 62/9
myopathy, 62/9
neuromuscular disease, 62/9
Reinnervation
motor unit, 62/62
muscle, see Muscle reinnervation
Relapsing aphthous uveitis, see Behçet syndrome
REM
limb girdle syndrome, 62/185
Renal dialysis
myoglobinuria, 62/556
Renal disease
periodic paralysis, 62/457
Renal dysfunction
Kearns-Sayre-Daroff-Shy syndrome, 62/309
mitochondrial DNA depletion, 62/511
progressive external ophthalmoplegia, 62/309
Renal failure
carnitine palmitoyltransferase deficiency, 62/494
childhood myoglobinuria, 62/561
myophosphorylase deficiency, 27/233, 62/485
Respiration, 1/650-680
see also Respiratory dysfunction and Respiratory
system
congenital myopathy, 62/334
Respiratory distress
childhood myoglobinuria, 62/561
congenital acetylcholine receptor deficiency/
synaptic cleft syndrome, 62/438
Respiratory dysfunction
see also Respiration and Respiratory system
adult acid maltase deficiency, 62/480
congenital fiber type disproportion, 62/356
congenital myopathy, 62/334, 349
late infantile acid maltase deficiency, 62/480
thyrotoxic myopathy, 62/528
Z band plaque myopathy, 62/344
Respiratory failure
cardiac, see Cardiorespiratory failure
Duchenne muscular dystrophy, 62/122, 133
Respiratory insufficiency
complex IV deficiency, 62/505

Respiratory system
   see also Respiration *and* Respiratory dysfunction
   myotonic dystrophy, 62/238
Retardations
   mental, *see* Mental deficiency
   psychomotor, *see* Psychomotor retardation
Retinal degeneration
   primary pigmentary, *see* Primary pigmentary
     retinal degeneration
   secondary pigmentary, *see* Secondary pigmentary
     retinal degeneration
Retinitis pigmentosa, *see* Primary pigmentary retinal
    degeneration *and* Secondary pigmentary retinal
    degeneration
Retinopathy
   complex V deficiency, 62/506
   facioscapulohumeral muscular dystrophy, 62/168
   Kearns-Sayre-Daroff-Shy syndrome, 62/311
   pigmentary, *see* Pigmentary retinopathy
   progressive external ophthalmoplegia, 62/305, 311
Retrovirus
   polymyositis, 62/381
Reversible ischemic neurological deficit, *see*
    Transient ischemic attack
Reye syndrome, 29/331-342
   medium chain acyl coenzyme A dehydrogenase
    deficiency, 62/495
Rhabdomyolysis
   see also Myoglobinuria
   myoglobinuria, 62/553
   myositis, 62/370
   thyrotoxic myopathy, 62/529
   toxic myopathy, 62/600
Rheumatoid arthritis, 38/479-499
   dermatomyositis, 62/374
   inclusion body myositis, 62/374
   myasthenia gravis, 41/240, 62/402
   polymyositis, 38/484, 41/77, 62/374
Ribosomal S6 protein
   muscle tissue culture, 62/90
Ribosome
   muscle fiber type, 62/4
Rice field fever, *see* Leptospirosis
Richardson-Steele-Olszewski syndrome, *see*
    Progressive supranuclear palsy
Rickets
   hyperparathyroid myopathy, 62/538
Rifampicin
   toxic myopathy, 62/597, 609
Rifampin, *see* Rifampicin
Rigid spine syndrome
   congenital fiber type disproportion, 62/355
   Emery-Dreifuss muscular dystrophy, 40/390,
    62/155
   EMG, 62/173
   family, 62/173
   heredofamilial myosclerosis, 62/173
   muscle fibrosis, 62/173
   myosclerosis, 62/173
   reducing body myopathy, 62/348
Rigidity

congenital myopathy, 62/335
Rimmed muscle fiber vacuole
   dermatomyositis, 62/376
   distal myopathy, 40/475, 62/204
   inclusion body myositis, 62/376
   oculopharyngeal muscular dystrophy, 40/253,
    62/297
   periodic paralysis, 40/29, 106, 219, 41/153, 160,
    62/465
   polymyositis, 62/204, 376
   Wohlfart-Kugelberg-Welander disease, 62/200, 204
Rimmed vacuole distal myopathy
   acid maltase deficiency, 62/481
   clinical course, 62/352
   congenital myopathy, 62/332
   course, 62/202
   creatine kinase, 62/204
   debrancher deficiency, 62/483
   distal myopathy, 62/199
   EMG, 62/204
   glycogen storage disease, 62/490
   histochemistry, 62/352
   histopathology, 62/204
   myophosphorylase deficiency, 62/485
   symptom, 62/202
RNA
   messenger, *see* mRNA
Rod myopathy, *see* Nemaline myopathy
Rohypnol, *see* Flunitrazepam
Ryanodine receptor
   malignant hyperthermia, 62/567, 569

Salbutamol
   hyperkalemic periodic paralysis, 62/472
   toxic myopathy, 62/597
Salicylic acid
   toxic myopathy, 62/601
Salmonella
   acquired myoglobinuria, 62/577
Sarcoidosis, 38/521-539
   polymyositis, 62/373
Sarcolemma
   see also Delta lesion
   muscle fiber, 62/31
   muscle fiber type, 62/3
   myotonic syndrome, 62/261
Sarcoplasmic body myopathy
   congenital myopathy, 62/332
   desmin intermediate filament, 62/345
Sarcoplasmic mass
   cap disease, 62/341
   muscle fiber, 40/27, 244, 497, 62/44
Sarcoplasmic reticulum
   muscle fiber, 62/37
   muscle fiber type, 62/3
Sarcoplasmic reticulum adenosine triphosphate
    deficiency syndrome
   malignant hyperthermia, 62/570
Sarcotubular myopathy
   congenital myopathy, 41/21, 62/332
   rimmed muscle fiber vacuole content, 62/18

rimmed muscle fiber vacuole type, 62/18
Scapula alata, *see* Winged scapula
Scapular winging, *see* Winged scapula
Scapulohumeral muscular atrophy (Vulpian-
    Bernhardt), *see* Scapulohumeral spinal muscular
    atrophy
Scapulohumeral muscular dystrophy
    facioscapulohumeral muscular dystrophy type,
        62/170
    winged scapula, 62/170
Scapulohumeral muscular dystrophy (Erb)
    see also Limb girdle syndrome
    features, 62/183
    limb girdle syndrome, 62/179
Scapulohumeral spinal muscular atrophy
    EMG, 62/171
    muscle biopsy, 62/171
Scapulohumeral syndrome, *see* Facioscapulohumeral
    syndrome
Scapuloperoneal amyotrophy, *see* Scapuloperoneal
    spinal muscular atrophy
Scapuloperoneal muscular dystrophy
    age at onset, 62/169
    autosomal dominant, 62/168
    EMG, 42/99, 62/169
    facioscapulohumeral muscular dystrophy, 62/169
    inflammatory, *see* Inflammatory scapuloperoneal
        muscular dystrophy
    muscle computerized assisted tomography, 62/169
    serum creatine kinase, 62/169
    sporadic, 62/168
Scapuloperoneal neuropathy, *see* Scapuloperoneal
    spinal muscular atrophy
Scapuloperoneal spinal muscular atrophy
    areflexia, 62/170
    autosomal dominant, 62/169, 170
    demyelination, 62/170
    EMG, 62/170
    hereditary motor and sensory neuropathy type I,
        62/170
    hereditary motor and sensory neuropathy type II,
        62/170
    hypertrophic interstitial neuropathy, 62/170
    hypertrophic nerve, 62/170
    motor conduction velocity, 62/170
    nerve biopsy, 62/170
    pes cavus, 62/170
    recessive subtype, 62/170
    symptom, 62/170
Scapuloperoneal syndrome, 22/57-63, 40/423-428
    differentiation, 62/161
    Emery-Dreifuss muscular dystrophy, 40/392, 426,
        62/171
    nosology, 62/168, 170
Schimke syndrome
    chorea, 62/304
    hearing loss, 62/304
    mental deficiency, 62/304
    progressive external ophthalmoplegia, 62/304
Schwartz-Jampel syndrome, *see* Chondrodystrophic
    myotonia

Scleroderma, 39/355-372
    dermatomyositis, 62/374
    inclusion body myositis, 62/374
    polymyositis, 62/374
Sclerosis
    amyotrophic lateral, *see* Amyotrophic lateral
        sclerosis
    systemic, *see* Systemic sclerosis
Scoliosis
    see also Lordosis
    Duchenne muscular dystrophy, 40/362, 62/119
SDH, *see* Succinate dehydrogenase
Secondary pigmentary retinal degeneration, 13/213-
    225
    see also Primary pigmentary retinal degeneration
    adenosine triphosphatase 6 mutation syndrome,
        62/510
Seizure, *see* Epilepsy
Senile ptosis
    progressive external ophthalmoplegia, 62/302
    progressive external ophthalmoplegia
        classification, 62/290
Sensorimotor polyneuropathy
    amiodarone myopathy, 62/608
    perhexiline myopathy, 62/608
Sensorineural deafness
    see also Deafness
    complex III deficiency, 62/504
    Kearns-Sayre-Daroff-Shy syndrome, 43/142,
        62/508
Sensory loss
    see also Sensory neuropathy
    adult polyglucosan body disease, 62/484
Sensory neuropathy
    see also Sensory loss
    adenosine triphosphatase 6 mutation syndrome,
        62/510
    complex I deficiency, 62/504
    complex III deficiency, 62/504
Serologic test
    Eaton-Lambert myasthenic syndrome, 62/425
Serous encephalitis
    acute, *see* Reye syndrome
Serum alanine aminotransferase
    dermatomyositis, 62/375
    facioscapulohumeral muscular dystrophy, 62/166
    inclusion body myositis, 62/375
    polymyositis, 62/375
Serum aspartate aminotransferase
    dermatomyositis, 62/375
    inclusion body myositis, 62/375
    polymyositis, 62/375
Serum creatine kinase
    scapuloperoneal muscular dystrophy, 62/169
Serum creatine phosphokinase, *see* Serum creatine
    kinase
Serum potassium
    myotonic hyperkalemic periodic paralysis, 62/267
Seven day fever, *see* Leptospirosis
Sex linked neurodegenerative disease
    monilethrix, *see* Trichopoliodystrophy

Sex ratio
  carnitine palmitoyltransferase deficiency, 62/494
  childhood myoglobinuria, 62/561
  inclusion body myositis, 62/204
  Leber hereditary optic neuropathy, 62/509
  myasthenia gravis, 43/157, 62/399
  myoglobinuria, 62/559
  myositis, 62/370
  myotonia congenita, 62/266
  steroid myopathy, 62/535
SGOT, see Serum aspartate aminotransferase
SGPT, see Serum alanine aminotransferase
Sharp disease, see Mixed connective tissue disease
Shawl sign
  dermatomyositis, 62/371
Shigella
  acquired myoglobinuria, 62/577
  childhood myoglobinuria, 62/561
Short chain acyl coenzyme A dehydrogenase
  metabolic myopathy, 62/492
Short chain acyl coenzyme A dehydrogenase
    deficiency
  lipid metabolic disorder, 62/495
  psychomotor retardation, 62/495
Short chain 3-hydroxyacyl coenzyme A
    dehydrogenase
  childhood myoglobinuria, 62/563
  myoglobinuria, 62/558
Short chain 3-hydroxyacyl coenzyme A
    dehydrogenase deficiency
  cardiomyopathy, 62/495
  lipid metabolic disorder, 62/495
  metabolic encephalopathy, 62/495
  metabolic myopathy, 62/495
  myoglobinuria, 62/495
  Reye like episode, 62/495
Short stature
  complex III deficiency, 62/504
  Kearns-Sayre-Daroff-Shy syndrome, 43/142,
    62/508
Sicca syndrome, see Sjögren syndrome
Sickle cell anemia, 38/33-44
  acquired myoglobinuria, 62/572
Sideroblastic anemia
  Pearson syndrome, 62/508
Signs
  Chvostek, see Chvostek sign
  Gowers, see Gowers sign
  lower motoneuron, see Lower motoneuron sign
  shawl, see Shawl sign
  upper motoneuron, see Upper motoneuron sign
Single fiber electromyography
  Eaton-Lambert myasthenic syndrome, 62/79
  Emery-Dreifuss muscular dystrophy, 62/150
  familial infantile myasthenia, 62/427
  limb girdle syndrome, 62/186
  motor unit, 62/49, 53
  myasthenia gravis, 41/128, 62/77, 79, 414
  myopathy, 62/49, 53
Sjögren syndrome
  dermatomyositis, 62/374

Eaton-Lambert myasthenic syndrome, 62/421
  inclusion body myositis, 62/374
  myasthenia gravis, 39/430, 41/240, 62/402
  polymyositis, 8/17, 62/374
Skeletal deformity
  congenital myopathy, 62/334
Skin
  temperature, see Thermoregulation
SLE, see Systemic lupus erythematosus
Sleep, 1/488-491, 3/80-105
  see also Hypersomnia and Lethargy
  myotonic dystrophy, 62/235
Sleep apnea
  myotonic dystrophy, 62/217, 235
Slow channel syndrome
  congenital, see Congenital slow channel syndrome
  congenital myopathy, 62/332
Snake venom
  acquired myoglobinuria, 62/577
  toxic myopathy, 62/601, 610
Sodium channel
  myotonia, 62/278
  myotonic dystrophy, 62/224
  myotonic hyperkalemic periodic paralysis, 62/224
Sodium valproate, see Valproic acid
Spasm
  torsion, see Dystonia musculorum deformans
Spastic ataxia
  progressive external ophthalmoplegia, 62/304
  ptosis, 62/304
Spastic paraparesis
  progressive external ophthalmoplegia, 62/305
Spastic paraplegia, 42/168-170
  familial, see Familial spastic paraplegia
  progressive external ophthalmoplegia, 13/307, 309,
    62/304
Spasticity
  MERRF syndrome, 62/509
Spectrin
  dystrophin, 62/126
  muscle fiber, 62/35
Spheroid body myopathy
  congenital myopathy, 62/332
  histopathology, 62/345
Spinal muscular atrophy
  infantile, see Infantile spinal muscular atrophy
  juvenile, see Wohlfart-Kugelberg-Welander
    disease
  juvenile proximal, see Wohlfart-Kugelberg-
    Welander disease
  ragged red fiber, 62/292
Spinal muscular atrophy type III, see Wohlfart-
    Kugelberg-Welander disease
Spine syndrome
  rigid, see Rigid spine syndrome
Spondylitis
  ankylosing, see Ankylosing spondylitis
Spontaneous activity
  motor unit, 62/50, 67
  myopathy, 62/50, 67
Sporadic progressive external ophthalmoplegia with

ragged red fiber
  age at onset, 62/508
  clinical features, 62/508
  cytochrome-c oxidase negative, 62/508
  differential diagnosis, 62/508
  mitochondrial DNA defect, 62/508
  muscle biopsy, 62/508
  muscle weakness, 62/508
  oculopharyngeal muscular dystrophy, 62/508
  ophthalmoplegia, 62/508
  ptosis, 62/508
Staphylococcus
  acquired myoglobinuria, 62/577
Staphylococcus aureus
  tropical polymyositis, 62/373
Stark-Kaeser syndrome, see Scapuloperoneal spinal
      muscular atrophy
Startle
  autosomal recessive generalized myotonia, 62/266
  fall, 62/266
Stature
  short, see Short stature
Status asthmaticus
  acquired myoglobinuria, 62/573
Status epilepticus
  see also Epilepsy
  acquired myoglobinuria, 62/573
Steatorrhea
  idiopathic, see Celiac disease
Steele-Richardson-Olszewski syndrome, see
      Progressive supranuclear palsy
Steely hair, see Trichopoliodystrophy
Steinert disease, see Myotonic dystrophy
Steroid myopathy
  atrophy fiber type II, 40/161, 62/384
  corticosteroids, 41/250, 62/384
  creatine kinase, 62/535, 605
  creatinuria, 62/535
  Cushing syndrome, 40/162, 41/250-253, 62/535
  dermatomyositis, 62/384
  diagnosis, 62/535
  dose-response relationship, 62/605
  EMG, 62/535
  inclusion body myositis, 62/384
  muscle fiber type II atrophy, 62/605
  muscle fiber type IIb atrophy, 62/535
  muscle ultrastructure, 62/535
  muscular atrophy, 62/535
  pathogenesis, 62/536
  polymyositis, 62/384
  sex ratio, 62/535
  treatment, 62/536
Stiff-man syndrome
  myotonia, 62/274
  myotonic syndrome, 62/274
  symptom, 62/274
Storage disease
  glycogen, see Glycogen storage disease
  lipid, see Lipid storage disease
  muscle biopsy, 62/17
  myopathy, 62/16

polysaccharide, see Polysaccharide storage disease
  triglyceride, see Triglyceride storage disease
Storage myopathy
  lipid, see Lipid storage myopathy
Streptococcus
  tropical polymyositis, 62/373
Streptococcus pneumoniae
  acquired myoglobinuria, 62/577
Stretch reflex, 1/63-68
  muscle, see Muscle stretch reflex
Striatopallidodentate calcification, 6/703-720
  see also Intracerebral calcification
  MELAS syndrome, 62/510
Stroke
  see also Brain infarction
  heat, see Heat stroke
Strümpell-Lorrain disease, see Familial spastic
      paraplegia
Strychnine
  acquired myoglobinuria, 62/573
  toxic myopathy, 62/601
Subacute cerebellar degeneration
  Eaton-Lambert myasthenic syndrome, 62/421
Subacute necrotizing encephalomyelopathy, 28/349-
      360
  adenosine triphosphatase 6 mutation syndrome,
      62/510
  childhood myoglobinuria, 62/563
  complex IV deficiency, 62/505
  mitochondrial disease, 41/212, 62/497
  pyruvate decarboxylase deficiency, 62/502
Subcortical argyrophilic dystrophy, see Progressive
      supranuclear palsy
Succinate cytochrome-c reductase
  complex II deficiency, 62/504
Succinate dehydrogenase
  mitochondrial disease, 62/497
Succinylcholine
  malignant hyperthermia, 40/546, 41/268, 270,
      62/568
Suppurative myositis, see Inflammatory myopathy
Supranuclear palsy
  progressive, see Progressive supranuclear palsy
Suxamethonium
  toxic myopathy, 62/597, 601
Suxamethonium induced muscle rigidity
  malignant hyperthermia, 62/569
Swamp fever, see Leptospirosis
Sweating
  oxidation-phosphorylation coupling defect, 62/506
Swineherd disease, see Leptospirosis
Synaptic vesical paucity syndrome
  congenital, see Congenital synaptic vesical paucity
      syndrome
Syncope, 11/532-547
  Emery-Dreifuss muscular dystrophy, 62/153
System degeneration
  heterogeneous, see Progressive supranuclear palsy
  multiple, see Multiple neuronal system
      degeneration
Systemic angiopathy, see Polymyositis

Systemic lupus erythematosus
    dermatomyositis, 62/374
    inclusion body myositis, 62/374
    myasthenia gravis, 62/402
    polymyositis, 62/374
Systemic sclerosis
    dermatomyositis, 62/371, 374
    inclusion body myositis, 62/374
    polymyositis, 62/374
Systems
    central nervous, *see* CNS

T-lymphocyte
    CD8+ cytotoxic, *see* CD8+ cytotoxic T-
      lymphocyte
T tubule
    hereditary cylindric spirals myopathy, 62/348
    muscle fiber, 62/23, 37
    muscle fiber type, 62/3
T tubule system
    myotonia, 62/277
T3 suppression test
    thyroid associated ophthalmopathy, 62/531
T4, *see* Thyrotropic hormone
Tachycardia
    malignant hyperthermia, 43/117, 62/567
    oxidation-phosphorylation coupling defect, 62/506
Taipoxin
    toxic myopathy, 62/610
Talipes cavus, *see* Pes cavus
Tapetoretinal degeneration, *see* Primary pigmentary
      retinal degeneration *and* Secondary pigmentary
      retinal degeneration
Tardive myopathy
    adult acid maltase deficiency, 62/480
Tarui disease, *see* Phosphofructokinase deficiency
Taurine
    myotonia, 62/276
Telodendrion, *see* Motor end plate
Temperature
    body, *see* Thermoregulation
Temperature regulation, *see* Thermoregulation
Temporomandibular ankylosis
    congenital myopathy, 62/335
    nemaline myopathy, 62/342
Tendon reflex, *see* Muscle stretch reflex
Tenotomy
    central core myopathy, 62/334
Tensilon test, *see* Edrophonium test
Terbutaline
    acquired myoglobinuria, 62/577
Tertiary Lyme disease, *see* Lyme disease
Testosterone
    myotonic dystrophy, 62/248
Tests
    edrophonium, *see* Edrophonium test
    serologic, *see* Serologic test
    T3 suppression, *see* T3 suppression test
Tetanus toxin
    central core myopathy, 62/338
Tetany

hypoparathyroid myopathy, 62/541
Tetracycline
    toxic myopathy, 62/609
Tetraplegia
    mitochondrial DNA depletion, 62/511
Tetrazolium reductase
    mitochondrial disease, 62/497
Theophylline
    acquired myoglobinuria, 62/573, 577
    toxic myopathy, 62/601
Thermoregulation
    see also Heat stroke
    myoglobinuria, 62/557
Thomsen disease, *see* Myotonia congenita
Thymectomy
    myasthenia gravis, 41/134, 62/419
Thymoma
    myasthenia gravis, 41/361, 43/157, 62/402
Thymopoietin
    myasthenia gravis, 62/406
Thymus gland
    myasthenia gravis, 62/405
Thyroid associated ophthalmopathy
    classification, 62/530
    clinical features, 62/530
    computer assisted tomography, 62/531
    diagnostic triad, 62/531
    diplopia, 62/531
    endocrine myopathy, 62/528
    Hashimoto thyroiditis, 62/530
    immunosuppression, 62/532
    lid lag, 62/530
    NOSPECS system, 62/530
    nuclear magnetic resonance, 62/531
    orbital myositis, 62/298
    pathogenesis, 62/531
    progressive external ophthalmoplegia
      classification, 62/290
    protirelin stimulation test, 62/531
    T3 suppression test, 62/531
    treatment, 62/531
Thyroid stimulating hormone, *see* Thyrotropic
      hormone
Thyroiditis
    autoimmune, *see* Hashimoto thyroiditis
    Hashimoto, *see* Hashimoto thyroiditis
Thyrotoxic myopathy
    amyotrophic lateral sclerosis, 62/529
    aspartate aminotransferase, 62/529
    biochemistry, 62/530
    creatine kinase, 62/529
    dysphagia, 62/528
    dysphonia, 62/528
    EMG, 62/529
    endocrine myopathy, 41/235-240, 62/528
    fructose bisphosphate aldolase, 62/529
    muscle biopsy, 62/529
    muscle cramp, 62/528
    muscle stretch reflex, 62/528
    muscle weakness, 62/528
    myalgia, 62/528

myoglobinuria, 62/529
myokymia, 62/528
neuropathy, 62/528
respiratory dysfunction, 62/528
rhabdomyolysis, 62/529
weakness distribution, 62/528
Thyrotoxic ocular myopathy, *see* Thyroid associated
   ophthalmopathy
Thyrotoxic periodic paralysis
   Asian race, 62/532
   attack, 62/461
   clinical features, 62/532
   endocrine myopathy, 41/165, 241-243, 62/528
   genetics, 62/461
   HLA haplotype, 62/532
   inheritance, 62/532
   precipitating factor, 62/532
   prevalence, 62/461
   treatment, 62/532
Thyrotoxicosis, *see* Hyperthyroidism
Thyrotropic hormone
   orbital myositis, 62/300
Thyrotropin, *see* Thyrotropic hormone
Tiopronin
   toxic myopathy, 62/597, 609
Titin
   limb girdle syndrome, 62/187
   muscle fiber, 62/26
Tocainide
   myotonia, 62/275
Toe walker
   late infantile acid maltase deficiency, 27/228,
     62/480
Tolosa-Hunt syndrome
   orbital myositis, 62/301
Toluene
   toxic myopathy, 62/601
Toluene intoxication
   toxic myopathy, 62/613
Tomé body, *see* Tomé-Fardeau body
Tomé-Fardeau body
   inclusion body, 62/103
   muscle tissue culture, 62/103
   oculopharyngeal muscular dystrophy, 62/103
Tomography
   computer assisted, *see* Computer assisted
     tomography
Torsion dystonia, *see* Dystonia musculorum
     deformans
Torsion dystonia (Mendel), *see* Dystonia
     musculorum deformans
Torsion spasm, *see* Dystonia musculorum deformans
Tortipelvis, *see* Dystonia musculorum deformans
Toxic agent
   acquired myoglobinuria, 62/575
Toxic encephalopathy
   acute, *see* Reye syndrome
Toxic myopathy
   alcohol, 62/601
   ε-aminocaproic acid, 62/598, 601
   amiodarone, 62/606

amoxapine, 62/601
amphetamine, 62/601
amphotericin B, 62/601
antihistaminic agent, 62/601
autophagia, 62/606
barbiturate, 62/601
basic mechanism, 62/595
beta adrenergic receptor blocking agent, 62/604
bezafibrate, 62/601
black widow spider venom, 62/611
botulism, 62/612
brown spider venom, 62/601, 611
captopril, 62/597
carbenoxolone, 62/601
carbon monoxide, 62/601
cardiac glycoside, 62/598
chloralose, 62/601
chloroquine, 62/606
chlorpromazine, 62/601
cimetidine, 62/597, 609
clofibrate, 62/597, 601
clofibride, 62/597
Clostridium welchii, 62/612
cocaine, 62/601
colchicine, 62/597, 600, 606
copper sulfate, 62/601
corticosteroids, 62/605
creatine kinase, 62/596
crotamine, 62/610
crotoxin, 62/610
cytotoxic agent, 62/597
danazol, 62/597
diamorphine, 62/601
diamorphine addiction, 62/600
diazepam, 62/601
dihydrocodeine, 62/601
diuretic agent, 62/597, 601
emetine, 62/600
enalapril, 62/597
eosinophilia myalgia syndrome, 62/604
ethchlorvynol, 62/597, 609
ethylene glycol, 62/601
experimental myotonia, 62/597
fasciculation, 62/596
fenfluramine, 62/601
fenofibrate, 62/601
flunitrazepam, 62/601
fluphenazine, 62/601
focal myopathy, 62/611
gasoline, 62/601
gasoline intoxication, 62/613
gemfibrozil, 62/601
glutethimide, 62/601
glycyrrhiza, 62/601, 603
gold, 62/597
Haff disease, 62/601, 612
haloperidol, 62/601
hornet venom, 62/601, 612
hypocholesterolemic agent, 62/598
hypokalemia, 62/603
inflammatory myopathy, 62/604

isoetarine, 62/597
isoniazid, 62/601
labetalol, 62/597
lindane, 62/601
lithium, 62/597, 601
loxapine, 62/601
lysergide, 62/601
malignant hyperthermia, 62/601
meprobamate, 62/601
mercuric chloride, 62/601
metal fume intoxication, 62/613
metaldehyde, 62/601
methadone, 62/601
metolazone, 62/597
mevinolin, 62/601
monensin intoxication, 62/612
morphine, 62/601
mulgatoxin a, 62/610
muscle cramp, 62/596
myalgia, 62/596
myokymia, 62/596
myotonia, 62/597
myotoxin a, 62/610
necrotizing myopathy, 62/598
needle, 62/611
nifedipine, 62/597
notechis II-5, 62/610
notexin, 62/610
opiate addiction, 62/598
organophosphate intoxication, 62/612
oxprenolol, 62/601
paracetamol, 62/601
penicillamine, 62/597
pentamidine, 62/601
perhexiline, 62/606
phencyclidine, 62/601
phencyclidine addiction, 62/600
phenelzine, 62/601
phenformin, 62/601
phenmetrazine, 62/601
phenylenediamine, 62/601
phenylpropanolamine, 62/601
phospholipase A, 62/610
pindolol, 62/597
plasmocid, 62/600
procainamide, 62/597
2-propanol, 62/601
quail ingestion, 62/601
quail myopathy, 62/612
rhabdomyolysis, 62/600
rifampicin, 62/597, 609
salbutamol, 62/597
salicylic acid, 62/601
snake venom, 62/601, 610
strychnine, 62/601
suxamethonium, 62/597, 601
taipoxin, 62/610
tarantula spider venom, 62/611
tetracycline, 62/609
theophylline, 62/601
tiopronin, 62/597, 609

toluene, 62/601
toluene intoxication, 62/613
trifluoperazine, 62/601
tryptophan, 62/597
vasopressin, 62/601
vidarabine, 62/609
vincristine, 62/600, 606
wasp venom, 62/601, 612
zidovudine, 62/597, 600, 602
zimeldine, 62/597
zinc phosphide, 62/601
Toxic shock syndrome
    acquired myoglobinuria, 62/574, 577
    malignant hyperthermia, 62/574
    metabolic encephalopathy, 62/574
    myoglobinuria, 62/574
    symptom, 62/574
Toxin
    acquired myoglobinuria, 62/576
    myoglobinuria, 41/271, 62/557
Toxoplasma gondii
    inflammatory myopathy, 62/373
    polymyositis, 62/373
Transient cerebral ischemia, see Transient ischemic
    attack
Transient global ischemia, see Transient ischemic
    attack
Transient ischemic attack
    Emery-Dreifuss muscular dystrophy, 62/154
Transient neonatal myasthenia gravis
    disease duration, 62/401
    symptom, 62/401
Trauma
    acquired myoglobinuria, 62/572
Tremor, 6/809-824
    congenital myopathy, 62/335
    fatigue, 6/818, 62/52
    motor unit, 62/52
    multiple mitochondrial DNA deletion, 62/510
Tricarboxylic acid cycle deficiency
    mitochondrial disease, 62/503
Trichinella
    inflammatory myopathy, 62/373
    polymyositis, 62/373
Trichopoliodystrophy
    mitochondrial disease, 62/497
Triglyceride storage disease
    congenital ichthyosis, 62/496
    lipid metabolic disorder, 62/496
Trilaminar myopathy
    congenital myopathy, 41/22, 62/332
    double ring myopathy, 62/341
    facial weakness, 62/334
    histopathology, 62/341
Triperinol intoxication
    myotonic syndrome, 62/264
Trisymptom complex, see Behçet syndrome
Trisymptomatic syndrome, see Behçet syndrome
Tropical myositis, see Tropical polymyositis
Tropical polymyositis
    acquired immune deficiency syndrome, 62/373

nuclear magnetic resonance, 62/373
Staphylococcus aureus, 62/373
Streptococcus, 62/373
Yersinia, 62/373
Trypanosoma
inflammatory myopathy, 62/373
polymyositis, 62/373
Tryptophan
toxic myopathy, 62/597
TSH, see Thyrotropic hormone
Tubular aggregate
periodic paralysis, 40/229, 41/154, 162, 62/465
Tubular aggregate myopathy
clinical features, 62/353
congenital myopathy, 62/332
enzyme histochemistry, 62/352
facial weakness, 62/334
Tubulofilament
dermatomyositis, 62/377
inclusion body myositis, 62/377
polymyositis, 62/377
Tubulomembranous inclusion myopathy
congenital myopathy, 62/332
facial weakness, 62/334
rimmed muscle fiber vacuole variant, 62/352
Tularemia
acquired myoglobinuria, 62/577
Tumor
hypothalamic, see Hypothalamic tumor
Tumor necrosis factor-α, see Cachectin
Twin
congenital fiber type disproportion, 62/356
Type I fiber, see Muscle fiber type I
Typhoid fever
acquired myoglobinuria, 62/577

Ubidecarenone
ataxia, 62/504
creatine kinase, 62/504
epilepsy, 62/504
exercise intolerance, 62/504
lactic acidosis, 62/504
multiple neuronal system degeneration, 62/504
muscle biopsy, 62/504
muscle weakness, 62/504
myoglobinuria, 62/504
Ubiquinone deficiency
myoglobinuria, 62/558
Uniform muscle fiber type I myopathy
congenital myopathy, 62/332
facial weakness, 62/334
muscle fiber type I predominance, 62/354
symptom, 62/354
Upper motoneuron sign
adult polyglucosan body disease, 62/484

Vacuole
autophagic, see Autophagic vacuole
rimmed muscle fiber, see Rimmed muscle fiber
vacuole
Valium, see Diazepam

Valproic acid
carnitine deficiency, 62/494
Valproic acid intoxication
lipid metabolic disorder, 62/493
Varicella zoster virus
congenital, see Chickenpox
Vascular disease
collagen, see Collagen vascular disease
Vasculitis
polymyositis, 62/373
Vasopressin
toxic myopathy, 62/601
Velocity
motor conduction, see Motor conduction velocity
Venous brain infarction, see Brain infarction
Verhaart progressive supranuclear palsy, see
Progressive supranuclear palsy
Vidarabine
toxic myopathy, 62/609
Vincristine
toxic myopathy, 62/600, 606
Vincristine myopathy
spheromembranous body, 62/609
Viral disease
myositis, 62/369
Viral infection
acquired myoglobinuria, 62/572
myasthenia gravis, 62/404
Viral myositis
dermatomyositis, 62/381
inclusion body myositis, 62/381
polymyositis, 62/381
Virus
Coxsackie, see Coxsackie virus
ECHO, see ECHO virus
enteric cytopathic human organ, see ECHO virus
Epstein-Barr, see Epstein-Barr virus
picorna, see Picornavirus
retro, see Retrovirus
Vision
blurred, see Visual impairment
α-oxoglutarate dehydrogenase deficiency, 62/503
Visual impairment
myotonic hyperkalemic periodic paralysis, 62/462
Vitamin D deficiency
limb girdle syndrome, 40/438, 62/188
Vitamin E deficiency
limb girdle syndrome, 62/188
Vitamin PP
acquired myoglobinuria, 62/577
Vitamins
nicotinic acid, see Vitamin PP
Vitiligo
Eaton-Lambert myasthenic syndrome, 62/421
Vomiting
α-oxoglutarate dehydrogenase deficiency, 62/503
Von Gräfe syndrome, see Progressive external
ophthalmoplegia
Von Strümpell-Lorrain disease, see Familial spastic
paraplegia
Vulpian-Bernhardt disease, see Scapulohumeral

spinal muscular atrophy

Wasp venom
   toxic myopathy, 62/601, 612
Wasting
   muscle, *see* Muscular atrophy
Weakness
   muscle, *see* Muscle weakness
Weil disease, *see* Leptospirosis
Welander distal muscular dystrophy, *see* Distal
   muscular dystrophy
Welander distal myopathy, *see* Distal myopathy
Werdnig-Hoffmann disease, *see* Infantile spinal
   muscular atrophy
Winged scapula
   Emery-Dreifuss muscular dystrophy, 62/146
   facioscapulohumeral muscular dystrophy, 40/417,
     62/162
   scapulohumeral muscular dystrophy, 62/170
Wohlfart-Kugelberg-Welander disease, 22/67-77,
   42/91-93
   atypical case, 62/200
   distal myopathy, 62/199
   inclusion body myositis, 62/204
   limb girdle syndrome, 40/433, 62/188
   muscle biopsy, 62/200
   pathology, 62/200
   progressive external ophthalmoplegia
     classification, 62/290
   rimmed muscle fiber vacuole, 62/200, 204
   symptom, 62/199

X-linked copper malabsorption, *see*
   Trichopoliodystrophy
X-linked humeroperoneal neuromuscular disease, *see*
   Emery-Dreifuss muscular dystrophy
X-linked muscular dystrophy
   Becker muscular dystrophy, 40/387-389, 62/117
   Duchenne muscular dystrophy, 40/277, 349-391,
     41/408, 415, 62/117
   Emery-Dreifuss muscular dystrophy, 40/389,
     62/117
   X-linked myotubular myopathy, 62/117
   X-linked neutropenic cardioskeletal myopathy,
     62/117
X-linked myotubular myopathy
   asphyxia, 62/156

chromosome Xq28, 62/156
   congenital myopathy, 62/117
   fatal course, 62/156
   X-linked muscular dystrophy, 62/117
X-linked neuromuscular dystrophy
   Charcot-Marie-Tooth disease, 62/117
   Foster Kennedy syndrome, 62/117
X-linked neutropenic cardioskeletal myopathy
   carnitine, 62/156
   chromosome Xq28, 62/156
   congenital myopathy, 62/117
   X-linked muscular dystrophy, 62/117
X-linked recessive disorder
   phosphoglycerate kinase deficiency, 62/489
X-linked recessive transmission (Menkes), *see*
   Trichopoliodystrophy
X-linked scapuloperoneal syndrome, *see* Emery-
   Dreifuss muscular dystrophy

Yersinia
   tropical polymyositis, 62/373

Z band
   muscle fiber, 62/7, 26
   myopathy, 62/7
   neuromuscular disease, 62/7
Z band plaque myopathy
   congenital myopathy, 62/332
   respiratory dysfunction, 62/344
Zebra body
   honeycomb myopathy, 62/353
Zebra body myopathy
   congenital myopathy, 41/21, 62/332
Zidovudine
   mitochondrial DNA depletion, 62/511
   polymyositis, 62/372, 382
   toxic myopathy, 62/597, 600, 602
Zidovudine myopathy
   ragged red fiber, 62/383
Ziehen-Oppenheim syndrome, *see* Dystonia
   musculorum deformans
Zimeldine
   toxic myopathy, 62/597
Zimelidine, *see* Zimeldine
Zinc phosphide
   toxic myopathy, 62/601